Symbol	Concept	Dimensions
M	Momentum	MLT^{-1}
$\dot{M}$	Momentum flux	MLT^{-2}
m	Mass	M
$\dot{m}$	Mass flow rate	MT^{-1}
$\mathcal{M}$	Moment	ML^2T^{-2}
$\mathbf{M}$	Mach number	1
N	Rotational speed (rpm)	T^{-1}
n	Coordinate normal to streamline (Chapter 4)	L
n	Number	$-$
n	Manning's n (roughness factor) (Chapter 11)	1
$\hat{n}$	Unit vector normal (perpendicular) to area	1
p	Pressure	$ML^{-1}T^{-2}$
p_T	Total pressure	$ML^{-1}T^{-2}$
p_0	Stagnation pressure	$ML^{-1}T^{-2}$
Δp	Pressure difference	$ML^{-1}T^{-2}$
Δp_d	Pressure drop	$ML^{-1}T^{-2}$
Δp_L	Pressure loss	$ML^{-1}T^{-2}$
P	Perimeter	L
Q	Volume flow rate	L^3T^{-1}
R	Radius or gas constant	L
R_0	Universal gas constant	$ML^2T^{-2}\theta^{-1}(\text{mole})^{-1}$
R_h	Hydraulic radius	L
r	Radial coordinate (cylindrical coordinates)	L
r'	Radial coordinate (spherical coordinates)	L
$\mathbf{R}$	Reynolds number	1
S	Specific gravity	1
S	Projected area for force coefficients (Chapters 8, 9, 10)	L^2
s	Streamline coordinate (Chapter 4)	L
$\hat{s}$	Unit vector in s-direction	1
$\tilde{s}$	Specific entropy	$L^2T^{-2}\theta^{-1}$
$\mathbf{S}$	Strouhal number	1
T	Temperature	θ
t	Time	T
$\mathcal{T}$	Torque	ML^2T^{-2}
U	Reference or given velocity	LT^{-1}
u	Velocity in x-direction	LT^{-1}
$u_{\max}$	Velocity at pipe centerline	LT^{-1}
$\tilde{u}$	Specific internal energy	L^2T^{-2}
V	Velocity	LT^{-1}
V_r	Relative velocity	LT^{-1}
$\mathcal{V}$	Volume	L^3

FUNDAMENTALS OF
FLUID
MECHANICS

FUNDAMENTALS OF
FLUID
MECHANICS

Philip M. Gerhart | University of Evansville
Richard J. Gross | University of Akron

ADDISON-WESLEY PUBLISHING COMPANY
Reading, Massachusetts ■ Menlo Park, California ■ Don Mills, Ontario
Wokingham, England ■ Amsterdam ■ Sydney ■ Singapore
Tokyo ■ Mexico City ■ Bogotá ■ Santiago ■ San Juan

Tom Robbins   Sponsoring Editor
Ann DeLacey   Manufacturing Supervisor
Barbara G. Flanagan   Production/Copy Editor
Melinda Grosser   Text Designer
Hannus Design Associates, Richard Hannus   Cover Designer
Martha K. Morong   Production Manager
Dick Morton   Art Editor

Library of Congress Cataloging in Publication Data

Gerhart, Philip M.
    Fundamentals of fluid mechanics.

    Bibliography: p.
    1. Fluid mechanics.   I. Gross, Richard.   II. Title.
TA357.G46 1985   532   84–9217
ISBN 0–201–11410–0

Reprinted with corrections, July 1985

CDEFGHIJ–RN–89876

Preface

This book was written to help prospective engineers learn fluid mechanics. Accordingly, it is primarily intended for use as a textbook in a sophomore or junior level course. The course may be taught by, and the students may be intending to become, civil engineers, mechanical engineers, or aerospace engineers. The assumption that the primary objective of the course is the preparation of engineers has strongly influenced our writing. We emphasize applications of the science of fluid mechanics to the student's chosen profession. The variety of applications presented shows that the subject of fluid mechanics crosses the boundaries of all branches of engineering.

This book, like any other textbook, is one corner of a triangle in the learning process. The other corners are the instructor and the student. The only corner of this triangle that does not suffer the pressures of limited time is the textbook. We have thus taken pains to provide complete and careful discussions of complex phenomena and have given special attention to several points that always seem to trouble students. This attention to detail, as well as the wide variety of topics, has resulted in a book that is somewhat larger than customary. We are confident that the book's flexibility and completeness qualify it as a valuable member of the learning triangle.

Objectives

The objectives of the book are the following:

- To help students develop a physical feel for the phenomena of fluid motion

- To present the fundamental laws that govern all fluid behavior and impress on students that all fluid mechanics problems are solved by application of these fundamental laws
- To develop practical methodologies for solution of engineering flow problems
- To give balanced emphasis to the differential and finite control volume approaches
- To illustrate the extremely wide variety of fluid-related phenomena in everyday life and in modern technology
- To prepare students to enter professional practice or to pursue graduate study
- To provide a valuable reference that can be used throughout an engineering career

Pre- and Corequisites

We assume that students have successfully completed courses in single- and multivariable calculus, statics, and dynamics. We also assume that most students have completed courses in mechanics of solids and ordinary differential equations or will be taking such courses concurrently with fluid mechanics. We have not assumed either prior or concurrent study of thermodynamics. Any necessary information from thermodynamics is introduced in the text. Students familiar with thermodynamics can skim over this material.

Fluid mechanics can be a very mathematical subject. We have not hesitated to introduce and use appropriate mathematical formulations; however our priorities have always placed physical understanding and practical application ahead of mathematical formulation and manipulation.

Features

Style. Our writing style is casual and informal. We hope to create a friendly atmosphere between the book and the reader and to promote the feeling of learning from experienced engineers.

Applications. We discuss both everyday life experiences that involve fluid mechanics principles (such as snowflakes carried in a wind gust or the aerodynamics of baseball pitching) and engineering applications of fluid mechanics. The latter applications are intended to show the importance of fluid mechanics in design, operation, maintenance, and other engineering functions. Applications are selected from many areas of civil, mechanical, and aerospace engineering.

Examples. The book contains 115 completely solved example problems reflecting a wide variety of applications. Many illustrate the need to make assumptions or obtain data from outside sources. Several involve iterative solution or evaluation of several alternatives. A degree of realism not often found in other texts is maintained. All examples are presented in a logical format. Careful attention is devoted to units in calculations. Discussions following each example comment on accuracy and generality of results, suggest alternative ways of solving the problem, and clarify tricky steps in the solution.

Problems. There are 967 end-of-chapter problems. Problems are organized into four types:

- extension and generalization
- lower-order application
- higher-order application
- comparison

Problems are drawn from all application areas of fluid mechanics and vary in difficulty. Many are based on real-world applications. An effective problem-solving method is described in Chapter 1.

Dimensions and Units. The text gives equal emphasis to SI units and English engineering (EE) units. As much as we might wish that SI units be used universally, U.S. industry still makes wide use of the older "foot, pound, inch" system. We feel that an engineering text written exclusively in either system does not adequately prepare students for the real world. Roughly half of the examples and problems are in each unit system.

Appendixes. An extensive set of appendixes is provided. Data on fluid properties, geometric characteristics of areas, and pipe and tubing sizes are given in both SI and EE units. An extensive list of conversion factors in a unique format is presented. Other items in the appendixes include compressible flow functions, a list of films, governing equations in cylindrical and spherical coordinates, and an exhaustive list of symbols. The appendixes are more extensive than customary because we believe that they add to the long-term usefulness of the book. We have found ourselves returning again and again to one or another of our old textbooks because it has an extensive appendix.

Solutions Manual. A comprehensive solutions manual is available to instructors from Addison-Wesley. Solutions to the problems are presented in the same format as the example problems. We have not included any answers in the text because we believe that the choice of making answers available to students should be up to the instructor. The solutions manual contains separate lists of answers to odd- and even-numbered problems for instructors who wish to use them.

Text Format. The text size is larger than customary to provide an open, inviting display for the reader. The typographical design reflects a concern with readability and comprehension. A second color is used in both type and art to highlight and emphasize important definitions and concepts and generally to focus the reader's attention. The effort of the production team has thus resulted in a book that is easy to read and pleasing to look at.

Organization. The organization of the book is broadly typical of most fluid mechanics texts. We have delayed introduction of most flow concepts until after the material on nonflowing fluids. Our experience indicates that introductions of

Eulerian and Lagrangian descriptions, velocity fields, streamlines, and other ideas can be intimidating in an introductory chapter. A special chapter is dedicated to fundamental flow concepts. A simple outline of the book is as follows:

- Introduction (Chapter 1)
- Fluid Statics and Rigid Body Motion (Chapter 2)
- Fundamental Flow Concepts (Chapter 3)
- Basic Techniques of Flow Analysis (Chapters 4–6)
- Applications of Flow Analysis (Chapters 7–11)

Chapters 1–6 should be studied in order (Chapter 2 can be moved or even omitted). Chapters 7–11 can be taken in any order or omitted.

We have tried to give balanced emphasis to the differential equation approach and the finite control volume approach. Since control volume analysis is more easily mastered by beginners, it is usually emphasized in an introductory course. There is nothing wrong with this; however, it often takes engineers (and graduate students) a long time to "integrate" introductory control volume analysis with "advanced" techniques based on the Navier-Stokes or Euler equations into a comprehensive analytical capability.

Course Planning

We have included far more material than can be covered in a single semester (assumed to include 45 class periods) or quarter (assumed to include 40 class periods). By selecting and omitting material, instructors can construct courses specifically tailored to the needs of civil, mechanical, and aerospace engineers. There is sufficient material for a second-semester or second-quarter course in "advanced" or "applied" fluid mechanics. Some instructors may wish to assign portions of the text that are not covered in class. Students may even read portions of the text that are not assigned (for one, that section on aerodynamics in sports looks interesting). We hope that students working on senior projects as well as practicing engineers will return to the book to find information not covered in their fluid mechanics course.

The following are suggested sections of the text to be included in courses for various purposes.

1. Single-semester or single-quarter introductory course for all types of engineers:
   Chapter 1
   Sections 2.1–2.4
   Sections 3.1–3.6
   Section 4.1
   Sections 5.1–5.4
   Chapter 6
   Sections 7.1–7.2.6, 7.3.1
   Sections 9.1, 9.3.1, 9.3.5
   Other material selected at instructor's option

2. Single-semester or single-quarter course for mechanical engineers:
   Chapter 1
   Sections 2.1–2.4
   Sections 3.1–3.6 (Omit 3.5.1)
   Section 4.1
   Chapter 5 (Omit 5.3.2, 5.3.3)
   Chapter 6
   Sections 7.1–7.2.6
   Sections 8.1–8.2
   Sections 10.2–10.4

3. Single-semester or single-quarter course for civil engineers:
   Chapter 1
   Sections 2.1–2.5
   Sections 3.1–3.6 (Omit 3.4.4, 3.5.1)
   Sections 4.1.1–4.1.3
   Sections 5.1–5.4 (Omit 5.3.2, 5.3.3)
   Chapter 6
   Sections 7.1–7.2.6, 7.3
   Section 8.1
   Sections 11.1–11.4

4. Single-semester or single-quarter course for aerospace engineers:
   Chapter 1
   Sections 2.1–2.4
   Chapter 3
   Chapter 4
   Sections 5.1–5.4
   Chapter 6
   Sections 8.1–8.3
   Chapter 9
   Sections 10.1–10.3, 10.6

5. Applied or advanced fluid mechanics (elective):
   Chapter 3 (review)
   Section 4.1 (review)

Sections 4.2–4.3
Sections 7.2.8–7.4
Sections 8.3–8.5
Chapter 9⎫
Chapter 10⎬ Various emphasis depending
Chapter 11⎭ on instructor and department

These are only suggestions. The specific interests and expertise of an instructor can be combined with the textbook to produce a wide variety of equally interesting and useful courses.

Acknowledgments

On completing a great task, it is customary to thank those who offered encouragement and help. In our task of writing this book we received aid from many people. Our colleagues Dr. John Hochstein and Dr. Jerry Drummond participated in many helpful discussions. Our department head at the University of Akron, Dr. R. J. Scavuzzo, offered much encouragement, as did Deans Joe Edminster and Lewis Hill, Jr. Our students kept us honest.

The manuscript was reviewed by Parker Lamb at the University of Texas, P. McDonald at North Carolina State University, and R. Warder at the University of Missouri. Their comments were helpful and encouraging.

The many drafts were typed and corrected by Mrs. Debbie Ganyard and Mrs. Dorothy Gulliams. Their contributions were extraordinary! We also received much help and encouragement from our department secretaries, Mrs. Irene Calvaruso and Mrs. Sandy Collins.

Our two wives and seven children displayed incredible patience and understanding. At the top of our current priority list is becoming reacquainted with them!

Evansville, Indiana   **P. M. G.**
Akron, Ohio   **R. J. G.**
November 1984

Contents

11

Applications of Flow Analysis: Liquid Flow in Open Channels 759

Appendixes 821

Appendix A Physical Properties 821

Appendix B Properties of Common Geometric Areas 828

Appendix C Conversion Factors 830

Introduction

1

1.1 Fluid Mechanics in Engineering

Fluid mechanics is the branch of engineering science that is concerned with forces and energies generated by fluids at rest and in motion. The study of fluid mechanics involves application of the fundamental principles of mechanics and, to a lesser degree, of thermodynamics to develop a qualitative understanding and quantitative analysis techniques that an engineer can apply to design or evaluation of equipment and processes that involve fluids.*

You are probably reading this book because fluid mechanics is a required course in your curriculum. The study of fluid mechanics is included in most engineering curricula because the principles and methods of fluid mechanics find many technological applications. To illustrate a few applications, we will consider the following fields:

- Fluid transport
- Energy generation
- Environmental control
- Transportation

Fluid transport is movement of a fluid from one place to another so that the fluid may be used or processed. Examples include home and city water supply systems, cross-country oil, natural gas and agricultural chemical pipelines, and chemical plant piping. Engineers involved in fluid transport might design systems involving pumps, compressors, pipes, valves, and a host of other components all directly involved with motion of fluids. In addition to the design of new systems, engineers may evaluate the adequacy of existing systems to meet new demands or they may maintain or upgrade existing systems.

In the field of energy generation, we find that, with the exception of chemical batteries and direct energy-conversion devices such as solar

---

* This discussion assumes that you already have some idea of what a fluid is. A formal definition will be presented shortly. The most common engineering fluids are air (a gas), water (a liquid), and steam (a vapor).

cells, no useful energy is generated without fluid movement. Typical energy-conversion devices such as steam turbines, reciprocating engines, gas turbines, hydroelectric plants, and windmills involve many complicated flow processes. In all of these devices, energy is extracted from a fluid, usually in motion. The thermodynamic cycles of these devices usually require fluid machinery such as pumps or compressors to do work on the fluid. Auxiliary equipment such as oil pumps, carburetors, fuel-injection systems, boiler draft fans, and cooling systems also depends on fluid motion.

Environmental control involves fluid motion. An estimated 75 percent of American homes are heated by forced-air systems, in which the home air is continually recirculated to transport heat from the combustion of fuel. In air-conditioning systems the circulating air is cooled by a flowing refrigerant. Similar processes occur in automobile engine cooling systems, in machine tool cooling systems, and in the cooling of electronic components by ambient air.

With the exception of space travel, all transportation takes place within a fluid medium (the atmosphere or a body of water). The relative motion between the fluid and the transportation device is responsible for the generation of a force that opposes the desired motion. This force can be minimized by the application of fluid mechanics to vehicle design. The fluid often contributes in a positive way, such as by floating a ship or generating lift by air motion over airplane wings. In addition, ships and airplanes derive propulsive force from propellers or jet engines that interact with the surrounding fluid.

These examples are by no means exhaustive. Other engineering applications of fluid mechanics include the design of canals and harbors as well as dams for flood control. The design of large structures must account for the effects of wind loading. In the relatively new fields of environmental engineering and biomedical engineering, engineers must deal with naturally occurring flow processes in the atmosphere and lakes, rivers, and seas or within the human body. Although it is not usually considered an engineering discipline, the phenomena of fluid motion are central to the field of meteorology and weather forecasting.

Few engineers can function effectively without at least a rudimentary knowledge of fluid mechanics. Large numbers of engineers are primarily involved with processes, devices, and systems in which a knowledge of fluid mechanics is essential to intelligent design, evaluation, maintenance, or decision making. You probably cannot foresee exactly what problems you will be called on to solve in your professional career; therefore, you would be wise to obtain a firm grasp of the fundamentals of fluid mechanics.

These fundamentals include a knowledge of the nature of fluids and the properties used to describe them, the physical laws that govern fluid behavior, the ways in which these laws may be cast into mathematical form, and the various methodologies (both analytical and experimental) that may be used to solve engineering problems. This book has been written to help you learn these fundamentals.

1.2 Fluids and the Continuum Hypothesis

All matter exists in one of two states: solid or fluid. You probably have at least a qualitative understanding of the difference between the two. A solid does not easily change its shape, but a fluid changes shape with relative ease. This concept of a fluid encompasses both liquids, which easily change shape but not volume, and gases, which easily change both shape and volume.

To develop a formal definition of a fluid, let us consider imaginary chunks of both a solid and a fluid (Fig. 1.1). The chunks are fixed along one edge, and a shear force is applied at the opposite edge.* First consider the solid. A short time after the application of the force, the solid assumes a deformed shape, which can be measured by the angle ϕ_1. If we maintain this force and examine the solid at a later time, we find that the deformation is exactly the same, that is, $\phi_2 = \phi_1$. On application of a shear force, a solid assumes a certain deformed shape and retains that shape as long as the force is applied. For most solids, the magnitude of the deformation is proportional to the magnitude of the force, and the solid will return to the original undeformed shape if the force is removed, at least if the magnitude of the shear and deformation are below certain limits.

Consider the response of the fluid to the applied shear force. A short time after application of the force, the fluid assumes a deformed shape, as indicated by the angle ϕ_1'. At a later time, the deformation is greater, $\phi_2' > \phi_1'$; in fact, the fluid continues to deform as long as the force is

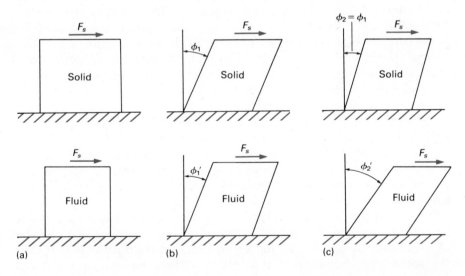

Figure 1.1 Response of samples of solid and fluid to applied shear force: condition (a) corresponds to the instant of application of the force; condition (b) to a short time after the application of the force; and (c) to a later time.

* Note that it is easy to imagine actually carrying out this experiment with a solid like steel or rubber but that it would be impossible to actually set a chunk of water on the table and carry out this experiment since the water would flow all over the table. This is exactly the nature of a fluid that we are seeking to define!

applied. If the force is removed, the fluid will not return to its undeformed shape but will retain whatever shape it had when the force was
removed. We can define a fluid:

> A fluid is a substance that deforms continuously under the action of
> an applied shear force or stress.

The process of continuous deformation is called *flowing*. A fluid is thus
a substance that is able to flow.

Because a fluid will always flow if a shear stress is applied, it is not
possible to analyze or discuss fluid behavior in terms of relations between stress and deformation as is done in solid mechanics. It is necessary to consider the relation between the applied stress and the *time
rate* of deformation. If a shear stress is applied to a fluid, the fluid will
deform at a rate related to the applied stress. The fluid attains a state
of "dynamic equilibrium" in which the applied stress is balanced by
the resisting stress. Thus, an alternative definition of a fluid is the
following:

> A fluid is a substance that can resist shear only in motion.

The distinction between solid and fluid seems rather simple, and
most substances can be easily classified; however, there are some substances that can fool you if you are not careful. Cold tar seems to act
like a solid; however, under the action of an applied shear it exhibits
a slow, continuous rate of deformation. And what about toothpaste?

The "flowing" property of fluids leads to some difficulty when we
begin to apply the laws of mechanics to an analysis of fluid motion. In
solid mechanics, particularly dynamics, it is a relatively simple matter
to identify a sample of matter (a ball, a beam, an engine connecting rod)
and to describe the motion and deformation that occur to that matter
because of the applied forces. This approach does not work well for
fluids because it is not possible to isolate (except in an imaginary way)
a particular sample of the fluid or to refer to some initial undeformed
state. Accordingly, we introduce a somewhat simplified *model* of a fluid
by means of the continuum hypothesis.

As far as we know, all substances are composed of an extremely
large number of discrete particles called molecules. In a pure substance
such as water, all molecules are identical; other substances such as air
are mechanical mixtures of different types of molecules. The molecules
of a substance interact with each other via collisions and intermolecular forces. The *phase* of a sample of matter, solid, liquid, or gas is a
consequence of the molecular spacing and intermolecular forces. The
molecules of a solid are relatively close (spacing on the order of a molecular diameter) and exert large intermolecular forces; the molecules
of a gas are far apart (spacing of an order of magnitude larger than
the molecular diameter) and exert relatively weak intermolecular
forces. Because intermolecular forces are weak, a gas easily changes its
volume as well as its shape. The stronger intermolecular forces in a
solid cause it to maintain both its volume and its shape. In a liquid,
the intermolecular forces are sufficiently strong to maintain volume
but not shape.

In principle it would be possible to describe a sample of fluid in terms of the dynamics of the individual molecules that compose the sample; in practice this is impossible because of the extremely large numbers of molecules involved. For most cases of practical interest, it is possible to ignore the molecular nature of matter and to assume that matter is continuous. This is called the *continuum hypothesis* and may be applied to both solids and fluids. The continuum hypothesis states that molecular structure is so small relative to the dimensions involved in problems of practical interest that we can ignore it.

If we adopt the continuum hypothesis, we describe a fluid in terms of its *properties,* which represent average characteristics of the molecular structure. As an example, we use the mass per unit volume or *density* rather than the number of molecules and the molecular mass. If matter were truly a continuum, the properties would be continuous mathematical functions of time and space. When a fluid is in motion, a fluid velocity exists at each point in the field. The fluid velocity is the velocity of the center of mass of the collection of molecules in the immediate vicinity of a given point.* Because the fluid properties and velocity are continuous functions, we can use calculus to analyze a continuum rather than applying discrete mathematics to each molecule.

For the continuum hypothesis to be valid, the smallest sample of matter of practical interest must contain a large number of molecules so that meaningful averages can be calculated. For air at sea-level conditions,[†] a volume of 10^{-9} mm^3 contains approximately 3×10^7 molecules. In most engineering problems, a volume of 10^{-9} mm^3 is very small, so the continuum hypothesis is valid. However, for certain cases, such as very-high-altitude flight, the molecular spacing becomes so large that a small volume contains only a few molecules, and the continuum hypothesis fails. For all the situations encountered in this book, the continuum hypothesis will be assumed to be valid.

It should be emphasized that, on adopting the continuum hypothesis, we seek the solution to a fluid mechanics problem in terms of the variation of a few fluid properties and the fluid velocity throughout the fluid. These properties and velocities are continuous functions of space and time, defined for every *point* within the fluid. Our notion of a "point" must be expanded to include a very small volume of space that is nevertheless large enough to contain a large number of molecules. Within these limits, it is possible to have discontinuous "jumps" in fluid properties and velocities at the interface between two fluids or across a shock wave in a compressible fluid (discussed in Chapter 10).

1.3 Fluid Properties

Fluid properties are used to characterize the state or condition of the fluid and are essentially the macroscopic (or continuum) representa-

---

* This fluid velocity, usually called simply *velocity,* is distinct from the random molecular velocity.

[†] By international agreement, sea-level conditions in the standard atmosphere correspond to a temperature of 15°C (59°F) and a pressure of 101,330 N/m^2 (14.696 lb/in^2).

tions of the microscopic (or molecular) structure and motion. You should already be familiar with some properties such as pressure and temperature. In the following discussion, we divide the properties into *mechanical properties, thermal properties,* and *auxiliary properties.*

1.3.1 Mechanical Properties of Fluids

The mechanical properties of a fluid involve the mass or weight of the fluid and the forces acting on or exerted by the fluid. These properties are always pertinent in fluid mechanics problems.

Density and Related Properties. The *density* of a fluid is its mass per unit volume. Density has a value at each point in a continuum and may vary from one point to another. Suppose that we pick an arbitrary volume ΔV in a fluid. We let Δm represent the mass of fluid contained in this volume. The average density ($\bar{\rho}$) is

$$\bar{\rho} \equiv \frac{\Delta m}{\Delta V}.$$

This density might depend on the exact size of the chosen volume. For example, consider a glass half full of water. If the chosen volume coincides exactly with the volume of water in the bottom half of the glass, we would obtain a value of about 1000 kg/m³ (62.4 lbm/ft³) for the average density; however, if our chosen volume was the entire glass, the average density would be only half as large because the top half of the glass contains no water. An exact definition of the density must involve a limit. Our natural inclination would be to take the limit of $\bar{\rho}$ as ΔV approaches zero; if we do, however, ΔV will become so small that the molecular structure of the fluid will become apparent and the density will vary, as indicated in Fig. 1.2. The irregularity at very small values

Figure 1.2 Dependence of (average) density on sample volume for small volumes.

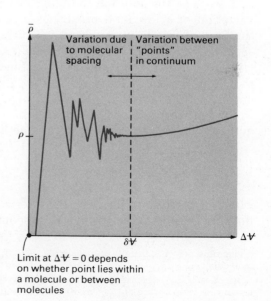

of ΔV is due to the molecular structure. As ΔV becomes larger, the density varies in a continuous manner because a large number of molecules are included in ΔV. The smallest volume (δV) for which the density variation is smooth is the limit of the continuum hypothesis. Thus we must define the fluid density by

$$\rho \equiv \lim_{\Delta V \to \delta V} \frac{\Delta m}{\Delta V}.$$

The small volume δV represents the size of a typical "point" in the continuum. For fluids near atmospheric pressure and temperature, δV is of the order of 10^{-9} mm^3. All of the fluid properties should be interpreted as representing an average of the fluid molecular structure over this small volume.

Fluid density varies widely between fluids. At atmospheric conditions, the density of air is about 1.22 kg/m^3 (0.076 lbm/ft^3), the density of water is about 1000 kg/m^3 (62.4 lbm/ft^3) and the density of mercury is about 13,500 kg/m^3 (846 lbm/ft^3). For a particular fluid, density varies with temperature and pressure. This variation is quite strong for gases but fairly weak for liquids. If the density of a fluid does not vary with pressure, the fluid is said to be *incompressible*.* In fluid mechanics, it is customary to treat liquids as incompressible fluids. It is often possible to treat gases as incompressible fluids if the speed† of the gas is low. Numerical values of density of several common fluids are presented in Appendix A, Tables A.3 to A.6.

Several other fluid properties are directly related to density. The *specific volume* (v), defined by

$$v \equiv \frac{1}{\rho},$$

is of considerable use in thermodynamics but is seldom used in fluid mechanics.

The *specific weight* (γ) is defined as the weight of the fluid per unit volume; thus

$$\gamma \equiv \rho g,$$

where g is the local acceleration of gravity.

The *specific gravity* of a fluid is the ratio of the density of the fluid to the density** of pure water at standard conditions.†† The defining equation is

$$S \equiv \frac{\rho}{\rho_{H_2O}}.$$

---

* Note that incompressibility is not concerned with the variation of density with temperature.

† Actually, the product of gas density and the square of the speed divided by the gas pressure must be small ($\rho V^2/p \ll 1$; see Chapter 10).

** Some authors define specific gravity as the ratio of specific weights. This requires the specification of a standard value of the acceleration of gravity to determine the specific weight of water.

†† Standard conditions for determining the density of pure water vary somewhat. We have selected 4°C (39.2°F) and 101,330 N/m^2 (14.696 lb/in^2).

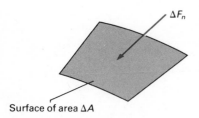

Surface of area ΔA

Figure 1.3 Illustration for definition of pressure.

In this equation, the specific weight of water, $\rho_{\text{H}_2\text{O}}$, is 1000 kg/m³ (62.4 lbm/ft³). Since the specific volume, specific weight, and specific gravity are all directly related to the density, they are all constant if the density is constant.

Pressure. *Pressure* is a fluid property of utmost importance. Most fluid mechanics problems involve prediction of the pressure distribution in the fluid or with the integrated effects of pressure over some surface or surfaces in contact with the fluid. Unlike density, which is usually one of the "known" quantities in a fluid mechanics problem, pressure is usually one of the "unknown" quantities to be determined by analysis or experiment.

Pressure is defined as the normal force per unit area acting on a real or imaginary surface in the fluid. Consider a small surface area ΔA within a fluid, as shown in Fig. 1.3. A force ΔF_n acts normal to the surface. Tangential forces may also be present but are not relevant to the definition of pressure. The pressure is defined by

$$p \equiv \lim_{\Delta A \to \delta A} \frac{\Delta F_n}{\Delta A}.$$

The limiting value δA represents the lower bound of the continuum assumption $(\delta A \approx (\delta \mathcal{V})^{2/3})$. The pressure can vary from point to point in a fluid. From a microscopic viewpoint, pressure represents molecular momentum and intermolecular forces within the fluid. Pressure is defined positively for compression.

1.3.2 Thermal Properties of Fluids

The thermal properties of a fluid are those related to the temperature and flow of heat within the fluid. From a microscopic viewpoint, they represent the energy contained in the molecular structure of the material. The thermal properties are not always pertinent in fluid mechanics problems. In problems involving heat transfer or in gas flows where compressibility is significant, the thermal properties are directly involved. In "incompressible" fluid problems, the involvement of the thermal properties is usually limited to the need to know a typical value of the fluid temperature so that other necessary properties (principally density and viscosity) can be found. The study of the thermal properties and their relations to the mechanical properties and to each other is the subject of thermodynamics. The following discussion of thermal properties is only a brief introduction. If you are interested in more information, consult standard textbooks on thermodynamics, such as [1, 2].

Temperature. *Temperature* (T) is a property that is familiar to everyone but rather difficult to define with the same exactness as density or pressure. We normally associate temperature with the degree of "hotness" or "coldness," but this is hardly a precise definition. Temperature is often defined as that property whose difference gives rise to the transfer of heat. Since heat is usually defined as flow of energy due to a difference of temperature, the definition is somewhat circular. An exact definition of temperature can be developed using the second law of thermodynamics, but this definition is of little use for introductory purposes. We will define the temperature of a fluid as a measure of (not "equal to") the energy contained in the molecular motions of the fluid.

Numerical values of temperature can be expressed with respect to a variety of scales. The major difference between them is that some scales are absolute (temperature expressed as a value relative to the lowest possible temperature, "absolute zero") and some scales are not absolute (temperatures expressed relative to an arbitrary datum). The most common nonabsolute temperature scales are the Fahrenheit scale and the Celsius scale. The corresponding (i.e., equal except for selection of datum) absolute scales are the Rankine and Kelvin scales, respectively. The relations between these scales are given by

$$T(\text{Rankine}) = T(\text{Fahrenheit}) + 459.67$$
$$T(\text{Kelvin}) = T(\text{Celsius}) + 273.15$$
$$T(\text{Rankine}) = 1.80\ T(\text{Kelvin})$$

Internal Energy, Enthalpy, Specific Heat. The *internal energy* $(\tilde{U})$ indicates the energy (both kinetic and potential) contained in random molecular motions and intermolecular forces. The *specific internal energy* $(\tilde{u})$ is the internal energy per unit mass. For single-phase fluids (liquids or gases), the internal energy is primarily a function of temperature. For fluids undergoing a change of phase (evaporating or condensing), the internal energy is also a function of the relative amounts of liquid and vapor in the two-phase mixture.

A property closely related to internal energy is the *specific enthalpy* $(\tilde{h})$, defined by

$$\tilde{h} \equiv \tilde{u} + \frac{p}{\rho}.$$

Enthalpy is a mixture of thermal and mechanical properties. It is an important property in the analysis of a compressible fluid because of the coupling between mechanical and thermal forms of energy; however, enthalpy is seldom used for incompressible fluids.

The relationship between the internal energy or enthalpy of a fluid and the temperature of the fluid is expressed in terms of *specific heat*. Heuristically, the specific heat of a substance is defined as the amount of heat that must be transferred to a unit mass of the substance to raise its temperature by one degree. If the substance is a compressible fluid, the specific heat depends on the nature of the heat transfer process, such as heat transfer at constant volume or heat transfer at constant pressure. The corresponding specific heats are the *specific heat at constant volume* (c_v) and the *specific heat at constant pressure* (c_p). It is shown in thermodynamics textbooks that a more general definition of the specific heats is given by

$$c_v \equiv \frac{\partial \tilde{u}}{\partial T}\bigg)_v \quad \text{and} \quad c_p \equiv \frac{\partial \tilde{h}}{\partial T}\bigg)_p.$$

For an incompressible fluid, there is essentially no difference between heat addition at constant volume and at constant pressure, so the specific heats are equal. The specific heat of an incompressible fluid is defined by

$$c \equiv \frac{d\tilde{u}}{dT}.$$

1.3.3 Property Relations, the Ideal Gas, and the Incompressible Fluid

Fluid properties are not independent. For a particular fluid, definite relations among the various properties exist. For most common fluids, specification of only two of the properties allows us to determine values for the other properties. Relations among the properties may be presented in graphical, tabular, or equation form. Tables A.3 to A.7 are examples of a limited property relation that gives the density and other properties at particular conditions of temperature and pressure.

Many gases have a property relation that can be closely approximated by the ideal gas law

$$p = \rho RT. \tag{1.1}$$

In this equation, the temperature and pressure must be expressed in absolute units.* R is the specific gas constant that is equal to the universal gas constant (R_0) divided by the molecular weight (M.W.) of the gas:

$$R = \frac{R_0}{\text{M.W.}} = \frac{8314}{\text{M.W.}} \, \text{N} \cdot \text{m/kg} \cdot \text{K} = \frac{1545}{\text{M.W.}} \, \text{ft} \cdot \text{lb/lbm} \cdot {}^\circ\text{R}.$$

For an ideal gas, the internal energy is a function of temperature only. Thus for the ideal gas[†]

$$\tilde{u} = \tilde{u}(T) = \int c_v(T) \, dT$$

and

$$\tilde{h} = \tilde{u} + p/\rho = \tilde{u}(T) + RT = \tilde{h}(T) = \int c_p(T) \, dT.$$

Often the specific heats are assumed to be constant. In this book, it will be assumed that all gases can be modeled by the ideal gas law and constant specific heats.

Liquids exhibit slight variation of density with temperature and pressure. The specific heat of liquids also varies slightly with temperature and, to a lesser degree, pressure. No simple, exact equations are available for properties of liquids. For most practical purposes, liquids can be treated as incompressible.

Incompressible fluids are modeled by particularly simple property relationships. The essence of the incompressible fluid is that the mechanical and thermal properties are not related; thus the incompressible fluid would obey the equations

$$\rho \equiv \text{constant} \quad \text{and} \quad d\tilde{u} = c \, dT$$

---

* Absolute pressures are defined in Section 2.3.

[†] The dashed parentheses ( ) mean "a function of."

or, approximately, for constant c,

$$\tilde{u} - \tilde{u}_{\text{ref}} \approx c(T - T_{\text{ref}}).$$

Note that the truly incompressible fluid has no pressure-density-temperature relation. The pressure is determined by purely mechanical phenomena, while the temperature is determined by thermal phenomena. The density is known and does not vary. If you choose to model a fluid as incompressible, the characteristic of incompressibility must be maintained even if the fluid is a gas. The following example illustrates this idea.

EXAMPLE 1.1 Illustrates the ideal gas law and the incompressible fluid assumption

An engineer is designing a system to provide combustion air for a boiler, as shown in Fig. E1.1. A fan draws air from a large room through an inlet duct. The fan increases the air pressure and discharges it into an outlet duct. The absolute pressure in the inlet duct is 98,700 N/m^2 and the temperature is 30.6°C. The absolute pressure in the outlet duct is 101,000 N/m^2. Because the air velocities are small, the engineer decides to assume that the air behaves as an incompressible fluid and that air is an ideal gas. Consistent with these assumptions, what can the engineer properly say about the following:

a) The density of the air at the fan inlet?
b) The density and temperature of the air at the fan outlet?

Solution

Given
Fan inlet pressure $= 0.987 \times 10^5$ N/m^2
Fan outlet pressure $= 1.01 \times 10^5$ N/m^2
Fan inlet temperature $= 30.6$°C
Air treated as incompressible and as an ideal gas

Find
a) The density of the air at the fan inlet
b) The density and temperature of the air at the fan outlet

Solution
a) Since the pressure and temperature of the air at the fan inlet are given, the air density may be found from the ideal gas law, Eq. (1.1)

$$\rho_{\text{in}} = \frac{p_{\text{in}}}{RT_{\text{in}}},$$

where the pressure and temperature must be expressed in absolute units. From Table A.3 in Appendix A, $R = 287$ N·m/kg·K for air. Then

$$\rho_{\text{in}} = \frac{0.987 \times 10^5 \text{ N/m}^2}{(287 \text{ N·m/kg·K})(30.6 + 273.2)\text{K}}$$

$$\boxed{\rho_{\text{in}} = 1.13 \text{ kg/m}^3.}$$

Answer ◀

b) Consistent with the engineer's assumption to treat the air as an incompressible fluid, we must have

$$\boxed{\rho_{\text{out}} \approx \rho_{\text{in}} = 1.13 \text{ kg/m}^3.}$$

Answer ◀

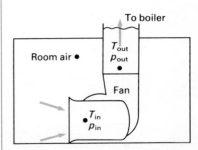

Figure E1.1 Fan setup for Example 1.1.

| Nothing can be said about the outlet temperature. | **Answer** ◀ |

Discussion

The ideal gas law was used to find the density of the air in the fan inlet duct. This same ideal gas law could be used to determine the density, pressure, or temperature in the fan outlet duct *if* the other *two* are known. The data in this example come from an actual fan where the outlet temperature is 35.6°C. In that case, the ideal gas law gives

$$\rho_{out} = \frac{p_{out}}{RT_{out}} = \frac{(1.01 \times 10^5 \text{ N/m}^2)}{(287 \text{ N·m/kg·K})(35.6 + 273.2)\text{K}} = 1.14 \text{ kg/m}^3.$$

If we use the actual outlet pressure and the *approximate* outlet density to calculate the outlet temperature, we get

$$T_{out} = \frac{p_{out}}{\rho R} = \frac{1.01 \times 10^5 \text{ N/m}^2}{(1.13 \text{ kg/m}^3)(287 \text{ N·m/kg·K})}$$

$$= 311.4\text{K}, \quad \text{or} \quad T_{out} = 38.2°\text{C}.$$

The temperature error of 2.6°C may be quite significant in energy-balance calculations; however, the density error of 0.01 kg/m³ is quite insignificant.

1.3.4 Auxiliary Properties of Fluids

The auxiliary properties of a fluid are used to express certain specific phenomena that arise when the fluid is subjected to particular conditions. By far the most important of these properties is the viscosity, which relates the stress generated in a fluid to the rate of strain of the fluid.

Viscosity. We learned earlier that a fluid is a substance that undergoes continuous deformation when subjected to a shear stress. For all fluids, the shear stress that resists this deformation is a function of the *rate* of deformation. For many common fluids, the shear stress is proportional to the rate of deformation. The constant of proportionality is called the *viscosity* and is a fluid property. To develop the defining equation for viscosity, we must first obtain an expression for the rate of deformation (i.e., the rate of shear strain) of a fluid particle. We do this by considering a flow in which the x-direction velocity of the fluid varies with y, as indicated in Fig. 1.4.

Since the fluid is assumed to be continuous, the fluid velocity is a continuous function (a smooth curve). A plot like Fig. 1.4, with the fluid velocity at various values of y indicated by arrows whose tips are connected by a smooth curve, is called a *velocity profile*.

Imagine that we can isolate a small rectangular chunk of fluid from somewhere near the center of the figure. The top of the chunk will move faster than the bottom, so if it is square at time t, it will be sheared at time $t + \delta t$, as shown in Fig. 1.5. The shear deformation is measured by the angle $\delta\phi$, which we can relate to the fluid velocity as follows. Since

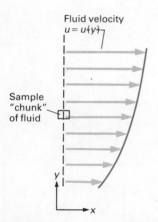

Fluid velocity
$u = u(y)$

Sample "chunk" of fluid

Figure 1.4 Typical fluid velocity variation.

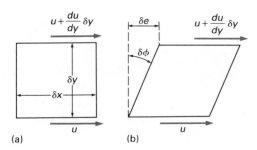

Figure 1.5 Fluid chunk (a) at time t and (b) at time $t + \delta t$.

the chunk is small, the velocity of the top *relative to the bottom* is

$$u_{\text{rel}} \approx \left(u + \frac{du}{dy} \delta y \right) - u = \frac{du}{dy} \delta y.$$

The distance, δe, that the top moves relative to the bottom is

$$\delta e = u_{\text{rel}} \delta t = \frac{du}{dy} \delta y \, \delta t.$$

Since $\delta\phi$ is small for small δt, we find

$$\delta\phi \approx \tan(\delta\phi) = \frac{\delta e}{\delta y} = \frac{du}{dy} \delta t.$$

The shear strain rate is the rate of change* of $\delta\phi$:

$$\frac{\delta\phi}{\delta t} = \frac{du}{dy}.$$

If the shear stress (τ) in the fluid is proportional to the strain rate, we can write

$$\tau = \mu \frac{du}{dy}. \tag{1.2}$$

The coefficient μ in this equation is the *viscosity*. Equation (1.2) is called *Newton's law of viscosity* since it was first suggested by Sir Isaac Newton. Fluids that obey this particular law and its generalization to multi-dimensional flow are called *Newtonian fluids*.

Not all fluids follow the Newtonian stress-strain relation. Some fluids, such as ketchup, are "shear-thinning," that is, the coefficient of resistance decreases with increasing strain rate (it all comes out of the bottle at once!). Others are "shear-thickening." Some fluids behave in a manner termed "plastic"; they do not begin to flow until a finite stress has been applied (this is the case for the toothpaste mentioned in Section 1.2). Fluids that do not follow the Newtonian relation are called *non-Newtonian*. Some possible types of stress–strain rate relations are illustrated in Fig. 1.6. If a viscosity were defined for a non-Newtonian fluid, it would be a function of the strain rate. For Newtonian fluids, the viscosity is independent of the strain rate.

---

* If u also varies with x, we would have to write $\partial u / \partial y$.

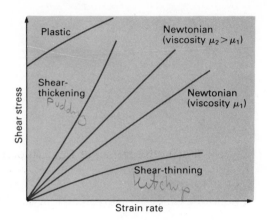

Figure 1.6 Various types of stress–strain rate relation.

Many common fluids such as air and other gases, water, and most simple solutions are Newtonian. Most petroleum products are Newtonian. Solutions containing long-chain polymers as well as slurries and suspensions are usually non-Newtonian. Blood is a non-Newtonian fluid. In this book, we limit our study to Newtonian fluids.

The viscosity of Newtonian fluids varies considerably from fluid to fluid. For a particular fluid, viscosity varies rather strongly with temperature but only slightly with pressure. Viscosity variation with pressure is usually neglected in practical calculations. Viscosities of several common fluids can be found in Tables A.3 to A.7 in Appendix A.

We can gain some insight into the mechanism of viscosity and its variation with temperature *in a gas* by considering the following very simplified model of molecular behavior. Suppose that the velocity distribution sketched in Fig. 1.4 represents a flowing gas. We further "magnify" the picture in Fig. 1.7 so that we can see two molecules of the gas at different *y*-locations. In part (a) of the figure, each molecule is moving to the right, with the fluid velocity appropriate to the molecule's particular location. Superimposed on its fluid velocity, each molecule also has a random molecular velocity and frequently collides with other gas molecules.* Suppose that the two molecules exchange places as a result of such collisions. We assume that immediately after changing places, each molecule retains the fluid velocity appropriate to its original position. This condition is represented in Fig. 1.7(b).

A result of this interchange of position is a net change of momentum on either side of plane *A-A* passing between the two molecules. Since this process, which has been illustrated for two molecules, actually happens continuously, there is a continuous momentum exchange due to random molecular motions in the gas. If we adopt the continuum hypothesis, we must imagine that a shear stress exists along the plane *A-A* between the molecules. According to Newton's second law, this shear stress (force per unit area) must equal the *rate* of momentum ex-

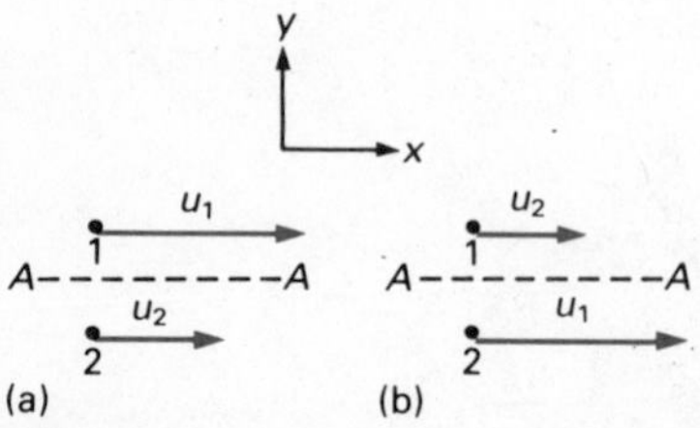

Figure 1.7 Representative gas molecules in a region of variable velocity.

* For air at standard conditions, approximately 5×10^9 collisions per cubic meter occur each second. A typical molecule travels a distance of 3×10^{-8} m between collisions. This distance is called the *mean free path*.

change per unit area. Notice that for this momentum exchange to take place, two things must occur:

- The molecules must have different fluid velocities at different y-locations.
- The molecules must be carried across the separating plane by random molecular motion.

In Newton's law of viscosity (Eq. 1.2), the velocity gradient, du/dy, represents the difference of fluid velocities, and the viscosity, μ, represents the molecular exchange process. Since molecular agitation increases with temperature, we would expect the viscosity of a gas to increase with temperature, and this is exactly the case!

The molecular momentum exchange model is quite accurate for gases; however, it does not explain the viscosity of liquids. The mechanisms of viscous stress generation in liquids are not completely understood. Intermolecular forces contribute more to liquid viscosity than does molecular momentum transfer. Since intermolecular forces generally decrease with temperature, the viscosity of most liquids decreases with temperature.

Viscosity provides a way to quantify the adjectives *thick* and *thin* as applied to liquids. "Thick" liquids have high viscosity and do not flow easily; the opposite is true for "thin" liquids. Fluid density has little to do with "thickness." Motor oil is usually considered a "thick" liquid, especially when cold, but it is less dense than water ($S_{oil} \approx 0.8$).

The ratio of viscosity to density often appears in the equations describing fluid motion. This ratio is given the name *kinematic viscosity* and is usually denoted by the symbol ν:

$$\nu \equiv \frac{\mu}{\rho}.$$

The kinematic viscosity is so named because it contains only length and time dimensions. Tables A.3 to A.7 in Appendix A give the kinematic viscosity of several common fluids as a function of temperature. When there is a need to carefully distinguish the types of viscosity, the viscosity μ is called the *dynamic viscosity* or *absolute viscosity*.

An effect usually associated with fluid viscosity is the *no-slip condition*. It is experimentally observed that whenever a fluid is in contact with a solid surface, the velocity of the fluid at the surface is equal to the velocity of the surface, that is, the fluid "sticks" to the surface and does not "slip" relative to it. This is true regardless of the type of fluid, type of surface, or surface roughness, so long as the continuum hypothesis is valid. Whenever a fluid flows over a surface, the variation of fluid velocity relative to the surface must be as sketched in Fig. 1.8. There will always be a velocity gradient near a surface when a fluid flows over the surface and, according to Newton's law of viscosity, there will always be shear stress in the fluid in this region.

The equations of motion for flowing fluids become very complicated when shear stresses are included. For this reason, simplified analyses are sometimes made by neglecting shear stresses. An easy way to introduce the assumption of zero shear stress is to pretend that the fluid has

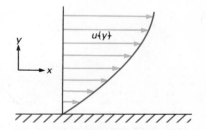

Figure 1.8 Variation of fluid velocity relative to a wall; note $u_{rel} = 0$ at the wall.

zero viscosity. According to Eq. (1.2), $\mu = 0$ gives $\tau = 0$. A fluid with zero viscosity is called an *inviscid fluid*. Although no real fluid has zero viscosity, the viscosity of many fluids is small, so shear stresses in these fluids will be small *if du/dy is not large*. Since this is often true except in the immediate vicinity of solid boundaries (as in Fig. 1.8), the assumption of an inviscid fluid is often useful for analyzing flow remote from solid boundaries.

The incompressible fluid and inviscid fluid models are combined in the so-called *ideal fluid*. The ideal fluid has $\mu \equiv 0$ and $\rho \equiv$ constant. Although this is a highly idealized fluid, many elegant mathematical solutions of the equations of motion of the ideal fluid can be found. You must be careful to distinguish between the ideal fluid defined here and the ideal gas discussed in Section 1.3.3.

EXAMPLE 1.2 Illustrates the calculation of shear stress from a velocity profile

Figure E1.2 shows a liquid flowing down an inclined surface. This may represent a number of physical phenomena, one being the flow of gasoline down the walls of an automobile carburetor. We will use this flow to illustrate the calculation of the shear stress from a velocity profile. The liquid is a Newtonian fluid and we are asked to find the shear stress at the fluid-solid interface ($y = 0$), at $y = Y/2$, and at the "free surface" ($y = Y$). The velocity profile is given by

$$u(y) = U\left[2\left(\frac{y}{Y}\right) - \left(\frac{y}{Y}\right)^2\right]$$

where U is a constant and Y is the thickness of the liquid layer.

Solution

    Given

Velocity profile of a Newtonian fluid:

$$u(y) = U\left[2\left(\frac{y}{Y}\right) - \left(\frac{y}{Y}\right)^2\right]$$

U and Y are constants.

    Find

The shear stress at $y = 0$, $Y/2$, and Y

    Solution

For a Newtonian fluid,

$$\tau = \mu\frac{du}{dy},$$

where μ is the liquid viscosity. Substituting the velocity u gives

$$\tau = \mu U\frac{d}{dy}\left[2\left(\frac{y}{Y}\right) - \left(\frac{y}{Y}\right)^2\right] = \frac{\mu U}{Y}\left[2 - 2\left(\frac{y}{Y}\right)\right].$$

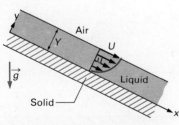

Figure E1.2 Liquid flowing down an inclined surface.

The shear stresses at the various y-locations are

$$\left. \tau \right)_{y=0} = 2\mu \frac{U}{Y}, *$$

Answer ◀

$$\left. \tau \right)_{y=Y/2} = \frac{\mu U}{Y},$$

Answer ◀

$$\left. \tau \right)_{y=Y} = 0.$$

Answer ◀

Discussion

The zero shear stress at the "free surface" is a common boundary condition for any type of liquid, as the air above the liquid exerts a negligible force on the liquid.

---

* For gasoline at 100°F with $Y = 0.05$ in. and $U = 13.4$ ft/sec, the shear stress at $y = 0$ would be 0.0025 lb/in², much smaller than the stresses normally encountered in solids.

EXAMPLE 1.3 Illustrates a common assumption for the velocity profile in small clearances and the development of a "working formula"

A concentric cylinder viscometer is a device used to measure the absolute viscosity of fluids. Figure E1.3 illustrates the details of this viscometer. The fluid is contained between a fixed outer cylinder and an inner cylinder that is free to rotate. The application of a torque $\mathscr{T}$ causes the inner cylinder to rotate at a constant speed ω. The viscometer has height H, and the width of the gap h is quite small compared to either the radius R_1 or R_2. Because of the small gap, the fluid velocity V in the tangential direction may be assumed to vary linearly across the gap. If the fluid is a Newtonian fluid with a constant viscosity μ, develop a formula for the viscosity μ in terms of the torque $\mathscr{T}$, speed ω, and the geometric parameters of the viscometer.

Solution

Given
Concentric cylinder viscometer shown in Fig. E1.3(a)
Enlarged view of a portion of the fluid, Fig. E1.3(b)

Find
A formula relating μ to the applied torque $\mathscr{T}$ and rotational speed ω

Solution
We see that μ appears in Newton's law of viscosity:

$$\tau = \mu \frac{du}{dy}, \quad \text{so} \quad \mu = \frac{\tau}{\left(\dfrac{du}{dy}\right)}.$$

To obtain the desired formula, we must express τ and du/dy in terms of the torque $\mathscr{T}$, speed ω, and other known quantities. Since the velocity varies

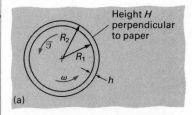
(a)

F_1 = Applied force on fluid by inner cylinder

F_2 = Resisting force on fluid by fixed outer cylinder

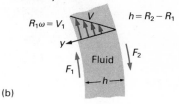
(b)

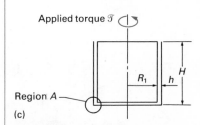
(c)

Figure E1.3 (a) Concentric cylinder viscometer; (b) enlarged view of portion of the fluid in concentric cylinder viscometer; (c) side view of concentric cylinder viscometer.

linearly over the gap h,

$$\frac{du}{dy} = \frac{V_1}{h} = \frac{R_1 \omega}{h}.$$

To express τ in terms of $\mathcal{T}$, we note that the force F_1 is the shear stress at the inner wall times the wall area ($2\pi R_1 H$):

$$F_1 = 2\pi R_1 H \tau)_{y=h}, \qquad \text{or} \qquad \tau)_{y=h} = \frac{F_1}{2\pi R_1 H}.$$

However, the force F_1 times the radius R_1 equals the applied torque on the inner cylinder:

$$\mathcal{T} = F_1 R_1.$$

Substituting gives

$$\tau)_{y=h} = \frac{\mathcal{T}}{2\pi R_1^2 H}.$$

Substituting into Newton's law of viscosity gives

$$\mu = \frac{\tau)_{y=h}}{\left(\dfrac{du}{dy}\right)_{y=h}} = \left(\frac{\mathcal{T}}{2\pi R_1^2 H}\right)\left(\frac{h}{R_1 \omega}\right),$$

$$\mu = \frac{\mathcal{T} h}{2\pi H \omega R_1^3}, \qquad \begin{array}{l} \text{Newtonian fluid,} \\ \text{constant viscosity,} \\ \text{constant speed,} \\ \text{small gap } (h \ll R_1). \end{array} \qquad \textbf{Answer} \quad \blacktriangleleft$$

Discussion

Note that the conditions under which the "working formula" is valid are given alongside the formula. These limitations arise from the assumptions explicit in the problem statement; however, additional limitations exist that may not be quite so evident. Referring to Fig. E1.3(c), we observe that the applied torque is resisted by the shear at the bottom of the inner cylinder as well as the shear of the fluid on the side of the cylinder. Since we have neglected the bottom torque, our equation is applicable only if $H \gg R_1$. To remove this restriction we could express the resisting torque on the bottom of the inner cylinder in terms of μ and develop a more accurate "working formula." A further limitation arises because we have neglected "end effects" on the gap velocity profile. Near the bottom of the viscometer (region A in Fig. E1.3c) the presence of the nonmoving bottom wall will have an effect on the gap velocity distribution. This effect will be negligible if $h \ll H$. Our formula is therefore subject to the additional restriction $H \gg R_1$ and $H \gg h$.

Bulk Modulus of Elasticity. Although we often assume that a fluid is incompressible, all real fluids exhibit some degree of compressibility; that is, the fluid density changes with the application of a compressive force (pressure). The *bulk modulus of elasticity* is a fluid property that indicates the degree of compressibility of a particular fluid.

Consider a fluid particle of volume δV. If the particle is subjected to a pressure dp, a volume change $d(\delta V)$ results. The density change is inversely proportional to the volume change since the mass of the fluid particle is constant. The bulk modulus of elasticity is defined by

$$E_v \equiv -\frac{dp}{\dfrac{d(\delta V)}{\delta V}} = -\delta V \frac{dp}{d(\delta V)}.$$

The negative sign is necessary because an increase in pressure results in a decrease in volume. Since the mass of the fluid particle is fixed, E_v can also be expressed in terms of the specific volume or the density:

$$E_v = -v\frac{dp}{dv} = \rho\frac{dp}{d\rho}.$$

The bulk modulus of elasticity is roughly equivalent to Young's modulus from solid mechanics. The units of E_v are those of pressure. For water at atmospheric conditions, E_v is approximately 2.1×10^9 N/m^2 (3×10^5 lb/in^2).*

If we consider a gas, the pressure-density relation also involves temperature, that is,

$$p = p(\rho, T);$$

thus

$$dp = \left(\frac{\partial p}{\partial \rho}\right)_T d\rho + \left(\frac{\partial p}{\partial T}\right)_\rho dT.$$

The exact relation between density change and pressure depends on temperature. If we assume constant temperature, we define the *isothermal bulk modulus*

$$E_{v,T} \equiv \rho\left(\frac{\partial p}{\partial \rho}\right)_T.$$

For an ideal gas,

$$p = \rho RT, \quad \text{so} \quad E_{v,T} = \rho\frac{\partial}{\partial \rho}(\rho RT) = \rho RT = p.$$

The isothermal bulk modulus of a gas is equal to the pressure of the gas. Air at sea-level conditions is some 20,000 times more compressible than water!

Coefficient of Thermal Expansion. Changing the temperature of a fluid can also cause a change of density. Most fluids expand (that is, density decreases) as temperature increases. The *coefficient of thermal expansion* is defined by considering a fluid particle of volume δV. Increasing the temperature of the fluid particle by an amount dT causes

---

* Water is thus 100 times more compressible than steel, which has a value of Young's modulus approximately 2.0×10^{11} N/m^2 (30×10^6 lb/in^2).

an increase in the fluid volume equal to $d(\delta V)$. The coefficient of thermal expansion is

$$\alpha_T \equiv \frac{\dfrac{d(\delta V)}{\delta V}}{dT} = \frac{1}{\delta V}\frac{d(\delta V)}{dT}.$$

Since the mass of the fluid particle is constant, α_T can also be expressed in terms of specific volume or density:

$$\alpha_T = \frac{1}{v}\frac{dv}{dT} = -\frac{1}{\rho}\frac{d\rho}{dT}.$$

For water at atmospheric conditions, α_T is approximately 1.53×10^{-4} K^{-1} (9.09×10^{-5} °R^{-1}).

In general, the temperature-density relation also involves pressure, so we must write

$$d\rho = \frac{\partial \rho}{\partial T}\bigg)_p dT + \frac{\partial \rho}{\partial p}\bigg)_T dp.$$

If we specify that the process of thermal expansion takes place at constant pressure, then

$$\alpha_{T,p} \equiv -\frac{1}{\rho}\frac{\partial \rho}{\partial T}\bigg)_p.$$

For an ideal gas with

$$\rho = \frac{p}{RT},$$

we find

$$\alpha_{T,p} = -\frac{1}{\rho}\frac{\partial}{\partial T}\left(\frac{p}{RT}\right) = \frac{1}{T}.$$

For sea-level conditions $\alpha_{T,p} = 3.470 \times 10^{-3}$ K^{-1} (1.928×10^{-3} °R^{-1}) for air, 23 times the value for water.

Surface Tension.   *Surface tension* (σ) is a property used to describe certain phenomena that are observed at interfaces between a gas and a liquid (e.g., water-air), two liquids, or one or more liquids, a gas, and a solid (e.g., water-air-glass). Two fairly familiar surface tension phenomena are the supporting of a small object of larger density (such as a steel sewing needle) by a water surface if the object is placed carefully on the surface and the beading of water into drops on the hood of a freshly waxed automobile.

Experimentally we observe that liquid surfaces resist deformation. This resistance can be explained by considering the intermolecular forces within the liquid. Molecules deep within the liquid are attracted by similar molecules on all sides evenly; however, a surface molecule is attracted differently by molecules on one side of the surface than by the molecules of a different type on the other. There is thus a net force on the surface molecules.

It is convenient to characterize the force on surfaces by defining a *surface tension*. Consider the simple apparatus sketched in Fig. 1.9. A loop of wire is dipped in a soap solution so that a soap film is formed. This film, although only a few molecules thick, has two surfaces, one on each side. To move the slide (or even hold it fixed) we must apply a force F. This force is proportional to the length ℓ, so we write

$$F = 2\sigma\ell.$$

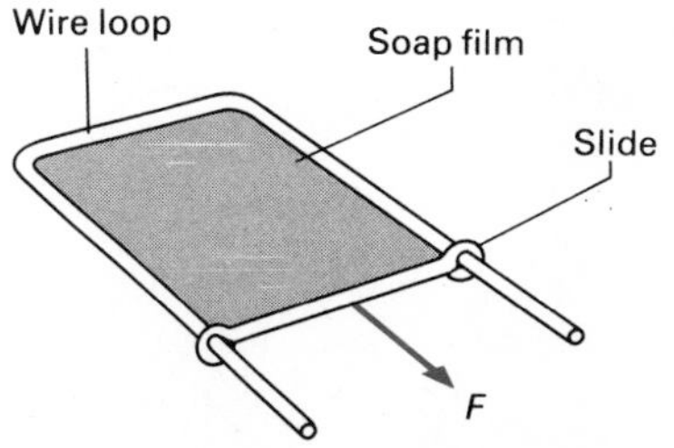

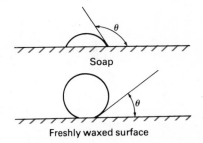

Figure 1.9 Wire loop and soap film.

The constant of proportionality, σ, is the surface tension of the film. The factor 2 is included to account for the fact that there are two surfaces. Surface tension has the dimensions of force per unit length.

It is sometimes more convenient to think of surface energy rather than surface tension. If we move the slide in Fig. 1.9 a distance Δx to the right, we generate more surface. The work required to generate this surface is

$$W = F(\Delta x) = 2\sigma\ell(\Delta x).$$

The product of Δx and ℓ is the amount of new surface created, ΔA, so

$$W = 2\sigma\Delta A.$$

This equation shows that surface tension can be interpreted as the energy per unit area required to create or maintain a surface.

In fluid mechanics, surface tension plays an important role in problems involving the formation of bubbles in liquids, the breakup of liquid jets into drops, and the determination of shapes of masses of liquid under conditions of "weightlessness" (zero-g).

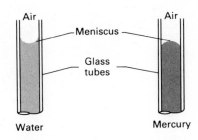

Figure 1.10 Drops of water on soap and freshly waxed surface.

The shapes of liquid surfaces that are in contact with solids and other fluids (usually air) are determined by the relative values of the surface energies. Specific liquid-solid combinations are classified as wetting or nonwetting depending on the contact angle between the liquid and solid surfaces. Figure 1.10 shows drops of water on soap and on a freshly waxed surface. If the contact angle θ is greater than 90°, the liquid *wets* the surface; if it is less than 90°, the liquid is *nonwetting*. Wetting behavior is determined by the particular liquid-solid combination. A situation where wetting behavior is important can be found in the field of manometry. A manometer is a liquid-filled glass tube used to measure pressure.* Figure 1.11 shows the liquid surface (called the *meniscus*) for water and mercury manometers. The height reading of the liquid in the manometer depends on the choice of reading the top, bottom, or center of the meniscus.

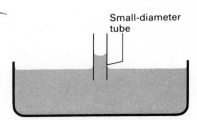

Figure 1.11 Shape of meniscus for water and mercury manometers.

A final phenomenon associated with surface tension is *capillarity*. Notice how the water in Fig. 1.11 inside the glass tube "climbs" the wall. If a very-small-diameter glass tube is inserted in a pan of water, the water will climb the walls of the tube until the upward force due to surface tension is balanced by the downward pull of the weight of the fluid column. See Fig. 1.12. In a nonwetting situation, the liquid in the tube is depressed below the level of liquid in the pan. Fluids can be lifted to great heights by capillary action in very small spaces. Capillary

Figure 1.12 Illustration of capillary rise of liquid in small tube.

* Manometers are discussed in detail in Chapter 2.

action is the mechanism by which water seeps through cloth, moving through the gaps in the weave of the cloth. When cloth is "water-proofed," it is treated with chemicals that change the water-cloth surface from wetting to nonwetting; thus the water beads up on the surface rather than seeping through.

As interesting as the effects of surface tension may be, they usually occur in small-scale phenomena. In most engineering fluid mechanics problems, forces associated with pressure, gravity, viscosity, and so on are much larger than those associated with surface tension. Accordingly, surface tension is usually neglected in this book; however, values of σ for some common liquids are presented in Appendix A.

1.4 Dimensions and Systems of Units

All engineering problems deal with physical entities that require quantitative description. We express the magnitude of the entity by a number and an associated unit of measurement. We may talk of a car with a mass of 1000 kilograms moving at a speed of 50 kilometers per hour or of a 2-inch-diameter pipe with a flow of 100 gallons per minute of water with a temperature of 150 degrees Fahrenheit. We recognize that "kilograms," "kilometers," "inches," "hours," "minutes," and "degrees Fahrenheit" are units of measurement for physical entities such as mass, length, time, and temperature. If we consider the situation carefully, we recognize two other facts. We know that the pipe diameter could also have been expressed as 0.16667 feet or 5.08 centimeters or 3.16×10^{-5} miles or any of several other combinations of numbers and units; however, the diameter of the pipe is a single entity that is independent of the choice of units used to express it. The pipe diameter is a characteristic of the pipe that can be expressed in a variety of units, so long as the units are of the same type.

We also recognize that the units of certain quantities are combinations of other units. The units for velocity are a combination of length and time units. The units for water flow are a combination of units of volume (gallons) and time; however, since a gallon contains 231 cubic inches, water flow could also be expressed in terms of units of length and time. We can generalize these observations by the following two statements:

- We can introduce the concept of a *dimension,* which is a sort of generalization of the concept of units. The dimensions of a physical entity tell us what kind of units must be involved in a quantitative statement of magnitude of the entity. Diameter has the dimension of length, velocity has the dimension of length divided by time, and so on.
- It is possible to select a set of independent *fundamental dimensions.* The dimensions of all physical entities can be expressed in terms of these fundamental dimensions.

The number of fundamental dimensions is surprisingly small, usually six: *mass, length, time, temperature, electric current,* and *luminous intensity.* The dimensions of all other physical entities can be derived from defining equations or from physical laws. Let us consider dimensions for velocity, energy, and force. Since velocity is the rate of change of distance with time, it has dimensions of length divided by time. We will use the symbol [ ] to represent "the dimension of." The symbols $[M]$, $[L]$, $[T]$, and $[\theta]$ will represent the fundamental dimensions of mass, length, time, and temperature. Using this notation, we write the previous statement as an equation

$$[V] = [L]/[T] = [L/T].*$$

This is read "The dimensions of velocity are length divided by time."

The dimensions of energy, $[E]$, can be determined by considering kinetic energy

$$E_K = \frac{1}{2} m V^2.$$

The factor $\frac{1}{2}$ is a pure number with no dimensions, so

$$[E] = [E_K] = [M][V]^2 = [ML^2/T^2].$$

Finally, we consider the dimension of force. Newton's second law ($F = ma$) shows that we can express force in dimensions of mass, length, and time:

$$[F] = [ML/T^2].$$

The selection of the number and types of fundamental dimensions is not unique. It is possible to select force as a fundamental dimension rather than mass or to select both force and mass as fundamental dimensions, increasing the number of fundamental dimensions from six to seven.[†] If we choose to do this, Newton's second law will require a conversion factor.

Turning to the practical side of the question, various systems of units have been developed to enable us to express the dimensions of physical entities quantitatively. While the number of fundamental dimensions is small, the number of units and unit systems is unnecessarily large. Unit systems have been developed in two ways. In the "scientific" method, a unit is defined for each fundamental dimension. The units of all physical entities are expressed as appropriate combinations of the units that have been assigned to the fundamental dimensions. Within this system, it is permissible to assign specific names to certain combinations of units, such as the name "newton" for the combination kilogram·meter/second².

In the "evolution" method, units were assigned to physical entities as the need arose. Units were often based on purely local standards or on specific physical phenomena. The "foot" was originally the length of the king's foot, while the BTU (British thermal unit) was defined as the amount of heat required to raise the temperature of one pound of

---

* The forms $[L]/[T]$ and $[L/T]$ are equivalent.

† We can easily increase the number of "fundamental" dimensions beyond six, but it is very difficult to decrease the number below six.

water by one Fahrenheit degree. This method often resulted in the assigning of different units to physical entities with the same dimensions, for example, the expression of heat in BTUs and work in foot·pounds although both heat and work are energy and have the dimensions $[ML^2/T^2]$ (or $[FL]$). When using such unit systems, various "conversion factors" must be employed. These conversion factors often had to be derived from experiments such as Joule's experiment that demonstrated the equivalence of heat and work and determined that 1 BTU = 778.16 foot·pounds.

It is obvious that the "scientific" method is preferable from the standpoint of simplicity and preciseness; however, it is equally obvious that the need for workable unit systems in commerce led to widespread adoption of evolved systems of units long before any scientific systems were worked out. Today, all nations of the world have officially adopted the SI* system of units as standard; however, in the United States, considerable use of other units persists, both in commerce and in engineering practice.

A major source of difficulty in most unit systems is the confusion between weight (a force) and mass and their units. We hope that as a result of your physics and elementary engineering mechanics courses, *you* know the difference between weight and mass; however, the distinction is by no means clear to everyone. If you go into the local supermarket to buy a "pound" of hamburger, are you buying weight or mass? If you fill out a professional résumé, do you list your "weight" as so many kilograms? The different approaches to the "weight-mass problem" can be illustrated by three unit systems: the SI system, the British gravitational (BG) system, and the English engineering (EE) system.

SI System.   The SI system is a "scientific" or defined system. The unit of mass is the kilogram (kg), the unit of length is the meter (m), the unit of time is the second (s) and the unit of temperature is the kelvin (K). The basis for the definition of these units is treated elsewhere [3] and will not be repeated here. The unit of force, the newton (N), is derived from Newton's second law:

$$1 \text{ N} \equiv 1 \text{ kg·m/s}^2.$$

The weight[†] of a mass is given by

$$W = mg,$$

where g is the local acceleration of gravity. By international agreement, the standard acceleration of gravity on the earth is 9.807 m/s², so a kilogram has a weight of 9.807 newtons.

The SI unit for energy is derived from

$$[E] = [ML^2/T^2] = [FL].$$

The energy unit is named a joule (J) and is defined by

$$1 \text{ J} \equiv 1 \text{ kg·m}^2/\text{s}^2 = 1 \text{ N·m}.$$

---

* SI stands for Système International or le Système International d'Unités.
† In the SI system the terms "force of gravity on" or "gravity force on" are preferred over "weight."

The unit of power is called a watt (W) and is defined by

$$1 \text{ W} \equiv 1 \text{ J/s} = 1 \text{ N·m/s} = 1 \text{ kg·m}^2/\text{s}^3.$$

British Gravitational System.   The British gravitational (BG) system adopts force rather than mass as a fundamental dimension. The unit of length is the foot (ft). The foot is defined with respect to the meter; 1 ft = 0.3048 m. The unit of time is the second (sec).* The unit of temperature is either the degree Fahrenheit (°F) or the degree Rankine (°R). The unit of force is the pound (lb). If we rewrite Newton's second law,

$$m = F/a,$$

the dimensions of mass become

$$[m] = [F]/[a] = \left[\frac{FT^2}{L}\right].$$

The unit of mass is the slug (slug), defined by

$$1 \text{ slug} = 1 \text{ lb·sec}^2/\text{ft}.$$

Since earth standard gravity is 32.174 ft/sec² in BG units, a mass of 1 slug weighs 32.174 lb under standard gravity. The unit of energy is the ft·lb, and the unit of power is the ft·lb/sec.

English Engineering System.   The English engineering (EE) system is similar to the BG system in most respects; however, the mass-weight problem is solved by defining units for *both* mass and weight such that 1 unit of mass has a weight of 1 unit of force under conditions of *standard gravity*. The length, time, and temperature units are the same as in the BG system (i.e., foot, second, degree Fahrenheit or degree Rankine). The mass unit is the pound mass (lbm), and the force unit is the pound force (lb).† The pound force (in the EE system) and the pound (in the BG system) are equivalent units. *One pound mass has a weight of one pound force under conditions of standard gravity.* One pound mass has a weight other than one pound force at values of gravity other than "earth standard." Using Newton's second law,

$$W = mg,$$

we can calculate the weight of 1 lbm under conditions of standard gravity, which, *by definition,* is 1 lb:

$$1 \text{ lb} = 1 \text{ lbm} \times 32.174 \text{ ft/sec}^2.$$

We seem to have trouble with this equation because the units are not the same on both sides and 1 does not equal 32.174. We correct the problem by dividing the right side by a *conversion factor* equal to

---

* Note that "second" is abbreviated "s" in SI units and "sec" in BG and EE units.

† In the EE system, the symbol lb_f is usually employed for the pound force. Since we use both the EE and BG systems in this book, we use lbm for the pound mass and lb for the pound force.

$32.174 \text{ ft} \cdot \text{lbm/lb} \cdot \text{sec}^2$, and we write:

$$1 \text{ lb} = \frac{1 \text{ lbm} \times 32.174 \text{ ft/sec}^2}{32.174 \text{ ft} \cdot \text{lbm/lb} \cdot \text{sec}^2}.$$

The conversion factor is assigned the symbol g_c, and is a *constant* factor that can be used *at any time* to relate pounds force and pounds mass. We *could* write Newton's second law as

$$F = \frac{ma}{g_c};$$

however, since g_c is only a conversion factor, it need not be written in equations. It would be possible to introduce g_c into the SI and BG systems:

$$g_c \text{ (SI)} = 1.0 \text{ m} \cdot \text{kg/N} \cdot \text{s}^2 \qquad \text{and} \qquad g_c \text{ (BG)} = 1.0 \text{ ft} \cdot \text{slug/lb} \cdot \text{sec}^2.$$

In the SI and BG systems, the combinations $\text{m} \cdot \text{kg/s}^2$ and $\text{lb} \cdot \text{s}^2/\text{ft}$ are named "newton" and "slug" rather than "converted" to them, certainly a simpler procedure; however, in these systems, a unit of mass does not have a weight equal to a unit of force. When making engineering calculations you should remember that

$$g_c = 32.174 \text{ ft} \cdot \text{lbm/lb} \cdot \text{sec}^2 = 1.$$

Since you can multiply any term in any equation by 1 at any time, you can multiply or divide by g_c any time you need to convert between units involving mass and units involving force.

Other units of measurement can be obtained by subdividing or combining the basic units. Examples include

- Length

$$1 \text{ inch} = 1/12 \text{ ft (BG, EE)}$$
$$1 \text{ yard} = 3 \text{ ft (BG, EE)}$$

- Volume

$$231 \text{ in}^3 = 1 \text{ gallon (BG, EE)}$$

- Pressure, stress

$$1 \text{ pascal (Pa)} = 1 \text{ N/m}^2 \text{ (SI)}$$
$$1 \text{ lb/in}^2 \text{ (psi)} = 144 \text{ lb/ft}^2 \text{ (BG, EE)}$$

- Energy

$$1 \text{ BTU} = 778.16 \text{ ft} \cdot \text{lb (BG, EE)}$$

- Power

$$1 \text{ horsepower (hp)} = 550 \text{ ft} \cdot \text{lb/sec (BG, EE)}$$

Table 1.1 lists dimensions, basic units, and other common units for several quantities of interest in fluid mechanics. Appendix C contains conversion factors, both between unit systems and between alternative units within a given system.

Table 1.1 Various quantities, their dimensions, and units

Quantity	Dimensions*	SI Units	BG Units	EE Units	Other Common Units
Area	$[L]^2$	m^2	ft^2	ft^2	in^2
Volume	$[L]^3$	m^3	ft^3	ft^3	in^3 gallon (gal) liter (l)
Velocity	$[L/T]$	m/s	ft/sec	ft/sec	—
Acceleration	$[L/T^2]$	m/s^2	ft/sec^2	ft/sec^2	—
Pressure or stress	$[M/LT^2]$	Pa (pascal) $(1\ Pa = 1\ N/m^2)$	lb/ft^2	lb/ft^2	$(lb/in^2$ (psi) kPa
Angular velocity	$[1/T]$	1/s	rad/sec	rad/sec	revolutions per minute (rpm)
Energy, work, heat	$[ML^2/T^2]$	J	ft·lb	ft·lb	BTU
Power	$[ML^2/T^3]$	W	ft·lb/sec	ft·lb/sec	BTU/sec, horsepower (hp)
Density	$[M/L^3]$	kg/m^3	$slug/ft^3$	lbm/ft^3	lbm/in^3
Viscosity	$[M/LT]$	Pa·s	slug/ft·sec	$lb·sec/ft^2$	poise (P) $(1\ P = 1\ Pa·s)$
Specific heat, gas constant	$[L^2/T^2\theta]$	J/kg·K	ft·lb/slug·°R	ft·lb/lbm·°R	BTU/lbm·°R BTU /slug·°R
Modulus of elasticity	$[M/LT^2]$	Pa	lb/ft^2	lb/ft^2	psi kPa
Surface tension	$[M/T^2]$	N/m	lb/ft	lb/ft	lb/in.

* Based on selection of mass rather than force as fundamental.

EXAMPLE 1.4 Illustrates the use of the conversion factor g_c

Bernoulli's equation for an ideal fluid flowing in the nozzle shown in Fig. E1.4 is

$$\frac{p}{\rho} + \frac{V^2}{2} + gz = \text{Constant.}$$

In the SI and EE systems, respectively, the description and units of each term are

p = Pressure, N/m^2 (lb/ft^2)
ρ = Density, kg/m^3 (lbm/ft^3)
V = Velocity, m/s (ft/sec)
g = Local acceleration of gravity, m/s^2 (ft/sec^2)
z = Elevation above some datum, m (ft).

Verify that each term on the left side of this equation can be expressed in units of N·m/kg in the SI system and of ft·lb/lbm in the EE system. What are the units of the constant?

Figure E1.4 Nozzle.

Solution

Given

$$\frac{p}{\rho} + \frac{V^2}{2} + gz = \text{Constant},$$

with units of

$$\text{EE:} \quad \frac{\text{lb/ft}^2}{\text{lbm/ft}^3} + (\text{ft/sec})^2 + (\text{ft/sec}^2)\text{ft}$$

$$\text{SI:} \quad \frac{\text{N/m}^2}{\text{kg/m}^3} + (\text{m/s})^2 + (\text{m/s}^2)\text{m}$$

Find

That each term on the left has units of $\text{N} \cdot \text{m/kg}$ (SI system) and $\text{ft} \cdot \text{lb/lbm}$ (EE system)

Solution

By inspection, the p/ρ term already has the proper units. The velocity and elevation terms have dimensions of $[L^2/T^2]$. Apparently, we need a conversion factor; we can try g_c and see if it helps us. Since the units of g_c are

$$\text{kg} \cdot \text{m/N} \cdot \text{s}^2 \quad \text{and} \quad \text{ft} \cdot \text{lbm/lb} \cdot \text{sec}^2$$

and we wish to express the terms of Bernoulli's equation with units of force in the numerator and units of mass in the denominator, we *divide* by g_c. In the SI system,

$$(\text{m/s})^2(\text{N} \cdot \text{s}^2/\text{kg} \cdot \text{m}) = |\ \text{N} \cdot \text{m/kg.}\ | \qquad \textbf{Answer} \ \blacktriangleleft$$

In the EE system,

$$(\text{ft/s})^2(\text{lb} \cdot \text{sec}^2/\text{ft} \cdot \text{lbm}) = |\ \text{ft} \cdot \text{lb/lbm.}\ | \qquad \textbf{Answer} \ \blacktriangleleft$$

$$\left|\ \begin{matrix} \text{The units of the constant are} \\ \text{N} \cdot \text{m/kg or ft} \cdot \text{lb/lbm.} \end{matrix}\ \right| \qquad \textbf{Answer} \ \blacktriangleleft$$

Discussion

In this example, we had to divide by g_c; note that we can either multiply or divide by g_c because it equals unity. The units of each term in the Bernoulli equation and the constant could be expressed in $(\text{m/s})^2$ or $(\text{ft/sec})^2$ if the pressure-density term were multiplied by the conversion factor g_c rather than the other terms being divided by g_c.

We must note that it is not customary to use g_c in the SI system; however, in this example, the use of g_c was more straightforward than attempting to manipulate the units of m^2/s^2 into $\text{N} \cdot \text{m/kg}$ by substituting the defining equation for a newton ($1\ \text{N} = 1\ \text{kg} \cdot \text{m/s}^2$).

1.5 Scope of Fluid Mechanics

Now that we have introduced the properties of fluids and have some idea of how a fluid behaves in certain sets of circumstances, we can put our study of fluid mechanics into perspective by developing a classification system for the various areas of fluid mechanics. This classification system will be based primarily on types of problems and types

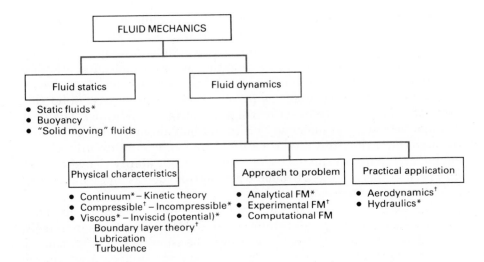

Figure 1.13 Specialty areas of fluid mechanics; areas marked with asterisk (*) are given special emphasis in this book; areas marked with dagger (†) are also covered.

of approaches to problem solution rather than on engineering applications. A few of the possible application areas of fluid mechanics were discussed in Section 1.1.

The major subdivisions of fluid mechanics are summarized in Fig. 1.13. Fluid mechanics is divided into the two main areas of *fluid statics* and *fluid dynamics*.

1.5.1 Fluid Statics

As the name implies, fluid statics deals with the mechanics of fluids that are not moving (strictly, fluids that are not flowing). In fluid statics, forces acting on the fluid are due to pressure and gravity only. No shear forces due to fluid deformation are present. Although they are not strictly static situations, we also include in this category those cases in which a fluid undergoes "rigid body motion," that is, the body of fluid moves without deformation. A liquid confined in a container rotating at constant angular velocity is an example of rigid body motion. The forces are still due to gravity and pressure and can be related easily to acceleration, so the problem is more like a fluid statics problem than a fluid dynamics (flow) problem. Fluid statics is an exact science; the only difficulties in practice arise from specific geometries of systems to be analyzed. Our discussions of fluid statics in Chapter 2 should prepare you to solve any fluid statics problem of practical interest. Examples of such problems include the calculation of forces on dams that impound large quantities of water, the calculation of buoyant forces on objects submerged in liquids, and the stability of floating bodies such as ships, in addition to the examples already cited for fluids in rigid body motion.

1.5.2 Fluid Dynamics

Fluid dynamics deals with the mechanics of fluids undergoing arbitrary motions. In addition to forces due to gravity (which are often negligible in fluid dynamics) and pressure, shear forces due to deformation of fluid particles may be significant. The fluid inertia usually must be taken into account. It is possible to formulate mathematically most fluid dynamics problems of practical interest by the use of a few basic laws; however, exact solutions are available for only a relatively few simple cases. Three items provide the major obstacles to obtaining solutions for most flow problems: geometry, viscosity, and turbulence.

Many engineering problems involve flow within or over solid surfaces of complex shape or with irregular surface conditions. The flow passage between the underside of an automobile and the highway or through the carburetor, intake manifold, and valves in the automobile's engine are familiar examples. Unfortunately, our ability to make exact flow predictions is limited to simple geometries such as pipes, flat plates, cylinders, and spheres.

The exact equations of fluid motion involving viscosity are nonlinear, second-order, partial differential equations that are extremely difficult to solve. If we discard viscosity, the equations can be reduced to a much simpler set; however, the solutions do not accurately reflect the real flow in many cases.

In addition to complicating the mathematics of flow analysis, viscosity may have a destabilizing effect on fluid flow that ultimately leads to the phenomenon we call *turbulence*. Turbulence is manifested by random, chaotic fluctuations in fluid velocity and represents a fluid motion so complex that the exact details of the motion are indeterminant except in a statistical or average sense. Most fluid flows of engineering interest include turbulence in at least some part of the flow.

Because of the difficulty (if not impossibility) of finding exact solutions of most flow problems by purely mathematical procedures, engineering fluid mechanics involves a mixture of analysis and experiment. The analysis is usually based on simplified approaches that do not require knowledge of the complete details of the flow (for example, the "control volume" approach described in Chapter 5), on simplified models of fluid behavior (such as the inviscid fluid assumption), or on simplified models of the physical system involved (such as modeling the flow in an automobile engine as flow in a network of pipes). Most practical prediction methods require the inclusion of information based on experimental measurements within the "theory"; in fact, there are absolutely no analytical methods for turbulent flow that do not require input from experiments. Some engineering fluid dynamics problems are solved entirely by experiment.

The digital computer has greatly extended our ability to make flow calculations. Computer-based methods usually predict the flow at a finite number of points in the field of interest by solving "finite-difference" approximations to the exact differential equations. Although computer-based methods are very powerful, it is still not possible to solve problems for all geometric configurations. If turbulence is involved, some experimental information must be included to find a solution.

Because of its complexity, fluid dynamics is often divided into several specialty areas. The division may be based on the particular assumptions made about the *physical characteristics* of the fluid or flow problem, on the type of *method(s)* used to solve the problem, or on the type of *engineering application.* Following is a description of a few of these areas.

Physical Characteristics.   Most fluid dynamics is based on the continuum hypothesis and may be called *continuum fluid dynamics;* however, one field, *kinetic theory,* takes the molecular nature of the fluid into account and applies statistical concepts together with the laws of particle dynamics for prediction of gas flows.

In *compressible fluid mechanics* (often called *gas dynamics*), problems associated with fluid compressibility are of primary importance. Such problems are often solved by assuming that the fluid obeys the ideal gas law.

The field of *incompressible flow* (often called *hydrodynamics*)* neglects fluid compressibility (Section 1.3.3). The methods of gas dynamics are necessary in high-speed gas flows, while the methods of hydrodynamics are valid in low-speed gas flows as well as liquid flows.

The field of *inviscid flow theory* deals with an imaginary fluid with zero viscosity. This area is often called *potential flow theory* since, in the absence of viscosity, the velocity can be obtained from a scalar potential function. If we also introduce the incompressible fluid assumption, we obtain the ideal fluid theory that is the most fully developed mathematical approach to flow prediction. A large number of exact solutions to the "ideal fluid" equations are known. These solutions are sometimes quite useful but at other times they fail miserably.

The area of viscous flow analysis is subdivided into several "specialties," including *boundary layer theory,* in which viscous effects are assumed confined to thin layers near walls (see Fig. 1.8); *lubrication,* in which flow in small clearances is considered; and *turbulence,* which seeks to develop information and models concerning turbulent flow.

Approach to Problems.   Approaches to the solution of fluid dynamics problems can be classified as analytical, experimental, and computational. In *analytical fluid mechanics,* one uses the tools of mathematics, physical modeling, and physical insight to develop solutions to engineering problems.

In *experimental fluid mechanics,* the tools include wind and water tunnels as well as a considerable amount of measurement apparatus. In experimental fluid mechanics, the fluid "solves" the problem by assuming a certain flow state; the challenge becomes measuring the flow and interpreting the results. This task is not as easy as you may think, especially since many experiments involve scale models rather than full-size apparatus. The experimental approach may be used directly (such as measuring the response of an airplane to wind gusts or measuring the rate of flow through a steam turbine in a power plant) or indirectly to provide basic information for use in mathematical models (such as studying the nature of turbulent flow near a rough wall). Several ex-

---

* The prefix *hydro* comes from the Greek word for "water"; however, hydrodynamics can be applied to all types of fluids, even air!

cellent references will provide more information on experimental fluid mechanics [4–7].

Probably the fastest-growing area in fluid mechanics is *computational fluid mechanics*. In this field, the digital computer is used to solve approximate forms of differential equations of fluid motion. These approximations involve the use of finite-difference or finite-element representations of the equations. Engineers in this field must combine knowledge of the physics of fluid flow with skills in numerical analysis and computer programming. For more details, consult [8–10].

Area of Application. The intended area of application provides the basis for the division between *aerodynamics* and *hydraulics*. Aerodynamics is concerned with the evaluation of forces generated between surfaces and large bodies of fluid in which they are moving. The flow may be compressible or incompressible and may be assumed viscous or inviscid. The ultimate application may be the design of aircraft, automobiles, or submarines. Hydraulics is concerned with the flow of fluids in pipes and channels as well as in process or power-generating equipment or in fluid machinery such as pumps or fans. The ultimate application may involve the design of a water supply system, chemical plant, power station, or hydraulic control system for an industrial machine.

Whichever specialty area of fluid mechanics is of most interest to you, you will need to apply the fundamental principles of fluid mechanics to solve problems. Even if you do not become a fluid mechanics specialist, a knowledge of the fundamental principles of fluid mechanics will be very helpful, if not essential, in your career. As you read this book and study the subject of fluid mechanics, make every effort to be sure you grasp the basic ideas, definitions, and methodologies. A firm understanding of these fundamentals is essential whether you use the computer or wind tunnel and whether you design an airplane or select a pump. The purpose of this book is to help you learn the fundamentals rather than to "short-circuit" you directly to a particular specialty.

1.6 Solving Fluid Mechanics Problems (A Note to the Student)

If you wish to learn the fundamentals of fluid mechanics (or to obtain a good grade in your fluids course!) you must do more than simply read this book. Your ultimate goal should be to obtain a *working* knowledge of the concepts and methodologies of fluid mechanics and to include this working knowledge in your professional "tool kit." Such a working knowledge can be obtained only by formulating and solving a variety of fluid mechanics problems.

In addition to learning to solve problems specific to the field of fluid mechanics, you should concentrate on developing an effective general approach to problem solution. The following two sections discuss the types of problems you will encounter in this book and propose an effective method for solving problems in fluid mechanics and other areas of engineering.

1.6.1 Types of Fluid Mechanics Problems

The problems at the end of each chapter are grouped according to the following classification.

Extension and Generalization (Derivation). Students often call problems of this type "theory" problems. Such problems typically involve the development of a "working formula" or the extension of a basic concept to a more general situation.

Example 1.3 is a typical problem involving the development of a working formula. The general relations involving viscosity, linear and angular velocity, stress, and torque were applied to a specific geometric configuration. The resulting equation, expressing the fluid viscosity as a function of torque, angular velocity, and the geometric parameters of the viscometer, can be used to obtain values of the viscosity from measured values of the torque and rotational speed for any specific fluid and viscometer so long as the fluid is Newtonian and the viscometer geometry fits the assumptions introduced in the example. With this equation available it is not necessary to start from "ground zero" each time a different-size viscometer or a different fluid is encountered. The equation is a working formula for small-gap concentric cylinder viscometers.

To illustrate the extension of a basic concept to a more general situation, let's investigate the development of a suitable equation to relate shear stress to rate of strain for a plastic material that has the characteristic shown in Fig. 1.14. We note that the material does not begin to deform until a finite stress (τ_0) has been applied. After deformation begins, the stress–strain rate relation is identical to that for a Newtonian fluid (i.e., it is linear). We can express this relation by the first-order algebraic equation

$$\tau = \tau_0 + \mu_p\left(\frac{du}{dy}\right),$$

where μ_p is a constant. This is a more general version of Newton's law of viscosity and is the desired result.

Lower-Order Application. In a lower-order application, the fluid mechanics problem to be solved is quite clear and is not disguised by the physical situation. To solve this type of problem, you must identify the basic physical phenomena, select the appropriate working equation(s), manipulate the equations as necessary, and substitute numbers and units to obtain the desired result. The problem will usually exactly fit the equations without the introduction of additional assumptions. The amount of given information will exactly match the amount needed.

As a simple illustration of this type of problem, let's consider the application of a horizontal force of 16.1 lb to a mass of 1.0 lbm. The mass slides along a frictionless horizontal surface and is in a vacuum. We wish to calculate the acceleration of the mass. We identify the phenomenon as the response of a mass to an applied force, a topic studied

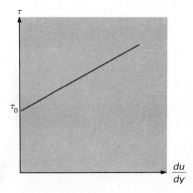

Figure 1.14 Stress–strain rate relation for a plastic.

in dynamics. We look into our professional "tool kit" and find that Newton's second law ($F = ma$) exactly "fits" our problem. The net force F is equal to the given applied force because air drag and sliding friction have been eliminated by the problem statement. Manipulating the equation and substituting numbers with their given units give

$$a = \frac{F}{m} = \frac{16.1 \text{ lb}}{1.0 \text{ lbm}} = 16.1 \frac{\text{lb}}{\text{lbm}}.$$

The answer does not have units of acceleration. Therefore, we again look into our "tool kit" and find the conversion factor g_c (Section 1.4). Noting that we can express g_c as

$$g_c = 32.2 \frac{\text{ft} \cdot \text{lbm}}{\text{lb} \cdot \text{sec}^2},$$

we observe that multiplying our answer by g_c eliminates both lb and lbm and leaves us with units of ft/sec² (acceleration!). Carrying out this operation, we have

$$a = 16.1 \frac{\text{lb}}{\text{lbm}} \left(\frac{32.2 \text{ ft} \cdot \text{lbm}}{\text{lb} \cdot \text{sec}^2} \right)$$

$$= 519 \text{ ft/sec}^2.$$

(Note only three significant figures in our answer.) Admittedly, this is a very simple example; however, examples like these were used in physics and dynamics to engrave Newton's second law into your mind so that it became a part of your professional knowledge.

You will find many lower-order application problems in this book. By working several of these problems you will increase your skills in the following areas:

- Correlating physical phenomena with the fundamental laws of nature that describe them
- Selecting and applying appropriate mathematical forms of these laws
- Properly handling units in numerical calculations, including the proper application of conversion factors, and carrying units through an entire calculation

Higher-Order Applications. Higher-order applications can include a variety of problems. In some cases it is obvious that the problem involves fluid mechanics, but it might be difficult to determine the applicable fundamental principle and exactly where the principle should be applied to "solve" the problem. In extreme cases, it might not even be clear that the problem involves fluid mechanics at all. In addition, the "standard" equations might not fit the problem without some approximations or simplifying assumptions, some of the given information may be superfluous, or some of the information may be missing and you might have to find it yourself. This type of problem more realistically represents fluid mechanics problems found in engineering practice and in everyday life.

A few examples will illustrate higher-order applications. Suppose we have been asked to design the keel of a small sailboat. If we were unfamiliar with sailing, we might assume that the only purpose of the keel is to improve the boat's stability by lowering its center of gravity. This assumption would be incorrect. The main purpose of the keel can be identified with the aid of Fig. 1.15. The force F represents the sideways component of the total wind force on the boat, and F_1 represents the sideways force of the water on the keel. The force F_1 keeps the sideways motion of the sailboat to a minimum. The keel is designed to keep the sideways motion at an acceptable value. The fluid mechanics problem thus involves designing the keel to generate a certain fluid dynamic force rather than produce a particular location of the boat's center of gravity. Of course the effect of the keel location on the center of gravity is important in the design of the sailboat; however, the main point of this example is that there are really two effects present, with the more important effect the less obvious. In other problems we may likewise be misled or we may have a difficult time identifying the real fluid mechanics problem. Once the real fluid mechanics problem has been identified, we can complete the solution in a straightforward manner.

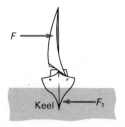

Figure 1.15 Aft view of simple sailboat.

Suppose we are assigned to choose a gasoline engine for a snow blower. The snow blower is designed to clear a 16-in.-wide path in 6-in.-deep snow. The direction of snow exiting the blower is known. We are given the performance characteristics of a number of engines. Before we can select one of the engines, we must determine the other important factors that must be considered in our selection process. We need to answer the following questions:

- Are the fundamentals of fluid mechanics applicable to this problem and are they adequate to obtain a solution? In other words, is this a fluid mechanics problem?
- How dense is the snow? Is it very dry and light or is it wet and heavy? Should we consider the worse case of the heavier snow?
- How fast will the person using the snow blower push it? Is this important?
- At what speed will the snow blower blades rotate? Will this speed be constant or will it change with the type of snow?

You may think of additional questions to be answered and may decide that some of the above questions are not appropriate to the situation. The answers to these questions are not contained in the problem statement. Exact answers to all of them are really not possible. If we wish to proceed with the design of the snow blower, we will be forced to assume approximate values for some of the data. It is important that we recognize the limitations these approximations introduce and that we fully document our assumptions. An investigation into the effects of different assumptions might also be valuable.

This particular problem will be solved as an example problem later in this book and will be used to demonstrate a technique to solve such problems.

Comparison. A comparison problem is actually two lower-order applications tied together. Such problems are intended to accomplish one of the following objectives:

- Illustrate how changing the fluid, the system geometry, or other parameters can alter the characteristics of a flow situation
- Illustrate the nature and properties of specific fluids
- Illustrate alternative ways to solve a problem and the differences in the results
- Illustrate the limitations of a particular equation or calculated result by evaluating the error introduced by assumptions or approximations
- Illustrate the added complexity of solving certain fluid mechanics problems without the usual assumptions and approximations

The following are examples of comparison problems that meet these objectives:

- Estimating the distances traveled by a dimpled golf ball and a smooth golf ball of the same diameter when hit with the same velocity from a tee
- Determining why oil is a better lubricant than water
- Estimating the error due to the assumptions of constant density and zero viscosity in calculating the air flow rate through a carburetor.

Considerable effort has gone into the selection of the problems at the ends of chapters; each problem has a specific purpose. After you complete each problem assigned by your instructor, review the problem and ask yourself how you can use it to equip your professional "tool kit."

1.6.2 A Systematic Approach to Problem Solving

Like engineering itself, problem solving is a combination of art and science. The art aspect of problem solving can be developed only by extensive practice; this is why engineering professors assign so many homework problems. On the other hand, problem solving must be approached in an orderly, or scientific, manner. Every engineer or engineering student can greatly increase his or her problem-solving capability by developing a systematic approach to problem solving. We suggest such an approach below. You may adopt our approach or you may develop your own approach. Whatever the case, we suggest that you label each step as you carry out a problem solution, at least until a systematic approach becomes natural. Our approach is as follows (see also Fig. 1.16).

Step 1 Make a sketch to help you visualize the physical situation. On the sketch indicate the given information in a descriptive manner so that you can recognize what you know about the problem. For example, a force or a flow rate may be indicated by an arrow in the direction of the flow, with the magnitude of the force or flow rate indicated near the arrow. This sketch is analogous to the free-body diagram used in statics or dynamics. *Do not neglect this step.*

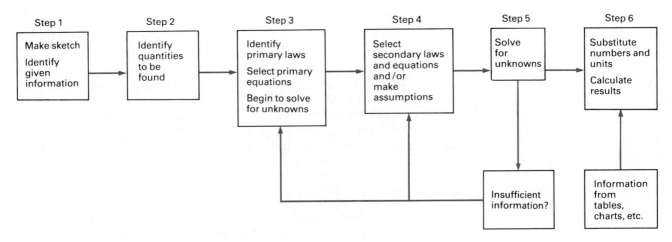

Figure 1.16 Schematic of problem-solving approach.

Step 2   List specifically what is to be found. This task is more difficult
for higher-order application problems. It amounts to identifying
the real fluid mechanics problem and requires an understanding
of the physical phenomena involved.

Step 3   Identify the *primary** physical law or laws that apply to the
particular problem. As illustrations of this step, consider the
viscometer problem of Example 1.3 and the acceleration problem
cited as an example of a lower-order application in Section 1.6.1.
The primary law applicable to the viscometer problem is the
stress–strain rate relation for fluids. The primary law in the
acceleration problem is Newton's second law of motion. *After* you
have identified the primary laws, select the appropriate mathematical forms of the laws. Be sure that any assumptions or
restrictions introduced in the development of the equations you
select are valid for your particular problem.

Step 4   In many cases, the primary laws and equations will not be sufficient to determine what you wish to know. It will be necessary to
introduce *secondary*† equations and/or to make some assumptions
to fit the equations to the problem. The secondary equations may
be based on other physical laws or on kinematic or geometric relations. Secondary equations in Example 1.3 are the equation
relating torque, force and radius; the equation relating stress and

---

* The primary physical law is the one that most directly applies to the physical
phenomenon. The quantity or quantities to be found usually appear explicitly in
the mathematical forms of the primary laws.

† Secondary equations do not usually contain the quantity to be found but involve
other variables that appear in the primary equations. The status of a particular law or
equation as "primary" or "secondary" depends on the particular problem. The same law
or equation may be primary in one problem and secondary (or irrelevant) in another
problem.

area; the equation relating angular velocity, radius, and linear velocity; and the equation relating the surface area of a cylinder to radius and height. As in Step 3, be sure that any inherent assumptions or restrictions in the secondary equations are valid.

If you must make assumptions, be careful that you do not "assume away the problem." Remember that any assumptions you make will probably affect the accuracy of your result. You should note the assumptions you introduce and also note any restrictions on their validity. The use of a linear velocity distribution for the fluid in the viscometer gap in Example 1.3 is an assumption. This velocity distribution will become a progressively worse approximation as the ratio of the gap width to cylinder radius increases.

Step 5 After you have selected the equations and introduced assumptions, manipulate the equations to solve for the desired unknown quantities in terms of given quantities. In carrying out this step, you usually come up against a roadblock. You might not have enough information to proceed past a certain point (did you count the number of equations and the number of unknowns?). At this point, you must return to Step 4 (or even to Step 3) and introduce more equations and/or assumptions. The inexperienced problem solver may have to repeat the Steps 3–4–5 cycle many times before proceeding to Step 6. The more experienced problem solver can do a better job of selecting equations and making proper assumptions in the first pass through Steps 3 and 4, and needs less feedback from Step 5. Only practice can lead to improvement in the selection/assumption process.

Step 6 After you have carried the solution as far as possible in terms of symbols, introduce numbers and units (at the same time) into the problem and complete the solution. At this step, it may be necessary to refer to tables or charts to find material (fluid) properties or conversion factors. You may also have to refer to charts to find other parameters.* Note any limitations due to assumptions, accuracy of given information, or information taken from tables or charts. Clearly label your answer.

The examples in this book have been worked, insofar as possible, by following this approach. We suggest that you study the examples and identify the above steps in their solution.

Before leaving the topic of problem solution, it is necessary to comment on the accuracy of numerical answers to engineering problems. (See Step 6.) In general, you should not give a numerical answer that appears to be more accurate than the information used to obtain the answer (no matter how many decimal places you have on your calculator!). To illustrate this point, let's say we want to calculate the force F due to a uniform pressure p acting on an area A by

$$F = pA.$$

---

* See, for example, the Moody chart, Fig. 7.11.

The pressure was measured as 91 psi by a gage with a maximum scale reading of 200 psi. The gage manufacturer stipulates the accuracy of the gage as $\pm 1.0\%$ of maximum scale reading.* The area is 61.0 in^2, with an uncertainty of ± 0.2 in^2. To calculate the uncertainty in F, we first calculate the uncertainty in the pressure:

$$\text{Uncertainty in } p = \pm 1\% \text{ of } 200 \text{ psi} = \pm 2.0 \text{ psi.}$$

The range of possible values of F is

$$F_{max} = (91 + 2) \text{ lb/in}^2 \times (61.0 + 0.2) \text{ in}^2 = 5691.6 \text{ lb,}$$
$$F_{min} = (91 - 2) \text{ lb/in}^2 \times (61.0 - 0.2) \text{ in}^2 = 5411.2 \text{ lb.}$$

Using the nominal values of pressure and area, we have

$$F = 91 \text{ lb/in}^2 \times (61 \text{ in}^2) = 5551.0 \text{ lb;}$$

however, uncertainty considerations expand the range in which the answer could lie:

$$5411.2 \leq F \leq 5691.6.$$

Therefore, we are not justified in writing our answer to more than two significant figures:

$$F = 5500 \text{ lb.}$$

There is an uncertainty of one or two units in the rightmost significant figure (the second 5). We usually shortcut this elaborate procedure by saying that the *percent* uncertainty in F is no less than the percent uncertainty in the most unreliable piece of given information. In this case it is 2 psi out of 91 psi, or about 2.2% of 5551 (± 122). Therefore we immediately write

$$F = 5500 \text{ lb,}$$

with a small uncertainty in the second 5. Note that the idea is to retain only one significant figure with an uncertainty in it.

In addition to uncertainties in given data, numerical results have limited accuracy because of assumptions introduced in obtaining the results. There is probably no quick and reliable way to estimate the inaccuracy due to such assumptions. The best way would be to repeat the calculation without the assumptions. This is usually impractical, or why did we need the assumptions in the first place? Another way would be to carry out a highly accurate experiment for comparison. This is usually impractical as well. Practice in doing problems and experience will help give the observant student a feeling for the errors introduced by assumptions. Sometimes we can at least bracket the correct answer on both the high side and the low side by making conservative assumptions for both extremes.

---

* Another way to state this is "The pressure has an uncertainty of $\pm 1.0\%$ of the maximum scale reading."

Problems

Extension and Generalization

1. Figure P1.1 shows a device known as a fluid drive. The driver rotates at an angular velocity ω_1, and the driven shaft rotates at a lower angular velocity ω_2. For steady-state conditions, ω_1 and ω_2 are constants. A torque $\mathcal{T}$ is transmitted by the driver to the driven shaft. Develop an expression for the slip $(\omega_1 - \omega_2)$ in terms of the applied torque $\mathcal{T}$, the spacing h of the two discs, and the absolute viscosity μ of the fluid enclosed between the two discs. Assume that the fluid is Newtonian. Then develop an expression for the efficiency, η, of the fluid drive in terms of ω_1, D, h, and $\mathcal{T}$. The efficiency is defined by

$$\eta = \frac{\omega_2 \mathcal{T}}{\omega_1 \mathcal{T}} = \frac{\omega_2}{\omega_1}.$$

Note all assumptions used.

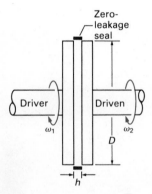

Figure P1.1

2. Oil of viscosity μ fills the gap of width t in Fig. P1.2. The truncated cone rotates at a constant rotational speed ω. Neglecting surface tension, develop an expression for the fluid viscosity μ as a function of the torque $\mathcal{T}$, the rotational speed ω, and the given dimensions.

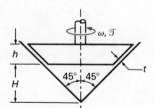

Figure P1.2

3. A glass tube is placed in a pan of water. Water rises in the tube to a height h above the level in the pan, as shown in Fig. P1.3. Show that the contact angle θ is given by

$$\theta = \cos^{-1}\left(\frac{\gamma h d}{4\sigma}\right),$$

where γ is the specific weight, d the inside diameter of the tube, and σ the surface tension.

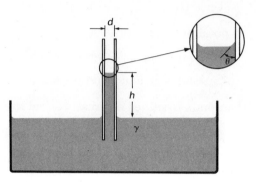

Figure P1.3

4. The circular disc of radius R in Fig. P1.4 is separated from a solid boundary by a distance h ($h \ll R$). The space is filled with a fluid of viscosity μ. The disc is rotated at angular velocity ω by a torque $\mathcal{T}$ applied to the shaft. Derive a relation between $\mathcal{T}$, μ, R, and ω.

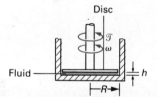

Figure P1.4

Lower-Order Application

5. An engineer is using a positive displacement pump to achieve very low pressure levels in an air tank and wants to check if the continuum assumption is valid at the pump inlet. The air in the tank is at 50°F and 0.25 lb/in². The inlet valve to the pump is 1 in. in diameter and opens a maximum of 1/8 in. The continuum assumption is valid if the smallest dimension of the physical problem is at least 100 times the mean free path λ. The mean free path λ for gas molecules is given by

$$\lambda = \frac{m}{\sqrt{2}\pi d^2 \rho},$$

where, for air,

$$m = \text{Mass of an air molecule}$$
$$= 4.8 \times 10^{-26} \text{ kg}$$
$$d = \text{Diameter of an air molecule}$$
$$= 3.7 \times 10^{-10} \text{ m}$$
$$\rho = \text{Mass density of air.}$$

Is our engineer justified in using the continuum assumption?

6. With the exception of the 410 bore (see Problem 7), the gauge of a shotgun barrel indicates the number of round lead balls, each having the bore diameter of the barrel, that are needed to weigh 1 lb. For example, a shotgun is called a 12-gauge shotgun if a 1/12-lb lead ball fits the bore of the barrel. Find the diameter of a 12-gauge shotgun barrel in inches and millimeters.

7. A 410-bore shotgun has a bore diameter of 0.410 in. What is its gauge? (See Problem 6 for an explanation of the gauge of a shotgun.)

8. Does a 10-gauge or a 20-gauge shotgun have a larger barrel bore diameter? (See Problem 6 for an explanation of the gauge of a shotgun.) Is the diameter of one twice the diameter of the other? If not, find the ratio of the two diameters.

9. Wine has a specific gravity of 1.15. A winery chemist decides to dilute wine with water to produce a specific gravity of 1.10. What percent of the new volume is the added water?

10. The aerodynamic wing used on race cars is designed to give a maximum "down-thrust" and a minimum drag. If a race car travels at speeds up to 200 mph and in temperatures up to 110°F, would you be justified in assuming the air flow was incompressible? No (at low speed) and (low temp)

11. A tank of fixed volume initially contains air at 30 psig and 70°F. Ten pounds of air are added to the tank by an air compressor. The final conditions in the tank are 65 psig and 75°F. Find the volume of the tank.

12. A compressed air tank in a service station has a volume of 10 ft³. It contains air at 70°F and 150 psia. How many tires can it fill to 30 psig at 70°F if each tire has a volume of 1.5 ft³ and the compressed air tank is not refilled?

13. Classify the following substances as plastic, shear-thickening, Newtonian, or shear-thinning.

Butter	Latex paint	Mustard
Gelatin	Margarine	Oil-base paint
Grease (warm)	Mayonnaise	Shampoo
Ketchup		

14. Explain why house paint should be a shear-thinning fluid.

15. A hydraulic lift in a service station has a 32.50-cm diameter ram that slides in a 32.52-cm diameter cylinder. The annular space is filled with SAE 10W oil at 10°C having a specific gravity of 0.85. The ram is traveling upward at the rate of 10 cm/s. Find the frictional force when 3.0 m of the ram is engaged in the cylinder.

16. The oil between the journal bearing and the journal (shaft) in Fig. P1.5 is a Newtonian fluid. The "power loss" is the product of the angular velocity of the shaft and the torque between the oil and the shaft. Calculate the power loss.

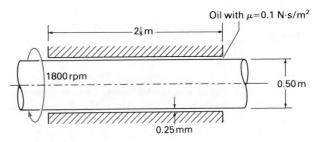

Figure P1.5

17. A journal bearing consists of a 6.000-in.-diameter shaft (or journal) rotating at an angular velocity of 60 rpm in a 6.030-in. sleeve (or bearing) that is 3.0 in. long. The annular space between the shaft and the sleeve is filled with SAE 30 oil at 100°F. Find the resisting torque. Assume the shaft and the sleeve remain concentric during rotation.

18. The viscosity of liquid water is tabulated in Table A.5 from 0°C to 100°C for atmospheric pressure. The following equation is proposed to fit the data:

$$\ln\left(\frac{\mu}{\mu_0}\right) \approx a + b\left(\frac{T_0}{T}\right) + c\left(\frac{T_0}{T}\right)^2.$$

T is the absolute temperature in K, $T_0 = 273.16\text{K}$, and $\mu_0 = \mu(T_0)$. Using the values of viscosity at 0°C, 50°C, and 100°C, evaluate the constants a, b, and c and find the maximum percent error of this curve fit for the viscosity of water. Write the approximating equation somewhere near Table A.5.

19. The viscosity of liquid water is tabulated in Table A.5 from 0°C to 100°C for atmospheric pressure. The following equation is proposed to fit the data:

$$\mu = ae^{-bT}.$$

T is the absolute temperature in K. Plot the viscosity versus the absolute temperature on semi-log paper and find suitable values for the constants a and b.

20. Figure P1.6 shows a plate sliding down a plane inclined at an angle of 30° with the horizontal. The plate weighs 40 lb, measures 20 in. by 40 in., and has a velocity of 0.5 in./sec. Find the thickness of the SAE 10W oil between the plate and the plane if the oil temperature is 0°F.

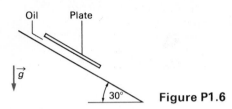

Figure P1.6

21. At atmospheric pressure and 60°F, a container holds 100 lbm of water. Additional water is forced into the container to increase the pressure to 10,000 psia while the temperature is held constant. If the container volume increases by 0.5% and water has an isothermal bulk modulus of elasticity of 3×10^5 lb/in², find the mass of water added to the container. What is the volume of the water at atmospheric pressure and 60°F?

22. If the isothermal bulk modulus of elasticity is constant, how does the density of a liquid change with pressure for constant temperature? Is the variation a linear, parabolic, or other function?

23. If the coefficient of thermal expansion is constant, how does the density of a liquid change with temperature for constant pressure? Is the variation a linear, parabolic, or other function?

24. A rectangular swimming pool measuring 20 m by 20 m has a depth of 2.0 m of 15°C water. The sun comes out and heats the water to an average temperature of 30°C. How much will the water level rise? Assume that the pool does not expand and no water is lost by evaporation.

25. A closed rigid container is filled with water at 55°F. Calculate the increase in pressure if the temperature is increased to 120°F.

26. A method used to determine the surface tension of a liquid is to find the force necessary to raise a wire ring as shown in Fig. P1.7. What is the value of the surface tension if a force of 0.015 N is required to raise a 4-cm-diameter ring?

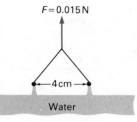

Figure P1.7

27. A soda straw with an inside diameter of 0.125 in. is inserted into a pan of water at 60°F. The water in the straw rises to a height of 0.150 in. above the water surface in the pan. Using the result of Problem 3, find the angle θ shown in Fig. P1.3.

28. Calculate the pressure difference between the inside and outside of a spherical water droplet having a diameter of 1/32 in. and a temperature of 50°F.

29. The momentum flux (discussed in Chapter 5) is given by the product $\dot{m}\vec{V}$, where $\dot{m}$ is mass flow rate and $\vec{V}$ is velocity. If mass flow rate is given in units of mass per unit time, show that the momentum flux can be expressed in units of force.

30. The air drag ($\mathscr{D}$) on a baseball thrown by a pitcher at a speed V in still air is given by

$$\mathscr{D} = \mathbf{C_D}\rho S V^2/2,$$

where

$\mathbf{C_D}$ = Dimensionless drag coefficient,

ρ = Air density,

S = Area of the baseball as seen by the catcher (also called the projected area or the maximum cross-sectional area).

Show that the drag can have units of force.

31. The Reynolds number, defined by

$$R = \frac{\rho V L}{\mu},$$

is a useful parameter in fluid mechanics. ρ is the fluid density, μ the fluid viscosity, V the fluid velocity, and L a relevant length dimension of an object in the flow stream or of a channel in which the fluid is flowing. Show that the Reynolds number is dimensionless.

32. A commercial advertisement shows a pearl falling in a bottle of shampoo. If the diameter D of the pearl is quite small and the shampoo sufficiently viscous, the drag $\mathscr{D}$ on the pearl is given by Stokes' law

$$\mathscr{D} = 3\pi\mu V D,$$

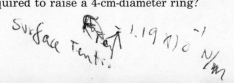

where V is the speed of the pearl and μ is the fluid viscosity. Show that the term on the right side of Stokes' law has units of force.

33. Air leaving a nozzle has a pressure of 14.7 lb/in^2, a temperature of 50°F, and a speed of 100 ft/sec. Calculate
a) the kinetic energy per unit mass of the air;
b) the kinetic energy per unit volume of the air.
Note all assumptions used.

34. Air leaving a nozzle has a pressure of 100 kPa, a temperature of 20°C, and a speed of 30 m/s. Calculate
a) the kinetic energy per unit mass of the air;
b) the kinetic energy per unit volume of the air.
Note all assumptions used.

35. Using the conversion factors between newtons and pounds force and between meters and inches, convert 14.7 lb/in^2 to pascals.

36. The pressure difference between points A and B in the manometer shown in Fig. P1.8 can be calculated by

$$p_A - p_B = (\rho_{H_2O} - \rho_{air})gh,$$

where $\rho_{H_2O} = 62.4$ lbm/ft^3, $\rho_{air} = 0.075$ lbm/ft^3, g is the local acceleration of gravity, and h is 12.0 in. Find $(p_A - p_B)$ in lb/in^2 at sea level.

$$P = 13.92 \; lb/in^2$$

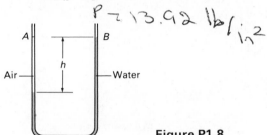

Figure P1.8

37. The axial velocity distribution for the flow of a viscous, Newtonian fluid in the circular pipe of inside radius R in Fig. P1.9 is

$$u = \frac{\Delta p_d}{4\mu L}\left[1 - \left(\frac{r}{R}\right)^2\right],$$

where the radius $r = 0$ at the centerline of the pipe, μ is the absolute viscosity, and Δp_d is the pressure drop in pipe length L. Find both the axial fluid velocity and the shear stress at $r = 0$ and at $r = R$.

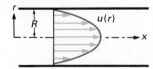

Figure P1.9

38. Air enters the converging nozzle shown in Fig. P1.10 at $T_1 = 70°F$ and a velocity of $V_1 = 50$ ft/sec. The velocity V_2 at the exit of the nozzle is given by

$$V_2 = \sqrt{V_1^2 + c_p(T_1 - T_2)},$$

where $c_p = 187$ ft·lb/lbm·°F and T_2 is the air temperature at the exit of the nozzle. Find the temperature T_2 for which $V_2 = 1000$ ft/sec.

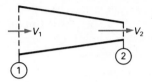

Figure P1.10

39. Explain why

$$g_c = 32.174 \text{ ft·lbm/lb·sec}^2 = 1$$

and can be substituted into any equation as needed.

40. A bag of salt weighs 2.2 newtons at sea level. What is its mass in kilograms?

41. For a mass m and gravitational acceleration g, explain why

$$mg = \frac{mg}{g_c}.$$

42. In Example 3.6 a velocity v_r is given by

$$v_r = \frac{C}{r},$$

where C is a constant, v_r has units of meters/second, and r has units of meters. What are the units of C?

43. Sir Isaac Newton is reported to have said the concept of gravity came to him while drinking tea in a garden and seeing an apple fall. What is the weight in pounds of an apple that weighs 1.0 N?

44. A correctly calibrated spring scale indicates the weight of a 10-lbm body as 8.0 lb. What is the local acceleration of gravity at the spring scale location?

45. Consider a mass of 1.0 lbm. What is its weight at sea level and at an altitude of 10,000 ft above sea level?

46. The orbiter of the NASA space shuttle *Columbia* can carry payloads of up to 29,000 kg (65,000 lbm) to orbits of up to 370 km (230 mi) altitude. Determine the weight of the payload at sea level and at the orbital altitude where the gravitational acceleration is 8.75 m/s^2 (28.743 ft/sec^2). Why is the payload said to be "weightless" in orbit? [You may wish to use the earth's radius of 6380 km (3960 mi) to show that the gravitational acceleration of the earth is 8.75 m/s² at this altitude.]

47. The specific gravity of a gas is sometimes defined as the ratio of the density of the gas to the density of dry *air* at standard conditions. If standard conditions are taken as 4°C (39.2°F) and 101,330 N/m² (14.696 lb/in²), find the specific gravity of dry air, carbon monoxide, and hydrogen at 20°C (68°F). If carbon monoxide were known to be present in a room, would you expect it near the floor, near the ceiling, or uniformly dispersed throughout the room? Would it depend on how long the carbon monoxide had been in the room?

48. An object is to be "weighed" on a beam balance using standard weights calibrated at sea level as a "counterweight". Will the beam balance give the mass or the force of gravity on the object being "weighed"? Does your answer depend on whether the object is "weighed" at sea level?

49. How high above sea level must an object be raised for the object's weight to be decreased by 1.0%?

50. In Chapter 3, we will show that fluid acceleration along a streamline is given by

$$a_s = \frac{\partial V}{\partial t} + V\frac{\partial V}{\partial s},$$

where V is velocity, t time, and s distance. Show that each term on the right side has dimensions of acceleration.

51. Convert the universal gas constant, $\bar{R} = 1545$ ft·lb/lbm·°R, to SI units.

52. From home plate, a baseball diamond measures 300 ft down the two foul lines and 400 ft to deep center field. Convert these dimensions to meters.

53. Show that each term in the following equation has units of lb/ft³. Consider u a velocity, y a length, x a length, p a pressure, and μ an absolute viscosity.

$$0 = -\frac{\partial p}{\partial x} + \mu\frac{\partial^2 u}{\partial y^2}$$

Higher-Order Application

54. Assume that the air volume in a small automobile tire is equal to the volume between two concentric cylinders 13 cm high with diameters of 33 cm and 52 cm. The air in the tire is initially at 25°C and 202 kPa. Immediately after air is pumped into the tire, the temperature is 30°C and the pressure is 303 kPa. What mass of air was added to the tire? What would be the air pressure after the air has

cooled down to a temperature of 0°C? List the assumptions used in your solution.

55. How many pounds of air must be pumped into a regulation basketball to inflate it to the proper playing pressure?

56. Use a hand pump to increase the air pressure in a size P225/75 R15 tire from 15 lb/in² to about 35 lb/in². Adjust your pace to average 30 strokes per minute and count the number of strokes, N, required. Then measure the inside diameter d, stroke L, and clearance ℓ of the hand pump. See Fig. P1.11. [*Author's note:* This is an experiment your instructor might perform for the class while the students help him or her maintain the proper pace.]

After completing the experiment, do a mathematical analysis and develop a formula to express the number of strokes, N, required to increase the tire pressure by an amount Δp in terms of d, L, ℓ, and the tire volume V. Obtain the numerical values of d, L, ℓ, and V for your hand pump and the tire and calculate N. Compare the two values of N and explain any differences. Did you assume the air temperature was constant while being compressed in the pump? If so, did this assumption increase or decrease your estimate of N?

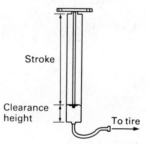

Stroke

Clearance height

To tire

Figure P1.11

57. A compressed air tank has a capacity of 30 gal and a maximum allowable pressure of 140 lb/in². How many size LR tires can be charged to a pressure of 43 lb/in² absolute at the ambient temperature without recharging the compressed air tank? List all assumptions used.

58. The concentric cylinder viscometer in Fig. P1.12 uses a falling weight W to produce a constant rotational speed of the inner cylinder. Find the required weights if the inner cylinder rotates at 30 rpm and the fluid in the viscometer is
a) tap water at 120°F,
b) gasoline at 120°F,
c) kerosene at 50°F,
d) SAE 10W 30 oil at 0°F,
c) SAE 20W oil at 0°F.

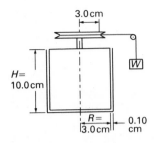

3.0 cm

H=
10.0 cm

R=
3.0 cm 0.10 cm   **Figure P1.12**

59. The closed-loop solar hot water system shown in Fig. P1.13 consists of copper tube, a tube bank in the solar collector, a tube bank in the heat exchanger, a pump to circulate water, and a surge tank. The surge tank prevents a dangerous pressure buildup as the water expands when heated. An enlarged view of the surge tank shows it to be a metal tank with a flexible bladder separating the water from an air space in the upper portion. The air in the upper portion can be pressurized to prevent boiling. In a particular system, there is 150 ft of 3/4-in. and 100 ft of 1/2-in. copper tube. The water is initially at 70°F and 20.0 lb/in² absolute pressure with no extension of the flexible bladder. Find the increase in volume of the water and in the air pressure if the average water temperature is increased to 120°F. The surge tank water volume and air volume are both initially 1.0 ft³ and the pump is not operating. List all assumptions used.

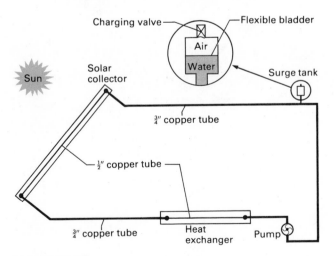

Figure P1.13

60. A 55-gal drum is completely filled with 40°F water and tightly sealed. The drum is then taken into a warm room at 75°F. Find the pressure in the drum when the drum and all the water reaches the room's

temperature. Recall that the drum and the water both expand with an increase in temperature.

61. Explain how sweat soldering of copper pipe works from a fluid mechanics viewpoint.

Comparison

62. The viscosity μ of common liquids decreases as temperature increases. A plot of the common logarithm of viscosity ($\log_{10} \mu$) versus the reciprocal of the absolute temperature ($1/T$) is approximately a straight line. Consider:
a) an SAE 10W 30 engine oil with viscosities of 1800 centipoise at 0°F and 4.7 centipoise at 210°F,
b) an SAE 10W 40 engine oil with viscosities of 1800 centipoise at 0°F and 20.0 centipoise at 210°F.
Plot $\log_{10} \mu$ versus $1/T$ and describe why a 10W 30 oil is better than a 10W 40 oil for starting an engine in 0°F to 100°F weather.

63. Explain why an SAE 10W 30 oil is superior to water for lubrication.

64. Compare the viscosity of an SAE 10W 30 with an SAE 30W oil at 0°F and 212°F. What would be the advantage, if any, of using an SAE 10W 30 oil in a cold engine and in an operating engine rather than the SAE 30W oil?

65. Calculate the change in volume of 2.0 liters of air and of 2.0 liters of water if both are initially at 20°C and 1 atmosphere pressure and the pressure of both is then increased to 100 atmospheres.

66. Consider 1 cubic meter each of water and air at 101 kPa and 20°C. What pressure change (in Pa) is required to produce the same volume change as a 40°C temperature change?

67. A fluid flows in a plane channel as shown in Fig. P1.14. The fluid viscosity is μ. Find expressions for the shear stress in the fluid at the wall ($y = 0$) for both laminar and turbulent flow in terms of V_{max}, h, and μ.

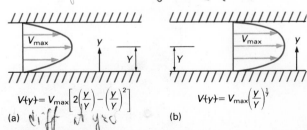

$$V(y) = V_{max}\left[2\left(\frac{y}{Y}\right) - \left(\frac{y}{Y}\right)^2\right]$$
(a)

$$V(y) = V_{max}\left(\frac{y}{Y}\right)^{\frac{1}{7}}$$
(b)

Figure P1.14 (a) Parabola; (b) power law.

References

1. Van Wylen, G. J., and R. E. Sonntag, *Fundamentals of Classical Thermodynamics* (2nd ed.), Wiley, New York, 1973.
2. Holman, J. P., *Thermodynammics* (3rd ed.), McGraw-Hill, New York, 1980.
3. ASME, "ASME Orientation and Guide for Use of SI (Metric) Units" (ASME Guide SI-1), American Society of Mechanical Engineers, New York, NY 10017.
4. Benedict, R. P., *Fundamentals of Temperature, Pressure and Flow Measurement* (2nd ed.), Wiley, New York, 1977.
5. Bradshaw, P., *Experimental Fluid Mechanics,* Macmillan, New York, 1964.
6. Bradshaw, P., *An Introduction to Turbulence and Its Measurement,* Pergamon Press, Oxford, 1971.
7. Dean, R. C., Jr., *Aerodynamic Measurements,* MIT Gas Turbine Lab Report, 1954, available from University Microfilms, Ann Arbor, MI.
8. Roache, P. J., *Computational Fluid Dynamics,* Hermosa Publishers, Albuquerque, 1976.
9. Chung, T. J., *Finite Element Analysis in Fluid Dynamics,* McGraw-Hill, New York, 1978.
10. Patankar, S. V., *Numerical Heat Transfer and Fluid Flow,* McGraw-Hill-Hemisphere, New York, 1980.

Mechanics of Nonflowing Fluids

2

In this chapter we will consider the mechanics of fluids that are not flowing; that is, the particles of the fluid are not experiencing any deformation. This condition is obviously satisfied for a static fluid. A second condition may also be classed as nonflow: "rigid body motion," in which the fluid moves without deformation. Examples of rigid body fluid motion include a tank of water experiencing uniform linear acceleration or rotating at constant angular velocity. Practical applications of rigid body motion exist but are not as numerous as those involving static fluids. Accordingly, most of this chapter concentrates on static fluids.

2.1 Pressure at a Point: Pascal's Law

If there is no deformation, then there are no shear stresses acting on the fluid. The forces on any fluid particle are due to gravity and pressure only.* We will be mainly concerned with calculation of the pressure distribution and with evaluating the resultant pressure forces at interfaces between a fluid and a solid.

In a nonflowing fluid, the pressure is a scalar quantity; that is, the pressure at a point in the fluid has a single value, independent of direction. This statement is known as *Pascal's law*. To prove this statement, we consider the equilibrium of forces for the small wedge of fluid shown in Fig. 2.1. For now, we assume that the fluid wedge is static.

The forces on the fluid are due to pressure, which we presume may be different for the y-, z-, and s-directions (we neglect the x-direction here for clarity), and gravity. The pressure force on any face of the wedge is the product of the pressure and the area of the face. Since the wedge is at rest, we can write

$$\sum F_y = p_y(\delta x\,\delta z) - p_s(\delta s\,\delta x)\sin\theta = 0$$

---

* We are neglecting surface tension and assuming that no electromagnetic forces act on the fluid.

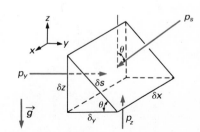

Figure 2.1 Wedge of fluid at rest.

and

$$\sum F_z = p_z(\delta x \, \delta y) - p_s(\delta s \, \delta x) \cos \theta - \gamma \left(\frac{\delta y \, \delta x \, \delta z}{2} \right) = 0$$

But

$$\delta y = \delta s \cos \theta \quad \text{and} \quad \delta z = \delta s \sin \theta,$$

so

$$p_y - p_s = 0 \quad \text{and} \quad p_z - p_s - \gamma \frac{\delta z}{2} = 0.$$

To evaluate this relation at a point, we take the limit as δx, δy, and δz approach zero, which results in

$$p_y = p_s \quad \text{and} \quad p_z = p_s.$$

Since the angle θ was arbitrary, these equations are valid for any angle. Now notice that the x- and y-axes are not unique in this derivation; if the wedge were rotated 90° about the z-axis, the x- and y-axes would be interchanged and we would conclude that $p_x = p_s$.

Consider what would happen if the wedge is not at rest. If there were no shear stresses, the only change to the analysis would be to equate the force sums to the mass of the wedge times the appropriate acceleration:

$$\sum F_y = ma_y = \rho \left(\frac{\delta x \, \delta y \, \delta z}{2} \right) a_y \quad \text{and} \quad \sum F_z = ma_z = \rho \left(\frac{\delta x \, \delta y \, \delta z}{2} \right) a_z.$$

When common terms are canceled, the terms

$$\rho \left(\frac{\delta z}{2} \right) a_y \quad \text{and} \quad \rho \left(\frac{\delta z}{2} \right) a_z$$

remain. When the limit is taken, these terms vanish just like $\gamma(\delta z/2)$ in the z-force equation. Therefore, pressure is also independent of direction in a moving fluid *if there are no shear stresses.*

If shear stresses were present in the fluid, they would have to be included in our force balance. Forces due to shear stresses are proportional to surface area, so stress terms would not vanish as the fluid wedge is "shrunk" to a point. In a fluid with shear stresses, the pressure is not necessarily independent of direction. In such cases, an average pressure is *defined* by

$$p \equiv \frac{1}{3} (p_x + p_y + p_z).$$

The difference between the average pressure and the three "directional" pressures is seldom significant.

2.2 Pressure Variation in a Static Fluid

We will now direct our attention specifically to static fluids. Even though the pressure at a point is the same in all directions in a static fluid, the pressure may vary from point to point. To evaluate the pres-

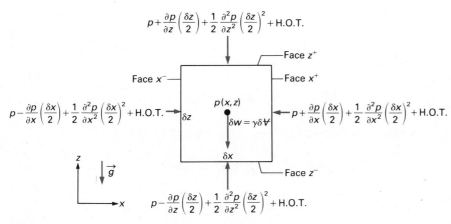

Figure 2.2 Rectangular fluid element in static fluid.

sure variation, consider a rectangular fluid element at rest (see Fig. 2.2). The fluid element is one length unit deep perpendicular to the paper. The sides of the fluid element are δx and δz, which are small but not necessarily infinitesimal quantities. The volume of the fluid element is

$$\delta V = (\delta x)(\delta z)(1).$$

The pressure at the center of the element is p. Since the fluid is continuous, the pressure at the faces of the element, $\pm \delta x/2$ and $\pm \delta z/2$ from the center, can be expressed in terms of the pressure and its derivatives at the center of the element by means of a Taylor series:

$$p_{z^+} = p + \frac{\partial p}{\partial z}\left(\frac{\delta z}{2}\right) + \frac{1}{2}\frac{\partial^2 p}{\partial z^2}\left(\frac{\delta z}{2}\right)^2 + \text{H.O.T.},$$

$$p_{z^-} = p - \frac{\partial p}{\partial z}\left(\frac{\delta z}{2}\right) + \frac{1}{2}\frac{\partial^2 p}{\partial z^2}\left(\frac{\delta z}{2}\right)^2 + \text{H.O.T.},$$

$$p_{x^+} = p + \frac{\partial p}{\partial x}\left(\frac{\delta x}{2}\right) + \frac{1}{2}\frac{\partial^2 p}{\partial x^2}\left(\frac{\delta x}{2}\right)^2 + \text{H.O.T.},$$

$$p_{x^-} = p - \frac{\partial p}{\partial x}\left(\frac{\delta x}{2}\right) + \frac{1}{2}\frac{\partial^2 p}{\partial x^2}\left(\frac{\delta x}{2}\right)^2 + \text{H.O.T.}$$

H.O.T. represents "higher-order terms" involving δz and δx. These series expansions are unnecessarily complicated, as we will see shortly. Since the fluid element is at rest, we can write equations expressing the equilibrium of forces:

$$\sum F_x = p_{x^-}(\delta z)(1) - p_{x^+}(\delta z)(1) = 0$$

and

$$\sum F_z = p_{z^-}(\delta x)(1) - p_{z^+}(\delta x)(1) - \gamma(\delta x)(\delta z)(1) = 0.$$

Substituting the series expressions for the pressure gives

$$\left[p - \frac{\partial p}{\partial x}\left(\frac{\delta x}{2}\right) + \frac{1}{2}\frac{\partial^2 p}{\partial x^2}\left(\frac{\delta x}{2}\right)^2 - p - \frac{\partial p}{\partial x}\left(\frac{\delta x}{2}\right) - \frac{1}{2}\frac{\partial^2 p}{\partial x^2}\left(\frac{\delta x}{2}\right)^2 + \text{H.O.T.}\right]\delta z(1) = 0$$

and

$$\left[p - \frac{\partial p}{\partial z}\left(\frac{\delta z}{2}\right) + \frac{1}{2}\frac{\partial^2 p}{\partial z^2}\left(\frac{\delta z}{2}\right)^2 - p - \frac{\partial p}{\partial z}\left(\frac{\delta z}{2}\right) - \frac{1}{2}\frac{\partial^2 p}{\partial z^2}\left(\frac{\delta z}{2}\right)^2 \right.$$
$$\left. + \text{H.O.T.} - \gamma(\delta z) \right](\delta x)(1) = 0.$$

Combining and canceling terms where possible, we get

$$\left(-\frac{\partial p}{\partial x} + \text{H.O.T.}\right)(\delta x)(\delta z)(1) = 0$$

and

$$\left(-\frac{\partial p}{\partial z} - \gamma + \text{H.O.T.}\right)(\delta x)(\delta z)(1) = 0.$$

We now divide by $(\delta x)(\delta z)(1)$, which, incidentally, is the volume of the fluid element, *and then* take the limit as (δx) and (δz) approach zero. This operation reduces the fluid element to a point. The H.O.T. terms contain positive powers of δx and δz so they vanish as we pass to the limit. Our equations become

$$\frac{\partial p}{\partial x} = 0 \tag{2.1}$$

and

$$\frac{\partial p}{\partial z} + \gamma = 0. \tag{2.2}$$

Before we discuss Eqs. (2.1) and (2.2), let's reflect on the mathematical procedure used to obtain them. Since the fluid was assumed continuous, the pressure in the vicinity of a point could be expressed by a Taylor series. Since at the end of the derivation we formally took the limit, shrinking the fluid element to a point, all terms involving derivatives higher than the first dropped out either by direct cancellation or because they included δx or δz to a positive power. When we repeat this type of derivation in the future, we will not include any terms beyond the first order in a series expansion because we know that they will drop out in the limiting process.* It would be good practice for you to repeat the derivation of Eqs. (2.1) and (2.2) using only the first-order terms of the series.

To return to the discussion of pressure variation in a static fluid: Eq. (2.1) shows that the pressure does not vary in a horizontal plane, perpendicular to the gravity vector (x could be any direction in the horizontal plane). Equation (2.2) shows that the pressure increases if we go "down" and decreases if we go "up." Since the pressure changes in only one direction, we can replace the partial derivative with an ordinary

---

* We used this shortcut method to find the strain rate of a fluid element in Section 1.3.4. You might try repeating that derivation by retaining higher-order velocity derivatives to see if you get the same result.

derivative,

$$\frac{dp}{dz} = -\gamma. \tag{2.3}$$

Equation (2.3) is the basic equation of fluid statics.

2.2.1 Pressure Variation in a Constant-Density Fluid

If the specific weight of the fluid is constant (as in a liquid), Eq. (2.3) can be easily integrated to give

$$p(z) = -\gamma z + p_0, \tag{2.4}$$

where p_0 is the pressure at $z = 0$.

Consider a body of liquid with a free surface, as shown in Fig. 2.3. Since the free surface is at constant pressure, it lies in a horizontal plane. We introduce the *depth* of the liquid, h, measured downward from the free surface. Since $z = -h$, we can write

$$p(h) = p_0 + \gamma h. \tag{2.5}$$

The pressure distribution implied by Eq. (2.5) is called a *hydrostatic pressure distribution*. The term γh is called the *hydrostatic pressure*.

Suppose that we have various containers of liquid as shown in Fig. 2.4. According to Eq. (2.5), the pressure at any point depends on the pressure at the free surface, the depth of the point, and the specific weight of the liquid, but *not* on the size or shape of the container. Thus the pressure at points A and B is the same, but the pressure at point C is different because C lies in a different fluid.

This principle is put to practical use in a hydraulic lift, illustrated in Fig. 2.5. You have probably seen such a system in a service station. Since elevation changes are usually small in such a system, the pressure at the piston is essentially equal to the pressure in the air tank. The weight of the automobile is supported by the pressure force on the piston. The cross-sectional area of the air tank and the pressure transmission line can be any size, larger or smaller than the piston area.

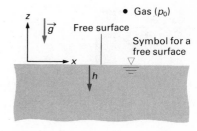

Figure 2.3 Body of liquid with a free surface.

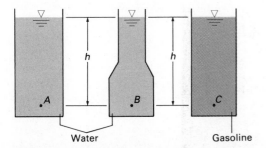

Figure 2.4 Liquids in different containers.

Figure 2.5 Hydraulic lift.

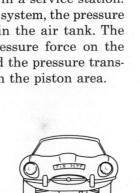

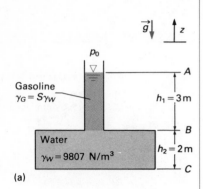

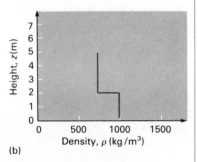

(b)

Figure E2.1 (a) Tank of water with column of gasoline; (b) variation of density with height.

EXAMPLE 2.1 Illustrates the calculation of pressure due to a liquid column

The tank of water shown in Fig. E2.1(a) has a 3-m column of gasoline ($S = 0.73$) above it. Find the pressure on the bottom of the tank. The atmospheric pressure is 101 kPa.

Solution

Given
Tank of water with 3-m column of gasoline ($S = 0.73$) above it
Atmospheric pressure 101 kPa

Find
Pressure at bottom of tank (2 m below interface)

Solution
Equation (2.5) is used to find the pressure at any point in a column of liquid if we know the pressure at another point and the difference in depth between the two points. We use the letters A, B, and C to represent the top of the gasoline, the interface between the gasoline and the water, and the bottom of the tank, respectively. Applying Eq. (2.5) separately to the gasoline and the water gives

$$p_B = p_A + \gamma_G h_1 \quad \text{and} \quad p_C = p_B + \gamma_W h_2.$$

Substituting the first equation into the second and setting $p_A = p_0$, we have

$$p_C = p_0 + \gamma_G h_1 + \gamma_W h_2.$$

The specific weight of gasoline $\gamma_G = S\gamma_W$, so

$$p_C = p_0 + \gamma_W(S h_1 + h_2).$$

Substituting numerical values, we obtain

$$p_C = 101 \text{ kPa} + 9807 \text{ N/m}^3 \, [0.73(3) + 2] \text{ m} \left(\frac{1 \text{ kPa} \cdot \text{m}^2}{1000 \text{ N}}\right),$$

$$\left| \; p_C = 142 \text{ kPa}. \; \right| \qquad \textbf{Answer}$$

Discussion
This example represents a simplified case of a variable-density fluid. Figure E2.1(b) shows how the density varies with distance above the bottom of the tank. If we use this density profile, Eq. (2.3) can be integrated to give

$$\int_{p_C}^{p_A} dp = -\int_{z_C}^{z_A} \gamma \, dz.$$

The integration is performed in two parts:

$$p_A - p_C = -\int_{z_C}^{z_B} \gamma_W \, dz - \int_{z_B}^{z_A} \gamma_G \, dz.$$

Since $p_A = p_0$,

$$p_0 - p_C = -\gamma_W(z_B - z_C) - \gamma_G(z_A - z_B).$$

Since $z_B - z_C = h_2$, $z_A - z_B = h_1$, and $\gamma_G = S\gamma_W$, we find

$$p_C = p_0 + \gamma_W(S h_1 + h_2),$$

the same result we obtained using Eq. (2.5) twice.

EXAMPLE 2.2 Illustrates that pressure, rather than force, is transmitted through a fluid and that force is the product of pressure and area

A simplified sketch of a hydraulic jack (or press) is shown in Fig. E2.2(a). A force $F_0 = 100$ N is applied at a distance $\ell_0 = 8.0$ cm from the pivot point A. Find the force F_E applied to the bottom of the larger piston due to the force F_0 at 0.

Figure E2.2 (a) Hydraulic jack (or press); (b) forces on hydraulic jack.

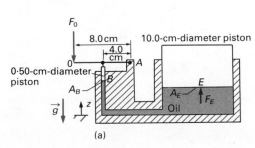

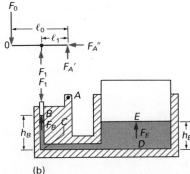

(a)   (b)

Solution

Given

Hydraulic jack with force $F_0 = 100$ N applied at 8.0 cm from pivot point A. Piston diameters 0.50 cm and 10.0 cm, with the smaller piston 4.0 cm from pivot point A
Figure E2.2(b)

Find

Force F_E applied to the bottom of the larger piston due to F_0

Solution

Since $p_E = F_E/A_E$, we must relate p_E to the applied force F_0. Referring to Fig. E2.2(b), we take moments about the pivot point A. Neglecting the bar's weight, we have

$$\sum \mathcal{M}_A = -F_0\ell_0 + F_1\ell_1 = 0.$$

Solving for F_1, we obtain

$$F_1 = \frac{F_0\ell_0}{\ell_1} = \frac{(100 \text{ N})(8.0 \text{ cm})}{4.0 \text{ cm}} = 200 \text{ N.}$$

If we neglect the weight of the smaller piston,

$$F_B = F_1 = 200 \text{ N.}$$

The pressure p_B is

$$p_B = \frac{F_B}{A_B} = \frac{200 \text{ N}}{\dfrac{\pi}{4}(0.5 \text{ cm})^2} = 1020 \text{ N/cm}^2 \ (=1.02 \times 10^4 \text{ kPa).}$$

Applying the hydrostatic pressure equation between levels B and C and between levels E and D gives

$$p_C = p_B + \gamma h_B \quad \text{and} \quad p_D = p_E + \gamma h_E.$$

Since points C and D are at the same elevation in a static fluid,

$$p_D = p_C.$$

Substituting for p_D and p_C gives

$$p_E + \gamma h_E = p_B + \gamma h_B \quad \text{or} \quad p_E = p_B + \gamma(h_B - h_E).$$

Assuming points B and E are at the same elevation, we have

$$p_E = p_B = 1020 \text{ N/cm}^2.$$

The force F_E is

$$F_E = p_E A_E = (1020 \text{ N/cm}^2)\left(\frac{\pi}{4}\right)(10.0 \text{ cm})^2,$$

$$F_E = 80{,}100 \text{ N}.$$

Since only two significant figures are justified,

$$| \; F_E = 80{,}000 \text{ N}. \; |$$

Answer
◀

Discussion

Note that we could directly argue that $p_E = p_B$ if they are at the same elevation in a static fluid.* Then

$$\frac{F_E}{A_E} = \frac{F_B}{A_B}$$

and

$$F_E = \frac{A_E}{A_B} F_B = \frac{\dfrac{\pi}{4}(10.0 \text{ cm})^2}{\dfrac{\pi}{4}(0.50 \text{ cm})^2} (200 \text{ N}),$$

$$F_E = 80{,}000 \text{ N}$$

with no significant-figure problems. (Why did this happen?)

Also note that the pressure error introduced by the assumption that points B and E are at the same elevation is $\gamma(h_B - h_E)$. For an oil with $\gamma = 8400 \text{ N/m}^3$ and $(h_B - h_E) = 4$ cm, the force F_E is changed by an amount

$$\Delta F = \gamma(h_B - h_E)A_E$$

$$= (8400 \text{ N/m}^3)(0.04 \text{ m})\left(\frac{\pi}{4}\right)\left(10.0 \times \frac{\text{m}}{100 \text{ cm}}\right)^2$$

$$= 2.6 \text{ N},$$

a negligible amount.

* Remember that pressure, not force, is transmitted through the fluid.

2.2.2 Pressure Variation in a Variable-Density Fluid and the Standard Atmosphere

If the specific weight is variable, we use Eq. (2.3) to determine the pressure distribution; however, it is necessary to relate the specific weight to the pressure and/or elevation before the equation can be integrated. A common case might involve an ideal gas for which

$$p = \rho RT. \tag{1.1}$$

For an ideal gas, Eq. (2.3) becomes

$$\frac{dp}{dz} = -\rho g = -\frac{pg}{RT}.$$

Solving for the pressure,

$$\int \frac{dp}{p} = \ln(p) = -\int \frac{g}{RT} dz + C. \tag{2.6}$$

To complete the integration on the right, we must know how the gas temperature varies with z.

The most useful application of Eq. (2.6) involves calculation of the variation of pressure with altitude in the earth's atmosphere. To carry out this calculation we must have information on the temperature-altitude relation for the atmosphere. Atmospheric temperature varies from day to day and season to season; however, calculations are usually based on the *U.S. Standard Atmosphere,* which has the temperature-altitude curve shown in Fig. 2.6(a). In the lower portion of the atmosphere, called the *troposphere,* the temperature decreases linearly with altitude. At an altitude of about 11 km (36,000 ft), the temperature becomes constant in the region called the *stratosphere.* The temperature remains constant to an altitude of about 20 km (66,000 ft) and then begins to increase linearly. In the outer region of the atmosphere, called the *ionosphere,* the temperature reaches very high values; however, the ionosphere is extremely rarified and cannot be modeled as a continuum.

The temperature variation in the troposphere is

$$T = T_0 - Bz, \tag{2.7}$$

where T_0 is the surface temperature and B is the *lapse rate.* For the U.S. Standard Atmosphere,

$$T_0 = 15°C = 288.15 \text{ K} = 59°F = 518.67°R$$

and

$$B = 0.00650 \text{ K/m} = 0.003566°R/\text{ft}.$$

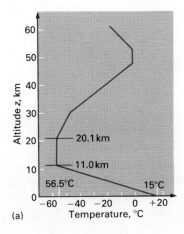

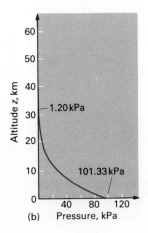

(a)

(b)

Figure 2.6 Temperature and pressure distribution in the U.S. Standard Atmosphere.

Substituting Eq. (2.7) into Eq. (2.6) and integrating we obtain

$$p = p_0 \left(1 - \frac{Bz}{T_0} \right)^{g/RB}.$$ (2.8)

The sea-level pressure is

$$p_0 = 101,330 \text{ Pa } (14.696 \text{ lb/in}^2).$$

In the stratosphere, the temperature is

$$T = T_c \ (T_c = 216.7 \text{ K } (-69.6°\text{F})),$$

and Eq. (2.6) becomes

$$\ln (p) = -\frac{g}{RT_c} \int dz + C.$$

If we let the pressure be p_c at the lower edge of the stratosphere z_c, then

$$p = p_c \exp\left(-\frac{g(z - z_c)}{RT_c} \right).$$ (2.9)

For the U.S. Standard Atmosphere,

$$z_c = 11 \text{ km } (36,100 \text{ ft}) \quad \text{and} \quad p_c = 22,600 \text{ Pa } (3.28 \text{ lb/in}^2).$$

Since the remainder of the atmosphere has either constant or linear temperature variation, the pressure for the entire atmosphere can be worked out. The results are shown in Fig. 2.6(b). The properties of the U.S. Standard Atmosphere are tabulated in Tables A.1 and A.2 in Appendix A.

2.3 Manometry and Pressure Measurement

2.3.1 Manometers

Consider the equation for the pressure in a static fluid of constant density:

$$p - p_0 = \gamma h.$$ (2.5)

A simple and effective way to measure pressure is to measure the height of a column of liquid that is supported by the pressure. A device for carrying this out is called a *manometer*. A simple manometer is sketched in Fig. 2.7. The pressure at point 1 is

$$p_1 = p_{\text{atm}} + \gamma_L h.$$

Since 1 and 2 are at the same level,

$$p_1 = p_2,$$

and since 2 is in pipe A,

$$p_2 = p_A, \quad \text{so} \quad p_A - p_{\text{atm}} = \gamma_L h.$$

Measurement of the height h together with knowledge of the specific weight of the manometer fluid provides a simple and accurate pressure measurement. In this example, the fluid in the manometer was identical to the fluid whose pressure was being measured. In many

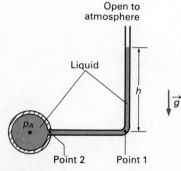

Open to
atmosphere

Liquid

h

$\vec{g}$

p_A

Point 2   Point 1

Figure 2.7 Simple manometer (piezometer).

practical cases, one end of the manometer is open to the atmosphere. This arrangement is called a *piezometer*. If the pressure in region A were very large and/or the fluid were very light (a gas, for example), a piezometer would be impractical because a very high column of fluid would be required. An obvious solution to this problem is to use a denser liquid in the manometer.

One of the most common types of manometer is the U-tube shown in Fig. 2.8. Since

$$p_1 = p_1,$$

we have

$$p_G + \gamma_L h' = p_{atm} + \gamma_L(h + h').$$

Canceling $\gamma_L h'$, we obtain

$$p_G - p_{atm} = \gamma_L h \qquad (2.10)$$

as the equation of this manometer. Note that equal columns of height h' cancel in the equation. This is true in general, so we can just "jump" across the tube to the other side, ignoring the equal columns whose pressures cancel from the equation. In this development, we have neglected the pressure due to the column of gas on the left of the tube and the column of air on the right. The relative magnitude of these pressures is illustrated in Example 2.3.

A manometer with several different fluids is shown in Fig. 2.9. We can find an expression for the pressure difference $p_A - p_B$ by dividing the manometer at the high point "1" and the low point "2."

For the part of the manometer between "1" and A, we write:

$$p_1 + \gamma_3 h_6 + \gamma_4 h_4 = p_A. \qquad (2.11a)$$

For the part between "1" and "2":

$$p_1 + \gamma_3 h_3 + \gamma_2(h_2 + h_7) = p_2. \qquad (2.11b)$$

And for the part between B and "2":

$$p_B + \gamma_1 h_1 + \gamma_2 h_7 = p_2. \qquad (2.11c)$$

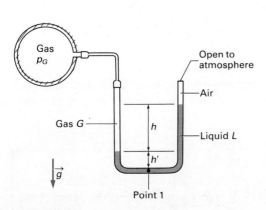

Figure 2.8 U-tube manometer.

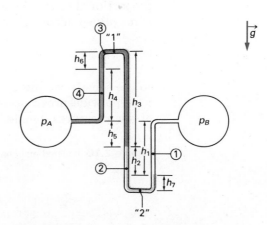

Figure 2.9 A complicated four-fluid manometer.

From Eqs. (2.11b) and (2.11c)

$$p_1 + \gamma_3 h_3 + \gamma_2 h_2 + \gamma_2 h_7 = p_B + \gamma_1 h_1 + \gamma_2 h_7,$$

so

$$p_1 = p_B + \gamma_1 h_1 - \gamma_3 h_3 - \gamma_2 h_2.$$

Substituting into Eq. (2.11a) and solving for $p_A - p_B$, we have

$$p_A - p_B = \gamma_1 h_1 - \gamma_2 h_2 - \gamma_3 (h_3 - h_6) + \gamma_4 h_4.$$

This development was not intended to produce a general result but to illustrate the principles by which the equation for any manometer can be developed. The essence of the method is the application of the hydrostatic pressure equation and the balance of pressure at a point.

We can avoid the slight complexity of solving simultaneous equations by following a three-step procedure (called the *manometer rule*):

Step 1   Begin at one end of the manometer and write the pressure there.

Step 2   Proceed through the manometer, *adding* the hydrostatic pressure if you are going *down* and *subtracting* if going *up* (note that going up and down equal distances in the same fluid cancels).

Step 3   At any point, the algebraic sum of pressures can be equated to the pressure at that point.

Applying this procedure to the manometer of Fig. 2.9:

$$p_A - \gamma_4 h_4 - \gamma_3 h_6 + \gamma_3 h_3 + \gamma_2 h_2 - \gamma_1 h_1 = p_B$$

so

$$p_A - p_B = \gamma_1 h_1 - \gamma_2 h_2 - \gamma_3 (h_3 - h_6) + \gamma_4 h_4. \tag{2.12}$$

The *sensitivity* (change in reading h per unit change in pressure) and maximum readable pressure can be changed by changing the liquid in the manometer. The most common manometer fluids are oil ($S \approx 0.8$), water ($S = 1.0$), and mercury ($S \approx 13.6$).

Sensitivity and accuracy can be increased by using an inclined manometer, as shown in Fig. 2.10. The pressure of the liquid column is proportional to the vertical distance h; however, the scale deflection ℓ is magnified according to

$$\ell = \frac{h}{\sin \theta},$$

so the sensitivity increases as θ is decreased.

The widespread use of the manometer, especially for measurement of low pressures, has led to the use of the "amount of length of fluid"

Figure 2.10 Inclined manometer.

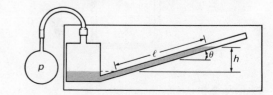

(e.g. "inch of water" or "millimeter of mercury") as a unit of pressure. These units can be converted to proper pressure units by

$$p\left[\frac{F}{L^2}\right] = \gamma\left[\frac{F}{L^3}\right]h[L].$$

EXAMPLE 2.3 Illustrates the manometer rule and compares pressures due to columns of gases and liquids

A gas pipe contains methane (CH_4) at a temperature of 70°F. A manometer is attached to the pipe as shown in Fig. E2.3. Calculate the pressure p in the pipe. The atmospheric pressure, p_{atm}, is 14.696 lb/in². The gas constant R equals 96.56 ft·lb/lbm·°R for methane.

Solution

Given

Gas pipe containing methane at 70°F and connected manometer with mercury and water as measuring fluids

Figure E2.3

Find

The pressure in the pipe

Solution

Applying the "manometer rule" gives

$$p = p_{atm} + \gamma_w(h_3 + h_2) + S_{Hg}\gamma_w(h_1 - h_2) - \gamma_M h_1.$$

Since a purpose of this example is to compare the pressures due to each column of fluid, we will compute each pressure individually. The water pressure is

$$\gamma_w(h_3 + h_2) = \frac{(62.4 \text{ lb/ft}^3)(12.0 \text{ in.})}{(1728 \text{ in}^3/\text{ft}^3)} = 0.433 \text{ lb/in}^2.$$

The mercury pressure is

$$S_{Hg}\gamma_w(h_1 - h_2) = \frac{13.6(62.4 \text{ lb/ft}^3)(12.0 \text{ in.})}{(1728 \text{ in}^3/\text{ft}^3)} = 5.89 \text{ lb/in}^2.$$

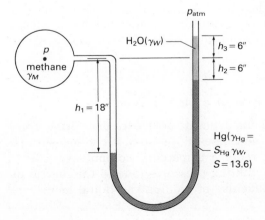

Figure E2.3 Gas pipe with manometer to measure methane pressure.

The methane contribution is $\gamma_M h_1$; however, the specific weight of the methane is not specified. Since the methane is at 70°F and its *approximate* pressure is 21.02 lb/in² (14.696 + 0.433 + 5.89), it is reasonable to assume that the methane is an ideal gas. Using the ideal gas law to compute ρ_M gives

$$\gamma_M h_1 = \rho_M g h_1 = \left(\frac{p}{RT}\right)g h_1.$$

The original equation then becomes

$$p = 21.02 \text{ lb/in}^2 - \frac{pgh_1}{RT}.$$

Solving for p, we have

$$p = \frac{21.02 \text{ lb/in}^2}{1 + \dfrac{gh_1}{RT}}.$$

Then

$$\frac{gh_1}{RT} = \frac{\left(32.2 \dfrac{\text{ft}}{\text{sec}^2}\right)(18 \text{ in.})\left(\dfrac{1 \text{ ft}}{12 \text{ in.}}\right)}{\left(96.56 \dfrac{\text{ft}\cdot\text{lb}}{\text{lbm}\cdot°\text{R}}\right)(459.67 + 70)°\text{R}\left(32.2 \dfrac{\text{ft}\cdot\text{lbm}}{\text{lb}\cdot\text{sec}^2}\right)}$$

$$= 2.93 \times 10^{-5},$$

a pure (dimensionless) number that is quite small compared to unity. Therefore

Answer

$$|\; p = 21.02 \text{ lb/in}^2. \;|$$

We can now calculate the pressure due to the column of methane:

$$\gamma_M h_1 = p\left(\frac{gh_1}{RT}\right)$$

$$= (21.02 \text{ lb/in}^2)(2.93 \times 10^{-5}) = 0.000616 \text{ lb/in}^2.$$

Discussion

The relative contributions of the various fluids are as follows:

Mercury 5.89 lb/in²
Water 0.433 lb/in²
Methane 0.000616 lb/in²

These numbers suggest that the pressure contribution of a gas column can usually be neglected compared to the pressure contribution of a liquid column unless the gas column is very high or a high degree of accuracy is desired.

2.3.2 Other Devices for Pressure Measurement

The manometer is a simple and accurate device for measuring pressures; however, its limits are the range of pressures it can measure (a pressure of 1000 lb/in² would require a mercury column over 165 ft high!), its slow response to fluctuating pressures (due to the inertia of the liquid column), and the difficulty of reading the liquid meniscus.

Alternative devices for pressure measurement generally are not based on the principles of fluid statics but are discussed here for the sake of completeness.

Most pressure-measuring devices operate on the principle that forces due to pressure produce deflection of an elastic element. Probably the most familiar and widely used pressure measurement device is the Bourdon tube gage, illustrated in Fig. 2.11. When a pressure is applied to the open end of the curved, flattened tube, it tends to straighten out, much like the paper noisemaker popular at New Year's Eve parties. The motion of the tube is converted to motion of a needle over the face of a dial by a linkage. The dial can be calibrated to read the pressure in any units desired. By varying the size and rigidity of the tube, one can construct a gage for pressures of almost any magnitude. By carefully controlling the quality of the tube material and its geometry, as well as by careful calibration, one can obtain a reasonably accurate gage (to, say, ± 0.1 percent of the gage's full scale reading).

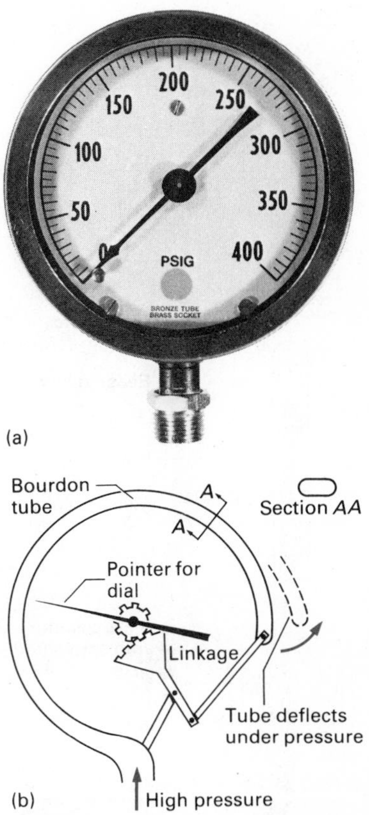

Electric or electronic pressure transducers* can be applied to an extremely wide range of situations requiring accurate pressure measurement with rapid response time, remote reading, and automatic recording. The most widely used type of pressure transducer is the strain-gage type sketched in Fig. 2.12. The application of pressure causes a deflection of the elastic diaphragm. This deflection is measured by the change of electrical resistance of a strain gage, a set of very fine wires whose electrical resistance changes when stretched. By calibration, the voltage output of a resistance-measuring bridge can be related to the applied pressure. The response of this type of gage to pressure changes is very rapid. The electrical output of the gage readily lends itself to automatic recording.

Similar characteristics are available with a piezoelectric pressure transducer, which uses a quartz crystal that generates a voltage when a pressure is applied. Piezoelectric transducers have extremely rapid response to varying pressure because no fluid volume need be present in the transducer.

This discussion is only a brief introduction to pressure measurement. Consult references [1] and [2] for more information.

2.3.3 Absolute, Gage, and Vacuum Pressures

All of the pressure-measuring devices we have considered measure a *difference* between two pressures. The manometer measures the difference between pressures at its two ends. The Bourdon tube responds to the pressure difference between the inside and outside of the tube. The diaphragm in the strain-gage transducer deflects only if there is a pressure difference across it.

In most applications, the pressure on one side of the gage is ambient atmospheric pressure. Pressure measured relative to local (that is, in

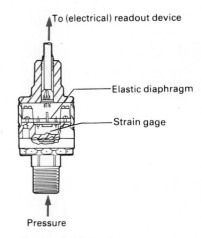

Figure 2.11 Bourdon tube pressure gage: (a) typical gage; (b) internal details.

Figure 2.12 Strain-gage pressure transducer.

---

* *Transducer* is the name for a device that has one type of signal (say, pressure) as its input and another type of signal (usually electric) as its output.

the vicinity of the gage) atmospheric pressure is called *gage pressure*. Pressure measured (or defined) relative to zero pressure is called *absolute pressure*. Gage and absolute pressures are related as follows:

(Absolute pressure) = (Gage pressure)

+ (Atmospheric pressure in vicinity of gage).

Pressures *below* local atmospheric pressure are sometimes called *vacuum pressures*. A vacuum pressure is expressed as a positive number, so

(Vacuum pressure) = (Atmospheric pressure) − (Absolute pressure)
= −(Gage pressure)

The subscripts "g" and "a" are sometimes used *with the units* of a pressure to indicate whether the pressure is gage or absolute. Thus 10 psig means "ten pounds per square inch, gage" and 10 psia means "ten pounds per square inch, absolute." The subscripts "abs" and "gage" are also used for the same purpose.

To convert from gage to absolute pressure, you must know the atmospheric pressure at the time and location of measurement. Often this information is not available. In such cases, we use a "standard" value of atmospheric pressure of 101,330 Pa (14.696 lb/in^2, abs; 29.92 in. Hg, abs). The actual atmospheric pressure can be measured with a barometer, one type of which is like a manometer with a vacuum on one side (Fig. 2.13).

Absolute pressures must always be used in the ideal gas law (Eq. 1.1). In most other situations that arise in fluid mechanics, we are more concerned with differences of pressure than with levels of pressure. Pressure differences are the same whether the pressures are considered as absolute or gage, *so long as all pressures are based on a common datum*. A pressure difference (say, between two locations in a pipeline) *is not* expressed in either "gage" or "absolute" units; these designations apply only to pressure levels.

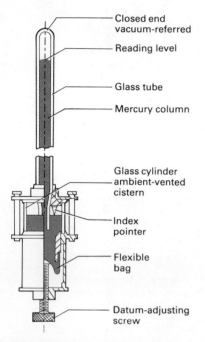

Closed end
vacuum-referred

Reading level

Glass tube

Mercury column

Glass cylinder
ambient-vented
cistern

Index
pointer

Flexible
bag

Datum-adjusting
screw

Figure 2.13 Typical (Fortin-type) barometer.

EXAMPLE 2.4 Illustrates the conversion of units of pressure

A tornado has a core pressure of 0.800×10^5 Pa. Express this pressure in units of lb/in^2, abs (psia); lb/in^2, gage (psig); in lb/in^2, vacuum; in H$_2$O, abs; and in mm Hg, abs. The barometer reads 675 mm Hg. Also express the core pressure in units relative to standard atmospheric pressure.

Solution

Given
Absolute core pressure, 0.800×10^5 Pa
Atmospheric pressure, 675 mm Hg

Find
Core pressure in units of lb/in^2, abs; lb/in^2, gage; lb/in^2, vacuum; H$_2$O, abs; and mm Hg, abs

Solution

Appendix C is used to convert from Pa to lb/in², abs, as follows:

$$p = (0.800 \times 10^5 \text{ Pa})\left(\frac{1 \text{ lb/in}^2}{6895 \text{ Pa}}\right),$$

$$\left| p = 11.6 \text{ lb/in}^2, \text{ abs (psia).} \right|$$ **Answer** ◀

In a barometer the atmospheric pressure supports the column of the barometer fluid. For our mercury-filled barometer,

$$p_{\text{atm}} = \gamma_{\text{Hg}}h = S_{\text{Hg}}\gamma_w h.$$

Using Tables A.5 and A.6 (the fluids are assumed to be at 4°C), we obtain

$$p_{\text{atm}} = \frac{\left(13.58 \times 62.43 \dfrac{\text{lb}}{\text{ft}^3}\right)(675 \text{ mm Hg})}{\left(\dfrac{1728 \text{ in}^3}{\text{ft}^3}\right)\left(\dfrac{25.4 \text{ mm}}{\text{in.}}\right)} = 13.0 \frac{\text{lb}}{\text{in}^2} \text{ abs.}$$

Then the core gage pressure is

$$p = (11.6 - 13.0) \text{ lb/in}^2,$$

$$\left| p = -1.4 \text{ lb/in}^2, \text{ gage (psig).} \right|$$ **Answer** ◀

The core vacuum pressure is

$$p = (13.0 - 11.6) \text{ lb/in}^2, \text{ vacuum,}$$

$$\left| p = 1.4 \text{ lb/in}^2, \text{ vacuum.} \right|$$ **Answer** ◀

The core pressure is converted to a column of liquid using Eq. (2.3)

$$h = \frac{p}{\gamma} = \frac{\left(11.6 \dfrac{\text{lb}}{\text{in}^2}\right)}{\left(62.4 \dfrac{\text{lb}}{\text{ft}^3}\right)\left(\dfrac{\text{ft}^3}{1728 \text{ in}^3}\right)},$$

$$\left| h = 321 \text{ in H}_2\text{O, abs.} \right|$$ **Answer** ◀

To obtain units of mm Hg, we use Eq. (2.3) to give

$$h = \frac{p}{S_{\text{Hg}}\gamma_W} = \frac{\left(11.6 \dfrac{\text{lb}}{\text{in}^2}\right)\left(1728 \dfrac{\text{in}^3}{\text{ft}^3}\right)\left(25.4 \dfrac{\text{mn}}{\text{in.}}\right)}{\left(13.59 \times 62.4 \dfrac{\text{lb}}{\text{ft}^3}\right)},$$

$$\left| h = 600 \text{ mm Hg, abs.} \right|$$ **Answer** ◀

The only pressures that can be referred to the standard atmospheric pressure are the gage and vacuum pressures. They are

$$p = (11.6 - 14.696) \text{ lb/in}^2,$$

$$\left| p = -3.1 \text{ lb/in}^2, \text{ gage (psig),} \right|$$ **Answer** ◀

and

$$p = (14.696 - 11.6) \text{ lb/in}^2,$$

$$\left| p = 3.1 \text{ lb/in}^2, \text{ vacuum.} \right|$$ **Answer** ◀

Discussion

We have chosen to measure the absolute pressures in the heights of columns of water and mercury at 4°C (39.2°F) and 101,330 N/m² (14.696 lb/in², abs). This specifies the density, but we must also specify the gravitational acceleration to obtain the specific weight. Our chosen value of g is the "sea-level standard value," 9.807 m/s² (32.174 ft/sec²). Choosing any other temperature, pressure, and gravitational acceleration would yield different values, so these conditions must be specified if very accurate results are necessary and the pressures are expressed as the height of a fluid column.

2.4 Pressure Forces on Surfaces

When a surface with a finite area is in contact with a fluid, a force is exerted on the surface because of the fluid pressure. This force is distributed over the surface; however, it is usually helpful in engineering calculations to replace the distributed force by a single resultant force. To completely specify the resultant force, we must determine its magnitude, line of action, and point of application. Since

$$p \equiv \frac{dF_n}{dA},$$

the resultant force can be found by integration

$$F_n = \int p \, dA.$$

In general, there are two difficulties to overcome in performing this calculation:

- The pressure and area must be expressed in terms of a common variable so that the integration can be performed.
- If the surface is curved, the normal direction varies from point to point on the surface, and F_n is meaningless because no single "normal" direction characterizes the entire surface. In this case, it is necessary to resolve the force at each point on the surface into components and then integrate each component.

In the following two sections, we consider calculation of forces due to hydrostatic pressure distributions. We first consider plane surfaces and then curved surfaces.

2.4.1 Forces on Plane Surfaces

Consider a horizontal plane surface submerged in a liquid, such as the bottom of the tank shown in Fig. 2.14. Since all of the surface is at uniform depth, it is at uniform pressure, and

$$F = \int p \, dA = pA = \gamma_L hA + p_{atm}A. \tag{2.13}$$

Note that in this case the force due to the liquid is equal to the weight of the liquid above the surface; however, the force would be the

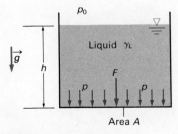

Figure 2.14 Tank of liquid.

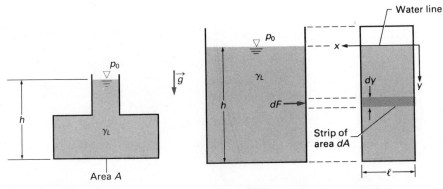

Figure 2.15 Another tank of liquid.

Figure 2.16 Tank of liquid showing elemental force on side.

same for the tank shown in Fig. 2.15, although the amount of liquid contained in the tank is less.*

Consider now the force due to pressure on a vertical surface, such as the side of the tank shown in Fig. 2.16. The force on a horizontal element of the wall is

$$dF = p\,dA = p\ell\,dy.$$

The pressure in the liquid varies according to Eq. (2.5), so, with y measured downward from the free surface,

$$dF = (p_{atm} + \gamma_L y)\ell\,dy.$$

Usually the outside of the tank experiences atmospheric pressure (p_{atm}), so

$$dF_{net} = (p_{atm} + \gamma_L y)\ell\,dy - p_{atm}\ell\,dy = \gamma_L \ell y\,dy.$$

Integrating, we have

$$F_{net} = \int_0^h \gamma\ell y\,dy = \frac{\gamma\ell h^2}{2} = \gamma\frac{h}{2}\,\ell h,$$

so

$$F_{net} = \gamma\frac{h}{2}\,A. \tag{2.14}$$

The net force on the rectangular area is the product of the average pressure (evaluated at the average depth) and the area. This result would also apply if the pressure in the tank were uniform, such as in a gas. In this case, the average pressure would just be the value of the pressure itself, and the force would be the product of the pressure and the area.

Finally, consider the general case of an inclined plane surface of arbitrary shape, shown in Fig. 2.17. The force on the strip of area dA is

$$dF = p\,dA.$$

* Can you explain this?

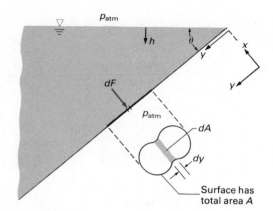

Figure 2.17 Arbitrary plane surface in contact with a liquid.

The pressure on the strip is the sum of two terms representing the hydrostatic pressure of the liquid and the pressure at the free surface. In many situations, the pressure at the free surface is the pressure of the atmosphere, and the surface on which we wish to calculate the force is exposed to the atmosphere on one side. The net force on such a surface is due to the hydrostatic pressure only. Even if this is not true, it is permissible to evaluate the force on any surface in two parts: the force due to γh and the force due to p_{atm}. The force on a submerged surface due to p_{atm} is

$$F_{atm} = p_{atm} A.$$

In the following analysis, we will consider only the force due to the hydrostatic pressure,

$$p = \gamma h,$$

where h is the depth from the free surface to the strip. We can relate h to y, the distance from the free surface to the strip *in the plane of the area,* by

$$h = y \sin \theta, \tag{2.15}$$

so the force due to hydrostatic pressure on the strip is

$$dF = \gamma y \sin \theta \, dA.$$

Integrating and noting that γ and θ are constant,* we obtain

$$F = \gamma \sin \theta \int y \, dA. \tag{2.16}$$

We can replace the integral on the right by noting that the *centroid* of an area is defined by

$$y_c \equiv \frac{\int y \, dA}{A}, \tag{2.17}$$

where y_c is the distance from the free surface to the centroid of area A, measured in the plane of area A. Substituting into Eq. (2.16), we have

$$F = \gamma y_c \sin \theta \, A.$$

* Note that θ is constant only if the surface is plane.

Using Eq. (2.15), we obtain

$$h_c = y_c \sin \theta$$

and

$$F = \gamma h_c A. \tag{2.18}$$

The net force on the surface is the product of the pressure at the centroid of the surface and the surface area. This result is independent of the shape of the surface or its angle of inclination. Equation (2.18) contains Eqs. (2.13) and (2.14) as special cases. The line of action of the resultant force is perpendicular to the surface and acts from the liquid toward the surface.

For plane surfaces, the major difficulty in calculating the net force is geometric, namely, locating the centroid. Appendix B gives information about the geometric characteristics of several common plane shapes.

EXAMPLE 2.5 Illustrates the calculation of hydrostatic force on a plane surface

A bathyscaph (Fig. E2.5) is a diving craft used for deep-sea observation; it consists of a thick-walled steel sphere attached to a large hull. The sphere protects the crew from the high pressure at more than 10 km below the surface of the sea. It has a flat, semicircular window with a 2.0-m diameter. The top of the window is 10 km below the surface of the sea. Find the magnitude of the hydrostatic force on the outside of the observation window. The specific gravity of seawater is 1.028.

Solution

Given

Inclined semicircular window with 2.0-m diameter making an angle of 45° with the horizontal

Top of window 10 km below sea level

Figure E2.5 (a) Bathyscaph with observation window; (b) pressure distributions on observation window along a radius in a vertical plane.

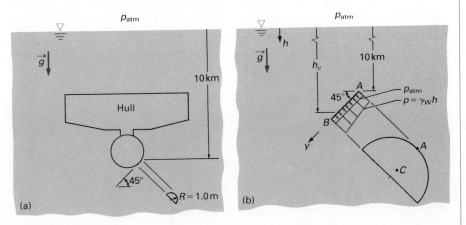

Find

Magnitude of the hydrostatic force on the outside of the window

Solution

Appendix B gives the location of the centroid of a semicircular area. Using Fig. E2.5(b), we have

$$h_c - h_A = \left(1 - \frac{4}{3\pi}\right) R \sin 45°.$$

Substituting numerical values gives

$$h_c - 10 \text{ km} = \left(1 - \frac{4}{3\pi}\right)(1 \text{ m}) \sin 45° = 0.41 \text{ m},$$

or

$$h_c = 10 \text{ km} + 0.41 \text{ m}$$

$$\approx 10 \text{ km}. \quad \text{(Note the units of km.)}$$

Assuming seawater has a constant specific weight, we have

$$F = \gamma h_c A = S \rho_w g h_c A = S \gamma_w h_c A$$

$$= (1.028)(9807 \text{ N/m}^3)(10^4 \text{ m})(\pi/2)(1.0 \text{ m})^2,$$

$$\left| \; F = 1.58 \times 10^8 \text{ N}. \; \right|$$

Answer

Discussion

The hydrostatic pressure distribution along a radius in a vertical plane is sketched in Fig. E2.5(b). The pressure increases linearly with depth; however, the slope of the hydrostatic pressure line is not noticeable because the hydrostatic pressure difference between points A and B is

$$\Delta p = p_B - p_A = S \gamma_w (h_B - h_A)$$

$$= 1.028(9807 \text{ N/m}^3)(1.0 \text{ m}) \sin 45° = 7.130 \text{ kPa},$$

compared to a pressure of

$$p_A = S \gamma_w h_A$$

$$= 1.028(9807 \text{ N/m}^3)(10^4 \text{ m}) = 1.01 \times 10^5 \text{ kPa}.$$

To complete the analysis of forces on plane surfaces, we must determine the point of application of the resultant force. This point is called the *center of pressure*. The principle that allows us to locate the center of pressure is the equality of the moment of the resultant force with the moment of the distributed (pressure) force. Since the resultant force can be calculated as the product of the area and the pressure at its centroid, we might be tempted to assume that the centroid is also the center of pressure; however, this is *not usually* the case, as the following discussion shows.

Consider once again a vertical rectangular surface, such as the one shown in Fig. 2.18. Suppose for now that the pressure is *uniform* over the entire surface. The resultant force would be simply

$$F = p_0 A,$$

where p_0 is the value of the uniform pressure.

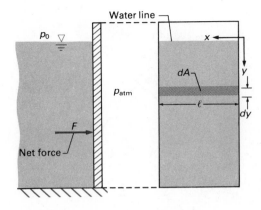

Figure 2.18 Vertical surface in contact with a liquid.

We determine the point of application of the resultant force by equating the moment of the resultant force to the moment of the distributed force. Taking moments about the x-axis, we have

$$Fy_p = \int_0^d yp\,dA = \int_0^d yp\ell\,dy,$$

where y_p is the y-location of the center of pressure and d is the maximum value of y (i.e., the depth of the bottom of the tank). Continuing to assume uniform pressure, we have

$$Fy_p = p_0\ell \int_0^d y\,dy = p_0\ell\,\frac{d^2}{2} = p_0(\ell d)\,\frac{d}{2}.$$

Since

$$F = p_0 A = p_0\ell d,$$

$$y_p = \frac{d}{2} \quad \begin{pmatrix}\text{location of point of application of} \\ \text{resultant force for } \textit{uniform} \text{ pressure}\end{pmatrix}. \qquad \textbf{(2.19)}$$

Of course $d/2$ is the location of the centroid of the rectangular surface.

Although this development considered a vertical rectangular surface, we can state the following general principle:

> If a surface is subjected to a *uniform* pressure, the center of pressure is at the centroid.*

Next consider a *hydrostatic* pressure variation. The pressure is not uniform but increases with depth. The force on an element of area near the bottom of the surface has a greater moment than a force on an element near the top, not only because the moment arm is greater but also because the force itself is greater. It follows that the center of pressure cannot be at the centroid of the area for a hydrostatic pressure distribution. Because the contribution to the moment is greater from locations in the deeper part of the liquid, the point of application of the resultant force must be shifted into the deeper portion. Thus we can state the general principle:

> The center of pressure of an area lies *below* the centroid of the area for hydrostatic pressure distribution.

---

* This is exactly the case for the free-surface pressure contribution to the force on a surface.

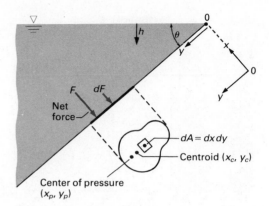

Figure 2.19 Arbitrary plane surface in contact with liquid; net force f acts at center of pressure.

To determine the center of pressure, consider an inclined plane surface in a liquid, as shown in Fig. 2.19. Taking moments about the x- and y-axes through point 0, we obtain

$$Fy_p = \int yp \, dA \tag{2.20}$$

and

$$Fx_p = \int xp \, dA. \tag{2.21}$$

First consider the location of y_p. The net force F is

$$F = \int p \, dA. \tag{2.22}$$

Substituting Eq. (2.22) into Eq. (2.20) gives

$$y_p = \frac{\int yp \, dA}{\int p \, dA}. \tag{2.23}$$

The pressure varies with y according to

$$p = \gamma y \sin \theta, \tag{2.24}$$

so

$$y_p = \frac{\int \gamma \sin \theta \, y^2 \, dA}{\int \gamma \sin \theta \, y \, dA}$$

Since γ and $\sin \theta$ are constants, they can be canceled. Using Eq. (2.17), we find

$$y_p = \frac{\int y^2 \, dA}{y_c A} = \frac{I_x}{y_c A}. \tag{2.25}$$

I_x is the moment of inertia of the area about the x-axis:

$$I_x \equiv \int y^2 \, dA.$$

Note that the x-axis lies on the free surface. It is usually more convenient to express y_p in terms of the moment of inertia of the area about its centroidal axis. The parallel axis theorem for moments of inertia is

$$I_x = I_{xc} + y_c^2 A,$$

where I_{xc} is the moment of inertia of the area about an axis parallel to the x-axis and passing through the centroid of the area. Substituting into Eq. (2.25) gives

$$y_p = y_c + \frac{I_{xc}}{y_c A}. \qquad (2.26)$$

The center of pressure lies *below* the centroid a distance equal to $I_{xc}/y_c A$. Since I_{xc} and A are constant for a given area, the center of pressure becomes closer to the centroid as the area is moved deeper. Remember that y_p is measured from the free surface in the plane of the area.

We can calculate x_p by similar steps. From Eqs. (2.21), (2.22), (2.24), and (2.17), we have

$$x_p = \frac{\int x(\gamma y \sin\theta)\, dA}{\gamma y_c \sin\theta A}.$$

Since γ and θ are constant,

$$x_p = \frac{\int xy\, dA}{y_c A} = \frac{I_{xy}}{y_c A}, \qquad (2.27)$$

where I_{xy} is the product of inertia. Using the parallel axis theorem for product of inertia results in

$$x_p = x_c + \frac{I_{xyc}}{y_c A}.$$

An easy way to compute x_p is to set up the x,y-coordinate system such that the y-axis passes through the centroid of the area so that $x = 0$ at the centroid. Then the product of inertia about the centroid can be used in Eq. (2.27). If the area is symmetric about either axis, the centroidal product of inertia is zero, and x_p coincides with the centroid. Moments of inertia of several plane shapes are tabulated in Appendix B.

EXAMPLE 2.6 Illustrates how the center of pressure changes with depth and how to calculate the moment of a force

Determine the locations of the centers of pressure and the moments of the hydrostatic forces about the base of the semicircular observation window in Example 2.5 if the top of the window is 10 m, 100 m, and 10 km below the surface of the sea.

Solution

Given

Inclined semicircular window with 2.0-m diameter making an angle of 45° with the horizontal

Top of window 10 m, 100 m, and 10 km below sea level

Figure E2.6

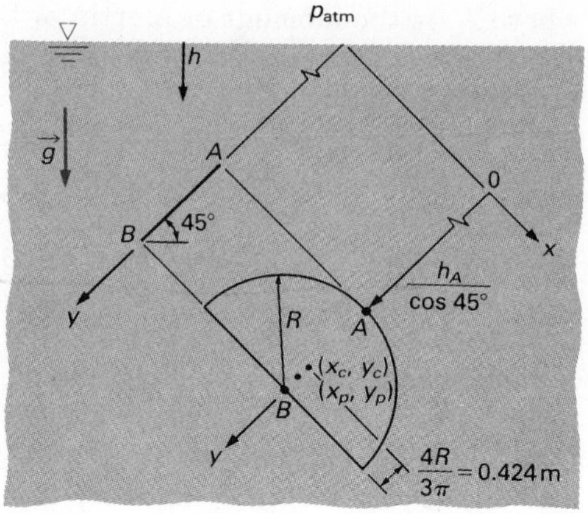

Figure E2.6 Centroid and center of pressure for a semicircular window.

Find
Centers of pressure and moments of the hydrostatic forces about the base of the semicircular window

Solution
We will first locate the centers of pressure (x_p, y_p). The y-coordinate of the center of pressure will be located relative to the centroid of the window, so we want to find $y_p - y_c$ in Fig. E2.6. Using Fig. E2.6 gives

$$y_c = \frac{h_A}{\cos 45°} + \left(R - \frac{4R}{3\pi}\right)$$

$$= 1.414h_A + 1.0 \text{ m} \left(1 - \frac{4}{3\pi}\right) = 1.414h_A + 0.576 \text{ m},$$

where h_A is in meters. For $h_A = 10$ m, using Eq. (2.26) and Table B.1 gives

$$y_c = 1.414(10 \text{ m}) + 0.576 \text{ m} = 14.7 \text{ m},$$

$$y_p - y_c = \frac{I_{xc}}{y_c A} = \frac{0.110R^4}{y_c \left(\dfrac{\pi R^2}{2}\right)}$$

$$= \frac{0.110(1.0 \text{ m})^4}{(14.7 \text{ m})\left(\dfrac{\pi}{2}\right)(1.0 \text{ m})^2} = 0.0048 \text{ m},$$

$$\left| \; y_p - y_c = 0.480 \text{ cm.} \; \right|$$ **Answer** ◄

For $h_A = 100$ m,

$$y_c = 1.414(100 \text{ m}) + 0.576 \text{ m} = 142.2 \text{ m}.$$

Then

$$y_p - y_c = \frac{I_{xc}}{y_c A} = \frac{0.110(1.0 \text{ m})^4}{(142.2 \text{ m})\left(\dfrac{\pi}{2}\right)(1.0 \text{ m})^2} = 0.00049 \text{ m},$$

$$\left| \; y_p - y_c = 0.0490 \text{ cm.} \; \right|$$ **Answer** ◄

For $h_A = 10$ km,

$$y_c = 1.414(10{,}000 \text{ m}) + 0.576 \text{ m} = 14{,}140 \text{ m},$$

$$y_p - y_c = \frac{I_{xc}}{y_c A} = \frac{0.110(1.0 \text{ m})^4}{(14{,}140 \text{ m})\left(\dfrac{\pi}{2}\right)(1.0 \text{ m})^2} = 0.0000050 \text{ m},$$

$$\left| \; y_p - y_c = 0.00050 \text{ cm}. \; \right|$$
Answer ◄

The symmetry of the area and of the hydrostatic pressure about a radius in the vertical plane places the x-coordinate of the center of pressure on the radius in the vertical plane (radius BA), so

$$\left| \; x_p - x_c = \frac{I_{xyc}}{y_c A} = 0. \; \right|$$
Answer ◄

The moment $\mathcal{M}$ of each resultant force about the base of the semicircular area is

$$\mathcal{M} = F\left[\frac{4R}{3\pi} - (y_p - y_c) \right]$$

$$= F\left[\frac{4(1.0 \text{ m})}{3\pi} - (y_p - y_c) \right]$$

$$= F[0.239 \text{ m} - (y_p - y_c)].$$

The force F is found from Eq. (2.18):

$$F = \gamma h A = \gamma y_c \sin 45° A$$
$$= (10{,}280 \text{ N/m}^3)(y_c)\, 0.707\pi\, (1.0 \text{ m})^2 = (22{,}800 \text{ N/m})\, y_c,$$

where y_c is in meters. For $h_A = 10$ m,

$$F = (22{,}800 \text{ N/m})(14.7 \text{ m}) = 3.40 \times 10^5 \text{ N}$$

and

$$\mathcal{M} = (3.4 \times 10^5 \text{ N})(0.239 \text{ m} - 0.0048 \text{ m}),$$

$$\left| \; \mathcal{M} = 79{,}600 \text{ N·m}. \; \right|$$
Answer ◄

For $h_A = 100$ m,

$$F = (22{,}800 \text{ N/m})(142.2 \text{ m}) = 3.24 \times 10^6 \text{ N}$$

and

$$\mathcal{M} = (3.24 \times 10^6 \text{N})(0.239 \text{ m} - 0.000490 \text{ m}),$$

$$\left| \; \mathcal{M} = 7.73 \times 10^5 \text{ N·m}. \; \right|$$
Answer ◄

For $h_A = 10$ km,

$$F = (22{,}800 \text{ N/m})(14{,}140 \text{ m}) = 3.22 \times 10^8 \text{ N}$$

and

$$\mathcal{M} = (3.22 \times 10^8 \text{ N})(0.239 \text{ m} - 0.0000050 \text{ m}),$$

$$\left| \; \mathcal{M} = 7.70 \times 10^7 \text{ N·m}. \; \right|$$
Answer ◄

Discussion

This example illustrates that the center of pressure moves closer to the centroid as the surface is submerged; however, it is always below the centroid. This example also illustrates a case where the area is symmetric with respect to the y-axis and x_p coincides with x_c. Finally, it illustrates how the resultant force is used in the calculation of moments.

You should note that although the formulas developed in this section (particularly Eqs. 2.18, 2.26, and 2.27) can be used only to calculate forces due to hydrostatic pressure on plane surfaces, the general principles of force and moment equality are applicable to any type of pressure distribution, as the following example demonstrates.

EXAMPLE 2.7   Illustrates how resultant forces are related to pressure distribution for other than hydrostatic cases

One of the twin towers of the World Trade Center in New York City is 110 stories high. It measures 241 ft square and 1350 ft high. The pressure distribution on this tower due to a 15 mph wind is approximated by the equation

$$p = Az^4 + Bz,$$

where

$A = -2.26 \times 10^{-11} \text{ lb/ft}^6,$
$B = 0.056 \text{ lb/ft}^3,$
z is measured in feet above the ground.

Find the magnitude and location of the resultant wind force on the windward side of the building.

Solution

Given
World Trade Center tower (Fig. E2.7a)
Pressure distribution $p = Az^4 + Bz$

Find
Magnitude and location of the resultant wind force on the building

Solution
Figure E2.7(b) shows the resultant wind force on the building. Assuming the pressure is uniform across the width, W, of the building, Eq. (2.13) gives the force as

$$F = \int p\,dA = \int_0^H pW\,dz.$$

Substituting for p gives

$$F = W \int_0^H (Az^4 + Bz)\,dz.$$

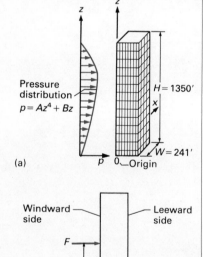

Figure E2.7 (a) Pressure distribution on the windward side of a World Trade Center tower; (b) wind force involved on a World Trade Center tower.

Integrating, we obtain

$$F = W\left(\frac{Az^5}{5} + \frac{Bz^2}{2}\right)_0^H$$

$$= WH^2\left(\frac{AH^3}{5} + \frac{B}{2}\right).$$

The numerical values give

$$F = (241 \text{ ft})(1350 \text{ ft})^2 \left[\frac{\left(-2.26 \times 10^{-11}\,\frac{\text{lb}}{\text{ft}^6}\right)(1350 \text{ ft})^3}{5} + \frac{0.0556\,\frac{\text{lb}}{\text{ft}^3}}{2}\right],$$

$$\boxed{F = 7.33 \times 10^6 \text{ lb.}}$$ **Answer** ◀

The vertical location of the resultant force is found by taking moments about the base of the building. Using Eq. (2.20), we have

$$Fz_p = \int zp\, dA = \int_0^H zp W\, dz.$$

Solving for z_p, we obtain

$$z_p = \frac{W \int_0^H z(Az^4 + Bz)\, dz}{F}.$$

Integrating gives

$$z_p = \frac{W}{F}\left(\frac{Az^6}{6} + \frac{Bz^3}{3}\right)_0^H = \frac{WH^3}{F}\left(\frac{AH^3}{6} + \frac{B}{3}\right).$$

The numerical values gives

$$z_p = \frac{(241 \text{ ft})(1350 \text{ ft})^3}{(7.33 \times 10^6 \text{ lb})}\left[\frac{\left(-2.26 \times 10^{-11}\,\frac{\text{lb}}{\text{ft}^6}\right)(1350 \text{ ft})^3}{6} + \frac{0.0556\,\frac{\text{lb}}{\text{ft}^3}}{3}\right],$$

$$\boxed{z_p = 750 \text{ ft.}}$$ **Answer** ◀

Since the pressure is assumed uniform across the width of the building, the x-location of the resultant force is one-half the distance from either side:

Answer

$$\boxed{x_p = 120.5 \text{ ft.}}$$ ◀

Discussion

In designing such a building, civil engineers must determine the necessary anchor force required to withstand the wind loads. Moments are taken about a lower edge of the building and equated to the moments of the anchoring force and the building's weight about the same line to find the necessary anchoring forces.

In addition to the pressure on the windward side, a vacuum exists on the leeward side, resulting in a larger total wind force on the tower.

Suppose we tried to determine the resultant force by calculating the pressure at the *centroid* of the windward side and multiplying this pressure by the area of the windward side. This gives

$$F = p_{y=675'} A$$

$$= \left[\left(-2.26 \times 10^{-11} \frac{\text{lb}}{\text{ft}^6} \right) (675 \text{ ft})^4 + \left(0.0556 \frac{\text{lb}}{\text{ft}^3} \right) (675 \text{ ft}) \right] (241 \text{ ft})(1350 \text{ ft}),$$

$$F = 10.7 \times 10^6 \text{ lb},$$

a value with about 50% error. The reason for the incorrect answer is that the average presssure does not occur at the centroid of the area. The average pressure occurs at the centroid only when the pressure is uniform or increases linearly with distance.

EXAMPLE 2.8   Illustrates how hydrostatic and atmospheric forces are handled in the same problem

In the Apollo spacecraft, a 5.0 lb/in², abs, pure oxygen atmosphere was used. It has been proposed to use a similar atmosphere in the bathyscaph of Example 2.5. As part of the analysis to determine the structural integrity of the bathyscaph, it is necessary to determine both the magnitude and the location of the net force on its semicircular window. Consider the case where the top of the window is 100 m below the surface of the water. The atmospheric pressure above the water surface is 100 kPa.

Solution

Given

Inclined semicircular window making an angle of 45° with the horizontal
Top of window 100 m below the surface of the water
Atmospheric pressure above the free surface 100 kPa
Pressure inside the bathyscaph 5.0 psia

Find

Magnitude and location of the net force acting on the window

Solution

The three forces acting on the window are shown in Fig. E2.8(a): the force F_0 due to the atmospheric pressure p_{atm} above the free surface acts on the outside at the centroid of the window, the force F'_0 due to the atmospheric pressure p_0 inside the bathyscaph acts on the inside at the centroid of the window, and the force F due to the hydrostatic pressure acts on the outside and at the center of pressure. The net uniform pressure force is

$$F_0 - F'_0 = p_{\text{atm}} A - p_0 A = (p_{\text{atm}} - p_0) A,$$

where

$$p_0 = (5.0 \text{ lb/in}^2) \left(6.895 \frac{\text{kPa}}{\text{lb/in}^2} \right) = 34.5 \text{ kPa}.$$

Then

$$F_0 - F'_0 = [(100 - 34.5)] \text{ kPa} \left(\frac{\pi}{2} \right) (1.0 \text{ m})^2 \left(1000 \frac{\text{N/m}^2}{\text{kPa}} \right)$$

$$= 1.03 \times 10^5 \text{ N}.$$

The force $(F_0 - F'_0)$ acts at the centroid of the window.

Using Fig. E2.8b and equating moments about a horizontal axis passing through the centroid to the moment of the resultant force F_R, we obtain

$$F_R (y_R - y_c) = F(y_p - y_c).$$

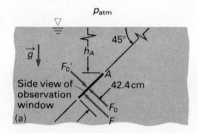

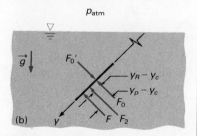

Figure E2.8 (a) Forces on the bathyscaph observation window; (b) forces and locations on the bathyscaph observation window.

Using

$$F_R = F + (F_0 - F'_0)$$

and solving for $(y_R - y_c)$, we have

$$(y_R - y_c) = \frac{F(y_p - y_c)}{F + (F_0 - F'_0)}.$$

The magnitude of F and $(y_p - y_c)$ is taken from Example 2.6. For $h_A = 100$ m,

$$F = 3.24 \times 10^6 \text{ N},$$

$$y_p - y_c = 4.90 \times 10^{-4} \text{ m},$$

$$F_R = (3.24 \times 10^6 \text{ N} + 1.03 \times 10^5 \text{ N}),$$

$$\left| \, F_R = 3.34 \times 10^6 \text{ N}. \, \right|$$ **Answer** ◀

$$y_R - y_c = \frac{(3.24 \times 10^6 \text{ N})(4.90 \times 10^{-4} \text{ m})}{(3.34 \times 10^6 \text{ N})},$$

$$\left| \, y_R - y_c = 4.66 \times 10^{-4} \text{ m}. \, \right|$$ **Answer** ◀

Since the pressure is uniform along a horizontal element of the window and since the semicircular window is symmetric about a radius in a vertical plane passing through the centroid, we conclude that the x-location, x_R, of the resultant force F_R is along the radius of symmetry, or

$$\left| \, x_R = x_c. \, \right|$$ **Answer** ◀

Discussion
Moments were taken about a horizontal axis to obtain the y-location of the resultant force as $(y_R - y_c)$. It is permissible to use any other parallel horizontal axis; however, the force $(F_0 - F'_0)$ would then have to be included.

2.4.2 Forces on Curved Surfaces

Consider a curved surface with hydrostatic pressure acting on it, as shown in Fig. 2.20. For simplicity, we will consider a two-dimensional curved surface. Note that y is now a vertical coordinate.

Since pressure always acts normal to area, the forces on each infinitesimal area have different directions. The force on each element

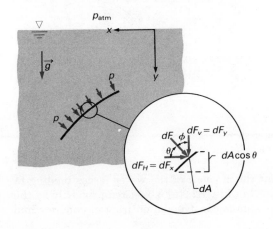

Figure 2.20 Curved surface in a liquid force on element of surface is broken into horizontal and vertical components.

of area may be broken into horizontal and vertical components. We can sum (integrate) the horizontal and vertical components separately.

We consider first the horizontal (x) component:

$$dF_x = dF \cos \theta = p \, dA \cos \theta.$$

But $dA \cos \theta$ is dA_x, the horizontal projection of dA, so we can write

$$F_x = \int p \, dA_x.$$

We can calculate the horizontal component of force on a curved area by calculating the force on the horizontal projection of the area:

$$F_x = \gamma h_{c,x} A_x. \qquad (2.28)$$

By the same reasoning, the line of action of the horizontal force passes through the center of pressure of the projected area.

Next we consider the vertical component of a pressure force on a curved surface. The elemental vertical force is

$$dF_v = p \, dA \cos \phi.$$

For a surface submerged in a liquid with a free surface,

$$p = \gamma h, \qquad \text{so} \qquad F_v = \int dF_v = \int \gamma h \cos \phi \, dA.$$

Now $\int h \cos \phi \, dA$ is V', the volume between the curved surface and the free surface, so the vertical force is equal to the weight of fluid above the surface:

$$F_v = \gamma V' \qquad (2.29)$$

You should be able to show that the line of action of the vertical component of the force is through the centroid of the volume V'. These concepts are illustrated in Fig. 2.21.

Figure 2.21 Curved surface: (a) Horizontal projection of area; (b) volume above curved area.

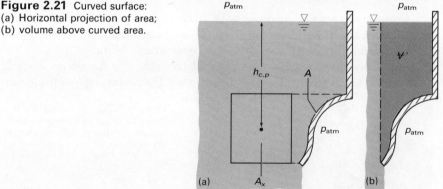

EXAMPLE 2.9 Illustrates the calculation of forces on a curved surface

A gravity dam is designed so that the moment of the water forces tending to topple the dam over about the toe B in Fig. E2.9(a) is resisted primarily by the moment of the dam's weight. The dam is constructed entirely of concrete and

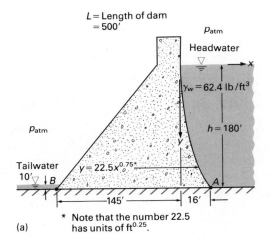

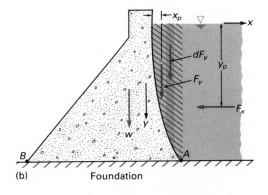

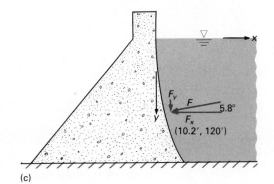

Figure E2.9 (a) Solid gravity dam; (b) forces acting on upstream side of dam; (c) resultant of upstream hydrostatic forces on dam.

is 190 ft high and 500 ft long. It maintains a headwater height of 180 ft and a tailwater height of 10 ft. Find the magnitudes and locations of the lines of action of the horizontal and vertical forces acting on the upstream side of the dam.

Solution

Given

Dam in Fig. E2.9(a), 500 ft long

Find

Forces F_x and F_y and their locations y_p and x_p, respectively, as shown in Fig. E2.9(b)

Solution

The horizontal force F_x is found from Eq. (2.28):

$$F_x = \gamma_w h_{c,x} A_x = (62.4 \text{ lb/ft}^3)(90 \text{ ft})(180 \text{ ft} \times 500 \text{ ft}),$$

$$\boxed{F_x = 5.05 \times 10^8 \text{ lb.}}$$

Answer

The vertical force F_y is found from Eq. (2.29):

$$F_y = \gamma_w \mathcal{V}.$$

The volume ($\mathcal{V}$) must be obtained by integration. Referring to Fig. E2.9(b), we have

$$F_y = \gamma_w L \int_0^{16} y \, dx,$$

where $L = 500$ ft. Substituting

$$y = 22.5 x^{0.75}$$

gives

$$F_y = \gamma_w L \int_0^{16} 22.5x^{0.75}\, dx$$

$$= 22.5\gamma_w L \left(\frac{x^{1.75}}{1.75}\right)_0^{16} = 1646 \text{ ft}^2\, \gamma_w L$$

$$= (1646 \text{ ft}^2)(62.4 \text{ lb/ft}^3)(500 \text{ ft}),$$

$$\left| \; F_y = 5.13 \times 10^7 \text{ lb.} \; \right|$$ **Answer**

The location y_p of F_x is found in the same manner as finding the vertical location of a horizontal force on a vertical plane measuring $h \times L$. Using a z-axis into the paper and Appendix B for the moment of inertia gives

$$y_p = \frac{I_z}{y_c A} = \frac{\dfrac{1}{3}Lh^3}{\left(\dfrac{h_1}{2}\right)(h \times L)} = \frac{2h}{3}$$

$$= \frac{2(180 \text{ ft})}{3},$$

$$\left| \; y_p = 120 \text{ ft.} \; \right|$$ **Answer**

The location of F_y is the centroid of the volume V'. The x-coordinate of the centroid is found from

$$x_p = \frac{\int x\, dV'}{V'} = \frac{\int x\, dV'}{\int dV'}$$

$$= \frac{\int_0^{16} x(Ly\, dx)}{\int_0^{16} Ly\, dx}.$$

Substituting $y = 22.5x^{0.75}$ gives

$$x_p = \frac{\int_0^{16} L(22.5x^{1.75})\, dx}{\int_0^{16} L(22.5x^{0.75})\, dx}.$$

Canceling constants and integrating, we obtain

$$x_p = \frac{\left(\dfrac{x^{2.75}}{2.75}\right)_0^{16}}{\left(\dfrac{x^{1.75}}{1.75}\right)_0^{16}} = \frac{1.75}{2.75}\left[\frac{(16 \text{ ft})^{2.75}}{(16 \text{ ft})^{1.75}}\right],$$

$$\left| \; x_p = 10.2 \text{ ft.} \; \right|$$ **Answer**

Discussion

Note that we did not find the z-coordinates (into the paper) of F_x and F_y. Both are located halfway across the dam.

The magnitude of the hydrostatic force F on the upstream side of the dam is

$$F = \sqrt{F_x^2 + F_y^2}$$
$$= \sqrt{(5.05 \times 10^8)^2 + (5.13 \times 10^7)^2} \text{ lb}$$
$$= 5.08 \times 10^8 \text{ lb}.$$

The line of action of F passes through the point

$$(x_p, y_p) = (10.2 \text{ ft}, 120 \text{ ft})$$

and makes an angle θ with the horizontal given by

$$\theta = \tan^{-1} \frac{F_y}{F_x}$$

$$= \tan^{-1} \frac{5.13 \times 10^7}{5.05 \times 10^8} = 5.8°$$

These results are shown in Fig. E2.9(c). Note that the point (10.2, 120) is not on the surface of the dam.

Also note that the constant 22.5 has units of $\text{ft}^{0.25}$ and that terms with fractional powers of units were properly used to produce the correct units of force or distance.

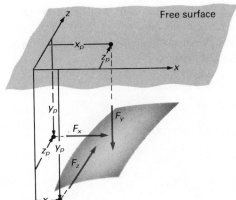

Figure 2.22 Components of a force on a three-dimensional curved surface. [*Note:* x_p for F_y may not equal x_p for F_z, etc.]

Pressure forces on three-dimensional curved surfaces can be found in the same way as we found pressure forces on two-dimensional curved surfaces. The main difference is that there are *two* horizontal components and one vertical component. Figure 2.22 illustrates the three components of pressure force on a three-dimensional curved surface.

In all of our discussions and examples so far, the liquid has been above the surface. There may be no liquid directly above the surface, as in Fig. 2.23. The pressure at each point on the surface is still calculated from the specific weight of the liquid and the depth below the free surface; that is, the pressure is the same as if there were liquid above the surface. The force is calculated by integrating the pressure over the area; in particular, the net vertical force will still have a magnitude

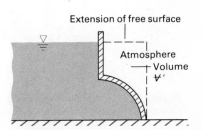

Figure 2.23 Curved surface with no liquid above it.

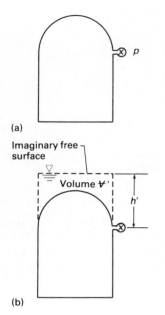

(a)

Imaginary free surface

Volume V'

h'

(b)

Figure 2.24 (a) Tank filled with pressurized fluid; (b) imaginary free surface above curved top of tank.

equal to the weight of liquid that would be contained in the volume V' between the curved surface and an extension of the free surface, so Eq. (2.29) can still be used. The only difference is that the vertical force will act *upward* rather than downward. You should always remember that pressure force acts *from* the liquid *toward* the surface.

If we are clever enough, we even can use this approach with a problem that has no free surface at all, like the pressurized container shown in Fig. 2.24(a). If we know the pressure at some level in the container, we can construct an imaginary free surface above this point (Fig. 2.24b). This "free surface" lies above the point of known pressure a distance h', where

$$h' = p/\gamma.$$

The net pressure force (from the fluid inside the tank) is equal to $\gamma V'$.* Notice that this force is upward.

2.5 Mechanics of Submerged and Floating Bodies

2.5.1 Buoyancy

When a body is either fully or partially submerged in a fluid, a net upward force called the *buoyant force* is exerted on the body. This force is caused by the imbalance of pressures on the upper and lower surfaces.

Consider a body submerged in a fluid as shown in Fig. 2.25. We wish to calculate the net vertical force on the body due to pressure. Since the body is closed, each "lower" area dA_1 has a corresponding "upper" area dA_2. The net pressure force on the two areas is

$$dF = p_1 \, dA_{1,y} - p_2 \, dA_{2,y}.$$

The vertical projections of dA_1 and dA_2 are equal, so

$$dF = (p_1 - p_2) \, dA_y = \gamma \ell \, dA_y.$$

Integrating, we obtain

$$F = \int \gamma l \, dA_y = \gamma V. \tag{2.30}$$

This equation shows that

- The bouyant force on a body submerged in a fluid is equal to the *weight* of the fluid displaced by the body.

We can also show that

- The line of action of the buoyant force passes through the centroid of the displaced volume. This centroid is called the *center of buoyancy.*

These principles also apply if the body is only partly submerged; however, you must be careful to consider only the part of the body that is in the fluid.

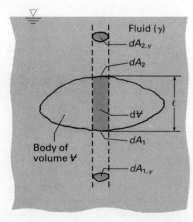

Fluid (γ)

$dA_{2,y}$

dA_2

ℓ

dV

dA_1

Body of volume V

$dA_{1,y}$

Figure 2.25 Arbitrary closed body submerged in fluid.

---

* Does γ have to be the specific weight of the fluid in the tank? What if h' does not lie above the top of the tank?

This information about the magnitude and line of action of buoyant forces is called Archimedes first principle of buoyancy because it was apparently discovered by Archimedes in 220 B.C.

A body will *float* if its average density is less than the density of the fluid in which it is placed. For a floating body (Fig. 2.26),

$$W = \gamma_f V_s, \tag{2.31}$$

where γ_f is the specific weight of the fluid and V_s is the submerged volume. This is Archimedes' second principle of buoyancy:

■ A floating body displaces a volume of fluid equivalent to its own weight.

In solving problems involving closed bodies in fluids, we can always replace the action of hydrostatic pressure by the buoyant force.

A useful application of the principles of buoyancy is the construction of a *hydrometer* to measure specific gravity of a liquid. Consider a hydrometer floating in pure water ($S = 1.0$) and in another liquid, with $S > 1.0$, as illustrated in Fig. 2.27. If $S > 1.0$, the buoyant force per unit submerged volume is greater, so less volume of submergence is necessary to balance the hydrometer's weight. In the water,

$$W = \gamma_{\mathrm{H_2O}} V_0, \tag{2.32}$$

while in the other fluid,

$$W = S\gamma_{\mathrm{H_2O}}(V_0 - A\,\Delta h).$$

Therefore

$$\Delta h = \frac{V_0}{A}\left(\frac{S-1}{S}\right). \tag{2.33}$$

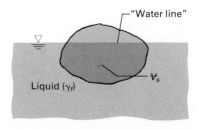

Figure 2.26 Floating body.

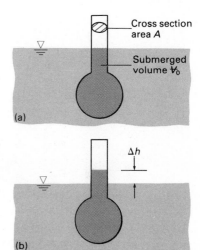

Figure 2.27 Hydrometer in two liquids: (a) Water, $S = 1.0$; (b) other liquid, $S > 1.0$.

EXAMPLE 2.10 Illustrates the equality of the buoyant force and the weight of a floating body.

A *spar buoy* does not heave and roll when hit by ocean waves as much as other buoys do, and it is well suited for use in scientific experimental stations. A particular spar buoy is shown in Fig. E2.10. A small radio sensor is mounted on top of the buoy. The buoy has a circular cross section and is supposed to float with 6 ft protruding above the water. Calculate the total height of the buoy if it is made from (a) aluminum and (b) steel. The other dimensions of the buoy are shown in the figure. The specific weights of seawater, aluminum, and steel are 64 lb/ft³, 169 lb/ft³, and 490 lb/ft³, respectively.

Solution

Given
Spar buoy, OD = 12.75 in., ID = 12.25 in., to float 6 ft out of water, constructed of (a) aluminum and (b) steel

Find
The total height of each buoy

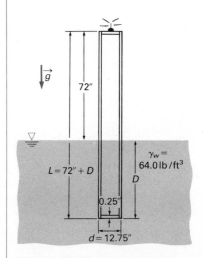

Figure E2.10 Spar buoy.

Solution

The primary equation is Eq. (2.31), expressing that the weight W of the floating buoy must be equal the buoyant force F_B $(=\gamma_w V_s)$. The weight W of the buoy is

$$W = \gamma_B(V_{sides} + V_{end\ plates})$$

$$= \gamma_B\left[\frac{\pi}{4}((12.75)^2 - (12.25)^2)L + 2\frac{\pi}{4}(12.25)^2(0.25)\right]in^3\left(\frac{ft^3}{1728\ in^3}\right)$$

$$= 0.00568\gamma_B L + 0.0341\gamma_B,$$

where γ_B is in lb/ft^3 and L is in inches, giving W in lb. The buoyant force F_B is

$$F_B = \gamma_w V_s$$

$$= (64\ lb/ft^3)\left[\frac{\pi}{4}(12.75)^2(L - 72)\right]in^3\left(\frac{ft^3}{1728\ in^3}\right)$$

$$= 4.73L - 340,$$

where L is in inches. Equating W and F_B, we have

$$0.00568\gamma_B L + 0.0341\gamma_B = 4.73L - 340,$$

or

$$L = \frac{340 + 0.0341\gamma_B}{4.73 - 0.00568\gamma_B}.$$

For aluminum,

$$L = \frac{340 + 0.0341(169)}{4.73 - 0.00568(169)},$$

$$\left|\ L = 91.7\ in.\ \right|$$ **Answer** ◄

For steel,

$$L = \frac{340 + 0.0341(490)}{4.73 - 0.00568(490)},$$

$$\left|\ L = 183\ in.\ \right|$$ **Answer** ◄

Discussion

For use in a later problem, note that the aluminum buoy is submerged a depth of $(91.7 - 72)$, or 19.7, in. and the steel buoy is submerged a depth of $(183 - 72)$, or 111, in.

2.5.2 Stability of Submerged and Floating Bodies

Buoyant force on a body always acts through the centroid of the displaced volume, while the weight always acts through the center of gravity. These facts can make a partially or totally submerged body either *stable* or *unstable*. The concept of stability is the next step beyond equilibrium. An object is in stable equilibrium if, on being slightly displaced from its equilibrium position, forces or moments are generated that restore the object to its original position. An object is in unstable equilibrium if, on being slightly displaced, forces or moments are generated that further displace the object. An object is in neutral equilib-

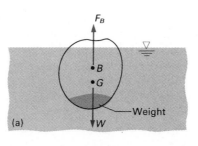

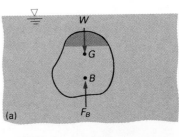

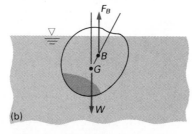

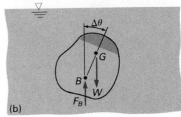

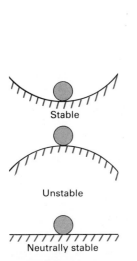

Figure 2.28 Ball-on-surface illustration of the concept of stability.

Figure 2.29 Floating body with center of gravity below center of buoyancy: (a) Equilibrium position; (b) tipped position.

Figure 2.30 Submerged body with center of gravity (G) above center of buoyancy (B): (a) Equilibrium position; (b) tipped position.

rium if no forces or moments are generated on displacement. Stability is often illustrated by the analogy of a ball on a curved surface, as in Fig. 2.28.

A completely or partially submerged (floating) body is in stable equilibrium if its center of gravity (G) lies *below* its center of buoyancy (B), as illustrated in Fig. 2.29. If the body is rotated, a moment is set up to right the body and return it to its original position with G directly below B.

If the center of gravity of a totally submerged body is above the center of buoyancy, the body is in unstable equilibrium since an unbalancing moment arises when the body is rotated (see Fig. 2.30).

If the center of gravity of a floating body is above the center of buoyancy, the body can be stable *or* unstable because the center of buoyancy shifts as the object is rotated. To illustrate, consider two floating blocks of wood, a short, wide one (Fig. 2.31a and b) and a tall,

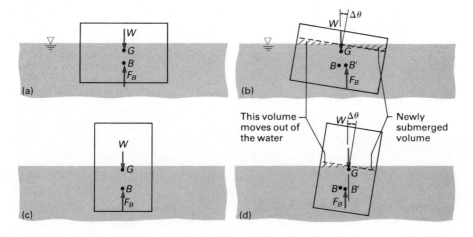

Figure 2.31 Comparison of stability of short, wide and tall, narrow blocks: (a) and (c) "Equilibrium" position; (b) and (d) tipped position.

narrow one (Fig. 2.31c and d). For both blocks, the center of buoyancy shifts to the right as some of the left side of the block moves above the surface. For the same angle of rotation, the center of buoyancy of the short, wide block moves farther to the right than does the center of buoyancy of the tall, narrow block. The result is that the new center of buoyancy (B') of the short, wide block now lies to the right of the center of gravity, while the new center of buoyancy of the tall, narrow block still lies to the left of the center of gravity. The short, wide block has a restoring moment and is stable, and the tall, narrow block has an upsetting moment and is unstable.

Engineers involved in the design of floating objects like ships and barges must evaluate the stability of such objects. The following analysis will illustrate how such an evaluation can be done for a simple situation with small angles of disturbance and no extraneous forces such as wind.

Consider a ship floating in equilibrium (Fig. 2.32a). The center of gravity G lies above the center of buoyancy B. The weight acts downward through G, and the buoyant force acts upward through B.

Now suppose that the ship is tipped to the right by small angle $\Delta\theta$, the *angle of heel*. The center of gravity retains its position, but the center of buoyancy shifts to B'. A new point, M, shown in Fig. 2.32(b), is the *metacenter,* the intersection of the line of action of the buoyant force and the line connecting the previous center of buoyancy B with the center of gravity.

In its tipped position, the ship experiences a restoring moment $\mathscr{M}_R$, given by the following equation (recall $W = F_B$):

$$\mathscr{M}_R = W(m + n) \sin (\Delta\theta). \tag{2.34}$$

Assuming that $\Delta\theta$ is small, Eq. (2.34) can be written

$$\mathscr{M}_R \approx W(m + n) \Delta\theta. \tag{2.35}$$

For the situation in Fig. 2.32(b), the ship is stable. Now suppose that the new center of buoyancy is at B'' (Fig. 2.32c). In this case, the moment acts to tip the ship; it is unstable. Equation (2.35) can still be used to calculate the moment if m is considered negative. The difference be-

Figure 2.32 Ship with center of gravity (G) above the center of buoyancy (B): (a) Equilibrium position; (b) tipped position, stable; (c) tipped position, unstable.

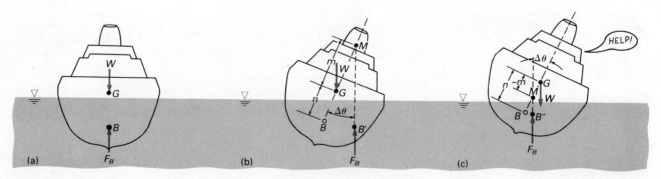

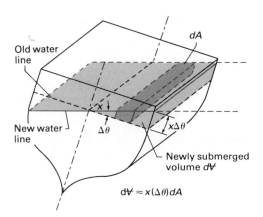

d$V \approx x(\Delta \theta) dA$

Figure 2.33 Volume (dV) that becomes submerged when ship tips.

tween the stable and unstable situations is that point M lies above G (m positive) for the stable case and point M lies below G (m negative) for the unstable case. We can summarize these facts in the following statement:

> A floating object is stable if the metacenter lies above the center of gravity and unstable if the metacenter lies below the center of gravity.

To develop an equation for the metacentric height m, we calculate the moment induced by the portion of the ship that goes under the water and the portion on the other side that comes out of the water when the ship heels. Figure 2.33 shows a portion of the ship's volume that becomes submerged when the ship is rotated $\Delta\theta$ about its longitudinal symmetry axis. The moment due to the buoyant force on this volume is

$$d\mathcal{M}_R = x\gamma\,dV = x\gamma(x\Delta\theta)\,dA.$$

Summing moments for all such volume elements and noting that γ and $\Delta\theta$ are constant, we obtain

$$\mathcal{M}_R = \gamma\Delta\theta \int x^2\,dA,$$

where the integral is taken over the right half of the ship; however, for each volume on the right that goes under water, an equal volume comes out of the water on the left, so the total restoring moment is

$$\mathcal{M}_R = \gamma\Delta\theta \int_{A_w} x^2\,dA = \gamma\Delta\theta I_x, \tag{2.36}$$

where I_x is the moment of inertia of the "plan view" area of the ship *at the waterline* (A_w).

We equate $\mathcal{M}_R$ from Eq. (2.35) and $\mathcal{M}_R$ from Eq. (2.36) and note that

$$W = F_B = \gamma V_s,$$

where V_s is submerged volume. Then we solve for m:

$$m = \frac{I_x}{V_s} - n. \tag{2.37}$$

The ship is stable if m is positive and unstable if m is negative.

Equation (2.37) is valid only for small angles of heel. For large angles of heel, I_x could change appreciably as larger portions of the ship become submerged.

EXAMPLE 2.11 Illustrates how to determine if a floating body is stable

Determine if the steel and aluminum spar buoys sized as in Example 2.10 are stable in an upright position.

Solution

Given
Two spar buoys shown in Fig. E2.11

Find
Whether the buoys are stable

Solution
Stability is determined by the sign of the metacentric height m, which is evaluated by Eq. (2.37). Moment of inertia I_x for both buoys is that of a circle about its diameter. Using Appendix B, we find that

$$I_x = \frac{\pi d^4}{64} = \frac{\pi}{64}(12.75 \text{ in})^4 = 1297 \text{ in}^4.$$

To find the values of n, we need the depth of submergence D, which is

$$D = 91.7 - 72 = 19.7 \text{ in.} \qquad \text{for aluminum,}$$
$$D = 183 - 72 = 111 \text{ in.} \qquad \text{for steel.}$$

Since the center of buoyancy lies at the centroid of the displaced volume and since n represents the distance from the center of gravity (at the center of the spar buoy) to the center of buoyancy,

$$n = \frac{L}{2} - \frac{D}{2} = \frac{1}{2}(L - D)$$

$$= \frac{1}{2}(91.7 - 19.7) \text{ in.} = 36 \text{ in.} \qquad \text{for aluminum}$$

$$= \frac{1}{2}(183 - 111) \text{ in.} = 36 \text{ in.} \qquad \text{for steel.}$$

Figure E2.11 Aluminum and steel spar buoys.

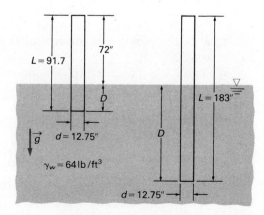

The volumes displaced are

$$V_s = \frac{\pi}{4}d^2 D$$

$$= \frac{\pi}{4}(12.75 \text{ in.})^2(19.7 \text{ in.}) = 2515 \text{ in}^3 \qquad \text{for aluminum,}$$

$$V_s = \frac{\pi}{4}(12.75 \text{ in.})^2(111 \text{ in.}) = 14{,}200 \text{ in}^3 \qquad \text{for steel.}$$

Then

$$m = \frac{I_x}{V_s} - n = \frac{1297 \text{ in}^4}{2515 \text{ in}^3} - 36 \text{ in.} = -35.5 \text{ in.}$$

Answer ◀

| Aluminum buoy is unstable. |

And

$$m = \frac{1297 \text{ in}^4}{14{,}200 \text{ in}^3} - 36 \text{ in.} = -35.9 \text{ in.}$$

Answer ◀

| Steel buoy is unstable. |

Discussion

How can these buoys be made stable? Assume that it is permissible for more than 6 ft of the buoy to protrude out of the water.

2.6 Mechanics of Fluids in Rigid Body Motion

Thus far we have considered the mechanics of static fluids. Since there was no motion of fluid relative to other fluid or to solid surfaces, there was no fluid shear deformation and hence no shear stresses. This is true in another class of situations: rigid body motion in which the fluid moves like a solid with no shear deformation.

To find the governing equation for rigid body motion, consider the infinitesimal fluid element shown in Fig. 2.34. The pressure at the center of the element is p. The pressures on the z- and y-faces of the element are as shown. The x-direction is omitted for clarity; however, it is similar to the y-direction. The fluid element is experiencing a known acceleration $\vec{a}\ (= a_y\hat{j} + a_z\hat{k})$. There are no shear stresses. Applying Newton's

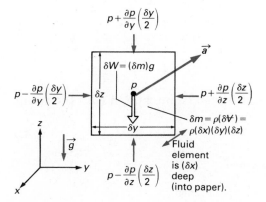

Figure 2.34 Fluid element experiencing uniform acceleration $\vec{a}$.

second law in the y-direction, we have

$$\sum F_y = ma_y.$$

The only y-forces are due to pressure, so

$$\left[p - \frac{\partial p}{\partial y}\left(\frac{\delta y}{2}\right) \right]\delta x\,\delta z - \left[p + \frac{\partial p}{\partial y}\left(\frac{\delta y}{2}\right) \right]\delta x\,\delta z = \rho(\delta x\,\delta y\,\delta z)a_y.$$

Canceling terms and taking the limit,* we obtain

$$\frac{\partial p}{\partial y} = -\rho a_y. \tag{2.38}$$

Considering the z-direction, we have

$$\sum F_z = ma_z.$$

The z-forces include gravity as well as pressure, so

$$\left[p - \frac{\partial p}{\partial z}\left(\frac{\delta z}{2}\right) \right]\delta x\,\delta y - \left[p + \frac{\partial p}{\partial z}\left(\frac{\delta z}{2}\right) \right]\delta x\,\delta y - \gamma\,\delta x\,\delta y\,\delta z = \rho(\delta x\,\delta y\,\delta z)a_z.$$

Canceling terms and taking the limit, we obtain

$$\frac{\partial p}{\partial z} = -\gamma\left(1 + \frac{a_z}{g}\right). \tag{2.39}$$

Since the x-direction is similar to the y-direction,

$$\frac{\partial p}{\partial x} = -\rho a_x. \tag{2.40}$$

The total differential of pressure is

$$dp = \frac{\partial p}{\partial x}\,dx + \frac{\partial p}{\partial y}\,dy + \frac{\partial p}{\partial z}\,dz.$$

Substituting Eqs. (2.38), (2.39) and (2.40) gives

$$dp = -\rho a_x\,dx - \rho a_y\,dy - \gamma\left(1 + \frac{a_z}{g}\right)dz. \tag{2.41}$$

Integration of this equation will yield the pressure distribution in the fluid. The following two sections illustrate particular cases.

2.6.1 Uniform Linear Acceleration

Consider a fluid moving in a plane, with a_x and a_z constant and a_y zero. For this case, Eq. (2.41) reduces to

$$dp = -\rho a_x\,dx - \gamma\left(1 + \frac{a_z}{g}\right)dz.$$

Since a_x and a_z are constant, this can be integrated with the result

$$p = -\rho a_x x - \gamma\left(1 + \frac{a_z}{g}\right)z + p_0, \tag{2.42}$$

where p_0 is the pressure at $x = z = 0$.

* If you are uneasy about the procedure, review Section 2.2.

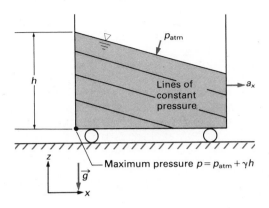

Maximum pressure $p = p_{atm} + \gamma h$

Figure 2.35 Container of liquid with constant acceleration in x-direction.

An item of considerable interest is the shape of constant-pressure surfaces. If p is constant,

$$-\rho a_x x - \gamma \left(1 + \frac{a_z}{g} \right) z = \text{Constant},$$

so surfaces of constant pressure are planes with slope equal to $-a_x/(a_z + g)$.

A particularly simple and interesting case occurs if a_x is constant and a_y and a_z are zero (see Fig. 2.35). In this case, the free surface is a plane with slope $-a_x/g$ and the pressure distribution in the z-direction is hydrostatic by Eq. (2.39).

2.6.2 Fluid Rotating About a Vertical Axis

Consider a container of liquid rotating at constant angular velocity ω (see Fig. 2.36). If the motion has been going on for a long time, the fluid rotates like a rigid body. The pressure distribution for this situation could be obtained by integrating Eq. (2.41); however, it is more convenient to set up the pressure equations in polar coordinates before integrating.

For rigid body rotation, the velocity of each fluid particle is

$$V_r = V_z = 0 \qquad \text{and} \qquad V_\theta = r\omega.$$

We note from Eq. (2.41) that in rigid body motion, the pressure gradient in any direction perpendicular to the gravity vector is proportional to acceleration in that direction. In this case,

$$a_z = a_\theta = 0 \qquad \text{and} \qquad a_r = -r\omega^2.$$

Thus

$$\frac{\partial p}{\partial \theta} = 0,$$

$$\frac{\partial p}{\partial z} = -\gamma \qquad \text{(hydrostatic pressure)},$$

and

$$\frac{\partial p}{\partial r} = -\rho a_r = \rho r\omega^2.$$

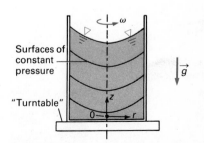

Figure 2.36 Tank of liquid rotating at constant angular velocity (ω).

The total differential of pressure is

$$dp = \frac{\partial p}{\partial z}\, dz + \frac{\partial p}{\partial r}\, dr = -\gamma\, dz + \rho\omega^2 r\, dr.$$

Integrating, we obtain

$$p = \frac{\rho\omega^2}{2}\, r^2 - \gamma z + p_0. \qquad (2.43)$$

If p is constant, Eq. (2.43) becomes

$$z = \frac{\omega^2}{2g}\, r^2 + \text{Constant}. \qquad (2.44)$$

Surfaces of constant pressure, including the free surface, are paraboloids, as sketched in Fig. 2.36.

EXAMPLE 2.12 Illustrates rigid body motion of a fluid rotating at a constant angular velocity

A child buys a 12-fl-oz cup of a soft drink, drinks one-fourth of it, and then places it directly in the center of a merry-go-round. The merry-go-round is then rotated by several children at 15 rpm. Will any soft drink spill? The merry-go-round has a diameter of 10 ft; the cup has an inside diameter of 2.5 in.

Solution

Given

12-oz cup with 9 fl oz of soft drink, rotating at 15 rpm
Inside diameter of cup 2.5 in.
Diameter of merry-go-round 10 ft.
Figure E2.12a, which shows the stationary cup and the rotating cup

Find

Height h_R of the soft drink when the cup is rotating at 15 rpm
Compare with height of the cup

Figure E2.12 (a) Stationary and rotating cup of soft drink; (b) volumes Ψ_1 and Ψ_2 of soft drink.

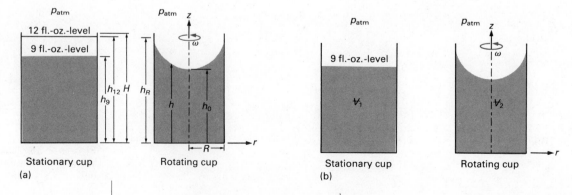

Solution
The height H is found using Table C.4 and noting the height h_{12}:

$$h_{12} = \frac{\text{Volume}}{\text{Area}}$$

$$= \frac{(12\text{ fl oz})\left(\dfrac{\text{gal}}{128\text{ fl oz}}\right)\left(\dfrac{231\text{ in}^3}{\text{gal}}\right)}{\dfrac{\pi}{4}(2.5\text{ in.})^2}$$

$$= 4.41\text{ in.}$$

The difference between H and h_{12} is not given and must be estimated. Assuming the top of a 12-oz cup is about $\frac{1}{4}$ in. above the 12-oz fluid level, we have

$$H \approx h_{12} + \frac{1}{4}\text{ in.}$$

$$\approx 4.66\text{ in.}$$

The height h_R is next found from Eq. (2.43) by noting that $p = p_{\text{atm}}$ at $z = h$ and that $z = h_0$ at $r = 0$ at the free surface. Then

$$h = h_0 + \frac{\rho\omega^2 r^2}{2\gamma}.$$

Recognizing $\gamma = \rho g$, we find that

$$h = h_0 + \frac{\omega^2 r^2}{2g}.$$

Since $h = h_R$ at $r = R$, we have

$$h_R = h_0 + \frac{\omega^2 R^2}{2g}.$$

We can find h_0 by recognizing that the volume V_1 shown in Fig. E2.12(b) and given by

$$V_1 = \pi R^2 h_9$$

must equal the volume V_2, also in Fig. E2.12(b), if no soft drink is spilled. We can calculate V_2 by

$$V_2 = \int_0^R 2\pi r h\, dr.$$

Substituting for h, we have

$$V_2 = 2\pi \int_0^R r\left(h_0 + \frac{\omega^2 r^2}{2g}\right)dr.$$

Integrating for constant h_0, ω, and g gives

$$V_2 = 2\pi\left(h_0\frac{r^2}{2} + \frac{\omega^2 r^4}{8g}\right)_0^R = \pi R^2\left(h_0 + \frac{\omega^2 R^2}{4g}\right).$$

Equating V_1 and V_2, we have

$$\pi R^2 h_9 = \pi R^2\left(h_0 + \frac{\omega^2 R^2}{4g}\right), \quad\text{or}\quad h_0 = h_9 - \frac{\omega^2 R^2}{4g}.$$

Then

$$h_R = h_9 + \frac{\omega^2 R^2}{4g}.$$

The numerical values give

$$h_9 = \frac{\text{Volume}}{\text{Area}} = \frac{(9 \text{ fl oz})\left(\dfrac{\text{gal}}{128 \text{ fl oz}}\right)\left(\dfrac{231 \text{ in}^3}{\text{gal}}\right)}{\dfrac{\pi}{4}(2.5 \text{ in.})^2} = 3.31 \text{ in.},$$

$$h_R = 3.31 \text{ in.} + \frac{\left(\dfrac{15 \text{ rev}}{\text{min}}\right)^2 \left(2\pi \dfrac{\text{rad}}{\text{rev}}\right)^2 (1.25 \text{ in.})^2}{4\left(386 \dfrac{\text{in.}}{\text{sec}^2}\right)\left(60 \dfrac{\text{sec}}{\text{min}}\right)^2}$$

$$= 3.31 \text{ in.} + 0.002 \text{ in.} = 3.33 \text{ in.}$$

Since $H > h_R$, we conclude:

Answer
◀

| No soft drink will spill from the cup. |

Discussion
The value of $\frac{1}{4}$ inch assumed for $(H - h_{12})$ was a realistic value but was not a significant factor in the determination of whether any soft drink was spilled.

Some students have difficulty in recognizing that

$$V_2 = \int_0^R 2\pi r h \, dr$$

in a problem of this nature. If you have such difficulty, a careful examination of Fig. E2.12(b) should be helpful.

EXAMPLE 2.13   Illustrates how to investigate approximations in mathematically difficult problems and illustrates liquids moving with constant linear acceleration

Another child has 8 fl oz of a soft drink in a 12-oz cup. She places it on some sticky bubble gum at the outer edge of the merry-go-round from Example 2.12, which is rotating at 15 rpm. Will any soft drink spill from the cup?

Solution

Given
Cup containing 8 fl oz of soft drink, placed on outer edge of 10-ft-diameter
merry-go-round rotating at 15 rpm
Inside diameter of cup 2.5 in.
Height of cup 4.66 in.
Figures E2.13(a) and (b)

Find
Height h_R of soft drink at outer edge of cup

Solution
We will start the solution in a manner similar to that used in Example 2.12. The height h is found using Eq. (2.44) and by noting that $z = h$ at $r = r_M$ with $z = h_R$ at $r = R_M$ at the free surface. Then

$$h = h_R - \frac{\omega^2}{2g}(R_M^2 - r_M^2).$$

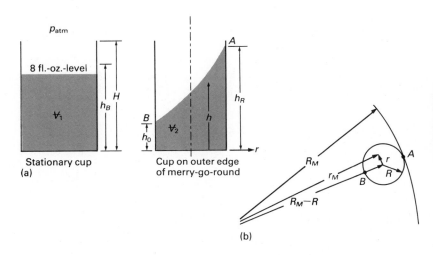

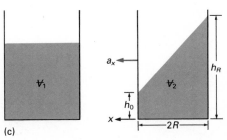

Stationary cup
(a)

Cup on outer edge
of merry-go-round

(b)

(c)

Figure E2.13 (a) Stationary cup and cup on outer edge of merry-go-round; (b) top view of cup of soft drink on merry-go-round; (c) stationary and linearly accelerating cup of soft drink.

An equation for h_R is found by recognizing that the volume V_1 ($= \pi R^2 h_8$) in Fig. E2.13(a) must equal the volume V_2 if no soft drink is spilled. The volume V_2 is

$$V_2 = \int_0^R 2\pi h r \, dr.$$

Substituting for h and equating V_1 and V_2 gives

$$R^2 h_8 = 2 \int_0^R r \left[h_R - \frac{\omega^2}{2g} (R_M^2 - r_M^2) \right] dr.$$

Referring to Fig. E2.13(b) shows that we will have to relate r_M to r by some tricky geometry, eventually obtaining an equation that can be solved for h_R (either in closed form or numerically).

At this point we ask ourselves if there is an easier way to determine if any soft drink will be spilled. There is, if we are willing to accept an approximate solution. Engineers should be willing to use a simpler solution technique if there is a high probability that it will lead to a correct conclusion. We will explore the easier approach.

We start by calculating the radial acceleration at point A, where $r_M = 5$ ft:

$$a = -\omega^2 r_M)_{r_M = 5 \, \text{ft}}$$

$$= -\left[\left(15 \frac{\text{rev}}{\text{min}} \right) \left(\frac{\text{min}}{60 \, \text{sec}} \right) \left(2\pi \frac{\text{rad}}{\text{rev}} \right) \right]^2 (5 \, \text{ft}) = -12.3 \, \text{ft/sec}^2.$$

The radial acceleration at point B, where $r_M = R_M - R = 4.79$ ft, is

$$a_r = -\omega^2 r_M)_{r_M = 4.79 \, \text{ft}} = -\left[\left(15 \frac{\text{rev}}{\text{min}} \right) \left(\frac{\text{min}}{60 \, \text{sec}} \right) \left(2\pi \frac{\text{rad}}{\text{rev}} \right) \right]^2 (4.79 \, \text{ft})$$

$$= -11.8 \, \text{ft/sec}^2.$$

The acceleration experienced by the liquid in the cup varies by only 4%. It would not be unrealistic to assume that the acceleration is approximately constant. If we assume that the acceleration is constant, then the free surface of the liquid is a plane, and the volume V_2 will be much easier to calculate. We have several reasonable choices for a constant value for the acceleration, among them (1) the acceleration at the outer edge of the cup (point A); (2) the acceleration at the inner edge of the cup (point B); and (3) the mean acceleration $(a_A + a_B)/2$. Which approximation we select depends on what we want to do. If we desire the most accurate approximation to the shape of the liquid surface, we would use the average acceleration. On the other hand, there may be advantages in using one of the limiting accelerations. To illustrate these advantages we will estimate the height h_R for each acceleration.

Using Fig. E2.13(c) and applying Eq. (2.41) along the free surface of the linearly accelerating cup, with $p = p_{atm} =$ Constant, $a_z = 0$, and $a_y = 0$, we obtain

$$\frac{dz}{dx} = -\frac{a_x}{g} = \frac{h_R - h_0}{2R}.^*$$

Thus

$$h_R - h_0 = -\frac{2a_x R}{g}.$$

Another equation involving h_R and h_0 comes from recognizing that the volume $\mathcal{V}_1$ must equal the volume $\mathcal{V}_2$, both illustrated in Fig. E2.13(c)

$$\mathcal{V}_2 = \text{(Base area)(Average height)} = \pi R^2 \left(\frac{h_0 + h_R}{2} \right),$$

so

$$\pi R^2 h_8 = \frac{\pi R^2}{2} (h_0 + h_R), \qquad \text{or} \qquad h_0 + h_R = 2h_8.$$

Adding the two equations involving h_0 and h_R gives

$$h_R = h_8 - \frac{a_x R}{g},$$

where

$$h_8 = \frac{\text{Volume}}{\text{Area}} = \frac{(8 \text{ fl oz}) \left(\dfrac{\text{gal}}{128 \text{ fl oz}} \right) \left(\dfrac{231 \text{ in}^3}{\text{gal}} \right)}{\dfrac{\pi}{4} (2.5 \text{ in.})^2} = 2.94 \text{ in.}$$

For $a_x = a_{r,B} = -11.8 \text{ ft/sec}^2$, we have

$$h_R = 2.94 \text{ in.} - \frac{(-11.8 \text{ ft/sec}^2)(1.25 \text{ in.})}{(32.2 \text{ ft/sec}^2)} = 3.40 \text{ in.}$$

For the average acceleration,

$$a_x = \left(\frac{r_{r,A} + a_{r,B}}{2} \right) = -\left(\frac{11.8 + 12.3}{2} \right) \text{ ft/sec}^2 = -12.0 \text{ ft/sec}^2,$$

$$h_R = 2.94 \text{ in.} - \frac{(-12.0 \text{ ft/sec}^2)(1.25 \text{ in.})}{(32.2 \text{ ft/sec}^2)} = 3.41 \text{ in.}$$

For

$$a_x = a_{r,B} = -12.3 \text{ ft/sec}^2,$$

$$h_R = 2.94 \text{ in.} - \frac{(-12.3 \text{ ft/sec}^2)(1.25 \text{ in.})}{(32.2 \text{ ft/sec}^2)} = 3.42 \text{ in.}$$

Example 2.12 gives $H = 4.66$ in. Since $H > h_R$ for all three accelerations, we conclude:

Answer

| Soft drink will not spill from the cup. |

◀

---

* Note that a_x is negative as x is radially outward.

Discussion

Since h_R increases as a_x increases, we could have been more efficient in our solution and calculated h_R for only the maximum acceleration. As long as this value of h_R is less than H, we are certain that no soft drink will spill out of the cup.

This type of approach to a problem is called a "conservative approach" or "making a conservative estimate." The general procedure is as follows:

- Hypothesize a possible occurrence. In this problem we could hypothesize either that some soft drink will spill or that no soft drink will spill. We used the hypothesis that some soft drink will spill.
- Proceed with calculations. When values of parameters are selected, select those values that are *least likely* to cause the hypothesized occurrence. In this case, we selected the minimum possible acceleration.
- If the hypothesized occurrence would occur under the least likely conditions, it will certainly occur under the actual conditions.
- If the hypothesized occurrence would not occur (i.e., soft drink will not spill) under these least likely conditions, it could still occur with more likely values of the parameters. In this case, *it is necessary to choose the opposite hypothesis* (i.e., some soft drink will spill) and proceed with the calculations using the parameter values least likely to produce this newly hypothesized occurrence. It is *not* sufficient to retain the original hypothesis and make calculations with the values of the parameters most likely to cause the hypothesized occurrence.

The key to the procedure is that you must always work with the least likely values of the parameters. If the answer is yes for the least likely values of the parameters, you can be sure that it is yes in the real case. If the answer is no for the least likely values of the parameters, you cannot make a conclusion about the real case.

Problems

Extension and Generalization

1. Prove that the static pressure at a point is the same in all three directions in a static fluid.

2. Find the center of pressure of an elliptical area of minor axis a and major axis b where axis a is vertical and axis b is horizontal. The center of the ellipse is a vertical distance h below the surface of the water ($h > a$).

3. Will the center of pressure of the ellipse in Problem 2 change if the water is replaced by another constant-density fluid? Explain.

4. Will the center of pressure of the ellipse in Problem 2 change if the vertical axis is tilted back an angle α from the vertical? Explain.

5. The U-tube manometer in Fig. P2.1 is used to measure the acceleration of the cart on which it sits. Develop an expression for the acceleration of the cart in terms of the liquid height h, the liquid density ρ, the local acceleration of gravity g, and the length ℓ.

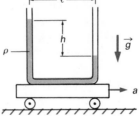

Figure P2.1

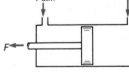

Figure P2.2

Lower-Order Application

6. The hydraulic cylinder in Fig. P2.2, with a 4-in.-diameter piston, is advertised as being capable of providing a force of $F = 16$ tons. If the piston has a design pressure (the maximum pressure at which the cylinder should safely operate) of 2500 lb/in², gage, can the cylinder safely provide the advertised force?

7. Express a pressure of 14.696 psia in (a) inches of mercury, (b) feet of water, and (c) millimeters of mercury.

8. Express the pressure in the tires of your car in pascals and in millimeters of mercury.

9. Even though women generally weigh less than men, why are the heel imprints of women's high-heel shoes more often found in soft floors than are the heel imprints of men's shoes?

10. A 12-in.-diameter, 10-ft-long piston of a hydraulic lift for automobiles is made of solid steel and is used to raise automobiles weighing up to 6000 lb. What gage pressure is necessary at the bottom of the piston to support this load?

11. Why is it dangerous to extinguish burning oil, grease, or gasoline with water?

12. A glass of liquid refreshment has a mixture, by volume, of 25% bourbon ($S \approx 0.78$) and 75% water. If the maximum suction that a person can develop is 100 mm Hg, what is the longest length of a straw of 1/4-inch inside diameter that this person can use and finish the drink?

13. Denver, Colorado, is called the "mile-high city" because its state capitol stands on land 1 mi above sea level. Assuming the standard atmosphere exists, what is the pressure and temperature of the air in Denver?

14. The interior of a 200-m-high building is maintained at 25°C, while the outside air is at 0°C. The inside and outside static pressures are equal at the building midheight. Find the difference of the inside and outside static pressures at the top of the building if there are no open windows to equalize the two pressures. Which pressure is higher?

15. Pilots use the rule of thumb that the air temperature drops 3.5°F for every 1000 ft increase in altitude. How does this rule compare with the standard atmosphere?

16. The Empire State Building is 380 m high. Estimate the magnitude of the atmospheric pressure in pascals at the top of the building if the atmospheric pressure at the bottom is 760 mm Hg, abs, and the atmospheric temperature is (a) $-18°C$ and (b) 30°C.

17. A tree has long tubes called *xylem tubes* to carry sap from the roots to the branches. The sap rises in the tubes partly due to surface tension and partly due to the higher pressure in the roots than in the branches. If sap has a surface tension of 0.0015 lb/ft and a density of 50 lbm/ft³, what is the maximum height due to surface tension to which the sap will rise in xylem tubes of the following diameters: 500 Å, 1000 Å, and 2000 Å?

18. A long glass tube is filled with light machine oil ($S = 0.834$), inverted, and then placed in a can of the same oil. Find the column height if the atmospheric pressure and temperature are 14.696 psia and 70°F, respectively.

19. Find the pressure p_T at the top of the water pipe in Fig. P2.3.

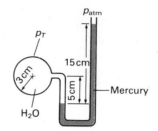

Figure P2.3

20. A U-tube manometer is used to check the pressure of natural gas entering a furnace. One side of the manometer is connected to the gas inlet line, and the water level in the other side rises 3 in. What is the gage pressure of the natural gas in the inlet line in inches of water and in lb/in², gage? List all assumptions used.

21. The container in Fig. P2.4 holds water and air as shown. Find the absolute pressures at locations A, B, and C.

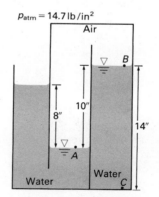

Figure P2.4

22. A commercial manometer was not available, so a U-tube manometer was hastily constructed from two pieces of straight glass tubing with the same outside diameter and a piece of flexible tubing. When a pressure difference was applied to the manometer, the left water column rose twice as much as the right column fell. Explain the probable cause for the unequal change in the two water columns. Find the applied pressure difference if the left column of 40°F water rose 4.0 in.

23. The U-tube manometer in Fig. P2.5 has two fluids, water and oil ($S = 0.80$). Find the height differ-

ence between the free water surface and the free oil surface with no applied pressure difference.

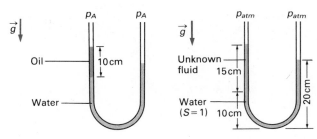

Figure P2.5   **Figure P2.6**

24. What is the specific gravity of the liquid in the left leg of the U-tube manometer in Fig. P2.6?

25. Determine the difference in pressure between points A and B in Fig. P2.7. Assume $h = 20.3$ in.

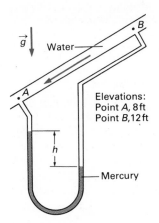

Elevations:
Point A, 8 ft
Point B, 12 ft

Figure P2.7

26. The difference in the two levels of the mercury manometer in Fig. P2.8 is 15 mm. Find the pressure difference between points A and B to which the manometer is connected.

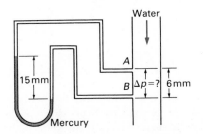

Figure P2.8

27. A small pocket of air is trapped in the horizontal leg of the top U-bend of the manometer in Fig. 2.9. For the same pressure difference $(p_A - p_B)$, will h_2 be different than if no air bubble were present?

28. The cistern shown in Fig. P2.9 has a diameter D that is 10 times the diameter d of the inclined tube. Find the drop in the fluid level in the cistern and the pressure difference $(p_A - p_B)$ if the liquid in the inclined tube rises $\ell = 20$ in. The angle θ is 20°. The fluid specific gravity is 0.85.

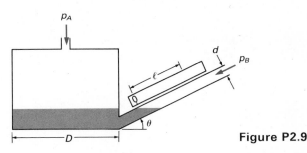

Figure P2.9

29. Consider the cistern manometer in Fig. P2.9, but with $\theta = 30°$ and the fluid 70°F water. Assume the fluid aligns with the zero scale reading when $p_A = p_B$. Find the length of the fluid in the tube (from the zero mark) when $(p_A - p_B) = 2.0$ psi.

30. A little Dutch boy puts his thumb in a hole in a dike. If the sea level is 3 m above the boy's thumb, estimate the force in newtons that the boy must exert to keep the water from coming out the hole. The boy's thumb has a cross-sectional area of 3.0 cm².

31. The container in Fig. P2.10 has square cross sections. Find the horizontal force on one of the four inclined surfaces $ABCD$. Also find the net vertical force on the bottom EF. Is the vertical force equal to the weight of the water in the container?

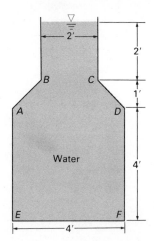

Figure P2.10

32. Assume that the container in Fig. P2.10 has circular cross sections. Find the vertical force on the inclined surface. Also find the net vertical force on the bottom *EF*. Is the vertical force equal to the weight of the water in the container?

33. The concrete gravity dam in Fig. P2.11 has a very pervious foundation such that water leaks under the dam to produce the pressure distribution shown and an uplift. The dam is 20 ft wide at the top and 120 ft wide at the base. Will the dam topple over? The concrete has a specific weight of 140 lb/ft^3.

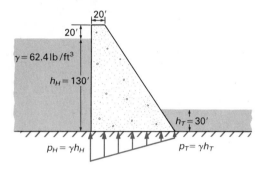

Figure P2.11

34. A 10-ft-long log is stuck against a dam as shown in Fig. P2.12. Find (a) the magnitude and direction of the force of the water on the log and (b) the force of gravity on the log.

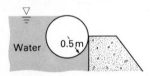

Figure P2.12

35. Determine the magnitude and direction of the force that must be applied to the bottom of the gate in Fig. P2.13 to keep the gate closed.

36. Find the maximum permissible water level so the gravity dam in Fig. P2.14 will not tip over. Assume that no hydrostatic pressure is exerted on the bottom of the dam.

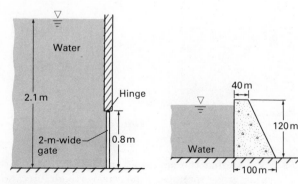

Figure P2.13   **Figure P2.14**

37. The side of the pool in Fig. P2.15 has a strut *ST* every 5 ft. Determine the magnitude of the force in the strut. Is this a compressive force or a tensile force?

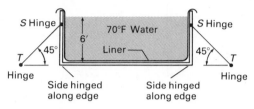

Figure P2.15

38. Estimate the "surface tension" in a 0.25-m-diameter balloon if the inside pressure exceeds the ambient pressure by 1.0 psi.

39. A small airplane has a mass of 700 kg and a wing area of 36 m^2. The average pressure on the lower wing surface is 70 kPa. What must be the average pressure on the upper wing surface for the plane to fly level?

40. What must be the average pressure on the upper wing surface if the plane in Problem 39 is to climb at an angle of 20° with the horizontal?

41. A yield point of 25,000 psi and a factor of safety of 2.0 are used in the design of a thin-walled cylindrical tank. The tank has inside diameter 18 in. and length 24 in. The tank is used for storing compressed air up to pressures of 150 psig. Find the minimum permissible wall thickness.

42. Find the force per unit width of the gate on the gear teeth in Fig. P2.16 to hold the gate in position.

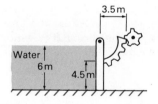

Figure P2.16

43. Figure P2.17 shows an approximate representation of the Keswick gravity dam in California. The dam is made of concrete having a density of 150 lbm/ft^3. The hydrostatic uplift pressure on the base of the dam varies linearly from 67% of the headwater pressure at the upstream toe of the dam to 100% of the tailwater pressure at the downstream toe. Determine the hydrostatic uplift pressure at both the upstream toe and the downstream toe of the dam, the uplift force per unit length of the dam (perpendicular to the paper), and the

location of the centers of pressure of the head-water and tailwater. Is this dam likely to overturn?

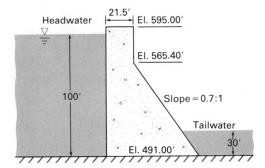

Figure P2.17

44. The coefficient of sliding friction μ between the base of the dam and the foundation is 0.65 for the Keswick gravity dam in Problem 43. Is this dam likely to slide downstream? Define an appropriate factor of safety for sliding for the dam and calculate its value for the Keswick dam.

45. Repeat Problem 43 for the approximate representation of the Altus gravity dam in Oklahoma, shown in Fig. P2.18.

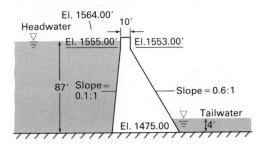

Figure P2.18

46. Repeat Problem 44 for the approximate representation of the Altus gravity dam shown in Fig. P2.18.

47. The completely enclosed hemispherical dome in Fig. P2.19 is built at the bottom of a deep lake for observations. Calculate the hydrostatic force on the dome.

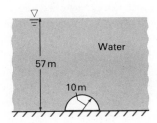

Figure P2.19

48. A vertical cylindrical water tank has a diameter of 20 ft and is filled to a depth of 30 ft. Find the hydrostatic force on the bottom of the tank.

49. A cylindrical water tank has a diameter of 7 m and a height of 20 m. Find the force F indicated in Fig. P2.20.

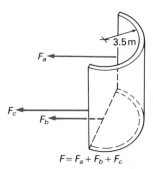

$F = F_a + F_b + F_c$ **Figure P2.20**

50. Find the force required to raise the top of the hemispherical dome in Problem 47 to the water surface. The hemispherical shell is clear plastic with thickness 2.5 cm and density 2300 kg/m³. Consider two cases: (a) the hemispherical dome has no base and (b) the hemispherical dome has a 2.5-cm-thick plastic base.

51. While building a high, tapered concrete wall, builders used the wooden forms in Fig. P2.21. If concrete has a specific gravity of about 2.5, find the total force on each of the three side sections of the wooden forms (neglect any restraining force of the two ends of the forms).

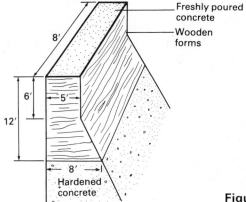

Figure P2.21

52. A tank contains 6 in. of oil ($S = 0.82$) above 6 in. of water ($S = 1.00$). Find the pressure on the bottom of the tank in psig. See Fig. P2.22.

53. Find the magnitude and location of the resultant force on the circular arc in Fig. P2.23. The arc is 25 m long.

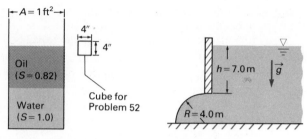

Figure P2.22

Figure P2.23

54. Find the magnitude and location of the resultant force on the hemispherical dome in Fig. P2.24.

55. Figure P2.25 shows a cross section of a submerged tunnel used by automobiles to travel under a river. Find the magnitude and location of the resultant force on the circular roof of the tunnel. The tunnel is 4 mi long.

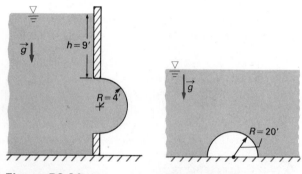

Figure P2.24

Figure P2.25

56. The dam in Fig. P2.26 is 200 ft long and is made of concrete with a specific gravity of 2.2. Find the magnitude and y-coordinate of the line of action of the net horizontal force.

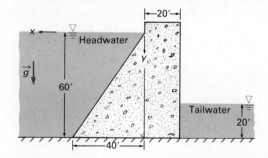

Figure P2.26

57. Repeat Problem 56, but find the magnitude and x-coordinate of the line of action of the vertical force on the dam due to the water.

58. An automobile has just dropped into a river. The car door is approximately a rectangle, measures 36 in. wide and 40 in. high, and has hinges on a vertical side. The water level inside the car is up to the midheight of the door, and the air inside the car is at atmospheric pressure. Find the force required to open the door if the force is applied 24 in. from the hinge line. See Fig. P2.27. (The driver did not have the presence of mind to open the window to escape.)

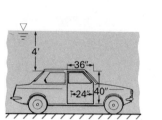

Figure P2.27

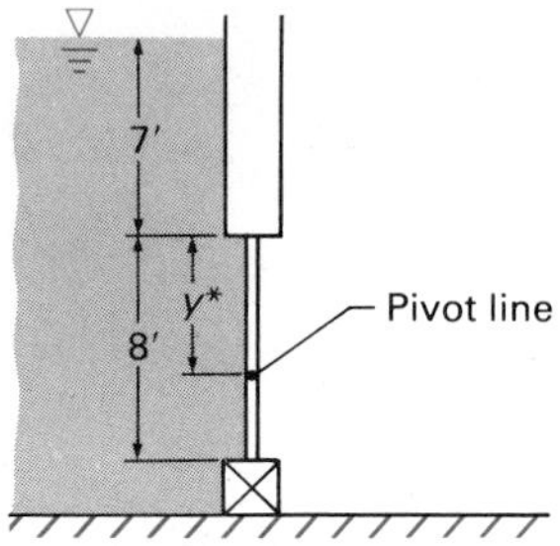

Figure P2.28

59. The gate in Fig. P2.28 is 10 ft wide and is exposed to freshwater on its left side. Find the magnitude of the force on the gate and the location y^* of the pivot line below the top of the gate so there is no moment tending to open the gate for the water depth shown.

60. A step-in viewing window having the shape of a half-cylinder is built into the side of a large aquarium. See Fig. P2.29. Find the magnitude, direction, and location of the net horizontal and net vertical forces on the viewing window.

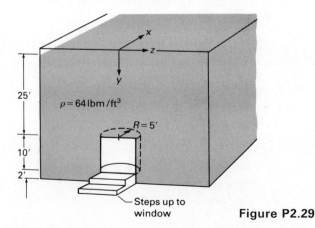

Figure P2.29

61. A hydrometer is used to measure the specific gravity of liquids. Figure P2.30 shows a hydrometer with a scale etched on the surface. Would the higher specific gravities be near the top or the bottom of the hydrometer?

Figure P2.30

62. The float in Fig. P2.31 is used to maintain a constant water level in a tank found in many homes. The valve closes, decreasing the flow rate as the float rises. The float and its support rod can be slid horizontally so as to move the float closer to or farther away from the pivot. If the water level at which a particular float stops the flow is too low, should you move the rod and float to the right or left? Could the same result be obtained by bending the rod near the float upward or downward?

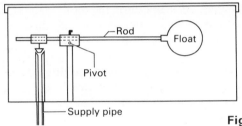

Figure P2.31

63. A floating 6-in.-thick piece of ice sinks 1 in. with a 500-lb polar bear in the center of the ice. What is the area of the ice in the plane of the water level?

64. Secret Agent 007 is set adrift in a hollow steel sphere with outside diameter 2 m and a 0.5-cm-thick shell. The agent weighs 800 N. How deep will the sphere float in sea water?

65. An empty 100,000-lb barge has a flat bottom and vertical sides. The barge is 20 ft wide, 60 ft long, and 10 ft high. The sides are made of $\frac{3}{4}$-inch steel plate. Find the draft of the barge. (The draft is the depth of the bottom of the barge below the water surface.)

66. A spherical balloon filled with helium at 40°F and 20 psia has a 25-ft diameter. What load can it support in atmospheric air at 40°F and 14.696 psia?

67. A not-too-honest citizen is thinking of making bogus gold bars by first making a hollow iridium ($S = 22.5$) ingot and plating it with a thin layer of gold ($S = 19.3$) of negligible weight. The bogus bar is to have a mass of 100 lbm. What must be the volumes of the bogus bar and of the air space inside the iridium so that an inspector would con-

clude it was real gold after weighing it in air and water to determine its density? Could lead ($S = 11.35$) or platinum ($S = 21.45$) be used instead of iridium?

68. What is the minimum volume of a life preserver ($S = 0.2$) to support a 170-lb person in freshwater, with 80% of the person's body submerged in the water? The person's specific gravity is about 0.94.

69. Find the weight of a ship that is said to "displace 100,000 tons" (i.e., is submerged a volume of water weighing 100,000 tons).

70. Will a boat displace more water at 1000 ft above sea level or at sea level? Assume the body of water at each elevation has no salt content.

71. Will the scale reading in Fig. P2.32 read the same or differently when the aluminum ball is in the water? The ball has a circumference of 12 in. and a weight at sea level of 6.5 oz, and it is submerged halfway. Does the scale reading depend on the depth to which the ball is immersed?

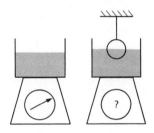

Figure P2.32

72. Will the buoyant force on a submarine be the same, smaller, or larger if the submarine moves from a floating position 20 fathoms below sea level to a floating position 40 fathoms below sea level? How does the submarine achieve a floating position at each level?

73. What volume fraction of an iceberg floats above seawater? Recall that an iceberg is frozen freshwater.

74. A supertanker is used to transport oil. It has an unloaded weight of 150 million lb, a length of 1000 ft, a width of 150 ft, and the cross section shown in Fig. P2.33. Find the depth the bottom of

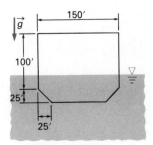

Figure P2.33

the tanker is below the water level if the tanker carries 8 million gal of oil with a specific gravity of 0.9.

75. Is the supertanker of Problem 74 stable?

76. A tall drinking glass is inverted and placed in a relatively large pan of water such that the glass floats in a vertical position. The pan is covered with a clear, thin plastic sheet securely fastened by rubber bands as shown in Fig. P2.34. If you push down on the plastic sheet, will the glass rise out of the water, sink into the water, or remain stationary? Explain.

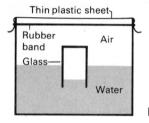

Figure P2.34

77. Check all correct answers: A completely submerged body is stable if
_____ (a) the center of buoyancy is directly above the center of gravity;
_____ (b) the center of buoyancy coincides with the center of gravity;
_____ (c) the center of buoyancy is directly below the center of gravity;
_____ (d) the center of buoyancy coincides with the centroid of the submerged volume.

78. An object weighs 35 lb in water and 55 lb in air. Find the object's specific gravity. Would the object be partially submerged or totally submerged while in water?

79. Find the volume of the object in Problem 78.

80. It is proposed to build a gate as sketched in Fig. P2.35 to control the water level in a reservoir. The spring extends (or is compressed) 1 in. for each 100 lb of tension (or compression) applied to it. Find the length of extension or compression of the

spring if the counterweight floats half-submerged in water.

81. How much must the water level in Fig. P2.35 rise (or fall) for the gate to move down 6 in.?

82. Assume that the tank with oil and water in Problem 52 is 18 in. high. What will be the pressure on the bottom of the tank if a cube having a volume of 64 in^3 and a specific gravity of 0.75 floats in the oil? See Fig. P2.22 on page 102.

83. Thirty cubic meters of wine having a specific gravity of 1.05 is fortified by removing 5 m^3 of wine and adding an equal volume of ethanol having a specific gravity of 0.85. How much deeper will a hydrometer float if it has a weight of 0.020 N and the stem protruding out of the wine has a diameter of 0.25 cm?

84. Find the specific gravity of the fortified wine in Problem 83.

85. The float in Fig. P2.36 is used to control the water level in the tank. Find the diameter d to maintain a water depth D in the tank. Neglect the mass and volume of the plug and the connecting string.

86. If the mass m and height h of the float in Fig. P2.36 were unchanged, would increasing the float diameter d above that found in Problem 85 cause the float to remain open or remain closed?

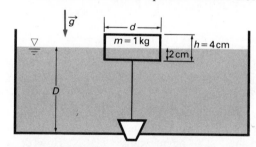

Figure P2.36

87. Consider the baron's hot air balloon in Fig. P2.37. The balloon, basket, burner, and butane weigh a total of 50 lb. The baron weighs 150 lb. What is the required temperature inside the balloon to lift the balloon, basket, burner, butane, and baron into the blue?

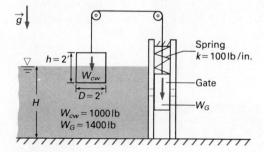

Figure P2.35

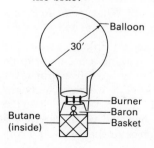

Figure P2.37

88. In drilling for oil in the Gulf of Mexico, some divers have to work at a depth of 400 m (1300 ft). (a) Assuming that seawater has a constant density of 64 lb/ft³, compute the pressure at this depth. (b) Estimate the change in density that this pressure would cause. The divers breathe a mixture of helium and oxygen, which is stored in cylinders as shown in Fig. P2.38 at a pressure of 3000 psia. (c) Calculate the force that the weld must resist and that tends to blow the end cap off while the diver is using the cylinder at 400 m. (d) After emptying a tank, a diver releases it. Will the tank rise or fall and what is its initial acceleration?

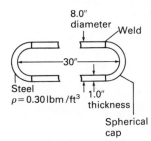

Figure P2.38

89. A right circular cylinder of specific gravity S has radius R and height H. For what range of values of R/H will the cylinder float stably in water if the circular ends are horizontal?

90. A cube of specific gravity S has sides of length L. For what range of values of S will the cube float stably in water?

91. A drinking glass has a wall thickness of 3 mm, an inside diameter of 6.0 cm, and an inside depth of 18.0 cm. The glass has a specific gravity of 2.5. Is the drinking glass in stable or unstable equilibrium if placed upright in a pan of 60°F water?

92. A solid pine ($S = 0.50$) spar buoy has a lead ($S \approx 11.3$) weight attached as shown in Fig. P2.39. Find the equilibrium position of the spar buoy in seawater (i.e., find d). Is this spar buoy stable or unstable?

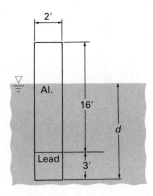

Figure P2.39

93. Find the minimum length of lead attached to the spar buoy in Fig. P2.39 such that the spar buoy will be stable when floating in seawater. How much of the spar buoy will then be floating out of the water?

94. How deep will an empty 1-gal paint can float in water? Will the paint can be stable or unstable? Use measured values of the outside diameter and weight of the paint can and the water density. Conduct an experiment with the paint can floating in water and compare the results of your analysis with those of the experiment.

95. Find the angle θ at which a wooden meter stick, as shown in Fig. P2.40, will float in the water. It will be necessary to obtain the weight and volume of a meter stick. Compare your calculated value of θ with that obtained by an experiment.

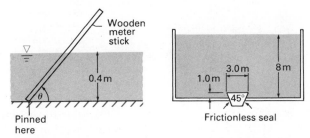

Figure P2.40   Figure P2.41

96. Will the conical stopper in Fig. P2.41 remain closed?

97. Find the minimum weight of the platform in Fig. P2.42 for which the floating oil platform is stable about an x,x-axis passing through two of the cylinders.

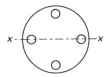

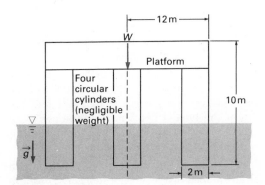

Figure P2.42

98. The circular buoy in Fig. P2.43 is to float in fresh-water to warn boaters. Will the buoy do its intended job? Is it necessary to connect a chain or rope from the bottom of the buoy to the bottom of the lake?

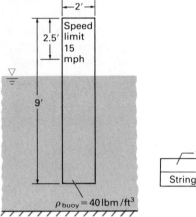

Figure P2.43

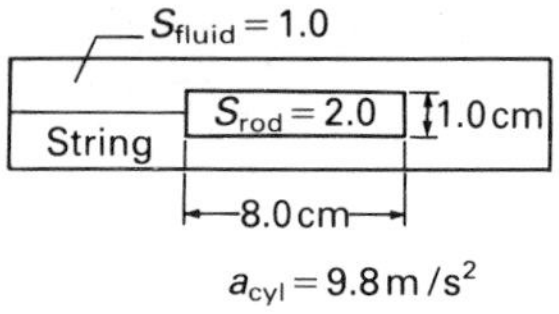

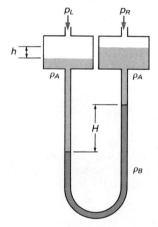

Figure P2.44

99. The cylinder in Fig. P2.44 accelerates to the left at the rate of 9.80 m/s². Find the tension in the string connecting the rod to the cylinder. The volume between the rod and the cylinder is completely filled with water.

100. Find the tension in the string of Problem 99 if the rod has a specific gravity $S = 0.75$.

101. Repeat Problems 99 and 100 if the cylinder is accelerated to the right at the rate of 9.80 m/s².

102. When an automobile brakes, the fuel gage indicates a fuller tank than when the automobile is traveling at a constant speed on a level road. Is the sensor for the fuel gage located near the front or rear end of the fuel tank? Assume a constant deceleration.

103. How fast must a 12-cm-diameter centrifuge rotate to separate blood plasma from heavier red blood cells? Plasma makes up about 55% of the volume of blood. Red blood cells have a specific gravity of 1.09, and plasma has a specific gravity of 1.03.

104. How fast must a bucket of water be rotated in a horizontal plane so that no water falls out of the bucket? Assume the bucket is half full and the bottom of the bucket is 2.0 m from the center of rotation. Is it necessary to know how much water is in the bucket?

105. A glass is filled to within $\frac{1}{2}$ in. of the top with water and placed in a level spot in a car. The glass has an inside diameter of $2\frac{1}{2}$ in. How fast can the driver of the car accelerate along a level road without spilling any of the water?

Higher-Order Application

106. The sensitivity S of the micromanometer in Fig. P2.45 is defined as

$$S = \frac{H}{p_L - p_R}.$$

Find the sensitivity of the micromanometer in terms of the densities ρ_A and ρ_B. How can the sensitivity be increased?

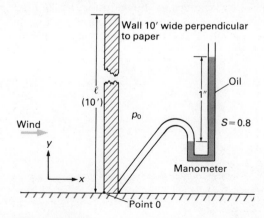

Figure P2.45

107. Consider the micromanometer in Fig. P2.45. For a given size, does increasing the sensitivity affect the pressure range $(p_L - p_R)$ that can be measured?

108. Wind blows against the large vertical wall shown in Fig. P2.46. Find the pressure p_0 at the bottom of the wall in psig. The pressure over the face of

Figure P2.46

the wall is given by

$$p = p_0 \left[1 - \left(\frac{y}{\ell}\right)^2 \right],$$

and the pressure over the back of the wall is atmospheric. Both p and p_0 are measured in psig. Find the resultant force on the wall and its line of action.

109. Consider the cistern manometer in Fig. P2.47. The scale is set up on the basis that the cistern area A_1 is infinite. However, A_1 is actually 50 times the internal cross-sectional area A_2 of the inclined tube. Find the percent error (based on the scale reading) in using this scale.

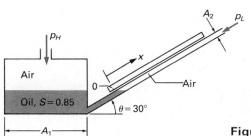

Figure P2.47

110. The submarine in Fig. P2.48 floats in seawater ($\gamma = 64.0$ lb/ft^3). The pressure inside the submarine is atmospheric. For shallow depths, the mercury manometer shown in the figure is used to indicate the depth. For the manometer levels shown, calculate the depth of the manometer tap in the submarine wall. Then find the minimum force required to open the submarine hatch at this depth. Neglect the weight of the hatch.

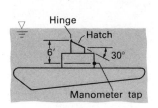

Figure P2.48

111. The Stickney gate in Fig. P2.49 is used for crest-level control. The gate is hinged at A and has two legs B and C. When the headwater level is below a limiting height h_1, the pressure on leg C keeps the gate closed in the position shown in the figure. As the headwater level reaches this limiting height, the pressure on leg B causes the gate to fall. A particular Stickney gate has a weight of 20,000 N/m of length perpendicular to the paper and a

center of gravity 1 m to the right of and 5 m above the hinge. Find the limiting height h_1 for this particular gate.

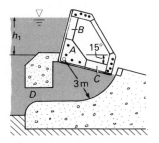

Figure P2.49

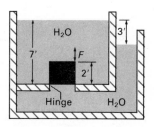

Figure P2.50

112. The 300-lb gate in Fig. P2.50 is a 3-ft-diameter cylinder hinged on the left so it can swing counterclockwise. Determine the force F to just open or to just hold the gate closed. Indicate the direction of F as well as its magnitude. Do Archimedes' principles apply?

113. A steel can with a $\frac{1}{16}$-in. wall thickness has an inside diameter of 2.5 in. and height 4.0 in. The can is inverted, placed in a large pan of water, and allowed to come to an equilibrium position. See Fig. P2.51. Assuming isothermal compression of the air in the can, find the depth of submergence H of the can.

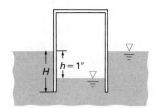

Figure P2.51

114. The polar bear in Problem 63 slowly moves from the center of the ice toward one of the sides. How far can the bear move before the top surface of the ice becomes submerged?

115. A submarine is modeled as a cylinder with a length of 300 ft and a diameter of 50 ft and with a conning tower as shown in Fig. P2.52. The submarine can dive a distance of 50 ft from the floating position in about 30 sec. Diving is accomplished by taking water into the ballast tank so the submarine will sink. When the submarine reaches the desired depth, some of the water in the ballast tank is discharged so the submarine will be in "neutral buoyancy" (i.e., it will neither rise nor sink). A simplified version of the ballast tank is shown in the figure. For the conditions illus-

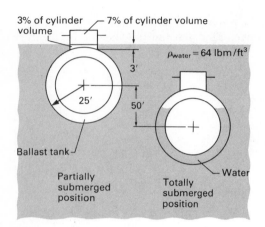

Figure P2.52

trated, find (a) the weight of the submarine; (b) the volume (or mass) of the water that must be in the ballast tank when the submarine is in neutral buoyancy. [*Notes of interest (not necessary to solve problem*): The intake and discharge of water in the ballast tank is accomplished by changing the air pressure above the water in the ballast tank. Since the submarine would most likely be moving as it submerges, it is equipped with diving planes (on the conning tower) to aid in descending and to help it level out at the desired depth.]

116. The wooden ($S = 0.5$) meter stick in Fig. P2.53 is floating vertically in water such that there is no tension in the string connecting it to the bottom of the container. Water is added to increase the water level by 4.0 cm. The string is then cut. Develop an expression for the depth of submergence of the meter stick as a function of time.

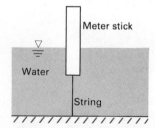

Figure P2.53

117. A wooden buoy is 16 ft long and floats in the vertical position with one-third of the buoy out of the water. The buoy is hit by a series of waves (a sinusoidal variation of the water height). The buoy is constrained to move only in the vertical direction. The waves are moving horizontally at a speed of 10 mph, the distance between crests is 15 ft, and the height between crests and valleys is 10 feet. Find the motion of the buoy as a function of time.

118. Is the meter stick in Problem 116 stable or unstable before the 4.0 cm of water is added? (Assume that the meter stick is 2.54 cm wide and 0.5 cm thick.)

119. Fir ($\gamma \approx 34$ lb/ft³) and oak ($\gamma \approx 55$ lb/ft³) trees felled by lumberjacks are floating down a river. Find how deep each floats in the river.

120. A tall drinking glass has a 3.0-mm wall thickness, a 6.0-cm inside diameter, an 18.0-cm inside depth, and a specific gravity of 1.2. Compare the depths of submergence and the stability of the glass if it is placed upright and then upside down in 60°F water.

121. A student wants to make a floating-ball hydrometer for an ethylene glycol–water mixture. See Fig. P2.54(a). The number of floating balls indicates the freezing temperature of the mixture. Find the density of each of four balls given the following data.

Number of Floating Balls	Freezing Temperature of Ethylene Glycol–Water Mixture
4	0°C
3	−10°C
2	−20°C
1	−30°C

The relation of the density of the ethylene glycol mixture to its freezing point is shown in Fig. P2.54(b).

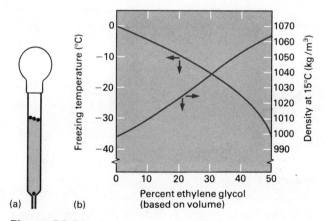

Figure P2.54

122. The U-tube in Fig. P2.55 rotates at 2.0 rev/sec. Find the absolute pressures at points C and B if the atmospheric pressure is 14.696 psia. Recalling that 70°F water evaporates at an absolute pressure of 0.363 psia, find the absolute pressures at points C and B if the U-tube rotates at 6.0 rev/sec.

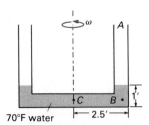

70°F water ⊢— 2.5′ —⊣   **Figure P2.55**

Comparison

123. Find the percent difference in the readings of the two identical U-tube manometers in Fig. P2.56. Manometer 90 uses 90°C water and manometer 35 uses 35°C water. Both have the same applied pressure difference. Does this percentage change with the magnitude of the applied pressure difference? Can the difference between the two readings be ignored?

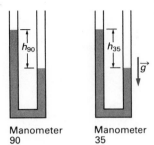

Manometer   Manometer
90   35   **Figure P2.56**

References

1. Beckwith, T., N. Buck, and R. Marangoni, *Mechanical Measurements* (3rd ed.), Addison-Wesley, Reading, MA, 1981.
2. Benedict, R. P., *Fundamentals of Temperature, Pressure, and Flow Measurements* (2nd ed.), Wiley, New York, 1977.

Fundamental Principles and Concepts of Flow Analysis

3

Although there are many engineering applications of fluid statics and a few applications of rigid body motion of fluids, most engineering fluid mechanics problems are concerned with fluid flow. We covered most methods of analysis of nonflowing fluids in a single chapter, but it will require several chapters to develop the basic methods of flow analysis.

This chapter sets down the fundamental concepts of flow analysis, states the basic laws of nature that apply to fluid motion, and discusses the ways in which these laws can be expressed mathematically and used to solve flow problems. In a broad sense, we will formulate the problem of flow analysis. A clear understanding of this chapter is absolutely essential for further progress; study this material carefully before proceeding to later chapters. We recommend that you reread Sections 1.2 and 1.5 before beginning this chapter.

3.1 Typical Flow Situations

An extremely wide variety of flow situations is important in engineering. Figures 3.1 and 3.2 illustrate a few typical flows of two broad types: *internal* or *confined* flows and *external* or *unconfined* flows. For the internal flows of Fig. 3.1, engineers are usually interested in the rate of fluid flow through the passage, changes of the fluid's energy during the flow, and, possibly, the forces exerted on the confining walls. The fluid is entirely confined by the walls of the passage and, in the case of the open channel flow, the free surface of the channel. The fluid involved may be gas or liquid. The flow may vary with time, as during the start-up of a pump, or may be invariant with time, as when a pump runs constantly.

In the external flows of Fig. 3.2, there is relative motion between an object and a large mass of fluid. Engineers are usually interested in the forces exerted on the object by the fluid. They may also need to know the details of the fluid motion near the object's surface so they can evaluate the rate of heat transfer to or from the surface. In this case, the fluid is not contained in a limited region; in fact, the fluid is often assumed to extend all the way to "infinity." The fluid may be gas or liquid or

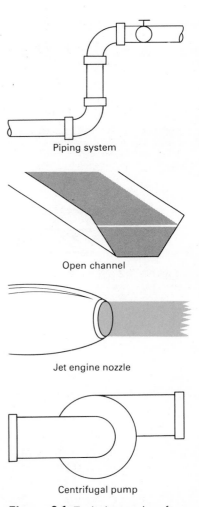

Piping system

Open channel

Jet engine nozzle

Centrifugal pump

Figure 3.1 Typical examples of internal flows.

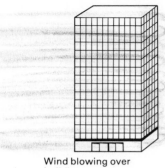

Wind blowing over
building

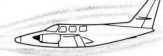

Flying airplane

Sailboat

Moving automobile

Figure 3.2 Typical examples of
external flows.

both, as with the sailboat. The flow may vary with time, such as during takeoff and climb of the airplane, or it may be invariant with time, as when the airplane cruises at constant speed and altitude in a still atmosphere.*

There are many differences between these two types of flows; nevertheless, *fluid motion in both types is governed by the same physical laws.* Engineers may use different *mathematical forms* of these laws and may use different *solution techniques,* but the physical laws are the same. In formulating an approach to flow problems, we will be as general as possible at the beginning, so that you do not mistakenly conclude that there are two "kinds" of fluid mechanics. There is only one kind of fluid mechanics, based on a rather small number of universally valid physical laws. What may later seem to be different kinds of fluid mechanics are just different formulations from a mathematical, but not physical, point of view.

The division between internal and external flows was introduced primarily to show the widely different types of engineering flow problems. Engineers often model flows of one type by exactly the opposite type. The flow over an airplane is often studied by placing a scale model of the airplane in a wind tunnel. If the model is small compared to the wind tunnel cross-sectional area, the flow near the model is similar to the flow near the real airplane in the unconfined atmosphere. Some flows are a mixture of both types. As an automobile moves along the road, the flow between the road surface and the underside of the auto is confined, while the flow over the sides and top is unconfined.

3.2 Fundamental Laws for Flow Analysis

Experience has shown that all fluid motion must be consistent with the following fundamental laws of nature.

1. *The law of conservation of mass.* Mass can be neither created nor destroyed; it can only be transported from place to place or stored.
2. *Newton's three laws of motion.*
   - A mass remains in a state of equilibrium, that is, at rest or in motion with constant speed and direction, unless acted on by an unbalanced force.
   - The rate of change of momentum of a mass is proportional[†] to the net force acting on the mass.
   - Any force action has an equal (in magnitude) and opposite (in direction) force reaction.
3. *The first law of thermodynamics (law of conservation of energy).* Energy, like mass, can be neither created nor destroyed. Energy can be transported, changed in form, or stored.

---

* This flow is time invariant only in a coordinate system fixed to the airplane, a trick often used in analysis of external flows.

† The "constant of proportionality" involves only the conversion between units of force, mass, length, and time.

4. *The second law of thermodynamics* (*law of degradation of energy*). This law is familiar to almost everyone by common experience but is formally studied only in thermodynamics. Although energy is conserved, it can change form. Some forms of energy (for example, kinetic and potential energies) are "better" than other forms (such as heat and internal energy). Although the "better" forms can be completely converted to the "lesser" forms, the "lesser" forms can be only partially converted to the "better" forms.* In thermodynamic language, the "better" forms are said to be more *available* than the "lesser" forms.

5. *The state postulate* (*law of property relations*). The various properties of a fluid are related. If a certain minimum number (usually two) of a fluid's properties are specified, the remainder of the properties can be determined (see Section 1.3).

The important thing to remember about these laws is that they apply to all situations.† They do not depend on the nature of the fluid, the geometry of the boundaries, or anything else. As far as we know, they have always been true and will continue to be true unless suspended by the Creator of the universe. It follows that we can firmly base analysis of *any* flow on these laws.

In addition to these universal laws, several lesser laws apply in restricted circumstances. An example is Newton's law of viscosity (see Section 1.3.4.):

> The shear stress in a fluid is proportional to the rate of deformation of the fluid.

This law is true for only some fluids and does not apply at all to solids. Such limited laws are better termed "constitutive relations." To solve most flow problems, we must use some constitutive relations, but we must select them carefully to match the particular problem.

3.3 Mathematical Formulation of the Fundamental Laws

The fundamental laws are the basis of our understanding of fluid (and solid) motion. Besides understanding, an engineer needs predictive capability. To design a piece of equipment or a structure, an engineer must know in advance how it will affect and be affected by fluids that

---

* This principle is easily illustrated by some familiar examples. Suppose a heavy block slides down an inclined plane. Because of friction, the block and the plane are both heated. If we start with the block at the bottom of the plane and heat the block and the plane with a blowtorch, we cannot make the block slide up the plane. Rubbing your hands together generates "heat," but heating your hands does not make them rub. A cup of hot coffee will transfer heat to the air in the kitchen and the coffee will get cold, but the cold coffee will never soak up heat from the kitchen air and get hotter. By now you should have the idea. See if you can come up with some other examples.

† In the case of nuclear reactions, mass *can* be changed into energy. In this case, laws 1 and 3 are combined, with mass simply a form of energy. We will not consider such cases in this book.

it contacts. An airplane's engines must be sufficiently powerful to overcome air resistance at the desired flight speed. A pipe must be large enough to convey water at a desired flow rate. A building must be strong enough to resist wind loads.

To obtain a predictive capability, we formulate the fundamental laws as mathematical expressions. We assign symbols to fluid velocity, pressure, and density, to the wind force on a building, and to the rate of flow in a pipe. We describe the proposed (or existing) equipment or structure (airplane, building, pipe) by certain geometric relations. Application of the fundamental laws permits us to write equations involving the symbols. By manipulation of the symbols according to the rules of mathematics, we can "solve" the problem, that is, we can predict the velocity or pressure or wind force that we need to know. It is unfortunate that while we can formulate a mathematical representation of the fundamental laws for almost any flow situation, our ability to solve the resulting mathematical problem is rather limited. In this section, we examine the various approaches to mathematical formulation of the fundamental laws. The remainder of the book is devoted to formulating the laws for specific situations and examining techniques for solving flow problems.

3.3.1 Field Representation: Lagrangian and Eulerian Approaches

Consider two of the flows illustrated in Figs. 3.1 and 3.2: the flow of water in a pipe and the flow of air around a moving automobile. In the pipe flow, the fluid of interest is confined within the pipe; in the flow around the automobile, the entire atmosphere is the fluid of interest. In both cases, however, the fluid is continuous. We imagine that at each point in the fluid region there is a tiny mass that we call a *fluid particle*. There are an infinite number of particles in both cases (but there are more in the automobile flow). Since a fluid is deformable, the velocity of each fluid particle may be different. In addition, the velocity of any fluid particle may change over time (due to forces on the particle). A complete description of fluid motion requires specification of the velocity, pressure, and so on of all particles at all times. Since a fluid is continuous, this can be accomplished only by mathematical functions that express the velocity and fluid properties for all particles at all times. Such a representation is called a *field representation,* and the dependent variables (such as velocity and pressure) are called *field variables*. The fluid region of interest (the water in the pipe or the air around the automobile) is called the *flow field*. You might try to identify the flow field for the other flow situations in Figs. 3.1 and 3.2.

To describe a flow field, we can adopt either of two approaches. The first approach, called the *Lagrangian description* (after the French scientist J. L. de Lagrange), identifies each particular fluid particle and describes what happens to it as time progresses. Mathematically we write the fluid velocity

$$\vec{V} = \vec{V}(\text{particle identity}, t).*$$

* Remember that the dashed parentheses ( ) mean "is a function of."

The independent variables are particle identity and time. The Lagrangian approach is very popular in solid mechanics. In your study of dynamics, you undoubtedly calculated the velocities of a thrown ball, a projectile, a particular block attached to a pulley system, and so forth. A Lagrangian description is attractive if a very small number of particles is involved, if all particles move together as a rigid body, or if all particles are displaced only a small amount from their initial or equilibrium position. In a flowing fluid, it is virtually impossible to identify and keep track of various particles.* Further complications arise because a typical fluid particle often experiences a large displacement. For these reasons, a Lagrangian description is not very useful in fluid mechanics.

The second approach, called the *Eulerian description* (after the Swiss mathematician L. Euler), focuses attention on a particular point (or region) in space and describes happenings at that point (or within and on the boundaries of the region) as time progresses. The properties of a fluid particle depend on the location of the particle in space and time. Mathematically, the velocity field is expressed by:

$$\vec{V} = \vec{V}(x, y, z, t).$$

The independent variables are location in space, represented above by the rectangular coordinates (x, y, z), and time. We might talk about the fluid velocity at the outlet of the pipe 3 seconds after the flow started or the air pressure 3 inches ahead of the hood ornament of the automobile. At each instant of time, a different fluid particle will probably be occupying those positions, but that would not matter to us. Since it is usually easier to identify fixed points in space rather than individual pieces of fluid, the Eulerian description is most often used in fluid mechanics. Solving a fluid flow problem then becomes tantamount to determining the velocity, pressure, and so on as functions of space coordinates and time. The function

$$\vec{V}(x, y, z, t) \qquad \text{or} \qquad p(x, y, z, t)$$

can then be used to find velocity or pressure anywhere within the field at any time simply by substituting values for x, y, z, and t.

The Eulerian description is especially suited to fluid mechanics problems because it does not involve what happens to any particular fluid particle. The engineering application of a flow analysis seldom is concerned with what happens to a specific piece of fluid but usually is concerned with the effects of the fluid motion on certain objects, such as the blades of a pump or the windows of a building. The pressure on the window, not the effect of the window on any particular fluid particle, is important to the building designer.

An amusing illustration of the Eulerian and Lagrangian descriptions of fluid motion as applied to traffic on a freeway has been suggested by White [1]. Engineers who design the freeway are concerned with the number of cars that must pass over the road, the movement of traffic as a whole, and the number of cars that may be expected to

---

* The only practical method identifies particles by their positions in space at a particular instant in time, for example, the initial or equilibrium position (x_0, y_0, z_0).

enter or leave the freeway at each ramp. An Eulerian description is perfectly suited because the freeway designer has no interest in any particular car and its characteristics—whether it is red or blue, manual or automatic transmission, and so on. A police officer patrolling the freeway, however, is interested in identifying those particular cars that are breaking the law and in giving the drivers tickets. Only a Lagrangian approach would stand up in court!

EXAMPLE 3.1   Illustrates a velocity field given by an Eulerian description and some characteristics of fluid flow

A constant-density fluid flows in the converging, two-dimensional channel shown in Fig. E3.1(a). The width W (perpendicular to the paper) is quite large compared to the channel height. The velocity in the z-direction is zero. The channel half-height, Y, and the fluid x-velocity, u, are given by*

$$Y = \frac{Y_0}{\left(1 + \dfrac{x}{\ell}\right)}, \qquad u = u_0\left(1 + \frac{x}{\ell}\right)\left[1 - \left(\frac{y}{Y}\right)^2\right],$$

where x, y, Y, and ℓ are in meters, u is in m/s, $u_0 = 1.0$ m/s, and $Y_0 = 1.0$ m. Tabulate and plot the velocity distribution $u(y)$ at $x/\ell = 0$, 0.5, and 1.0. Use y/Y values of 0, ± 0.2, ± 0.4, ± 0.6, ± 0.8, and ± 1.0.

Solution

  Given
$Y(x)$ and $u(x)$ as specified
Figure E3.1(a)

Figure E3.1 (a) Two-dimensional channel; (b) plot of the velocity $u(x, y)$ as a function of the transverse coordinate y at various axial locations.

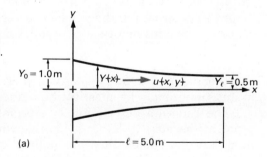

(a)

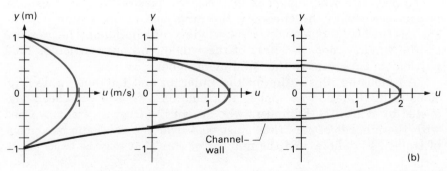

(b)

* Note that u does not depend on time.

Find

u for specified values of x and y

Solution

The solution of this problem simply involves substitution of numerical values of x/ℓ and y/Y into the equation for $u(x, y)$.

| The results are given in Table E3.1 | **Answer**
| and plotted in Fig. E3.1(b). | ◀

Discussion

With the exception of the fluid in contact with the wall, the fluid velocity u increases as the channel height $2Y$ decreases. The fluid in contact with the wall has zero velocity, a characteristic of any viscous fluid flowing inside a pipe or channel. An inviscid fluid flowing in a pipe or channel would not have zero velocity at the wall.

Table E3.1 Tabulated values of the velocity $u(x, y)$ as a function of the length ratio x/ℓ, the height ratio y/Y, and the height y for the two-dimensional channel.

Length Ratio x/ℓ	Height Ratio y/Y	Height y (m)	Fluid Velocity u (m/s)
0.0	0.00	0.00	1.00
	± 0.20	± 0.20	0.96
	± 0.40	± 0.40	0.84
	± 0.60	± 0.60	0.64
	± 0.80	± 0.80	0.36
	± 1.00	± 1.00	0.00
0.5	0.00	0.00	1.50
	± 0.20	± 0.13	1.44
	± 0.40	± 0.27	1.26
	± 0.60	± 0.40	0.95
	± 0.80	± 0.53	0.54
	± 1.00	± 0.67	0.00
1.0	0.00	0.00	2.00
	± 0.20	± 0.10	1.92
	± 0.40	± 0.20	1.68
	± 0.60	± 0.30	1.28
	± 0.80	± 0.40	0.72
	± 1.00	± 0.50	0.00

3.3.2 Visualizing the Velocity Field: Flow Lines

To visualize the velocity field in a complex flow, we can construct various flow lines either on paper or, sometimes, in the laboratory.* These lines are given the names *streamline, pathline, streakline,* and *timeline.* The first two are primarily analytical concepts; the last two are primarily laboratory concepts.

* Flow lines are illustrated very well in the film *Flow Visualization* [2].

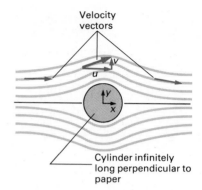

Figure 3.3 Typical streamlines for inviscid flow over circular cylinder.

A *streamline* is an imaginary line that is everywhere tangent to the fluid velocity vector. A few streamlines for the flow of an inviscid fluid over a circular cylinder are shown in Fig. 3.3. The entire flow field could be covered with streamlines. For any flow field, the streamlines can be expressed as a family of curves:

$$\Psi(x,\, y,\, z,\, t) = C,$$

where C is a constant. The equations of the streamline curves can (in principle) be obtained from the condition that the streamline be tangent to the velocity vector; thus

$$\left.\frac{dy}{dx}\right)_{\Psi} = \frac{v}{u}, \tag{3.1a}$$

$$\left.\frac{dy}{dz}\right)_{\Psi} = \frac{v}{w}, \tag{3.1b}$$

$$\left.\frac{dz}{dx}\right)_{\Psi} = \frac{w}{u}, \tag{3.1c}$$

where u, v, and w are the instantaneous x-, y-, and z-components* of the fluid velocity vector $\vec{V}$. Since the fluid velocity (and therefore its components) may change with time, the streamlines may change with time. The streamline curves are better represented by

$$\Psi(x,\, y,\, z) = C(t). \tag{3.2}$$

If the velocity field does not change with time, the streamlines are fixed curves in space. These curves are loosely called the *flow pattern*.

The streamline is an analytical concept. In the laboratory there is no way of marking curves that are instantaneously tangent to the fluid velocity, especially if the velocity varies with time.

A *pathline* is the curve marked out by the trajectory of a particular fluid particle as it moves through the flow field. Any fluid particle actually travels along its own pathline. The pathline is a Lagrangian concept; if we could analytically identify the pathline for all fluid particles, we could apply a Lagrangian method of description.

* An alternative notation is V_x, V_y, V_z; but the u, v, w notation is more popular in fluid mechanics.

It is possible to determine a pathline in the laboratory. We simply need to "mark" a fluid particle at some instant and then take a time-exposure photograph of its subsequent motion. We might mark a particle of water flowing in a channel by injecting dye with a hypodermic needle or by electrolyzing some of the water into hydrogen and oxygen bubbles with a pulse of current through a tiny wire in the water. The oxygen bubbles will quickly disappear, but the hydrogen bubbles will remain, marking the fluid.

The equation of the pathline of a specific particle is

$$f(x, y, z) = C(t), \tag{3.3}$$

where

$$\frac{dx}{dt} = u_{\text{particle}}(t), \tag{3.4a}$$

$$\frac{dy}{dt} = v_{\text{particle}}(t), \tag{3.4b}$$

$$\frac{dz}{dt} = w_{\text{particle}}(t), \tag{3.4c}$$

If the velocity field is not a function of time, then since any particle is always traveling in the direction of its own velocity, which is the same for all times at any point, a pathline and a streamline through the same point are identical. An alternative way of stating this is simply that, if the velocity is not a function of time, any fluid particle always travels along a unique streamline.

A *streakline* is a fluid line (that is, a line composed of fluid particles) made up of all particles that have passed a certain point. The streakline is essentially a laboratory concept. A streakline can be constructed in the laboratory by continuously marking all fluid particles that pass a certain point, say by the continuous injection of dye in water or smoke in air. See Fig. 3.4. The marker material forms a streak downstream of the injection point. If the velocity field is not a function of time, then all particles passing a certain point (the injection point) follow exactly the same path downstream of the point and always move along the streamline passing through the point. A streakline is therefore identical to a pathline and a streamline. Streaklines are used in the laboratory

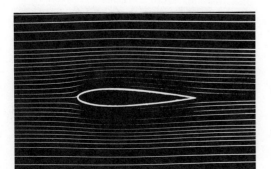

Figure 3.4 Smoke marks streaklines in flow over an airfoil in a wind tunnel (from the "NCFMF Book of Film Notes," 1974, The MIT Press with Education Development Center, Inc., Newton, Mass.).

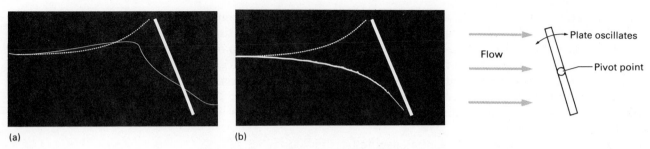

(a) (b)

Figure 3.5 Difference between streamlines, streaklines, and pathlines in unsteady flow over an oscillating plate: (a) Streamline (dotted) and pathline; (b) streamline and streakline. (From the "NCFMF Book of Film Notes," 1974, The MIT Press with Education Development Center, Inc., Newton, Mass.).

to mark streamlines in steady flows so that complicated flow patterns can be studied.

In a flow with a time-varying velocity, streamlines, pathlines, and streaklines are all different (see Fig. 3.5) and none are very helpful for visualization of flow.

Streamlines, pathlines, and streaklines, although different in concept, reduce to identical lines in most cases where they are of any use at all. A timeline is distinctly different from the other three. A *timeline* is a line of fluid particles that have been marked at a particular instant of time. The usual marking technique is to electrolyze a line of hydrogen bubbles in water by pulsing current through a wire. Timelines are usually marked such that they are (initially) perpendicular to the direction of flow across the line. Each piece of a timeline subsequently travels at the velocity appropriate to its location at the time of marking. A photograph of the timeline taken at a later time will show how far the various elements of the line have traveled in the elapsed time. See Fig. 3.6. Since the distance traveled by each element is proportional to the velocity of the element, the distorted timeline is a picture of the velocity profile at the instant and location of marking.

Figure 3.6 Timelines marked by hydrogen bubbles in water flow through a diffuser (from the "NCFMF Book of Film Notes," 1974, The MIT Press with Education Development Center, Inc., Newton, Mass.).

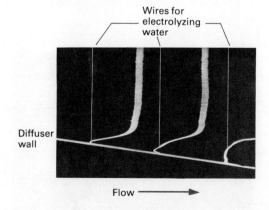

EXAMPLE 3.2   Illustrates the difference between streamlines, streaklines, and pathlines

A small water storage tank is punctured by a bullet. A civil engineering student who is studying fluid mechanics sees the water flowing out of the hole and thinks this might be an illustration of several flow concepts. Grabbing her camera, she takes several photographs at the rate of one per second. After developing the photographs, she makes a sketch that superimposes several of the photographs. The sketch is shown in Fig. E3.2(a). Sketch the streamline, pathline, and streakline at the instant of the last photograph shown on the sketch.

Solution

Given

Water flowing out of water storage tank as sketched in Fig. E3.2(a).

Find

Streamline, pathline, and streakline in Fig. E3.2(a) at the instant of the last photograph

Solution

A streamline is an imaginary line tangent to the fluid velocity vector. The velocity vectors for the instant of the last photograph are drawn in Fig. E3.2(b); the streamline at this instant is coincident with the water stream at the instant of the last photograph.

The pathline is a curve marked out by the trajectory of a particular fluid particle as it moves through the flow field. This means that although we can identify a pathline at a particular instant of time, the pathline must be developed or mapped out during a certain period of time and we must consider the path followed by a particular particle. We will consider the particle identified as A at the instant of the last photograph. We cannot precisely identify A's path after leaving the tank, but we can give a reasonable representation of how A got from the hole in the tank to the location shown in the sketch. This representation of A's path is given in Fig. E3.2(b).

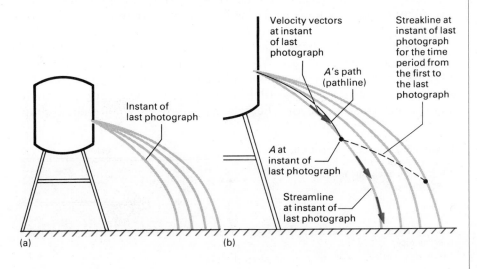

Velocity vectors at instant of last photograph

Streakline at instant of last photograph for the time period from the first to the last photograph

A's path (pathline)

Instant of last photograph

A at instant of last photograph

Streamline at instant of last photograph

(a)   (b)

Figure E3.2 (a) Superposition of several photographs of flow out of a tank; (b) illustration of a streamline, pathline, and streakline for the flow out of a tank.

A streakline could be called the present locus of all the particles that passed through a particular point during a particular time period. We will consider the particular point as the hole in the tank and the particular time period as the time from the instant of the first photograph to the instant of the last photograph. Again, we cannot precisely identify the exact streakline, but we can give a reasonable representation of it. This representation is also illustrated in Fig. E3.2(b).

Discussion
Can a timeline be illustrated in Fig. E3.2(b)? How, or why not?

3.3.3 Classifying the Velocity Field: Dimensionality, Directionality, and Steadiness

Dimensionality, directionality, and steadiness are used to classify the mathematical and, to some extent, physical nature of the velocity field. The fluid velocity in a flow field is a *vector* function of four independent variables: three space coordinates and time. If rectangular coordinates are used, we write

$$\vec{V} = \vec{V}(x, y, z, t)$$

or, in terms of components,

$$\vec{V} = u(x, y, z, t)\hat{i} + v(x, y, z, t)\hat{j} + w(x, y, z, t)\hat{k}$$

The *dimensionality* of a flow field is the number of *independent space coordinates* necessary to specify the velocity.* From another point of view, dimensionality is the number of directions in which the velocity can vary. The dimensionality can vary from zero to three. A flow with zero dimensionality is called a *uniform parallel flow.*

The *directionality* of a flow field is the number of nonzero velocity components present in the field, that is, the number of nonzero *dependent* variables in the velocity function. Directionality can vary from one[†] to three. It may seem that dimensionality and directionality are the same thing, but a few examples will show that they are really quite different.

First consider the flow of a viscous fluid between two large parallel plates, as shown in Fig. 3.7. We assume that the plates are infinite in all directions and that there is no flow perpendicular to the paper. The streamlines are straight, parallel lines. There is no velocity perpendicular to the streamlines, and the velocity profile does not change shape

Figure 3.7 Flow of a viscous fluid between parallel plates: Flow is one-dimensional in *y* and one-directional in *x*.

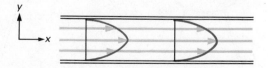

* A slightly more general definition would involve all fluid properties rather than just the velocity.
[†] If directionality is zero, there is no flow!

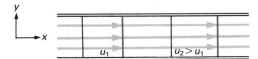

Figure 3.8 This flow is one-dimensional and one-directional in x.

as the flow proceeds downstream. This flow has $v = 0$ and $w = 0$, and it is one-directional in x. The single velocity, u, varies in a direction perpendicular to the plates, and the flow is one-dimensional in y (but not in x).

Next consider the flow shown in Fig. 3.8. The streamlines are straight, and $v = w = 0$. The velocity u varies in the x-direction but not in the y-direction. In this case, the single dimension and the single direction are aligned.

Finally, consider the flow over a circular cylinder shown in Fig. 3.3 on page 118. If the cylinder is very long perpendicular to the paper and if the fluid approaches the cylinder in a direction exactly perpendicular to the axis of the cylinder, then there is no velocity along the cylinder and the flow is identical in all planes perpendicular to the cylinder's axis. This flow is two-dimensional and two-directional. If we were to superimpose a uniform fluid velocity perpendicular to the paper on the entire flow field (equivalent to moving the cylinder parallel to its axis), the flow would become three-directional but would still be two-dimensional.*

If we want to perform an analysis, we would like both the dimensionality and the directionality to be as low as possible because the dimensionality determines whether we must solve (at least in principle) ordinary (one-dimensional) or partial (two- or three-dimensional) differential equations, and the directionality determines the number of equations we have to solve. We have some control over these aspects of a particular problem by our selection of a coordinate system, as the following illustration shows.

Consider the flow of water into a drain. Far from the drain, the water swirls around the drain in almost circular streamlines, with very little motion toward the drain. The streamline pattern and the velocity vectors are approximately as sketched in Fig. 3.9. The swirling velocity increases as the flow moves toward the center. If we choose to analyze this flow in a rectangular (x, y) coordinate system, the flow is two-dimensional and two-directional. If we choose a polar (r, θ) coordinate system, the flow is one-dimensional (in r) and one-directional (in θ).

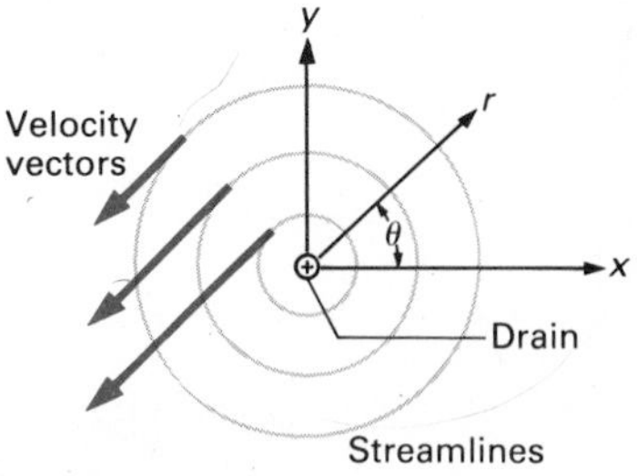

We can often greatly simplify calculations with only a small sacrifice in accuracy by using average velocities to replace a velocity distribution in one or two directions. The two-dimensional (and two-directional) flow in the diverging passage of Fig. 3.10(a) can be approximated by the one-dimensional, one-directional flow of Fig. 3.10(b) by

Figure 3.9 Approximate streamlines of swirling flow around a drain.

---

* Unfortunately, the distinction between dimensionality and directionality is seldom clearly made. "Dimensionality" is usually used as a catch-all for both ideas. It is usually implied that a "two-dimensional" flow is also two-directional *and* that the two directions line up with the two dimensions. At least two of the preceding examples would violate one of these implications.

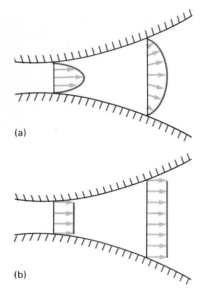

(a)

(b)

Figure 3.10 One-dimensional and one-directional flow approximation for diverging passage: (a) Actual two-dimensional and two-directional flow; (b) approximate flow.

replacing the velocity distribution $u(x, y)$ by a uniform average velocity given by, say,*

$$\bar{u}(x) = \frac{1}{Y(x)} \int_0^{Y(x)} u(x, y)\, dy.$$

By cleverly choosing a coordinate system and using average velocities, we can reduce some very complicated flows to rather simple problems.

The concept of flow steadiness or unsteadiness is closely associated with dimensionality in that it deals with one of the independent variables (time). A flow is *steady* if the velocity and other flow properties at all points in the flow field do not vary with time. A flow is *unsteady* if the velocity or any other property varies with time at any point.

Since time is an independent variable, steady flows (with one fewer variable) are easier to analyze than unsteady ones. Many flows that are unsteady with respect to one coordinate system are steady with respect to another coordinate system. If we observe a passing boat from the bank of a river, the flow of water around the boat is unsteady, even if the boat travels at constant speed. If we watch the water from the bow of the boat, we see a steady flow of the water relative to the boat. In such a case, we can simply attach a coordinate system to the boat and perform a steady-flow analysis. Since our analysis invariably makes use of Newton's second law of motion somewhere, this trick works only if the coordinate system is moving at constant velocity. The trick fails at other times too. If the boat is approaching a pier, the motion is unsteady from both the boat and the pier, even if the boat is moving at constant velocity.

---

* There are several possibilities for defining an average, depending on exactly what you want to do in your analysis [3]. The point here is that an average velocity can be defined and that it reduces the dimensionality of the problem.

EXAMPLE 3.3 Illustrates the difference between dimensionality and directionality and illustrates the concept of streamlines

Determine the dimensionality and the directionality of the velocity field of Example 3.1. Then sketch some typical streamlines of the flow field.

Solution

Given
Velocity field illustrated in Fig. E3.1(b) on page 116

Find
Dimensionality and directionality of the field in Fig. E3.1(b)
Several streamlines for the flow

Solution
The dimensionality of a velocity field is the number of independent space variables necessary to specify the velocity. The velocity u in Example 3.1 depends on both x and y, so the velocity field is two-dimensional.

The directionality of a velocity field is the number of nonzero velocity components in the field. We definitely have one nonzero velocity component, u. The problem statement of Example 3.1 says that the z velocity is zero. What about the y velocity component, v? Since the walls of the channel converge as the fluid flows from left to right, we would expect them to "push" the fluid toward the center of the channel. This gives rise to a velocity v directed toward the center. Since v is then nonzero, we have two nonzero velocities (u and v) and the flow is two-directional.   **Answer**

| Two-dimensional, two-directional |

A streamline is an imaginary line that is tangent to the fluid velocity vector. The total velocity of each particle is directed toward the center and toward the outlet of the channel.   **Answer**

| The streamlines must be as shown in Fig. E3.3. |

Discussion

We will see later how to mathematically determine v by knowing only $u(x, y)$. We will see that v is definitely nonzero. Incidentally, this flow is steady because the velocity field does not vary in time.

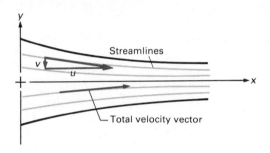

Figure E3.3 Typical streamlines for a flow field.

3.3.4 Differential versus Finite Control Volume Formulations

We consider again the possibilities for mathematical formulation of the fundamental laws. The laws govern the fluid motion at every point in an entire flow field; they are independent of the size of the region or of the amount of fluid to which they are applied. The laws must also be satisfied in any finite region that contains many points and a finite fluid mass. We can express the laws either in a form that applies to each point or in a form that applies to a finite volume of space. The best choice depends on how much information we are given and on what we need to know.

Consider a steady flow of air in a variable area duct (Fig. 3.11). Plane 1 is the duct inlet and plane 2 is the duct outlet. For certain applications we may need to determine the fluid velocity and pressure at every point in the volume between plane 1 and plane 2. To determine the flow in such detail, we must apply the fundamental laws to each point in the field. This will result in a set of differential equations for the fluid velocity and pressure. The solution of these differential equations, together with appropriate boundary conditions representing the shape

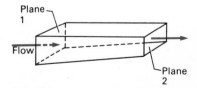

Figure 3.11 Air flow in a variable-area duct.

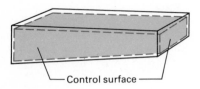

Figure 3.12 Control volume for variable-area duct.

and size of the duct, the fluid velocity at the inlet, and so forth, will be two functions, $\vec{V}(x, y, z)$ and $p(x, y, z)$, that can tell us the velocity and pressure at every point. We might call this a *local* description because it gives the detailed information at any locality.

Often we do not need to know all of the details of a flow field. If our duct is part of an air-conditioning system, we might want to know only the air flow rate in the duct (how many kg/s of air are passing through the duct) and the change in pressure and temperature of the air between the inlet and the outlet. An effective approach in this case is to apply the fundamental laws to a finite control volume. A *control volume* is an imaginary region of space, selected for the purposes of analysis. The boundary surface of a control volume is called the *control surface*. Fluid flows into or out of a control volume with no resistance from the control surface. A control volume for the duct flow in this discussion is shown in Fig. 3.12. Note that no fluid flows across the sides of the control volume because the sides are adjacent to the duct walls.

The fundamental laws apply to the finite volume enclosed by the control surface and to the fluid in that volume. For example, the principle of conservation of mass states that nothing can occur within the control volume that results in a *net* creation or destruction of mass. Control volume equations deal with the relations between the fluid properties and the velocity at the control volume boundaries, without revealing the details of what is going on inside the volume.* Since most of the fundamental laws are conservation laws, control volume equations deal with a balance between the amount of a quantity that enters the control volume, the amount of the quantity that leaves the control volume, the amount of the quantity stored inside the control volume, and the net creation or destruction of the quantity† within the control volume. Control volume equations sometimes require integration of the fluid velocity and property distributions over the control surface, but algebraic expressions often are satisfactory. The control volume approach might be called a global approach.

Both the finite control volume approach and the differential approach are important in fluid mechanics. Both have their advantages and disadvantages. A comparison of the two approaches is provided in Table 3.1.

The difference between the differential and the finite control volume approaches sometimes leads students and engineers to conclude that there are two different kinds of fluid mechanics, as we mentioned in Section 3.1. We try to give equal emphasis to both approaches in this book and, in particular, to show how both approaches may work together in the solution of engineering fluid mechanics problems.**

---

* Nothing in the control volume formulation prevents the existence of, say, 100 points in the control volume that create mass, so long as there are a sufficient number of points that destroy mass so that no net creation or destruction occurs.

† Net creation of mass and energy is zero; however, momentum is created or destroyed by force.

** Students and engineers usually prefer the control volume approach because it involves less complicated mathematics. White [1] says that the approaches are equal but that the control volume approach is more equal.

Table 3.1 Comparison of differential and finite control volume formulations

	Advantages	**Disadvantages**
Differential formulation	1. Reveals all details of the flow 2. Forces fluid to obey fundamental laws at all points 3. Solves problem with minimum of input information (boundary conditions)	1. Produces differential equations that are often difficult or impossible to solve 2. Often requires that equations be solved by computer, which can be expensive 3. May give more information than we really need
Finite control volume formulation	1. Simpler mathematics 2. Much less sensitive to approximations and assumptions; often yields quite useful approximate information with very crude assumptions 3. Pencil-and-paper method requiring about one hour's work 4. Often reveals only the information we really need	1. Does not reveal all details of flow; does not force fluid to obey fundamental laws at every point 2. Often yields *only* approximate answers 3. Requires more input information, such as velocity distribution at convenient boundaries 4. Often cannot tell us everything we need to know

3.4 Illustration of Approaches to Formulation of the Fundamental Laws: Conservation of Mass

The purpose of this chapter is to lay the foundation for flow analysis rather than to derive the equations that describe flow. Nevertheless, we will illustrate both the differential and finite control volume formulations by applying them to the law of conservation of mass. Although our primary purpose for including these derivations here is illustration, the result will be some very useful equations.

3.4.1 Differential Formulation: The Differential Continuity Equation

There are two reasons to begin with the differential formulation. The differential approach is the more general of the two approaches* because it reveals all details of the flow and requires the fluid at each point to obey the particular law of interest. While the finite control volume approach is very useful in analyzing confined flows, it is usually not helpful in analyzing unconfined flows. The differential formulation

---

* Historically, the differential approach was the first to be developed, formally beginning with Euler and John and Daniel Bernoulli in the eighteenth century.

is applied to the confined or the unconfined case in exactly the same way and is as likely to yield useful answers in an unconfined flow as in a confined flow. The second reason for beginning with the differential approach may seem strange: It will be used less in our study of fluid mechanics than the control volume approach. This may seem inconsistent with the claim that the differential formulation is more general. The unfortunate fact is that the differential approach often leads to intractable mathematics, especially for a junior engineering student. Because we want you to gain an appreciation of the importance of the differential approach in the entire field of fluid mechanics, we consider it first so that it does not seem to be tacked on after seemingly more useful material based on the control volume approach.

Consider a two-dimensional, two-directional, unsteady flow. We omit the third direction and dimension for clarity. Choose a rectangular coordinate system* and consider a rectangular fluid particle of dimensions δx by δy by δz located at an arbitrary point (x, y, z) in the field at a specific instant of time (t). In Fig. 3.13, the solid lines represent the boundaries of the fluid particle. They are not fixed in the coordinate system but move as the particle moves. If we adopt an Eulerian description, the fluid density and the velocity components are functions of x, y, and t. The faces of the fluid particle are not exactly at (x, y), so they have the velocities indicated in the figure. Higher-order terms have been omitted from these expressions because they will later vanish anyway (recall Section 2.2).

Assuming that the density at the center of the particle is an appropriate average, the mass of the particle is

$$\delta m = \rho(\delta \mathcal{V}) = \rho(\delta x)(\delta y)(\delta z). \tag{3.5}$$

The law of conservation of mass states that the mass of this fluid particle cannot be created or destroyed; that is, the total time rate of

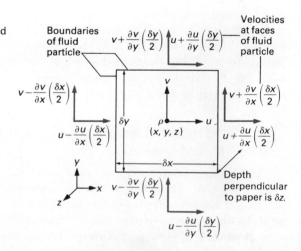

Figure 3.13 Rectangular fluid particle at point (x, y, z) in a two-dimensional, two-directional flow field; particle is at (x, y, z) at time t; w is zero and z is constant.

---

* We will consistently use rectangular (x, y, z) coordinates in the development of the differential approach. The development can also be done in other (cylindrical or spherical) coordinates, or the resulting equations can be transformed into other coordinate systems. Appendix H lists the important differential equations of fluid mechanics in cylindrical and spherical coordinates.

change of the particle's mass must be zero:

$$\frac{d(\delta m)}{dt} = \frac{d}{dt}[\rho(\delta x)(\delta y)(\delta z)] = 0. \tag{3.6}$$

Using the product rule for derivatives, we have

$$(\delta x)(\delta y)(\delta z)\frac{d\rho}{dt} + \rho\left[(\delta y)(\delta z)\frac{d(\delta x)}{dt} + (\delta x)(\delta z)\frac{d(\delta y)}{dt} + (\delta x)(\delta y)\frac{d(\delta z)}{dt}\right] = 0. \tag{3.7}$$

We will examine each derivative in this equation. First consider the density derivative, $d\rho/dt$. Since we are using an Eulerian description of the field, the density of any particular fluid particle depends on the particle's location and on the specific time that the particle is at that location.

We will *temporarily* use the subscript p to remind us that we are talking about a specific fluid particle. Mathematically we can write the density for the two-dimensional flow as a function:

$$\rho_p = \rho_p(x_p, y_p, t),$$

where x_p and y_p give the particle's location at time t.

The total time rate of change of ρ_p must include the fact that ρ_p changes as the particle moves from point to point in the field over time as well as the fact that the density at a fixed point may change with time.* Since ρ_p is a function of three variables, its total derivative with respect to time is

$$\frac{d\rho_p}{dt} = \frac{d}{dt}[\rho_p(x_p, y_p, t)] = \frac{\partial \rho_p}{\partial x_p}\left(\frac{dx_p}{dt}\right) + \frac{\partial \rho_p}{\partial y_p}\left(\frac{dy_p}{dt}\right) + \frac{\partial \rho_p}{\partial t}, \tag{3.8}$$

Note that $\partial \rho_p/\partial t$ is the rate of change of density at a fixed location and is not the same as $d\rho_p/dt$, which also includes the change of density with position. Now notice that x_p and y_p are the position coordinates of the fluid particle, so

$$\frac{dx_p}{dt} = u_p \quad \text{and} \quad \frac{dy_p}{dt} = v_p.$$

Substituting these relations into Eq. (3.8), we have

$$\frac{d\rho_p}{dt} = \frac{\partial \rho_p}{\partial t} + u_p\left(\frac{\partial \rho_p}{\partial x_p}\right) + v_p\left(\frac{\partial \rho_p}{\partial y_p}\right).$$

We now drop the subscript p, so the density derivative that we need for Eq. (3.7) is

$$\frac{d\rho}{dt} = \frac{\partial \rho}{\partial t} + u\left(\frac{\partial \rho}{\partial x}\right) + v\left(\frac{\partial \rho}{\partial y}\right). \tag{3.9}$$

This form of derivative is very important in fluid mechanics, so we will discuss it briefly before we continue our derivation of the differential equation of conservation of mass. Recall that $d\rho/dt$ is the total rate of change of the density of a fluid particle as that particle moves

---

* If a particle of air moves upward in the atmosphere, for example, its density decreases with elevation even though the density of the air at any particular elevation may be constant with time. If the same particle of air were fixed near the ground, its density might change with time as the atmosphere cools off at night or heats up during the day.

from point to point. Equation (3.9) reflects the fact that the density of a particle can change with location (the second and third term) or with time at a fixed location (the first term). The complicated form of Eq. (3.9), requiring three terms to express a single total time derivative, is a consequence of the selection of the Eulerian method of description in which fluid properties depend on both location and time. This type of derivative is variously called the *Eulerian derivative* (since it is necessary in an Eulerian description), the *material derivative* (since it expresses the rate of change of a property of a particle of the material), the *substantial derivative* (substituting the word *substance* for *material*), or the *total derivative*. As a reminder that it is a special type of derivative, we often write it

$$\frac{d}{dt} = \frac{D}{Dt} \equiv \frac{\partial}{\partial t} + u\left(\frac{\partial}{\partial x}\right) + v\left(\frac{\partial}{\partial y}\right) + w\left(\frac{\partial}{\partial z}\right), \tag{3.10}$$

where the derivative has been written as a mathematical operator and has been extended to three directions and three dimensions. A shorthand notation using vectors is

$$\frac{D}{Dt} \equiv \frac{\partial}{\partial t} + \vec{V} \cdot \nabla, \tag{3.10a}$$

where $\vec{V}$ is the fluid velocity vector and ∇ is the gradient operator.

Returning to the differential equation of conservation of mass, consider the next derivative in Eq. (3.7), namely, $d(\delta x)/dt$. We have δx as the distance between the left and right faces of the fluid particle; $d(\delta x)/dt$ is the relative x-velocity between the two faces:

$$\frac{d(\delta x)}{dt} = u)_{\text{right face}} - u)_{\text{left face}}$$

or

$$\frac{d(\delta x)}{dt} = \left[u + \frac{\partial u}{\partial x}\left(\frac{\delta x}{2}\right)\right] - \left[u - \frac{\partial u}{\partial x}\left(\frac{\delta x}{2}\right)\right].$$

Simplifying, we get

$$\frac{d(\delta x)}{dt} = \frac{\partial u}{\partial x}(\delta x). \tag{3.11}$$

By a similar argument,

$$\frac{d(\delta y)}{dt} = \frac{\partial v}{\partial y}(\delta y) \tag{3.12}$$

and

$$\frac{d(\delta z)}{dt} = \frac{\partial w}{\partial z}(\delta z). \tag{3.13}$$

Equation (3.13) gives $d(\delta z)/dt$ for the general case. But since we have assumed two-dimensional, two-directional flow, $w \equiv 0$, so $\partial w/\partial z = 0$ and, for our specific case,

$$\frac{d(\delta z)}{dt} = 0. \tag{3.14}$$

Now we substitute Eqs. (3.9), (3.11), (3.12), and (3.14) into Eq. (3.7) to get

$$(\delta x)(\delta y)(\delta z)\left[\frac{\partial \rho}{\partial t} + u\left(\frac{\partial \rho}{\partial x}\right) + v\left(\frac{\partial \rho}{\partial y}\right)\right] + \rho(\delta x)(\delta y)(\delta z)\left(\frac{\partial u}{\partial x} + \frac{\partial v}{\partial y}\right) = 0.$$

We now divide by $(\delta x)(\delta y)(\delta z)$ and take the limit of the expression, letting δx, δy, δz all approach zero, shrinking our fluid particle to a point. The result is the differential equation of conservation of mass at the point

$$\frac{\partial \rho}{\partial t} + u\left(\frac{\partial \rho}{\partial x}\right) + v\left(\frac{\partial \rho}{\partial y}\right) + \rho\left(\frac{\partial u}{\partial x}\right) + \rho\left(\frac{\partial v}{\partial y}\right) = 0. \qquad \textbf{(3.15)}$$

A shorter form can be written by combining the second and fourth terms and the third and fifth terms in accordance with the product rule for derivatives:

$$\frac{\partial \rho}{\partial t} + \frac{\partial(\rho u)}{\partial x} + \frac{\partial(\rho v)}{\partial y} = 0. \qquad \textbf{(3.16)}$$

Equation (3.16) is called the "differential continuity equation" or the "continuity equation in differential form." Before discussing this equation and its use, we can find some other forms of the equation by inspection.

For three-dimensional, three-directional flow, the equation is

$$\frac{\partial \rho}{\partial t} + \frac{\partial(\rho u)}{\partial x} + \frac{\partial(\rho v)}{\partial y} + \frac{\partial(\rho w)}{\partial z} = 0. \qquad \textbf{(3.17)}$$

Using vector notation, we can write the equation as

$$\frac{\partial \rho}{\partial t} + \nabla \cdot (\rho \vec{V}) = 0. \qquad \textbf{(3.18)}$$

For steady flow ($\partial \rho/\partial t = 0$), the equation becomes

$$\frac{\partial(\rho u)}{\partial x} + \frac{\partial(\rho v)}{\partial y} + \frac{\partial(\rho w)}{\partial z} = 0.$$

For flow of an incompressible fluid ($\rho = $ constant), the equation becomes

$$\frac{\partial u}{\partial x} + \frac{\partial v}{\partial y} + \frac{\partial w}{\partial z} = 0 \qquad \textbf{(3.19)}$$

for steady *or* unsteady flow.

If you are like most students, you are probably somewhat overwhelmed by all of the mathematics of this derivation and are wondering what the use of all this is. You can see the value of the differential continuity equation by considering one of its simplest forms, namely the equation for two-dimensional, two-directional flow of an incompressible fluid:

$$\frac{\partial u}{\partial x} + \frac{\partial v}{\partial y} = 0. \qquad \textbf{(3.19a)}$$

This differential equation relates the velocity components u and v to the space coordinates x and y. Only those velocity fields that satisfy this equation obey the law of conservation of mass. If we could solve (that is, integrate) this differential equation to find u and v as functions of x and y, we could determine a velocity field that is consistent with the law of conservation of mass. Of course, we can't solve the equation by itself to predict u and v because it is only one equation and has two unknowns. The point is that we could use this equation together with differential equations representing other fundamental laws to predict an entire velocity field. All we need to know about the field are values of the velocity (and, possibly, fluid properties) on the boundaries of the field so that we can evaluate the "constants of integration" in the solution of the differential equations. If we could solve such differential equations, we could predict the complete details of all the flows illustrated in Figs. 3.1 and 3.2. We would simply need to specify such things as the shape of the pipe and the pressure at its inlet and outlet or the shape of the airplane, its velocity of flight, and the properties of the ambient air.

EXAMPLE 3.4   Illustrates the differential continuity equation

Find the transverse velocity $v(x, y)$ for the fluid flowing in the converging channel of Example 3.1

Solution

   Given

Constant density fluid flowing in a converging channel with half-height

$$Y = \frac{Y_0}{\left(1 + \dfrac{x}{\ell}\right)}$$

Longitudinal velocity, given by

$$u = u_0\left(1 + \frac{x}{\ell}\right)\left[1 - \left(\frac{y}{Y}\right)^2\right],$$

   with $u_0 = 1.0$ m/s
Figure E3.1(a) on page 116

   Find
$v(x, y)$

   Solution
   The appropriate form of the differential continuity equation is Eq. (3.19a). We rewrite it as

$$\frac{\partial v}{\partial y} = -\frac{\partial u}{\partial x}.$$

The equations for $u(x, y)$ and $Y(x)$ give

$$u = u_0\left(1 + \frac{x}{\ell}\right)\left[1 - \frac{y^2}{Y_0^2}\left(1 + \frac{x}{\ell}\right)^2\right].$$

Substituting this equation for u into the differential continuity equation gives

$$\frac{\partial v}{\partial y} = -\frac{\partial}{\partial x}\left\{u_0\left(1+\frac{x}{\ell}\right)\left[1-\frac{y^2}{Y_0^2}\left(1+\frac{x}{\ell}\right)^2\right]\right\}$$

$$= -\frac{u_0}{\ell}\left[1-\frac{3y^2}{Y_0^2}\left(1+\frac{x}{\ell}\right)^2\right].$$

Multiplying by dy and integrating, we obtain

$$v = -\int \frac{u_0}{\ell}\,dy + \int \frac{3y^2u_0}{\ell Y_0^2}\left(1+\frac{x}{\ell}\right)^2 dy$$

$$= -\frac{u_0 y}{\ell} + \frac{u_0 y^3}{\ell Y_0^2}\left(1+\frac{x}{\ell}\right)^2 + C(x),$$

where $C(x)$ is a "constant" of integration. Since we are integrating with respect to y, then $C(x)$ cannot be a function of y but could possibly be a function of the other coordinate, x. The fluid immediately adjacent to the wall has zero velocity; therefore, $v = 0$ at $y = Y$ for all x.

We can use either boundary to evaluate $C(x)$. Using the upper boundary where $Y = Y_0/(1 + x/\ell)$ gives

$$0 = -\frac{u_0 Y_0}{\ell\left(1+\frac{x}{\ell}\right)} + \frac{u_0 Y_0^3}{\ell Y_0^2\left(1+\frac{x}{\ell}\right)} + C(x), \qquad \text{and} \qquad C(x) = 0.$$

Then

$$\boxed{\; v(x, y) = -\frac{yu_0}{\ell} + \frac{y^3 u_0}{\ell Y_0^2}\left(1+\frac{x}{\ell}\right)^2. \;}$$

Answer ◀

Discussion

The velocity $v(x, y)$ is tabulated in Table E3.4. It is zero at the wall (satisfying the boundary condition), zero along the center of the channel, negative for

Table E3.4 Tabulated values of the velocity $v(x, y)$ for the two-dimensional channel of Example 3.1

Longitudinal Coordinate x (m)	Height* y (m)	Transverse Velocity v (m/s)*
0.0	0.00	0.000
0.0	± 0.20	∓ 0.038
0.0	± 0.40	∓ 0.067
0.0	± 0.60	∓ 0.077
0.0	± 0.80	∓ 0.058
0.0	± 1.00	0.000
2.5	0.00	0.000
2.5	± 0.13	∓ 0.024
2.5	± 0.27	∓ 0.045
2.5	± 0.40	∓ 0.051
2.5	± 0.67	0.000
5.0	0.00	0.000
5.0	± 0.10	∓ 0.019
5.0	± 0.20	∓ 0.034
5.0	± 0.30	∓ 0.078
5.0	± 0.40	∓ 0.029
5.0	± 0.50	0.000

* Top sign for top half of channel, lower sign for lower half of channel.

positive y, and positive for negative y. This pattern indicates that fluid is moving toward the center of the channel, a quite reasonable result since we would expect the converging walls to "push" the fluid toward the center.

Recall the discussion in Example 3.1 that the fluid in contact with the wall has zero velocity ($u = v = 0$). This is due to the no-slip and impermeability conditions, which require

$$V_{\text{Tangent at wall}} = 0 \qquad \text{and} \qquad V_{\text{Normal at wall}} = 0.$$

The no-slip condition was introduced in Section 1.3.4. Wall boundary conditions for velocity are discussed further in Section 4.2.1.

The velocity u was given in units of m/s. The velocity v would have the same units. This statement can be verified by noting that u_0 appears in the equation for v.

3.4.2 Finite Control Volume Formulation: The Integral Continuity Equation

Now we will apply the law of conservation of mass to a finite control volume. Consider a three-dimensional, three-directional, unsteady flow field. As shown in Fig. 3.14, we superimpose a finite control volume on the flow. The shape of the control volume is arbitrary. The volume enclosed by the control volume is V_{cv}. The surface that bounds V_{cv} is called the control surface, whose area is A_{cv}. The control surface is completely permeable, offering no resistance to flow. The control volume is not attached to the fluid. We will specify that the control volume remain fixed in space, although it is sometimes useful to let the control volume move at some specified velocity.*

Now we look at the flow through the control volume at some particular instant of time. Fluid flows into the control volume at some locations and out of the control volume at other locations. A certain amount of mass is contained (stored) inside the control volume. The law of conservation of mass requires that regardless of what occurs in the control volume, there cannot be a net creation or destruction of mass. If more mass flows into the control volume than flows out of the control volume in some time δt, the excess inflow must be stored inside the control volume. We can write a word equation representing this principle:

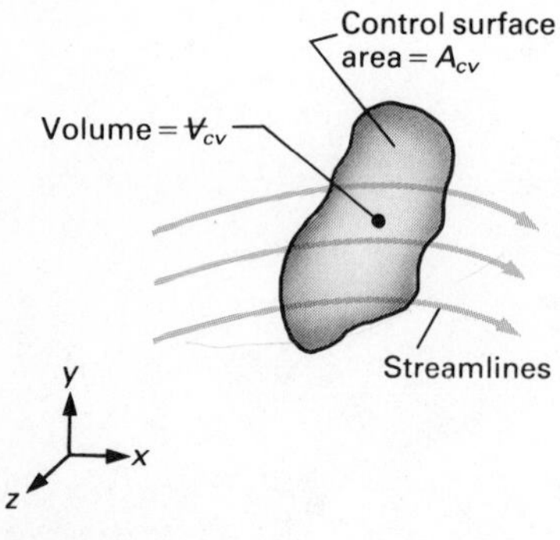

$$\begin{pmatrix} \text{Rate of mass flow} \\ \text{into control volume} \end{pmatrix} - \begin{pmatrix} \text{Rate of mass flow} \\ \text{out of control volume} \end{pmatrix} = \begin{pmatrix} \text{Rate of change of mass} \\ \text{stored in control volume} \end{pmatrix}.$$

Introducing the symbols $\dot{m}$ to represent mass flow rate and m_{cv} to represent mass stored in the control volume, we can rearrange the equation and write

$$\frac{d}{dt}(m_{cv}) + \dot{m}_{\text{out}} - \dot{m}_{\text{in}} = 0. \tag{3.20}$$

Sometimes Eq. (3.20) is quite useful just as it is. If you pull up to a gasoline pump and pump 0.25 kg/s of gasoline into the tank ($\dot{m}_{\text{in}}$), the amount of gasoline in the tank (m_{cv}) increases at a rate of 0.25 kg/s,

Control surface area = A_{cv}

Volume = V_{cv}

Streamlines

Figure 3.14 Finite control volume superimposed onto flow field.

---

* We will consider moving control volumes in Chapter 5.

at least if your engine is turned off and your gas tank doesn't leak ($\dot{m}_{\text{out}} = 0$).

We can relate the terms in Eq. (3.20) to the fluid's velocity and properties and to the control volume geometry. The mass contained in the control volume at any instant is

$$m_{cv} = \int_{V_{cv}} \rho \, dV,$$

where dV is a differential element of volume within the control volume and ρ is the density of the fluid at the center of dV. Since it is possible for the density to vary from point to point within the control volume, m_{cv} is written as an integral. The rate of change of stored mass is

$$\frac{d}{dt}(m_{cv}) = \frac{d}{dt}\int_{V_{cv}} \rho \, dV. \qquad \textbf{(3.21)}$$

Next consider the rate of flow of mass out of the control volume. Figure 3.15 shows an enlargement of a small portion of the control surface. Mass flows out of the control volume through the area dA. Part (a) of the figure shows the fluid velocity at dA at a particular time t; $\hat{n}$ is a unit vector drawn perpendicular to dA and pointing outward from the control volume. In general, the fluid velocity is not perpendicular to the control surface, so the angle θ between $\vec{V}$ and $\hat{n}$ is not zero. Part (b) of the figure shows the fluid that has left the control volume through dA in time δt. The fluid that was at dA at time t has now moved to dA'. All the fluid in the volume $d(\delta V)$ between dA and dA' has flowed out of the control volume through dA in time δt. We can write:*

$$d(\delta m)_{\text{out}} = \rho d(\delta V). \qquad \textbf{(3.22)}$$

From Fig. 3.15(b), we have

$$d(\delta V) = (V\delta t) \cos \theta \, dA. \qquad \textbf{(3.23)}$$

Figure 3.15 Details of outflow through a small piece of the control surface (dA): (a) Velocity and unit outdrawn normal vector at surface; (b) fluid particles that were at dA at t are at (dA') at $t + \delta t$; fluid in volume $d(\delta V)$ has left control volume through dA.

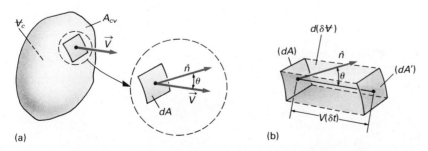

(a)   (b)

---

* In this development, we have used both $d(\ )$ and $\delta(\ )$ to indicate small quantities. The $d(\ )$ is generally associated with the smallness of the area considered and the $\delta(\ )$ is generally associated with the smallness of the time considered.

To obtain an expression for the rate of outflow through dA, we substitute Eq. (3.23) into Eq. (3.22), divide by δt, and take the limit as $\delta t \to 0$. The result is

$$dm_{\mathrm{out}} = \rho V \cos \theta \, dA.$$

The total rate of outflow is obtained by integrating over all of the outflow area:

$$\dot{m}_{\mathrm{out}} = \int_{A_{\mathrm{out}}} \rho V \cos \theta \, dA. \tag{3.24a}$$

If you are wondering how to identify the outflow area, notice that fluid flows outward across the control surface at any point where $0° \leq \theta \leq 90°$. In vector notation, we can write

$$\dot{m}_{\mathrm{out}} = \int_{A_{\mathrm{out}}} \rho \vec{V} \cdot \hat{n} \, dA. \tag{3.24b}$$

Remember that the magnitude of vector $\hat{n}$ is unity.

Finally, consider the rate of flow of mass into the control volume. Figure 3.16 illustrates a portion of the control surface through which mass enters the control volume. Part (a) of the figure shows the velocity vector $\vec{V}$ and the unit outward normal vector $\hat{n}$ at the area dA at time t. Note that for inflow, the angle θ lies between $90°$ and $180°$. The angle θ' is the complement of the angle θ. Part (b) of the figure shows the situation at time $t + \delta t$. The fluid that was at dA at time t has now moved to dA', inside the control volume. All the fluid in volume $d(\delta V)$ has flowed into the control volume across dA in time δt. We can write

$$d(\delta m)_{\mathrm{in}} = \rho \, d(\delta V), \tag{3.25}$$

where

$$d(\delta V) = (V \, \delta t) \cos \theta' \, dA = -V \, \delta t \cos \theta \, dA.$$

Dividing by δt and taking the limit as $\delta t \to 0$ gives

$$d\dot{m}_{\mathrm{in}} = -\rho V \cos \theta \, dA.$$

Figure 3.16 Details of inflow through a small piece of control surface: (a) Velocity and unit outdrawn normal vector at surface; (b) fluid particles that were at (dA) at time t are at (dA') at time $t + \delta t$; fluid in volume $d(\delta V)$ has entered the control volume.

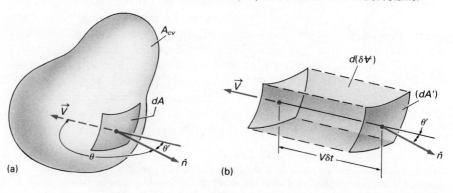

Integrating over all of the inflow surface, we have

$$\dot{m}_{\text{in}} = -\int_{A_{\text{in}}} \rho V \cos \theta \, dA. \qquad \textbf{(3.26a)}$$

In vector notation,

$$\dot{m}_{\text{in}} = -\int_{A_{\text{in}}} \rho \vec{V} \cdot \hat{n} \, dA. \qquad \textbf{(3.26b)}$$

We can now substitute Eqs. (3.21), (3.24a) and (3.26a) into Eq. (3.20) to obtain the integral equation of conservation of mass for the control volume:

$$\frac{d}{dt} \int_{\Psi_{cv}} \rho \, dV + \int_{A_{\text{out}}} \rho V \cos \theta \, dA + \int_{A_{\text{in}}} \rho V \cos \theta \, dA = 0. \qquad \textbf{(3.27)}$$

$$(0 \leq \theta \leq 90°) \qquad (90° \leq \theta \leq 180°)$$

We can combine the two area integrals into a single integral over the whole control surface if we note that each piece of the control surface has either outflow, inflow, or $\theta = 90°$ ($\cos \theta = 0$), so we can write

$$\frac{d}{dt} \int_{\Psi_{cv}} \rho \, dV + \oint_{A_{cv}} \rho V \cos \theta \, dA = 0, \qquad \textbf{(3.28)}$$

where A_{cv} represents the area of the entire control surface. In vector notation, Eq. (3.28) is

$$\frac{d}{dt} \int_{\Psi_{cv}} \rho \, dV + \oint_{A_{cv}} \rho (\vec{V} \cdot \hat{n}) \, dA = 0. \qquad \textbf{(3.29)}$$

Remember that $\hat{n}$ is the unit outward normal vector.

Either of Eqs. (3.28) and (3.29) is called the *integral continuity equation* or the *continuity equation for a control volume,* or sometimes simply the *continuity equation.* If these equations look rather frightening to you because of the mathematical symbols, remember that they simply express the word equation of the law of conservation of mass:

(Rate of mass in) − (Rate of mass out) = (Rate of mass stored).

It is interesting at this point to compare Eqs. (3.17) and (3.18) with Eqs. (3.28) and (3.29). Both pairs of equations represent the same physical law, conservation of mass, but, at least to the untrained observer, they do not look very much alike.* The dissimilarity is due to the different viewpoints adopted for derivation of the equations; certainly it is not a result of different physics of the flow in each case. These equations each find different areas of application. The application of the differential equation was discussed in Section 3.4.1 Several applications of the control volume equation will be discussed in the next section, and applications of both equations will reappear throughout this book. In a later section we will also illustrate the equivalence between the differential and control volume approaches by using the control volume continuity equation to rederive the differential continuity equation.

---

* Note that you can see a term involving the time derivative of density and a term involving $\rho \vec{V}$ in both pairs of equations.

EXAMPLE 3.5 Illustrates the integral continuity equation

Show that the flow field in Example 3.1 satisfies the integral continuity equation.

Solution

Given

Converging channel in Fig. E3.1(a) on page 116, with

$$Y = \frac{Y_0}{\left(1 + \dfrac{x}{\ell}\right)}, \qquad u = u_0\left(1 + \frac{x}{\ell}\right)\left[1 - \left(\frac{y}{Y}\right)^2\right]$$

from Example 3.1 and

$$v = -\frac{yu_0}{\ell} + \frac{y^3 u_0}{\ell Y_0^2}\left(1 + \frac{x}{\ell}\right)$$

from Example 3.4

Constant density fluid, two-dimensional channel, and steady flow

Find

That the given velocity field satisfies

$$\frac{d}{dt}\int_{\mathcal{V}_{cv}} \rho\, d\mathcal{V} + \oint_{A_{cv}} \rho(\vec{V}\cdot\hat{n})\, dA = 0.$$

Solution

The first term of the equation

$$\frac{d}{dt}\int_{\mathcal{V}_{cv}} \rho\, d\mathcal{V} = 0,$$

because the flow is steady. Since ρ is constant,

$$\int_{A_{cv}} \rho(\vec{V}\cdot\hat{n})\, dA = \rho\int_{A_{cv}} (\vec{V}\cdot\hat{n})\, dA.$$

For the control volume shown in Fig. E3.5, we note that $\vec{V}\cdot\hat{n} = 0$ along the walls of the channel because $u = v = 0$ at the wall (see the discussion for Example 3.4). However, $\vec{V}\cdot\hat{n} \neq 0$ at the exit or at the inlet. Since the flow is two-directional,

$$\vec{V} = u\hat{i} + v\hat{j}.$$

At the outlet plane, the unit outward normal vector is $+\hat{i}$. At the inlet plane, the unit outward normal vector is $-\hat{i}$. Therefore

$$\int_{A_{cv}} (\vec{V}\cdot\hat{n})\, dA = \int_{A_{\text{out}}} (u\hat{i} + v\hat{j})\cdot(\hat{i})\, dA + \int_{A_{\text{in}}} (u\hat{i} + v\hat{j})\cdot(-\hat{i})\, dA$$

$$= \int_{A_{\text{out}}} u\, dA - \int_{A_{\text{in}}} u\, dA.$$

Substituting for u at the exit, we have

$$\int_{A_{\text{out}}} u\, dA = \int_{A_{\text{out}}} u_0\left(1 + \frac{\ell}{\ell}\right)\left[1 - \left(\frac{y}{Y_\ell}\right)^2\right] dA$$

$$= W u_0 \int_{-Y_\ell}^{+Y_\ell} 2\left[1 - \left(\frac{y}{Y_\ell}\right)^2\right] dy,$$

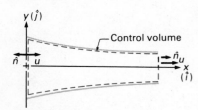

Figure E3.5 Control volume for converging two-dimensional channel.

$y\,(\hat{j})$

Control volume

$\hat{n}$ u $\hat{n}$ u

x

$(\hat{i})$

where Y_ℓ is the magnitude of the channel half-height at $x = \ell$ and W is the width of the channel (perpendicular to the paper). Integrating, we have

$$\int_{A_{\text{out}}} u\, dA = 2Wu_0 \left[y - \frac{y^3}{3Y_\ell^2} \right]_{-Y_\ell}^{+Y_\ell} = \frac{8Wu_0 Y_\ell}{3}.$$

Since $Y_\ell = 0.5$ m and $u_0 = 1.0$ m/s,

$$\int_{A_{\text{out}}} u\, dA = \frac{4}{3}\, W.$$

Similarly,

$$\int_{A_{\text{in}}} u\, dA = \int_{A_{\text{in}}} u_0(1) \left[1 - \left(\frac{y}{Y_0} \right)^2 \right] dA$$

$$= Wu_0 \int_{-Y_0}^{+Y_0} \left[1 - \left(\frac{y}{Y_0} \right)^2 \right] dy,$$

where Y_0 is the magnitude of the channel half-height at $x = 0$. Integrating, we obtain

$$\int_{A_{\text{in}}} u\, dA = Wu_0 \left[y - \frac{y^3}{3Y_0^2} \right]_{-Y_0}^{+Y_0}$$

$$= \frac{4Wu_0 Y_0}{3}.$$

Since $Y_0 = 1.0$ m and $u_0 = 1.0$ m/s,

$$\int_{A_{\text{in}}} u\, dA = \frac{4}{3}\, W.$$

Then

$$\oint_{A_{cv}} (\vec{V} \cdot \hat{n})\, dA = \frac{4}{3}\, W - \frac{4}{3}\, W = 0.$$

Thus

$$\left| \frac{d}{dt} \int_{V_{cv}} \rho\, dV + \oint_{A_{cv}} \rho(\vec{V} \cdot \hat{n})\, dA = 0. \right|$$ **Answer** ◄

Discussion
Although the mathematical form of the integral continuity equation looks rather formidable, evaluation of the terms was not too difficult. Note that the unit outward normal vector was $+\hat{i}$ at the outlet and $-\hat{i}$ at the inlet. This vector is determined by the control volume geometry, not by the flow direction.

3.4.3 Simplified Forms and Applications of the Integral Continuity Equation

The integral continuity equation is one of the most useful equations in fluid mechanics, but we seldom use the equation in its most general form, Eqs. (3.28) and (3.29). In this section we will develop some simplified forms of the equation and illustrate their application to a variety of flow problems. Although the simplified forms are usually easier and quicker to use, remember that the more general form, Eq. (3.28) or (3.29), will always produce identical results if carefully applied.

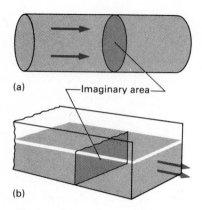

(a)

Imaginary area

(b)

Figure 3.17 Areas for calculating flow rate in (a) a closed conduit and (b) an open channel.

Flow Rate. The first item is not a form of the continuity equation but of one of the terms in it. Consider a fluid flowing in a closed conduit (Fig. 3.17a) or an open channel (Fig. 3.17b). An important parameter in many engineering applications is the fluid flow rate in the passage. To evaluate the flow rate in any passage, we consider an imaginary surface that spans the passage. This surface is usually selected so that it is perpendicular to the axis of the conduit or channel. We imagine a unit vector $\hat{n}$ perpendicular to the surface and pointing in the direction of flow ($0 < \theta < 90°$). Since we are considering a single surface, $\hat{n}$ is neither inward nor outward; however, Eq. (3.24a) can be applied to calculate the rate of flow of mass across the surface by simply dropping the subscript "out":

$$\dot{m} = \int_A \rho V \cos\theta \, dA \qquad (3.30)$$

or, in vector notation,

$$\dot{m} = \int_A \rho(\vec{V} \cdot \hat{n}) \, dA. \qquad (3.31)$$

A third way of writing the equation is

$$\dot{m} = \int_A \rho V_n \, dA, \qquad (3.32)$$

where V_n is the component of the fluid velocity in a direction perpendicular to the area.

These equations for mass flow rate will work for any choice of the cross-sectional area. As the analyst, you get to select the area that you like best. The simplest choice is an area perpendicular to the duct axis or an area perpendicular to the velocity vector. In many cases these reduce to the same area, but not always. (See Example 3.6.)

Sometimes engineers wish to express flow rate in terms of volume rather than mass, as in rating pump flow in gallons per minute delivered or fan flow in cubic feet per minute. The volume flow rate, Q, crossing a surface can be computed by deleting the density from the mass flow integral:

$$Q = \int V \cos\theta \, dA \qquad (3.33)$$

or

$$Q = \int (\vec{V} \cdot \hat{n}) \, dA \qquad (3.34)$$

or

$$Q = \int V_n \, dA. \qquad (3.35)$$

If the fluid density is uniform over the area of integration,

$$\dot{m} = \rho Q. \qquad (3.36)$$

If the density and velocity are both uniform over the cross-sectional area and if the area is perpendicular to the velocity, Eqs. (3.30)–(3.32) all reduce to

$$\dot{m} = \rho V A. \qquad (3.37)$$

If the velocity is uniform and perpendicular to the area, Eqs. (3.33)–(3.35) all reduce to

$$Q = VA. \qquad (3.38)$$

EXAMPLE 3.6 Illustrates that the complexity involved in calculating a volume flow rate depends on the area selected

Figure E3.6(a) shows an inviscid, constant-density fluid flowing steadily in a two-dimensional diffuser of constant width W (perpendicular to the paper). The radial velocity $V_r(r, \theta)$ is given by

$$V_r(r, \theta) = \frac{C}{r},$$

where C is a constant, r is measured in meters, and V_r is measured in meters per second. The circumferential velocity $V_\theta = 0$. Calculate the volume flow rate for the two surfaces shown in Fig. E3.6(b):

1. $r = r_1 = $ Constant,
2. $r_1 \cos \theta_{max} = x_1 = $ Constant.

Solution

Given

Inviscid, constant-density fluid flowing through the two-dimensional diffuser
of width W in Fig. E3.6(a)
Radial velocity v_r:

$$V_r(r, \theta) = \frac{C}{r}.$$

Find

Volume flow rate for two surfaces:

1. $r = r_1 = $ Constant,
2. $x = x_1 = $ Constant

Solution
The volume flow rate is found using Eq. (3.35):

$$Q = \int V_n \, dA.$$

Figure E3.6 (a) Plane diffuser; (b) plane diffuser with two surfaces to calculate volume flow rate; (c) relation of velocity and unit normal vectors.

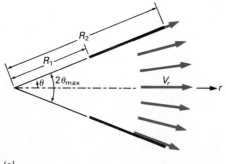

(a)

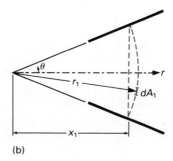

(b)

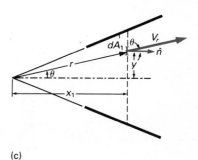

(c)

For the surface $r = r_1 =$ Constant, V_r is perpendicular to each differential area dA. Then

$$V_n = V_r \quad \text{and} \quad dA = Wr_1\,d\theta.$$

Using the symmetry about the centerplane of the diffuser, we have

$$Q = 2 \int_0^{\theta_{max}} V_r(Wr_1\,d\theta).$$

Substituting for V_r gives

$$Q = 2 \int_0^{\theta_{max}} \frac{C}{r_1} Wr_1\,d\theta$$

$$= 2WC[\theta]_0^{\theta_{max}}$$

$$\left| \; Q = 2WC\theta_{max}. \; \right|$$

Answer ◄

For the surface $x = x_1 =$ Constant, Fig. E3.6(c) shows

$$V_n = V_r \cos\theta, \quad dA = W\,dy, \quad r = \sqrt{x_1^2 + y^2}.$$

Using the symmetry about the centerplane of the diffuser, we have

$$Q = 2 \int_{y=0}^{y_{max}} V_r \cos\theta(W\,dy)$$

Substituting for V_r and r gives

$$Q = 2W \int_{y=0}^{y_{max}} \frac{C \cos\theta}{\sqrt{x_1^2 + y^2}}\,dy.$$

We can calculate $\cos\theta$ from

$$\cos\theta = \frac{x_1}{r} = \frac{x_1}{\sqrt{x_1^2 + y^2}},$$

so

$$Q = 2WCx_1 \int_{y=0}^{y_{max}} \frac{dy}{x_1^2 + y^2}.$$

Integrating, we have

$$Q = 2WCx_1 \left[\frac{1}{x_1} \tan^{-1}\left(\frac{y}{x_1}\right) \right]_{y=0}^{y_{max}}$$

$$= 2WC \tan^{-1}\left(\frac{y_{max}}{x_1}\right).$$

However, the angle whose tangent is y_{max}/x_1 is θ_{max}, so

$$\left| \; Q = 2WC\theta_{max}, \; \right|$$

Answer ◄

the same answer as before.

Discussion

The units of r_1 were given as meters and V_r as meters per second. Therefore the units of C must be

$$[C] = [V_r \cdot r]$$

$$= [\text{m/s} \cdot \text{m}] = [\text{m}^2/\text{s}].$$

The brackets indicate that we are considering only the units of the equation. Can you show that the units of Q are cubic meters per second?

This example showed that a careful choice of a surface can result in a simpler calculation. In the same way, we carefully choose a control volume whenever we use one to solve a problem.

It is seldom true that the fluid velocity is precisely uniform over the cross section of a pipe or channel. In Section 3.3.3 we found that it is often desirable to approximate such flows, which are in fact multidimensional and multidirectional, as one-dimensional and one-directional by the use of average velocities. For the approximate one-dimensional flow, we require Eq. (3.38) to be valid, so we define an average velocity (V) by

$$Q = \int V_n \, dA = \bar{V} A.$$

Therefore

$$\bar{V} = \int \frac{V_n \, dA}{A}. \tag{3.39}$$

The overbar ($\bar{\ }$) represents an average quantity. We will not normally use the overbar when dealing with one-dimensional flow, but we have used it here to distinguish the average velocity from the true velocity. Figure 3.18 shows real and average velocities for flow in a pipe or channel.

We would also like the mass flow equation, Eq. (3.37), to apply to our approximate one-dimensional flow:

$$\dot{m} = \int \rho V_n \, dA = \bar{\rho} \bar{V} A. \tag{3.40}$$

To use this equation, we need an average density as well as an average velocity. This is no problem if the fluid is incompressible because $\bar{\rho} = \rho$. If the fluid's density varies from point to point in the cross section,* we find the average density by substituting Eq. (3.39) into Eq. (3.40), with the result

$$\bar{\rho} = \frac{\int \rho V_n \, dA}{\int V_n \, dA}. \tag{3.41}$$

Notice that the proper average density is a "velocity-weighted" average. An average density computed by

$$\bar{\rho}' = \frac{\int \rho \, dA}{A}, \tag{3.42}$$

similar to the average velocity, would not be the same as $\bar{\rho}$, and the

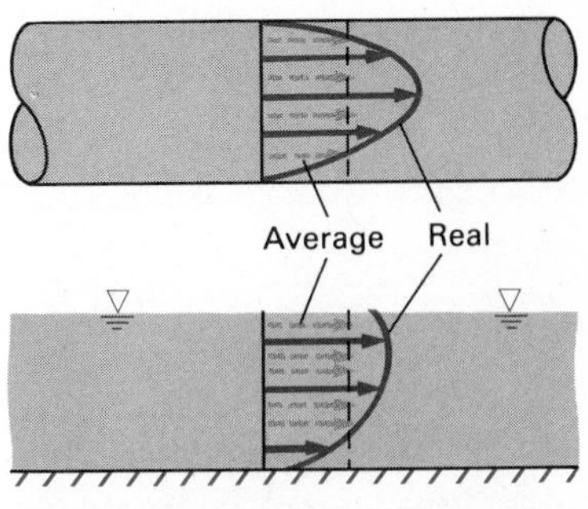

Average   Real

Figure 3.18 Comparison of real and average velocities for pipe and open channel flow.

---

* This is not as rare as you might think. Air flowing in a pipe with a heated wall will have a lower density at the hot wall than at the cooler center of the pipe.

mass flow rate would not be correctly calculated by using $\bar{\rho}'$ in Eq. (3.40). You can work Problem 51 at the end of this chapter to convince yourself of this. In the remainder of this book, we assume that proper averages are implied when we use the one-dimensional flow approximation.

EXAMPLE 3.7 Illustrates how to compute average velocity

Compute the average velocity $\bar{V}$ for the flow in Example 3.1.

Solution

Given
Fluid velocity

$$u = u_0\left(1 + \frac{x}{\ell}\right)\left[1 + \left(\frac{y}{Y}\right)^2\right]$$

for the two-dimensional channel flow in Fig. E3.1(a)

Find
Average velocity $\bar{V}$

Solution
The problem is not adequately defined. Are we to calculate the average velocity at $x = 0$, $x = \ell$, or at all values of x? Since u is known as a function of x and y, we will calculate $\bar{V}(x)$, the average velocity as a function of x. An average velocity is found using Eq. (3.39):

$$\bar{V} = \frac{\int V_n \, dA}{A},$$

where V_n is the component of the velocity normal to dA. Using u as the x-component of the total velocity and normal to each area dA, as shown in Fig. E3.7, we have

$$\bar{V}(x) = \frac{\int u(x, y) \, dA}{A}.$$

Using the channel width W, we obtain

$$dA = W \, dy,$$

where dA is considered a positive quantity. Because $u(x, y)$ is symmetric with respect to the centerplane of the channel, we can eliminate errors of signs by using

$$\bar{V}(x) = \frac{2 \int_0^Y u(x, y) \, dA}{A},$$

with

$$dA = W \, dy \quad \text{and} \quad A = 2WY.$$

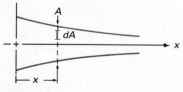

Figure 3.E3.7 Flow area for calculating the average velocity $\bar{v}(x)$.

Substituting for $u(x, y)$, A, and dA gives

$$\bar{V}(x) = \frac{\int_0^Y u_0 \left(1 + \frac{x}{\ell}\right)\left[1 - \left(\frac{y}{Y}\right)^2\right] W \, dy}{2WY}$$

$$= \frac{u_0 \left(1 + \frac{x}{\ell}\right)\int_0^Y \left[1 - \left(\frac{y}{Y}\right)^2\right] dy}{Y}$$

$$= \frac{u_0 \left(1 + \frac{x}{\ell}\right)\left[y - \frac{y^3}{3Y^2}\right]_0^Y}{Y}$$

$$= \frac{u_0 \left(1 + \frac{x}{\ell}\right)\left[Y - \frac{1}{3}Y\right]}{Y},$$

$$\boxed{\bar{V}(x) = \frac{2}{3}u_0 \left(1 + \frac{x}{\ell}\right).}$$

Answer ◀

Discussion

The first three columns of Table E3.7 give $\bar{V}$ and A as a function of the axial coordinate x/ℓ for a channel of width $W = 1.0$ m. The fourth column gives the product $\bar{V}A$ as a function of x/ℓ. Note that

$$(\bar{V}A)_{\text{inlet } (x=0)} = (\bar{V}A)_{\text{any } x\text{-location}}$$

Do you know why this is true?

The transverse velocity $v(x, y)$ is nonzero for this flow (Example 3.4) but does not influence the calculation of $\bar{V}$ because v is parallel to the cross-sectional area.

Table E3.7 Values of $\bar{V}$, A, and $\bar{V}A$ as a function of x/ℓ for the flow in Example 3.1

Length Ratio x/ℓ	Average Velocity $\bar{V}$ (m/s)	Area A (m²)	$\bar{V}A$ (m³/s)
0.0	0.667	2.000	1.334*
0.2	0.800	1.667	1.334
0.4	0.933	1.429	1.333
0.6	1.067	1.250	1.334
0.8	1.200	1.111	1.333
1.0	1.333	1.000	1.333

* Note the small error due to rounding off $\bar{V}$ and A to four significant figures.

Integral Continuity Equation for Steady One-Dimensional Flow.

Consider the flow through the apparatus shown in Fig. 3.19. This may represent a water storage tank with multiple inlets and outlets, a steam turbine with steam extraction at various points, or some other piece of equipment. We assume that the flow is steady (there is no accumulation of fluid inside the apparatus) and that the flow is one-dimensional and

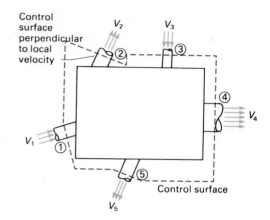

Figure 3.19 Apparatus with several inlets and outlets; flow is locally one-dimensional and one-directional at each inlet/outlet.

one-directional at each inlet and outlet. To formally apply the integral continuity equation, we select a control volume that encloses the entire apparatus, as shown in Fig. 3.19.

The integral continuity equation in the form of Eq. (3.27) is

$$\frac{d}{dt}\int_{\Psi_{cv}}\rho\,dV + \int_{A_{\text{out}}}\rho V\cos\theta\,dA + \int_{A_{\text{in}}}\rho V\cos\theta\,dA = 0.$$

The first term is zero by the steady-flow condition. Recall that for all the inflow surfaces, the third term is negative, so using the angle θ' instead of θ at the inlet (refer to Fig. 3.16),

$$\int_{A_{\text{in}}}\rho V\cos\theta'\,dA = \int_{A_{\text{out}}}\rho V\cos\theta\,dA.$$

It is not necessary to evaluate integrals since mass flows in and out at a finite number of locations and the velocity and density are uniform at each location; instead we can write the integrals as finite sums:

$$\sum(\rho V_n A)_{\text{in}} = \sum(\rho V_n A)_{\text{out}}. \tag{3.43}$$

An alternate form is

$$\sum \dot{m}_{\text{in}} = \sum \dot{m}_{\text{out}}. \tag{3.44}$$

For the apparatus of Fig. 3.19,

$$\rho_1 V_1 A_1 + \rho_3 V_3 A_3 = \rho_2 V_2 A_2 + \rho_4 V_4 A_4 + \rho_5 V_5 A_5.$$

If the density of the fluid is the same at all points (a good assumption if the apparatus is a water tank but a poor one if the apparatus is a steam turbine), we can use volume flows instead of mass flows:

$$\sum(V_n A)_{\text{in}} = \sum(V_n A)_{\text{out}} \qquad \text{(constant density only)} \tag{3.45}$$

or

$$\sum Q_{\text{in}} = \sum Q_{\text{out}}. \tag{3.46}$$

For a flow in a conduit or channel with only one inlet and one outlet, as shown in Fig. 3.20, we can drop the summations:

$$\rho_1 V_1 A_1 = \rho_2 V_2 A_2 \tag{3.47}$$

or, if $\rho_1 = \rho_2$,

$$V_1 A_1 = V_2 A_2 \qquad \text{(constant density only)} \tag{3.48}$$

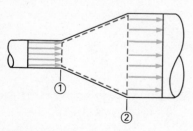

Figure 3.20 Conduit with one inlet and one outlet.

EXAMPLE 3.8 Illustrates the steady, one-dimensional integral continuity equation

Water at 60°F flows steadily with an average velocity of 160 ft/min through a 6-in. (nominal) diameter, type L copper pipe. Some distance downstream, the water flows through a reducer and into a 4-in. (nominal) diameter, type L copper pipe. Find the average velocity in the 4-in. pipe.

Solution

Given
Figure E3.8

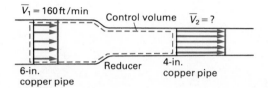

Figure E3.8 Steady flow of water through copper piping.

Find
Downstream velocity $\bar{V}_2$

Solution
Apply the one-dimensional integral continuity equation to the control volume shown in Fig. E3.8 For steady conditions, Eq. (3.20) gives

$$\dot{m}_{\text{out}} - \dot{m}_{\text{in}} = 0.$$

Using Eq. (3.40) for $\dot{m}$ gives

$$\rho_2 \bar{V}_2 A_2 - \rho_1 \bar{V}_1 A_1 = 0.$$

Solving for $\bar{V}_2$, we have

$$\bar{V}_2 = \left(\frac{\rho_1}{\rho_2}\right)\left(\frac{A_1}{A_2}\right)\bar{V}_1.$$

The areas are

$$A_1 = \frac{\pi}{4} D_1^2 \quad \text{and} \quad A_2 = \frac{\pi}{4} D_2^2.$$

Considering the water density as constant, we find

$$\bar{V}_2 = \left(\frac{D_1}{D_2}\right)^2 \bar{V}_1.$$

Appendix D gives the dimensions of various schedules of steel pipe and types of copper tube. For type L tube, $D_1 = [6.125 - 2(0.140)] = 5.845$ in. and $D_2 = [4.125 - 2(0.110)] = 3.905$ in. Then

$$\bar{V}_2 = \left(\frac{5.845 \text{ in.}}{3.905 \text{ in.}}\right)^2 (160 \text{ ft/min})$$

$$\left| \ \bar{V}_2 = 358 \text{ ft/min.} \ \right|$$ **Answer**

Discussion
See the discussion for Example 3.10.

EXAMPLE 3.9 Illustrates the steady, one-dimensional integral continuity equation

A *hydraulic jump* is an open channel flow phenomenon in which there is an abrupt increase in the depth of the flow stream. Engineers may design a dam spillway (a passage for excess water to run around or under the dam) such that a hydraulic jump occurs in the spillway to reduce the water velocity entering the river. Figure E3.9a shows a hydraulic jump occurring at the base of a dam. Water flows at a steady rate with an average velocity $\bar{V}_1 = 11$ m/s. The depth increases from $y_1 = 2$ m to $y_2 = 6$ m. Find the average velocity downstream of the hydraulic jump.

Solution

Given

Hydraulic jump shown in Fig. E3.9(a)

$\bar{V}_1 = 11$ m/s, $y_1 = 2$ m, $y_2 = 6$ m

Find

Average velocity $\bar{V}_2$

Solution

Apply the one-dimensional integral continuity equation to the control volume shown in Fig. E3.9(b). For steady flow conditions, Eq. (3.20) gives

$$\dot{m}_{\text{out}} - \dot{m}_{\text{in}} = 0.$$

Using Eq. (3.40) for $\dot{m}$ gives

$$\rho_2 \bar{V}_2 A_2 - \rho_1 \bar{V}_1 A_1 = 0.$$

Solving for $\bar{V}_2$, we have

$$\bar{V}_2 = \left(\frac{\rho_1}{\rho_2}\right)\left(\frac{A_1}{A_2}\right)\bar{V}_1.$$

Each area is the product of the water depth and the width of the water. We have no given information on the widths W_1 and W_2 of the river; however, we expect the ratio W_1/W_2 to be close to 1.* Therefore

$$\frac{A_1}{A_2} = \frac{y_1 W_1}{y_2 W_2} \approx \frac{y_1}{y_2}$$

Figure E3.9 (a) Hydraulic jump at the base of a dam; (b) control volume.

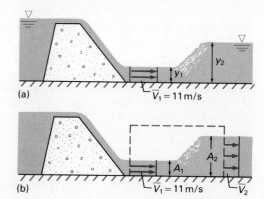

* If you were to observe a hydraulic jump at the base of a dam, it would occur within about 20 m of the base of the dam, and the width of the river would not change appreciably over that distance.

and

$$\bar{V}_2 = \left(\frac{y_1}{y_2}\right)\bar{V}_1$$

$$= \left(\frac{2\,\mathrm{m}}{6\,\mathrm{m}}\right)(11\,\mathrm{m/s})$$

$$\left|\ \bar{V}_2 = 3.67\,\mathrm{m/s.}\ \right|$$

Answer ◀

Discussion
See the discussion for Example 3.10.

EXAMPLE 3.10 Illustrates the steady, one-dimensional integral continuity equation

A shock is a flow phenomenon that can occur when a compressible fluid flows at a speed greater than the speed of sound. There is an abrupt increase in the fluid pressure and temperature and a decrease in the fluid velocity across the shock. Shocks are so thin (thickness of only a few mean free paths) that they can be assumed to have zero thickness. Figure E3.10(a) shows air flowing through a normal (perpendicular to the fluid velocity) shock in a constant-area duct. The pressure, temperature, and velocity upstream of the shock are $p_1 = 80$ kPa, $T_1 = 20°$C, and $V_1 = 500$ m/s. Downstream of the shock the pressure $p_2 = 180$ kPa and the temperature $T_2 = 97°$C. Find the downstream velocity V_2.

Solution

Given
Figure E3.10(a)

Find
Velocity V_2

Solution
Apply the one-dimensional integral continuity equation to the control volume shown in Fig. E3.10(b). For steady flow conditions, Eq. (3.20) gives

$$\dot{m}_{\mathrm{out}} - \dot{m}_{\mathrm{in}} = 0.$$

Figure E3.10 (a) Normal shock in a constant-area duct; (b) control volume.

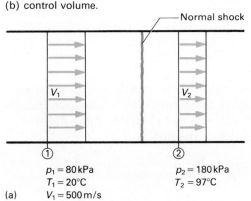

(a)
$p_1 = 80$ kPa
$T_1 = 20°$C
$V_1 = 500$ m/s

$p_2 = 180$ kPa
$T_2 = 97°$C

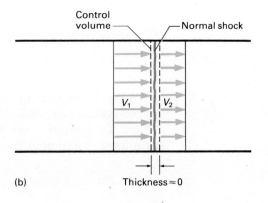

(b)

Thickness ≈ 0

Using Eq. (3.40) for $\dot{m}$, we have

$$\rho_2 V_2 A_2 - \rho_1 V_1 A_1 = 0.$$

The density is assumed uniform over both area A_1 and area A_2. Noting $A_1 = A_2$ and solving for V_2 give

$$V_2 = \left(\frac{\rho_1}{\rho_2}\right) V_1.$$

Air may be assumed to be an ideal gas at the temperatures and pressures involved. The ideal gas law, Eq. (1.1), gives

$$\frac{\rho_1}{\rho_2} = \left(\frac{p_1}{RT_1}\right)\left(\frac{RT_2}{p_2}\right) = \left(\frac{p_1}{p_2}\right)\left(\frac{T_2}{T_1}\right).$$

Substituting into the integral continuity equation, we have

$$V_2 = \left(\frac{p_1}{p_2}\right)\left(\frac{T_2}{T_1}\right) V_1.$$

The numerical values give

$$V_2 = \left(\frac{80\text{ kPa}}{180\text{ kPa}}\right)\left(\frac{(273 + 97)\text{K}}{(273 + 20)\text{K}}\right)(500\text{ m/s}),$$

$$\left| \; V_2 = 280 \text{ m/s}. \; \right|$$

Answer ◀

Discussion

Examples 3.8, 3.9, and 3.10 involved applications of the steady, one-dimensional integral continuity equation. Examples 3.8 and 3.9 involved a liquid, so it was realistic to assume constant density; Example 3.10 involved a gas with pressure and temperature changes large enough that density changes must be considered (compare to Example 1.1).

The term "average velocity" was used in Examples 3.8 and 3.9, while the term "uniform velocity" was used in Example 3.10. Many internal flows are assumed to have a uniform velocity profile to obtain a reasonable estimate for engineering purposes.

One-Dimensional, Nonsteady Flow. In a nonsteady flow, mass accumulates in the control volume, so

$$\frac{d}{dt}(m_{cv}) = \frac{d}{dt}\int_{\Psi_{cv}} \rho\, d\Psi \neq 0.$$

For a finite number of inlets and outlets, the continuity equation is

$$\frac{d}{dt}\int_{\Psi_{cv}} \rho\, d\Psi = \sum \dot{m}_{\text{in}} - \sum \dot{m}_{\text{out}}. \tag{3.49}$$

We know that the mass flow rates can be evaluated from density-velocity-area products, so we can concentrate on the mass accumulation. To simplify our analysis, we assume that the fluid density is uniform throughout the volume at any instant, so

$$\frac{d}{dt}\int_{\Psi_{cv}} \rho\, d\Psi = \frac{d}{dt}(\rho \Psi_{cv}).$$

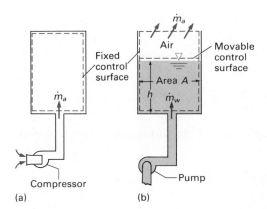

Figure 3.21 Two tank-filling operations: (a) A compressor fills a closed tank with air; (b) a pump fills an open tank with water.

Either the volume or the density or both may be time-dependent. To see these possibilities, consider the two cases shown in Fig. 3.21. In case (a) a compressor fills a rigid closed tank with air. In case (b) a pump fills an open tank with water.

In case (a), the air occupies the entire tank. As more air enters the tank, the density increases. The only sensible choice for a control volume is a volume that encloses the entire tank. Since this volume is constant, the continuity equation becomes

$$\frac{d}{dt} \int_{V_{cv}} \rho \, dV = V_{cv} \left(\frac{d\rho_a}{dt} \right) = \dot{m}_{\text{in}}.$$

Consider case (b), the tank of water. Since the water has constant density, the water level must rise. Suppose that we choose a control volume that encloses the entire tank, as for case (a). At any instant of time, this control volume must be divided into two parts, one containing water and the other containing air (above the water). Then

$$\frac{d}{dt} \int_{V_{cv}} \rho \, dV = \frac{d}{dt} \left(\rho_w V_w + \rho_a V_a \right)$$

$$= \rho_w \frac{dV_w}{dt} + \rho_a \frac{dV_a}{dt}.$$

Since the air above the water is unconfined, we have assumed that ρ_a is uniform and constant in time. Applying the continuity equation, we have

$$\rho_w \frac{dV_w}{dt} + \rho_a \frac{dV_a}{dt} = \dot{m}_{w,\text{in}} - \dot{m}_{a,\text{out}}.$$

Since the air and water don't mix, we have

$$\rho_w \frac{dV_w}{dt} = \dot{m}_{w,\text{in}} \quad \text{and} \quad \rho_a \frac{dV_a}{dt} = -\dot{m}_{a,\text{out}}.$$

Since $V_w = hA$ (refer to Fig. 3.21b), we get

$$\frac{dh}{dt} = \frac{\dot{m}_{w,\text{in}}}{\rho_w A} = \frac{Q_{w,\text{in}}}{A}.$$

A great deal of manipulation and thinking was involved to arrive at the final result, especially the part dealing with the air.* Suppose that we choose our control volume such that it always exactly coincides with the water in the tank. Then

$$\frac{d}{dt} \int_{\mathcal{V}_{cv}} \rho \, d\mathcal{V} = \frac{d}{dt} (\rho_w \mathcal{V}_{cv}) = \rho_w \frac{d\mathcal{V}_{cv}}{dt}.$$

Since $\mathcal{V}_{cv} = hA$, substituting into the continuity equation and solving for dh/dt gives

$$\frac{dh}{dt} = \frac{\dot{m}_{w,in}}{\rho_w A} = \frac{Q_{w,in}}{A}.$$

The choice of a deformable control volume saved us the trouble of considering the air in the tank at all. You should always be on the alert for clever choices of control volumes that will save you a lot of work in the long run.

* Think how much more complicated it would have been if the top of the tank were closed so that the air was compressed rather than allowed to escape.

EXAMPLE 3.11   Illustrates how to apply the integral continuity equation to a nonsteady problem

Water is being added to a storage tank at the rate of 500 gal/min. Water also flows out of the bottom through a 2.0-in.-inside-diameter pipe with an average velocity of 60 ft/sec, as shown in Fig. E3.11(a). The storage tank has an inside diameter of 10.0 ft. Find the rate at which the water level is rising or falling.

Solution

   Given
Figure E3.11(a)
Storage tank with inside diameter 10.0 ft
Water with an average velocity of 60 ft/sec in discharge pipe

Figure E3.11   (a) Water storage tank; (b) control volume.

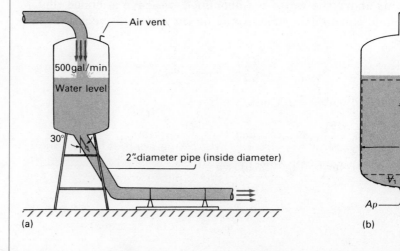

(a)   (b)

Find

Rate at which water level rises or falls

Solution

We choose a control volume that exactly coincides with the water in the tank. Assuming the water density is constant, the integral continuity equation is

$$\rho_w \frac{dV_{cv}}{dt} = \sum \dot{m}_{in} - \sum \dot{m}_{out},$$

where V_{cv} is the volume of the control volume (or of the water in the tank). Equations (3.36) and (3.40) give, respectively,

$$\sum \dot{m}_{in} = \rho_w Q_{in} \quad \text{and} \quad \sum \dot{m}_{out} = \rho_w (\bar{V}A)_{out}.$$

Choosing the control volume in Fig. E3.11(b), we have

$$V_{cv} = V_1 + V_2$$
$$= V_1 + A_T h.$$

Substituting into the integral continuity equation gives

$$\rho_w \frac{d}{dt} (V_1 + A_T h) = \rho_w Q_{in} - (\rho_w \bar{V}A)_{out}.$$

Since ρ_w, V_1, A_T, and A_{out} are constant,

$$A_T \frac{dh}{dt} = Q_{in} - (\bar{V}A)_{out}, \quad \text{or} \quad \frac{dh}{dt} = \frac{Q_{in} - (\bar{V}A)_{out}}{A_T}.$$

The numerical values give

$$\frac{dh}{dt} = \frac{\left(500 \frac{\text{gal}}{\text{min}}\right)\left(\frac{\text{ft}^3}{7.48 \text{ gal}}\right) - \left(60 \frac{\text{ft}}{\text{sec}}\right)\left(\frac{\pi}{4}\right)\left(\frac{2}{12} \text{ft}\right)^2 \left(\frac{60 \text{ sec}}{\text{min}}\right)}{\frac{\pi}{4} (10.0 \text{ ft})^2}$$

$$\left| \frac{dh}{dt} = -0.149 \text{ ft/min.} \right| \qquad \textbf{Answer} \blacktriangleleft$$

The water level is falling, indicated by the negative sign.

Discussion

Note that mass flow rate is calculated two ways in this problem:

$$\dot{m} = \rho_w Q \quad \text{and} \quad \dot{m} = \rho_w \bar{V}A.$$

Both are for a constant-density fluid.

3.4.4 Derivation of the Differential Continuity Equation from the Integral Continuity Equation

Since we have shown you many interesting applications of the integral continuity equation, you have probably pushed the differential continuity equation to the back of your mind or joined the ranks of those who assume that the differential continuity equation belongs to another kind of fluid mechanics. To reemphasize the equivalence between the differential and finite control volume approaches, we will now use the control volume approach to derive the differential continuity equation. We will do this by applying the control volume formulation to a very small control volume.

Consider a two-dimensional, two-directional, unsteady flow field. (By now you should realize that the two-dimensional assumptions are only for clarity in presentation.) Figure 3.22 shows a point in the field. A control volume of dimensions δx, δy, δz is superimposed around the point. The fluid velocities at the faces of the control volume are shown. This figure is very much like Fig. 3.13 except that in Fig. 3.22 the dotted lines refer to the boundaries of a fixed control volume rather than the boundaries of a specific fluid particle.*

We assume that the velocities are uniform across the faces of the small control volume and that the fluid density at the center of the control volume is an appropriate average. The integral continuity equation, Eq. (3.49), becomes

$$\frac{d}{dt}(\rho\,\delta V) + \sum \dot{m}_{\text{out}} - \sum \dot{m}_{\text{in}} = 0.$$

The volume δV of the control volume is constant, so

$$\frac{d}{dt}(\rho\,\delta V) = \delta V \frac{d\rho}{dt} = (\delta x)(\delta y)(\delta z)\frac{d\rho}{dt}.$$

Since the control volume is fixed in space, the time derivative of density is taken at a fixed point and should be written $\partial\rho/\partial t$, so

$$\frac{d}{dt}(\rho\,\delta V) = (\delta x)(\delta y)(\delta z)\frac{\partial\rho}{\partial t}.$$

Noting that mass flows into the control volume through the left and bottom faces and out through the right and top faces and that only the

Figure 3.22 Rectangular control volume enclosing point (x, y) in a two-dimensional flow field; velocities and densities at control volume faces included.

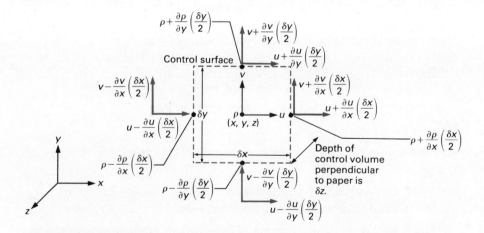

velocity component perpendicular to an area carries mass through the area, we compute the mass flow rates:

$$\sum \dot{m}_{in} = \left[\rho - \frac{\partial \rho}{\partial x}\left(\frac{\delta x}{2}\right)\right]\left[u - \frac{\partial u}{\partial x}\left(\frac{\delta x}{2}\right)\right](\delta y)(\delta z)$$

$$+ \left[\left(\rho - \frac{\partial \rho}{\partial y}\left(\frac{\delta y}{2}\right)\right)\left[v - \frac{\partial v}{\partial y}\left(\frac{\delta y}{2}\right)\right]\right](\delta x)(\delta z)$$

and

$$\sum \dot{m}_{out} = \left[\rho + \frac{\partial \rho}{\partial x}\left(\frac{\delta x}{2}\right)\right]\left[u + \frac{\partial u}{\partial x}\left(\frac{\delta x}{2}\right)\right](\delta y)(\delta z)$$

$$+ \left[\rho + \frac{\partial \rho}{\partial y}\left(\frac{\delta y}{2}\right)\right]\left[v + \frac{\partial v}{\partial y}\left(\frac{\delta y}{2}\right)\right](\delta x)(\delta z).$$

Substituting into the continuity equation and canceling terms where possible give

$$(\delta x)(\delta y)(\delta z)\left(\frac{\partial \rho}{\partial t}\right) + (\delta x)(\delta y)(\delta z)\left[\rho\left(\frac{\partial u}{\partial x}\right) + u\left(\frac{\partial \rho}{\partial x}\right) + \rho\left(\frac{\partial v}{\partial y}\right) + v\left(\frac{\partial \rho}{\partial y}\right)\right] = 0.$$

We now divide by $(\delta x)(\delta y)(\delta z)$ and take the limit as δx, δy, and δz approach zero, shrinking the control volume to a point. The equation becomes

$$\frac{\partial \rho}{\partial t} + u\left(\frac{\partial \rho}{\partial x}\right) + \rho\left(\frac{\partial u}{\partial x}\right) + v\left(\frac{\partial \rho}{\partial y}\right) + \rho\left(\frac{\partial v}{\partial y}\right) = 0.$$

Combining terms in accordance with the product rule for derivatives, we get

$$\frac{\partial \rho}{\partial t} + \frac{\partial(\rho u)}{\partial x} + \frac{\partial(\rho v)}{\partial y} = 0,$$

which is Eq. (3.16), the differential continuity equation for two-dimensional, two-directional, unsteady flow.

You have probably noticed that this development of the differential continuity equation from the integral equation was a little shorter than our earlier approach, which involved consideration of a particle of constant mass. Since this is generally true, the control volume forms of the fundamental laws are often used to derive the differential forms. This derivation can be done by a completely formal mathematical procedure using theorems from vector integral calculus. If you are interested in this approach, consult [1, 4, 5]. It is also possible to derive the finite control volume forms of the fundamental laws from the differential forms by integration.

3.5 Some Characteristics of Flows and Flow Fields

So far in this chapter we have concentrated primarily on introducing the *methods* that are used to formulate flow problems. We will now consider some of the characteristics of flows and flow fields that must be recognized and dealt with in engineering.

3.5.1 Kinematics of a Fluid Particle: Acceleration, Rotation, Deformation

Kinematics is the study of motion; as such it involves velocities and accelerations, both linear and angular, and the relation between them. Analysis of the forces that cause motion involves kinematics since force and acceleration are related by Newton's second law. In our consideration of flow kinematics, we will also examine the deformation of fluid particles, since stress in fluids is related to rate of deformation.

We will consider a two-dimensional, two-directional, unsteady flow. Figure 3.23 shows a rectangular fluid particle at point (x, y) at time t. Notice that the situation is exactly the same as in Fig. 3.13. Using an Eulerian description, the velocity field is

$$\vec{V} = \vec{V}(x, y, t) = u(x, y, t)\hat{i} + v(x, y, t)\hat{j}.$$

Acceleration. The acceleration of the particle is the total time rate of change of its velocity. We reintroduce the subscript p to remind us that we are considering a specific fluid particle, and we write the velocity as

$$\vec{V}_p = \vec{V}_p(x_p, y_p, t).$$

The acceleration is

$$\vec{a}_p = \frac{d}{dt}(\vec{V}_p) = \frac{d}{dt}(\vec{V}_p(x_p, y_p, t)).$$

We can write the velocity and acceleration in x- and y-components:

$$a_{p,x} = \frac{d}{dt}(u_p(x_p, y_p, t)) \quad \text{and} \quad a_{p,y} = \frac{d}{dt}(v_p(x_p, y_p, t)).$$

Considering $a_{p,x}$, we have

$$a_{p,x} = \frac{\partial u_p}{\partial x_p}\left(\frac{dx_p}{dt}\right) + \frac{\partial u_p}{\partial y_p}\left(\frac{dy_p}{dt}\right) + \frac{\partial u_p}{\partial t}.$$

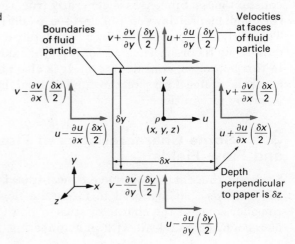

Figure 3.23 Rectangular fluid particle in two-dimensional, two-directional velocity field; particle center is at point (x, y) with velocity (u, v) at time t; the particle's z-coordinate is arbitrary.

However,

$$\frac{dx_p}{dt} = u_p \quad \text{and} \quad \frac{dy_p}{dt} = v_p,$$

so

$$a_{p,x} = \frac{\partial u_p}{\partial t} + u_p \frac{\partial u_p}{\partial x_p} + v_p \frac{\partial u_p}{\partial y_p}.$$

At the instant in question, the fluid particle is at $x_p = x$, $y_p = y$, with $u_p = u$ and $v_p = v$, so

$$a_x = \frac{\partial u}{\partial t} + u\left(\frac{\partial u}{\partial x}\right) + v\left(\frac{\partial u}{\partial y}\right). \tag{3.50}$$

Note that we have dropped the subscript p. By a similar development,

$$a_y = \frac{\partial v}{\partial t} + u\left(\frac{\partial v}{\partial x}\right) + v\left(\frac{\partial v}{\partial y}\right). \tag{3.51}$$

Notice that the derivatives on the right of both acceleration equations are the Eulerian derivatives from Eq. (3.10) (in two dimensions), so

$$a_x = \frac{Du}{Dt} \quad \text{and} \quad a_y = \frac{Dv}{Dt}.$$

These expressions can be extended to three dimensions and three directions:

$$a_x = \frac{Du}{Dt} = \frac{\partial u}{\partial t} + u\left(\frac{\partial u}{\partial y}\right) + v\left(\frac{\partial u}{\partial y}\right) + w\left(\frac{\partial u}{\partial z}\right), \tag{3.52}$$

$$a_y = \frac{Dv}{Dt} = \frac{\partial v}{\partial t} + u\left(\frac{\partial v}{\partial x}\right) + v\left(\frac{\partial v}{\partial y}\right) + w\left(\frac{\partial v}{\partial z}\right), \tag{3.53}$$

$$a_z = \frac{Dw}{Dt} = \frac{\partial w}{\partial t} + u\left(\frac{\partial w}{\partial x}\right) + v\left(\frac{\partial w}{\partial y}\right) + w\left(\frac{\partial w}{\partial z}\right). \tag{3.54}$$

Or they can be written in a single equation using vector notation:

$$\vec{a} = \frac{D\vec{V}}{Dt} = \frac{\partial \vec{V}}{\partial t} + (\vec{V} \cdot \nabla)\vec{V}. \tag{3.55}$$

These equations for acceleration are somewhat complicated. In most cases in this book, several of the terms in these equations will vanish.

Remember the basic idea that, for any property of a fluid particle, we can write

$$\begin{bmatrix} \text{Total rate of change} \\ \text{of property of a} \\ \text{fluid particle} \end{bmatrix} = \begin{bmatrix} \text{Rate of change} \\ \text{at a fixed location } \left(\frac{\partial}{\partial t}\right) \end{bmatrix} + \begin{bmatrix} \text{Velocity times} \\ \text{derivative with} \\ \text{respect to space} \\ \text{coordinates } (\vec{V} \cdot \nabla) \end{bmatrix}$$

The first term on the right in this word equation is called the *local unsteady term*. The second term on the right is called the *convective rate of change* since it reflects that the fluid is convected (carried) about in the field. So far we have applied this rate of change equation to density (in Eq. 3.9) and velocity (in Eqs. 3.52–3.55).

Angular Velocity and Vorticity. The acceleration of the fluid particle of Fig. 3.23 was obtained by a purely mathematical procedure. The particle's angular velocity cannot be obtained quite so rigorously. To understand why, consider a similar particle of a rigid material (like steel), as shown in Fig. 3.24(a). Two perpendicular lines are marked on the particle. Since the particle is rigid, both lines rotate at the same angular velocity. If the angular velocity is not zero, a short time later the lines will appear as in part (b) of the figure. Next consider a rectangular fluid particle with two perpendicular lines marked on it (Fig. 3.24c). Since the fluid is deformable, a short time later the fluid and the lines will appear as shown in part (d). Due to deformation, the two lines may have rotated different amounts. To obtain a single value for the angular velocity of the fluid particle, we must introduce a definition:

> The instantaneous angular velocity (ω) of a fluid particle is the average of the instantaneous angular velocities of two mutually perpendicular lines on the fluid particle. Usually these lines are parallel to the coordinate axes.

Figure 3.25 shows the fluid particle with two lines A-A' and B-B'. By definition

$$\omega \equiv \frac{1}{2}(\omega_{AA'} + \omega_{BB'}), \tag{3.56}$$

where

$$\omega_{AA'} = \frac{v_{A'} - v_A}{\delta x} = \frac{\left[v + \frac{\partial v}{\partial x}\left(\frac{\delta x}{2}\right)\right] - \left[v - \frac{\partial v}{\partial x}\left(\frac{\delta x}{2}\right)\right]}{\delta x} = \frac{\partial v}{\partial x} \tag{3.57}$$

and

$$\omega_{BB'} = -\frac{u_{B'} - u_B}{\delta y} = -\frac{\left[u + \frac{\partial u}{\partial y}\left(\frac{\delta y}{2}\right)\right] - \left[u - \frac{\partial u}{\partial y}\left(\frac{\delta y}{2}\right)\right]}{\delta y} = -\frac{\partial u}{\partial y}, \tag{3.58}$$

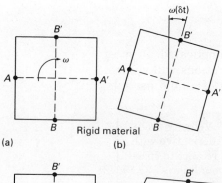

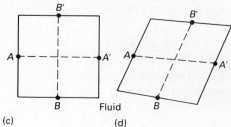

Figure 3.24 Comparison of rotation of rectangular particles of rigid material at (a) time t and (b) time $t + \delta t$; and of fluid at (c) time t and (d) time $t + \delta t$.

Figure 3.25 Rectangular fluid particle with two instantaneous perpendicular lines AA' and BB'; velocities perpendicular to AA' and BB' are also shown.

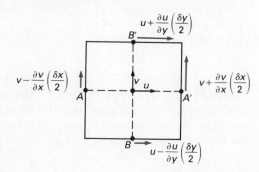

so

$$\omega_z = \frac{1}{2}\left(\frac{\partial v}{\partial x} - \frac{\partial u}{\partial y}\right). \tag{3.59}$$

We have added the subscript z to ω to remind us that this is actually the angular velocity about the z-axis (perpendicular to the paper). In general, the angular velocity is a vector ($\vec{\omega}$) with three components.

The $\frac{1}{2}$ is somewhat inconvenient, so instead of the angular velocity ω_z we often use the vorticity ζ_z, defined by

$$\zeta_z \equiv 2\omega_z = \frac{\partial v}{\partial x} - \frac{\partial u}{\partial y}. \tag{3.60}$$

A flow in which angular velocity and vorticity are zero is called an *irrotational flow*. The definition of angular velocity and vorticity can be extended to three dimensions, but we will not do that here. An irrotational three-dimensional flow must have all three components of $\vec{\omega}$ and $\vec{\zeta}$ equal to zero.

Rate of Volumetric Strain (Stretching). A fluid particle can be "stretched" and "squeezed" as illustrated in Fig. 3.26. This process can cause a change in the volume of the fluid particle. The rate of change of volume, divided by the volume itself, is called the "rate of volumetric strain":

$$\left(\frac{1}{\delta V}\frac{d(\delta V)}{dt}\right).$$

In Section 3.4.1 we considered the rates of stretching in each direction and found them to be

$$\frac{d(\delta x)}{dt} = \frac{\partial u}{\partial x}\,\delta x, \tag{3.10}$$

$$\frac{d(\delta y)}{dt} = \frac{\partial v}{\partial y}\,\delta y. \tag{3.11}$$

The rate of volumetric change for a two-dimensional particle is

$$\frac{d(\delta V)}{dt} = (\delta z)(\delta y)\frac{d(\delta x)}{dt} + (\delta z)(\delta x)\frac{d(\delta y)}{dt},$$

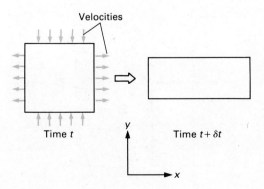

Velocities

Time t

y

Time $t + \delta t$

x

Figure 3.26 Fluid particle stretched in x-direction and squeezed in y-direction.

so the rate of volumetric strain is

$$\frac{1}{\delta \mathcal{V}} \frac{d(\delta \mathcal{V})}{dt} = \frac{\partial u}{\partial x} + \frac{\partial v}{\partial y}. \tag{3.61}$$

Rate of Shear Deformation (Shear Strain Rate). In addition to rotation and volumetric strain, a fluid particle in a flow field also experiences shear deformation. Next to acceleration, the rate of shear deformation is the most important kinematic property of a flow because the shear stresses in the fluid are related to this strain rate (see Sections 1.2 and 1.3.4).

Our earlier discussions of shear deformation in fluids (see Figs. 1.1, 1.4, and 1.5) used a somewhat simplified picture, with the upper and lower edges of a rectangular fluid particle always remaining parallel to the x-coordinate. A more general definition of strain rate can be developed with the aid of Fig. 3.27. Part (a) of the figure shows an initially rectangular fluid particle with two perpendicular lines on it. Part (b) of the figure shows the deformed fluid particle. The average shear deformation, ϕ, is defined by

$$\phi \equiv \frac{1}{2}(\phi_1 + \phi_2),$$

with ϕ_1 and ϕ_2 positive as shown. The rate of shear deformation, $\dot{\phi}$, is

$$\dot{\phi} = \frac{d\phi}{dt} = \frac{1}{2}\left(\frac{d\phi_1}{dt} + \frac{d\phi_2}{dt}\right) \tag{3.62}$$

Borrowing results from our recently completed discussion of angular velocity, we have

$$\frac{d\phi_1}{dt} = \omega_{AA'} \quad \text{and} \quad \frac{d\phi_2}{dt} = -\omega_{BB'}.$$

Using Eqs. (3.57) and (3.58), Eq. (3.62) becomes

$$\dot{\phi} = \frac{1}{2}\left(\frac{\partial u}{\partial y} + \frac{\partial v}{\partial x}\right). \tag{3.63}$$

This can be extended to a three-dimensional fluid particle, where there are six component rates of shear strain, but we will not do that here.

We can use the strain rate equation, Eq. (3.63), to make some comments about the situation in Figs. 1.4 and 1.5, where we have assumed that

$$\frac{\partial v}{\partial x} = 0.$$

Figure 3.27 Deformation of fluid particles; deformation angles ϕ_1 and ϕ_2 are positive as shown.

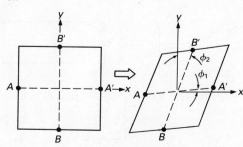

In flows of the type sketched in Figs. 1.4 and 1.5 it is usually true that

$$\frac{\partial u}{\partial y} \gg \frac{\partial v}{\partial x}.$$

EXAMPLE 3.12   Illustrates the kinematic characteristics of a flow field

Calculate the linear acceleration, rotation, vorticity, rate of volumetric strain, and rate of shear deformation for the flow field in Example 3.1.

Solution

Given
Velocity distribution

$$u = u_0 \left(1 + \frac{x}{\ell} \right) \left[1 - \left(\frac{y}{Y} \right)^2 \right]$$

from Example 3.1 and

$$v = u_0 \left[\frac{y^3}{\ell Y_0^2} \left(1 + \frac{x}{\ell} \right)^2 - \frac{y}{\ell} \right]$$

from Example 3.4, both valid for the flow in Example 3.1

Find
Linear acceleration
Rotation
Vorticity
Rate of volumetric strain
Rate of shear deformation

Solution
There are two linear accelerations, a_x and a_y. Equations (3.52) and (3.53) give

$$a_x = \frac{Du}{Dt} = \frac{\partial u}{\partial t} + u \left(\frac{\partial u}{\partial x} \right) + v \left(\frac{\partial u}{\partial y} \right),$$

$$a_y = \frac{Dv}{Dt} = \frac{\partial v}{\partial t} + u \left(\frac{\partial v}{\partial x} \right) + v \left(\frac{\partial v}{\partial y} \right).$$

Since Example 3.1 is a steady-flow situation,

$$\frac{\partial u}{\partial t} = \frac{\partial v}{\partial t} = 0.$$

Before performing the derivative operations, we simplify the mathematics by first substituting

$$Y = \frac{Y_0}{\left(1 + \dfrac{x}{\ell} \right)}$$

into the equation for u. This gives

$$u = u_0 \left(1 + \frac{x}{\ell} \right) \left[1 - \frac{y^2}{Y_0^2} \left(1 + \frac{x}{\ell} \right)^2 \right] = u_0 \left[\left(1 + \frac{x}{\ell} \right) - \frac{y^2}{Y_0^2} \left(1 + \frac{x}{\ell} \right)^3 \right].$$

Then

$$\frac{\partial u}{\partial x} = u_0\left[\frac{1}{\ell} - \frac{3y^2}{\ell Y_0^2}\left(1 + \frac{x}{\ell}\right)^2\right],$$

$$\frac{\partial u}{\partial y} = u_0\left[-\frac{2y}{Y_0^2}\left(1 + \frac{x}{\ell}\right)^3\right],$$

$$\frac{\partial v}{\partial x} = u_0\left[\frac{2y^3}{\ell^2 Y_0^2}\left(1 + \frac{x}{\ell}\right)\right],$$

$$\frac{\partial v}{\partial y} = u_0\left[\frac{3y^2}{\ell Y_0^2}\left(1 + \frac{x}{\ell}\right)^2 - \frac{1}{\ell}\right].$$

Substituting and simplifying, we obtain

$$\left| \; a_x = \frac{u_0}{\ell}\left[\left(1 + \frac{x}{\ell}\right) - 2\left(\frac{y}{Y_0}\right)^2\left(1 + \frac{x}{\ell}\right)^3 + \left(\frac{y}{Y_0}\right)^4\left(1 + \frac{x}{\ell}\right)^5\right] \; \right| \qquad \textbf{Answer} \;\blacktriangleleft$$

and

$$\left| \; a_y = \frac{u_0^2}{\ell^2}\left[y - 2\frac{y^3}{Y_0^2}\left(1 + \frac{x}{\ell}\right)^2 + \frac{y^5}{Y_0^4}\left(1 + \frac{x}{\ell}\right)^4\right]. \; \right| \qquad \textbf{Answer} \;\blacktriangleleft$$

The rotation or angular velocity is

$$\omega_z = \frac{1}{2}\left(\frac{\partial v}{\partial x} - \frac{\partial u}{\partial y}\right).$$

Substituting for $\partial v/\partial x$ and $\partial u/\partial y$, we have

$$\left| \; \omega_z = \frac{yu_0}{Y_0^2}\left(1 + \frac{x}{\ell}\right)\left[\frac{y^2}{\ell^2} + \left(1 + \frac{x}{\ell}\right)^2\right]. \; \right| \qquad \textbf{Answer} \;\blacktriangleleft$$

The vorticity is

$$\zeta_z = 2\omega_z,$$

so

$$\left| \; \zeta_z = \frac{2yu_0}{Y_0^2}\left(1 + \frac{x}{\ell}\right)\left[\frac{y^2}{\ell^2} + \left(1 + \frac{x}{\ell}\right)^2\right]. \; \right| \qquad \textbf{Answer} \;\blacktriangleleft$$

The rate of volumetric strain (or stretching) is

$$\frac{1}{\delta V}\frac{d(\delta V)}{dt} = \left(\frac{\partial u}{\partial x} + \frac{\partial v}{\partial y}\right).$$

Substituting for $\partial u/\partial x$ and $\partial v/\partial y$, we have

$$\left| \; \frac{1}{\delta V}\frac{d(\delta V)}{dt} = 0. \; \right| \qquad \textbf{Answer} \;\blacktriangleleft$$

The rate of shear deformation is

$$\dot{\phi} = \frac{1}{2}\left(\frac{\partial u}{\partial y} + \frac{\partial v}{\partial x}\right).$$

Substituting for $\partial u/\partial y$ and $\partial v/\partial x$, we obtain

$$\left| \; \dot{\phi} = \frac{yu_0}{Y_0^2}\left(1 + \frac{x}{\ell}\right)\left[-\left(1 + \frac{x}{\ell}\right)^2 + \frac{y^2}{\ell^2}\right] \; \right| \qquad \textbf{Answer} \;\blacktriangleleft$$

Discussion

It would be informative if we could observe the rotation, stretching, and deformation of a fluid particle as it moved through a flow field. Even though we can mathematically separate each quantity, the rotation, stretching, and deformation occur simultaneously and one cannot be isolated from the others. It would be instructive for you to calculate a few numerical values of $\partial u/\partial y$ and $\partial v/\partial x$ and compare them.

Note that the continuity equation requires that the rate of volumetric strain be zero for this incompressible flow.

3.5.2 Flow Regimes: Inviscid Flow, Laminar Flow, Turbulent Flow

Our discussion of the various mathematical means of describing flow may have led you to believe that fluid motion is always orderly and predictable. All of the velocity distributions that we have drawn have been smooth curves (see Figs. 1.4, 3.7, 3.10, and 3.18) and our streamlines, pathlines, and streaklines have also been shown as smooth curves (see Figs. 3.2, 3.3, 3.4, and 3.9). Unfortunately, in a great many real-world flows, smooth velocity distributions and streamlines are the exception rather than the rule.

There are three regimes of flow: inviscid, laminar, and turbulent. In the inviscid and laminar flow regimes, the fluid motion is orderly, but in the turbulent regime the motion is quite disorderly. Figure 3.28 is a photograph of water flowing in a diverging channel. The flow is from left to right and is made visible by generating hydrogen bubbles continuously from a wire stretched across the channel. Notice that the flow is quite orderly in the center of the channel and reasonably orderly near the upper wall; however, a disorderly motion appears near the lower wall. The motion near the center is (nearly) inviscid, the motion near the upper wall is laminar, and the motion near the lower wall is turbulent. We will consider the characteristics of each regime in detail.

Inviscid flow is an imaginary condition that involves a fictitious fluid with zero viscosity. Such a flow would be free of shear stress. Since there are no real fluids with zero viscosity, there are no truly inviscid flows; however, if the rate of shear deformation of the fluid particles is small and the viscosity is not too large, then an almost inviscid flow results. Equation (3.63) shows that the shear strain rate, $\dot{\phi}$, is large only where the velocity changes appreciably over a short distance perpendicular to the velocity itself (that is, only if $\partial u/\partial y$ or $\partial v/\partial x$ is large).

Figure 3.28 Water flowing in a diverging channel; flow is made visible by generating hydrogen bubbles from a wire spanning the channel (from the "NCFMF Book of Film Notes," 1974, The MIT Press with Education Development Center, Inc., Newton, Mass.).

Figure 3.29 Velocity profiles at three locations in water flow in a diverging channel (from the "NCFMF Book of Film Notes," 1974, The MIT Press with Education Development Center, Inc., Newton, Mass.).

Figure 3.29 shows velocity profiles for the diverging channel flow. The profiles are marked by generating bubbles from three wires stretched across the channel. Note the large value of $\partial u/\partial y$ near the wall and the almost-flat velocity profile ($\partial u/\partial y \approx 0$) near the center. For this flow, v and $\partial v/\partial x$ are small everywhere, so the flow near the center of the channel behaves as if it were inviscid.

It is possible to show that flow of a fluid with zero viscosity would be irrotational, that is, the vorticity and angular velocity of all fluid particles would be zero.* Since the word *inviscid* refers to an unreal, imaginary situation, many engineers prefer to call a flow like that in the center of the channel irrotational since the condition of zero vorticity (angular velocity) is possible in the real world.

In a *laminar flow,* at least one of the velocity gradients ($\partial u/\partial y$ or $\partial v/\partial x$) is not zero, but the velocity profile is a smooth curve. The fluid moves along smooth streamlines. The flow is called laminar because it looks like a series of thin sheets of fluid (laminae) sliding over one another (Fig. 3.30). In laminar flow, fluid particles move along fixed streamlines with very little mixing across streamlines. The concept of "friction" in the fluid gives a good picture of shear stress in laminar flow even though the shear stress is actually the result of molecular momentum transfer and/or intermolecular forces (see Section 1.3.4 on viscosity).

Inviscid (irrotational) and laminar flows are in sharp contrast to the disorderly fluid motion that we call turbulent flow. A *turbulent flow* is characterized by random fluctuations in fluid velocity and by intense mixing of the fluid on the macroscopic level. The disorderly pattern of bubbles near the lower wall of the channel of Fig. 3.28 is due to the mixing action of the turbulent flow there. Everyone has seen turbulent flow in everyday life. The water from a kitchen faucet usually flows

Figure 3.30 Laminar flow visualized as thin sheets (laminae) of fluid in relative sliding motion.

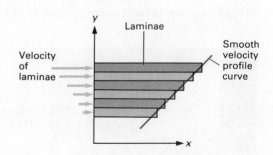

* This will be done in Section 9.2.1.

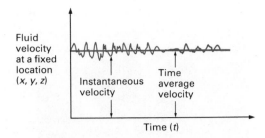

Fluid velocity at a fixed location (x, y, z)

Instantaneous velocity

Time average velocity

Time (t)

Figure 3.31 Velocity-time history for a fixed point in a turbulent flow.

turbulently. A plume of smoke or steam from an industrial stack displays turbulent motion. If you live in the northern half of the United States or in Canada, you have probably driven a car along a highway on a snowy winter night and watched "clouds" of snowflakes billowing aimlessly about the highway. The snowflakes serve as markers for turbulent wind gusts. In addition to these familiar occurrences, turbulent flow occurs in a great many engineering applications, so we must be prepared to deal with it.

Suppose we inserted a velocity-measuring instrument at a particular point near the lower wall of the channel of Fig. 3.28 and recorded the velocity as a function of time. If the instrument has a fast response time,* it would show a velocity-versus-time plot similar to that of Fig. 3.31. The flow is hopelessly unsteady; the velocity apparently changes randomly with time. We can, however, define a time average velocity and define a steady turbulent flow as one in which this average velocity does not vary with time at any point in the flow field.

Next consider a velocity distribution curve. Figure 3.32 shows two instantaneous velocity profiles for flow near a solid flat plate. These profiles are actually timelines generated shortly before the photograph was taken. The flow on the upper side of the plate is turbulent and the flow on the lower side of the plate is laminar. If we generated instantaneous velocity profiles at another instant of time at the same location, the shape of the turbulent profile would be different but the shape of

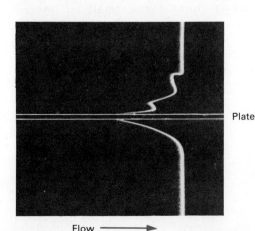

Plate

Flow ⟶

Figure 3.32 Instantaneous velocity profiles in laminar (lower) and turbulent (upper) flow over a solid plate (from the "NCFMF Book of Film Notes," 1974, The MIT Press with Education Development Center, Inc., Newton, Mass.).

---

* Response time characterizes how rapidly the instrument responds to changes in velocity.

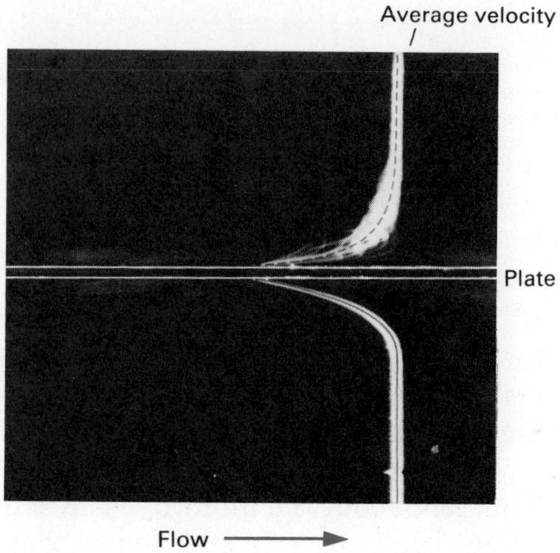

Average velocity

Plate

Flow ⟶

Figure 3.33 Superposition of a large number of laminar (lower) and turbulent (upper) instantaneous velocity profiles; lines indicate time average velocity profile (from the "NCFMF Book of Film Notes," 1974, The MIT Press with Education Development Center, Inc., Newton, Mass.).

the laminar profile would be the same. Figure 3.33 shows the superposition of a large number of velocity profiles. The laminar (lower) profile is identical to the instantaneous profile of Fig. 3.32; however, the turbulent (upper) profile is a wide "smear" because of the different instantaneous profiles. Notice that it is still possible to define an average velocity profile for the turbulent flow.

Engineering analysis of turbulent flow is based on the idea of an average velocity distribution on which the turbulent fluctuations are superimposed. This concept is illustrated in Fig. 3.34. The average velocity profile is considered to be deterministic, that is, reproducible for the same boundary and initial conditions; however, the fluctuations (loosely called "the turbulence") are random and can be dealt with only statistically. The details of turbulent flow are only partly understood. No theory is capable of predicting the details of the velocity-time trace of Fig. 3.31.

The mixing action of turbulent flow is very important in engineering applications. Because of the velocity fluctuations, small but macroscopic "lumps" of fluid (called *eddies*) are thrown about in the flow. Because these lumps carry mass, momentum, and energy, the rate of mixing of dissimilar fluids, "apparent" shear stress (due to momentum transfer), and "apparent" heat transfer are considerably larger than in

Figure 3.34 Instantaneous velocity profile in pipe flow; fluctuating velocity is superimposed on time average velocity.

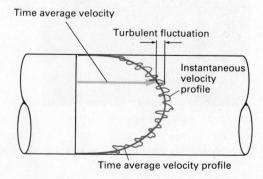

Time average velocity

Turbulent fluctuation

Instantaneous velocity profile

Time average velocity profile

laminar flow. This enhanced mixing is sometimes harmful, as in the increase of pressure drop in a fluid flowing in a pipe, and sometimes helpful, as in the rapid mixing of cream into hot coffee when it is stirred (compared to the mixing of paint, usually a laminar process).

A laminar flow can change to a turbulent flow in a process called *transition*. Transition from a laminar to a turbulent flow state can be seen in a plume of smoke rising from a cigarette* (Fig. 3.35). The flow is initially laminar near the tip of the cigarette. Higher up, the laminar flow becomes unstable. By mechanisms that are not fully understood, these instabilities grow and the flow becomes turbulent. The transition in the smoke plume occurs naturally without outside disturbances. Transition can be promoted by introducing disturbances (e.g., by roughening a surface in contact with the fluid). The flow near the lower wall in Fig. 3.28 and the flow over the top of the plate in Fig. 3.32 have been made turbulent by roughening the wall upstream of the location shown in the figures.

Engineering approaches to turbulent flow and transition are based on a combination of theory and experiment. Theory suggests how turbulent flows behave qualitatively and suggests possible forms for mathematical description, but experimental data are necessary to make the results quantitative.

A parameter called the *Reynolds number* is used to decide if the flow is laminar or turbulent. The Reynolds number is defined by

$$\mathbf{R} \equiv \frac{\rho V \ell}{\mu},$$

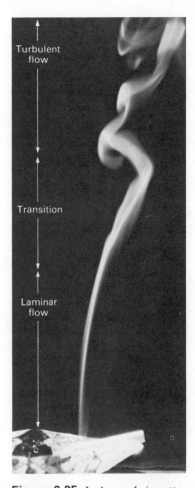

Figure 3.35 A plume of cigarette smoke illustrates laminar flow, transition, and turbulent flow.

where ρ is the fluid density, V is a typical (average) fluid velocity, ℓ is a relevant length of the problem at hand (diameter of a pipe, depth of a channel, length of a plate), and μ is the fluid viscosity. The Reynolds number is dimensionless, that is, if all parameters are expressed in the same units (say SI units), all units cancel and $\mathbf{R}$ is a pure number. Experimental evidence indicates that if the Reynolds number is small, the flow will be laminar, but if it is large, the flow will be turbulent. The definition of "large" and "small" must be determined by experiment; Reynolds numbers below about 10^3 are usually "small" and Reynolds numbers above about 10^4 are usually "large."

The Reynolds number does not help in deciding if a flow is inviscid. A flow will behave as if it were inviscid at any Reynolds number if velocity gradients are zero. In particular, a high Reynolds number due to low viscosity usually means turbulent flow in regions where the velocity gradients are appreciable and inviscid flow only where velocity gradients are very small.

3.6 Possible Approaches for the Solution of Flow Problems

Although the main purposes of this chapter were to describe the possibilities for formulating flow problems and to describe some of the phenomena that are associated with fluid flow, it would be improper to

* Illustrating transition is one of the few beneficial uses of a cigarette.

close the chapter without a discussion of how engineers go about solving flow problems.

The solution of flow problems is a combination of good news and bad news. If you are like many engineering students and are somewhat uncomfortable with a lot of mathematics, you may look over the material we have just covered and conclude that it is all bad news; however, even if you are a mathematical whiz, you are still in for some bad news. In spite of our ability set up mathematical problems representing complicated flows and our clever approaches to formulation of the fundamental laws, detailed exact solutions of the resulting equations usually cannot be found. In some cases, the system geometry is too complicated. If viscosity is involved, the (differential) equations of the flow are non-linear,* and nonlinear differential equations are notoriously difficult to solve. Any turbulent flow is so complicated in its fine details that an exact solution is utterly impossible.

A great deal of the bad news deals with our inability to find detailed, exact solutions for arbitrary flow situations using mathematical methods. The good news is in three parts:

- For engineering purposes, we may not need detailed, exact solutions. Reliable estimates of a few key parameters and gross features of the flow may be sufficient.
- Detailed, exact solutions are known for several special cases.
- Methods other than mathematical ones are available and can be put to good use in fluid mechanics.

We will consider each point in detail.

Engineering estimates (accurate to, say, $\pm 20\%$) of important flow parameters often can be obtained by application of the control volume method, by modeling a complicated flow in terms of simple cases for which exact (or approximate) solutions are already available, or by rule-of-thumb estimates based on years of experience. You cannot gain years of experience by studying this book, but we can cover the other two methods. The powerful control volume approach has already been illustrated in connection with the law of conservation of mass. The control volume formulation of the remaining fundamental laws will be developed in Chapter 5, and a great many applications will be illustrated.

Flow modeling involves the judicious application of those solutions that we do have and the introduction of key assumptions that reduce a complicated flow to a simpler, though approximate, case. If we wish to predict the flow in the channel shown in Fig. 3.28, we might model the flow in the central region of the channel as inviscid, the flow near the top wall as a laminar boundary layer,† and the flow near the bottom wall as a turbulent boundary layer. The complicated flow in the entire

---

* The differential equations are nonlinear even without viscosity, but if there is no viscosity or compressibility, the equations have an alternative linear form. We consider this special case in Chapter 9.

† A *boundary layer* is a thin region of flow where shear stresses are important. Since such layers are thin, the change of pressure across them is usually neglected (a simplification). Boundary layers will be discussed further in Chapter 8 and considered extensively in Chapter 9.

channel is calculated by "patching" together the three simpler flows. Chapters 7, 9, 10, and 11 will deal with some simplified flows and simplified flow models that can form the basis for analysis of more general cases.

Exact solutions to the flow equations are known for some special cases. The most general flow involves compressibility, viscosity, and turbulence and takes place in a complex, three-dimensional geometry. The motion of a fluid particle depends on the balance between "inertia force" (using the d'Alembert principle), shear forces due to viscosity and turbulence, pressure forces, elastic forces, and gravity forces. Seldom are all of these forces of comparable magnitude. Many simplified cases exist in which only two or three forces are relevant.

If we neglect shear stress and compressibility, a very general solution to the equation of motion (Newton's second law) called Bernoulli's equation can be found. This equation can be applied to any geometry whatsoever.

Flows with shear stresses are usually highly dependent on the specific geometry involved. Of course many complicated geometric arrangements can occur in engineering flow problems, but several simple geometries are also of great engineering interest. An outstanding example is flow in a pipe. The geometry, although quite relevant for engineering applications, is very simple. In most cases, the flow in a pipe is determined by considering a balance between pressure, shear, and gravity forces, with fluid acceleration and compressibility negligible. We have an exact mathematical solution for laminar flow in a pipe and it agrees beautifully with experimental data. Although we do not have an exact solution for turbulent flow in a pipe, we do have a reliable mathematical model of the flow based on experimental measurements.

We will consider several exact solutions for simplified cases in this book. Chapter 4 illustrates the application of the differential formulation of the fundamental laws and develops Bernoulli's equation as well as an exact solution for a simple viscous flow. Chapter 7 deals with a simple case, namely incompressible flow in closed conduits, Chapter 9 considers inviscid flows and simple boundary layer flows, Chapter 10 considers the effects of compressibility in simple situations, and Chapter 11 considers simple open channel flows.

Finally, remember that there are approaches other than mathematical ones for solving flow problems. One of these alternative approaches is numerical approximation. In this method we approximate the differential equations describing a flow by finite-difference or finite-element methods. The result is always a very large number of simultaneous *algebraic* equations for the fluid properties and velocities at a finite number of discrete points in the field. These equations are usually solved by an electronic computer. Numerical solutions, similar to analytical solutions, often become very difficult for complicated geometries. Many numerical solutions make use of a great deal of modeling of a complicated flow in terms of simple components. We will not consider numerical methods very much in this book. We feel that the best way to think about numerical fluid mechanics is as an extension of analytical fluid mechanics, at least in the context of an introductory course in fluid mechanics.

We are left at last with experimental methods of solution of flow problems. In experimental methods, either we set up and investigate a flow in a laboratory environment (Fig. 3.36) or we make measurements in a real flow in the field (Fig. 3.37). A laboratory investigation may involve either a scale model of a piece of equipment or the actual equipment. In experimental fluid mechanics, the fluid "solves the problem" for us and we have to extract the answers using measuring instruments. Experimental fluid mechanics may be applied on two levels. Probably the most obvious application is measurement of the quantities of direct interest for a particular engineering application. If we were designing a truck and wanted to know the aerodynamic drag force on the cab, we could build a scale model of the cab, put it in a wind tunnel, and measure the drag force directly. Similarly, if we want to know the flow rate in a certain pipe in a plant, we could attach instruments and measure it directly. While these direct approaches are very important in fluid mechanics, they are costly and sometimes impractical. If 100 possible alternative designs of the truck cab were proposed, a model of each would have to be built and tested. On the other hand, measurement of the flow in a particular pipe presumes that the pipe has already been built and that there is flow in it. Such measurements are necessarily after the fact and are obviously of no use for plant design.

A second application of experimental methods involves the investigation of basic phenomena. Detailed measurements of turbulent flow in a particular circular pipe can form the basis for a model of turbulent flow for all circular pipes (and maybe for noncircular pipes as well). In fact, since no theory is capable of predicting turbulent flow, all of our information about turbulent flow is ultimately based on correlations of experimental data obtained from careful experiments.

Figure 3.36 Experimental fluid mechanics: The engineer is using a laser Doppler anemometer system to measure fluid velocities between the blades of a high-speed propeller. The propeller is mounted in a wind tunnel (photo courtesy of NASA Lewis Research Center).

Figure 3.37 Solving a problem by measurements in the field: These engineers are measuring the flow rate through a large fan that exhausts flue gases from a boiler. The fan is driven by two 3000-hp electric motors. Gas velocity is measured at several points at the fan inlet plane and outlet plane. The flow rate is then calculated by equations like Eq. (3.33).

There are two key areas of concern in experimental fluid mechanics. The most obvious of these deals with measurement techniques. Our treatment of measurement techniques in this book will be cursory; we will discuss them only when they illustrate the application of some particular principle or equation of fluid mechanics (see the discussion of pressure measurement in Chapter 2, for example). Several excellent books deal with measurement techniques in general [6, 7] and experimental fluid mechanics in particular [8, 9, 10].

A less obvious but extremely important aspect of experimental fluid mechanics deals with efficient presentation and extrapolation of data. If we were performing an experimental study of various truck cab designs, we would probably find ourselves making measurements on scale-model cabs rather than on full-sized cabs. We expect that the data from the models tell us something about the full-sized cabs, but what do they tell us? How do we find out? Similarly, we hope that detailed measurements of the turbulent flow in a 6-in.-diameter circular pipe tell us something about the flow in a 4-in. pipe, but how do we find out? These problems are directly addressed by the technique of *dimensional analysis,** the key to "getting the most for your money" in experimental fluid mechanics. It is also the only way that the vast amount of experimental information available can be organized and passed on to students and engineers.

We will formally consider the techniques of dimensional analysis in Chapter 6. Extremely useful experimental data about pipe flow and

---

* Dimensional analysis is also important in numerical fluid mechanics. You don't want to rerun an expensive computer program for every case that comes along if you already have the solution in some form.

external flow over simple shapes will be presented in Chapters 7 and 8, respectively, using approaches suggested by dimensional analysis.

Fluid mechanics problems can be formulated and solved in a bewildering array of possible approaches. While it may seem difficult to put all of this in perspective right now, as you proceed through the remaining chapters, sinking your teeth into some practical (and some not so practical) fluid mechanics problems, you will begin to fit everything in its place.

Problems

Extension and Generalization

1. Derive the differential continuity equation in polar coordinates.

2. Derive the differential continuity equation in spherical coordinates.

3. A tank has a horizontal cross-sectional area A and a height H. Water flows into the tank at a steady rate Q_i and leaves at a rate of

$$Q_e = A_e\sqrt{2gh},$$

where A_e is the exit area of a nozzle in the side of the tank, g the local acceleration of gravity, and h the water height above the centerline of the nozzle. Develop the differential equation for the water height h as a function of time if the water height is h_0 at time $t = 0$.

4. A particle experiences a drag of kV, where k is a constant and V is the velocity. The particle is fired vertically upward with an initial velocity of V_0. Find the velocity of the particle as a function of position and time.

5. A fluid stream leaves a nozzle with velocity V_0 and cross-sectional area A_0. The stream is directed vertically upward. Find the velocity V and cross-sectional area A as a function of the vertical position y if the product AV is constant.

6. The surface velocity of a river is measured at several locations x and can be reasonably represented by the equation

$$V = V_0 + \Delta V(1 - e^{-ax}),$$

where V_0, ΔV, and a are constants. Find the velocity of a fluid particle flowing along the surface if its velocity is V_0 at $x = 0$ with time $t = 0$.

7. Find an equation for the nozzle flow area as a function of the coordinate x for the flow of a constant-density fluid in Problem 17.

8. A tank contains a volume V_0 of pure water at time $t = 0$. An entering stream of volume flow rate Q_i has a steady salt concentration c_i, and the salt is well mixed in the tank. Develop an expression for the salt concentration c in the tank as a function of time.

9. A constant-density fluid flows through a converging section whose area A is given by

$$A = \frac{A_0}{1 + x/\ell},$$

where A_0 is the area at $x = 0$. Find the velocity and acceleration of the fluid in Eulerian form and then the velocity and acceleration of a fluid particle in Lagrangian form. The velocity is V_0 at $x = 0$ when $t = 0$.

Lower-Order Application

10. Water leaves a nozzle with a downward velocity of 10 ft/sec. Find the velocity of the stream in both Eulerian and Lagrangian descriptions.

11. Water leaves a nozzle with a velocity of 10 ft/sec. The nozzle makes an angle θ above the horizontal. Find the velocity in both Eulerian and Lagrangian descriptions.

12. The parametric equations for the horizontal and vertical position of a particle are

$$x = x_0 + V_{0_x}t \quad \text{and} \quad y = y_0 + V_{0_y}t - \frac{1}{2}gt^2.$$

Find the locations (x, y) of the point of maximum velocity and of maximum distance equal to $\sqrt{(x - x_0)^2 + (y - y_0)^2}$ from the initial position (x_0, y_0).

13. A two-dimensional, unsteady velocity field is given by

$$u = 5x(1 + t) \quad \text{and} \quad v = 5y(-1 + t),$$

where u is the x-velocity component and v the y-velocity component. Find $x(t)$ and $y(t)$ if $x = x_0$ and $y = y_0$ at $t = 0$.

14. Do the velocity components given in Problem 13 represent an Eulerian description or a Lagrangian description?

15. A car accelerates from rest to a final constant velocity V_f and a police officer records the following velocities at various locations x along the highway:

$$
\begin{array}{ll}
x = 0 & V = 0\,\text{mph} \\
 = 100\,\text{ft} & = 34.8\,\text{mph} \\
 = 200\,\text{ft} & = 47.6\,\text{mph} \\
 = 300\,\text{ft} & = 52.3\,\text{mph} \\
 = 400\,\text{ft} & = 54.0\,\text{mph} \\
 = 1000\,\text{ft} & = 55.0\,\text{mph} = V_f.
\end{array}
$$

Find a mathematical expression for the velocity V traveled by the car as a function of the final velocity V_f of the car and time t if $t = 0$ at $V = 0$. [*Hint:* Try an exponential fit to the data.]

16. A tornado has the following velocity components in polar coordinates:

$$
V_r = -\frac{C_1}{r} \quad \text{and} \quad V_\theta = -\frac{C_2}{r}.
$$

Note that the air is spiraling inward. Find an equation for the streamlines of the air.

17. The velocity field in a nozzle is both one-dimensional and one-directional. The axial velocity is given by

$$
u = U_0 e^{x/4\ell},
$$

where $x = 0$ represents the inlet and the inlet area is 0.25 m². For constant-density flow, find the exit area where $x = \ell$.

18. Rain is falling vertically downward with a velocity of 3 ft/sec. There are 100 $\frac{1}{8}$-in.-diameter water droplets per cubic foot. How long will it take for a small vessel measuring 1 in. on a side and 4 in. high to fill up with water?

19. Repeat Problem 18 if the rain is falling at an angle of 60° with the horizontal.

20. A hypodermic syringe is being used to inject oil into a machine part. The inside diameter of the glass tube is 1.0 cm and the plunger moves at the rate of 0.5 cm/s. How long will it take to discharge 12 cc of oil?

21. A stationary tube is used to inject oil into bearings moving on the belt shown in Fig. P3.1. The bearings

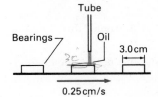

Bearings · Tube · Oil · 3.0 cm · 0.25 cm/s · **Figure P3.1**

have a diameter of 3.0 cm and the belt moves at a speed of 0.25 cm/s. What must be the minimum flow rate out of the tube to provide each bearing with at least 3.0 cc of oil?

22. Find the depth (h) of rainwater flowing down a hill making an angle of 30° with the horizontal. The rain is falling vertically downward with a velocity of 3.0 ft/sec and there are 50 $\frac{1}{8}$-in.-diameter water droplets per cubic foot of rain. The water flowing down the hill has the velocity profile

$$
u = U\frac{y}{Y},
$$

where u is the fluid velocity parallel to the hill and Y is the water depth perpendicular to the hill at a distance x from the top. Assume $Y = 0$ at the top, where $x = 0$ and $U = 1.0$ ft/sec.

23. Rainwater is flowing down a street having an incline of 30° with the horizontal. A scrap of paper on the surface of the water moves at the rate of 1 ft/sec. The street is 20 ft wide and the water depth 1/2 in. Estimate the flow rate of rainwater down the street.

24. A river is approximately 30 ft wide and has an average depth of 7 ft. It is estimated that the surface water flows at the rate of 6 ft/sec. Make a reasonable estimate of the river flow rate if the fluid velocity varies linearly with depth.

25. A tank contains 300 m³ of pure water. Brine containing 3 kg/m³ of salt flows into the tank at the rate of 3 m³/s. The tank is well mixed and the mixture runs out of the tank at the rate of 2 m³/s. When will the tank contain 100 kg of salt?

26. Brine flows through a pipe of inside radius $R = 1.5$ in. with an axial velocity of

$$
u = U_0\left[1 - \left(\frac{r}{R}\right)^2\right],
$$

where $U_0 = 1.5$ ft/sec. The salt concentration is

$$
c = c_0(r/R),
$$

where $c_0 = 0.05$ lbm/ft³. Find the mass flow rate of salt through the pipe. Is it equal to the average brine velocity times the average brine concentration? Explain.

27. A 1-mi-long tunnel has an unlimited number of cars passing through it daily. Each car is 15 ft long and each driver maintains a distance of $1.5V$ ft from the car ahead. V is the car speed in miles per hour. Find the speed the cars should travel so that the maximum number of cars will pass through the tunnel each hour.

28. Kerosene is pumped into a 1000-gal storage tank at the rate of 10 gal/min. However, 100 gal of kerosense is drained out of the tank before the tank is filled. If the tank is initially empty, find the time required to fill the tank.

29. Kerosene has an average velocity of 1.2 m/s through a 0.025-m-inside-diameter pipe. Calculate the volume flow rate.

30. Oil flows into a tank at the rate of 10 m³/s and leaves through a 0.5-m² hole in the side of the tank with a velocity

$$V = \sqrt{19.6\,h},$$

where V is in meters per second and h, the vertical distance from the centerline of the hole to the free surface of the oil, is in meters. Find the steady-state value of h.

31. Oil flows into an initially empty tank of 10.0 m² cross section at the rate of 10 m³/s and leaves through a 0.25-m² hole in the side of the tank with a velocity

$$V = \sqrt{19.6\,h},$$

where V is in meters per second and h is in meters. Find $h(t)$ where h is the vertical distance from the centerline of the hole to the free surface of the oil.

32. A lawn sprinkler has 25 nozzles. Nozzle number 1 (closest to the faucet) has an exit velocity of 16 ft/sec, nozzle number 25 has an exit velocity of 6.32 ft/sec, and the exit velocity V_n varies as

$$V_n = \sqrt{265.0 - 9.0n},$$

where n is the nozzle number. Find the volume flow rate Q in cubic feet per second leaving the lawn sprinkler if each velocity is perpendicular to the nozzle area. Each nozzle area is 0.15 in².

33. Laminar flow in a circular pipe has the following velocity profile:

$$u = u_{max}\left[1 - \left(\frac{r}{R}\right)^2\right],$$

where u_{max} is the maximum fluid velocity and occurs at the pipe centerline ($r = 0$) and R is the inside radius of the pipe. Find the volume flow rate passing through the pipe in terms of R and u_{max}.

34. The axial fluid velocities given in Fig. P3.2 were measured at the locations indicated by the plus signs in the circular duct. Estimate the volume flow rate and the average velocity.

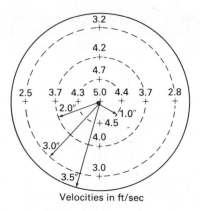

Velocities in ft/sec   **Figure P3.2**

35. The cross-sectional area of a rectangular duct is divided into 16 smaller and equal rectangular areas, as shown in Fig. P3.3. The axial fluid velocity measured in feet per second in each smaller area is given in the figure. Estimate the volume flow rate and average axial velocity.

20.0″			
3.0	3.4	3.6	3.1
3.7	4.0	3.9	3.8
3.9	4.6	4.5	4.2
3.7	4.4	4.3	3.9

16.0″

Velocities in ft/sec   **Figure P3.3**

36. A square duct measures 6 in. on a side and has a volume flow rate of 20 gal/min. What diameter must a circular pipe have for the same average velocity as the square duct if the velocity in the circular pipe is given by

$$u = U_0\left[1 - \left(\frac{r}{R}\right)^2\right],$$

where R is the pipe's inside radius, r the radial coordinate, and U_0 the centerline velocity.

37. The velocity components u and v of a two-dimensional flow are given by

$$u = ax + \frac{bx}{x^2y^2} \quad \text{and} \quad v = ay + \frac{by}{x^2y^2},$$

where a and b are constants. Calculate the vorticity, the rate of volumetric strain, and the rate of shear deformation.

38. Is the flow of Problem 37 irrotational? Justify your answer.

39. Find the accelerations a_x and a_y of the flow in Problem 37.

40. The velocity at any point in a flow is given by the curl of the velocity vector $\vec{V}$,

$$\vec{\zeta} = \nabla \times \vec{V},$$

where ∇ is the gradient operator. Show that this reduces to Eq. (3.60) for two-dimensional flow in Cartesian coordinates.

41. The vorticity at any point in a flow field is given by the curl of the velocity vector, or

$$\vec{\zeta} = \nabla \times \vec{V},$$

where ∇ is the gradient operator. In two-dimensional cylindrical coordinates (r, θ),

$$\nabla = \hat{r}\frac{1}{r}\left(\frac{\partial(r)}{\partial r}\right) + \hat{\theta}\frac{1}{r}\left(\frac{\partial}{\partial \theta}\right) \quad \text{and} \quad \vec{V} = \hat{r}V_r + \hat{\theta}V_\theta,$$

where $\hat{r}$ and $\hat{\theta}$ are the respective unit vectors. Show that

$$\zeta_x = \frac{1}{r}\left(\frac{\partial}{\partial r}\right)(rV_\theta) - \frac{1}{r}\left(\frac{\partial V_r}{\partial \theta}\right).$$

 42. Consider the flow field

$$V_r = \frac{a}{r^2}\cos\theta, \qquad V_\theta = \frac{a}{r^2}\sin\theta.$$

Use the result of Problem 41 to determine the vorticity ζ_x. Is this flow irrotational?

43. Consider the velocity field

$$V_\theta = br,$$

where b is a constant. Use the result of Problem 41 to determine the vorticity. Is this flow irrotational?

 44. Consider the velocity field

$$V_\theta = \frac{b}{r},$$

where b is a constant. Use the result of Problem 41 to determine the vorticity. Is this flow irrotational?

Higher-Order Application

45. The axial velocity u in the entrance region of a pipe is given by

$$u = u_0\left(1 - \frac{r^2}{R^2}\right)^{x/\ell_e},$$

where u_0 is a constant, r the radial coordinate, R the pipe inside radius, x the axial coordinate, and ℓ_e an entrance length. Find the acceleration of a fluid particle as a function of r, x, and time t.

46. Show that the flow field described by velocity components

$$u = \frac{-2xyz}{(x^2 + y^2)^2}, \qquad v = \frac{(x^2 - y^2)z}{(x^2 + y^2)^2},$$

and

$$w = \frac{y}{x^2 + y^2}$$

is a possible incompressible fluid flow. What is the vorticity, rate of volumetric strain, and rate of shear deformation? Find the components of the acceleration of a fluid particle as functions of x, y, z, and t.

47. List some beneficial and some harmful aspects of turbulence. Consider both engineering applications and everyday occurrences.

48. Water is leaving a hole in the vertical side of a tank with a horizontal velocity of 20 ft/sec. The hole is 25 ft above the ground. How far will the water travel horizontally before it hits the ground?

49. A student has 7 min to go from one building to another through a moderate rain. Would the student get wetter by running or by walking? Assume that the rain is falling straight downward.

Comparison

50. Show that $\bar{\rho}$ in Eq. (3.41) equals $\bar{\rho}'$ in Eq. (3.42) for constant density.

51. Show that $\bar{\rho}$ in Eq. (3.41) does not necessarily equal $\bar{\rho}'$ in Eq. (3.42) for a variable-density fluid. You may choose to consider a particular velocity profile and a particular density profile.

References

1. White, F. M., *Fluid Mechanics*, McGraw-Hill, New York, 1979.
2. Kline, S. J., *Flow Visualization*, National Committee for Fluid Mechanics Films, distributed by Encyclopaedia Britannica Educational Corporation.

3. Gerhart, P. M., "Averaging Methods for Determining the Performance of Large Fans from Field Measurements," *Transactions of the American Society of Mechanical Engineers: Journal of Engineering for Power,* April 1981.

4. Fox, R. W., and A. T. MacDonald, *Introduction to Fluid Mechanics* (2nd ed.), Wiley, New York, 1978.

5. Mironer, A., *Engineering Fluid Mechanics,* McGraw-Hill, New York, 1979.

6. Beckwith, T., N. Buck, and R. Marangoni, *Mechanical Measurements* (3rd ed.), Addison-Wesley, Reading, MA, 1981.

7. Benedict, R. P., *Fundamentals of Temperature, Pressure, and Flow Measurements* (2nd ed.), Wiley, New York, 1977.

8. Bradshaw, P., *Experimental Fluid Mechanics*, Macmillan, New York, 1964.

9. Ower, E., and R. Pankhurst, *The Measurement of Air Flow* (5th ed.), Pergamon Press, Oxford, 1977.

10. Dean, R. C., Jr. (Ed.), *Aerodynamic Measurements,* MIT Gas Turbine Lab Report, 1954, available from University Microfilms, Ann Arbor, MI.

The Differential Approach to Flow Analysis

4

In Chapter 3 we pointed out that all fluid flows are governed by a rather small number of fundamental laws and that these laws can be expressed mathematically using either the differential approach or the finite control volume approach. Both approaches were illustrated by applying them to the law of conservation of mass. The control volume approach produced some very useful equations for flow analysis, while the differential approach produced some rather formidable-looking partial differential equations. If you were to base a judgment of the practicality of the two approaches only on the evidence of Chapter 3, you would probably discard the differential approach altogether. If you did, you would make a serious error.

The purpose of this chapter is to develop further the differential approach to flow analysis. We still have other fundamental laws to formulate, especially Newton's laws of motion. In this chapter, we will be primarily concerned with applying Newton's second law to a flowing fluid. We will consider both inviscid and laminar viscous flows.

A complete application of the differential approach to a flow problem involves both the formulation of the differential equations and their solution (integration). We will find that in some cases we can easily integrate the differential equations and obtain general relations, usually in the form of algebraic equations. In other cases, a general solution of the differential equations is not possible. This chapter will deal with both situations. We will develop what is probably the most widely used algebraic equation of fluid mechanics, Bernoulli's equation. We will also develop the general differential equations of viscous flow, the Navier-Stokes equations, which defy exact solution in most cases and are the basis of most advanced work in fluid mechanics.

4.1 Inviscid Flow: Euler and Bernoulli Equations

We will first consider flow of an inviscid fluid, that is, we will assume that the fluid's viscosity is zero. We will also assume that there is no turbulence in the flow.* These two assumptions imply that the fluid experiences no shear stresses. This zero shear stress condition provides the desired simplifications in the analysis. Of course no fluid has zero viscosity or is immune to turbulence; however, if the viscosity is small and if the rate of strain is also small, then forces due to shear stress will be negligible compared to forces due to pressure and gravity. Under these conditions, Newton's second law for a fluid particle can be expressed as

$$\begin{pmatrix} \text{Mass of} \\ \text{particle} \end{pmatrix} \times \begin{pmatrix} \text{Acceleration} \\ \text{of particle} \end{pmatrix} = \begin{pmatrix} \text{Net pressure} \\ \text{force on particle} \end{pmatrix} + \begin{pmatrix} \text{Net gravity} \\ \text{force on particle} \end{pmatrix}.$$

(4.1)

Our objective is to express this relation in terms of the fluid velocity and fluid properties and the geometry of the flow field.

4.1.1 Streamline Coordinates

Since Newton's second law is a fundamental law of nature, it can be expressed in any (inertial) coordinate system we choose. After choosing a coordinate system, we must express the fluid velocity and acceleration and the forces on the fluid in that coordinate system. We then substitute these expressions into Eq. (4.1) to obtain the equations of motion in that coordinate system. Since Eq. (4.1) is a vector equation, it may represent as many as three separate component equations. The complexity of the resulting equations is highly dependent on the choice of a coordinate system. Our choice usually lies between a simple coordinate system (like rectangular (x, y, z) or cylindrical (r, θ, x) coordinates) in which the velocity field is complicated and a complicated coordinate system in which the velocity field is simple.[†]

One example of the latter type of coordinate is the *streamline coordinate system* in which the coordinates are the flow streamlines and a set of lines normal (perpendicular) to them. Figure 4.1 illustrates a streamline coordinate system for a two-dimensional, two-directional flow. The coordinate lines are the streamlines (s) and the normal lines (n). The n-lines are perpendicular to the streamlines and point toward their center of curvature. Note that although the s-lines and n-lines are always perpendicular to each other, the s- and n-directions are variable because the streamlines are not straight.

The situation illustrated in Fig. 4.1 presumes that the streamlines always lie in a single plane. This is true only for a two-dimensional, two-directional flow. In a three-dimensional, three-directional flow, the streamlines are curves in space. At any point on a streamline, the unit vectors $\hat{s}$ and $\hat{n}$ define a plane and the flow is locally two-directional in

---

* Strictly speaking, a fluid with zero viscosity could not have turbulence in it; however, since no real fluid has zero viscosity, we will treat the absence of turbulence as an additional assumption.

[†] Sometimes, we get lucky and a particular flow yields to analysis in a simple coordinate system in which the velocity field is simply expressed as well. In such cases, we can usually obtain exact solutions to the equations.

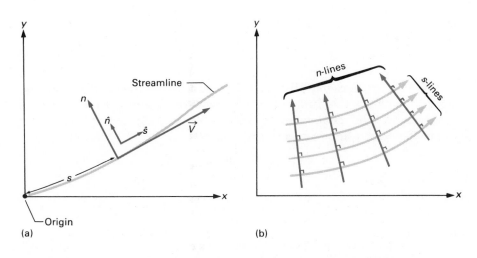

Figure 4.1 Streamline coordinate system: (a) The coordinates for a single streamline; (b) a grid of s,n-lines in a plane.

the plane. The plane could have a different orientation for each point on the streamline.

The major advantage of the s,n-coordinate system is that the velocity at any point is always parallel to the s direction. We can write the velocity vector

$$\vec{V} = V_s \hat{s} + V_n \hat{n} = V_s \hat{s}$$

because $V_n = 0$. Since the fluid velocity is always parallel to the (local) s-direction, any fluid particle is instantaneously moving along the s-line passing through it. If the flow is steady, any fluid particle always moves along the same s-line.[*]

Consider the acceleration of a fluid particle in the s,n-coordinate system. We can write the velocity components[†] as

$$V_s = V_s(s, n, t) \qquad \text{and} \qquad V_n = V_n(s, n, t).$$

The s-direction acceleration is

$$a_s = \frac{dV_s}{dt} = \frac{DV_s}{Dt} = \frac{\partial V_s}{\partial t} + \frac{\partial V_s}{\partial s}\left(\frac{ds}{dt}\right) + \frac{\partial V_s}{\partial n}\left(\frac{dn}{dt}\right),$$

but

$$\frac{ds}{dt} = V_s \qquad \text{and} \qquad \frac{dn}{dt} = V_n = 0,$$

so

$$a_s = \frac{\partial V_s}{\partial t} + V_s\left(\frac{\partial V_s}{\partial s}\right). \qquad\qquad \textbf{(4.2)}$$

Next consider the n-direction. Note that although $V_n = 0$ at every point at any instant, it does not follow that a_n is zero because the n-direction can change with time or with movement along a streamline

[*] The streamline coordinate system is the same as the path coordinate system or the tangential-and-normal coordinate system in elementary dynamics.

[†] You may think that we need not consider V_n at all since it is always zero but that would be a mistake, as you will soon see.

Streamline at t Streamline at $t + \delta t$

(a)

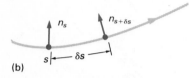

(b)

Figure 4.2 Changes of the n-direction: (a) Streamline changes with time; $n_{t+\delta t}$ is different from n_t; (b) changes with movement along a streamline; $n_{s+\delta s}$ is different from n_s.

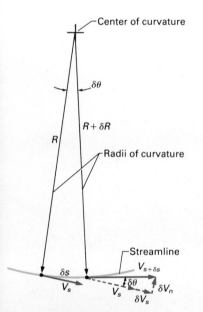

Figure 4.3 Changes of velocity along a streamline.

(see Fig. 4.2). The n-direction acceleration is

$$a_n = \frac{dV_n}{dt} = \frac{DV_n}{Dt} = \frac{\partial V_n}{\partial t} + \frac{\partial V_n}{\partial s}\left(\frac{ds}{dt}\right) + \frac{\partial V_n}{\partial n}\left(\frac{dn}{dt}\right) = \left(\frac{\partial V_n}{\partial t}\right) + V_s\left(\frac{\partial V_n}{\partial s}\right). \quad (4.3)$$

We can further simplify Eq. (4.3) by examining the geometry of Fig. 4.3, which shows the fluid velocity at two points along a streamline in *steady* flow. The change of normal velocity, δV_n, due to movement along the streamline from s to $s + \delta s$ is

$$\delta V_n \approx V_s \tan (\delta\theta) \approx V_s (\delta\theta).$$

We can also write

$$\delta V_n \approx \frac{\partial V_n}{\partial s} (\delta s).$$

From the geometry of the figure, we have

$$\delta\theta \approx \frac{\delta s}{R},$$

where R is the local radius of curvature of the streamline. Combining these three equations gives

$$\frac{\partial V_n}{\partial s} \approx \frac{V_s}{R}.$$

If we use this equation, the second term in Eq. (4.3) becomes

$$V_s\left(\frac{\partial V_n}{\partial s}\right) \approx V_s\left(\frac{V_s}{R}\right) = \frac{V_s^2}{R}.$$

If we take the limit of this expression as $\delta s \to 0$, it becomes exact, and the normal acceleration is given by

$$a_n = \frac{\partial V_n}{\partial t} + \frac{V_s^2}{R}. \quad (4.4)$$

Summarizing, the accelerations in a streamline coordinate system are

$$a_s = \frac{\partial V_s}{\partial t} + V_s\left(\frac{\partial V_s}{\partial s}\right) \quad (4.2)$$

and

$$a_n = \frac{\partial V_n}{\partial t} + \frac{V_s^2}{R}. \quad (4.4)$$

Recall that n points toward the center of curvature.

4.1.2 Euler's Equations of Motion in Streamline Coordinates

Because the fluid velocity and acceleration are expressed rather simply in streamline coordinates, the equations of motion in streamline coordinates have a correspondingly simple form. To develop these equations, consider Fig. 4.4, which shows a rectangular fluid particle at a particular point on a streamline. The velocity and pressure fields are

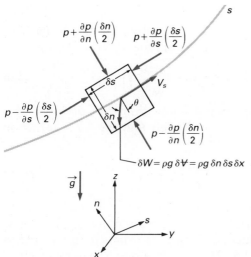

Figure 4.4 A rectangular fluid particle in streamline coordinates; pressures and gravity force acting on the particle are shown. Depth perpendicular to paper is δ_x.

expressed in the s,n-coordinate system. A rectangular (x, y, z) coordinate system is also shown for reference. Note that the z-coordinate is parallel to the gravity vector $(\breve{g})$ and points upward. We assume that the flow is two-dimensional and two-directional so we need not consider the x-direction. We must develop separate equations of motion for the s- and n-directions.

First consider the s (streamwise) direction. Newton's second law for the particle is

$$\sum \delta F_s = (\delta m_p) a_s, \tag{4.5}$$

where $\sum \delta F_s$ represents the sum of the streamwise components of all forces acting on the particle and δm_p is the mass of the particle,

$$\delta m_p = \rho \delta V = \rho (\delta s)(\delta n)(\delta x). \tag{4.6}$$

This equation, together with Eq. (4.2), allows us to write

$$(\delta m_p) a_s = \rho \left[\frac{\partial V_s}{\partial t} + V_s \left(\frac{\partial V_s}{\partial s} \right) \right] (\delta s)(\delta n)(\delta x). \tag{4.7}$$

Now consider the forces on the particle. Since we have decided to assume there is no shear stress, the only forces are due to pressure and gravity. With the aid of Fig. 4.4, we find

$$\sum \delta F_s = \underbrace{\left[p - \frac{\partial p}{\partial s} \left(\frac{\delta s}{2} \right) \right] (\delta n)(\delta x) - \left[p + \frac{\partial p}{\partial s} \left(\frac{\delta s}{2} \right) \right] (\delta n)(\delta x)}_{\text{net pressure force}} - \underbrace{\delta W \sin \theta.}_{\substack{\text{gravity} \\ \text{force}}}$$

The particle's weight is

$$\delta W = g(\delta m_p) = g\rho(\delta s)(\delta n)(\delta x),$$

and the force term can be simplified so that

$$\sum \delta F_s = \left(-\frac{\partial p}{\partial s} - \rho g \sin \theta \right)(\delta n)(\delta s)(\delta x). \tag{4.8}$$

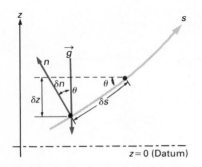

Figure 4.5 Illustration of relation between movement along a streamline (δs) and vertical displacement (δz).

We obtain the streamwise equation of motion at the point (s, n) by substituting Eqs. (4.7) and (4.8) into Eq. (4.5), dividing by $(\delta n)(\delta s)(\delta x)$, and then taking the limit as δn, δs, and δx approach zero. The result is

$$-\frac{\partial p}{\partial s} - \rho g \sin \theta = \rho \left[\frac{\partial V_s}{\partial t} + V_s \left(\frac{\partial V_s}{\partial s} \right) \right]. \tag{4.9}$$

A more popular form of this equation is obtained by dividing by ρ and writing the acceleration terms on the left and the force terms on the right:

$$\frac{\partial V_s}{\partial t} + V_s \frac{\partial V_s}{\partial s} = -\frac{1}{\rho} \left(\frac{\partial p}{\partial s} \right) - g \sin \theta. \tag{4.10}$$

To prepare for integration of this equation, it is convenient to replace the second term on the right by an expression involving a derivative with respect to s. Consider the relation between the s- and z-directions illustrated in Fig. 4.5. Movement along the streamline a distance δs results in an upward movement δz, where

$$\delta z \approx \delta s \sin \theta.$$

Dividing by δs and taking the limit as $\delta s \to 0$, we have

$$\frac{dz}{ds} = \sin \theta.$$

This should really be written as a partial derivative because we were moving along a streamline. Making this change and substituting into Eq. (4.10), we get

$$\frac{\partial V_s}{\partial t} + V_s \left(\frac{\partial V_s}{\partial s} \right) + \frac{1}{\rho} \left(\frac{\partial p}{\partial s} \right) + g \left(\frac{\partial z}{\partial s} \right) = 0. \tag{4.11}$$

This equation represents Newton's second law of motion along a streamline in the absence of shear stress.

Now consider the equation of motion for the n-direction. We should be able to do this a little faster since the procedure is identical to that used for the s-direction. Newton's second law for the n-direction for the particle of Fig. 4.4 is

$$\sum \delta F_n = (\delta m_p) a_n. \tag{4.12}$$

The mass of the particle is given by Eq. (4.6), and the n-direction acceleration is given by Eq. (4.4). The net pressure force in the n-direction is

$$\delta F_{n,\text{pressure}} = \left[\left(p - \frac{\partial p}{\partial n} \frac{\delta n}{2} \right) - \left(p + \frac{\partial p}{\partial n} \frac{\delta n}{2} \right) \right] \delta s \, \delta x$$

$$= -\frac{\partial p}{\partial n} (\delta s)(\delta n)(\delta x),$$

and the n-component of the gravity force is

$$\delta W_n = -\delta W \cos \theta = -\rho g \cos \theta \, (\delta s)(\delta n)(\delta x).$$

There is no shear force by assumption. Now we substitute the expressions for δm_p, a_n, and the forces into Eq. (4.12), divide by $(\delta s)(\delta n)(\delta x)$, and take the limit. The result is

$$-\frac{\partial p}{\partial n} - \rho g \cos \theta = \rho\left(\frac{\partial V_n}{\partial t} + \frac{V_s^2}{R}\right).$$

A more popular form is

$$\frac{\partial V_n}{\partial t} + \frac{V_s^2}{R} = -\frac{1}{\rho}\left(\frac{\partial p}{\partial n}\right) - g\cos\theta. \tag{4.13}$$

We can replace $\cos \theta$ by noting from Fig. 4.5 that

$$\frac{\partial z}{\partial n} = \lim_{\substack{\delta n \to 0, \\ s\ \text{fixed}}} \frac{\delta z}{\delta n} = \cos\theta.$$

Equation (4.13) can then be written

$$\frac{\partial V_n}{\partial t} + \frac{V_s^2}{R} + \frac{1}{\rho}\frac{\partial p}{\partial n} + g\left(\frac{\partial z}{\partial n}\right) = 0. \tag{4.14}$$

This equation represents Newton's second law in a direction normal to a streamline in the absence of shear stresses. Equations (4.11) and (4.14) are called *Euler's equations of motion in streamline coordinates*. For steady flow, Euler's equations become

$$V_s\frac{\partial V_s}{\partial s} + \frac{1}{\rho}\left(\frac{\partial p}{\partial s}\right) + g\left(\frac{\partial z}{\partial s}\right) = 0 \tag{4.15}$$

and

$$\frac{V_s^2}{R} + \frac{1}{\rho}\left(\frac{\partial p}{\partial n}\right) + g\left(\frac{\partial z}{\partial n}\right) = 0. \tag{4.16}$$

EXAMPLE 4.1 Illustrates Euler's equation along a streamline

Figure E4.1(a) shows an ideal fluid (zero viscosity and constant density) flowing through a planar converging nozzle that lies in a horizontal plane. The pressure and velocity are uniform over both section 1 and section 2. Is the pressure p_2 greater or less than the pressure p_1 for steady flow?

Solution

  Given
Figure E4.1(a)
Uniform pressure and velocity over sections 1 and 2

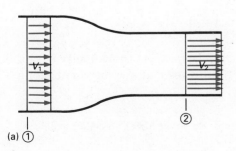

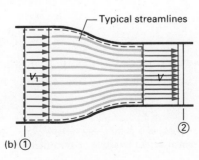
Typical streamlines

Figure E4.1 (a) Planar nozzle; (b) planar nozzle with control volume and typical streamlines.

Find

Is p_2 greater than or less than p_1?

Solution

Euler's equation along a streamline for steady flow is

$$V_s \frac{\partial V_s}{\partial s} + \frac{1}{\rho}\left(\frac{\partial p}{\partial s}\right) + g\left(\frac{\partial z}{\partial s}\right) = 0.$$

Figure E4.1(b) shows typical streamlines along which this equation can be applied and for which $\partial z/\partial s = 0$. Solving the above equation for $\partial p/\partial s$, we have

$$\frac{\partial p}{\partial s} = -\rho V_s \frac{\partial V_s}{\partial s}.$$

Applying the integral continuity equation Eq. (3.47) to the control volume in Fig. E4.1(b) gives

$$\rho_1 V_1 A_1 = \rho V A, \qquad \text{or} \qquad V = \left(\frac{A}{A_1}\right) V_1.$$

The velocity V is greater than V_1 because A is less than A_1. Therefore

$$\frac{\partial V_s}{\partial s} > 0$$

because s increases in the direction of flow. Therefore

$$\frac{\partial p}{\partial s} < 0$$

along any streamline in the nozzle. This leads to the conclusion

Answer

$$|\ p_2 < p_1.\ |$$

◀

4.1.3 Integration of the Streamline Equation of Motion: Bernoulli's Equation

After we obtain differential equations of fluid motion, the next step is to integrate them. Usually we can integrate the equations only for a specific geometry, that is, only for a particular flow problem. An exception to this is that it is possible to integrate the *streamwise* Euler equation once and for all without reference to any specific geometry. The result of this integration is one of the most widely used (and misused!) equations of fluid mechanics.

Consider the streamwise Euler equation for steady flow:*

$$V\left(\frac{\partial V}{\partial s}\right) + \frac{1}{\rho}\left(\frac{\partial p}{\partial s}\right) + g\left(\frac{\partial z}{\partial s}\right) = 0.$$

The subscript s has been dropped from V_s because the velocity is in the s-direction. Next we multiply by an infinitesimal displacement along a

---

* Have you been keeping track of assumptions? So far we have steady flow and no shear stress (viscous or turbulent).

streamline, ds:

$$V\left(\frac{\partial V}{\partial s}\right)ds + \frac{1}{\rho}\left(\frac{\partial p}{\partial s}\right)ds + g\left(\frac{\partial z}{\partial s}\right)ds = 0. \tag{4.17}$$

In general, the total differential of any parameter of the flow field (say pressure p) is given by

$$dp = \frac{\partial p}{\partial s}(ds) + \frac{\partial p}{\partial n}(dn)$$

because p is a function of both s and n. If we restrict ourselves to *remain on the same streamline,* then

$$dn)_{\text{on streamline}} = 0 \qquad \text{and} \qquad dp)_{\text{on streamline}} = \frac{\partial p}{\partial s}(ds).$$

Similar relations hold for other properties.

With the restriction of staying on the same streamline, Eq. (4.17) becomes

$$V\,dV + \frac{dp}{\rho} + g\,dz = 0 \tag{4.18}$$

and can be integrated *along the streamline*

$$\frac{V^2}{2} + \int \frac{dp}{\rho} + \int g\,dz = C,$$

where C is a constant.

In most flows of engineering interest, the acceleration of gravity does not vary throughout the field, so we can write

$$\int \frac{dp}{\rho} + \frac{V^2}{2} + gz = C. \tag{4.19}$$

In many (but by no means all) flows, the density is constant. If we assume constant density, the equation becomes

$$\frac{p}{\rho} + \frac{V^2}{2} + gz = C. \tag{4.20}$$

This is *Bernoulli's equation.* Before we discuss its meaning and use, take careful note of the following assumptions and restrictions on its validity.

- Bernoulli's equation applies only to flow with no shear stress.
- Bernoulli's equation applies only to steady flow.
- Bernoulli's equation applies only along a particular streamline.
- Bernoulli's equation applies only if the density (and gravity) are constant.

There is a further restriction on the applicability of Bernoulli's equation that is actually contained in the four listed above but is rather subtle, so it is usually listed separately:

- Bernoulli's equation applies only if there is no mechanical work done on or by the fluid due to outside agencies (e.g., pumps, turbines).

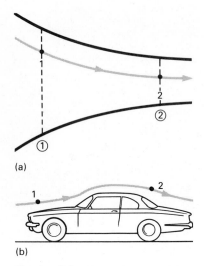

(a)

(b)

Figure 4.6   Two points 1 and 2 on the same streamline: (a) Flow in a nozzle; (b) flow over an automobile.

An attempt to use Bernoulli's equation in any situation where any one of these five restrictions is violated will usually result in an error.

The way to relax at least one of these restrictions is obvious. If the density is not constant but can be related to the pressure, then Eq. (4.19), which can be called "Bernoulli's equation for compressible flow," can be applied with $\int dp/\rho$ evaluated from the actual pressure-density relation. The result of this process *will not* be Eq. (4.20).

If you are satisfied that all of the conditions on the applicability of Bernoulli's equation are valid, it is extremely useful because it relates pressure, velocity, and elevation for flow along a streamline. Consider flow between some point 1 and some other point 2 on the same streamline. Figure 4.6(a) shows such a streamline for a flow in a nozzle, and Fig. 4.6(b) shows such a streamline for flow over an automobile. If the restrictions on Bernoulli's equation are valid, the constant C on the right side of Eq. (4.20) is the same at both 1 and 2, and we can write

$$\frac{p_1}{\rho} + \frac{V_1^2}{2} + gz_1 = \frac{p_2}{\rho} + \frac{V_2^2}{2} + gz_2. \qquad (4.21)$$

Note that 1 and 2 could be any two points on the streamline. Equation (4.21) is the most useful form of Bernoulli's equation.

The usefulness of Bernoulli's equation can be greatly extended by combining it with the one-dimensional flow assumption. Consider the nozzle flow of Fig. 4.6(a). If all of the streamlines passing plane ① containing point 1 have the same pressure, velocity, and elevation and all of the streamlines passing plane ② containing point 2 have the same pressure, velocity, and elevation, then the Bernoulli equation for any streamline between planes ① and ② gives us the relation between the flow parameters for both planes. By combining the one-dimensional flow assumption, Bernoulli's equation, Eq. (4.21), and the continuity equation, Eq. (3.48), we can calculate a great many flows of engineering interest. The following examples will illustrate some applications.

EXAMPLE 4.2   Illustrates Bernoulli's equation

A storage tank has an inside diameter of 6.77 in. and a 1.0-in.-diameter nozzle in its side. At a particular instant, the water level is 4 ft above the center of the hole. Find the velocity of the water leaving the hole at this instant.

Solution

Given
Figure E4.2
$D = 6.77$ in., $d = 1.0$ in., $h = 4.0$ ft

Find
Velocity of the jet leaving the nozzle, V_2, in Fig. E4.2

Solution
Since the cross-sectional area of the tank is somewhat greater than the cross-sectional area of the nozzle, we expect the water level in the tank to drop

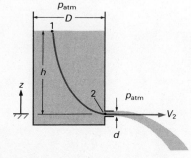

Figure E4.2   Storage tank and nozzle with exiting jet of water.

relatively slowly. We will obtain an approximate solution by assuming a steady-flow condition at each instant of time. We also assume constant density and gravity and an inviscid fluid. Applying Bernoulli's equation between points 1 and 2 gives

$$\frac{p_1}{\rho} + \frac{V_1^2}{2} + gz_1 = \frac{p_2}{\rho} + \frac{V_2^2}{2} + gz_2 .$$

We want to solve for V_2. Because the water level drops slowly,

$$V_1 \approx 0.$$

We establish a reference level for elevation by setting

$$z_2 = 0, \quad \text{so} \quad z_1 = h.$$

Since the tank is open to the atmosphere,

$$p_1 = p_{atm}.$$

Since the nozzle discharges to the atmosphere,

$$p_2 = p_{atm}.$$

(Do you think p_2 is $(p_{atm} + \gamma h)$? Do you understand why it isn't?) We can now solve for V_2 and get

$$V_2^2 = 2\left[\frac{p_1 - p_2}{\rho} + gh\right]$$

or, since $p_1 = p_2$,

$$V_2 = \sqrt{2gh}.$$

The numerical values give

$$V_2 = \sqrt{2(32.2 \text{ ft/sec}^2)(4 \text{ ft})},$$

$$\left| \; V_2 = 16.0 \text{ ft/sec.} \; \right| \qquad\qquad \textbf{Answer}$$

◀

Discussion
Note that the fluid density does not appear in the resulting equation for V_2; therefore, this analysis suggests that V_2 is independent of the type of fluid. The mass flow rate from the nozzle can be calculated by Eq. (3.37),

$$\dot{m} = \rho A_2 V_2,$$

by assuming that V_2 is uniform over the area A_2 (all streamlines in area A_2 originate at the free surface in the tank and all streamlines have $z \approx 0$). For 50°F water, the numerical values give

$$\dot{m} = \left(62.4 \frac{\text{lbm}}{\text{ft}^3}\right)\left(\frac{\pi}{4}\right)\left(1.0 \text{ in.} \times \frac{\text{ft}}{12 \text{ in.}}\right)^2\left(16.0 \frac{\text{ft}}{\text{sec}}\right)$$

$$= 5.45 \text{ lbm/sec.}$$

The above approach is called a "quasi-steady" analysis. It assumes that neglecting the term $(\partial V_s/\partial t)$ in Euler's equation will not lead to a serious error in the jet velocity V_2.

EXAMPLE 4.3 Illustrates the error introduced by assumptions

Estimate the error introduced into the answer of Example 4.2 by neglecting the velocity V_1. Retain the quasi-steady assumption.

Solution

Given
Figure E4.2
$D = 6.77$ in., $d = 1.0$ in., $h = 4.0$ ft

Find
Estimate of the error due to neglecting the velocity V_1

Solution
Apply the integral continuity equation, Eq. (3.20), to the control volume in Fig. E4.3(a). Assuming that density is constant and that V_1 and V_2 are uniform over the areas A_1 and A_2, respectively, gives

$$A_2 V_2 = A_1 V_1, \qquad \text{or} \qquad V_1 = \frac{A_2 V_2}{A_1}.$$

In terms of the diameters,

$$V_1 = \left(\frac{d}{D}\right)^2 V_2.$$

Using this velocity V_1 in Eq. (4.21) gives

$$\frac{p_1}{\rho} + \left(\frac{d}{D}\right)^4 \left(\frac{V_2^2}{2}\right) + gz_1 = \frac{p_2}{\rho} + \frac{V_2^2}{2} + gz_2.$$

Solving for V_2, we obtain

$$V_2 = \sqrt{\frac{2g(z_1 - z_2)}{1 + \left(\dfrac{d}{D}\right)^4}}.$$

The numerical values give

$$V_2 = \sqrt{\frac{2\left(32.2\,\dfrac{\text{ft}}{\text{sec}^2}\right)(4\text{ ft})}{1 + \left(\dfrac{1.0\text{ in.}}{6.77\text{ in.}}\right)^4}}$$

$$= 16.0\text{ ft/sec},$$

the same answer to three significant figures. The percent error is

$$\text{Error} = \left(\frac{16.0 - 16.0}{16.0}\right) \times 100,$$

$$\boxed{\text{Error} = 0.00\%}$$

Answer ◀

Discussion
It is informative to estimate the error as a function of the diameter ratio d/D. The percent error estimate is

$$\% \text{ error} = \frac{\sqrt{2gh} - \sqrt{\dfrac{2gh}{1 + (d/D)^4}}}{\sqrt{\dfrac{2gh}{1 + (d/D)^4}}} \times 100$$

$$= 100\left[\sqrt{1 + (d/D)^4} - 1\right].$$

The percent error is plotted in Fig. E4.3(b).

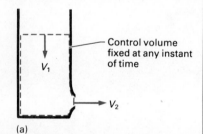

Control volume fixed at any instant of time

V_1

V_2

(a)

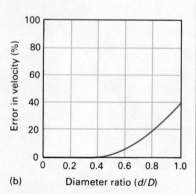

(b)   Diameter ratio (d/D)

Figure E4.3 (a) Control volume for applying the integral continuity equation; (b) estimated error in exit velocity, due to neglect of free surface velocity.

EXAMPLE 4.4 Illustrates the simultaneous application of the integral continuity and Bernoulli equations

Water at 10°C enters the horizontal Venturi tube in Fig. E4.4(a) with a uniform and steady velocity of 2.0 m/s and an inlet pressure of 150 kPa. Find the pressures at the throat 2 and the exit 3; $D = 6.0$ cm, $d = 3.0$ cm.

Solution

Given
Figure E4.4(a)
$D = 6.0$ cm, $d = 3.0$ cm
$V_1 = 2.0$ m/s, $p_1 = 150$ kPa

Find
p_2 and p_3

Solution
Assume constant density and uniform velocity over planes 1 and 2. Applying the continuity equation between these two planes gives

$$V_1 A_1 = V_2 A_2, \quad \text{or} \quad V_2 = \left(\frac{A_1}{A_2}\right) V_1 = \left(\frac{D}{d}\right)^2 V_1.$$

Assuming an inviscid fluid and applying Bernoulli's equation to a streamline connecting cross sections 1 and 2, we have

$$\frac{p_1}{\rho} + \frac{V_1^2}{2} + gz_1 = \frac{p_2}{\rho} + \frac{V_2^2}{2} + gz_2.$$

Assuming $z_1 \approx z_2$ and solving for p_2, we have

$$p_2 = p_1 + \frac{\rho}{2}(V_1^2 - V_2^2).$$

Substituting for V_2 gives

$$p_2 = p_1 + \frac{\rho}{2}\left[1 - \left(\frac{D}{d}\right)^4\right] V_1^2.$$

Figure E4.4 (a) Horizontal Venturi with entering uniform velocity; (b) typical static pressure distribution in a real Venturi.

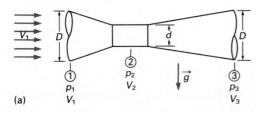

(a)

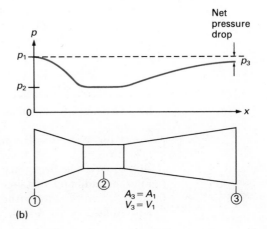

(b)

Using numerical values from Table A.5, we obtain

$$p_2 = 150 \text{ kPa} + \frac{(10^3 \text{ kg/m}^3)}{2} \left[1 - \left(\frac{6.0 \text{ cm}}{3.0 \text{ cm}} \right)^4 \right] \left(2.0 \frac{\text{m}}{\text{s}} \right)^2 \left(\frac{\text{N} \cdot \text{s}^2}{\text{kg} \cdot \text{m}} \right)$$

$$= 150 \text{ kPa} - 30{,}000 \text{ N/m}^2, \qquad \qquad \textbf{Answer}$$

$$\boxed{p_2 = 120 \text{ kPa.}} \qquad \blacktriangleleft$$

In a similar manner, applying the continuity equation, Eq. (3.48), between planes 1 and 3 gives

$$V_1 A_1 = V_3 A_3.$$

Since $A_3 = A_1$,

$$V_3 = V_1.$$

Bernoulli's equation between planes 1 and 3 gives

$$\frac{p_1}{\rho} + \frac{V_1^2}{2} + g z_1 = \frac{p_3}{\rho} + \frac{V_3^2}{2} + g z_3.$$

Assuming $z_1 \approx z_3$ and using $V_3 = V_1$, we have

$$\frac{p_1}{\rho} = \frac{p_3}{\rho},$$

or
$$\qquad \qquad \textbf{Answer}$$

$$\boxed{p_3 = p_1 = 150 \text{ kPa.}} \qquad \blacktriangleleft$$

Discussion

The water is assumed to be inviscid and to have a constant density. The integral continuity equation shows that the water velocity increases in the converging section and decreases in the diverging section. Bernoulli's equation then shows that the static pressure decreases in the converging section and increases in the diverging section. The diverging section is called a diffuser. A *diffuser* increases the static pressure by decreasing the fluid kinetic energy. The converging section acts as a nozzle. A *nozzle* increases the fluid velocity (kinetic energy) by decreasing the static pressure. We encountered nozzles previously in Examples 1.5 and 3.1.

Since the water is assumed inviscid, the static pressure drop $(p_1 - p_2)$ was fully recovered in the diffuser section by decreasing the fluid velocity to V_1. Full pressure recovery would not occur in a real Venturi. Viscous effects would produce a net pressure drop between sections 1 and 3. See Fig. E4.4(b).

A Venturi is often used to measure flow rate in a pipe. We will discuss Venturis and other flow-measuring devices further in Section 7.4.

EXAMPLE 4.5 Illustrates Bernoulli's equation and velocities as vectors

A city has a fire truck whose pump and hose can deliver 1000 gal/min with a nozzle velocity of 120 ft/sec. The tallest building in the city is 100 ft high and 40 ft square. The firefighters hold the nozzle at an angle of 75° with the ground. Find the minimum distance the firefighters must stand from the building to put out a fire on the roof without the aid of a ladder. Assume the water velocity is not reduced by air resistance.

Solution

Given
Fire hose that delivers 1000 gal/min
Nozzle velocity 120 ft/sec
Tallest building 100 ft high and 40 ft square
Nozzle inclined at 75° angle
No ladders used

Find
Minimum distance firefighters must stand from the building to put out a
  fire on the roof

Solution
A sketch of the physical situation is shown in Fig. E4.5. Assuming constant
density and gravity, Bernoulli's equation is applied between point 1 and any
other point on the streamline:

$$\frac{p_1}{\rho} + \frac{V_1^2}{2} + gz_1 = \frac{p}{\rho} + \frac{V^2}{2} + gz$$

At any point, $p = p_1 = p_{atm}$. Taking $z_1 = 0$ and solving for V^2 give

$$V^2 = V_1^2 - 2gz.$$

Since air resistance is negligible, there is no force on the fluid in the x (horizontal) direction and so

$$V_x = V_1 \cos \theta.$$

Bernoulli's equation can be written

$$V_x^2 + V_z^2 = V_1^2(\cos^2 \theta + \sin^2 \theta) - 2gz.$$

Since $V_x^2 = V_1^2 \cos^2 \theta$, we get

$$V_z^2 = V_1^2 \sin^2 \theta - 2gz, \qquad \text{or} \qquad V_z = \sqrt{V_1^2 \sin^2 \theta - 2gz}.$$

Now

$$V_z = \frac{dz}{dt}, \qquad \text{so} \qquad \frac{dz}{dt} = \sqrt{V_1^2 \sin^2 \theta - 2gz}.$$

Separating the variables and integrating, we have

$$\int_{t=0}^{t} dt = \int_{z=0}^{z} \frac{dz}{\sqrt{V_1^2 \sin^2 \theta - 2gz}}$$

or

$$t = \frac{1}{g}(V_1 \sin \theta - \sqrt{V_1^2 \sin^2 \theta - 2gz}), \tag{E4.1}$$

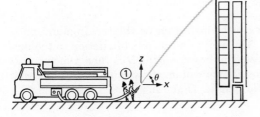

Figure E4.5 Firefighters in action.

where t is the time for the water to reach elevation z. Now

$$V_x = \frac{dx}{dt} = V_1 \cos \theta.$$

Integrating gives

$$x = V_1 \cos \theta \, t.$$

Substituting t from Eq. (E4.1) gives

$$x = \frac{V_1 \cos \theta}{g} \left(V_1 \sin \theta - \sqrt{V_1^2 \sin^2 \theta - 2gz} \right).$$

Note that x and z are coordinates of the water stream. Some rearrangement gives

$$x = \frac{V_1^2 \sin 2\theta}{2g} \left(1 - \sqrt{1 - \frac{2gz}{V_1^2 \sin^2 \theta}} \right).$$

We now assume that the firefighters hold the nozzle 5 ft above the ground. Substituting $z = 95$ ft, $\theta = 75°$, and $V_1 = 120$ ft/sec gives the minimum distance the firefighters must stand from the building:

$$x = \frac{\left(120 \, \dfrac{\text{ft}}{\text{sec}} \right)^2 (\sin 150°)}{2 \left(32.2 \, \dfrac{\text{ft}}{\text{sec}^2} \right)} \left[1 - \sqrt{1 - \frac{2 \left(32.2 \, \dfrac{\text{ft}}{\text{sec}^2} \right) (95 \text{ ft})}{\left(120 \, \dfrac{\text{ft}}{\text{sec}} \right)^2 (\sin^2 75°)}} \right],$$

$$\left| \; x = 29.3 \text{ ft.} \; \right| \qquad \textbf{Answer} \; \triangleleft$$

Discussion

If we used a nozzle velocity less than approximately 81 ft/sec, the quantity

$$\left(1 - \frac{2gz}{V_1^2 \sin^2 \theta} \right)$$

would then be negative. What would this mean?

Bernoulli's equation was our primary equation to solve this problem. The secondary equations were fundamental kinematic relations taught in a basic physics course. The solution of some problems in fluid mechanics may not have all the tools clearly laid out in this book and may require that you occasionally go back to some of the tools you learned previously. A good student must be able to draw on past learning experiences.

Do not be disheartened if you think you would have difficulty doing this problem on your own. One of the authors was hung up on it for a while himself!

EXAMPLE 4.6 Illustrates Bernoulli's equation and its application to an open channel flow measuring device

The flow rate in an open channel can be measured by the rectangular sharp-crested weir shown in Fig. E4.6(a). The weir presents an obstruction to the flow and raises the upstream liquid level. The liquid height H is then a measure of the flow rate. Develop an expression for the volume flow rate through the weir of Fig. E4.6(a) in terms of the liquid heights H and h.

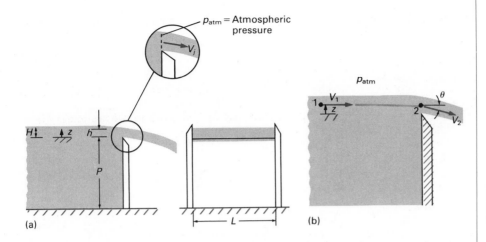

Figure E4.6 (a) Rectangular weir; (b) streamline used for a rectangular weir.

Solution

Given
Rectangular sharp-crested weir in Fig. 4.6(a)

Find
Expression for the volume flow rate as a function of H and h

Solution
Assume constant density and gravity and an inviscid fluid. Applying Bernoulli's equation between point 1 and point 2 in Fig. E4.6(b) gives

$$\frac{p_1}{\rho} + \frac{V_1^2}{2} + gz_1 = \frac{p_2}{\rho} + \frac{V_2^2}{2} + gz_2,$$

where

$$p_2 = p_{atm}.$$

Next apply Euler's normal equation between point 1 and the free surface immediately above it. Equation (4.16) gives

$$\frac{V_s^2}{R} + \frac{1}{\rho}\left(\frac{\partial p}{\partial n}\right) + g\left(\frac{\partial z}{\partial n}\right) = 0.$$

Since the streamlines are straight, $R \to \infty$. Multiplying by $\rho\, dn$ and integrating gives

$$p_{atm} - p_1 + \rho g(H - z_1) = 0.$$

Solving for p_1, we have

$$p_1 = p_{atm} + \rho g(H - z_1) = 0.$$

Substituting this expression for p_1 into Bernoulli's equation gives

$$\frac{p_{atm}}{\rho} + g(H - z_1) + \frac{V_1^2}{2} + gz_1 = \frac{p_{atm}}{\rho} + \frac{V_2^2}{2} + gz_2.$$

Solving for V_2, we find

$$V_2 = \sqrt{2g(H - z_2) + V_1^2}.$$

Referring to Fig. E4.6(b), we see that the volume flow rate Q through the weir (width L) is

$$Q = \int_0^h V_2 \cos\theta \, L \, dz_2$$

$$= \int_0^h L\sqrt{2g(H - z_2) + V_1^2} \, \cos\theta \, dz_2,$$

where V_2 and θ vary with z_2. To carry out the integration, we assume

$$V_1^2 \ll V_2^2 \quad \text{and} \quad \cos\theta \approx 1.$$

This gives

$$Q \approx \int_0^h L\sqrt{2g(H - z_2)} \, dz_2 = -\frac{2}{3}\sqrt{2g}\,L[(H - h)^{3/2} - H^{3/2}],$$

$$\left| \; Q \approx \frac{2\sqrt{2g}}{3}\,L[H^{3/2} - (H - h)^{3/2}]. \; \right| \qquad \textbf{Answer}$$

Discussion

In any measurement device it is simpler to record one quantity than two; so the equation for Q above is simplified by assuming $(H - h) \ll H$ and using a coefficient of discharge $\mathbf{C_d}$ to account for this simplification ($\mathbf{C_d}$ also accounts for the two previous assumptions, $V_1 \ll V_2$ and $\cos\theta \approx 1$). The result is

$$Q = \frac{2}{3}\sqrt{2g}\,L\,\mathbf{C_d}H^{3/2}.$$

The coefficient $\mathbf{C_d}$ must be evaluated by experiment. Typical values for a rectangular sharp-crested weir vary from about 0.60 to about 0.75. The exact value depends on the values of H, P, and L in Fig. E4.6(a), the local acceleration of gravity g, and the physical properties μ, ρ, and σ of the fluid. Once $\mathbf{C_d}$ has been determined for a particular application, the weir is said to be calibrated.

Now that you have seen a few possible applications of Bernoulli's equation, we will consider some interpretations and alternative forms of the equation. Although it is derived from Newton's second law of motion, Bernoulli's equation is actually an energy equation; to be specific, it is a *mechanical energy* equation. When Eq. (4.15), which involves forces and acceleration, is multiplied by ds, a displacement along a streamline, the result (Eqs. 4.17 and 4.18) is an energy equation. This can be more clearly seen by rewriting Eq. (4.18) in the following form:

$$-\frac{dp}{\rho} = V \, dV + g \, dz.$$

The three terms in this equation are interpreted as follows:

$-(dp/\rho)$ is the work per unit mass done by pressure on a particle of fluid as it moves a distance ds. This can be verified by noting that $-(\partial p/\partial s)$ is the net pressure force per unit *volume*; $-(1/\rho)(\partial p/\partial s)$ is the net pressure force per unit *mass*; and (Force per unit mass) $\times$ (Displacement) = (Work per unit mass).

$V\,dV\,[=d(V^2/2)]$ is the change of kinetic energy per unit mass of a
   fluid particle as it moves a distance ds.
$g\,dz$ is most usually interpreted as the change of gravitational potential
   energy per unit mass of a fluid particle as it moves a distance ds;
   $-g\,dz$ can also be interpreted as the work done on the fluid particle
   by the gravity force.

The equivalence between work done by the gravitational force and
the change of gravitational potential energy is sometimes extended to
pressure, and dp/ρ is interpreted as "change of potential energy due to
pressure." The integrated form, Bernoulli's equation,

$$\frac{V^2}{2} + \frac{p}{\rho} + gz = C, \tag{4.20}$$

shows that the sum of kinetic energy, gravitational potential energy,
and pressure work* is constant if the fluid is incompressible, there is
no shear stress, and no mechanical work is done on or by the fluid
(except that done by pressure).

We might call Eqs. (4.20) and (4.21) the energy form of Bernoulli's
equation since each term has the dimensions and units of energy per
unit mass (refer to Example 1.4). Other forms of Bernoulli's equation
are the pressure form, obtained by multiplying the equation by density,
ρ:

$$p + \frac{1}{2}\rho V^2 + \gamma z = C; \tag{4.22}$$

and the elevation form or head form, obtained by dividing the equation
by the acceleration of gravity, g:

$$\frac{p}{\gamma} + \frac{V^2}{2g} + z = C. \tag{4.23}$$

In the pressure form, Eq. (4.22), each term has the dimensions and
units of pressure. The terms are often given names that imply that they
actually are pressures. The true fluid pressure, p, is called the *static
pressure* because it is the pressure that would be measured by an instrument that is static *with respect to the fluid*. Of course, if the instrument were static with respect to the fluid, it would have to move
with the fluid. The term $\frac{1}{2}\rho V^2$, which is actually kinetic energy per
unit volume, is called the *velocity pressure* or, sometimes, the *dynamic
pressure*. This is of course not really a pressure at all; however, if a fluid
particle is brought to rest (zero velocity) in the absence of shear stress
and external work, the pressure of the particle will rise by an amount
equal to $\frac{1}{2}\rho V^2$. The kinetic energy of the fluid particle will have been
"converted" into a pressure rise. The term γz, actually the potential
energy per unit volume, is called the *hydrostatic pressure,* by obvious
analogy with Eq. (2.5). Note that γz is not actually a pressure but could

---

* In thermodynamics, dp/ρ is called "flow work." We use "pressure work" here because
the fluid pressure is the agent that performs the work. In Chapter 5, we use the term
"flow work."

be converted to pressure if the fluid were brought to zero elevation. The constant on the right side of Eq. (4.22) also has dimensions and units of pressure and is called the *total pressure* (p_T) or, sometimes, the *stagnation pressure* (p_0). The total pressure is the maximum pressure that could be attained by a fluid particle unless mechanical work were done on it. To attain this pressure, the particle would have to be brought to rest and to zero elevation so that the "velocity pressure" and "hydrostatic pressure" could be converted into true pressure.*

The pressure form of Bernoulli's equation is put to use in a Pitot-static tube, a device for measuring fluid velocity. Figure 4.7 shows a Pitot-static tube and a schematic of the tube in a fluid stream. The flow is assumed to be locally one-dimensional, so streamlines A and B have the same value of p_T at plane 1. Since the inner tube is closed off by the pressure gage at $\textcircled{A}$, the fluid particles on streamline A are brought to rest at point 2. Applying Eq. (4.22), we obtain

$$p_1 + \frac{1}{2}\rho V_1^2 + \gamma z_1 = p_T = p_2 + \frac{1}{2}\rho V_2^2 + \gamma z_2.$$

If we select $z = 0$ at the tube centerline and note that $V_2 = 0$, this equation becomes

$$p_1 + \frac{1}{2}\rho V_1^2 = p_2 = p_T. \tag{4.24}$$

Gage $\textcircled{A}$ measures the total pressure of the fluid on streamline A.[†]

For streamline B,

$$p_{1'} + \frac{1}{2}\rho V_{1'}^2 + \gamma z_{1'} = p_3 + \frac{1}{2}\rho V_3^2 + \gamma z_3.$$

If the tube is small and the open holes are located far enough from the nose of the tube (usually eight tube diameters is sufficient), then $V_{1'} = V_3$, so

$$p_3 + \gamma z_3 = p_{1'} + \gamma z_{1'}. \tag{4.25}$$

Figure 4.7 Pitot-static tube.

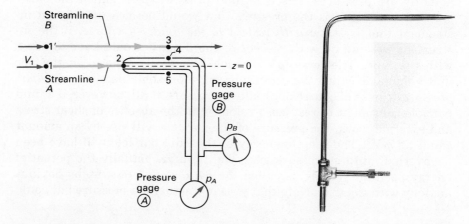

---

* Recall that this entire discussion is applicable only if the fluid density is constant.

[†] Note that the stagnation process between points 1 and 2 converts the velocity pressure at 1 into a true pressure at 2.

Now we apply Eq. (4.16) *across streamlines* between 3 and 4. Noting that the streamlines are straight, we get

$$p_4 + \gamma z_4 = p_3 + \gamma z_3. \tag{4.26}$$

Substituting Eq. (4.26) into Eq. (4.25) gives

$$p_{1'} + \gamma z_{1'} = p_4 + \gamma z_4.$$

Since the streamlines are straight at plane 1,

$$p_{1'} + \gamma z_{1'} = p_1 + \gamma z_1 = p_1, \quad \text{so} \quad p_4 = p_1 + \gamma z_4.$$

If we consider point 5 on the lower side of the tube, we find

$$p_5 = p_1 + \gamma z_5 = p_1 - \gamma z_4.$$

Most Pitot-static tubes have several static holes equally spaced around the tube so that the gage (B) measures an average pressure. Considering only holes 4 and 5, we have

$$\bar{p} = \frac{1}{2}(p_4 + p_5) = p_1;$$

that is, the pressure measured by the static holes is the static pressure on the tube centerline far in front of the tube. Because p_2 is measured by gage (A) and $\bar{p}$ is measured by gage (B),* we can compute the fluid velocity from

$$V_1 = \sqrt{2\left(\frac{p_2 - \bar{p}}{\rho}\right)} = \sqrt{2\left(\frac{p_{(A)} - p_{(B)}}{\rho}\right)}. \tag{4.27}$$

Note that the pressure difference $p_{(A)} - p_{(B)}$ is equal to the fluid velocity pressure ($\frac{1}{2}\rho V_1^2$). If a manometer or other differential pressure gage were connected directly between the locations of gages (A) and (B), we could say that the manometer directly measures the velocity pressure, although the manometer really measures the difference between the total pressure (a true pressure at 2) and the static pressure (the average of true pressures at 4 and 5).[†]

---

* Be careful to correct the gage readings for the difference in elevation between the gage and the tube centerline ($z = 0$).

† In this analysis of the Pitot-static tube, we have been very careful in our treatment of elevation changes. In practice, these are usually negligible. If this is so, they can be ignored.

EXAMPLE 4.7   Illustrates Bernoulli's equation for Pitot-type tubes

One difficulty with using the Pitot-static tube in Fig. 4.7 is the possible misalignment of the stagnation hole with the streamlines. This difficulty can be partially overcome by using the Fechheimer probe shown in Fig. E4.7(a). It has a cylindrical body, a leading stagnation hole, and two off-center holes serving for alignment in one plane and as a static pressure reference. The Fechheimer probe is properly aligned in one plane when $p_1 = p_3$. Find θ_0 in Fig. E4.7(b) such that $p_1 = p_3 = p_\infty$ if the velocity over the cylindrical surface is given by

$$V_\theta = 2V_\infty \sin \theta \qquad (0 \le \theta \le 180°).$$

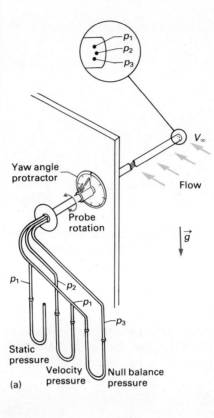

Figure E4.7 (a) Fechheimer probe; (b) flow over cylindrical body of Fechheimer probe.

Solution

Given
Fechheimer probe in Fig. E4.7(a)
Velocity distribution over cylindrical body: $V_\theta = 2V_\infty \sin\theta$

Find
Angle θ_0 such that $p_1 = p_3 = p_\infty$

Solution

The following analysis assumes constant density and gravity and an inviscid fluid; the resulting value of θ_0 is valid only for situations approximating these conditions.

Apply Bernoulli's equation to a streamline connecting points ∞ and 2 in Fig. E4.7(b). If the probe is properly aligned, half the particles moving along the ∞-2 streamline will pass over point 1 and the other half will pass over point 3. Applying Bernoulli's equation along the streamline from ∞ to 1 and along the streamline from ∞ to 3 gives

$$\frac{p_\infty}{\rho} + \frac{V_\infty^2}{2} + gz_\infty = \frac{p_1}{\rho} + \frac{V_1^2}{2} + gz_1 \quad \text{and} \quad \frac{p_\infty}{\rho} + \frac{V_\infty^2}{2} + gz_\infty = \frac{p_3}{\rho} + \frac{V_3^2}{2} + gz_3,$$

where $z_\infty \approx z_1 \approx z_3$. Solving for $p_1 - p_\infty$ and $p_3 - p_\infty$ gives

$$p_1 - p_\infty = 0 = \frac{\rho}{2}(V_\infty^2 - V_1^2), \qquad p_3 - p_\infty = 0 = \frac{\rho}{2}(V_\infty^2 - V_3^2).$$

Substituting for V_1 or V_3 gives the same result:

$$0 = V_\infty^2[1 - 4\sin^2\theta_0].$$

Solving for $\sin \theta_0$ gives

$$\sin \theta_0 = \frac{1}{2},$$

so

$$\boxed{\theta_0 = 30°}$$

Answer ◀

Discussion

The probe is properly aligned with the flow when the null balance pressure manometer gives a pressure difference $(p_1 - p_3)$ of zero. The direction of the flow is measured by the yaw angle protractor.

Next we consider the head form of Bernoulli's equation, Eq. (4.23). Each term has the dimensions and units of length. In the equation, p/γ is called the *pressure head* or *static head*, $V^2/2g$ is called the *velocity head*, and z is called the *elevation head*. The sum of static and elevation heads $(p/\gamma + z)$ is called the *piezometric head* because this is the height to which fluid would rise in a piezometer attached to a pipe with flowing fluid (Fig. 4.8). The constant on the right side of Eq. (4.23) is called the *total head* and is assigned the symbol h_T. The concept of "head" is quite useful when we consider flow systems where gravity and elevation provide the driving potential for the flow. For this reason the head form of Bernoulli's equation is often preferred by civil engineers.

The head form lends itself to a graphical interpretation of the mechanical energy balance of a flowing fluid. We define two grade lines as follows (see Fig. 4.9):

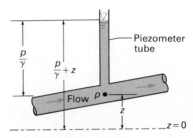

Figure 4.8 Piezometric head for flow in a closed pipe.

- The *energy grade line* is a line drawn above the $z = 0$ datum that shows the total head of the fluid.
- The *hydraulic grade line* is a line drawn above the $z = 0$ datum that shows the piezometric head of the fluid.

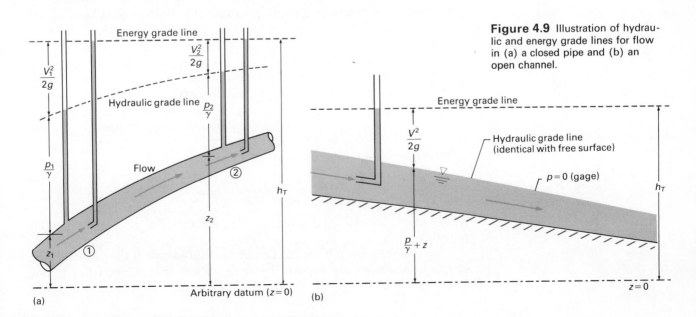

Figure 4.9 Illustration of hydraulic and energy grade lines for flow in (a) a closed pipe and (b) an open channel.

At any point, the distance between these two lines equals the fluid velocity head. The height of the energy grade line can be measured with a Pitot tube (a total head [pressure] tube with no static holes), and the height of the hydraulic grade line can be measured by a piezometer tube. The grade line representation is especially helpful in open channel flow since the free surface of the fluid is identical with the hydraulic grade line.

Whether you choose the energy form of Bernoulli's equation, the pressure form, or the head form, you should remember that the sum of the three terms is a measure of the total mechanical energy of the fluid. This sum will be constant if there is no friction (shear stress) and no external work but will change if there is either friction or external work along the streamline.

4.1.4 Integration of the Equation of Motion Normal to Streamlines

Although Bernoulli's equation is very useful, our derivation restricts the equation to flow along a single streamline. Although the one-dimensional flow assumption allows us to extend the results for one streamline to many identical streamlines, it is not always a valid assumption. To relate pressure, velocity, and elevation changes between streamlines in the general case, it is necessary to apply the equation of motion for the normal (cross-stream) direction. If we assume a steady flow, the appropriate equation is Eq. (4.16). Following the procedure used in deriving Bernoulli's equation, we multiply by dn and integrate *in the direction normal to the streamlines:*

$$\int \frac{V^2}{R}\, dn + \int \frac{dp}{\rho} + \int g\, dz = C. \qquad \textbf{(4.28)}$$

The pressure and elevation integrals can be evaluated in the same way as in Bernoulli's equation if we assume constant density and gravity; however, it is not possible to evaluate the first integral because, in general, we don't know how V and R vary in the n-direction. The normal equation can be integrated only for specific cases in which V and R are known as functions of n. One such case arose in our discussion of the Pitot-static tube. If the flow streamlines are straight and parallel, then R, the radius of curvature, is infinite, and

$$\int \frac{V^2}{R}\, dn = 0.$$

In this case (assuming constant ρ and g):

$$\frac{p}{\rho} + gz = \text{Constant}$$

from streamline to streamline if the streamlines are straight and parallel.

For other cases, $\int (V^2/R)\, dn$ can be evaluated only if the velocity and streamline shapes are known. In many cases, this information is not known; however, Eqs. (4.16) and (4.28) can still help us make a qualitative estimate of the pressure distribution. You should remember that the normal equation *must* be applied across streamlines and that *applying Bernoulli's equation across streamlines may lead to considerable errors.*

EXAMPLE 4.8 Illustrates a common misapplication of Bernoulli's equation

Figure E4.8 shows two simple flow fields called a *free vortex* and a *forced vortex*. The circumferential velocities in a free vortex and a forced vortex, respectively, are

$$V = \frac{C}{r} \quad \text{and} \quad V = r\omega.$$

In this case $C = 4.0 \text{ m}^2/\text{s}$, $\omega = 1.0 \text{ rad/s}$, r is in meters, and V is in meters per second. Both flows have been introduced previously but not named. The drain flow of Fig. 3.9 is a free vortex, and the fluid motion in the rotating tank in Section 2.6.2 is a forced vortex.

Apply Bernoulli's equation and Euler's normal equation between point A where $r_A = 2.0$ m and point B where $r_B = 8.0$ m to calculate the pressure difference. $(p_B - p_A)$ for both types of vortex. Both flows involve 20°C water and lie in a horizontal plane.

Solution

Given
Figure E4.8
Free vortex with

$$V = \frac{(4.0 \text{ m}^2/\text{s})}{r}$$

Forced vortex with

$$V = (1.0 \text{ rad/s})r,$$

where V is in meters per second and r is in meters
20°C water
Flows lying in a horizontal plane

Find
Pressure difference $(p_B - p_A)$ for both flows, where $r_A = 2.0$ m and $r_B = 8.0$ m

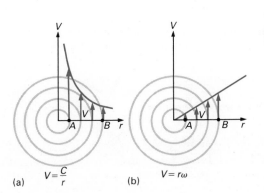

Figure E4.8 (a) Free vortex and (b) forced vortex with typical velocity distributions.

Solution

To apply Bernoulli's equation, we must assume constant density. Bernoulli's equation is

$$\frac{V_1^2}{2} + \frac{p_1}{\rho} + gz_1 = \frac{V_2^2}{2} + \frac{p_2}{\rho} + gz_2. \tag{4.21}$$

In Fig. E4.8, we identify point A as 1 and point B as 2, and we note that $z_A = z_B$, so

$$\frac{V_A^2}{2} + \frac{p_A}{\rho} = \frac{V_B^2}{2} + \frac{p_B}{\rho}, \quad \text{or} \quad p_B - p_A = \frac{\rho}{2}(V_A^2 - V_B^2).$$

For the free vortex,

$$p_B - p_A = \frac{\rho C^2}{2}\left(\frac{1}{r_A^2} - \frac{1}{r_B^2}\right).$$

Table A.5 in Appendix A gives $\rho = 998 \text{ kg/m}^3$. Then

$$p_B - p_A = \frac{\left(998\,\dfrac{\text{kg}}{\text{m}^3}\right)\left(4.0\,\dfrac{\text{m}^2}{\text{s}}\right)^2}{2}\left[\frac{1}{(2.0\text{ m})^2} - \frac{1}{(8.0\text{ m})^2}\right]$$

$$= 1{,}870 \text{ kg}\cdot\text{m/s}^2\cdot\text{m}^2,$$

$$\boxed{\; p_B - p_A = 1.87 \text{ kPa.} \;}$$

Bernoulli's equation, free vortex

Answer ◄

For the forced vortex,

$$p_B - p_A = \frac{\rho}{2}(r_A^2\omega^2 - r_B^2\omega^2)$$

$$= \frac{\rho\omega^2}{2}(r_A^2 - r_B^2)$$

$$= \frac{\left(998\,\dfrac{\text{kg}}{\text{m}^3}\right)\left(1.0\,\dfrac{\text{rad}}{\text{s}}\right)^2}{2}[(2.0\text{ m})^2 - (8.0\text{ m})^2]$$

$$= -31{,}900 \text{ kg}\cdot\text{m/m}^2\cdot\text{s}^2,$$

$$\boxed{\; p_B - p_A = -31.9 \text{ kPa.} \;}$$

Bernoulli's equation, forced vortex

Answer ◄

Next we apply Euler's normal equation between points A and B of both vortexes. Integrating Eq. (4.28) with $dz = 0$ and constant density gives

$$p_B - p_A = -\rho\int_A^B \frac{V^2}{R}\,dn.$$

Since the normal coordinate points toward the center of curvature, $dn = -dr$. Also $R = r$, so

$$p_B - p_A = +\rho\int_A^B \frac{V^2}{r}\,dr.$$

For the free vortex,

$$p_B - p_A = \rho \int_A^B \frac{C^2}{r^3}\, dr$$

$$= \frac{\rho C^2}{2}\left(\frac{1}{r_A^2} - \frac{1}{r_B^2}\right)$$

$$= \frac{\left(998\,\frac{kg}{m^3}\right)\left(4.0\,\frac{m^2}{s}\right)^2}{2}\left[\frac{1}{(2.0\ m)^2} - \frac{1}{(8.0\ m)^2}\right]$$

$$= 1{,}870\ kg \cdot m/m^2 \cdot s^2,$$

$$\boxed{p_B - p_A = 1.87\ kPa.}$$

Normal equation,
free vortex

Answer
◀

This is the same answer obtained from Bernoulli's equation.

For the forced vortex,

$$p_B - p_A = \rho \int_A^B \frac{r^2 \omega^2}{r}\, dr$$

$$= \frac{\rho \omega^2}{2}\left(r_B^2 - r_A^2\right)$$

$$= \frac{\left(998\,\frac{kg}{m^3}\right)\left(1.0\,\frac{rad}{s}\right)^2}{2}\left[(8.0\ m)^2 - (2.0\ m)^2\right]$$

$$= 29{,}900\ kg \cdot m/m^2 \cdot s^2,$$

$$\boxed{p_B - p_A = 29.9\ kPa.}$$

Normal equation,
forced vortex

Answer
◀

This is *not* the same answer obtained from Bernoulli's equation.

Discussion

Euler's normal equation was properly applied in both cases and gave us the correct answer. Bernoulli's equation was not properly applied since we violated one of the assumptions used in the development of the equation, namely, that the equation must be applied *along* a streamline. We applied it normal to streamlines, so we should not expect the correct answers from Bernoulli's equation. Surprisingly, the calculations showed that Bernoulli's equation did give us the correct answer for the free vortex even though we violated one of the assumptions implicit in the equation.*

The misapplication of Bernoulli's equation in this example illustrates an important point:

Applying an equation to a situation where all of the assumptions used in the development of that equation are not satisfied usually leads to incorrect answers.

---

* This curious fact is explained in Section 4.1.6.

Also note that Bernoulli's equation erroneously indicated that the pressure decreased with radius for the forced vortex. Recall that a mass particle will move in a straight line unless acted on by a force; in this case each particle tends to move in the tangential direction. The particle can only follow a circular path if the pressure at its larger radius exceeds the pressure at its inner radius and forces the particle to continually change its direction to form the circular path.

You might often be tempted to misapply Bernoulli's equation because it is an algebraic (rather than differential or integral) equation, because it does not require that the exact streamlines be known, and because it is an easy equation to apply (or misapply). On the other hand, Euler's normal equation can be integrated only if the streamlines and their radii of curvature can be identified. In this example, the streamlines are very simple shapes and Euler's normal equation can in fact be integrated. Other problems are not so simple. Students are more likely to use an easy algebraic equation than one that is difficult to integrate and use. But you must use equations that are applicable to your problem!

4.1.5 Euler's Equations in "Ordinary" Coordinates

The equations of motion in streamline coordinates, Eqs. (4.11) and (4.14), and the streamwise integral, Bernoulli's equation, Eq. (4.20), are extremely useful in many cases, but they have one serious drawback. To make calculations, we must know the location and, sometimes, the shape of the streamlines. In some cases, such as flow in a pipe, we can determine the streamlines by inspection; in other cases, such as flow in a converging passage, we can guess or approximate the streamlines reasonably well; however, in many cases we simply do not know the shapes of the streamlines. When this happens, we must select some coordinate system other than streamline coordinates. We usually pick one of the "ordinary" coordinate systems—rectangular, cylindrical, or spherical—since these are simple and we are reasonably familiar with them. If the flow is three- or even two-dimensional, the equations of motion will be rather difficult to solve, but we really have no other choices available. In this section, we develop the differential equations of motion for an inviscid fluid (Euler's equations) for the simplest ordinary coordinate system, rectangular coordinates.

For geometric simplicity, we will consider a two-dimensional, two-directional flow field. We obtain the equations of motion by applying Newton's second law to a typical fluid particle at a point (x, y) in the field. Figure 4.10 shows such a fluid particle and the pressures and gravity force that act on it. By assumption, there are no shear stresses, so Newton's second law for the x-direction is

$$\delta m_p a_x = \sum \delta F_x = \delta F_{x,\text{pressure}} + \delta F_{x,\text{gravity}}.$$

The mass of the particle is

$$\delta m_p = \rho \, \delta V = \rho(\delta x)(\delta y)(\delta z).$$

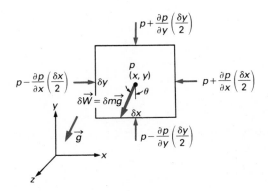

Figure 4.10 Rectangular fluid particle in two-dimensional, two-directional flow showing pressures and gravity force; depth of particle (perpendicular to paper) is δz.

The x-component of the particle's acceleration is given by Eq. (3.52), with $w = 0$ and $\partial u/\partial z = 0$ for two-dimensional, two-directional flow:

$$a_x = \frac{\partial u}{\partial t} + u\left(\frac{\partial u}{\partial x}\right) + v\left(\frac{\partial u}{\partial y}\right).$$

The net pressure force in the x-direction is

$$\delta F_{x,\text{pressure}} = \left[p - \frac{\partial p}{\partial x}\left(\frac{\delta x}{2}\right)\right](\delta y)(\delta z) - \left[p + \frac{\partial p}{\partial x}\left(\frac{\delta x}{2}\right)\right](\delta y)(\delta z)$$

$$= -\frac{\partial p}{\partial x}(\delta x)(\delta y)(\delta z).$$

The x-component of the gravity force is

$$\delta F_{x,\text{gravity}} = \delta W_x = (\delta m_p)g_x = \rho(\delta x)(\delta y)(\delta z)g_x,$$

where g_x is the component of the acceleration of gravity vector in the (positive) x-direction.

If we substitute the forces and acceleration, Newton's second law becomes

$$\rho(\delta x)(\delta y)(\delta z)\left[\frac{\partial u}{\partial t} + u\left(\frac{\partial u}{\partial x}\right) + v\left(\frac{\partial u}{\partial y}\right)\right] = -\frac{\partial p}{\partial x}(\delta x)(\delta y)(\delta z) + \rho g_x(\delta x)(\delta y)(\delta z).$$

We divide by $\rho(\delta x)(\delta y)(\delta z)$ and take the limit, shrinking the fluid particle to the point (x, y). The result is

$$\frac{\partial u}{\partial t} + u\left(\frac{\partial u}{\partial x}\right) + v\left(\frac{\partial u}{\partial y}\right) = -\frac{1}{\rho}\left(\frac{\partial p}{\partial x}\right) + g_x. \tag{4.29}$$

Next we consider Newton's second law for the y-direction:

$$\delta m_p a_y = \delta F_{y,\text{pressure}} + \delta F_{y,\text{gravity}},$$

with δm_p the same as before. We have a_y given by Eq. (3.53) with $w = 0$ and $\partial v/\partial z = 0$:

$$a_y = \frac{\partial v}{\partial t} + u\left(\frac{\partial v}{\partial x}\right) + v\left(\frac{\partial v}{\partial y}\right).$$

The net pressure force is

$$\delta F_{y,\text{pressure}} = -\frac{\partial p}{\partial y}(\delta y)(\delta x)(\delta z),$$

and the gravity force is

$$\delta F_{y,\text{gravity}} = \rho(\delta x)(\delta y)(\delta z)g_y.$$

Substituting into Newton's second law, we have

$$\rho(\delta x)(\delta y)(\delta z)\left[\frac{\partial v}{\partial t} + u\left(\frac{\partial v}{\partial x}\right) + v\left(\frac{\partial v}{\partial y}\right)\right] = -\frac{\partial p}{\partial y}(\delta x)(\delta y)(\delta z) + g_y(\delta x)(\delta y)(\delta z).$$

Dividing by $\rho(\delta x)(\delta y)(\delta z)$ and taking the limit, we get

$$\frac{\partial v}{\partial t} + u\left(\frac{\partial v}{\partial x}\right) + v\left(\frac{\partial v}{\partial y}\right) = -\frac{1}{\rho}\left(\frac{\partial p}{\partial y}\right) + g_y. \tag{4.30}$$

Equations (4.29) and (4.30) are Euler's equations for two-dimensional, two-directional flow. The equations for three-dimensional, three-directional flow are quite similar. The x- and y-equations must contain the acceleration terms $w\,\partial u/\partial z$ and $w\,\partial v/\partial z$, respectively. The z-direction equation is similar to the equations for the x- and y-directions, with a_z appearing on the left and the z-direction pressure gradient and gravity vector component appearing on the right. Euler's equations for three-dimensional, three-directional, unsteady flow are

$$\frac{\partial u}{\partial t} + u\left(\frac{\partial u}{\partial x}\right) + v\left(\frac{\partial u}{\partial y}\right) + w\left(\frac{\partial u}{\partial z}\right) = -\frac{1}{\rho}\left(\frac{\partial p}{\partial x}\right) + g_x, \tag{4.31}$$

$$\frac{\partial v}{\partial t} + u\left(\frac{\partial v}{\partial x}\right) + v\left(\frac{\partial v}{\partial y}\right) + w\left(\frac{\partial v}{\partial z}\right) = -\frac{1}{\rho}\left(\frac{\partial p}{\partial y}\right) + g_y, \tag{4.32}$$

$$\frac{\partial w}{\partial t} + u\left(\frac{\partial w}{\partial x}\right) + v\left(\frac{\partial w}{\partial y}\right) + w\left(\frac{\partial w}{\partial z}\right) = -\frac{1}{\rho}\left(\frac{\partial p}{\partial z}\right) + g_z. \tag{4.33}$$

The equations for special cases (steady, two-dimensional flow) can be obtained from these equations by deleting the appropriate terms.

These equations look rather formidable, and indeed they are. If we are interested in predicting a three-dimensional, three-directional flow (one of the flows in Fig. 3.2, for instance), we need to find the three velocities u, v, and w and the pressure p as functions of three space coordinates and time:

$$p = p(x, y, z, t), \qquad u = u(x, y, z, t), \qquad \text{etc.}$$

The three Euler equations (4.31), (4.32), and (4.33) together with the differential continuity equation (3.17) are four equations for the four unknowns u, v, w, and p, provided that the fluid density is known. The Euler equations represent the other equations that we said we needed in the discussion of the differential equation at the end of Section 3.4.1.*

---

* Actually, the Euler equations are the other equations only if we assume an inviscid flow.

All we need to do is solve the differential equations! This might seem to be an impossible task, but remember that we have already found one solution (Bernoulli's equation). Other analytical solutions can be found for special cases (we will illustrate one in the next section). Numerical solutions can be found to specific problems by using a digital computer.

Euler's equations can be written in all sorts of coordinate systems. We have already seen them in rectangular coordinates and in the very special streamline coordinates. The equations in other coordinates (say, cylindrical coordinates) can be derived by applying Newton's second law to a representative fluid particle in that coordinate system or they can be derived by mathematical transformation between coordinate systems. Finally, since the Euler equations represent Newton's second law, they can be written as a single vector equation:

$$\frac{D\vec{V}}{Dt} = \frac{\partial \vec{V}}{\partial t} + (\vec{V} \cdot \nabla)\vec{V} = -\frac{\nabla p}{\rho} + \vec{g}. \tag{4.34}$$

Euler's equations in cylindrical and spherical coordinates can be obtained from Appendix H. These forms are particularly useful when a flow has cylindrical or spherical symmetry since many terms drop out.

4.1.6 Integration of Euler's Equations for Irrotational Flow

For the special case of irrotational flow, Euler's equations can be integrated once and for all. To illustrate this, we will consider a two-dimensional, two-directional steady flow. In this case Euler's equations are

$$u\frac{\partial u}{\partial x} + v\frac{\partial u}{\partial y} = -\frac{1}{\rho}\frac{\partial p}{\partial x} + g_x$$

and

$$u\frac{\partial v}{\partial x} + v\frac{\partial v}{\partial y} = -\frac{1}{\rho}\frac{\partial p}{\partial y} + g_y.$$

First, consider the gravity term. For this derivation, we will adopt the coordinate system shown in Fig. 4.11. A z-coordinate is defined that

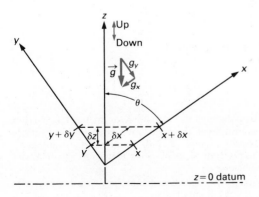

Figure 4.11 x, y-coordinate system with z-coordinate (elevation) pointing upward; movement from x to $(x + \delta x)$ or from y to $(y + \delta y)$ results in elevation change δz.

is parallel to the gravity vector and points upward. We call z the *elevation*.* From the figure

$$g_x = -|\vec{g}| \cos \theta = -g \cos \theta$$

and

$$g_y = -|\vec{g}| \sin \theta = -g \sin \theta.$$

If the fluid particle moves from x to $x + \delta x$, it moves "up" an amount δz where

$$\delta z = \delta x \cos \theta, \quad \text{so} \quad \cos \theta = \frac{\delta z}{\delta x} = \frac{\partial z}{\partial x}.$$

Similarly,

$$\sin \theta = \frac{\partial z}{\partial y},$$

so substituting gives

$$g_x = -g\left(\frac{\partial z}{\partial x}\right) \quad \text{and} \quad g_y = -g\left(\frac{\partial z}{\partial y}\right).$$

Substituting the g-components into the right side of the Euler equations gives[†]

$$u\left(\frac{\partial u}{\partial x}\right) + v\left(\frac{\partial u}{\partial y}\right) = -\frac{1}{\rho}\left(\frac{\partial p}{\partial x}\right) - g\left(\frac{\partial z}{\partial x}\right) \tag{4.35}$$

and

$$u\left(\frac{\partial v}{\partial x}\right) + v\left(\frac{\partial v}{\partial y}\right) = -\frac{1}{\rho}\left(\frac{\partial p}{\partial y}\right) - g\left(\frac{\partial z}{\partial y}\right). \tag{4.36}$$

Next we introduce the condition of irrotationality. An irrotational flow has angular velocity (ω) and vorticity ($\zeta = 2\omega$) equal to zero. From Eq. (3.60),

$$\zeta = \frac{\partial v}{\partial x} - \frac{\partial u}{\partial y} = 0.$$

Thus

$$\frac{\partial u}{\partial y} = \frac{\partial v}{\partial x}.$$

If we use this relationship, Eq. (4.35) becomes

$$u\left(\frac{\partial u}{\partial x}\right) + v\left(\frac{\partial v}{\partial x}\right) = \frac{\partial}{\partial x}\left(\frac{u^2}{2} + \frac{v^2}{2}\right) = -\frac{1}{\rho}\left(\frac{\partial p}{\partial x}\right) - g\left(\frac{\partial z}{\partial x}\right),$$

---

* It seems that we are having a little trouble with the z-coordinate. In the last section, x, y, and z were ordinary rectangular coordinates. In this section (and also in Sections 4.1.3 and 4.1.4) z is the elevation and is not perpendicular to x and y. This unfortunate confusion results from a long-standing tradition of using z for elevation in integrated equations of motion like Bernoulli's equation.

[†] Note that the replacement of g_x, g_y, and g_z by $-g[\partial(\text{Elevation})/\partial x(y \text{ or } z)]$ can also be done in the general Euler equations (4.31), (4.32), and (4.33).

and Eq. (4.36) becomes

$$u\left(\frac{\partial u}{\partial y}\right) + v\left(\frac{\partial v}{\partial y}\right) = \frac{\partial}{\partial y}\left(\frac{u^2}{2} + \frac{v^2}{2}\right) = -\frac{1}{\rho}\left(\frac{\partial p}{\partial y}\right) - g\left(\frac{\partial z}{\partial y}\right).$$

The magnitude of the velocity vector $\vec{V}$ is

$$|\vec{V}|^2 = V^2 = u^2 + v^2,$$

so substituting and rearranging (constant g is assumed), we obtain

$$\frac{\partial}{\partial x}\left(\frac{V^2}{2} + gz\right) + \frac{1}{\rho}\left(\frac{\partial p}{\partial x}\right) = 0 \qquad (4.37)$$

and

$$\frac{\partial}{\partial y}\left(\frac{V^2}{2} + gz\right) + \frac{1}{\rho}\left(\frac{\partial p}{\partial y}\right) = 0. \qquad (4.38)$$

Multiplying Eq. (4.37) by dx and Eq. (4.38) by dy, adding them and using the fact that, for any quantity,

$$d(\ ) = \frac{\partial(\ )}{\partial x}(dx) + \frac{\partial(\ )}{\partial y}(dy),$$

we get

$$d\left(\frac{V^2}{2}\right) + d(gz) + \frac{dp}{\rho} = 0.$$

Integrating (g is assumed constant) gives

$$\frac{V^2}{2} + gz + \int\frac{dp}{\rho} = C, \qquad (4.39)$$

where C is a constant. If ρ is a constant, we get

$$\frac{V^2}{2} + \frac{p}{\rho} + gz = C. \qquad (4.40)$$

This equation is identical to Bernoulli's equation, Eq. (4.20), but there are some important differences. Bernoulli's equation applies only along a streamline but contains no restrictions about irrotationality. Equation (4.40) applies between any two points in a flow, even across streamlines, if the flow is irrotational. Since Eq. (4.40) is identical in form to Bernoulli's equation, we will not treat it as a different equation but simply relax one of the restrictions on Bernoulli's equation:*

> Bernoulli's equation can be applied across streamlines if the flow is irrotational.

Although we carried out this development for a two-dimensional, two-directional flow, it is also valid for a three-dimensional, three-directional flow, so long as the flow is irrotational, with all three components of vorticity equal to zero.

---

* Now can you explain why we were able to obtain correct results from the application of Bernoulli's equation across streamlines for the free vortex but not for the forced vortex in Example 4.8?

4.2 Viscous Flow

Euler's and Bernoulli's equations apply only if there is no shear stress acting on the fluid. Although there are many cases where reasonably accurate calculations can be made by neglecting shear stress, many flow phenomena are closely tied to the existence of shear stresses in the fluid. It seems that it ought to be easy enough to include shear stresses; Newton's second law for a fluid particle would become

$$
\begin{pmatrix} \text{Mass} \\ \text{of} \\ \text{particle} \end{pmatrix} \times \begin{pmatrix} \text{Acceleration} \\ \text{of} \\ \text{particle} \end{pmatrix} = \begin{pmatrix} \text{Net pressure} \\ \text{force on} \\ \text{particle} \end{pmatrix} + \begin{pmatrix} \text{Net gravity} \\ \text{force on} \\ \text{particle} \end{pmatrix} + \begin{pmatrix} \text{Net stress} \\ \text{force on} \\ \text{particle} \end{pmatrix}.
$$

$$(4.41)$$

This equation is very similar to Eq. (4.1). We have already formulated all of the terms except the last one, so it seems that all we have to do is add the stress force term and carry on. In principle, this is true enough. The operation of adding this force term *in terms of stress* can be accomplished by a few lines of algebra. Relating the stress to the flow field is more complicated; in fact, we will not even do it in a completely rigorous manner in this book. We will, however, show the resulting equations.

When we try to solve the differential equations for flow with shear stress (where the real payoff is), we won't get very far. In inviscid flow, the introduction of streamline coordinates allowed us to find a very useful general solution to the differential equations of motion. The use of streamline coordinates in a viscous flow *does not* allow us to reduce the equations to an easily integrable set because the net shear stress force depends on the second derivative of velocity normal to a streamline. At the least, a $(\partial^2 V_s/\partial n^2)$ term will appear in the streamline equation of motion, and the streamline and normal equations must be solved simultaneously. No general solutions (like Bernoulli's equation) exist for viscous flow and so each problem must be solved a priori.

We will begin our investigation of the differential approach to viscous flow analysis by considering a very simple flow for which we can obtain a solution of the differential equation. Then we will describe and discuss the general equations that describe viscous flow.

4.2.1 A Simple Viscous Flow: Fully Developed Flow Between Parallel Plates

Consider flow of a viscous fluid between two very large, parallel, flat plates spaced a distance $2Y$ apart (Fig. 4.12). The plates are so large that we pretend they are infinite.* The fluid flows from left to right. We assume that the flow is steady and that the fluid is incompressible. Since the fluid has viscosity, it will not "slip" at the surface of the plates. We expect the velocity profile to appear roughly as shown in Fig. 4.12(b). The velocity increases from the lower plate to the centerline and then

---

* Formally, all dimensions of the plates are very much larger than Y.

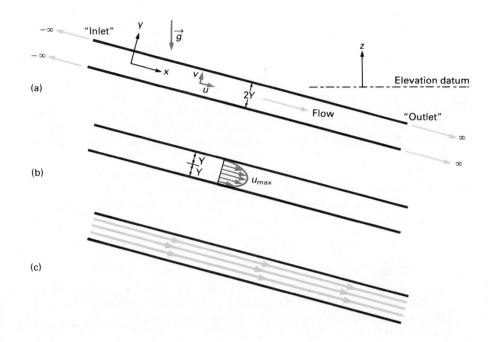

Figure 4.12 Viscous flow between two large (infinite) parallel plates: (a) Basic geometry and coordinate system; (b) expected velocity profile (u-velocity); (c) typical streamlines.

decreases from the centerline to the upper plate. Figure 4.12(c) shows some streamlines (remember, a flow does not have to be inviscid to have streamlines). We assume that these streamlines are straight and parallel to the plates. It is possible that the streamlines are not straight and parallel far upstream where the fluid first enters the space between the plates, but in the region we are considering, the streamlines are straight and parallel.

To analyze the flow, we select an x, y-coordinate system as shown in Fig. 4.12(a). We assume that the flow is identical in all planes parallel to the paper, so we have at most a two-dimensional flow. Before we apply Newton's second law, we will see what we can learn from the continuity equation. Since the flow is steady, two-dimensional, and incompressible, the appropriate continuity equation is Eq. (3.19), with $\partial w/\partial z$ dropped:

$$\frac{\partial u}{\partial x} + \frac{\partial v}{\partial y} = 0.$$

Since the streamlines are straight and parallel to the x-direction,

$$v = 0 \quad \text{(everywhere)},$$

so

$$\frac{\partial v}{\partial x} = \frac{\partial v}{\partial y} = 0. \tag{4.42}$$

If we use Eq. (4.42), the continuity equation becomes

$$\frac{\partial u}{\partial x} = 0,$$

which tells us that the velocity does not change with x (it can, and does, change with y). The velocity profile is identical at all x's. The velocity field is given by

$$v = 0, \qquad u = u(y).$$

The flow is one-dimensional (in y) and one-directional (the single velocity is in the x-direction). This type of flow, with the velocity profile unchanging as the fluid moves downstream, is called *fully developed flow*.

To find the velocity distribution, we now apply Newton's second law to a fluid particle. Figure 4.13 shows a fluid particle at some point (x, y) together with the pressures and shear stresses acting on the particle. For the x-direction, Newton's second law is

$$(\delta m_p)a_x = \delta F_{x,\text{pressure}} + \delta F_{x,\text{gravity}} + \delta F_{x,\text{shear}}.$$

The pressure and gravity forces should be familiar by now; they are*

$$\delta F_{x,\text{pressure}} = -\frac{\partial p}{\partial x}(\delta x)(\delta y)(1) \qquad \text{and} \qquad \delta F_{x,\text{gravity}} = \rho g_x(\delta x)(\delta y)(1).$$

The shear force is

$$\delta F_{x,\text{shear}} = \left[\tau + \frac{\partial \tau}{\partial y}\left(\frac{\delta y}{2}\right)\right]\delta x(1) - \left[\tau - \frac{\partial \tau}{\partial y}\left(\frac{\delta y}{2}\right)\right]\delta x(1)$$

$$= \frac{\partial \tau}{\partial y}(\delta x)(\delta y)(1).$$

Figure 4.13 Pressure, shear stresses, and gravity force acting on particle between parallel plates.

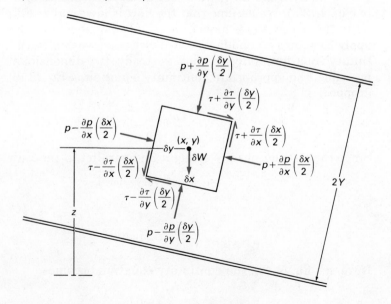

---

* The fluid particle is one unit (1) deep perpendicular to the paper.

The acceleration of the particle is

$$a_x = \frac{\partial u}{\partial t} + u\left(\frac{\partial u}{\partial x}\right) + v\left(\frac{\partial u}{\partial y}\right) = 0 + u(0) + 0\left(\frac{\partial u}{\partial y}\right) = 0.$$

Newton's second law reduces to a simple force balance

$$0 = \left(-\frac{\partial p}{\partial x} + \rho g_x + \frac{\partial \tau}{\partial y}\right)(\delta x)(\delta y)(1).$$

Dividing by $(\delta x)(\delta y)(1)$ and taking the limit for δx and δy approaching zero, we have

$$-\frac{\partial p}{\partial x} + \rho g_x + \frac{\partial \tau}{\partial y} = 0.$$

Introducing a z-coordinate pointing upward, opposite the gravity vector, we obtain

$$-\frac{\partial p}{\partial x} - \rho g\left(\frac{\partial z}{\partial x}\right) + \frac{\partial \tau}{\partial y} = 0.$$

For ρ and g constant, we get

$$-\frac{\partial}{\partial x}(p + \gamma z) + \frac{\partial \tau}{\partial y} = 0. \qquad (4.43)$$

In a similar fashion, Newton's second law for the y-direction gives

$$-\frac{\partial}{\partial y}(p + \gamma z) + \frac{\partial \tau}{\partial x} = 0.$$

Since the velocity profile does not change in the x-direction,

$$\frac{\partial \tau}{\partial x} = 0, \quad \text{so} \quad \frac{\partial}{\partial y}(p + \gamma z) = 0.$$

The pressure distribution in the y-direction is hydrostatic. The term $(p + \gamma z)$ is a function of x only, and the partial derivative with respect to x in Eq. (4.43) can be replaced by an ordinary derivative:

$$-\frac{d}{dx}(p + \gamma z) + \frac{\partial \tau}{\partial y} = 0. \qquad (4.44)$$

As it stands, this equation applies to turbulent flow as well as laminar flow since we have not specified how the shear stress is related to the fluid velocity. To proceed further, we specify that the flow is laminar, so

$$\tau \sim \text{Rate of shear deformation of fluid particle } (\dot{\phi}).$$

From Eq. (3.63),

$$\dot{\phi} = \frac{1}{2}\left(\frac{\partial u}{\partial y} + \frac{\partial v}{\partial x}\right) = \frac{1}{2}\left(\frac{\partial u}{\partial y}\right).$$

According to Newton's law of viscosity, Eq. (1.2), the constant of proportionality must be twice the fluid viscosity, 2μ:

$$\tau = 2\mu\dot{\phi} = \mu\left(\frac{\partial u}{\partial y}\right).$$

Substituting into Eq. (4.44), we have

$$-\frac{d}{dx}(p + \gamma z) + \frac{\partial}{\partial y}\left[\mu\left(\frac{\partial u}{\partial y}\right)\right] = 0.$$

If we assume that the viscosity is constant, we get

$$-\frac{d}{dx}(p + \gamma z) + \mu\frac{\partial^2 u}{\partial y^2} = 0.$$

For this fully developed flow, u is a function of y only, so

$$\frac{\partial^2 u}{\partial y^2} = \frac{d^2 u}{dy^2},$$

and we get

$$\mu\frac{d^2 u}{dy^2} = \frac{d}{dx}(p + \gamma z). \tag{4.45}$$

Examining this equation, we can conclude that each term, the one on the right and the one on the left, must be equal to a constant because the left term is a function of only y and the right term is a function of only x. If the equality is to be true for all x's and y's, the terms must be constant. To remind us of this, let

$$\tilde{p} = \frac{d}{dx}(p + \gamma z).$$

To obtain the velocity distribution, we must solve

$$\frac{d^2 u}{dy^2} = \frac{\tilde{p}}{\mu}. \tag{4.46}$$

Integrating (4.46) twice, we obtain

$$u = \frac{\tilde{p}}{2\mu}(y^2) + C_1 y + C_2.$$

We can evaluate the constants of integration from the conditions

$$u = 0 \qquad \text{at } y = \pm Y,$$

giving

$$C_1 = 0, \quad C_2 = -\frac{\tilde{p}Y^2}{2\mu}.$$

The velocity distribution is given by

$$u = -\frac{\tilde{p}Y^2}{2\mu}\left[1 - \left(\frac{y}{Y}\right)^2\right].$$

The maximum velocity is

$$u_{\max} = -\frac{\tilde{p}Y^2}{2\mu} \qquad \text{at } y = 0.$$

The average velocity, $\bar{V}$, is

$$\bar{V} = \frac{1}{2Y}\int_{-Y}^{Y} u\, dy = -\frac{2}{3}\frac{\tilde{p}Y^2}{2\mu} = \frac{2}{3}u_{\max}.$$

We reintroduce the definition of $\tilde{p}$, and our results are

$$u = -\frac{Y^2}{2\mu}\frac{d(p+\gamma z)}{dx}\left[1-\left(\frac{y}{Y}\right)^2\right], \tag{4.47}$$

$$u_{max} = -\frac{Y^2}{2\mu}\frac{d(p+\gamma z)}{dx}, \tag{4.48}$$

and

$$\bar{V} = \frac{2}{3}u_{max} = -\frac{Y^2}{3\mu}\frac{d(p+\gamma z)}{dx}. \tag{4.49}$$

To maintain flow in the x-direction, the piezometric head ($p/\gamma + z$) must continually decrease. From a physical point of view, we must have either an elevation difference or a pressure difference (possibly generated by a pump) to overcome the resistance due to fluid "friction." Notice also that Bernoulli's equation does not apply to this flow because that would require ($p + \gamma z$) to be constant if $\rho u^2/2$ were constant; $\rho u^2/2$ is indeed constant along any streamline since $\partial u/\partial x = 0$ for all streamlines, but ($p + \gamma z$) is definitely *not* constant so long as there is any flow at all.

4.2.2 General Equations of Motion for a Viscous Fluid: Cauchy and Navier-Stokes Equations

The success of our analysis of viscous flow between parallel plates is strongly dependent on the fact that all of the acceleration (or "inertia force") terms vanish from the equations of motion. Although there are a few other flow geometries that result in this simplification (the most notable and useful being fully developed flow in a circular pipe), most flow situations encountered in engineering do not lend themselves to such simple mathematical treatment.

The objective of this section is to develop the general equations that describe viscous fluid flow. Our development will not be completely rigorous since it is primarily for the purpose of introduction. We will consider only incompressible fluids with constant viscosity and concentrate on two-dimensional, two-directional flow. Although the use of a coordinate system that is well matched to the flow geometry is a very powerful tool, we will consider only rectangular coordinates. Even with all these restrictions, solution of the resulting equations is well beyond the scope of this book (unless of course the flow is very simple, such as the flow between infinite parallel plates). These equations do, however, form the basis of some very successful approximate methods of flow analysis, such as the boundary layer model, that we will discuss in Chapter 9.

We begin by developing the differential equations of motion in terms of stress. Newton's second law for a fluid particle that experiences forces due to pressure, gravity, and stress has already been presented as Eq. (4.41). Using symbols similar to those used for Euler's equations, we can write the equation for both the x- and the y-direction:

$$(\delta m_p)a_x = \delta F_{x,pressure} + \delta F_{x,gravity} + \delta F_{x,stress} \tag{4.50}$$

and

$$(\delta m_p)a_y = \delta F_{y,\text{pressure}} + \delta F_{y,\text{gravity}} + \delta F_{y,\text{stress}}. \tag{4.51}$$

All terms in these equations, with the exception of the forces due to stress, have already been found in our previous work. The already derived terms are

$$\delta m_p = \rho(\delta x)(\delta y)(\delta z),$$

$$a_x = \frac{\partial u}{\partial t} + u\left(\frac{\partial u}{\partial x}\right) + v\left(\frac{\partial u}{\partial y}\right)$$

$$a_y = \frac{\partial v}{\partial t} + u\left(\frac{\partial v}{\partial x}\right) + v\left(\frac{\partial v}{\partial y}\right)$$

$$\delta F_{x,\text{pressure}} = -\frac{\partial p}{\partial x}(\delta x)(\delta y)(\delta z)$$

$$\delta F_{y,\text{pressure}} = -\frac{\partial p}{\partial y}(\delta x)(\delta y)(\delta z)$$

$$\delta F_{x,\text{gravity}} = \rho g_x(\delta x)(\delta y)(\delta z)$$
$$\delta F_{y,\text{gravity}} = \rho g_y(\delta x)(\delta y)(\delta z)$$

To determine the net stress force acting on a fluid particle, consider Fig. 4.14, which shows a rectangular fluid particle at point (x, y) in a two-dimensional, two-directional flow field. The force due to gravity and the fluid velocities have been omitted for clarity. In general, there may be two types of stress acting on the fluid, tangential (shear) stress and normal stress. Up to this point, we have assumed that the normal stress in fluids is completely represented by pressure. In Section 2.1 we showed that the normal stress (pressure) is indeed equal in all directions for a static fluid or a moving inviscid fluid, but we also noted that the normal stress might be different in different directions for a viscous fluid. This possible difference in normal stresses is due to volumetric strain of the fluid particle. Figure 4.14 shows the total normal stress in two parts: the pressure p (now defined as the *average* of the normal stresses at a point) and the normal viscous stresses σ_x and σ_y, with

Figure 4.14 Pressures and stresses acting on a rectangular fluid particle.

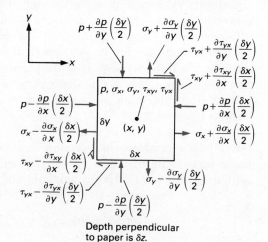

Depth perpendicular to paper is δz.

p, σ_x, and σ_y the values at the point (x, y). Values at the faces of the rectangular fluid particle are shown in the figure.

At the center of the particle, the shear stresses are τ_{xy} and τ_{yx}. The shear stresses on the faces of the rectangular particle are also shown. The subscripts are interpreted as follows. The first subscript identifies the coordinate direction perpendicular to the face of the particle on which the particular shear stress acts. The second subscript identifies the direction in which the particular shear stress acts. In a three-dimensional flow, an x-face would have stresses τ_{xy} and τ_{xz} acting on it.

The forces on the particle due to stresses other than pressure are

$$\delta F_{x,\text{stress}} = \left[\sigma_x + \frac{\partial \sigma_x}{\partial x}\left(\frac{\delta x}{2}\right) \right](\delta y)(\delta z) - \left[\sigma_x - \frac{\partial \sigma_x}{\partial x}\left(\frac{\delta x}{2}\right) \right](\delta y)(\delta z)$$
$$+ \left[\tau_{yx} + \frac{\partial \tau_{yx}}{\partial y}\left(\frac{\delta y}{2}\right) \right](\delta x)(\delta z) - \left[\tau_{yx} - \frac{\partial \tau_{yx}}{\partial y}\left(\frac{\delta y}{2}\right) \right](\delta x)(\delta z)$$

and

$$\delta F_{y,\text{stress}} = \left[\sigma_y + \frac{\partial \sigma_y}{\partial y}\left(\frac{\delta y}{2}\right) \right](\delta x)(\delta z) - \left[\sigma_y - \frac{\partial \sigma_y}{\partial y}\left(\frac{\delta y}{2}\right) \right](\delta x)(\delta z)$$
$$+ \left[\tau_{xy} + \frac{\partial \tau_{xy}}{\partial x}\left(\frac{\delta x}{2}\right) \right](\delta y)(\delta z) - \left[\tau_{xy} - \frac{\partial \tau_{xy}}{\partial x}\left(\frac{\delta x}{2}\right) \right](\delta y)(\delta z).$$

These may be simplified to

$$\delta F_{x,\text{stress}} = \left(\frac{\partial \sigma_x}{\partial x} + \frac{\partial \tau_{yx}}{\partial y} \right)(\delta x)(\delta y)(\delta z)$$

and

$$\delta F_{y,\text{stress}} = \left(\frac{\partial \sigma_y}{\partial y} + \frac{\partial \tau_{xy}}{\partial x} \right)(\delta x)(\delta y)(\delta z).$$

Next we substitute the particle's mass and acceleration and all of the forces into Eqs. (4.50) and (4.51), divide by $(\delta x)(\delta y)(\delta z)$, and take the limit, shrinking our particle to the point (x, y). The results are

$$\rho\left[\frac{\partial u}{\partial t} + u\left(\frac{\partial u}{\partial x}\right) + v\left(\frac{\partial u}{\partial y}\right) \right] = -\frac{\partial p}{\partial x} + \rho g_x + \frac{\partial \sigma_x}{\partial x} + \frac{\partial \tau_{yx}}{\partial y} \tag{4.52}$$

and

$$\rho\left[\frac{\partial v}{\partial t} + u\left(\frac{\partial v}{\partial x}\right) + v\left(\frac{\partial v}{\partial y}\right) \right] = -\frac{\partial p}{\partial y} + \rho g_y + \frac{\partial \sigma_y}{\partial y} + \frac{\partial \tau_{xy}}{\partial x}. \tag{4.53}$$

These equations, the equations of motion in terms of stress, are called the *Cauchy equations*.* The Cauchy equations must replace Euler's equations if we are considering a viscous fluid. Together with the continuity equation,

$$\frac{\partial u}{\partial x} + \frac{\partial v}{\partial y} = 0, \tag{3.19a}$$

---

* At this point, we would normally extend the equations to three dimensions; however, that would add very little to this discussion.

they describe the two-dimensional, two-directional motion of an incompressible viscous fluid. Consider what would happen if we tried to solve these equations (we realize that this seems like a very tall order, and you'll soon see that it's even taller than you think). First we count the equations: (4.52), (4.53), (3.19a); three. Next we count the unknowns:[*] u, v, p, σ_x, σ_y, τ_{xy}, τ_{yx}; seven. Not even your professor can solve three equations for seven unknowns!

The way out of this dilemma is to relate the stresses to the velocity field using, say, Newton's law of viscosity. We have to generalize this law to include normal viscous stresses (σ_x and σ_y) as well as shear stresses. The basic idea is that viscous stresses are proportional to the rates of strain of the fluid. The shear stresses are proportional to the shear rate of strain[†]

$$\tau_{xy} = \tau_{yx} = 2\mu\dot{\phi} = \mu\left(\frac{\partial u}{\partial y} + \frac{\partial v}{\partial x}\right), \tag{4.54}$$

and the normal viscous stresses are proportional to the stretching or volumetric strain rate (see Section 3.5.1):

$$\sigma_x = 2\mu\frac{\partial u}{\partial x} \tag{4.55}$$

and

$$\sigma_y = 2\mu\frac{\partial v}{\partial y}. \tag{4.56}$$

These relations must be altered slightly for a compressible fluid. References [1] and [2] present a complete development of the relations between stresses and strain rates for a Newtonian fluid.

If we use Eqs. (4.54) and (4.55), the last two terms on the right side of Eq. (4.52) become

$$\frac{\partial \sigma_x}{\partial x} + \frac{\partial \tau_{yx}}{\partial y} = \frac{\partial}{\partial x}(2\mu)\left(\frac{\partial u}{\partial x}\right) + \frac{\partial}{\partial y}(\mu)\left(\frac{\partial u}{\partial y} + \frac{\partial v}{\partial x}\right).$$

If we assume constant viscosity,

$$\frac{\partial \sigma_x}{\partial x} + \frac{\partial \tau_{yx}}{\partial y} = 2\mu\left(\frac{\partial}{\partial x}\right)\left(\frac{\partial u}{\partial x}\right) + \mu\left(\frac{\partial}{\partial y}\right)\left(\frac{\partial u}{\partial y}\right) + \mu\left(\frac{\partial}{\partial y}\right)\left(\frac{\partial v}{\partial x}\right).$$

Interchanging the order of differentiation of the last term, we obtain

$$\frac{\partial \sigma_x}{\partial x} + \frac{\partial \tau_{yx}}{\partial y} = \mu\left(\frac{\partial^2 u}{\partial x^2} + \frac{\partial^2 u}{\partial y^2}\right) + \mu\frac{\partial}{\partial x}\left[\frac{\partial u}{\partial x} + \frac{\partial v}{\partial y}\right].$$

The term in the square brackets is zero by Eq. (3.19a). Substituting the remaining terms into Eq. (4.52), we get

$$\rho\left[\frac{\partial u}{\partial t} + u\left(\frac{\partial u}{\partial x}\right) + v\left(\frac{\partial u}{\partial y}\right)\right] = -\frac{\partial p}{\partial x} + \rho g_x + \mu\left(\frac{\partial^2 u}{\partial x^2} + \frac{\partial^2 u}{\partial y^2}\right).$$

---

[*] The unknowns are the dependent variables. We don't count x, y, or t.

[†] The equality of τ_{xy} and τ_{yx} can be demonstrated by considering the rotational equilibrium of the fluid particle.

Similarly, Eq. (4.53) becomes

$$\rho\left[\frac{\partial v}{\partial t} + u\left(\frac{\partial v}{\partial x}\right) + v\left(\frac{\partial v}{\partial y}\right)\right] = -\frac{\partial p}{\partial y} + \rho g_y + \mu\left(\frac{\partial^2 v}{\partial x^2} + \frac{\partial^2 v}{\partial y^2}\right).$$

If we extend the equations to three-dimensional, three-directional flow, they become

$$\rho\left[\frac{\partial u}{\partial t} + u\left(\frac{\partial u}{\partial x}\right) + v\left(\frac{\partial u}{\partial y}\right) + w\left(\frac{\partial u}{\partial z}\right)\right] = -\frac{\partial p}{\partial x} + \rho g_x + \mu\left(\frac{\partial^2 u}{\partial x^2} + \frac{\partial^2 u}{\partial y^2} + \frac{\partial^2 u}{\partial z^2}\right),$$

$$\tag{4.57}$$

$$\rho\left[\frac{\partial v}{\partial t} + u\left(\frac{\partial v}{\partial x}\right) + v\left(\frac{\partial v}{\partial y}\right) + w\left(\frac{\partial v}{\partial z}\right)\right] = -\frac{\partial p}{\partial y} + \rho g_y + \mu\left(\frac{\partial^2 v}{\partial x^2} + \frac{\partial^2 v}{\partial y^2} + \frac{\partial^2 v}{\partial z^2}\right),$$

$$\tag{4.58}$$

$$\rho\left[\frac{\partial w}{\partial t} + u\left(\frac{\partial w}{\partial x}\right) + v\left(\frac{\partial w}{\partial y}\right) + w\left(\frac{\partial w}{\partial z}\right)\right] = -\frac{\partial p}{\partial z} + \rho g_z + \mu\left(\frac{\partial^2 w}{\partial x^2} + \frac{\partial^2 w}{\partial y^2} + \frac{\partial^2 w}{\partial z^2}\right).$$

$$\tag{4.59}$$

These are the *Navier-Stokes equations** for an incompressible, constant viscosity fluid. Together with the differential continuity equation, Eq. (3.19), they describe the motion of a viscous, incompressible fluid. Note that the four equations have only four unknowns (u, v, w, p) if ρ and μ are known. The equations can be derived in or transformed to other coordinates such as cylindrical or spherical (see Appendix H). They do not become any prettier in these systems.

At this point, you may be ready to give up on fluid mechanics, but these equations are not for you to solve, at least not yet. If you could solve them, for any arbitrary geometry, you would be able to do something no one else has ever done! We *can* solve them for some specific cases; in fact, we have already done one case—flow between infinite parallel plates. For the flow geometry shown in Fig. 4.12, Eq. (4.59) goes away altogether, and Eqs. (4.57) and (4.58) reduce to Eq. (4.45), which is quite simple.

The Navier-Stokes equations are more or less the ultimate equations of fluid mechanics. They even apply to turbulent flow, if we recognize that turbulent flows are randomly unsteady, even with steady boundary conditions. They don't apply to flow of non-Newtonian fluids, the mechanics of which are beyond the scope of this book. The Navier-Stokes equations form the basis of most advanced studies in fluid mechanics, serving as the point of departure for approximations like the boundary layer model, for numerical (computer) flow calculations, and for statistical models of turbulent flow. Notice too that the Navier-Stokes equations reduce to the Euler equations by simply putting $\mu = 0$. The inviscid flow model is thus seen to be just one possible approximation to a real flow.

Our main purpose for presenting the Navier-Stokes equations in this introductory book is to assure you that flow problems can be completely formulated, if not solved, from a mathematical point of view and

---

* Named for the Frenchman L. Navier and the Englishman Sir G. G. Stokes.

to show you that there are plenty of problems left for clever engineers like yourself in basic and advanced fluid mechanics. Analytical and numerical techniques for solving the equations are discussed in [1, 3, 4, 5, 6].

EXAMPLE 4.9   Illustrates how to simplify the Navier-Stokes equations

Simplify the three-dimensional, three-directional Navier-Stokes equations, Eqs. (4.57), (4.58), and (4.59), to produce Eq. (4.45) for the steady, one-directional, one-dimensional, fully developed flow between the two very large parallel flat plates of Fig. 4.12 on page 211.

Solution

Given
Equations (4.57), (4.58), and (4.59)
Steady flow between parallel plates

Find
Equation (4.45) by properly simplifying Eqs. (4.57), (4.58), and (4.59)

Solution
Equations (4.57), (4.58), and (4.59) are rewritten below. The z'-direction is perpendicular to the x- and y-directions for these equations, and the z-direction is parallel to the gravity vector. See Fig. E4.9.

$$\rho\left[\frac{\partial u}{\partial t} + u\left(\frac{\partial u}{\partial x}\right) + v\left(\frac{\partial u}{\partial y}\right) + w\left(\frac{\partial u}{\partial z'}\right)\right] = -\frac{\partial p}{\partial x} + \rho g_x + \mu\left(\frac{\partial^2 u}{\partial x^2} + \frac{\partial^2 u}{\partial y^2} + \frac{\partial^2 u}{\partial z'^2}\right)$$

(4.57a)

$$\rho\left[\frac{\partial v}{\partial t} + u\left(\frac{\partial v}{\partial x}\right) + v\left(\frac{\partial v}{\partial y}\right) + w\left(\frac{\partial v}{\partial z'}\right)\right] = -\frac{\partial p}{\partial y} + \rho g_y + \mu\left(\frac{\partial^2 v}{\partial x^2} + \frac{\partial^2 v}{\partial y^2} + \frac{\partial^2 v}{\partial z'^2}\right)$$

(4.58a)

Figure E4.9 Channel orientation.

z' is perpendicular to x- and y-directions.

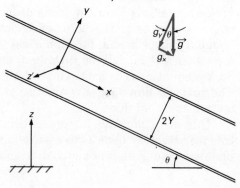

and

$$\rho\left[\frac{\partial w}{\partial t} + u\left(\frac{\partial w}{\partial x}\right) + v\left(\frac{\partial w}{\partial y}\right) + w\left(\frac{\partial w}{\partial z'}\right)\right] = -\frac{\partial p}{\partial z'} + \rho g_{z'} + \mu\left(\frac{\partial^2 w}{\partial x^2} + \frac{\partial^2 w}{\partial y^2} + \frac{\partial^2 w}{\partial z'^2}\right)$$

(4.59a)

These three equations are applicable to a constant-density, constant-viscosity, Newtonian fluid. We must now use each specification in our problem statement to simplify these equations.

The first specification is steady flow. This gives

$$\frac{\partial u}{\partial t} = \frac{\partial v}{\partial t} = \frac{\partial w}{\partial t} = 0.$$

The next specification is one-directional flow. Figure 4.13 shows

$$u \neq 0, \qquad v = w = 0.$$

Since v and w are zero *everywhere*, all of their derivatives are also zero. The one-dimensional specification means u is a function of only one space variable. The problem statement rules out z', and the fully developed flow specification gives

$$\frac{\partial u}{\partial x} = 0, \qquad \frac{\partial v}{\partial x} = 0, \qquad \text{and} \qquad \frac{\partial w}{\partial x} = 0$$

and rules out x. Therefore

$$u = u(y).$$

Equations (4.57a), (4.58a), and (4.59a) now give

$$\rho\left[0 + u(0) + 0\left(\frac{\partial u}{\partial y}\right) + (0)(0)\right] = -\frac{\partial p}{\partial x} + \rho g_x + \mu\left[0 + \frac{\partial^2 u}{\partial y^2} + 0\right],$$

$$\rho[0 + u(0) + (0)(0) + (0)(0)] = -\frac{\partial p}{\partial y} + \rho g_y + \mu[(0) + (0) + (0)],$$

and

$$\rho[0 + u(0) + (0)(0) + (0)(0)] = -\frac{\partial p}{\partial z'} + \rho g_{z'} + \mu[(0) + (0) + (0)].$$

These simplifications give

$$0 = -\frac{\partial p}{\partial x} + \rho g_x + \mu\frac{d^2 u}{dy^2}$$

$$0 = -\frac{\partial p}{\partial y} + \rho g_y$$

$$0 = -\frac{\partial p}{\partial z'} + \rho g_{z'}.$$

The channel orientation in Fig. E4.9 shows

$$g_{z'} = 0, \qquad g_y = -g\cos\theta, \qquad \text{and} \qquad g_x = g\sin\theta.$$

Referring to Fig. E4.9 gives

$$\cos\theta = \frac{\partial z}{\partial y} \qquad \text{and} \qquad \sin\theta = -\frac{\partial z}{\partial x}.$$

Then

$$g_y = -g\left(\frac{\partial z}{\partial y}\right) \quad \text{and} \quad g_x = -g\left(\frac{\partial z}{\partial x}\right).$$

If we use $\gamma = \rho g$, our three equations are

$$0 = -\frac{\partial p}{\partial x} - \gamma\left(\frac{\partial z}{\partial x}\right) + \mu\left(\frac{d^2 u}{dy^2}\right), \quad 0 = -\frac{\partial p}{\partial y} - \gamma\frac{\partial z}{\partial y}, \quad \text{and} \quad 0 = -\frac{\partial p}{\partial z'}.$$

The second of these equations shows that

$$0 = -\frac{\partial}{\partial y}(p + \gamma z),$$

or $p + \gamma z$ is not a function of y. Since $p + \gamma z$ is also not a function of z', it is a function of only x. Therefore, all that remains of the three Navier-Stokes equations is

$$0 = -\frac{d}{dx}(p + \gamma z) + \mu\left(\frac{d^2 u}{dy^2}\right). \qquad \textbf{Answer}$$

Discussion
Note that we have developed two other equations:

$$\frac{\partial p}{\partial z} = -\gamma \quad \text{and} \quad \frac{\partial p}{\partial z'} = 0.$$

The first shows that the static pressure decreases with elevation and is identical to Eq. (2.3). However, Eq. (2.3) applies to a static fluid while the above equation applies to a specific flow situation. The second equation simply states that the static pressure is constant in a horizontal plane if there is no flow in that direction.

4.3 Boundary and Initial Conditions for Flow Problems

If we intend to solve differential equations to predict the details of a particular flow, we must specify *boundary conditions*. If the flow is unsteady, we must also specify *initial conditions*. Boundary conditions are values of the fluid velocity and properties at the boundaries (walls, inlets, outlets) of the region in which we are trying to calculate the flow. Initial conditions are the values of the fluid velocity and properties at all points within the region at the initial instant of time. Mathematically, boundary and initial conditions are necessary to evaluate any constants of integration in the solutions of the differential equations.

In this discussion, we limit our consideration to incompressible fluids with constant viscosity. In this case, we need boundary and initial conditions for pressure and velocity. Compressible fluids or fluids with variable viscosity require not only additional boundary and initial conditions for the temperature and density but also a differential equation representing the first law of thermodynamics, an equation of state, and, usually, a constitutive relation to relate heat transfer to temperature. Since we have not yet considered these equations,* we will not consider the necessary boundary and initial conditions either.

Consider the problem of predicting the velocity and pressure field for flow in a nozzle, shown in Fig. 4.15. Suppose that we are interested only in the flow in the region between the inlet and the outlet, planes 1 and 2. The fluid motion in this region is governed by the continuity and Navier-Stokes equations. If the flow is unsteady, the initial conditions are

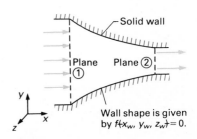

Figure 4.15 Flow through a nozzle.

$$\vec{V}(x, y, z, t = 0) = \vec{V}_i(x, y, z)$$

and

$$p(x, y, z, t = 0) = p_i(x, y, z),$$

where V_i and p_i are known functions. It is customary to set $t = 0$ at the beginning of the calculation. Initial conditions are usually relatively easy to determine. If we were interested in evaluating the start-up of flow from rest, we would probably put $V_i = 0$ everywhere; p_i might be uniform or there might be a hydrostatic pressure distribution.

Next we consider the boundary conditions. The boundaries of the flow field are the inlet (plane 1), the outlet (plane 2), and the side walls of the nozzle. The mathematical character of the governing equations is such that we must specify values of the velocity at all points on the entire boundary. At the solid walls, there can be no component of velocity perpendicular to the wall:

$$V_{\text{perpendicular to wall}} = V_n = 0.$$

Also, if the fluid is viscous, there is no slip at the wall,[†] that is,

$$V_{\text{tangent to wall}} = V_t = 0.$$

Together these conditions give

$$\vec{V}(x_w, y_w, z_w, t) = 0,$$

where x_w, y_w, z_w are the coordinates of the nozzle wall and satisfy some relation that describes the nozzle geometry, that is,

$$f(x_w, y_w, z_w) = 0.$$

At the inlet and outlet planes, we must specify the pressure as well as the velocity:

$$\vec{V}_{\text{at plane 1}} = \vec{V}_1(x_1, y_1, z_1, t), \qquad p_{\text{at plane 1}} = p_1(x_1, y_1, z_1, t)$$

and

$$\vec{V}_{\text{at plane 2}} = \vec{V}_2(x_2, y_2, z_2, t), \qquad p_{\text{at plane 2}} = p_2(x_2, y_2, z_2, t).$$

---

* We will never get around to considering the differential forms of these equations in this book since this book is about fluid mechanics rather than heat transfer. We will consider one-dimensional flow of compressible fluids in Chapter 10.

[†] See Section 1.3.4 for a discussion of the no-slip condition.

These conditions are not specific to our particular nozzle; in fact, we can generalize them to all flows:

- At a solid boundary the normal component of velocity relative to the boundary is zero. **(4.60)**
- At a solid boundary the tangential component of velocity relative to the boundary is zero. **(4.61)**
- At an inlet the fluid pressure and velocity distributions must be known as functions of time. **(4.62)**
- At an outlet the fluid pressure and velocity distributions must be known as functions of time. **(4.63)**

Notice that we have inserted the phrase "relative to the boundary" in the first two conditions. This accounts for moving solid boundaries as well as fixed solid boundaries.

Some special situations must be mentioned. First, suppose that we are interested in predicting the flow around an airplane cruising at constant speed V_∞ through the atmosphere, as shown in Fig. 4.16(a). It is possible to consider this as a steady flow if we attach a coordinate system to the airplane. In this system, the airplane is at rest and the air approaches the airplane at constant speed, V_∞, as shown in Fig. 4.16(b). For this flow, the solid surfaces of the airplane are obviously one boundary, but where are the "inlet" and "outlet" boundaries? These boundaries are far upstream and far downstream. Mathematically, we put them at $\pm \infty$. There are also boundaries far above and far below the airplane. Mathematically, we handle these conditions by writing

$$\lim_{x,y,z \to \infty} \vec{V}(x, y, z) = V_\infty \hat{i} \quad \text{and} \quad \lim_{x,y,z \to \infty} p(x, y, z) = p_\infty.$$

Next suppose that we decide to analyze a particular flow using the inviscid fluid assumption. The governing equations are now the Euler equations rather than the Navier-Stokes equations. This immediately leads to a mathematical problem. The Euler equations are of lower order than the Navier-Stokes equations because the Euler equations do not contain any second derivatives. Since the Euler equations are of lower order, it is impossible to enforce all of the boundary conditions, Eqs. (4.60–4.63). We must drop one of these conditions. The only sensible condition to drop is Eq. (4.61), the no-slip condition, since the non-slip of fluid at a solid boundary is actually associated with viscosity. For

Figure 4.16 Flow around an airplane. (a) Actual case: Airplane moves at constant speed V_∞ through still air. (b) Flow from coordinate system fixed to airplane: Air approaches airplane at constant speed V_∞.

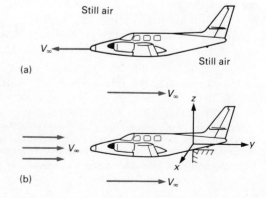

an inviscid fluid, only conditions (4.60), (4.62), and (4.63) are applied. The tangential (slip) velocity at a solid surface will be calculated as part of the solution and *will not* be zero.

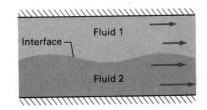

Figure 4.17 Flow of two imiscible fluids that form an interface between them.

If there is an internal boundary in the flow field, we must pay special attention to boundary conditions. Consider the flow of two immiscible fluids that form an interface between them (Fig. 4.17). This may represent a layer of oil over water or a free surface with air over water. The flow fields of the two fluids must be patched together at the interface by means of boundary conditions, which must express the conditions of continuity of mass and balance of forces at the interface. In words, the conditions are as follows:

- The velocities of each fluid normal to the interface at the interface must be equal. **(4.64)**
- The shear stresses in each fluid must be equal at the interface. **(4.65)**
- The pressures in each fluid at the interface must be equal except for surface tension effects. **(4.66)**

These can be very complicated conditions, and we will not consider them further here* except for two special cases. The first is an air-water interface in an open channel (Fig. 4.18). In this case we usually simplify the conditions to

$$\left.\begin{array}{r} \tau = 0 \\ p_{\text{liquid}} = p_{\text{air}} \end{array}\right\} \quad \text{at the interface.} \qquad \begin{array}{l} \textbf{(4.65a)} \\ \textbf{(4.66a)} \end{array}$$

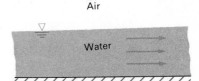

Figure 4.18 Flow in an open channel with an air-water interface.

The second special case involves the issue of a jet into a large quiescent body of fluid (see Fig. 4.19). This may represent water issuing from a hose into still air, air escaping from an air hose into the atmosphere, or water issuing from a pipe submerged in a lake or ocean. Consider the pressure of the fluid in the jet immediately at the plane of issue. Condition (4.66) says that the pressure just inside the jet must equal the pressure just outside the jet; that is, the pressure inside the jet must equal the pressure of the quiescent body of fluid, except for surface tension effects. There is an exception to this condition. If the fluid in the jet is compressible and moving at a speed greater than the speed of sound, then it is possible for the jet pressure to adjust to the exhaust pressure by means of compression or expansion waves. In most cases encountered in engineering practice, surface tension effects are negligible and the fluid moves at a speed less than the speed of sound, so we usually are justified in simply setting the pressure just inside the jet equal to the pressure of the exhaust region. If the streamlines in the exiting jet are straight, then the pressure is uniform (or hydrostatic) across the entire jet and equal to the exhaust region pressure. Note that this condition applies irrespective of whether the jet and exhaust region fluids are similar or dissimilar (i.e., air-air or water-air). We can formally write this jet exit condition as

$$p_e = p_E. \qquad \textbf{(4.66b)}$$

Note that we have already used this condition in Example 4.3.

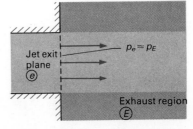

Figure 4.19 Jet of fluid issuing into a large quiescent "exhaust region."

---

* See [1] if you are interested in further discussion.

When we introduced the need for boundary (and initial) conditions, our justification was primarily mathematical; however, if we consider the facts expressed by the boundary conditions, we see that these conditions are necessary to make the mathematical solutions to the differential equations representative of the true physical behavior of the fluid. Boundary conditions perform another very important task for us. If we consider, for example, the Navier-Stokes equations, we notice that they are exactly the same for all flow geometries. These same equations apply to the nozzle flow of Fig. 4.15, the airplane flow of Fig. 4.16, and the two-fluid flow of Fig. 4.17 (with different values of density and viscosity for each fluid), and all of the different flows of Figs. 3.1 and 3.2. Only the boundary conditions distinguish one flow from another. For the nozzle flow (Fig. 4.15), different nozzles have different functions $f(x_w, y_w, z_w)$ describing the nozzle contour, and this information is fed into the problem only by means of the boundary conditions. Specifying proper boundary conditions for a flow problem is as important as using the right governing equations.

EXAMPLE 4.10   Illustrates how to write boundary conditions

Write the necessary boundary conditions for the steady, two-directional, two-dimensional flow of a constant-density, Newtonian fluid flowing in the entrance region between the two wide and parallel flat plates in Fig. E4.10.

Solution

   Given

Two wide, parallel, flat plates with $u(x, y)$ and $v(x, y)$
Figure E4.10
Constant-density, Newtonian fluid

   Find

Necessary boundary conditions to find $u(x, y)$ and $v(x, y)$.

   Solution

   Boundary conditions are necessary for both u and v at the inlet 1 and the outlet 2, at the upper solid boundary, and at the lower solid boundary. See Eqs. (4.60) to (4.63). At the inlet

Figure E4.10 Entrance region between two wide and parallel flat plates.

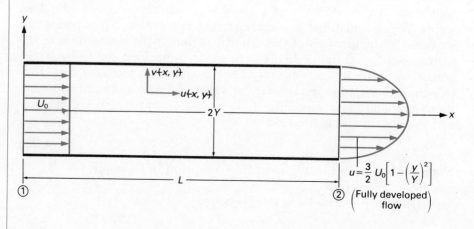

$$\begin{vmatrix} u(0, y) = U_0 \\ v(0, y) = 0 \\ p(0, y) = p_0 - \gamma y \end{vmatrix} \quad inlet$$

Answer ◀

where p_0 is the static pressure at $y = 0$.

At the outlet

$$\begin{vmatrix} u(L, y) = \dfrac{3}{2} U_0\left[1 - \left(\dfrac{y}{Y}\right)^2\right] \\ v(L, y) = 0 \\ p(L, y) = p_L - \gamma y \end{vmatrix} \quad outlet$$

Answer ◀

where p_L is the static pressure at $y = 0$.

At the upper solid boundary

$$\begin{vmatrix} u(x, Y) = 0 \\ v(x, Y) = 0. \end{vmatrix} \quad top$$

Answer ◀

At the lower solid boundary

$$\begin{vmatrix} u(x, -Y) = 0 \\ v(x, -Y) = 0. \end{vmatrix} \quad bottom$$

Answer ◀

Discussion

If the fluid were inviscid, the boundary conditions at the upper and lower boundaries would be $v(x, Y) = 0$ and $v(x, -Y) = 0$. There would be no boundary conditions on u at the upper and lower boundaries.

The flow downstream of section 2 does not change with x and is fully developed. The region of flow investigated in Fig. 4.12 was also considered fully developed; there was a similar entrance region upstream of this fully developed region.

The requirement that the pressure p_L be known at $x = L$ is rather stringent. This requirement is often replaced by the fully developed flow condition

$$\frac{\partial p}{\partial x} = \text{Constant} \qquad \text{for } x \geq L$$

in problems of this type.

Problems

Extension and Generalization

1. Integrate Bernoulli's equation for compressible flow, Eq. (4.9), for an ideal gas undergoing an isothermal process.

2. Integrate Bernoulli's equation for compressible flow, Eq. (4.9), for a fluid whose pressure p and density ρ obey the equation

$$\frac{p}{\rho^n} = C,$$

where C and n are constants.

3. Start with Newton's second law for a fluid element and derive an equation for the velocity profile of a viscous fluid entrained between two very wide, very long, horizontal, parallel flat plates. The plates are separated by a distance $2Y$. The upper plate moves with a constant horizontal velocity U_∞, and the lower plate is fixed. [*Hint:* Assume fully developed flow with an axial pressure gradient of zero.]

4. Derive an equation for the velocity profile of a viscous fluid entrained between two very long concentric cylinders. The inner radius is R_i and the outer radius is R_0. The inner cylinder is stationary and the outer cylinder rotates with a constant peripheral speed of U_0.

5. Show that the velocity $u(y)$ for the simple model of the slipper bearing in Fig. P4.1 is given by

$$u = \frac{1}{2\mu} \frac{dp}{dx}(y^2 - hy) + U\left(\frac{y}{h}\right).$$

Assume the bearing is infinitely wide. Find an expression for the load L per unit width that the bearing in Fig. P4.1 can support.

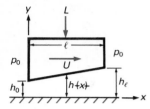

Figure P4.1

6. A liquid of constant density ρ and constant viscosity μ flows down a very wide, very long inclined flat plate. The plate makes an angle θ with the horizontal. The velocity components do not change in the direction of the plate, and the fluid depth h normal to the plate is constant. There is negligible shear stress by the air on the fluid. Find the velocity profile $u(y)$, where u is the velocity parallel to the plate and y is measured perpendicular to the plate. Find an expression for the volume flow rate per unit width of the plate.

7. An infinitely wide, long flat plate is immersed in a still fluid of constant viscosity μ and constant density ρ. At time $t = 0$, the plate is made to move with a constant velocity U in its own plane. Find the velocity $u(y, t)$ of the fluid surrounding the plate, where y is the coordinate perpendicular to the plate and u is the velocity parallel to the plate. The boundary conditions are

$$u(0, t) = U \quad \text{for } t > 0,$$

and

$$v(0, t) = 0 \quad \text{for } t > 0,$$

where v is the velocity component in the y-direction. The initial condition is

$$u(y, 0) = 0 \quad \text{for } y > 0.$$

[*Hint:* The resulting partial differential equation can be converted to an ordinary differential equation by defining an independent variable η by

$$\eta = \frac{y}{\sqrt{2vt}}.$$

8. In Example 1.3 in Chapter 1, a working formula for a concentric cylinder viscometer was devel-

oped. The result was

$$\mu = \frac{\mathcal{T}h}{2\pi H\omega R_1^3}.$$

Torque $\mathcal{T}$ causes the inner cylinder to rotate at a constant angular velocity ω. The viscometer has a height H, a gap width h, and radii R_1 and R_2, where $h \ll R_1$ or R_2. Develop an improved working formula relating the fluid viscosity μ and the torque $\mathcal{T}$ by considering the viscous friction between the inner cylinder and the fluid along both the side and the bottom of the cylinder. The thickness of the viscous fluid below the inner cylinder is h_1.

Lower-Order Application

9. A 15-mph wind is blowing against a closed door that opens outward. If the door measures 3 ft by 7 ft, how much force must be applied to start opening the door? Assume the air pressure on the inside of the door is equal to the local atmospheric pressure.

10. The steady axial velocity of flow through a nozzle is given by

$$u = \frac{U_0}{\left(1 - \dfrac{x}{2\ell}\right)^2},$$

where U_0 is the entrance velocity at $x = 0$ and ℓ is the length of the nozzle. The flow is approximately incompressible and irrotational. Gravity acts perpendicular to the axial or x-direction. Find the pressure drop across the nozzle (i.e., from $x = 0$ to $x = \ell$) for $U_0 = 5$ m/s and $\ell = 3.0$ m.

11. The velocity of the water leaving the faucet on the first floor of a home is 1.2 ft/sec. What is the maximum possible velocity from an identical faucet 10 ft higher on the second floor?

12. The pressure and average velocity at point A in the pipe in Fig. P4.2 are 16.0 psia and 4.0 ft/sec, respectively. Find the height h and the pressure and average velocity at point B.

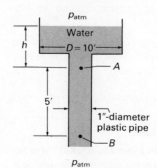

Figure P4.2

13. The pressure and average velocity at point A in the pipe in Fig. P4.3 are 120 kPa and 10 m/s, respectively. Find the average velocity at point B.

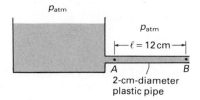

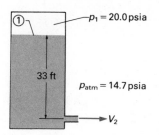

Figure P4.3

14. A gate valve holds back water in a dam whose water level is 45 m above the valve. Find the water pressure on the closed gate valve. What is the maximum velocity through the valve if the valve is wide open?

15. What pressure, p_1, is necessary to double the flow rate leaving the tank in Fig. P4.4?

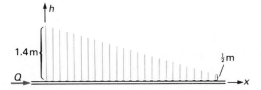

Figure P4.4

16. Fortified wine containing five parts wine with a specific gravity of 1.05 and one part ethanol with a specific gravity of 0.85 flows through a Venturi having an inlet diameter of 2.0 cm and a throat diameter of 0.50 cm. Find the pressure drop between the inlet and throat of the Venturi.

17. What is the minimum height for an oil ($S = 0.75$) manometer to measure airplane speeds up to 30 m/s at altitudes up to 1500 m? The manometer is connected to a Pitot-static tube.

18. The water leaving the lawn sprinkler hose in Fig. P4.5 reaches the heights shown. Each of the 25 nozzles has an exit area of 1.5 cm². Find the volume flow rate of water entering the sprinkler.

Figure P4.5

19. Figure P4.6 shows a duct for testing a centrifugal fan. Air is drawn from the atmosphere ($p_{atm} = 14.7$

psia, $T_{atm} = 70°F$). The inlet box is 4 ft by 2 ft. At section 1, the duct is 2.5 ft square. At section 2, the duct is circular with a 3-ft diameter. A water manometer in the inlet box measures a static pressure of -2.0 in of water. Calculate the volume flow rate of air into the fan and the average fluid velocity at both sections 1 and 2. Assume constant density.

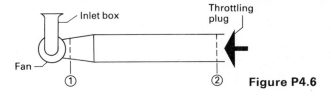

Figure P4.6

20. Find the water height h_B in tank B of Fig. P4.7.

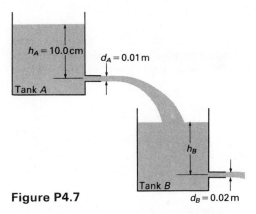

Figure P4.7

21. Water is leaving a hole in the side of a tank with a horizontal velocity of 20 ft/sec. The hole is 25 ft above the ground. How far will the water travel horizontally before it hits the ground?

22. Consider the Pitot-static tube in Fig. 4.7 on page 196; $p_A - p_B = 1.4$ psi, and the flowing fluid is water. Find the fluid velocity V_1.

23. Consider the two tanks in Fig. P4.8. Tank A initially contains salt water with 35 grams of salt per

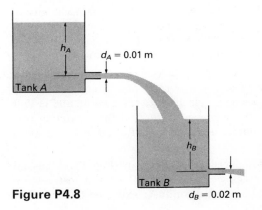

Figure P4.8

liter solution and $h_A = 10.0$ m. Tank B initially contains pure water with $h_B = 10.0$ m. Both tanks have a diameter of 8.0 m. Find the salt concentration of tank B in grams per liter of solution at 10 s after the initial time $t = 0$. Assume the contents of both tanks are well mixed, so the salt is uniformly distributed throughout each tank.

24. Find the rate at which oil ($S = 0.85$) flows out of the tank in Fig. P4.9.

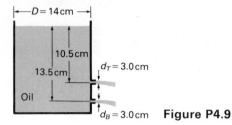

Figure P4.9

25. The hovercraft shown in Fig. P4.10 takes in air at the top and discharges it at the bottom. Find the maximum possible weight of the hovercraft for $Q = 40$ m³/min.

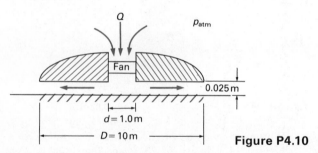

Figure P4.10

26. Would the hovercraft in Problem 25 work if the fan were reversed and the air drawn out from underneath the hovercraft and up the center tube? Would the maximum possible weight of the hovercraft be larger or smaller? Explain your answer.

27. A fire hose has a nozzle outlet velocity of 30 mph. What is the maximum height the water can reach?

28. Water leaves a pump at 700 kPa and a velocity of 12 m/s. It then enters a diffuser to increase its pressure to 1400 kPa. What must be the ratio of the outlet area to the inlet area of the diffuser?

29. A sump pump is submerged in 60°F ordinary water that vaporizes at a pressure of 0.256 psia. The pump inlet has an inside diameter of 2.067 in. and is 15 ft below the water surface. Find the maximum possible flow rate before cavitation (vaporization) occurs.

30. The low-speed wind tunnel in Fig. P4.11 has an inlet cross-sectional area of 10 ft², a test section area of 5 ft², and a diffuser exit area of 10 ft². The

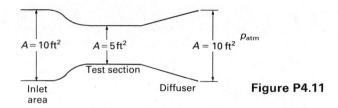

Figure P4.11

atmospheric pressure is 14.690 psia. Find the following for a test section speed of 120 mph:
a) inlet area velocity,
b) diffuser exit area velocity,
c) test section absolute pressure,
d) inlet area absolute pressure.

31. An airplane wing is placed in a wind tunnel and the pressures indicated in Fig. P4.12 are measured. Find the corresponding velocities if the upstream air velocity and pressure are 170 km/hr and 101,200 Pa.

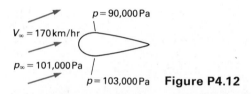

Figure P4.12

32. The cap is removed from the fire hydrant in Fig. P4.13 and water shoots out a horizontal distance of 70 ft before hitting the ground. The vertical distance between the ground and the bottom of the 6-in.-inside-diameter pipe is 14 in. Estimate the distance ℓ' and find the volume flow rate leaving the hydrant.

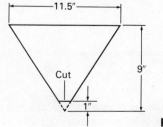

Figure P4.13

33. A newspaper is folded into the shape of a conical cup as shown in Fig. P4.14. The cup is then filled with water and the bottom of the cup cut as shown.

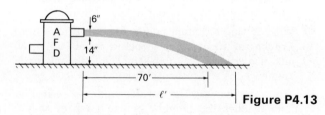

Figure P4.14

Find the time for all of the water to drain out of the cup.

34. Someone siphoned 15 gal of gasoline from a gas tank in the middle of the night. The gas tank measures 12 in. wide, 24 in. long, and 18 in. high and was full when the thief started. If the siphoning tube has an inside diameter of 1/2 in., find the minimum amount of time needed to siphon the 15 gal out of the tank. Assume that at any instant of time the steady-state equations are adequate to predict the gasoline velocity in the siphon tube. Also assume that the end of the siphon tube outside the gas tank is at the same level as the bottom of the tank. You may consider the gasoline inviscid.

35. The water clock (clepsydra) shown in Fig. P4.15 is an ancient device for measuring time by the fall of the water level in a large glass container. The water slowly drains out through a small hole in the bottom. Determine the approximate shape $R(z)$ of a container of circular cross section such that the water level falls at a constant rate of 5 cm/hr if the drain hole is 1.5 mm in diameter. What size container (h and R_0) permits the clock to run for 24 hr without refilling?

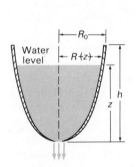

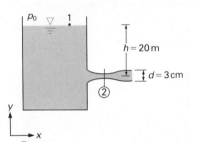

Figure P4.15   **Figure P4.16**

36. The sealed tank in Fig. P4.16 contains water with air at atmospheric pressure above the water. Assuming inviscid water flow and constant temperature of the air, find the amount of water that will flow out of the tank.

37. Water enters a pump through a 20-cm-diameter pipe and leaves the pump through a 12-cm-diameter pipe, as shown in Fig. P4.17. If the mercury column

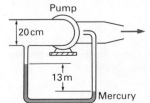

Mercury   **Figure P4.17**

difference is 13 m, what flow rate is delivered by the pump? Note that upstream of the pump the manometer is connected to a static-pressure tap and downstream to a stagnation probe. The downstream static pressure is 10 m of mercury higher than the upstream static pressure.

38. A flow of water occurs in the system shown in Fig. P4.18. Assume frictionless flow.
    a) Calculate the rate Q at which water must be added at the inlet to maintain the 16-ft head.
    b) Calculate the height h in feet of water in the static-pressure tube.

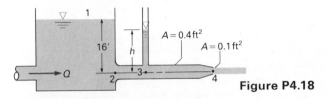

Figure P4.18

39. Water flows from the storage tank through the converging-diverging passage and to the atmosphere as shown in Fig. P4.19. The water is at 20°C and boils at a pressure of 2.34 kPa. Find the area A_2 such that the water boils at A_2.

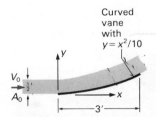

Figure P4.19

40. A water jet flows up the vane shown in Fig. P4.20. The inlet area A_0 and velocity V_0 are 12 in^2 and 12 in./sec, respectively. Assuming inviscid flow, find the exit area A_e and exit velocity V_e if the jet width remains constant at 2 in.

Curved vane with $y = x^2/10$

Figure P4.20

41. Find the force exerted on the vane by the water jet in Problem 40. Use the equation normal to flow to find the pressure distribution on the vane. Assume a uniform velocity over each area normal to flow.

42. A liquid jet has a velocity of V_0 and makes an angle θ with the horizontal. Find the value of the angle θ for the liquid to travel the maximum horizontal distance along level ground.

43. A liquid jet has a velocity of 100 ft/sec and makes an angle of 45° with the horizontal. Find the maximum possible height that can be attained by the fluid.

44. Water flowing over a waterfall has a velocity of 4 m/s at its crest and a drop of 30 m. Find the maximum possible velocity of the water.

45. A U-tube manometer indicates the difference in the stagnation pressures across a pump. The water flow rate through the pump is 100 gal/min, the inlet pipe is $2\frac{1}{2}$-in.-nominal-diameter, type L copper pipe, the discharge pipe is a 4-in.-nominal-diameter, type L copper pipe, and the pump has a static pressure rise of 45 psi. Find the difference in the mercury levels of the U-tube manometer. There is no elevation difference across the pump.

46. An airship flies at a speed of 30 km/hr into a 15 km/hr oncoming wind. Find the stagnation pressure at the nose of the airship if the air temperature is 20°C and the ambient static pressure is 100 kPa.

47. The water nozzle in Fig. P4.21 is directed downward. Find the water flow rate.

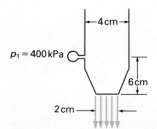

Figure P4.21

48. The water from a fire hose reaches a peak of 90 ft above the nozzle with a velocity of 20 ft/sec. Find the velocity of the water leaving the nozzle and the radius of curvature of the water stream at its peak.

49. A stream of grain and air leave a combine with a velocity of 10 ft/sec through an area of 15 ft². It is desired to reduce the area to increase the velocity to 15 ft/sec at the exit where the pressure is the atmospheric pressure. Assuming a constant density of 40 lbm/ft³ for the grain and air, find the exit area and the pressure at the 15-ft² area.

50. Water flows at the rate of 1200 ft³/sec through the aqueduct in Fig. P4.22. The aqueduct is 10 ft wide

at point A and 15 ft wide at point B. Find the height and average velocity at point B.

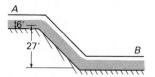

Figure P4.22

51. Consider the fluid motion that forms above an open drain as a free vortex (Fig. E4.8 on page 201). For $V_A = 0.50$ m/s, $V_B = 0.10$ m/s, $r_A = 0.20$ m, and $r_B = 1.0$ m, find the difference in the elevation of the surface between A and B.

52. Consider a cyclone as a free vortex. The velocity is 7.5 km/hr at a radius of 30 km from the center, where the atmospheric pressure is 105 kPa. Find the atmospheric pressure at a radius of 3 km from the center of the cyclone for the same elevation.

Higher-Order Application

53. The velocity of the water leaving the faucet on the first floor of a home is 1.2 ft/sec. What would be the maximum possible velocity from an identical faucet on the second floor? What about an identical faucet on the third floor?

54. A puck for an air hockey game weighs 0.060 lb and has a diameter of 4.0 in. The air space maintained between the hockey puck and the table is 0.005 in. Make a reasonable estimate of the air flow rate per hole to support the puck. The holes are spaced 1/2 in. apart.

55. A beer keg roughly measures 27 in. high by 16 in. in diameter. It contains $15\frac{1}{2}$ gal of beer and weighs 172 lb full and 18 lb empty. At a party, the keg is sitting upright on the floor; a 6-ft-tall person holds the spigot 5 ft above the ground and starts to fill a 1/2-gal pitcher. Estimate the air pressure required in the keg for beer to flow out at the rate of 1/4 gal/min.

56. Sludge, which is formed from foreign impurities and oil, is denser than oil. A rotating cylinder is used as a centrifugal oil separator. Where on the bottom of the rotating cylinder should the drain pipes be located to remove the sludge? Where should the drain pipes be located to remove the oil?

57. Water drains from the higher of the two large tanks and through the Venturi in the discharge line in Fig. P4.23. Find the range of throat diameter d such that water will flow from the lower tank into the Venturi for the given heights H and h.

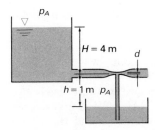

Figure P4.23

58. A commercial spray jar used to spray lawns and gardens is shown in Fig. P4.24. Find the maximum permissible pressure at the throat of the Venturi to empty the contents of the jar. Then estimate the shortest time to empty the contents of the jar if this throat pressure is maintained.

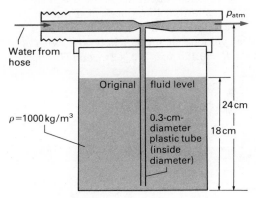

Figure P4.24

59. A centrifugal water pump is advertised to have a shutoff head of 95 m. (The shutoff head occurs at zero flow rate.) The pump impeller has a radius of 1.0 m and rotates at a speed of 900 Hz. Is the advertisement correct?

60. Lubricating oil at 20°C fills the space between the two concentric cylinders in Fig. P4.25. The outer cylinder is fixed, and the inner cylinder rotates at 100 Hz. Neglect end effects and find $v(r)$.

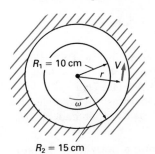

Figure P4.25

61. The Navier-Stokes equations for steady-flow, constant-density, and constant-viscosity conditions are given in Appendix H. Simplify these equations

to determine the differential equation for both the rotational velocity $V_\theta(r)$ and the axial velocity $V_x(r)$ for the setup in Fig. P4.26. Assume a fully developed ($\neq f(x)$) axial flow.

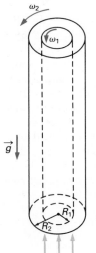

Q = Volume flow rate   **Figure P4.26**

62. A Bingham plastic is a fluid in which the stress τ is related to the rate of strain du/dy by

$$\tau = \tau_0 + \mu\left(\frac{du}{dy}\right),$$

where τ_0 and μ are constants. Consider the flow of a Bingham plastic between two fixed, horizontal, infinitely wide, flat plates. For fully developed flow with $\partial u/\partial x = 0$ and $\partial p/\partial x$ constant, find $u(y)$.

63. Consider the journal bearing shown in Fig. P4.27. R is the journal radius, $h(x)$ the oil film thickness, U the peripheral velocity of the journal, and b the bearing width. Assume $h \ll R$, constant fluid properties, and negligible gravity and use Cartesian coordinates. Also assume constant fluid density and viscosity and negligible gravity. Develop the

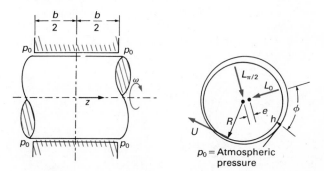

p_0 = Atmospheric pressure

Figure P4.27

Reynolds equation

$$\frac{1}{6\mu}\left(\frac{\partial}{\partial x}\right)h^3\left(\frac{\partial p}{\partial x}\right) + \frac{1}{6\mu}\left(\frac{\partial}{\partial z}\right)h^3\left(\frac{\partial p}{\partial z}\right) = U\left(\frac{dh}{dx}\right),$$

where $x = R\phi$.

64. The bearing in Fig. P4.28 consists of two parallel discs of radius R separated from each other by a small distance h ($h \ll R$). The lower disc is made of a porous material, and an incompressible, viscous fluid is pumped through it and into the gap so that the space between the discs is completely filled. The pores of the lower disc are very closely spaced so that the velocity of the fluid as it leaves the surface of the porous disc may be regarded as a uniform value w_0 that is very small compared to the mean radial velocity. Find the radial velocity u and the load L that the bearing can support as a function of w_0, R, and h.

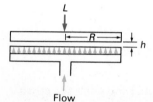

Figure P4.28

65. Consider a spherical bubble or balloon in an infinite Newtonian liquid where the radial motion of the surface of the bubble or balloon $R(t)$ is known as a function of time. Find the liquid velocity as a function of time and the radial position from the center of the bubble or balloon.

66. A viscosity motor/pump is shown in Fig. P4.29. The rotor is concentric within a stationary housing. The clearance h between the housing and the rotor is very small compared to the width w and radius R of the rotor. A seal divides the clearance space as shown. When the device operates as a motor, the viscous liquid is introduced at the high pressure side of the seal and leaves at the low pressure side of the seal. This flow causes the rotor to turn and produce power. For $R = 1.0$ ft, $h = 0.50$ in., $w = 1.0$ ft, and $\omega = 10$ rpm, find the power output of

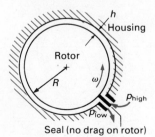

Seal (no drag on rotor)   **Figure P4.29**

the rotor for a flow rate of 1.0 ft³/min. The fluid is 0°F, SAE 10W crankcase oil.

67. Consider operating the viscosity motor/pump in Problem 66 as a pump. The input power to the rotor is $\dot{W} = 0.25$ hp, $w = 1.0$ ft, $\omega = 10$ rpm, $R = 1.0$ ft, and $h = 0.50$ in. The developed pressure rise is 15.0 psi, and the fluid is 0°F, SAE 10W crankcase oil. Find the oil volume flow rate Q and the useful power ($Q\Delta p$) transmitted to the fluid, where Δp is the pressure rise. How would the overall efficiency of this pump ($Q\Delta p/\dot{W}$) compare with that of a conventional pump?

68. A liquid is rotating at level A in a stationary, flat-bottom cylinder as shown by the solid concentric circles in Fig. P4.30. The fluid below this level follows a path somewhat like the dashed lines in the figure. Find the fluid velocity below level A if the tangential velocity at level A is $v = r\omega$, with $\omega = 0.50$ rad/s, $R = 0.125$ m, $h = 0.002$ m, and $H = 0.25$ m.

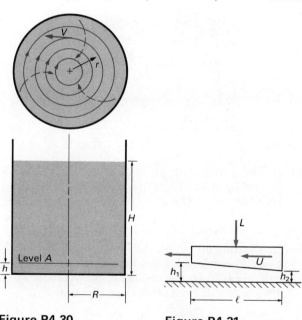

Figure P4.30   **Figure P4.31**

69. The simplified slipper bearing in Fig. P4.31 has a velocity $U = 4.0$ m/s, length $\ell = 10.0$ in., width $w = 10.0$ cm, $h_1 = 0.5$ cm, and $h_2 = 0.25$ cm. SAE 20W crankcase oil at $-18°$C separates the pad from the lower surface. Find the load L that can be supported by the pad.

70. A flat block is pulled along a horizontal flat surface by a horizontal string perpendicular to one of the sides. The block measures 1.0 m by 1.0 m, has a mass of 10.0 kg and a constant velocity of 1.0 m/s, and is separated from the flat surface by a 0.10-cm-

thick oil layer of $-18°C$ SAE 20W crankcase oil. Find the coefficient of friction for the block. Does this coefficient of friction change with the velocity?

71. Consider the velocity field

$$u = A(x^2 - y^2), \qquad v = -2Axy, \qquad w = 0,$$

with $g_z = -g$. Determine the conditions under which the Navier-Stokes equations are satisfied. Consider constant viscosity μ.

72. The disc shown in Fig. P4.32 has a diameter D of 1 m and a rotational speed of 1800 rpm. It is positioned 4 mm from a solid boundary. The spacing is filled with SAE 30W oil at $-18°C$. Find the torque required to overcome the frictional resistance of the oil on the disc.

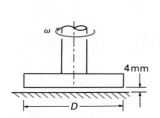

Figure P4.32

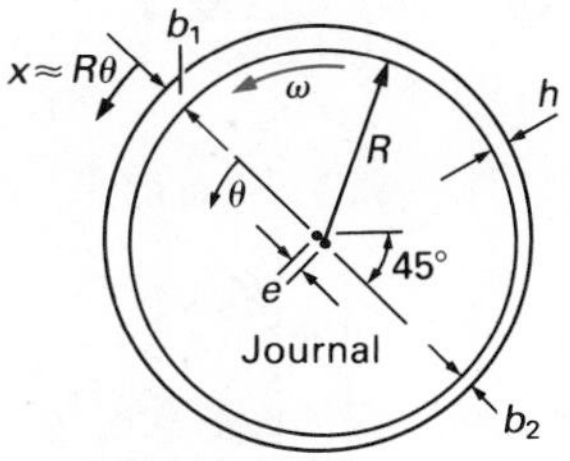

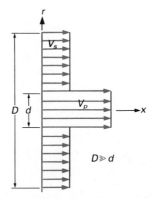

Figure P4.33

73. The spacing h between a rotating journal and its bearing varies around the circumference of the bearing as shown in Fig. P4.33. Assume that this spacing may be taken as a linear relationship

$$h = b_1 - \left(\frac{b_1 - b_2}{\ell}\right)R\theta \qquad R\theta > 0$$

$$= b_1 + \left(\frac{b_1 - b_2}{\ell}\right)R\theta \qquad R\theta < 0,$$

where

$$b_1 = b_2 + 2e \qquad \text{and} \qquad \ell = \pi R.$$

Take e as the eccentricity of the journal relative to the center of the bearing, R the journal radius, $x \approx R\theta$, and $h \ll R$. Find the velocity $u(y)$ for a 210°F, SAE 50W oil separating the journal and the bearing. Next find the load L that the journal can support if the pressure at $\theta = 0$ is 120,000 N/m², $e = 0.005$ mm, $b_2 = 0.002$ mm, $R = 4.0$ cm, and $\omega = 1000$ Hz.

74. Find the torque per unit length required to rotate the shaft against the resistance of the oil in the journal bearing in Problem 73.

75. The basic equations for the flow of a constant-density, constant-viscosity fluid in the entrance region of a circular pipe are

$$V_x\left(\frac{\partial V_x}{\partial x}\right) + V_r\left(\frac{\partial V_x}{\partial r}\right) = \frac{1}{\rho}\left(\frac{\partial p}{\partial x}\right) + v\left[\frac{\partial^2 V_x}{\partial r^2} + \frac{1}{r}\left(\frac{\partial V_x}{\partial r}\right)\right]$$

and

$$\frac{\partial V_x}{\partial x} + \frac{1}{r}\left(\frac{\partial(rV_r)}{\partial r}\right) = 0.$$

Referring to Fig. P4.34, write the appropriate boundary conditions.

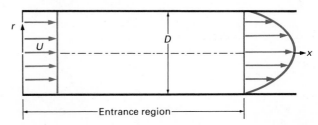

Figure P4.34

76. Figure P4.35 shows a jet of diameter d and velocity V mixed with the slower-velocity stream of outer diameter D and velocity v_s. An incompressible Navier-Stokes equation and the continuity equation are used to find the jet velocity as it mixes with the slower stream. These equations are

$$V_x\left(\frac{\partial V_x}{\partial x}\right) + V_r\left(\frac{\partial V_x}{\partial r}\right) = \frac{v}{r}\left(\frac{\partial}{\partial r}\right)\left(\frac{\partial V_x}{\partial r}\right)$$

and

$$\frac{\partial V_x}{\partial x} + \frac{1}{r}\frac{\partial(rV_r)}{\partial r} = 0.$$

Find the appropriate boundary conditions.

Figure P4.35

77. Natural convection occurs in a fluid when gravitational forces are present and density variations arise due to temperature variations. For many

problems, pressure variations do not significantly change the density, and the density variations can be accounted for by using the coefficient of thermal expansion α_T. The corresponding equations for the problem described by Fig. P4.36 are

$$u\left(\frac{\partial u}{\partial x}\right) + v\left(\frac{\partial u}{\partial y}\right) = v\left(\frac{\partial^2 u}{\partial y^2}\right) + \alpha_T g_x(T - T_\infty),$$

$$\frac{\partial u}{\partial x} + \frac{\partial v}{\partial y} = 0,$$

$$u\left(\frac{\partial T}{\partial x}\right) + v\left(\frac{\partial T}{\partial y}\right) = \frac{k}{\rho c_p}\frac{\partial^2 T}{\partial y^2},$$

where k is called the thermal conductivity and T is the fluid temperature. Find the appropriate boundary conditions.

Comparison

78. Which of the two tanks in Fig. P4.37 will be completely drained first?

79. Repeat Problem 66 for an SAE 50W oil at 210°F. Is there any advantage to using this oil rather than the SAE 10W oil?

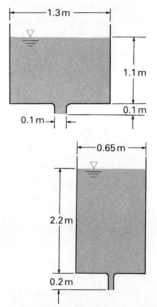

Figure P4.37

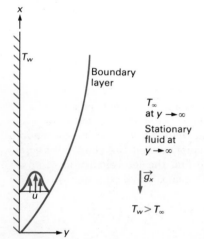

Figure P4.36

References

1. White, F. M., *Viscous Fluid Flow,* McGraw-Hill, New York, 1974.
2. Daily, J. W., and D. R. F. Harleman, *Fluid Dynamics,* Addison-Wesley, Reading, MA, 1966.
3. Hughes, W. F., *Introduction to Viscous Flow,* McGraw-Hill-Hemisphere, New York, 1979.
4. Schlichting, H., *Boundary Layer Theory* (7th ed.), McGraw-Hill, New York, 1980.
5. Roache, P. J., *Computational Fluid Dynamics,* Hermosa, Albuquerque, NM, 1976.
6. Patankar, S. V., *Numerical Heat Transfer and Fluid Flow,* McGraw-Hill-Hemisphere, New York, 1980.
7. White, F. M., *Fluid Mechanics,* McGraw-Hill, New York, 1979, pp. 230–235.

The Finite Control Volume Approach to Flow Analysis

5

In Chapter 3 we introduced a trio of approaches that engineers use to solve flow problems. Chapter 4 dealt with the most mathematical and exact method, the differential approach. The differential approach claims some outstanding successes (such as Bernoulli's equation); however, when applied to viscous flows or complicated geometries, the differential approach often promises more than it can deliver.

In this chapter, we consider the alternative analytical approach to flow problems, the finite control volume approach. The advantage of the finite control volume approach is that it can overlook the details of a particular flow and give relations between various gross features of the flow. Instead of detailed density and velocity predictions, we deal directly with flow rate. Instead of point-by-point prediction of fluid pressure, we deal with the net forces due to pressure. This "lumped" or "global" feature of the finite control volume approach is its greatest strength. Given a relatively small amount of information, the control volume approach can produce valuable results.

Very often it is necessary to introduce some simplifying assumptions (such as the one-dimensional, uniform-flow assumption) to complete a flow analysis. The control volume approach is much less sensitive to such assumptions than is the differential approach. The control volume approach does not have nearly as high a failure rate as the differential approach. With a clever choice of control volume, careful attention to the use of given information, and by introducing some reasonable assumptions and approximations, it is almost always possible to obtain at least a partial answer. In many cases, the approach produces the complete answer with a high degree of accuracy.

The finite control volume approach is a very important tool for flow analysis. Like any other single tool, it will not solve all your problems, but in many cases it is unquestionably the best tool available. By carefully studying this chapter and by solving a lot of the problems at the end of the chapter, you will learn how to use this tool and come to appreciate its advantages and limitations.

5.1 Control Volumes and Systems

The finite control volume approach applies the fundamental laws of mechanics and thermodynamics (Section 3.2) to a finite region of space (the control volume) containing a finite amount of fluid. If the fundamental law in question is a conservation principle such as conservation of mass or conservation of energy, all we need to do is make a balance between the amounts of some entity (mass or energy) entering the control volume, leaving the control volume, and stored in the control volume. We applied such a balance technique to develop the integral continuity equation in Section 3.4.2.

Not all the fundamental laws are conservation principles. Newton's laws of mechanics, for example, are not conservation principles, so we cannot simply write control volume balance statements for them. Newton's laws of mechanics apply to *a specific mass of fixed identity*. If you reexamine all of the formulations of Newton's second law in Chapter 4, you will note that we were always concentrating on a specific fluid particle, identified by its location in space, and equating the sum of forces on that particle to its rate of change of momentum.

Recall from your study of dynamics that Newton's laws can be applied to a finite collection of mass particles (a lump of mass) as well as to a single particle. In fluid mechanics, we call a lump of mass with fixed identity a *system*. Newton's laws of motion can be applied directly to a system. All the other fundamental laws can also be applied to systems. By definition, the mass of a system is constant, so the law of conservation of mass simply states that system mass is constant. The law of conservation of energy requires that any net energy transfer to a system results in an equal increase of the system's energy. The laws of degradation of energy and property relations are intrinsically system laws.

In fluid mechanics, we are interested in rate processes, that is, processes that involve time and motion. The system forms of the laws of conservation of mass and energy and Newton's second law of motion can be stated in terms of rates of change, as follows:

Conservation of mass: The rate of change of a system's mass is zero. Mathematically,*

$$\frac{dm_{\text{sys}}}{dt} = 0. \tag{5.1}$$

Newton's second law of motion: The rate of change of a system's momentum is equal to the sum of all forces acting on the system. Mathematically,

$$\frac{d\vec{M}_{\text{sys}}}{dt} = \sum \vec{F}_{\text{sys}}. \tag{5.2}$$

Conservation of energy: The rate of change of a system's intrinsic energy is equal to the net rate of energy transfer to the system by

---

* In Section 3.4.1, where we derived the differential continuity equation, we applied this equation to a single mass particle.

heat and work. Mathematically,

$$\frac{dE_{sys}}{dt} = \frac{dQ}{dt} - \frac{dW}{dt}.$$

(5.3)

According to thermodynamic sign convention, heat added *to* a system is positive and work done *on* a system is negative.

Note that all these equations require calculation of the rate of change of some system property.

Before we can use Eqs. (5.1)–(5.3), we have to figure out how to identify a specific system in a moving, deforming body of fluid and how to calculate the rate of change of properties of this system. This is where a control volume comes in. Although fluid flows through a control volume continuously, a control volume contains a specific mass of fluid at any instant. We designate our system as the fluid mass that instantaneously occupies the control volume. A different system occupies the control volume at each instant.* Since the equations deal with rates of change, we can apply them at any particular instant. At every instant, we can apply the rate of change form of the fundamental laws to whatever system occupies the control volume *at that instant*. The control volume becomes a way to identify a specific system, even if only for a very short time.

5.2 The Control Volume Rate-of-Change Equation

The key to the control volume approach is the use of a control volume to instantaneously identify a system and to use information at the control volume boundaries to determine rates of change of system properties. Since we have at least three equations requiring rates of change of different system properties (mass, momentum, and energy), we might have to repeat our analysis three times. We can save ourselves a lot of work by deriving a general relation for the rate of change of an arbitrary system property in terms of control volume parameters. In this section, we will do just that.

5.2.1 Derivation

Figure 5.1 shows a system passing through a control volume. Part (a) shows the situation at time t, when the system exactly fills the control volume. Part (b) shows the situation at a later time, $t + \delta t$, when a portion of the system has left the control volume. V_I is the portion of the control volume that has been vacated by the system in time δt; V_{II} is the portion of the control volume that contained part of the system at time t and still contains part of the system at time $t + \delta t$; and V_{III}

---

* This idea is somewhat like taking motion pictures. We point the camera at a particular location (say a busy street corner) and take pictures. We record frame by frame the pictures of people who occupy the location. When we run the film at regular speed, we see a continuous flow of people; but each frame shows a particular person occupying the location.

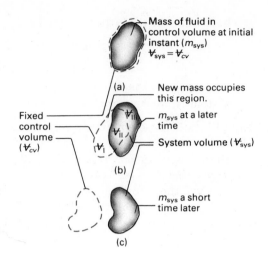

Figure 5.1 A system flows through a control volume.

lies outside the control volume and contains the portion of the system that has left the control volume in time δt.*

We can calculate the rate of change of an arbitrary property B of this system. B could be mass, momentum, energy, or any other property. B may be nonuniformly distributed throughout the system, so let b be the per-unit-mass equivalent of B. If B represents system momentum ($\vec{M} = m\vec{V}$), then b is velocity ($\vec{V}$). If B represents system kinetic energy, then b is $V^2/2$. The total amount of B in the system is

$$B_{\text{sys}} = \int_{m_{\text{sys}}} b \, dm = \int_{\mathcal{V}_{\text{sys}}} b\rho \, d\mathcal{V}, \tag{5.4}$$

where m_{sys} is the system mass and $\mathcal{V}_{\text{sys}}$ is the system volume.

Now consider the rate of change of B_{sys}. By definition,

$$\frac{dB_{\text{sys}}}{dt} = \lim_{\delta t \to 0} \frac{B_{\text{sys}})_{t+\delta t} - B_{\text{sys}})_t}{\delta t}.$$

Substituting from Eq. (5.4), we have

$$\frac{dB_{\text{sys}}}{dt} = \lim_{\delta t \to 0} \frac{\int_{\mathcal{V}_{\text{sys}}} \rho b \, d\mathcal{V})_{t+\delta t} - \int_{\mathcal{V}_{\text{sys}}} \rho b \, d\mathcal{V})_t}{\delta t}. \tag{5.5}$$

At time t,

$$\mathcal{V}_{\text{sys}})_t = \mathcal{V}_{cv},$$

and at time $t + \delta t$,

$$\mathcal{V}_{\text{sys}})_{t+\delta t} = \mathcal{V}_{\text{II}} + \mathcal{V}_{\text{III}} = \mathcal{V}_{cv} - \mathcal{V}_{\text{I}} + \mathcal{V}_{\text{III}},$$

so

$$\frac{dB_{\text{sys}}}{dt} = \lim_{\delta t \to 0} \frac{\int_{\mathcal{V}_{cv}} \rho b \, d\mathcal{V})_{t+\delta t} - \int_{\mathcal{V}_{\text{I}}} \rho b \, d\mathcal{V})_{t+\delta t} + \int_{\mathcal{V}_{\text{III}}} \rho b \, d\mathcal{V})_{t+\delta t} - \int_{\mathcal{V}_{cv}} \rho b \, d\mathcal{V})_t}{\delta t}.$$

* If the fluid is incompressible, $\mathcal{V}_{\text{III}} = \mathcal{V}_{\text{I}}$; this fact is irrelevant to the present discussion.

This can be rewritten as

$$\frac{dB_{sys}}{dt} = \underbrace{\lim_{\delta t \to 0} \frac{\int_{V_{cv}} \rho b\, dV)_{t+\delta t} - \int_{V_{cv}} \rho b\, dV)_t}{\delta t}}_{\text{Term A}} + \underbrace{\lim_{\delta t \to 0} \frac{\int_{V_{III}} \rho b\, dV)_{t+\delta t}}{\delta t}}_{\text{Term B}}$$

$$- \underbrace{\lim_{\delta t \to 0} \frac{\int_{V_I} \rho b\, dV)_{t+\delta t}}{\delta t}}_{\text{Term C}}$$

Each of the three terms has a separate meaning. Term Ⓐ looks very similar to the right side of Eq. (5.5). There is one important difference: In term Ⓐ the integration is over the *control* volume, while in Eq. (5.5), the integration is over the *system* volume. Using the definition of a derivative, we find

$$\text{Term } Ⓐ = \frac{d}{dt} \int_{V_{cv}} \rho b\, dV = \frac{d}{dt}(B_{cv}),$$

which represents the time rate of change of the amount of B contained in the control volume. Note that this amount of B could change due to changes in V_{cv} (a moving or deforming control volume) as well as to changes of ρ or b at points within the control volume. Term Ⓐ is sometimes called the *rate of accumulation*.

The integral in the numerator of term Ⓑ represents the amount of B that has flowed out of the control volume between time t and time $t + \delta t$. Dividing by δt, we get the average rate of outflow, and passing to the limit, we get the instantaneous rate of outflow

$$\text{Term } Ⓑ = \dot{B}_{out}.$$

The integral in term Ⓒ is the amount of B that has flowed into the control volume between time t and time $t + \delta t$. (Note that this B is not in the original system.) Dividing by δt, we get the average rate of inflow, and passing to the limit, we get the instantaneous rate of inflow

$$\text{Term } Ⓒ = \dot{B}_{in}.$$

Terms Ⓑ and Ⓒ are sometimes called the *flux terms* or the *B flux terms*. Collecting terms Ⓐ, Ⓑ, and Ⓒ, we get

$$\frac{dB_{sys}}{dt} = \frac{d}{dt} \int_{V_{cv}} \rho b\, dV + \dot{B}_{out} - \dot{B}_{in}. \tag{5.6}$$

In words:

> The rate of change of any property of the system that occupies a control volume at any particular instant is equal to the instantaneous rate of accumulation of the property inside the control volume plus the difference between the instantaneous rates of outflow and inflow of the property. This latter term is the net rate of outflow of the property across the control surface.

Sometimes Eq. (5.6) can be used directly, but it is usually more convenient to relate $\dot{B}_{out}$ and $\dot{B}_{in}$ to the fluid velocity. We can do this very

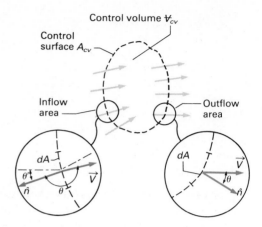

Figure 5.2 Control volume with detail of inflow and outflow areas.

rapidly if we borrow a few results from our discussion of the integral continuity equation in Section 3.4.2. Consider the control volume shown in Fig. 5.2. The rate of mass flow out of this control volume is

$$\dot{m}_{\text{out}} = \int_{A_{\text{out}}} \rho V \cos\theta \, dA = \int_{A_{\text{out}}} \rho(\vec{V} \cdot \hat{n}) \, dA. \qquad \textbf{(3.24a,b)}$$

Each unit of mass leaving the control volume carries b units of B out with it. Since b may not be uniform, we write

$$d\dot{B}_{\text{out}} = b(d\dot{m})_{\text{out}}.$$

Since

$$\dot{B}_{\text{out}} = \int d\dot{B}_{\text{out}} = \int b(d\dot{m})_{\text{out}},$$

we get

$$\dot{B}_{\text{out}} = \int_{A_{\text{out}}} \rho b V \cos\theta \, dA = \int_{A_{\text{out}}} \rho b(\vec{V} \cdot \hat{n}) \, dA. \qquad \textbf{(5.7)}$$

By similar arguments, using Eqs. (3.26a,b), we find

$$\dot{B}_{\text{in}} = -\int_{A_{\text{in}}} \rho b V \cos\theta \, dA = -\int_{A_{\text{in}}} \rho b(\vec{V} \cdot \hat{n}) \, dA. \qquad \textbf{(5.8)}$$

Using Eqs. (5.7) and (5.8), we can write Eq. (5.6) in the following forms:

$$\frac{dB_{\text{sys}}}{dt} = \frac{d}{dt}\int_{\mathcal{V}_{cv}} \rho b \, d\mathcal{V} + \int_{A_{\text{out}}} \rho b V \cos\theta \, dA + \int_{A_{\text{in}}} \rho b V \cos\theta \, dA, \qquad \textbf{(5.9a)}$$

$$\frac{dB_{\text{sys}}}{dt} = \frac{d}{dt}\int_{\mathcal{V}_{cv}} \rho b \, d\mathcal{V} + \int_{A_{\text{out}}} \rho b V_n \, dA - \int_{A_{\text{in}}} \rho b V_n \, dA, \qquad \textbf{(5.9b)}$$

and

$$\frac{dB_{\text{sys}}}{dt} = \frac{d}{dt}\int_{\mathcal{V}_{cv}} \rho b \, d\mathcal{V} + \oint_{A_{cv}} \rho b(\vec{V} \cdot \hat{n}) \, dA. \qquad \textbf{(5.9c)}$$

We will call Eq. (5.9), in any of its forms, the *control volume rate-of-change equation*. The equation is also called the *Reynolds transport theorem*, after the English scientist and engineer Osborne Reynolds. Although it may not be apparent, this equation is the finite control volume equivalent of the Eulerian derivative, Eq. (3.10).

5.2.2 Illustrative Application: The Integral Continuity Equation

We use the control volume rate-of-change equation to find rates of change of properties for use in the system formulation of the fundamental laws. We illustrate this idea by using the equation together with the system formulation of the law of conservation of mass to derive the integral continuity equation.

The system formulation of the law of conservation of mass is

$$\frac{dm_{\text{sys}}}{dt} = 0. \tag{5.1}$$

We calculate the rate of change of m_{sys} by using the control volume rate-of-change equation with $B = m_{\text{sys}}$. The property b, the mass per unit mass, is

$$b = \frac{m_{\text{sys}}}{m_{\text{sys}}} = 1.$$

Using Eq. (5.9a), we have

$$\frac{dm_{\text{sys}}}{dt} = \frac{d}{dt}\int_{V_{cv}} \rho\, dV + \int_{A_{\text{out}}} \rho V \cos\theta\, dA + \int_{A_{\text{in}}} \rho V \cos\theta\, dA = 0.$$

This of course is just the integral continuity equation that we have already derived and thoroughly discussed in Chapter 3, so we needn't pursue it any further here.

It may seem that this illustration is trivial, and to a certain extent it is. In fact, we already did most of the real work associated with the derivation of the control volume rate-of-change equation in Section 3.4.2, and there is no need to rederive the continuity equation this way. The real value of the control volume rate-of-change equation will become apparent soon when we deal with the laws of mechanics and thermodynamics.

5.2.3 The Control Volume Rate-of-Change Equation for Some Special Cases

In the forms of Eqs. (5.9a,b,c) the control volume rate-of-change equation can be applied to any control volume, fixed or moving, rigid or deformable. Likewise, the flow can be three-dimensional, three-directional, and unsteady. Most applications do not involve all of these complexities, and we will examine some special cases involving some combinations of them.

Moving, Deforming Control Volume.   If the control volume moves and deforms, we must be very careful in interpreting the first term of the equation,

$$\frac{d}{dt}\int_{V_{cv}} \rho b\, dV.$$

For a moving, deforming control volume, V_{cv} is time-dependent, so the limits of the integral are time-dependent. Of course, ρ and b may be time-dependent as well. If this is the case, it is necessary to evaluate the integral first and then carry out the differentiation. Section 3.4.3 discusses a deformable control volume.

If the control volume moves and deforms, we must also be careful in calculating the flux terms. Since these terms represent the rate that fluid crosses the control surface, we must use the fluid velocity relative to the control surface:

$$\dot{B}_{\text{out}} = \int_{A_{\text{out}}} \rho b V_r \cos\theta \, dA \quad \text{and} \quad \dot{B}_{\text{in}} = \int_{A_{\text{in}}} \rho b V_r \cos\theta' \, dA,$$

where V_r is the velocity relative to the control surface. This requirement can be relaxed if the entire control volume moves at constant velocity because the constant velocity will not contribute to the net flux integral. In this case, we can use either the relative velocity or the absolute velocity. This idea is discussed further in Section 5.3.2.

Fixed, Rigid Control Volume. If the control volume is fixed and rigid, then it does not change with time, and we can write

$$\frac{d}{dt} \int_{\mathcal{V}_{cv}} \rho b \, d\mathcal{V} = \frac{\partial}{\partial t} \int_{\mathcal{V}_{cv}} \rho b \, d\mathcal{V} = \int_{\mathcal{V}_{cv}} \frac{\partial}{\partial t} (\rho b) \, d\mathcal{V};$$

that is, we can perform differentiation and integration in any order.

Steady-Flow, Fixed, Rigid Control Volume. In this case, the first term can be dropped altogether:

$$\frac{d}{dt} \int_{\mathcal{V}_{cv}} \rho b \, d\mathcal{V} \equiv 0.$$

One-Dimensional Uniform Flow at a Fixed Number of Inlets and Outlets. In this case, the flux term(s) can be simplified:

$$\oint_{A_{cv}} \rho b (\vec{V} \cdot \hat{n}) \, dA = \sum_{\text{out}} b(\rho A V) - \sum_{\text{in}} b(\rho A V),$$

or

$$\oint_{A_{cv}} \rho b (\vec{V} \cdot \hat{n}) \, dA = \sum_{\text{out}} b(\dot{m}) - \sum_{\text{in}} b(\dot{m}).$$

5.2.4 Selection of Appropriate Control Volumes

We have just demonstrated how the mathematical complexity of the control volume rate-of-change equation can be strongly affected by the choice of control volume. You might have concluded that you always want to use fixed, rigid control volumes. This would be a bad idea, as a quick review of the tank-filling problem discussed in Section 3.4.3 will show. Figure 5.3 illustrates some control volumes especially tailored to particular flow problems.

Having decided on control volume type (fixed, moving, rigid, etc.) you still have to select the size, shape, and boundary location. In many problems, the choice is reasonably obvious. In other problems, it doesn't really make much difference—any of several choices will do. A good control volume selection has the following characteristics:

- The control volume fits the problem (try a rigid control volume for the piston/cylinder or a moving control volume for the pipe flow of Fig. 5.3 and see what happens!).
- The quantities you want to calculate (flows, forces, pressures, etc.) appear explicitly in your formulation. This usually requires that these quantities be at the boundaries of the control volume.
- The control surface should be perpendicular to the local fluid velocity.
- No unnecessary mathematical complexity is introduced by your choice.

Skill at selecting good control volumes can be developed only by practice. Studying the examples in this chapter with an eye toward the specific choice of control volume will also be valuable.

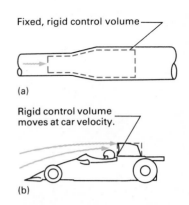

(a)

(b)

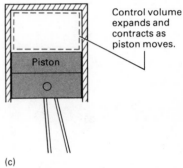

(c)

Figure 5.3 Situations requiring different control volume choices: (a) Flow in pipe with diffuser section. (b) flow into race car air scoop; (c) reciprocating air compressor.

5.3 Newton's Second Law: The Linear Momentum Theorem

By combining the control volume rate-of-change equation with Newton's second law of motion, we obtain a very useful result known as the *linear momentum theorem*. Consider a flow field with a superimposed control volume (Fig. 5.4). Newton's second law for the system enclosed by the control volume is

$$\frac{d\vec{M}_{\text{sys}}}{dt} = \sum \vec{F}_{\text{sys}}. \tag{5.2}$$

To apply the control volume rate-of-change equation, note that the arbitrary property B is the system momentum, $\vec{M}_{\text{sys}} (= m_{\text{sys}} \vec{V}_{\text{sys}})$. The per unit mass property b is the fluid velocity $\vec{V}$. Using Eq. (5.9c), we obtain

$$\frac{d}{dt} \int_{V_{cv}} \rho \vec{V} \, dV + \oint_{A_{cv}} \rho \vec{V} (\vec{V}_r \cdot \hat{n}) \, dA = \sum \vec{F}. \tag{5.10}$$

Equation (5.10), called the *linear momentum equation,* is the mathematical expression of the general principle called the linear momentum theorem:*

> The rate of accumulation of linear momentum within a control volume plus the net rate of momentum flow out of the control volume (momentum out minus momentum in) equals the sum of the forces acting on the mass inside the control volume.

Before investigating simplified forms and possible uses of this equation, note that any simplified form of the linear momentum theorem must reflect the following important characteristics.

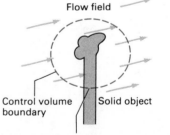

Figure 5.4 Flow around solid object with superimposed control volume.

---

* Equation (5.10) is sometimes called the principle or equation of conservation of linear momentum; however, this is a misnomer because momentum is not conserved but created or destroyed by forces.

- The velocity $\vec{V}$ in the first integral and the first velocity $\vec{V}$ in the second integral represent the momentum and must be measured relative to an inertial frame of reference. The second velocity in the second integral (written as $\vec{V}_r$) represents the flow rate passing through the control surface and must be measured relative to the control surface. $\vec{V}_r$ need not be measured relative to an inertial frame of reference.

- $\sum \vec{F}$ is the vector sum of all forces acting *on* the mass within the control volume. If the control volume contains any solid material, as in Fig. 5.4, the forces on this mass are included. $\sum \vec{F}$ is sometimes called "the sum of forces acting on the control volume," omitting the words "mass within." Note that forces from all sources (gravitational, pressure, magnetic, etc.) are included in $\sum \vec{F}$.

- Equation (5.10) is a vector equation. As such, it represents three separate component (x, y, z or r, θ, x, etc.) equations. The vector nature of the equation is embodied in $\sum \vec{F}$ and in $\vec{V}$ in the integrals (but *not* in the $\vec{V}_r$ that is dotted with $\hat{n}$).

5.3.1 Evaluating the Terms in the Linear Momentum Equation

To apply the linear momentum theorem to a particular flow problem, we must evaluate the terms in Eq. (5.10). This equation can take many different forms, depending on the specific application. In this section, we examine various forms of the primary terms in the equation. As we proceed, it will be helpful to refer to the following symbolic form of Eq. (5.10):

$$\frac{d}{dt}(\vec{M}_{cv}) + \dot{\vec{M}}_{out} - \dot{\vec{M}}_{in} = \sum \vec{F}. \tag{5.11}$$

In this equation $\vec{M}$ represents momentum and $\dot{\vec{M}}$ represents momentum flux. The momentum flux integral has been split into two terms, $\dot{\vec{M}}_{out}$ and $\dot{\vec{M}}_{in}$, where

$$\dot{\vec{M}}_{out} = \int_{A_{cv,out}} \rho \vec{V}(\vec{V}_r \cdot \hat{n})\, dA = \int_{A_{cv,out}} \rho \vec{V}(V_r \cos \theta)\, dA \tag{5.12a}$$

and

$$\dot{\vec{M}}_{in} = -\int_{A_{cv,in}} \rho \vec{V}(\vec{V}_r \cdot \hat{n})\, dA = \int_{A_{cv,in}} \rho \vec{V}(V_r \cos \theta')\, dA. \tag{5.12b}$$

We will discuss separately the evaluation of the momentum accumulation term, the momentum flux term(s), and the sum-of-forces term.

Momentum Accumulation Term. The momentum accumulation term,

$$\frac{d}{dt}(\vec{M}_{cv}) = \frac{d}{dt} \int_{V_{cv}} \rho \vec{V}\, dV,$$

may be important if there is accumulation or depletion of mass in the control volume or if the control volume is accelerating. Accelerating control volumes introduce the special problem of noninertial coordinates. Analysis of accelerating control volumes will be considered in a later section.

If the control volume is fixed and rigid, the accumulation term is

$$\frac{d}{dt}(\vec{M}_{cv}) = \frac{\partial}{\partial t}\int_{\Psi_{cv}} \rho \vec{V} d\Psi.$$

If the flow is steady, the momentum accumulation term vanishes:

$$\frac{d}{dt}(\vec{M}_{cv}) = 0. \quad \text{Steady flow}$$

The condition of steady flow is probably the most widely used simplification of the linear momentum equation. For steady flow, the linear momentum equation becomes

$$\dot{\vec{M}}_{\text{out}} - \dot{\vec{M}}_{\text{in}} = \sum \vec{F}. \quad \text{Steady flow} \qquad \textbf{(5.13)}$$

Momentum Flux Term. If the fluid velocity is nonuniformly distributed over the control surface (Fig. 5.5), we must use Eqs. (5.12a,b), to evaluate the momentum flux and we will be in for some integration. In other cases, it is possible to bypass the integration by taking advantage of certain characteristics of a particular problem or by making some simplifying assumptions. We will discuss these special cases, but we hope you will not conclude that it is always possible to avoid integration in evaluation of momentum flux.

Often we choose a control volume with inflow and outflow at only a finite number of locations and we assume that the flow is locally uniform and one-directional at these locations (see Fig. 5.6). In this case, the momentum flux can be calculated by

$$\dot{\vec{M}}_{\text{out}} - \dot{\vec{M}}_{\text{in}} = \sum_{\text{out}} (\rho V_n A)\vec{V} - \sum_{\text{in}} (\rho V_n A)\vec{V}, \qquad \textbf{(5.14a)}$$

where

$$V_n = |\vec{V}_r \cdot \hat{n}| = |V_r \cos\theta|.$$

The summations are taken over the number of outlets and inlets, respectively. Since at any outlet or inlet,

$$\dot{m} = \rho V_n A,$$

Equation (5.14a) can also be written as

$$\dot{\vec{M}}_{\text{out}} - \dot{\vec{M}}_{\text{in}} = \sum_{\text{out}} \dot{m}\vec{V} - \sum_{\text{in}} \dot{m}\vec{V}. \qquad \textbf{(5.14b)}$$

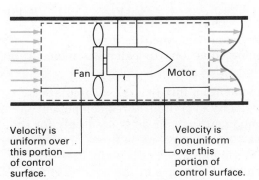

Figure 5.5 Control volume for analysis of flow through a fan in a duct illustrating nonuniform velocity over part of the control surface.

Fan Motor

Velocity is uniform over this portion of control surface.

Velocity is nonuniform over this portion of control surface.

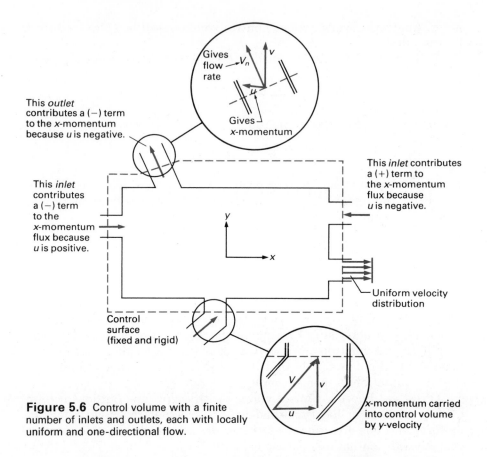

Figure 5.6 Control volume with a finite number of inlets and outlets, each with locally uniform and one-directional flow.

Since the momentum flux is a vector, it can be evaluated in terms of its separate components. For the flow situation shown in Fig. 5.6, the three component momentum fluxes are

$$\dot{M}_{x,\text{out}} - \dot{M}_{x,\text{in}} = \sum_{\text{out}} (\rho V_n A)u - \sum_{\text{in}} (\rho V_n A)u, \tag{5.15a}$$

$$\dot{M}_{y,\text{out}} - \dot{M}_{y,\text{in}} = \sum_{\text{out}} (\rho V_n A)v - \sum_{\text{in}} (\rho V_n A)v, \tag{5.15b}$$

$$\dot{M}_{z,\text{out}} - \dot{M}_{z,\text{in}} = \sum_{\text{out}} (\rho V_n A)w - \sum_{\text{in}} (\rho V_n A)w. \tag{5.15c}$$

Even if the flow is not locally uniform, the momentum flux can be evaluated in terms of components. The x-momentum flux is

$$\dot{M}_{x,\text{out}} - \dot{M}_{x,\text{in}} = \int_{A_{cv,\text{out}}} \rho u(V_r \cos \theta)\, dA - \int_{A_{cv,\text{in}}} \rho u(V_r \cos \theta')\, dA, \tag{5.16}$$

with similar expressions for the y- and z-momentum fluxes.

When using these component equations to calculate momentum flux, you must obey the following rules:

- The mass flow part of the momentum flux $(\rho V_n A)$ is calculated using the relative normal velocity (V_n) in all three component equations. The velocity component $(u, v, \text{ or } w)$ is used only for the momentum per unit mass part of the momentum flux.
- You must pick a positive direction for x, y, and z. Velocities (u, v, w) are entered as positive numbers if they are in the positive direc-

tion and as negative numbers if they are opposite this direction. As a result, any of the momentum flux terms can have a positive or negative sign. The sign is not determined only by the label "in" or "out." See Fig. 5.6.

■ The components of the momentum flux are combined with components of the momentum accumulation (if accumulation is not zero) and equated to the components of the sum of forces, for example,

$$\frac{d}{dt} M_{x,cv} + \dot{M}_{x,\text{out}} - \dot{M}_{x,\text{in}} = \sum F_x. \tag{5.17}$$

The force terms must obey the same sign convention as the momentum flux and accumulation terms.

In many practical cases, the fluid velocity is locally one-directional but not uniform. Figure 5.7 illustrates this situation for flow in a circular pipe. At plane 1 the flow is one-directional (parallel to the pipe axis) but nonuniform. The momentum flux at plane 1 is*

$$\dot{M}_1 = \int \rho \vec{V}[V \cos(0)]\, dA = \left(\int \rho V^2\, dA\right)\hat{n}_v. \tag{5.18}$$

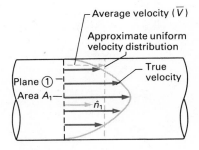

Figure 5.7 One-directional flow in a pipe with nonuniform velocity distribution.

To avoid a tedious integration, it would be desirable to replace the nonuniformly distributed velocity with an average velocity. We assume that the fluid is incompressible. To ensure that the uniformly distributed velocity carries the same flow rate as the real distribution, we use the average velocity $\vec{V}$

$$\bar{V} = \frac{\int V_n\, dA}{A}. \tag{3.39}$$

In terms of the average velocity, the mass flow rate at any cross section is

$$\dot{m} = \rho \bar{V} A;$$

however, the approximating uniform flow does not have the correct momentum flux, because the real V is nonuniform. In general $\dot{M}_1 > (\rho \bar{V}^2 A)\hat{n}_v$.

We can evaluate the true momentum flux using a *momentum correction factor, β,* defined by

$$\beta \equiv \frac{\int \rho V^2\, dA}{\rho \bar{V}^2 A}. \tag{5.19a}$$

For constant density,

$$\beta = \frac{\int V^2\, dA}{\bar{V}^2 A}. \tag{5.19b}$$

The momentum flux across plane 1 is then calculated by

$$\dot{M} = \beta_1 (\rho \bar{V}_1^2 A_1)\hat{n}_v. \tag{5.20}$$

---

* Since we have only a single plane, not an entire control surface, we don't know whether $\dot{M}_1$ is "in" or "out." The unit vector $\hat{n}_v$ points in the same direction as $\vec{V}$.

For a control volume with a finite number of inlets and outlets with locally one-directional flow,

$$\dot{M}_{out} - \dot{M}_{in} = \sum_{out} \beta(\rho \bar{V}^2 A)\hat{n}_v - \sum_{in} \beta(\rho \bar{V}^2 A)\hat{n}_v. \tag{5.21}$$

The vector nature of the momentum flux is indicated by the unit vector $\hat{n}_v$, which is parallel to the *local* velocity vector.

For fully developed flow in a circular pipe, numerical values of β range from 1.33 for laminar flow to as low as 1.02 for turbulent flow. Since β is often approximately equal to 1.0, momentum correction factors are often ignored in engineering calculations.

EXAMPLE 5.1 Illustrates the concepts of momentum flux and momentum correction factor

Figure E5.1 shows a constant-density fluid flowing at a steady rate in the entrance region of a circular pipe of inside diameter $2R$. The velocity profile is uniform across the inlet and parabolic at a cross section 60 diameters downstream of the inlet. Find the momentum flux and momentum correction factor at each cross section.

Solution

Given
Constant-density fluid flowing at a steady rate
Figure E5.1
Velocity uniform at inlet and parabolic 60 diameters downstream

Find
Momentum flux and momentum correction factor at each cross section

Solution
The momentum flux is

$$\dot{M} = \int_A \rho \vec{V}(\vec{V}_r \cdot \hat{n})\, dA.$$

Density is constant and $\hat{n}$ is taken as positive to the right. See Fig. E5.1. At the inlet,

$$V = V_r = U_0 \quad \text{and} \quad \dot{M} = \rho \int_A U_0^2 \hat{n}\, dA$$

$$= \rho U_0^2 A \hat{n}.$$

Figure E5.1 Velocity distribution at two cross sections in a circular pipe.

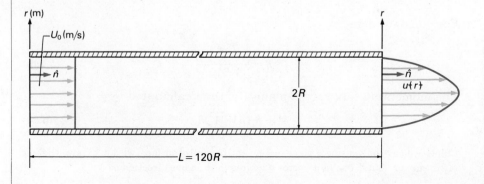

Since the flow area is circular,

Answer

$$\boxed{\dot{M}_{\text{inlet}} = \pi R^2 \rho U_0^2 \hat{n}.}$$

The momentum correction factor β for a constant density fluid is defined by Eq. (5.19b). At the inlet,

$$\beta \equiv \frac{\int V^2 \, dA}{\bar{V}^2 A}$$

$$= \frac{\int U_0^2 \, dA}{U_0^2 A},$$

or

Answer

$$\boxed{\beta_{\text{inlet}} = 1.}$$

The momentum flux 60 diameters downstream is

$$\dot{M} = \rho \int u^2 \hat{n} \, dA.$$

Since the velocity profile is parabolic,

$$u = u_{\max}\left[1 - \left(\frac{r}{R}\right)^2\right],$$

with $u_{\max}$ found from the integral continuity equation, Eq. (3.29):

$$\rho U_0 \pi R^2 = \rho \int u \, dA = \rho \int_0^R u_{\max}\left[1 - \left(\frac{r}{R}\right)^2\right] 2\pi r \, dr.$$

Integrating gives

$$\rho U_0 \pi R^2 = 2\pi \rho u_{\max}\left(\frac{r^2}{2} - \frac{r^4}{4R^2}\right)_0^R$$

$$= \frac{\pi \rho u_{\max} R^2}{2}.$$

Solving for $u_{\max}$ gives

$$u_{\max} = 2U_0 \quad \text{and} \quad u = 2U_0\left[1 - \left(\frac{r}{R}\right)^2\right].$$

The momentum flux is then

$$\vec{M} = \rho \int_0^R (2U_0)^2 \left[1 - \left(\frac{r}{R}\right)^2\right]^2 \hat{n} 2\pi r \, dr.$$

Integrating gives

$$\vec{M} = 8\pi U_0^2 \rho \hat{n} \int_0^R \left[1 - 2\left(\frac{r}{R}\right)^2 + \left(\frac{r}{R}\right)^4\right] r \, dr$$

$$= 8\pi U_0^2 \rho \hat{n} \left(\frac{r^2}{2} - \frac{r^4}{2R^2} + \frac{r^6}{6R^4}\right)_0^R$$

Answer

$$\boxed{\dot{M}_{60} = \frac{4}{3}\pi R^2 \rho U_0^2 \hat{n}.}$$

The momentum correction factor is

$$\beta \equiv \frac{\int \rho V^2 \, dA}{\rho \bar{V}^2 A},$$

where, for our case, $\bar{V} = U_0$ and $\int \rho V^2 \, dA$ is the magnitude of the momentum flux. Substituting gives

$$\beta \equiv \frac{\frac{4}{3}\pi R^2 \rho U_0^2}{\rho U_0^2 \pi R^2},$$

$$\boxed{\; \beta_{60} = \frac{4}{3}. \;}$$

Answer

Discussion

Since both locations have the same average velocity and $\beta_{60} > \beta_{\text{inlet}}$, then $\dot{M}_{60} > \dot{M}_{\text{inlet}}$. Equation (5.11) shows that the net force acting on the fluid between the inlet and outlet must be positive.

Sum-of-Forces Term. The sum-of-forces term is $\sum \vec{F}$. A control volume momentum analysis is usually carried out to calculate $\sum \vec{F}$ or a part of it. Considering the mass within a control volume as a free body, as in Fig. 5.8, we can split the force into three parts,

$$\sum \vec{F} = \sum \vec{F}_{\text{pressure}} + \sum \vec{F}_{\text{shear}} + \sum \vec{F}_{\text{gravity}}. \tag{5.22}$$

Gravity force is a *body force*. It acts on the mass as a whole. The pressure and shear forces are *surface forces* because they act on the mass at the control surface where it interfaces with material outside of the control volume. All surface forces between adjacent masses inside the control volume (say between the fluid and the portion of the solid object enclosed in the control volume) cancel because they occur in equal and opposite pairs, by Newton's third law.

The surface forces include those where the control surface "slices" through solid objects that protrude into the fluid. These forces are usually left in a single lump, so we write

$$\sum \vec{F} = \sum \vec{F}_{\text{presssure}} + \sum \vec{F}_{\text{shear (in fluid)}} + \sum \vec{F}_{\text{gravity}} + \vec{F}_R.$$

The lump force is labeled $\vec{F}_R$ to indicate that it is a reaction to the force exerted on the portion of the solid object inside the control volume by the fluid in contact with it.

The pressure force can be evaluated by

$$\sum \vec{F}_{\text{pressure}} = -\oint_{A_{cv}} p\hat{n} \, dA. \tag{5.23}$$

The direction of the pressure force that acts on an element of area dA is opposite to the outdrawn unit normal vector $\hat{n}$ because pressure is perpendicular to area and is positive for compression. Evaluation of the pressure integral is often easier than it looks. We can drop any constant uniform pressure from consideration because the integral of a constant over a closed surface is zero. This is illustrated in Fig. 5.9. Using this trick, a large portion of the pressure integral can often be reduced to zero.

The shear forces acting on the fluid can be evaluated by

$$\sum \vec{F}_{\text{shear (on fluid)}} = \int_{A_{cv},\text{fluid}} \vec{\tau} \, dA, \tag{5.24}$$

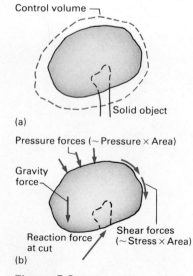

(a)

(b)

Figure 5.8 Forces acting on mass within a control volume: (a) Mass in a control volume; (b) free body of mass in control volume.

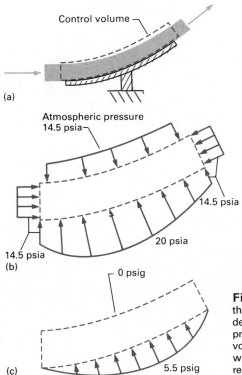

(a)

(b)

(c)

Figure 5.9 Removing part of the pressure integral: (a) Jet deflection by a fixed vane; (b) pressure distribution on control volume; (c) pressure distribution with atmospheric pressure removed.

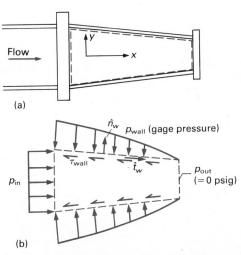

(a)

(b)

Figure 5.10 Pressure and shear stress on fluid flowing in a fire hose nozzle: (a) Nozzle and control volume; (b) pressure and shear stress on fluid.

where $A_{cv,fluid}$ is the portion of the control surface that is in contact with fluid and $\vec{\tau}$, the stress vector, is defined as the shear force per unit area. Because it is dependent on the specific orientation of the elemental area* dA, $\vec{\tau}$ is much more complicated than the pressure. In problems involving the linear momentum equation, the shear stress force is often evaluated as a single lump rather than by integration. This trick of leaving part of the force as a lump is also commonly applied to the pressure integral, which is often split up between parts that are evaluated by integration and parts that are left as lumps. The various lump forces from the pressure and shear stress integrals can be combined into a single force. Often, calculation of this force is object of an analysis.

To illustrate these ideas consider the force on the walls of the fire hose nozzle shown in Fig. 5.10. We select a control volume as shown, and the sum of forces is

$$\sum \vec{F} = \sum \vec{F}_{\text{pressure}} + \sum \vec{F}_{\text{shear}} + \sum \vec{F}_{\text{gravity}}.$$

The gravity force is simply the weight W of the water contained in the nozzle. The pressure force on the water is

$$\sum \vec{F}_{\text{pressure}} = \left(\int_{A_{\text{in}}} p \, dA \right) \hat{i} - \left(\int_{A_{\text{out}}} p \, dA \right) \hat{i} - \int_{A_w} p \hat{n}_w \, dA.$$

---

* If you are familiar with the concept of the stress tensor, you may know that $\vec{\tau}$ is the dot product of the stress tensor and the unit normal vector $\hat{n}$.

The shear force is due to the shear stress exerted on the water by the nozzle walls,

$$\int_{A_w} \vec{\tau}\, dA = -\int_{A_w} \tau_w \hat{t}_w\, dA,$$

where $\hat{t}_w$ is a unit vector tangent to the wall, pointing in the flow direction.

Substituting the expressions for pressure, shear, and gravity forces, we have

$$\sum \vec{F} = \left(\int_{A_{\text{in}}} p\, dA\right)\hat{i} - \left(\int_{A_{\text{out}}} p\, dA\right)\hat{i} - W\hat{j} - \int_{A_w} \tau \hat{t}_w\, dA - \int_{A_w} p\hat{n}_w\, dA.$$

Each of the last two terms would be very difficult to evaluate, requiring detailed flow predictions to find wall pressure and shear stress; however, together these terms represent the net force that the nozzle wall exerts on the fluid, so we can combine them:

$$-\int_{A_w} \tau \hat{t}_w\, dA - \int_{A_w} p\hat{n}_w\, dA = \vec{F}_w.$$

Substituting this result, we find that the sum of forces is

$$\sum \vec{F} = \left(\int_{A_{\text{in}}} p\, dA - \int_{A_{\text{out}}} p\, dA\right)\hat{i} - W\hat{j} + \vec{F}_w$$

In a typical problem, the inlet and outlet pressures and areas and the weight of fluid in the nozzle would be known, so these terms can be moved to the momentum side of the linear momentum equation, and the unknown wall force can be calculated as a single lump. If we assume that the flow is one-directional and that the pressure is uniform at the inlet and outlet, the resulting equation for $\vec{F}_w$ is*

$$\vec{F}_w = (\beta_{\text{out}}\rho_{\text{out}}\bar{V}_{\text{out}}^2 A_{\text{out}} - \beta_{\text{in}}\rho_{\text{in}}\bar{V}_{\text{in}}^2 A_{\text{in}} + p_{\text{out}}A_{\text{out}} - p_{\text{in}}A_{\text{in}})\hat{i} + W\hat{j}.$$

Note that $\vec{F}_w$ *thus calculated is the force that the nozzle wall exerts on the fluid.* An engineer would more likely want to calculate the force that the fluid exerts on the nozzle. This force can be obtained from Newton's third law:

$$\vec{F}(\text{Fluid on nozzle}) = -\vec{F}_w(\text{Nozzle on fluid}). \qquad (5.25)$$

A final point to consider in connection with the force term is the vector nature of $\sum \vec{F}$. If you are using component equations, remember that $\sum \vec{F}$ has x-, y-, and z-components $\sum F_x$, $\sum F_y$, and $\sum F_z$. These components have signs; they are positive if the force acts in the positive coordinate direction and negative if the force acts in the negative coordinate direction. If you are solving the linear momentum equation for an unknown force and do not know its direction (sign), simply assume that it is positive when writing the momentum equation. If the calculated force is positive, you made the right assumption. If the calculated force is negative, it acts in the negative coordinate direction.

---

* Do you know where all of the terms in this equation are from?

EXAMPLE 5.2 Illustrates calculation of the force terms for the linear
momentum equation

Mercury at 0°C flows upward through a vertical pipe with an inside diameter of
0.50 cm and a length of 1.0 m. The flow is fully developed, with an axial velocity profile given by

$$u = u_0\left[1 - \left(\frac{r}{R}\right)^2\right],$$

where R is the pipe radius and $u_0 = 1.0$ m/s. For the control volume shown in
Fig. E5.2, find the net force acting on the fluid.

Solution

Given

Mercury at 0°C flowing upward through a vertical pipe
Pipe inside diameter 0.50 cm, length 1.0 m
Axial velocity given by

$$u = u_0\left[1 - \left(\frac{r}{R}\right)^2\right],$$

where R is the pipe radius and $u_0 = 1.0$ m/s
Steady, fully developed flow
Figure E5.2

Find

Net force on the mercury in the control volume

Solution

In this problem we are concerned with the force in the axial direction, $\sum F_x$.
This force consists of three parts:

- Pressure force $(F_x)_p$
- Shear force $(F_x)_s$
- Gravity force $(F_x)_g$

The pressure force is

$$(F_x)_p = \int_{A_{in}} p\, dA - \int_{A_{out}} p\, dA.$$

If we assume angular symmetry ($\partial/\partial\theta = 0$) and no circumferential flow ($V_\theta = 0$),
the pressure is uniform over each horizontal plane for fully developed flow. Then
the pressure force is

$$(F_x)_p = (p_1 - p_2)\pi R^2.$$

The numerical values give

$$(F_x)_p = (404{,}000 - 271{,}000)\frac{\text{N}}{\text{m}^2}\frac{\pi(0.50\text{ cm})^2}{\left(\dfrac{100\text{ cm}}{\text{m}}\right)^2}$$

$$= 10.4\text{ N}.$$

The shear force is

$$(F_x)_s = \int_{A_w} \tau_w\, dA.$$

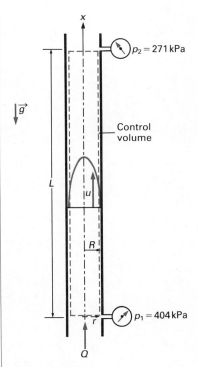

Figure E5.2 Flow of mercury
upward through a vertical pipe.

Since the flow is fully developed,

$$\frac{\partial \tau_w}{\partial x} = 0, \qquad \text{or} \qquad \tau_w = \text{Constant}.$$

Integrating, we have

$$(F_x)_s = \tau_w(2\pi RL).$$

Since mercury is a Newtonian fluid,

$$\tau_w = \mu\left(\frac{du}{dr}\right)\bigg|_{r=R}$$

Substituting for u gives

$$\tau_w = \mu u_0\left(\frac{-2r}{R^2}\right)_{r=R}, \qquad \text{or} \qquad \tau_w = -\frac{2\mu u_0}{R}.$$

The negative sign means that there is a downward force on the fluid in the control volume due to the stress at the pipe wall. Substituting τ_w into $(F_x)_s$ gives

$$(F_x)_s = \frac{-2\mu u_0}{R}(2\pi RL) = -4\pi\mu u_0 L.$$

Table A.5 of Appendix A gives $\mu = 1.68 \times 10^{-3}$ N·s/m². Then

$$(F_x)_s = -4\pi\left(1.68 \times 10^{-3}\frac{\text{N·s}}{\text{m}^2}\right)\left(1.0\frac{\text{m}}{\text{s}}\right)(1.0\text{ m})$$

$$= -0.0211 \text{ N}.$$

The gravity force is

$$(F_x)_g = -\int_{V_{cv}} \gamma dV = -\gamma V_{cv}$$

where γ is the specific weight and assumed constant and $V_{cv} = \pi R^2 L$. Thus

$$(F_x)_g = -\gamma\pi R^2 L.$$

If we assume the mercury is at sea level, Table A.5 gives

$$\gamma = \rho g = 13.60\left(1000\frac{\text{kg}}{\text{m}^3}\right)\left(9.807\frac{\text{m}}{\text{s}^2}\right)$$

$$= 133{,}000 \text{ N/m}^3,$$

and

$$(F_x)_g = \frac{-\left(133{,}000\frac{\text{N}}{\text{m}^3}\right)\pi(0.5\text{ cm})^2(1.0\text{ m})}{\left(\frac{100\text{ cm}}{\text{m}}\right)^2} = -10.4 \text{ N}.$$

The net axial force $\sum F_x$ is

$$\sum F_x = (F_x)_p + (F_x)_s + (F_x)_g$$
$$= (10.4 \text{ N}) + (-0.0212 \text{ N}) + (-10.4 \text{ N}),$$

$$\boxed{\sum F_x \approx -0.02 \text{ N.}}$$

Answer

Discussion
Since the flow is fully developed, should the net axial force be zero or just a small value? Is our answer accurate?

In this problem, the pressure and gravity forces are much greater than the stress force. This is not always the case.

A good exercise for you would be to apply the conditions

■ fully developed flow $\left(\dfrac{\partial}{\partial x} = 0 \text{ except } \dfrac{\partial p}{\partial x}\right)$,

■ angular symmetry $\left(\dfrac{\partial}{\partial \theta} = 0\right)$,

■ no rotational flow $(V_\theta = 0)$, and
■ $g_x = -g, \quad g_r = 0, \quad \text{and} \quad g_\theta = 0$

to the differential continuity equation and the Navier-Stokes equations (see Appendix H) and show that $\partial p/\partial r = 0$, which implies that the pressure is uniform in each horizontal plane.

In this section we have considered several specialized forms of the various terms in the linear momentum equation. We have deliberately not substituted all of these specialized terms into the general equation to obtain various specialized equations. We believe that each problem should be formulated by first writing the general expression for the linear momentum theorem (Eq. 5.10 or 5.11) and then introducing appropriate specialized forms of the terms as needed. By carefully studying the following examples and working several problems, you can become adept at applying the linear momentum theorem.

EXAMPLE 5.3 Illustrates how to calculate the force associated with a momentum change and the choice of alternative control volumes

The firefighters from Example 4.5 on page 190 are directing their fire hose against the 4 ft × 6 ft rectangular target shown in Fig. E5.3(a). The water leaves the fire hose at a rate of 1000 gal/min and strikes the target with a velocity of 120 ft/sec. Find the horizontal force that the supporting pipe exerts on the target. Find this force for each of the control volumes shown in Fig. E5.3(b).

Solution

Given
1000-gal/min water stream with 120-ft/sec velocity
Water stream hitting a 4 ft × 6 ft rectangular target
Figure E5.3(a)

Find
Horizontal force F_p of supporting pipe on target using each of the control volumes in Fig. E5.3(b)

Solution
We first use control volume 1 in Fig. E5.3(b). A free-body diagram of the target in Fig. E5.3(c) shows the horizontal force, F_p, of the supporting pipe and the horizontal force, F_w, of the water. Applying Newton's first law to the target

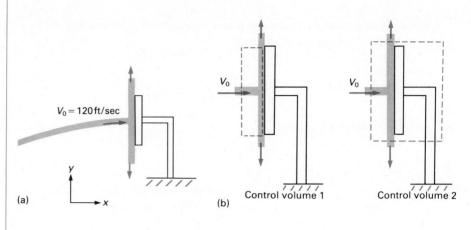

Figure E5.3 (a) Fire hose directed at target; (b) control volumes used to find the force of the supporting pipe on the target; (c) forces on control volume 1, target, and supporting pipe. W_T = gravity force on target; W_p = gravity force on pipe. Bending moments not shown.

gives

$$F_p = F_w.$$

This result shows that we can find F_p by finding F_w. If we assume that the fluid density and velocity are uniform over each jet stream, Eq. (5.17) is the appropriate form of the linear momentum equation for the control volume:

$$\dot{M}_{x,\text{out}} - \dot{M}_{x,\text{in}} = \sum F_x.$$

Equation (5.15a) gives

$$\dot{M}_{x,\text{out}} - \dot{M}_{x,\text{in}} = (\rho V_n A)_{\text{out}} u_{\text{out}} - (\rho V_n A)_{\text{in}} u_{\text{in}}.$$

If we assume 60°F water, Table A.6 gives $\rho = 62.4$ lbm/ft³. Equation (3.38) gives

$$A_{\text{in}} = \frac{Q}{(V_n)_{\text{in}}} = \frac{\left(1000\ \dfrac{\text{gal}}{\text{min}}\right)\left(\dfrac{\text{min}}{60\ \text{sec}}\right)}{\left(120\ \dfrac{\text{ft}}{\text{sec}}\right)\left(\dfrac{7.48\ \text{gal}}{\text{ft}^3}\right)}$$

$$= 0.0186\ \text{ft}^2,$$

where $(V_n)_{\text{in}} = u_{\text{in}} = V_0$. Neither the exit area A_{out} nor the exit velocity $(V_n)_{\text{out}}$ are known, and they need not be found since $u_{\text{out}} = 0$. Therefore

$$\dot{M}_{x,\text{out}} - \dot{M}_{x,\text{in}} = 0 - \rho V_0 A_{\text{in}}(V_0) = -\rho A_{\text{in}} V_0^2.$$

Since F_w on the control volume is assumed to act in the positive x-direction (see Fig. E5.3c),

$$\sum F_x = +F_w.$$

The linear momentum theorem for control volume 1 gives

$$F_w = -\rho A_{\text{in}} V_0^2 \quad \text{and} \quad F_p = -\rho A_{\text{in}} V_0^2.$$

The numerical values give

$$F_p = -\frac{\left(\dfrac{62.4 \text{ lbm}}{\text{ft}^3}\right)(0.0186 \text{ ft}^2)\left(120 \dfrac{\text{ft}}{\text{sec}}\right)^2}{\left(32.2 \dfrac{\text{ft} \cdot \text{lbm}}{\text{lb} \cdot \text{sec}^2}\right)},$$

$$\left| \; F_p = -519 \text{ lb.} \; \right|$$

Answer ◀

The negative sign means that F_p acts opposite to the assumed direction.

We now use control volume 2 in Fig. E5.3(b). A free-body diagram of the supporting pipe in Fig. E5.3(c) shows the horizontal force F_p' (internal to the pipe) and the horizontal force F_p of the target, both acting on the upper portion of the supporting pipe. Newton's first law gives

$$F_p = +F_p'.$$

This result shows that we can find F_p by finding F_p'.

Similar to the previous analysis,

$$\dot{M}_{x,\text{out}} - \dot{M}_{x,\text{in}} = 0 - \rho A_{\text{in}} V_0(V_0) \quad \text{and} \quad \sum F_x = F_p'.$$

The linear momentum theorem gives

$$F_p' = -\rho A_{\text{in}} V_0^2 \quad \text{or} \quad F_p = -\rho A_{\text{in}} V_0^2,$$

the same result as before. The numerical result is then

$$\left| \; F_p = -519 \text{ lb.} \; \right|$$

Answer ◀

Discussion

There is also a vertical force and a moment acting between the target and the supporting pipe and at the cut section of the support pipe. We were not asked to find these in this example problem.

The water was assumed to be at 60°F. Repeat the calculations for 40°F water and compare the magnitude of the answer with the answer to this example. Did the temperature difference lead to a significant difference in the force?

EXAMPLE 5.4   Illustrates simultaneous application of Bernoulli's equation, the continuity equation, and the linear momentum theorem

Water at 20°C flows through the nozzle in Fig. E5.4(a) at a steady rate of 12.0 m³/min. Find the force required to hold the nozzle. Neglect gravitational force.

Solution

Given
Nozzle in Fig. E5.4(a)
Water flow rate 12.0 m³/min.
Neglect the gravitational force

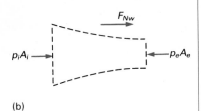

(a)

(b)

Figure E5.4 (a) Nozzle; (b) control volume for water in nozzle.

Find

Force F required to hold nozzle

Solution

Apply Newton's first law to the nozzle:

$$2\left(\frac{F}{2}\right) + F_{wN} = 0.$$

F_{wN} is the force of the water on the nozzle. Then*

$$F = -F_{wN} = F_{Nw},$$

which says that the force required to hold the nozzle can be found by finding the force of the nozzle on the water (F_{Nw}). Figure E5.4(b) shows a control volume enclosing the fluid. Note that F_{Nw} is assumed positive. Its actual direction will be determined by the algebraic sign of the resulting value.

If we assume the fluid density and velocity are uniform over each flow area, Eqs. (5.13) and (5.15a) are the appropriate forms for the linear momentum equation. For the x-direction,

$$\dot{M}_{x,\text{out}} - \dot{M}_{x,\text{in}} = \sum F_x,$$

where

$$\dot{M}_{x,\text{out}} - \dot{M}_{x,\text{in}} = \sum_{\text{out}} (\rho V_n A)u - \sum_{\text{in}} (\rho V_n A)u.$$

For the control volume in Fig. E5.4(b), the force terms are

$$\sum F_x = p_i A_i + F_{Nw},$$

where p_i is the gage pressure since we can drop any constant pressure that acts around the entire control volume. A_i is known, and p_i is found by assuming constant density and applying Bernoulli's equation to a streamline connecting the inlet and outlet

$$\frac{p_i}{\rho} + \frac{V_i^2}{2} + gz_i = \frac{p_e}{\rho} + \frac{V_e^2}{2} + gz_e.$$

Since the force of gravity is to be neglected, the elevation terms are dropped. Since p_i is gage pressure, $p_e = 0$. The velocities V_i and V_e are found from Eq. (3.38):

$$V_i = \frac{Q}{A_i} \quad \text{and} \quad V_e = \frac{Q}{A_e}.$$

Solving Bernoulli's equation for p_i gives

$$p_i = \frac{\rho Q^2}{2}\left(\frac{1}{A_e^2} - \frac{1}{A_i^2}\right).$$

Substituting this expression for p_i into the force term gives

$$\sum F_x = \frac{\rho Q^2}{2}\left(\frac{1}{A_e^2} - \frac{1}{A_i^2}\right)A_i + F_{Nw}.$$

The terms in the momentum flux are

$$A_{\text{out}} = A_e, \quad (V_n)_{\text{out}} = u_{\text{out}} = V_e,$$
$$A_{\text{in}} = A_i, \quad (V_n)_{\text{in}} = V_i.$$

* If F_{wN} is positive, then F_{Nw} must be negative for the directions assumed for F_{wN} and F_{Nw}.

Then

$$\dot{M}_{x,\text{out}} - \dot{M}_{x,\text{in}} = \rho A_e V_e^2 - \rho A_i V_i^2$$

$$= \rho A_e \left(\frac{Q}{A_e}\right)^2 - \rho A_i \left(\frac{Q}{A_i}\right)^2$$

$$= \rho Q^2 \left(\frac{1}{A_e} - \frac{1}{A_i}\right).$$

Equating the force terms and momentum flux terms, we have

$$\frac{\rho Q^2}{2}\left(\frac{1}{A_e^2} - \frac{1}{A_i^2}\right)A_i + F_{Nw} = \rho Q^2\left(\frac{1}{A_e} - \frac{1}{A_i}\right),$$

and solving for F_{Nw},

$$F_{Nw} = \frac{\rho Q^2}{A_e}\left(1 - \frac{A_i}{2A_e}\right).$$

The numerical values give

$$F_{Nw} = -\frac{\left(998\,\dfrac{\text{kg}}{\text{m}^3}\right)\left(12.0\,\dfrac{\text{m}^3}{\text{min}}\right)^2\left(1 - \dfrac{0.0075\ \text{m}^2}{2(0.015\ \text{m}^2)} - \dfrac{0.015\ \text{m}^2}{2(0.0075\ \text{m}^2)}\right)\left(\dfrac{\text{min}}{60\ \text{s}}\right)}{(0.0075\ \text{m}^2)}$$

or

$$F_{Nw} = -133\ \text{kg}\cdot\text{m/s}^2.$$

$$\left| F_{Nw} = -133\ \text{N.} \right|$$

Answer ◀

F acts to the left since it was assumed positive to the right and the numerical value is negative.

Discussion

If the gravity force were included in Bernoulli's equation, the pressure p_i would change with elevation. An average pressure $\bar{p}_i$ would then be used for the pressure force term in the linear momentum equation.

Since we did not know the direction of F_{Nw} at the beginning of the problem, we assumed it to be positive (to the right or in the positive x-direction). The resulting negative value means that F_{Nw} acts to the left, or the negative x-direction.

EXAMPLE 5.5   Illustrates the vector nature of the linear momentum theorem

A carbon steel, 14-in., schedule 30, 90° elbow is to be butt welded to a pipe carrying water at 4000 gal/min. Find the force in the weld to support the elbow. Carbon steel has a density of 490 lbm/ft^3. See Fig. E5.5(a).

Solution

Given
Figure E5.5(a)
Carbon steel, 14-in., schedule 30, 90° elbow for butt welding
Water flow rate 4000 gal/min
Carbon steel density 490 lbm/ft^3

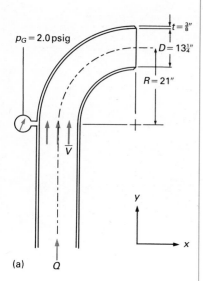

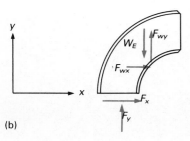

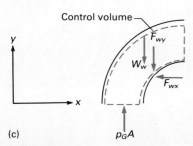

(a)

(b)

Control volume

(c)

Figure E5.5 (a) Water flowing through 90° elbow butt welded to 14-in. pipe; (b) forces on 90° elbow; (c) forces on water in control volume.

Find

Force in the butt weld to support the elbow

Solution

A free-body diagram of the elbow is given in Fig. E5.5(b). The following are the forces involved.

F_y = Vertical component of the force of the weld on the elbow
F_x = Horizontal component of the force of the weld on the elbow
W_E = Force of gravity on the elbow
F_{wy} = Vertical component of the force of the water on the elbow
F_{wx} = Horizontal component of the force of the water on the elbow

Applying Newton's first law in the horizontal (x) direction gives

$$F_x = -F_{wx}.$$

Applying Newton's first law in the vertical (y) direction gives

$$F_y = W_E - F_{wy}.$$

We must now evaluate F_{wx}, W_E, and F_{wy}. W_E is the product of the carbon steel density (ρ_{cs}) and the volume (V_E) of the elbow. V_E is approximated by*

$$V_E \approx \frac{\pi}{4}[(D + 2t)^2 - D^2]\left(\frac{\pi}{2}R\right).$$

Using values from Table D.1 for 14-in., schedule 30 pipe, we have

$$V_E \approx \frac{\pi}{4}[(14.0 \text{ in.})^2 - (13.25 \text{ in.})^2]\left(\frac{\pi}{2}\right)(21 \text{ in.})$$

$$= 529 \text{ in}^3\left(\frac{\text{ft}^3}{1728 \text{ in}^3}\right)$$

$$= 0.306 \text{ ft}^3.$$

Then

$$W_E = \gamma_{cs}V_E \approx \left(490 \frac{\text{lb}}{\text{ft}^3}\right)(0.306 \text{ ft}^3) = 150 \text{ lb.}^\dagger$$

F_{wy} and F_{wx} are evaluated by applying the linear momentum theorem to a control volume enclosing the water in the elbow. See Fig. E5.5(c). For steady flow, the x- and y-components of Eq. (5.13) give

$$\dot{M}_{x,\text{out}} - \dot{M}_{x,\text{in}} = \sum F_x \qquad \text{and} \qquad \dot{M}_{y,\text{out}} - \dot{M}_{y,\text{in}} = \sum F_y.$$

By Newton's third law, the force terms are

$$\sum F_x = -F_{wx} \qquad \text{and} \qquad \sum F_y = p_G A - W_w - F_{wy},$$

where p_G is the gage pressure. The uniform atmospheric pressure acting on the material in the control volume is neglected (see Section 5.3.1). W_w is the force of gravity on the water in the elbow.

---

* The theorem of Pappus states that the volume developed by rotating an area about an axis not intersecting the area is the product of the area and the length of the path described by the centroid of the area [1].

† Actual weight of this type of elbow is about 159 lb. The elbow is not exactly the shape of a torus as we assumed.

We now assume that the velocity profile is uniform at the inlet and at the outlet ($\beta_{in} = \beta_{out} = 1$). With this assumption, Eqs. (5.15a) and (5.15b) give

$$\dot{M}_{x,out} - \dot{M}_{x,in} = (\rho V_n A u)_{out} - (\rho V_n A u)_{in}$$

and

$$\dot{M}_{y,out} - \dot{M}_{y,in} = (\rho V_n A v)_{out} - (\rho V_n A v)_{in}.$$

The continuity equation gives

$$(\rho V_n A)_{in} = (\rho V_n A)_{out}.$$

Assuming constant water density and noting that

$$A_{in} = A_{out} = A,$$

we find

$$(V_n)_{in} = (V_n)_{out} = \bar{V},$$

The momentum flux terms are then

$$\dot{M}_{x,out} - \dot{M}_{x,in} = (\rho \bar{V} A)(\bar{V}) - (\rho \bar{V} A)(0) = \rho \bar{V}^2 A$$

and

$$\dot{M}_{y,out} - \dot{M}_{y,in} = (\rho \bar{V} A)(0) - (\rho \bar{V} A)(\bar{V}) = - \rho \bar{V}^2 A.$$

Substituting into the x-momentum equation gives

$$F_{wx} = -\rho \bar{V}^2 A,$$

so

$$F_x = \rho \bar{V}^2 A.$$

The numerical values give

$$A = \frac{\pi D^2}{4} = \frac{\pi (13.25 \text{ in.})^2}{4\left(144 \dfrac{\text{in}^2}{\text{ft}^2}\right)} = 0.958 \text{ ft}^2.$$

and

$$\bar{V} = \frac{Q}{A} = \frac{\left(4000 \dfrac{\text{gal}}{\text{min}}\right)\left(\dfrac{\text{ft}^3}{7.48 \text{ gal}}\right)\left(\dfrac{\text{min}}{60 \text{ sec}}\right)}{(0.958 \text{ ft}^2)} = 9.31 \text{ ft/sec}.$$

If we assume 60°F water, Table A.6 gives $\rho = 62.3 \text{ lbm/ft}^3$. Then

$$F_x = \frac{\left(62.4 \dfrac{\text{lbm}}{\text{ft}^3}\right)\left(9.31 \dfrac{\text{ft}}{\text{sec}}\right)^2 (0.958 \text{ ft}^2)}{\left(32.2 \dfrac{\text{ft}\cdot\text{lbm}}{\text{lb}\cdot\text{sec}^2}\right)},$$

and

Answer

$$\left| \; F_x = 161 \text{ lb.} \; \right|$$

F_x acts to the right since it is positive.

Substituting the y-momentum flux term and the y-forces into the y-linear momentum equation gives

$$-\rho \bar{V}^2 A = p_G A - W_w - F_{wy},$$

or

$$F_{wy} = \rho \bar{V}^2 A + p_G A - W_w$$

and

$$F_y = W_E - \rho \bar{V}^2 A - p_G A + W_w.$$

The water gravity force term is approximately*

$$W_w = \gamma_w \mathcal{V}_w \approx \gamma_w \left(\frac{\pi D^2}{4}\right)\left(\frac{\pi R}{2}\right) \approx \frac{\pi^2 \gamma_w D^2 R}{8}.$$

The numerical values are

$$W_w \approx \frac{\pi^2 \left(62.4 \dfrac{\text{lb}}{\text{ft}^3}\right)(13.25 \text{ in.})^2 (21 \text{ in.})}{8\left(\dfrac{1728 \text{ in}^3}{\text{ft}^3}\right)} = 164 \text{ lb}.$$

The y-momentum flux term is

$$\rho \bar{V}^2 A = \frac{\left(\dfrac{62.4 \text{ lbm}}{\text{ft}^3}\right)\left(9.31 \dfrac{\text{ft}}{\text{sec}}\right)^2 (0.958 \text{ ft}^2)}{\left(32.2 \dfrac{\text{ft} \cdot \text{lbm}}{\text{lb} \cdot \text{sec}^2}\right)} = 161 \text{ lb}.$$

F_y is now computed to be

$$F_y = 150 \text{ lb} - 161 \text{ lb} - \frac{\left(2.0 \dfrac{\text{lb}}{\text{in}^2}\right)(0.958 \text{ ft}^2)}{\left(\dfrac{\text{ft}^2}{144 \text{ in}^2}\right)} + 164 \text{ lb},$$

$$\left| F_y = -123 \text{ lb}. \right|$$

Answer ◀

F_y acts downward since it is negative.

Discussion

Note that a force acts in the direction assumed if calculated to be positive and acts opposite to the direction assumed if calculated to be negative. It is a good idea always to assume an unknown force to be in the *positive* coordinate direction.

Also note that the linear momentum equation is a vector equation, has three components, and can be written in any of the three directions. You might try writing the linear momentum equation in the z-direction and showing $F_z = 0$. Another interesting exercise would be to include realistic values for the momentum correction factor (say $\beta \approx 1.05$ for turbulent flow) in the analysis.

EXAMPLE 5.6   Illustrates the linear momentum theorem applied to a nonfluid mechanics problem

Figure E5.6(a) shows a 400-m-long mechanized walkway in an airport. The walkway moves at a speed of 4 km/hr and transports people and their luggage from one location to another. As many as 2000 persons per hour use the walkway at

---

* The theorem of Pappus again.

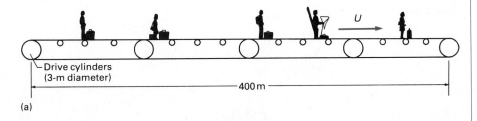

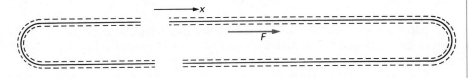

(b) F = Force on walkway by drive cylinders

Figure E5.6 (a) Mechanized walkway for an airport; (b) control volume for mechanized walkway. (Drawings not to scale.)

peak traffic times. Estimate the force that must be supplied to the walkway by the five drive cylinders.

Solution

Given
400-m mechanized walkway moving at 4 km/hr
Peak traffic rate 2000 persons per hour
Figure E5.6(a)

Find
Force that must be supplied to the walkway by the five drive cylinders

Solution
Assume that the people enter the walkway and leave the walkway at the same rate such that a steady-flow situation exists at the peak traffic period.

The linear momentum theorem is applied to a control volume enclosing the walkway and written in its direction of travel (the x-direction) (see Fig. E5.6b):

$$\dot{M}_{x,\text{out}} - \dot{M}_{x,\text{in}} = \sum F_x.$$

But

$$\dot{M}_{x,\text{out}} - \dot{M}_{x,\text{in}} = \sum_{\text{out}} \dot{m}u - \sum_{\text{in}} \dot{m}u.$$

We assume that the people and the luggage moving onto the walkway have no x-velocity component and that any people and luggage leaving the walkway have an x-velocity component equal to that of the walkway. Therefore

$$u_{\text{out}} = U \quad \text{and} \quad u_{\text{in}} = 0.$$

We take $\dot{m}_{\text{in}}$ and $\dot{m}_{\text{out}}$ to represent the rate of mass entering and leaving the control volume, specifically the mass of the people and their luggage moving onto or off the walkway in unit time. Assuming an average mass of 70 kg per person and 2000 persons per hour gives

$$\dot{m}_{\text{out}} = \dot{m}_{\text{in}} \approx \left(70\,\frac{\text{kg}}{\text{person}}\right)\left(2000\,\frac{\text{persons}}{\text{hr}}\right)\left(\frac{\text{hr}}{3600\ \text{s}}\right) = 38.9\ \text{kg/s}.$$

The linear momentum theorem gives

$$\dot{m}(U) - \dot{m}(0) = F, \quad \text{or} \quad F = \dot{m}U.$$

The numerical values give

$$F = \left(38.9\,\frac{\text{kg}}{\text{s}}\right)\left(4\,\frac{\text{km}}{\text{hr}}\right)\left(\frac{10^3\,\text{m}}{\text{km}}\right)\left(\frac{\text{hr}}{3600\,\text{s}}\right)$$
$$= 43.2\,\text{kg}\cdot\text{m/s}^2,$$

or

$$|\;F = 43.2\,\text{N}.\;|$$

Answer ◀

Discussion

Although this was not a fluid mechanics problem, the principles of fluid mechanics (developed from Newton's second law) were applicable. This is not uncommon; you will uncover similar situations in your professional career.

The calculated force is a very small value. The actual force would be considerably larger due to friction in the system and aerodynamic drag on the people being transported. The calculated force represents the force required to accelerate the passengers to the walkway's speed.

EXAMPLE 5.7   Illustrates the linear momentum theorem with nonuniform pressure distribution

The gate in a small lock is stuck open as shown in Fig. E5.7(a). Water flows through the lock at the steady rate of 500 ft³/sec. Find the force on the partially open gate.

Solution

Given

Water flow rate 500 ft³/sec through partially opened gate
Figure E5.7(a)

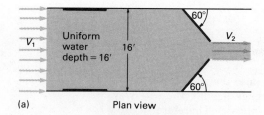

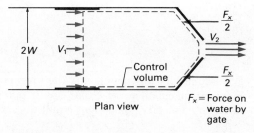

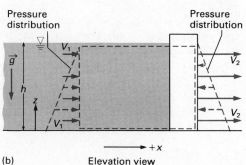

(a)   Plan view

Figure E5.7 (a) Lock; (b) control volume.

Find

Force on the partially opened gate

Solution

Plan and elevation views of the water and the lock are shown in Fig. E5.7(b).

The linear momentum theorem is written in the direction of flow for the control volume shown in Fig. E5.7(b):

$$\dot{M}_{x,\text{out}} - \dot{M}_{x,\text{in}} = \sum F_x.$$

We assume uniform velocity over both inlet and outlet, and the momentum flux terms are given by Eq. (5.14b):

$$\dot{M}_{x,\text{out}} - \dot{M}_{x,\text{in}} = \dot{m}_{\text{out}} V_{x,\text{out}} - \dot{m}_{\text{in}} V_{x,\text{in}}.$$

For steady flow, the integral continuity equation (3.44) gives

$$\dot{m}_{\text{in}} = \dot{m}_{\text{out}} = \dot{m},$$

and for constant density, Eq. (3.43) gives

$$V_{\text{in}} A_{\text{in}} = V_{\text{out}} A_{\text{out}}$$

or

$$V_1 A_1 = V_2 A_2 \quad \text{and} \quad V_2 = \left(\frac{A_1}{A_2}\right) V_1.$$

The momentum flux terms become

$$\dot{M}_{x,\text{out}} - \dot{M}_{x,\text{in}} = \dot{m} V_1 \left(\frac{A_1}{A_2} - 1\right).$$

The force term includes the forces F_x of the gate on the water and the force due to pressure at the inlet and outlet* of the control volume. The equation normal to a streamline, Eq. (4.16), is

$$\frac{V^2}{R} + \frac{1}{\rho}\left(\frac{\partial p}{\partial n}\right) + g\left(\frac{\partial z}{\partial n}\right) = 0.$$

Assuming straight streamlines at the inlet and outlet ($R \to \infty$) and integrating gives

$$p = C - \gamma z,$$

where $\gamma = \rho g$. The constant C is evaluated from the boundary condition

$$p = 0 \quad \text{at} \quad z = h, \quad C = \gamma h.$$

The pressure distribution is

$$p = \gamma(h - z),$$

and the force term is then

$$\sum F_x = -F_x + \int_0^h p(2W)\,dz - \int_0^h p(2)(W - W \sin 60°)\,dz.$$

Substituting for p gives

$$\sum F_x = -F_x + 2W \sin 60° \, \gamma \int_0^h (h - z)\,dz.$$

Integrating, we have

$$\sum F_x = -F_x + \gamma W h^2 \sin 60°.$$

---

* Remember that the uniform atmospheric pressure cancels out.

Substituting the momentum flux terms and the force terms into the momentum theorem gives

$$F_x = -\dot{m}V_1\left(\frac{A_1}{A_2} - 1\right) + \gamma Wh^2 \sin 60°.$$

Referring to Fig. E5.7(b), we have

$$\left(\frac{A_1}{A_2} - 1\right) = \left[\frac{2Wh}{2(W - W\sin 60°)h}\right] - 1 = 6.46.$$

If we assume 60°F water, Table A.6 and Eq. (3.36) give

$$\dot{m} = \rho Q$$

$$= \left(\frac{62.4\ \text{lbm}}{\text{ft}^3}\right)\left(500\ \frac{\text{ft}^3}{\text{sec}}\right) = 31,200\ \frac{\text{lbm}}{\text{sec}}.$$

Equation (3.39) gives

$$V_1 = \frac{Q}{A_1} = \frac{Q}{2Wh}$$

$$= \frac{500\ \dfrac{\text{ft}^3}{\text{sec}}}{(16\ \text{ft})(16\ \text{ft})} = 1.95\ \text{ft/sec}.$$

The force F_x is

$$F_x = -\frac{\left(31,200\ \dfrac{\text{lbm}}{\text{sec}}\right)\left(1.95\ \dfrac{\text{ft}}{\text{sec}}\right)(6.46)}{\left(32.2\ \dfrac{\text{ft}\cdot\text{lbm}}{\text{lb}\cdot\text{sec}^2}\right)} + \left(\frac{62.4\ \text{lb}}{\text{ft}^3}\right)(8\ \text{ft})(16\ \text{ft})^2 \sin 60°$$

$$= (-12,200 + 110,700)\ \text{lb},$$

$$F_x = 98,500\ \text{lb}.$$

This force acts to the *left* (Fig. E5.7b). The force of the water on the lock is the opposite of F_x:

$$\left| \begin{array}{c} F(\text{water on lock}) = 98,500\ \text{lb} \\ \text{Acts to the right.} \end{array} \right.$$

Answer ◄

Discussion

The pressure distribution is hydrostatic, so we could have borrowed results from Chapter 2 to calculate the pressure force on the inlet and outlet areas:

$$F_{\text{pressure}} = \gamma(\text{Area})\left(\frac{\text{Depth}}{2}\right).$$

5.3.2 Linear Momentum Equation for Moving and Deforming Control Volumes

The general form of the linear momentum equation, Eq. (5.10), is applicable to moving or deforming control volumes if the velocities are measured with respect to an inertial reference frame.

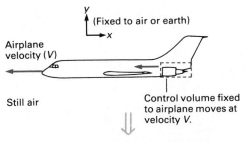

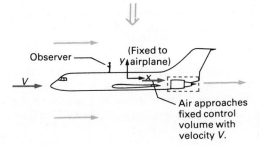

Figure 5.11 Transforming moving control volume to fixed control volume.

The most common moving control volume application involves a control volume with constant velocity. Suppose that an engineer needs to analyze the flow through an airplane's jet engines. The logical control volume choice is one that encloses the engine and travels along with it (Fig. 5.11). If the airplane is cruising at constant velocity, then the control volume also moves with constant velocity. One way to handle this problem is to add a uniform velocity equal and opposite to the airplane velocity at every point in the field, which brings the airplane, and hence the control volume, to rest. This is equivalent to attaching a coordinate system to the control volume (airplane). In such a coordinate system, the control volume is fixed, so the simplified methods of the last section can be employed. The forces acting on the material in the control volume are unaffected by this transformation. This approach is called a *Galilean transformation*.

Another way to handle this problem is to imagine an observer riding on the airplane. This observer sees the air approaching the airplane with a velocity that is equal in magnitude and opposite in direction to the airplane's velocity. The observer naturally chooses a coordinate system that is fixed to the airplane, draws a control volume around the engine, and proceeds to analyze the flow through a fixed control volume. If the airplane is not accelerating, the observer can use Eq. (5.10), or any of its simplified forms, with the velocities defined with respect to the airplane-based coordinates. The forces acting on the material in the control volume are unaffected by a Galilean transformation or by the motion of the observer.

The most common application of a deforming control volume involves situations where the control volume must grow or shrink to completely contain the fluid. The control volume inside the piston-cylinder shown in Fig. 5.3 is a typical example. Deforming control volumes usually have accumulation of momentum within them* so the first term of the linear momentum equation

$$\left(\frac{d}{dt} \int_{V_{cv}} \rho \, \vec{V} \, dV \right)$$

must be carefully evaluated.

---

* Deforming control volumes sometimes have accumulation of mass also.

EXAMPLE 5.8   Illustrates application of the linear momentum theorem to moving control volume

The curved vane in Fig. E5.8(a) moves with a constant absolute velocity U along a horizontal frictionless surface. Water flows out of the nozzle with a constant absolute velocity V and impinges on the vane. Develop an expression for the force $\vec{F}_R$ resisting the motion of the vane. Neglect the force of gravity. Consider two approaches:

a) The control volume moves with the vane, but the observer (and the coordinate system) is fixed on the nozzle (or ground).
b) The control volume, the observer, and the coordinate system all move with the vane.

Solution

Given
Curved vane in Fig. E5.8(a)
Absolute velocity U along horizontal frictionless surface
Water with absolute horizontal velocity V impinging on vane

Find
Expression for the force $\vec{F}_R$ resisting the motion of the vane by using a control volume moving with the vane
Use two different observers (coordinate systems):
a) observer (and coordinate system) fixed on ground,
b) observer (and coordinate system) moving with vane.

Solution
The control volume moves with the vane, as shown in Fig. E5.8(b). For this control volume the uniform atmospheric pressure acting on the entire control surface may be removed from consideration. The velocity V is assumed to be greater than the velocity U, so the vane does not outrun the water. The water density is assumed constant, and fluid velocities are assumed uniform over each flow area.

The appropriate principle is the linear momentum theorem. We will first consider an observer XY fixed on the ground. This observer uses Eq. (5.11):

$$\frac{d}{dt} (\vec{M}_{cv}) + \vec{M}_{out} - \vec{M}_{in} = \sum \vec{F}.$$

Since V and U are constant and the vane does not outrun the water, the observer sees the identical flow situation inside the control volume at every in-

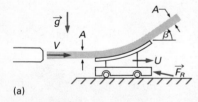

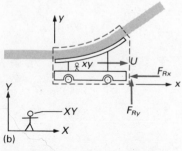

Figure E5.8 (a) Water impinging on a moving vane; (b) control volume moving with vane and observers *xy* and *XY* fixed on their respective coordinate systems.

stant of time and concludes

$$\frac{d}{dt}(\vec{M}_{cv}) = 0.$$

The linear momentum equations in the x- and y-directions are

$$\sum_{out}(\rho V_n A)u - \sum_{in}(\rho V_n A)u = -F_{Rx}$$

and

$$\sum_{out}(\rho V_n A)v - \sum_{in}(\rho V_n A)v = F_{Ry},$$

where ρ, A_{in}, and A_{out} are given information (either in the problem statement or in Fig. E5.8a).

Experience shows that it is useful to apply the continuity equation before evaluating the rest of the terms in the linear momentum theorem. Equation (3.43) gives

$$\sum_{in}(\rho V_n A) = \sum_{out}(\rho V_n A).$$

The inlet and outlet have the same density and flow area, so

$$(V_n)_{in} = (V_n)_{out}.$$

$(V_n)_{in}$ is the inlet normal velocity *relative to the control surface*,

$$(V_n)_{out} = (V_n)_{in} = V - U.$$

Evaluating the terms in the linear momentum theorem for the observer on the ground, we have

$$u_{in} = V,$$
$$v_{in} = 0,$$
$$u_{out} = U + (V_n)_{out}\cos\beta = U + (V - U)\cos\beta,$$

and

$$v_{out} = 0 + (V_n)_{out}\sin\beta = (V - U)\sin\beta.$$

Substituting these expressions into the x-component of the linear momentum theorem, we have

$$\rho(V - U)A[U + (V - U)\cos\beta] - \rho(V - U)AV = -F_{Rx},$$

and, solving for F_{Rx},   **Answer**

$$\left| \; F_{Rx} = \rho A(V - U)^2(1 - \cos\beta). \; \right| \quad \blacktriangleleft$$

Substituting the appropriate expressions into the y-component of the linear momentum theorem gives

$$\rho(V - u)A(V - U)\sin\beta - \rho(V - U)A(0) = F_{Ry},$$

so   **Answer**

$$\left| \; F_{Ry} = \rho A(V - U)^2\sin\beta. \; \right| \quad \blacktriangleleft$$

We now consider an observer xy moving with the vane. This observer sees a steady flow, so

$$\frac{\partial}{\partial t}(\vec{M}_{CV}) = 0.$$

The observer then uses Eqs. (5.15a) and (5.15b) to give

$$\sum_{out}(\rho V_n A)u - \sum_{in}(\rho V_n A)u = -F_{Rx}$$

and

$$\sum_{\text{out}} (\rho V_n A)v - \sum_{\text{in}} (\rho V_n A)v = +F_{Ry},$$

where ρ, A_{in}, and A_{out} are given information. The observer on the moving vane sees

$$(V_n)_{\text{in}} = V - U,$$
$$(V_n)_{\text{out}} = V - U,$$
$$u_{\text{in}} = V - U,$$
$$v_{\text{in}} = 0,$$
$$u_{\text{out}} = (V - U)\cos\beta, \qquad \text{and}$$
$$v_{\text{out}} = (V - U)\sin\beta.$$

The force terms are independent of the observer (or the coordinate system). The x- and y-components are again denoted by F_{Rx} and F_{Ry} and shown in Fig. E5.8(b).

Substituting these expressions into the x-component of the linear momentum theorem gives

$$\rho(V - U)A(V - U)\cos\beta - \rho(V - U)A(V - U) = -F_{Rx},$$

or

Answer

$$|\ F_{Rx} = \rho A(V - U)^2(1 - \cos\beta),\ |$$

the same answer we obtained previously.

Substituting the appropriate expressions into the y-component of the linear momentum theorem gives

$$\rho(V - U)A(V - U)\sin\beta - \rho(V - U)A(0) = F_{Ry},$$

or

Answer

$$|\ F_{Ry} = \rho A(V - U)^2\sin\beta,\ |$$

also the same answer we obtained previously.

Discussion

It should not be surprising that both approaches gave the same expression for F_{Rx} and F_{Ry}. The magnitude of a force is determined by the physical problem. Any method of analysis used to evaluate the force should give the same result if both approaches make use of identical assumptions (as done in this problem).

Did you notice that the analysis was much simpler using the moving observer?

Notice that the vane would accelerate to the right if the resisting force were not present, as there would be an unbalanced force acting on the vane.

EXAMPLE 5.9 Illustrates a higher-order application using the linear momentum theorem

A 3/4-ton pickup truck is traveling at 40 km/hr and is cleaning snow from the street. The blade on the front of the truck is 2.0 m wide and the snow is 0.2 m deep. Find the power required to clean the snow from the street. See Fig. E5.9(a). Neglect friction between the street and the blade and assume the snow follows the contour of the blade.

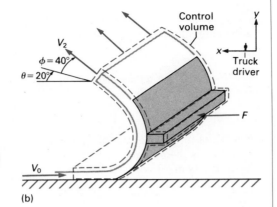

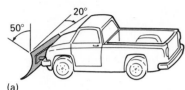

(a)

(b)

Figure E5.9 (a) Pickup truck cleaning snow from street; (b) control volume to find the required force on the blade.

Solution

Given
3/4-ton pickup traveling at 40 km/hr and removing a swath of snow 2.0 m wide and 0.2 m deep.
Neglect friction between the street and the blade
Assume the snow follows the contour of the blade
Figure E5.9(a)

Find
The power required to clean the snow from the streets.

Solution
Since the truck is moving at a constant velocity, we will use the driver as the observer and a coordinate system fixed to the truck. Consider a control volume fixed to the truck as shown in Fig. E5.9(b). Snow enters the control volume with velocity V_0 at the bottom front and leaves with a velocity V_2 at the top. V_2 is normal to the exit area and has been deflected an angle of 20° to the right of the direction of travel and an angle of 40° above the horizontal. If we assume steady conditions (constant truck speed), the horizontal component of Eq. (5.13) is

$$\dot{M}_{x,\text{out}} - \dot{M}_{x,\text{in}} = \sum F_x.$$

The force terms are F and the atmospheric pressure. As in Example 5.8, the uniform atmospheric pressure acting over the entire control surface can be removed from consideration. Therefore

$$\sum F_x = F.$$

The momentum flux terms are evaluated using Eq. (5.15a):

$$\dot{M}_{x,\text{out}} - \dot{M}_{x,\text{in}} = \sum_{\text{out}} (\rho \, V_n A)u - \sum_{\text{in}} (\rho \, V_n A)u.$$

The snow density is unknown. Measurements taken during a recent winter in Akron, Ohio, indicate a range of 96 kg/m³ (6 lbm/ft³) to 320 kg/m³ (20 lbm/ft³). We will use the larger value to obtain a conservatively large value of F. Then

$$\rho_{\text{in}} = \rho_{\text{out}} = 320 \text{ kg/m}^3.$$

The outlet flow area is also unknown. We assume that it is equal to the inlet flow area:

$$A_{\text{out}} = A_{\text{in}} = (2.0 \text{ m})(0.2 \text{ m}) = 0.4 \text{ m}^2.$$

V_n is the magnitude of the normal component of the inlet velocity relative to the control volume. Therefore

$$(V_n)_{in} = V_0.$$

We also have u_{in} as the x-component of the inlet velocity relative to the observer (the truck driver):

$$u_{in} = -V_0$$

since V_0 is in the negative x-direction.

$(V_n)_{out}$ is the magnitude of the normal component of the outlet velocity relative to the control volume. It is found by applying the continuity equation

$$(\rho V_n A)_{out} = (\rho V_n A)_{in}.$$

Since

$$\rho_{out} = \rho_{in} \quad \text{and} \quad A_{out} = A_{in},$$

we conclude

$$(V_n)_{out} = (V_n)_{in} = V_0.$$

Now u_{out} is the x-component of the outlet velocity relative to the truck. Figure E5.9(b) gives

$$u_{out} = V_2 \cos \phi \cos \theta.$$

Since

$$V_2 = (V_n)_{out}, \quad \text{then} \quad V_2 = V_0.$$

Substituting the above into the linear momentum theorem gives

$$\rho_{in} A_{in}[(V_0)(V_0 \cos \phi \cos \theta) - (V_0)(-V_0)] = F,$$

or

$$F = \rho_{in} A_{in} V_0^2 (1 + \cos \phi \cos \theta).$$

The numerical values give

$$F = \frac{\left(320 \, \dfrac{kg}{m^3}\right)(0.4 \, m^2)\left(40 \, \dfrac{km}{hr}\right)^2 (1 + \cos 40° \cos 20°)}{\left(\dfrac{10^{-3} \, km}{m}\right)^2 \left(\dfrac{3600 \, s}{hr}\right)^2}$$

$$= 27{,}200 \, N.$$

F acts to the left since it is negative.

The required power is

$$\dot{W} = FV_0$$

$$= (27{,}200 \, N)\left(40 \, \frac{km}{hr}\right)\left(\frac{10^3 \, m}{km}\right)\left(\frac{hr}{3600 \, s}\right)$$

$$= 302{,}000 \, \frac{N \cdot m}{s} = 302{,}000 \, W,$$

or

Answer

$$\left| \, \dot{W} = 302 \, kW. \, \right|$$

Discussion

Several assumptions were made in solving this example. One was that the snow density is constant; however, it is likely that the snow would be compacted as the blade is pushed into the snow, and the snow may break up into lumps as it passes through the blade. A second assumption is that the exit flow area equals the inlet flow area. Both assumptions would be hard to justify without conducting a test. If we conducted several tests, the results may not even be repeatable as the snow may break up in one test and not in another. The extent to which the snow is compacted may vary, with wet snow being compacted more than dry snow.

However, we need not run tests to determine the effect of these assumptions on the result. We write the momentum flux terms as

$$\dot{M}_{x,\text{out}} - \dot{M}_{x,\text{in}} = (\rho A)_{\text{out}}(V_n u)_{\text{out}} - (\rho A)_{\text{in}}(V_n u)_{\text{in}}.$$

The integral continuity equation gives

$$(V_n)_{\text{out}} = \frac{(\rho A)_{\text{in}}}{(\rho A)_{\text{out}}} (V_n)_{\text{in}} = \frac{(\rho A)_{\text{in}}}{(\rho A)_{\text{out}}} V_0.$$

Also

$$u_{\text{in}} = -V_0,$$

$$u_{\text{out}} = (V_n)_{\text{out}} \cos\phi \cos\theta = \frac{(\rho A)_{\text{in}}}{(\rho A)_{\text{out}}} (V_0) \cos\phi \cos\theta.$$

Substituting into the linear momentum equation gives

$$F = (\rho A)_{\text{in}} V_0^2 \left[1 + \frac{(\rho A)_{\text{in}}}{(\rho A)_{\text{out}}} \cos\phi \cos\theta \right].$$

The power requirement is

$$\dot{W}' = (\rho A)_{\text{in}} V_0^3 \left[1 + \frac{(\rho A)_{\text{in}}}{(\rho A)_{\text{out}}} \cos\phi \cos\theta \right].$$

The effect of the constant-density, constant-area assumption is given by the ratio of $\dot{W}'$ to $\dot{W}$ as calculated in the solution:

$$R = \frac{\dot{W}'}{\dot{W}} = \frac{1 + \dfrac{(\rho A)_{\text{in}}}{(\rho A)_{\text{out}}} (\cos\phi \cos\theta)}{(1 + \cos\phi \cos\theta)}.$$

You may wish to plot values of R as a function of

$$r = \frac{(\rho A)_{\text{in}}}{(\rho A)_{\text{out}}}$$

for values of r from 0.25 to 2.0. The result should give you an appreciation of the effect of the assumed constant-density and equal inlet and exit area assumptions.

You may wish to also solve the following two problems:

■ Find the y- and z-components of the force acting between the blade and the truck. Note that we found only the x-component in this example.
■ Solve for F using an observer and coordinate system fixed on the street.

Application of the linear momentum theorem to accelerating control volumes is sometimes necessary. An engineer who designs pumps often uses a control volume fixed to a rotating impeller. In the analysis of meteorological flows for weather prediction, it is usually necessary to account for the fact that a control volume fixed to the surface of the earth has a (small) acceleration with respect to the fixed stars due to the earth's rotation. Momentum analysis of fluid motion through an accelerating control volume is complicated by the fact that a coordinate system fixed with respect to the control volume is noninertial.

An engineer may select a control volume that is fixed in a noninertial coordinate system because the flows into and out of the control volume can be expressed conveniently in this system. Figure 5.12 shows a control volume that is fixed with respect to a noninertial reference frame, xyz. The noninertial frame moves with respect to an inertial reference frame, XYZ. We can apply Newton's second law only in the XYZ-frame. For the system that instantaneously occupies the control volume,

$$\frac{d}{dt}(\vec{M}_{\text{sys}})_{XYZ} = \sum \vec{F}.$$

To derive an expression for the rate of change of system momentum, we first consider a single mass particle within the control volume. The momentum of this particle is

$$\delta \vec{M}_{XYZ} = (\delta m)\vec{V}_{XYZ}.$$

From elementary dynamics, we have

$$\vec{V}_{XYZ} = \vec{V}_{xyz} + \frac{d\vec{R}}{dt} + \vec{\Omega} \times \vec{r}. \tag{5.26}$$

The rate of change of the particle's momentum is

$$\frac{d}{dt}(\delta \vec{M}_{XYZ}) = \frac{d}{dt}(\delta m \, \vec{V}_{XYZ}) = \delta m \left(\frac{d\vec{V}_{XYZ}}{dt}\right).$$

Figure 5.12 Control volume in non-inertial reference frame.

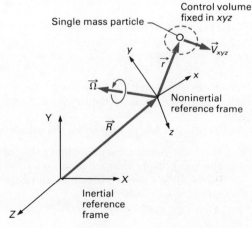

Borrowing another result from elementary dynamics, we find

$$\frac{d\vec{V}_{XYZ}}{dt} = \frac{d\vec{V}_{xyz}}{dt} + \frac{d^2\vec{R}}{dt^2} + \frac{d\vec{\Omega}}{dt} \times \vec{r} + 2\vec{\Omega} \times \vec{V}_{xyz} + \vec{\Omega} \times \vec{\Omega} \times \vec{r}. \qquad (5.27)$$

The rate of change of system momentum is obtained by integrating over all particles that make up the system:

$$\frac{d}{dt}\vec{M}_{XYZ} = \int_{\Psi_{sys}} \frac{d\vec{V}_{XYZ}}{dt}(dm) = \int_{\Psi_{sys}} \rho \frac{d\vec{V}_{XYZ}}{dt}(d\Psi)$$

$$= \int_{\Psi_{sys}} \rho \left(\frac{d\vec{V}_{xyz}}{dt} + \frac{d^2\vec{R}}{dt^2} + \frac{d\vec{\Omega}}{dt} \times \vec{r} + 2\vec{\Omega} \times \vec{V}_{xyz} + \vec{\Omega} \times \vec{\Omega} \times \vec{r} \right) d\Psi.$$

The integral

$$\int_{\Psi_{sys}} \rho \left(\frac{d\vec{V}_{xyz}}{dt} \right) d\Psi$$

represents the rate of change of the system's *xyz*-momentum. An observer in the *xyz*-frame would calculate this by using the control volume rate-of-change equation for a fixed control volume:

$$\int_{\Psi_{sys}} \rho \left(\frac{d\vec{V}_{xyz}}{dt} \right) d\Psi = \frac{\partial}{\partial t} \int_{\Psi_{cv}} \rho \vec{V}_{xyz}\, d\Psi + \oint_{A_{cv}} \rho \vec{V}_{xyz}(\vec{V}_{xyz} \cdot \hat{n})\, dA.$$

The remainder of the terms that make up the rate of change of the system's *XYZ*-momentum are not affected by the flow as viewed from the *xyz*-frame. Since the system of interest instantaneously fills the control volume, the integration of these terms can be performed over the control volume Ψ_{cv}. Newton's second law for a control volume that is fixed with respect to a noninertial reference frame thus becomes

$$\frac{\partial}{\partial t} \int_{\Psi_{cv}} \rho \vec{V}_{xyz}\, d\Psi + \oint_{A_{cv}} \rho \vec{V}_{xyz}(\vec{V}_{xyz} \cdot \hat{n})\, dA + \int_{\Psi_{cv}} \rho \left[\frac{d^2\vec{R}}{dt^2} + \left(\frac{d\vec{\Omega}}{dt} \times \vec{r} \right) \right.$$

$$\left. + (2\vec{\Omega} \times \vec{V}_{xyz}) + (\vec{\Omega} \times \vec{\Omega} \times \vec{r}) \right] d\Psi = \sum \vec{F}. \qquad (5.28)$$

This equation is the linear momentum equation for a noninertial rigid control volume. The first two terms on the left are the "ordinary" momentum accumulation and momentum flux terms, evaluated by an observer in the noninertial (moving) reference frame. To an observer in this frame, the additional momentum terms appear to be forces. Often these terms are transferred to the force side of the equation and treated as additional forces. An observer in the moving reference frame would use the following form of the equation:

$$\frac{\partial}{\partial t} \int_{\Psi_{cv}} \rho \vec{V}\, d\Psi + \oint_{A_{cv}} \rho \vec{V}(\vec{V}_r \cdot \hat{n})\, dA$$

$$= \sum \vec{F} - \int_{\Psi_{cv}} \rho \left[\frac{d^2\vec{R}}{dt} + \left(\frac{d\vec{\Omega}}{dt} \times \vec{r} \right) + (2\vec{\Omega} \times \vec{V}) + (\vec{\Omega} \times \vec{\Omega} \times \vec{r}) \right] d\Psi. \qquad (5.29)$$

The last two terms in the brackets under the integral on the right contribute the Coriolis force and centrifugal force. The *xyz*-subscript has been omitted from the velocity $\vec{V}$ since this equation is written from the viewpoint of an observer in the *xyz*-system.

5.3.3 Connection Between Control Volume and Differential Approaches: Derivation of Euler's Equations Using the Linear Momentum Theorem

As a demonstration of the essential equivalence between the differential and control volume approaches, we will now derive Euler's equations starting from the control volume formulation. In the interest of simplicity, we will consider a two-dimensional, two-directional flow. To arrive at Euler's equations, we assume that the fluid is inviscid. Since Euler's equations are differential equations, it is necessary to analyze the flow through a differentially small control volume. Figure 5.13 shows the details of the flow in the vicinity of a point (x, y) in a two-directional, two-dimensional flow. A rectangular control volume of dimensions δx, δy, and δz encloses the point. The control volume is fixed and rigid, so the linear momentum equation is

$$\frac{\partial}{\partial t} \int_{V_{cv}} \rho \vec{V} \, dV + \int_{A_{\text{out}}} \rho \vec{V}(V_n \, dA) - \int_{A_{\text{in}}} \rho \vec{V}(V_n \, dA) = \sum \vec{F}.$$

First consider the x-component of this equation. Since the control volume is small, we can evaluate the integrals by assuming that velocities and densities are uniformly distributed across each face of the control volume. The x-momentum equation is

$$\frac{\partial}{\partial t} (\rho u)(\delta x)(\delta y)(\delta z) + \sum_{\text{out}} (\rho V_n \delta A) u - \sum_{\text{in}} (\rho V_n \delta A) u$$

$$= \sum \delta F_x = \delta F_{x,\text{pressure}} + \delta F_{x,\text{gravity}}.$$

Figure 5.13 Differentially small control volume in two-directional, two-dimensional flow field.

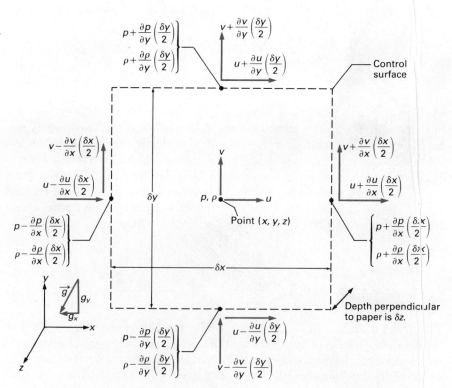

Note that mass and momentum flow into the control volume at the left and bottom faces and out of the control volume at the right and top faces. Substituting the appropriate velocities and expressions for the pressure and gravity forces, we get

$$
\frac{\partial}{\partial t}(\rho u)(\delta x)(\delta y)(\delta z) + \left[\rho + \frac{\partial \rho}{\partial x}\left(\frac{\delta x}{2}\right)\right]\left[u + \frac{\partial u}{\partial x}\left(\frac{\delta x}{2}\right)\right](\delta y)(\delta z)\left[u + \frac{\partial u}{\partial x}\left(\frac{\delta x}{2}\right)\right]
$$

$$
+ \left[\rho + \frac{\partial \rho}{\partial y}\left(\frac{\delta y}{2}\right)\right]\left[v + \frac{\partial v}{\partial y}\left(\frac{\delta y}{2}\right)\right](\delta x)(\delta z)\left[u + \frac{\partial u}{\partial y}\left(\frac{\delta y}{2}\right)\right]
$$

$$
- \left[\rho - \frac{\partial \rho}{\partial x}\left(\frac{\delta x}{2}\right)\right]\left[u - \frac{\partial u}{\partial x}\left(\frac{\delta x}{2}\right)\right](\delta y)(\delta z)\left[u - \frac{\partial u}{\partial x}\left(\frac{\delta x}{2}\right)\right]
$$

$$
- \left[\rho - \frac{\partial \rho}{\partial y}\left(\frac{\delta y}{2}\right)\right]\left[v - \frac{\partial v}{\partial y}\left(\frac{\delta y}{2}\right)\right](\delta x)(\delta z)\left[u - \frac{\partial u}{\partial y}\left(\frac{\delta y}{2}\right)\right]
$$

$$
= \left[p - \frac{\partial p}{\partial x}\left(\frac{\delta x}{2}\right)\right](\delta y)(\delta z) - \left[p + \frac{\partial p}{\partial x}\left(\frac{\delta x}{2}\right)\right](\delta y)(\delta z) + \rho g_x(\delta x)(\delta y)(\delta z).
$$

Carrying out the indicated multiplications, canceling terms where possible, and using the product rule for derivatives in the first term on the left, we obtain

$$
\left[\rho\left(\frac{\partial u}{\partial t}\right) + u\left(\frac{\partial \rho}{\partial t}\right) + 2\rho u\left(\frac{\partial u}{\partial x}\right) + u^2\left(\frac{\partial \rho}{\partial x}\right) + \rho v\left(\frac{\partial u}{\partial y}\right) + \rho u\left(\frac{\partial v}{\partial y}\right) + uv\left(\frac{\partial \rho}{\partial y}\right)\right]
$$

$$
\times (\delta x)(\delta y)(\delta z) = -\frac{\partial p}{\partial x}(\delta x)(\delta y)(\delta z) + \rho g_x(\delta x)(\delta y)(\delta z).
$$

We now divide by $(\delta x)(\delta y)(\delta z)$ and take the limit, shrinking the control volume to a point. We also group the second, fourth, sixth, and seventh terms on the left together with one of the two $[\rho u(\partial u/\partial x)]$ terms in the third term on the left to obtain

$$
\rho\left(\frac{\partial u}{\partial t}\right) + \rho u\left(\frac{\partial u}{\partial x}\right)\rho v\left(\frac{\partial u}{\partial y}\right) + u\left[\frac{\partial \rho}{\partial t} + u\left(\frac{\partial \rho}{\partial x}\right) + v\left(\frac{\partial \rho}{\partial y}\right) + \rho\left(\frac{\partial u}{\partial x} + \frac{\partial v}{\partial y}\right)\right]
$$

$$
= -\frac{\partial p}{\partial x} + \rho g_x.
$$

The differential continuity equation for two-dimensional, two-directional flow, Eq. (3.16), shows that the sum of the terms enclosed in the [ ] is zero, so the final form of the momentum equation is (on dividing by ρ)

$$
\frac{\partial u}{\partial t} + u\left(\frac{\partial u}{\partial x}\right) + v\left(\frac{\partial u}{\partial y}\right) = -\frac{1}{\rho}\left(\frac{\partial p}{\partial x}\right) + g_x.
$$

This is Euler's equation of motion for the x-direction, derived earlier as Eq. (4.29).

If we apply the linear momentum theorem in the y-direction, we eventually obtain Euler's equation of motion for the y-direction, earlier derived as Eq. (4.30):

$$
\frac{\partial v}{\partial t} + u\left(\frac{\partial v}{\partial x}\right) + v\left(\frac{\partial v}{\partial y}\right) = -\frac{1}{\rho}\left(\frac{\partial p}{\partial y}\right) + g_y.
$$

We will leave that derivation to you as an exercise. If you decide to try it, remember:

- y-momentum flows across all four faces of the control volume.
- Be sure to remove the terms involved with the differential continuity equation to get the simplest form of the result (Eq. 4.30).

5.4 The Laws of Thermodynamics: The General Energy Equation and the Mechanical Energy Equation

We now turn to consideration of the energy generated by and transported with moving fluids. We will begin by combining the control volume rate-of-change equation with the first law of thermodynamics (conservation of energy) to obtain the general energy equation.

The first law of thermodynamics requires that if a net energy transfer to a system occurs, the energy contained (stored) in the system must increase by an amount equal to the energy transferred. Energy can be transferred to a system by two different processes, *heat* and *work*. Heat transfer (Q) is energy transfer across a system boundary due to difference in temperature between the system and its surroundings. Work (W) is energy transfer across a system boundary due to the action of a force through a distance or any other energy transfer that could be replaced by action of a force through a distance.*

We call energy stored in a system (that is, energy contained in the system's mass) *intrinsic energy*. There are five forms of intrinsic energy:

1. *kinetic energy,* the energy of mass in motion;
2. *potential energy,* the energy of position in a force field (usually a gravitational field but sometimes an electric or magnetic field);
3. *internal energy,* the energy of molecular structure and motion;
4. *chemical energy,* the energy associated with the rearrangement of molecules (released only in a chemical reaction); and
5. *nuclear energy,* the energy associated with the structure of atoms (released only by fission or fusion).

In many problems in fluid mechanics, changes in nuclear and chemical energy[†] do not occur; accordingly, we will neglect these forms of energy in subsequent discussions.

The intrinsic energies are proportional to the amount of mass in the system. The intrinsic energies can also be related to properties of the system (see Section 1.3). It is a fundamental postulate of thermodynamics that internal energy is a system property and, as such, is related to other system properties, primarily mass, pressure, and temperature.

---

* In fluid mechanics we are almost always concerned with work that actually is the action of a force through a distance; however, the flow of electric current into a system is uaually also considered work since a mechanism could be devised (a frictionless electric motor) that completely converts the current flow to the action of a force through a distance (for example, using the motor to raise a weight).

† Probably the most common example of an engineering flow with changes of chemical energy is combustion.

We often deal with *specific energies,* which are energies divided by system mass, that is, energy per unit mass. Specific intrinsic energies are not proportional to system mass since system mass has been divided out. Specific kinetic energy is one-half the square of the fluid velocity ($V^2/2$). Specific potential energy is the acceleration of gravity multiplied by elevation above a specific datum (gz). Specific internal energy* ($\tilde{u}$) is a property that depends only on other specific properties, primarily temperature.[†] We also define specific work, w, and specific heat transfer, q, as work and heat transfer divided by the system mass; however, *specific work and specific heat transfer are not system properties.*

For later discussion, we can classify various energies as either *thermal* or *mechanical.* Thermal energies are associated with temperature, molecular structure, and heat transfer. Mechanical energies are associated with force and motion. Neglecting chemical and nuclear energies, we can classify the other forms as follows:

Mechanical Energies	**Thermal Energies**
Work	Heat
Kinetic energy	Internal energy
Potential energy	

The law of conservation of energy in rate-of-change form for a system is

$$\begin{pmatrix} \text{Rate of} \\ \text{heat transfer} \\ \text{to system} \end{pmatrix} + \begin{pmatrix} \text{Rate of work} \\ \text{done on} \\ \text{system} \end{pmatrix} = \begin{bmatrix} \text{Rate of increase of intrinsic} \\ \text{(kinetic + potential + internal)} \\ \text{energy of system} \end{bmatrix} \quad \textbf{(5.30)}$$

In this equation, there is no distinction between the thermal and mechanical forms of energy; a given energy transfer may result in an increase of the system's mechanical energy, thermal energy, or both.

Introducing symbols and sign conventions customary in thermodynamics and fluid mechanics, we have

$$\frac{dQ}{dt} - \frac{dW}{dt} = \frac{dE_{\text{sys}}}{dt},$$

where

$$E_{\text{sys}} = \int_{\mathcal{V}_{\text{sys}}} \rho \left(\tilde{u} + \frac{V^2}{2} + gz \right) d\mathcal{V};$$

heat transferred to a system is positive; and
work done by (that is, transferred from) a system is positive.

We can obtain the control volume form of this equation by using the control volume rate-of-change equation with $b = \tilde{u} + V^2/2 + gz$. The

---

* The symbol u for internal energy per unit mass is almost universal in thermodynamics. In fluid mechanics, it might be confused with the x-direction velocity. To avoid confusion, we will use the slightly more cumbersome symbol $\tilde{u}$ to represent internal energy per unit mass.

† Similar to specific potential energy, a zero datum must be selected for internal energy. We will specify that $\tilde{u} = 0$ at the absolute zero of temperature.

result is

$$\dot{Q} - \dot{W} = \frac{d}{dt} \int_{V_{cv}} \rho \left(\tilde{u} + \frac{V^2}{2} + gz \right) dV + \oint_{A_{cv}} \rho \left(\tilde{u} + \frac{V^2}{2} + gz \right) (\vec{V}_r \cdot \hat{n}) \, dA. \quad \textbf{(5.31)}$$

It is customary to use the overdot (˙) to indicate rates of heat transfer and work. Rate of work is also called *power*.

Before we consider simplified working forms of this equation, we will examine the work term. If we neglect electrical and other "equivalent" forms of work, there are three types of work that might be done on or by the fluid inside the control volume (see Fig. 5.14):

- Shaft work (W_{shaft}), transmitted by a rotating shaft such as a pump drive shaft or a turbine output shaft that is "cut" by the control surface. This work is done by shear stresses in the "cut" shaft, so it is somewhat similar to shear work. Shaft work is sometimes called "pump work" (W_p), if work is done on the fluid, or "turbine work" (W_t), if work is delivered to the surroundings.
- Shear work (W_{shear}), done by the shear stresses *in the fluid* acting on the boundaries of the control volume.
- Pressure work (W_{pressure}), done by the fluid pressure acting on the boundaries of the control volume.

The total work is the sum of these three types:

$$W = W_{\text{shaft}} + W_{\text{shear}} + W_{\text{pressure}}.$$

Equation (5.31) requires the rate of work, which is simply

$$\dot{W} = \dot{W}_{\text{shaft}} + \dot{W}_{\text{shear}} + \dot{W}_{\text{pressure}}.$$

We can develop equations for calculating each of the work rates. The shear work rate is evaluated from the product of shear stress, area, and fluid velocity in the direction of the stress force:

$$\dot{W}_{\text{shear}} = \oint_{A_{cv}} (\vec{\tau} \cdot \vec{V}) \, dA, \quad \textbf{(5.32)}$$

where $\vec{\tau}$ is the shear stress vector discussed earlier.

Figure 5.14 Various forms of work exchange with moving fluid.

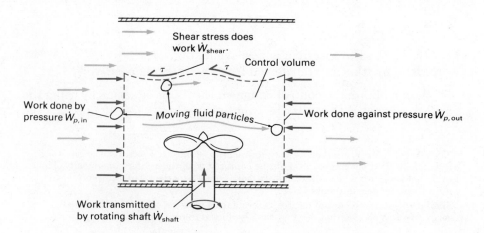

We often choose a control volume with control surfaces lying adjacent to solid boundaries so that the fluid velocity is zero there (Fig. 5.15a). If this is the case, there is no shear work even though there may be a shear stress. If for some reason we choose a control surface that lies within a moving fluid (Fig. 5.15b), there will be shear work unless the fluid is inviscid.

The shaft work rate is equal to the shaft torque multiplied by the shaft rotational speed:

$$\dot{W}_{\text{shaft}} = \mathscr{T}\omega.$$

The net shaft torque can be calculated by integrating the shear stress over the shaft cross section, so

$$\dot{W}_{\text{shaft}} = \omega \int_{A,\text{shaft}} \tau r \, dA. \tag{5.33}$$

Rather than calculating shaft work by this equation, it is usually left as a lumped term $\dot{W}_{\text{shaft}}$ (or possibly as two lumps, $\dot{W}_p$ and $\dot{W}_t$).

The effects of shaft work and shear work are quite similar. We will usually find it convenient to combine $\dot{W}_{\text{shaft}}$ and $\dot{W}_{\text{shear}}$ into a single work $\dot{W}_s$,

$$\dot{W}_s = \dot{W}_{\text{shaft}} + \dot{W}_{\text{shear}}. \tag{5.34}$$

W_s is work done on the fluid inside the control volume by forces other than those due to pressure.

Work done by pressure acting at the control volume boundary is of two types, *flow work* and *work of (control volume) deformation*. Flow work must always be included in the energy equation for moving fluids. Work of deformation occurs only if a deforming control volume is selected.

We can develop an expression for flow work by examining outflow from and inflow to the control volume. Figure 5.16 shows a portion of the control volume where fluid is flowing outward across the control surface. To leave the control volume, the fluid must "push" the fluid on

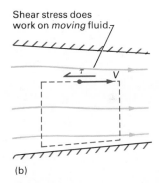

Fluid is at rest at solid boundary.

(a)

Shear stress does work on *moving* fluid.

(b)

Figure 5.15 Control volumes (a) without and (b) with shear work and (a) with and (b) without shaft work.

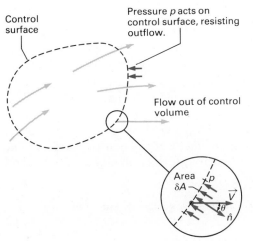

Control surface

Pressure p acts on control surface, resisting outflow.

Flow out of control volume

Area δA

Figure 5.16 Detail of flow outward from control volume for calculation of pressure work.

the outside of the control volume out of the way. This push is accomplished by the fluid pressure (the force per unit area) at the control surface.

As the outside fluid is pushed out of the way, work is done on it by the fluid behind it pushing its way out. At every point of outflow, the fluid leaving the control volume is doing work on the fluid outside. According to the sign convention, this is positive work. Similarly, at every point where fluid flows into the control volume, the fluid entering does work on the fluid inside as it pushes its way in. According to the sign convention, this is negative work.

Consider the detail of the outflow process shown in Fig. 5.16. Since

$$\text{Work} = (\text{Force}) \times (\text{Distance moved parallel to force}),*$$

then

$$\delta \dot{W}_{\text{flow}} = \delta F \times (\text{Rate of motion parallel to force}).$$

The pressure force is

$$\delta F = p \, \delta A,$$

and the rate of motion of the fluid parallel to the force is

$$(\text{Rate of motion in direction of force}) = V_n = V \cos \theta = \vec{V} \cdot \hat{n}.$$

Since we are calculating the work done by fluid leaving the control volume on fluid outside the control volume, we really should be using the velocity relative to the control surface ($\vec{V}_r$). The rate of flow work done at area δA is

$$\delta \dot{W}_{\text{flow}} = p(\vec{V}_r \cdot \hat{n}) \, \delta A. \tag{5.35a}$$

If we consider an inflow area, the flow work is calculated by the same expression except that ($\vec{V}_r \cdot \hat{n}$) is negative because the angle θ lies between 90° and 180°. This is exactly as we would like it because outflow work is positive and inflow work is negative. The expression of Eq. (5.35a) is capable of computing either outflow or inflow work. To obtain the total flow work, all we need to do is integrate over the entire control surface:

$$\dot{W}_{\text{flow}} = \oint_{A_{cv}} p(\vec{V}_r \cdot \hat{n}) \, d\hat{A} = \oint_{A_{cv}} p V_{r,n} \, dA. \tag{5.35b}$$

Work of deformation (W_D) occurs if the control volume is deforming. Consider the process of blowing up a balloon. We choose a control volume that exactly coincides with the outside of the balloon. As we blow air in, the balloon and the control volume expand. The material inside the control volume must do work to push the atmosphere out of its way. This work is calculated by the product of the pressure acting on the control surface and the displacement of the control surface. The *rate* of work is the product of pressure and the velocity of the control

---

* Algebraic signs are neglected in this relation since we already know that flow work is positive for outflow regions.

surface. The total rate of deformation work is

$$\dot{W}_D = \oint_{A_{cv}} p(\vec{V}_c \cdot \hat{n})\, dA, \tag{5.36}$$

where $\vec{V}_c$ is the velocity of the control surface.

The total pressure work is

$$\dot{W}_{\text{pressure}} = \dot{W}_{\text{flow}} + \dot{W}_D$$

$$= \oint_{A_{cv}} p(\vec{V}_r \cdot \hat{n})\, dA + \oint_{A_{cv}} p(\vec{V}_c \cdot \hat{n})\, dA. \tag{5.37}$$

We can now collect the various types of work,

$$\dot{W} = \dot{W}_{\text{shaft}} + \dot{W}_{\text{shear}} + \oint_{A_{cv}} p(\vec{V}_r \cdot \hat{n})\, dA + \oint_{A_{cv}} p(\vec{V}_c \cdot \hat{n})\, dA,$$

and substitute into the control volume energy equation, Eq. (5.31), to get

$$\dot{Q} - \dot{W}_{\text{shaft}} - \dot{W}_{\text{shear}} - \oint_{A_{cv}} p(\vec{V}_r \cdot \hat{n})\, dA - \oint_{A_{cv}} p(\vec{V}_c \cdot \hat{n})\, dA$$

$$= \frac{d}{dt} \int_{V_{cv}} \rho\left(\tilde{u} + \frac{V^2}{2} + gz\right) dV + \oint_{A_{cv}} \left(\tilde{u} + \frac{V^2}{2} + gz\right)(\vec{V}_r \cdot \hat{n})\, dA.$$

Now notice the similar form of the flow work integral on the left and the energy flux integral on the right. If we multiply and divide by the density ρ inside the flow work integral, move it to the right side, and replace $\dot{W}_{\text{shaft}} + \dot{W}_{\text{shear}}$ by $\dot{W}_s$, we get

$$\dot{Q} - \dot{W}_s - \int_{A_{cv}} p(\vec{V}_c \cdot \hat{n})\, dA = \frac{d}{dt} \int_{V_{cv}} \rho\left(\tilde{u} + \frac{V^2}{2} + gz\right) dV$$

$$+ \oint_{A_{cv}} \rho\left(\tilde{u} + \frac{p}{\rho} + \frac{V^2}{2} + gz\right)(\vec{V}_r \cdot \hat{n})\, dA. \tag{5.38}$$

We will call this the *general energy equation*. This equation is central to the study of thermodynamics, but it plays a surprisingly minor role in most problems in (incompressible) fluid mechanics. In thermodynamics, the sum $\tilde{u} + p/\rho$ is replaced by its equivalent, the fluid enthalpy $\tilde{h}$. Note that the enthalpy represents a combination of an intrinsic energy ($\tilde{u}$) and an energy transfer (flow work).

5.4.1 Some Simplified Forms of the General Energy Equation

Like the integral continuity equation and the linear momentum equation, the general energy equation is seldom used in its most general mathematical form, Eq. (5.38). In this section we describe some simplified forms and illustrate their use by examples.

Since the general energy equation is a scalar equation, it is not necessary that velocities be measured with respect to an inertial reference frame. If you use a noninertial reference frame, you must be sure to include any work due to imaginary "forces" such as Coriolis forces or centrifugal forces.

Equation (5.38) applies for a moving, deforming control volume. If the control volume is fixed and rigid, then

$$\frac{d}{dt} \int_{V_{cv}} \left(\tilde{u} + \frac{V^2}{2} + gz \right) dV = \frac{\partial}{\partial t} \int_{V_{cv}} \rho \left(\tilde{u} + \frac{V^2}{2} + gz \right) dV,$$

$$\oint_{A_{cv}} p(\vec{V}_c \cdot \hat{n}) \, dA = 0, \quad \text{and} \quad \vec{V}_r = \vec{V}.$$

A great many applications involve steady flow through a fixed, rigid control volume. In this case, $\partial/\partial t = 0$ and the general energy equation becomes

$$\dot{Q} - W_s = \oint_{A_{cv}} \rho \left(\tilde{u} + \frac{p}{\rho} + \frac{V^2}{2} + gz \right) (\vec{V}_r \cdot \hat{n}) \, dA. \tag{5.39}$$

A very common simplification is the assumption that the control volume has inflow and outflow at a finite number of inlets and outlets and that the flow is locally uniform and one-directional at these locations (see Fig. 5.6). Assuming that the control volume is fixed and rigid, the general energy equation is

$$\dot{Q} - \dot{W}_s = \frac{\partial}{\partial t} \int_{V_{cv}} \rho \left(\tilde{u} + \frac{V^2}{2} + gz \right) dV + \sum_{\text{out}} (\rho A V_n) \left(\tilde{u} + \frac{p}{\rho} + \frac{V^2}{2} + gz \right)$$

$$- \sum_{\text{in}} (\rho A V_n) \left(\tilde{u} + \frac{p}{\rho} + \frac{V^2}{2} + gz \right), \tag{5.40}$$

or

$$\dot{Q} - \dot{W}_s = \frac{\partial}{\partial t} \int_{V_{cv}} \rho \left(\tilde{u} + \frac{V^2}{2} + gz \right) dV + \sum_{\text{out}} \dot{m} \left(\tilde{u} + \frac{p}{\rho} + \frac{V^2}{2} + gz \right)$$

$$- \sum_{\text{in}} \dot{m} \left(\tilde{u} + \frac{p}{\rho} + \frac{V^2}{2} + gz \right). \tag{5.41}$$

If the flow is steady, the first term on the right is zero. In these equations, all terms under the summation signs are positive.* They are not vector equations, so the labels "in" and "out" serve to determine the algebraic signs.

* Unless z or $\tilde{u}$ is below an arbitrarily selected zero datum.

EXAMPLE 5.10 Illustrates the general energy equation

A compressor compresses 6 kg/s of air from inlet conditions of $T_1 = 300\text{K}$ and $p_1 = 90$ kPa to discharge conditions of $T_2 = 390\text{K}$ and $p_2 = 310$ kPa. See Fig. E5.10(a). The air in the inlet pipe has a uniform velocity profile. The air in the discharge pipe has a parabolic velocity profile given by

$$u = u_{\max} \left[1 - \left(\frac{r}{R_2} \right)^2 \right],$$

where R_2 is the inside radius of the discharge pipe. Elevation changes are negligible, and the internal energy change of the air is given by

$$(\tilde{u}_2 - \tilde{u}_1) = c_v(T_2 - T_1), \quad \text{with} \quad c_v = 720 \text{ J/kg·K}.$$

Assuming steady flow and negligible heat transfer between the air in the compressor and pipe and the ambient, find the power input to the compressor.

Solution

Given
Air compressor in Fig. E5.10(a) operating at a steady rate, with

$$\dot{m} = 6 \text{ kg/s} \qquad R_1 = R_2 = 0.25 \text{ m}$$
$$T_1 = 300 \text{K} \qquad T_2 = 390 \text{K}$$
$$p_1 = 90 \text{ kPa} \qquad p_2 = 310 \text{ kPa}$$

Uniform velocity profile at inlet and parabolic velocity profile at discharge
Internal energy change of air given by

$$\tilde{u}_2 - \tilde{u}_1 = (720 \text{ J/kg} \cdot \text{K})(T_2 - T_1)$$

Negligible heat transfer and elevation changes

Find
Power input to the compressor

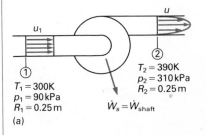

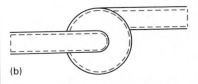

Figure E5.10 (a) Air compressor with inlet and discharge pipes; (b) control volume enclosing air in the compressor and piping.

Solution
We apply the general energy equation, Eq. (5.38), to a control volume enclosing the air in the compressor and piping (see Fig. E5.10b):

$$\dot{Q} - \dot{W}_{\text{shaft}} - \dot{W}_{\text{shear}} - \int_{A_{cv}} p(\vec{V}_c \cdot \hat{n})\, dA$$

$$= \frac{d}{dt}\int_{\mathcal{V}_{cv}} \left(\rho + \frac{V^2}{2} + gz\right) d\mathcal{V} + \int_{A_{cv}} \rho\left(\tilde{u} + \frac{p}{\rho} + \frac{V^2}{2} + gz\right)(\vec{V}_r \cdot \hat{n})\, dA.$$

We want to solve for the shaft power $\dot{W}_{\text{shaft}}$. The problem statement gives

$$\dot{Q} = 0.$$

The shear stress work is given by Eq. (5.33):

$$\dot{W}_{\text{shear}} = \int_{A_{cv}} (\vec{\tau} \cdot \vec{V})\, dA,$$

where $\vec{V}$ is the fluid velocity at the control surface. The only shear stress along the control surface is between the fluid and the pipe wall or compressor outer casing, where $\vec{V} = 0$. Therefore

$$\dot{W}_{\text{shear}} = 0.$$

Since the control volume is not moving, $\vec{V}_c = 0$, and

$$\oint_{A_{cv}} p(\vec{V}_c \cdot \hat{n})\, dA = 0.$$

Assuming steady flow, the time derivative term is zero. The remaining integral in the general energy equation is simplified by the negligible elevation change condition to give

$$-\dot{W}_{\text{shaft}} = \oint_{A_{cv}} \rho\left(\tilde{u} + \frac{p}{\rho} + \frac{V^2}{2}\right)(\vec{V}_r \cdot \hat{n})\, dA.$$

Assuming uniform temperature and pressure over both the inlet and the exit of the control surface and recognizing that

$$\int_{A_{\text{in}}} \rho(\vec{V}_r \cdot \hat{n})\, dA = -\dot{m} \qquad \text{and} \qquad \int_{A_{\text{out}}} \rho(\vec{V}_r \cdot \hat{n})\, dA = \dot{m},$$

we have

$$-\dot{W}_{\text{shaft}} = \dot{m}\left[(\tilde{u}_2 - \tilde{u}_1) + \left(\frac{p_2}{\rho_2} - \frac{p_1}{\rho_1}\right)\right] - \dot{m}\frac{u_1^2}{2} + \int_{A_{\text{out}}}\rho\left(\frac{u^3}{2}\right)dA.$$

Substituting for the discharge velocity $u(r)$ gives

$$\int_{A_{\text{out}}}\rho\left(\frac{u^3}{2}\right)dA = 2\pi\int_0^{R_2}\frac{\rho u_{\max}^3}{2}\left[1 - \left(\frac{r}{R_2}\right)^2\right]^3 r\,dr.$$

We now assume that air is an ideal gas, so the uniform temperature and pressure assumption implies that the density is uniform over the exit area. Integration gives

$$\int_{A_{\text{out}}}\rho\frac{u^3}{2}dA = \pi\rho_2 u_{\max}^3\int_0^{R_2}\left[1 - 3\left(\frac{r^2}{R_2^2}\right) + 3\left(\frac{r^4}{R_2^4}\right) - \frac{r^6}{R_2^6}\right]r\,dr$$

$$= \frac{\pi\rho_2 u_{\max}^3 R_2^2}{8}.$$

For this parabolic velocity profile in a circular pipe, Example 5.1 shows that

$$\bar{V} = \frac{u_{\max}}{2}.$$

Making use of the expressions for $(\tilde{u}_2 - \tilde{u}_1)$, $u_{\max}$, the area integral, and the ideal gas law, $p = \rho RT$, we get

$$-\dot{W}_{\text{shaft}} = \dot{m}c_v(T_2 - T_1) + \dot{m}R(T_2 - T_1) - \dot{m}\frac{u_1^2}{2} + \pi\rho_2\bar{V}_2^3 R_2^2.$$

Recognizing

$$\dot{m} = \pi\rho_2\bar{V}_2 R_2^2$$

gives

$$-\dot{W}_{\text{shaft}} = \underbrace{mc_v(T_2 - T_1)}_{\substack{\text{internal}\\\text{energy}\\\text{change}}} + \underbrace{\dot{m}R(T_2 - T_1)}_{\text{flow work}} + \underbrace{\dot{m}\left(\bar{V}_2^2 - \frac{u_1^2}{2}\right)}_{\substack{\text{kinetic}\\\text{energy}\\\text{change}}}.$$

Using Eqs. (1.1), (3.36), and (3.39) gives

$$u_1 = \frac{\dot{m}}{\rho_1 A_1} = \frac{\dot{m}RT_1}{p_1 A_1} \quad\text{and}\quad \bar{V}_2 = \frac{\dot{m}}{\rho_2 A_2} = \frac{\dot{m}RT_2}{p_2 A_2}.$$

R, the gas constant, is obtained from Table A.3:

$$R = 287\ \text{N}\cdot\text{m/kg}\cdot\text{K}.$$

Also

$$A_1 = A_2 = \pi R_1^2 = \pi(0.25\ \text{m})^2 = 0.196\ \text{m}^2,$$

so

$$u_1 = \frac{\left(6\ \dfrac{\text{kg}}{\text{s}}\right)\left(287\ \dfrac{\text{N}\cdot\text{m}}{\text{kg}\cdot\text{K}}\right)(300\text{K})}{\left(90{,}000\ \dfrac{\text{N}}{\text{m}^2}\right)(0.196\ \text{m}^2)} = 29.3\ \text{m/s},$$

and

$$\bar{V}_2 = \frac{\left(6\,\dfrac{\text{kg}}{\text{s}}\right)\left(287\,\dfrac{\text{N}\cdot\text{m}}{\text{kg}\cdot\text{K}}\right)(390\text{K})}{\left(310{,}000\,\dfrac{\text{N}}{\text{m}^2}\right)(0.196\text{ m}^2)} = 11.1\text{ m/s}.$$

Then

$$\dot{W}_{\text{shaft}} = -\left(6\,\frac{\text{kg}}{\text{s}}\right)\left(720\,\frac{\text{J}}{\text{kg}\cdot\text{K}}\right)(390 - 300)\text{K}$$

$$-\left(6\,\frac{\text{kg}}{\text{s}}\right)\left(287\,\frac{\text{N}\cdot\text{m}}{\text{kg}\cdot\text{K}}\right)(390 - 300)\text{K}$$

$$-\left(6\,\frac{\text{kg}}{\text{s}}\right)\left[\left(11.1\,\frac{\text{m}}{\text{s}}\right)^2 - \frac{1}{2}\left(29.3\,\frac{\text{m}}{\text{s}}\right)^2\right]$$

$$= (-388{,}800 - 156{,}000 + 1800)\text{ J/s}$$

$$\ \underset{\substack{\text{internal} \\ \text{energy}}}{} \quad \underset{\substack{\text{flow} \\ \text{work}}}{} \quad \underset{\substack{\text{kinetic} \\ \text{energy}}}{}$$

$$= -546{,}600\text{ J/s},$$

$$\left|\ \dot{W}_{\text{shaft}} = -546.6\text{ kW}.\ \right|$$

Answer ◀

Discussion

Note that a large portion of the compressor input work appears as an increase in the thermal (internal) energy and the mechanical "pressure energy" of this compressible fluid. The kinetic energy change is much smaller. The negative sign means work is done on the air by the compressor.

Usually, the uniform, one-directional flow assumption is only an approximation, and we use average property values to represent the flow of energy at the various inlet and outlet planes. For extremely accurate calculations, the averages must be carefully defined [2] so that they truly represent the associated energy flows. In most cases, appropriate average values of $\tilde{u}$, p, ρ, and z are readily apparent since these properties are often closely uniform across the cross section. Since the fluid velocity is usually nonuniform, representing the kinetic energy flux in terms of an average velocity is a little more complicated.

Consider fluid flowing in a pipe with the velocity profile shown in Fig. 5.7. The flux of kinetic energy across plane 1 is

$$\dot{E}_{K,1} = \int_{A_1} \rho\left(\frac{V^2}{2}\right)(V_n\,dA). \tag{5.42}$$

The average velocity at plane 1 is

$$\bar{V}_1 = \frac{\int_{A_1} V_n\,dA}{A_1}. \tag{3.39}$$

However, since V is not uniformly distributed,

$$\dot{E}_{k,1} \neq \rho_1 \frac{(\bar{V}_1)^3}{2} A_1.$$

To evaluate kinetic energy flux in terms of the average velocity, a kinetic energy correction factor, α, is defined by

$$\alpha \equiv \frac{\frac{1}{2}\int \rho V^3\, dA}{\left(\frac{1}{2}\right)\bar{\rho}\bar{V}^3 A}. \tag{5.43}$$

The true kinetic energy flux across plane 1 is

$$E_{K,1} = \alpha_1 \left(\frac{1}{2}\rho_1 \bar{V}_1^3\right) A_1. \tag{5.44}$$

With the kinetic energy correction factor, the general energy equation (for steady flow and a fixed, rigid control volume) is

$$\dot{Q} - \dot{W}_s = \sum_{out} \dot{m}\left[\bar{u} + \frac{\bar{p}}{\bar{\rho}} + \alpha\left(\frac{\bar{V}^2}{2}\right) + \bar{g}\bar{z}\right] - \sum_{in} \dot{m}\left[\bar{u} + \frac{\bar{p}}{\bar{\rho}} + \alpha\left(\frac{\bar{V}^2}{2}\right) + \bar{g}\bar{z}\right]. \tag{5.45}$$

For flow in a circular pipe, α ranges from 2.0 for fully developed laminar flow to about 1.05 for fully developed turbulent flow. The value of α for the flow at the compressor discharge in Example 5.10 is 2.0.

A very important special form of the general energy equation is the energy equation for flow along a single streamline. Figure 5.17(a) shows a single streamline in a steady flow. Figure 5.17(b) shows a small steamtube that surrounds the particular streamline. A *streamtube* is an imaginary tube whose surface is composed entirely of streamlines. Since the fluid velocity is always parallel to a streamline, there is no velocity component perpendicular to the walls of a streamtube, so no fluid enters or leaves a streamtube and the mass flow rate is constant.

We select a control volume that is coincident with the streamtube. The control surface is perpendicular to the central streamline at the ends of the streamtube (Fig. 5.17c). If we select the central streamline as a coordinate line, the flow is one-dimensional and one-directional. Since we have assumed that the flow is steady, the general energy equation for the streamtube control volume is

$$\delta\dot{Q} - \delta\dot{W}_s = \delta\dot{m}_2\left(\tilde{u}_2 + \frac{p_2}{\rho_2} + \frac{V_2^2}{2} + gz_2\right) - \delta\dot{m}_1\left(\tilde{u}_1 + \frac{p_2}{\rho_2} + \frac{V_1^2}{2} + gz_1\right).$$

Since no fluid leaves the sides of the streamtube and the flow is steady, conservation of mass requires that

$$\delta\dot{m}_1 = \delta\dot{m}_2 = \delta\dot{m}.$$

Substituting $\delta\dot{m}$ for $\delta\dot{m}_1$ and $\delta\dot{m}_2$ and then dividing by $\delta\dot{m}$, we obtain

$$q - w_s = \tilde{u}_2 - \tilde{u}_1 + \frac{p_2}{\rho_2} - \frac{p_1}{\rho_1} + \frac{V_2^2}{2} - \frac{V_1^2}{2} + gz_2 - gz_1,$$

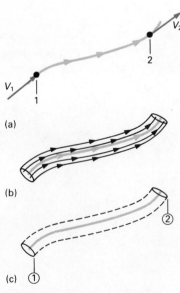

Figure 5.17 Flow along streamlines: (a) A single streamline; (b) streamtube with the streamline at the center; (c) control volume surrounding the streamtube.

where

$$q = \frac{\delta \dot{Q}}{\delta \dot{m}}$$

is the heat transfer per unit mass of moving fluid, and

$$w_s = \frac{\delta \dot{W}_s}{\delta \dot{m}}$$

is the work per unit mass of moving fluid. If we take the limit of this expression, shrinking the streamtube and collapsing it around the central streamline, we get the steady-flow general energy equation along a streamline:

$$q - w_s = \tilde{u}_2 - \tilde{u}_2 + \frac{p_2}{\rho_2} - \frac{p_1}{\rho_1} + \frac{V_2^2}{2} - \frac{V_1^2}{2} + gz_2 - gz_1. \qquad \textbf{(5.46)}$$

In steady flow, any particular mass particle always moves along a fixed streamline. Equation (5.46) relates the changes of specific intrinsic energies and the specific energy transfers for all fluid particles that move from point 1 to point 2 on this streamline. We might interpret this equation as an energy balance for a specific particle that moves from position 1 to position 2 on a particular streamline. Then $(\tilde{u}_2 - \tilde{u}_1) + (V_2^2/2 - V_1^2/2) + (gz_2 - gz_1)$ is the net gain of the particle's intrinsic energy, $(-p_1/\rho_1)$ and (p_2/ρ_2) represent flow work done on the particle as it begins its trip from 1 to 2 and flow work done by the particle as it finishes its trip, and $(q - w_s)$ is the net energy transferred to the particle as it travels between 1 and 2. If we rewrite Eq. (5.46) as

$$\left(\tilde{u}_1 + \frac{p_1}{\rho_1} + \frac{V_1^2}{2} + gz_1 \right) + (q - w_s) = \left(\tilde{u}_2 + \frac{p_2}{\rho_2} + \frac{V_2^2}{2} + gz_2 \right), \qquad \textbf{(5.47)}$$

we might interpret it as follows:

$$\left(\begin{array}{c} \text{Initial energy of} \\ \text{fluid particle} \end{array} \right) + \left(\begin{array}{c} \text{Net energy transferred} \\ \text{to fluid particle} \end{array} \right) = \left(\begin{array}{c} \text{Final energy of} \\ \text{fluid particle} \end{array} \right). \qquad \textbf{(5.48)}$$

This interpretation is very helpful but not strictly correct because it implies that p/ρ is energy of the fluid mass (intrinsic energy) rather than energy transferred to the fluid mass (work done by pressure forces).

If we have a steady flow in a *finite* streamtube (the inside of a pipe, for example) and if flow conditions are identical for all streamlines inside the streamtube, Eq. (5.46) applies to any particular streamline. If we multiply the equation by $\dot{m}$, the mass flow rate in the finite streamtube, we get

$$\dot{Q} - \dot{W}_s = \dot{m} \left(\tilde{u}_2 - \tilde{u}_1 + \frac{p_2}{\rho_2} - \frac{p_1}{\rho_1} + \frac{V_2^2}{2} - \frac{V_1^2}{2} + gz_2 - gz_1 \right). \qquad \textbf{(5.49)}$$

This is the general energy equation for steady, one-directional, *uniform* flow with a single inlet and a single outlet. The total heat transfer rate and total rate of work done (that is, total power) are related to the specific rates by $\dot{Q} = \dot{m}q$ and $\dot{W}_s = \dot{m}w_s$.

5.4.2 Mechanical Energy and the Steady-Flow Mechanical Energy Equation

Equation (5.47), the general energy equation for flow along a streamline, bears a striking resemblance to Bernoulli's equation in the form of Eq. (4.21). Both equations are listed together for comparison:

$$\frac{p_1}{\rho} + \frac{V_1^2}{2} + gz_1 \qquad = \qquad \frac{p_2}{\rho} + \frac{V_2^2}{2} + gz_2 \qquad \textbf{(4.21)}$$

$$\tilde{u}_1 + \frac{p_1}{\rho_1} + \frac{V_1^2}{2} + gz_1 + (q - w_s) = \tilde{u}_2 + \frac{p_2}{\rho_2} + \frac{V_2^2}{2} + gz_2. \qquad \textbf{(5.47)}$$

The obvious differences are the internal energy, heat, and work terms. A more subtle difference is the inclusion of the subscripts 1 and 2 on the density in Eq. (5.47). Table 5.1 summarizes the assumptions and conditions applicable to each equation. Notice that Bernoulli's equation contains many more restrictions (no stress, work, or compressibility) than the general energy equation. Notice also that even if we introduced the incompressibility ($\rho_1 = \rho_2 = \rho$) and no work ($w_s = 0$) assumptions into the general energy equation, we would still need to assume that ($q = \tilde{u}_2 - \tilde{u}_1$) to obtain Bernoulli's equation. Examining the list of assumptions and conditions in Table 5.1, we seem to find that assuming that ($q = \tilde{u}_2 - \tilde{u}_1$) is equivalent to assuming that there are no shear stresses, at least for an incompressible fluid. We will investigate this situation in more detail.

In Section 4.1.3 we pointed out that Bernoulli's equation is a *mechanical* energy equation, although a restricted one because no work or shear stresses are present. Consider a fluid particle moving along a streamline from point 1 to point 2 *in a steady flow*. Following the pattern of Eq. (5.48) we can write a balance of *mechanical* energy for the particle:*

$$\begin{bmatrix} \text{Initial mechanical} \\ \text{energy of particle} \\ \text{(at point 1)} \end{bmatrix} + \begin{pmatrix} \text{Mechanical energy} \\ \text{transferred to} \\ \text{particle between} \\ \text{1 and 2} \end{pmatrix} + \begin{pmatrix} \text{Amount of thermal} \\ \text{energy converted to} \\ \text{mechanical energy} \\ \text{between 1 and 2} \end{pmatrix}$$

$$- \begin{pmatrix} \text{Amount of mechanical} \\ \text{energy converted to} \\ \text{thermal energy between} \\ \text{1 and 2} \end{pmatrix} = \begin{bmatrix} \text{Final mechanical} \\ \text{energy particle} \\ \text{(at point 2)} \end{bmatrix}. \qquad \textbf{(5.50)}$$

This mechanical energy balance might seem to imply a "law of conservation of mechanical energy," but it does not! It is simply a special way of writing the first law of thermodynamics (Section 3.2), which states that energy can be transported, *changed in form* (that is, converted), or stored (in the fluid particle in this case), but not created or destroyed.[†]

---

* Notice that we have switched to the "following the fluid" approach rather than a strict control volume approach for this discussion.

† As an exercise, try writing a "thermal energy balance equation." Then show that adding your "thermal energy balance" to the "mechanical energy balance" results in the total or general energy balance, Eq. (5.47).

Table 5.1 Comparison of assumptions and conditions for the streamline form of the general energy equation and Bernoulli's equation

General Energy Equation (5.47)	Bernoulli's Equation (4.21)
Derived from first law of thermodynamics	Derived from Newton's second law of motion
Valid along a streamline	Valid along a streamline
Steady flow	Steady flow
Compressible or incompressible fluid	Incompressible fluid only
	No shear stress
	No shaft work
Contains mechanical and thermal energies	Contains only mechanical energies

In this equation, the mechanical energy of the particle is represented by its specific kinetic and potential energies ($V^2/2$ and gz, respectively) and the mechanical energy transfer to the particle is the *net* work done on the particle by outside agencies (pump/turbine blades), shear stress, and pressure. The net work done on the particle by outside agencies and shear stress is $-w_s$. According to our earlier development of Bernoulli's equation (see Section 4.1.3), the work done by pressure on a fluid particle as it moves from point 1 to point 2 along a streamline is*

$$(\text{Net work done on a particle by pressure}) = -\int_1^2 \frac{dp}{\rho}.$$

Next we consider the conversions between mechanical and thermal energies. The second law of thermodynamics (Section 3.2) has something to say about such conversions. Rather than invoke this law, we will rely on some relatively common occurrences to illustrate the thermal-mechanical energy conversion process. It is well known that mechanical energy can be easily converted into thermal energy. On a cold day, you might rub your hands together to warm them up, converting kinetic energy into thermal energy. The brakes on a car bring it to a stop by converting the car's kinetic energy into thermal energy. As a heavy block slides down an inclined plane, friction between the block and the plane causes both to heat up. In this case, the block's potential energy is converted to kinetic energy, which is then converted to thermal energy by friction. There are two common elements in all of these

---

* If you are unsure about this, note that $-\partial p/\partial s$ is the pressure force per unit volume, $(-1/\rho)(\partial p/\partial s)$ is the pressure force per unit mass, $(-1/\rho)(\partial p/\partial s)ds = -dp/\rho$ is the work per unit mass done in moving a distance ds, and so $\int_1^2 dp/\rho$ is the total work done in moving along the streamline from 1 to 2.

processes:

- Friction causes a conversion of mechanical (kinetic) energy to thermal energy.
- The process cannot be run backward (reversed), converting thermal energy into mechanical energy.

In each case, friction has brought about a loss of mechanical energy.

"Friction" (actually shear stresses due to viscosity and turbulence) in moving fluids is responsible for similar conversions of mechanical energy into thermal energy. If a viscous fluid like oil is pumped through a long pipeline, the mechanical energy is continually converted to thermal energy, resulting in an increase in the oil's temperature. The mechanical energy loss is manifested as a pressure drop. Pumps are placed at intervals along the pipeline to transfer mechanical energy to the fluid from outside sources, increasing the pressure, which then supplies the mechanical energy to the oil to keep it moving to the next pump station.

The amount of mechanical energy converted to thermal energy by friction is sometimes called "lost energy" or simply *losses,* since this energy cannot be converted back from the thermal form to the mechanical form. These names are somewhat misleading since (total) energy is conserved; it is *mechanical* energy that is lost. The term *mechanical energy loss* would be much better, but it is not very popular. Probably the most popular term is *head loss,* a term that refers to the head form of Bernoulli's equation (Eq. 4.23). Head loss, assigned the symbol h_L, expresses the lost mechanical energy in terms of elevation. The lost mechanical energy is gh_L.

So far we have considered mechanical-to-thermal energy conversions that cannot be reversed. What about the process of thermal-to-mechanical energy conversion? The answer to the question is this: The only mechanism for conversion of thermal energy to mechanical energy is associated with fluid pressure.* If increasing the thermal energy of a fluid causes an increase in its pressure, then the forces generated by the increased pressure are capable of doing work on the fluid, increasing its mechanical energy. The pressure forces are also capable of doing work on the surroundings. This is exactly the way a gasoline or diesel engine works (Fig. 5.18). Combustion of fuel in air (equivalent to heating the air) generates a high temperature and pressure. The force due to the pressure acting on the moving piston does work on the piston. This work is transmitted to a driven load (say a car or truck or lawnmower).

In any case, we can see that *the process of thermal-to-mechanical energy conversion requires fluid compressibility.*† If the fluid is incompressible, increasing its thermal energy (increasing its temperature) does not cause a pressure increase and does not increase the potential

---

* We are assuming that there are no electric or magnetic effects in the fluid.
† The fluid's bulk modulus (β_T) must be finite and its coefficient of thermal expansion (α_T) must be nonzero.

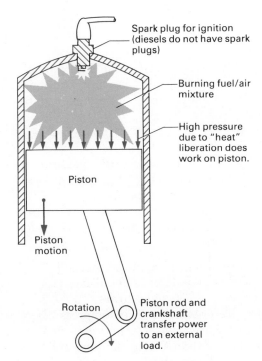

Spark plug for ignition (diesels do not have spark plugs)

Burning fuel/air mixture

High pressure due to "heat" liberation does work on piston.

Piston

Piston motion

Rotation

Piston rod and crankshaft transfer power to an external load.

Figure 5.18 Gasoline and diesel engines generate power from high pressure due to rapid temperature rise in a gas.

of the fluid for doing work. In an incompressible fluid, the only possible mechanical-to-thermal energy conversion process is the "frictional" conversion, resulting in a complete "loss" of the converted mechanical energy.

For an incompressible fluid, the mechanical energy balance, Eq. (5.47), becomes

$$\frac{V_1^2}{2} + gz_1 + \underbrace{\left(-w_s - \int_1^2 \frac{dp}{\rho} \right)}_{\substack{\text{Work done} \\ \text{on particle} \\ \text{by pressure}}} + \underbrace{(0)}_{\substack{\text{Conversion} \\ \text{of thermal} \\ \text{to mechanical} \\ \text{energy}}} - \underbrace{(gh_L)}_{\substack{\text{Conversion} \\ \text{of mechanical} \\ \text{to thermal} \\ \text{energy}}} = \left(\frac{V_2^2}{2} + gz_2 \right).$$

Since the fluid was assumed incompressible, ρ is constant, so the integral can be evaluated and the equation becomes

$$\frac{V_1^2}{2} + gz_1 - w_s - \frac{p_2 - p_1}{\rho} - gh_L = \frac{V_2^2}{2} + gz_2.$$

Regrouping terms, we have

$$\frac{p_1}{\rho} + \frac{V_1^2}{2} + gz_1 - w_s = \frac{p_2}{\rho} + \frac{V_2^2}{2} + gz_2 + gh_L. \qquad \textbf{(5.51)}$$

Equation (5.33) is called the "mechanical energy equation for incompressible flow along a streamline," or, often, simply the "incompressible energy equation." If we subtract Eq. (5.51) from the general energy equation, Eq. (5.47), we get

$$\tilde{u}_2 - \tilde{u}_1 = q + gh_L, \tag{5.52}$$

which tells us that the increase of internal energy of a fluid particle equals the net heat transferred to the particle plus the net "frictional heating," the conversion of mechanical energy into thermal energy. By rewriting Eq. (5.52) as

$$gh_L = (\tilde{u}_2 - \tilde{u}_1) - q, \tag{5.53}$$

we see that Eq. (5.51) can be used as a replacement for the general energy equation of an incompressible fluid. This is why Eq. (5.51) is sometimes called simply "the incompressible energy equation."

Equation (5.51) is almost always used rather than Eq. (5.47) when we consider flow of an incompressible fluid. There are two reasons for this:

- For incompressible fluids, we often don't care about thermal energies, so we lump them together. We do care about friction, pressure change, and work, so we keep them separate.
- When thermal energy changes are present in an incompressible flow, for example, flow through a water heater, q and $\tilde{u}_2 - \tilde{u}_1$ are often orders of magnitude larger than the mechanical energy terms and Eq. (5.52) is approximately true without gh_L. To investigate the mechanical aspects of such a flow (pressure drop, pump sizing, etc.), we must handle mechanical and thermal effects separately.

If we wish to use Eq. (5.51), we must evaluate the head loss by considering the action of friction in the fluid* because $(\tilde{u}_2 - \tilde{u}_1)$ and q are usually either so large that their difference is meaningless or so small that they cannot be calculated or measured with confidence.

* This is a very complicated problem. There are very few analytical solutions for head losses, but there is a fair amount of reliable measured data. This problem will be considered in Chapter 7.

EXAMPLE 5.11 Illustrates proper energy equations for the flow of an incompressible, inviscid fluid

A Boy Scout troop builds a camp shower with a 55-gal drum, as shown in Fig. E5.11. The Boy Scouts use "cold" 55°F water. The troop leaders build a fire to heat their water to 100°F. The two pipe lines are identical, and constant-density, frictionless flow occurs in both. Find the flow rate to each shower stall when the water level in the drum is 3 ft above the pipe discharges.

The internal energy change of the water is given by

$$\tilde{u}_2 - \tilde{u}_1 = 1.0(T_2 - T_1),$$

where $\tilde{u}$ is in BTU/lbm and T is in degrees Fahrenheit. Find the heat added to each pound of water in each pipe line. Assume no heating of the water by the sun or the air.

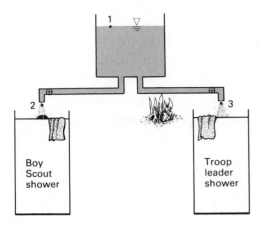

Figure E5.11 Boy Scout shower.

Solution

Given
55-gal drum with two identical discharge pipe lines
Constant-density, frictionless flow
55°F water in the Scouts' line
100°F water in the leaders' line

Find
The flow rate to each shower stall for a water level height of 3 ft above
the pipe discharges
The heat added to each pound of water in each pipe line

Solution
We define point 1 as the water surface in the drum, point 2 as the Scouts'
end of the pipe, and point 3 as the leaders' end of the pipe. We write Bernoulli's
equation for streamlines connecting points 1 and 2 and points 1 and 3:

$$\frac{p_1}{\rho_c} + \frac{V_1^2}{2} + gz_1 = \frac{p_2}{\rho_c} + \frac{V_2^2}{2} + gz_2,$$

$$\frac{p_1}{\rho_H} + \frac{V_1^2}{2} + gz_1 = \frac{p_3}{\rho_H} + \frac{V_3^2}{2} + gz_3.$$

In both cases $V_1 \approx 0$. Also

$$p_1 = p_2 = p_3 = p_{atm},$$

and we take

$$z_2 = z_3 = 0,$$

so the discharge velocities are equal:

$$V_2 = V_3 = \sqrt{2gz_1}$$

The numerical values give

$$V_2 = V_3 = \sqrt{2(32.2 \text{ ft/sec})(3 \text{ ft})}$$ **Answer**
$$|\ V_2 = V_3 = 13.9 \text{ ft/sec.}\ |$$ ◀

The thermal energy equation for incompressible flow, Eq. (5.52), can be applied to both flows:

$$q_{1-2} = \tilde{u}_2 - \tilde{u}_1 - gh_L)_{1-2}$$

and

$$q_{1-3} = \tilde{u}_3 - \tilde{u}_1 - gh_L)_{1-3}.$$

In both cases $gh_L = 0$ by assumption. The internal energy changes are

$$\tilde{u}_2 - \tilde{u}_1 = 1.0(T_2 - T_1) \quad \text{and} \quad \tilde{u}_3 - \tilde{u}_1 = 1.0(T_3 - T_1).$$

Substitution gives

$$q_{1-2} = 1.0(T_2 - T_1) \quad \text{and} \quad q_{1-3} = 1.0(T_3 - T_1).$$

The numerical values are

$$q_{1-2} = \left(1.0 \, \frac{\text{BTU}}{\text{lbm}°\text{F}}\right)(55 - 55)°\text{F},$$

$$\left| \; q_{1-2} = 0, \; \right| \qquad \qquad \textbf{Answer}$$

and

$$q_{1-3} = \left(1.0 \, \frac{\text{BTU}}{\text{lbm}°\text{F}}\right)(100 - 55)°\text{F},$$

$$\left| \; q_{1-3} = 45 \text{ BTU/lbm.} \; \right| \qquad \textbf{Answer}$$

Discussion

This example illustrates that Bernoulli's equation and the thermal energy equation can be applied to the same problem if the flow is incompressible and frictionless. Why? Note that the heat transfer in the leaders' shower pipe had no effect on the mechanical energy equation.

We still have at least one loose end to tie up: a mathematical expression of the mechanical energy equation for a *compressible* fluid. For a compressible fluid, both thermal-to-mechanical energy conversions and mechanical-to-thermal conversions that do not result in lost mechanical energy* are possible; however, the essential parameter in these conversions, the pressure and its ability to do work, already appears explicitly in the pressure work term. Since a direct conversion of thermal energy to kinetic or potential energy is not possible, the mechanical energy equation for a compressible fluid is

$$\frac{V_1^2}{2} + gz_1 + (-w_s) - \int_1^2 \frac{dp}{\rho} + \underbrace{(0)} - \underbrace{(gh_L)} = \frac{V_2^2}{2} + gz_2.$$

Includes thermal-to-mechanical and mechanical-to-thermal energy conversion due to pressure changes

Thermal-to-mechanical energy conversion other than via pressure (not possible)

Mechanical-to-thermal energy conversion other than via pressure

---

* An example of this is the internal energy (temperature) rise that occurs when air is compressed. The associated high pressure is capable of doing work.

In simpler form the mechanical energy equation for a compressible fluid is:

$$-w_s - gh_L = \int_1^2 \frac{dp}{\rho} + \left(\frac{V_2^2}{2} - \frac{V_1^2}{2}\right) + (gz_2 - gz_1). \qquad (5.54)$$

Note that the pressure-density term must be left as an integral. Different pressure-density relations will produce different forms of the equation.

Equation (5.54) can also be derived by multiplying the Navier-Stokes equations, Eqs. (4.57, 4.58, 4.59), by the fluid velocity and integrating them over a control volume. Of course steady flow must be assumed. This development is very tedious mathematically, and we will not present it here. The approach does provide us with an expression for evaluating the mechanical energy loss in terms of the shear stress. It turns out that the loss is proportional to terms like $\mu(\partial u/\partial y)^2$, the product of the fluid viscosity with the square of a velocity gradient.

5.4.3 Summary and Comparison of Various Forms of the Energy Equation

Students often have a great deal of trouble deciding which energy equation to use. Sometimes they use too many energy equations. This is particularly true if they are familiar with thermodynamics as well as fluid mechanics. This section is intended to help you select the correct equation for your needs. We will first consider steady-flow streamline forms of the energy equations. Suppose you wish to analyze a flow in a pipe system like that shown in Fig. 5.19. You might consider any one of the following three equations for the streamline between points 1 and 2.

Bernoulli's equation:

$$\frac{p_1}{\rho} + \frac{V_1^2}{2} + gz_1 = \frac{p_2}{\rho} + \frac{V_2^2}{2} + gz_2 \qquad (4.21)$$

General energy equation:

$$\tilde{u}_1 + \frac{p_1}{\rho_1} + \frac{V_1^2}{2} + gz_1 + q - w_s = \tilde{u}_2 + \frac{p_2}{\rho_2} + \frac{V_2^2}{2} + gz_2 \qquad (5.47)$$

Incompressible energy equation:

$$\frac{p_1}{\rho} + \frac{V_1^2}{2} + gz_1 - w_s = \frac{p_2}{\rho} + \frac{V_2^2}{2} + gz_2 + gh_L \qquad (5.51)$$

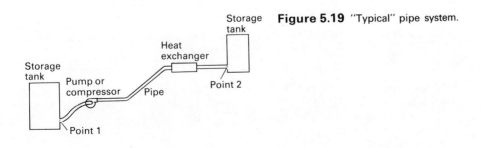

Figure 5.19 "Typical" pipe system.

Storage tank

Heat exchanger

Storage tank

Pump or compressor

Pipe

Point 2

Point 1

The most important thing to remember is this:

You can *never* use more than one of these equations at the same time.

This is because either:

■   the assumptions and restrictions on a particular equation make it inconsistent with the others *or*
■   the equations are in fact identical.

Table 5.2 summarizes the assumptions, restrictions, and conditions on these equations. We will consider some specific circumstances. First suppose that the flow in the pipe is compressible. We can't use the Bernoulli or incompressible energy equation; only the general energy equation applies. Next suppose we assume that the flow is incompressible but has shear stresses. The Bernoulli equation does not apply, but the general energy and incompressible energy equations both do; however, if the flow is incompressible, $\rho_2 = \rho_1$ and

$$gh_L = (\tilde{u}_2 - \tilde{u}_1) - q, \tag{5.53}$$

so the general energy and incompressible energy equations are not separate equations; they are actually the same. Finally, suppose that the flow is incompressible and there is no friction and no work. We know immediately that the Bernoulli equation applies. A check of Table 5.2 shows that the other two equations are also good; however, if there is no friction, then

$$gh_L = 0 \quad \text{and} \quad \tilde{u}_2 - \tilde{u}_1 = q.$$

Since we also assumed that $w_s = 0$, the general energy and incompressible energy equations reduce identically to the Bernoulli equation and do not produce independent information.

You are probably wondering when to assume that a fluid is incompressible and when you have to treat it as compressible.* Obviously, liquids usually behave as if they were strictly incompressible. Sometimes even gas flow can be treated as incompressible. The general energy equation gives us a clue about this problem. If the thermal and mechanical energies are of different orders of magnitude, then the

Table 5.2 Assumptions and conditions for the Bernoulli, general energy, and incompressible energy equations

	Bernoulli	**General Energy**	**Incompressible Energy**
Steady/unsteady	Steady only	Steady only	Steady only
Compressibility	Incompressible	Compressible or incompressible	Incompressible
Friction	None	Allowed	Allowed
Heat transfer	Allowed	Allowed	Allowed
Shaft work	None	Allowed	Allowed

---

* Refer to Section 1.3.1 for a few preliminary ideas.

larger forms of energy will dominate the general energy equation. To examine the smaller form of energy, it becomes necessary to separate it out into its own equation. The incompressible energy equation is one of these separated-out equations. We must use the incompressible energy equation to examine the mechanical energies if the ratio

$$\frac{\text{Typical mechanical energy}}{\text{Typical thermal energy}}$$

is very small. The assumption of an incompressible fluid is justified if the ratio of a typical mechanical energy ($V_1^2/2$) to a typical thermal energy ($\tilde{u}_1$) is small. For air flowing at 50 m/s (164 ft/s), $V^2/2$ is 1250 m²/s². If the air's temperature is 20°C, its internal energy ($\tilde{u}$) is approximately 210,000 m²/s², so the ratio

$$\frac{V^2/2}{\tilde{u}} \approx 0.006.$$

Since this is quite small, the air can be treated as if it were incompressible.

All three of the streamline energy equations can be converted to forms appropriate for a finite streamtube control volume (like the inside of a pipe or nozzle) by multiplying by mass flow rate and integrating over all of the streamlines inside the streamtube. The resulting equations are written in terms of appropriate average fluid properties at each inlet and outlet and, if the velocity distribution is nonuniform at inlet and outlet, kinetic energy correction factors. The equations are the following.

General energy:

$$\dot{Q} - \dot{W}_s = \dot{m}\left[\tilde{u}_2 - \tilde{u}_1 + \frac{\bar{p}_2}{\rho_2} - \frac{\bar{p}_1}{\rho_1} + \alpha_2\left(\frac{\bar{V}_1^2}{2}\right) - \alpha_1\left(\frac{\bar{V}_2^2}{2}\right) + \overline{gz}_2 - \overline{gz}_1\right] \quad \textbf{(5.55)}$$

Incompressible energy:

$$\dot{m}\left(\frac{\bar{p}_1}{\rho} + \alpha_1\frac{\bar{V}_1^2}{2} + \overline{gz}_1\right) - \dot{W}_s = \dot{m}\left[\frac{\bar{p}_2}{\rho} + \alpha_2\left(\frac{\bar{V}_2^2}{2}\right) + \overline{gz}_2\right] + \dot{m}(\overline{gh}_L), \quad \textbf{(5.56)}$$

where $\overline{gh}_L$ is the mass flow averaged mechanical energy loss,[*]

$$\overline{gh}_L = \frac{\displaystyle\int_{\substack{\text{All} \\ \text{streamlines}}} (gh_L)\, d\dot{m}}{\dot{m}}$$

The product $\dot{m}(\overline{gh}_L)$ can also be written as a total mechanical energy loss rate, $\dot{L}$:

$$\dot{m}(\overline{gh}_L) = \dot{L}.$$

Bernoulli's equation cannot be applied to a finite streamtube unless uniform flow is assumed.

---

[*] A more proper definition of gh_L is

$$\overline{gh}_L \equiv \frac{\int_{\Psi_{cv}} (\text{Loss per unit of volume})\, d\Psi}{\dot{m}} = \frac{\dot{L}}{\dot{m}}.$$

In general, losses are not uniformly distributed over the control volume.

Equations (5.55) and (5.56) can be divided by the mass flow to produce specific energy equations:

Specific general energy equation:

$$q - w_s = \bar{\bar{u}}_2 - \bar{\bar{u}}_1 + \frac{\bar{p}_2}{\rho_2} - \frac{\bar{p}_1}{\rho_1} + \alpha_2\left(\frac{\bar{V}_1^2}{2}\right) - \alpha_1\left(\frac{\bar{V}_2^2}{2}\right) + \overline{gz}_2 - \overline{gz}_1 \qquad \textbf{(5.57)}$$

Specific incompressible energy equation:

$$\frac{\bar{p}_1}{\rho} + \alpha_1\left(\frac{\bar{V}_1^2}{2}\right) + \overline{gz}_1 - w_s = \frac{\bar{p}_2}{\rho} + \alpha_2\left(\frac{\bar{V}_2^2}{2}\right) + \overline{gz}_2 + \overline{gh}_L \qquad \textbf{(5.58)}$$

By subtraction, we find a specific thermal energy equation:

$$\overline{gh}_L = \bar{\bar{u}}_2 = \bar{\bar{u}}_1 - q. \qquad \textbf{(5.59)}$$

These equations are very similar to the streamline forms. They apply in an *average* sense for an entire streamtube; the streamline forms, Eqs. (5.47) and (5.51), apply only to a particular streamline.

The general energy and incompressible energy equations can be extended to control volumes with multiple inlets and outlets (Fig. 5.20). The resulting general energy equation is Eq. (5.45). The incompressible energy equation is

$$\sum_{\text{in}} \dot{m}\left(\frac{\bar{p}}{\rho} + \alpha\frac{\bar{V}^2}{2} + g\bar{z}\right) - \dot{W}_s = \sum_{\text{out}} \dot{m}\left(\frac{\bar{p}}{\rho} + \alpha\frac{\bar{V}^2}{2} + g\bar{z}\right) + \dot{L}.$$

In this case, $\dot{L}$, the total rate of mechanical energy loss, must be calculated from a volume integral involving shear stresses inside the control volume. We *cannot* calculate $\dot{L}$ from

$$\dot{L} = \sum \dot{m}(\overline{gh}_L) \qquad \text{(Improper equation)}$$

because all streamlines mix inside the control volume.

Extensions of the general energy and incompressible energy equations to unsteady cases require addition of the term

$$\frac{\partial}{\partial t} \int_{\mathcal{V}_{cv}} \left(\tilde{u} + \frac{V^2}{2} + gz\right) d\mathcal{V}$$

to the right side of the equation (see Eq. 5.40). The explicit appearance

Figure 5.20 Control volume with multiple inlets and outlets.

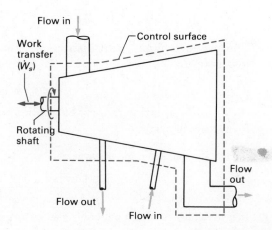

of the internal energy under the integral sign tends to destroy the usefulness of the incompressible energy equation, unless $\tilde{u}$ is constant.

The streamline form or the specific form of the incompressible energy equation can be written in head form (see Section 4.1.3) by dividing by the acceleration of gravity:

$$\frac{p_1}{\gamma} + \frac{V_1^2}{2g} + z_1 - \frac{w_s}{g} = \frac{p_2}{\gamma} + \frac{V_2^2}{2g} + z_2 + h_L \tag{5.60}$$

and

$$\frac{\bar{p}_1}{\gamma} + \alpha_1 \left(\frac{\bar{V}^2}{2g}\right) + \bar{z}_1 - \frac{w_s}{g} = \frac{\bar{p}_2}{\gamma} + \alpha_2 \left(\frac{\bar{V}^2}{2g}\right) + z_2 + h_L. \tag{5.61}$$

The dimensions of all terms are length. Proponents of the head form (typically civil engineers or mechanical engineers involved in pump and piping engineering) usually prefer to substitute turbine and pump work $(w_t - w_p)$ for w_s. In a flow system that contains a pump or turbine, part of the head loss actually occurs in the pump or turbine; that is, the rise of mechanical energy across a pump is less than the amount of work, w_p, that the pump does on the fluid, and the mechanical energy drop across a turbine is greater than the amount of work, w_t, delivered by the turbine. The total head loss in a system can be divided between the head loss in pumps and turbines ($h_{L,p}$ and $h_{L,t}$) and the head loss that occurs in the rest of the system. A *pump head* is defined by

$$h_p \equiv \frac{w_p}{g} - h_{L,p}, \tag{5.62}$$

and a *turbine head* is defined by

$$h_t \equiv \frac{w_t}{g} + h_{L,t}. \tag{5.63}$$

In terms of pump head and turbine head, Eqs. (5.60) and (5.61) are

$$\frac{p_1}{\gamma} + \frac{V_1^2}{2g} + z_1 + h_p = \frac{p_2}{\gamma} + \frac{V_2^2}{2g} + z_2 + h_t + h_L \tag{5.64}$$

and

$$\frac{\bar{p}_1}{\gamma} + \alpha_1 \left(\frac{\bar{V}^2}{2g}\right) + \bar{z}_1 + h_p = \frac{\bar{p}_2}{\gamma} + \alpha_2 \left(\frac{\bar{V}^2}{2g}\right) + \bar{z}_2 + h_t + \bar{h}_L, \tag{5.65}$$

where $\bar{h}_L$ is the head loss that occurs in all parts of the flow system except the pump(s) and turbine(s). Equations (5.64) and (5.65) have the easily remembered form

$$\begin{matrix} \text{Initial} \\ \text{head} \end{matrix} + \begin{matrix} \text{Head added} \\ \text{(by pumps)} \end{matrix} = \begin{matrix} \text{Final} \\ \text{head} \end{matrix} + \begin{matrix} \text{Head removed} \\ \text{(by turbines)} \end{matrix} + \begin{matrix} \text{Head} \\ \text{loss} \end{matrix}. \tag{5.66}$$

If you use the head form of the incompressible energy equation, a plot or sketch of the energy grade line will help you visualize energy gains and "losses."* Figure 5.21 shows the energy grade line for water

---

* The term "energy losses" is well established. Remember that only *mechanical* energy, not total energy, is lost.

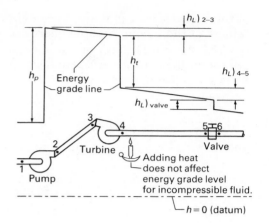

Figure 5.21 Energy grade line for a flow system with mechanical energy addition, extraction, and loss.

flowing in a piping system. Note that the energy grade line rises abruptly by h_p at a pump, drops abruptly by h_t at a turbine, and gradually drops as the flow passes along a run of pipe, reflecting the continuous loss of mechanical energy due to friction in the pipe.

EXAMPLE 5.12 Illustrates the mechanical energy equation for incompressible flow and various methods for calculating mechanical energy loss

An incompressible viscous fluid flows between two horizontal parallel plates (Fig. E5.12). Pressure gages are placed at the locations shown. The plates are spaced 0.5 cm apart and are very wide perpendicular to the paper. The flow is laminar and fully developed. Calculate the average mechanical energy loss $\overline{gh_L}$ between the two pressure gages. Then show that the mechanical energy loss also satisfies the equations

$$\overline{gh_L} = \frac{\int_{\mathcal{V}_{cv}} \mu\left(\frac{\partial u}{\partial y}\right)^2 dV}{\dot{m}} \tag{E5.1}$$

and

$$\overline{gh_L} = f(\mathbf{R}) \frac{L}{Y}\left(\frac{\bar{V}^2}{2}\right), \tag{E5.2}$$

where

$$\mathbf{R} = \frac{\rho \bar{V} Y}{\mu}.$$

The fluid density is 850 kg/m³.

Figure E5.12 Fully developed flow between two very wide, horizontal, flat plates.

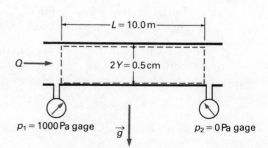

Solution

Given
Flow between parallel plates in Fig. E5.12
Fully developed, laminar flow
Fluid density 850 kg/m^3

Find
Average mechanical energy loss $\overline{gh}_L$
Show that

$$\overline{gh}_L = \frac{\int_{V_{cv}} \mu \left(\frac{\partial u}{\partial y}\right)^2 dV}{\dot{m}} \qquad \textbf{(E5.1)}$$

and

$$\overline{gh}_L = f\left(\frac{\rho \bar{V} Y}{\mu}\right) \frac{L}{Y} \frac{\bar{V}^2}{2}. \qquad \textbf{(E5.2)}$$

Solution
We select a control volume enclosing the fluid between the plates and the inlet and outlet planes. Since the fluid is incompressible, we use the specific incompressible energy equation, Eq. (5.58):

$$\frac{\bar{p}_1}{\rho} + \alpha_1 \left(\frac{\bar{V}_1^2}{2}\right) + \overline{gz}_1 - w_s = \frac{\bar{p}_2}{\rho} + \alpha_2 \left(\frac{\bar{V}_2^2}{2}\right) + \overline{gz}_2 + \overline{gh}_L.$$

Since the plates are horizontal,

$$\overline{gz}_1 = \overline{gz}_2.$$

There is no work done on the fluid by pumps or stress at the boundary of the control volume, so $w_s = 0$. From the continuity equation,

$$\bar{V}_1(2Y) = \bar{V}_2(2Y), \quad \text{so} \quad \bar{V}_1 = \bar{V}_2.$$

Since the flow is fully developed, the shape of the velocity profile is identical at planes 1 and 2, so

$$\alpha_1 = \alpha_2.$$

With these simplifications, the energy equation gives

$$\overline{gh}_L = \frac{\bar{p}_1 - \bar{p}_2}{\rho} = \frac{p_1 - p_2}{\rho}. \qquad \textbf{(E5.3)}$$

Substituting numbers, we have

$$\overline{gh}_L = \frac{1000 \, \dfrac{N}{m^2}}{850 \, \dfrac{kg}{m^3}},$$

$$\left| \; \overline{gh}_L = 1.18 \text{ J/kg.} \; \right| \qquad \textbf{Answer} \; \blacktriangleleft$$

We demonstrate the validity of Eq. (E5.1) by showing that the expression on the right of Eq. (E5.1) reduces to

$$\frac{p_2 - p_1}{\rho}.$$

For fully developed laminar flow between parallel plates, Eq. (4.48) gives the velocity profile as

$$u = -\frac{Y^2}{2\mu}\left(\frac{dp}{dx}\right)\left[1 - \left(\frac{y}{Y}\right)^2\right].$$

Differentiating, we obtain

$$\frac{\partial u}{\partial y} = \frac{y}{\mu}\left(\frac{dp}{dx}\right).$$

Substituting into the right side of Eq. (E5.1) and denoting the plate width by W, we have

$$\frac{\int_{V_{cv}} \mu\left(\frac{\partial u}{\partial y}\right)^2 dV}{\dot{m}} = \frac{2\mu\int_0^Y \left[\frac{y}{\mu}\left(\frac{dp}{dx}\right)\right]^2 WL\,dy}{\rho\,\bar{V}(2Y)W}$$

$$= \frac{L\left(\frac{dp}{dx}\right)^2}{\rho\mu\bar{V}Y}\int_0^Y y^2\,dy = \frac{Y^2 L}{3\rho\mu\bar{V}}\left(\frac{dp}{dx}\right)^2.$$

The average velocity is

$$\bar{V} = \frac{2\int_0^Y u\,dy}{2Y} = \frac{-2\int_0^Y \left(\frac{Y^2}{2\mu}\right)\left(\frac{dp}{dx}\right)\left[1 - \left(\frac{y}{Y}\right)^2\right]dy}{2Y}.$$

Integrating and simplifying give

$$\bar{V} = -\frac{1}{3}\left(\frac{Y^2}{\mu}\right)\frac{dp}{dx} = \frac{2}{3}u_{max}.$$

Substituting the average velocity, we obtain

$$\frac{\int_{V_{cv}} \mu\left(\frac{\partial u}{\partial y}\right)^2 dV}{\dot{m}} = -\frac{1}{\rho}\left(\frac{dp}{dx}\right)L.$$

The pressure gradient is constant, so

$$\frac{dp}{dx}(L) = p_2 - p_1 \quad \text{and} \quad \frac{\int_{V_{cv}} \mu\left(\frac{\partial u}{\partial y}\right)^2 dV}{\dot{m}} = \frac{p_1 - p_2}{\rho} = \overline{gh_L}.$$

| Equation (E5.1) is verified. | **Answer** ◀

To verify Eq. (E5.2), note from the average velocity equation that

$$-\frac{dp}{dx} = \frac{3\mu\bar{V}}{Y^2}.$$

Multiplying the right side by $2\bar{V}/2\bar{V}$ and both sides by L/ρ, we obtain

$$-\frac{dp}{dx}\left(\frac{L}{\rho}\right) = \frac{6\mu}{\rho\bar{V}Y}\left(\frac{L}{Y}\right)\frac{\bar{V}^2}{2}.$$

But

$$-\frac{dp}{dx}(L) = p_1 - p_2 \quad \text{and} \quad \frac{\rho\bar{V}Y}{\mu} = \mathbf{R},$$

so

$$\frac{p_1 - p_2}{\rho} = \overline{gh_L} = \frac{6}{\mathbf{R}}\left(\frac{L}{Y}\right)\frac{\bar{V}^2}{2}.$$

Since $6/\mathbf{R}$ is clearly a function of $\mathbf{R}$,

| Eq. (E5.2) is verified. |

Discussion

The easiest calculation of $\overline{gh_L}$ was from the incompressible energy equation itself; however, we could make this calculation only because all other terms were known. Usually, we need to evaluate $\overline{gh_L}$ from independent information so that we can use the incompressible energy equation to calculate pressure or work.

We have certainly not *derived* Eqs. (E5.1) and (E5.2); we have only *verified* them for a particular case. Equations (E5.1) and (E5.2) are, in fact, more general than Eq. (E5.3). Equation (E5.1) can be used to calculate mechanical energy loss for laminar flow if the velocity distribution is known. Equation (E5.2) is used to calculate mechanical energy loss in any fully developed pipe or channel flow, laminar or turbulent; however, $f(\mathbf{R})$ must usually be obtained by experiments.

EXAMPLE 5.13   Illustrates energy equations for the flow of an incompressible, viscous fluid

A 15-m-diameter storage tank contains fuel oil at 0°C. The fuel oil is pumped from the storage tank, through a steam heater, and through a 50-m-long pipe with an average velocity of 2.5 m/s. The pipe has a 4.0-cm inside diameter and an average discharge pressure of 210 kPa. The fuel oil is heated to 70°C in the steam heater and has an internal energy change given by

$$\tilde{u}_2 - \tilde{u}_1 = 2000(T_2 - T_1),$$

where $\tilde{u}$ is measured in J/kg and T in °C. The fuel oil head loss in the pipe is expressed by

$$\bar{h}_L = f\left(\frac{L}{D}\right)\frac{\bar{V}^2}{2g}.$$

where $f = 0.02$ and $\bar{V}$ is the average velocity. The fuel oil density is 850 kg/m^3. Assume that gravity is constant and that the tank, pump, and pipeline are all thermally insulated from the ambient air. Find the pump work per unit mass of the flowing fluid, the temperature rise of the oil in the pipe, the heater discharge pressure, and the heat transfer to the fuel oil in the heater.

Solution

Given
Fuel oil pumped from storage tank, through steam heater, and through pipe
   to a pressure of 210 kPa
Pipe inside diameter 4.0 cm
Pipe length 50 m
Internal energy change of

$$\tilde{u}_2 - \tilde{u}_1 = 2000(T_2 - T_1),$$

with

$$T_2 = 70°C = 343.2K \quad \text{and} \quad T_1 = 0°C = 273.2K$$

Head loss of

$$\bar{h}_L = 0.02 \left(\frac{L}{D}\right)\frac{\bar{V}^2}{2g},$$

with

$$\bar{V} = 2.5 \text{ m/s} \quad \text{and} \quad \rho = 850 \text{ kg/m}^3 = \text{Constant}$$

Pipeline thermally insulated
Constant gravity
Figure E5.13

Find

Pump work per unit mass
Temperature rise of oil in the pipe
Pressure at the heater outlet
Heat transfer to the fuel oil in the heater

Solution

Since the density is considered constant, apply the mechanical energy equation, Eq. (5.58), to a streamline connecting points 1 and 3:

$$\frac{\bar{p}_1}{\rho} + \alpha_1\left(\frac{\bar{V}^2}{2}\right) + \bar{g}\bar{z}_1 - w_s = \frac{\bar{p}_3}{\rho} + \alpha_3\left(\frac{\bar{V}^2}{2}\right) + \bar{g}\bar{z}_3 + \bar{g}\bar{h}_L.$$

To find w_s we use the "quasi-steady" analysis introduced in Example 4.2. We assume the oil level in the tank drops very slowly, so $\bar{V}_1 \approx 0$. We next substitute the given expression for $\bar{h}_L$ and solve for w_s to get

$$w_s = \frac{\bar{p}_1 - \bar{p}_3}{\rho} - \alpha_3\left(\frac{\bar{V}_3^2}{2}\right) + g(\bar{z}_1 - \bar{z}_3) - f\left(\frac{L}{D}\right)\frac{\bar{V}^2}{2}.$$

Assuming a uniform velocity profile in the pipe gives

$$\bar{V}_3 = \bar{V} \quad \text{and} \quad \alpha_3 = 1.0.$$

Since $\bar{z}_1 - \bar{z}_3 = h$, substituting gives

$$w_s = \frac{\bar{p}_1 - \bar{p}_3}{\rho} + gh - \left[f\left(\frac{L}{D}\right) + 1\right]\frac{\bar{V}^2}{2}.$$

Assuming sea-level elevation, the numerical values give

$$w_s = \frac{(101 - 210) \text{ kPa}\left(1000 \frac{\text{N/m}^2}{\text{kPa}}\right)}{850 \frac{\text{kg}}{\text{m}^3}}$$

$$+ \left(9.807 \frac{\text{m}}{\text{s}^2}\right)(2.0 \text{ m})\left(\frac{\text{N}\cdot\text{s}^2}{\text{kg}\cdot\text{m}}\right)$$

$$- \left[0.02\left(\frac{50 \text{ m}}{0.04 \text{ m}}\right) + 1\right]\frac{\left(2.5 \frac{\text{m}}{\text{s}}\right)^2}{2},$$

$$w_s = (-128.2 + 19.6 - 81.3)\frac{\text{N}\cdot\text{m}}{\text{kg}},$$

$$\boxed{w_s = -189.9 \text{ J/kg.}}$$

Answer ◀

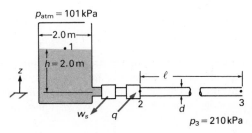

Figure E5.13 Flow from storage tank through pump, heater, and pipeline.

The temperature rise of the oil in the pipeline is found by applying Eq. (5.59) to a streamline connecting points 2 and 3:

$$\tilde{u}_3 - \tilde{u}_2 = q + \overline{gh}_L.$$

The problem statement gives $q = 0$. Substituting for $(\tilde{u}_3 - \tilde{u}_2)$ and $\bar{h}_L$ gives

$$2000(\bar{T}_3 - \bar{T}_2) = f\left(\frac{L}{D}\right)\frac{\bar{V}^2}{2}, \quad \text{or} \quad \bar{T}_3 - \bar{T}_2 = \frac{f}{2000}\left(\frac{L}{D}\right)\frac{\bar{V}^2}{2}.$$

The numerical values give

$$\bar{T}_3 - \bar{T}_2 = \frac{0.02(50 \text{ m})}{\left(2000 \dfrac{\text{J}}{\text{kg} \cdot \text{K}}\right)(0.04 \text{ m})}\left[\frac{\left(2.5 \dfrac{\text{m}}{\text{s}}\right)^2}{2}\right],$$

$$\left| \; \bar{T}_3 - \bar{T}_2 = 0.039 \text{K}. \; \right| \qquad \textbf{Answer}$$

The heater discharge pressure is found by applying the mechanical energy equation to a streamline connecting points 2 and 3:

$$\frac{\bar{p}_2}{\rho} + \alpha_2\left(\frac{\bar{V}_2^2}{2}\right) + g\bar{z}_2 - w_s = \frac{\bar{p}_3}{\rho} + \alpha_3\left(\frac{\bar{V}^2}{\rho}\right) + g\bar{z}_3 + gh_L.$$

There is no shaft work or elevation change, so

$$w_s = 0 \quad \text{and} \quad \bar{z}_2 = \bar{z}_3.$$

The constant-density condition and the integral continuity equation (3.47) give

$$\bar{V}_2 = \bar{V}_3 = \bar{V}.$$

Assuming $\alpha_2 = \alpha_3$ and solving for $\bar{p}_2$ give

$$\bar{p}_2 = \bar{p}_3 + \rho\overline{gh}_L.$$

Substituting for $\bar{h}_L$, we have

$$p_2 = p_3 + \rho f\left(\frac{L}{D}\right)\frac{\bar{V}^2}{2}.$$

The numerical values give

$$\bar{p}_2 = 210 \text{ kPa} + 0.02\left(850 \frac{\text{kg}}{\text{m}^3}\right)\left(\frac{50 \text{ m}}{0.04 \text{ m}}\right)\frac{\left(2.5 \dfrac{\text{m}}{\text{s}}\right)^2}{2}\left(\frac{\text{kPa} \cdot \text{m} \cdot \text{s}^2}{1000 \text{ kg}}\right),$$

$$\left| \; \bar{p}_2 = 276 \text{ kPa}. \; \right| \qquad \textbf{Answer}$$

The heat transferred to the oil in the steam heater is found by applying the thermal energy equation for incompressible flow to a streamline connecting points 1 and 2:

$$\overline{gh}_L = \bar{\bar{u}}_2 - \bar{\bar{u}}_1 - q, \quad \text{or} \quad q = \bar{\bar{u}}_2 - \bar{\bar{u}}_1 - \overline{gh}_L = 2000(\bar{T}_2 - \bar{T}_1) - \overline{gh}_L,$$

where $\overline{gh}_L$ is the mechanical energy loss that occurs in the inlet pipe, pump, and heater. Our calculations above showed that such losses, represented by the frictional loss in the discharge pipe, produce very small temperature changes. To make a conservative estimate of the mechanical energy loss, we will assume that it is equal to one-half of the pump work:

$$\overline{gh}_L \approx 95 \text{ J/kg}.$$

Substituting numerical values, we obtain

$$q = \left(2000 \frac{\text{J}}{\text{kg} \cdot \text{K}}\right)(343.2\text{K} - 273.2\text{K}) - 95 \frac{\text{J}}{\text{kg}}$$

$$= 140{,}000 \text{ J/kg} - 95 \text{ J/kg} \approx 140{,}000 \text{ J/kg}.$$

Since

$$\dot{m} = \rho \bar{V} A = 850 \frac{\text{kg}}{\text{m}^3}\left(2.5 \frac{\text{m}}{\text{s}}\right)\frac{\pi}{4}(0.04)^2 \text{ m}^2 = 2.67 \text{ kg/s},$$

$$\dot{Q} = \dot{m}q = 2.67 \frac{\text{kg}}{\text{s}}\left(140{,}000 \frac{\text{J}}{\text{kg}}\right), \quad \text{and}$$

$$\left| \dot{Q} = 374 \text{ kW.} \right|$$

Answer
◀

Discussion

Note that the pressure drop in the pipe ($\bar{p}_2 - \bar{p}_3 = 66$ kPa) is quite significant, while the temperature rise ($\bar{T}_2 - \bar{T}_3 = 0.039$ K) is very small for this incompressible fluid; the relative temperature change is several orders of magnitude smaller than the relative pressure change. Compare with the temperature and pressure changes of the compressible fluid across the compressor in Example 5.10. Do you expect temperature and pressure changes of the same or significantly different orders of magnitude for the flow of a compressible fluid in a pipeline?

The kinetic energy changes both in this example and in Example 5.12 were quite small compared to the magnitude of the changes of other quantities. Can you think of a physical problem where the kinetic energy change is a significant factor?

Also note that the heat transfer was several orders of magnitude larger than the work input or the potential energy of the oil in the reservoir. Our assumed value for mechanical energy loss in the inlet pipe and pump had a negligible effect on the heat transfer calculation.

5.5 The Angular Momentum Theorem

5.5.1 Derivation of the Angular Momentum Equation

Our final application of the control volume equation is the angular momentum theorem. This theorem relates the net moment or torque acting on the fluid mass in a control volume to the rate of change of the angular momentum of the mass. The starting point is the angular momentum law for a system:

$$\sum(\vec{\mathcal{M}}_0)_s = \frac{d}{dt}\int_{V_{sys}} \rho(\tilde{r}_0 \times \vec{V})\,dV, \qquad (5.67)$$

where

$$\sum(\vec{\mathcal{M}}_0)_s = \sum(\tilde{r}_0 \times \vec{F})_{sys}$$

is the sum of moments about an arbitrary point 0 of the forces acting on the system and $\tilde{r}_0$ is the radius vector measured from the inertial point 0.

Equation (5.67) does not represent an independent fundamental law of nature; it can be derived from Newton's second law of motion. Consider a single mass particle as a system. Figure 5.22 shows the particle, its velocity $\vec{V}$, the net force $\vec{F}$ acting on the particle, and the radius vector from an arbitrary point 0 to the particle, $\tilde{r}_0$. Newton's second law for the particle is

$$\vec{F} = \frac{d}{dt}(m_p\vec{V}).$$

Taking the vector (cross) product of $\tilde{r}_0$ with this equation, we have

$$\tilde{r}_0 \times \vec{F} = \tilde{r}_0 \times \frac{d}{dt}(m_p\vec{V}). \qquad (5.68)$$

From the definition of the moment of a force,

$$\vec{\mathcal{M}}_0 \equiv \tilde{r}_0 \times \vec{F}. \qquad (5.69)$$

Consider the term

$$\frac{d}{dt}[\tilde{r}_0 \times (m_p\vec{V})].$$

According to the product rule for derivatives,

$$\frac{d}{dt}[\tilde{r}_0 \times (m_p\vec{V})] = \frac{d\tilde{r}_0}{dt} \times (m_p\vec{V}) + \tilde{r}_0 \times \frac{d}{dt}(m_p\vec{V}).$$

The derivative of the position vector $\tilde{r}_0$ is the particle velocity $\vec{V}$, so

$$\frac{d}{dt}[\tilde{r}_0 \times (m_p\vec{V})] = \vec{V} \times (m_p\vec{V}) + \tilde{r}_0 \times \frac{d}{dt}(m_p\vec{V}).$$

For the first term on the right,

$$\vec{V} \times (m_p\vec{V}) = m_p(\vec{V} \times \vec{V}) = 0$$

since the vector product of identical vectors is zero. Then

$$\frac{d}{dt}[\tilde{r}_0 \times (m_p\vec{V})] = \tilde{r}_0 \times \frac{d}{dt}(m_p\vec{V}).$$

With this result and Eq. (5.69), Eq. (5.68) becomes

$$\vec{\mathcal{M}}_0 = \frac{d}{dt}[\tilde{r}_0 \times (m_p\vec{V})]. \qquad (5.70)$$

If the single particle is part of a finite system, Eq. (5.70) applies to each particle of the system. If the system is continuous, Eq. (5.70)

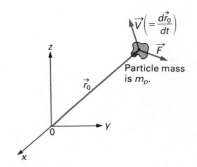

Figure 5.22 Mass particle with velocity $\vec{V}$ at position $\tilde{r}_0$; force $\vec{F}$ is acting on the particle.

applies to each differential piece of the system with

$$m_p = dm = \rho\, d\mathcal{V}.$$

Integrating the individual equations for each differential mass yields Eq. (5.67).

The control volume form of the angular momentum law, called the *angular momentum theorem,* is obtained by applying the control volume rate of change equation with $b = \vec{r}_0 \times \vec{V}$:

$$\sum \vec{M}_0 = \frac{d}{dt} \int_{\mathcal{V}_{cv}} \rho(\vec{r}_0 \times \vec{V})\, d\mathcal{V} + \oint_{A_{cv}} \rho(\vec{r}_0 \times \vec{V})(\vec{V}_r \cdot \hat{n})\, dA. \qquad \textbf{(5.71)}$$

In this equation, $\vec{V}_r$ is the velocity relative to the control surface, and $\vec{V}$ must be measured with respect to an inertial coordinate system. $\sum \vec{M}_0$ is the vector sum of the moments of all forces acting on the mass inside the control volume. Extension to a noninertial coordinate system requires addition of the term

$$-\int_{\mathcal{V}_{cv}} \rho\left(\vec{r}_0 \times \left[\frac{d^2\vec{R}}{dt^2} + \frac{d\vec{\Omega}}{dt} \times \vec{R} + 2\vec{\Omega} \times \vec{V} + \vec{\Omega} \times \vec{\Omega} \times \vec{R}\right]\right) d\mathcal{V}$$

to the left side of Eq. (5.71) (see Fig. 5.11).

Equation (5.71) is in general form, applicable to nonsteady, multidimensional flow through a moving, deformable control volume. Simplified forms exist for several common cases, similar to simplified forms of the other control volume equations. For example, the moment of momentum equation for steady flow through a control volume with a finite number of inlets and outlets, each with locally one-directional and uniform flow is

$$\sum \vec{M}_0 = \sum_{\text{out}} (\rho V_n A)(\vec{r}_0 \times \vec{V}) - \sum_{\text{in}} (\rho V_n A)(\vec{r}_0 \times \vec{V})$$

$$= \sum_{\text{out}} \dot{m}(\vec{r}_0 \times \vec{V}) - \sum_{\text{in}} \dot{m}(\vec{r}_0 \times \vec{V}). \qquad \textbf{(5.72)}$$

Like the linear momentum theorem, the angular momentum theorem is a vector equation, so it has three separate component equations.

EXAMPLE 5.14 Illustrates the angular momentum theorem and how to solve higher-order application problems

Your company is designing a snow blower similar to that shown in Fig. E5.14(a). The primary design requirement is that the machine must clear a 16-in.-wide path in 6-in.-deep snow. The vanes are driven by a small gasoline engine. The drive pulley and belt system reduce the vane rotational speed to one-fourth of the engine speed. The company currently manufactures two gasoline engines, models 100A and 100B. The performance characteristics of the two engines at wide-open throttle (WOT) conditions are shown in Fig. E5.14(b). The company wishes to use the smaller and cheaper model 100A in the snow blower. Is it possible to use model 100A in the snow blower? If not, is model 100B acceptable?

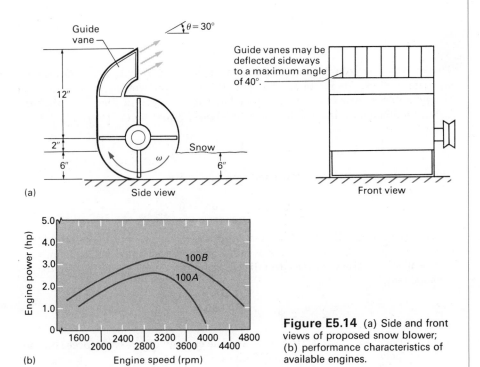

(a)   Side view

Guide vane

$\theta = 30°$

12″

2″

6″

Snow

6″

ω

Guide vanes may be deflected sideways to a maximum angle of 40°.

Front view

(b)

Figure E5.14 (a) Side and front views of proposed snow blower; (b) performance characteristics of available engines.

Solution

Given
Snow blower and gasoline engine characteristics shown in Figs. E5.14(a) and E5.14(b), respectively
Snow blower vanes' rotational speed is 1/4 engine speed
Requirement that snow blower clear a path 16 in. wide and 6 in. deep

Find
Is engine model 100A acceptable for the snow blower?
If not, is engine model 100B acceptable?

Solution
This is a complex problem, and you may not immediately see a method of solution until you have seen a few examples of this type. We will follow the systematic approach to problem solving outlined in Section 1.6.2. In fact, complex problems are more effectively solved by following such a systematic approach.

We have already started to use this systematic approach since we already have made a sketch of the physical system and noted the given information. The second step is to identify the quantity to be found. The quantity to be found is rather ill defined at this point, but we hope it will become clearer as the solution progresses.

The third step is to identify the appropriate fundamental law(s) or principle(s), select primary equations, and start to solve for the unknowns. The fundamental principle is that, neglecting losses in the pulley system, the power output of the engine must equal the power input to the vanes. Apparently, the required vane power input can be considered the quantity to be found. The power input to the vanes is the product of a torque $\mathcal{T}$ and a rotational speed

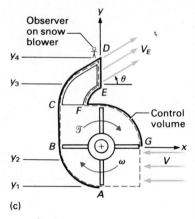

Figure E5.14 (c) Control volume for analysis of snow blower.

ω, or*

$$\dot{W}_s = \mathcal{T}\omega.$$

This tells us we need relations and/or equations to find $\mathcal{T}$ and ω. An appropriate equation to find $\mathcal{T}$ is Eq. (5.71). To apply this equation, we choose a coordinate system and observer fixed on the snow blower, as shown in Fig. E5.14(c). The control volume used is also shown in the figure. Equation (5.71) gives[†]

$$\sum \vec{\mathcal{M}}_0 = \frac{d}{dt}\int_{\mathcal{V}_{cv}} \rho(\vec{r}_0 \times \vec{V})\, d\mathcal{V} + \oint_{A_{cv}} \rho(\vec{r}_0 \times \vec{V})(\vec{V}_r \cdot n)\, dA.$$

The observer sees a steady-flow condition and writes

$$\frac{d}{dt}\int_{\mathcal{V}_{cv}} \rho(\vec{r}_0 \times \vec{V})\, d\mathcal{V} = 0.$$

(This step and the following development of this equation and the power equation are steps 3 and 5. No secondary equations and assumptions are obvious at this point.) The observer then specializes the angular momentum theorem to the z-component:

$$\sum \mathcal{M}_z = \oint_{A_{cv}} \rho(\vec{r}_0 \times \vec{V})_z(\vec{V}_r \cdot \hat{n})\, dA.$$

Neglecting friction between the snow and the casing and pressure forces on the material inside the control volume** gives

$$\sum \mathcal{M}_z = -\mathcal{T}.$$

Uniform velocity is now assumed to exist over both the inlet and outlet areas (another assumption). The inlet velocity V is normal to the inlet area, while the outlet velocity V_E makes an angle of 30° with the normal to the area. The snow leaving the blower is assumed not to be deflected sideways (another assumption). At the inlet,

$$(\vec{V}_r \cdot \hat{n}) = -V \qquad \text{and} \qquad (\vec{r}_0 \times \vec{V})_z\, dA = -yVb\, dy,$$

where b is the vane width. At the outlet,

$$(\vec{V}_r \cdot \hat{n}) = V_E \cos\theta$$

and

$$(\vec{r}_0 \times \vec{V})_z\, dA = -yV_E \sin\left(\frac{\pi}{2} - \theta\right)b\, dy = -yV_E \cos\theta\, b\, dy.$$

Substituting into the angular momentum theorem gives

$$-\mathcal{T} = \int_{y_1}^{y_2} \rho(-yVb\, dy)(-V) + \int_{y_3}^{y_4} \rho(-yV_E \cos\theta\, dy)(V_E \cos\theta).$$

Assuming that density is constant and that both V and V_E are uniform over their respective areas (more assumptions), we integrate and obtain

$$-\mathcal{T} = \rho V^2 b\left(\frac{y^2}{2}\right)_{y_1}^{y_2} - \rho V_E^2 \cos^2\theta\, b\left(\frac{y^2}{2}\right)_{y_3}^{y_4}$$

---

* Our first primary equation.
[†] Our second primary equation. The equation for ω will also be a primary equation.
** We have begun to introduce simplifying assumptions (step 4).

and

$$-\mathcal{T} = \frac{\rho V^2 b}{2}(y_2^2 - y_1^2) - \frac{\rho V_E^2 \cos^2\theta \, b}{2}(y_4^2 - y_3^2).$$

We next apply the integral continuity equation* to the control volume. For constant density,

$$Vb(y_2 - y_1) = V_E \cos\theta \, b(y_4 - y_3),$$

where V_E and y_3 are both unknown. V_E is assumed equal to the vane tip velocity[†] given by**

$$V_E = V_{\text{tip}} = R\omega,$$

where $R = -y_1$. The integral continuity equation becomes

$$Vb(y_2 - y_1) = R\omega \cos\theta \, b(y_4 - y_3),$$

or

$$y_4 - y_3 = \frac{V}{R\omega \cos\theta}(y_2 - y_1) \quad \text{and} \quad y_3^2 = \left[y_4 - \frac{V(y_2 - y_1)}{R\omega \cos\theta}\right]^2.$$

Then

$$y_4^2 - y_3^2 = y_4^2 - \left[y_4^2 - \frac{2y_4 V(y_2 - y_1)}{R\omega \cos\theta} + \left(\frac{V(y_2 - y_1)}{R\omega \cos\theta}\right)^2\right]$$

$$= \frac{2Vy_4(y_2 - y_1)}{R\omega \cos\theta} - \left(\frac{V(y_2 - y_1)}{R\omega \cos\theta}\right)^2.$$

The required torque $\mathcal{T}$ is then

$$-\mathcal{T} = \frac{\rho V^2 b}{2}(y_2^2 - y_1^2)$$

$$- \frac{\rho R^2 \omega^2 \cos^2\theta \, b}{2}\left[\frac{2Vy_4(y_2 - y_1)}{R\omega \cos\theta} - \left(\frac{V(y_2 - y_1)}{R\omega \cos\theta}\right)^2\right]$$

Simplifying gives

$$\mathcal{T} = \rho b V(y_2 - y_1)(y_4 R\omega \cos\theta + y_1 V).$$

Since $R = -y_1$,

$$\mathcal{T} = \rho b VR(y_2 + R)(y_4 \omega \cos\theta - V).$$

The power input to the vanes is

$$\dot{W}_s = \mathcal{T}\omega, \quad \text{or} \quad \dot{W}_s = \rho b VR\omega(y_2 + R)(y_4 \omega \cos\theta - V).$$

The known numerical values are[‡]

$$b = 16 \text{ in.} = 1.33 \text{ ft},$$
$$R = 8 \text{ in.} = 0.667 \text{ ft},$$
$$y_2 = -2 \text{ in.} = -0.167 \text{ ft},$$
$$y_4 = 12.0 \text{ in.} = 1.0 \text{ ft},$$
$$\cos\theta = \cos 30° = 0.866.$$

---

* This equation is a secondary equation (step 4).
 [†] Step 4 (an assumption).
** Step 3 (a primary equation for ω).
 [‡] We have started to use step 6, continuing steps 3, 4, and 5.

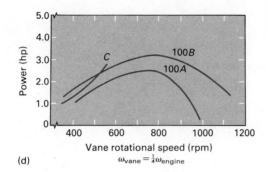

(d)

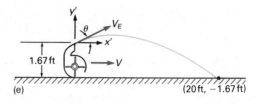

Figure E5.14 (d) Engine output power and required vane power as a function of vane rotational speed; (e) trajectory of snow leaving snow blower.

The snow density* is taken as that of wet snow. This is a conservative assumption, which gives us a maximum power requirement for the engine. Example 5.9 gives

$$\rho = 20 \text{ lbm/ft}^3.$$

The vane input power is

$$\dot{W}_s = \frac{\left(20 \dfrac{\text{lbm}}{\text{ft}^3}\right)(1.33 \text{ ft})\,V(0.667 \text{ ft})\omega(0.167 \text{ ft} + 0.667 \text{ ft})}{\left(32.2 \dfrac{\text{ft} \cdot \text{lbm}}{\text{lb} \cdot \text{sec}^2}\right)} \times [(1.0 \text{ ft})\omega(0.866) - V]$$

$$= 0.275\,V\omega(0.866\omega - V)\ \text{ft} \cdot \text{lb/sec}.$$

To use this equation, V must be in feet per second and ω in radians per second.

The power required to rotate the vanes depends on the vane rotational speed and the velocity at which the snow blower is being pushed. The pushing velocity is unspecified, so we must assume a reasonable value. We assume that $V = 2$ ft/sec. The vane power requirement becomes

$$\dot{W}_s = 0.275(2.0)\omega(0.866\omega - 2.0)\ \frac{\text{ft} \cdot \text{lb}}{\text{sec}}$$

$$= (0.476\omega^2 - 1.10\omega)\ \frac{\text{ft} \cdot \text{lb}}{\text{sec}}\left(\frac{\text{sec} \cdot \text{hp}}{550 \text{ ft} \cdot \text{lb}}\right)$$

$$= (8.66\omega^2 - 20\omega) \times 10^{-4} \text{ hp},$$

where ω is still in rad/sec. Converting ω to rpm gives

$$\dot{W}_s = \left[8.66\omega^2\left(\frac{2\pi \text{ rad}}{\text{rev}}\right)^2\left(\frac{\text{min}}{60 \text{ sec}}\right)^2 - 20\omega\left(\frac{2\pi \text{ rad}}{\text{rev}}\right)\left(\frac{\text{min}}{60 \text{ sec}}\right)\right] \times 10^{-4} \text{ hp}$$

$$= (0.095\omega^2 - 2.09\omega) \times 10^{-4} \text{ hp.}^{\dagger}$$

Using our previous assumption of negligible power loss in the pulley system, the vane input power must equal the engine output power. The above equation then represents the power that must be supplied by an engine to rotate the vanes at a rotational speed ω. This equation is plotted in Fig. E5.14(d) as curve C. The actual power output of the two engines is also plotted in the figure. Since there is no intersection of curve C and engine curve 100A, it is not possible for a snow blower equipped with engine 100A to remove 6 in. of wet snow while being pushed at 2.0 ft/sec with the engine operating at WOT. The

---

*Step 4 again.

† Note how these conversion factors were used: ω has units of revolutions per minute in these equations.

conclusion:

| Engine 100A is not an acceptable engine to meet the design specifications. | **Answer** ◀

Note that since the engine 100A is not acceptable at WOT conditions, it is not acceptable at part-throttle conditions.

Note that curve C and engine curve 100B intersect at about 515 rpm. This means the vanes would rotate at 515 rpm if engine 100B is installed in the snow blower and operated at wide-open throttle in 6-in.-deep wet snow. Although the snow blower would work with engine 100B, we cannot be sure that it is acceptable. To be useful, the snow blower must throw the snow a substantial distance. We now specify that the snow blower must throw the snow a minimum distance of 20 ft. This implies that the velocity of the snow leaving the vanes (V_E) must be large enough to carry the snow 20 ft. We have already assumed that the exit velocity of the snow is equal to the vane tip velocity; thus the minimum acceptable vane rotational speed is

$$\omega_{\min} = \frac{V_{E,\min}}{R},$$

where $V_{E,\min}$ is the velocity required to carry the snow 20 ft.

Recalling that V_E is the exit velocity relative to the snow blower and referring to Fig. E5.14(e) and Example 4.5, we find the trajectory of the snow (x', y') as a parametric function of time:

$$x' = (V_E \cos\theta + V)t \quad \text{and} \quad y' = V_E \sin\theta\, t - 1/2 g t^2.$$

Solving the first equation for t, substituting into the second, and rearranging, we have the following quadratic equation for V_E:

$$(y' \cos\theta - x' \sin\theta)V_E^2 + V(2y' - x' \tan\theta)V_E + \left(\frac{gx'^2}{2} - V^2 y'\right)\sec\theta = 0.$$

From Fig. E5.14(e),

$$x' = 20 \text{ ft}, \quad y' = -1.67 \text{ ft}, \quad \theta = 30°, \quad g = 32.2 \text{ ft/sec}^2,$$

so the equation becomes

$$-11.45 V_E^2 - 29.78 V_E + 6447 = 0.$$

Using the quadratic formula and discarding the unrealistic negative value give

$$V_E = 25.06 \text{ ft/sec}.$$

This is $V_{E,\min}$ since a 20-ft throw was assumed. The minimum acceptable rotational speed is

$$\omega_{\min} = \frac{V_{E,\min}}{R} = \frac{25.06 \dfrac{\text{ft}}{\text{sec}}}{0.667 \text{ ft}} = 37.6 \text{ rad/sec}.$$

Converting to rpm,

$$\omega_{\min} = 37.6 \frac{\text{rad}}{\text{sec}} \left(\frac{\text{rev}}{2\pi \text{ rad}}\right)\left(\frac{60 \text{ sec}}{\text{min}}\right) \approx 360 \text{ rpm}.$$

The wide-open throttle value of ω with engine 100B is greater than this minimum acceptable value, so

| Engine model 100B is an acceptable engine to meet the design specifications.* | **Answer** ◀

---

* Assuming the snow blower is being pushed at 2.0 ft/sec in wet snow and a 20-ft throw is acceptable.

Discussion

Since the vane rotational speed at wide-open throttle is larger than the minimum acceptable speed, the snow would be thrown farther than 20 ft if the engine is operated at wide-open throttle. In practice, the snow blower engine would be operated at a lower throttle setting. Commercially available snow blowers often employ a governing device that automatically adjusts the throttle position to maintain a fixed rotational speed.

The assumptions in this higher-order application problem fall into three categories:

- Neglect of items that are considered minor
- Conservative assumptions
- Unavailable information that must be made up

Examples of the first type of assumptions are the following:

- Neglect of friction between the snow and the casing
- Uniform velocity over both the inlet and exit areas and snow exit velocity equal to the vane tip velocity
- Constant snow density
- Negligible power loss in the pulley system

Examples of the second type of assumption are the following:

- Use of the density of wet snow rather than of dry snow
- Assumption that the snow leaving the snow blower was not deflected sideways
- Neglect of the pressure forces of the casing on the material in the control volume

Examples of the third type of assumption are the following:

- Snow density of 20 lbm/ft^3
- 20-ft throw
- Speed of snow blower 2.0 ft/sec

We now discuss each assumption in detail. Consider the first type.

- The frictional force between the snow and the casing was expected to be small. However, this assumption can be validated only by the methods outlined in Chapter 9; such an analysis would be quite cumbersome.
- The uniform inlet velocity is reasonable, as all the snow enters the machine at the speed at which a person pushes the machine. The uniform exit velocity would have to be checked, but it appears reasonable that all the snow would have a velocity close to the vane tip velocity as it leaves the machine. The snow must last be in contact with the vane tip as it leaves the vanes.
- The negligible power loss in the pulley system is reasonable. Pulley systems transmit power at an efficiency of about 95%.*

The second type of assumption is conservative for the following reasons:

- The maximum snow density, that of wet snow, was used because the largest snow density, ρ, gives the maximum value of the power requirement. (See the equation for $\dot{W}_s$.)

---

* This efficiency could easily have been incorporated into the analysis.

- The snow leaving the snow blower was assumed not to be deflected sideways. If it were deflected an angle ϕ (sideways and upward such that $\phi > \theta$), the power equation is

$$\dot{W}_s = \rho b \, VR\omega(y_2 + R)(y_4 \, \omega \cos \phi - V).$$

Since

$$y_4 = y_3 + \frac{V}{R\omega \cos \phi} (y_2 - y_1),$$

then

$$\dot{W}_s = \rho b \, VR\omega(y_2 + R) \left[y_3 \omega \cos \phi + \frac{V(y_2 - y_1)}{R} - V \right].$$

We see that the power decreases as ϕ increases; hence we have considered the case of the maximum power requirement by using $\phi = \theta$, as θ is the minimum value of ϕ.

- Neglect of the pressure forces of the casing on the snow is partly justified because the pressure along the circular surface AB in Fig. E5.14(c) is directed toward the center of rotation of the vanes. Since moments were taken about this line of rotation, the moment of the pressure force along AB is zero. The pressure forces along surfaces BC, CD, and EF are not directed toward the center of rotation. However, the pressure along CD is expected to exceed the pressure along EF (pressure increases as we move away from the center of curvature, Eq. 4.16). In addition, area CD exceeds EF, and the clockwise moment $\mathscr{M}_{CD}$ of the pressure forces along CD exceeds the counterclockwise moment $\mathscr{M}_{EF}$ of the pressure force along EF. Taking moments gives

$$\sum \mathscr{M}_z = -\mathscr{T} - \mathscr{M}_{CD} + \mathscr{M}_{EF}.$$

Solving for $\mathscr{T}$ in the angular momentum theorem gives a smaller value of $\mathscr{T}$ if we include these two pressure forces. Hence neglect of these two pressure forces (and of that along BC) gives larger torque and power estimates.

The third type of assumptions involve values for information that was not given and had to be postulated to solve the problem. The density of wet snow was the best available, the 20-ft throw is the minimum throw we would expect of a snow blower, and the 2.0-ft/sec velocity of pushing the snow blower is a reasonable estimate. Design engineers often must make such reasonable assumptions to carry out an analysis and complete their design.

5.5.2 An Application of the Angular Momentum Theorem: Turbomachinery

The angular momentum theorem is useful in the analysis of turbomachinery. A *turbomachine* is a device that uses a moving rotor, carrying a set of *blades* or *vanes,* to accomplish work transfers to or from a moving stream of fluid. If the work is done on the fluid by the rotor, the machine is called a *pump* or *compressor;* if the fluid delivers work to the rotor, the machine is called a *turbine.* Figure 5.23 illustrates some typical turbomachines.

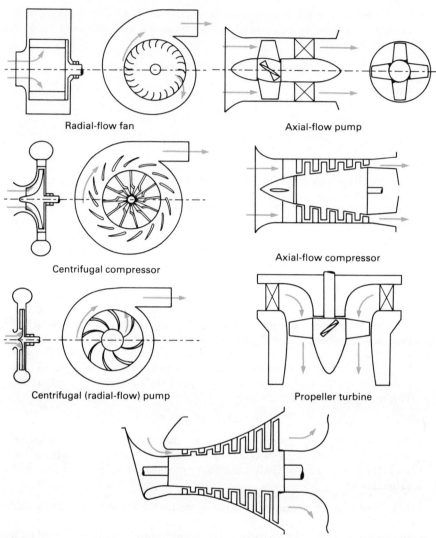

Figure 5.23 Sketches of typical turbomachines.

The principle of turbomachinery operation is not difficult to understand. In Example 5.8 we considered the force generated when a jet of fluid is deflected by a moving vane. The product of vane force and vane speed is the rate of work done on the vane. Note that vane motion is required in order for work to be done. A simple turbine would be made up of several vanes (Fig. 5.24). Instead of moving in a straight path, the vanes move in a circular path because they are connected to the rotor. If we run the rotor "backwards," driving it with a motor and pushing the blades into the fluid, then the blade would do work on the fluid and the fluid's energy would increase. This is the principle of pump operation.

Turbomachines are classified as *axial flow, radial flow,* or *mixed flow,* depending on the direction of fluid motion with respect to the rotor's axis of rotation as the fluid passes over the blades (see Fig. 5.23).

In an axial-flow rotor, the fluid maintains an essentially constant radial position as it flows from rotor inlet to rotor outlet. In a radial-flow rotor, the fluid moves primarily radially from rotor inlet to outlet, although the fluid may be moving in the axial direction at the machine inlet or outlet. In a mixed-flow rotor, the fluid has both axial and radial velocity components as it passes through the rotor. For a given diameter and rotational speed, radial- and mixed-flow rotors are capable of exchanging more work with the moving fluid than axial-flow rotors because the "centrifugal forces" associated with the radius change do work.

To apply the angular momentum theorem to a turbomachine flow, consider a control volume that lies just outside the rotor (Fig. 5.25). The control volume is fixed. We can use the angular momentum theorem to compute the moment (torque) about any axis. The most useful computation is the torque about the driveshaft axis (z) since this will contain the driving torque. For moments about the z-axis, Eq. (5.71) becomes

$$\sum M_z = \left[\frac{d}{dt} \int_{\forall_{cv}} \rho(\vec{r} \times \vec{V}) \, dV \right] \cdot \hat{k} + \left[\oint_{A_{cv}} \rho(\vec{r} \times \vec{V})(\vec{V}_r \cdot \hat{n}) \, dA \right] \cdot \hat{k}, \quad \text{(5.73)}$$

where $\hat{k}$ is the unit vector in the z-direction. Taking the dot product with $\hat{k}$ produces the z-component on the right side. We now introduce the following assumptions:

■ The only moment acting on the material in the control volume is the driveshaft torque $\mathcal{T}_z$. We neglect moments due to the weight of the material in the control volume and the pressure and shear forces in the fluid at the control surface.*

■ There is no accumulation of angular momentum inside the control volume.

■ The flow is uniform at the inlet (radius r_1) and at the outlet (radius r_2).

With these assumptions, the angular momentum equation becomes

$$\mathcal{T}_z = \dot{m}_2(r_2 V_{\theta 2}) - \dot{m}_1(r_1 V_{\theta 1}),$$

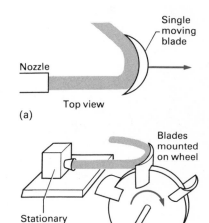

Figure 5.24 A simple turbine made of several moving vanes for fluid jet deflection.

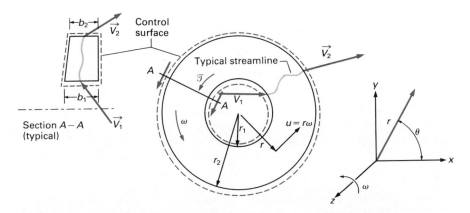

Figure 5.25 Control volume enclosing a generalized turbomachine rotor.

* Note that moments due to pressure would be zero for a circular control volume.

where V_θ is the tangential component of the fluid velocity (see Fig. 5.25). The mass flow rates, $\dot{m}_2$ and $\dot{m}_1$, come from the integration of the product of density and the normal component of velocity over the control volume inlet and outlet areas, the rV_θ-terms come from $(\vec{r} \times \vec{V}) \cdot \vec{k}$, and $\mathcal{T}_z$ is the torque required to drive the rotor.

Application of the continuity equation to the control volume of Fig. 5.25 gives

$$\dot{m}_1 = \dot{m}_2 = \dot{m},$$

so we can write

$$\mathcal{T}_z = \dot{m}(r_2 V_{\theta 2} - r_1 V_{\theta 1}). \tag{5.74}$$

The power transferred to the fluid by the rotor (that is, the rate at which the rotor does work on the fluid) is equal to the product of the driveshaft torque and the rotational speed ω:

$$-\dot{W}_s = \mathcal{T}_z \omega = \dot{m}\omega(r_2 V_{\theta 2} - r_1 V_{\theta 1}).$$

The minus sign is added to $\dot{W}_s$ according to the customary sign convention for work. Since the control volume lies just outside of the rotor, r_1 and r_2 are the radii of the rotor at inlet and outlet respectively.* The product of these rotor radii with the rotor angular velocity equals the *linear* velocity of the rotor at points 1 and 2. In turbomachinery analysis, it is customary to use U to represent the linear speed of the rotor at a point and V_u to represent the tangential ("U-wise") component of the fluid velocity, so we write

$$-\dot{W}_s = \dot{m}(U_2 V_{u2} - U_1 V_{u1}). \tag{5.75}$$

This equation is called Euler's pump and turbine equation. It applies to turbines (for which $U_2 V_{u2}$ must be less than $U_1 V_{u1}$, so that $\dot{W}_s$ is positive) as well as to pumps (for which $U_2 V_{u2} > U_1 V_{u1}$).

Analysis and design of turbomachines is facilitated by drawing *velocity diagrams* to represent the fluid and rotor velocities at various points in the machine. The fluid velocity can be viewed from two coordinate systems, one fixed to the rotor and one fixed to the machine casing. The velocity as viewed from a coordinate system fixed to the rotor is called the *relative velocity* and is assigned the symbol $\vec{W}$. Since the relative coordinate system moves at the rotor velocity $\vec{U}$ (written here as a vector), the relation between the absolute velocity, $\vec{V}$, and the relative velocity is

$$\vec{V} = \vec{U} + \vec{W}. \tag{5.76}$$

Velocity diagrams reflect this vector relation. Figure 5.26 shows velocity diagrams for flow entering and leaving a blade in a radial-flow turbomachine. In terms of the geometry of the velocity diagram, the power absorbed by the rotor and transferred to the fluid is

$$-\dot{W}_s = \dot{m}(U_2 V_2 \cos \alpha_2 - U_1 V_1 \cos \alpha_1). \tag{5.77}$$

---

* Note carefully that 1 is rotor *inlet* and 2 is rotor *outlet*. In this discussion 1 is the smaller radius and 2 is larger; in some machines the inlet and outlets may be reversed, with fluid entering at the larger radius and exiting at the smaller.

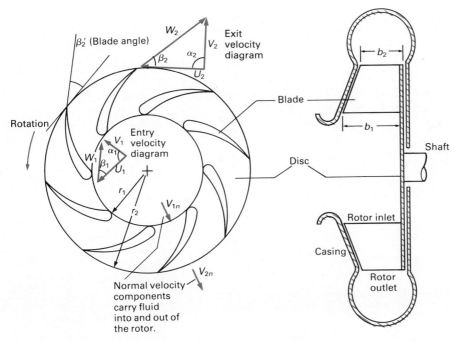

Figure 5.26 Details of flow entering and leaving a radial-flow rotor.

The mass flow rate through the rotor can be calculated as the flow across the inlet or outlet:*

$$\dot{m} = \rho_1 V_1 \sin \alpha_1 (2\pi r_1 b_1) = \rho_2 V_2 \sin \alpha_2 (2\pi r_2 b_2). \tag{5.78}$$

The angles in the velocity diagrams are related to the angles of the blades; a particularly simple assumption is

$$\beta_2 \text{ (velocity diagram angle)} = \beta_2' \text{ (blade angle).} \tag{5.79}$$

Simple preliminary analysis and design of turbomachines can be based on Eqs. (5.77), (5.78), and (5.79), together with consideration of the geometry of the velocity diagrams.

---

* This equation is applicable only to radial-flow rotors because of the expression used to calculate the areas ($2\pi rb$). A similar equation can be developed for axial-flow rotors where the flow area is an annulus.

EXAMPLE 5.15   Illustrates the analysis of a simple turbomachine

The axial-flow hydraulic turbine sketched in Fig. E5.15(a) has a water flow rate of 75 m³/s, an outer radius $R = 5.0$ m, and a blade height $h = 0.5$ m. Assume uniform properties and velocities over both the inlet and the exit. The water temperature is 20°C, and the turbine rotates at 60 rpm. The relative velocities W_1 and W_2 make angles of 30° and 10°, respectively, with the normal to the flow area. Find the output torque and power developed by the turbine.

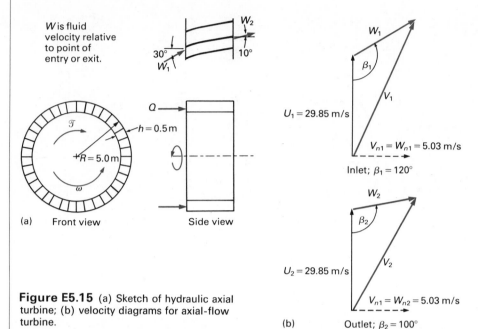

W is fluid velocity relative to point of entry or exit.

(a) Front view   Side view

$U_1 = 29.85$ m/s

$V_{n1} = W_{n1} = 5.03$ m/s

Inlet; $\beta_1 = 120°$

$U_2 = 29.85$ m/s

$V_{n1} = W_{n2} = 5.03$ m/s

Outlet; $\beta_2 = 100°$

Figure E5.15 (a) Sketch of hydraulic axial turbine; (b) velocity diagrams for axial-flow turbine.

(b)

Solution

Given
Turbine in Fig. E5.15(a)
Volume flow rate 75 m³/s
Outer radius 5.0 m
Blade height 0.5 m
Rotational speed 60.0 rpm
Fluid 60°C water

Find
Output torque and power developed by turbine

Solution
The output torque is found using Eq. (5.75):

$$\mathcal{T} = \dot{m}(r_2 V_{u2} - r_1 V_{u1}).$$

Using Table A.5 of Appendix A, Eq. (3.36) gives

$$\dot{m} = \rho Q = \left(998\ \frac{\text{kg}}{\text{m}^3}\right)\left(75\ \frac{\text{m}^3}{\text{s}}\right) = 74{,}850\ \text{kg/s}.$$

For an axial-flow machine where the blade height h is small compared to R,

$$r_2 = r_1 \approx R - \frac{1}{2}h = 5.0\ \text{m} - \frac{1}{2}(0.5\ \text{m}) = 4.75\ \text{m}.$$

The tangential components of the absolute velocity are found from the velocity triangles in Fig. E5.15(b), where V represents the fluid absolute velocity, W the fluid velocity relative to the turbine wheel, and U the wheel peripheral velocity.

In this figure,

$$U_1 = r_1\omega = (4.75 \text{ m})\left(60 \frac{\text{rev}}{\text{min}}\right)\left(2\pi \frac{\text{rad}}{\text{rev}}\right)\left(\frac{\text{min}}{60 \text{ s}}\right) = 29.85 \text{ m/s},$$

$$U_2 = r_2\omega \approx r_1\omega = 29.85 \text{ m/s},$$

$$V_{n1} = \frac{Q}{A_1} = \frac{75 \text{ m}^3/\text{s}}{\pi[(5.0 \text{ m})^2 - (4.5 \text{ m})^2]} = 5.03 \text{ m/s}, \quad \text{and}$$

$$V_{n2} = \frac{Q}{A_2} = 5.03 \text{ m/s}.$$

V_{n1} and V_{n2} represent the normal components of the absolute velocities V_1 and V_2 (or of the relative velocities W_1 and W_2), respectively.

Referring to Fig. E5.15(b), we calculate the tangential components of the absolute velocities as

$$V_{u1} = U_1 + W_1 \cos(\pi - \beta_1) = U_1 - W_1 \cos \beta_1$$

and

$$V_{u2} = U_2 + W_2 \cos(\pi - \beta_2) = U_2 - W_2 \cos \beta_2.$$

Since

$$W_{n1} = W_1 \sin(\pi - \beta_1) = W_1 \sin \beta_1 \quad \text{and} \quad W_{n2} = W_2 \sin(\pi - \beta_2) = W_2 \sin \beta_2,$$

we have

$$V_{u1} = U_1 - \frac{W_{n1}}{\tan \beta_1} \quad \text{and} \quad V_{u2} = U_2 - \frac{W_{n2}}{\tan \beta_2}.$$

The numerical values give

$$V_{u1} = 29.85 \text{ m/s} - \frac{5.03 \text{ m/s}}{\tan 120°} = 32.75 \text{ m/s}$$

and

$$V_{u2} = 29.85 \text{ m/s} - \frac{5.03 \text{ m/s}}{\tan 100°} = 30.74 \text{ m/s}.$$

The output torque is

$$\mathcal{T} = \left(74,850 \frac{\text{kg}}{\text{s}}\right)\left[(4.75 \text{ m})\left(30.74 \frac{\text{m}}{\text{s}}\right) - (4.75 \text{ m})\left(32.75 \frac{\text{m}}{\text{s}}\right)\right] = -7.15 \times 10^5 \text{ N} \cdot \text{m}.$$

The significance of the negative sign is that the torque is in the direction opposite that assumed to be positive (counterclockwise for front view of Fig. E5.15a). The numerical value is merely

$$\left| \mathcal{T} = 7.15 \times 10^5 \text{ N} \cdot \text{m}. \right|$$

Answer ◀

The output power is

$$\dot{W} = \mathcal{T}\omega$$

$$= (7.15 \times 10^5 \text{ N} \cdot \text{m})\left(60 \frac{\text{rev}}{\text{min}}\right)\left(2\pi \frac{\text{rad}}{\text{rev}}\right)\left(\frac{\text{min}}{60 \text{ s}}\right),$$

$$\dot{W}_s = 4.49 \times 10^6 \text{ N} \cdot \text{m/s},$$

$$\left| \dot{W}_s = 4.49 \text{ MW}. \right|$$

Answer ◀

Discussion

We assumed that the relative velocity was aligned with the rotor blade at both the inlet and the outlet. This means that the fluid would be swirling in the direction of rotor motion prior to entering onto the blades ($V_{u1} \neq 0$; see Fig. E5.15b). A set of nozzles or deflector vanes must be provided to direct the flow as it enters the turbine.

More extensive considerations of flow in turbomachines are beyond the scope of this book. The major effort in this field is directed along two paths:

- Relating the flow over the blades to the blade shapes. Crudely, we may think of this analysis as an improvement to the "perfectly guided fluid" assumption embodied in Eq. (5.79) and the one-dimensional flow assumption embodied in Eqs. (5.77) and (5.78). This requires analysis of fluid motion in a coordinate system fixed to the rotor, usually a noninertial reference frame.
- Calculating the "losses" associated with the flow through the turbomachine. Although Eq. (5.77) applied to a pump, for example, accurately calculates the work transferred between the fluid and the rotor (within the limits of the one-dimensional flow assumption), not all of this work results in an increase in the fluid's mechanical energy. If flow conditions are very poor, well over half of the work done on the fluid may be converted into "losses."

The fluid mechanics of turbomachinery is currently a very active area of research. If you are interested in further reading in this area, consult references [3–6].

Problems

Extension and Generalization

1. Find the acceleration of the cart in Fig. P5.1 as a function of time. The initial total mass is m_0 and the fluid density is ρ_0. Assume frictionless bearings and a frictionless surface.

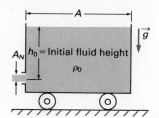

Figure P5.1

2. Show that the magnitude of the pressure pulse Δp that travels upstream at the speed of sound c due to the rapid closure of a valve in a liquid piping system is given by

$$\Delta p = \rho V c,$$

where ρ is the fluid density, V the fluid velocity before closure, and c the speed of sound in the liquid.

3. Water with a constant absolute velocity V strikes the cart moving with velocity U in Fig. E5.8(a). Find the cart velocity U such that the power developed by the cart ($\dot{W} = UF_{Rx}$) is a maximum for fixed values of V, A, ρ, and β.

Lower-Order Application

4. The water from a fire hose having a volume flow rate of $1.0 \text{ m}^3/\text{s}$ and an exit velocity of 100 m/s strikes a cart having a circular area of 2.0 m^2, as shown in Fig. P5.2. If the cart's mass is 1.0 kg and its initial velocity is zero, find its initial acceleration.

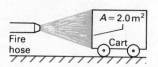

Figure P5.2

5. Water leaves the nozzle in Fig. P5.3 with a velocity of 5 ft/sec. The nozzle exit area is 0.725 in². The water falls a distance of 60 ft before hitting the flat plate. Find the force on the flat plate by the water stream.

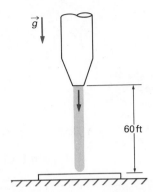

Figure P5.3

6. Civil engineering students are participating in a concrete canoe race. The canoe shown in Fig. P5.4 springs a leak in the bottom of the canoe, and the velocity of the water through the hole is 10 ft/sec. The students set up a hand pump to bail out the canoe. The head losses in the pump line system are given by $gh_L = 1.5V^2$, where V is the water velocity in the pipe. Find the rate at which water is filling or emptying the canoe.

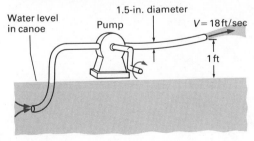
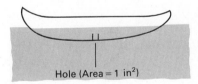

Figure P5.4

7. For the students in Problem 6 to slow down the canoe behind them, they direct their water jet from the hand pump against the bow of the trailing canoe. See Fig. P5.5. Compute the force exerted on the trailing canoe by the water jet. Will this action increase the velocity of the canoe with the hand pump?

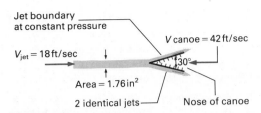

Figure P5.5

8. Consider the jet of water hitting the cart in Fig. E5.8(a). The cart has a mass of 5.0 kg, the velocity V is 1.0 m/s, the area A is 0.10 m², $\beta = 45°$, and $U = 0$ at time $t = 0$. Find the velocity $U(t)$. Assume no air resistance to the motion of the cart. Next find the velocity $U(t)$ assuming the air resistance on the cart can be expressed as

$$F_R = KU,$$

where $K = 20.0 \text{ N·s/m}$.

9. A two-engine airplane weighing 150,000 lb touches down on a runway at 125 mph. The thrust reversers in Fig. P5.6 are activated at 120 mph. How long must the thrust reversers be operated to slow the airplane down to 30 mph? The mass flow rate out of each engine is 200 lbm/sec with a density of 0.02 lbm/ft³ and a velocity of 2500 ft/sec.

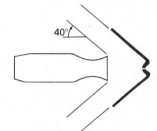

Figure P5.6

10. An unmanned space vehicle has a velocity of 7500 m/s and a mass of 1500 kg and is moving in outer space with negligible gravitational attraction. A rocket on the engine fires at a rate of 5 kg/s with an exit velocity of 1000 m/s for 2.0 s to accelerate the rocket. Find the new velocity of the space vehicle.

11. The rocket-powered sled in Fig. P5.7 is traveling at 600 mph and is slowed down by lowering a scoop on the sled into water. Find the shortest distance in which the sled can be stopped. The sled weighs 1000 lb, and the scoop is 6 ft wide. Neglect air resistance.

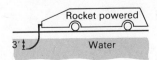

Figure P5.7

12. The rocket sled in Problem 11 is brought up to speed by burning fuel at a constant rate of 250 lbm/sec. The exhaust gases have the properties of air at 1600°F and 14.7 psia. The exit area is 2.0 ft². Find the time for the sled to reach 600 mph.

13. A 4-ton coal hopper carrying 8 tons of coal is rolling freely down a 10° incline at a constant speed of 1.0 mph. A chute opens in the bottom of the hopper and the coal empties out of the hopper at a constant rate in 20 sec. Find the velocity of the coal hopper after the 20 sec if air drag $\mathscr{D}$ is CV^2, where $C = 0.030 \ \text{lb} \cdot \text{hr}^2/\text{mi}^2$.

14. A 10-ton semitrailer carrying 12 tons of coal is moving up a 5° incline at a speed of 50 mph. The air drag on the trailer is equal to CV^2, where $C = 0.040 \ \text{lb} \cdot \text{hr}^2/\text{mi}^2$. Assume that a constant driving force is applied to the road through the tires. The tailgate of the trailer comes loose and coal falls out at the rate of 100 lbm/sec through an area of 4 ft². Find the speed of the semitrailer after 5 min.

15. Find the vertical and horizontal components of force required to hold the tank in Fig. P4.19 in place.

16. Two streams mix as shown in Fig. P5.8. Find the pressure change $(p_3 - p_1)$ due to the mixing. Neglect friction between the fluid and the wall.

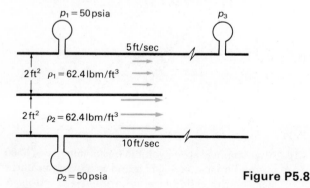

Figure P5.8

17. Find the pressure change $(p_2 - p_1)$ for the jet pump in Fig. P5.9. The fluid is 20°C water. Assume negligible friction between the fluids and the walls and uniform pressure over each flow area. See Problem 122 in this chapter for a further description of a jet pump.

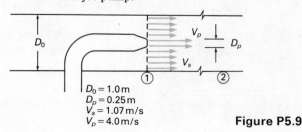

$D_0 = 1.0 \, \text{m}$
$D_p = 0.25 \, \text{m}$
$V_s = 1.07 \, \text{m/s}$
$V_p = 4.0 \, \text{m/s}$

Figure P5.9

18. Find the x- and y-components of the force required to hold the bend in Fig. P5.10 in place.

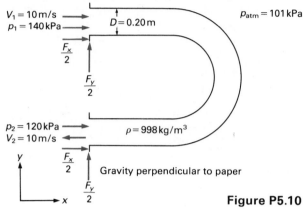

Gravity perpendicular to paper

Figure P5.10

19. Water flows through the nozzle in Fig. P5.11 and discharges to the atmosphere. Find the force F necessary to hold the nozzle assembly onto the support.

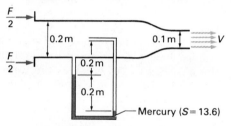

Mercury $(S = 13.6)$

Figure P5.11

20. How much force must the firefighters exert to hold the nozzle in Fig. P5.12? Assume that no force is exerted on the nozzle by the hose.

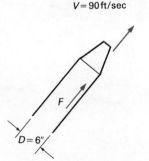

$Q = 100 \, \text{gal/min}$
$V = 90 \, \text{ft/sec}$

F

$D = 6''$

Figure P5.12

21. Fortified wine having a specific gravity of 1.03 flows through a 2-in., 90° standard elbow with a pressure drop of 0.50 psi and a flow rate of 40 gal/min. Find all the forces on the elbow if the elbow is in a horizontal plane and the elbow inlet pressure is 12 psig.

22. Water flows between the fixed vanes as shown in Fig. P5.13. Find the x- and y-components of the force on the vanes by the water. The total volume flow rate is 1000 m^3/s, the pressure $p_1 = 150$ kPa, and the pressure $p_2 = 101$ kPa.

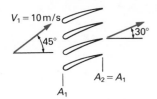

Figure P5.13

23. A moving jet engine draws in air at 15 kg/s with an inlet relative velocity of 150 m/s and an exit relative velocity of 700 m/s. Find the engine thrust if atmospheric pressure exists around the engine.

24. The nozzle of a rocket has the velocities and pressures shown in Fig. P5.14. Find the force F required to hold the nozzle in place.

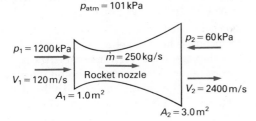

Figure P5.14

25. A stationary jet engine has an inlet velocity of 400 ft/sec, an inlet density of 0.075 lbm/ft^3, and an inlet area of 1.0 ft^2. The exit velocity is given by

$$u = \left(4000 \, \frac{\text{ft}}{\text{sec}}\right)\left[1 - \left(\frac{r}{R}\right)^2\right],$$

with $R = 1.0$ ft. The exit density is 0.040 lbm/ft^3. Find the thrust on the plane by the engine.

26. A youngster places her finger in the stream of a wide jet of water. The water stream is deflected as shown in Fig. P5.15. Find the force required to hold her finger in the water stream.

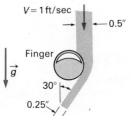

Figure P5.15

27. The hydraulic brake in Fig. P5.16 consists of a movable piston in a cylinder of sligthly larger diameter. In a particular case, the piston velocity V is constant. Find the velocity V_j of the jet escaping between the piston and the cylinder. Then find the pressure p_H on the head end of the cylinder; $D_p = 1.0$ cm, $D_c = 1.10$ cm, $V = 10$ m/s, and $x = 9.0$ cm. $S = 0.90$.

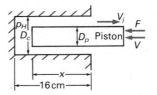

Figure P5.16

28. Water at 60°F and at a flow rate of 80 gal/min flows through a 6-in. schedule 40 pipe, through a reducer, and then through a 4-in. schedule 40 pipe. The pressure at the reducer inlet is 40 psig and at the reducer outlet is 34 psig. Find the force of the water on the reducer.

29. Find the forces F_x and F_y required to hold the tank steady in Fig. P5.17. The pipe diameter D is that required for a constant water level. Assume inviscid flow.

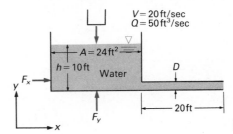

Figure P5.17

30. Find the diameter D of the jet to hold a plate in suspension, as shown in Fig. P5.18.

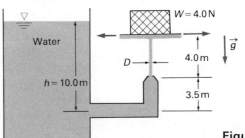

Figure P5.18

31. The cart in Fig. P5.19 is hit by the water jet as shown. The cart has frictionless wheels and rolls along a frictionless, horizontal surface. The cart has a mass of 4.0 kg and can hold up to 2.0 m^3 of

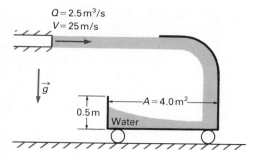

Figure P5.19

water. Find the velocity of the cart as a function of time if the cart initially has no water and a velocity of zero when first hit by the water jet.

32. The centrifugal pump in Fig. P5.20 delivers water at the rate of 10 ft³/sec. The water is an ideal fluid, and the pump, piping, and internal fluid weigh 500 lb. Each bolt holding the pump to the floor can withstand a shear force of 500 lb and a tensile force of 1000 lb. Find the minimum number of bolts required to hold the pump to the floor.

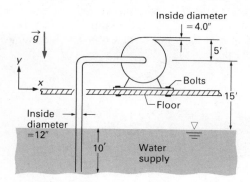

Figure P5.20

33. Consider the 0.10-m-diameter water stream hitting the conical deflector in Fig. P5.21. Find the force F required to hold the deflector stationary.

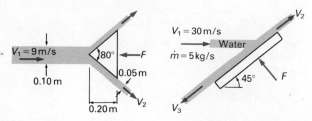

Figure P5.21   **Figure P5.22**

34. Find the fraction of the mass $\dot{m}$ diverted down the plate in Fig. P5.22. Also find the magnitude of the force F required to hold the plate stationary. Neglect friction between the water and the plate.

35. Find the spacing h required to lift the disc in Fig. P5.23. The plate weighs 10.0 N. Assume the 20°C air density is constant.

36. Find the spacing h to lift the disc in Fig. P5.23 if the direction of the air flow is reversed. The plate weighs 10.0 N. Assume the 20°C air density is constant.

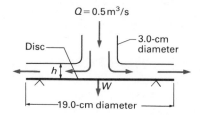

Figure P5.23

37. A fireboat is traveling at 35 mph and is discharging water toward the rear at the rate of 390 gal/min through a 6-in.-diameter nozzle and at a 45° angle above the horizontal. How much faster would the fireboat travel if the water were directed at a fixed, solid wall and the water stream were turned 90° to the right or left?

38. The locomotive in Fig. P5.24 is plowing through 0.2-m-deep snow while traveling at 20 km/hr. It is estimated that the locomotive has an extra 25 hp to remove the snow. What angle α should the plow at the front of the locomotive make with the direction of travel for the locomotive to remove the snow at this speed?

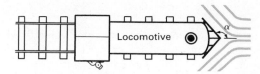

Figure P5.24

39. Figure P5.25 shows coal dropped from a hopper onto a conveyor belt at a constant rate of 25 ft³/sec. The coal has a specific gravity ranging from 1.12 to 1.50. The belt has a speed of 5.0 ft/sec

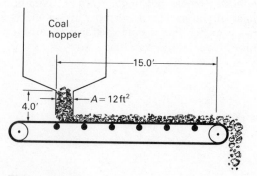

Figure P5.25

and a coal length of 15.0 ft. Estimate the torque required to turn the drive pulley of the conveyor belt. The drive pulley diameter is 2.0 ft.

40. When a particular airplane flies at 700 mph it takes in air at the rate of 1000 lbm/sec, has an air-fuel ratio of 30:1 on a mass basis, and an exhaust gas velocity of 3000 ft/sec. The jet plane weighs 40,000 lb and experiences a drag force $\mathscr{D}$ equal to

$$\mathscr{D} = 100 \ V^2,$$

where $\mathscr{D}$ is in pounds and V is the absolute airplane velocity in miles per hour. Find the thrust required for the airplane to travel both horizontally and vertically upward at 700 mph.

41. A commercial airliner with four jet engines is traveling at a constant speed of 600 mph. Each engine has an inlet area of 1.0 ft^2, an exit area of 0.30 ft^2, an air-fuel ratio of 35:1 on a mass basis, an exhaust velocity of 2000 ft/sec relative to the plane, and an exhaust pressure equal to the local atmospheric pressure. The ambient temperature and pressure are $-65°$F and 4.0 psia, respectively. Find the drag on the airplane.

42. The dredger in Fig. P5.26 delivers a mixture of sand and water at the rate of 4000 gal/min to the barge. Find the forces tending to separate the dredger and barge. Neglect the resistance of the river water to the motion of the dredger and barge. The sand-and-water mixture has a specific gravity of 1.7. Both the dredger and barge weigh 200,000 lb at the instant shown.

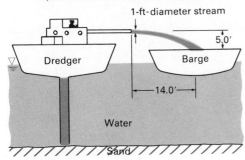

Figure P5.26

43. The pipeline in Fig. P5.27 has a water flow rate of 600 gal/min. Find the force on the elbow due

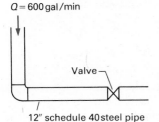

12″ schedule 40 steel pipe   **Figure P5.27**

to the sudden closure of the valve. The speed of sound in water is about 5000 ft/sec.

44. Using the result of Problem 2, find the magnitude of the pressure pulse for a water flow rate of 4.0 m^3/min in a 0.30-m-inside-diameter steel pipe due to the sudden closure of a valve. The speed of sound in water is about 1700 m/s.

45. A boat weighs 100,000 lb and is traveling at 30 knots. How much less horsepower is required to maintain this constant speed if the boat has two guns shooting 6000 rounds per minute to the rear with a muzzle velocity of 3000 ft/sec? Each round weighs 0.125 lb.

46. An air-lift pump is shown in Fig. P5.28. Compressed air is mixed with water, and the mixture has a specific gravity of 0.6. Assume that the air and water have the same velocity V at the exit and that pressure is uniform across planes 1 and 2. Find the water flow rate.

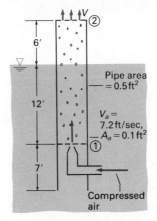

Figure P5.28

47. The stationary jet engine in Fig. P5.29 produces a thrust of 200,000 N. Find the mass flow rate through the engine.

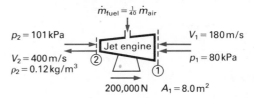

Figure P5.29

48. Air enters a stationary jet engine at a uniform velocity of 200 m/s and leaves at a uniform velocity of 600 m/s. The fuel mass flow rate is 1/40 the air mass flow rate into the engine. The inlet area is 0.25 m^2, the exit area 0.10 m^2, the inlet air density 0.75 kg/m^3, and the exit pressure equal to the atmospheric pressure. Find the thrust developed by the engine.

49. Air flows through a 90° miter bend at the rate of 400 m³/s. The duct cross section is rectangular, measuring 0.67 m by 0.50 m. The air is at 15°C and 101 kPa. Find the force on the elbow due to the turning of the air.

50. Find the angle α between the sail and the absolute wind velocity for a sailboat to sail "across the wind" at the fastest speed. "Sailing across the wind," "sailing with the wind abeam," or "reaching" occurs when the sailboat sails in a direction perpendicular to the wind. For simplicity, assume the sail is flat. Neglect water resistance to the motion of the sailboat.

51. Find the angle α between the sail and the absolute wind velocity for a sailboat to "run with the wind" at the fastest speed. "Running with the wind" is sailing in the direction of the wind.

52. The drag force on an object can be measured by placing the object in a horizontal circular pipe, causing a fluid to flow in the pipe, neglecting the wall shear stress on the fluid, and measuring the fluid pressure and velocity profiles both upstream and downstream of the object. In a particular experiment of 70°F water flowing over an object, the upstream pressure is 75.0 psig, the downstream pressure is 71.3 psig, the upstream velocity is uniform at 2.17 ft/sec and the downstream velocity is given by

$$u = (4.34 \text{ ft/sec})\left(1 - \frac{r^2}{R^2}\right).$$

Find the drag on the submerged object.

53. A two-dimensional object is placed in a 2.0-m-wide water channel, as shown in Fig. P5.30. Neglecting friction between the walls and the water, find the drag per unit length of the object.

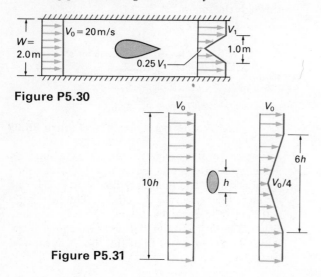

Figure P5.30

Figure P5.31

54. Figure P5.31 shows a two-dimensional object placed in an infinite stream. The velocity profiles are as shown. Find the drag force per unit length of the fluid on the object.

55. Find the force components F_x and F_y required to hold the box in Fig. P5.32 stationary. The fluid is oil with a specific gravity of 0.85.

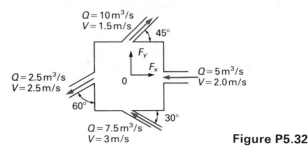

Figure P5.32

56. A water jet is aimed at the hole in the wall of Fig. P5.33; 25% of the water escapes through the hole, and the remaining water is turned 90° to the water jet. Find the force on the wall by the jet.

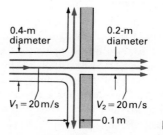

Figure P5.33

57. Consider the inviscid flow of water out of the tank in Fig. P5.34. Find the spring scale reading as a function of the water height h if the spring scale is set at zero when the tank is full and no water is leaving the tank. The balance beam remains approximately horizontal.

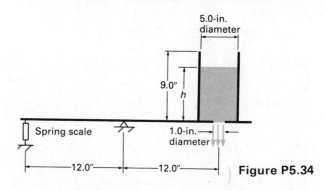

Figure P5.34

58. Consider the gate in the small lock of Fig. E5.7(a). Find the angle α (60° in Example 5.7) that produces the greatest force on the gate. Assume the

59. A stationary 100-kg object is hit by a 200-mg bullet having a velocity of 700 m/s. Find the velocity of the object as a function of time, assuming the bullet is embedded in the object.

60. Water flows over the 100-ft-long, 25-ft-high dam in Fig. P5.35. The water velocity V_1 some distance upstream of the dam is uniform at 10 ft/sec, and the water velocity V_2 some distance downstream is also uniform. Assume inviscid flow and a constant river width and find the force on the dam by the water.

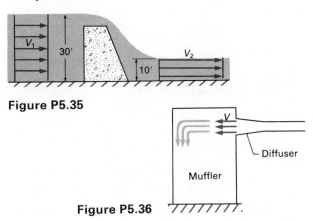

Figure P5.35

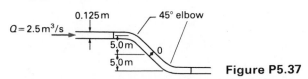

Figure P5.36

61. The muffler shown in Fig. P5.36 was built to reduce the noise level of a supersonic wind tunnel. The air coming out of the diffuser is moving at 20 mph through an area of 25 in². The exit temperature and pressure are $-70°$F and 25 psia. Find the horizontal force on the muffler wall when the air stream is deflected 90°.

62. Two straight pipes are connected by the two 45° elbows shown in Fig. P5.37. Find the moment required about point 0 to prevent rotation of the two pipes. Water is flowing.

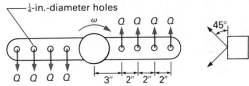

Figure P5.37

63. A tube containing gunpowder is wrapped around a wheel. When the gunpowder is ignited, the exhaust gases are expelled, forcing the wheel to rotate about its axis. The wheel mass is 10 kg, its radius of gyration is 1.0 m, and its radius is 2.0 m. The velocity of the exhaust gases relative to the wheel rim is 70 m/s. The gunpowder has an initial

mass of 1.0 kg and is burned at the rate of 0.025 kg/s. Find the angular speed of the wheel as a function of time.

64. The radial-flow pump in Fig. 5.26 has a uniform inlet radial velocity of 10 m/s, a uniform outlet radial velocity of 2.5 m/s, an inlet radius $R_1 = 1.0$ m, and an outlet radius of 4.0 m. The flow rate is 100 m³/s, $\alpha_1 = 90°$, and $\alpha_2 = 120°$. Find the power input to the fluid by the rotor blades and sketch the inlet and outlet velocity diagrams.

65. A lossless motor drives the fan in Fig. P5.38 at 4000 Hz. The power input to the motor is 16 amps at 110 volts. For the geometry shown, find the flow rate of 15°C air through the fan. Assume the tangential component of the velocity leaving the impeller is equal to that of the impeller at that point.

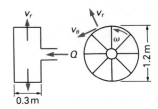

Figure P5.38

66. Figure P5.39 shows a simplified sketch of a dishwasher water supply manifold. Find the manifold's rotational speed ω. Neglect air resistance and bearing friction and take $Q = 0.25$ gal/min.

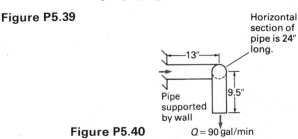

Figure P5.39

67. Water is flowing through the 2-in. schedule 40 steel pipe in Fig. P5.40 at the rate of 90 gal/min. Find the torque developed at the base where the pipe is supported. Neglect the pipe and water weights.

68. Find the torque required about point 0 of the box in Fig. P5.41 to keep the box from rotating. The

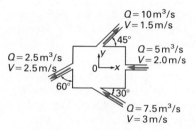

Figure P5.41

top and bottom are each 1.0 m long, and the two sides are 0.8 m long. Use clockwise moments as positive.

69. Water is pumped steadily through the apparatus shown in Fig. P5.42. The velocity, pipe area, and gage pressure are shown for both outlet sections 1 and 2. Assuming water as a frictionless and incompressible fluid, compute the horsepower input to the pump.

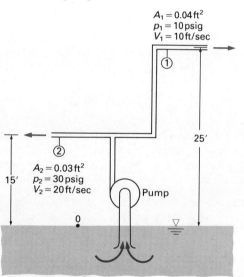

Figure P5.42

70. The flow rate through the pump in Fig. P5.43 is 2 ft³/sec. Find the pump head $(h_2 - h_1)$ in feet; $d_1 = d_2 = 4$ in., and $d_3 = 0.5$ in.

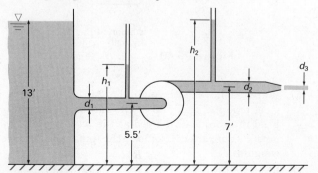

Figure P5.43

71. The subsonic wind tunnel in Fig. P5.44 consists of a converging nozzle, test section, diffuser, fan, and connecting ductwork. The test section has a flow area of 1.0 m³ and a maximum velocity of 120 km/hr of 10°C air. The ductwork has a flow area of 9.0 m². The mechanical energy loss in the wind tunnel is given by $KV^2/2$, where $K = 15$. Find the pressure drop across the nozzle and the power input to the air by the fan.

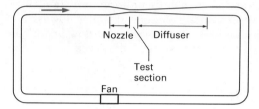

Figure P5.44

72. A motor-driven fan has a flow rate of 200 m³/min and a pressure rise of 20 cm of water. The fan inlet and outlet velocities are equal. The motor has an output power of 10.0 kW and draws a steady current of 60 amps at 220 volts. Find the fan and motor efficiencies. The efficiency η is defined as

$$\eta = \frac{\text{Power output}}{\text{Power input}}.$$

73. The pump in Fig. P5.45 has the performance characteristics shown. The mechanical energy loss in the piping and fittings in the suction and discharge piping is given by

$$gh_L = K\left(\frac{V^2}{2}\right),$$

where V is the average fluid velocity in the pipe and $K = 10$. Find the flow rate through the pump and the power supplied by the pump to the fluid.

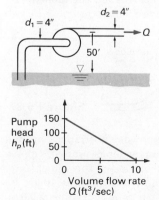

Figure P5.45

74. The hydraulic turbine in Fig. P5.46 has an efficiency of 90% (ratio of output power to available

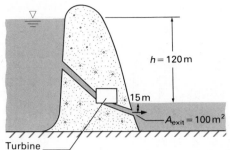

Turbine
generator   **Figure P5.46**

power). The flow rate is 10,000 m³/min. Find the
turbine output power in kilowatts.

75. A horizontal oil $(S = 0.80)$ pipeline has a flow
rate of 1000 gal/min and a pressure drop of 3.0 psi
per 100 ft. Find the power input to the oil by a
pump installed every 3 mi. The pump has no
losses.

76. A fluid flows through a nozzle without heat loss
or gain. The inlet enthalpy $(\tilde{u}_1 + p_1/\rho_1)$ is 1000
BTU/lbm, while the exit enthalpy $(\tilde{u}_2 + p_2/\rho_2)$
is 500 BTU/lbm. The inlet velocity is 10 ft/sec.
Find the exit velocity. Neglect potential energy
changes.

77. Consider the fan test setup in Problem 19 in Chapter 4. The static pressure is 6.0 in. of water (gage)
at section 1 and 5.0 in. of water at section 2. Calculate the energy loss between sections 1 and 2.
Where did this energy loss go? The throttling plug
is a cone of 120° included angle. Estimate the
force required to hold the plug in place.

78. The sump pump in Fig. P5.47 has an input current of 20 amps and an input voltage of 110 volts.
What is the maximum possible height h that the
60°F water can be raised if the pump has a capacity
of 100 gal/min?

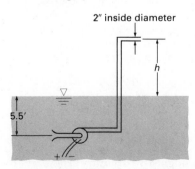

Figure P5.47

79. A pump delivers 10 m³/s through a vertical pipeline 300 ft high. Find the minimum horsepower
input to the pump. Assume that atmospheric pressure exists at both the inlet and the outlet of the
pipeline and that kinetic energy changes are negligible.

80. When the pump in Fig. P5.48 is stopped, there is
no flow through the system and the spring force
is zero. With the pump running, a 6-in.-diameter
jet leaves the pipe, and the spring force is 420 lb.
The water surface elevation in the tank can be
considered constant. Find (a) the water flow rate
and (b) the pump horsepower required to maintain
the flow. The flow is frictionless.

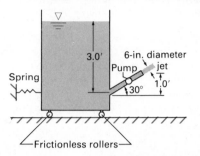

Figure P5.48

81. Oil $(S = 0.85)$ at 40°F flows through a long horizontal pipeline and has a pressure drop of 150 psi.
The mechanical energy loss is given by

$$gh_L = 0.0075\left(\frac{L}{D}\right)V^2,$$

where L is the pipe length, D its inside diameter,
and V the average fluid velocity. For $L = 3000$ ft
and $D = 1.5$ ft, find the internal energy change of
the fluid if the pipe is thermally insulated.

82. The internal energy change of the fluid in Problem 81 is related to its temperature change by

$$\tilde{u}_2 - \tilde{u}_1 = (2000 \text{ J/kg·°C})(T_2 - T_1).$$

Find the temperature change of the fluid for the
conditions given.

83. Liquid water at 40°F flows down a vertical, thermally insulated, 2-in. schedule 40 steel pipe at the
rate of 100 gal/min. The temperature change of the
water is related to its internal energy change by

$$\tilde{u}_2 - \tilde{u}_1 = 1.0 \text{ BTU/lbm·°F } (T_2 - T_1).$$

Find the temperature change of the water per
100 ft of drop if the pressure change $(p_1 - p_2)$
per 100 ft is 12.0 psi and

$$gh_L = 0.0065\left(\frac{L}{D}\right)V^2.$$

84. The axial velocity $u(r)$ for the flow in the entrance
region of a circular pipe of radius R is

$$u = u_0\left(1 - \frac{r^2}{R^2}\right)^{x/L_e},$$

where L_e is the entrance length. Find the pressure
drop due to wall friction and to the momentum

change of the fluid in this entrance region. Assume constant fluid density ρ, $u_0 = 1.0$ m/s, $R = 0.25$ m, $L_e = 15$ m, and $\rho = 990$ kg/m^3.

85. The density in Problem 84 changes according to the equation

$$\rho = \rho_0(1 - x/\ell_e).$$

Find the pressure drop due to wall friction and to the momentum change; $\rho_0 = 990$ kg/m^3.

86. Fuel oil is heated as it flows upward in a 0.2-m-diameter vertical tube. The tube is 25 m long. At the entrance, the velocity is uniform at 6.0 m/s, the pressure is 180 kPa, and the fluid density is 850 kg/m^3. At the exit the velocity is uniform at 7.0 m/s, and the density is uniform. The mechanical energy loss is given by

$$d(gh_L) = \frac{0.02\,V^2}{2D}\,dx,$$

where V is the average velocity at each cross section and D is the tube diameter. The fluid velocity and density change linearly with distance along the tube. Find the pressure drop along the tube. What is the change in the fluid's internal energy over and above that due to heat transfer $(\Delta\tilde{u} - q)$?

87. Liquid water at 40°F flows upward through a vertical, thermally insulated, 2-in. schedule 40 steel pipe at the rate of 100 gal/min. The water temperature change is related to its internal energy change by the relation in Problem 83. Find the temperature change of the water if the pressure change $(p_1 - p_2)$ is 12.0 psi and

$$gh_L = 0.0065\left(\frac{L}{D}\right)V^2.$$

88. The wall shear stress between planes 1 and 2 are assumed negligible in Problem 16. Why can't we apply Bernoulli's equation between planes 1 and 2?

89. Air enters a vacuum pump at 7.0 kPa and 15°C with a velocity of 30 m/s and through an inlet area of 0.004 m^2. The air leaves the vacuum pump at 101 kPa and a negligible velocity. The input power to the pump is 16 kW, and the pump is thermally insulated from the surroundings. What is the air discharge temperature if the elevation changes are negligible?

90. Calculate the torque required to drive the pump in Fig. P5.49 at 300 Hz and to deliver water at 3.0 m^3/min.

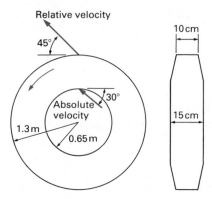

Figure P5.49

91. Figure P5.50 shows an open-circuit wind tunnel. The two manometers indicate the pressure in the test section and at the fan discharge. Assuming air as an ideal fluid, steady-state conditions, and negligible elevation changes, find the power input to the fan in horsepower. The mechanical energy loss is given by $KV^2/2$, where $K = 4.0$.

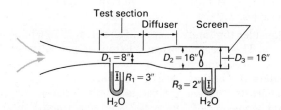

Figure P5.50

92. A diver is using a flexible hose to suck up a mixture of water and sand with a specific gravity of 1.5. The inside diameter of the hose is 10 cm. Find the amount of sand that can be picked up per minute by a 1.5-kW pump. The end of the flexible hose is 6.0 m below the water surface, and the open end of the hose connected to the pump discharge is at the water surface. Neglect all friction losses.

93. A centrifugal pump has an input power of 75 kW and a flow rate of 1.0 m^3/min. Neglecting elevation changes, what is the maximum pressure rise that can be generated by the pump at this flow rate?

94. Water flows through a hydraulic turbine at the rate of 10 m^3/s. The inlet pressure and velocity are 250 kPa gage and 10 m/s, respectively, and the discharge pressure and velocity are -90 kPa gage and 1.0 m/s, respectively. Find the maximum power delivered to the turbine by the water.

95. A simplified diagram of a hydraulic turbine is shown in Fig. P5.51. Water from the lake flows down the penstock and through the turbine. Calculate the power output of the turbine. Neglect mechanical energy losses in the penstock and turbine.

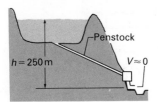

Figure P5.51

96. Air enters a converging nozzle at 40 psia, 150°F, and a velocity of 30 ft/sec. The inlet area is 0.25 ft^2 and the exit pressure is 35 psia. Find the exit area and the exit temperature if the air internal energy change is given by

$$\tilde{u}_2 - \tilde{u}_1 = (0.17 \text{ BTU/lbm} \cdot °\text{F})(T_2 - T_1)$$

and the nozzle is thermally insulated from the ambient air.

97. Air enters a small turbine at 50 psia, 150°F, and 300 ft/sec and leaves at 13 psia, 80°F, and 50 ft/sec. The shaft work delivered by the turbine is 12,000 ft·lb per lbm of air passing through the turbine. Find the mechanical energy loss, gh_L, of the turbine.

98. A diffuser is used to reduce the velocity of a flow stream. A particular diffuser has inlet conditions of 250 m/s, 202 kPa, and 300°C and exit conditions of 50 m/s and 101 kPa. The diffuser is thermally insulated from the ambient air, and the internal energy change of the air is given by

$$\tilde{u}_2 - \tilde{u}_1 = (750 \text{ J/kg} \cdot °\text{C})(T_2 - T_1).$$

Find the exit temperature. Elevation changes are negligible.

99. An oil having a specific gravity of 0.85 is pumped at a rate of 0.25 m^3/s from a pressure of 80 kPa to a pressure of 160 kPa. The inlet area is 0.025 m^2 and the exit area is 0.10 m^2. Find the minimum power input to the pump for negligible elevation changes.

100. A pump delivers 10,000 lbm/hr of 60°F water from inlet conditions of 15 psia and zero elevation to exit conditions of 15 psia and 100 ft elevation with negligible kinetic energy changes. Find the minimum power input to the pump.

101. When the spigot in a home is opened, water flows from a 24-in. water main outside the home to the 1/2-in. outlet at the spigot. The outlet is 40 ft above the water main. What is the minimum water-main pressure for water to flow at 40 gal/min?

102. Find the height h of the nozzle in Fig. P5.52 such that the water will travel the longest horizontal distance ℓ. The nozzle has an energy loss given by

$$gh_L = 0.5 V^2.$$

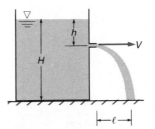

Figure P5.52

103. A fan draws in 20°C air from the atmosphere through a 1.0-m-diameter duct. The rounded entrance has a mechanical energy loss given by

$$gh_L = 0.10 V^2,$$

where V is the average air velocity in the duct. The static pressure at the fan inlet is 3.0 cm H$_2$O vacuum, and the fan discharges to the atmosphere. Find the air volume flow rate at the fan inlet and the fan input power.

104. A 10-ft-diameter water storage tank has 2500 lbm of water below 40 lbm of air at 60 psia and 70°F. A 1-in.-diameter plug is pulled out of the bottom, and the flowing water is used to produce work. Find the maximum work that can be developed by the water from the tank.

105. A pump is used to supply 100 ft^3/sec of 60°F water from a reservoir at atmospheric pressure through a 4-in. schedule 40 steel pipe to an open tank on top of a 400-ft-high building. Kinetic energy changes are negligible. Find the minimum pump work required.

106. Air enters a turbine at 700 kPa, 300°C, and a flow rate of 500 kg/min. The turbine power output is 200 kW and there is no heat transfer between the turbine and the ambient air. Kinetic energy and elevation changes are negligible. The air's internal energy change is given by

$$\tilde{u}_2 - \tilde{u}_1 = (750 \text{ J/kg} \cdot °\text{C})(T_2 - T_1).$$

Find the exit temperature.

107. Water flows at the rate of $1.0 \text{ m}^3/\text{s}$ through the contraction in Fig. P5.53. The contraction has an energy loss given by

$$gh_L = 0.1 V^2.$$

Find the manometer reading h.

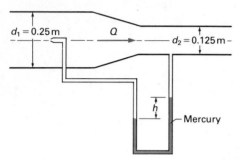

Figure P5.53

108. An exhaust fan in a wall of a building has a pressure rise of 3.0 in. of water and a volume flow rate of $5000 \text{ ft}^3/\text{min}$. Find the minimum power input to the fan.

109. Water is pumped through a horizontal pipeline having a pressure difference of 202 kPa and a flow rate of $100 \text{ m}^3/\text{min}$. What is the minimum power input to a pump installed in this pipeline such that this pressure difference is brought to zero?

110. Determine the volume flow rate and minimum power input to the water pump in Fig. P5.54.

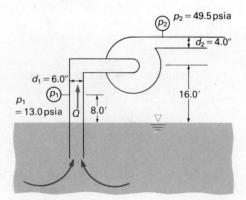

Figure P5.54

111. The pump in Fig. P5.55 has a constant input power of 10 hp and no energy losses. Find the time to fill up the storage tank. Assume negligible energy losses.

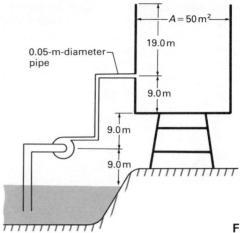

Figure P5.55

112. Calculate the kinetic energy correction factor for each of the following velocity profiles for a circular pipe:

a) $u = u_{\max}\left(1 - \dfrac{r}{R}\right)$

b) $u = u_{\max}\left(1 - \dfrac{r^2}{R^2}\right)$

c) $u = u_{\max}\left[1 - \left(\dfrac{r}{R}\right)^{1/9}\right]$

R is the pipe radius and r is the radial coordinate.

113. Find the total flow rate through the 2-in. schedule 40 steel pipe in Fig. P5.56. Neglect friction between the fluid and the pipe walls.

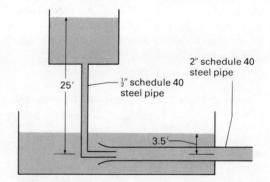

Figure P5.56

114. A reciprocating pump has a constant volume flow rate of $Q = 100 \text{ gal/min}$ and a static pressure rise Δp. The pump has a constant efficiency (output power $Q\Delta p$ divided by the input power) of 0.75. Find the energy input required to fill the 20,000-gal storage tank through pipeline A in Fig. P5.57.

The fluid has a specific gravity of 0.78. Assume inviscid flow.

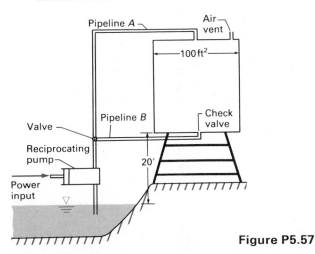

Figure P5.57

115. Repeat Problem 114 using pipeline *B*. The check valve permits fluid to enter the storage tank from pipeline *B* but will stop any fluid from flowing out of the tank. Compare your result with that of Problem 114 and draw any worthwhile conclusions.

Higher-Order Application

116. A simplified schematic of the carburetor of a gasoline engine is shown in Fig. P5.58. The throat area is 0.5 in². The running engine draws air downward through the carburetor Venturi and maintains a throat pressure of 14.0 psia. The low throat pressure draws fuel from the float chamber and into the airstream. The energy losses in the 0.06-in.-diameter fuel metering line and valve are given by

$$gh_L = \frac{KV^2}{2g},$$

where $K = 3.0$. Assume:
a) The air is an ideal fluid with constant density

$$\rho_A = 0.075 \text{ lbm/ft}^3.$$

b) The atmospheric pressure is 15 psia.

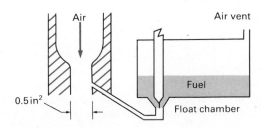

Figure P5.58

c) Elevation changes are negligible.
d) The fuel specific gravity is 0.75.
Find the air-fuel ratio (A/F) (the ratio of the air mass flow rate to the fuel mass flow rate).

117. A young man blows up a spherical balloon to a pressure of 200 mm Hg gage and a diameter of 0.25 m. The surface tension in the balloon increases linearly with diameter and is 10.7 N/m at a diameter of 0.25 m. The mass of the balloon is 0.01 kg. The young man lets go of the neck of the balloon and the balloon moves in a straight path. Find the differential equation relating the distance traveled as a function of time if the neck area is 0.001 m² and the air follows the relation $p\mathcal{V}^{1.4} = $ Constant. The atmospheric pressure is 750 mm Hg.

118. The small toy rocket in Fig. P5.59 is released from its launching pad at time $t = 0$. Find its velocity as a function of time if it moves vertically upward. The rocket mass is 0.25 lbm and the rocket contains 1.0 lbm of water under a pressure of 10.0 psig when launched. Neglect air resistance. Assume frictionless flow inside the rocket.

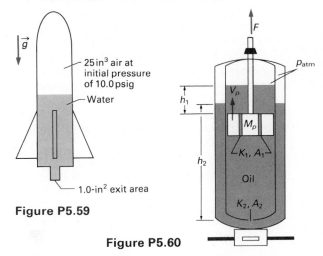

Figure P5.59

Figure P5.60

119. Consider the simplified shock absorber shown in Fig. P5.60. Find an expression for the force F as a function of the piston velocity V_p, piston mass M_p, orifice areas A_1 and A_2, and the loss coefficients K_1 and K_2 for the orifices, where

$$(gh_L)_{\text{orifice 1}} = \frac{K_1 V^2_{\text{fluid/piston}}}{2}$$

and

$$(gh_L)_{\text{orifice 2}} = \frac{K_2 V^2_{\text{fluid/casing}}}{2}.$$

Note that $V_{\text{fluid/piston}}$ is the fluid velocity in the orifice relative to the piston.

120. The empty 1-gal paint can of weight W in Fig. P5.61 is punctured with a 1/2-in.-diameter hole in the bottom. The paint can is then placed on top of a large body of water and allowed to sink. Assuming an ideal fluid, develop an equation for the depth h_0 as a function of time.

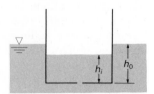

Figure P5.61

121. Repeat Problem 120 if the paint can is empty but the lid has been securely put on prior to being placed in the water. The air temperature inside the can is assumed to remain constant.

122. Figure P5.62 shows a water supply system that uses a jet pump. A small amount of water is pumped down the primary line at a high pressure. Inside the jet pump, the water is passed through the nozzle, producing a high velocity jet at ①. This jet mixes with the water from the well, dragging the well water along with it. The mixing takes place between ① and ②; at ② the water moves with a uniform velocity. Assume the pressure is uniform across plane ① and plane ②, the primary jet velocity is uniform at ①, the secondary velocity is uniform at ①, the delivery velocity is uniform at ②, the wall shear stress is negligible between ① and ②, and the primary and secondary line flows are frictionless. Calculate:
a) the water velocity in the delivery pipe;
b) the required pressure at ②;

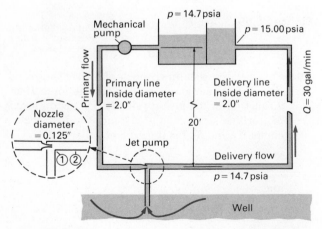

Figure P5.62

c) the primary and secondary velocities at ① if the secondary velocity is 1/100 the primary velocity at ①;
d) the pressure at plane ①; and
e) the mechanical pump input power.

123. Repeat Problem 122 where the primary and secondary line energy losses are given by

$$gh_L = 15\left(\frac{V^2}{2}\right), \qquad \text{primary}$$

$$gh_L = 7\left(\frac{V^2}{2}\right), \qquad \text{secondary}$$

where V is in feet per second. Both the primary and discharge lines are 120 ft long.

124. Find the temperature rise of the water in the discharge line of Problem 123 if the internal energy change of water is given by

$$\tilde{u}_2 - \tilde{u}_1 = \left(\frac{1.0 \text{ BTU}}{\text{lbm} \cdot {}^\circ\text{F}}\right)(T_2 - T_1)$$

and the discharge line is perfectly insulated.

125. Superman pushes off the earth's surface and rises vertically upward to a height of 2 km, where his velocity is zero. Neglect air resistance and find the force he must exert against the earth to rise to this height.

Comparison

126. A steel cylinder has a volume of 1.0 m^3 and an inside diameter of 0.25 m. It is fitted with a 0.0005-m-thick, 0.30-m-diameter end cap that is glued to the cylinder. The glue's holding strength is to be tested in a pressure test with a fluid at 700 kPa and 100°C. The test engineers decide to use either water or air. They are concerned with the safety of test personnel and have asked you to determine the maximum velocity of the end cap should the glue fail with either test fluid. The air may be assumed to expand isothermally, and the water follows the relation

$$dp = -(2.07 \times 10^9 \text{ N/m}^2)\, dv/v,$$

where v is the water specific volume and p the water pressure.

127. Combustion products at 300°C of a furnace flow up the chimney in Fig. P5.63. The gases may be assumed to have constant density, a molecular weight of 29.0, and a flow rate of 1000 kg/hr. Compute the pressure p_1 at the bottom of the

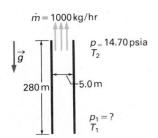

$\dot{m} = 1000\,\text{kg/hr}$

$\vec{g}$

$p = 14.70\,\text{psia}$
T_2

280 m — 5.0 m

$p_1 = ?$
T_1

Figure P5.63

is given by

$$gh_L = 0.28\,V^2,$$

where V is the velocity.

128. Combustion products at 300°C from a furnace flow up the chimney in Fig. P5.63. The gases have a molecular weight of 29.9 and a flow rate of 1000 kg/hr. The flow is frictionless. Compute the pressure p_1 at the bottom of the chimney if the density is constant, and compare with the pressure p_1 if the temperature drops linearly from $T_1 = 325°C$ to $T_2 = 275°C$.

chimney if the flow is frictionless, and compare with the pressure p_1 if the mechanical energy loss

References

1. Thomas, G., *Calculus and Analytic Geometry,* Addison-Wesley, Reading, MA.
2. Gerhart, P. M., "Averaging Methods for Determining the Performance of Large Fans from Field Tests," *ASME Journal of Engineering for Power,* October 1981.
3. Shepherd, D. G., *Principles of Turbomachinery,* Macmillan, New York, 1956.
4. Dixon, S. L., *Fluid Mechanics: Thermodynamics of Turbomachinery* (3rd ed.), Pergamon Press, Oxford, 1981.
5. Stepanoff, A. J., *Centrifugal and Axial Flow Pumps,* Wiley, New York, 1948.
6. Daugherty, R. L., and J. B. Franzini, *Fluid Mechanics* (7th ed.), McGraw-Hill, New York, 1977.

Organizing the Experimental Approach to Flow Problems: Dimensional Analysis

6

The differential approach of Chapter 4 and the finite control volume approach of Chapter 5 are analytical tools for the solution of engineering flow problems. If you have valid information about every term in the differential or control volume equations and if you are able to perform the mathematical operations* required to solve the equations, you can use the information gained from your flow analysis with confidence. The unfortunate fact is that some rather serious gaps exist in our ability to account for every term in the equations and to solve them. Consider the problem of evaluating the mechanical energy (head) loss for turbulent flow in a pipe. Nearly one hundred years of effort has not produced a complete, workable theory. You would think that such a basic problem with obvious engineering relevance would have been solved long ago. Of course the status of flow prediction in more complicated geometric configurations is no better.

How then do engineers deal with complicated flow problems? How can airplanes, ships, harbors, and piping systems be successfully designed and operated? The answer, in a single word, is *experimentation*. Engineering fluid mechanics is a combination of theory and experiment, much more so than, say, structural mechanics or electric and electronic circuit analysis. For turbulent pipe flow, the "theoretical" incompressible energy equation must be supplemented by information gained from measurements of energy loss. Theoretical analysis suggests tentative designs for an airplane or a harbor. The candidate designs are evaluated and refined by extensive testing. An introduction to fluid mechanics would be incomplete without considering the contributions of experimentation.

This chapter deals with the place of experimentation in fluid mechanics. It does not deal with the how-to of experimental techniques; in fact, we will not even present many experimental results. We will discuss a method by which experimental information can be packaged and transmitted efficiently from an experimenter to other engineers.

---

* We include numerical approximation and computer programming in these mathematical operations.

Although almost all engineering fluid mechanics is, to some degree, based on experimental information, only a small percentage of fluids engineers are actually involved in performing experiments. The work of these engineers is frequently used by others. Our objective is to help you understand the forms in which experimental data are usually packaged.

The specific topic of this chapter is *dimensional analysis*. Dimensional analysis is a packaging or compacting technique used to reduce the complexity of experimental programs and at the same time increase the generality of experimental information. Before we consider the subject in detail, note the following points about dimensional analysis:

- Dimensional analysis is sometimes very confusing to beginners. The basic principle involved seems obvious, almost trivial. The execution of the technique, however, often seems almost magical.
- The confusion arises because we usually expect theories to be complete, that is, to begin at the beginning and carry through to the end. Dimensional analysis stops far short of providing a final answer to a particular engineering problem; it simply leads to a more efficient organization of the variables. The ultimate answer comes only in the laboratory or from analytical methods.
- Dimensional analysis tells us that we *can* organize variables efficiently and suggests possible ways that the organization can be carried out; however, it does not allow us to find unique results.

Two analogies will help illustrate the utility of dimensional analysis. Dimensional analysis is similar to a trash compactor. Trash thrown into the compactor is crushed into a smaller space so it is easier to handle. The compactor does not solve the trash problem nor does it ultimately dispose of any of the trash put into it; it only reduces it to a more convenient package. Dimensional analysis is also similar to packing a suitcase for a trip. The traveler must select the items to take. Any item that is not selected will not be in the suitcase. The traveler can organize the items in the suitcase in one of several equally convenient arrangements. Different travelers prefer different arrangements, but that does not make any arrangements superior. After the suitcase is packed, it is up to the traveler to carry it to the ultimate destination. The suitcase won't go on its own, and just packing it doesn't accomplish the objective of the trip.

6.1 The Need for Dimensional Analysis

Each experiment carried out in a laboratory is unique. The apparatus being tested has a unique geometry and a specific size. Only one fluid at a specific flow rate can be used in a particular experimental run. The unknown parameters whose evaluation is the object of the experiment (e.g., energy loss, pressure distribution, or resistance force) have unique numerical values. Information about the effect of varying any "controllable" parameters such as flow rate or fluid viscosity can be obtained only by repeated experiments. Trends can be established only by systematic variation of the controllable parameters. Since experimentation

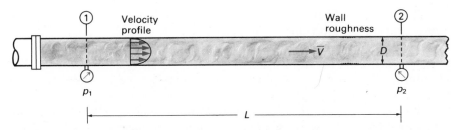

Figure 6.1 Incompressible turbulent flow in a circular pipe.

is costly and time consuming, it is a good idea to document experimental results so that a new experiment is not required the next time a similar problem arises. We would also like to extend and generalize any information that is obtained from a particular experiment.

We will consider two different flow problems that must be solved, at least in part, by experiment. The first is evaluation of mechanical energy (head) loss for incompressible turbulent flow in a pipe, illustrated in Fig. 6.1. The second problem is evaluation of the resistance force (drag) on a two-person submarine being designed for oceanographic research, illustrated in Fig. 6.2.

The first problem is quite general since pipes appear in countless engineering applications. A generalization of experimental data that allows evaluation of energy loss in an arbitrary pipe flow would be extremely useful. Suppose we are to carry out a systematic evaluation of the effects of various parameters on pipe flow energy loss. First we establish that the energy loss can indeed be measured. Using Eq. (5.58), the energy loss is

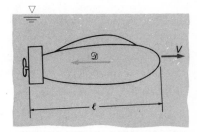

Figure 6.2 Drag force on small two-person submarine.

$$gh_L = \frac{\bar{p}_1 - \bar{p}_2}{\rho} + \frac{\alpha_1 \bar{V}_1^2 - \alpha_2 \bar{V}_2^2}{2} + g(\bar{z}_1 - \bar{z}_2) - w_s. \tag{6.1}$$

If we construct the experimental apparatus as shown in Fig. 6.1, then there is no work between stations 1 and 2. A convenient setup might be a horizontal pipe with $\bar{z}_1 = \bar{z}_2$. If the pipe is long enough, the flow will be fully developed ($\alpha_1 = \alpha_2$). We are interested in steady flow, so conservation of mass requires that $\bar{V}_1$ and $\bar{V}_2$ be equal because density and pipe cross-sectional area are fixed. Under these conditions, Eq. (6.1) reduces to*

$$gh_L = \frac{p_1 - p_2}{\rho}. \tag{6.2}$$

The energy loss can be evaluated by measuring the pressure difference between stations 1 and 2, together with knowledge of the fluid density. Satisfied that energy loss can be measured, we will consider what parameters affect it. A little thought leads to the following list:

1. Length of pipe between 1 and 2 (L),
2. Pipe diameter (D),
3. Fluid velocity (V),
4. Fluid density (ρ),
5. Fluid viscosity (μ).

---

* Overbars ($^-$) representing averages are dropped for simplicity.

This list should have been reasonably obvious. A final, somewhat subtle, parameter that might be missed is

6. Pipe wall roughness height (ε).

Of course the shape of the pipe is important, but we are limiting this discussion to straight pipes with circular cross section.

We might indicate that we expect *all* of these parameters to have an effect on the loss by writing the following *functional relation:*

$$gh_L = f(L, D, V, \rho, \mu, \varepsilon). \tag{6.3}$$

It is extremely important that you understand what this relation means. If we change *any one* of the arguments of f, we expect the energy loss to change, even if all other arguments remain fixed. The functional relation indicates only that there is interdependence between gh_L and the arguments of f. The form of the function f is not known; in fact, discovering the function is the objective of the experimental program. It need not be an analytical function; in fact, experiments simply produce tables of data (which can be presented as graphs and charts). Tables and charts are perfectly acceptable forms of f. So are curve fits of numerical data.

Now we ask, What do we need to do to discover the relation implied by Eq. (6.3), and how can we present the results? To discover the relation, we must systematically vary each parameter on the right side. First we fix D, V, ρ, μ, and ε and systematically vary L. This would require, say, 10 runs (for 10 different L's) and produce results like those indicated in Fig. 6.3. These runs establish the dependence of gh_L on L only for the specific values of D, V, ρ, μ, and ε that we used in the 10 runs. Next we change to a new value of D, keep the same values of V, ρ, μ, and ε, and repeat the runs for all 10 values of L. We repeat this for, say, 10 values of D. Next we run through all combinations of L and D for, say, 10 different values of V. Figure 6.4 shows a few of the results we might expect. We proceed through the list of variables, each time repeating all combinations of the variables preceding the current one. A

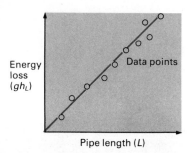

Figure 6.3 Typical results: Pipe flow energy loss versus pipe length for fixed values of D, V, ρ, μ, and ε.

Figure 6.4 A few of the 10^4 plots necessary to present the results of a complete investigation of turbulent pipe flow.

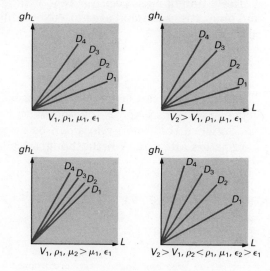

complete investigation would require about 10^6 different experimental runs. The cost of such a program would be astronomical, not to mention that we would have to find 100 different fluids since we need 10 different ρ's and 10 different μ's!

To present the results of all our work, we would need about 10^5 plots like those of Fig. 6.3. We could save some paper by preparing parametric plots like those of Fig. 6.4; this would reduce the required number of plots to "only" 10^4!* The compilation of results would be so unwieldy that it would be useless.†

Dimensional analysis to the rescue! According to the fundamental theorem of dimensional analysis (to be discussed in Section 6.3), the functional relation of Eq. (6.3) can be replaced by a *completely equivalent* relation involving *groups* of variables, as follows:

$$\frac{gh_L}{\frac{1}{2}V^2} = F\left(\frac{L}{D}, \frac{\rho VD}{\mu}, \frac{\varepsilon}{D}\right). \tag{6.4}$$

This is still a general functional relation, with the form of the *new* function F unspecified; however, it contains *all* of the information implicit in Eq. (6.3). Instead of six controllable parameters, we now have only three. Even if no other simplifications could be found (and they can), we would need "only" 10^3 experiments. Only one fluid would be needed because the second parameter on the right ($\rho VD/\mu$) can be varied by changing V instead of ρ and μ. Only 10 pages would be required to present the results. *Note, however, that someone still must run the 10^3 experiments and fill the 10 pages.*** The problem has been greatly simplified but still not solved. We will describe how to do this reduction in a later section.

Now we consider the second fluid mechanics problem, finding the drag force ($\mathscr{D}$) on a small submarine (Fig. 6.2). Since the shape of the submarine is rather complex and we would naturally expect the flow to be at least partially turbulent, accurate calculation of the drag force is very difficult and we must usually have recourse to experiment. Similar to the pipe problem, we first list the parameters that we expect to affect the drag force:

1. The size of the submarine. We can choose any single length dimension to characterize the size; we use the length (ℓ).
2. The speed of the submarine (V).
3. The fluid density (ρ). (Even though the submarine will operate only in water, saltwater and freshwater have different densities.)
4. The fluid viscosity (μ).

---

* According to Kline [1], one independent parameter requires one line; two, a page; three, a book; and four, a library! We have six.

† According to "Murphy's law," any particular case you need doesn't exactly correspond to one of the 10^6 specifically documented, so you would have to do a six-parameter interpolation between cases.

** In the case of pipe flow losses, the experiments have been run and the information is available for general use. This information will be discussed in Chapter 7.

The force will, of course, depend on the specific shape of the submarine, but we suppose that we are investigating a specific design, so the shape is fixed. In functional notation we write

$$\mathcal{D} = f(\ell, V, \rho, \mu). \tag{6.5}$$

To determine f, we must systematically vary ℓ, V, ρ, and μ.

If we compare this problem to the pipe energy loss investigation, we seem to be in better shape because we have fewer variables. A complete investigation of submarine drag would require "only" about 10^4 experiments. This case has problems of its own, however. Supposedly, this submarine is only in the design phase. We may wish to compare several different shapes (designs). Building and testing several different designs would be very costly. There is an even greater problem; if we build the submarines full size, there is no laboratory where they can be tested.* A full-scale test program is out of the question. The only possible approach is to test a small replica (model) of the actual submarine. It seems reasonable that if the model is built *exactly* like the full-size (prototype) submarine, then measurements of forces on the model could be extrapolated to the prototype—but how?

Once again, dimensional analysis comes to the rescue. If we apply dimensional analysis to the functional relation of Eq. (6.5), it reduces to the equivalent form

$$\frac{\mathcal{D}}{\frac{1}{2}\rho V^2 \ell^2} = F\left(\frac{\rho V \ell}{\mu}\right). \tag{6.6}$$

In Eq. (6.6), the size of the submarine has been combined with the other parameters in such a way that the relation is explicitly independent of size. If the model and prototype are exactly similar geometrically, then the function F for the prototype will be identical to the function F for the model. Determining the latter experimentally gives us the former. Notice also that the parameters have once again been "compacted," so only a few experimental runs and a single curve are required.

These two examples should have served to whet your appetite for dimensional analysis. While these examples may have overstated the case, they are representative of what can be done with dimensional analysis, from the viewpoints of both experimenters and users of experimental results.

6.2 The Foundations of Dimensional Analysis

Engineers are concerned with physical phenomena. The parameters of interest to engineers (such as force, velocity, energy) can be expressed in terms of a numerical magnitude and associated units of measurement. All physical entities have certain dimensions; that is, a given physical entity always has the same dimensions, even if the units employed to express the dimensions change.[†]

* This is especially true for large aircraft, supertankers, skyscrapers, and the like.
† Reread Section 1.4 if you need to brush up on the concepts of units and dimensions.

To deal with physical entities, engineers use the language and methods of mathematics. We write equations that allow us to calculate expected values of pressure, velocity, flow rate, and so forth. Some typical equations in fluid mechanics are Bernoulli's equation, the linear momentum equation, the hydrostatic pressure formula, and the Navier-Stokes equations. These equations are not abstract mathematical expressions; they represent the physical laws of nature. There are certain rules, other than mathematical, that these equations must obey. One of these rules is the *principle of dimensional homogeneity* (PDH).

An equation is *dimensionally homogeneous* if it is valid regardless of the units of measurement assigned to the terms. An example of a dimensionally homogeneous equation is Bernoulli's equation:

$$\frac{p_1}{\rho} + \frac{V_1^2}{2} + gz_1 = \frac{p_2}{\rho} + \frac{V_2^2}{2} + gz_2. \tag{4.21}$$

This equation is valid if we use SI units, BG units, EE units, or any other units. Note that the numerical values of the terms will change if we change units, but the validity of the equation will not be affected. Two things are required for an equation to be dimensionally homogeneous:

- The dimensions of the right side must be the same as the dimensions of the left side.
- The dimensions of all additive terms must be the same.*

If we write only the dimensions of Bernoulli's equation, using mass $[M]$, length $[L]$, and time $[T]$ as fundamental dimensions, we get

$$\frac{[M/LT^2]}{[M/L^3]} + [L/T]^2 + [L/T^2]L = \frac{[M/LT^2]}{[M/L^3]} + [L/T]^2 + [L/T^2][L].$$

Simplifying, we get

$$[L^2/T^2] = [L^2/T^2].$$

As we claimed, Bernoulli's equation is dimensionally homogeneous.

The principle of dimensional homogeneity can be stated as follows:

> Any equation that completely describes a physical phenomenon must be dimensionally homogeneous.

The PDH provides the entire foundation of dimensional analysis. In this sense it is analogous to the fundamental physical laws of Section 3.2. All further developments of dimensional analysis are essentially mathematical.

The PDH is a very helpful check of mathematical operations. If, after a lengthy derivation, you find that the dimensions on the left side of the equation are different from the dimensions on the right side, you made an error somewhere. A comparison of the dimensions often helps you find the source of the error.

The PDH applies to derivatives and integrals as well as algebraic terms. The d or ∂ of a derivative and the $\int$ and d of an integral do not have dimensions, but the independent and dependent variables do. The

---

* How many times have you heard the saying "you can't add apples and oranges"?

following dimensional equations should illustrate how this works:

$$\left[\frac{dV}{dt}\right] = \frac{[V]}{[t]} = \frac{[L/T]}{[T]} = [L/T^2],$$

$$\left[\frac{d^2p}{dx^2}\right] = \frac{[p]}{[x^2]} = \frac{[M/LT^2]}{[L^2]} = \left[\frac{M}{L^3T^2}\right], \quad \text{and}$$

$$\left[\int p\, dA\right] = [p][A] = \left[\frac{M}{LT^2}\right]L^2 = \left[\frac{ML}{T^2}\right].^*$$

You should be aware that there are some equations that do not satisfy the PDH, such as the following:

- Manning's formula for open channel flow (English units):

$$V = \frac{1.49}{n}\, R_h^{2/3} S^{1/2},$$

where V is velocity, R_h is radius, S is slope (dimensionless), and n is a number taken from tables;

- Formula for flow over a $90°\,V$-notch weir:

$$Q = 2.5(h)^{2.5},$$

where Q is volume flow rate and h is height of fluid.

These equations are true only for a specific set of units. In Manning's formula, R_h must be in feet and V must be in feet per second. In the weir formula, h must be in feet and Q must be in gallons per minute. Such equations might be called *dimensionally inhomogeneous equations* or *unit-specific equations*. Such equations usually arise from curve fits of experimental data. One possible way of looking at these equations is to assign units to the "constants" (1.49 and 2.5) and/or to the numerical value n. This is a very tricky business since 1.49 would have units only if it appeared in Manning's equation.

If an equation satisfies the PDH, only three types of quantities should appear in it:

- *Dimensional variables.* These are the parameters that vary in real life. They usually include those parameters that we are trying to evaluate and those that represent known material properties or flow conditions. In Bernoulli's equation, p, ρ, V, and z are dimensional variables. In an experiment, dimensional variables are either measured or controlled.

- *Dimensional constants.* These are constants that are associated with fundamental phenomena of nature, such as the gravitational constant G in Newton's law of universal gravitation

$$F = G\frac{m_1 m_2}{r^2},$$

R_0, the universal gas constant, and c_0, the speed of light in a vacuum.

---

* If we use force $[F]$ instead of mass $[M]$ in our fundamental set of dimensions, the dimensions of this term are just $[F]$.

■ *Pure numbers.* Pure numbers have no dimensions, never did, and never will. They usually result from mathematical operations. The 1/2 in Bernoulli's equation is a pure number, having arisen in the integration of $V\,dV$. The exponent 2 on the velocity is also a pure number. Other pure numbers include e and π.

The distinction between dimensional variables and dimensional constants is a little fuzzy at times. Consider the acceleration of gravity, g, in Bernoulli's equation. We know that g is indeed variable, especially the farther we go from the earth. On the other hand, most engineers who use Bernoulli's equation are interested only in happenings on the earth's surface. Certainly few experimenters plan to change g during the course of an experiment. In most cases, g would be considered a dimensional constant. We will use the name *dimensional parameters* to refer to both dimensional variables and dimensional constants.

A fourth type of quantity that may sometimes appear in a dimensionally homogeneous equation is a *conversion factor*. Conversion factors are used to convert units and, if they appear in an equation, must be treated as pure numbers (equal to 1). In our opinion, conversion factors (including g_c) should not be written in equations but should be introduced as necessary in numerical calculations.

EXAMPLE 6.1 Illustrates how to determine if an equation is dimensionally homogeneous

The frictional pressure loss, Δp_L, in a certain pipe can be calculated by the following equation

$$\Delta p_L = 0.03\left(\frac{L}{D}\right)\left(\frac{\rho V^2}{2}\right),$$

where L and D are in meters, ρ in kilograms per cubic meters, and V in meters per second. Is this equation dimensionally homogeneous or dimensionally inhomogeneous?

Solution

Given
Equation:

$$\Delta p_L = 0.03\left(\frac{L}{D}\right)\left(\frac{\rho V^2}{2}\right)$$

Units:
L in meters
D in meters
ρ in kilograms per cubic meters
V in meters per second

Find
Whether the equation is dimensionally homogeneous

Solution
L, D, ρ, and V are expressed in SI units. Therefore we anticipate that ΔP_L will be expressed in SI units of pressure, N/m². We try to determine if the units

of

$$\frac{L}{D}\left(\frac{\rho V^2}{2}\right)$$

are N/m². The units of this term are

$$\left[\frac{L}{D}\left(\frac{\rho V^2}{2}\right)\right]^* = \left[\frac{m}{m}\left(\frac{kg}{m^3}\right)\left(\frac{m}{s}\right)^2\right] = \left[\frac{kg \cdot m}{s^2}\left(\frac{1}{m^2}\right)\right] = \left[\frac{N}{m^2}\right],$$

which are the units of pressure in the SI system. Therefore:

| The equation is dimensionally homogeneous. |

Answer ◄

Discussion
This example could also be solved using dimensions as follows:

$$\left[\frac{L}{D}\left(\frac{\rho V^2}{2}\right)\right] = \left[\frac{L}{L}\left(\frac{M}{L^3}\right)\left(\frac{L}{T}\right)^2\right] = \left[\frac{ML}{T^2}\left(\frac{1}{L^2}\right)\right] = \left[\frac{F}{L^2}\right],$$

which are the dimensions of pressure.

* Note that these brackets mean we are writing the units of the enclosed terms.

6.3 Dimensionless Parameters and the Pi Theorem

Now we will consider how to use dimensional analysis to organize (experimental) information and reduce the complexity of an investigation. Consider once again the examples of Section 6.1. The various groups in Eqs. (6.4) and (6.6) are combinations of dimensional parameters. The dimensions of the groups are not obvious. We will evaluate the dimensions of these groups with respect to an $[M, L, T, \theta]$ set of fundamental dimensions.* The dimensions of the groups in Eq. (6.4) are

$$\frac{gh_L}{\frac{1}{2}V^2} = \frac{[L/T^2][L]}{[L/T]^2} = \frac{[L/T]^2}{[L/T]^2} = [1],$$

$$[L/D] = \frac{[L]}{[L]} = [1],$$

$$\left[\frac{\rho VD}{\mu}\right] = \frac{[M/L^3][L/T][L]}{[M/LT]} = \frac{[M/LT]}{[M/LT]} = [1], \quad \text{and}$$

$$[\varepsilon/D] = \frac{[L]}{[L]} = [1],$$

where [1] means dimensionless.

The dimensions of the groups in Eq. (6.6) are

$$\frac{\mathscr{D}}{\frac{1}{2}\rho V^2\ell^2} = \frac{[F]}{[M/L^3][L/T]^2[L^2]} = \frac{[ML/T^2]}{[ML/T^2]} = [1]$$

* See Table 1.1 in Chapter 1 for dimensions of the various quantities.

and

$$\frac{\rho V \ell}{\mu} = \frac{[M/L^3][L/T][L]}{[M/LT]} = [1].$$

All of the parameters in the reduced equations are dimensionless.* When two or more dimensional parameters are combined into a dimensionless group, the group is called a *dimensionless parameter*. Perhaps the combination of the dimensional parameters in Eqs. (6.3) and (6.5) into the dimensionless parameters in Eqs. (6.4) and (6.6) is the key to the resulting simplification.

To demonstrate (not derive) how this might occur, we consider a flow for which we have a theory. Figure 6.5 shows a horizontal converging nozzle. An incompressible fluid flows steadily through the nozzle. We assume that shear stresses in the flow are negligible. Suppose we wish to investigate the relation between the pressure drop $\Delta p_d (\Delta p_d \equiv p_1 - p_2)$, the fluid velocity V_1, and the nozzle areas A_1 and A_2. Of course Bernoulli's equation applies to this flow, so we can write immediately

$$\frac{p_1}{\rho} + \frac{V_1^2}{2} = \frac{p_2}{\rho} + \frac{V_2^2}{2}.$$

Solving for the pressure drop, we obtain

$$\Delta p_d = \frac{1}{2} \rho (V_2^2 - V_1^2).$$

Since the flow is steady, the fluid velocities are related by the continuity equation

$$V_1 A_1 = V_2 A_2,$$

so the pressure drop can be computed by

$$\Delta p_d = \frac{1}{2} \rho V_1^2 \left(\frac{A_1^2}{A_2^2} - 1 \right). \tag{6.7}$$

Equation (6.7) has the functional form

$$\Delta p_d = f(V_1, \rho, A_2, A_1). \tag{6.8}$$

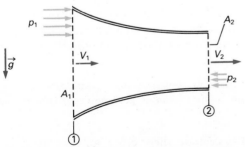

Figure 6.5 Flow in horizontal converging nozzle.

* You might try calculating the dimensions of these parameters using a force-length-time-temperature $[F, L, T, \theta]$ system to verify that they are still dimensionless.

Equation (6.7) and its generalized functional form, Eq. (6.8), are dimensionally homogeneous equations involving five dimensional parameters. Is there a corresponding dimensionless form, and does this dimensionless form involve fewer parameters? The answer to both questions is yes. We divide Eq. (6.7) by $\frac{1}{2}\rho V_1^2$ and simplify the area-ratio term to get

$$\frac{\Delta p_d}{\frac{1}{2}\rho V_1^2} = \left(\frac{A_1}{A_2}\right)^2 - 1. \tag{6.9}$$

Now we check the dimensions of the terms of this equation:

$$\left[\frac{\Delta p_d}{\frac{1}{2}\rho V_1^2}\right] = \frac{[M/LT^2]}{[M/L^3][L/T]^2} = [1] \quad \text{and} \quad [A_1/A_2] = [L^2]/[L^2] = [1].$$

Equation (6.9) has the general functional form

$$\frac{\Delta p_d}{\frac{1}{2}\rho V_1^2} = F(A_1/A_2). \tag{6.10}$$

Clearly, Eq. (6.9) and its generalization, Eq. (6.10), are in every way equivalent to Eq. (6.7) and its generalization, Eq. (6.8); however, Eq. (6.10) contains only two dimensionless parameters, while Eq. (6.7) contains five dimensional parameters. Notice also that the reduction in the number of parameters by three is exactly equal to the number of fundamental dimensions (M, L, and T) involved in the list of dimensional parameters.

The value of this reduction in the number of parameters is not apparent in this example because we knew about the Bernoulli and continuity equations and could develop a complete theory for the nozzle. Equation (6.7) is a perfectly adequate relation, and the dimensionless form, Eq. (6.9), is only a curiosity. Note, however, that if we did not have a complete theory, it would be much easier to discover the two-parameter relation implied by Eq. (6.10) than the five-parameter relation implied by Eq. (6.8), provided that we knew how to group the dimensional parameters of Eq. (6.8) into the dimensionless parameters of Eq. (6.10).

We are guaranteed that it is possible in every case to reduce the number of independent parameters in a problem by compacting the dimensional parameters into a set of dimensionless parameters. The basis of this guarantee is the following theorem of dimensional analysis, called the *pi theorem*:

If it is known that a physical process is governed by a dimensionally homogeneous relation involving *n dimensional* parameters, such as

$$x_1 = f(x_2, x_3, \ldots, x_n),$$

where the x's are dimensional variables, there exists an equivalent relation involving a smaller number, $(n - k)$, of *dimensionless* parameters, such as

$$\Pi_1 = F(\Pi_2, \Pi_3, \ldots, \Pi_{n-k}),$$

where the Π's are dimensionless groups constructed from the x's. The reduction, k, is usually equal to, but never more than, the number of fundamental dimensions involved in the x's.

The name of Buckingham [2] is usually associated with this theorem, although it was apparently independently discovered by others. The name pi theorem is due to the assigning of the symbol Π (capital Greek pi) to the dimensionless parameters.

The pi theorem is a mathematical theorem, provable by strictly mathematical manipulations. The only "physics" involved is the principle of dimensional homogeneity, which is part of the "if" statement of the theorem. The proof of the theorem is too involved to give here, but the idea is not too complicated. The basic idea is that the general function involving the dimensional parameters is *constrained* by the principle of dimensional homogeneity—not every relation among the x's will do; only those relations that satisfy the principle of dimensional homogeneity are allowed. This leads to a relation among a smaller number of Π's. A proof of the theorem is given in [3, 4, 5].

The pi theorem is the entire basis for dimensional analysis. It only tells us that we *can* do the reduction. It does not tell us how to do it. In fact, there is no unique reduction of the n x's to the $(n - k)$ Π's. Having the pi theorem is like being told that an airplane can fly and then being placed in the cockpit and told to take it up.* Three questions come to mind immediately:

■ How does one combine the x's into a set of Π's, making sure to get just the right number?

■ If k is only usually equal to the number of fundamental dimensions involved, how does one know when it is not? If it is not, what is it?

■ If a complicated physical phenomenon is under investigation, how does one even know what dimensional variables are involved?

We regret to tell you that there is no single answer to the first and last questions. Dimensional analysis depends on experience, taste, and custom. In the next section, we present a way to do dimensional analysis, which will provide the answer to the second question. But our way is not the only way.

6.4 Putting the Pi Theorem to Work: A Prescription for Dimensional Analysis

The pi theorem tells us that it is possible to construct a set of dimensionless parameters, thereby reducing the complexity of a problem by reducing the number of independent parameters. The construction of the dimensionless parameters from the dimensional parameters is up to the analyst. The only help the pi theorem offers is to tell the analyst how many dimensionless parameters to construct.

---

* You might take encouragement from the fact that engineers simply use the results of dimensional analysis done by others far more frequently than they make their own dimensional analysis.

The dimensionless parameters are constructed as products of powers of the dimensional parameters. Three of the dimensionless parameters that we introduced in Section 6.1 are

$$\frac{gh_L}{\frac{1}{2}V^2}, \qquad \frac{L}{D}, \qquad \text{and} \qquad \frac{\rho VD}{\mu}.$$

These can be written as

$$2(gh_L)(V)^{-2}, \qquad (L)(D)^{-1}, \qquad \text{and} \qquad (\rho)(V)(D)(\mu)^{-1},$$

that is, as products of powers of the dimensional parameters.*

The following stepwise method can be used to construct a set of dimensionless parameters. Each step will be illustrated by carrying out a complete dimensional analysis for the pipe flow mechanical energy loss problem discussed in Section 6.1.

Step 1   Write a functional expression for the dimensional relation under investigation. Be sure to include all relevant dimensional parameters.

Writing a functional relation is easy enough, but how do we know which dimensional parameters are relevant and which are not? There is no simple answer. The best guide is experience, but that is something that you, a student, don't have much of yet. A simple theory describing the phenomenon in question, no matter how crude, is very helpful.

Dimensional analysis itself does not begin until after Step 1, so it is not much help. If a relevant parameter is overlooked, it will be forever lost and experimental data will be confusing because the parameter will not be under control. On the other hand, if an irrelevant parameter is included, it will appear in at least one of your dimensionless groups. Several experiments may be required to demonstrate that the parameter (and the group that it is in) is irrelevant to the particular phenomenon.

There is one very helpful check. Any dimension must appear in at least two parameters or it would not be possible to obtain dimensionless groups. Whenever a dimension is represented in only one parameter, either that parameter is irrelevant or another parameter that also includes the unique dimension must be included. Note that dimensional constants (G, g, R_0) can be used for this purpose.

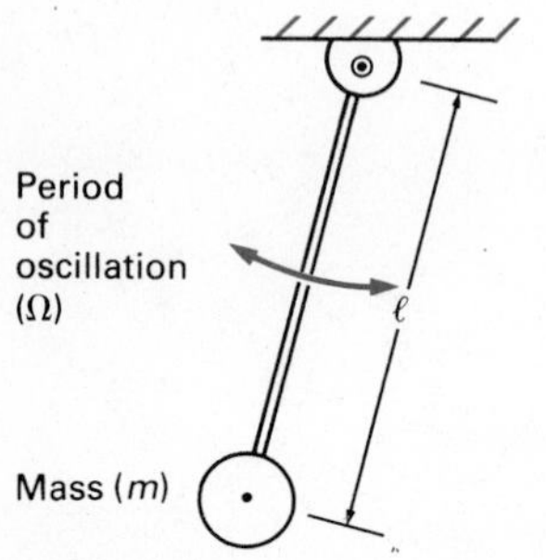

An excellent illustration of these points is the problem of determining the period (Ω) of a simple pendulum (Fig. 6.6). With no knowledge of the physics of pendulums, we would probably propose a dimensional relation of the form:

$$\Omega = f\{m, \ell\}.$$

This relation has serious trouble since each fundamental dimension, mass, length, and time, appears only once. We must add something and possibly remove something as well. We might add the acceleration of gravity g (a dimensional constant) or the weight W. Since weight and

Period
of
oscillation
(Ω)

Mass (m)

Figure 6.6 Simple pendulum.

---

* The factors 1/2 and 2 are irrelevant to the discussion of dimensions. They are included because we know that $\frac{1}{2}V^2$ is kinetic energy per unit mass.

mass are always related by $W = mg$, it would be redundant to include all three. Our new trial functional relation is

$$\Omega = f(m, g, \ell).$$

We still have a problem because the dimension of mass appears in only one parameter. Including the weight will not help because the only dimensionless parameter that could be formed from W, m, and g is mg/W, which is always unity. We conclude that the pendulum's mass is irrelevant and drop m from the functional relation, so its final form is

$$\Omega = f(g, \ell).$$

Returning to the field of fluid mechanics, we find that the functional relation for our pipe flow example is

$$gh_L = f(L, D, V, \rho, \mu, \varepsilon). \tag{6.3}$$

Note that g and h_L are not considered separate parameters; gh_L is just our way of writing a single parameter, mechanical energy loss.

It is sometimes helpful to write a *dimension equation* that exhibits only the dimensions of the parameters involved in the postulated dimensional relation. The dimension equation corresponding to Eq. (6.3) is

$$\left[\frac{L^2}{T^2}\right] = f\left([L], [L], [L/T], \left[\frac{M}{L^3}\right], \left[\frac{M}{LT}\right], [L]\right). \tag{6.11}$$

Step 2 Determine the number of dimensionless parameters you need to construct.

This number is equal to the number of dimensional parameters (n) in the functional relation minus a number (k) that is equal to *the maximum number of dimensional parameters that cannot form a dimensionless group (pi) among themselves*. The number k is usually equal to the number of fundamental dimensions (m) involved in the dimensional parameters; it is never greater than the number of fundamental dimensions.

To find k, initially assume that it is equal to m and try to find a set of m of your dimensional parameters that cannot be formed into a pi group. If you succeed in finding any set, $k = m$, so stop. If you do not find such a set, reduce k by 1 and try again. *Hint:* Try to find parameters that have only one dimension or are fundamentally of a given dimensional type.

In Eq. (6.3), there are seven dimensional parameters (don't forget to count the parameter on the left side). The number of fundamental dimensions is three ($[M]$, $[L]$, $[T]$). So we hope that $k = 3$, and we look for three parameters that cannot be formed into a dimensionless group. If we pick L, ρ, and V, we note that mass appears only in ρ and time only in V, so no combination of L, ρ, and V could be dimensionless, and k is equal to 3. The number of dimensionless parameters that we wish to construct is four:

$$n - k = 7 - 3 = 4.$$

The rare cases where k is less than the number of fundamental dimensions involved in the dimensional parameters occur when certain dimensions (usually mass and time) occur only in fixed combinations in the dimensional parameters. In stress analysis problems, for example, mass and time occur only in the combination $[M/T^2]$. If we were to analyze a stress problem in an $[F, L, T, \theta]$-dimension system instead of an $[M, L, T, \theta]$-dimension system, we would find only two dimensions occurring, namely $[F]$ and $[L]$. A helpful check in determining k is to express the parameters of the problem in terms of both $[M, L, T, \theta]$ and $[F, L, T, \theta]$ systems and see if the number of fundamental dimensions involved is the same in both systems. If not, then k will be the smaller number of fundamental dimensions.

Step 3 Select k of the dimensional parameters that contain among them all of the fundamental dimensions. Combine these parameters with the remaining $(n - k)$ dimensional parameters to form the required number, $(n - k)$, of dimensionless parameters. This is done by selecting the remaining $(n - k)$ parameters one at a time and multiplying by appropriate powers of the k repeating variables so that the result is dimensionless.

Although it sounds difficult, this part is easy. Anyone can do it by following a few simple steps. First we consider the pipe energy loss problem. We must select three of the parameters as repeating variables and use them with the remaining four parameters to form four dimensionless groups. We select ρ, V, D as the repeating variables. We now take the remaining four parameters $(gh_L, L, \mu, \varepsilon)$ and combine them with ρ, V, D to form four dimensionless products (Π's).

We begin with (gh_L). A dimensionless product involving gh_L, ρ, V, and D has the general form

$$\Pi_1 = (gh_L)\rho^a V^b D^c, \tag{6.12}$$

where a, b, and c are exponents to be selected such that Π_1 is dimensionless. The dimensional equation equivalent to Eq. (6.12) is

$$[\Pi_1] = [gh_L][\rho]^a[V]^b[D]^c.$$

Substituting the appropriate dimensions, we have

$$[\Pi_1] = \left[\frac{L^2}{T^2}\right]\left[\frac{M}{L^3}\right]^a\left[\frac{L}{T}\right]^b[L]^c.$$

Combining exponents, we obtain

$$[\Pi_1] = [M]^a[L]^{2-3a+b+c}[T]^{-2-b}. \tag{6.13}$$

Now Π_1 is *dimensionless,* that is

$$[\Pi_1] = [M]^0[L]^0[T]^0.$$

We must therefore equate the *net* exponent of any dimension on the right side of Eq. (6.13) to zero. This gives the three equations

$$a = 0 \quad \text{(from } M\text{)},$$
$$2 + b = 0 \quad \text{(from } T\text{)},$$
$$2 - 3a + b + c = 0 \quad \text{(from } L\text{)}.$$

It is easily verified that the solution to these equations is

$$a = 0, \quad b = -2, \quad c = 0.$$

Substituting back into Eq. (6.12), we find

$$\Pi_1 = \frac{gh_L}{V^2}. \tag{6.14}$$

The next dimensional parameter to be formed into a Π is L. Using the formal procedure just described, we have

$$\Pi_2 = (L)\rho^a V^b D^c. \tag{6.15}$$

Note that a, b, and c are "new" values for this equation. The dimensional equation is

$$[\Pi_2] = [L]\left[\frac{M}{L^3}\right]^a\left[\frac{L}{T}\right]^b [L]^c,$$

or

$$[M]^0[L]^0[T]^0 = [M]^a[L]^{1-3a+b+c}[T]^{-b}.$$

This implies

$$a = 0,$$
$$b = 0,$$
$$1 - 3a + b + c = 0.$$

The solution of these equations is

$$a = 0, \qquad b = 0, \qquad c = -1,$$

so

$$\Pi_2 = \frac{L}{D}. \tag{6.16}$$

Possibly, Π_2 was obvious to you at the start. If a dimensionless group is obvious, there is no need to go through all of the mathematical manipulation; just write $\Pi = $ (whatever) and go on to the next parameter. Note that our Π_2 is a ratio of two lengths. The most obvious dimensionless parameters are ratios of like quantities.

To obtain the third Π, we combine powers of ρ, V, D with μ:

$$\Pi_3 = (\mu)\rho^a V^b D^c. \tag{6.17}$$

The dimensional equation is

$$[\Pi_3] = \left[\frac{M}{LT}\right]\left[\frac{M}{L^3}\right]^a\left[\frac{L}{T}\right]^b [L]^c.$$

This implies

$$1 + a = 0,$$
$$-1 - 3a + b + c = 0,$$
$$-1 - b = 0.$$

The solution to these three equations is

$$a = -1, \qquad b = -1, \qquad c = -1,$$

so

$$\Pi_3 = \frac{\mu}{\rho VD}. \tag{6.18}$$

The fourth and final Π is obtained from ρ, V, D, and ε. The obvious parameter is

$$\Pi_4 = \frac{\varepsilon}{D}. \qquad \textbf{(6.19)}$$

Finally we collect all of our results into a dimensionless functional relation:

$$\frac{gh_L}{V^2} = F\left(\frac{L}{D}, \frac{\mu}{\rho VD}, \frac{\varepsilon}{D}\right). \qquad \textbf{(6.20)}$$

This finishes the dimensional analysis. Any further information must come from theories or experiments. We will offer an optional Step 4 shortly.

The mathematical manipulations required to form the pi's are rather simple. The only tricky element is selecting the dimensional parameters to be used as repeating variables. This selection is rather important because the parameters may appear in every one of the dimensionless groups. There is some room for choice in this selection. The following rules should help:

- *Do not* select the primary *dependent* parameter (in our example, gh_L) as a repeating variable.
- *Do not* select any parameter whose influence on the problem is questionable or important only part of the time. (In the pipe flow example, for instance, viscosity and roughness are only sometimes relevant, as we will see in Chapter 7.)
- It is easier if you select parameters whose dimensions are as close to "pure" as possible. In our case, pipe diameter has the pure dimension of length, density is as close as we can get to pure mass, and velocity is as close as we can get to pure time. If we are able to choose a set of parameters with pure dimensions, the various Π groups can usually be found by inspection.

In case of conflict between these rules, they are listed in order of importance. Application of these rules to our pipe flow problem leaves only the choice between D and L. For fully developed pipe flow, D is more relevant than L, so we selected it.

Step 4   Rearrange the Π groups to please yourself or to correspond with customary usage.

This is an optional step, although you will be very unpopular with your professional colleagues if you insist on using nontraditional types of dimensionless parameters. You may have noticed that the dimensionless pipe energy loss functional form derived as Eq. (6.20) does not correspond to the earlier stated form, Eq. (6.4). The differences are slight: a factor of 2 in Π_1 and a reciprocal in Π_3. The fact is that Eqs. (6.4) and (6.20) are completely equivalent from a theoretical standpoint. The parameters of Eq. (6.4) are traditional in fluid mechanics, while those of Eq. (6.20) are not.

Recall that the pi theorem itself tells us only that we can do a dimensional analysis. Any set of $(n - k)$ independent dimensionless parameters formed from the n dimensional parameters is acceptable from the point of view of the theorem. A set of dimensionless parameters is independent if none of the parameters can be obtained from the others by multiplying parameters together or by combining powers of the parameters. The set of parameters

$$\left(\frac{L}{D}, \frac{\rho VD}{\mu}, \frac{\varepsilon}{D}, \frac{\rho VL}{\mu}\right)$$

is not an independent set since the last parameter can be obtained by multiplying the first two.

According to the pi theorem, any set of independent dimensionless parameters is as good as any other set. The first three steps of the procedure will always produce a set of independent dimensionless parameters (independence is guaranteed since each of the $(n - k)$ nonrepeating dimensional parameters appears in only one Π). After this set is found, any member of the set can be replaced by a different dimensionless parameter formed in the following way:

- By multiplying the parameter by a constant,
- By raising the parameter to *any* positive or negative power,
- By multiplying any power of the parameter with any power of any of the other parameters, or
- By any combination of the above.

To reconcile Eqs. (6.4) and (6.20), note that

$$\frac{gh_L}{\frac{1}{2}V^2} = 2\left(\frac{gh_L}{V^2}\right) \quad \text{and} \quad \frac{\rho VD}{\mu} = \left(\frac{\mu}{\rho VD}\right)^{-1}.$$

In a similar fashion, the functional form

$$\frac{gh_L}{\frac{1}{2}V^2} = F\left(\frac{L}{D}, \frac{\rho VD}{\mu}, \frac{\varepsilon}{D}\right) \tag{6.4}$$

is equivalent* to

$$\frac{D\rho gh_L}{V\mu} = \tilde{F}\left(\frac{\rho VL}{\mu}, \frac{\rho VD}{\mu}, \frac{\rho V\varepsilon}{\mu}\right). \tag{6.21}$$

The dimensionless parameters in Eq. (6.21) are obtained by the following multiplications:

$$\frac{D\rho gh_L}{V\mu} = \frac{1}{2}\left(\frac{gh_L}{\frac{1}{2}V^2}\right)\left(\frac{\rho VD}{\mu}\right), \quad \frac{\rho VL}{\mu} = \left(\frac{L}{D}\right)\left(\frac{\rho VD}{\mu}\right), \quad \text{and}$$

$$\frac{\rho V\varepsilon}{\mu} = \left(\frac{\varepsilon}{D}\right)\left(\frac{\rho VD}{\mu}\right).$$

The "function" in Eq. (6.21) is written as $\tilde{F}\{\ \}$ to emphasize that the *form* of the function will be different. Likewise, the $F\{\ \}$ of Eq. (6.4) will

---

* Note the word *equivalent,* meaning "carries the same information," as opposed to the word *equal* or *identical.*

be different from the $F\{\ \}$ of Eq. (6.20). Whichever form you choose for the dimensionless parameters, dimensional analysis ends with the functional statements of Eqs. (6.4), (6.20), and (6.21). Finding the specific functional relations requires experiments or theories.

Although our approach to dimensional analysis in this chapter has implied that the functional relation is found by experimentation, dimensional analysis is equally applicable to theoretical work; in fact, the appropriate dimensionless parameters will be involved in our theory whether we like it or not.* Many useful functional relations are determined by a combination of theory and experiment.

The relevance of dimensional analysis is not limited to fluid mechanics or even to engineering. The pi theorem is completely general and applies to any physical relation. Dimensional analysis is very important in heat transfer and has even been applied in the fields of meteorology, astrophysics, and economics. Whatever type of problem is involved, dimensional analysis proceeds in the same manner and yields the same benefits.

The four-step method for applying the pi theorem to a particular problem can be summarized as follows:

Step 1 Write the functional dependence between the dimensional parameters.

Step 2 Determine the number of dimensionless parameters to be constructed from the dimensional parameters.

Step 3 Select a set of *repeating variables* from the dimensional parameters that will be used to form each dimensionless group. Use the remaining dimensional parameters one at a time to combine with the repeating variables and form dimensionless groups. This may be done by analysis of the dimensional equation or, in simple cases, by inspection.

Step 4 Rearrange the dimensionless groups found in Step 3 to correspond to customary usage, to suit yourself, or to facilitate an experimental program.

* Sometimes the parameters will be hidden, and a little work will be required to uncover them. Compare Eqs. (6.7), (6.8), (6.9), and (6.10).

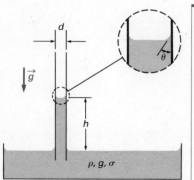

Figure E6.2 Straw inserted in liquid.

EXAMPLE 6.2 Illustrates the procedure to determine dimensionless parameters

A soda straw of inside diameter d is placed in a pan of liquid of density ρ as shown in Fig. E6.2. The water rises to height h in the straw due to the surface tension σ between the water and the straw and makes a contact angle θ at the water-straw interface. Find the appropriate dimensionless parameters describing this phenomenon.

Solution

Given

Figure E6.2

Dimensional parameters d, ρ, g, h, σ, θ

Find

Appropriate dimensionless parameters

Solution

Step 1: Write the dimensional relation. Considering h as the dependent parameter gives

$$h = f(d, \rho, g, \sigma, \theta).$$

Step 2: Determine the maximum number of dimensionless parameters. There are six parameters, but θ is dimensionless since it can be expressed in radians. Therefore $n = 5$:

$$[d] = [L],$$
$$[\rho] = [M/L^3],$$
$$[g] = [L/T^2],$$
$$[h] = [L],$$
$$[\sigma] = [F/L] = [M/T^2].$$

There are three fundamental dimensions, so we expect $k = 3$; ρ, d, and σ cannot form a dimensionless group, so $k = 3$ and the maximum number of dimensionless parameters is $(5 - 3) = 2$.

Step 3: Select k dimensional parameters that contain all the fundamental dimensions. We select d, ρ, and σ. Combining d, ρ, and σ with h gives

$$\Pi_1 = h d^a \rho^b \sigma^c.$$

The dimensional equation is

$$[h][d]^a[\rho]^b[\sigma]^c = [M]^0[L]^0[T]^0$$

or

$$[L][L]^a\left[\frac{M}{L^3}\right]^b\left[\frac{M}{T^2}\right]^c = [M]^0[L]^0[T]^0.$$

Equating powers of each dimension gives

$$M: \quad b + c = 0,$$
$$L: \quad 1 + a - 3b = 0,$$
$$T: \quad -2c = 0.$$

Solving, we obtain

$$c = 0, \quad b = 0, \quad a = -1.$$

The first dimensionless parameter is

$$\left| \; \Pi_1 = \frac{h}{d}. \; \right|$$

Answer ◄

Combining d, ρ, and σ with g gives

$$[g][d]^a[\rho]^b[\sigma]^c = [M]^0[L]^0[T]^0$$

or

$$\left[\frac{L}{T^2}\right][L]^a\left[\frac{M}{L^3}\right]^b\left[\frac{M}{T^2}\right]^c = [M]^0[L]^0[T]^0.$$

Equating powers of each dimension, we have

$$M: \quad b + c = 0$$
$$L: \quad 1 + a - 3b = 0$$
$$T: \quad -2 - 2c = 0.$$

Solving, we obtain

$$c = -1, \qquad b = -c = 1, \qquad a = -1 + 3b = -1 + 3(1) = 2.$$

The second dimensionless parameter is

$$\Pi_2 = gd^2\rho^1\sigma^{-1},$$

or

$$\left| \; \Pi_2 = \frac{\rho g d^2}{\sigma}. \; \right| \qquad\qquad \textbf{Answer} \; \blacktriangleleft$$

Step 4: Rearrange the Π groups. The Π groups are acceptable.

Recalling that θ also influences the phenomenon, the complete dimensionless equation is

$$\left| \; \frac{h}{d} = F\!\left(\frac{\rho g d^2}{\sigma}, \theta \right). \; \right| \qquad\qquad \textbf{Answer} \; \blacktriangleleft$$

Discussion

As a check on the maximum number of dimensionless parameters, we list the fundamental dimensions of each dimensional parameter in the $[F, L, T, \theta]$ system:

$$[d] = [L],$$
$$[\rho] = [M/L^3] = [FT^2/L^4],$$
$$[g] = [L/T^2],$$
$$[h] = [L],$$
$$[\sigma] = [F/L].$$

Again we have three fundamental dimensions and five dimensional parameters, so $(n - k) = (5 - 3) = 2$, or two dimensionless parameters.

An interesting result is that the terms ρ and g appear only as the product ρg, which is the specific weight γ. This suggests that we could have started with the single dimensional parameter γ rather than the two separate parameters ρ and g. It would be instructive for you to start with the four dimensional parameters d, γ, h, and σ and develop the dimensionless parameters using both the $[M, L, T, \theta]$ and the $[F, L, T, \theta]$ system.

EXAMPLE 6.3 Illustrates the procedure to determine dimensionless parameters for a case where $k \neq m$

A cantilever beam of length ℓ and second moment of area I has a transverse force P applied at its free end. The bending stress σ at a distance c from its neutral axis is given by*

$$\sigma = f(\ell, c, x, I, P),$$

where x is defined in Fig. E6.3. Determine the dimensionless parameters describing the beam stress using

a) fundamental dimensions $[F, L, T, \theta]$,
b) fundamental dimensions $[M, L, T, \theta]$.

---

* From strength of materials,

$$\sigma = \frac{P(\ell - x)c}{I}.$$

Solution

Given
Cantilever beam in Fig. E6.3
Length ℓ
Second moment of area I
Transverse force P
Bending stress σ at distance c from neutral axis at distance x from fixed end

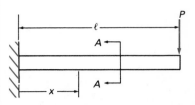

Find
The dimensionless parameters using
Fundamental dimensions $[F, L, T, \theta]$
Fundamental dimensions $[M, L, T, \theta]$

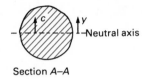

Figure E6.3 Cantilever beam.

Solution using $[F, L, T, \theta]$ system
Step 1: Write the dimensional relation. Considering σ as the dependent variable gives

$$\sigma = f(\ell, c, x, I, P).$$

Step 2: Determine the maximum number of dimensionless parameters. There are six dimensional parameters, so $n = 6$. Their fundamental dimensions are

$$[\sigma] = [F/L^2], \qquad [x] = [L],$$
$$[\ell] = [L], \qquad [I] = [L^4],$$
$$[c] = [L], \qquad [P] = [F].$$

There are two fundamental dimensions in the $[F, L, T, \theta]$ system, so we expect $k = 2$. The maximum number of dimensionless parameters is then $(n - k) = (6 - 2) = 4$.

Step 3: Select k dimensional parameters that contain all the fundamental dimensions. Since $k = 2$, we select ℓ and P. Combining ℓ and P with σ gives

$$[\sigma][\ell]^a[P]^b = [F]^0[L]^0,$$

or

$$\left[\frac{F}{L^2}\right][L]^a[F]^b = [F]^0[L]^0.$$

Equating powers of each dimension gives

$$F: \quad 1 + b = 0, \qquad L: \quad -2 + a = 0.$$

Solving, we obtain

$$b = -1, \qquad a = 2.$$

The first dimensionless parameter is

$$\left| \ \Pi_1 = \frac{\sigma \ell^2}{P}. \ \right| \qquad \textbf{Answer} \ \blacktriangleleft$$

Combining ℓ and P with c gives the obvious parameter

$$\left| \ \Pi_2 = \frac{c}{\ell}. \ \right| \qquad \textbf{Answer} \ \blacktriangleleft$$

We next combine ℓ and P with x. By inspection we get

$$\left| \ \Pi_3 = \frac{x}{\ell}. \ \right| \qquad \textbf{Answer} \ \blacktriangleleft$$

Combining ℓ and P with I gives

$$[I][\ell]^a[P]^b = [F]^0[L]^0, \quad \text{or} \quad [L]^4[L]^a[F]^b = [F]^0[L]^0.$$

Equating powers of each dimension gives

$$F: \quad b = 0, \qquad L: \quad 4 + a = 0.$$

Solving, we obtain

$$b = 0, \qquad a = -4.$$

The fourth dimensionless parameter is

$$\left| \; \Pi_4 = \frac{I}{\ell^4}. \; \right| \qquad \qquad \textbf{Answer}$$

Step 4: Rearrange the Π groups. The Π groups are acceptable.

Solution using $[M, L, T, \theta]$ system

Step 1: Write the dimensional relation. Considering σ as the dependent variable gives

$$\sigma = f(\ell, c, x, I, P).$$

Step 2: Determine the maximum number of dimensionless parameters. There are six dimensional parameters, so $n = 6$. Their fundamental dimensions are

$$[\sigma] = [F/L^2] = [M/LT^2], \qquad [x] = [L],$$
$$[\ell] = [L], \qquad\qquad\qquad\; [I] = [L^4],$$
$$[c] = [L], \qquad\qquad\qquad\; [P] = [F] = [ML/T^2].$$

There are three fundamental dimensions in the $[M, L, T, \theta]$ system, so we expect $k = 3$. If this is true, the maximum number of dimensionless parameters is $(n - k)$, or $(6 - 3) = 3$.

This value is not equal to that predicted by the $[F, L, T, \theta]$ system. We ignore this for now and proceed with the $[M, L, T, \theta]$ dimensional analysis.

Step 3: Select k dimensional parameters that contain all the fundamental dimensions. Since $k = 3$, we select ℓ, I, and P. Combining ℓ, I, and P with σ gives

$$[\sigma][\ell]^a[I]^b[P]^c = [M]^0[L]^0[T]^0,$$

or

$$\left[\frac{M}{LT^2} \right][L]^a[L^4]^b \left[\frac{ML}{T^2} \right]^c = [M]^0[L]^0[T]^0.$$

Equating powers of each dimension gives

$$M: \quad 1 + c = 0,$$
$$L: \quad -1 + a + 4b + c = 0,$$
$$T: \quad -2 - 2c = 0.$$

Note that the first and last equation both give $c = -1$; a and b both must be calculated from the second equation. Unfortunately, we have two unknowns and only one equation; it is impossible to solve for both a and b. Apparently, we cannot complete the dimensional analysis; we must have made an error.

It is not difficult to locate our error. At Step 2, we did not verify that $k = 3$. Apparently, we have uncovered a case where k is less than the number of fundamental dimensions. Remember that k is equal to the maximum number of dimensional parameters that cannot form a dimensionless group. Some three-parameter combinations of our six dimensional parameters are as follows:

σ, ℓ, c ℓ, c, x c, x, I
σ, ℓ, x ℓ, c, I c, x, P
σ, ℓ, P

It is possible to form a dimensionless group from *any* of these combinations; for example,

σ, ℓ, c form ℓ/c, ℓ, c, x form $c/\ell, \ell/x$,
σ, ℓ, x form ℓ/x, c, x, P form c/x,

and so on.

All possible three-parameter combinations could form a dimensionless group, so k cannot equal three. We now assume $k = 2$ and examine possible two-parameter combinations. We immediately see that σ and ℓ *cannot* form a dimensionless group, so $k = 2$. Then $(n - k) = (6 - 2) = 4$, and we must find four dimensionless parameters. This conclusion agrees with that from the $[F, L, T, \theta]$ system. We must now repeat Step 3. We select ℓ and P as repeating variables. Combining ℓ and P with σ gives

$$[\sigma][\ell]^a[P]^b = [M]^0[L]^0[T]^0,$$

or

$$\left[\frac{M}{LT^2}\right][L]^a\left[\frac{ML}{T^2}\right]^b = [M]^0[L]^0[T]^0.$$

Equating powers of each dimension gives

$$M: \quad 1 + b = 0,$$
$$L: \quad -1 + a + b = 0,$$
$$T: \quad -2 - 2b = 0.$$

Solving,* we have

$$b = -1, \quad a = 2.$$

The first dimensionless parameter is

$$\boxed{\Pi_1' = \frac{\sigma\ell^2}{P}.}$$ **Answer** ◀

Combining ℓ and P with c gives the obvious parameter

$$\boxed{\Pi_2' = \frac{c}{\ell}.}$$ **Answer** ◀

Combining ℓ and P with x gives our third dimensionless parameter by inspection:

$$\boxed{\Pi_3' = \frac{x}{\ell}.}$$ **Answer** ◀

Combining ℓ and P with I gives

$$[I][\ell]^a[P]^b = [M]^0[L]^0[T]^0,$$

or

$$[L]^4[L]^a\left[\frac{ML}{T^2}\right]^b = [M]^0[L]^0[T]^0.$$

---

* Note that we have two *independent* equations and two unknowns.

Equating powers of each dimension gives

$$M: \quad b = 0,$$
$$L: \quad 4 + a + b = 0,$$
$$T: \quad 2b = 0.$$

Solving, we have

$$b = 0, \quad a = -4.$$

The fourth dimensionless parameter is

$$\left| \; \Pi'_4 = \frac{I}{\ell^4}. \; \right|$$

Answer ◀

Discussion

The dimensionless parameters developed by each method are as follows:

$$\Pi_1 = \frac{\sigma \ell^2}{P} \qquad \Pi'_1 = \frac{\sigma \ell^2}{P}$$

$$\Pi_2 = \frac{c}{\ell} \qquad \Pi'_2 = \frac{c}{\ell}$$

$$\Pi_3 = \frac{x}{\ell} \qquad \Pi'_3 = \frac{x}{\ell}$$

$$\Pi_4 = \frac{I}{\ell^4} \qquad \Pi'_4 = \frac{I}{\ell^4}$$

Both methods give the same dimensionless parameters.

This example illustrates that we must be careful in determining k. Since k is usually equal to m (the number of fundamental dimensions), it is very tempting to set $k = m$ and bypass the lengthier procedure to check k. If you do this and k is actually less than m for your problem, you will invariably come up with an indeterminant set of equations in Step 3. In mechanics problems, a quick check for the value of k is to list the fundamental dimensions in both the $[M, L, T, \theta]$ and the $[F, L, T, \theta]$ system. If the number of fundamental dimensions involved is different for the two systems, k can be no larger than the smaller number of fundamental dimensions.

6.5 Customary Dimensionless Parameters in Fluid Mechanics

Generally it is not necessary to begin each new problem in fluid mechanics with a formal dimensional analysis. Only a limited number of dimensional parameters can be relevant (length, velocity, density, viscosity, etc.). Although it is possible to form an infinite number of dimensionless groups from a finite number of dimensional parameters, there is a fixed, finite number of independent groups that can be found. Certain dimensionless groups have become standard in fluid mechanics, and all engineers are more or less expected to use these parameters when conducting and reporting experimental investigations or proposing theories. It is more important for beginning engineers to know and understand the standard dimensionless parameters than to be able to perform a rigorous dimensional analysis each time a new fluid mechanics problem crops up. In this section, we will carry out a complete dimensional analysis of fluid mechanics problems and introduce you to the customary dimensionless parameters.

Figures 6.7 and 6.8 illustrate two typical problems in fluid mechanics: an external flow and an internal flow. For now we will assume that the flow in each case is steady. A general functional relation for these flows is

$$\text{Dimensional parameter of primary interest} = f\left\{\begin{array}{l}\text{Size of object (or passage),}\\ \text{shape of object, fluid velocity,}\\ \text{fluid properties, dimensional constants}\end{array}\right\}. \tag{6.22}$$

In general, the *type* of parameters in the function on the right changes very little from problem to problem; however, the parameter on the left can be of several types. For external flow, the parameter of interest is probably the force exerted on the object by the fluid. Instead of the total force, we may be interested in components in the streamwise direction (called *drag force*) and the direction normal to the stream (called *lift force*). Rather than the vector components of the force, we may be interested in separating the force into pressure and shear stress contributions. For the internal flow, the parameter of interest may be the wall force or it may be the mechanical energy loss. For the purpose of discussion, we will assume that we are interested in drag force ($\mathscr{D}$) for the external flow and mechanical energy loss (gh_L) for the internal flow.

Notice that the size and shape of the object have been separated on the right side of Eq. (6.22). The shape of an object is determined by a set of *dimensionless* parameters, essentially length ratios and angles. For a given shape, the size of an object is specified by a *single* length dimension. Since shape is defined by a (typically very large) set of already dimensionless parameters, it is usually dropped from the general functional equation (6.22). If shape is not shown explicitly, the relation is restricted to a specific shape, such as forces on cubes or losses in conical nozzles. Sometimes simple shape parameters like the length-to-diameter ratio for pipes are retained in the equation.

Since velocity is a vector, it may be necessary to specify its direction as well as speed. In internal flow problems, velocity parallel to the axis

Figure 6.7 External flow problem.

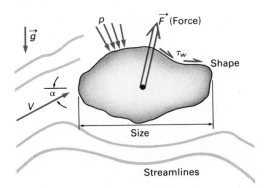

Figure 6.8 Internal flow problem.

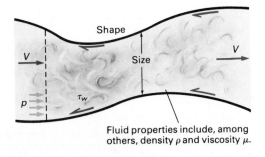

of the channel is usually understood; in external flow problems, the angle between the approaching fluid velocity and an axis of the object must be specified. This is angle α in Fig. 6.7. Since angles are already dimensionless,* they do not enter into the dimensional analysis (but they propagate directly from the dimensional equation to the dimensionless equation).

The fluid properties that are important in fluid mechanics problems include density, viscosity, modulus of elasticity, and surface tension. If the flow is highly compressible so that mechanical-thermal energy exchanges are important, then specific heats are also important. Note that pressure and temperature have not been included. This is because pressure is usually a parameter of primary interest and so is included on the left side rather than the right. For compressible fluids, pressure is also included on the right side in the modulus of elasticity because

$$E_{v,T} = p,$$

at least for an ideal gas. In incompressible flow, temperature itself is not relevant unless we are considering a heat transfer problem. In compressible flow, specification of density, pressure, and specific heats (and so, indirectly, gas constant) is equivalent to specification of temperature. In other words, for a given gas, we need to list any two of pressure, density, or temperature, but not all three.

If there are no electric, magnetic, or relativistic effects in the flow, the only relevant dimensional constant is the acceleration of gravity g. The dimensional functional relations for the flows of Figs. 6.7 and 6.8 become[†]

$$\mathscr{D} = f(\ell, V, \rho, \mu, E_v, \sigma, c_p, c_v, g) \tag{6.23}$$

and

$$gh_L = f(\ell, V, \rho, \mu, E_v, \sigma, c_p, c_v, g). \tag{6.24}$$

Since only the left-side parameter is different in these two equations, we continue with only Eq. (6.23). The dimensional equation is

$$\left[\frac{ML}{T^2}\right] = f\left([L], \left[\frac{L}{T}\right], \left[\frac{M}{L^3}\right], \left[\frac{M}{LT}\right], \left[\frac{M}{LT^2}\right], \left[\frac{M}{T^2}\right], \left[\frac{L^2}{T^2\theta}\right], \left[\frac{L^2}{T^2\theta}\right], \left[\frac{L}{T^2}\right]\right).$$

There are ten dimensional parameters and four fundamental dimensions ($[M]$, $[L]$, $[T]$, $[\theta]$). Since ρ, V, ℓ, and c_p cannot form a dimensionless product, $k = 4$ and we can form six dimensionless parameters. We select ρ, V, ℓ and c_p as repeating variables. Performing the dimensional analysis and putting the resulting dimensionless groups in standard form gives

$$\frac{\mathscr{D}}{\frac{1}{2}\rho V^2 \ell^2} = F\left(\frac{\rho V\ell}{\mu}, \frac{V}{\sqrt{E_v/\rho}}, \frac{\rho V^2 \ell}{\sigma}, \frac{V}{\sqrt{g\ell}}, \frac{c_p}{c_v}\right). \tag{6.25}$$

---

* Angles should be expressed in radians for this purpose.
† The angle α has been omitted since it is already dimensionless. It must be reinserted in the final dimensionless relation.

Each of the dimensionless parameters on the right side of Eq. (6.25) is a "standard" dimensionless parameter in fluid mechanics. Each parameter focuses on a particular effect in the flow. If the effect represented by a parameter is not important in the flow, the dimensionless parameter approaches a limiting value, usually zero or infinity. With one exception, the parameters are named after famous persons from the history of fluid mechanics. The parameters, their symbols, names, and significance are as follows:

$\mathbf{R} \equiv \rho V\ell/\mu$ is the *Reynolds number,* probably the most important dimensionless parameter in fluid mechanics. $\mathbf{R}$ represents the effects of viscosity on the flow. Flows with large values of $\mathbf{R}$ $(\mathbf{R} \to \infty)$ do not depend on viscosity but rather are turbulent (see Section 3.5.2).

$\mathbf{M} \equiv V/\sqrt{E_v/\rho}$ is the *Mach number,* which is important in compressible flows. It can be shown (see Chapter 10) that the speed of sound (c) in a fluid is given by

$$c^2 = \frac{E_v}{\rho},$$

so the Mach number is the ratio of the fluid speed to the speed of sound in the fluid. Flows with low values of Mach number $(\mathbf{M} \approx 0)$ do not exhibit compressibility effects.

$\mathbf{W} \equiv \rho V^2\ell/\sigma$ is the *Weber number;* it represents the effects of surface tension. Since surface tension is important only if there is an interface, a flow depends on Weber number only if there is an interface between two fluids. If the passage in Fig. 6.8 is filled with a single fluid or if the object in Fig. 6.7 is far away from an interface, the Weber number can be dropped from the parameter list. Even if there is a two-fluid interface present, surface tension effects are not relevant unless the length scale (size) of the problem is very small.

$\mathbf{F} \equiv V/\sqrt{g\ell}$ is the *Froude number*,* which represents free-surface effects. If there is no free surface, the Froude number can be dropped from the parameter list. Note that the Froude number does not deal with hydrostatic pressure, even though it is the only parameter that contains g. This is because the dimensionless parameters deal with dynamic effects while the hydrostatic pressure is a static effect.

$k \equiv c_p/c_v$ is the *specific heat ratio; k* is relevant only in compressible flow problems. For an incompressible fluid, k may be dropped from the list.

The Reynolds, Mach, Weber, and Froude numbers and the specific heat ratio are the independent dimensionless parameters and will appear in all problems unless the effect specifically represented by the parameter is absent.

---

* Some authors define the Froude number as $V^2/g\ell$.

Next we consider the parameters that could appear on the left side of the functional relation. The dimensional parameters that usually appear are the following:

- Drag force ($\mathscr{D}$) or
- Lift force ($\mathscr{L}$) or
- Pressure difference ($\Delta p \equiv p - p_{\text{ref}}$)* or
- Wall shear stress (τ_w) or
- Mechanical energy loss (gh_L).

If we combine these dimensional parameters with ρ, V, and L (we don't need c_p), we get the following dimensionless parameters:

$$\mathbf{C_D} \equiv \frac{\mathscr{D}}{\frac{1}{2}\rho V^2 \ell^2} \qquad \text{drag coefficient}$$

$$\mathbf{C_L} \equiv \frac{\mathscr{L}}{\frac{1}{2}\rho V^2 \ell^2} \qquad \text{lift coefficient}$$

$$\mathbf{C_p} \equiv \frac{\Delta p}{\frac{1}{2}\rho V^2} \qquad \text{pressure coefficient (or Euler number)}$$

$$\mathbf{C_f} \equiv \frac{\tau_w}{\frac{1}{2}\rho V^2} \qquad \text{skin friction coefficient}$$

$$\mathbf{K} \equiv \frac{gh_L}{\frac{1}{2}\rho V^2} \qquad \text{loss coefficient}$$

Since these represent the dependent parameters, any one of these coefficients can be represented as a function of the five independent-variable dimensionless parameters. For example, an appropriate dimensionless relation for the internal flow of Fig. 6.8 is

$$\mathbf{K} = F(\mathbf{R}, \mathbf{M}, \mathbf{W}, \mathbf{F}, k).$$

Our list of relevant dimensional parameters and corresponding dimensionless parameters is not quite complete. There are certain special cases that involve a few more dimensional variables and so require a few more dimensionless parameters.

In the pipe flow illustration of Sections 6.1 and 6.4, we included the roughness of the pipe wall, ε, in the list of dimensional parameters and so the parameter ε/D appeared in the dimensionless parameter list. Since ε/D (more generally ε/ℓ in terms of an arbitrary length ℓ) is a ratio of two lengths, we might be tempted to include it with the shape parameters and drop it from the list. This is not a good idea because roughness is a characteristic of the type of surface rather than the shape of a passage or object. Laminar flow does not depend on surface roughness, but turbulent flow does. If we are considering a turbulent flow, then ε/ℓ, the

---

* p_{ref} is a reference pressure.

relative roughness, should be included with the independent dimensionless parameters.

Many flow problems are unsteady. Suppose that the airfoil shown in Fig. 6.9 is oscillating with frequency ω. The frequency of oscillation will definitely affect the flow. The reciprocal of the frequency is a characteristic time of the problem. If ω is added to the list of independent dimensional variables, then the dimensionless parameter

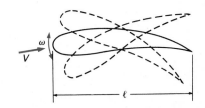

Figure 6.9 Airfoil oscillating with frequency ω.

$$\mathbf{S} \equiv \frac{\omega\ell}{V}$$

will appear in the list of independent dimensionless parameters. $\mathbf{S}$ is called the *Strouhal number.* Sometimes an oscillatory phenomenon is the problem under investigation, that is, the frequency is the dependent parameter rather than an independent parameter. This would be the case with flow-induced vibration of heat exchanger tubes or suspension bridge cables. In this instance, the Strouhal number would appear as a dependent parameter rather than an independent parameter, typically in an equation of the form

$$\mathbf{S} = f(\mathbf{R}, \mathbf{M}).$$

A similar situation occurs when there are two relevant velocities in a problem rather than just one. With flow in a centrifugal pump rotor, both the fluid velocity and the rotor velocity would be important dimensional parameters. The rotor linear velocity U (or, twice its value, ND, where N is rotational speed and D is rotor diameter) would have to be included in the list of independent dimensional parameters. The corresponding dimensionless parameter would be

$$\frac{V}{U} \quad \text{or} \quad \frac{V}{ND}.$$

The second of these parameters is actually a type of reciprocal Strouhal number. A similar parameter would be required to investigate a flow using a coordinate system that moves with velocity U (see Example 5.8).

It would be possible to extend this list of special cases indefinitely. The important thing to remember is that each extra *dimensional* parameter that is important will introduce a *dimensionless* parameter to represent it. Valid dimensionless parameters that represent these extra parameters can usually be expressed in terms of length or velocity ratios.

Table 6.1 is a comprehensive list of the dimensionless parameters discussed in this section. Using this list, you do not need to do a dimensional analysis for each situation you encounter. All you need to do is select the relevant parameters from the table and substitute them into the general functional relation

$$\begin{matrix}\text{Dimensionless parameter} \\ \text{containing relevant} \\ \text{dependent variables}\end{matrix} = f \begin{pmatrix} \text{Relevant dimensionless} \\ \text{parameters from} \\ \text{independent parameters list} \end{pmatrix} \qquad \textbf{(6.26)}$$

A parameter is irrelevant if the effect it represents (e.g., surface tension for $\mathbf{W}$, compressibility for $\mathbf{M}$ and k) is not important in the particular

Table 6.1 Customary dimensionless parameters of fluid mechanics

Parameter	Symbol	Name	Relevance
Independent Variables			
$\dfrac{\rho V \ell}{\mu}$	**R**	Reynolds number	Laminar/turbulent flow
$\dfrac{V}{\sqrt{E_v/\rho}}$	**M**	Mach number	Compressible flow
$\dfrac{\rho V^2 \ell}{\sigma}$	**W**	Weber number	Flow with two-fluid interface
$\dfrac{V}{\sqrt{g\ell}}$	**F**	Froude number	Flow with a free surface
c_p/c_v	k	Specific heat ratio	Compressible flow
$\dfrac{\omega \ell}{V}$	**S**	Strouhal number	Oscillating boundary
$\dfrac{V}{U}$	–	Velocity ratio	Two relevant velocities
ℓ'/ℓ	–	Length ratio	Two relevant lengths
$\dfrac{\varepsilon}{\ell}$	–	Relative roughness	Turbulent flow
α	–	Angle of attack	External flow over objects
Dependent Variables			
$\dfrac{\mathscr{D}}{\frac{1}{2}\rho V^2 \ell^2}$	$\mathbf{C_D}$	Drag coefficient	External flow
$\dfrac{\mathscr{L}}{\frac{1}{2}\rho V^2 \ell^2}$	$\mathbf{C_L}$	Lift coefficient	External flow
$\dfrac{\Delta p}{\frac{1}{2}\rho V^2}$	$\mathbf{C_p}$	Pressure coefficient	All flows
$\dfrac{\tau_w}{\frac{1}{2}\rho V^2}$	$\mathbf{C_f}$	Skin friction coefficient	All flows with shear stress
$\dfrac{g h_L}{\frac{1}{2}V^2}$	**K**	Loss coefficient	Internal flows
$\dfrac{\omega \ell}{V}$	**S**	Strouhal number	Flows with oscillatory output, such as vortex shedding

problem. If there is a compelling reason to use a different set of dimensionless parameters to describe a particular flow, the different parameters can be constructed from the standard parameters by multiplying powers of the standard parameters.

EXAMPLE 6.4   Illustrates an alternative procedure for identifying dimensionless parameters

Figure E6.4 shows a centrifugal pump with a rotor of diameter D that rotates at speed N (rpm). Liquid flows through the pump with volume flow rate Q. The pressure of the liquid increases by Δp as it passes through the pump. Suggest a useful dimensionless functional relation to express the pump's performance.

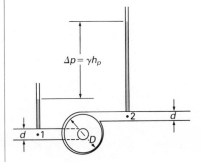

Figure E6.4 Centrifugal pump and its head rise h_p.

Solution

   Given
Centrifugal pump in Fig. E.6.4

   Find
A useful dimensionless functional relation to express pump performance

   Solution
   We can deduce a useful set of dimensionless parameters with the help of Table 6.1. We select the pressure rise as the primary dependent parameter. The primary *dimensionless* dependent parameter would then be a pressure coefficient:

$$C_p = \frac{\text{Pressure rise}}{(\text{Density})(\text{Velocity})^2}.$$

We expect the pressure rise to depend on the liquid's viscosity. This implies that we should include the Reynolds number as an important dimensionless parameter:

$$R = \frac{(\text{Velocity})(\text{Length})}{\text{Kinematic viscosity}}.$$

There are two relevant velocities in a centrifugal pump: a typical fluid velocity V and the pump rotor tip velocity U. The *velocity ratio*

$$\frac{V}{U}$$

is surely an important dimensionless parameter.
   Effects of compressibility, surface tension, and surface gravity waves are irrelevant in a liquid pump, so we expect a dimensionless performance law of the form

$$C_p = f\!\left(\frac{V}{U},\, R\right).$$

We now want to identify useful expressions for each dimensionless parameter. The pressure coefficient is

$$C_p = \frac{\Delta p}{\rho(\text{Velocity})^2}.$$

The most obvious choice for the velocity is a typical fluid velocity. We can find a characteristic fluid velocity from

$$V = \frac{\text{Volume flow rate}}{\text{Area}},$$

thus

$$V \propto \frac{Q}{D^2}.$$

If we use V in the pressure coefficient, $\mathbf{C_p}$ will contain two parameters that vary while the pump operates, namely Δp and Q. Centrifugal pumps are often driven by constant-speed AC electric motors, so the pump rotor velocity U is a more convenient velocity for the pressure coefficient. Since

$$U \propto ND,$$

the most convenient pressure coefficient is

$$\mathbf{C_p} = \frac{\Delta p}{\rho U^2},$$

or, ignoring constants of proportionality,

$$\mathbf{C_p} = \frac{\Delta p}{\rho N^2 D^2}.$$

We can proceed one step further by recognizing that the pressure rise Δp can be expressed as

$$\Delta p = \rho g h_p,$$

where h_p is the pump head. The pressure coefficient can be expressed in terms of pump head as

$$\mathbf{C_p} = \frac{\rho g h_p}{\rho N^2 D^2},$$

or

$$\left| \; \mathbf{C_p} = \frac{g h_p}{N^2 D^2}. \; \right| \qquad\qquad \textbf{Answer} \; \blacktriangleleft$$

This special form of pressure coefficient is called the *head coefficient* and is often denoted by Ψ.

Although Reynolds number is usually based on a fluid velocity, the rotor velocity (ND) is more convenient for expressing Reynolds number in a pump. Using the rotor diameter as the relevant length in the Reynolds number gives

$$\left| \; \mathbf{R} = \frac{ND^2}{\nu}. \; \right| \qquad\qquad \textbf{Answer} \; \blacktriangleleft$$

The final dimensionless parameter, velocity ratio, is given by

$$\frac{V}{U} = \frac{\text{Fluid velocity}}{\text{Rotor velocity}} \propto \frac{Q}{D^2(ND)}.$$

Ignoring constants of proportionality, we find the dimensionless parameter

$$\left| \; C_Q = \frac{Q}{ND^3}. \; \right| \qquad\qquad \textbf{Answer} \; \blacktriangleleft$$

This parameter is called the *flow coefficient* and is often denoted by ϕ.

Collecting all our results, we can write a useful dimensionless performance law for a centrifugal pump as

$$\left| \frac{gh_P}{N^2D^2} = f\left(\frac{Q}{ND^3}, \frac{ND^2}{\nu}\right). \right|$$ **Answer** ◀

Discussion

Using V instead of U in the pressure coefficient and Reynolds number gives

$$\frac{gh_P D^4}{Q^2} = f\left(\frac{Q}{ND^3}, \frac{Q}{\nu D}\right).$$

an alternative, but not quite as convenient, performance law. The recommended parameters can be derived from this set by multiplying the pressure coefficient by $(Q/ND^3)^2$ and dividing the Reynolds number by (Q/ND^3).

EXAMPLE 6.5 Illustrates an alternative method for obtaining dimensionless parameters by using the equations that describe a flow problem

The two appropriate differential equations to study the boundary layer* in incompressible flow over the flat plate shown in Fig. E6.5 are

$$\rho\left[u\left(\frac{\partial u}{\partial x}\right) + v\left(\frac{\partial u}{\partial y}\right)\right] = \mu\left(\frac{\partial^2 u}{\partial y^2}\right)$$

and

$$\frac{\partial u}{\partial x} + \frac{\partial v}{\partial y} = 0,$$

where x and y are coordinates and u and v are the corresponding velocity components, ρ is the density, and μ the fluid viscosity. Convert the equations to dimensionless form and identify any familiar parameters.

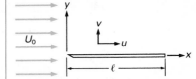

Figure E6.5 Flow of a fluid over a flat surface.

Solution

Given
The Prandtl boundary layer equations for a flat plate

Find
Dimensionless form of the equations
Any familiar parameters

Solution
In this approach, we select a reference velocity U_0, a reference length ℓ, and define dimensionless velocities and lengths as follows:[†]

$$U = \frac{u}{U_0}, \qquad V = \frac{v}{U_0}, \qquad X = \frac{x}{\ell}, \qquad Y = \frac{y}{\ell}.$$

Substituting for u, v, x, and y in terms of U, V, X, and Y gives

$$\rho\left[U_0 U\left(\frac{\partial(U_0 U)}{\partial(\ell X)}\right) + U_0 V\left(\frac{\partial(U_0 U)}{\partial(\ell Y)}\right)\right] = \mu\left(\frac{\partial}{\partial(\ell Y)}\right)\left(\frac{\partial(U_0 U)}{\partial(\ell Y)}\right)$$

---

* The boundary layer is the fluid region very near a solid surface.

[†] The reference velocity U_0 could be the approach velocity, and the reference length ℓ could be the length of the flat plate in the direction of flow.

or, since U_0 and ℓ are constant,

$$\frac{\rho U_0^2}{\ell}\left[U\left(\frac{\partial U}{\partial X}\right) + V\left(\frac{\partial U}{\partial Y}\right)\right] = \frac{\mu U_0}{\ell^2}\left(\frac{\partial^2 U}{\partial Y^2}\right).$$

Similarly,

$$\frac{U_0}{\ell}\frac{\partial U}{\partial X} + \frac{U_0}{\ell}\frac{\partial V}{\partial Y} = 0.$$

Rearranging and simplifying give

$$U\left(\frac{\partial U}{\partial X}\right) + V\left(\frac{\partial U}{\partial Y}\right) = \frac{\mu}{\rho U_0 \ell}\frac{\partial^2 U}{\partial Y^2} \quad \text{and} \quad \frac{\partial U}{\partial X} + \frac{\partial V}{\partial Y} = 0.$$

The quantity $(\rho U_0 \ell / \mu)$ is a Reynolds number:

$$\frac{\rho U_0 \ell}{\mu} = \frac{(\text{Density})(\text{Velocity})(\text{Length})}{\text{Viscosity}} = \mathbf{R}.$$

The Prandtl boundary layer equations are written as follows in dimensionless form

$$\left| U\left(\frac{\partial U}{\partial X}\right) + V\left(\frac{\partial U}{\partial Y}\right) = \frac{1}{\mathbf{R}}\left(\frac{\partial^2 U}{\partial Y^2}\right) \right| \qquad \textbf{Answer} \blacktriangleleft$$

and

$$\left| \frac{\partial U}{\partial X} + \frac{\partial v}{\partial Y} = 0, \right| \qquad \textbf{Answer} \blacktriangleleft$$

where

$$\left| \mathbf{R} = \frac{\rho U_0 \ell}{\mu}. \right| \qquad \textbf{Answer} \blacktriangleleft$$

Discussion

The solution of the dimensionless equations would exhibit a dependence on Reynolds number. Obviously, Reynolds number is a relevant dimensionless parameter for this type of flow.

This example demonstrates that relevant dimensionless parameters appear *naturally* when the equations describing a physical phenomenon are put in dimensionless form. This method can be used instead of the pi theorem to obtain dimensionless parameters; however, the method is limited to those phenomena for which we have appropriate equations available.

It is important to note that the boundary conditions may also introduce relevant dimensionless parameters.

You may be curious about why the particular dimensionless parameters of Table 6.1 have come to be preferred. You may also be wondering if any physical meaning can be attached to these parameters or if they are just mathematical curiosities. The key to answering both of these questions is that each dimensionless parameter can be interpreted as a ratio of *typical* values of two physical entities. These entities are usually velocities or forces or energies. Sometimes it is possible to interpret a particular dimensionless parameter in two or more ways, usually as either a ratio of typical forces or a ratio of typical energies.

The physical relevance of the various dimensionless parameters can be demonstrated by identifying and estimating the magnitude of the various forces on the fluid. If we adopt the viewpoint advocated by d'Alembert, a moving fluid particle can be imagined as being in a state dynamic equilibrium, influenced by an inertia force ($\vec{F}_I = -m\vec{a}$) as well as the actual forces due to pressure, shear stress, surface tension, and so on. We can estimate typical magnitudes of some of these forces in a flow.

We can estimate the inertia force by

$$\text{Inertia force} \sim \text{"Dynamic Pressure"} \times \text{Area}.$$

A typical value of the dynamic pressure is the value corresponding to the reference velocity, namely $\frac{1}{2}\rho V^2$. A typical value of area is the square of the reference length dimension, ℓ^2. Using these estimates, we have

$$\text{Inertia force} \sim \rho V^2 \ell^2.$$

An estimate of the pressure force can be obtained from a typical pressure difference and a typical area

$$\text{Pressure force} \sim \text{Pressure difference} \times \text{Area} \sim (\Delta p)\ell^2.$$

An estimate of the viscous shear force is

$$\text{Shear force} \sim \text{Shear stress} \times \text{Area}.$$

The shear stress is given by

$$\tau = \mu\left(\frac{\partial u}{\partial y}\right).$$

We can estimate the velocity derivative by assuming that the velocity changes from zero to the typical value V over an appropriate length ℓ:

$$\frac{\partial u}{\partial y} \approx \frac{\Delta u}{\Delta y} \sim \frac{V - 0}{\ell} = \frac{V}{\ell}.$$

Using this derivative estimate, the viscous force estimate is

$$\text{Viscous force} \sim \mu\left(\frac{V}{\ell}\right)\ell^2 = \mu V\ell.$$

Now we will form some ratios from our force estimates:

$$\frac{\text{Inertia force}}{\text{Viscous force}} \sim \frac{\rho V^2 \ell^2}{\mu V\ell} = \frac{\rho V\ell}{\mu} = \mathbf{R},$$

$$\frac{\text{Pressure force}}{\text{Inertia force}} \sim \frac{(\Delta p)\ell^2}{\frac{1}{2}\rho V^2 \ell^2} = \frac{\Delta p}{\frac{1}{2}\rho V^2} = \mathbf{C_p}.$$

The dimensionless parameters $\mathbf{R}$ and $\mathbf{C_p}$ can be interpreted as ratios of typical forces in the flow field. This procedure can be extended to other dimensionless groups.

Not all of the standard dimensionless parameters are force ratios. The Strouhal number and velocity ratio are more readily interpreted as

$$\mathbf{S} = \frac{\omega \ell}{V} = \frac{(\ell/V)}{\left(\dfrac{1}{\omega}\right)} \sim \frac{\text{Characteristic flow time}}{\text{Characteristic oscillation time}}$$

and

$$\frac{V}{U} \sim \frac{\text{Fluid velocity}}{\text{Velocity of coordinate system}}.$$

It is also possible to interpret many of the dimensionless parameters in terms of energy ratios. For example,

$$\frac{\text{Work done by pressure}}{\text{Fluid kinetic energy}} \sim \frac{\dfrac{\Delta p}{\rho}}{\dfrac{1}{2}V^2} = \frac{\Delta p}{\dfrac{1}{2}\rho V^2} = \mathbf{C_p}.$$

Table 6.2 gives interpretations of various dimensionless parameters.

We close this discussion with three notes: First, the choice of ρ, V, and L as repeating variables in the dimensional analysis was tantamount to choosing to compare all forces to the inertia force. Second,

Table 6.2 Physical interpretation of common dimensionless parameters

Parameter	Interpretation	Alternative Interpretation
R	$\dfrac{\text{Inertia force}}{\text{Viscous force}}$	$\dfrac{\text{Flow of kinetic energy}}{\text{Rate of mechanical energy loss}}$
M^2	$\dfrac{\text{Inertia force}}{\text{Elastic force resisting compression}}$	$\dfrac{\text{Kinetic energy}}{\text{Thermal energy}}$
W	$\dfrac{\text{Inertia force}}{\text{Surface tension force}}$	$\dfrac{\text{Kinetic energy}}{\text{Surface energy}}$
F^2	$\dfrac{\text{Inertia force}}{\text{Gravity force}}$	$\dfrac{\text{Kinetic energy}}{\text{Potential energy}}$
S	$\dfrac{\text{Time scale of flow}}{\text{Time scale of boundary}}$	$\dfrac{\text{Speed of boundary}}{\text{Speed of flow}}$
$\dfrac{V}{U}$	$\dfrac{\text{Speed of flow}}{\text{Speed of coordinate system}}$	$\dfrac{\text{Time scale of coordinate system}}{\text{Time scale of flow}}$
C$_D$, C$_L$	$\dfrac{\text{Resultant force on object}}{\text{Inertia force}}$	—
C$_p$	$\dfrac{\text{Pressure force}}{\text{Inertia force}}$	$\dfrac{\text{Work done by pressure}}{\text{Kinetic energy}}$
K	$\dfrac{\text{Mechanical energy loss}}{\text{Kinetic energy}}$	—

whenever a particular force (e.g., surface tension) is known to have a negligible effect on the flow when compared to the effect of other forces, notably inertia, the dimensionless parameter representative of that force can be dropped. Third, we can immediately obtain parameters representing other force ratios by multiplying or dividing the standard parameters. For example,

$$\frac{\text{Pressure force}}{\text{Viscous force}} = \frac{\text{Inertia force}}{\text{Viscous force}} \times \frac{\text{Pressure force}}{\text{Inertia force}} \sim \mathbf{R} \times \mathbf{C_p}$$

$$\sim \left(\frac{\rho V \ell}{\mu}\right) \times \left(\frac{\Delta p}{\rho V^2}\right) = \frac{(\Delta p)\ell}{\mu V}.$$

EXAMPLE 6.6   Illustrates how dimensional analysis is used to determine the form of an equation and that "customary" dimensionless parameters are not always the best choice

Consider external flow (Fig. 6.7). At very low Reynolds numbers, the drag force ($\mathscr{D}$) is expected to depend on viscous effects only. At very high Reynolds numbers, the drag force is expected to depend on inertia effects only. Determine the form of the equations for the drag on the object for both cases. What conclusions can be drawn?

Solution

Given
Flow over an object
At very low Reynolds numbers, drag depends on viscous effects only
At very high Reynolds numbers, drag depends on inertia effects only

Find
Form of:
$\mathscr{D} = f\{\text{Inertia effects}\}$ at very high $\mathbf{R}$
$\mathscr{D} = f\{\text{Viscous effects}\}$ at very low $\mathbf{R}$

Solution
Recall from the text that

$$\text{Inertia force} \sim \frac{1}{2}\rho V^2 \ell^2.$$

$$\text{Viscous force} \sim \mu V \ell$$

At very high Reynolds numbers, the dimensional relation involves ρ but not μ and is

$$\mathscr{D} = f_H\{\rho, V, \ell\}.$$

The dimensions of these parameters are

$$[\mathscr{D}] = ML/T^2, \qquad [V] = L/T,$$
$$[\rho] = M/L^3, \qquad [\ell] = L.$$

There are four dimensional parameters and three fundamental dimensions; therefore we expect one dimensionless parameter. This single parameter would have the form

$$\frac{\text{Drag force}}{\text{Intertia force}} \sim \frac{\mathcal{D}}{\frac{1}{2}\rho V^2 \ell^2}.$$

This is the drag coefficient C_D. Since there is only one dimensionless parameter,

$$C_D = \text{Constant},$$

and the drag at high Reynolds number can be easily calculated from

$$\left| \; \mathcal{D} = C_D\left(\frac{1}{2}\rho V^2 \ell^2\right) = \text{Constant}\left(\frac{1}{2}\rho V^2 \ell^2\right). \; \right|$$ **Answer** ◀

At very low Reynolds numbers, the dimensional relation involves μ but not ρ and is

$$\mathcal{D} = f_L(\mu, V, \ell).$$

The dimensions of these parameters are

$$[\mathcal{D}] = ML/T^2, \qquad [V] = L/T$$
$$[\mu] = M/LT, \qquad [\ell] = L.$$

There are four dimensional parameters and three fundamental dimensions; therefore we expect no more than one dimensionless parameter. The form of this parameter is

$$\frac{\text{Drag force}}{\text{Viscous force}} \sim \frac{\mathcal{D}}{\mu V \ell}.$$

We will call this the *viscous drag coefficient* C_D'. Since there is only one dimensionless parameter,

$$C_D' = \text{Constant},$$

and the drag at very low Reynolds number can be calculated from **Answer**

$$\left| \; \mathcal{D} = C_D'(\mu V \ell) = \text{Constant }(\mu V \ell). \; \right|$$ ◀

C_D and C_D' are related by

$$C_D = \frac{\mathcal{D}}{\frac{1}{2}\rho V^2 \ell^2} = \frac{\mathcal{D}}{\mu V \ell}\left(\frac{\mu V \ell}{\frac{1}{2}\rho V^2 \ell^2}\right)$$

Simplifying, we obtain

$$C_D = \left(\frac{\mathcal{D}}{\mu V \ell}\right)\left(\frac{2\mu}{\rho V \ell}\right), \qquad \text{or} \qquad C_D = \frac{2C_D'}{R}.$$

The "ordinary" drag coefficient is very convenient in flow with very high Reynolds number because it is constant. The ordinary drag coefficient is not as convenient as C_D' in very low Reynolds number flow because C_D' is constant and C_D depends on the exact value of the Reynolds number. We conclude:

The ordinary drag coefficient

$$C_D = \frac{\mathcal{D}}{\frac{1}{2}\rho V^2 \ell}$$

is convenient for flow with high Reynolds number, but an alternative "drag coefficient"

$$\mathbf{C'_D} \equiv \frac{\mathscr{D}}{\mu V \ell}$$

is more convenient for flow with very low Reynolds number.

Discussion

Two important questions are (1) How low is a "very low Reynolds number"? and (2) How high is a "very high Reynolds number"? In general, a very low Reynolds number is less than one (say 10^{-1}) and a very high Reynolds number is greater than about 10^4. Between Reynolds numbers of 10^{-1} and 10^4, the drag is proportional to some power(s) of velocity between one and two. At these intermediate Reynolds numbers, both $\mathbf{C_D}$ and $\mathbf{C'_D}$ depend on the exact value of Reynolds number, and neither is especially convenient. Custom prefers the use of $\mathbf{C_D}$ instead of $\mathbf{C'_D}$ at intermediate Reynolds numbers.

6.6 An Application of Dimensional Analysis: Model Testing and Similitude

Many engineering fluid mechanics problems must be solved by experimentation. A partial list of fluid mechanics problems for which there is no complete analytical solution includes aerodynamic drag on aircraft and road vehicles, surface wave resistance on ship hulls, and flow and erosion patterns in rivers and harbors. Engineering designs of airborne, land-based, and sea going vehicles, river and harbor improvements, and large dams are based on extensive experimental testing. Obviously construction and testing of several alternative designs, all built to full scale and operated under all expected conditions of wind speed, water flow, and so forth, is completely out of the question. The only possible approach is to test smaller-scale models.* Dimensional analysis provides the key to relating the data obtained from a model experiment to the expected values for the full-scale object (called the *prototype*). Dimensional analysis also allows us to determine the conditions that must be maintained in the model test if the conditions under which the prototype must operate are known.

Suppose a group of engineers is designing a two-person submarine for oceanographic research. A design has been proposed for the submarine hull. The engineers need to determine the drag on the hull at various cruising speeds to size the submarine's propulsion system. Model tests are to be carried out to study the drag. Figure 6.10 shows the proposed (prototype) hull shape and a model of the hull.† The first, almost obvious, requirement of model testing is that the model look exactly like the prototype. This is called *geometric similarity*. Every

---

* Sometimes, larger-scale models are used. This would occur if the actual device were too small to test conveniently.

† In this example, the prototype exists only on paper; it has not actually been built. Nevertheless it will prove helpful to think of the prototype as actually having been built.

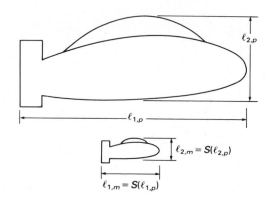

Figure 6.10 Prototype submarine and a scale model.

model length dimension is related to the corresponding prototype length dimension by a constant scale factor (S):

$$\ell_{1,m} = S\ell_{1,p},$$
$$\ell_{2,m} = S\ell_{2,m}, \qquad \text{and so on.}$$

Model areas are S^2 times corresponding prototype areas, and the model volume is S^3 times the prototype volume.

Applying what we know about drag and dimensional analysis, we can state the drag coefficient for the prototype submarine as

$$\mathbf{C_{D},_p} = F_p(\mathbf{R}_p, \mathbf{M}_p, \mathbf{W}_p, \mathbf{F}_p, k_p, \alpha_p),$$

where α, called the "angle of attack," is the angle between a reference axis on the submarine and the approach velocity vector. Since the prototype will operate in water, compressibility effects will no doubt be negligible, so a simpler relation is

$$\mathbf{C_{D},_p} = F_p(\mathbf{R}_p, \mathbf{W}_p, \mathbf{F}_p, \alpha_p). \tag{6.27}$$

When the model is tested, its drag will obey a relation of the same form:

$$\mathbf{C_{D},_m} = F_m(\mathbf{R}_m, \mathbf{W}_m, \mathbf{F}_m, \alpha_m). \tag{6.28}$$

Since explicit effects of size, fluid properties, and velocity have been removed by the use of dimensionless parameters, the functions F_p and F_m are determined solely by the shape of the prototype and model. Since the prototype and model have the same shape, F_p and F_m are the same function. The relation

$$\mathbf{C_{D}} = F(\mathbf{R}, \mathbf{W}, \mathbf{F}, \alpha) \tag{6.29}$$

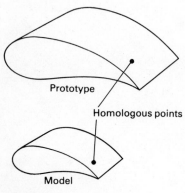

Prototype

Homologous points

Model

Figure 6.11 Corresponding points on model and prototype are called homologous points.

applies to both model and prototype. A function F determined from model tests can be used to calculate the prototype drag.

This principle applies point by point as well as for the entire function; if the model and prototype have the same Reynolds number, Weber number, Froude number, and orientation relative to the flow, they will have the same drag coefficient. They will also have the same lift coefficient. Any pair of corresponding geometric points on the model and prototype (called *homologous points;* see Fig. 6.11) will have the same pressure coefficient and skin friction coefficient.

If the model and prototype were of a high-speed airplane, then Mach and Reynolds numbers and specific heat ratio would replace Froude and Weber numbers in the independent parameter list, but the principle would be the same. In general:

If a model and its prototype have equal values of all of the relevant independent-variable dimensionless parameters, then the model and prototype have equal values of all of the dependent-variable dimensionless parameters.

This principle can be used to scale up model test results and to determine the conditions (density, velocity, etc.) necessary in a model test to "duplicate" prototype operating conditions.

EXAMPLE 6.7 Illustrates how dimensional analysis applies to model testing

A one-tenth-scale model of a soap box derby car is tested in a wind tunnel. The air speed in the wind tunnel is 70 m/s, the air drag on the model car is 240 N, and the air temperature and pressure are identical to those expected when the prototype car is racing. The important dimensionless parameters are the Reynolds number and the drag coefficient. Find the corresponding racing speed in still air and the drag on the car.

Solution

Given
One-tenth-scale model test of soap box derby car
Air speed 70 m/s
Air temperature and pressure identical to actual racing conditions

Find
Corresponding racing speed in still air
Drag on the car

Solution
The functional relation is

$$\mathbf{C_D} = f(\mathbf{R}),$$

where

$$\mathbf{R} = \frac{\rho V \ell}{\mu},$$

ρ is the air density, V the car velocity, ℓ a characteristic length of the car, and μ the air viscosity. The drag coefficient is

$$\mathbf{C_D} = \frac{\mathscr{D}}{\frac{1}{2}\rho V^2 \ell^2},$$

where $\mathscr{D}$ is the drag on the car.
For the wind tunnel test to simulate the actual (prototype) car, the Reynolds number of the model car must equal that of the actual car. Then

$$\mathbf{R}_{\text{model}} = \mathbf{R}_{\text{actual}},$$

or

$$\frac{\rho_m V_m \ell_m}{\mu_m} = \frac{\rho_p V_p \ell_p}{\mu_p}.$$

The problem statement shows

$$\rho_m = \rho_p \quad \text{and} \quad \mu_m = \mu_p.$$

Then

$$V_p = V_m \frac{\ell_m}{\ell_p}$$

$$= (70 \text{ m/s})\left(\frac{1}{10}\right),$$

$$\boxed{V_p = 7.0 \text{ m/s.}}$$

Answer

With this prototype velocity, the independent dimensionless parameter has the same value for both the model and the actual car; then the dependent dimensionless parameter has the same value. Therefore

$$\mathbf{C_{D,model}} = \mathbf{C_{D,prototype}}, \quad \text{or} \quad \frac{\mathscr{D}_m}{\frac{1}{2}\rho_m V_m^2 \ell_m^2} = \frac{\mathscr{D}_p}{\frac{1}{2}\rho_p V_p^2 \ell_p^2}.$$

The problem statement shows

$$\rho_m = \rho_p, \quad \text{so} \quad \mathscr{D}_p = \mathscr{D}_m \left(\frac{\ell_p V_p}{\ell_m V_m}\right)^2.$$

The numerical values give

$$\mathscr{D}_p = 240 \text{ N}\left[\left(\frac{10}{1}\right)\left(\frac{7 \text{ m/s}}{70 \text{ m/s}}\right)\right]^2 = 240 \text{ N}\left(\frac{70 \text{ m/s}}{70 \text{ m/s}}\right)^2,$$

or

$$\boxed{\mathscr{D}_p = 240 \text{ N.}}$$

Answer

Discussion

Even though this problem contained one dependent and one independent dimensionless parameter, other problems may contain more than one dependent and/or independent parameter. The same principle holds; each dependent dimensionless parameter has the same value for the model test and the prototype if each independent dimensionless parameter has the same value for the model test and the prototype.

The equality of model and prototype drag in this example is an interesting but nongeneral result.

The principles underlying model testing and the equality of the various dimensionless parameters between model and prototype are illustrated by the concept of *similitude*. To understand this concept, we must introduce three types of similarity: *geometric similarity, kinematic similarity,* and *dynamic similarity.* Consider the deflection of a jet of liquid by a turning vane attached to a moving cart. Figure 6.12 shows both a prototype and a model of this flow. Geometric similarity, which we have already discussed, requires that the shape of the model and the shape of the prototype be exactly the same. Formally, the requirement of *geometric similarity* is as follows:

> All lengths of the model and prototype must be in the same ratio. All corresponding angles must be equal.

Strictly speaking, the attainment of geometric similarity is essential for any meaningful model test.

Kinematic similarity requires that the ratio of all relevant velocities be the same for the model and prototype. In the moving vane example, the ratio of cart speed to fluid jet speed must be the same for the model and the prototype. Kinematic similarity also requires that the orientation of the flow with respect to the object be the same. If there are unsteady effects, the ratio of flow time scale to boundary time scale (Strouhal number) must be the same. Formally, the requirement of *kinematic similarity* is as follows:

> Ratios of fluid velocity and other relevant velocities must be the same for the model and the prototype. The orientation of the flow with respect to the object must be the same. Ratios of flow time scale and boundary time scale must be the same.

If two flows are geometrically and kinematically similar, they will have geometrically similar streamline patterns.

Dynamic similarity involves the similarity of the force polygons for the model and prototype flows. Figure 6.12 shows the polygons of the forces acting on fluid particles at homologous points in the model and prototype flows. In the specific case of the cart, inertia "forces," pressure forces, viscous forces, and surface tension forces are assumed to be present. Similarity of force polygons requires that the ratios of all

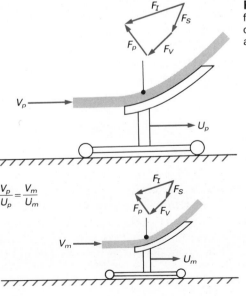

Figure 6.12 Prototype and model flow: Fluid jet deflection by a moving cart; force polygons for typical point are shown.

forces in one polygon are equal to the ratios of all forces in the other polygon:

$$\frac{\text{Inertia force in model flow}}{\text{Viscous force in model flow}} = \frac{\text{Inertia force in prototype flow}}{\text{Viscous force in prototype flow}},$$

$$\frac{\text{Pressure force in model flow}}{\text{Inertia force in model flow}} = \frac{\text{Pressure force in prototype flow}}{\text{Inertia force in prototype flow}},$$

and

$$\frac{\text{Surface tension force in model flow}}{\text{Inertia force in model flow}} = \frac{\text{Surface tension force in prototype flow}}{\text{Inertia force in prototype flow}}.$$

The requirements are equivalent to

$$\mathbf{R}_m = \mathbf{R}_p, \qquad \mathbf{C}_{\mathbf{p},m} = \mathbf{C}_{\mathbf{p},p}, \qquad \text{and} \qquad \mathbf{W}_m = \mathbf{W}_p.$$

If we have geometric and kinematic similarity, then the angles in the force polygons will be equal for the model and the prototype. Given this, one requirement for dynamic similarity can be relaxed. If

$$\mathbf{R}_m = \mathbf{R}_p \qquad \text{and} \qquad \mathbf{W}_m = \mathbf{W}_p,$$

then it must be true that

$$\mathbf{C}_{\mathbf{p},m} = \mathbf{C}_{\mathbf{p},p}.$$

This fact is stated by equations like Eqs. (6.27) and (6.28). Formally, the complete *principle of similitude* is as follows:

> If model and prototype flows have geometric similarity and kinematic similarity in all relevant parameters and if there is dynamic similarity in all relevant independent-variable parameters, then there will be complete dynamic similarity, that is, all dependent-variable parameters will be equal.

There is some debate concerning which parameters are geometric, kinematic, and dynamic. The angle of orientation of flow with respect to the object could be considered geometric rather than kinematic. Froude and Mach numbers can be considered either kinematic or dynamic parameters (see Table 6.2). It really makes little difference in the end.

Extensive and costly facilities have been constructed for regular testing of models of aircraft, spacecraft, ships, dams, harbors, and rivers. Figures 6.13 to 6.16 illustrate some models and model-testing facilities. Sometimes, special one-of-a-kind model-testing facilities are constructed to obtain information about a specific piece of equipment. Figure 6.17 illustrates a model test of this type.

The rather well developed theory of dimensional analysis and its interpretation in terms of similitude may lead you to believe that model testing is quite simple and straightforward. Unfortunately, this is not true, as the following situation demonstrates. Suppose we want to determine the drag on a large ship hull by performing tests on a smaller-scale model ship (Fig. 6.13). The drag is due to viscous shear stress and surface wave resistance, so the drag coefficient depends on Reynolds number and Froude number:

$$\mathbf{C_D} = F(\mathbf{R}, \mathbf{F}).$$

Figure 6.13 Ship model in towing tank facility (courtesy of U.S. Navy David Taylor Ship Research and Development Center).

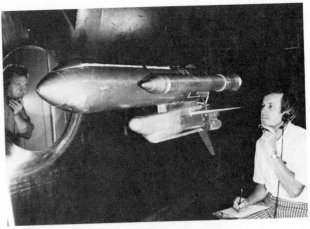

Figure 6.14 Model of the U.S. Space Shuttle in a supersonic wind tunnel (courtesy of NASA Lewis Research Center).

Figure 6.15 Hydraulic model of a dam and spillway (U.S. Army Corps of Engineers photograph).

Figure 6.16 Hydraulic model used to determine effects of improvements to a river channel (U.S. Army Corps of Engineers photograph).

Figure 6.17 Scale-model test of a power plant induced draft fan, electrostatic precipitator, and stack (reproduced from Dynatech R/D Company Report No. 1928, Figure 2.2, page 22 of "An Experimental Gas Flow Model Study on the Precipitator, Fan and Silencer for the Michigan City Plant Boiler 12 of the Northern Indiana Public Service Company," completed under the direction of the Stone & Webster Engineering Corporation and NIPSCO).

Suppose we decide to use a 1/25-scale model for the test. For similarity, we must have

$$\mathbf{F}_m = \mathbf{F}_p \quad \text{and} \quad \mathbf{R}_m = \mathbf{R}_p.$$

From the Froude number requirement,

$$\frac{V_m}{\sqrt{g_m \ell_m}} = \frac{V_p}{\sqrt{g_p \ell_p}}.$$

Since undoubtedly $g_m = g_p$, this equation requires that

$$\frac{V_m}{V_p} = \sqrt{\frac{\ell_m}{\ell_p}} = \sqrt{S}.$$

Since $S = 1/25$,

$$\frac{V_m}{V_p} = \frac{1}{5}.$$

Next we consider the Reynolds number requirement:

$$\frac{\rho_m V_m \ell_m}{\mu_m} = \frac{\rho_p V_p \ell_p}{\mu_p}.$$

The necessary kinematic viscosity ratio is

$$\frac{v_m}{v_p} = \frac{V_m}{V_p}\frac{\ell_m}{\ell_p} = \sqrt{\frac{\ell_m}{\ell_p}}\left(\frac{\ell_m}{\ell_p}\right) = \left(\frac{\ell_m}{\ell_p}\right)^{3/2} = (S)^{3/2}.$$

In this case, with $S = 1/25$, $v_m/v_p = 1/125$. Since the prototype ship will undoubtedly operate in water, we must perform the model test in a fluid with 1/125 the kinematic viscosity of water. No such fluid exists. In fact, the only practical test fluid is also water. It is not possible to simultaneously match Reynolds number and Froude number in the scale-model test.

The same type of problem arises when testing high-speed aerodynamic vehicles. In this case, both Reynolds and Mach numbers must be matched. Assuming that the fluid is an ideal gas with* $E_{v,T} = p$, the Mach number requirement is

$$\frac{\sqrt{\rho_m}\,V_m}{\sqrt{p_m}} = \frac{\sqrt{\rho_p}\,V_p}{\sqrt{p_p}}.$$

Using the ideal gas law, we have

$$\frac{p}{\rho} = RT,$$

and so

$$\frac{V_m}{\sqrt{R_m T_m}} = \frac{V_p}{\sqrt{R_p T_p}}, \quad \text{or} \quad \frac{V_m}{V_p} = \sqrt{\frac{R_m T_m}{R_p T_p}}.$$

---

* The proper elastic modulus is actually not $E_{v,T}$ but is the adiabatic modulus, equal to kp (see Chapter 10).

The Reynolds number requirement is

$$\frac{\rho_m V_m \ell_m}{\mu_m} = \frac{\rho_p V_p \ell_p}{\mu_p}.$$

The necessary viscosity ratio is

$$\frac{\mu_m}{\mu_p} = \frac{\rho_m V_m}{\rho_p V_p}\left(\frac{\ell_m}{\ell_p}\right) = \frac{p_m R_p T_p}{p_p R_m T_m}\sqrt{\frac{R_m T_m}{R_p T_p}}\left(\frac{\ell_m}{\ell_p}\right) = \frac{p_m}{p_p}\sqrt{\frac{R_p T_p}{R_m T_m}}\left(\frac{\ell_m}{\ell_p}\right).$$

The prototype gas is air. The model test gas is most likely air also, so the gas constants are equal ($R_m = R_p$). The viscosity of a gas can be varied by changing its temperature. For air, an approximate equation for viscosity ratio is

$$\frac{\mu_m}{\mu_p} \approx \left(\frac{T_m}{T_p}\right)^{0.75},$$

so

$$\frac{\ell_m}{\ell_p} = S \approx \frac{p_p}{p_m}\left(\frac{T_m}{T_p}\right)^{1.25}.$$

For a small-scale-model test (S small), the pressure in the test gas must be high and the temperature must be low. This would require prohibitively large compressor power and air cooler costs. Similar to the ship model test, a completely similar model test is impractical. In most cases when two independent dimensionless parameters must be matched, the only way to do it is to have the model be the same size as the prototype, which is not a model test at all.

If it is impractical to match two (or more) independent similarity parameters, information must be obtained from a series of model tests or a combination of model test results with theoretical calculations. Consider the case of determining aircraft drag. A series of tests may be made at very low values of Mach number (maybe even using water as the test fluid). These tests can establish the influence of Reynolds number on drag. This information can then be used as a basis for extrapolation of high Mach number (but low Reynolds number) test results to higher values of Reynolds number.

Alternatively, the low Mach number test results may be used to construct a theory that permits calculation of the drag due to viscous forces. If it is assumed that viscous and compressibility effects on drag are independent, the theory can be used to "remove" viscous drag from model test data by calculating the viscous drag for the model and subtracting it. The remaining "compressibility drag" can be converted to compressibility drag for the prototype by similarity considerations. The theory can then be used to calculate the viscous drag for the prototype. This type of procedure is also used to separate viscous and wave drag for ship model tests. Unfortunately, such a procedure is approximate at best.

There are other, less obvious, problems in model testing. One of these has to do with surface roughness. Since surface roughness is a geometric parameter, complete similarity requires that the relative roughness (ε/ℓ) be the same for a model and a prototype. In many cases, model roughness cannot be made small enough for complete similarity, and a "roughness correction" must be made. These corrections are essentially extrapolations based on a few studies of the effects of roughness.

The problems that we have discussed so far involve cases where it is impossible to properly model some effect that is actually present in the prototype flow. Another type of problem occurs when certain influences that are not relevant in the prototype flow occur in the model flow. When modeling harbors or rivers (Fig. 6.16), a very small scale factor must be used. The water depth in the scale model becomes so small that surface tension forces, entirely irrelevant in the prototype, become important in the model flow. To eliminate this effect, the model must use deeper water, and so complete similarity is violated. The data must be dealt with using a theory of dissimilar models, the consideration of which is beyond the scope of this book.

In any multiparameter model test, a hierarchy of the dimensionless parameters should be followed if it is not possible to match all of them. Table 6.3 lists such a hierarchy for most types of tests. The parameters indicated as "primary" must be matched for the given type of test. The "secondary" parameters are also important but often cannot be matched simultaneously with the primaries in a scale-model test. The secondary parameters must be addressed by extrapolation of data, theoretical calculation, or, at the minimum, they must be shown to be irrelevant.* The parameters indicated as "normally irrelevant" are irrelevant only for the particular type of test indicated.

Table 6.3 Similarity parameters and their typical importance in various types of model tests

Type of Test	Primary Similarity Parameter (Must be matched)	Secondary Similarity Parameter(s)	Normally Irrelevant Parameter(s)
Low-speed flow, no free surface (e.g., pipes,valves, and fittings, low-speed aerodynamics)	$\mathbf{R}, \alpha$	ε/ℓ	$\mathbf{W}, \mathbf{F}, \mathbf{M}, k$
Surface vessels, ships	$\mathbf{F}$	$\mathbf{R}, \varepsilon/\ell$	$\mathbf{W}, \mathbf{M}, k$
Dam spillways and runoffs, harbors, rivers	$\mathbf{F}$	$\mathbf{R}, \varepsilon/\ell$	$\mathbf{W}, \mathbf{M}, k$
High-speed flows	$\mathbf{M}, k, \alpha$	$\mathbf{R}, \varepsilon/\ell$	$\mathbf{F}, \mathbf{W}$
Fluid machinery	V/U	$\mathbf{R}, \varepsilon/\ell, \mathbf{M}$	$\mathbf{W}, \mathbf{F}, k$
Unsteady flows	$\mathbf{S}$	Various	Various

* For example, at high values of $\mathbf{R}$, drag forces are independent of the exact value of $\mathbf{R}$.

EXAMPLE 6.8 Illustrates the use of dimensionless parameters in extrapolating test data and some consequences of a multiparameter similarity law

Performance test data for a centrifugal pump is tabulated in Table E6.8a. The pump rotor diameter is 2.772 ft. The data were obtained at pump speeds of 2000 rpm and 3000 rpm. Find the head developed by the pump at a speed of 3500 rpm and a volume flow rate of 5000 gal/min. The fluid is water.

Table E6.8a Test data for a centrifugal pump

2000 rpm		3000 rpm	
Q (gal/min)	h_p (ft)	Q (gal/min)	h_p (ft)
0	115	0	259
400	122	600	280
800	122	1200	280
1200	118	1800	266
1600	112	2400	250
2000	105	3000	232
2400	96	3600	215
2800	87	4200	194
3200	78	4800	170

Solution

Given
Test data in Table E6.8a for a centrifugal pump
Pump rotor diameter 2.772 ft
Rotor rotational speeds 2000 rpm and 3000 rpm
Fluid is water
See also Fig. E6.4

Find
Head developed by the pump at 3500 rpm and volume flow rate of 5000 gal/min

Solution
Example 6.4 shows that the appropriate dimensionless performance function for a centrifugal pump is

$$\Psi = f(\phi, \mathbf{R}),$$

where

$$\Psi = \text{Head coefficient} = \frac{gh_p}{N^2 D^2},$$

$$\phi = \text{Flow coefficient} = \frac{Q}{ND^3},$$

$$\mathbf{R} = \text{Reynolds number} = \frac{ND^2}{\nu}.$$

We want to find a value of the pump head at a certain speed and flow. The head can be calculated by

$$h_p = \frac{\Psi(N^2 D^2)}{g}.$$

Since N, D, and g are known, we must find Ψ, which depends on ϕ and $\mathbf{R}$; ϕ and $\mathbf{R}$ can be calculated from the specified speed (3500 rpm), the specified flow (5000 gal/min), the pump rotor diameter, and the water viscosity. To establish the relation between Ψ, ϕ, and $\mathbf{R}$, we must use the test data of Table E6.8a. We calculate ϕ and Ψ from the test data by

$$\phi = \frac{Q}{ND^3} = \frac{Q\left(\dfrac{\text{gal}}{\text{min}}\right)\left(\dfrac{\text{ft}^3}{7.48\ \text{gal}}\right)}{N\left(\dfrac{\text{rev}}{\text{min}}\right)(2.772\ \text{ft})^3\left(2\pi\ \dfrac{\text{rad}}{\text{rev}}\right)} = \frac{Q}{1000N},$$

where Q is in gal/min and N in rpm, and

$$\Psi = \frac{gh_p}{N^2D^2} = \frac{\left(32.2\ \dfrac{\text{ft}}{\text{sec}^2}\right)h_p\left(60\ \dfrac{\text{sec}}{\text{min}}\right)^2}{N^2\left(\dfrac{\text{rev}}{\text{min}}\right)^2(2.772\ \text{ft})^2\left(2\pi\ \dfrac{\text{rad}}{\text{rev}}\right)^2} = 382\ \frac{h_p}{N^2},$$

where h_p is in feet and N in rpm.

The Reynolds number is calculated by

$$\mathbf{R} = \frac{ND^2}{\nu} = \frac{N\left(\dfrac{\text{rev}}{\text{min}}\right)(2.772\ \text{ft})^2\left(2\pi\ \dfrac{\text{rad}}{\text{rev}}\right)}{\left(1.22 \times 10^{-5}\ \dfrac{\text{ft}^2}{\text{sec}}\right)\left(\dfrac{60\ \text{sec}}{\text{min}}\right)} = (6.60 \times 10^4)N,$$

where N is in rpm and 60°F water is assumed.

Of the parameters appearing in the Reynolds number, only pump speed is different in the two tests; therefore there are only two different values of Reynolds number in the data.

Table E6.8b shows the values of the dimensionless parameters calculated from the test data. Figure E6.8 is a plot of the dimensionless parameters. The table and the plot represent the functional relation

$$\Psi = f(\phi, \mathbf{R})$$

for this particular pump.

From the plot and Table E6.8b, it is clear that Ψ depends on ϕ; however, it seems that Ψ is independent of Reynolds number. Experience shows that if the Reynolds number is high enough, flow phenomena become independent of

Table E6.8b Dimensionless parameters for Example 6.8

2000 rpm ($\mathbf{R} \approx 1.31 \times 10^8$)		3000 rpm ($\mathbf{R} \approx 1.96 \times 10^8$)	
ϕ	Ψ	ϕ	Ψ
0	0.0110	0	0.0110
0.0002	0.0117	0.0002	0.0119
0.0004	0.0117	0.0004	0.0119
0.0006	0.0113	0.0006	0.0113
0.0008	0.0107	0.0008	0.0106
0.0010	0.0100	0.0010	0.0098
0.0012	0.0092	0.0012	0.0091
0.0014	0.0083	0.0014	0.0082
0.0016	0.0075	0.0016	0.0072

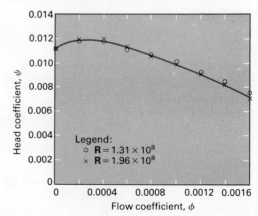

Figure E6.8 Head coefficient versus flow coefficient for two different Reynolds numbers.

the exact value of Reynolds number. The data for this particular pump indicate that the range of Reynolds number independence has been reached, and we can assume

$$\Psi = f(\phi \text{ only}),$$

at least at Reynolds numbers equal to or greater than the lowest value achieved in the test ($\mathbf{R} = 1.31 \times 10^8$).

At the new pump speed of 3500 rpm and flow of 5000 gal/min, the independent dimensionless parameters are

$$\phi = \frac{Q}{1000N} = \frac{5000 \text{ gal/min}}{1000(3500 \text{ rpm})} = 0.00143,$$

and

$$\mathbf{R} = 6.60 \times 10^4(N) = (6.60)(3500) \times 10^4 = 2.31 \times 10^8.$$

Since $\mathbf{R} > 1.31 \times 10^8$, Ψ is independent of Reynolds number. The value of Ψ at $\phi = 0.00143$ is read from Fig. E6.8 and is

$$\Psi = 0.0082.$$

The head at this operating point is

$$h_p = \frac{\Psi N^2}{382} = \frac{0.0082(3500 \text{ rpm})^2}{382},$$

$$| \ h_p = 263 \text{ ft.} \ |$$

Answer ◀

Discussion

Note that the volume flow rates in the data were such that the flow coefficients in Table E6.8b were identical; this is normally not the case. If it is not, then a plot of Ψ versus ϕ such as Fig. E6.8 is useful.

Strictly speaking, this is not a model test because the same pump was considered throughout. We might think of the low-speed pump as a model of the high-speed pump. At any rate, this example demonstrates the principles of similarity.

If the speed of the pump is reduced below 2000 rpm, the effects of Reynolds number may become important. If this occurs, the simple dependence of Ψ on ϕ only will not be true. This would typically occur if $\mathbf{R}$ drops below about 10^5.

Approaches similar to the approach used in this problem would be used in the following situations:

- Finding the rotor diameter of a geometrically similar pump to deliver a given flow rate for a given pump speed,
- Finding a substitute liquid to test a pump when Reynolds number effects must be considered,
- Finding a pump speed for a geometrically similar pump to deliver a specific flow rate at a specific head, and many others.

You might want to try solving the three cases listed above.

Problems

Lower-Order Application

1. Find appropriate dimensionless parameters for the flow over the weir in Example 4.6.

2. Experiments are conducted on a washing machine agitator. The relevant dimensional parameters are the driving torque $\mathcal{T}$, the oscillation frequency f, the angular velocity ω, the number of paddles N, the

paddle height H, and the paddle width w. Specify the relevant fluid properties and find the appropriate dimensionless parameters.

3. The input power $\dot{W}$ to a large industrial fan depends on the fan impeller diameter D, fluid viscosity μ, fluid density ρ, blade rotational speed ω, and fan efficiency η. Find appropriate dimensionless parameters.

4. The magnitude of a pressure pulse that travels through a liquid due to the quick closure of a valve depends on the isothermal bulk modulus of elasticity, the density, the fluid velocity prior to the closure of the valve, and the speed of sound in the fluid. Find the dimensionless groups.

5. Find appropriate dimensionless parameters for the jet pump of Fig. P5.9. The relevant dimensional parameters are the diameters D_0 and D_p, the velocities V_0, V_p, and V_s, and the pressures p_1 and p_2. Don't forget to include relevant fluid properties.

6. Find appropriate dimensionless parameters for the maximum static deflection δ of a simply supported beam under the load P. The other relevant dimensional parameters are the beam dimensions of length ℓ, width w, and thickness h, and the modulus of elasticity E.

7. Find appropriate dimensionless parameters for the period τ of transverse vibration of a turbine rotor of mass m connected to a shaft of stiffness k (N/m) and length ℓ (m). Other relevant dimensional parameters are the eccentricity $\varepsilon(m)$ of the center of mass of the rotor, the speed of rotation N (rpm) of the shaft, and the amplitude A (m) of the vibration.

8. Consider the force F (N) on an electrical conductor carrying a current I (amps). A length ℓ (m) of the conductor experiences a magnetic field of flux density B (webers/m^2 or N/amp·m). Find the appropriate dimensionless parameter(s).

9. Consider the radial air flow between the two parallel discs in Problem 35 or 36 in Chapter 5. The upper disc is fixed, and the lower disc of weight W is movable. Other relevant dimensional parameters are the air flow rate Q, the disc radius R, and the disc spacing h. Find appropriate dimensionless parameter(s).

10. Assume that the lower disc in Problem 35 or 36 in Chapter 5 is fixed and the disc spacing $(h) \ll$ disc radius (R). In this case the fluid viscosity μ and density become relevant. Find appropriate dimensionless parameter(s).

11. The volume flow rate Q flowing out of the tank in Example 4.2 depends on the diameters D and d, the gravitational acceleration g, and the water height $h(t)$. Find the dimensionless functional relation.

12. Consider the standpipe system of Example 7.2 and Fig. E7.2. Determine the relevant dimensional parameters. Find an appropriate dimensionless functional relation and compare with the result of Example 7.2.

13. Seawater tests are run on a 1/50-scale-model submarine traveling on the surface and also when deeply submerged. Specify appropriate dimensionless parameter(s) for each situation.

14. Consider model testing of a hydrofoil. Is it practical to satisfy both the Reynolds number and the Froude number for the hydrofoil when it is operating near the water surface? Support your decision.

15. Consider a vapor bubble rising in a liquid. The relevant dimensional parameters are the liquid density ρ_f, vapor density ρ_g, bubble velocity V, bubble diameter d, surface tension σ, and liquid viscosity μ_f. Find appropriate dimensionless parameters.

16. Determine the relevant dimensional parameters for the journal bearing in Fig. P4.27 and find appropriate dimensionless parameters.

17. Determine the relevant dimensional parameters for the viscosity motor/pump in Fig. P4.29 and find appropriate dimensionless parameters.

18. Perform a dimensional analysis for the clepsydra for Fig. P4.15.

19. The relevant dimensional parameters for a centrifugal compressor delivering an ideal gas are the absolute inlet pressure p_i, the absolute discharge pressure p_d, the absolute inlet temperature T_i, the absolute discharge temperature T_d, the specific gas constant R, the dimensionless specific heat ratio k (see Chapter 10), the inlet gas density ρ, the inlet gas velocity V, the impeller rotational speed N, the impeller diameter D, the inlet volume flow rate Q_i, the discharge volume flow rate Q_d, the input power $\dot{W}$ to the gas, and the gas specific heat c_p. Perform a dimensional analysis on the compressor.

20. The speed of deep ocean waves depends on the wave length. Find appropriate dimensionless parameters.

21. The frequency of a vertically oscillating buoy depends on the gravitational acceleration g, the buoy density ρ_b, the water density ρ_w, and the volume V of the buoy. Find the appropriate functional relation in dimensionless form.

22. A screw propeller has the following relevant dimensional parameters: axial thrust F, propeller diameter D, fluid kinematic viscosity ν, fluid density ρ, gravitational acceleration g, advance velocity V, and rotational speed N. Find appropriate dimensionless parameters to present the data taken in a test.

23. At very low velocities, the pressure drop Δp_d in a horizontal pipe of length L depends on the fluid viscosity μ, the volume flow rate Q, and the pipe inside diameter d. Find appropriate dimensionless parameter(s).

24. Develop the Froude number by starting with expressions for the inertia and gravitational forces.

25. Develop the Weber number by starting with expressions for the inertia and surface tension forces.

26. Consider the steady flow of water in the sewer pipe of Example 11.2. The relevant dimensional parameters are the pipe diameter D, the water depth d, the volume flow rate Q, the average fluid velocity V, the hydraulic radius R_h, the slope S_0 of the pipe in degrees, the fluid kinematic viscosity v, the pipe absolute roughness ε (in feet), and the gravitational acceleration g. Find the appropriate dimensionless parameters.

27. Find appropriate dimensionless parameters for the oscillatory motion of a mass m suspended by a spring with damping. The other relevant dimensional parameters are spring stiffness k in N/m, the damping coefficient c in N·s/m, the period of oscillation τ, and the amplitude of oscillation A.

28. A concentric cylinder viscometer to measure fluid absolute viscosity was analyzed in Example 1.3. The viscometer had a height H, and the gap width h was quite small compared to R_1 and R_2. The fluid in the viscometer was Newtonian with constant viscosity μ. A torque $\mathscr{T}$ applied to the inner cylinder caused it to rotate with angular velocity ω. The outer cylinder was fixed. Determine the dimensionless parameters for this viscometer.

29. Using the four-step procedure outlined in this chapter, use the dimensional parameters density ρ, reference velocity U_0, reference length ℓ, and the fluid viscosity v to develop the dimensionless parameter appropriate to the Prandtl boundary layer equations for the flow over a flat surface.

30. The relevant dimensional parameters for the rectangular weir in Example 4.6 are the volume flow rate Q, the width L, the water heights H, h, and P, the velocity V_j, the gravitational acceleration g, and the volume flow rate Q. Find the dimensionless functional relation.

31. The power output $\dot{W}$ of a hydraulic turbine depends on the fluid density ρ, the height h of the water surface above the turbine, the gravitational acceleration g, the volume flow rate Q, the turbine wheel angular velocity ω, and the turbine efficiency η. Develop the dimensionless functional relation.

32. Discuss the physical significance of the ratio of the Froude number squared to the Weber number.

33. The basic equation that describes the motion of the fluid above a large oscillating flat plate is
$$\frac{\partial u}{\partial t} = \mu\left(\frac{\partial^2 u}{\partial y^2}\right),$$
where u is the fluid velocity component parallel to the plate, t is time, y is the spatial coordinate perpendicular to the plate, and μ is the fluid viscosity. Find appropriate dimensionless parameters.

34. The appropriate equation for fully developed flow in a circular pipe of radius R is
$$\frac{\partial^2 u}{\partial r^2} + \frac{1}{r}\left(\frac{\partial u}{\partial r}\right) = \frac{1}{\mu}\left(\frac{dp}{dx}\right).$$
Find appropriate dimensionless parameters.

35. A 1/50-scale model of a boat's propeller is tested in a towing tank at 1000 rpm and requires a power input of 0.0031 hp. For negligible Reynolds number effects, find the corresponding prototype values.

36. A 1/40-scale model of a dam spillway is to be tested using Froude number scaling. The prototype flow and velocity are 180 m³/s and 4.0 m/s, respectively. Find the corresponding model flow and velocity.

37. A dam spillway measures 40 ft long and has a fluid velocity of 10 ft/sec. Considering Weber number effects as minor, find the corresponding model fluid velocity for a model length of 5 ft.

38. A 1/10-scale model of an automobile is tested in a wind tunnel under dynamically similar conditions. The prototype speed is 75 km/hr and the model drag is 300 N. Find the corresponding prototype drag and the prototype horsepower required to overcome this drag.

39. A torpedo 12 ft long has a velocity of 50 ft/sec. Find the corresponding velocity for a 1/10-scale model.

40. Assume the pump performance characteristics shown in Fig. E7.12(b) are for a pump running at 1200 rpm. Assuming that the pump head h_p, the pump impeller diameter D, and the flow rate Q through the pump are the other relevant parameters, find the pump performance characteristics at speeds of 1800 rpm and 2400 rpm. Neglect viscous effects.

41. Air at 70°F and 100 ft³/min and 14.7 psia flows through a $2\frac{1}{2}$-in. schedule 40 steel pipe. A model test is to be run using 60°F water with a maximum flow rate of 10 gal/min. What is the standard size steel pipe that should be used for dynamic similarity?

42. A 1/5-scale model of an automobile is tested in water. What should be the velocity at which the car is towed through the water to ensure dynamic similarity between the model and the prototype traveling in air at 55 mph?

43. A ship's propeller operates close to the water surface. A particular ship measuring 200 m long will have a propeller diameter of 3.0 m that rotates at 100 Hz. A force of 100 N is required to tow a 1/200-scale model of the ship through a water tank. What should be the model's propeller speed? What is the drag force on the prototype ship?

44. A 1/200-scale model is made of a spillway. Assume that both the model and prototype flows are weakly dependent on the Reynolds number. The fluid velocity and flow rate at a particular point in the model spillway were measured to be 2.0 m/s and 8.0 m³/s. Find the prototype values at the corresponding location of the spillway.

45. We wish to model a 150-ft-long airship traveling at 120 mph in the standard atmosphere at a 2000-ft altitude. The model is to be towed through 60°F water at 2 ft/sec. Find the necessary length of the model. What precautions should we take in the water test?

46. A 1/10-scale model of an airplane is tested in a wind tunnel at 70°F and 14.40 psia. The model test results are shown in the following chart.

Velocity (mph)	0	50	100	150	200
Drag (lb)	0	5	21	46	85

Find the corresponding airplane velocities and drags if only fluid compressibility is important and the airplane is flying in the U.S. Standard Atmosphere at 30,000 ft. Assume the air is an ideal gas.

47. Can the corresponding velocities and drags be found in Problem 46 if both compressibility and viscous effects were considered equally important? Support your decision.

48. A 1/400-scale model of a beach is made to study tides. What length of time in the model would correspond to one hour in the prototype?

49. We wish to test a 1/20-scale model of an airplane at the same Reynolds number as the prototype. The air in both cases is at the same temperature and pressure. Is the wind tunnel velocity practical for an airplane speed of 500 mph?

50. A 1/20-scale model of a spillway measures 2 ft high and 1 ft wide. The scale model flow rate is 1.0 ft³/sec under a head of 0.50 ft. What is the prototype flow rate for a 400-ft-long spillway?

51. A company manufactures geometrically similar airplane propellers up to 4.0 m in diameter. Wind tunnel tests are run on geometrically similar propellers up to 0.33 m in diameter. In the test of a 0.33-m model of a 4.0-m propeller, the air velocity was 50 m/s, the propeller rotational speed was 2000 Hz, the thrust 100 N, and the propeller input torque 20 N·m. Find the corresponding prototype rotational speed, thrust, and input torque for an airplane speed of 100 m/s.

52. Calculate the model wind velocity to provide dynamic similarity between a suspension bridge experiencing 100-mph winds and a 1/50-scale model. Both the bridge and the model have air flows of the same temperature and pressure.

53. Two geometrically similar reciprocating air compressors operate side by side. One is one-half the size of the other ($S = 1/2$). The smaller air compressor is operating at twice the reciprocating speed of the larger one. Find the ratios of their volume flow rates and absolute discharge pressures.

54. Two irrigation canals have the same Reynolds number and the same Froude number. One canal is two-thirds the size of the other. Both canals experience atmospheric pressure. What are acceptable ratios of their absolute temperatures and absolute viscosities?

55. It is intended to pump 60°F hydrogen through a 2-ft-inside-diameter, 6000-ft-long pipeline. To determine information about the flow in this pipeline, tests will be conducted in a geometrically similar pipe with an inside diameter of 0.02 ft. Do the following:
a) Consider the physical quantities as mass density ρ, pressure drop Δp, average fluid velocity V, pipe inside diameter D, and kinematic viscosity v. Find the dimensionless groups pertinent to this problem.
b) If water is used in the smaller pipe, find the temperature of the water (by first determining the water kinematic viscosity) so that dynamic similarity exists between the tests run in the smaller pipe and the actual flow conditions in the larger pipe. The average fluid velocity is the same in both the smaller pipe and the larger pipe.

56. A torpedo 3 m long is expected to travel at a speed of 7 m/s. A model of the torpedo is tested in a towing tank at a maximum speed of 30 m/s. What is the minimum length of the scale model?

57. Is it more important to match the relative roughness or the Reynolds number when trying to obtain dynamic similarity of a flow with complete turbulence? Explain.

58. We wish to run a wind tunnel model test of a rocket capable of speeds up to 2000 mph. Is it more important to match Mach number or Reynolds number?

59. A 1/20-scale model of a 20-ft-diameter fan is tested in a wind tunnel at 70°F and an inlet pressure of 14.40 psia. The model fan has an input power of 1.24 hp and a flow rate of 220 ft³/min while rotating at 1800 rpm. Find the corresponding input power, flow rate, and rotational speed of the prototype fan.

60. Lubricating oil at 100°C flows through a 0.25-m-inside-diameter pipe at the rate of 0.74 m/s. What must be the flow rate of 15°C water through the same pipe for dynamic similarity of the two flows?

61. A 1/5-scale model of a centrifugal pump runs at 900 rpm and has an input power of 2.5 hp, an impeller diameter of 0.5 ft, a water flow rate of 1200 gal/min, and a head rise of 20 ft. Find the flow rate Q, head rise h_p, and input power $\dot{W}$ for the prototype pump, which runs at 450 rpm.

62. Calculate the air velocity that must be used with a 1/50-scale model to simulate a submarine traveling at 25 knots in seawater.

63. It is desired to use a 1/50-scale model to study the air drag characteristics of a large tractor-trailer traveling at 55 mph. Suggest a practical fluid and fluid velocity for the model test.

64. An engineer wants to model the flight of an airplane flying at 30,000 ft where the air is at −58°F and 4.4 psia. The engineer tests a 1/20-scale model in a wind tunnel where the air is at 70°F and 14.40 psia. What wind tunnel air velocity is needed to simulate an airplane speed of 500 mph?

65. If the test of Problem 64 showed a drag force at 120 lb, what would be the drag force on the airplane?

66. A breakwater is a wall built around a harbor so the incoming waves dissipate their energy against it. The significant dimensionless parameters are the Froude number **F** and the Reynolds number **R**. A particular breakwater measuring 450 m long and 20 m deep is hit by waves 5 m high and velocities up to 30 m/s. In a 1/100-scale model of the breakwater, the wave height and velocity can be controlled. Can similarity be obtained using water for the model test?

67. An oil tanker measures 700 ft long and travels the ocean at 12 knots. Find the speed at which a 7-ft model of the tanker must be towed in 60°F freshwater to study the wave resistance on the tanker.

68. A model airplane is 1/25 the size of a jet fighter plane. The model is to be tested in a wind tunnel at the same speeds as the jet fighter. The wind tunnel's air temperature and viscosity are the same as those encountered by the jet fighter. What must the air pressure in the wind tunnel be?

69. A torpedo measures 6 m long and travels at 35 knots below the ocean surface. What must be the length of a model torpedo being tested in 15°C air?

70. The flow of 500 gal/min of 68°F benzene through a 14-in. schedule 80, nominal-diameter steel pipe is modeled by the flow of 60°F water through a 1-in. schedule 40 steel pipe. Find the flow rate through the 1-in. steel pipe.

71. Tests are run to determine the torque required on the inner of the two cylinders in Fig. P4.25. The results for $R_1 = 10$ cm and $R_2 = 15$ cm are shown in the following chart.

Fluid	Viscosity (N·s/m²)	Rotational Speed (Hz)	Torque/length (N·m/m)
Lubricating oil			
20°C	1.31×10^{-2}	10.0	1.29
40°C	6.81×10^{-3}	10.0	6.72×10^{-1}
60°C	4.18×10^{-3}	10.0	4.12×10^{-1}
Water			
10°C	1.31×10^{-3}	5.0	6.46×10^{-2}
20°C	1.00×10^{-3}	5.0	4.93×10^{-2}

Find the torque required to rotate the inner cylinder at 10 Hz with SAE 20W oil at −18°C. Are all the data in this table necessary? Support your answer.

72. An airplane travels at 160 mph in air at 14.7 psia and 0°F. A 1/10-scale model of the airplane is tested in a wind tunnel at 70°F. What must be the air pressure and velocity in the wind tunnel so that both the Reynolds number and the Mach number are satisfied?

73. The following data were taken for the flow of 60°C benzene in a 120-cm-inside-diameter, 120-m-long, horizontal steel pipe.

Pressure drop (kPa)	52.5	210	840	3350
Flow rate (m³/s)	10.0	20.0	40.0	80.0

What would be the pressure drop in the same pipe for the flow of 0.05 m³/s of 0°C lubricating oil?

74. It is desired to know the drag $\mathscr{D}$ of 20°C kerosene flowing at 10.5 m/s over an object. For safety reasons, tests are to be run using air or water. The maximum available air and water velocities are 25 m/s and 2.0 m/s, respectively, and both can be heated to 90°C or cooled to 0°C. Which fluid, if any, can be used to determine the drag $\mathscr{D}$?

75. A very small needle valve is used to control the flow of air in an 1/8-in. air line. The valve has a pressure drop of 4.0 psi at a flow rate of 0.005 ft³/sec of 60°F air. Tests are performed on a large, geometrically similar valve and the results are used to predict the performance of the smaller valve. How many times larger can the model valve be if 60°F water is used in the test and the water flow rate is limited to 70 gal/min?

76. We wish to know the drag on an airship traveling at 60 mph in air at 60°F and 14.7 psia. We perform a test on a 1/100-scale model in 60°F water at 14.7 psia. What is the necessary water velocity to simulate the 60 mph speed in air?

77. A centrifugal pump operates at 300 Hz to deliver 20°C lubricating oil. A 1/10-size, geometrically similar pump delivering 15°C water is used to model the larger pump. How fast should the smaller pump run?

78. Consider the two concentric cylinders in Fig. P4.25. The relevant dimensional parameters are the radii R_1 and R_2, the torque $\mathcal{T}$ applied to the inner cylinder, the rotational speed ω, the oil viscosity μ, and the length ℓ of the inner cylinder. Show (a) how the torque is changed if ω is halved, (b) how the torque is changed if R_1 and R_2 are both halved, and (c) how the torque is changed if the temperature of the lubricating oil is raised from 20°C to 60°C.

Higher-Order Application

79. A ship measures 250 m long and has a wetted area of 8000 m². A 1/100-scale model is tested in a towing tank with the prototype fluid, and the results are tabulated below. Find the prototype drag at 7.5 m/s and 12.0 m/s. The Froude number is the same for the model and the ship.

| Model velocity (m/s) | 0.57 | 1.02 | 1.40 |
| Model drag (N) | 0.50 | 1.02 | 1.65 |

80. Consider the drag on a ship where the appropriate dimensionless parameters are the Froude number and the Reynolds number. Experimental testing on the model at various Reynolds numbers produced the data tabulated below for the drag coefficient. Estimate the drag coefficients C_D for the prototype at a Reynolds number of 5×10^8. The Froude number is the same for the model and the prototype.

R	4.4×10^4	1×10^5	4×10^5	1×10^6
C_D	0.44	0.55	0.61	0.67

R	4×10^6	1×10^7	2×10^7	6×10^7
C_D	0.69	0.72	0.72	0.72

81. Consider the flight of a rocket at atmospheric pressure and 20°C where the appropriate dimensionless parameters are the Reynolds number and the Mach number. A model is 1/10 the size of the prototype. Model experiments are also to be performed at atmospheric pressure. Suggest appropriate fluids for the model test for equality of both the Reynolds number and the Mach number.

82. Assume the fluids for the model testing in Problem 81 are unacceptable (too expensive or potentially dangerous) and the model testing is performed in air at atmospheric pressure. The test results for the drag coefficient on the rocket are tabulated below. Find the drag coefficient C_D for the rocket at a Reynolds number of 2×10^9. Both the model and the rocket have the same Mach number.

R	4.2×10^5	4.2×10^6	4×10^7	8×10^7	1.8×10^8
C_D	0.53	0.34	0.27	0.26	0.25

Comparison

83. A straw of inside diameter d is placed in a pan of liquid of specific gravity γ, as shown in Fig. E6.1. The water rises a height h in the straw due to surface tension σ between the water and the straw and makes a contact angle θ at the water-straw interface. Start with the dimensional parameters h, d, γ, and σ and develop the appropriate dimensionless parameters describing the phenomena. Use both the $[F, L, T, \theta]$ system and the $[M, L, T, \theta]$ system. Note and explain the differences that occur in the two methods.

References

1. Kline, S. J., *Similitude and Approximation Theory*, McGraw-Hill, New York, 1965.
2. Buckingham, E., "On Physically Similar Systems: Illustrations of the Use of Dimensional Equations", *Phys. Rev.*, vol. 4, no. 4, pp. 345–376, 1914.
3. Bridgman, P. W., *Dimensional Analysis* (rev. ed.), Yale University Press, New Haven, 1931.
4. Langhaar, H. L., *Dimensional Analysis and Theory of Models*, Wiley, New York, 1951.
5. Ipsen, E. C., *Units, Dimensions, and Dimensionless Numbers*, McGraw-Hill, New York, 1960.

Applications of Flow Analysis: Steady Incompressible Flow in Pipes and Ducts

Now that we are familiar with the basic tools of flow analysis (differential and control volume formulations and dimensional analysis), we can consider some specific applications. This chapter deals with incompressible flow in closed conduits. If a conduit has a circular cross section of constant area, it is called a pipe. Constant-area conduits with other cross sections (the most common being rectangular and annular) are called ducts. We will consider only cases where the flow is steady and the conduit is running full.*

Fluid transport systems composed of pipes or ducts have many practical applications. Chemical plants and oil refineries seem to be a maze of pipes. Power plants contain many pipes and ducts for transporting fluids involved in the energy-conversion process. City water supply and waste-water removal systems consist of many kilometers of pipe. Many machines are controlled by hydraulic systems with control fluid transported in hoses or tubes. Chemical, civil, and mechanical engineers are often involved in designing, maintaining, and operating these and other types of flow systems.

Piping systems are constructed by piecing together components like those shown in Fig. 7.1. These components transport the fluid, change its direction, control its flow rate, divide it up, speed it up, or slow it down. In addition to conveying fluid, most of the components shown are responsible for a mechanical energy loss. Only the pump does not contribute a net loss† but instead increases the fluid's mechanical energy. Analysis of pipe and duct systems consists of relating flow variables, such as energy loss and flow rate, to pipe/duct system parameters such as size, shape, length, and number and location of fittings. Other important items include evaluating reaction forces that result from fluid momentum changes. This evaluation can be accomplished by applying

---

* Partially full conduits are open channels and must be dealt with using the methods described in Chapter 11.

† There are losses even in a pump. The net increase of the fluid's mechanical energy is less than the work done on the fluid; however, the *net* change of the fluid's mechanical energy is positive.

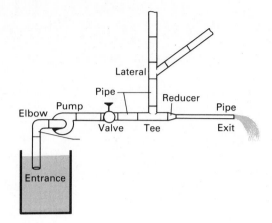

Figure 7.1 Piping system constructed by piecing together various components.

the linear or angular momentum theorem after fluid velocities and pressures have been established.

In this chapter we consider flow in straight pipes and ducts, in fittings and valves, and in complete systems. We begin with flow in straight pipes and ducts.

7.1 Regimes of Flow in a Straight Pipe or Duct

Before we can discuss or analyze the flow in a particular pipe, we must know if the flow is *laminar* or *turbulent* and if it is *developing* or *fully developed*.

7.1.1 Laminar Flow and Turbulent Flow

The nature of laminar and turbulent flows was introduced in Section 3.5.2. If the flow in a pipe is laminar, the fluid moves along smooth streamlines. The fluid velocity at any point in the pipe is virtually constant in time. If the flow is turbulent, a rather violent mixing of the fluid occurs, and the fluid velocity at a point varies randomly with time (Figs. 3.31 and 3.34).

The differences between laminar and turbulent pipe flows were first clarified by Osborne Reynolds in 1883. Reynolds performed a series of experiments in which dye was injected into water flowing in a glass pipe. Figure 7.2 illustrates Reynolds's observations. At low flow rates, the dye streak remained smooth and regular as the dye flowed downstream. At higher flow rates, the dye seemed to explode, mixing rapidly throughout the entire pipe. A modern spark photograph of the mixing dye would reveal a very complex flow pattern, not discernible by Reynolds's experiments.

Reynolds's experiments and his analytical work showed that the nature of pipe flow depends on the Reynolds number,

$$\mathbf{R} = \frac{\rho \bar{V} D}{\mu},$$

in which $\bar{V}$ is the average velocity and D is the pipe inside diameter.* Reynolds found that if the value of **R** was below about 2000, the flow was always laminar, while at higher Reynolds numbers the flow was turbulent. The exact value of Reynolds number that defined the boundary between laminar and turbulent flow depended on the experimental conditions. If the water in the inlet reservoir was very still and there was no vibration in the equipment, Reynolds found that laminar flow could be maintained at Reynolds numbers much higher than 2000. Reynolds also found that if he began at a high value of **R** with turbulent flow in the pipe and then decreased **R**, the flow became laminar at a value of **R** of about 2000.

Although it is possible to obtain laminar flow at higher Reynolds numbers in the laboratory, most engineering situations do not qualify as "undisturbed." In engineering practice, the upper limit of Reynolds number for laminar flow in a pipe is usually taken to be

$$\mathbf{R} \approx 2300 = \text{Maximum for laminar flow in a pipe}$$

For Reynolds numbers between 2300 and about 4000, the flow is unpredictable, sometimes pulsating or switching back and forth between laminar and turbulent. If the Reynolds number is above 4000, the flow is usually turbulent:

$$\mathbf{R} \approx 4000 = \text{Minimum for stable turbulent flow in a pipe.}$$

The randomly fluctuating velocities and violent mixing action of turbulent flow make it significantly different from laminar flow. In turbulent flow, we must deal with time average velocities, defined by

$$\langle V \rangle \equiv \int_0^T V \, dt. \tag{7.1}$$

(See Fig. 7.3.) Note carefully that $\langle V \rangle$ is a *time* average of the instantaneous velocity at a point and is different from $\bar{V}$, the *area* average. The

Figure 7.2 Schematic of Reynolds's observations of laminar and turbulent pipe flow.

Figure 7.3 Illustration of the definition of time average velocity in turbulent flow.

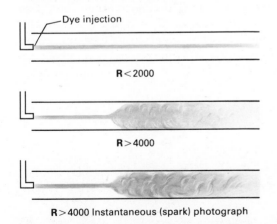

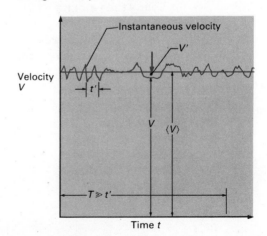

* Note that for steady flow of an incompressible, constant-viscosity fluid, **R** is constant all along the pipe. In his experiments, Reynolds changed **R** by changing $\bar{V}$, that is, by changing the flow rate.

$\langle \ \rangle$ notation is cumbersome and will be used only when it is necessary to distinguish between instantaneous and time average velocities (or other properties) for turbulent flow. The instantaneous velocity is related to the time average according to

$$V = \langle V \rangle + V', \qquad (7.2)$$

where V' is the fluctuating component of the velocity. By definition,

$$\langle V' \rangle = 0. \qquad (7.3)$$

This relation also holds for the velocity components in a two- or three-directional flow:

$$u = \langle u \rangle + u', \qquad (7.4a)$$

$$v = \langle v \rangle + v', \qquad (7.4b)$$

$$w = \langle w \rangle + w'. \qquad (7.4c)$$

Most engineering considerations of turbulent flow are concerned only with evaluation of the time average velocity. The time average velocity characterizes the net flow of mass, momentum, and energy. When we refer to the fluid velocity in turbulent flow, we usually mean the time average velocity and simply write V for local time average velocity and $\bar{V}$ for area-and-time average velocity.

If we limit consideration to the time average flow, we must account for additional *apparent stresses* in the fluid due to turbulent motion. The existence of these apparent stresses can be demonstrated by simple physical arguments. In turbulent flow, macroscopic "lumps" of fluid are thrown about by the turbulent motion. Figure 7.4 shows two such lumps. The time average velocity distribution is also shown. The lumps are moving with the time average velocities $\langle u \rangle$ and $\langle u \rangle + \delta\langle u \rangle$ appropriate to their locations. Now assume that the two lumps exchange places due to turbulent mixing. The slower lump arrives at its new location with velocity $\langle u \rangle$, and the faster lump arrives at its new location with velocity $\langle u \rangle + \delta\langle u \rangle$. Now neither lump has the velocity appropriate for its location. At $y + \delta y$, the velocity perturbation is $u' \approx -\delta\langle u \rangle$, and at y the velocity perturbation is $u' \approx +\delta\langle u \rangle$. The time average velocities at y and $y + \delta y$ are still $\langle u \rangle$ and $\langle u \rangle + \delta\langle u \rangle$. The interchange of the two lumps has resulted in an increase of momentum at location y and a decrease in momentum at $y + \delta y$. Since this type of interchange occurs continuously in turbulent flow, there is a net rate of momentum transfer

Figure 7.4 Two lumps of fluid with different mean velocities exchange places in a turbulent flow.

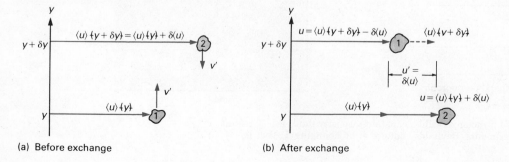

(a) Before exchange

(b) After exchange

between y and $y + \delta y$ due to the velocity fluctuations. From the time average point of view, this momentum transfer is equivalent to a force or, on a per unit area basis, a stress.* The momentum represented by u' is transported between y and $y + \delta y$ by the velocity perturbation v'. The momentum flux (per unit area) is $\rho u'v'$. If we treat this momentum flux as a stress, according to d'Alembert's principle, its instantaneous value is

$$\tau_{\text{app}} = -\rho u'v',$$

where τ_{app} is the instantaneous apparent stress due to turbulent momentum transport.[†] The time average of τ_{app} is the apparent stress that we must include in the mechanics of the time average flow

$$\langle \tau_{\text{app}} \rangle = -\rho \langle u'v' \rangle. \tag{7.5}$$

This analysis shows that the apparent stress is related to the gradient of the time average velocity since we assumed

$$u' \approx \pm\delta\langle u \rangle \approx \pm\frac{\partial\langle u \rangle}{\partial y}\,\delta y. \tag{7.6}$$

The apparent stresses are also called *Reynolds stresses* or *turbulent stresses*. When we consider the mechanics of turbulent flow, it is convenient to think of the fluid as experiencing extra stresses due to turbulence. These extra stresses *cannot* be evaluated from the viscosity and velocity gradient. To date, the only way to evaluate these stresses has been the use of ad hoc models based on experimental measurements.

The concept of momentum exchange by macroscopic lumps of fluid is the basis of the mixing length model, one of many approximate models for the apparent stresses. The mixing length model assumes that the velocity fluctuations can be calculated by

$$u' \approx -\ell\left(\frac{\partial\langle u \rangle}{\partial y}\right) \quad \text{and} \quad v' \approx -u' \approx \ell\left(\frac{\partial\langle u \rangle}{\partial y}\right),$$

so

$$\langle \tau_{\text{app}} \rangle = -\rho \langle u'v' \rangle = \rho\ell^2\left(\frac{\partial\langle u \rangle}{\partial y}\right)^2. \tag{7.7}$$

The parameter ℓ, representing the typical lateral distance moved by a lump of fluid, is called the *mixing length*. The mixing length corresponds to δy in Eq. (7.6) and must be determined by experiment.

In turbulent flow, apparent stresses may be many orders of magnitude larger than laminar (viscous) stresses. It is instructive to interpret this in terms of what we know about the Reynolds number. In Section 6.5, we pointed out that the Reynolds number represents the ratio of inertia and viscous forces for a typical fluid particle. If there are no velocity gradients (i.e., no fluid deformation), then there are no stresses

---

* The strong parallel between this discussion and the discussion of viscosity in a gas presented in Section 1.3.4 is no accident. The salient difference is that viscosity is due to molecular momentum transport and is a fluid property, while turbulent momentum transport is a *flow characteristic* rather than a *fluid characteristic*.

[†] Note that positive values of v' (upward movement) generate negative values of u'.

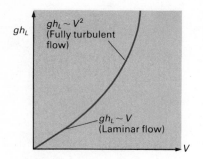

Figure 7.5 Variation of mechanical energy loss with velocity for flow of a specific fluid (ρ, μ fixed) in a specific pipe (D, L fixed).

and the Reynolds number is irrelevant. If there are velocity gradients, then experiments reveal that the flow is turbulent at high Reynolds numbers. In this case, the inertia forces in the Reynolds number are due to turbulent stresses. The high value of **R** then means

$$\langle \tau_{\text{app}} \rangle \gg \langle \tau_{\text{viscous}} \rangle.$$

The high stress levels in turbulent flow are due to the intense mixing of turbulence. These high stress levels cause much higher mechanical energy loss in turbulent pipe flow than in laminar pipe flow. Figure 7.5 shows typical trends of experimental measurements of energy loss in pipe flow. For a particular pipe of given diameter and for a particular fluid, increasing velocity corresponds to increasing Reynolds number. Note that at low velocities (low **R**), energy loss varies as

$$gh_L \sim V,$$

which indicates that energy loss is due to viscous forces (see Section 6.5). At high velocities (high **R**), energy loss varies as

$$gh_L \sim V^2,$$

which indicates that energy loss is due to inertia effects (that is, turbulence).

7.1.2 Developing and Fully Developed Flow

The flow in a constant-area pipe or duct is said to be *fully developed* if the shape of the velocity profile is the same at all cross sections. We encountered fully developed flow in Section 4.2.1. Fully developed flow is relatively simple to analyze because it is one-dimensional (the velocity varies with distance across the pipe but not along it) and one-directional (there is no velocity perpendicular to the pipe axis).

It might seem that fully developed flow would occur only under very unique circumstances, but, fortunately, many pipe flows are fully developed. We can understand this by considering the flow from a large reservoir into a pipe (Fig. 7.6). If the fluid in the reservoir is very tranquil and the junction between the pipe and the reservoir is nicely rounded, the velocity of the fluid entering the pipe at cross section 1 will be nearly uniform; however, fluid in contact with the pipe wall has zero velocity (remember the no-slip condition). The velocity gradient near the pipe wall is associated with a retarding shear stress on the

Figure 7.6 Developing flow.

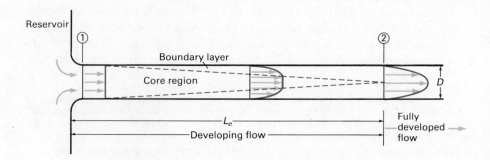

fluid. A region of slower-moving fluid near the wall is established.* This region of slower-moving fluid grows wider as the flow moves downstream. The fluid in the central region of the pipe is called the *core flow*. Since the fluid near the wall is moving more slowly, the core flow must accelerate as it moves downstream to maintain the same average velocity (constant mass flow) at all cross sections. Figure 7.6 also shows some velocity profiles for the entrance or developing region. Clearly, the velocity profiles change as the fluid moves downstream.

Note that the distinction between laminar and turbulent flow in a pipe applies only to the fully developed flow condition. In the development zone, the core flow is irrotational and thus neither laminar nor turbulent. The flow in the boundary layer is laminar or turbulent. If the fully developed pipe flow is laminar ($\mathbf{R} < 2300$), the boundary layer will be laminar, but if the fully developed pipe flow is turbulent ($\mathbf{R} \geq 4000$), then the boundary layer will be laminar near the pipe entrance, will undergo transition, and will be turbulent as it approaches the fully developed condition.

Obviously, the boundary layer cannot continue to grow indefinitely; eventually it fills the entire pipe. This occurs at plane 2 in Fig. 7.6. Downstream of plane 2 the velocity profile does not change and the flow is fully developed. The flow between planes 1 and 2 is said to be *developing*. The region between 1 and 2 is called the *entrance region*. The length L_e is called the *entrance length* or the *development length*. By dimensional analysis,

$$\frac{L_e}{D} = F(\mathbf{R}).$$

The function F is different for laminar and turbulent flow. Many analytical and experimental investigations have demonstrated the validity of the following correlations:

$$\frac{L_e}{D} \approx 0.06 \, \mathbf{R}, \qquad \text{Laminar flow} \tag{7.8}$$

$$\frac{L_e}{D} \approx 4.4 \, (\mathbf{R})^{1/6}. \qquad \text{Turbulent flow} \tag{7.9}$$

The longest practical development length corresponds to laminar flow with $\mathbf{R} \approx 2300$:

$$\left(\frac{L_e}{D}\right)_{max} \approx 140.$$

In many engineering pipe flows, $\mathbf{R}$ is between 10^4 and 10^5. For this case, the flow is turbulent and

$$\frac{L_e}{D} \approx 25. \qquad \text{Typical engineering flow}$$

Since many pipes in engineering applications are hundreds or thousands of diameters long, the flow is fully developed over most of their length.

---

* The region of slow-moving fluid near the wall is called a *boundary layer*. Boundary layers are discussed in Chapters 8 and 9.

This discussion has concentrated on flow entering a pipe from a large reservoir. Any disturbance such as a valve, elbow, or change in pipe diameter will cause the flow to depart from fully developed conditions; however, the flow returns to fully developed conditions several diameters downstream. Generally speaking, development lengths for other disturbances are shorter than for a reservoir entrance.

Aside from the analytical advantage of one-dimensionality and one-directionality, fully developed flow has several other interesting features:

- The streamlines are straight and parallel to the pipe axis.
- The shear stress at the pipe wall is the same at all locations.
- The momentum and kinetic energy correction factors β and α are constant.

Can you explain these facts based on the condition that the shape of the velocity profile is constant?

EXAMPLE 7.1 Illustrates developing flow

A wind tunnel has a 3.0-m diameter, a 4.0-m-long test section, and a flow rate of 300 m³/s, as shown in Fig E7.1. Estimate the size of the inviscid core at the end of the test section.

Solution

Given

Wind tunnel having 3.0-m diameter, 4.0-m length, and 300-m³/s flow rate. Figure E7.1

Find

Size of the inviscid core at end of test section

Solution

The diameter d_{IC} of the inviscid core is related to the wind tunnel diameter D by

$$d_{IC} = D - 2\delta,$$

where δ is the boundary layer thickness. This thickness increases along the direction of flow, so the inviscid core diameter decreases along the direction of flow. The limit of this process is reached when the boundary layer fills the entire wind tunnel, that is, when the flow becomes fully developed.

Figure E7.1 Expected boundary layer development in a wind tunnel.

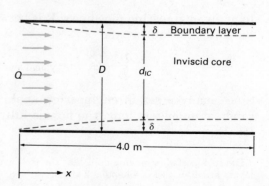

We first calculate the Reynolds number and determine whether to use Eq. (7.8) or (7.9) for the development length. The average velocity is found from Eq. (3.38):

$$\bar{V} = \frac{Q}{A} = \frac{\left(300\ \dfrac{m^3}{s}\right)}{\dfrac{\pi}{4}(3.0\ m)^2}$$

$$= 42.4\ m/s.$$

Assuming the air is at 15°C, Table A.3 gives

$$v = 1.49 \times 10^{-5}\ m^2/s,$$

and so

$$\mathbf{R} = \frac{\bar{V}D}{v} = \frac{\left(42.4\ \dfrac{m}{s}\right)(3.0\ m)}{\left(1.49 \times 10^{-5}\ \dfrac{m^2}{s}\right)}$$

$$= 8.5 \times 10^6.$$

Since $\mathbf{R} > 2300$, we use Eq. (7.9), which gives

$$L_e \approx 4.4(8.5 \times 10^6)^{1/6}D \approx 63(3.0\ m) \approx 190\ m.$$

The flow would require a length of 190 m to become fully developed. Assuming the boundary layer thickness increases linearly from a value of zero at $x = 0$ to a value of $\delta = D/2$ at $x = 190$ m, the thickness after 4 m is estimated as

$$\delta \approx \frac{4.0\ m}{190\ m}\left(\frac{3.0\ m}{2}\right) = 0.032\ m.$$

The inviscid core diameter is

$$d_{IC} \approx 3.0\ m - 2(0.032\ m), \qquad \textbf{Answer}$$
$$|\ d_{IC} \approx 2.94\ m.\ |$$

Discussion

In Chapter 9 we will discover that the boundary layer growth is not linear but instead can be approximated by

$$\delta \sim x^{4/5},$$

where x is measured from the wind tunnel inlet and along the direction of flow. For our wind tunnel with a 1.5-m radius,

$$\delta \approx 1.5\ m\left(\frac{4.0}{190}\right)^{4/5},$$

$$\approx 0.068\ m.$$

A more accurate estimate of d_{IC} is

$$d_{IC} \approx 3.0\ m - 2(0.068\ m)$$
$$\approx 2.86\ m.$$

Wind tunnel test sections are kept short so that a large region of uniform flow (the inviscid core) is available to simulate "free-flight" conditions in the atmosphere.

7.2 Analysis of Fully Developed Flow in Pipes and Ducts

We now consider fully developed flow in a straight pipe or duct of constant cross section. Consideration will be limited to steady flow of incompressible fluids with constant viscosity. Items of interest include relations between energy loss and flow rate, pressure and shear stress distributions, and characteristics of velocity profiles. By using all of the tools that we have available (control volume approach, differential approach, and experimental results expressed in the language of dimensional analysis), we can obtain a good working knowledge of pipe flow. Although there are significant differences between fully developed laminar and turbulent flows, we will carry the discussion as far as possible before introducing these differences.

7.2.1 Application of Energy Equation

Figure 7.7 illustrates fully developed flow in a pipe or duct. Since the fluid is assumed incompressible, Eq. (5.56) can be used to obtain

$$\frac{\bar{p}_1}{\rho} + \alpha_1\left(\frac{\bar{V}_1^2}{2}\right) + g\bar{z}_1 = \frac{\bar{p}_2}{\rho} + \alpha_2\left(\frac{\bar{V}_2^2}{2}\right) + g\bar{z}_2 + \overline{gh}_L. \tag{7.10}$$

Since the flow is fully developed and steady and the cross-sectional area is constant, $\alpha_1 = \alpha_2$ and $\bar{V}_1 = \bar{V}_2$. The energy equation becomes

$$\frac{\bar{p}_1}{\rho} + g\bar{z}_1 = \frac{\bar{p}_2}{\rho} + g\bar{z}_2 + \overline{gh}_L.$$

The overbars (¯) representing averages are somewhat cumbersome, so we will omit them unless they are necessary to clarify the difference between a local quantity and its area average. Since the streamlines are straight and parallel for fully developed flow, the variation of pressure over the cross section is hydrostatic. The sum of averages at any cross section $(\bar{p}/\rho + g\bar{z})$ is equal to the sum $(p/\rho + gz)$ at any point in the cross section. To avoid confusion, we will assume that p and z are evaluated at the pipe centerline.

Dropping the overbar and solving for the energy loss, we get

$$gh_L = \frac{p_1 - p_2}{\rho} + g(z_1 - z_2). \tag{7.11}$$

Figure 7.7 Fully developed flow in duct of arbitrary cross section.

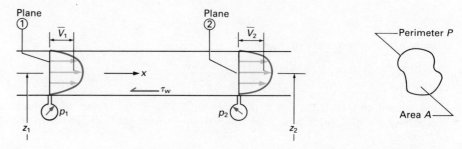

Equation (7.11) reveals a very interesting feature about fully developed flow: friction *does not* cause a decrease in fluid velocity (kinetic energy) along the pipe.

It is sometimes convenient to express energy loss in terms of an equivalent potential energy loss or an equivalent pressure difference. The equivalent potential energy loss is the head loss, given by

$$h_L = \frac{(gh_L)}{g}.$$

For fully developed pipe flow,

$$h_L = \frac{p_1 - p_2}{\gamma} + z_1 - z_2. \qquad (7.12)$$

Energy loss can also be expressed by a *pressure loss,* defined by

$$\Delta p_L \equiv \rho(gh_L). \qquad (7.13)$$

Pressure loss is actually energy loss per unit volume. A more exact name for pressure loss is *stagnation pressure loss* (see Eq. 4.22 and the discussion following it). For fully developed flow,

$$\Delta p_L = (p_1 - p_2) + \gamma(z_1 - z_2). \qquad (7.14)$$

It is important to recognize the difference between pressure change, pressure drop, and pressure loss. *Pressure change* between two points in a flow is

$$\Delta p = p_2 - p_1.$$

Pressure drop is the negative of pressure change,

$$\Delta p_d = p_1 - p_2.$$

In fully developed flow, pressure change is usually negative and pressure drop is usually positive. Pressure loss is not necessarily equal to a difference between two pressures but is the energy loss expressed in units of pressure. For *fully developed* flow in a *horizontal pipe,* pressure drop and pressure loss are equal. In a nonhorizontal pipe or in a flow with acceleration, pressure drop and pressure loss are not equal. Unfortunately, engineers sometimes use the term "pressure drop" when they mean "pressure loss." Since the change in elevation between two points in a pipe is usually known, relating pressure drop and pressure loss is usually simple.

All of the results of this section apply to either laminar or turbulent flow.

7.2.2 Application of Linear Momentum Equation

Next we apply the linear momentum equation to fully developed flow. Consider a control volume with inlet at plane 1, outlet at plane 2, and coincident with the inner wall of the duct (Fig. 7.7). The forces on the fluid in the control volume are due to pressure at the ends and along the wall, shear stress between the wall and the fluid, and gravity. To simplify the analysis, we assume that the pipe axis is horizontal.

The linear momentum theorem for the axial (x) direction gives

$$\sum \dot{M}_{x,\text{out}} - \sum \dot{M}_{x,\text{in}} = \sum F_x.$$

If we use various results from Section 5.3.1, the momentum equation becomes

$$\rho \beta_2 \bar{V}_2^2 A_2 - \rho \beta_1 \bar{V}_1^2 A_1 = \bar{p}_1 A_1 - \bar{p}_2 A_2 - \int_{A_w} \tau_w \, dA.$$

For fully developed flow,

$$\beta_2 = \beta_1 \quad \text{and} \quad \bar{V}_2 = \bar{V}_1.$$

The pipe area is constant, so

$$A_1 = A_2 = A.$$

With these simplifications, the axial momentum equation gives

$$\bar{p}_1 - \bar{p}_2 = \frac{1}{A} \int_{A_w} \tau_w \, dA. \tag{7.15}$$

The "wetted area" on which the shear stress acts is

$$dA_w = (dx)(P),$$

where P is the perimeter of the duct. Let $\bar{\tau}_w$ be the average shear stress over the perimeter at any x-location. Then

$$\int_{A_w} \tau_w \, dA = P \int_{x_1}^{x_2} \bar{\tau}_w \, dx.$$

For fully developed flow, the wall shear stress does not vary with x, so

$$\int_{A_w} \tau_w \, dA = P\bar{\tau}_w \int_{x_1}^{x_2} dx = P\tau_w(x_2 - x_1).$$

If we define

$$\Delta x \equiv x_2 - x_1.$$

Equation (7.15) becomes

$$\bar{p}_1 - \bar{p}_2 = \frac{P\Delta x}{A} \bar{\tau}_w. \tag{7.16}$$

This equation can be rearranged and generalized. Although we assumed a horizontal pipe, we can extend this result to a non-horizontal pipe by replacing $\bar{p}$ by $(\bar{p} + \gamma \bar{z})$. Making this change and dropping the overbars, we have

$$(p_1 + \gamma z_1) - (p_2 + \gamma z_2) = \frac{P\tau_w}{A} \Delta x. \tag{7.17}$$

Combining Eqs. (7.11), (7.13), and (7.12) with Eq. (7.17) gives

$$\Delta p_L = \tau_w \left(\frac{P}{A}\right) \Delta x, \tag{7.18}$$

$$gh_L = \frac{\tau_w}{\rho} \left(\frac{P}{A}\right) \Delta x. \tag{7.19}$$

and

$$h_L = \frac{\tau_w}{\gamma}\left(\frac{P}{A}\right)\Delta x \qquad\qquad (7.20)$$

for fully developed flow.

The most common duct cross section is circular. For a circular cross section (pipe),

$$\frac{P}{A} = \frac{\pi D}{\frac{\pi D^2}{4}} = \frac{4}{D},$$

and Eqs. (7.17)–(7.20) become

$$(p_1 + \gamma z_1) - (p_2 + \gamma z_2) = \Delta p_L = 4\tau_w\left(\frac{\Delta x}{D}\right), \qquad\qquad (7.21)$$

$$gh_L = 4\frac{\tau_w}{\rho}\left(\frac{\Delta x}{D}\right), \qquad\qquad (7.22)$$

and

$$h_L = 4\frac{\tau_w}{\gamma}\left(\frac{\Delta x}{D}\right). \qquad\qquad (7.23)$$

Equations (7.11), (7.17), and (7.19) provide relations between three of the four most important parameters in pipe and duct flow: energy loss, pressure (plus elevation) change, and wall shear stress. Given any one of these parameters, together with fluid density and duct geometry, we can determine the other two for fully developed flow. The remaining problem is to relate any one of these parameters to the fourth parameter, flow rate. Since we have "used up" all of the applicable control volume equations* (momentum, continuity, energy), it is necessary to turn to a differential analysis *or* experimental data to complete the analysis. The differential approach can solve the remainder of the problem if the flow is laminar; however, the turbulent flow problem can be solved only by referring to experimental data. These detailed considerations also require that we be specific about the particular geometry involved. For now we will limit discussion to flow in circular pipes.

7.2.3 Application of Dimensional Analysis: Friction Factors and Loss Coefficients

Since we anticipate the need to refer to experimental data for further information about pipe flow, it is appropriate to consider dimensional analysis before we proceed. Dimensional analysis of energy loss

---

* This situation clearly points up the most common failure of the finite control volume approach. While the approach can provide valuable relations between key flow parameters, at least one of these parameters must often be evaluated by the differential or experimental approach.

in pipe flow was discussed extensively in Chapter 6 as the example introducing dimensional analysis and the pi theorem. The result of the dimensional analysis was*

$$\frac{gh_L}{\frac{1}{2}V^2} = F\left(\frac{L}{D}, \frac{\rho VD}{\mu}, \frac{\varepsilon}{D}\right),$$ (6.4)

or

$$\mathbf{K} = F\left(\frac{L}{D}, \mathbf{R}, \frac{\varepsilon}{D}\right),$$ (7.24)

where $\mathbf{K}$ is the loss coefficient. If $\mathbf{K}$ is known, the energy loss can be calculated by

$$gh_L = \mathbf{K}\left(\frac{V^2}{2}\right),$$ (7.25a)

and the head loss can be calculated by

$$h_L = \mathbf{K}\left(\frac{V^2}{2g}\right).$$ (7.25b)

Using Eq. (7.13), we find that the pressure loss is

$$\Delta p_L = \mathbf{K}\left(\frac{\rho V^2}{2}\right)$$ (7.26)

To use Eqs. (7.24)–(7.26), we must know the dependence of $\mathbf{K}$ on length to diameter ratio, Reynolds number, and relative roughness. If $\mathbf{K}$ must be determined completely by experiment, about 10^3 runs would be required because $\mathbf{K}$ depends on three parameters.

The problem of determining $\mathbf{K}$ can be simplified by noting that, according to Eq. (7.22), energy loss in a pipe is directly proportional to the length-diameter ratio. This is because the wall shear stress is constant for fully developed flow. If we replace the general length Δx with the pipe length L, the general function F in Eq. (7.24) must have the special form

$$F\left(\frac{L}{D}, \mathbf{R}, \frac{\varepsilon}{D}\right) = \left(\frac{L}{D}\right)\tilde{F}\left(\mathbf{R}, \frac{\varepsilon}{D}\right).$$

The dimensionless parameter defined by

$$f \equiv \tilde{F}\left(\mathbf{R}, \frac{\varepsilon}{D}\right)$$ (7.27)

is called the *Darcy friction factor* (f is sometimes called the *Moody friction factor* for reasons that will become apparent later). The Darcy friction factor is related to the loss coefficient by

$$\mathbf{K} = f\left(\frac{L}{D}\right).$$ (7.28)

---

* $F(\ )$ is used for the functional notation rather than $f(\ )$ to avoid confusion with the friction factor, to be introduced shortly.

In terms of the Darcy friction factor, the energy loss, head loss, and pressure loss are calculated by

$$gh_L = f\left(\frac{L}{D}\right)\frac{V^2}{2},$$ (7.29a)

$$h_L = f\left(\frac{L}{D}\right)\frac{V^2}{2g},$$ (7.29b)

and

$$\Delta p_L = f\left(\frac{L}{D}\right)\left(\frac{\rho V^2}{2}\right).$$ (7.30)

Equation (7.29b) is called the *Darcy-Weisbach equation.* This name could also be extended to the other two equations.

To determine f experimentally, we would need only about 10^2 experiments. This shows how even a simple, incomplete theory (in this case, Eq. 7.22) can reduce the complexity of an investigation.

The Darcy friction factor is related to the skin friction coefficient. According to Eq. (7.22) with $\Delta x = L$,

$$gh_L = 4\frac{\tau_w}{\rho}\left(\frac{L}{D}\right).$$

Evaluating gh_L in terms of the Darcy friction factor from Eq. (7.29a), we have

$$f\left(\frac{L}{D}\right)\frac{V^2}{2} = 4\frac{\tau_w}{\rho}\left(\frac{L}{D}\right).$$

Rearranging gives

$$\frac{\tau_w}{\frac{1}{2}\rho V^2} = \mathbf{C_f} = \frac{f}{4}.$$ (7.31)

The skin friction coefficient ($\mathbf{C_f}$) for fully developed pipe flow is sometimes called the *Fanning friction factor;* We will not use this terminology, however.

The problem of predicting losses or pressure drop can be reduced to the problem of determining any one of the three dimensionless parameters $\mathbf{K}$, f, or $\mathbf{C_f}$ because they are related by Eqs. (7.28) and (7.31). Whichever we select, we expect it to depend on the Reynolds number and relative roughness. To proceed further, we must consider laminar and turbulent flow separately.

7.2.4 Laminar Flow in a Circular Pipe

For fully developed laminar flow in a circular pipe, analytical methods produce an exact solution. The key is finding an expression for the velocity profile (see Fig. 7.8):

$$u = u\left(r, R, \mu, \frac{dp}{dx}\right).$$

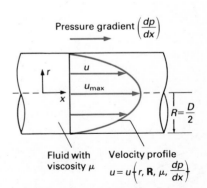

Figure 7.8 Schematic of velocity profile for fully developed pipe flow.

After we obtain the velocity profile, we can find the flow rate and average velocity by integration. The fluid viscosity and pressure gradient will remain as parameters, so a relation between pressure drop, pipe length and diameter, fluid viscosity, and flow rate can be found.

There are two ways to obtain a differential equation for the velocity distribution. One way is to start with the Navier-Stokes equations written in the appropriate coordinate system (cylindrical coordinates for a circular pipe) and simplify them for fully developed flow. We will leave this approach for you to try. An alternative approach is to derive the differential equation from a force balance on a fluid element. We will use this approach because it will allow us to delay the specification of laminar flow.

Figure 7.9 shows a cylindrical fluid element of radius r and length δx. The fluid element is coaxial with the pipe, which we assume to be horizontal. Pressure and shear stress on the fluid element are shown. Since mechanical energy loss is always positive, Eq. (7.14) tells us that the pressure drops in the flow direction. It is convenient to work with pressure drop δp as a positive quantity, so the pressure at $x + \delta x$ is $p - \delta p$.

Newton's second law for the fluid element is

$$(\delta m)a_x = \sum F_x.$$

Since the flow is steady and fully developed, the fluid element has no acceleration, so

$$\sum F_x = 0.$$

Substituting appropriate expressions for shear and pressure forces, we have

$$(p)\pi r^2 - (p - \delta p)\pi r^2 - \tau(2\pi r \, \delta x) = 0.$$

This simplifies to

$$\tau = \frac{r}{2}\frac{\delta p}{\delta x}. \tag{7.32}$$

For fully developed flow, $\delta p/\delta x$ is a constant, given by

$$\frac{\delta p}{\delta x} = \left|\frac{dp}{dx}\right|,$$

Figure 7.9 Cylindrical fluid element in pipe flow. Flow is from left to right.

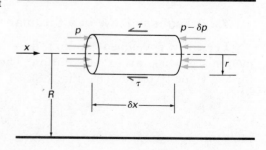

and Eq. (7.32) shows that the shear stress increases linearly with radius. Putting $\tau = \tau_w$ at $r = R$ gives

$$\tau_w = \frac{R}{2}\left(\frac{\delta p}{\delta x}\right) = \frac{R}{2}\left(\frac{\Delta p_d}{\Delta x}\right),$$

where Δp_d is the pressure drop in length Δx. This expression is equivalent to Eq. (7.16). Equation (7.32) is valid for both laminar and turbulent flow.

To proceed further, we now specify that the flow is laminar and write Newton's law of viscosity as

$$\tau = -\mu\left(\frac{du}{dr}\right).$$

The minus sign is necessary because τ has been treated as a positive quantity, while du/dr is negative (Fig. 7.8). Substituting into Eq. (7.32) gives

$$\mu\left(\frac{du}{dr}\right) = -\frac{r}{2}\left(\frac{\delta p}{\delta x}\right).$$

Separating variables and integrating give

$$u = -\frac{1}{4\mu}\left(\frac{\delta p}{\delta x}\right)r^2 + \text{Constant}.$$

The constant of integration is evaluated from the condition

$$u = 0 \quad \text{at} \quad r = R.$$

The resulting velocity distribution is

$$u = \frac{1}{4\mu}\left(\frac{\delta p}{\delta x}\right)(R^2 - r^2). \tag{7.33}$$

The maximum velocity, $u_{\max}$, occurs at $r = 0$ (Fig. 7.8) and is

$$u_{\max} = u)_{r=0} = \frac{R^2}{4\mu}\left(\frac{\delta p}{\delta x}\right). \tag{7.34}$$

The average velocity is

$$\bar{V} = \frac{\displaystyle\int V_n\,dA}{A} = \frac{2\pi\displaystyle\int_0^R ur\,dr}{\pi R^2} = \frac{2\pi\displaystyle\int_0^R \frac{1}{4\mu}\left(\frac{\delta p}{\delta x}\right)(R^2 - r^2)r\,dr}{\pi R^2}.$$

Evaluating the integral, we have

$$\bar{V} = \frac{R^2}{8\mu}\left(\frac{\delta p}{\delta x}\right) = \frac{1}{2}u_{\max}. \tag{7.35}$$

The volume flow rate is

$$Q = \bar{V}A = \frac{\pi R^4}{8\mu}\left(\frac{\delta p}{\delta x}\right). \tag{7.36}$$

Equation (7.36) is the desired relation between pressure drop and flow rate. Since the pressure gradient is constant, δp and δx need not be infinitesimally small.

We can extend this result to nonhorizontal pipes by replacing p by $p + \gamma z$. Since the flow is fully developed, we can use Eqs. (7.13) and (7.14) and write

$$(p_1 + \gamma z_1) - (p_2 + \gamma z_2) = \Delta p_L = \rho g h_L = \frac{8\mu L}{\pi R^4} Q, \qquad (7.37)$$

where Δp_L and gh_L are the pressure loss and energy loss that occur in a pipe of length L. Equation (7.37) is called the *Hagen-Poiseuille formula* for pressure loss in laminar pipe flow. Note that the equation predicts that the loss is directly proportional to flow rate and hence directly proportional to the fluid's (average) velocity. This is because the loss is due to viscous stress only.

The friction factor for laminar pipe flow can be found from the Hagen-Poiseuille formula, which can be written*

$$gh_L = \frac{8vL}{\pi R^4} Q.$$

The average velocity[†] is introduced by

$$Q = VA = V(\pi R^2),$$

so

$$gh_L = \frac{8v \, VL}{R^2}.$$

From Eq. (7.29a), we have

$$f = \frac{gh_L}{\frac{1}{2} V^2} \left(\frac{D}{L}\right).$$

Substituting ($R = D/2$) and simplifying give

$$f = \frac{64\mu}{\rho VD} = \frac{64}{\mathbf{R}}. \qquad (7.38)$$

The loss coefficient and skin friction coefficient are

$$\mathbf{K} = \left(\frac{D}{L}\right) f = \frac{64}{\mathbf{R}} \left(\frac{D}{L}\right) = \frac{64}{\mathbf{R}_L} \quad \text{and} \qquad (7.39)$$

$$\mathbf{C}_f = \frac{f}{4} = \frac{16}{\mathbf{R}}. \qquad (7.40)$$

Equation (7.38) implies that the friction factor for laminar flow (and hence the loss) does not depend on pipe wall roughness. This surprising fact is verified by experimental results, as we will see in the next section.

Before closing this discussion of laminar pipe flow, we note that the momentum and kinetic energy correction factors can be calculated by substituting the parabolic velocity profile given by Eqs. (7.33) and

---

* Generally the kinematic viscosity (v) is more convenient with energy loss, and the dynamic viscosity (μ) is more convenient with pressure loss.
[†] The overbar is dropped for convenience.

Assuming the development length is a small fraction of the total length of the pipe, Eq. (7.37) gives

$$\rho g h_L = \frac{8\mu L Q}{\left(\dfrac{\pi d^4}{16}\right)} = \frac{128\mu L Q}{\pi d^4}, \qquad \text{and}$$

$$p_2 - p_3 = \frac{\rho V_2^2}{2} + \frac{128\mu L Q}{\pi d^4}.$$

Substituting the expressions obtained for $(p_1 - p_2)$ and $(p_2 - p_3)$ into the equation for $(p_1 - p_3)$ gives

$$p_1 - p_3 = \left(\frac{\rho V_2^2}{2} - \rho g H\right) + \left(\frac{\rho V_2^2}{2} + \frac{128\mu L Q}{\pi d^4}\right).$$

Since $p_1 = p_3$, we have

$$\rho V_2^2 - \rho g H + \frac{128\mu L Q}{\pi d^4} = 0.$$

The average velocity at 2 is

$$V_2 = \frac{Q}{A_3} = \frac{4Q}{\pi d^2}.$$

Substituting this result into the previous equation gives

$$\rho\left(\frac{4Q}{\pi d^2}\right)^2 - \rho g H + \frac{128\mu L Q}{\pi d^4} = 0.$$

Solving for the viscosity μ gives

$$\left|\ \mu = \frac{\pi d^4 \rho g H}{128 L Q} - \frac{\rho Q}{8\pi L}.\ \right| \qquad \textbf{Answer}$$

This result is the answer because it shows that viscosity can be calculated from known values of d and L and measured values of ρ, H, and Q. The density can be measured with a hydrometer, and Q can be obtained by collecting a specified volume of fluid in a measured time period. The ratio of tank cross-sectional area to pipe cross-sectional area should be large so that the height H does not change appreciably during the time period required to collect the fluid.

Discussion
Note that the above result is valid only for the following conditions:

- Laminar flow ($\mathbf{R} \le 2300$)
- Velocity $V_1 \approx 0$
- Constant-density fluid
- Negligible energy loss in tank and in entrance nozzle
- Fully developed flow at pipe exit
- Development length short compared to overall pipe length
- Horizontal pipe

You should make sure you understand the reason(s) each of these conditions must be satisfied.

Some numerical values would be instructive. An experiment was conducted in our laboratory using $H = 1.0$ ft, $D = 6.77$ in., $d = 0.125$ in., $L = 48$ in., and $\rho = 62.3$ lbm/ft^3. The fluid was 70°F water. Q was measured as 1.74×10^{-4} ft^3/sec.

(7.35) into the defining equations to obtain

$$\beta = \frac{\int u^2 \, dA}{V^2 A} = \frac{4}{3} \qquad \text{Laminar fully developed} \tag{7.41}$$

$$\text{flow in a circular pipe}$$

$$\text{and} \qquad \alpha = \frac{\int u^3 \, dA}{V^3 A} = 2. \tag{7.42}$$

(See Example 5.1.) Equations (7.37)–(7.42) apply only to laminar flow, so the Reynolds number must be less than 2300 to use them.

EXAMPLE 7.2 Illustrates laminar flow in a circular pipe

It has been suggested that the standpipe system in Fig. E7.2 can be used as a viscometer. Evaluate this proposal. Neglect energy loss in the entrance nozzle.

Solution

> Given
>
> Standpipe system in Fig. E7.2
>
> Find
>
> Whether the standpipe system can be used as a viscometer

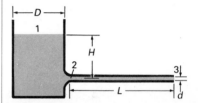

Figure E7.2 Standpipe system used as a viscometer.

> Solution
>
> The flow from point 1 to point 2 is assumed inviscid, but viscous effects are important from point 2 to point 3. We will analyze each section differently and therefore break the pressure drop $(p_1 - p_3)$ into two parts:

$$p_1 - p_3 = (p_1 - p_2) + (p_2 - p_3).$$

Overbars denoting average quantities are dropped. Assuming constant density, we apply the mechanical energy equation, Eq. (5.58), to a streamline between points 1 and 2 to give

$$\frac{p_1}{\rho} + \alpha_1 \left(\frac{V_1^2}{2} \right) + gz_1 - w_s = \frac{p_2}{\rho} + \alpha_2 \left(\frac{V_2^2}{2} \right) + gz_2 + gh_L.$$

Now $V_1 = 0$, $w_s = 0$, and $z_1 - z_2 = H$. Neglecting losses between points 1 and 2, we have

$$gh_L = 0 \qquad \text{and} \qquad \alpha_2 = 1.0.$$

Then the mechanical energy equation gives

$$p_1 - p_2 = \rho \frac{V_2^2}{2} - \rho g H.$$

Next we apply the mechanical energy equation between points 2 and 3 to give

$$\frac{p_2}{\rho} + \alpha_2 \left(\frac{V_2^2}{2} \right) + gz_2 - w_s = \frac{p_3}{\rho} + \alpha_3 \left(\frac{V_3^2}{2} \right) + gz_3 + gh_L.$$

Now $w_s = 0$, $\alpha_2 = 1.0$, $z_2 - z_1 = 0$, and Eq. (3.40) gives $V_2 = V_3$. We assume that the flow between 2 and 3 is laminar. Equation (7.42) shows that $\alpha_3 = 2.0$ for fully developed laminar flow. The mechanical energy equation gives

$$p_2 - p_3 = \rho \left(\frac{\alpha_3 V_3^2}{2} \right) - \rho \left(\frac{\alpha_2 V_2^2}{2} \right) + \rho g h_L = \frac{\rho V_2^2}{2} + \rho g h_L.$$

The viscosity is then calculated to be

$$\mu = \frac{\pi\left(0.125 \text{ in.} \times \dfrac{1 \text{ ft}}{12 \text{ in.}}\right)^4 \left(62.3 \dfrac{\text{lbm}}{\text{ft}^3}\right)\left(32.2 \dfrac{\text{ft}}{\text{sec}^2}\right)(1.0 \text{ ft})}{128\left(48 \text{ in.} \times \dfrac{1 \text{ ft}}{12 \text{ in.}}\right)\left(1.74 \times 10^{-4} \dfrac{\text{ft}^3}{\text{sec}}\right)\left(32.2 \dfrac{\text{ft} \cdot \text{lbm}}{\text{lb} \cdot \text{sec}^2}\right)}$$

$$- \frac{\left(62.3 \dfrac{\text{lbm}}{\text{ft}^3}\right)\left(1.74 \times 10^{-4} \dfrac{\text{ft}^3}{\text{sec}}\right)}{8\pi\left(48 \text{ in.} \times \dfrac{\text{ft}}{12 \text{ in.}}\right)\left(32.2 \dfrac{\text{ft} \cdot \text{lbm}}{\text{lb} \cdot \text{sec}^2}\right)}$$

$$= 2.26 \times 10^{-5} \frac{\text{lb} \cdot \text{sec}}{\text{ft}^2},$$

which compares reasonably well with a value of 2.08×10^{-5} lb·sec/ft² from Table A.6.

If we use the tabulated value of the kinematic viscosity from Table A.6, the Reynolds number for the experiment was

$$\mathbf{R} = \frac{Vd}{\nu} = \frac{Q}{\dfrac{\pi}{4}d^2}\left(\frac{d}{\nu}\right) = \frac{4Q}{\pi\nu d}$$

$$= \frac{4\left(1.74 \times 10^{-4} \dfrac{\text{ft}^3}{\text{sec}}\right)}{\pi\left(2.08 \times 10^{-5} \dfrac{\text{ft}^2}{\text{sec}}\right)\left(\dfrac{1}{8} \text{ in.} \times \dfrac{\text{ft}}{12 \text{ in.}}\right)} = 1022.$$

Therefore, the flow was laminar. Would you expect better agreement between the experimental value and the Table A.6 value if the pipe length were longer or shorter? Explain.

7.2.5 Turbulent Flow in a Circular Pipe: Experimental Information and the Moody Chart

Since our differential analysis of laminar pipe flow was so successful, we might try to analyze turbulent flow in the same way. The procedure from the beginning of the analysis to Eq. (7.32) is completely general. We encounter a problem when we must substitute an expression for the shear stress into Eq. (7.32) because in turbulent flow the shear stress is composed of both viscous and turbulent parts. The viscous stress is expressed by Newton's law of viscosity, but there is no similar relation for obtaining turbulent shear stress. All available information about turbulent stress is based on experimental data, a great deal of which was obtained from pipe flow experiments. Since we must use experimental data for the turbulent stress, we might just as well rely on experimental data for the friction factor itself.

Pipe flow experiments have been conducted from as early as the mid-1800s. With hindsight, we can condense the available information; more important, we can combine physical reasoning with experimental evidence to obtain a physical "feel" as well as some working information about turbulent pipe flow.

According to Eq. (7.27), we expect the friction factor for turbulent pipe flow to depend on Reynolds number and relative roughness. Early experiments, including those of Reynolds, verified the roughness effect. At a given Reynolds number, rough pipes have higher energy loss than smooth pipes. This is easily explained if we recognize that roughness disturbs the flow, promoting more turbulence.

Experiments have shown that Eq. (7.27) has two limiting forms. For smooth pipes ($\varepsilon/D \approx 0$),

$$f = F_s(\mathbf{R} \text{ only}). \qquad \text{Smooth pipe} \qquad\qquad (7.43)$$

For very rough pipes, we expect the flow to be highly turbulent, at least if the Reynolds number is not too small. The turbulent stress would be much larger than the viscous stress and the loss would depend only on inertia forces. Since $\mathbf{R}$ is the ratio of inertia force to viscous force and since viscous force is irrelevant for highly turbulent flow, the Reynolds number becomes irrelevant. For a very rough pipe,

$$f = F_r(\varepsilon/D \text{ only}). \qquad \text{Very rough pipe} \qquad\qquad (7.44)$$

A critical question we must answer in dealing with rough pipes is how to determine the surface roughness (ε). A commercially rough surface does not have uniform roughness. Pipes made of riveted steel or corrugated sheet metal have regular patterns of roughness. Pipes made of cast iron or concrete have random roughness distributions. It is far from clear how to quantify all types of roughness by a single parameter ε.

Most of our detailed, reliable information about friction factor in rough pipes is based on experiments performed by J. Nikuradse in 1933 [1]. Nikuradse solved the problem of quantifying the roughness by using artificially roughened pipes. By a screening process, Nikuradse obtained and classified sand grains of various sizes. He glued these sand grains to the inside of various pipes, producing a pipe wall roughness very similar to sandpaper. Since the sand grain size was known, the parameter

$$\frac{\varepsilon_s}{D}$$

could be accurately determined. The subscript s reminds us that the roughness elements are sand grains of uniform size.

Some results of Nikuradse's experiments are shown in Fig. 7.10. The friction factor f is plotted versus the Reynolds number $\mathbf{R}$ with ε_s/D as a parameter. Logarithmic coordinates are used because f and $\mathbf{R}$ change by several orders of magnitude. The straight line labeled H-P represents the Hagen-Poiseuille equation

$$f = \frac{64}{\mathbf{R}}.$$

The straight line labeled B represents an early attempt by H. Blasius to correlate data on turbulent friction factor in smooth pipes. His equation,

$$f_{\text{smooth}} \approx 0.3164\, \mathbf{R}^{-1/4}, \qquad\qquad (7.45)$$

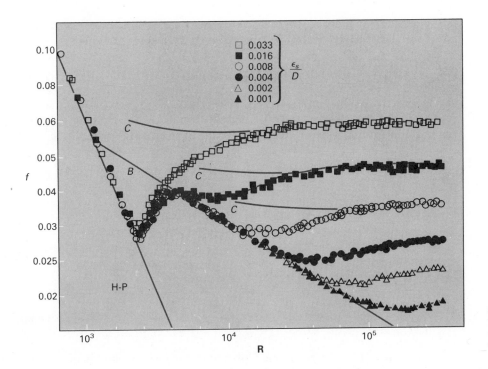

Figure 7.10 Friction factor for sand-roughened circular pipes as a function of Reynolds number and relative roughness. Based on the tests of Nikuradse [1].

is reasonably accurate for Reynolds numbers between 4000 and 10^5. The experimental data provide excellent verification of several points already discussed:

- For low Reynolds numbers (laminar flow), the Hagen-Poiseuille equation is accurate, and pipe roughness has no effect on f.
- For rough pipes, f is constant over a wide range of Reynolds number.
- For smooth pipes (ε_s/D small) f depends on Reynolds number but not on ε_s/D.

The experimental data also show that the smoothness of a pipe depends on Reynolds number. Up to a Reynolds number of about 4000, all pipes follow the same curve, indicating that they behave as if they were smooth. As the Reynolds number increases, pipes with different roughness break away from the group at various Reynolds numbers. Rougher pipes break away at lower Reynolds number.

One rather puzzling aspect of Nikuradse's data is the dip in f between Reynolds numbers of 2000 and 4000. This dip is characteristic of the uniform sand roughness used by Nikuradse and is not observed in commercial pipes. The curves labeled C in Fig. 7.10 are characteristic of commercial pipes.

Knowledge gained from careful experiments like those of Nikuradse has aided development of semi-empirical theories* and working charts

---

* A semi-empirical theory is a theory that, while based on sound physical reasoning, requires experimental data to complete the theory or fill in gaps.

for pipe friction factors. In 1935, L. Prandtl deduced the following formula for the friction factor in turbulent flow in smooth pipes:

$$\frac{1}{\sqrt{f}} = 2.0 \log \left(\mathbf{R}\sqrt{f} \right) - 0.8. \qquad (7.46)$$

At about the same time, T. von Karman derived the following relation for friction factor for completely turbulent (very rough pipe) flow:

$$\frac{1}{\sqrt{f}} = -2.0 \log \left(\frac{\varepsilon/D}{3.7} \right). \qquad (7.47)$$

Equations (7.46) and (7.47) are the limiting forms implied by Eqs. (7.43) and (7.44). Both equations will be discussed in Section 7.2.7.

In 1939, C. F. Colebrook [2] devised a formula that combines the smooth and rough limits, Eqs. (7.46) and (7.47), and also covers the transition range in between. The *Colebrook formula* is

$$\frac{1}{\sqrt{f}} = -2.0 \log \left(\frac{\varepsilon/D}{3.7} + \frac{2.51}{\mathbf{R}\sqrt{f}} \right). \qquad (7.48)$$

The main difficulty with the Colebrook formula is that it is implicit in f. To overcome this difficulty, L. F. Moody [3], in 1944, published a chart of f versus $\mathbf{R}$ and ε/D based on the Colebrook formula. This chart, reproduced in Fig. 7.11, is called a *Moody chart*. Before the days of computers and calculators, the Moody chart was the primary source of information for friction factor, but the engineer of today and tomorrow might just as easily use the Colebrook formula or one of the modifications that will be developed in the next section.

There is one very big gap in the pipe flow information we have developed so far. The Nikuradse data and the von Karman and Colebrook formulas are based on a uniform size and randomly distributed sand grain roughness ε_s. But the engineer needs information on the roughness of commercial pipes. Colebrook and Moody have provided the necessary information. Figure 7.12 reproduces a second chart from Moody's paper. This chart gives the equivalent relative roughness (ε/D) of various types of commercial pipe and is based on information published by Colebrook.

Table 7.1 develops equivalent absolute roughness (ε) for various materials. The roughness values listed in Table 7.1 and implied in Fig. 7.12 have little to do with the actual *geometric* roughness of the surface. The given roughnesses are the values that make the Colebrook equation work; they were obtained by "backing out" a value of ε/D from pressure drop data on commercial pipes. The values of ε/D apply to clean, new pipe only. Pipe that has been in service for a long time will experience corrosion or scale buildup and may have an equivalent roughness that is an order of magnitude larger than the value implied by Table 7.1 and Fig. 7.12. Because of uncertainty in evaluating equivalent roughness, values of f from the Moody charts (Figs. 7.11 and 7.12) are accurate to

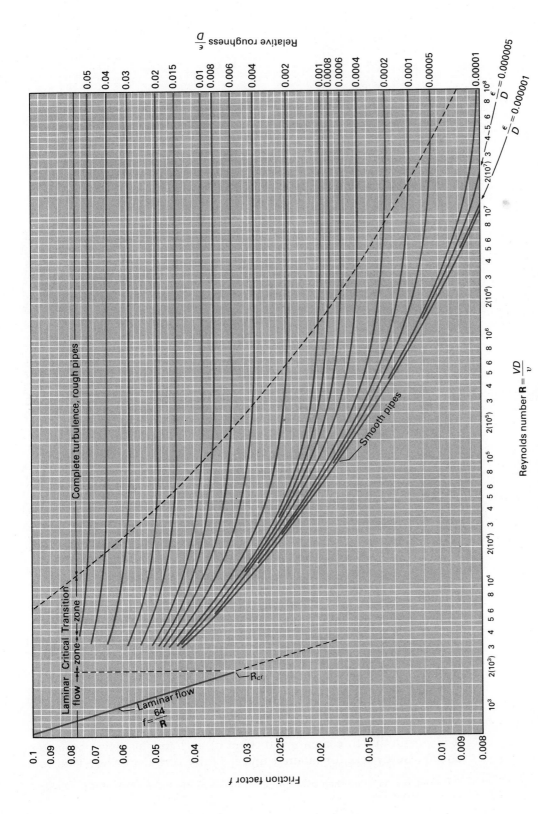

Figure 7.11 The Moody chart for friction factor (from [3]).

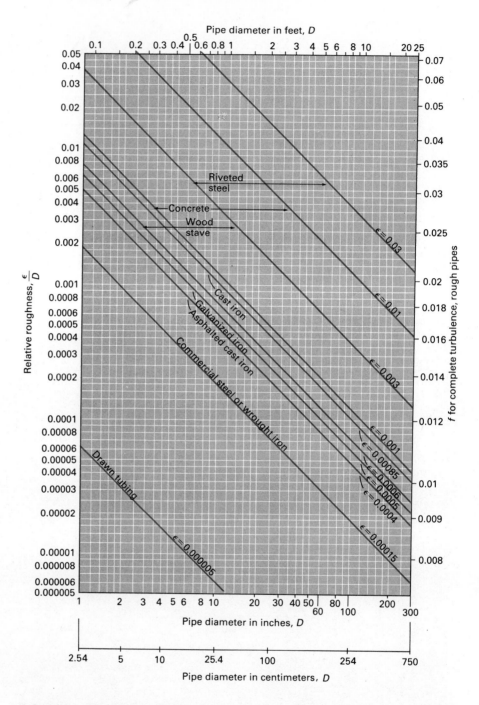

Figure 7.12 Moody's original chart for relative roughness of commercial pipes (from [3]).

only about 10 percent. The use of several significant figures in pipe flow calculations is accordingly *unjustified*.*

Now that we have discussed the friction factor for turbulent flow, it is instructive to compare the effect of various pipe flow parameters on the energy loss for laminar and for fully turbulent flow. This com-

---

* For this reason, more than two significant figures in the calculation of **R** is usually not justified either.

Table 7.1 Equivalent roughness of new commercial pipe materials (after Colebrook [2] and Moody [3])

Material	ε ft	ε mm
Riveted steel	0.003–0.03	0.9–9.0
Concrete	0.001–0.01	0.3–3.0
Wood stave	0.0006–0.003	0.18–0.9
Cast iron	0.00085	0.26
Galvanized iron	0.0005	0.15
Asphalted cast iron	0.0004	0.12
Commercial steel or wrought iron	0.00015	0.046
Drawn tubing	0.000005	0.0015
Glass, plastic	Smooth	Smooth

parison is presented in Table 7.2 and is based on the following formulas, adapted from Eqs. (7.29a), (7.38), and (7.47):

$$gh_L = \frac{64}{\mathbf{R}}\left(\frac{L}{D}\right)\left(\frac{V^2}{2}\right) = \frac{32\,\mu VL}{\rho D^2}, \qquad \text{Laminar flow}$$

$$gh_L = \left(\frac{0.5}{\log\left(\dfrac{3.7}{\varepsilon/D}\right)}\right)^2 \left(\frac{L}{D}\right)\left(\frac{V^2}{2}\right). \qquad \text{Completely turbulent flow}$$

The most interesting effects are those of V and D. Doubling V doubles the loss in laminar flow but quadruples it in turbulent flow. The effect of diameter is opposite. Halving the diameter quadruples the loss in laminar flow, but in turbulent flow halving the diameter slightly more than doubles the loss. Note also that energy loss depends on both density and viscosity in laminar flow but is independent of both in fully turbulent flow. If we consider pressure loss instead of energy loss, we find that pressure loss depends on viscosity but not on density in laminar flow and on density but not on viscosity in fully turbulent flow.

We will briefly discuss velocity profiles in fully developed turbulent pipe flow. Figure 7.13 shows velocity profiles for laminar and turbulent pipe flow. In part (a) the profiles have the same *average* velocity, while in part (b) the profiles have the same *maximum* velocity. The turbulent velocity profiles are much fuller than the laminar profiles and are

Table 7.2 Comparison of energy loss dependence in laminar and completely turbulent pipe flow

Parameter	Laminar Flow	Completely Turbulent Flow
Velocity (V)	V	V^2
Density (ρ)	$1/\rho$	No dependence
Viscosity (μ)	μ	No dependence
Diameter (D)	$1/D^2$	$\approx 1/D$
Length (L)	L	L
Roughness (ε)	No dependence	Depends on ε/D

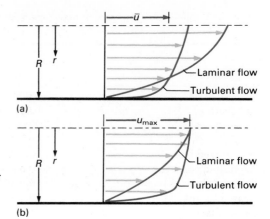

Figure 7.13 Comparison of laminar and turbulent velocity profiles for pipe flow. (a) Same average velocity; (b) same maximum velocity.

relatively flat over most of the pipe, falling rapidly to zero velocity as the wall is approached.

Turbulent velocity profiles can be represented by a rather complicated logarithmic function (see Section 7.2.7) that has a firm physical basis. An alternative approximation for turbulent flow velocity profiles is the *power law formula:*

$$\frac{u}{u_{\max}} \approx \left(\frac{y}{R}\right)^{1/n},\tag{7.49}$$

where y is the distance measured from the wall toward the centerline. In terms of the radial coordinate r,

$$\frac{u}{u_{\max}} \approx \left(\frac{R-r}{R}\right)^{1/n}.\tag{7.50}$$

The power law formula fits most of the profile but is unsatisfactory near the wall and the centerline. The value of n is often taken as 7; however, a single value of n cannot represent all conditions of Reynolds number and roughness. Table 7.3 gives values of n for various Reynolds numbers for smooth pipe. A correlation that is reasonably accurate for smooth or rough pipes is

$$n \approx \frac{1}{\sqrt{f}}.\tag{7.51}$$

The power law profile can be used to evaluate momentum and kinetic energy correction factors for fully developed turbulent pipe flow. Substitution of Eq. (7.49) into the equations

$$\bar{V} = \frac{\int u\,dA}{A}, \qquad \beta = \frac{\int u^2\,dA}{\bar{V}^2 A},$$

Table 7.3 Approximate values of power law profile parameter (n) for smooth pipes

R	4×10^3	2.3×10^4	1.1×10^5	1.1×10^6	2×10^6	3.2×10^6
n	6.0	6.6	7.0	8.8	10	10

and

$$\alpha = \frac{\int u^3 \, dA}{\bar{V}^3 A}$$

gives

$$\frac{\bar{V}}{u_{max}} = \frac{2n^2}{(n+1)(2n+1)}, \qquad \text{(7.52)}$$

$$\beta = \frac{(n+1)(2n+1)^2}{4n^2(n+2)}, \qquad \text{(7.53)}$$

and

$$\alpha = \frac{(n+1)^3(2n+1)^3}{4n^4(n+3)(2n+3)}. \qquad \text{(7.54)}$$

Table 7.4 gives values of α and β for various values of n.

7.2.6 Three Types of Pipe Flow Problems

The three types of pipe flow calculations that occur in engineering practice are the following:

- The energy/pressure loss problem: Given pipe diameter (D), length (L), surface roughness (ε), fluid properties (ρ, μ), and velocity (V) or flow rate (Q), find the energy loss (gh_L) or pressure loss (Δp_L).
- The velocity/flow rate problem: Given pipe diameter, length, roughness, fluid properties, and energy or pressure loss, find the resulting velocity or flow rate.
- The pipe sizing problem: Given pipe length and roughness, fluid properties, energy or pressure loss, and velocity *or* flow rate (but not both), select an appropriate pipe diameter.

These problems can be solved by applying the Darcy-Weisbach equation,

$$gh_L = f\left(\frac{L}{D}\right)\frac{V^2}{2}, \qquad \text{(7.29a)}$$

with friction factors evaluated from Eq. (7.38) for laminar flow or the Colebrook formula, Eq. (7.48), for turbulent flow. The graphical equivalent of the friction factor formulas, the Moody chart, can also be used. For laminar flow, friction factor calculations can be bypassed by using the Hagen-Poiseuille formula.

Table 7.4 Values of momentum and kinetic energy correction factors for various values of power law profile parameter (n)

n	5	6	7	8	9
β	1.037	1.027	1.020	1.016	1.013
α	1.106	1.077	1.058	1.046	1.037

Flow rate and velocity are related by

$$Q = VA = \frac{\pi D^2}{4} V. \tag{7.55}$$

The three different problems are all very simple if the flow is laminar because Eqs. (7.37) and (7.55) can be combined and solved for any single variable in terms of the others. The resulting equations are as follows:

■ Energy/pressure loss problem:

$$gh_L = \frac{\Delta p_L}{\rho} = \frac{32\mu L V}{\rho D^2} = \frac{128\nu L Q}{\pi D^4} \tag{7.56}$$

■ Velocity/flow rate problem:

$$V = \frac{\Delta p_L D^2}{32\mu L} = \frac{(gh_L)D^2}{32\nu L} \qquad \text{or} \tag{7.57}$$

$$Q = \frac{\pi \Delta p_L D^4}{128\mu L} = \frac{\pi(gh_L)D^4}{128\nu L}. \tag{7.58}$$

■ Pipe sizing problem:

$$D = \left(\frac{32\mu L V}{\Delta p_L}\right)^{1/2} = \left(\frac{32\nu L V}{gh_L}\right)^{1/2} \qquad \text{or} \tag{7.59}$$

$$D = \left(\frac{128\mu L Q}{\pi \Delta p_L}\right)^{1/4} = \left(\frac{128\nu L Q}{\pi gh_L}\right)^{1/4}. \tag{7.60}$$

These equations all apply if $\mathbf{R} < 2300$.

If the flow is turbulent, the three problems are more difficult because two of the three key parameters, namely V and D, appear explicitly in the friction factor equations and the Moody chart. Both V and D are in the Reynolds number, which is supposedly an independent variable. D also appears in the relative roughness. In addition to these complications, the Colebrook formula is implicit in f. The latter fact has led to the popularity of the Moody chart because it gives f directly if $\mathbf{R}$ and ε/D are known, but using the Colebrook formula requires iteration. This is not as serious a drawback as it used to be because now calculators and computers can make iterative calculations quickly. The following discussion is based on use of the Colebrook formula rather than the Moody chart.

Energy/Pressure Loss Problem. The Reynolds number is calculated from

$$\mathbf{R} = \frac{VD}{\nu} = \frac{4Q}{\pi \nu D},$$

and the relative roughness is calculated with the aid of Table 7.1 or read from Fig. 7.12. The friction factor can be read from Fig. 7.11 and the energy or pressure loss calculated from Eqs. (7.29a)–(7.30) as appropriate. If desired, the pressure drop can be calculated by

$$\Delta p_d = \Delta p_L - \gamma(z_1 - z_2). \tag{7.61}$$

If you wish to bypass the Moody chart, the friction factor can be evaluated by successive approximation using the following scheme based on the Colebrook formula:

- Guess a value of f. The "fully turbulent" value for the given ε/D is usually a good guess if the pipe is rough. The Blasius formula, Eq. (7.45), gives a good first guess for smooth pipes.
- Calculate an improved value of f from

$$f_{\text{new}} = \frac{0.25}{\left[\log \left(\dfrac{\varepsilon/D}{3.7} + \dfrac{2.51}{\mathbf{R}\sqrt{f_{\text{old}}}} \right) \right]^2}.$$

- Repeat the iteration until you are satisfied that f_{new} is close enough to f_{old} (usually two iterations is sufficient).

EXAMPLE 7.3   Illustrates the energy/pressure loss problem for turbulent flow

A horizontal commercial steel pipe has an inside diameter of 1.0 m. Water at 20°C flows through the pipe at the rate of 1.0 m³/s. Find the pressure drop and the energy loss in a 75-m length of the pipe. Assume fully developed flow.

Solution

Given
Commercial steel pipe
Inside diameter 1.0 m
Length 75 m
Water flow rate 1.0 m³/s
Water temperature 20°C

Find
Pressure drop
Energy loss

Solution
We will solve this problem two ways. The first method will use the Moody chart and the second method the Colebrook formula. In both methods the energy loss is found from the Darcy-Weisbach formula:

$$gh_L = f\left(\frac{L}{D}\right)\left(\frac{V^2}{2}\right).$$

The relative roughness for the pipe found using Fig. 7.12 is

$$\frac{\varepsilon}{D} = 0.00007.$$

Obtaining the kinematic viscosity from Table A.5, we have a Reynolds number of

$$\mathbf{R} = \frac{4Q}{\pi v D} = \frac{4\left(1.0\,\dfrac{\text{m}^3}{\text{s}}\right)}{\pi\left(1.00\times 10^{-6}\,\dfrac{\text{m}^2}{\text{s}}\right)(1.0\text{ m})} = 1.27\times 10^6.$$

The Moody chart gives $f = 0.0127$. Next we use Eq. (3.37a) to give

$$V = \frac{Q}{A} = \frac{4Q}{\pi D^2} = \frac{4\left(1.0 \, \frac{m^3}{s}\right)}{\pi (1.0 \, m)^2} = 1.27 \, m/s.$$

The Darcy-Weisbach formula gives

$$gh_L = 0.0127 \left(\frac{75 \, m}{1.0 \, m}\right) \frac{(1.27 \, m/s)^2}{2},$$

$$| \; gh_L = 0.77 \, m^2/s^2. \; |$$

Answer ◄

The pressure drop is

$$\Delta p_d = \Delta p_L - \gamma(z_1 - z_2),$$

where

$$\Delta p_L = \rho g h_L.$$

Since the pipe is horizontal, $z_1 = z_2$ and $\Delta p_d = \rho(gh_L)$. Table A.5 gives $\rho = 998 \, kg/m^3$, so

$$\Delta p_d = \left(998 \, \frac{kg}{m^3}\right)\left(0.77 \, \frac{N \cdot m}{kg}\right) = 770 \, N/m^2,$$

$$| \; \Delta p_d = 770 \, Pa. \; |$$

Answer ◄

To use the Colebrook formula, we first calculate the relative roughness and the Reynolds number, obtaining

$$\frac{\varepsilon}{D} = 0.00007, \qquad \mathbf{R} = 1.27 \times 10^6.$$

Since the flow is turbulent, the Blasius formula can be used to make a first guess for f:

$$f \approx \frac{0.3164}{\mathbf{R}^{0.25}} = \frac{0.3164}{(1.27 \times 10^6)^{0.25}} = 0.0094.$$

Iterating with Eq. (7.48) gives

$$f_{new} = \frac{0.25}{\left[\log\left(\dfrac{0.00007}{3.7} + \dfrac{2.51}{1.27 \times 10^6 \sqrt{0.0094}}\right)\right]^2}$$

$$= 0.0129.$$

Using this value of f on the right side gives

$$f_{new} = \frac{0.25}{\left[\log\left(\dfrac{0.00007}{3.7} + \dfrac{2.51}{1.27 \times 10^6 \sqrt{0.0129}}\right)\right]^2}$$

$$= 0.0127,$$

and, repeating the process,

$$f_{new} = \frac{0.25}{\left[\log\left(\dfrac{0.00007}{3.7} + \dfrac{2.51}{1.27 \times 10^6 \sqrt{0.0127}}\right)\right]^2},$$

$$= 0.0127.$$

Convergence has occurred. Using this value of f (the same value we obtained from the Moody chart) with the Darcy-Weisbach and pressure drop equations, we have **Answer**

$$| \; gh_L = 0.77 \text{ N} \cdot \text{m/kg} \; |$$

and **Answer**

$$| \; \Delta p_d = 770 \text{ Pa.} \; |$$

Discussion

Note that pressure *drop* and pressure *loss* are equal because the pipe is horizontal and the flow is fully developed. How would the pressure drop and pressure loss change if the pipe were vertical ($z_1 - z_2 = L$)?

Velocity/Flow Rate Problem. This problem turns out to be very easy using the Colebrook formula but more difficult with the Moody chart. Recalling that gh_L, ε, and D are given, we define a new Reynolds number,

$$\mathbf{R}_f \equiv \sqrt{f} \mathbf{R},$$

and write the Colebrook formula,

$$f = \frac{0.25}{\left[\log \left(\dfrac{\varepsilon/D}{3.7} + \dfrac{2.51}{\mathbf{R}_f} \right) \right]^2}. \tag{7.62}$$

Evaluating $\mathbf{R}_f$ in terms of dimensional parameters, we have

$$\mathbf{R}_f = \sqrt{\frac{gh_L}{\dfrac{1}{2} V^2} \left(\frac{D}{L} \right)} \left(\frac{\rho V D}{\mu} \right) = \left[\frac{2(gh_L) D^3}{v^2 L} \right]^{1/2}. \tag{7.63}$$

$\mathbf{R}_f$ does not contain velocity and can be calculated from the given data. We can solve the velocity/flow rate problem as follows:

- Compute $\mathbf{R}_f$ by substituting given values of gh_L, D, v, and L into Eq. (7.63).
- Compute f from Eq. (7.62).
- Compute V from the Darcy-Weisbach equation,

$$V = \sqrt{\frac{2(gh_L) D}{fL}}, \tag{7.64}$$

or from

$$V = \frac{v \mathbf{R}_f}{D \sqrt{f}}. \tag{7.65}$$

- If desired, calculate Q from Eq. (7.55).

Note that no iteration is required if the Colebrook formula is used. Iteration is required if the Moody chart is used because $\mathbf{R}$ contains the unknown velocity.

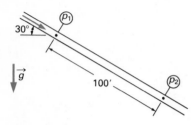

Figure E7.4 Flow through inclined pipe.

EXAMPLE 7.4 Illustrates the velocity/flow rate problem for turbulent flow

Water flows steadily down the inclined, 3/4-in.-inside-diameter copper pipe shown in Fig. E7.4. Two pressure gages 100 ft apart indicate identical pressures. Find the volume flow rate. Assume the flow is fully developed.

Solution

Given
Figure E7.4
Fully developed flow in an inclined, 3/4-in.-inside-diameter copper pipe
$p_1 = p_2$

Find
Volume flow rate

Solution
We will solve this problem two ways. The first method will use the Moody chart and the second method the Colebrook formula. Both methods start with the steady-state mechanical energy equation between 1 and 2.

$$\frac{\bar{p}_1}{\rho} + \alpha_1\left(\frac{\bar{V}_1^2}{2}\right) + g\bar{z}_1 - w_s = \frac{\bar{p}_2}{\rho} + \alpha_2\left(\frac{\bar{V}_2^2}{2}\right) + g\bar{z}_2 + \overline{gh_L}.$$

In this equation, $\bar{p}_1 = \bar{p}_2$, $\alpha_1 = \alpha_2$, $\bar{V}_1 = \bar{V}_2$, and $w_s = 0$. Then we have

$$gh_L = g(z_1 - z_2),$$

where overbars are dropped for convenience.

Both methods also use the relative roughness for the copper pipe (drawn tubing). Using Table 7.1, we have

$$\frac{\varepsilon}{D} = \frac{0.000005 \text{ ft}}{0.75 \text{ in.}}\left(\frac{12 \text{ in.}}{\text{ft}}\right) = 8.0 \times 10^{-5}.$$

Using the Moody chart method, we use Eq. (7.29a) for gh_L to give

$$f\left(\frac{L}{D}\right)\left(\frac{V^2}{2}\right) = g(z_1 - z_2).$$

Solving for V gives

$$V = \sqrt{\frac{2g(z_1 - z_2)D}{fL}}.$$

Since f is unknown, iteration is necessary. Guessing $f = 0.009$, we calculate V to be

$$V = \sqrt{\frac{2\left(32.2\,\frac{\text{ft}}{\text{sec}^2}\right)(100 \text{ ft})\sin 30° (0.75 \text{ in.})}{0.009(100 \text{ ft})\left(\dfrac{12 \text{ in.}}{\text{ft}}\right)}} = 15.0 \text{ ft/sec.}$$

Assuming 60°F water, Table A.6 gives $\nu = 1.22 \times 10^{-5}$ ft²/sec. The Reynolds number is

$$\mathbf{R} = \frac{VD}{\nu} = \frac{\left(15.0\,\dfrac{\text{ft}}{\text{sec}}\right)(0.75 \text{ in.})}{\left(1.22 \times 10^{-5}\,\dfrac{\text{ft}^2}{\text{sec}}\right)\left(\dfrac{12 \text{ in.}}{\text{ft}}\right)} = 7.7 \times 10^4.$$

The Moody chart, Fig. 7.11, gives

$$f = 0.019.$$

Two more iterations give

$$V = 9.88 \text{ ft/sec}, \quad \mathbf{R} = 5.1 \times 10^4, \quad \text{and} \quad f = 0.0208.$$

The volume flow rate is

$$Q = VA = \left(9.88 \frac{\text{ft}}{\text{sec}}\right) \frac{\pi(0.75 \text{ in.})^2}{4\left(\dfrac{144 \text{ in}^2}{\text{ft}^2}\right)},$$

Answer

$$\mid Q = 0.030 \text{ ft}^3/\text{sec.} \mid$$

Using the Colebrook formula to solve the problem, we substitute

$$gh_L = g(z_1 - z_2)$$

into Eq. (7.63) and obtain

$$\mathbf{R}_f = \left[\frac{2g(z_1 - z_2)D^3}{v^2 L}\right]^{1/2}.$$

Substituting the known values gives

$$\mathbf{R}_f = \left[\frac{2\left(32.2 \dfrac{\text{ft}}{\text{sec}^2}\right)(100 \text{ ft}) \sin 30° (0.75 \text{ in.})^3}{\left(1.22 \times 10^{-5} \dfrac{\text{ft}^2}{\text{sec}}\right)^2 (100 \text{ ft})\left(\dfrac{12 \text{ in.}}{\text{ft}}\right)^3}\right]^{1/2} = 7270.$$

Equation (7.48) gives the friction factor:

$$f = \frac{0.25}{\left[\log\left(\dfrac{0.00008}{3.7} + \dfrac{2.51}{7270}\right)\right]^2} = 0.0212.$$

Equation (7.64) gives the velocity:

$$V = \sqrt{\frac{2g(z_1 - z_2)D}{fL}}$$

$$= \sqrt{\frac{2\left(32.2 \dfrac{\text{ft}}{\text{sec}^2}\right)(100 \text{ ft}) \sin 30° (0.75 \text{ in.})}{0.0212(100 \text{ ft})\left(\dfrac{12 \text{ in.}}{\text{ft}}\right)}} = 9.74 \text{ ft/sec.}$$

The volume flow rate is

$$Q = VA = \left(9.74 \frac{\text{ft}}{\text{sec}}\right) \frac{\pi(0.75 \text{ in.})^2}{4\left(\dfrac{144 \text{ in}^2}{\text{ft}^2}\right)}$$

Answer

$$\mid Q = 0.030 \text{ ft}^3/\text{sec.} \mid$$

Discussion

The answers are equal since the two methods are equivalent. However, the Moody chart method requires iteration. The Colebrook method doesn't, is shorter, and is therefore more advantageous for solving this velocity/flow rate problem.

Note that there is no net pressure drop in the direction of flow. This intriguing phenomenon can be explained from either energy considerations or force considerations.

In terms of energy, the work required to overcome the mechanical energy loss is provided by the decrease in potential energy. In terms of forces, the gravitational force exactly balances the resisting force due to shear stress at the pipe walls.

In this example the static pressures at planes 1 and 2 are equal. It is possible that the pressure could either increase or decrease from plane 1 to plane 2. If the pressure increases, the resisting force at the walls is less than the gravitational force. If the pressure decreases, then both static pressure drop and the gravitational force are needed to overcome the wall resistance.

Pipe Sizing Problem. If we use the Moody chart, the pipe sizing problem is the most difficult of all since the diameter appears in both Reynolds number and relative roughness. In addition, the diameter is required to determine the friction factor from the energy loss. In both the energy loss problem and the velocity/flow rate problem, we could treat velocity and flow rate almost interchangeably because of Eq. (7.55); however, if the diameter is unknown, then either velocity or flow rate, but not both, may be known. In the following discussion we assume that Q is given. The key to direct solution of the pipe sizing problem is to rearrange the Colebrook formula by introducing new dimensionless parameters in such a way that the diameter appears in only one of them. Consider the following dimensionless parameters:

$$\mathbf{R} = \frac{VD}{v} = \frac{4Q}{\pi v D}, \tag{7.66}$$

$$f' \equiv f(\mathbf{R})^5 = \left(\frac{gh_L}{8Q^2}\right)\left(\frac{\pi^2 D^5}{L}\right)\left(\frac{4Q}{\pi v D}\right)^5 = \frac{128(gh_L)Q^3}{\pi^3 L v^5}, \tag{7.67}$$

$$e = \frac{4}{\pi}\left(\frac{\varepsilon}{D}\right)\left(\frac{1}{\mathbf{R}}\right) = \frac{\varepsilon v}{Q}. \tag{7.68}$$

Note that f' and e can be calculated from given parameters and neither contains D. Substituting

$$f = f'\mathbf{R}^{-5} \qquad \text{and} \qquad \frac{\varepsilon}{D} = \frac{\pi \mathbf{R} e}{4}$$

into the Colebrook formula and rearranging give

$$\mathbf{R} = \left[-2.0\sqrt{f'}\,\log\left(\frac{\pi e \mathbf{R}}{14.8} + \frac{2.51}{\sqrt{f'}}\mathbf{R}^{1.5}\right)\right]^{0.4}. \tag{7.69}$$

This horrendous-looking equation is implicit in $\mathbf{R}$, the only parameter that contains the unknown diameter, so it seems that we have to solve the problem by iteration, in which case all of the advantages that we hoped to gain by reformulating the Colebrook formula would be lost. Fortunately, White [4] has shown that the relation implied by Eq. (7.69) is only weakly dependent on the pipe roughness and can be approximated by*

$$\mathbf{R} \approx 1.43(f')^{0.208}. \tag{7.70}$$

---

* White also presents alternative forms of the Moody chart for the velocity/flow rate and pipe sizing problems.

This equation may be used directly or it may be substituted into the *right side* of Eq. (7.69) to produce the complicated but explicit equation

$$\mathbf{R} \approx \left[-2.0\sqrt{f'} \log \left(\frac{\pi e}{10.35} [f']^{0.208} + 4.29[f']^{-0.188} \right) \right]^{0.4}. \qquad \textbf{(7.71)}$$

Direct solution of the pipe sizing problem proceeds according to the following steps:

- Substitute given values of gh_L, Q, v, and L into Eqs. (7.67) and (7.68) to calculate f' and e.
- Compute $\mathbf{R}$ using Eq. (7.70) for a "good" value or Eq. (7.71) for a "better" value.*
- Calculate the required pipe size from

$$D = \frac{4Q}{\pi v \mathbf{R}}. \qquad \textbf{(7.72)}$$

Commercial pipe and tubing come in standard sizes (see Appendix D). It will seldom happen that the pipe diameter calculated according to the procedure above exactly corresponds to a standard value. It will usually be more economical to specify a standard pipe than a custom fabrication just to match your carefully computed diameter. It is customary to specify the standard pipe size that has the next largest diameter. Note that the practice of rounding upward implies that there is little need to be extremely accurate in pipe sizing calculations (the "good" calculation using Eq. 7.70 is usually adequate). Note also that the energy loss and/or flow rate will not equal the originally specified values when the larger pipe is used, so you might have to do an energy loss or flow rate calculation for the pipe size you choose.

---

* The "best" value can be obtained by solving Eq. (7.71) by iteration, but this is usually not worth the trouble.

EXAMPLE 7.5   Illustrates the pipe sizing problem for turbulent flow

Find the minimum acceptable pipe size for a schedule 80, commercial steel pipe to permit the flow of 175 gal/min of 60°F water with a maximum pressure drop of 1.2 psi per 100 ft of pipe. Assume the pipe is horizontal.

Solution

    Given
Horizontal, schedule 80, commercial steel pipe
Flow of 175 gal/min
60°F water
Maximum pressure drop 1.2 psi per 100 ft of pipe

    Find
Minimum acceptable pipe size

    Solution
    This problem could be solved in two ways. One method uses the Colebrook formula and the second uses the Moody chart. The Colebrook method is much simpler and direct and will be used here.

Using the Colebrook method, the energy loss is calculated for a 100-ft length of pipe using Eq. (7.11):

$$gh_L = \frac{p_1 - p_2}{\rho}$$

$$= \frac{\left(1.2\,\dfrac{\text{lb}}{\text{in}^2}\right)\left(\dfrac{144\ \text{in}^2}{\text{ft}^2}\right)}{\left(62.4\,\dfrac{\text{lbm}}{\text{ft}^3}\right)\left(\dfrac{\text{lb}\cdot\text{sec}^2}{32.2\ \text{ft}\cdot\text{lbm}}\right)} = 89.2\ \text{ft}^2/\text{sec}^2.$$

Equation (7.67) is used to calculate f':

$$f' = \frac{128(gh_L)Q^3}{\pi^3 L\nu^5}$$

$$= \frac{128\left(89.2\,\dfrac{\text{ft}^2}{\text{sec}^2}\right)\left(175\,\dfrac{\text{gal}}{\text{min}}\right)^3\left(\dfrac{\text{ft}^3}{7.48\ \text{gal}}\right)^3}{\pi^3(100\ \text{ft})\left(1.22\times10^{-5}\,\dfrac{\text{ft}^2}{\text{sec}}\right)^5\left(\dfrac{60\ \text{sec}}{\text{min}}\right)^3} = 8.08\times10^{23}.$$

For commercial steel pipe, Table 7.1 gives $\varepsilon = 0.00015$ ft. Then, from Eq. (7.68),

$$e = \frac{\varepsilon\nu}{Q} = \frac{(0.00015\ \text{ft})\left(1.22\times10^{-5}\,\dfrac{\text{ft}^2}{\text{sec}}\right)}{\left(175\,\dfrac{\text{gal}}{\text{min}}\right)\left(\dfrac{\text{ft}^3}{7.48\ \text{gal}}\right)\left(\dfrac{\text{min}}{60\ \text{sec}}\right)} = 4.69\times10^{-9}.$$

The approximate formula, Eq. (7.70), gives

$$\mathbf{R} = 1.43(f')^{0.208} = 1.43(8.08\times10^{23})^{0.208} = 1.34\times10^5.$$

The diameter D is found using Eq. (7.72):

$$D = \frac{4Q}{\pi\nu\mathbf{R}}$$

$$= \frac{4\left(175\,\dfrac{\text{gal}}{\text{min}}\right)\left(\dfrac{\text{ft}^3}{7.48\ \text{gal}}\right)\left(\dfrac{\text{min}}{60\ \text{sec}}\right)}{\pi\left(1.22\times10^{-5}\,\dfrac{\text{ft}^2}{\text{sec}}\right)(1.34\times10^5)} = 0.303\ \text{in.}$$

Table D.1 shows that this inside diameter corresponds to a schedule 80, 1/4-in., commercial steel pipe.

Answer

| Use 1/4-in. (nominal) pipe. |

◀

Discussion

Using the more exact Eq. (7.71) for the Reynolds number gives

$$\mathbf{R} = \left[-2.0\sqrt{8.08\times10^{23}}\ \log\left(\frac{\pi(4.69\times10^{-9})}{10.35}\ (8.08\times10^{23})^{0.208}\right.\right.$$

$$\left.\left. + 4.29(8.08\times10^{23})^{-0.188}\right)\right]^{0.4} = 1.33\times10^5,$$

compared to the approximate value of 1.34×10^5.

You may wish to solve this pipe sizing problem using the Moody chart. See Example 7.11 as a guide.

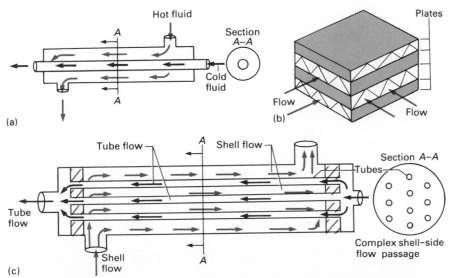

Figure 7.14 Some internal flow passages of complex shape: (a) Double-pipe heat exchanger; (b) compact "plate fin" heat exchanger; (c) shell and tube heat exchanger.

7.2.7 Fully Developed Flow in Noncircular Ducts

The Hagen-Poiseuille, Darcy-Weisbach, and Colebrook equations and the Moody chart provide adequate solutions for flow in *circular* pipes. Unfortunately, not all internal flows occur in circular pipes. Air- and gas-handling systems such as home heating systems or power plant air and flue gas ducts commonly employ rectangular ducts. A double-pipe heat exchanger (Fig. 7.14a) has flow in a concentric annulus, and a "compact" plate-fin heat exchanger (Fig. 7.14b) has flow in triangular passages. Flow in shell and tube heat exchangers (Fig. 7.14c) or in nuclear fuel-rod bundles occurs in passages of very complex shape. If the flow is laminar, it is sometimes possible to obtain an exact solution similar to the Hagen-Poiseuille equation; however, these solutions involve lengthy mathematical manipulations that are beyond the scope of this book.* If the flow is turbulent, no purely analytical solutions exist. Available experimental data are not nearly so abundant as for circular pipes.

Practical analysis of flow in noncircular ducts is based on the idea of finding an "equivalent" circular pipe flow. The main question is how to determine the equivalence. The primary geometric characteristic of a circle is that it can be completely specified by a single characteristic dimension, the diameter.[†] Maybe there is some way to define an equivalent single characteristic dimension for an arbitrary duct that can replace the circular pipe diameter in flow equations. The first such characteristic dimension that comes to mind is the *equivalent diameter*

---

* Two exceptions to this are laminar flow in a very wide rectangular duct (see Section 4.2.1) and laminar flow in a concentric annulus (see Problem 2 at the end of this chapter).

† On purely geometric grounds, the radius is preferable to the diameter; however, pipe flow data and analysis are based on diameter.

(D_{eq}) defined by

$$\frac{\pi}{4}(D_{eq})^2 = A_{NCD} \tag{7.73a}$$

or

$$D_{eq} \equiv \sqrt{\frac{4A_{NCD}}{\pi}}, \tag{7.73b}$$

where A_{NCD} is the cross-sectional area of the noncircular duct and D_{eq} is the diameter of the circle that has the same area as the noncircular duct. For a given flow rate the average velocity in a pipe with diameter D_{eq} is the same as the average velocity in the noncircular duct.

A second possible characteristic dimension can be defined from the linear momentum equation for fully developed flow,

$$(p_1 + \gamma z_1) - (p_2 + \gamma z_2) = \frac{P\tau_w}{A}(\Delta x). \tag{7.18}$$

For a circular pipe, the equation becomes

$$(p_1 + \gamma z_1) - (p_2 + \gamma z_2) = \left(\frac{4}{D}\right)\tau_w(\Delta x). \tag{7.21}$$

Comparing Eqs. (7.18) and (7.21), we find that the *hydraulic diameter* (D_h), defined by

$$D_h \equiv \frac{4A}{P} = \frac{4(\text{Duct cross-sectional area})}{\text{Wetted perimeter}}, \tag{7.74}$$

is a characteristic dimension whose definition depends directly on the dynamics of the flow. Evaluating the wall shear stress from Eq. (7.31),

$$\tau_w = C_f\left(\frac{1}{2}\rho V^2\right) = \frac{f}{4}\left(\frac{1}{2}\rho V^2\right),$$

using Eqs. (7.13), (7.14) and (7.21), and substituting L for Δx, we find

$$gh_L = f\frac{L}{D_h}\left(\frac{V^2}{2}\right). \tag{7.75}$$

The Darcy-Weisbach equation can be generalized by using the hydraulic diameter. For any duct shape except circular, the hydraulic diameter and the equivalent diameter are not equal. Expressions for calculating the equivalent diameter and hydraulic diameter of a few common shapes are given in Table 7.5. Since the hydraulic diameter is required for the Darcy-Weisbach equation, it seems logical to choose the hydraulic diameter as the characteristic dimension of an equivalent circular pipe flow.

Changing from D to D_h in the Darcy-Weisbach equation is simple enough. The real question is "How do we evaluate the friction factor f?" Consider calculation of energy loss in a duct of arbitrary cross section. If we choose the hydraulic diameter as the relevant length dimension for the cross section, the energy loss would obey an equation of the form

$$gh_L = F\{V, L, D_h, \rho, \mu, \varepsilon\}.$$

Table 7.5 Equivalent diameter and hydraulic diameter of common geometric shapes

Shape	Equivalent Diameter D_{eq}	Hydraulic Diameter D_h
Circle	D	D
Rectangle	$\dfrac{2}{\sqrt{\pi}}\sqrt{ab}$	$\dfrac{ab}{2(a+b)}$
Isosceles triangle	$\left(\dfrac{4a^2b^2 - a^4}{4\pi^2}\right)^{1/4}$	$\dfrac{\sqrt{4a^2b^2 - a^4}}{a + 2b}$
Ellipse	$\sqrt{ab}$	$\approx \dfrac{2\sqrt{2}ab}{\sqrt{a^2 + b^2}}$
Concentric annulus	$\sqrt{D_2^2 - D_1^2}$	$D_2 - D_1$

In terms of dimensionless parameters,

$$\frac{gh_L}{\frac{1}{2}V^2} = F\left(\frac{L}{D_h}, \frac{\rho V D_h}{\mu}, \frac{\varepsilon}{D_h}\right).$$

With the aid of Eq. (7.75), this can be reduced to

$$\frac{gh_L}{\frac{1}{2}V^2} = \left(\frac{L}{D_h}\right)\tilde{F}\left(\mathbf{R}_h, \frac{\varepsilon}{D_h}\right),$$

where

$$\mathbf{R}_h = \frac{\rho V D_h}{\mu}. \tag{7.76}$$

The friction factor for the arbitrary duct is defined by

$$f \equiv \tilde{F}\left(\mathbf{R}_h, \frac{\varepsilon}{D_h}\right).$$

This equation is a generalization of Eq. (7.27).

The function $\tilde{F}$ might be different for different duct shapes. We have rather complete information about $\tilde{F}$ for one particular duct shape (circular). This information consists of the Hagen-Poiseuille equation, the Colebrook equation, and the Moody chart. Using the "equivalent" circular pipe concept, we assume that the function $\tilde{F}$ for circular pipes can be used as the general function for any duct simply by replacing the circular pipe diameter wherever it occurs in the circular pipe $\tilde{F}$ by the hydraulic diameter. According to this assumption, the friction factor for a noncircular duct can be approximated by

$$f \approx \frac{64}{\mathbf{R}_h}, \qquad \mathbf{R}_h < 2300, \tag{7.77}$$

or

$$\frac{1}{\sqrt{f}} \approx -2.0 \log\left(\frac{\varepsilon/D_h}{3.7} + \frac{2.51}{\mathbf{R}_h\sqrt{f}}\right), \qquad \mathbf{R}_h > 4000, \tag{7.78}$$

or, reading f from the Moody chart with

$$\mathbf{R} = \mathbf{R}_h, \qquad \frac{\varepsilon}{D} = \frac{\varepsilon}{D_h}.$$

There is no theoretical basis for this assumption, although it seems reasonable. The proof of an assumption is comparison with more exact theories or experimental data. It turns out that the hydraulic diameter approach works remarkably well, being accurate to $\pm10\%$ in turbulent flow. Since the Colebrook equation and Moody chart are accurate to only about $\pm10\%$ for circular pipe flow, the hydraulic diameter approach is quite acceptable for turbulent flow calculations.

The following summarizes the hydraulic diameter approach for the energy/pressure loss problem:

- The duct shape is assumed to be known. Calculate the hydraulic diameter from Eq. (7.74) or use Table 7.5.
- Calculate $\mathbf{R}_h$ and ε/D_h.
- Using $\mathbf{R}_h$ and ε/D_h, calculate f from Eq. (7.77) *or* Eq. (7.78) or read f from the Moody chart.
- Calculate energy losses from Eq. (7.75).

Analysis of the flow rate/velocity problem or the pipe (duct) sizing problem is very similar to the procedures outlined in the previous section (with D replaced by D_h of course), except for one twist. The average velocity and flow rate are not related by the hydraulic diameter, so Eq. (7.55) doesn't apply. The proper equation is

$$Q = \frac{\pi}{4} V(D_{eq})^2 = \frac{\pi}{4} V(D_h)^2 \left(\frac{D_{eq}}{D_h}\right)^2. \tag{7.79}$$

For a given cross-section shape, D_{eq}/D_h is a constant. We will leave the modification of the procedures for the flow rate/velocity and pipe (duct) sizing problems as an exercise.

Although it is widely used, the hydraulic diameter approach is not very accurate for laminar flow. Also, while the accuracy is acceptable for turbulent flow, it could be better. Can we find a better equivalent circular pipe dimension? The answer is yes. We reconsider how the equivalent diameter D_{eq} and the hydraulic diameter D_h were defined. The equivalent diameter makes the continuity equation work exactly. The hydraulic diameter makes the momentum equation work exactly. Can we find a new characteristic dimension (we will call it the *effective diameter, D_{eff}*) that makes the circular pipe friction factor equation work exactly? We can for laminar flow at least.

To begin we need an exact solution for the friction factor in a non-circular duct. We have one such solution in Section 4.2.1. The flow considered there, fully developed flow between infinite parallel plates, can be thought of as flow in a rectangular duct whose dimension perpendicular to the x,y-plane is very large (see Fig. 7.15). The hydraulic diameter of this duct is

$$D_h = \lim_{W \to \infty} \frac{4(2Y)(W)}{2(2Y + W)} = 4Y.$$

The friction factor for this flow is found using Eq. (7.75) and a result of Example 5.12 and is*

$$f = \frac{24\mu}{\rho VY} = \frac{96\mu}{\rho VD_h} = \frac{96}{\mathbf{R}_h}. \tag{7.80}$$

This is an *exact* expression for the friction factor in terms of the hydraulic diameter. If we use the hydraulic diameter in the Hagen-Poiseuille equation, we get

$$f \approx \frac{64}{\mathbf{R}_h}.$$

This value of f is 33% too low. If, however, we introduce an *effective diameter* defined by

$$D_{eff} \equiv \frac{64}{96} D_h$$

and the effective Reynolds number

$$\mathbf{R}_{eff} = \frac{\rho VD_{eff}}{\mu},$$

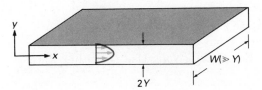

Figure 7.15 Flow in a very wide rectangular duct.

---

* Remember that V is the average velocity.

then

$$f = \frac{64}{\mathbf{R}_{\text{eff}}},$$

exactly.

We can extend this idea to laminar flow in other geometries. Although it is beyond the scope of this book, exact solutions for the laminar friction factor have been obtained for many geometries [5]. All these solutions are of the general form

$$f\mathbf{R}_h = \text{Constant}. \tag{7.81}$$

For the circular pipe, the constant is 64, and for the very wide rectangular duct it is 96. The form of these equations is deducible from dimensional reasoning. The product

$$f\mathbf{R}_h \sim \left(\frac{\text{Force due to pressure drop}}{\text{Inertia force}} \right) \left(\frac{\text{Inertia force}}{\text{Viscous force}} \right)$$

$$\sim \frac{\text{Force due to pressure drop}}{\text{Viscous force}}.$$

In fully developed laminar flow, pressure drop is due to viscous effects *only*, so the product $f\mathbf{R}_h$ should be constant. The constant is different for different geometries. Table 7.6 gives values of $f\mathbf{R}_h$ for several common geometries.

Using Eq. (7.81), we can define an effective diameter for any shape by

$$D_{\text{eff}} = \frac{64}{(f\mathbf{R}_h)} D_h. \tag{7.82}$$

You have probably concluded that there would be little point in going to all the trouble to calculate an effective diameter for laminar flow because f can be calculated directly from Eq. (7.81) and Table 7.6. If so, you are right. Since the effective diameter approach considers the details of the flow, it may prove to be more accurate than the hydraulic diameter approach even in turbulent flow. Experimental measurements by Jones [6] indicate that this is indeed the case. The "laminar effective diameter," defined by Eq. (7.82), is the most accurate replacement for the circular pipe diameter D when evaluating friction factor in both laminar and turbulent flows. The procedure for calculating losses in noncircular duct flow using the effective diameter approach is as follows:

- The duct shape is assumed to be known. Calculate the effective diameter from Eq. (7.82) and Table 7.6 (or other information on the laminar flow friction factor for your particular duct shape).
- Calculate $\mathbf{R}_{\text{eff}}$ and $\varepsilon/D_{\text{eff}}$.
- Using $\mathbf{R}_{\text{eff}}$ and $\varepsilon/D_{\text{eff}}$, calculate f from the Hagen-Poiseuille or Colebrook equations or read f from the Moody chart.
- Calculate energy loss from Eq. (7.75). (*Note:* Use D_h rather than D_{eff} in L/D_h.)

Solution of the velocity/flow rate and pipe (duct) sizing problems is similar to the procedure using hydraulic diameter.

Table 7.6 Laminar friction constant for various duct shapes

Duct Shape	Laminar Friction Constant $f\,R_h$
Ellipse	64
Equilateral triangle	53.33
Concentric annulus $\left(d=\dfrac{D_1}{D_2}\right)$	$64\left[\dfrac{(1-d)^2}{1+d^2+\dfrac{1-d^2}{\ln d}}\right]$
Rectangle	$\dfrac{b}{a}=\begin{cases}0.0 & 96.00\\0.05 & 89.91\\0.1 & 84.68\\0.125 & 82.34\\0.167 & 78.81\\0.25 & 72.93\\0.4 & 65.47\\0.5 & 62.19\\0.75 & 57.89\\1.0 & 56.91\end{cases}=f\mathbf{R_h}$

The effective diameter approach is not in widespread use. The hydraulic diameter approach is much more common. For turbulent flow, either approach is satisfactory, although the effective diameter approach is slightly more accurate. The effective diameter approach is superior for laminar flow.

EXAMPLE 7.6 Compares the difference in using the hydraulic and effective diameters to solve the energy/pressure loss problem for laminar flow

Machine oil with a specific gravity of 0.85 and a kinematic viscosity of 1.3×10^{-4} m²/s flows through a horizontal, annular channel between concentric tubes at a rate of 0.0010 m³/s. The channel has an inner diameter of 25 cm, an outer diameter of 27 cm, and a length of 1.0 m. Find the pressure loss.

Solution

Given
Machine oil with $S = 0.85$ and $v = 1.3 \times 10^{-4}$ m²/s.
Volume flow rate 0.0010 m³/s through annular channel between concentric
   drawn tubing

Inner diameter 25 cm
Outer diameter 27 cm
Length 1.0 m

Find

Pressure loss

Solution

We assume that the flow is fully developed. We will solve the problem first using the hydraulic diameter and then the effective diameter. In both cases the pressure loss is

$$\Delta p_L = \rho g h_L = \rho f \left(\frac{L}{D_h}\right)\left(\frac{V^2}{2}\right).$$

The velocity V is

$$V = \frac{Q}{A} = \frac{4Q}{\pi(D_2^2 - D_1^2)},$$

where D_1 is the inner diameter and D_2 the outer diameter. Substituting given values, we have

$$V = \frac{4\left(0.0010 \, \frac{m^3}{s}\right)\left(\frac{100 \, cm}{m}\right)^2}{\pi[(27 \, cm)^2 - (25 \, cm)^2]} = 0.122 \, m/s.$$

With the hydraulic diameter approach, Table 7.5 gives

$$D_h = D_2 - D_1 = 27 \, cm - 25 \, cm = 2.0 \, cm.$$

The corresponding Reynolds number is calculated from Eq. (7.76):

$$\mathbf{R}_h = \frac{VD_h}{\nu}$$

$$= \frac{\left(0.122 \, \frac{m}{s}\right)(2.0 \, cm)\left(\frac{m}{100 \, cm}\right)}{\left(1.3 \times 10^{-4} \, \frac{m^2}{s}\right)} = 18.8.$$

Since we have laminar flow, Eq. (7.77) gives the friction factor

$$f = \frac{64}{\mathbf{R}_h} = \frac{64}{18.8} = 3.4.$$

The head loss is

$$gh_L = 3.4\left(\frac{1.0 \, m}{2.0 \, cm}\right)\frac{\left(0.122 \, \frac{m}{s}\right)^2}{2}\left(\frac{100 \, cm}{m}\right)$$

$$= 1.265 \, \frac{m^2}{s^2} = 1.265 \, \frac{N \cdot m}{kg}.$$

Using water density at 4°C, we have machine oil density of

$$\rho = (0.85)(1000 \, kg/m^3) = 850 \, kg/m^3.$$

The pressure loss is

$$\Delta p_L = \rho g h_L = \left(850 \, \frac{kg}{m^3}\right)\left(1.265 \, \frac{N \cdot m}{kg}\right),$$

Answer

$$| \; \Delta p_L = 1075 \, Pa. \; |$$

With the effective diameter approach, Table 7.5 gives

$$d = \frac{D_1}{D_2} = \frac{25 \text{ cm}}{27 \text{ cm}} = 0.926,$$

$$f\mathbf{R}_h = 64 \left[\frac{(1 - 0.926)^2}{1 + 0.926^2 + \dfrac{1 - 0.926^2}{\ln 0.926}} \right] = 64(1.489).$$

Equation (7.82) gives the effective diameter,

$$D_{\text{eff}} = \frac{64}{f\mathbf{R}_h} D_h = \frac{64(2.0 \text{ cm})}{64(1.489)} = 1.34 \text{ cm},$$

and the friction factor,

$$f = \frac{64}{\mathbf{R}_h} \left(\frac{D_h}{D_{\text{eff}}} \right) = \left(\frac{64}{18.8} \right) \left(\frac{2.0 \text{ cm}}{1.34 \text{ cm}} \right) = 5.1.$$

The head loss is (using D_h in L/D)

$$gh_L = 5.1 \left(\frac{1.0 \text{ m}}{2.0 \text{ cm}} \right) \frac{\left(0.122 \dfrac{\text{m}}{\text{s}} \right)^2}{2} \left(\frac{100 \text{ cm}}{\text{m}} \right) = 1.898 \text{ N·m/kg}.$$

The pressure loss is

$$\Delta p_L = \left(850 \frac{\text{kg}}{\text{m}^3} \right) \left(1.898 \frac{\text{N·m}}{\text{kg}} \right),$$

Answer

$$\mid \Delta p_L = 1613 \text{ Pa.} \mid$$

Discussion

The effective diameter approach provides a more accurate estimate of the pressure loss. The "effective" friction factor would also have been computed as

$$f = \frac{64}{\mathbf{R}_{\text{eff}}} = \frac{64 \, v}{V D_{\text{eff}}}.$$

This example is a comparison problem because it shows two approaches to solve a problem and the difference in the results.

7.2.8 Detailed Consideration of Fully Developed Turbulent Flow in a Circular Pipe

In this section, we fill in some of the background for the Colebrook formula and the Moody chart. We investigate the nature of turbulent flow near a surface and develop a semi-empirical theory for the velocity profile. As was the case with laminar flow, integration of the velocity profile yields a useful relation between friction factor, average velocity, and other key parameters. The information about turbulent flow that we present here is a condensation of many years of research by many people. Using the benefit of hindsight, we will present only the correct assumptions and deductions, disregarding all of the blind alleys that were pursued in developing the information.

First we consider turbulent flow near a smooth surface such as a pipe wall. Figure 7.16(a) shows the (time average) velocity profile. If

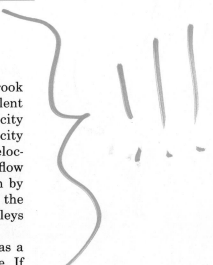

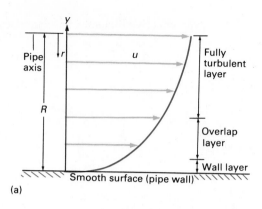

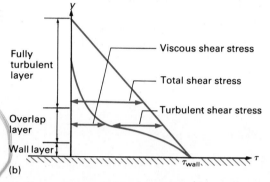

Figure 7.16 Velocity profile and shear stress distributions in turbulent flow near a smooth surface.

this profile represents a pipe flow, Eq. (7.32) tells us that the shear stress distribution is linear. The contributions of viscous stress and turbulent stress to the total stress are shown in Fig. 7.16(b). We can divide the flow into three regions:

- An "inner" *wall layer* where turbulent stress is very small and viscous stress is large;
- An "outer" *fully turbulent layer* where viscous stress is negligible and turbulent stress is large;
- An "*overlap*" *layer* where both stresses are significant.

This multilayer structure is the basis of most mathematical modeling and analysis of turbulent flow near a surface. A very accurate theory describing the velocity profile can be constructed by examining each layer in detail and then patching the layers together. Although there are several possible approaches, we will present the approach based on dimensional reasoning that was pioneered by L. Prandtl, T. von Karman, and C. Millikan.

The presence of the inner layer can be understood by recalling that turbulent stress results from velocity fluctuations, which must vanish very near the wall. The thickness of this layer has been magnified in Fig. 7.16; it typically occupies about 2% of the entire layer. Since turbulence effects are small in the inner layer, it is sometimes called the *viscous sublayer* or the *laminar sublayer*. Since the inner layer is very thin, Prandtl assumed that the flow behaves as if it knew nothing about the outer flow; that is, there is no dependence on happenings near the pipe axis ($y = R$) or on the maximum (or average) velocity. Accordingly, the velocity in the inner layer is given by a function of the form

$$u = f\{\mu, \tau_w, \rho, y\}.$$

Applying dimensional analysis, we obtain

$$\frac{u\sqrt{\rho}}{\sqrt{\tau_w}} = F\left(\frac{y\sqrt{\tau_w}}{\sqrt{\rho}\,\mu}\right). \qquad (7.83)$$

Introducing the definitions

$$u_\tau \equiv \sqrt{\frac{\tau_w}{\rho}}, \qquad (7.84)$$

$$u^+ \equiv u/u_\tau, \qquad (7.85)$$

and

$$y^+ \equiv \frac{yu_\tau}{\nu}, \qquad (7.86)$$

we can write Eq. (7.83) as

$$u^+ = F\{y^+\}. \qquad (7.87)$$

The parameter u_τ is called the *friction velocity* and has the dimensions $[L/T]$. The dimensionless velocity u^+ is related to the skin friction coefficient $u^+ = (u/V)\sqrt{2/C_f}$. The dimensionless distance y^+ is a type of Reynolds number. Equation (7.87) is called the *law of the wall*. Determination of the function F requires a specific theory or experimental data. A very simple theory assumes that the stress in the inner layer is entirely due to viscosity and is constant. According to these assumptions,

$$\tau = \mu\left(\frac{du}{dy}\right) = \tau_w.$$

Integrating and using the no-slip condition give

$$u = \frac{\tau_w}{\mu}\,y,$$

or

$$u^+ = y^+. \qquad (7.88)$$

Equation (7.88) is valid only *very* near the wall.

Next we consider the outer layer. In this layer, viscous stress is negligible, so viscosity is not important. Likewise, the details of the flow near the wall are not important; the only thing that the outer layer "knows" about the wall is that there is shear stress there. The velocity distribution in the outer flow does depend on the total thickness of the layer (the radius R in pipe flow) and on the maximum velocity, u_{max}. We assume that the velocity in the outer layer obeys a relation of the form

$$u = g\{u_{max}, R, \tau_w, \rho, y\}. \qquad (7.89)$$

Von Karman deduced that Eq. (7.89) can be written in the special form

$$(u_{max} - u) = \tilde{g}\{R, \tau_w, \rho, y\}, \qquad (7.90)$$

where $(u_{max} - u)$ is called the *velocity defect* and is treated as a *single* parameter. Equation (7.90) can be reduced to the dimensionless form

$$u^+ = \frac{u_{max}}{u_\tau} + G\left(\frac{y}{R}\right). \tag{7.91}$$

Equation (7.91) is called the *outer law* or the *velocity defect law*. Evaluation of the function G requires a specific theory or experimental data.

Finally we consider the overlap layer. In this layer, both velocity distributions, Eqs. (7.87) and (7.91), must apply. Equating u^+ from both equations, we have

$$\frac{u_{max}}{u_\tau} + G\left(\frac{y}{R}\right) = F(y^+).$$

We can introduce y^+ on the left also:

$$\frac{u_{max}}{u_\tau} + G\left(\frac{yu_\tau}{v}\left(\frac{v}{Ru_\tau}\right)\right) = \frac{u_{max}}{u_\tau} + G\left(y^+\left(\frac{v}{Ru_\tau}\right)\right) = F(y^+). \tag{7.92}$$

This equation is rather cryptic, but we can deduce some very valuable information from it. The parameters u_{max}, u_τ, R, and v do not vary with y, and so the equation has the form

$$G(\text{Constant}_1 y^+) = F(y^+) + \text{Constant}_2$$

In the overlap layer, G and F have the property that a function of a constant times the argument equals a function of the argument plus another constant. Only one type of function has this property; that is the logarithm,

$$\ln (Cy^+) = \ln y^+ + \ln C.$$

In the overlap layer, both F and G must be logarithmic. A general form of Eq. (7.87), with F a logarithmic function, is

$$\frac{u}{u_\tau} = \frac{1}{\kappa} \ln \left(\frac{yu_\tau}{v}\right) + B, \tag{7.93}$$

where $1/\kappa$ and B are constants. The logarithmic velocity profile in the overlap layer is probably the most important item of information available about turbulent flow. Notice that this law does not apply all the way to the wall ($\ln 0$ is undefined); it must be patched to an inner law such as Eq. (7.88).

You are probably thinking that it took some very fancy "footwork" to derive the logarithmic velocity profile. Is it valid? We can answer this question by comparing the logarithmic velocity profile with data. Comparison with experimental data is also necessary to establish the constants κ and B.

Figure 7.17 shows typical velocity profile data in semilogarithmic coordinates. Both the logarithmic law and the linear (near wall) law are verified by the data. Note that the difference between the outer layer velocities and the logarithmic law is very small for pipe flow even though the logarithmic law is not supposed to apply in the outer region. The logarithmic scale for y^+ can be deceiving. In the figure, it looks as if the wall layer takes up about a third of the pipe, but in fact the frac-

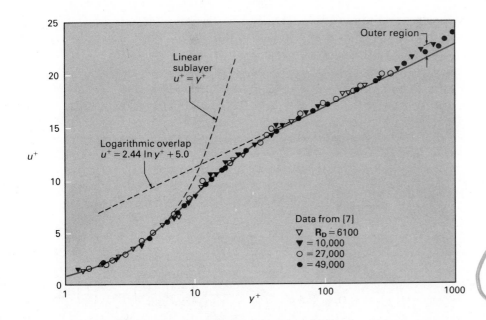

Figure 7.17 Comparison of log law and linear wall law velocity profiles with experimental data.

tion taken up by the wall layer is

$$\frac{y^+ \text{ (Edge of wall layer)}}{R^+} \approx \frac{20}{1000} = 0.02.$$

By fitting a straight line to the overlap region data, the constants κ and B are found to be

$$\kappa \approx \frac{1}{2.44} = 0.41 \quad \text{and} \quad B \approx 5.0.$$

The need to determine these constants from experiments emphasizes that any theory of turbulent flow requires experimental data to be complete.

At this point, we will compare the information we have about turbulent flow velocity profiles and our information about laminar profiles. In laminar flow, the velocity profile is related to viscosity, pipe radius, and pressure drop (Eq. 7.33). By integrating the velocity distribution across the pipe, we were able to relate pressure drop, viscosity, pipe radius, and average velocity. The ultimate result was the Hagen-Poiseuille equation for friction factor in terms of Reynolds number.

Now consider the velocity profile for turbulent flow. The logarithmic law is accurate over most of the cross section, failing badly only in the region very near the wall ($y^+ < 5$). Using the logarithmic profile for the whole cross section and noting that

$$y = R - r,$$

we have

$$u \approx 2.44 u_\tau \ln\left[\frac{u_\tau(R-r)}{\nu}\right] + 5.0 u_\tau. \tag{7.94}$$

Integrating this velocity distribution across the pipe from $y = 0$ to $y = R$ will yield a relation between friction velocity, average velocity, density, viscosity, and pipe radius. This can be transformed into a relation involving friction factor and Reynolds number.

Carrying out this procedure, we obtain the average velocity:*

$$V = \frac{Q}{A} = \frac{2\pi \int_0^R ur\,dr}{\pi R^2}.$$

Substituting the logarithmic velocity profile gives

$$V = \frac{2}{R^2} \int_0^R \left[2.44u_\tau \ln\left(\frac{u_\tau(R-r)}{\nu}\right) + 5.0u_\tau \right] r\,dr.$$

Performing the integration, we get

$$\frac{V}{u_\tau} = 2.44 \ln\left(\frac{u_\tau R}{\nu}\right) + 1.34. \tag{7.95}$$

Now

$$\frac{V}{u_\tau} = \sqrt{\frac{\rho V^2}{\tau_w}} = \sqrt{\frac{2}{C_f}} = \sqrt{\frac{8}{f}} \quad \text{and} \quad \frac{u_\tau R}{\nu} = \left(\frac{u_\tau}{V}\right)\left(\frac{VD}{2\nu}\right) = \frac{1}{2}\sqrt{\left(\frac{f}{8}\right)}(\mathbf{R}).$$

Substituting into Eq. (7.95) and changing to common logarithms, we get

$$\frac{1}{\sqrt{f}} = 1.99 \log\left(\mathbf{R}\sqrt{f}\right) - 1.02.$$

This equation was first derived by Prandtl in 1935. Upon comparing it with Nikuradse's data for smooth pipes, Prandtl found it to be slightly inaccurate due to the approximation of the wall and outer layers by the logarithmic law. Prandtl obtained a better fit by adjusting the constants slightly:

$$\frac{1}{\sqrt{f}} = 2.0 \log\left(\mathbf{R}\sqrt{f}\right) - 0.8. \tag{7.46}$$

This is the smooth pipe friction factor formula presented earlier.

The theory above applies only to smooth walls. A theory for rough walls can be developed by appropriate modifications. Roughness disturbs the flow in the wall region. If roughness elements are large enough, they completely break up the nearly laminar flow in the wall region. We can obtain a clue about the effects of roughness from Fig. 7.17. If the surface roughness is completely buried in the sublayer such that

$$\varepsilon^+ = \frac{\varepsilon u_\tau}{\nu} < 5,$$

then the pipe wall will appear smooth. If the roughness is large enough to protrude into the overlap layer such that

$$\varepsilon^+ > 70,$$

then the sublayer will be completely broken up and the flow will be

---

* The overbar is dropped for convenience.

completely turbulent. For values of roughness given by

$$5 < \varepsilon^+ < 70,$$

the flow is *transitionally rough* (see the Moody chart).

Von Karman developed a theory for the velocity profile in fully turbulent flow. If the flow is fully turbulent, the wall region velocity depends on roughness height but not on viscosity:

$$\frac{u}{u_\tau} = F\left(\frac{y}{\varepsilon}\right). \tag{7.96}$$

For fully turbulent flow, this equation replaces Eqs. (7.83) and (7.87). Since the flow in the outer layer does not depend on the flow details near the wall, Eqs. (7.89)–(7.92) still apply. By the same arguments that led to Eq. (7.93), we find that the velocity profile in the overlap layer for fully rough flow is given by

$$\frac{u}{u_\tau} = C_1 \ln\left(\frac{y}{\varepsilon}\right) + C_2,$$

where C_1 and C_2 are constants.

Comparison of this equation with experimental data for fully rough pipe flow shows that the logarithmic form is valid and gives values for the constants:

$$C_1 = \frac{1}{\kappa} = \frac{1}{0.41}, \qquad \text{Same as smooth wall}$$

$$C_2 = 8.5. \qquad \text{Different from smooth wall}$$

Using these constants, we find the velocity profile for fully turbulent flow:

$$u \approx 2.44 u_\tau \ln\left(\frac{y}{\varepsilon}\right) + 8.5 u_\tau. \tag{7.97}$$

A comparison of smooth wall and fully rough wall velocity profiles is shown in Fig. 7.18. The rough wall velocity profile can be integrated to find the average velocity. The resulting equation is rearranged to give a friction factor equation. The procedure is exactly the same as that which led to Eq. (7.46). The result is

$$\frac{1}{\sqrt{f}} = -2.0 \log\left(\frac{\varepsilon/D}{3.7}\right), \tag{7.47}$$

valid for fully rough flow.

The Colebrook formula is constructed from the smooth and fully rough friction factor expressions. The smooth wall formula can be written

$$\frac{1}{\sqrt{f}} = 2.0 \log\left(\mathbf{R}\sqrt{f}\right) - \log\left(10^{0.8}\right)$$

$$= -2.0\left[\log\left(10^{0.4}\right) - \log\left(\mathbf{R}\sqrt{f}\right)\right]$$

$$= -2.0 \log\left(\frac{2.51}{\mathbf{R}\sqrt{f}}\right). \tag{7.98}$$

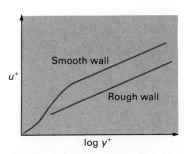

u^+

Smooth wall

Rough wall

log y^+

Figure 7.18 Comparison of smooth wall and rough wall velocity distribution in semi-logarithmic coordinates.

The rough wall formula can be rearranged as

$$\frac{1}{\sqrt{f}} = -2.0 \log \left[\frac{\left(\frac{\varepsilon u_\tau}{v} \right) \left(\frac{v}{u_\tau D} \right)}{3.7} \right] = -2.0 \log \left[\frac{\varepsilon^+}{3.7 \frac{u_\tau}{V} \mathbf{R}} \right].$$

Since

$$\frac{u_\tau}{V} = \sqrt{\frac{f}{8}},$$

$$\frac{1}{\sqrt{f}} = -2.0 \log \left(\frac{\varepsilon^+}{1.31 \mathbf{R} \sqrt{f}} \right).$$

Colebrook constructed his formula by combining the smooth wall and the rough wall and the rough wall parameters in the same logarithm:

$$\frac{1}{\sqrt{f}} = -2.0 \log \left[\frac{\varepsilon^+}{1.31 \mathbf{R} \sqrt{f}} + \frac{2.51}{\mathbf{R} \sqrt{f}} \right].$$

This equation covers both limits since if ε^+ is small ($\varepsilon^+ \ll 3.3$), the smooth wall equation results and if ε^+ is large ($\varepsilon^+ \gg 3.3$), the rough wall equation results.*

* Note that $(1.31)(2.51) \approx 3.3$.

EXAMPLE 7.7 Illustrates calculations with turbulent velocity profiles

Water at 20°C flows through a horizontal, smooth pipe with an average velocity of 0.807 m/s and a Reynolds number of 2.3×10^4. Compare the dimensionless velocity (u^+) calculated from the following:

- the linear law of the wall, the power law equation, and Fig. 7.17 for $y^+ = 5$,
- the logarithmic velocity equation, the power law equation, and Fig. 7.17 for $y^+ = 100$ and $y^+ = 500$.

Solution

Given
20°C water flowing through horizontal, smooth pipe
Average velocity 0.807 m/s
Reynolds number 2.3×10^4

Find
Comparison of dimensional velocities from linear law of the wall, power law equation, logarithmic velocity equation, and Fig. 7.17 for $y^+ = 5$, $y^+ = 100$, and $y^+ = 500$

Solution
The primary equations are

$$u^+ = y^+, \qquad u^+ = \frac{1}{0.41} \ln y^+ + 5.0, \qquad \text{and}$$

$$u = u_{max} \left(\frac{y}{R} \right)^{1/n},$$

where R is the pipe radius and y the distance measured from the pipe wall. Using Eqs. (7.85) and (7.86) gives the power law formula as

$$u^+ = \frac{u_{\max}}{u_\tau}\left(\frac{y^+}{R^+}\right)^{1/n},$$

where

$$R^+ = \frac{Ru_\tau}{v}.$$

To find the dimensionless velocities, we will need values for n, u_τ, v, R^+, and $u_{\max}$. For $\mathbf{R} = 2.3 \times 10^4$, Table 7.3 gives $n = 6.6$. For water at 20°C, Appendix A.5 gives

$$\mu = 1.00 \times 10^{-3}\ \text{N·s/m}^2, \qquad \rho = 998\ \text{kg/m}^3.$$

For smooth pipes, Eq. (7.45) gives

$$f = 0.3164(\mathbf{R})^{-0.25} = 0.3164(2.3 \times 10^4)^{-0.25} = 0.0257.$$

Equation (7.31) then gives

$$\frac{\tau_w}{\rho} = \frac{f}{8}V^2 = \frac{0.0257(0.807\ \text{m/s})^2}{8} = 0.00209\ \text{m}^2/\text{s}^2.$$

Next we have

$$u_\tau = \sqrt{\frac{\tau_w}{\rho}} = 0.0457\ \text{m/s}$$

and

$$v = \frac{\mu}{\rho} = \frac{\left(1.00 \times 10^{-3}\ \dfrac{\text{N·s}}{\text{m}^2}\right)}{\left(998\ \dfrac{\text{kg}}{\text{m}^3}\right)} = 1.00 \times 10^{-6}\ \text{m}^2/\text{s}.$$

The definition of the Reynolds number gives

$$R = \frac{D}{2} = \frac{v\mathbf{R}}{2V} = \frac{\left(1.00 \times 10^{-6}\ \dfrac{\text{m}^2}{\text{s}}\right)(2.3 \times 10^4)}{2\left(0.807\ \dfrac{\text{m}}{\text{s}}\right)} = 0.0143\ \text{m}.$$

R^+ is calculated as*

$$R^+ = \frac{Ru_\tau}{v} = \frac{(0.0143\ \text{m})\left(0.0457\ \dfrac{\text{m}}{\text{s}}\right)}{\left(1.00 \times 10^{-6}\ \dfrac{\text{m}^2}{\text{s}}\right)} = 654.$$

Equation (7.52) is now used to calculate the maximum velocity $u_{\max}$:

$$u_{\max} = \frac{(n+1)(2n+1)}{2n^2}(\bar{V}) = \frac{7.6(14.2)}{2(6.6)^2}\left(0.807\ \frac{\text{m}}{\text{s}}\right) = 1.0\ \text{m/s}.$$

We are now able to calculate the dimensionless velocities. For $y^+ = 5.0$, the law of the wall gives

$$u^+ = 5.0,$$

---

* R^+ could also be calculated from

$$R^+ = \frac{Ru_\tau}{v} = \frac{R}{2}\sqrt{\frac{f}{8}}.$$

the power law equation gives

$$u^+ = \left(\frac{1.00 \, \frac{m}{s}}{0.0457 \, \frac{m}{s}} \right) \left(\frac{5.0}{654} \right)^{1/6.6} = 10.5,$$

and Fig. 7.17 gives $u^+ = 5.0$.

For $y^+ = 100$, the logarithmic velocity equation gives

$$u^+ = \frac{1}{0.41}(\ln 100) + 5.0 = 16.2,$$

the power law equation gives

$$u^+ = \left(\frac{1.00 \, \frac{m}{s}}{0.0457 \, \frac{m}{s}} \right) \left(\frac{100}{654} \right)^{1/6.6} = 16.5,$$

and Fig. 7.17 gives $u^+ \approx 16.5$.

For $y^+ = 500$, the logarithmic velocity equation gives

$$u^+ = \frac{1}{0.41}(\ln 500) + 5.0 = 20.2,$$

the power law equation gives

$$u^+ = \left(\frac{1.00 \, \frac{m}{s}}{0.0457 \, \frac{m}{s}} \right) \left(\frac{500}{654} \right)^{1/6.6} = 21.0,$$

and Fig. 7.17 gives $u^+ \approx 21.5$.

A comparison summary of u^+ is contained in the following table.

y^+	Linear Law of Wall	Logarithmic Equation	Power Law	Figure 7.17
5	5.0	Not applicable	10.5	5.0
100	Not applicable	16.2	16.5	16.5
500	Not applicable	20.2	21.0	21.5

Answer
◀

Discussion

The Reynolds number of 2.3×10^4 used in this example falls in the range of Reynolds numbers used to plot Fig. 7.17. The power law equation with $n = 6.6$ gives a reasonably good approximation to the turbulent velocity profile of Fig. 7.17.

You might try to answer the following:

- Does the power law equation with $n = 6.6$ provide a better approximation for u^+ than either the law of the wall or the logarithmic velocity equation at $y^+ = 10$ and at $y^+ = 20$?
- Find a value of n other than 6.6 that gives a better fit to the data of Fig. 7.17.

7.3 Flow in Pipe and Duct Systems

Pipe and duct systems include many components in addition to pipes and ducts. We will now consider the flow in some of these components. As in our analysis of pipe/duct flow, our primary objective is to discuss relations between the component geometry, the flow rate through the component, and the mechanical energy loss generated by the component. After considering the components, we will examine complete systems made up by combining pipes or ducts with various components.

7.3.1 System Components and Local Losses

Figure 7.19 shows some auxiliary components that are used in a pipe or duct system. They include the following:

- *Transitions* for changing pipe size
- *Elbows* and *bends* for changing flow (pipe) direction
- *Tees* and *laterals* for dividing or mixing streams
- *Valves* for controlling flow
- *Entrances* and *exits,* special cases of transitions where the upstream (entrance) or downstream (exit) regions are considered infinite in extent

All of these components introduce disturbances that cause turbulence and mechanical energy loss in addition to that which occurs in the basic pipe flow. Figure 7.20 illustrates flow through a sudden enlargement, a rather crude transition. The flow is not able to turn the sharp corner, so it separates, generating a great deal of disorderly motion. The disorderly motion is often loosely called "turbulence"; however, the motion near the sudden enlargement is more orderly than true turbulence. The motion degenerates into true turbulence as the fluid proceeds downstream. The disturbance caused by the expansion persists for some distance downstream as the kinetic energy of the disorderly motion is

Figure 7.19 Some piping system components (valve diagrams courtesy of The Crane Co.).

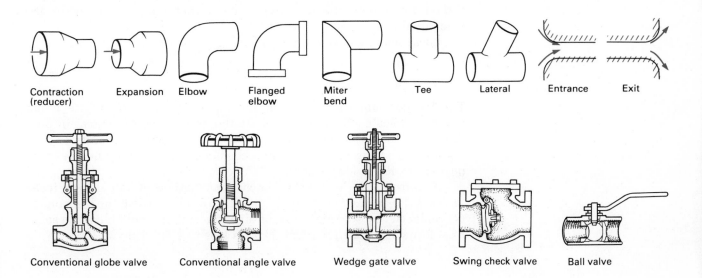

Contraction (reducer)   Expansion   Elbow   Flanged elbow   Miter bend   Tee   Lateral   Entrance   Exit

Conventional globe valve   Conventional angle valve   Wedge gate valve   Swing check valve   Ball valve

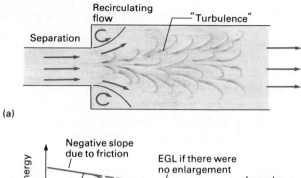

Figure 7.20 Flow detail and energy loss for flow through a sudden enlargement.

dissipated and the flow gradually returns to a fully developed condition. This flow behavior is not unique to the sudden expansion; other components generate similar disturbances. These disturbances, loosely called "turbulence," are responsible for energy loss in the region immediately downstream of the component. More profound disturbances persist further downstream and produce larger losses. A plot (Fig. 7.20) of the energy grade line (EGL) in the immediate vicinity of the component shows that the energy loss occurs over a finite distance; however, when viewed from the perspective of an entire pipe system, the energy losses are localized near the component. We will refer to such losses as *local losses*. Most authors and many engineers refer to such losses as *minor losses*. This is an unfortunate misnomer because local losses may be the dominant losses in some systems. Local losses are added to pipe friction losses to calculate the total energy loss in the system. Local loss is usually the *total* loss in the component. The length of the component should not be included in the pipe length when pipe friction losses are calculated.

Local losses are calculated by using a loss coefficient **K**:

$$gh_L = \mathbf{K}\left(\frac{V^2}{2}\right).\qquad(7.99)$$

The loss coefficient is a function of component geometry and Reynolds number. A scale-effect parameter similar to relative roughness in pipe flows may also be important. Since most components generate energy loss by promoting turbulence, loss coefficients are usually nearly independent of Reynolds number.

An alternative to the loss coefficient method of calculating local losses is the "equivalent length" method. In this method, the component is "replaced" by a straight run of pipe (duct) that would have the same loss. Equating the equivalent friction loss to local loss, we have

$$f\frac{L_{eq}}{D_h}\frac{V^2}{2} = \mathbf{K}\frac{V^2}{2}.$$

so the equivalent length is calculated by

$$L_{eq} = \frac{\mathbf{K}}{f} D_h,$$ (7.100)

where f is the friction factor in the pipe (duct) to which the component is attached. The equivalent length concept is reasonably straightforward for all components except transitions. For transitions, one must be careful to specify whether the equivalent length is added to the smaller or larger pipe (duct).

In practice, local loss coefficients or equivalent lengths are obtained from handbooks or, for best accuracy, from component manufacturers' specifications. These data have usually been obtained by experiment. References [8, 9, 10] contain extensive information about local loss coefficients and equivalent lengths. We present a brief discussion of loss generation in various types of components and indicate typical values of $\mathbf{K}$.

Transitions. The sudden expansion (Fig. 7.20) is the only configuration for which the loss coefficient can be derived by simple analytical methods. Figure 7.21 shows a control volume for this flow. We assume uniform flow enters the expansion at plane 1. Plane 2 is placed downstream a sufficient distance such that the flow is uniform there also. We also assume that the pressure is uniform across planes 1 and 2.

The continuity, linear momentum, and energy equations for the control volume are

$$V_1 A_1 = V_2 A_2,$$

$$(p_1 - p_2)A_2 + \tau_{avg} A_{wall} = \rho(V_2^2 A_2 - V_1^2 A_1), \quad \text{and}$$

$$\frac{p_1}{\rho} + \frac{V_1^2}{2} = \frac{p_2}{\rho} + \frac{V_2^2}{2} + gh_L,$$

respectively. For this particular flow, the average shear stress on the wall (τ_{avg}) can be neglected because the fluid motion near the wall is very sluggish and random. The continuity and momentum equations are solved for the pressure difference

$$p_1 - p_2 = \rho V_1^2 \left[\left(\frac{A_1}{A_2} \right)^2 - \frac{A_1}{A_2} \right].$$

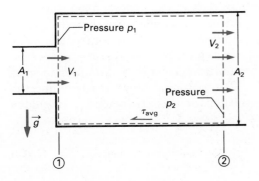

Figure 7.21 Control volume for analysis of flow in a sudden enlargement.

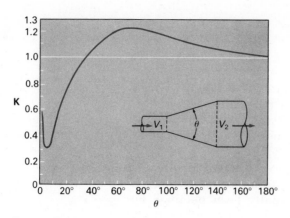

Figure 7.22 Typical values of loss coefficient for a conical gradual expansion.

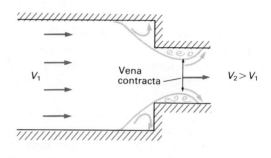

Figure 7.23 Sketch of the flow in a sudden contraction.

Substituting this result into the energy equation and eliminating V_2 by the continuity equation give

$$gh_L = \left(1 - \frac{A_1}{A_2}\right)^2 \left(\frac{V_1^2}{2}\right).$$

Comparing this equation with Eq. (7.99), we find

$$\mathbf{K} = \left(1 - \frac{A_1}{A_2}\right)^2. \qquad \text{Sudden expansion} \qquad \textbf{(7.101)}$$

Note that $\mathbf{K}$ is based on the smaller area (larger velocity). This is an excellent place to point out that minor losses are not always interchangeable with pressure drop. The pressure actually rises in a sudden expansion because of the velocity decrease. The pressure *loss* is positive; however, pressure loss is not interchangeable with pressure drop or pressure change.

Next consider a gradual expansion. Our ability to calculate $\mathbf{K}$ for the sudden expansion was based on the assumption of zero wall shear stress. This is not true for a gradual expansion; in fact, if the expansion is very gradual, the friction loss can be significant. Figure 7.22 shows typical loss coefficients for a gradual conical expansion as a function of included angle. An angle of 0° is a straight pipe, and an angle of 180° corresponds to a sudden enlargement. Note that for angles greater than about 40° it is better to use a sudden expansion. For $\theta > 40°$, separation losses are about the same as in the sudden expansion. Adding friction loss makes the net loss greater than that in the sudden expansion.

Contractions generally have lower loss coefficients than do expansions. Figure 7.23 is a sketch of the flow in a sudden contraction. The flow separates at the corner, and a region of flow disturbance exists downstream. The main flow streamlines converge to a minimum area and then diverge to fill the pipe. The minimum flow area is called the *vena contracta*. An empirical equation for loss coefficient in a sudden contraction in a circular pipe is

$$\mathbf{K} \approx 0.42 \left[1 - \left(\frac{D_2}{D_1}\right)^2\right]. \qquad \text{Sudden contraction} \qquad \textbf{(7.102)}$$

This loss coefficient is based on the larger velocity (V_2). Loss coefficients for gradual contractions are given in Table 7.7.

Bends. Figure 7.24 is a sketch of the flow in a smooth bend in a circular pipe. Losses in bends result from three effects:

- Ordinary friction losses, which are dependent on the length-diameter ratio of the bend and the relative roughness;
- Flow separation on the downstream side of the bend; and
- Secondary flow in the cross-sectional plane due to centrifugal forces.

Figure 7.25 gives coefficients for calculating losses due to separation and secondary flow in bends. Loss due to friction in the bend must be added. The total loss coefficient in a 90° bend is

$$\mathbf{K}_{\text{total}} = \mathbf{K}_{\text{Fig. 7.25}} + f\frac{\left(\dfrac{\pi R}{2}\right)}{D}. \qquad (7.103)$$

Miter bends are often used where space or construction costs must be kept to a minimum. Figure 7.26 gives equivalent lengths for miter bends. The addition of turning vanes can reduce loss by about 80%.

Entrances and Exits. Figure 7.27 shows three types of entrances with corresponding loss coefficients. Note that sharp corner entrances exhibit flow separation and a vena contracta. Loss coefficients for entrances only compute the local energy loss; they do not account for the

Table 7.7 Typical loss coefficients for gradual contractions

Contraction Cone Angle (degrees)	K
30	0.02
45	0.04
60	0.07

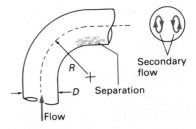

Figure 7.24 Flow in a bend in a circular pipe.

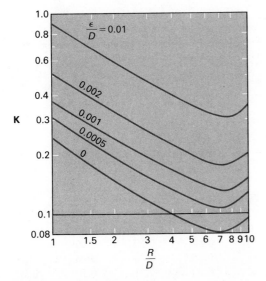

Figure 7.25 Loss coefficients for separation and secondary flow losses in circular pipe bends.

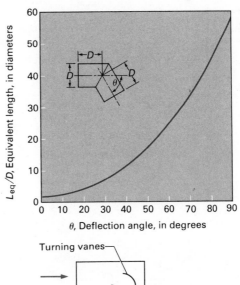

Figure 7.26 Equivalent length for miter bends without turning vanes.

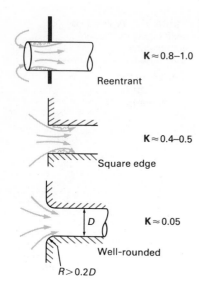

Figure 7.27 Loss coefficients for pipe entrances.

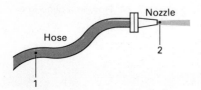

Figure 7.28 Flow through a fire hose and nozzle.

pressure drop that accompanies flow development. The energy represented by this pressure drop is still present at the end of the development length in the form of increased kinetic energy in the fully developed velocity profile* ($\alpha_{FD} > 1.0$). Similarly, the ordinary frictional energy loss due to the development length is not included. The total energy loss between the inlet and the fully developed location is

$$gh_L = \mathbf{K}\frac{V^2}{2} + f\left(\frac{L_e}{D}\right)\left(\frac{V^2}{2}\right),$$

where f is evaluated as if the flow were fully developed.

An exit is a sudden expansion with a very large downstream area. No matter what its shape, the loss coefficient for an exit is 1.0. A word of caution: Exit loss is just the kinetic energy of the flow at the exit. Consider calculation of flow through a fire hose and nozzle (Fig. 7.28). The energy equation for flow between points 1 and 2 is (uniform flow is assumed)

$$\frac{p_1}{\rho} + \frac{V_1^2}{2} + gz_1 = \frac{p_2}{\rho} + \frac{V_2^2}{2} + gz_2 + gh_L.$$

Since the exit kinetic energy has already been included as $V_2^2/2$, it must not be included in the losses. Exit loss is considered only when the downstream point in the energy equation lies beyond the exit a sufficient distance that the exit kinetic energy has been dissipated.

Commercial Pipe Fittings and Valves. Smaller pipes and tubing are assembled with commercial fittings screwed or soldered onto the pipe or bolted by flanges. Table 7.8 gives typical loss coefficients for commer-

Table 7.8 Typical loss coefficients for commercial pipe fittings

Nominal Diameter (in.)	Screwed or Soldered				Flanged				
	$\frac{1}{2}$	1	2	4	1	2	4	8	20
Valves (fully open)									
Globe	14	8.2	6.9	5.7	13	8.5	6.0	5.8	5.5
Gate	0.30	0.24	0.16	0.11	0.80	0.35	0.16	0.07	0.03
Swing check	5.1	2.9	2.1	2.0	2.0	2.0	2.0	2.0	2.0
Angle	9.0	4.7	2.0	1.0	4.5	2.4	2.0	2.0	2.0
Elbows									
45° standard	0.39	0.32	0.30	0.29					
45° long radius					0.21	0.20	0.19	0.16	0.14
90° standard	2.0	1.5	0.95	0.64	0.50	0.39	0.30	0.26	0.21
90° long radius	1.0	0.72	0.41	0.23	0.40	0.30	0.19	0.15	0.10
180° standard	2.0	1.5	0.95	0.64	0.41	0.35	0.30	0.25	0.20
180° long radius					0.40	0.30	0.21	0.15	0.10
Tees									
Line flow	0.90	0.90	0.90	0.90	0.24	0.19	0.14	0.10	0.07
Branch flow	2.4	1.8	1.4	1.1	1.0	0.80	0.64	0.58	0.41

* Once again, pressure drop and energy loss are not equivalent.

cial fittings. Note that there is a scale effect present, with larger fittings having lower loss coefficients.

Table 7.8 also contains loss coefficients for fully open valves. Valves control the flow rate by introducing large losses, especially when partly closed. Table 7.9 gives loss coefficient multipliers for partially closed valves.

Table 7.9 Loss coefficient multipliers for partially closed valves

| Condition | Ratio K/K (Open) | |
	Gate Valve	Globe Valve
Open	1.0	1.0
Closed		
25%	3.0–5.0	1.5–2.0
50%	12–22	2.0–3.0
75%	70–120	6.0–8.0

EXAMPLE 7.8 Illustrates calculation of local losses and the difference between pressure loss and pressure drop

Find the pressure loss and pressure drop between cross sections 1 and 2 for each of the two conical expansions in Fig. E7.8. Water flows at 2.0 m³/s, $D_1 = 1.0$ m, $D_2 = 2.0$ m, $L = 2.75$ m, and $\theta = 30°$.

Solution

Given
Two conical expansions in Fig. E7.8
Water flow rate 2.0 m³/s, $D_1 = 1.0$ m, $D_2 = 2.0$ m, $L = 2.75$ m, $\theta = 30°$

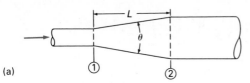

(a)

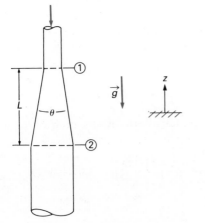

(b)

Figure E7.8 Conical gradual expansions: (a) Horizontal conical gradual expansion; (b) vertical conical gradual expansion.

Find

Pressure loss and pressure drop between sections 1 and 2 for both expansions

Solution

In either expansion, the pressure loss is

$$\Delta p_L = \mathbf{K}\frac{\rho V^2}{2},$$

where V is the inlet velocity. Assuming 20°C water, Table A.5 gives $\rho = 998$ kg/m³. For an expansion angle of 30°, Fig. 7.22 gives $\mathbf{K} = 0.93$ based on the velocity V_1. Equation (3.38) is next used to give

$$V_1 = \frac{Q}{A} = \frac{4Q}{\pi D_1^2} = \frac{4\left(2.0\,\dfrac{\text{m}^3}{\text{s}}\right)}{\pi(1.0\text{ m})^2} = 2.55\text{ m/s}.$$

The pressure loss is then

$$\Delta p_L = \frac{0.93\left(998\,\dfrac{\text{kg}}{\text{m}^3}\right)\left(2.55\,\dfrac{\text{m}}{\text{s}}\right)^2}{2} = 3020\,\frac{\text{kg}\cdot\text{m}}{\text{s}^2\cdot\text{m}^2},$$

$$\boxed{\Delta p_L = 3.02\text{ kPa,}}$$

Answer ◀

for both the horizontal and vertical conical expansion.

Assuming constant density, we find the pressure drop by applying the mechanical energy equation between sections 1 and 2 to give

$$\frac{p_1}{\rho} + \alpha_1\left(\frac{V_1^2}{2}\right) + gz_1 = \frac{p_2}{\rho} + \alpha_2\left(\frac{V_2^2}{2}\right) + gz_2 + gh_L.$$

(Overbars are omitted for clarity.) The continuity equation gives

$$V_2 = V_1\left(\frac{A_1}{A_2}\right) = V_1\left(\frac{D_1}{D_2}\right)^2.$$

Solving the mechanical energy equation for the pressure drop ($\Delta p_d = p_1 - p_2$) gives

$$p_1 - p_2 = \rho g(z_2 - z_1) + \rho\left(\frac{V_1^2}{2}\right)\left[\alpha_2\left(\frac{D_1}{D_2}\right)^2 - \alpha_1\right] + \rho g h_L.$$

Using Eq. (7.13), we have

$$p_1 - p_2 = \rho g(z_2 - z_1) + \rho\left(\frac{V_1^2}{2}\right)\left[\alpha_2\left(\frac{D_1}{D_2}\right)^2 - \alpha_1\right] + \Delta p_L,$$

which shows that the pressure drop ($p_1 - p_2$) depends on both the elevation change and the velocity profiles. Assuming uniform velocity at both sections 1 and 2, $\alpha_1 = \alpha_2 = 1.0$. Then

$$p_1 - p_2 = \rho g(z_2 - z_1) + \rho\left(\frac{V_1^2}{2}\right)\left[\left(\frac{D_1}{D_2}\right)^2 - 1.0\right] + \Delta p_L.$$

Substituting the numerical values for the horizontal expansion gives

$$p_1 - p_2 = \rho g(0) + \frac{\left(998\,\dfrac{\text{kg}}{\text{m}^3}\right)\left(2.55\,\dfrac{\text{m}}{\text{s}}\right)^2}{2}\left[\left(\frac{1.0\text{ m}}{2.0\text{ m}}\right)^2 - 1.0\right] + 3020\,\frac{\text{N}}{\text{m}^2}$$

$$= (-2430 + 3020)\text{ N/m}^2,$$

$$p_1 - p_2 = 590 \text{ Pa.}$$
Horizontal expansion

Answer
◀

This result illustrates a situation where the pressure drop due to viscous effects is partly offset by the pressure rise due to the decrease in velocity. Substituting the numerical values for the vertical expansion gives

$$p_1 - p_2 = \left(998 \frac{\text{kg}}{\text{m}^3} \right) \left(9.806 \frac{\text{m}}{\text{s}^2} \right) (-2.75 \text{ m})$$

$$+ \frac{\left(998 \frac{\text{kg}}{\text{m}^3} \right) \left(2.55 \frac{\text{m}}{\text{s}} \right)^2}{2} \left[\left(\frac{1.0 \text{ m}}{2.0 \text{ m}} \right)^2 - 1.0 \right] + 2430 \text{ Pa}$$

$$= (-26,900 - 2430 - 3020) \text{ N/m}^2,$$

$$p_1 - p_2 = -26 \text{ kPa.}$$
Vertical expansion

Answer
◀

This negative pressure drop (i.e., a pressure rise) indicates that the pressure increase due to the decrease in elevation and velocity exceeds the pressure drop due to viscous effects.

Discussion
The pressure loss is the same in both cases; however, the pressure drop is not the same because it reflects the difference in the average static pressure as would be measured by two pressure gages at sections 1 and 2.

Note that the pressure loss is always positive for a real fluid flowing through a fitting, valve, or pipe. This is true regardless of the orientation of the fitting, valve, or pipe.

The data in this section have two important limitations:

- With the exception of sudden expansions and exits, the values of **K** are based on experimental measurements and have an uncertainty of about 10%.
- The value of **K** for any particular component applies only to the component in isolation with either uniform or fully developed flow upstream. If two components are placed close together, the resulting losses usually will be greater than the sum of the losses of the components in isolation.

7.3.2 Analysis and Design of Pipe and Duct Systems

The analysis or design of a pipe or duct system is accomplished by patching together the information we have discussed about pipes, ducts and components. In an analysis problem, the system configuration is given and the engineer wants to calculate flow rate, energy loss, and pressure drop or required pumping power to maintain flow in the system. In a design problem, the engineer needs to select pipe sizes, specify number and location of pumps or fans, and select other system geometric and operating parameters to obtain a given flow rate (or flow rates if

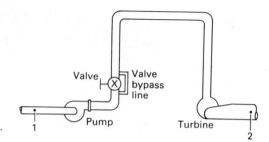

Figure 7.29 Pipe system.

a multiple branch system is required) at an economical cost. The mathematical tools at the engineer's disposal are the (incompressible flow) energy equation, the continuity equation, and the friction factor and loss coefficient information that we have developed.

Consider the pipe system shown in Fig. 7.29. The energy equation for flow between planes 1 and 2 is Eq. (5.56) (overbars denoting average values are dropped):

$$\dot{m}\left[\frac{p_1}{\rho} + \alpha_1\left(\frac{V_1^2}{2}\right) + gz_1\right] - \dot{W}_s = \dot{m}\left[\frac{p_2}{\rho} + \alpha_2\left(\frac{V_2^2}{2}\right) + gz_2\right] + \dot{m}(gh_L). \quad \textbf{(5.56)}$$

A very popular and useful form of the energy equation for pipe system calculations is the specific energy equation in head form, Eq. (5.65):

$$\frac{p_1}{\gamma} + \alpha_1\left(\frac{V_1^2}{2g}\right) + z_1 + h_p = \frac{p_2}{\gamma} + \alpha_2\left(\frac{V_2^2}{2g}\right) + z_2 + h_t + h_L. \quad \textbf{(5.65)}$$

This form is popular for two reasons:

- The head form provides a very helpful visualization using the energy grade line and hydraulic grade line (see Figs. 5.20 and 4.9).
- The losses that occur in pumps or turbines are combined with the shaft work in h_p and h_t. This permits separate treatment of the active components (pumps and turbines) and the passive components (the rest of the system).

It is customary to ignore the kinetic energy correction factors in Eq. (5.65). Some justifications of this simplification are that 1 and 2 often lie in reservoirs where the velocity is zero; that loss coefficients are often obtained from experiments that do not consider the effect of α, so it would be inconsistent to include such effects in the energy equation; and that the slight increase in accuracy that might be gained by using α's is completely offset by the uncertainty in typical values of loss coefficients and friction factors. The working equation for pipe/duct system analysis is thus

$$\frac{p_1}{\gamma} + \frac{V_1^2}{2g} + z_1 + h_p = \frac{p_2}{\gamma} + \frac{V_2^2}{2g} + z_2 + h_t + h_L. \quad \textbf{(7.104)}$$

We will examine how to deal with the terms in this equation and discuss certain practical aspects of pipe/duct flow calculations.

Evaluation of Losses. The head loss (h_L) that appears in Eq. (7.104) is the sum of all losses in the system except those that occur within any pumps or turbines and is calculated by

$$h_L = \sum h_{L,\text{friction}} + \sum h_{L,\text{local}} = \sum f\left(\frac{L}{D_h}\right)\left(\frac{V^2}{2g}\right) + \sum K\left(\frac{V^2}{2g}\right). \qquad \textbf{(7.105)}$$

If any local losses are evaluated by the equivalent length method, the equivalent lengths are included in L.

If all pipes have the same cross-sectional area, they all have the same average velocity, so Eq. (7.105) becomes

$$h_L = \left(f\frac{L}{D_h} + \sum K\right)\frac{V^2}{2g}. \qquad \textbf{(7.106)}$$

If the pipe/duct line has elements of different cross-sectional area, the flow rate may be a more convenient parameter than the velocity:

$$h_L = \left(\sum \frac{f}{A^2}\frac{L}{D_h} + \sum \frac{K}{A^2}\right)\frac{Q^2}{2g}, \qquad \textbf{(7.107)}$$

where A is the inside cross-sectional area of the segment of pipe or duct under consideration.

EXAMPLE 7.9 Illustrates the energy/pressure loss problem in a piping system

Heavy machine oil having a specific gravity of 0.85 and an absolute viscosity of 400×10^{-3} N·s/m² is pumped through 20 m of 0.052-m-inside-diameter PVC pipe. The pipe line is shown in Fig. E7.9 and contains one swing check valve, two gate valves, four 45° standard elbows, and a nozzle with a throat diameter of 0.026 m. A manometer connecting the inlet and throat of the nozzle reads 2.0 m of mercury. All fittings and valves have cemented connections. Find the pressure loss between points 2 and 3. Assume frictionless flow in the nozzle.

Figure E7.9 Piping system for Example 7.9.

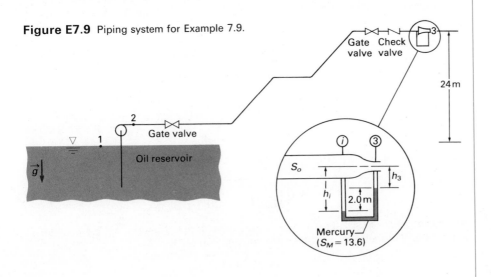

Solution

Given

Sketch of piping system in Fig. E7.9

Find

Pressure loss Δp_L from point 2 to point 3

Solution

The pressure loss is

$$\Delta p_L = \rho g h_L.$$

The total energy loss can be expressed as the sum of the individual energy losses:

$$gh_L = \left(\mathbf{K}\frac{V^2}{2}\right)_{\substack{\text{check} \\ \text{valve}}} + 2\left(\mathbf{K}\frac{V^2}{2}\right)_{\substack{\text{gate} \\ \text{valve}}} + 4\left(\mathbf{K}\frac{V^2}{2}\right)_{\substack{45° \\ \text{standard} \\ \text{elbow}}} + \left[f\left(\frac{L}{D}\right)\left(\frac{V^2}{2}\right)\right]_{\text{pipe}},$$

V, the average velocity, is presently unknown. Assuming constant density, the mechanical energy equation between the inlet and outlet of the nozzle gives

$$\frac{p_i}{\rho_0} + \alpha_i\left(\frac{V_i^2}{2}\right) + gz_i = \frac{p_3}{\rho_0} + \alpha_3\left(\frac{V_3^2}{2}\right) + gz_3 + gh_L,$$

where ρ_0 is the oil density. The nozzle is horizontal, so $z_i = z_3$. Using Eq. (3.48) gives

$$V_3 = V_i\left(\frac{A_i}{A_3}\right) = V_i\left(\frac{D_i}{D_3}\right)^2.$$

If we assume $\alpha_i = \alpha_3$, the mechanical energy equation gives

$$V_i^2\left[\left(\frac{D_i}{D_3}\right)^4 - 1\right] = \frac{p_i - p_3}{\rho_0}.$$

We next apply the manometer rule to the nozzle manometer to obtain

$$p_3 = p_i + S_0\gamma_w h_i - S_M\gamma_w(h_i - h_3) - S_0\gamma_w h_3,$$

where S_0 and S_M are the specific gravities of the oil and the mercury and γ_w is the specific weight of water at 4°C and 101,330 Pa. Simplifying, we obtain

$$p_i - p_3 = (h_i - h_3)\gamma_w(S_M - S_0),$$

or, since $V_i = V$, the mechanical energy equation gives

$$V = \sqrt{\frac{g(h_i - h_3)(S_M - S_0)}{S_0\left[\left(\frac{D_i}{D_3}\right)^4 - 1\right]}}.$$

From Table A.5, the numerical values give

$$V = \sqrt{\frac{\left(9.807\,\frac{\text{m}}{\text{s}^2}\right)(2.0\text{ m})(13.6 - 0.85)}{(0.85)\left[\left(\frac{0.052}{0.026}\right)^4 - 1\right]}} = 9.90\text{ m/s}.$$

Noting that $0.052\text{ m} \approx 2.0$ in. and assuming cemented connections are equivalent to screwed connections, we find values from Table 7.8:

$$\mathbf{K}_{\substack{\text{check} \\ \text{valve}}} = 2.1, \qquad \mathbf{K}_{\substack{\text{gate} \\ \text{valve}}} = 0.16, \qquad \mathbf{K}_{\substack{45° \\ \text{standard} \\ \text{elbow}}} = 0.30.$$

The oil kinematic viscosity is

$$v_0 = \frac{\mu_0}{\rho_0} = \frac{\left(400 \times 10^{-3} \, \frac{N \cdot s}{m^2}\right)}{(0.85)\left(1000 \, \frac{kg}{m^3}\right)} = 4.71 \times 10^{-5} \, m^2/s.$$

The pipe Reynolds number is

$$R = \frac{VD}{v} = \frac{\left(9.90 \, \frac{m}{s}\right)(0.052 \, m)}{\left(4.71 \times 10^{-5} \, \frac{m^2}{s}\right)} = 10{,}900.$$

Table 7.1 shows that plastic pipe has a zero relative roughness. Therefore the Moody chart gives $f = 0.030$. The total energy loss can now be calculated as follows:

$$gh_L = \left(\mathbf{K}_{\substack{check \\ valve}} + 2\mathbf{K}_{\substack{gate \\ valve}} + 4\mathbf{K}_{\substack{45° \\ standard \\ elbow}} + f\frac{L}{D}\right)\frac{V^2}{2}$$

$$= \left[2.1 + 2(0.16) + 4(0.30) + 0.30\left(\frac{20 \, m}{0.052 \, m}\right)\right]\frac{\left(9.90 \, \frac{m}{s}\right)^2}{2}$$

$$= 743 \, \frac{m^2}{s^2}\left(\frac{N \cdot s^2}{kg \cdot m}\right) = 743 \, \frac{N \cdot m}{kg}.$$

The pressure loss is

$$\Delta p_L = 0.85\left(1000 \, \frac{kg}{m^3}\right)\left(743 \, \frac{N \cdot m}{kg}\right),$$

Answer

$$|\ \Delta p_L = 631 \, kPa. \ |$$

◀

Discussion
Note that no energy loss term was included for the exit because the kinetic energy at the exit was included explicitly in the energy equation.

Can this problem be solved with the Colebrook formula as in Example 7.3? Would an iterative solution be necessary?

EXAMPLE 7.10   Illustrates the velocity/flow rate problem in a piping system

This example takes a closer look at the Boy Scout camp shower in Example 5.11. The Scouts used cold 55°F water while the leaders built a fire to heat their water to 100°F. Both showers used about 25 ft of 1/2-in. (inside diameter = 0.569 in.) copper pipe and had two 90° standard elbows. Find the volume flow rate to each shower when the water level in the drum is 3 ft above the pipe discharge. Assume the kinetic energy correction factor is unity.

Solution

Given
Camp showers in Fig. E5.11(a)
25 ft of 1/2-in. copper pipe to each shower with two 90° standard elbows

Scouts' water 55°F
Leaders' water 100°F

Find

Volume flow rate to each shower when drum water level is 3 ft above
pipe discharge

Solution

We start with the steady-state, mechanical energy equation in head form:

$$\frac{p_1}{\gamma} + \frac{V_1^2}{2g} + z_1 = \frac{p_2}{\gamma} + \frac{V_2^2}{2g} + z_2 + h_L \qquad \text{and}$$

$$\frac{p_1}{\gamma} + \frac{V_1^2}{2g} + z_1 = \frac{p_3}{\gamma} + \frac{V_3^2}{2g} + z_3 + h_L,$$

where p, V, and z represent average values at each location. $V_1 \approx 0$ for each
case. Also

$$p_1 = p_2 = p_3.$$

With the subscript 55 denoting the boys' shower and 100 denoting the leaders'
shower, the mechanical energy equation gives

$$V_{55}^2 = 2g(z_1 - z_2 - h_L) \qquad \text{and} \qquad V_{100}^2 = 2g(z_1 - z_3 - h_L).$$

In each case the total head loss is the sum of the individual head losses:*

$$h_L = \left(\mathbf{K}\frac{V^2}{2g}\right)_{\text{entrance}} + 2\left(\mathbf{K}\frac{V^2}{2g}\right)_{\substack{90° \\ \text{standard} \\ \text{elbow}}} + \left[f\left(\frac{L}{D}\right)\left(\frac{V^2}{2g}\right)\right]_{\text{pipe}}$$

$$= \left(\mathbf{K}_{\text{entrance}} + 2\mathbf{K}_{\substack{90° \\ \text{standard} \\ \text{elbow}}} + f\frac{L}{D}\right)\frac{V^2}{2g}.$$

Substituting into the mechanical energy equation, we obtain

$$V_{55} = \sqrt{\frac{2g(z_1 - z_2)}{1 + \mathbf{K}_{\text{entrance}} + 2\mathbf{K}_{\substack{90° \\ \text{standard} \\ \text{elbow}}} + f\frac{L}{D}}} \qquad \text{and}$$

$$V_{100} = \sqrt{\frac{2g(z_1 - z_3)}{1 + \mathbf{K}_{\text{entrance}} + 2\mathbf{K}_{\substack{90° \\ \text{standard} \\ \text{elbow}}} + f\frac{L}{D}}}.$$

Assuming a reentrant pipe entrance for the 1/2-in. copper pipe, we find from
Fig. 7.27 and Table 7.8:

$$\mathbf{K}_{\text{entrance}} \approx 1.0 \qquad \text{and} \qquad \mathbf{K}_{\substack{90° \\ \text{standard} \\ \text{elbow}}} = 2.0,$$

assuming crude field installation of the entrance and soldered or screwed connections at the elbow. The friction factor can be found using the relative roughness and the Reynolds number. Since the (unknown) velocity is involved in the
Reynolds number, iteration is necessary. For copper pipe (drawn tubing), Table

---

* Note that an exit energy loss term is not included. Do you know why?

7.1 gives

$$\frac{\varepsilon}{D} = \frac{0.000005 \text{ ft}}{0.569 \text{ in.}}\left(\frac{12 \text{ in.}}{\text{ft}}\right) = 0.00011.$$

Guessing $f = 0.013$ (the complete turbulence value from Fig. 7.10), we calculate each velocity as

$$V_{100} = V_{55} = \sqrt{\frac{2\left(32.2 \dfrac{\text{ft}}{\text{sec}^2}\right)(3 \text{ ft})}{\left[1 + 1.0 + 2(2.0) + 0.013\dfrac{(25 \text{ ft})}{(0.569 \text{ in.})}\left(\dfrac{12 \text{ in.}}{\text{ft}}\right)\right]}}$$

$$= 3.88 \text{ ft/sec.}$$

We next check our guess of the friction factor by calculating the Reynolds number. From Table A.6,

$$v_{55} = 1.33 \times 10^{-5} \text{ ft}^2/\text{sec} \quad \text{and} \quad v_{100} = 0.739 \times 10^{-5} \text{ ft}^2/\text{sec.}$$

The corresponding Reynolds numbers are

$$\mathbf{R}_{55} = \frac{VD}{v} = \frac{\left(3.88 \dfrac{\text{ft}}{\text{sec}}\right)(0.569 \text{ in.})}{\left(1.33 \times 10^{-5} \dfrac{\text{ft}^2}{\text{sec}}\right)}\left(\frac{\text{ft}}{12 \text{ in.}}\right) = 1.4 \times 10^4$$

and

$$\mathbf{R}_{100} = \frac{VD}{v} = \frac{\left(3.88 \dfrac{\text{ft}}{\text{sec}}\right)(0.569 \text{ in.})}{\left(0.739 \times 10^{-5} \dfrac{\text{ft}^2}{\text{sec}}\right)}\left(\frac{\text{ft}}{12 \text{ in.}}\right) = 2.5 \times 10^4.$$

Figure 7.11 gives $f_{55} = 0.0277$ and $f_{100} = 0.0247$.

Repeating the calculations with these new values of friction factor and iterating until friction factors and velocities do not change, we obtain

$$\mathbf{R}_{55} = 1.1 \times 10^4, \quad \mathbf{R}_{100} = 2.0 \times 10^4, \quad f_{55} = 0.030, \quad f_{100} = 0.026,$$
$$V_{55} = 2.98 \text{ ft/sec}, \quad V_{100} = 3.13 \text{ ft/sec.}$$

The flow rates are found using Eq. (3.38):

$$Q_{55} = V_{55}A = V_{55}\left(\frac{\pi D^2}{4}\right)$$

$$= \left(2.96 \frac{\text{ft}}{\text{sec}}\right)\frac{\pi}{4}\left(\frac{0.569 \text{ in.}}{\frac{12 \text{ in.}}{\text{ft}}}\right)^2\left(\frac{60 \text{ sec}}{\text{min}}\right),$$

Answer

$$\boxed{Q_{55} = 0.31 \text{ ft}^3/\text{min}}$$

and

$$Q_{100} = V_{100}A = V_{100}\left(\frac{\pi D^2}{4}\right)$$

$$= \left(3.11 \frac{\text{ft}}{\text{sec}}\right)\frac{\pi}{4}\left(\frac{0.569 \text{ in.}}{\frac{12 \text{ in.}}{\text{ft}}}\right)^2\left(\frac{60 \text{ sec}}{\text{min}}\right),$$

Answer

$$\boxed{Q_{100} = 0.33 \text{ ft}^3/\text{min.}}$$

Discussion
Notice the drastic change in the calculated exit velocity (and the flow rate) by including the resistance of the copper pipe, the entrance, and the two elbows. The comparison is given below:

Lossless Flow (Example 5.11)	**Flow with Losses (Example 7.10)**
$V_{55} = 13.9$ ft/sec	$V_{55} = 2.96$ ft/sec
$V_{100} = 13.9$ ft/sec	$V_{100} = 3.11$ ft/sec

This comparison also shows that heating the fluid has a slight effect on the fluid velocity as well as providing a more enjoyable shower.

This example used the head form of the mechanical energy equation, and the previous example used the energy form. The choice between the two is up to you. Often, there is no particular reason to choose one form over the other.

It would be instructive for you to investigate whether this problem could be solved in a noniterative fashion using the Colebrook formula, which provided a direct solution of the velocity/flow rate problem in Example 7.4.

EXAMPLE 7.11 Illustrates the pipe sizing problem in a piping system

Find the schedule 40 pipe size required to deliver at least 70 gal/min from the higher reservoir to the lower reservoir in Fig. E7.11 when the water levels are as shown. The pipe line contains 250 ft of straight galvanized iron pipe, three fully open globe valves, and six 90° standard elbows. All connections are flanged.

Solution

Given
Figure E7.11
Water flow rate 70 gal/min
Flanged connections
Schedule 40 pipe

Find
Required pipe size

Solution
We apply the mechanical energy equation,

$$\frac{p_1}{\rho} + \alpha_1\left(\frac{V_1^2}{2}\right) + gz_1 = \frac{p_2}{\rho} + \alpha_2\left(\frac{V_2^2}{2}\right) + gz_2 + gh_L.$$

Now $p_1 = p_2$ and $V_1 \approx V_2 \approx 0$. The mechanical energy equation becomes

$$gh_L = g(z_1 - z_2),$$

Figure E7.11 Flow between reservoirs.

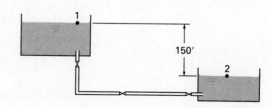

where $z_1 - z_2$ is 150 ft. Equations (7.29a) and (7.99) give*

$$gh_L = \left[\mathbf{K}_{\substack{\text{entrance}}} + 3\mathbf{K}_{\substack{\text{globe} \\ \text{valve}}} + 6\mathbf{K}_{\substack{90° \\ \text{standard} \\ \text{elbow}}} + \mathbf{K}_{\text{exit}} + \left(f\frac{L}{D} \right)_{\text{pipe}} \right] \frac{V^2}{2},$$

where V is the average fluid velocity in the pipe. The average velocity is

$$V = \frac{Q}{A} = \frac{4Q}{\pi D^2},$$

so

$$gh_L = \left[\mathbf{K}_{\substack{\text{entrance}}} + 3\mathbf{K}_{\substack{\text{globe} \\ \text{valve}}} + 6\mathbf{K}_{\substack{90° \\ \text{standard} \\ \text{elbow}}} + \mathbf{K}_{\text{exit}} + \left(f\frac{L}{D} \right)_{\text{pipe}} \right] \frac{8Q^2}{\pi^2 D^4}.$$

Equating the expressions for gh_L and rearranging give

$$\frac{\pi^2 D^4 g(z_1 - z_2)}{8Q^2} - \left(f\frac{L}{D} \right)_{\text{pipe}} = \mathbf{K}_{\substack{\text{entrance}}} + 3\mathbf{K}_{\substack{\text{globe} \\ \text{valve}}} + 6\mathbf{K}_{\substack{90° \\ \text{standard} \\ \text{elbow}}} + \mathbf{K}_{\text{exit}}$$

For a reentrant pipe, Fig. 7.27 gives $\mathbf{K}_{\text{entrance}} \approx 0.9$. The section "Entrances and Exits" gives $\mathbf{K}_{\text{exit}} = 1.0$. Substituting known numerical values gives

$$\frac{\pi^2 D^4 g(z_1 - z_2)}{8Q^2} = \frac{\pi^2 D^4 \left(32.2 \, \frac{\text{ft}}{\text{sec}^2} \right) (150 \text{ ft})}{8 \left(70 \, \frac{\text{gal}}{\text{min}} \right)^2 \left(\frac{\text{ft}^3}{7.48 \text{ gal}} \right)^2 \left(\frac{\text{min}}{60 \text{ sec}} \right)^2} = 2.45 \times 10^5 \left(\frac{D^4}{\text{ft}^4} \right)$$

and

$$f\left(\frac{L}{D} \right) = \frac{f(250 \text{ ft})}{D}.$$

Table 7.8 shows that loss coefficients for globe valves and standard elbows depend on the pipe diameter. Therefore

$$\mathbf{K}_{\substack{\text{entrance}}} + 3\mathbf{K}_{\substack{\text{globe} \\ \text{valve}}} + 6\mathbf{K}_{\substack{90° \\ \text{standard} \\ \text{elbow}}} + \mathbf{K}_{\text{exit}} = 1.9 + 3\mathbf{K}_{\substack{\text{globe} \\ \text{valve}}} + 6\mathbf{K}_{\substack{90° \\ \text{standard} \\ \text{elbow}}}.$$

The mechanical energy equation is then

$$2.45 \times 10^5 \left(\frac{D}{\text{ft}} \right)^4 - 250 f\left(\frac{\text{ft}}{D} \right) - \left[1.9 + 3\mathbf{K}_{\substack{\text{globe} \\ \text{valve}}} + 6\mathbf{K}_{\substack{90° \\ \text{standard} \\ \text{elbow}}} \right] = 0,$$

where D is measured in feet. D must now be found by trial and error. For each chosen value of D there is a corresponding value of f found from the Moody chart after computing the Reynolds number and relative roughness. As a first guess, we choose $D = 2.0$ in. $= 1/6$ ft. For 55°F water, Table A.6 gives $v = 1.33 \times 10^{-5}$ ft²/sec. Then

$$\mathbf{R} = \frac{VD}{v} = \frac{4Q}{\pi v D}$$

$$= \frac{4 \left(70 \, \frac{\text{gal}}{\text{min}} \right) \left(\frac{\text{ft}^3}{7.48 \text{ gal}} \right) \left(\frac{\text{min}}{60 \text{ sec}} \right)}{\pi \left(1.33 \times 10^{-5} \, \frac{\text{ft}^2}{\text{sec}} \right) \left(\frac{1}{6} \text{ ft} \right)} = 9.1 \times 10^4.$$

* Do you know why the exit loss coefficient is included here?

Using Table 7.1 for the relative roughness for galvanized pipe, we have

$$\frac{\varepsilon}{D} = \frac{0.00015 \text{ ft}}{1/6 \text{ ft}} = 0.0009.$$

The Moody chart gives $f = 0.022$. Table 7.8 gives

$$\mathbf{K}_{\substack{\text{globe} \\ \text{valve}}} = 8.5 \qquad \text{and} \qquad \mathbf{K}_{\substack{90° \\ \text{standard} \\ \text{elbow}}} = 0.30.$$

The mechanical energy equation gives

$$2.45 \times 10^5 \left(\frac{\frac{1}{6} \text{ ft}}{\text{ft}} \right)^4 - 250(0.022)\left(\frac{\text{ft}}{\frac{1}{6} \text{ ft}} \right) - [1.9 + 3(8.5) + 6(0.30)] = 126 \neq 0.$$

As a second guess, choose $D = 1.0$ in. $= 1/12$ ft. Repeating the calculations gives

$$\mathbf{R} = 1.8 \times 10^5, \quad \frac{\varepsilon}{D} = 0.0018, \quad f = 0.024, \quad \mathbf{K}_{\substack{\text{globe} \\ \text{valve}}} = 13, \quad \mathbf{K}_{\substack{90° \\ \text{standard} \\ \text{elbow}}} = 0.5.$$

The mechanical energy equation gives

$$2.45 \times 10^5 \left(\frac{\frac{1}{12} \text{ ft}}{\text{ft}} \right)^4 - 250(0.024)\left(\frac{\text{ft}}{\frac{1}{12} \text{ ft}} \right) - [1.9 + 3(13) + 6(0.5)] = -104 \neq 0.$$

Since this number is negative and the first trial yielded a positive number, D lies between 1.0 and 2.0 in. As a third guess, choose $D = 1.5$ in. Repeating the calculations gives

$$\mathbf{R} = 1.2 \times 10^5, \quad \frac{\varepsilon}{D} = 0.00124, \quad f = 0.022, \quad \mathbf{K}_{\substack{\text{globe} \\ \text{valve}}} = 10.0, \quad \mathbf{K}_{\substack{90° \\ \text{standard} \\ \text{elbow}}} = 0.44.$$

The mechanical energy equation gives

$$2.45 \times 10^5 \left(\frac{0.125 \text{ ft}}{\text{ft}} \right)^4 - 250(0.022)\left(\frac{\text{ft}}{0.125 \text{ ft}} \right) - [1.9 + 3(10) + 6(0.44)] = -19 \neq 0$$

The inside pipe diameter D must be slightly larger than 1.5 in. Table D.1 in Appendix D shows that a 1.5-in., nominal-diameter schedule 40 pipe, with a 1.610-in inside diameter, would be a good possibility. Repeating the calculations for a 1.610-in. inside diameter gives

$$\mathbf{R} = 1.1 \times 10^5, \quad \frac{\varepsilon}{D} = 0.0011, \quad f = 0.022, \quad \mathbf{K}_{\substack{\text{globe} \\ \text{valve}}} = 10.0, \quad \mathbf{K}_{\substack{90° \\ \text{standard} \\ \text{elbow}}} = 0.44.$$

The mechanical energy equation gives

$$2.45 \times 10^5 \left(\frac{\frac{1.610}{12} \text{ ft}}{\text{ft}} \right)^4 - 250(0.022)\left(\frac{\text{ft}}{\frac{1.610}{12} \text{ ft}} \right) - [1.9 + 3(10.0) + 6(0.44)] = 3.9 \approx 0.$$

We use

$$\left| \begin{array}{c} D_{\text{nominal}} = 1.50\text{-in.} \\ \text{schedule 40} \\ \text{galvanized pipe.} \end{array} \right.$$

Answer
◀

Discussion

Can this problem be solved directly using the Colebrook formula as in Example 7.5?

If you are familiar with sophisticated iteration techniques (such as Newton-Rhapson), you might be able to decrease the number of iterations to arrive at the final answer.

Pump (or Fan)-System Matching. Consider a system that contains one or more pumps. The head generated by a pump* (h_p) is not a constant but is a function of the speed of rotation of the pump shaft and the flow rate through the pump. Since pumps are frequently driven by AC electric motors, which have a nearly constant rotational speed, pump performance is usually presented graphically as pump head versus flow rate at constant rotational speed. Figure 7.30(a) shows a typical *pump curve*. Pump curves are usually determined by experiment and are supplied by the pump manufacturer.

When a pump is installed in a system, the flow rate through the pump must equal the flow rate through the system.† Likewise, the head supplied by the pump must equal the head "used" by the system. The *system head* is defined as the algebraic sum of all heads in Eq. (7.104) except the pump head (we will neglect turbine head for this discussion):

$$h_{\text{sys}} \equiv \frac{p_2 - p_1}{\gamma} + \frac{V_2^2 - V_1^2}{2g} + z_2 - z_1 + h_L, \qquad \textbf{(7.108)}$$

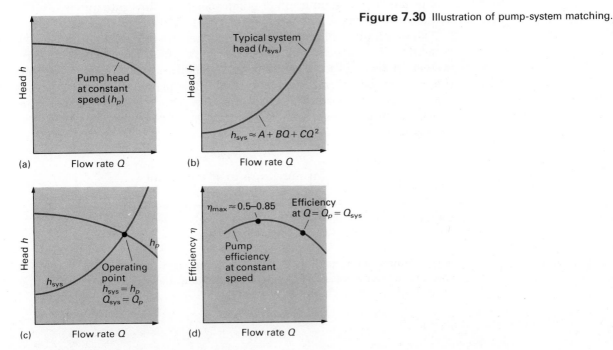

Figure 7.30 Illustration of pump-system matching.

---

* See Eq. (5.62) and the attendant discussion.
† This ignores leaks in the pump, which are usually very small.

where h_L is calculated by Eqs. (7.105)–(7.107). System head is a function of system configuration $(z_2 - z_1)$ and flow rate (Fig. 7.30b). A system head curve is a characteristic of the system. Such a curve is also called a *system resistance curve*.

When a pump is installed in a system, Eq. (7.104) requires

$$h_p = h_{\text{sys}}, \tag{7.109a}$$

while continuity requires

$$Q_{\text{pump}} = Q_{\text{sys}}. \tag{7.109b}$$

The pump-system combination operates at the head and flow rate that correspond to the intersection of the pump curve and system curve, as shown in Fig. 7.30(c). This intersection can be determined graphically or, if a curve fit of the pump curve is available, mathematically.

Knowing just the head and flow rate does not permit calculation of the (motor) power required to drive the pump; we must know something about the losses in the pump and also about the mechanical friction in the pump drive mechanism (shaft seals, bearings, etc.). Pump power is usually calculated from an *overall efficiency,* defined by

$$\eta = \frac{\text{Useful energy transfer to fluid}}{\text{Mechanical power input}}.$$

The useful energy transfer to the fluid is the *net* mechanical energy rise (pump head) multiplied by the flow rate, so

$$\eta = \frac{\dot{m}(gh_p)}{\dot{W}_{\text{shaft}}} = \frac{\rho Q g h_p}{\dot{W}_{\text{shaft}}} = \frac{\gamma Q h_p}{\dot{W}_{\text{shaft}}}. \tag{7.110}$$

Pump efficiency is a function of pump speed and flow rate and is usually available from manufacturer's curves (Fig. 7.30d).

The technique of matching a fan to an air- or gas-handling system is similar to the technique for matching a pump to a liquid-handling system, at least as long as air or gas velocities are small enough that the flow can be assumed incompressible. The primary difference is that fans are usually rated in terms of pressure rather than head, so system resistance must be expressed in terms of pressure rather than head. The equation

$$\Delta p_{\text{sys}} = \rho g h_{\text{sys}} \tag{7.111}$$

relates system resistance pressure to system head.

EXAMPLE 7.12 Illustrates the velocity/flow rate problem with a pump in the piping system

A centrifugal pump having the head characteristic shown in Fig. E7.12(a) is used in the piping system of Example 7.11 to pump water from the low reservoir to the high reservoir. Find the flow rate.

Solution

Given

Piping system in Example 7.11 modified as shown in Fig. E7.12(a)
Pump characteristics as in Fig. E7.12(b)

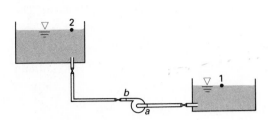

Figure E7.12 (a) Modification of Example 7.11 piping system.

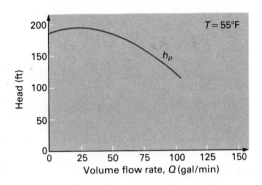

Figure E7.12 (b) Pump characteristics.

Find
Volume flow rate with pump installed

Solution
The system head is given by Eq. (7.108):

$$h_{sys} = \frac{p_2 - p_1}{\gamma} + \frac{V_2^2 - V_1^2}{2g} + (z_2 - z_1) + h_L,$$

where p, V, and z represent average values and a kinetic energy correction factor of unity is assumed.

As in Example 7.11, $p_1 - p_2 = 0$ and $V_1 \approx V_2 \approx 0$. The system head is then

$$h_{sys} = (z_2 - z_1) + h_L,$$

where $(z_2 - z_1)$ is 150 ft and h_L is found from Example 7.11:

$$h_L = \left(\mathbf{K}_{\substack{\text{entrance}}} + 3\mathbf{K}_{\substack{\text{globe} \\ \text{valve}}} + 6\mathbf{K}_{\substack{90° \\ \text{standard} \\ \text{elbow}}} + \mathbf{K}_{\text{exit}} + f\frac{L}{D} \right) \frac{V^2}{2g}.$$

A pump is installed in the system, so Eq. (7.109a) gives

$$h_{sys} = h_p.$$

Substituting for h_{sys} and h_L and solving for the velocity V, we obtain

$$V = \sqrt{\frac{2g[h_p - (z_2 - z_1)]}{\mathbf{K}_{\substack{\text{entrance}}} + 3\mathbf{K}_{\substack{\text{globe} \\ \text{valve}}} + 6\mathbf{K}_{\substack{90° \\ \text{standard} \\ \text{elbow}}} + \mathbf{K}_{\text{exit}} + \left(f\frac{L}{D} \right)_{\text{pipe}}}}.$$

The unknowns in this equation are the velocity V, the pump head h_p, and the friction factor f; h_p and f depend on the volume flow rate and V. Using the numerical values from Example 7.11 and for 55°F water (Table A.6) gives

$$V = \sqrt{\frac{2\left(32.2\,\dfrac{\text{ft}}{\text{sec}^2} \right)(h_p - 150\text{ ft})}{0.9 + 3(10.0) + 6(0.44) + f\left(\dfrac{250\text{ ft}}{1.610\text{ in.}} \right)\left(\dfrac{12\text{ in.}}{\text{ft}} \right)}}.$$

Simplifying, we have

$$V = \sqrt{\frac{2\left(32.2\,\dfrac{\text{ft}}{\text{sec}^2} \right)(h_p - 150\text{ ft})}{33.54 + 1863f}},$$

where h_p is in ft. We must now solve for V in this velocity/flow rate problem. The velocity/flow rate problem in Example 7.4 was most conveniently solved using the Colebrook formula since $\mathbf{R}_f \ (=\sqrt{f}\mathbf{R})$ could be conveniently grouped in the mechanical energy equation. This is not possible in this problem. The other method of solution is iterative, using the Moody chart. For 55°F water,

$$v = 1.33 \times 10^{-5} \text{ ft}^2/\text{sec} \qquad \text{and} \qquad \rho = 62.4 \text{ lbm/ft}^3.$$

We begin the iteration by choosing a value for Q.* Choosing $Q = 50$ gal/min and using Eq. (3.38), we obtain

$$V = \frac{4Q}{\pi D^2} = \frac{4\left(50 \dfrac{\text{gal}}{\text{min}}\right)\left(231 \dfrac{\text{in}^3}{\text{gal}}\right)}{\pi(1.610 \text{ in.})^2 \left(\dfrac{60 \text{ sec}}{\text{min}}\right)\left(\dfrac{12 \text{ in.}}{\text{ft}}\right)} = 7.88 \frac{\text{ft}}{\text{sec}}$$

and

$$\mathbf{R} = \frac{VD}{v} = \frac{\left(7.88 \dfrac{\text{ft}}{\text{sec}}\right)(1.610 \text{ in.})}{\left(1.33 \times 10^{-5} \dfrac{\text{ft}^2}{\text{sec}}\right)\left(\dfrac{12 \text{ in.}}{\text{ft}}\right)} = 7.9 \times 10^4.$$

With $\varepsilon/D = 0.0011$ from Example 7.11, the Moody chart gives $f = 0.0193$. For $Q = 50$ gal/min, Fig. E7.12(b) gives $h_p = 184$ ft. The mechanical energy equation gives

$$V = \sqrt{\frac{\left(64.4 \dfrac{\text{ft}}{\text{sec}^2}\right)(184 \text{ ft} - 150 \text{ ft})}{33.54 + 1863(0.0193)}} = 5.61 \text{ ft/sec.}$$

Since this velocity is not equal to the velocity found from the assumed flow rate of 50 gal/min, our guess is not correct. Repeating the calculations for $V = 5.61$ ft/sec gives

$$Q = \frac{\pi D^2}{4}(V) = \frac{\pi}{4} \frac{(1.610 \text{ in.})^2 \left(5.61 \dfrac{\text{ft}}{\text{sec}}\right)\left(\dfrac{7.48 \text{ gal}}{\text{ft}^3}\right)}{\left(\dfrac{144 \text{ in}^2}{\text{ft}^2}\right)\left(\dfrac{\text{min}}{60 \text{ sec}}\right)} = 35.6 \text{ gal/min}$$

and

$$\mathbf{R} = \frac{VD}{v} = \frac{\left(5.61 \dfrac{\text{ft}}{\text{sec}}\right)(1.610 \text{ in.})}{\left(1.33 \times 10^{-5} \dfrac{\text{ft}^2}{\text{sec}}\right)\left(\dfrac{12 \text{ in.}}{\text{ft}}\right)} = 5.7 \times 10^4.$$

This Reynolds number and flow rate are used to give $f = 0.0205$ and $h_p = 193$ ft, and the mechanical energy equation gives

$$V = \sqrt{\frac{\left(64.4 \dfrac{\text{ft}}{\text{sec}^2}\right)(193 \text{ ft} - 150 \text{ ft})}{33.54 + 1863(0.0205)}} = 6.21 \text{ ft/sec.}$$

Repeating the calculations until the velocity converges, we have

$$V = 6.02 \text{ ft/sec}, \qquad \mathbf{R} = 6.1 \times 10^4, \qquad f = 0.0202, \qquad h_p = 190,$$

$$\left| \ Q = 38.3 \text{ gal/min.} \ \right|$$

Answer ◀

---

* In a velocity/flow rate problem, the iteration often converges faster by selecting f rather than Q or V. Do you see why selecting f is not the better approach in this problem?

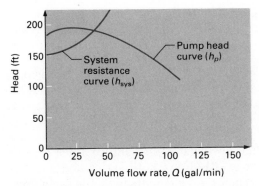

Figure E7.12 (c) Graphical determination of system operating point.

Discussion

An iteration of the type involved in this problem often converges faster if the new guess is taken as the average of the past trial value and the newly calculated value, that is,

$$Q_{\text{guess}} = \frac{1}{2}(Q_{\text{old}} + Q_{\text{new}}).$$

You might repeat the calculations using this idea and see how rapidly they converge.

An alternative method of finding Q is by graphical means. Note that the operating point is the intersection of a system resistance curve and the pump operating curve, as illustrated in Fig. 7.30(c). To find the system curve, we use Eq. (7.108), which gives

$$h_{\text{sys}} = (z_1 - z_2) + \left[\mathbf{K}_{\text{entrance}} + 3\mathbf{K}_{\underset{\substack{\text{valve}}}{\text{globe}}} + 6\mathbf{K}_{\underset{\substack{\text{standard} \\ \text{elbow}}}{90°}} + \mathbf{K}_{\text{exit}} + \left(f\frac{L}{D} \right)_{\text{pipe}} \right] \frac{V^2}{2g}.$$

Note that h_{sys} is the head that must be supplied to the system by the pump to raise the fluid from z_1 to z_2 and to overcome the losses associated with flow at velocity V. Using Eq. (3.38) gives

$$h_{\text{sys}} = (z_1 - z_2) + \left[\mathbf{K}_{\text{entrance}} + 3\mathbf{K}_{\underset{\substack{\text{valve}}}{\text{globe}}} + 6\mathbf{K}_{\underset{\substack{\text{standard} \\ \text{elbow}}}{90°}} + \mathbf{K}_{\text{exit}} + \left(f\frac{L}{D} \right)_{\text{pipe}} \right] \frac{8Q^2}{\pi^2 g D^4}.$$

The numerical values give

$$h_{\text{sys}} = 150 \text{ ft} + (0.0129 + 0.719f)Q^2,$$

where Q is in gal/min. We next choose several values of Q and plot h_{sys} versus Q in Fig. E7.12(c) along with the pump head curve. The intersection ($Q \approx 38$ gal/min, $h_p = h_{\text{sys}} \approx 190$ ft) represents the operating point.

The pump manufacturer usually supplies the pump head curve. It is obtained experimentally by measuring appropriate pressures, velocities, and elevation changes across the pump. Applying the mechanical energy equation, Eq. (5.65), across the pump gives

$$h_p = \frac{p_b - p_a}{\gamma} + \frac{V_b^2 - V_a^2}{2g} + (z_b - z_a),$$

where subscript a represents the pump inlet and subscript b the pump outlet.

The pump head is also equal to the work done on the fluid (w_p/g) minus any head losses inside the pump (Eq. 5.62). However, it is very difficult to measure both (w_p/g) and the losses.

The head form of the mechanical energy equation was especially useful in this example because the pump head was given via the performance curve.

Series and Parallel Lines.   Pipe systems often contain lines or elements in series or parallel. Figure 7.31 shows both series and parallel arrangements. In a series arrangement, all elements have the same flow, and the total loss is the sum of all the element losses:

$$Q_A = Q_B = \cdots = Q_N,$$
$$h_L = h_{L,A} + h_{L,B} + \cdots + h_{L,N}. \qquad \text{Series arrangement}$$

In a parallel arrangement, the total flow is the sum of the flows through all elements, and the loss is the same for each element. This is because there can be only one pressure and velocity at point 1 and only one pressure and velocity at point 2:*

$$Q = Q_A + Q_B + Q_C + \cdots + Q_N,$$
$$h_{L,A} = h_{L,B} = h_{L,C} = \cdots = h_{L,N}. \qquad \text{Parallel arrangement}$$

Equations (7.28) and (7.107) are directly applicable to series systems. In general, series calculations are quite simple if flow rate and geometry are given (the energy loss/pressure drop problem) and are more complicated if losses are given and either flow rate or pipe sizes are to be found. The equivalent length concept is quite helpful in these latter cases because the direct calculation methods of Section 7.2.5 can be used if friction is the only loss.

Parallel system calculations are easiest if the loss is given. With loss known, the flow rate (or pipe size, depending on the problem) can be separately calculated for each element in the system. The total flow is obtained by summing the element flows.

Figure 7.31 Pipes in series and parallel.

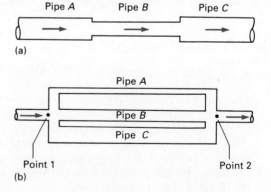

* A graphical interpretation is that the hydraulic grade line for each pipe must pass through a common point at 1 and 2.

A common problem for a parallel system involves determining the head loss and division of flow between parallel branches for given system geometry and total flow rate. Solution of this problem typically requires iteration. The procedure is as follows:

- Guess a flow rate for one branch of the system. A good rule of thumb is

$$Q_A \approx \left(\frac{A_A}{A_A + A_B + \cdots + A_N} \right) Q_{\text{total}}.$$

- Calculate the head loss for the selected branch using the guessed flow and the appropriate loss equations (energy/pressure loss problem).
- Since the loss is the same for all branches, calculate the flow in each of the remaining branches using the tentative loss value (flow rate/velocity problem).
- Sum the flow rates in all of the branches. If the sum equals the known total flow rate through the system, the calculations are complete.
- If the sum of the branch flows does not equal the total flow, adjust the trial flow rate according to

$$Q_{A,\text{new}} = \left[\frac{Q_{\text{total, given}}}{(Q_A + Q_B + \cdots + Q_N)_{\substack{\text{last} \\ \text{try}}}} \right] Q_{A,\text{old}}.$$

Return to the second step and repeat calculations until convergence.

This procedure will converge after a few cycles. Note that there may be internal iterations built into each step, such as calculating f for known Q and geometry using the Colebrook formula.

EXAMPLE 7.13 Illustrates how to solve a parallel piping system problem

Figure E7.13 shows a bypass line containing a globe valve in parallel with a cooling coil. Both the cooling coil and the bypass piping are 1.34-cm-inside-diameter copper pipe. The cooling coil contains 25 m of straight pipe and thirty-five 180° standard elbows. The bypass piping contains 5 m of straight pipe and a 50% closed globe valve. All connections are soldered. Find the flow rate through the cooling coil and the bypass for a total flow rate of 0.00040 m³/s. The fluid is water at 10°C.

Solution

Given
Cooling coil in parallel with bypass line
Cooling coil 25 m of 1.34-cm-inside-diameter copper pipe and thirty-five 180°
    standard elbows
Bypass piping 5 m of 1.34-cm-inside-diameter copper pipe and 50% closed
    globe valve
Soldered connections
Figure E7.13
Total flow rate 0.00040 m³/s

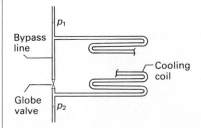

Figure E7.13 Parallel piping system.

Find
Flow rate through cooling coil and bypass

Solution
Assume constant density. As indicated in the "Series and Parallel Lines" section of the text,

$$Q_{\text{total}} = Q_{\text{bypass}} + Q_{\text{coil}} \quad \text{and} \quad h_{L,\text{bypass}} = h_{L,\text{coil}}.$$

The head loss in each line is

$$h_{L,\text{bypass}} = \left[\left(f\frac{L}{D} \right)_{\substack{\text{pipe}}} + 2\mathbf{K}_{\substack{\text{tee} \\ \text{line} \\ \text{flow}}} + \mathbf{K}_{\substack{\text{globe} \\ \text{valve}}} \right] \frac{8Q_{\text{bypass}}^2}{\pi^2 g D^4}$$

and

$$h_{L,\text{coil}} = \left[\left(f\frac{L}{D} \right)_{\substack{\text{pipe}}} + 2\mathbf{K}_{\substack{\text{tee} \\ \text{branch} \\ \text{flow}}} + 35\mathbf{K}_{\substack{90° \\ \text{standard} \\ \text{elbow}}} \right] \frac{8Q_{\text{coil}}^2}{\pi^2 g D^4}.$$

Noting that 1.34 cm $\approx$ 1/2 in. and assuming that the losses in soldered fittings are equal to those in screwed fittings, we find from Tables 7.8 and 7.9 that

$$h_{L,\text{bypass}} = \left[f\left(\frac{5 \text{ m}}{1.34 \text{ cm}} \right)\left(\frac{100 \text{ cm}}{\text{m}} \right) + 2(0.90) + (2.5 \times 14) \right] \frac{8Q_{\text{bypass}}^2 \left(\frac{100 \text{ cm}}{\text{m}} \right)^4}{\pi^2 \left(9.807 \frac{\text{m}}{\text{s}^2} \right)(1.34 \text{ cm})^4}$$

$$= 9.42 \times 10^7 (10.1f + 1)\left(\frac{\text{s}^2}{\text{m}^5} \right)(Q_{\text{bypass}}^2)$$

and

$$h_{L,\text{coil}} = \left[f\left(\frac{25 \text{ m}}{1.34 \text{ cm}} \right)\left(\frac{100 \text{ cm}}{\text{m}} \right) + 2(2.4) + 35(2.0) \right] \frac{8Q_{\text{coil}}^2 \left(\frac{100 \text{ cm}}{\text{m}} \right)^4}{\pi^2 \left(9.807 \frac{\text{m}}{\text{s}^2} \right)(1.34 \text{ cm})^4}$$

$$= 1.91 \times 10^8 (24.9f + 1)\left(\frac{\text{s}^2}{\text{m}^5} \right)(Q_{\text{coil}}^2).$$

Since each parallel path has the same head loss, we equate the two expressions for h_L to get

$$9.42 \times 10^7 (10.1f + 1)\left(\frac{\text{s}^2}{\text{m}^5} \right)(Q_{\text{bypass}}^2) = 1.91 \times 10^8 (24.9f + 1)\left(\frac{\text{s}^2}{\text{m}^5} \right)(Q_{\text{coil}}^2).$$

or

$$Q_{\text{coil}} = \sqrt{\frac{0.493(10.1f + 1)_{\text{bypass}}}{(24.9f + 1)_{\text{coil}}}} \, (Q_{\text{bypass}}).$$

The problem statement and the continuity equation give

$$Q_{\text{bypass}} = 0.00040 \text{ m}^3/\text{s} - Q_{\text{coil}}.$$

Substituting into the energy equation, we have

$$Q_{\text{coil}} = \sqrt{\frac{0.493(10.1f + 1)_{\text{bypass}}}{(29.4f + 1)_{\text{coil}}}} \, \left(0.0004 \frac{\text{m}^3}{\text{s}} - Q_{\text{coil}} \right).$$

Solving for Q_{coil}, we have

$$Q_{coil} = \frac{0.0004 \frac{m^3}{s} F}{1 + F},$$

where

$$F = \sqrt{\frac{0.493(10.1f + 1)_{bypass}}{(29.4f + 1)_{coil}}}$$

The factor F depends on Q_{coil} and Q_{bypass} through the friction factors, but this dependence is fairly weak. We can rapidly generate a solution by guessing values for the two f's and so on until convergence. Preparing for the friction factor calculations, we obtain the Reynolds number for either path:

$$\mathbf{R} = \frac{4Q}{\pi v D} = \frac{4\left(\frac{100\ cm}{m}\right)Q}{\pi\left(1.31 \times 10^{-6} \frac{m^2}{s}\right)(1.34\ cm)} = 7.25 \times 10^7\ Q.$$

The relative roughness is

$$\frac{\varepsilon}{D} = \frac{0.00015\ cm}{1.34\ cm} = 0.00011.$$

We now guess

$$f_{coil} = f_{bypass} = 0.03.$$

Then

$$F = \sqrt{\frac{(0.493)[10.1(0.03) + 1]}{(29.4)(0.03) + 1}} = 0.584,$$

$$Q_{coil} = \left(\frac{0.584}{1.584}\right)(0.0004\ m^3/s) = 0.000148\ m^3/s, \qquad \text{and}$$

$$Q_{bypass} = 0.0004\ m^3/s - 0.000148\ m^3/s = 0.000252\ m^3/s.$$

Then

$$\mathbf{R}_{coil} = 10,730 \qquad \text{and} \qquad \mathbf{R}_{bypass} = 18,270.$$

From the Moody chart,

$$f_{coil} \approx 0.031 \qquad \text{and} \qquad f_{bypass} \approx 0.027.$$

These values give

$$F = 0.573,$$

$$Q_{coil} = \frac{0.573}{1.573}(0.0004\ m^3/s) = 0.000146\ m^3/s,$$

$$Q_{bypass} = 0.0004\ m^3/s - 0.000146\ m^3/s = 0.000254\ m^3/s.$$

The Reynolds numbers are

$$\mathbf{R}_{coil} = 10,600 \qquad \text{and} \qquad \mathbf{R}_{bypass} = 18,400,$$

giving

$$f_{coil} \approx 0.031 \qquad \text{and} \qquad f_{bypass} \approx 0.027.$$

The calculations have converged, and

$$\begin{vmatrix} Q_{coil} = 0.000146 \text{ m}^3/\text{s}, \\ Q_{bypass} = 0.000254 \text{ m}^3/\text{s}. \end{vmatrix}$$ **Answer**

Discussion

Note how we constructed our iteration to isolate that portion of the equation with the weakest dependence on the unknown Q's. The trick was to substitute the continuity equation for Q_{bypass} into the energy equation. Had we not done this but instead solved the continuity and energy equations by straightforward iteration, at least twice as many trials would have been required to generate the same answers. You should be able to show that the pressure drop $p_1 - p_2$ is equal to 70.7 kPa.

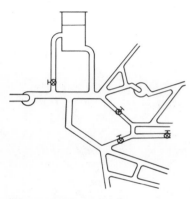

Figure 7.32 Complex multibranch pipe/duct network.

Pipe/Duct Networks. The ultimate problem in pipe/duct system analysis involves a complex network with many branches. Such a network is illustrated schematically in Fig. 7.32. Two examples of such networks are municipal water supply systems and air-handling ducts for large air-conditioning systems. Methods for calculating flow and losses in such systems are similar to methods for analyzing electrical networks. The following rules are the basis of any calculation procedure:

- The net flow into any junction must equal the net flow out of the junction. This is most conveniently handled by letting each flow have an algebraic sign so that the net flow into any junction can be set to zero.
- The sum of head (or pressure) increases, and losses around any loop must be zero. Alternative ways of stating this requirement are that the static pressure and velocity at any junction must be the same regardless of which path is followed to arrive at the junction and that the hydraulic grade lines for all pipes that come into a junction must intersect at the junction.
- All losses must satisfy the pipe friction equations or the local loss equations. All pumps must operate at a point on their pump curve.

The primary difference between electrical networks and hydraulic networks is that hydraulic networks are governed by nonlinear equations. Application of the three rules above to a hydraulic network produces a set of nonlinear algebraic equations (some of these "equations" may even be graphical, such as pump curves) for the flows and pressure at each junction. In the old days, these equations had to be solved by tedious hand calculations [11], but now the digital computer handles this type of problem in nothing flat [12, 13].

7.4 Flow Measurement

Up to this point, all of our discussions of internal flow have concentrated on calculation of energy loss or flow rate. An equally important practical problem is measurement of these quantities. Although this book is not intended to teach measurement methods, a brief discussion of internal flow measurements will illustrate an important application of the concepts and methods discussed in this chapter.

As we have previously pointed out, measurement of losses can be accomplished indirectly by measuring pressure, elevation, and, sometimes, velocity. Pressure measurement was discussed in Section 2.3. Elevation measurement is quite straightforward. At least one device for measuring velocity, the Pitot-static tube was discussed in Section 4.1.3. Measurement of losses need not be discussed any further.

Measurement of flow rate is required in many situations. Flow measurement is not only important in laboratory experiments but is also necessary to monitor plant operation or to provide information for process control. Many different devices and techniques are available for flow measurement. Generally a device for flow measurement is called a *flow meter* or a *fluid meter*. We will discuss the most common type of fluid meter, called the *obstruction type* or *differential pressure type*.

Figure 7.33 shows a pipe with a flow obstruction. Holes are tapped into the pipe wall upstream of the obstruction and downstream at the vena contracta, and instruments are attached to measure the pressures p_1 and p_2. Initially we assume that losses between the upstream point 1 and downstream point 2 in the vena contracta are negligible and apply Bernoulli's equation:

$$p_1 + \frac{\rho V_1^2}{2} + \gamma z_1 = p_2 + \frac{\rho V_2^2}{2} + \gamma z_2.$$

We also use the continuity equation and assume uniform flow at 1 and 2:

$$V_1 A_1 = V_2 A_2 = Q.$$

Combining these two equations gives

$$V_2 = \sqrt{\frac{(p_1 + \gamma z_1) - (p_2 + \gamma z_2)}{1 - (A_2/A_1)^2}}, \qquad \textbf{(7.112)}$$

and so

$$Q = V_2 A_2 = A_2 \sqrt{\frac{(p_1 + \gamma z_1) - (p_2 + \gamma z_2)}{\rho[1 - (A_2/A_1)^2]}}. \qquad \textbf{(7.113)}$$

According to these equations, all that is necessary for flow determination (and velocity determination as well) is the measurement of pressure differential and some geometric quantities. There are three

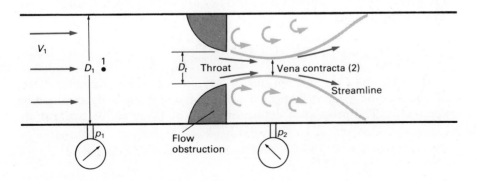

Figure 7.33 Generalized flow obstruction in a pipe.

V_1

D_1 • 1

D_t Throat

Vena contracta (2)

Streamline

p_1

Flow obstruction

p_2

problems with these equations. First, the area of the vena contracta is very difficult to determine. It would be much better to use the throat area of the obstruction (A_t). Second, since there are losses in the flow, the velocity will be lower than Eq. (7.112) indicates. Third, the exact location of the vena contracta is difficult to determine.

The first two problems are handled by defining a *contraction coefficient,*

$$C_c \equiv \frac{A_2}{A_t},$$

and a *velocity coefficient,*

$$C_v \equiv \frac{V_{2,\text{real}}}{V_{2,\text{ideal}}}.$$

Substituting these definitions into Eqs. (7.112) and (7.113), we have

$$V_2 = C_v \sqrt{\frac{\Delta p_z}{\rho\left[1 - \left(C_c \frac{A_t}{A_1}\right)^2\right]}} \quad \text{and} \quad Q = C_c C_v A_t \sqrt{\frac{\Delta p_z}{\rho\left[1 - \left(C_c \frac{A_t}{A_1}\right)^2\right]}},$$

where

$$\Delta p_z = (p_1 + \gamma z_1) - (p_2 + \gamma z_2).$$

The contraction coefficient and velocity coefficient are rather difficult to obtain experimentally and really represent more detail than we need. They are combined into a single *discharge coefficient* (C_d) so that the flow equation becomes

$$Q = C_d A_t \sqrt{\frac{\Delta p_z}{\rho\left[1 - \left(\frac{A_t}{A_1}\right)^2\right]}}. \tag{7.114}$$

Note that C_c is also included in C_d. We can also define a *flow coefficient** (C_Q) according to

$$C_Q \equiv \frac{C_d}{\sqrt{1 - (A_t/A_1)^2}} = \frac{C_d}{\sqrt{1 - \beta^4}}, \tag{7.115}$$

where

$$\beta = \sqrt{\frac{A_t}{A_1}} = \frac{D_t}{D_1}. \quad \text{For circular cross sections}$$

The term $\sqrt{1 - \beta^4}$ is called the *velocity of approach factor*. Using C_Q, we have for Eq. (7.114)

$$Q = C_Q A_t \sqrt{\frac{\Delta p_z}{\rho}} \tag{7.116}$$

Both C_d and C_Q are functions of Reynolds number, diameter ratio, and the specific geometry of the obstruction and must be determined by experiment. The process of determining C_d or C_Q is called *calibration*.

* The most common symbol for flow coefficient is **K**. We have used C_Q to avoid confusion with the loss coefficient.

Calibration is also the means by which the problem of locating the vena contracta can be avoided. The downstream pressure tap can be located at any convenient location near the obstruction. Calibration gives the value of C_Q for that particular pressure tap location.

Any type of obstruction can be used as a fluid meter. Each different type of obstruction requires calibration. Certain types of obstructions have been standardized, and calibration data for these types are available in the literature. Professional organizations, primarily the American Society of Mechanical Engineers (ASME) and the International Organization for Standardization (ISO), have established standard designs for obstruction-type flow meters. Their publications [14, 15] provide values of the various calibration coefficients that can be used for flow meters of standard design. If you choose a standard obstruction meter, specific calibration is not necessary unless a high degree of accuracy is required.

The desirable characteristics of a fluid meter are as follows:

- Reliable, repeatable calibration (that is, the published values of C_Q and C_d can be used with a high degree of certainty);
- Introduces small energy loss into the system;
- Inexpensive;
- Requires minimum space.

You might expect that no single type of obstruction fluid meter satisfies all these requirements. This is indeed the case, so there are three standard types of obstruction flow meters (Fig. 7.34). The *Venturi* meter meets the first two requirements and the *thin-plate orifice* meets the last two requirements. The *flow nozzle* represents a compromise between them. The pressure drop Δp_d is indicative of the energy loss due to the device.

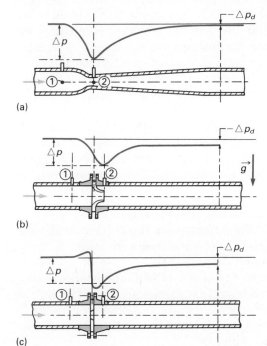

Figure 7.34 Three common types of fluid meters installed in a pipe.

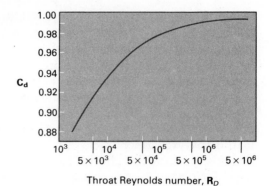

Figure 7.35 Discharge coefficient for ASME flow nozzles. Note that Reynolds number is based on throat diameter.

The Venturi is really a combined nozzle and diffuser. The problem of locating the vena contracta is not present in a Venturi since the maximum velocity and minimum pressure occur at the throat. The design of Venturi meters is not completely standardized because the "ideal" flow equations are reasonably accurate for a Venturi. Published calibration information for Venturis is usually given in terms of the discharge coefficient C_d. C_d is a weak function of Reynolds number and β. Typically,

$$0.94 < C_d < 0.98.$$

Consult references [14] and [15] for further details.

There are three standard designs for flow nozzles. ASME has published both a short-radius and a long-radius design. ISO has published a design called the "type 1932" flow nozzle. The ASME and ISO standards specify the location of the upstream pressure tap. The downstream tap is located at the nozzle exit. Since flow nozzles are smooth and relatively gradual contractions, the minimum pressure and maximum velocity occur at the nozzle exit and there is no problem in locating the vena contracta. Calibration data for flow nozzles are usually presented in terms of flow coefficient C_Q, which is a function of β and Reynolds number. Figure 7.35 shows a typical plot of C_d versus R. Curve fit equations for calculating C_Q and C_d are also available. Consult references [14] and [15] for details.

Thin-plate orifices are the cheapest type of flow meter. They are also very easy to install since the orifice can be simply clamped between pipe flanges. Orifices can be of sharp edge or square edge design. There are three standard locations for upstream and downstream pressure taps for orifices:

- Vena Contracta taps. The upstream tap is one pipe diameter upstream of the orifice, and the downstream tap is located at the plane of the vena contracta. Vena contracta location is determined from an experimental curve [14].
- D and $D/2$ taps. The upstream tap is one pipe diameter upstream of the orifice, and the downstream tap is one-half diameter downstream.
- Flange taps. The pressure taps are located one inch upstream and one inch downstream of the orifice.

Each pressure tap location requires a different calibration curve. Likewise, sharp edge and square edge orifices have different calibration curves. Figure 7.36 shows the variation of C_Q with β and **R** for a sharp edge orifice with D and $D/2$ taps. Further information, including tables and curve fit equations for flow coefficient and discharge coefficient, is available in [14, 15, 16]. These references also give information on eccentric orifices in which the hole is not concentric with the pipe.

If a fluid meter is part of a complete flow system, energy loss due to the meter must be included in the system resistance. Figure 7.37 shows typical loss coefficients (based on the throat velocity $V_t = Q/A_t$) for fluid meters.

If you use fluid meters of standard design, you can use the published information on calibration coefficients, but you must exercise some care. The published coefficients are valid only if the flow entering the meter is fully developed; accordingly, the fluid meter must be located a sufficient distance downstream of flow disturbances such as elbows, valves, and so on. Information on acceptable locations is given in [14]. Even if all restrictions are met, the calibration information is accurate only to about $\pm 2\%$ (Venturis have better accuracy and orifices may have worse). For extreme accuracy, a flow meter should be calibrated in place.

Many other flow measurement methods are available. Consult [14, 17, 18] for information. Two very basic methods are the "catch-and-weigh" technique and the velocity traverse method. In the catch-and-weigh method, the flow from the piping system is collected in a tank. The quantity of fluid collected in a given time period (measured with a stopwatch) is determined by weighing. The flow rate is the quantity

Figure 7.36 Flow coefficients for sharp-edge orifice plate with D and $D/2$ taps.

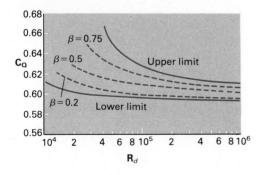

Figure 7.37 Typical loss coefficients for fluid meters. **K** is based on throat velocity.

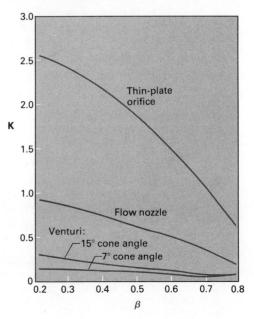

collected divided by the elapsed time. Obviously, this method works only for liquid flows.

In the velocity traverse method, a velocity-measuring instrument, typically a Pitot-static tube, is positioned at various locations in a cross-sectional plane. The measured velocity distribution can be integrated (graphically or numerically) over the cross section to compute the flow. Details of the velocity traverse method can be found in [18, 19, 20]. Both the catch-and-weigh and velocity traverse methods can be used to calibrate other types of fluid meters.

EXAMPLE 7.14 Illustrates considerations in selecting flow meters

You are an engineer for a small company and are to select an appropriate fluid meter from your warehouse stock to measure the water flow rate in a 6-in. schedule 40, horizontal commercial steel pipe. The fluid meter is needed immediately, so no time is available for machining or modification. The flow rate is estimated to be between 100 and 200 gal/min. A mercury manometer is to be used to measure the appropriate pressure difference to determine the flow rate.

Your instructions are to choose a fluid meter to determine the flow rate with a maximum uncertainty of 10% due to errors in reading the manometer. You estimate that the manometer can be read with an uncertainty of 0.05 in. The net pressure drop across the meter must not exceed 20.0 psi. The following meters are available:

Type of Fluid Meter	Throat Diameter
Venturi	5.7 in.
ASME long-radius flow nozzle	3.0 in.
Thin-plate orifice	3.0 in.

Solution

Given
Three fluid meters:
   ASME long-radius flow nozzle with 3.0-in. throat diameter
   Thin-plate orifice with 3.0-in. throat diameter
   Venturi with 5.7-in. throat diameter
Water flow rate estimated between 100 and 200 gal/min
Mercury manometer that can be read reliably to within 0.05 in.
6-in. schedule 40, horizontal commercial steel pipe
Net pressure drop (pressure loss) not to exceed 20.0 psi
Figure E7.14

Find
Which meter(s) will measure the flow rate with a maximum uncertainty
   due to manometer reading of 10% and maximum pressure loss of 20.0 psi

Solution
Each fluid meter and manometer combination must satisfy two requirements:

- A minimum manometer reading of at least 0.5 in. so that a 0.05-in. reading error represents no more than 10% of the reading. This reading must be for 100 gal/min since a 200-gal/min flow rate would give a higher reading.

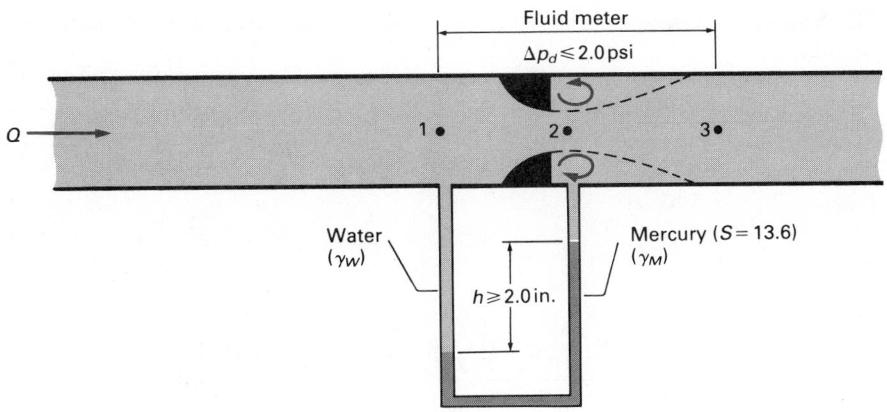

Figure E7.14 Manometer used with fluid meter.

- Net pressure drop cannot exceed 20.0 psi for the maximum flow rate of 200 gal/min.

The relation between the flow rate and the maximum pressure drop in a fluid meter is given by Eq. (7.114):

$$Q = C_d A_t \sqrt{\frac{\Delta p_z}{\rho \left[1 - \left(\dfrac{A_t}{A_1} \right)^2 \right]}}.$$

For a horizontal pipe, $\Delta p_z = p_1 - p_2 = \Delta p$. Solving for Δp gives

$$\Delta p = \rho \left[1 - \left(\frac{A_t}{A_1} \right)^2 \right] \left(\frac{Q}{C_d A_t} \right)^2.$$

Applying the manometer rule (Fig. E7.14) gives

$$p_2 = p_1 + (\gamma_W - \gamma_M)h, \quad \text{or} \quad \Delta p = p_1 - p_2 = (\gamma_M - \gamma_W)h.$$

Eliminating Δp gives the manometer reading h in terms of the flow rate Q. Introducing the diameters results in

$$h = \frac{\rho \left[1 - \left(\dfrac{D_t}{D_1} \right)^4 \right]}{(\gamma_M - \gamma_W)} \left[\frac{4Q}{C_d \pi D_t^2} \right]^2.$$

For 60°F water, Table A.6 gives

$$\rho = 62.4 \text{ lbm/ft}^3, \quad \gamma_W = 62.4 \text{ lb/ft}^3, \quad \text{and} \quad \gamma_M = 13.6 \,(62.4 \text{ lb/ft}^3) = 849 \text{ lb/ft}^3.$$

For a 6-in. schedule 40 commercial steel pipe, Table D.1 gives $D_1 = 6.065$ in. We now calculate the manometer reading for each fluid meter with a 100-gal/min flow rate. For a Venturi the text shows $0.94 < C_d < 0.98$. The maximum manometer reading for a Venturi-manometer system and 100 gal/min is obtained using $C_d = 0.94$:

$$h = \frac{\left(62.4 \dfrac{\text{lbm}}{\text{ft}^3} \right) \left[1 - \left(\dfrac{5.70 \text{ in.}}{6.065 \text{ in.}} \right)^4 \right] \left[4 \left(100 \dfrac{\text{gal}}{\text{min}} \right) \left(\dfrac{\text{ft}^3}{7.48 \text{ gal}} \right) \left(\dfrac{\text{min}}{60 \text{ sec}} \right) \right]^2}{(849 - 62.4) \dfrac{\text{lb}}{\text{ft}^3} \left[(0.94)\pi(5.70 \text{ in.})^2 \left(\dfrac{\text{ft}}{12 \text{ in.}} \right)^2 \right]^2}$$

$$= \frac{0.0312 \text{ lbm} \cdot \text{ft}^2}{\text{lb} \cdot \text{sec}^2} \left(\frac{12 \text{ in.}}{\text{ft}} \right) \left(\frac{\text{lb} \cdot \text{sec}^2}{32.2 \text{ ft} \cdot \text{lbm}} \right) = 0.0116 \text{ in.}$$

The Venturi is not acceptable because its maximum manometer reading is less than 0.5 in.

Next consider the ASME flow nozzle. The Reynolds number based on the throat diameter is first calculated for 60°F water; Table A.6 gives

$$\nu = 1.22 \times 10^{-5} \text{ ft}^2/\text{sec}.$$

Then

$$\mathbf{R} = \frac{V_t D_t}{\nu} = \frac{4Q}{\pi \nu D_t}$$

$$= \frac{4\left(100 \dfrac{\text{gal}}{\text{min}}\right)\left(\dfrac{\text{ft}^3}{7.48 \text{ gal}}\right)\left(\dfrac{\text{min}}{60 \text{ sec}}\right)}{\pi\left(1.22 \times 10^{-5} \dfrac{\text{ft}^2}{\text{sec}}\right)(3.0 \text{ in.})\left(\dfrac{\text{ft}}{12 \text{ in.}}\right)} = 9.3 \times 10^4.$$

Figure 7.35 gives $\mathbf{C_d} = 0.976$. The minimum manometer reading for the flow nozzle–manometer system is then

$$h = \frac{\left(62.4 \dfrac{\text{lbm}}{\text{ft}^3}\right)\left[1 - \left(\dfrac{3.0 \text{ in.}}{6.065 \text{ in.}}\right)^4\right]}{(849 - 62.4) \dfrac{\text{lb}}{\text{ft}^3}}\left[\frac{4\left(100 \dfrac{\text{gal}}{\text{min}}\right)\left(\dfrac{\text{ft}^3}{7.48 \text{ gal}}\right)\left(\dfrac{\text{min}}{60 \text{ sec}}\right)}{(0.976)\pi(3.0 \text{ in.})^2\left(\dfrac{\text{ft}}{12 \text{ in.}}\right)^2}\right]^2$$

$$= 1.61 \frac{\text{lbm} \cdot \text{ft}^2}{\text{lb} \cdot \text{sec}^2}\left(\frac{12 \text{ in.}}{\text{ft}}\right)\left(\frac{\text{lb} \cdot \text{sec}^2}{32.2 \text{ ft} \cdot \text{lbm}}\right) = 0.60 \text{ in.}$$

The flow nozzle provides a high enough reading. Now consider the orifice. The Reynolds number based on the throat diameter is the same as for the flow nozzle, so $\mathbf{R} = 9.3 \times 10^4$. Using the β ratio defined as

$$\beta = \frac{D_t}{D_1} = \frac{3.0 \text{ in.}}{6.065 \text{ in.}} = 0.495,$$

Figure 7.36 gives $\mathbf{C_d} = 0.612$. The minimum manometer reading for the orifice–manometer system is

$$h = \frac{\left(62.4 \dfrac{\text{lbm}}{\text{ft}^3}\right)\left[1 - \left(\dfrac{3.0 \text{ in.}}{6.065 \text{ in.}}\right)^4\right]}{(849 - 62.4) \dfrac{\text{lb}}{\text{ft}^3}}\left[\frac{4\left(100 \dfrac{\text{gal}}{\text{min}}\right)\left(\dfrac{\text{ft}^3}{7.48 \text{ gal}}\right)\left(\dfrac{\text{min}}{60 \text{ sec}}\right)}{(0.612)\pi(3.0 \text{ in.})^2\left(\dfrac{\text{ft}}{12 \text{ in.}}\right)^2}\right]^2$$

$$= 4.10 \frac{\text{lbm} \cdot \text{ft}^2}{\text{lb} \cdot \text{sec}^2}\left(\frac{12 \text{ in.}}{\text{ft}}\right)\left(\frac{\text{lb} \cdot \text{sec}^2}{32.2 \text{ ft} \cdot \text{lbm}}\right) = 1.53 \text{ in.}$$

The orifice also provides a high enough manometer reading. At this point, only the nozzle and the orifice satisfy the first requirement. We will now check if these two satisfy the second requirement that the flow meter must have a net pressure drop less than 20.0 psi. This net pressure drop is found by applying the mechanical energy equation between inlet 1 and outlet 3 in Fig. E7.14:

$$\frac{p_1}{\rho} + \alpha_1\left(\frac{V_1^2}{2}\right) + gz_1 = \frac{p_3}{\rho} + \alpha_3\left(\frac{V_3^2}{2}\right) + gz_3 + gh_L,$$

where p, V, and ρ represent average values. Since the fluid meter is horizontal,

$z_1 = z_3$. The pipe Reynolds number for 200 gal/min is

$$\mathbf{R} = \frac{V_1 D_1}{v} = \frac{4Q}{\pi v D_1}$$

$$= \frac{4\left(200\,\frac{\text{gal}}{\text{min}}\right)\left(\frac{\text{ft}^3}{7.48\,\text{gal}}\right)\left(\frac{\text{min}}{60\,\text{sec}}\right)}{\pi\left(1.22 \times 10^{-5}\,\frac{\text{ft}^2}{\text{sec}}\right)(6.065\,\text{in.})\left(\frac{\text{ft}}{12\,\text{in.}}\right)} = 9.2 \times 10^4.$$

The flow is turbulent. Noting that $D_1 = D_3$, we get, from the continuity equation, $V_1 = V_3$. If we assume $\alpha_1 = \alpha_3$, the mechanical energy equation results in $p_1 - p_3 = \rho g h_L$. Using Eq. (7.99) to express the energy loss, $g h_L$, in terms of a loss coefficient, $\mathbf{K}$, gives

$$p_1 - p_3 = \rho \mathbf{K} \frac{V_t^2}{2},$$

where $p_1 - p_3$ is the net pressure drop across the fluid meter. First consider the flow nozzle; its β-ratio is the same as that for the orifice:

$$\beta = \frac{D_t}{D_1} = 0.495.$$

Figure 7.37 gives $\mathbf{K} = 0.60$. The throat velocity is

$$V_t = \frac{Q}{A_t} = \frac{4Q}{\pi D_t^2}$$

$$= \frac{4\left(200\,\frac{\text{gal}}{\text{min}}\right)\left(\frac{\text{min}}{60\,\text{sec}}\right)}{\pi(3.0\,\text{in.})^2\left(\frac{\text{ft}}{12\,\text{in.}}\right)^2} = 67.9\,\text{ft/sec}.$$

Then

$$p_1 - p_3 = \frac{\left(62.4\,\frac{\text{lbm}}{\text{ft}^3}\right)(0.60)\left(67.9\,\frac{\text{ft}}{\text{sec}}\right)^2}{2\left(32.2\,\frac{\text{ft}\cdot\text{lbm}}{\text{lb}\cdot\text{sec}^2}\right)\left(\frac{144\,\text{in}^2}{\text{ft}^2}\right)} = 18.6\,\text{psi}.$$

The flow nozzle–manometer system satisfies both requirements, so it is acceptable. Next consider the orifice. It has the same β-ratio and pipe velocity as the flow nozzle. Using Fig. 7.37 gives $\mathbf{K} = 1.85$, and so

$$p_1 - p_3 = \frac{\left(62.4\,\frac{\text{lbm}}{\text{ft}^3}\right)(1.85)\left(67.9\,\frac{\text{ft}}{\text{sec}}\right)^2}{2\left(32.2\,\frac{\text{ft}\cdot\text{lbm}}{\text{lb}\cdot\text{sec}^2}\right)\left(\frac{144\,\text{in}^2}{\text{ft}^2}\right)} = 57.4\,\text{psi}.$$

The orifice is not acceptable because its net pressure drop is too high. Our conclusion is **Answer**

| use ASME flow nozzle. |

◀

Discussion

The actual uncertainty of the measured flow rate will be larger than the uncertainty due to manometer reading alone. One source of additional uncertainty is the discharge coefficient. See references [17] and [18] for discussion of uncertainty analysis.

We presented the maximum manometer reading for the Venturi and the minimum manometer readings for the flow nozzle and orifice. Only those calculations that were critical to a decision were presented in this solution. We have made a so-called worst-case analysis. In practice, we may not know beforehand which of several calculations are the critical ones and may make more than the necessary minimum.

Problems

Extension and Generalization

1. Consider the volume flow rate Q of blood in a circular tube of radius R. The blood cells concentrate and flow near the center of the tube, while the cell-free fluid (plasma) flows in the outer region. The center core of radius R_c has a viscosity μ_c, while the plasma has a viscosity μ_p. Assuming laminar, fully developed flow for both the core and plasma flows, show that an "apparent" viscosity defined by

$$\mu_{\text{app}} \equiv \frac{\pi R^4 \Delta p}{8LQ}$$

is given by

$$\mu_{\text{app}} = \frac{\mu_p}{1 - \left(\dfrac{R_c}{R}\right)^4 \left(1 - \dfrac{\mu_p}{\mu_c}\right)}.$$

2. Consider the horizontal, fully developed, laminar flow of a viscous liquid through an annulus with inner radius R_1 and outer radius R_2. Develop an expression for the axial velocity as a function of the radius r. Next develop an expression for the friction factor f in terms of the volume flow rate Q, the liquid properties, and the annulus dimensions.

3. The mixing length in the overlap region of a turbulent flow is given by $\ell = \kappa y$. Assume that the turbulent shear stress in the overlap region is constant and approximately equal to the wall shear stress and derive the logarithmic velocity distribution.

4. Approximate the velocity distribution in turbulent flow in a smooth annulus by a pair of logarithmic velocity profiles joined in the center of the annulus and develop an expression for the friction factor.

Lower-Order Application

5. A 250-ft-high building has a 6-in.-diameter standpipe and a 100-ft-long, $2\frac{9}{16}$-in.-diameter fire hose on each floor. The nearest fireplug is 100 ft from the standpipe's ground-level connection. Assume that firefighters connect a 6-in.-diameter fire hose from the fireplug to the fire truck and a 4-in.-diameter fire hose from the fire truck to the standpipe's ground-level connection. The National Fire Protection Association (NFPA) requires that a minimum pressure of 65 psig be maintained at the connection of the $2\frac{9}{16}$-in.-diameter hose and the standpipe while maintaining a flow rate of 500 gal/min through the fire hose. What pressure rise must the pump on the fire engine supply to satisfy the NFPA requirement for this building?

6. A horizontal commercial steel pipe is 0.5 m in diameter and has a total head loss of 12.0 m between cross sections 1 and 2, which are 5000 m apart. Find the value of the Moody friction factor if the average water velocity is 5.0 m/s.

7. Consider the standpipe system in Figure E7.2 on page 419. Find the flow rate for $H = 4.0$ ft, $D = 6.77$ in., $d = 0.125$ in., $L = 48$ in. The fluid is 70°F water. Assume steady-state conditions. Calculate the pressure loss Δp_L, the energy loss gh_L, and the pressure change between points 1 and 3.

8. The pressure and average velocity at point A in the pipe in Fig. P4.2 on page 228 are 16.0 psia and 4.0 ft/sec, respectively. Find the pressure and average velocity at point B. Compare your result with that of Problem 12 in Chapter 4.

9. Find the energy loss, pressure loss, and pressure drop for the contraction of Fig. 7.23. The volume flow rate is $2.0 \text{ m}^3/\text{s}$, and elevation changes are negligible. Assume the velocity is uniform over each flow area. The fluid is 20°C ethanol.

10. The pump in Fig. P7.1 delivers 80°F water at a steady rate of 30 ft³/min and a pump discharge pressure of 200 psig. Find the pressure at points B and C.

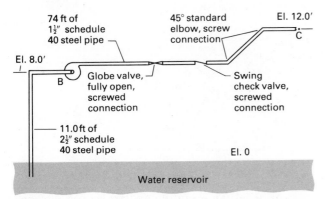

Figure P7.1

11. What air pressure is necessary to deliver 50 gal/min of 68°F ethanol through the piping system in Fig. P7.2?

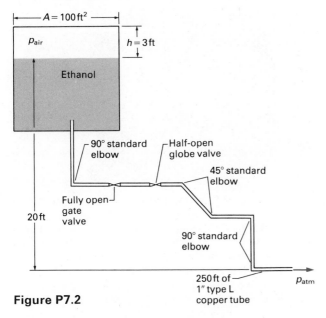

Figure P7.2

12. Methanol at 68°F flows through a 25-cm-inside-diameter, clear plastic tube at the rate of 1.2 m³/min. The tube is inclined upward in the direction of flow at an angle of 30° with the horizontal. Find the pressure drop, pressure loss, and energy loss for a 50-m-long section of the tube.

13. Find the pressure drop, pressure loss, and energy loss for a flow rate of 500 gal/min flowing upward in a vertical 6-in. schedule 80 pipe. The pipe is 400 ft long and has one swing check valve and three fully open globe valves.

14. Air flows in a horizontal 100-ft-long, 24-in.-by-24-in. duct at the rate of 5000 ft³/min. The air then flows through an expansion into 200 ft of 36-in. by 36-in. duct. The expansion has a loss coefficient of 0.80 based on the higher inlet velocity. Find the static pressure rise across the expansion and the static pressure drop across the entire duct system. The duct is made of galvanized sheet metal.

15. Crude oil having a specific gravity of 0.80 and a viscosity of 6.0×10^{-5} ft²/sec flows through a pumping station at a rate of 10,000 barrels/hr. The oil then flows through 120,000 ft of 24-in.-inside-diameter commercial steel pipe, enters another pumping station at 20 psig, and leaves at a discharge pressure equal to the discharge pressure of the previous pumping station. Find the discharge pressures and the power supplied to the crude oil at the second pumping station.

16. An aorta has an inside diameter of 1.0 cm and a length of 40 cm. The flowing blood has an absolute viscosity of 0.035 Poise and an average velocity of 50 cm/s. Find the pressure drop if the aorta is (a) horizontal and (b) vertical with downward flow. Express your answers in mm Hg.

17. A 24-in. schedule 40 commercial steel pipe has a flow rate of 77°F kerosene of 1200 ft³/min. At point A the static pressure is 50.6 psig and the elevation is 150 ft. At point B, 2000 ft downstream of point A, the static pressure is 24.4 psig and the elevation is 25 ft. The pipeline has no fittings or valves. Find the mechanical energy loss and the pressure loss between points A and B.

18. Liquid mercury at 20°C flows through the channel in Fig. P7.3. Assume fully developed flow and find the mechanical energy loss and pressure drop per unit length. The channel walls are made of copper plate.

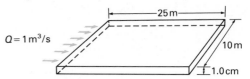

Figure P7.3

19. Methanol at 32°F is pumped through a 3-in. schedule 40 pipe having a length of 700 ft. Find the flow rate through the pipe if the pipe is sloped downward in the direction of flow and the static pressure is constant along the pipe.

20. Find the amount of water that will drain out of the tank in Fig. P7.4. The air volume above the water is 4 ft^3 at 14.696 psia and 70°F for the condition shown. The air temperature may be considered constant.

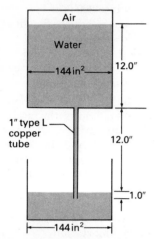

Air

Water

144 in^2

12.0″

1″ type L copper tube

12.0″

1.0″

144 in^2

Figure P7.4

21. Lubricating oil at 20°F flows through a horizontal, 2½-in. schedule 80 commercial steel pipe with a head loss (h_L) of 0.75 ft per 100 ft of pipe. Find the flow rate.

22. Methanol at 10°C flows upward through a 5.0-cm-inside-diameter commercial steel pipe with a pressure drop of 20 kPa per 100 m of pipe. The pipe is inclined upward at an angle of 20° with the horizontal. Find the flow rate.

23. After steady-state conditions were achieved the globe valve in Fig. P7.5 is suddenly moved from fully open to half-open. Find the flow rate as a function of time.

Figure P7.5

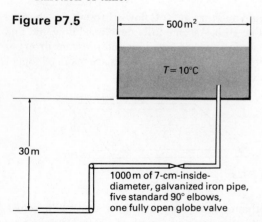

500 m^2

$T = 10°C$

30 m

1000 m of 7-cm-inside-diameter, galvanized iron pipe, five standard 90° elbows, one fully open globe valve

24. A centrifugal pump operates at 1000 rpm and has the characteristics shown in Fig. P7.6. The pump delivers water through a piping system where the pressure drop is 45 psi and the outlet

of the pipe is 100 ft above the inlet. Find the flow rate. Find the new flow rate if the pump speed is doubled. Assume the fluid viscosity does not alter the pump performance significantly. [*Hint:* Refer to Example 6.8.]

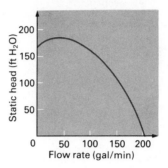

Figure P7.6

25. An axial fan operating at 1000 Hz has the characteristics shown in Fig. P7.7. It delivers 15°C air through a 15-cm-inside-diameter, galvanized sheet-metal duct having a length of 175 m and seven 90° long-radius elbows. Find the flow rate if the duct discharges to the atmosphere.

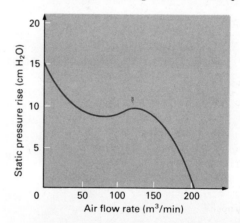

Figure P7.7

26. Water at 210°F and 14.7 psia flows out of a boiler through a blowdown line. The line is 150 ft of 2-in.-nominal-diameter schedule 40 commercial steel pipe. There are also twelve 90° standard elbows and two gate valves. The flow discharges to the atmosphere. The discharge is 60 ft below the inlet. Find the flow rate in this line.

27. Kerosene at 20°C enters the main header in Fig. P7.8 at $Q = 5.0$ m^3/min. Find the flow rate in each of the parallel lines. All lines are type K copper tube.

28. Find the flow rate of 20°C kerosene from tank A to tank B for the conditions shown in Fig. P7.9.

29. Estimate the time for the water depth in the reservoir in Fig. P7.5 to drop from a height of 25 m to 5 m.

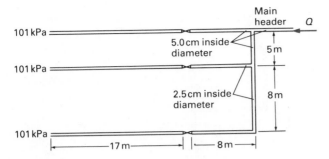

Figure P7.8

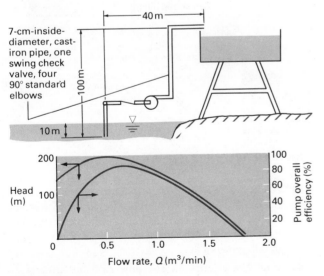

3-m-long, 2.5-cm-inside-
diameter, circular clear
plastic pipe

Figure P7.9

30. A pump has the head and efficiency characteristics shown in Fig. P7.10 and delivers water from
the lake to the reservoir. Find the average fluid
velocity in the piping and the input power to the
pump.

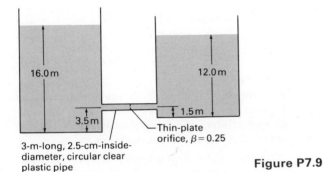

7-cm-inside-
diameter, cast-
iron pipe, one
swing check
valve, four
90° standard
elbows

Figure P7.10

31. An experienced fan engineer suggests that the
flow in the system of Fig. P7.11 can be increased
and the mechanical energy losses decreased by
rounding the inlet box entrance (giving $K = 0.05$)

and adding turning vanes in the miter bends
(giving $K = 0.33$ per bend). The pressure rise developed by the fan is given by

$$\Delta p = 17.0 - 0.125(Q \times 10^{-5})^2,$$

where Δp is in inches of water and Q in ft^3/min.
Without the modifications, the flow rate is 4×10^5 ft^3/min. Determine the flow rate with the suggested modifications.

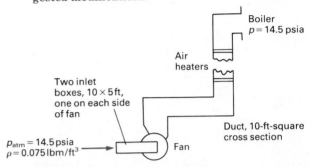

Figure P7.11

32. The flow rate of 60°F methanol through the piping system of Fig. P7.12 is 250 gal/min. Find the
division of flow between the two paths and the
pressure drop $(p_A - p_D)$.

33. The flow rates from the branches of the main
supply line of Fig. P7.13 are shown. Find the total
pressure drop $(p_A - p_D)$.

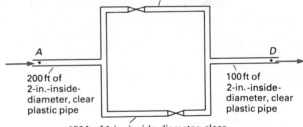

Figure P7.12

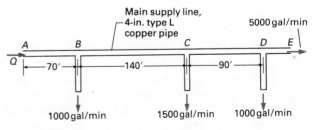

Figure P7.13

34. Assume corrosion and scaling has increased the absolute roughness of the cast-iron pipe in Problem 30 from 0.00085 ft to 0.0017 ft. Find the reduction in the flow rate.

35. Find the flow rate in the piping system of Fig. P7.14.

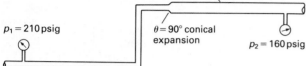

310 m of 0.15-m-inside-diameter, type L copper pipe, five fully open gate valves, five 90° street elbows, one fully open globe valve

$p_1 = 210$ psig $\theta = 90°$ conical expansion $p_2 = 160$ psig

120 m of 0.05-m-inside-diameter, type L copper pipe, four fully open gate valves, five 90° street elbows

Figure P7.14

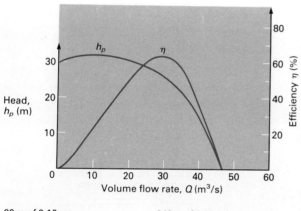

Head, h_p (m)

Volume flow rate, Q (m³/s)

Efficiency η (%)

60 m of 0.10-m-inside-diameter, commercial steel pipe, one swing check valve, four 90° standard elbows

240 m of 0.10-m-inside-diameter, commercial steel pipe, four gate valves, one fully open globe valve, twelve 90° standard elbows, four tees with flow through lines

40 m

Figure P7.15

36. A pump has the head and efficiency curves shown in Fig. P7.15. Determine the flow rate and the power input to the pump if it is connected to the piping system shown.

37. Find the water flow rate in the system of Fig. P7.16. The piping system includes four gate valves, two half-open globe valves, fourteen 90° standard elbows, and 250 ft of 2-in. commercial steep pipe.

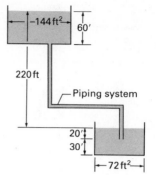

144 ft² 60'

220 ft

Piping system

20'
30'

72 ft² **Figure P7.16**

38. Initially the lower tank in Fig. P7.16 is empty and the upper tank has a water depth of 60 ft. Find the time for the water depth in the lower tank to reach a depth of 30 ft. See Problem 37 for details of the piping system.

39. Both the suction and discharge piping for the pump in Fig. P5.45 on page 334 consists of 4-in.-inside-diameter, 40-ft-long plastic pipe. Find the volume flow rate through the pump.

40. The pump in the piping system of Fig. P7.17 has the head curve shown in Fig. E7.12(b). Find the pump operating point. All fittings have screwed connections.

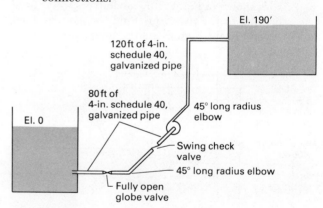

El. 190'

120 ft of 4-in. schedule 40, galvanized pipe

80 ft of 4-in. schedule 40, galvanized pipe

El. 0

45° long radius elbow

Swing check valve

45° long radius elbow

Fully open globe valve

Figure P7.17

41. A force of $F = 4.0$ N is applied to the hypodermic needle of Fig. P7.18. Find the flow rate Q. Assume the piston moves with a constant velocity. The fluid has a specific gravity of 0.93 and a viscosity of 6.0 centipoise.

42. Find the change in the flow rate out of or into each reservoir of Fig. P7.17 if each pipe is replaced by a plastic pipe of the same inside diameter. Compare your results with that of Problem 40.

43. Consider the pump and piping system in Fig. P5.55 on page 338. The inlet pipe is 0.10-m-inside-diameter galvanized steel and the discharge piping

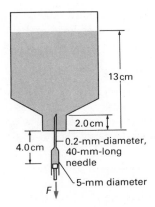

Figure P7.18

is 0.05-m-inside-diameter galvanized steel. Both pipes are 27 m long. Find the time to fill the storage tank, and compare your result with that of Problem 111 in Chapter 5.

44. A 2-in.-inside-diameter copper tube is used to bleed water from the open tank to the closed tank in Fig. P7.19. Find the amount of water bled to the lower tank. The air in the lower tank is initially at 14.7 psia and 70°F. Assume the air temperature remains constant at 70°F.

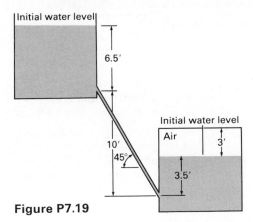

Figure P7.19

45. For very large tanks and steady-flow conditions, find the net rate of water flow out of each tank in Fig. P7.20.

46. A neighboring city has asked to tie a 45-in.-inside-diameter pipe into the galvanized pipe from the city storage reservoir to the city supply reservoir in Fig. P7.21. The tie-in point has been agreed to be 13 mi from the city storage reservoir, with two gate valves and twelve 90° standard elbows upstream of the tie-in point. The new pipeline will be galvanized pipe with two gate valves, twelve 90° standard elbows, and four 45° standard elbows and will tie into a reservoir having a water-level elevation of 250.0 ft. Find the flow rate to each city.

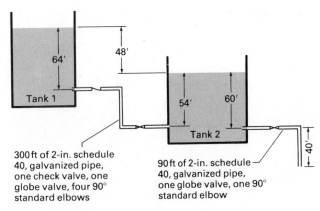

Figure P7.20

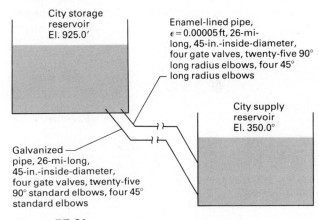

Figure P7.21

47. Find the maximum flow rate of wine out of the wine vat in Fig. P7.22 if a pump were not used.

48. Find the inside diameter D of the commercial steel pipe in Fig. P7.23 if $h = 12$ cm.

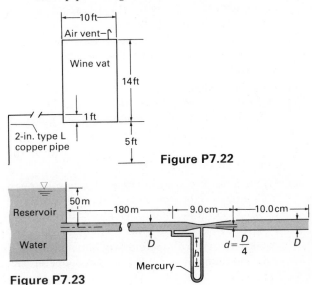

Figure P7.22

Figure P7.23

49. Suppose the two pipelines for the city water supply system in Problem 46 were extensively damaged by an earthquake. The city engineer has asked you to size a single pipeline that will supply the same flow rate as the two pipelines. Find an appropriate size for a galvanized pipe system.

50. It is necessary to deliver 270 ft^3/min of water from reservoir A to reservoir B in Fig. P7.24. The connecting piping consists of four fully open gate valves, twelve 90° elbows, one check valve, two fully open globe valves, and 3000 ft of commercial steel pipe. Find the minimum-diameter steel pipe needed to obtain this flow rate. All fittings and valves have flanged connections.

51. Reservoir B in Problem 50 is empty, and reservoir A has a water depth of 60 ft. Find the minimum schedule 40 pipe size necessary for a flow rate of 100 gal/min.

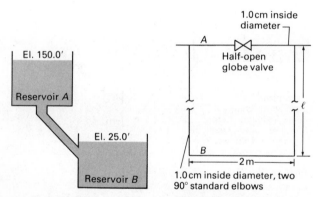

Figure P7.24   **Figure P7.25**

52. Find the length ℓ of pipeline B in Fig. P7.25 for equal flows through both path A and path B. Assume fully turbulent flow.

53. Find the length ℓ in Problem 52 if the flow is laminar. Does the length ℓ change with Reynolds number?

54. Find the minimum diameter necessary for the flow in Problem 13 to have a pressure loss less than 20.0 psi. Find the corresponding pressure drop.

55. Find the inside diameter of the two lower pipelines in Problem 27 so that the flow rate is the same in all three parallel lines.

56. Water is to flow at a rate of 200 gal/min through 1200 ft of straight pipe that has a downward slope of 20°. Find the minimum diameter of the schedule 80 pipe if the pressure drop is 40 psi.

57. A 5000-ft cast-iron pipeline originates in a reservoir and terminates in a 2-in.-outlet-diameter noz-

zle that discharges to the atmosphere. The nozzle is 200 ft below the reservoir water level and the pipeline has seventeen 90° standard elbows and two fully open gate valves. Find the pipe size required to deliver 10,000 ft^3/min.

58. A 3-in. schedule 40 commercial steel pipe carries 210°F SAE 10W crankcase oil at the rate of 6.0 gal/min. For the same pressure drop, what must the new pipe size be to carry the oil at 8.0 gal/min?

59. Water flows from the reservoir in Fig. P7.26 and through the pipeline shown. Find the steady-state water height h of the surge tank. All of the connecting piping is 1.0-cm-inside-diameter plastic pipe.

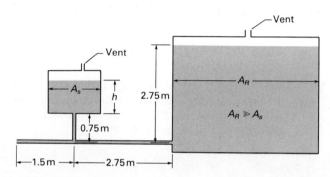

Figure P7.26

60. Repeat Problem 59 if the air vent on the surge tank is capped off and the surge tank is filled with 15°C air at atmospheric pressure prior to flow in the pipeline; $A_s = 2.5$ m^2.

61. Water at 20°C is to flow through a 3-cm-inside-diameter pipe at the rate of 1.0 m^3/s. Find the incline angle of the pipe so the static pressure is constant along the pipe.

62. Find the diameter of a wrought-iron pipe required to deliver 20°C kerosene from one tank to another at the rate of 2.0 m^3/s. The pipe is 1500 m long and has two fully open globe valves and six 45° standard elbows. The two free liquid surfaces in the tanks have an elevation difference of 120 m.

63. A 2½-in. schedule 80 commercial steel pipe is 120 ft long and carries 70°F benzene from one reservoir to another at the rate of 200 gal/min. Find the difference in the free-surface elevations of the two reservoirs.

64. Consider the forced draft fan and air duct shown in Fig. P7.10. The air flow rate through the fan at normal operating conditions is 4×10^5 ft^3/min. All of the losses in the inlet box, ducts, and heaters

are represented by

$$gh_L = \mathbf{K}\left(\frac{V^2_{\text{duct}}}{2}\right).$$

Find the value of **K**.

65. The test setup in Fig. P7.27 is used to determine the mechanical energy loss characteristics of a valve. The catch basin receives 10 liters of water in 50 s. What is the loss coefficient **K** for the valve? What would be the water flow rate if the water level were 30 m rather than 10 m?

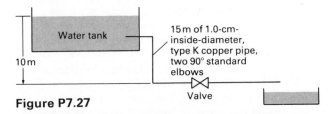

Figure P7.27

66. Air flows through the contraction shown in Fig. P7.28. The static pressure drop is given by

$$\Delta p = \frac{1.02\rho_{\text{air}}(V_2^2 - V_1^2)}{2}.$$

Find the value of the loss coefficient **K** and compare with that of Fig. 7.29.

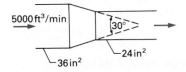

Figure P7.28

67. Water flows from a large reservoir to a storage tank as shown in Fig. P7.29. The pipe entrance at *B* is well rounded. The pipeline from *B* to *C* contains four gate valves, three standard 90° elbows, and one tee with flow through the main run. The pipeline from *D* to *E* contains four gate valves, six 90° standard elbows, two 45° standard elbows, and one tee with flow through the main run. The average fluid velocity in the pipeline is 12.0 ft/sec. Find the power input to the pump if it has an overall efficiency of 85%.

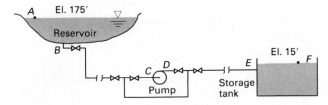

Figure P7.29

68. Wine having a specific gravity of 1.02 and a viscosity of 1.0 centipoise is pumped from the wine vat in Fig. P7.30 by a constant-volume pump at the rate of 100 gal/min. Find the power required to pump the wine through a 2-in. copper pipe that is 250 ft long, has 14 regular 90° elbows, four fully open gate valves, two half-open globe valves, and one check valve. The pump overall efficiency is 89%.

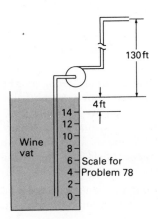

Figure P7.30

69. Air at 90°F flows through an 18-in. by 24-in. duct at the rate of 10,000 ft³/min. Find the head loss, pressure loss, and pressure drop in 100 ft of the duct. Also find the power that must be supplied by a fan to the air to overcome the duct resistance.

70. A pump at a lake delivers water at a maximum rate of 4.0 m³/min through a 15-cm-inside-diameter cast-iron pipe line to atmospheric pressure at an elevation of 150 m above the lake water surface. The pipe line is 2000 m long and has three flanged elbows and two flanged gate valves. Find the minimum power output of the motor driving the pump.

71. Kerosene at 15°C is pumped through 200 m of smooth 5-cm-inside-diameter pipe at a rate of 10 m³/min from zero velocity, atmospheric pressure, and an elevation of 150 m to zero velocity, atmospheric pressure, and an elevation of 540 m. Find the power input to the fluid by the pump.

72. A motor-driven centrifugal pump delivers water at the rate of 10 m³/min from a reservoir, through a 2500-m-long, 30-cm-inside-diameter plastic pipe, and to a second reservoir. The water level in the second reservoir is 40 m above the water level in the first reservoir. The pump efficiency is 75%. Find the motor output power.

73. An emergency flooding system for a nuclear reactor core is shown in Fig. P7.31. Find the power input required to flood the core at the rate of 5000 gal/min.

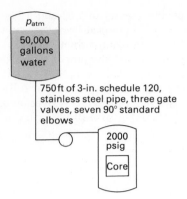

50,000 gallons water

p_{atm}

750 ft of 3-in. schedule 120, stainless steel pipe, three gate valves, seven 90° standard elbows

2000 psig

Core

Figure P7.31

74. A factory draws its water supply from a lake whose water surface is 200 ft above the pipe discharge. The 5-in. cast-iron supply pipe is 1000 ft long and has four gate valves, one globe valve, and twenty 90° standard elbows. Find the pump head required for a flow rate of 1000 gal/min if the pipe outlet is open to the atmosphere.

75. The water flow rate through the enamel-lined pipe of Fig. P7.21 must be 1000 gal/min. Is a pump necessary?

76. What horsepower is needed to pump 120 gal/min of 140°F benzene through 500 ft of 3-in. type K, horizontal copper pipe?

77. The Air Force is designing a new air-to-air missile. The full-scale missile is shown in Fig. P7.32(a). A 1/5-scale model of the missile is tested in a wind tunnel. To power the model's rocket engine,

kerosene (specific gravity = 0.75, $\mu = 5 \times 10^{-4}$ lb·sec/ft^2) and liquid oxygen (specific gravity = 1.1, $\mu = 5 \times 10^{-5}$ lb·sec/ft^2) are pumped to the model from outside storage tanks. The flow rates of kerosene and liquid oxygen are 10 lbm/sec and 30 lbm/sec, respectively. The fuel/oxidizer supply system is shown in Fig. P7.32(b). Do the following:

a) The pressure in the supply tanks is atmospheric and the pressure in the combustion chamber is 350 psia. Calculate the power required to pump the kerosene and liquid oxygen to the model.

b) The density of the combustion products at the exit of the rocket nozzle in the model is 0.03 lbm/ft^3. Calculate the exit velocity and the thrust.

c) The thrust (T) of a rocket is a function of the combustion chamber pressure (p_c), ambient air pressure (p_e), nozzle exit diameter (D_e), and the geometric properties of the nozzle. A dimensional analysis yields

$$\frac{T}{p_c D^2} = f\left(\frac{p_e}{p_c}, \text{geometry}\right).$$

Calculate the thrust developed by the full-scale missile if the combustion chamber pressure is 500 psia and p_e/p_c is the same as in the model test.

78. How much energy is required to lower the wine from the 14-ft level down to the 1-ft level for the system described in Fig. P7.30? The wine is pumped out at the rate of 100 gal/min.

79. Figure P7.33 shows a pumped storage system. At night when the electrical demand is low, electrical motor driven pumps transfer water from the lower reservoir to the higher reservoir. During the day when the electrical demand is higher than available generation capacity, water flows from the higher reservoir, through the turbine-gener-

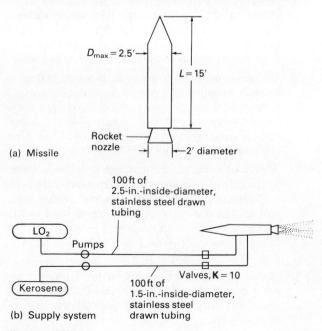

$D_{max} = 2.5'$

$L = 15'$

Rocket nozzle

2' diameter

(a) Missile

100 ft of 2.5-in.-inside-diameter, stainless steel drawn tubing

LO_2

Pumps

Valves, **K** = 10

100 ft of 1.5-in.-inside-diameter, stainless steel drawn tubing

Kerosene

(b) Supply system

Figure P7.32

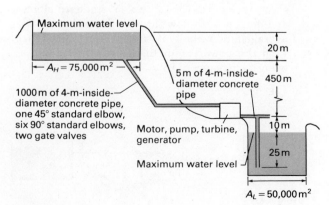

Maximum water level

20 m

$A_H = 75,000\,m^2$

5 m of 4-m-inside-diameter concrete pipe

450 m

1000 m of 4-m-inside-diameter concrete pipe, one 45° standard elbow, six 90° standard elbows, two gate valves

Motor, pump, turbine, generator

10 m

Maximum water level

25 m

$A_L = 50,000\,m^2$

Figure P7.33

ator set, and into the lower reservoir generating electricity. Assume the electrical motors, pump, turbine, and generator all have no energy losses. Find the mechanical energy loss in transferring one million cubic meters of water from the lower reservoir to the higher reservoir and back again.

80. A centrifugal pump has a suction pressure of 5.5 psig, a discharge pressure of 110 psig, a water flow rate of 120 gal/min, and an overall efficiency of 91.5%. Find the input power to the pump.

81. Water flows through the 2-in. schedule 40, 1200-ft-long commercial steel pipe in Fig. P7.34 at a rate of 200 gal/min. Is a pump required or can a turbine be installed in the line?

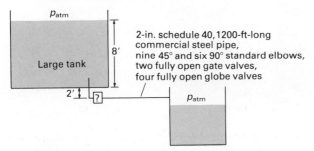

2-in. schedule 40, 1200-ft-long commercial steel pipe, nine 45° and six 90° standard elbows, two fully open gate valves, four fully open globe valves

Figure P7.34

82. Plot the hydraulic and energy grade lines for the two pipelines in Fig. P7.21.

83. Plot the hydraulic and energy grade lines for the piping system in Fig. 7.29.

84. Plot the hydraulic and energy grade lines for the piping system in Fig. P7.33.

85. A thin-plate orifice is to measure flow rates between 50 gal/min and 250 gal/min. The pressure drop is to be measured by a vertical mercury manometer 1 yd long. Recommend a β-ratio for the orifice.

86. A flow nozzle has a β-ratio of 0.5 and is to measure water flow rates between 50 gal/min and 250 gal/min. Recommend a suitable fluid for a vertical manometer that can be no longer than 1 yd long.

87. A thin-plate orifice with a throat diameter of 2.1 in. is used in a 3-in. type K copper pipe to measure the flow of 0°F SAE 20W crankcase oil. Find the pressure drop for a flow rate of 120 gal/min for D and $(1/2)D$ taps and for both upward flow and downward flow.

88. A thin-plate orifice with D and $(1/2)D$ taps and a 0.25 β-ratio is used in a piping system. The pressure drop across the orifice is to be measured by a mercury manometer 1 m long. What is the maximum flow rate that can be indicated?

89. A flow nozzle with a 0.25 β-ratio is used in a piping system. The pressure drop across the nozzle is to be measured by a mercury manometer 1 m long. What is the maximum flow rate that can be indicated?

90. Find the instantaneous mass flow rate of 60°F kerosene from tank A to tank B for the situation shown in Fig. P7.35.

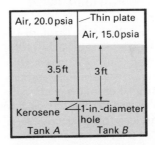

Figure P7.35

91. Air at 70°F and 20 psia flows through a 4-in. schedule 40 commercial steel pipe. A horizontal ASME long-radius flow nozzle having a β-ratio of 0.70 is located in the pipeline. A manometer connected between the inlet and throat of the nozzle indicates a level difference of 0.40 of the two legs of the manometer. The manometer contains 68°F lubricating oil. Find the air flow rate.

92. The minimum flow area A or vena contracta for the fluid leaving an orifice is given by

$$A = [0.62 + 0.38(A_2/A_1)^3]A_2,$$

where A_2 is the orifice throat area and A_1 is the pipe area. Find the maximum β-ratio such that the area ratio A/A_2 is at least 95%.

Higher-Order Application

93. Assume the tank in Fig. P7.2 is empty. How much work is required to fill the tank with 68°F ethanol pumped through the piping at the rate of 10 gal/min. The air temperature remains constant and the final air pressure is 40 psig.

94. Assume that all standpipes used to supply water to the upper floors of multistory buildings have a 6-in inside diameter and are capable of supplying a flow rate of at least 500 gal/min to the top floor. Can the tall buildings in your city satisfy this flow requirement?

95. Figure P7.36 shows a portion of a piping system where a differential pressure gage is used to measure the pressure drop across a flow nozzle. Valves C and D are normally open and the flow rate in the pipe line containing the flow nozzle is obtained by opening valves A and B, recording the pressure drop across the nozzle, and then using a calibra-

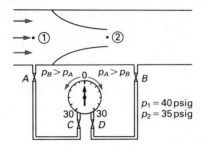

Figure P7.36

tion curve that gives the flow rate through the nozzle as a function of the pressure drop. Valves A and B are closed once the reading is taken. On a particular occasion, the pressure difference indicated by the gage continued to increase after the valves A and B were closed. Valve C was also observed to have a small leak from the valve stem. Is the leakage from valve C a possible explanation for the increasing pressure difference observed after valves A and B were closed? Give reasons for your answer. Valves A, B, C, and D are lever-handle cock valves.

96. A water line is to be constructed between a pumping station and a reservoir. The line is to be 4000 ft long and will have no elbows or fittings. The following costs for various pipe sizes have been determined:

Schedule 40 Pipe Size	Estimated Cost of Construction
12-in.	$ 75,000
14-in.	120,000
16-in.	150,000
18-in.	175,000

The pipe line is to have a 40-year life, and the money for the construction cost may be borrowed with an interest rate of 10% per year. Ignore depreciation and assume no salvage value. The pipe line will be used 8760 hours per year. The pumping cost is $0.06/hp·hr, the pump and drive motor are 100% efficient, and the pumping costs increase at the rate of 10% per year. Find the most economical pipe size.

97. Write a FORTRAN program to solve pipe sizing problems. Include loss coefficients for standard fittings and valves. As a test case, input the given information for Example 7.5 and attempt to get the correct answer.

98. Write a BASIC program to solve the velocity/flow rate problem. Include loss coefficients for standard fittings and valves. As a test case, use the given information for Example 7.4 and attempt to get the correct answer.

99. Estimate the calibration coefficient and the pressure loss for an ASME flow nozzle operating with reverse flow.

100. Infiltration of outside air into a building through cracks around doors and windows must often be estimated. To do this consider the crack between the swinging door and its frame in Fig. P7.37. For a static pressure difference of 0.01 in. water, estimate the air flow through a crack measuring 1/16 in. by 20 ft.

Figure P7.37

101. Find the flow rate from each water reservoir in Fig. P7.38.

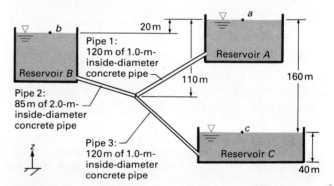

Figure P7.38

102. It is necessary to increase the flow rate in pipe 3 of Fig. P7.38. The pump whose head flow is shown in Fig. P7.39 is available for use. Is it more suitable to install the pump in pipe 1, pipe 2, or pipe 3? Support your answer.

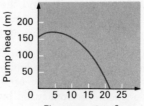

Figure P7.39

103. Two of the pumps in Fig. P7.15 are operated in parallel to supply water through the piping system. Determine the flow rate and total power input to the two pumps. Compare your results with those of Problem 36.

104. Repeat Problem 103 if the two pumps are operated in series rather than in parallel.

105. Consider a person donating a pint of blood. The pint bag is initially flat and contains no air. Neglect the initial mass of air in the 1/8-in.-inside-diameter, 4-ft-long plastic tube carrying blood to the pint bag. The average blood pressure in the vein is 100 mm Hg above atmospheric pressure. Estimate the time for the person to donate one pint of blood. Assume blood has a specific gravity of 1.06 and a viscosity of 1.0×10^{-4} lb·sec/ft². Donate a pint of blood and check your answer.

106. Make a more realistic estimate of the time required to siphon 15 gal of gasoline out of the tank in Problem 34 in Chapter 4. The siphon tube is 4 ft long.

107. Consider the horizontal, fully developed, laminar flow of 20°C lubricating oil through an annulus of inner radius $R_1 = 0.5$ cm and outer radius $R_2 = 0.75$ cm. The Reynolds number is 1000. Find the pressure drop per unit length of the channel using the hydraulic diameter approach and the equivalent diameter approach. Compare the answers. Which answer is more accurate?

108. Repeat Problem 107 for a Reynolds number of 30,000. Which of the two answers is more accurate?

Comparison

109. Von Karman suggested that the fully rough friction factor be expressed by the equation

$$f = \frac{1}{4\left(0.57 - \log_{10}\dfrac{\varepsilon}{D}\right)^2},$$

where ε is the absolute roughness of the pipe. Compare the values predicted by this equation and those indicated on the Moody chart.

110. For the same pressure rise, is more power required to pump 100 lbm of water or 100 lbm of air? Assume that the air density may be considered constant.

111. Assume that laminar flow of 60°F water has been maintained in a 1-in. schedule 160, horizontal, 100-ft-long commercial steel pipe at a Reynolds number of 25,000. Compare the mechanical energy loss, pressure loss, and pressure drop for this flow and that of turbulent flow for the same Reynolds number.

112. Repeat Problem 111 for a 0.815-in.-inside-diameter smooth pipe. Compare the answers to those of Problem 111 and make a conclusion about the effect of wall roughness on mechanical energy loss.

113. Kerosene at 0°F flows through a 2-in. schedule 40, horizontal, 100-ft-long commercial steel pipe at a rate of 200 gal/min. Find the increase in flow rate for the same pressure drop if the kerosene is heated to 80°F.

114. Consider laminar flow in a horizontal circular pipe. For a given pressure difference over a length ℓ, how will the flow rate change if the pipe inside diameter is doubled? Assume fully developed flow.

115. Consider fully turbulent flow in a horizontal, circular pipe. For a given pressure difference over a length ℓ, how will the flow rate change if the pipe inside diameter is doubled?

116. An SAE oil has a published viscosity of 76 centistokes at a temperature of 100°F. Experiments show that 352 sec are required for 60 ml of the 100°F fluid to drain out of the tube shown in Fig. P7.40. The fluid has a specific gravity of 0.92. Use the experimental result to determine the kinematic viscosity of the fluid and compare with the published value.

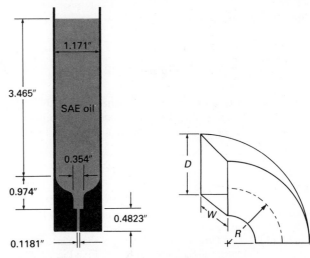

Figure P7.40   **Figure P7.41**

117. Consider the rectangular radius elbow in Fig. P7.41 with an air flow rate of 10,000 ft³/min. The duct is made of galvanized sheet metal. Find the length of straight duct that has the same static pressure drop as the elbow. $D = 12$ in., $W = 16$ in., and $R = 1.25D$. Compare your answer to the value of seven duct lengths published in duct design manuals.

118. Find the equivalent lengths of a 2-in., an 8-in., and a 20-in. fully open globe valve. Is it reasonable to assume the equivalent length is constant?

119. The static pressure drop Δp (inches H$_2$O) for air flowing in a horizontal circular duct of diameter D (in.) and length L (ft) with average velocity V (ft/min) is given by

$$\Delta p = 0.027\left(\frac{L}{D^{1.22}}\right)\left(\frac{V}{1000}\right)^{1.82}.$$

Plot Δp versus the average velocity V for a duct 100 ft long with a diameter of 24 in. Consider turbulent flow and a galvanized sheet-metal duct. Compare your result with that obtained using the Moody chart.

120. It is desired to build an experiment to demonstrate the development flow in the entrance region of a pipe. The experimental setup is shown in Fig. P7.42. Would you recommend using cold water (50°F), warm water (120°F), or kerosene (70°F) so fully developed flow will occur in as short a length as possible?

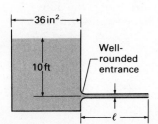

Figure P7.42

121. It is desired to pump heavy machine oil at 15.6°C through a 2.5-cm-inside-diameter, 1000-m-long pipeline at an average velocity of 1.5 m/s. The pipe has a relative roughness of 0.004. Which of the following would require less energy?
a) Pump the oil through the pipeline while keeping the oil temperature at 15.6°C.
b) Heat the oil to 37.8°C and then pump the oil. The oil has a specific heat at constant pressure of 2000 N·m/kg·K.

122. The following equation for the Moody friction factor f is sometimes used for computer calculations:

$$f = a + b/R^c,$$

where

$$a = 0.104\left(\frac{\varepsilon}{D}\right)^{0.225} + 0.532\left(\frac{\varepsilon}{D}\right),$$

$$b = 88\left(\frac{\varepsilon}{D}\right)^{0.44}, \qquad c = 1.62\left(\frac{\varepsilon}{D}\right)^{0.134}.$$

Compare this equation with the Moody chart and decide if it is an acceptable replacement for the Colebrook formula.

123. Compare the computer time required to obtain an answer using the equations of Problem 122 with an iteration scheme based on the Colebrook formula. Which is faster?

124. Observations have shown that the absolute roughness of pipes increases with time. Find the increase in pressure drop required to pump 100 gal/min of water through a 2-in. schedule 80, 100-ft-long pipe if the absolute roughness doubles over a period of 10 years.

125. Compare friction factors calculated from the formula derived in Problem 4 for an annulus with calculated values based on the hydraulic diameter and effective diameter models.

References

1. Nikuradse, J., "Stomungsgesetze in Rauhen Rohren," *VDI-Forschungsch*, no. 361 (1933). Also Available as NACA Tech Memo 1292.
2. Colebrook, C. F., "Turbulent Flow in Pipes with Particular Reference to the Transition Between the Smooth and Rough Pipe Laws," *J. Inst. Civ. Eng. Lond.*, vol. 11, 1939.
3. Moody, L. F., "Friction Factors for Pipe Flow," *Transactions of the ASME*, vol. 66, 1944.
4. White, F. M., *Fluid Mechanics*, McGraw-Hill, New York, 1979.
5. White, F. M., *Viscous Fluid Flow*, McGraw-Hill, New York, 1974.
6. Jones, O. C., Jr., "An Improvement in the Calculation of Turbulent Friction in Rectangular Ducts," *ASME Journal of Fluids Engineering*, June 1976.
7. Lindgren, E. R., Oklahoma State University Civil Engineering Dept., Report 1AD621071, 1965.
8. "Flow of Fluids Through Valves, Fittings, and Pipe," Technical Paper No. 410 (1978), The Crane Co. Available through The Crane Co., New York.

9. *ASHRAE Handbook of Fundamentals*, published every four years by the American Society of Heating, Refrigeration and Air-Conditioning Engineers.
10. Idel'chek, I. E., "Handbook of Hydraulic Resistance—Coefficients of Local Resistance and of Friction." Available from U.S. National Technical Information Service as AEC-TR-6630, 1960.
11. Cross, H. "Analysis of Flow in Networks of Conduits or Conductors," University of Illinois Bulletin No. 286, November 1936.
12. Jepson, R. W., *Analysis of Flow in Pipe Networks,* Ann Arbor Science Publishers, Ann Arbor, MI, 1976.
13. Sarakelle, S., "AQUA, Akron Water Distribution Model: Program Description," Department of Civil Engineering, The University of Akron, 1980.
14. Bean, H. S. (ed.), *Fluid Meters: Their Theory and Application* (6th ed.), American Society of Mechanical Engineers, New York, 1971.
15. "Measurement of Fluid Flow by Means of Orifice Plates, Nozzles, and Venturi Tubes Inserted in Circular Cross Section Conduits Running Full," International Standards Organization Report DIS-5167, 1976.
16. "The ISO-ASME Orifice Coefficient Equation," *Mechanical Engineering,* July 1981.
17. Beckwith, T., N. Buck, and R. Marangoni, *Mechanical Engineering Measurements* (3rd ed.), Addison-Wesley, Reading, MA, 1981.
18. Benedict, R. P., *Measurement of Temperature, Pressure, and Flow* (2nd ed.), Wiley, New York, 1977.
19. "Laboratory Methods of Testing Fans for Rating," AMCA Bulletin 210-74, ASHRAE Standard 210-75, a Joint Bulletin of the Air Moving and Conditioning Association and the American Society of Heating, Refrigeration and Air Conditioning Engineers, 1974.
20. "ASME Performance Test Code No. 11—Fans," American Society of Mechanical Engineers, New York, 1984.

Applications of Flow Analysis: External Flow

8

In this chapter we consider flow over bodies immersed in a stream of fluid. Such flows are called *external flows* because the fluid is external to a surface. Figure 8.1 shows a few external flows that are of interest to engineers. The fluid stream in which the body is immersed is usually considered to be infinite in extent. The items of interest are the forces that the fluid exerts on the body and, to a lesser extent, the details of the flow pattern near the body. The study and engineering application of external flow is often called *aerodynamics,* although this title is inappropriate if the fluid is not air.

Complete information on aerodynamic forces and flow patterns can usually be obtained only from experimental data. Current theories of flow over immersed bodies are capable of providing excellent qualitative explanations of flow phenomena. Accurate quantitative predictions are somewhat limited, however, because the differential approach must be used to reveal the detail necessary for accurate flow prediction. In principle, it is necessary to solve the Navier-Stokes equations (see Section 4.2.2), a nearly impossible task. Most practical prediction methods are based on a simplified model of the real flow using Prandtl's boundary layer hypothesis. In this chapter we will discuss the physical basis for this model and show how it provides the key to understanding most external flows that occur in engineering. The predictive aspects of the boundary layer model will be discussed in the next chapter.

Much research has been done in recent years on computer methods for computing external flows and predicting forces and flow patterns. Calculations for simple body shapes are fairly routine nowadays; nevertheless, much aerodynamic engineering is still based on experimental data. We will concentrate on examining such data and on understanding the physical behavior of external flow. We will also have a little fun as we consider applications of aerodynamics in sports.

8.1 Force on Bodies: Drag and Lift

A body immersed in a moving stream (or alternatively, a body moving through a still fluid) experiences a force. In this chapter we will be concerned only with those forces that are the result of relative motion

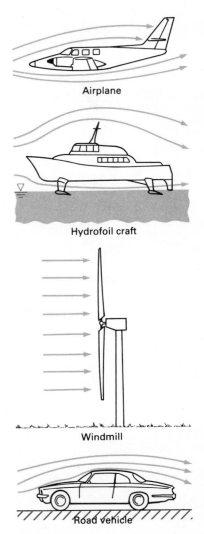

Figure 8.1 Some external flows.

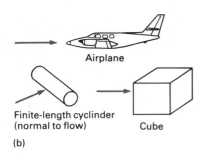

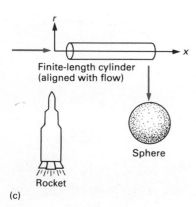

Figure 8.2 (a) Some two-dimensional bodies; (b) some three-dimensional bodies; (c) some axisymmetric bodies.

between the fluid and the body; buoyancy forces due to hydrostatic pressure are not included in this discussion.

8.1.1 Body Geometry and Force Components

In this chapter we refer to two-dimensional bodies, three-dimensional bodies, and axisymmetric bodies. A *two-dimensional body* has the same shape in all planes perpendicular to an axis. This axis is infinitely long. Figure 8.2(a) shows some two-dimensional bodies. The shape of the body is the same in all planes parallel to the paper.

A *three-dimensional body* has finite length. It may have the same shape in planes perpendicular to its length axis, or it may not. Figure 8.2(b) illustrates some three-dimensional bodies.

An *axisymmetric body* is a body of revolution with circular cross sections. An axisymmetric body has the same shape in all meridional (x, r) planes. Figure 8.2(c) illustrates some axisymmetric bodies.

When a body is immersed in a stream, the dimensionality and directionality of the flow are determined by the dimensionality of the body and the alignment between the approach flow and the body. Two-dimensional bodies produce two-dimensional, two-directional flow if there is no flow along the body's long axis. Three-dimensional bodies produce three-dimensional, three-directional flow. Even if the body shape is the same in all planes perpendicular to the width axis and the approaching flow is perpendicular to this axis, there will still be flow around the "ends" of the body. If a three-dimensional body is very wide, we may approximate the flow in the central portion as two-dimensional. The flow over an axisymmetric body with its axis parallel to the approaching flow is two-dimensional and two-directional in a meridional (x, r) plane. If the approach flow is not parallel to the body axis, the flow is three-dimensional and three-directional.

The aerodynamic force on a body can be broken into components along any arbitrary set of axes. First consider a two-dimensional or axisymmetric body (Fig. 8.3). For the two-dimensional body, the resultant force is in the plane of the body. For the axisymmetric body, the resultant force lies in the plane defined by the body axis and the approaching flow velocity vector. In either case, the force component in the direction of the approaching flow is called *drag* and the force component perpendicular to the approaching flow is called *lift*.

Figure 8.3 Aerodynamic forces on two-dimensional axisymmetric bodies.

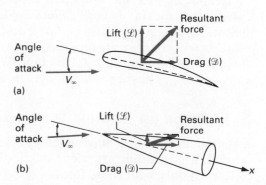

Experiments show that any body placed in a moving stream experiences drag. If a body moves relative to a still fluid, drag force resists the motion. The drag force vector always points downstream.

Lift forces are not necessarily present in all flows; they occur only if there is asymmetry. This asymmetry may be due to asymmetry of the body or it may be due to a misalignment between the body and the approaching flow. The angle of misalignment is called the *angle of attack*. This is shown in Fig. 8.3.

Now consider a three-dimensional body. Many three-dimensional bodies have at least one plane of symmetry (Fig. 8.4a). If the approach flow velocity vector is parallel to this symmetry plane, the resultant force on the body lies in the symmetry plane and therefore has only two components, still identified as drag and lift. If the flow is completely asymmetric, either because the three-dimensional body has no plane of symmetry (Fig. 8.4b) or because the approach flow is not parallel to the symmetry plane (Fig. 8.4c), then the resultant force has three components, called drag, lift, and *side force*. The drag axis is defined by the direction of the approaching flow, but the selection of lift and side force axes is somewhat arbitrary. If the body is in a gravity field, the lift and side force axes are selected such that the lift is vertical and the side force is horizontal. A car moving along a highway in the presence of a cross wind experiences all three forces. Drag resists the motion of the car and must be compensated by constant pressure on the gas pedal. Side force tends to move the car right or left and must be compensated by the steering wheel. The driver is generally unaware of lift at ordinary highway speeds, so long as the car's weight is greater than the lift!

The cause of both lift and side force is the same, flow asymmetry. The cause of drag is fundamentally different. In this book, we will concentrate on those cases in which there is at least one plane of symmetry, so the only forces present will be drag and lift.

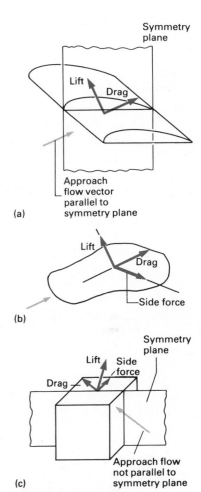

Figure 8.4 Flow over three-dimensional bodies.

8.1.2 Calculation of Drag and Lift: Force Coefficients

Drag and lift are the resultants of forces due to pressure and shear stress on the body surface. If the pressure and shear stress distributions over the surface can be determined, either by calculation or by experiment, the aerodynamic forces can be obtained by integration. Figure 8.5 illustrates typical pressure and shear stress distributions for an airfoil, a two-dimensional body designed to produce high lift and low drag. The magnitude of the pressure at a point on the airfoil surface is indicated

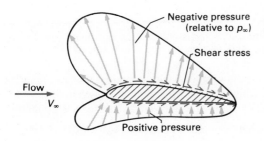

Figure 8.5 Pressure and shear stress distributions on an airfoil.

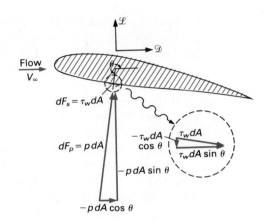

Figure 8.6 Geometry of elemental forces on an airfoil.

by the length of an arrow perpendicular to the surface. In Fig. 8.5, pressure is plotted relative to the pressure in the approaching (undisturbed) stream. Note that pressures on the top of the airfoil are negative (lower than free stream pressure) while pressures on the bottom are positive. The shear stresses at the airfoil surface are indicated by arrows parallel to the surface. The shear stresses all point in the downstream direction and are roughly the same magnitude.

The net force on the airfoil can be calculated by integrating the pressure and shear stress over the surface:

$$\vec{F} = -\oint p\hat{n}\,dA + \oint \tau_w \hat{t}\,dA, \tag{8.1}$$

where $\hat{n}$ and $\hat{t}$ are unit vectors perpendicular and tangent to the airfoil surface.

Drag and lift are the force components in the flow direction and perpendicular to it. With the aid of Fig. 8.6, we find*

$$\mathscr{D} = \oint (-p\cos\theta + \tau_w \sin\theta)\,dA \tag{8.2}$$

and

$$\mathscr{L} = \oint (-p\sin\theta + \tau_w \cos\theta)\,dA. \tag{8.3}$$

Since θ is approximately 90° over most of the top of the airfoil and approximately 270° over most of the bottom of the airfoil, the lift force is due mainly to pressure and the drag force is due mainly to shear stress.

---

* τ_w is positive if it is clockwise; thus τ_w is positive on the top of the airfoil and negative on the bottom.

EXAMPLE 8.1   Illustrates how pressure and shear stresses are used to calculate lift and drag forces

A highway sign has the shape of a half-cylinder with diameter 2.0 m and length 25.0 m. A rearward 10-km/hr wind is blowing over the sign, as shown in Fig. E8.1. The pressure and shear stress distributions on the circular surface are given by

$$p - p_\infty = 4.63(1 - 4\sin^2\theta)\ \text{N/m}^2$$

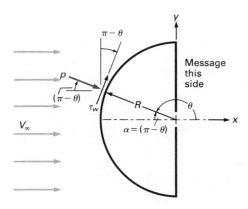

Figure E8.1 Rearward wind on sign.

and

$$\tau_w = [1.74 \times 10^{-3}\,(\pi - \theta) - 1.59 \times 10^{-5}\,(\pi - \theta)^2]\text{ N/m}^2,$$

where θ is in radians. The pressure on the flat face is constant and equal to the pressure at the corner. There are no shear stresses on the flat face. Find the lift and drag on the sign. Neglect end effects.

Solution

Given
Sign having shape of half-cylinder, shown in Fig. E8.1
Sign diameter 2.0 m, length 25.0 m
Rearward, 10-km/hr wind
Pressure and shear stress distribution on circular surface given by

$$p - p_\infty = 4.63(1 - 4\sin^2\theta)\text{ N/m}^2$$

and

$$\tau_w = [1.74 \times 10^{-3}\,(\pi - \theta) - 1.59 \times 10^{-5}\,(\pi - \theta)^2]\text{ N/m}^2,$$

where θ is in radians
Neglect end effects.

Find
Lift and drag on the sign

Solution
The directions of positive lift and drag are the positive y- and x-directions respectively. The drag is found using Eq. (8.2), with p and τ_w having units of N/m². Since

$$dA = bR\,d\theta,$$

where b is the sign length and R the radius, the drag is

$$\mathscr{D} = \int_0^{2\pi} (-p\cos\theta + \tau_w \sin\theta)bR\,d\theta.$$

Recognizing that the drag on the top half equals the drag on the bottom half gives

$$\mathscr{D} = 2\int_0^{\pi} (-p\cos\theta + \tau_w \sin\theta)bR\,d\theta.$$

Substituting for p and τ_w, recognizing that the pressure on the flat face is constant and equal to that at the corners, and noting that the shear stress is zero over the flat face, we have

$$\mathscr{D} = 2 \int_0^{\pi/2} -[p_\infty + 4.63(-3)] \cos \theta \, bR \, d\theta$$

$$+ 2 \int_{\pi/2}^{\pi} -[p_\infty + 4.63(1 - 4 \sin^2 \theta)] \cos \theta \, bR \, d\theta$$

$$+ 2 \int_{\pi/2}^{\pi} [1.74 \times 10^{-3} (\pi - \theta) - 1.59 \times 10^{-5} (\pi - \theta)^2] \sin \theta \, bR \, d\theta.$$

Integrating and noting that $\sin \theta = \sin (\pi - \theta)$, we obtain

$$\mathscr{D} = -2bR(p_\infty - 13.9)(\sin \theta)_0^{\pi/2}$$

$$- 2bR \left[p_\infty \sin \theta + 4.63 \sin \theta - 18.5 \left(\frac{\sin^3 \theta}{3} \right) \right]_{\pi/2}^{\pi}$$

$$+ 3.48 \times 10^{-3} \, bR \int_{\pi/2}^{\pi} (\pi - \theta) \sin (\pi - \theta) \, d\theta$$

$$- 3.18 \times 10^{-5} \, bR \int_{\pi/2}^{\pi} (\pi - \theta)^2 \sin (\pi - \theta) \, d\theta.$$

Substituting in the limits for the pressure terms and letting $\phi = \pi - \theta$ so that $d\phi = -d\theta$ gives

$$\mathscr{D} = -2bR(p_\infty - 13.9)(1 - 0)$$

$$- 2bR \left[p_\infty(0 - 1) + 4.63(0 - 1) - 18.5 \left(\frac{0 - 1}{3} \right) \right]$$

$$- 3.48 \times 10^{-3} \, bR \int_{\pi/2}^{0} \phi \sin \phi \, d\phi$$

$$+ 3.18 \times 10^{-5} \, bR \int_{\pi/2}^{0} \phi^2 \sin \phi \, d\phi$$

$$= 2bR(12.4) - 3.48 \times 10^{-3} \, bR[\sin \phi - \phi \cos \phi]_{\pi/2}^{0}$$

$$+ 3.18 \times 10^{-5} \, bR[2\phi \sin \phi - (\phi^2 - 2) \cos \phi]_{\pi/2}^{0}$$

$$= \left(24.8 \, \frac{\text{N}}{\text{m}^2} \right) bR = \left(24.8 \, \frac{\text{N}}{\text{m}^2} \right) (25.0 \text{ m})(1.0 \text{ m}),$$

Answer

$$| \ \mathscr{D} = 620 \text{ N}. \ |$$

The lift is found using Eq. (8.3). Recognizing that $dA = bR \, d\theta$ and that there is no lift component on the flat face, we have

$$\mathscr{L} = \int_{\pi/2}^{3\pi/2} (-p \sin \theta - \tau_w \cos \theta) bR \, d\theta.$$

Substituting for p and τ_w gives

$$\mathscr{L} = \int_{\pi/2}^{3\pi/2} -[p_\infty + 4.63(1 - 4 \sin^2 \theta)] \sin \theta \, bR \, d\theta$$

$$+ \int_{\pi/2}^{3\pi/2} [1.74 \times 10^{-3} (\pi - \theta) - 1.59 \times 10^{-5} (\pi - \theta)^2] \cos \theta \, bR \, d\theta.$$

We now use the trigonometric identity

$$\sin^2 \theta = 1 - \cos^2 \theta$$

in the first integral and $\cos \theta = -\cos (\pi - \theta)$, $\pi - \theta = \phi$, and $-d\theta = d\phi$ in the second integral. The lift equation is then

$$\mathscr{L} = bR \int_{\pi/2}^{3\pi/2} -[p_\infty + 4.63(-3 + 4 \cos^2 \theta)] \sin \theta \, d\theta$$

$$+ 1.74 \times 10^{-3} \, bR \int_{\pi/2}^{-(\pi/2)} \phi \cos \phi \, d\phi - 1.58 \times 10^{-5} \, bR \int_{\pi/2}^{-(\pi/2)} \phi^2 \cos \phi \, d\phi.$$

Integrating yields

$$\mathscr{L} = bR\left[(p_\infty - 13.9)\cos\theta + \frac{18.52\cos^3\theta}{3}\right]_{\pi/2}^{3\pi/2}$$
$$+ 1.74 \times 10^{-3}\, bR[\cos\phi + \phi\sin\phi]_{\pi/2}^{-(\pi/2)}$$
$$- 1.59 \times 10^{-5}\, bR[2\phi\cos\phi + (\phi^2 - 2)\sin\phi]_{\pi/2}^{-(\pi/2)}.$$

This reduces to

$$\mathscr{L} = (0)bR,$$

so **Answer**

$$|\ \mathscr{L} = 0.\ |$$

Discussion
The zero lift should not be surprising because the vertical component of the pressure or shear stress at any point on the top half of the sign is exactly balanced by an equal but oppositely directed vertical component of the pressure or shear stress on the bottom half.

Equations (8.2) and (8.3) are useful for calculation of drag and lift only if the pressure and shear stress distributions are known. Calculation of these distributions from the basic equations governing the flow, namely the Navier-Stokes equations (4.57)–(4.59), is very difficult. Sometimes computer solutions of these equations (or approximations of them) can be made. More frequently such information is determined by experiment. Since the engineer's ultimate interest is often the drag and lift forces rather than the pressure and stress distributions, experimental results are often obtained and presented directly as drag and lift. The data are usually presented in terms of dimensionless *drag coefficient* and *lift coefficient,* defined by

$$C_D \equiv \frac{\mathscr{D}}{\frac{1}{2}\rho V_\infty^2 S} \tag{8.4}$$

and

$$C_L \equiv \frac{\mathscr{L}}{\frac{1}{2}\rho V_\infty^2 S}, \tag{8.5}$$

where V_∞ is the velocity of the fluid relative to the object and S is a reference area. According to "pure" dimensional analysis (Chapter 6), the reference area should be taken as the square of a single reference length ($S = \ell^2$). In practice, various definitions are used for the area, the two most common being the *frontal area* (the area that you would see if you looked at the body from the direction of the approach flow) and the *planform area* (the area you would see if you looked at the body from above) (Fig. 8.7). For two-dimensional bodies, the area is based on a unit width ($L = 1$). From dimensional analysis, drag and lift coefficients for a given shape in steady flow are functions of the primary dimensionless parameters:

$$C_D = C_D(\alpha, \mathbf{R}, \mathbf{M}, \mathbf{W}, \mathbf{F}) \quad \text{and} \quad C_L = C_L(\alpha, \mathbf{R}, \mathbf{M}, \mathbf{W}, \mathbf{F}).$$

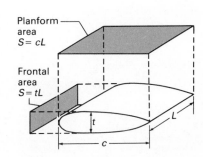

Planform area
$S = cL$

Frontal area
$S = tL$

Figure 8.7 Definition of frontal area and planform area.

In most flows of practical interest, surface tension (Weber number) effects are irrelevant. In almost all cases (except for hydrofoils operating near a free surface), gravity wave (Froude number) effects are also irrelevant. With these simplifications, we write

$$\mathbf{C_D} = \mathbf{C_D}(\alpha, \mathbf{R}, \mathbf{M}) \tag{8.6}$$

and

$$\mathbf{C_L} = \mathbf{C_L}(\alpha, \mathbf{R}, \mathbf{M}). \tag{8.7}$$

In this chapter, we will limit our attention to incompressible flow,* for which Mach number effects are irrelevant, and so

$$\mathbf{C_D} = \mathbf{C_D}(\alpha, \mathbf{R}) \tag{8.8}$$

and

$$\mathbf{C_L} = \mathbf{C_L}(\alpha, \mathbf{R}). \tag{8.9}$$

Drag and lift coefficients can be related to pressure and skin friction coefficients ($\mathbf{C_p}$ and $\mathbf{C_f}$) by equations similar to Eqs. (8.2) and (8.3):

$$\mathbf{C_D} = \oint (-\mathbf{C_p} \cos\theta + \mathbf{C_f} \sin\theta) \, d\left(\frac{A}{S}\right) \tag{8.10}$$

and

$$\mathbf{C_L} = \oint (-\mathbf{C_p} \sin\theta - \mathbf{C_f} \cos\theta) \, d\left(\frac{A}{S}\right). \tag{8.11}$$

The remainder of this chapter is devoted to presentation of typical values of $\mathbf{C_D}$ and $\mathbf{C_L}$ and development of a qualitative understanding of drag and lift.

* The effect of compressibility on drag and lift is discussed briefly in Chapter 10.

EXAMPLE 8.2 Illustrates lift and drag coefficients

Calculate the lift and drag coefficients for the sign in Example 8.1. What would be the lift and drag if the sign were reduced one-half size and the wind velocity doubled?

Solution

Given
Highway sign in Fig. E8.1
Drag $\mathscr{D} = 620$ N
Lift $\mathscr{L} = 0$

Find
Drag coefficient $\mathbf{C_D}$ and lift coefficient $\mathbf{C_L}$
Values of lift and drag on a similar sign of half size in a wind with double the original velocity

Solution
The drag coefficient is defined by Eq. (8.4):

$$\mathbf{C_D} = \frac{\mathscr{D}}{\frac{1}{2}\rho V_\infty^2 S}.$$

In our case, V_∞ is 10 km/hr and S is the frontal area,

$$S = 2R\ell = 2(1.0 \text{ m})(25.0 \text{ m}) = 50 \text{ m}^2.$$

Assuming 20°C air at standard atmospheric pressure, we get from Table A.3

$$\rho = 1.20 \text{ kg/m}^3.$$

The numerical values give

$$\mathbf{C_D} = \frac{620 \text{ N}}{\frac{1}{2}\left(1.20 \, \frac{\text{kg}}{\text{m}^3}\right)\left(\frac{10 \text{ km}}{\text{hr}}\right)^2 (50 \text{ m}^2)\left(\frac{1000 \text{ m}}{\text{km}}\right)^2 \left(\frac{\text{hr}}{3600 \text{ s}}\right)^2},$$

$$\boxed{\mathbf{C_D} = 2.68.}$$

Answer
◄

The lift coefficient is defined by Eq. (8.5):

$$\mathbf{C_L} = \frac{\mathscr{L}}{\frac{1}{2}\rho V_\infty^2 S}.$$

Since $\mathscr{L} = 0$,

$$\boxed{\mathbf{C_L} = 0.}$$

Answer
◄

With incompressible flow, Eqs. (8.8) and (8.9) show that the lift and drag coefficients are functions of the angle of attack α and the Reynolds number **R**. Assuming α is the same for both the full-size and half-size signs, we have

$$\mathbf{C_D} = \mathbf{C_D(R)} \quad \text{and} \quad \mathbf{C_L} = \mathbf{C_L(R)}.$$

An appropriate Reynolds number for the sign is

$$\mathbf{R} = \frac{V_\infty R}{\nu}.$$

Since

$$R_{\text{full-size}} = 2R_{\text{half-size}} \quad \text{and} \quad (V_\infty)_{\text{full-size}} = \frac{1}{2}(V_\infty)_{\text{half-size}},$$

both signs have the same **R**. Therefore, both signs have the same values of $\mathbf{C_D}$ and of $\mathbf{C_L}$. Solving for $\mathscr{D}$ and $\mathscr{L}$ for the half-size sign gives

$$\mathscr{D}_{\text{half-size}} = \left(\frac{1}{2}\mathbf{C_D}\rho V_\infty^2 S\right)_{\text{half-size}}$$

and

$$\mathscr{L}_{\text{half-size}} = \left(\frac{1}{2}\mathbf{C_L}\rho V_\infty^2 S\right)_{\text{half-size}}.$$

Substituting for the velocity and frontal area S gives

$$\mathscr{D}_{\text{half-size}} = \frac{1}{2}\mathbf{C_D}\rho\left[(2V_\infty)^2\left(\frac{1}{2}\right)(2Rb)\right] = 2\left[\frac{1}{2}\mathbf{C_D}\rho V_\infty^2(2Rb)\right]$$

and

$$\mathscr{L}_{\text{half-size}} = \frac{1}{2}\mathbf{C_L}\rho\left[(2V_\infty)^2\left(\frac{1}{2}\right)(2Rb)\right] = 2\left[\frac{1}{2}\mathbf{C_L}\rho V_\infty^2(2Rb)\right].$$

Then

$$\mathscr{D}_{\text{half-size}} = 2\mathscr{D}_{\text{full-size}},$$

or

$$| \ \mathscr{D}_{\text{half-size}} = 1240 \text{ N} \ |$$

Answer ◀

and

$$\mathscr{L}_{\text{half-size}} = 2\mathscr{L}_{\text{full-size}},$$

or

$$| \ \mathscr{L}_{\text{half-size}} = 0. \ |$$

Answer ◀

Discussion

We were able to calculate the drag and lift for the half-size sign only because both signs had the same Reynolds numbers and the same angles of attack. Unless the drag and lift coefficients are independent of Reynolds number, this would not have been possible if the Reynolds numbers were not equal.

8.2 Drag

The phenomenon of aerodynamic drag is familiar to anyone who has ever tried walking or bicycling into the wind or watched leaves flutter to the ground. As children, we learn from experience that "heavier" objects like rocks fall faster than "lighter" objects like paper wads. We are surprised when we learn in high school or college physics that an object's rate of fall in a vacuum is independent of its weight. Aerodynamic drag is the culprit in our childhood experience; heavier objects experience a larger acceleration because the *net* force to mass ratio (weight minus drag divided by mass) is larger than on a lighter object. In the same way, drag obscured the true laws of mechanics for centuries. The ancient Greeks believed that a force must be exerted on a body to move it at *constant* velocity. Isaac Newton in about 1670 realized that moving bodies experience a resisting force and deduced that force was necessary to *change* a body's velocity.

After the existence of aerodynamic drag was realized, it remained a confusing phenomenon. The fluid mechanics of Euler and Bernoulli was completely incapable of predicting any drag forces because their equations did not contain viscosity.* Not only were theories incapable of predicting drag, but no one was able to explain the fundamental mechanisms by which significant drag forces could arise in fluids of very small viscosity like air and water. It was not until the twentieth century that even a qualitative explanation of drag was developed. Quantitative a priori prediction of drag for any but the most simple shapes (specifically excluding such "simple" shapes as cylinders and spheres) is still beyond the capabilities of analytical fluid mechanics. In the following section we examine typical experimental data and offer a qualitative explanation of drag.[†]

---

* This is called d'Alembert's paradox and is quite general. Any theory of fluid flow that neglects viscosity is incapable of an a priori prediction of drag.

[†] An excellent presentation of the principles of drag is the film *The Fluid Dynamics of Drag* [1]. Two hours spent watching this film is worth about one week of reading assignments or classroom lectures.

8.2.1 Drag of Simple Shapes: Types of Drag

Equations (8.2) and (8.10) show that drag arises from two sources, pressure and shear stress. Drag due to pressure is called *form drag* because it depends on the shape or form of the body. Drag due to shear stress is called *skin friction drag* or *friction drag*. Friction drag depends primarily on the amount of surface in contact with the fluid.

The drag on most bodies is a combination of both form and friction drag. We begin our discussion by considering bodies that have only one type of drag. First, consider a very thin, flat plate that is aligned with the approach flow (Fig. 8.8a). For the present we will assume that the extent of the plate is infinite perpendicular to the paper. As long as the plate is parallel to the approach flow, the situation would not change appreciably if the plate had finite width. Since the plate is very thin, it has negligible area perpendicular to the flow and experiences no form drag. The fluid velocity at the plate surface is zero and increases to the free-stream velocity a short distance from the plate. Since the velocity gradient is not zero at the plate surface, there is a shear stress at the plate surface and the plate experiences friction drag.

Figure 8.8(b) shows the drag coefficient of a flat plate aligned with the flow as a function of Reynolds number and plate roughness. The reference area is the planform area. Notice the similarity between the flat plate drag coefficient curves and those for the friction factor for pipe flow (Fig. 7.11). At low Reynolds numbers, the flow near the plate is laminar and the drag coefficient is rather small. At high Reynolds numbers, the flow near the plate is turbulent over most of the plate length and the drag coefficient is higher because the shear stress is larger. The

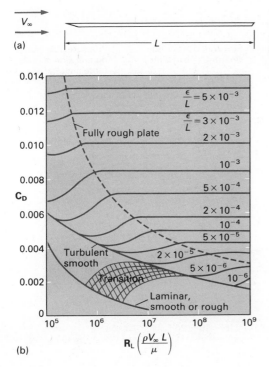

(a)

(b)

Figure 8.8 Flow over a thin flat plate aligned with the flow. (a) Geometry; (b) drag coefficient for smooth and rough plates.

transition between the laminar curve and the turbulent curve is not a single line but a family of lines that depend on plate roughness and amount of turbulence in the fluid above the plate. At any particular value of Reynolds number, a smooth parallel flat plate has the lowest drag coefficient of any shape.

If we rotate the plate so that it is perpendicular to the approach flow, we obtain a completely different flow pattern (Fig. 8.9). The fluid must turn as shown to flow over the plate. At the edges of the plate, the fluid is incapable of turning to flow over the back, so the streamlines *separate* at the corner. The region behind the plate, called the *wake,* contains low-pressure, low-velocity fluid that moves aimlessly.

Although there are large shear stresses on the front of the plate and also shear stresses on the back of the plate, the forces due to these stresses are perpendicular to the approaching flow and do not contribute to drag. The drag on a normal plate is due to pressure only; it is all form drag. Figure 8.9(b) shows a typical (measured) pressure distribution for flow over a two-dimensional normal flat plate. At the center of the plate (the stagnation point), the pressure is equal to the free-stream total pressure since the velocity is zero. Near the edges of the plate, the pressure falls rapidly, dropping to a value well below the free-stream pressure. The pressure over the back of the plate is constant and equal to the low pressure at the edge of the plate. The drag coefficient can be calculated by

$$\mathbf{C_D} = \int_{A_{\text{front}}} - \mathbf{C_p} \cos(180°) d\left(\frac{A}{S}\right) + \int_{A_{\text{back}}} - \mathbf{C_p} \cos(0°) d\left(\frac{A}{S}\right)$$

$$= \bar{\mathbf{C}}_{\text{p,front}} - \bar{\mathbf{C}}_{\text{p,back}}.$$

Figure 8.9 Flow over a flat plate normal to the flow. (a) Flow schematic; (b) pressure distribution on the plate; (c) effect of width on $\mathbf{C_D}$ for finite plate.

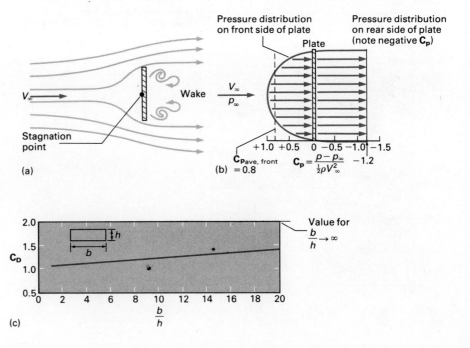

(a)

(b)

(c)

TABLE 8.1 Comparison of flat plate drag

	Plate Aligned with Flow	Plate Normal to Flow
Drag type	Friction drag only	Form drag only
Magnitude of C_D	Small	Large
Reynolds number dependence	Strong	Weak or none
Effect of plate roughness	Strong	None
Difference between two-dimensional and three-dimensional plate	Minor	Significant

Obtaining the average pressure coefficients from Fig. 8.9(b), we have

$$C_D = 0.80 - (-1.20) = 2.0.$$

This value of drag coefficient is valid for Reynolds numbers above about 10^3; that is, the drag coefficient is insensitive to Reynolds number.*

If the plate has finite width (a three-dimensional plate), the drag coefficient depends on the plate width-to-height ratio (Fig. 8.9c). Note that these drag coefficients are based on frontal area. Table 8.1 summarizes the differences in drag coefficients for flat plates aligned with and normal to the flow.

The nature of drag on flat plates aligned with and normal to the flow is relatively easy to understand because only one type of drag is present in each. Now we consider simple shapes for which both types of drag are present. Figure 8.10(a) illustrates flow over a smooth body of circular cross section. If the body is two-dimensional, it is a cylinder; if it is axisymmetric, it is a sphere. Both a cylinder and a sphere have equal amounts of area parallel and perpendicular to the flow and will experience both form drag and friction drag. Drag coefficients for smooth cylinders and spheres are plotted in Fig. 8.10(b). For both bodies, the Reynolds number is based on diameter and the drag coefficient is based on frontal area

$$S = \begin{cases} D(1), & \text{Cylinder} \\ \dfrac{\pi}{4} D^2. & \text{Sphere} \end{cases}$$

These curves are based almost entirely on experimental data. For hand calculations, drag coefficients can be read from the curves. White [2]

---

* Simple "logic" might imply that the maximum value for C_D should be 1.0, assuming that the plate experiences a force equal to the free-stream dynamic pressure times the area. Because the pressure on the back side of the plate is lower than the free-stream pressure, this "logic" would be in error by 100%!

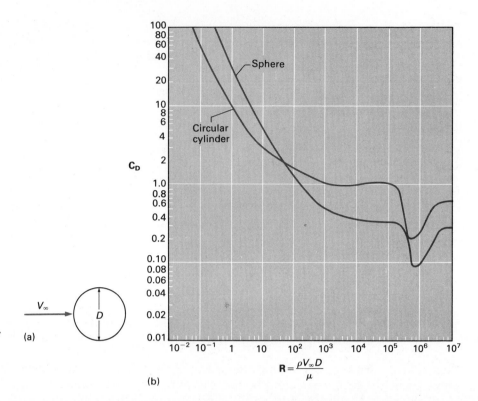

(a)

(b)

Figure 8.10 Drag of smooth body of circular cross section (cylinder or sphere). (a) Schematic; (b) drag coefficients as a function of Reynolds number.

has suggested the following curve fits for numerical work:

$$\mathbf{C_{D,cylinder}} \approx 1.0 + 10.0(\mathbf{R})^{-0.67}, \qquad \mathbf{R} < 2 \times 10^5, \qquad \textbf{(8.12a)}$$

$$\mathbf{C_{D,sphere}} \approx \frac{24}{\mathbf{R}} + \frac{6}{1 + \sqrt{\mathbf{R}}} + 0.4, \qquad \mathbf{R} < 2 \times 10^5. \qquad \textbf{(8.12b)}$$

It is instructive to compare the cylinder and sphere drag coefficients with those of the smooth parallel flat plate and the normal flat plate. The drag coefficients of cylinders and spheres have some striking similarities to those of the plates, but they also have some striking differences. At low Reynolds numbers, the cylinder/sphere drag coefficients vary with Reynolds number, decreasing as Reynolds number increases. This behavior is similar to that of the flat plate aligned with the flow. At a Reynolds number of about 10^3, the drag coefficient becomes essentially constant and the curve is flat. This behavior is similar to that of the normal flat plate. At a Reynolds number of about 5×10^5 the cylinder/sphere drag coefficient dips and then returns to a lower constant value. Note that Eqs. (8.12a) and (8.12b) do not apply in this region.

The sudden change in drag coefficient is similar to that which occurs for the aligned flat plate except that the cylinder/sphere coefficient *drops sharply* while the flat plate coefficient *rises gradually*. The effect of surface roughness on cylinder/sphere drag is illustrated schematically in Fig. 8.11. Roughness has very little effect until the dip occurs. Roughness causes the dip to occur at lower Reynolds numbers and causes higher drag beyond the dip.

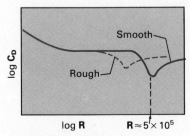

Figure 8.11 Effect of surface roughness on cylinder/sphere drag.

If we attempted to enter the drag characteristics of the cylinder/sphere in a table such as Table 8.1, we would not learn very much. Each space would contain several entries, based on the value of the Reynolds number. A complete explanation of the "strange" behavior of the cylinder/sphere drag coefficient requires separate consideration of high Reynolds number and low Reynolds number regimes. We take up this discussion in the next two sections.

EXAMPLE 8.3   Illustrates high Reynolds number drag and a simple method to measure a drag coefficient

The smooth sphere shown in Fig. E8.3(a) has weight W and hangs from a hook by a rigid and weightless wire of length L. A wind of velocity V_∞ blows over the sphere. Develop an expression relating the drag coefficient C_D of the sphere, the velocity V_∞, and the angle θ. The air density is ρ and the sphere diameter is D. Use the resulting expression to calculate the angle θ for a 6-in.-diameter sphere weighing 1.0 lb, a wind velocity of 35 mph, and a 24-in.-long wire. Assume the sphere diameter is very large compared to the diameter of the wire so that the air drag on the wire may be neglected.

Solution

Given
Figure E8.3(a)
$W = 1.0$ lb, $D = 6$ in., $V_\infty = 35$ mph, $L = 24$ in.

Find
Relation between angle θ, drag coefficient, and properties of the sphere and the wind
Calculate θ for a specific case.

Solution
We first apply Newton's first law in the horizontal (x) direction. Using the free-body diagram in Fig. E8.3(b), we get

$$\mathscr{D} = T \sin \theta,$$

where T is unknown but can be found by applying Newton's first law in the vertical (y) direction. Referring to Fig. E8.3(b), we see that

$$W = T \cos \theta, \quad \text{so} \quad T = \frac{W}{\cos \theta}.$$

Substituting this expression for T into the equation for $\mathscr{D}$ gives

$$\mathscr{D} = W \tan \theta.$$

Equation (8.4) gives

$$\mathscr{D} = \frac{1}{2} \rho C_D V_\infty^2 S.$$

Eliminating $\mathscr{D}$ from the last two equations gives

$$W \tan \theta = \frac{1}{2} \rho C_D V_\infty^2 S,$$

where S is the frontal area, $S = \pi D^2/4$.

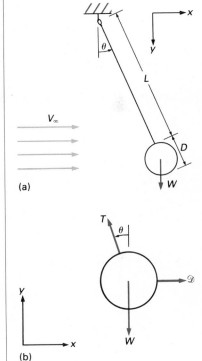

(a)

(b)

Figure E8.3 (a) Simple experiment to determine drag coefficients for a sphere; (b) free-body diagram of sphere.

Solving for θ gives

$$\left|\; \theta = \tan^{-1}\left(\frac{\rho C_D V_\infty^2 \pi D^2}{8W}\right). \;\right|$$

Answer

Assuming 60°F air, Table A.4 gives $\rho = 0.077$ lbm/ft³ and $\nu = 1.59 \times 10^{-4}$ ft²/sec. The Reynolds number is

$$\mathbf{R} = \frac{V_\infty D}{\nu} = \frac{\left(35\,\dfrac{\text{mi}}{\text{hr}}\right)\left(\dfrac{5280\ \text{ft}}{\text{mi}}\right)\left(\dfrac{\text{hr}}{3600\ \text{sec}}\right)\left(\dfrac{1}{2}\,\text{ft}\right)}{1.59 \times 10^{-4}\,\dfrac{\text{ft}^2}{\text{sec}}} = 1.6 \times 10^5.$$

Figure 8.10 gives $C_D = 0.28$. (Note that Eq. 8.12b does not apply.) The numerical values give

$$\theta = \tan^{-1}\left[\frac{\left(0.077\,\dfrac{\text{lbm}}{\text{ft}^3}\right)(0.28)\left(35\,\dfrac{\text{mi}}{\text{hr}}\right)^2\left(\dfrac{5280\ \text{ft}}{\text{mi}}\right)^2\left(\dfrac{\text{hr}}{3600\ \text{sec}}\right)^2 \pi(6\ \text{in.})^2}{8(1.0\ \text{lb})\left(\dfrac{32.2\ \text{ft}\cdot\text{lbm}}{\text{lb}\cdot\text{sec}^2}\right)\left(\dfrac{12\ \text{in}}{\text{ft}}\right)^2}\right],$$

$$\left|\; \theta = 9.8°. \;\right|$$

Answer

Discussion

This simple pendulum could be used to measure the drag coefficient for an object by solving the equation for C_D:

$$C_D = \frac{8W \tan\theta}{\rho V_\infty^2\, \pi D^2}.$$

For known values of W, ρ, V_∞, and D, measuring the angle θ will give the value of C_D.

Such a pendulum could also be used as a wind speed indicator by developing a calibration curve. The data point of $V_\infty = 35$ mph and $\theta = 9.8°$ is one of many points that would be used to develop this calibration curve for a sphere as the indicator.

EXAMPLE 8.4 Illustrates form drag and viscous drag at high
Reynolds numbers

Phil's Pizza Parlor decides to place a thin, rectangular, plastic sign on top of its delivery van. The sign measures 2 ft by 5 ft. Estimate the extra power required to drive the van in still air at 35 mph if the sign faces forward rather than sideways.

Solution

Given

Sign on top of van traveling at 35 mph in still air
Sign dimensions 2 ft by 5 ft

Find

Extra power required to drive van if sign faces forward rather than sideways

Solution

We assume the sign is far enough away from the van that the air flow over the sign is not influenced by the van, so we may use Figs. 8.8 and 8.9. We first consider the sign placed sideways. Assuming 60°F air at atmospheric pressure, we have from Table A.4 $v = 1.59 \times 10^{-4}$ ft²/sec. The Reynolds number is $\mathbf{R} = V_\infty L/v$, where $L = 5$ ft and $V_\infty = 35$ mph. Then

$$V_\infty = \frac{\left(35 \frac{\text{mi}}{\text{hr}}\right)\left(\frac{5280 \text{ ft}}{\text{mi}}\right)}{\left(\frac{3600 \text{ sec}}{\text{hr}}\right)} = 51.3 \text{ ft/sec} \quad \text{and} \quad \mathbf{R} = \frac{\left(51.3 \frac{\text{ft}}{\text{sec}}\right)(5 \text{ ft})}{1.59 \times 10^{-4} \frac{\text{ft}^2}{\text{sec}}} = 1.61 \times 10^6.$$

Table 7.1 shows that plastic has $\varepsilon = 0$ (smooth surface). Assuming a turbulent flow, from Fig. 8.8 we have $\mathbf{C_D} = 0.004$. Equation (8.4) is used to calculate the drag,

$$\mathscr{D} = \frac{1}{2} \rho \mathbf{C_D} V_\infty^2 S,$$

where S is the planform area. Using Table A.4, we have $\rho = 0.077$ lbm/ft³. Then

$$\mathscr{D} = \frac{1}{2}\left(0.077 \frac{\text{lbm}}{\text{ft}^3}\right)(0.004)\left(51.3 \frac{\text{ft}}{\text{sec}}\right)^2 (2 \text{ ft} \times 5 \text{ ft})$$

$$= \left(3.55 \frac{\text{ft} \cdot \text{lbm}}{\text{sec}^2}\right)\left(\frac{\text{lb} \cdot \text{sec}^2}{32.2 \text{ ft} \cdot \text{lbm}}\right) = 0.11 \text{ lb}.$$

The power required to overcome this air drag is the product of the drag force and the van speed:

$$\dot{W}_1 = \mathscr{D} \cdot V_\infty = \frac{(0.11 \text{ lb})\left(51.3 \frac{\text{ft}}{\text{sec}}\right)}{\left(550 \frac{\text{ft} \cdot \text{lb}}{\text{sec} \cdot \text{hp}}\right)} = 0.011 \text{ hp}.$$

We now consider the forward-facing sign. For 60°F air,

$$\mathbf{R} = \frac{V_\infty h}{v} = \frac{\left(51.3 \frac{\text{ft}}{\text{sec}}\right)(2 \text{ ft})}{1.59 \times 10^{-4} \frac{\text{ft}^2}{\text{sec}}} = 6.5 \times 10^5.$$

Figure 8.9 is then used with $b/h = 2.5$ to give $\mathbf{C_D} \approx 1.2$, and Eq. (8.4) gives

$$\mathscr{D} = \frac{1}{2} \rho \mathbf{C_D} V_\infty^2 S$$

$$= \frac{1}{2}\left(0.077 \frac{\text{lbm}}{\text{ft}^3}\right)(1.2)\left(51.3 \frac{\text{ft}}{\text{sec}}\right)^2 (2 \text{ ft} \times 5 \text{ ft})$$

$$= \left(1215 \frac{\text{ft} \cdot \text{lbm}}{\text{sec}^2}\right)\left(\frac{\text{lb} \cdot \text{sec}^2}{32.2 \text{ ft} \cdot \text{lbm}}\right) = 37.8 \text{ lb}.$$

The power required to overcome this air drag is

$$\dot{W}_2 = \mathscr{D} \cdot V_\infty = \frac{(37.8 \text{ lb})\left(51.3 \frac{\text{ft}}{\text{sec}}\right)}{\left(550 \frac{\text{ft} \cdot \text{lb}}{\text{sec} \cdot \text{hp}}\right)} = 3.53 \text{ hp}.$$

The extra power required to drive the van if the sign faces forward rather than sideways is

$$\dot{W}_2 - \dot{W}_1 = 3.53 \text{ hp} - 0.010 \text{ hp},$$

$$| \dot{W}_2 - \dot{W}_1 = 3.52 \text{ hp.} |$$

Answer ◀

Discussion

The values of 0.010 hp and 3.53 hp represent the power required to overcome air drag at 35 mph. Considering that the van's drive train efficiency might be as low as 85%, the corresponding engine output powers would be

$$\dot{W}_1' = \frac{0.010 \text{ hp}}{0.85} = 0.014 \text{ hp} \quad \text{and} \quad \dot{W}_2' = \frac{3.52 \text{ hp}}{0.85} = 4.15 \text{ hp.}$$

The extra *engine* power could then be as large as 4.14 hp.

8.2.2 Drag at High Reynolds Number: The Boundary Layer

Most flows of interest in engineering have rather large Reynolds numbers, owing to two factors: The size of the object of interest is significant (dimensions range from a few centimeters to several meters) and the (kinematic) viscosity of the fluids involved (usually air or water) is not large. Since Reynolds number is the ratio of inertia force to viscous force in the fluid (see Chapter 6), it would seem reasonable to assume that high Reynolds number flow is approximately inviscid. If we discard viscosity, we would have no mechanism for friction drag, but it would seem that the inviscid flow assumption might give us a good understanding of form drag.

If we neglect viscous forces, it is possible to calculate streamline patterns and pressure distribution for flow over simple bodies (discussed in Chapter 9). Figure 8.12 illustrates results of an inviscid calculation for flow over a circular cylinder (results for flow over a sphere are qualitatively similar). Both the flow pattern and the pressure distribution are symmetrical about the midplane of the cylinder. Consider the calculation of form drag from the pressure distribution shown in Fig. 8.12(b). Since the pressure distribution is symmetrical, each element of the surface on the front of the cylinder that experiences a "push" in the flow direction is balanced by an element on the back that experiences an equal push opposite the flow direction. All such elements occur in pairs, so the net drag force would be zero! The prediction of zero drag for an inviscid fluid even extends to the *normal* flat plate. Figure 8.13 shows streamlines for flow over a normal flat plate calculated using the inviscid fluid assumption. The predicted pressure distribution on the plate is the same on the front and back, so no drag would result. Apparently, the inviscid fluid assumption does not allow us to investigate form drag, even approximately. This fact is known as *d'Alembert's paradox*.

In 1904, Prandtl introduced the boundary layer hypothesis, which explains how both form drag and friction drag arise in a high Reynolds number flow. Consider the flow over a long, slender body like a parallel flat plate or airfoil (Fig. 8.14). Prandtl realized that no matter how small the viscosity, the fluid must still satisfy the no-slip condition at

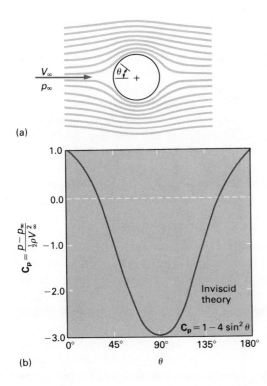

(a)

(b)

Figure 8.12 Calculated streamline pattern and pressure distribution for inviscid flow over a circular cylinder.

the surface. The fluid velocity must then rise from zero at the surface to a value that is the order of magnitude of the free-stream velocity a short distance above the surface. This region where the velocity increases from zero to the free-stream value is called the *boundary layer*. Prandtl noted that for large Reynolds numbers, the boundary layer is very thin; in fact, he showed that

$$\frac{\delta}{L} \sim \sqrt{\frac{v}{V_\infty L}} = \frac{1}{\sqrt{R_L}}.$$ (8.13)

Of course, the thinner the boundary layer, the larger the value of $\partial u/\partial y$ in the boundary layer. There are significant shear stresses in the boundary layer fluid, even if the viscosity is small.

Figure 8.13 Inviscid flow streamlines for flow over a normal flat plate.

Figure 8.14 Thin viscous boundary layer on a slender body in high Reynolds number flow.

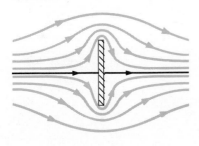

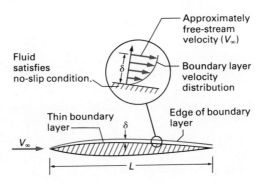

As long as the boundary layer is thin, its presence has little effect on the pressure distribution or the flow pattern near the body. For the flat plate or thin airfoil, the streamline shapes are essentially unchanged by the boundary layer. By analyzing the boundary layer, it is possible to calculate the shear stress and thereby calculate the drag on long, slender shapes like the parallel flat plate (we will consider this type of calculation in Chapter 9). Since the boundary layer is a region where the fluid experiences shear deformation, the flow may be laminar or turbulent depending on the value of the Reynolds number. Flow in the boundary layer near the leading edge of a plate is always laminar because the (local) Reynolds number is small; if the plate is long enough, transition to turbulence occurs and the flow is turbulent over the downstream portion of the plate.* If the flow is disturbed by surface roughness or turbulence in the stream outside the boundary layer, transition occurs nearer the leading edge. Because of enhanced mixing in turbulent flow, the turbulent boundary layer has larger shear stresses than the laminar boundary layer. Likewise, rough surfaces experience larger stress than do smooth ones (see Sections 7.1.1 and 7.2.7). These facts explain the behavior of the drag coefficient for a flat plate aligned with the flow (Fig. 8.8b).

The boundary layer model adequately explains flat plate drag, but what about cylinder drag? Consider the boundary layer on a circular cylinder (Fig. 8.15). The primary difference between the flat plate and cylinder flows is that the shape of the cylinder causes a pressure change in the flow. The pressure distribution has a profound effect on the boundary layer. The flow outside the boundary layer is determined by a balance between pressure force and fluid momentum. On the front of the cylinder, the fluid momentum increases and the pressure drops,

Figure 8.15 Boundary layer on a circular cylinder.

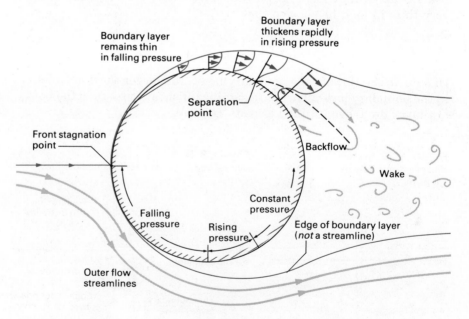

* Note that the same plate can have laminar flow over part of it and turbulent flow over another part. This is not true of *fully developed* pipe flow, which must be either laminar or turbulent.

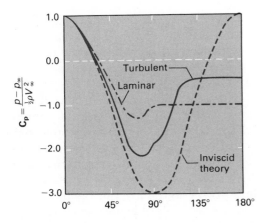

Figure 8.16 Pressure distribution on cylinder surface for laminar and turbulent boundary layers.

while on the back side the fluid exchanges momentum for increasing pressure. Now consider the boundary layer flow. In the boundary layer, the flow is determined by a balance between momentum and pressure *plus* viscous forces. Because of viscosity, the fluid inside the boundary layer has less momentum than the fluid outside the boundary layer. This situation is not too serious on the front side of the cylinder, where the falling pressure helps "push" the fluid in the boundary layer along. On the back side of the cylinder, the situation is different. As the pressure begins to rise, it is necessary for the fluid to exchange momentum for pressure. Since the boundary layer is deficient in momentum, it is unable to penetrate very far into the rising pressure. The pressure causes the fluid in the boundary layer to stop and, ultimately, reverse its direction. The boundary layer separates from the cylinder surface and a broad wake is formed behind the cylinder. The separating boundary layer pushes the flow streamlines outward and so alters the entire flow pattern and pressure distribution. The pressure on the rear of the cylinder, in the wake, is low and approximately constant. The pressure on the front of the cylinder is high, so there is a net pressure force (form drag) on the cylinder. The shear stress on the cylinder surface produces friction drag. At Reynolds numbers greater than about 10^3, form drag accounts for about 95% of the total drag, so the drag coefficient becomes independent of Reynolds number* except near the dip at a Reynolds number of about 5×10^5.

The dip in drag coefficient, as well as the resulting lower value of C_D after the dip, is the result of transition from laminar to turbulent flow in the boundary layer; however, the effect is opposite that in the flat plate flow, where turbulence causes a drag increase. To understand the effects of turbulence in the cylinder flow, it is necessary to note that a turbulent boundary layer contains more momentum than a laminar boundary layer because the velocity profile is fuller (see Fig. 7.13). Since the turbulent boundary layer contains more momentum, it can penetrate farther into the rising pressure on the rear of the cylinder, and separation is delayed. The wake is smaller, and the pressure in the wake is higher. Figure 8.16 shows pressure distributions on cylinders

---

* Reynolds number independence means that the drag depends on inertia force but not on viscous force.

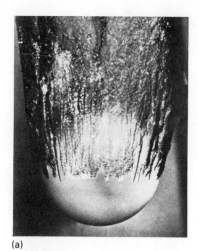

(a)

(b)

Figure 8.17 Effect of turbulence in boundary layer on delaying separation. An 8.5-in. bowling ball is dropped into water at a speed of 25 ft/sec. (a) The ball is smooth; (b) the ball has a patch of sandpaper on the nose that causes transition. (U.S. Navy photographs.)

with laminar and turbulent boundary layers. The higher average pressure in the back of the cylinder with a turbulent boundary layer results in lower form drag and a lower drag coefficient, even though the friction drag of the turbulent boundary layer is higher than the laminar layer.

The effects of turbulence in delaying separation are graphically illustrated in Fig. 8.17. The two spheres are identical except that one has a patch of sandpaper on the nose. This sandpaper causes the boundary layer to become turbulent, which delays the separation. Note that the wake is turbulent in both cases; there is no visible difference between the laminar and turbulent boundary layers on the spheres.

We are now in a position to better understand drag on a normal flat plate. Although there are boundary layers on the front of the plate, the flow is unable to undergo the infinite acceleration necessary to turn the sharp corner at the edge of the plate. The separation point is fixed at the plate edge for all Reynolds numbers, so the drag coefficient is independent of Reynolds number.

The above considerations help us understand the *mechanisms* of drag at high Reynolds number, but, except for the parallel plate, it is very difficult to analytically predict drag coefficients with confidence. It is possible to design low-drag shapes based on our qualitative understanding of drag. There are two important principles to follow if you wish to design a body with low drag:

- If the body is long and thin, like a parallel flat plate or an airfoil, the drag is primarily due to friction. This drag can be reduced by keeping the flow laminar as much as possible. This implies smooth surfaces and, if the body has thickness (like an airfoil), selecting the shape that delays transition as long as possible.
- If the body is blunt, the (high Reynolds number) drag is primarily form drag. This drag can be reduced by delaying separation as long as possible.

One way to delay boundary layer separation on a blunt (thick) body is to promote transition to a turbulent boundary layer (Fig. 8.17). A better method is *streamlining*, that is, elongating the rear portion of the body. It turns out that boundary layer separation depends on the rate of pressure rise, that is, on the pressure gradient dp/dx, where x is the distance along the surface.

Figure 8.18(a) shows a circular cylinder and a streamlined strut with the same frontal area. The streamlined strut will have a drag coefficient

Figure 8.18 Illustration of the effects of streamlining on drag at high Reynolds number. (a) Cylinder and strut with the same frontal area; $C_{D,\text{cylinder}} \approx 5C_{D,\text{strut}}$; (b) cylinder and strut with equal drag force at a given flow speed.

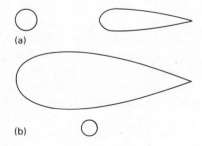

(a)

(b)

about one-fifth that of the cylinder. Alternatively, Fig. 8.18(b) shows a cylinder and strut that have equal drag for a given flow speed.

If the nose of a body is flat, nose rounding can also reduce the drag. Figure 8.19 illustrates the benefits of nose rounding and rearward taper. It is interesting to note that the round nose–rear tapered body would have higher drag if it were reversed with the tapered section pointing forward. The forward portion of a body can be quite blunt with a large (negative) pressure gradient because separation does not occur in falling pressure. It is the rearward (rising) pressure change that must be accomplished gradually.

There is a limit to the amount of drag reduction that can be accomplished by streamlining. As the body becomes more elongated, friction drag increases because there is more "wetted" surface (see Fig. 8.20). The minimum drag body provides the proper balance between friction drag and form drag.

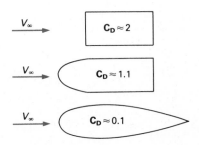

Figure 8.19 Effect of streamlining on drag reduction. Drag coefficients are based on frontal area. All bodies are two-dimensional.

8.2.3 Drag at Low Reynolds Number

Occasionally an engineer must deal with a low Reynolds number flow. Low Reynolds numbers are due to small body sizes (like dust or pulverized coal particles) or large fluid viscosities (oil, grease). The mechanisms of drag are fundamentally different in low and high Reynolds number flow. The boundary layer concept, so valuable in high Reynolds number flow, is invalid at low Reynolds number. At low Reynolds number, viscous forces are not confined to a thin region near the body but are significant everywhere. We may think of the boundary layer as being so thick that it distorts the streamline pattern everywhere. At very low Reynolds number (less than 1), inertia forces are negligible and only pressure and shear forces are present. Of course, low Reynolds number flow is always laminar.

Consider the flow over a sphere at very low Reynolds number. This flow was successfully analyzed by Sir G. G. Stokes in 1851. The streamline pattern and pressure distribution are shown in Fig. 8.21. The streamlines are nearly symmetric with only a very thin wake (no separation); however, the pressure distribution is highly asymmetric. Stokes

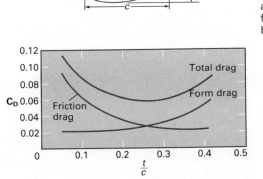

Figure 8.20 Total drag is the sum of form drag, which increases as a body becomes more blunt, and friction drag, which increases as a body becomes longer.

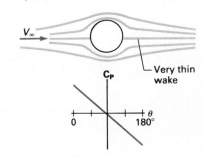

Figure 8.21 Streamline pattern and pressure distribution for very low Reynolds number flow over a sphere.

found that the drag on a sphere at very low Reynolds numbers is given by

$$\mathcal{D} = 3\pi\mu D V_\infty. \tag{8.14}$$

Equation (8.14) is called Stokes' law. The drag is not due to friction only; one-third of the drag is due to pressure and two-thirds is due to friction. The drag coefficient for a sphere at very low Reynolds number is*

$$C_D = \frac{\mathcal{D}}{\frac{1}{2}\rho V_\infty^2 \left(\frac{\pi D^2}{4}\right)} = \frac{24\mu}{\rho VD} = \frac{24}{R}. \tag{8.15}$$

Actually, the "usual" drag coefficient defined by Eq. (8.4) is not the best definition for low Reynolds number flow because the drag does not depend on fluid inertia (see Example 6.6). A *viscous drag coefficient*, defined by

$$C'_D \equiv \frac{\mathcal{D}}{\mu V_\infty \ell} \qquad \left(\sim \frac{\text{Drag force}}{\text{Viscous force}}\right), \tag{8.16}$$

is more useful. For the sphere, $\ell = D$ and

$$C'_D = 3\pi,$$

a constant.

The viscous drag coefficient for any body in a very low Reynolds number flow is a constant, dependent on only the body shape. Table 8.2 gives values of C'_D for various body shapes. Note that C'_D is roughly the same for all shapes except very long, slender ones. As a first approximation, C'_D of a sphere can be used for all blunt bodies in very low R flow.

Table 8.3 compares drag in high and low Reynolds number flows. These comparisons are based on the following approximate equations:

$$\mathcal{D}(\text{High } R) \approx (\text{Constant})\left(\frac{1}{2}\rho V_\infty^2\right)(\text{Area}),$$

$$\mathcal{D}(\text{Low } R) \approx (\text{Constant})\mu V_\infty(\text{Length}).$$

It is important to note that steps taken to reduce drag in high Reynolds number flow are ineffective or even counterproductive in low Reynolds number flow. In particular, streamlining usually results in a drag *increase* in low R flow because body length and wetted area are increased.

Drag in the intermediate Reynolds number range ($10 < R < 10^3$) is most difficult to analyze, even qualitatively, because the boundary layer concept is not valid but inertia forces are significant. Information on

---

* Note the similarity with the Hagen-Poiseuille equation for friction factor in laminar pipe flow. Also note the leading term of Eq. (8.12b).

Table 8.2 Viscous drag coefficients for low Reynolds number flow

Shape	Reference Length	C_D'
Sphere	D	3π
Hemisphere	D	8.7
Normal disc	D	8
Parallel disc	D	$\dfrac{16}{3}$
Normal rod	L	$\dfrac{4\pi}{\ln\left(\dfrac{2L}{D}\right) + 0.5}$
Parallel rod	L	$\dfrac{2\pi}{\ln\left(\dfrac{2L}{D}\right) - 0.72}$

Table 8.3 Dependence of drag on various parameters in high **R** and low **R** flow (compare to Table 7.2)

Parameter	High R Flow	Low R Flow
Velocity (V)	V^2	V
Density (ρ)	ρ	None
Viscosity (μ)	Slight to none	μ
Size (ℓ)	ℓ^2	ℓ

drag in this range is based almost entirely on experiment. Data are usually presented in terms of C_D. Both C_D and C_D' are functions of Reynolds number.*

* Note that $C_D' = C_D \times R$; C_D' is an alternative dimensionless group.

EXAMPLE 8.5   Illustrates low Reynolds number drag

The 1.0-mm-diameter steel ball in Fig. E8.5 is released from rest at the oil surface and reaches a constant velocity at point 1. Find the velocity of the steel ball and estimate the time for the ball to fall from position 1 to position 2. The fluid is light machine oil with an absolute viscosity of 0.10 N·s/m² and a specific weight of 8340 N/m³. Steel has a specific gravity of 7.8.

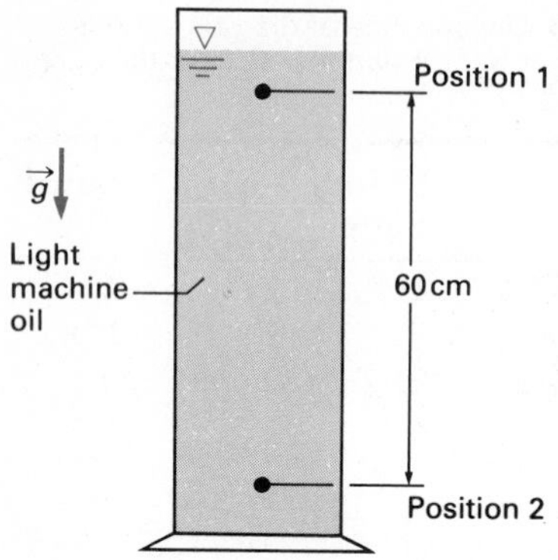

Figure E8.5 Steel ball falling in light machine oil.

Solution

Given

1.0-mm steel ball falling in light machine oil
Ball falling with constant velocity for a distance of 60 cm
Oil viscosity 0.10 N·s/m²
Oil specific weight 8340 N/m³
Figure E8.5
Steel specific gravity 7.8

Find

The time for the steel ball to fall 60 cm
Constant velocity during the time of the fall

Solution

We first apply Newton's second law in the vertical direction, with the positive direction taken as upward. Since the velocity is constant, there is no acceleration, and Newton's second law gives

$$\sum F = \mathscr{D} - W + F_B = 0,$$

where $\mathscr{D}$ is the viscous drag on the ball, W its weight, and F_B the buoyant force on the ball. Assuming the ball falls very slowly, Stokes' law (Eq. 8.14) gives

$$\mathscr{D} = 3\pi\mu D V_\infty.$$

The ball's weight is $W = \gamma_s \mathcal{V}$, where γ_s is the specific weight of steel and $\mathcal{V}$ is the ball's volume. Now

$$\gamma_s = S\gamma_w = 7.8(9810 \text{ N/m}^3) = 76,500 \text{ N/m}^3.$$

The buoyant force is $F_B = \gamma_0 \mathcal{V}$, where γ_0 is the oil specific weight.

Substituting for $\mathscr{D}$, W, and F_B gives

$$3\pi\mu D V_\infty - \gamma_s \mathcal{V} + \gamma_0 \mathcal{V} = 0.$$

Using $\mathcal{V} = \pi D^3/6$ gives

$$V_\infty = \frac{(\gamma_s - \gamma_0)D^2}{18\mu}.$$

The numerical values give

$$V_\infty = \frac{\left(76,500 \dfrac{\text{N}}{\text{m}^3} - 8340 \dfrac{\text{N}}{\text{m}^3}\right)(1.0 \text{ mm})^2}{18\left(0.10 \dfrac{\text{N·s}}{\text{m}^2}\right)\left(\dfrac{1000 \text{ mm}}{\text{m}}\right)^2},$$

$$\left| V_\infty = 0.0379 \text{ m/s}. \right|$$ **Answer** ◀

Before proceeding, we now check to see if our assumption of very low Reynolds number is correct. The oil kinematic viscosity is

$$\nu_0 = \frac{\mu_0}{\rho_0} = \frac{g\mu_0}{\gamma_0}.$$

Assuming sea-level elevation, the numerical values give

$$\nu_0 = \frac{\left(9.81 \dfrac{\text{m}}{\text{s}^2}\right)\left(0.10 \dfrac{\text{N·s}}{\text{m}^2}\right)}{\left(8340 \dfrac{\text{N}}{\text{m}^3}\right)} = 1.18 \times 10^{-4} \text{ m}^2/\text{s}$$

and

$$\mathbf{R} = \frac{V_\infty D}{\nu_0} = \frac{\left(0.0379 \dfrac{\text{m}}{\text{s}}\right)(1.0 \text{ mm})}{\left(1.18 \times 10^{-4} \dfrac{\text{m}^2}{\text{s}}\right)\left(\dfrac{1000 \text{ mm}}{\text{m}}\right)} = 0.321.$$

Since we are in the Reynolds number range where Stokes' law is valid, our velocity estimate is also valid. The time required for the sphere to fall 60 cm is

$$t = \frac{\text{Distance}}{\text{Velocity}} = \frac{L}{V_\infty}.$$

The numerical values give

$$t = \frac{(60 \text{ cm})\left(\dfrac{\text{m}}{100 \text{ cm}}\right)}{\left(0.0379 \dfrac{\text{m}}{\text{s}}\right)},$$

or   **Answer**

$$\mid t = 15.8 \text{ s.} \mid$$

◀

Discussion

We can combine the equations for V_∞ and t to give

$$\mu = \frac{(\gamma_s - \gamma_0)D^2 t}{18\,L}.$$

Could we use this experiment and equation to find the viscosity of fluids? Must the fluid be a particular type? Do the walls of the container have any effect?

8.2.4 Working Information on Drag Coefficients

With the exception of very low Reynolds number flow, working information on drag coefficients is obtained from experimental measurements. Figures 8.8(b), 8.9(c), 8.10(b), and 8.11 are examples of this type of information and may be used for drag calculations. Extensive information on drag is contained in [3]. Information on drag coefficients for a few simple two-dimensional, three-dimensional, and axisymmetric shapes is contained in Tables 8.2, 8.4, 8.5, and 8.6. Most of the geometries shown in the figures in the tables are rather angular, with (high Reynolds number) separation points fixed at sharp corners, so the quoted drag coefficients are valid for Reynolds numbers greater than about 10^4. You should be aware that these values are based on experiment and have uncertainty of about 5%. Values of $\mathbf{C_D}$ for situations not listed can sometimes be *estimated* from listed data. The drag coefficient for a very thin streamlined two-dimensional body might be approximated by the value for a flat plate. Values of $\mathbf{C_D}$ for cylinders of finite length with axis perpendicular to flow might be estimated by assuming that the ratio of $\mathbf{C_D}$ for the finite-length cylinder to $\mathbf{C_D}$ for the infinite cylinder is equal to the corresponding ratio of $\mathbf{C_D}$'s for a normal flat plate (Fig. 8.9c). Such estimates should be used with care.

Table 8.4 Drag of various two-dimensional bodies

Shape	Reynolds number	C_D
Parallel flat plate	$> 10^5$	See Fig. 8.8(b).
Normal flat plate	$> 10^3$	2.0
Circular cylinder	All	See Fig. 8.10.
Square rod	$> 10^4$	2.0
Square rod	$> 10^4$	1.50
Equilateral triangular rod	$> 10^4$	Sharp edge forward: 1.40 Flat face forward: 2.0
C-section	$> 10^4$	2.30
C-section	$> 10^4$	1.20
Airfoils	Various	See Reference [5]

Table 8.5 Drag of various three-dimensional objects

Shape	Reynolds number	C_D
Parallel flat plate (finite width)	$> 10^5$	Use Fig. 8.8(b).
Normal plate	$> 10^3$	See Fig. 8.9(c).
Cube	$> 10^4$	1.10
Cube	$> 10^4$	0.81

Table 8.6 Drag of various axisymmetric objects

Shape	Reynolds number	C_D
Sphere	All	See Fig. 8.10.
Cylinder/disc ($L = 0$)	$> 10^4$	$L/D = 0$ 1.17 $L/D = 0.5$ 1.15 $L/D = 1$ 0.90 $L/D = 2$ 0.85 $L/D = 4$ 0.87 $L/D = 8$ 0.99
Hemispherical cup	$> 10^4$	1.40
Hemispherical cup	$> 10^4$	0.40
60° cone	$> 10^4$	0.50
Parachute	$> 10^5$	1.2

EXAMPLE 8.6 Illustrates drag at high Reynolds numbers

A 9 × 18-ft rectangular sign rests atop a 1-ft-diameter, 150-ft-high pole. Find the bending moment at the bottom of the pole for a uniform wind velocity of 50 mph.

Solution

Given
9 × 18-ft rectangular sign supported by a 150-ft-high pole
50-mph wind

Find
Bending moment at bottom of the pole

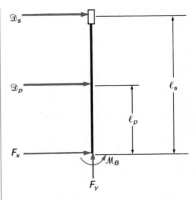

Figure E8.6 Forces and bending moment on sign and support pole.

Solution

The bending moment at the bottom of the pole is due to the drag force $\mathscr{D}_s$ on the sign and the drag $\mathscr{D}_p$ on the pole. Referring to Fig. E8.6, we see that the bending moment at the base, $\mathscr{M}_B$, is

$$\mathscr{M}_B = \ell_s \mathscr{D}_s + \ell_p \mathscr{D}_p.$$

$\mathscr{D}_s$ is found using Eq. (8.4):

$$\mathscr{D}_s = \frac{1}{2} \rho \mathbf{C_D} V_\infty^2 S,$$

where S is the frontal area. Assuming 60°F air, Table A.4 gives $\rho = 0.077$ lbm/ft^3 and $v = 1.59 \times 10^{-4}$ ft^2/sec. The Reynolds number is $\mathbf{R} = V_\infty h/v$, where h is the sign height and V_∞ is

$$V_\infty = \left(50 \frac{\text{mi}}{\text{hr}}\right)\left(\frac{5280 \text{ ft}}{\text{mi}}\right)\left(\frac{\text{hr}}{3600 \text{ sec}}\right) = 73.3 \text{ ft/sec}.$$

Then

$$\mathbf{R} = \frac{\left(73.3 \dfrac{\text{ft}}{\text{sec}}\right)(9 \text{ ft})}{\left(1.59 \times 10^{-4} \dfrac{\text{ft}^2}{\text{sec}}\right)} = 4.1 \times 10^6.$$

Figure 8.9(c) may now be used with $b/h = 2.0$ to give $\mathbf{C_D} = 1.15$. Then

$$\mathscr{D}_s = \frac{\left(0.077 \dfrac{\text{lbm}}{\text{ft}^3}\right)(1.15)\left(73.3 \dfrac{\text{ft}}{\text{sec}}\right)^2 (9 \text{ ft} \times 18 \text{ ft})}{2\left(32.2 \dfrac{\text{ft} \cdot \text{lbm}}{\text{lb} \cdot \text{sec}^2}\right)} = 1200 \text{ lb}.$$

$\mathscr{D}_p$ is also found using Eq. (8.4):

$$\mathscr{D}_p = \frac{1}{2} \rho \mathbf{C_D} V_\infty^2 S.$$

The Reynolds number is

$$\mathbf{R} = \frac{\left(73.3 \dfrac{\text{ft}}{\text{sec}}\right)(1.0 \text{ ft})}{\left(1.59 \times 10^{-4} \dfrac{\text{ft}^2}{\text{sec}}\right)} = 4.61 \times 10^5.$$

Figure 8.10 gives $C_D \approx 0.40$. Then

$$\mathscr{D}_p = \frac{\left(0.077 \dfrac{\text{lbm}}{\text{ft}^3}\right)(0.40)\left(73.3 \dfrac{\text{ft}}{\text{sec}}\right)^2 (1 \text{ ft} \times 150 \text{ ft})}{2\left(32.2 \dfrac{\text{ft} \cdot \text{lbm}}{\text{lb} \cdot \text{sec}^2}\right)} = 385 \text{ lb.}$$

Now $\ell_s = 154.5$ ft and $\ell_p = 75$ ft, so the bending moment at the base is

$$\mathscr{M}_B = (154.5 \text{ ft})(1200 \text{ lb}) + (75 \text{ ft})(385 \text{ lb}),$$

Answer

$$|\; \mathscr{M}_B = 2.14 \times 10^5 \text{ ft} \cdot \text{lb.} \;|$$

◀

Discussion

Note that the resultant drag on the pole acts at the midheight because the drag is assumed uniform along the pole. The drag on the sign is expected to act at the center of the sign unless the pole causes the flow to be asymmetric. The pole will produce some asymmetry, but its effect is assumed to be slight. Likewise, the pole will experience some asymmetry with the ground on the bottom and the sign at the top. This asymmetry is assumed to be slight. A more correct estimate of $\mathscr{D}_p$ can be made using a power law equation for the wind velocity profile.

Before ending the discussion of drag, it is useful to warn you that the qualitative dependence of drag *force* on *velocity* is different from the dependence of drag *coefficient* on *Reynolds number*. By examining the drag coefficients for cylinders and spheres (Fig. 8.10b), you might conclude that the drag force drops with increasing velocity. We can show how false this conclusion is by replotting the information in a different way. Consider a sphere of fixed diameter D in a fluid of given density and viscosity. In this case, the drag on the sphere will depend on the velocity only. The Reynolds number varies directly with velocity, but the drag coefficient varies with both drag and velocity. If we consider an alternative dimensionless parameter, defined by

$$\tilde{C}_D \equiv \frac{\pi}{8} C_D R^2, \tag{8.17a}$$

we find

$$\tilde{C}_D = \frac{\pi}{8} \frac{\mathscr{D}}{\dfrac{1}{2}\rho V_\infty^2 \left(\dfrac{\pi D^2}{4}\right)} \left(\frac{\rho V_\infty D}{\mu}\right)^2 = \frac{\rho \mathscr{D}}{\mu^2}. \tag{8.17b}$$

The parameter $\tilde{C}_D$ is directly proportional to drag for fixed ρ and μ. We can prepare a plot of $\tilde{C}_D$ versus R from Fig. 8.10(b) by picking values of R, reading C_D, and calculating $\tilde{C}_D$. The resulting plot is shown in Fig. 8.22. This plot clearly shows that drag continually increases with velocity except for a slight dip at transition.

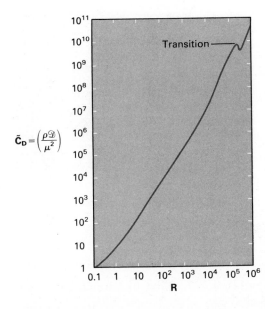

Figure 8.22 Dimensionless drag force ($\tilde{C}_D$) versus Reynolds number for a sphere. For fixed size and fluid properties, drag force clearly increases with velocity.

8.3 Lift

Drag is present in all external flows. Lift is present only if there is asymmetry. Usually (but not always), drag is undesirable, leading to increased fuel consumption in vehicles, wind loading on structures, and the like. Lift is usually beneficial; for example, aerodynamic lift on the wings of an airplane allows it to fly. Lift forces also do most of the useful work in flow over propeller, compressor, and turbine blades.

8.3.1 Mechanism of Lift Generation: The Simple Airfoil

To understand how lift is generated, we will consider a shape designed to produce high lift with low drag—a two-dimensional airfoil. An airfoil may be symmetric about its centerline (Fig. 8.23a) or asymmetric (Fig. 8.23b). In either case, a typical airfoil has a rounded nose and a sharp trailing edge. The free-stream flow approaches the airfoil at an angle of attack α.

Airfoil lift is primarily the result of surface pressure, so it would seem that fluid viscosity would have little effect on lift. This is almost true; measured lift coefficients show very little dependence on Reynolds number, and calculations of lift based on an assumed inviscid fluid are reasonably accurate provided that we take account of one extremely important "viscosity" effect, without which there would be no lift. If we assume an absolutely inviscid fluid and calculate the streamline pattern for flow over an airfoil at an angle of attack (we will discuss how this can be done in Chapter 9), the results are as shown in Fig. 8.24(a). The flow over the front half of the airfoil is reasonable, but the flow over the back half is not. Note that the flow on the lower side is expected to turn the sharp corner at the trailing edge and flow "backward" over the top surface to the rear stagnation point. If we calculate the pressure dis-

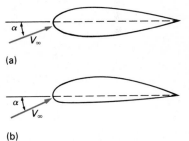

(a)

(b)

Figure 8.23 Two-dimensional airfoil profiles: (a) Symmetric; (b) asymmetric.

Figure 8.24 Streamline pattern for airfoil at angle of attack: (a) Theoretical pattern for inviscid flow; (b) more realistic trailing edge flow; (c) streamlines for real airfoil flow. (Part c from "NCFMF Book of Film Notes," 1974, The MIT Press with Education Development Center, Inc. Newton, MA.)

tribution on the airfoil surface and then calculate the lift using Eq. (8.3) (with $\tau_w = 0$), we would predict zero lift.

The obvious problem is the trailing edge flow. A real fluid could not turn the sharp corner but would separate at the trailing edge. The fluid on the upper surface at the trailing edge and the upper and lower streams would join smoothly. The real streamline pattern would be as sketched in Fig. 8.24(b). Figure 8.24(c) shows a photograph of streamlines (actually streaklines marked by smoke) of flow over an airfoil at a small angle of attack. Note the streamline pattern near the sharp trailing edge.

It is possible to produce the realistic streamline pattern of Figs. 8.24(b) and (c) from that of Fig. 8.24(a) by adding a clockwise circulating velocity pattern that increases the velocity on top of the airfoil and decreases the velocity on the bottom of the airfoil (Fig. 8.25). This is equivalent to adding a free vortex flow (see Example 4.8) to the flow of Fig. 8.24(a). Of course, the flow very near the center of the free vortex is quite unrealistic, so we "bury" that part of the vortex inside the airfoil. In this way, we are able to produce a realistic streamline pattern without actually considering the effects of viscosity and shear stress. When we calculate the lift, we note that, according to Bernoulli's equation for two-dimensional irrotational flow, the pressure on the top of the airfoil, where the velocity is larger, will be lower than the pressure on the bottom of the airfoil, where the velocity is smaller, and so a net upward (lift) force will be generated.

This explanation and method of calculating lift may seem artificial, but it has a firm physical basis. Figure 8.26 is a series of sketches of the processes that occur during the acceleration of an airfoil from rest to a uniform velocity. Immediately after start-up, the streamline pattern resembles that of Fig. 8.24(a). Separation of the lower surface boundary layer occurs at the trailing edge, and a starting vortex is formed. The flow on the upper surface fills the void, and the separation point on the upper surface moves to the trailing edge. The starting vortex, which is equal in strength and rotates in the opposite direction to the imaginary

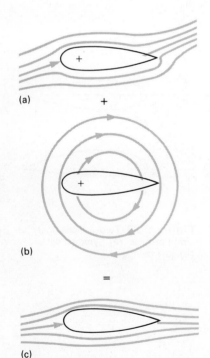

Figure 8.25 Realistic streamline pattern obtained by adding circulatory flow to theoretical inviscid flow over airfoil.

vortex that we place in the airfoil, remains at the starting location. The motion of the starting vortex decays slowly due to the action of viscosity.*

Increasing the angle of attack of the airfoil increases the flow asymmetry, resulting in higher velocities over the top of the airfoil and larger lift. For very thin symmetrical airfoils, theory predicts

$$\mathbf{C_L} \approx 2\pi\alpha, \tag{8.18}$$

where α is in radians. The increase of $\mathbf{C_L}$ with α predicted by Eq. (8.18) persists to an angle of about $15°$, when the airfoil stalls and the lift drops rapidly. *Stall* occurs when the upper separation point moves from the trailing edge toward the nose. Figure 8.27 shows an airfoil in stall at a high angle of attack.

8.3.2 The Kutta-Joukowsky Theorem

A theorem derived independently by W. M. Kutta in 1902 and N. E. Joukowsky[†] in 1906 permits calculation of lift in two-dimensional flows directly from the velocity field without calculating the pressure distribution. This theorem also shows the strong connection between a circulating flow pattern and lift. Since the theorem is derived by using the finite control volume linear momentum and continuity equations, it also shows that finite control volume formulations are incapable of providing complete solutions to external flow problems.

Figure 8.28(a) shows an airfoil in an infinite stream. The fluid velocity and pressure far from the airfoil are V_∞ and p_∞. We assume that the fluid is inviscid and the flow is irrotational (this is actually a very good assumption outside of the thin boundary layer near the airfoil surface and the downstream wake). Figure 8.28(b) shows a coordinate sys-

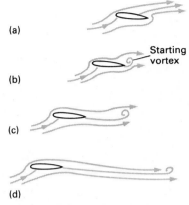

Figure 8.26 Sketches of the establishment of flow during acceleration of an airfoil from rest: (a) Instant of start-up; (b) lower surface boundary layer separates; (c) nearly steady flow; (d) steady flow pattern (starting vortex decays with time).

Figure 8.27 Airfoil in stall at high angle of attack; upper surface boundary layer separates near leading edge. (From "NCFMF Book of Film Notes," 1974, The MIT Press with Education Development Center, Inc., Newton, MA.)

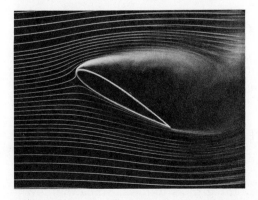

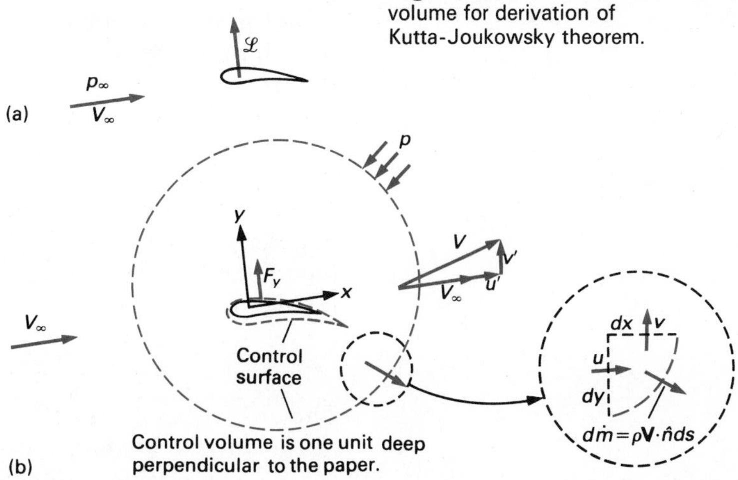

Figure 8.28 Airfoil and control volume for derivation of Kutta-Joukowsky theorem.

* Starting vortexes can be clearly seen by moving a spoon in a cup of coffee.

[†] Joukowsky is sometimes spelled Joukowski or Zoukowskii.

tem and control volume for analysis of the airfoil forces. The x-direction is parallel to the approach flow. The y-direction is perpendicular to the approach flow and is the direction of the lift force. Note that the airfoil is excluded from the control volume by drawing the control surface adjacent to the airfoil. The y-direction force that the airfoil exerts on the fluid at the inner portion of the control surface is the reaction to the lift force, that is

$$\mathscr{L} = -F_y. \tag{8.19}$$

No fluid flows across the inner boundary of the control volume because it is adjacent to the airfoil.

The fluid velocity at the outer boundary of the control volume is disturbed from the free-stream velocity V_∞ by the presence of the airfoil. The velocity components can be written as

$$u = V_\infty + u' \tag{8.20a}$$

and

$$v = v', \tag{8.20b}$$

where u' and v' are *pertubations* from the values of u and v far ahead of the airfoil. By selecting the outer boundary of the control surface very far from the airfoil, it is possible to make u' and v' very small compared to V_∞; that is, far from the airfoil,

$$u'/V_\infty \ll 1 \tag{8.21a}$$

and

$$v'/V_\infty \ll 1. \tag{8.21b}$$

We now apply the y-direction linear momentum equation to the control volume. Since the flow is two-dimensional and two-directional, we assume that the control volume is one unit deep perpendicular to the paper and calculate the force per unit width of airfoil. The forces are F_y at the inner boundary and the pressure force at the outer boundary. Momentum flows across the outer boundary only. The momentum equation is

$$F_y + \oint_{\substack{\text{outer} \\ \text{boundary}}} p\,dx(1) = \oint_{\substack{\text{outer} \\ \text{boundary}}} \rho v(\vec{V} \cdot \hat{n})\,ds(1).$$

We evaluate the integrals by moving counterclockwise around the boundary. The (1) in the integrals is the unit depth. In the pressure integral, $dx(1)$ is the projected area that results in a y-direction pressure force . Solving for F_y, we have

$$F_y = -\oint p\,dx + \oint \rho v(\vec{V} \cdot \hat{n})\,ds. \tag{8.22}$$

The unit depth has been dropped, and the integrals are to be evaluated at the outer boundary only.

We can evaluate the pressure integral as follows. Since the flow is assumed ideal and irrotational, Bernoulli's equation gives

$$p + \frac{1}{2}\rho V^2 + \gamma z = p_\infty + \frac{1}{2}\rho V_\infty^2 + \gamma z_\infty,$$

and so

$$p = p_\infty + \gamma(z_\infty - z) + \frac{1}{2}\rho(V_\infty^2 - V^2). \qquad (8.23)$$

Now

$$V^2 = u^2 + v^2 = (V_\infty + u')^2 + v'^2$$
$$= V_\infty^2 + 2u'V_\infty + u'^2 + v'^2$$
$$= V_\infty^2\left[1 + 2\left(\frac{u'}{V_\infty}\right) + \left(\frac{u'}{V_\infty}\right)^2 + \left(\frac{v'}{V_\infty}\right)^2\right]. \qquad (8.24)$$

By virtue of Eqs. (8.21a) and (8.21b), the last two terms are very small compared to the first two, so*

$$\oint p\,dx \approx -\rho\oint u'V_\infty\,dx. \qquad (8.25)$$

Next consider the momentum flux integral. If we note that the term $\rho(\vec{V}\cdot\hat{n})\,ds$ represents mass flow across the control surface, we can write it as (See Fig. 8.28b)

$$\rho(\vec{V}\cdot\hat{n})\,ds = \rho(u\,dy - v\,dx). \qquad (8.26)$$

Note that dy and dx have either positive or negative signs depending on whether we are moving in the direction of increase or decrease of that coordinate. If we use Eq. (8.26), the momentum flux integral is

$$\dot{M}_y = \oint \rho v(\vec{V}\cdot\hat{n})\,ds = \oint \rho v(u\,dy - v\,dx).$$

Substituting Eqs. (8.20a) and (8.20b), we have

$$\dot{M}_y = \oint \rho v'[(V_\infty + u')\,dy - v'\,dx]$$
$$= \oint \rho v'V_\infty\,dy + \oint \rho v'u'\,dy - \oint \rho v'^2\,dx.$$

By virtue of Eqs. (8.21a) and (8.21b), the second and third integrals are very much smaller than the first, so

$$\dot{M}_y \approx \oint \rho v'V_\infty\,dy. \qquad (8.27)$$

This approximation can be made as exact as we wish by selecting the outer control surface far from the airfoil.

We now substitute the momentum flux and pressure integrals into Eq. (8.22) to get

$$F_y \approx \rho V_\infty \oint (u'\,dx + v'\,dy). \qquad (8.28)$$

Since V_∞ is a constant,

$$\oint V_\infty\,dx = 0. \qquad (8.29)$$

Adding Eqs. (8.28) and (8.29), we have

$$F_y \approx \rho V_\infty \oint [(V_\infty + u')\,dx + v'\,dy]$$
$$= \rho V_\infty \oint (u\,dx + v\,dy).$$

---

* The integrals of p_∞ and $\gamma(z_\infty - z)$ around a closed curve are zero.

We now define the path length vector

$$d\vec{s} = dx\hat{i} + dy\hat{j},$$

so

$$\vec{V} \cdot d\vec{s} = (u\hat{i} + v\hat{j}) \cdot (dx\hat{i} + dy\hat{j}) = u\,dx + v\,dy.$$

Using Eq. (8.19), we obtain the following equation for the lift force:

$$\mathscr{L} \approx -\rho V_\infty \oint \vec{V} \cdot d\vec{s}. \tag{8.30}$$

The integral is called the *circulation* and is assigned the symbol Γ:

$$\Gamma \equiv \oint \vec{V} \cdot d\vec{s}.$$

It can be shown that in an inviscid fluid, the circulation is the same for *any* closed path that encloses the airfoil. If we evaluate the circulation on a path that is very far from the airfoil, Eq. (8.30) becomes exact; however, since the circulation is the same for any other path, the equation is exact for any other path as well. We can therefore write

$$\mathscr{L} = -\rho V_\infty \Gamma = -\rho V_\infty \oint \vec{V} \cdot d\vec{s}. \tag{8.31}$$

Equation (8.31) is the Kutta-Joukowsky theorem.

The Kutta-Joukowsky theorem shows that adding a free vortex flow with a clockwise circulation to transform the unrealistic streamline pattern of Fig. 8.24(a) into the realistic streamline pattern of Fig. 8.24(b) causes a positive lift to be developed. Note that we would not want to add a forced vortex to the flow since that would result in an infinite velocity far from the airfoil. The theorem also shows that it is necessary for the flow to move faster over the top of the airfoil (where $\vec{V} \cdot d\vec{s}$ is negative) than over the bottom of the airfoil (where $\vec{V} \cdot d\vec{s}$ is positive) to produce a positive (upward) lift force.

The Kutta-Joukowsky theorem shows that it is not possible for a control volume analysis alone to predict lift (or drag) force. To actually calculate lift, the circulation must be known. To calculate the circulation, the velocity field in the vicinity of the airfoil must be known. This velocity field can be calculated only from a differential analysis.

8.3.3 Lift on Spinning Cylinders and Spheres: The Magnus Effect

The Kutta-Joukowsky theorem indicates that lift will be produced in any flow in which there is circulation or flow asymmetry. In airfoil flows, the asymmetry is produced by airfoil geometry and/or angle of attack. It is possible to produce lift on a completely symmetrical body like a cylinder or sphere by spinning the body. Figure 8.29(a) illustrates the ideal streamline pattern for a stationary two-dimensional cylinder in a flow stream. If the cylinder is caused to spin with angular speed ω in a clockwise direction (Fig. 8.29b), the fluid at the cylinder surface will be forced to move at the surface velocity U ($U = R\omega$). The fluid near the surface will also have a clockwise velocity due to viscous effects.

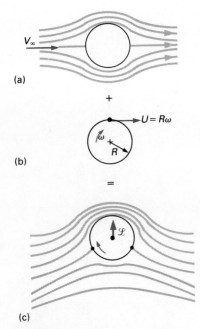

Figure 8.29 Asymmetric streamline pattern produced by spinning cylinder in a uniform stream.

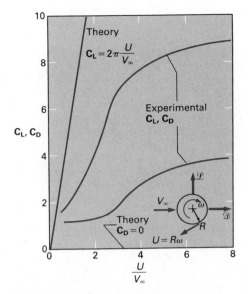

Figure 8.30 Lift and drag coefficients for spinning cylinders in uniform flow.

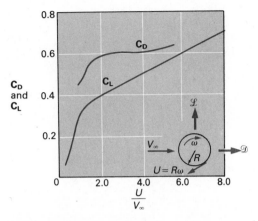

Figure 8.31 Lift and drag coefficients for spinning spheres in uniform flow.

The net result will be the establishment of a circulation around the cylinder. This flow can be modeled by a cylinder with a free vortex at the center (inside the cylinder) to produce the asymmetric streamline pattern shown in Fig. 8.29(c). Note that this is an inviscid flow model even though the actual circulation in the fluid is due to viscous action near the cylinder surface. The higher velocities on the top of the cylinder are associated with lower pressure, so a net lift force is generated. A theoretical analysis of the flow pattern of Fig. 8.29(c) gives the lift coefficient as

$$\mathbf{C_L} = 2\pi\left(\frac{U}{V_\infty}\right). \tag{8.32}$$

The ratio of surface speed to approach flow velocity, U/V_∞, is analogous to the angle of attack for airfoils. The inviscid flow model that leads to Eq. (8.32) is, of course, incapable of predicting drag.

This is fine theory, but does it work? Figure 8.30 shows measured lift and drag coefficients for a spinning cylinder. Although the high-lift (Eq. 8.32), zero-drag prediction is quantitatively incorrect, there is considerable lift generated on the spinning cylinder. It is interesting to note that roughening the surface of the cylinder produces *more* lift since it makes the surface more effective at causing circulation.

The phenomenon of lift generation by spinning bodies is called the *Magnus effect*. Spinning spheres are also capable of generating significant lift, as shown in Fig. 8.31. In engineering, the Magnus effect is mainly a curiosity, although a few devices that rely on the Magnus effect have been proposed or actually built. Magnus effect devices are attractive in applications where the approach flow speed is very low because they have very large lift and drag coefficients. Two of the more

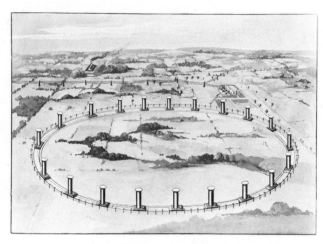

Figure 8.32 Artist's conception of Madaras rotor power plant.

interesting applications are the Flettner rotor ship and the Madaras wind energy plant.

The Flettner rotor ship, which was built in Germany in 1924, used rotating vertical cylinders to generate propulsive thrust from the wind. The ship employed two rotors 50 ft (15 m) high and 9 ft (3 m) in diameter that rotated at 750 rpm. The rotor ship never became popular because the era of cheap energy and efficiency improvements made engine/propeller propulsion systems more attractive.

A large plant for generating electric power from the wind was proposed by J. P. Madaras in 1927. The plant was to employ eighteen 90-ft (27-m) high by 23-ft (7-m) diameter vertical rotating cylinders, each mounted on a railroad car. The cars were to move around a closed 1500-ft (460-m) diameter track, propelled by the lift force generated by the wind (Fig. 8.32). Electric generators attached to the car wheels were to generate power. Several electric utility companies funded the development of the Madaras rotor plant concept, but it was never built. More recent studies [4] have indicated that in a high-energy-cost economy, commercial utility–size Madaras plants, with a power-generating capacity of over 230 megawatts are economically and technologically feasible.

8.3.4 Lift (and Drag) Information for Standard Airfoils

Theories based on the inviscid flow assumption and the Kutta-Joukowsky theorem are capable of reasonably accurate predictions of lift for thin airfoils at low angles of attack but are incapable of predicting airfoil drag or stall. In practice, lift and drag on airfoils are evaluated from experimental data. Several agencies, most notably the U.S. National Advisory Committee for Aeronautics (NACA) (the forerunner of the National Aeronautics and Space Administration, NASA), have produced a large number of standardized shapes of airfoils and have published information on their lift and drag coefficients established from wind tunnel tests. The amount of information is so extensive (see [5]) that only typical examples are presented here.

Figure 8.33 shows a typical airfoil shape. The size of the airfoil is measured by the *chord* (c), a straight line from the leading edge to the trailing edge. Angle of attack is measured with respect to the chord. The thickness of the airfoil is indicated by the maximum thickness (t). Airfoil thickness is usually expressed as a percentage of chord length. The *midline* of an airfoil is a line located halfway between the upper and lower surfaces. If the midline is curved, the airfoil is said to be *cambered*. Camber is measured by the distance h, usually expressed in percent chord. An airfoil is symmetric if the shapes above and below the camber line are identical. Cambered airfoils produce lift at zero angle of attack.

A (NACA) family of airfoil shapes is determined by specifying a thickness distribution and a midline shape. Various members of the family of airfoils are obtained by varying the camber, the location of the maximum camber, and the maximum thickness. For example, the NACA four-digit series of airfoils has a thickness distribution given by

$$\frac{y_t}{c} = \pm 5\left(\frac{t}{c}\right)\left[0.3\sqrt{(x/c)} - 0.13\left(\frac{x}{c}\right) - 0.35\left(\frac{x}{c}\right)^2 + 0.28\left(\frac{x}{c}\right)^3 - 0.28\left(\frac{x}{c}\right)^4\right].$$

The midline is given by

$$\frac{y_m}{c} = \frac{(m/c)}{1-(P/c)^2}\left[2\left(\frac{Px}{c^2}\right) - \left(\frac{x}{c}\right)^2\right], \qquad x < P,$$

and

$$\frac{y_m}{c} = \frac{(m/c)}{1-(P/c)^2}\left[1 - 2\left(\frac{P}{c}\right) + 2\left(\frac{Px}{c^2}\right) - \left(\frac{x}{c}\right)^2\right], \qquad x > P,$$

where x and y are measured along and are perpendicular to the chord line and P and m are the position and amount of the maximum deviation of the meanline from the chord. A typical member of this family, say a 2415 profile, has the following characteristics:

2	4	15
Maximum camber is 2% of chord.	Maximum camber occurs at 40% of chord.	Maximum thickness is 15% of chord.

In addition to the four-digit series, NACA developed other families of airfoil shapes. A family of particular interest is the 6-series, which was designed to produce laminar boundary layers over a large portion of the surface, resulting in low friction drag. This series also shows good performance at high Mach numbers. The designations of 6-series airfoil shapes are rather complicated, a typical designation being 63_2-615. The interpretation of these numbers will not be given here; see [5] if you are interested.

Performance information for standard airfoils is given as plots of lift and drag coefficients versus angle of attack. Figure 8.34 illustrates such plots for NACA 2415 and 63_2-615 airfoils as well as the shapes of the profiles themselves. Lift and drag coefficients are based on planform area (chord times length), not frontal area. Note that these nonsymmetrical airfoils produce lift at zero angle of attack. Note also that the slope of the lift curves ($dC_L/d\alpha$) is nearly 2π, except near and after stall.

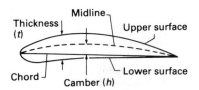

Figure 8.33 Typical airfoil shape.

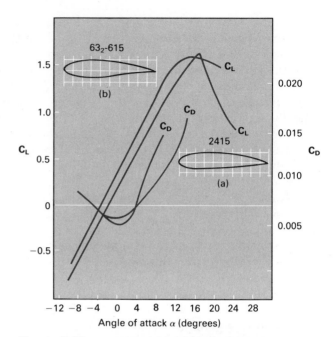

Figure 8.34 Lift and drag coefficients for two NACA airfoil shapes: (a) 2415 profile; (b) 63_2-615 profile. (Adapted from [5].)

The plots indicate that C_L and C_D are nearly independent of Reynolds number. This is quite true for C_L and approximately true for C_D, as long as the Reynolds number (based on airfoil chord) is high. The drag coefficient is sensitive to surface roughness and can increase up to 300% of its smooth surface value (the value indicated in Fig. 8.34) if the surface is very rough. This behavior is similar to that of the flat plate parallel to the flow stream, which should come as no surprise since an airfoil is a thin, elongated body.

EXAMPLE 8.7 Illustrates airfoil lift

A race car uses an inverted NACA 2415 airfoil to provide downthrust on its rear wheels (see Fig. E8.7). The airfoil has two end plates, so the flow may be assumed two-dimensional. The airfoil is 6 ft long and has a chord of 1 ft. Find the angle of attack to produce a downward thrust of 500 lb at a car speed of 170 mph.

Solution

   Given
Inverted NACA 2415 airfoil measuring 6 ft long with 1-ft chord
Two-dimensional flow

   Find
Angle of attack to produce a downward thrust of 500 lb at a speed of 170 mph

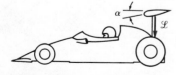

Figure E8.7 Inverted airfoil on race car to provide downthrust.

Solution

Equation (8.4) relates the downward thrust (or lift) $\mathscr{L}$ and the lift coefficient $\mathbf{C_L}$:

$$\mathscr{L} = \frac{1}{2}\mathbf{C_L}\rho V_\infty^2 S,$$

when S is the planform area

$$S = (1\text{ ft} \times 6\text{ ft}) = 6\text{ ft}^2.$$

Solving for the lift coefficient gives

$$\mathbf{C_L} = \frac{2\mathscr{L}}{\rho V_\infty^2 S}.$$

Assuming 60°F air, Table A.4 gives $\rho = 0.077$ lbm/ft^3. The numerical values give

$$\mathbf{C_L} = \frac{2(500\text{ lb})\left(\dfrac{32.2\text{ ft}\cdot\text{lbm}}{\text{lb}\cdot\text{sec}^2}\right)}{\left(0.077\dfrac{\text{lbm}}{\text{ft}^3}\right)\left(170\dfrac{\text{mi}}{\text{hr}}\right)^2\left(\dfrac{5280\text{ ft}}{\text{mi}}\cdot\dfrac{\text{hr}}{3600\text{ sec}}\right)^2(6\text{ ft}^2)} = 1.12.$$

Figure 8.35 shows the angle of attack α must be **Answer**

$$|\ \alpha = 9°.\ |$$

Discussion

Note that there is a second solution to this problem: $\alpha \approx 19.5°$. The smaller angle is preferable as it provides lower drag and better fuel economy. Any angle between 9° and 19.5° would produce a lift greater than 500 lb.

8.3.5 Practical Considerations: Lift Augmentation Devices and Three-Dimensional Effects

Consider the processes of takeoff and landing of an airplane. For takeoff, the wing lift must exceed the weight. In descent and landing, the weight exceeds the lift, but the difference between them must be small so that the airplane does not descend too rapidly. The lift *force* varies with the square of the velocity, which is typically low at takeoff and landing. To provide the required lift at low speeds, it is necessary to increase the lift coefficient. One way to increase the lift coefficient is to increase the angle of attack. During landing, this can be accomplished by raising the nose of the airplane and lowering the tail. This works only up to a point; if the angle of attack gets too high, the wing will stall with sudden loss of lift—usually resulting in a crash. To keep takeoff and landing speeds reasonable and avoid catastrophic stall, it is necessary to fit an airplane wing with lift augmentation devices to increase the lift coefficient at low speeds. The most common lift augmentation devices are trailing edge flaps and leading edge slats (Fig. 8.35). Flaps increase the effective angle of attack, increasing lift coefficient. Leading edge slats produce a fast-moving jet of air over the upper surface, giving a low pressure and at the same time energizing the boundary layer and delaying stall. These lift augmentation devices also produce high drag and so are used only during takeoff and landing.

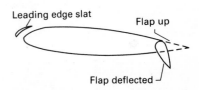

Figure 8.35 Airfoil with leading edge slat and trailing edge flap.

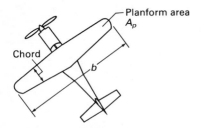

Figure 8.36 Wing of finite span, viewed from above.

Our previous discussion of lift has assumed two-dimensional, two-directional flow over a two-dimensional airfoil. Real wings have finite span and are three-dimensional bodies. Figure 8.36 shows a wing with span b. An important parameter of such a wing is the *aspect ratio, AR*, defined by

$$AR = \frac{b^2}{A_p},$$

where A_p is the planform area. For a nontapered wing (constant chord along the span) without sweepback, the aspect ratio reduces to the ratio of wing span to chord length.

Finite-span wings have a larger drag coefficient and a smaller lift coefficient than infinite-span wings (such as airfoils). This fact can be explained by considering the generation of tip vortices at the wingtips. Figure 8.37 is a sketch of the flow pattern near a wingtip. Since the pressure on the lower side of the wing is higher than the pressure on the upper side, the air near the wingtip flows from the lower surface to the upper surface, setting up a vortex. This tip vortex trails downstream, far behind the aircraft.* The tip vortexes (one from each end) cause the fluid behind the wing to have an induced downward velocity, called the *downwash velocity,*[†] w. The downwash causes the effective angle of attack of the wing sections to be lower. Referring to Fig. 8.38, we can see that this has two effects. Since lift is perpendicular to the effective angle of attack, the lift vector is rotated backward. The lift vector now has a component in the (true) downstream direction that adds to the drag of the wing. This is called *induced drag*. The second

Figure 8.37 Detail of flow near the tip of a finite wing.

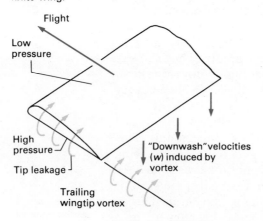

Figure 8.38 Downwash velocity reduces effective angle of attack and rotates lift force backward from true direction of travel.

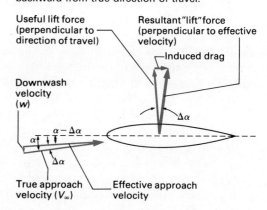

* Light planes must avoid the tip vortexes of large commercial airliners, which may extend 10 mi downstream of the aircraft. Theoretically, the tip vortexes extend all the way back to the takeoff point, where they join up with the starting vortex that was shed at takeoff. See Fig. 8.26. The imaginary bound vortex in the wing, the tip vortexes, and the starting vortex then form one giant closed loop.

[†] Note that there is an induced *upwash* velocity outboard of the wingtips. It has been suggested that migrating birds fly in a V formation so that the downstream birds can take advantage of the upwash of the upstream birds [6].

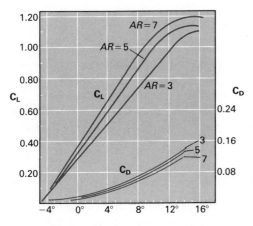

Figure 8.39 Effect of aspect ratio on lift and drag coefficient of finite wing. All wings have the same (airfoil) profile shape.

effect of the lowered effective angle of attack is lower lift (the lower lift can also be understood by noting that the average pressure on the lower wing surface is smaller due to the tip leakage). A theory due to Prandtl shows that if the shape of the spanwise lift distribution curve is elliptical (a reasonably good approximation for nonswept wings), the downwash velocity is

$$w = \frac{2\mathscr{L}}{\pi \rho V_\infty b^2}, \tag{8.33}$$

the reduction in effective angle of attack is

$$\Delta \alpha \approx -\frac{w}{V_\infty} = -\frac{2\mathscr{L}}{\pi \rho V_\infty^2 b^2} = -\frac{\mathbf{C_L}}{\pi AR}, \tag{8.34}$$

and the induced drag is

$$\mathscr{D}_{\text{ind}} = \frac{2\mathscr{L}^2}{\pi \rho V_\infty^2 b^2}. \tag{8.35}$$

The induced drag coefficient is

$$\mathbf{C_{D,ind}} = \frac{\mathscr{D}_{\text{ind}}}{\frac{1}{2}\rho V_\infty^2 A_p} = \frac{\mathbf{C_L^2}}{\pi AR}. \tag{8.36}$$

Problems of induced drag and lower effective angle of attack are especially critical during takeoff and landing because the lift coefficient is especially large. These effects can be reduced by increasing the aspect ratio of the wing. Figure 8.39 shows the effect of aspect ratio on the lift and drag of a finite wing. Wings that produce high lift and low drag must have large aspect ratio. The wing of a modern sailplane (glider) may have $AR \approx 40$. Birds that fly long distances such as seagulls have wings of large aspect ratio. Modern sport sailboats normally have tall, narrow sails.

In some applications, it is practical to fit endplates on a lifting surface to suppress formation of the tip vortex. Notice the endplates on the cylinders of the Madaras wind plant shown in Fig. 8.32.

8.4   Nonaviation Applications of Aerodynamics

The practical applications of aerodynamics are not limited to aircraft. Whenever an object moves through air or water or when wind or water flows over a stationary object, aerodynamic forces are generated. These aerodynamic forces are often of considerable importance in the operation, stability, and durability of engineering devices and systems. In what follows, we will consider three nonaviation applications of aerodynamics.

8.4.1   Road Vehicle Aerodynamics

Aerodynamic considerations are important in the design of road vehicles such as trucks and automobiles. The most important aerodynamic force on a road vehicle is drag. Consider a vehicle traveling at constant speed on a level road. The vehicle experiences two forces that resist its motion: rolling resistance and aerodynamic drag. *Rolling resistance* is due to the continuous deformation of the vehicle's tires. The sum of the aerodynamic drag and rolling resistance is called the *road load*. The engine must continuously supply power to overcome the road load. This power is equal to the product of the road load force and the vehicle velocity. Figure 8.40 shows the variation of road load power with vehicle velocity for a typical automobile. Rolling resistance power is nearly linear with velocity, while aerodynamic drag power varies with the cube of the velocity (the drag coefficient is approximately constant). The rolling resistance and aerodynamic drag curves intersect (each contributes equally to road load) at a speed between about 50 mph (80 km/hr) and 60 mph (96 km/hr). Above this speed, the power required to overcome aerodynamic drag increases rapidly and becomes the controlling factor in the vehicle's speed. Fuel economy depends strongly on the road load power curve; if the magnitude of road load can be reduced, fuel economy is improved. Since most road vehicles normally operate in the speed range from 35 mph to 70 mph, significant reductions in road load power can be realized by reducing either rolling

Figure 8.40 Variation of road load power and its components with vehicle velocity.

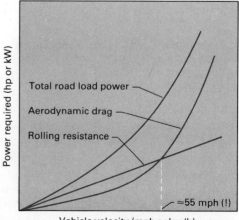

Total road load power

Aerodynamic drag

Rolling resistance

Power required (hp or kW)

≈55 mph (!)

Vehicle velocity (mph or km/h)

Figure 8.41 Semitrailer with an air deflector to reduce drag.

resistance or aerodynamic drag. The optimum design must address both road load components [7].

Since road vehicles operate at high Reynolds numbers (a typical five-passenger sedan traveling at 55 mph has a Reynolds number of about 10^7, based on length) and are rather blunt bodies, most of the drag on them is form drag. The primary method of reducing form drag is well known: streamlining. A second method, obvious but often overlooked by the beginner, is to reduce the size (frontal area) of the vehicle. Early (1920 vintage) automobiles had high profiles (large frontal area per unit car volume) and boxy shapes that produced drag coefficients of about 0.9–1.0. Modern automobiles have a much more streamlined shape with a lower profile. Current production cars have drag coefficients in the range of 0.45–0.65. The ultimate streamlined car would have a nearly circular cross section, a rounded blunt nose, and a long, pointed tail (like a teardrop shape) and might be expected to yield a value of C_D as low as 0.05–0.10; however, such a car would probably be unacceptable from cost, load-carrying volume, and aesthetic viewpoints.

Drag reduction for automobiles is complicated by problems introduced by underbody drag and interference drag. The underbody of a road vehicle is aerodynamically very complicated due to protruding surfaces such as oil pan, muffler, and suspension parts. Each of these surfaces experiences a relatively large drag force. Interference drag is caused by protrusions from the vehicle surface such as rearview mirrors, door handles, and luggage racks. These objects not only experience drag on themselves but, equally important, they disturb the flow over the basic body shape, influencing its drag as well. The concept of interference drag refers to the fact that the total drag due to a surface protrusion is usually larger than the sum of the drags of the basic body in isolation and the protrusion in isolation.

Aerodynamic drag on trucks and buses presents its own set of problems. Legal regulations set limits on the length, width, and height of these commercial vehicles. Truck and bus operators normally wish to maximize the quantity of goods or number of people that can be carried per trip, so these vehicles are designed with a basically rectangular shape. This limits the amount of drag reduction that can be realized; however, attention to a few details such as rounding of corners on forward portions of the vehicle (such as the front of a semitrailer or bus) can substantially reduce drag with little sacrifice in interior volume. In recent years, cab rooftop air deflectors have been installed on many semitractors to improve the flow pattern over the front of the trailer. See Fig. 8.41.

Although drag reduction is probably the most important aerodynamic problem in road vehicle design, it is by no means the only area

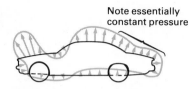

Figure 8.42 Typical pressure distribution on symmetry plane of a sedan: Pressure curve lying outside the contour indicates negative (suction) pressure; pressure curve lying inside the contour indicates positive pressure. Pressures are gage pressure.

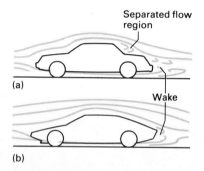

Figure 8.43 Streamline patterns for (a) sedan and (b) "fastback" automobiles.

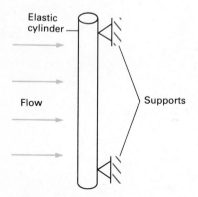

Figure 8.44 Flow perpendicular to axis of long, flexible elastic cylinder.

of concern. Other aerodynamic considerations in road vehicles include the following:

- Secondary aerodynamic forces such as lift and side force
- Provision of air flow to the interior of the vehicle for engine cooling and passenger compartment ventilation
- Provision of an acceptable flow pattern near the car surface to minimize dirt deposits on the vehicle and transport exhaust gases away from the vehicle
- Aerodynamically generated noise

Figures 8.42 and 8.43 show the pressure distribution and streamline pattern for a typical automobile. Note that the lower surface pressure is considerably larger (smaller negative value) than the upper surface pressure, resulting in a lift force. Upward lift is an undesirable phenomenon because it reduces traction and can adversely affect handling characteristics. Problems caused by lift forces are seldom significant for ordinary passenger and commercial vehicles but can be important for high-speed race cars. Sometimes race cars are fitted with "negative lift" devices (basically upside-down wings) to improve traction (Fig. E8.7).

Air flow into the vehicle usually occurs at two locations. Engine cooling and combustion air usually enter the front of the vehicle. The pressure drop of the air as it flows through the radiator contributes to the vehicle's drag. Passenger ventilation air is normally taken in at the high-pressure, low-velocity zone at the base of the windshield.

Transport of dirt and other particles by the air flow can result in deposits on vehicle windows, reducing visibility. This problem is especially severe in station wagons where the recirculating flow behind the vehicle picks up dirt or water from the road and deposits it on the rear windows. Station wagons are sometimes equipped with deflecting vanes to jet air from the top downward over the back window to keep it clean.

Aerodynamic considerations are an important part of road vehicle design, although, unlike in the case of aircraft, aerodynamics is not the primary consideration. References [8, 9, 10] provide excellent insight into this field.

8.4.2 Flow-Induced Vibration

We now consider a situation in which there is a dynamic interaction between the flow over an object and the object itself. Many practical applications involve flow perpendicular to the axis of a long flexible cylinder (Fig. 8.44). Important examples include wind blowing over power poles, power lines, suspension bridge cables, and smokestacks, water flow over bridge pilings, and liquid or gas flow over banks of tubes in a heat exchanger or boiler. Because of the cylinder's flexibility, the aerodynamic load causes a deflection (bending) of the cylinder, which is balanced by an elastic restoring force. If the aerodynamic load is removed, the cylinder will spring back to its original shape. If the aerodynamic load fluctuates, the cylinder will vibrate under the influence of the fluctuating load. The frequency and amplitude (maximum deflection) of the vibration depends on the frequency and magnitude of the fluctuating load and the elastic and support characteristics of the cylinder. All elastic structures possess one or more *natural frequencies,* which are the frequencies at which the structure would vibrate under the influence of its mass and elastic force conditions alone. If the structure is subjected to a fluctuating load whose frequency is equal to (or near) one of the natural frequencies, the amplitude of vibration becomes very large, often resulting in destruction of the structure.* Catastrophic failures have resulted from this type of phenomenon, the most dramatic being the failure of the Tacoma Narrows suspension bridge just a few weeks after its opening in 1940 [11]. Less drastic problems associated with flow-induced vibration include gradual fatigue of heat exchanger tubes and wear of tubes as they rub against their supports.

To understand some of the causes of flow-induced vibration, first consider the details of flow over an isolated circular cylinder. Experiments show that at Reynolds numbers greater than about 50, vortexes are shed on the downstream side of the cylinder. These vortexes are shed alternately from the top and the bottom of the cylinder with a definite frequency. The vortexes trail behind the cylinder in two rows (Fig. 8.45), called a *Karman vortex street.* The oscillating streamline pattern caused by the alternate vortex shedding causes a fluctuating pressure force on the cylinder and hence a time-varying load. The frequency of the fluctuating force is equal to the frequency of the vortex shedding. Dimensional analysis shows that the frequency of vortex shedding is governed by a relation of the form

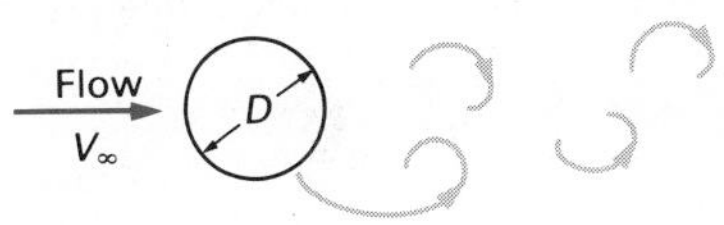

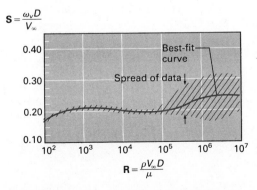

Figure 8.45 Vortex street shed from a circular cylinder.

$$\mathbf{S} = f(\mathbf{R}),\qquad (8.37)$$

where

$$\mathbf{S} = \frac{\omega_v D}{V_\infty},\qquad \text{Strouhal number of the vortex pattern}^\dagger$$

$$\mathbf{R} = \frac{V_\infty D}{v},\quad\text{and}\quad \omega_v = \text{Frequency of vortex shedding in Hz.}$$

The functional relation implied by Eq. (8.37) has been the subject of many experimental investigations. Figure 8.46 represents its typical

$\mathbf{S} = \frac{\omega_v D}{V_\infty}$

Figure 8.46 Strouhal number versus Reynolds number for vortex shedding from a single circular cylinder.

*This concept can be illustrated by the action of pushing a child on a swing. If the swing is pushed each time it instantaneously comes to a stop at the high point of its travel, it will keep going higher and higher. This once-per-cycle push is exactly in time with the "natural frequency" of the swing.

†Note the appearance of Strouhal number as a dependent variable.

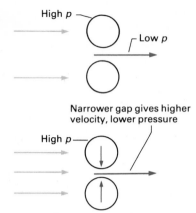

Figure 8.47 Flow in the gap between two circular cylinders. Narrowing the gap causes increased velocity and lower pressure and tends to narrow the gap further.

behavior. A structural engineer must be sure that the natural frequency of the structure is not too close to the vortex-shedding frequency. This type of vortex shedding is responsible for wires "singing" in the wind, a phenomenon extensively investigated by Strouhal.

If more than one cylinder is present in the flow, other mechanisms can induce vibration. Obviously, if a second circular cylinder were placed in the wake of the cylinder in Fig. 8.45, the shed vortexes would generate a fluctuating flow over the second cylinder. In a bank of tubes, the Karman vortex pattern is rapidly broken up into turbulence. Large amounts of turbulence in the flow cause fluctuation and can lead to flow-induced vibration.

A third mechanism for generation of flow-induced vibration is *hydroelastic instability*. To understand this phenomenon, consider flow between two circular cylinders as shown in Fig. 8.47. Suppose that due to some unknown perturbation, the two cylinders are deflected in such a way that the gap between them decreases. The narrowing of the gap requires a higher fluid velocity in the gap by virtue of conservation of mass. According to Bernoulli's equation, the higher velocity in the gap is accompanied by a lower pressure. The lower pressure on the gap side causes the cylinders to deflect inward, narrowing the gap and further decreasing the pressure, which causes more deflection. Eventually, the elastic forces in the deflected cylinders act to restore the cylinders to their initial position; however, if the cylinders overshoot their equilibrium position, the process of gap narrowing will start on the other side. At some flow velocities, it is possible that the frequency of vibration induced by the process matches the natural frequency of the cylinders, and the resulting vibration can become quite large.

A great deal of research is currently being conducted in the field of flow-induced vibrations. Consult [12] for further details.

EXAMPLE 8.8 Illustrates vortex shedding and its relation to a structural natural frequency

The 150-ft-high sign pole in Example 8.6 is a steel pipe with an inside diameter $d = 11.94$ in. and an outside diameter $D = 12.75$ in. Its natural frequency is

$$\omega = \frac{\pi^2}{4} \sqrt{\frac{EI}{m'\ell^4}} \, ,$$

where E is the pipe's modulus of elasticity, m' its mass per unit length, I its second moment of area, and ℓ its length; $E = 30 \times 10^6$ psi for steel and $I = \pi(D^4 - d^4)/64$. Find the critical wind velocity for vortex shedding if the wind velocity is uniform along the pole. Steel density is 490 lbm/ft^3.

Solution

Given
Pole in Fig. E8.8
Pole inside diameter $d = 11.94$ in., outside diameter $D = 12.75$ in.

Natural frequency $\omega = \dfrac{\pi^2}{4} \sqrt{\dfrac{EI}{m'\ell^4}}$

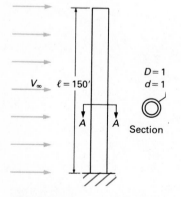

Figure E8.8 Given information for Example 8.8.

Find

Wind velocity at which the pole vibrates at its resonant frequency due to vortex shedding

Solution

Figure 8.46 relates the Strouhal number and the Reynolds number. Assuming 60°F air, Table A.4 gives $v = 1.7 \times 10^{-4}$ ft²/sec, and the Reynolds number is

$$\mathbf{R} = \frac{V_\infty D}{v} = \frac{V_\infty(12.75 \text{ in.})\left(\dfrac{\text{ft}}{12 \text{ in.}}\right)}{\left(1.7 \times 10^{-4} \dfrac{\text{ft}^2}{\text{sec}}\right)} = \left(6250 \frac{\text{sec}}{\text{ft}}\right) V_\infty.$$

The second moment of area of the pipe is

$$I = \frac{\pi}{64}(D^4 - d^4) = \frac{\pi}{4}[(12.75 \text{ in.})^4 - (11.94 \text{ in.})^4] = 4793 \text{ in}^4.$$

The mass per unit length of the pipe is

$$m' = \rho A = \rho \frac{\pi}{4}(D^2 - d^2)$$

$$= \frac{\left(490 \dfrac{\text{lbm}}{\text{ft}^3}\right)\dfrac{\pi}{4}(12.75 \text{ in}^2 - 11.94 \text{ in}^2)}{\left(\dfrac{144 \text{ in}^2}{\text{ft}^2}\right)} = 53.4 \text{ lbm/ft}.$$

The pipe's natural frequency is

$$\omega = \frac{\pi^2}{4}\sqrt{\frac{\left(30 \times 10^6 \dfrac{\text{lb}}{\text{in}^2}\right)(4793 \text{ in}^4)\left(\dfrac{32.2 \text{ ft} \cdot \text{lbm}}{\text{lb} \cdot \text{sec}^2}\right)}{\left(53.4 \dfrac{\text{lbm}}{\text{ft}}\right)(150 \text{ ft})^4\left(\dfrac{144 \text{ in}^2}{\text{ft}^2}\right)}} = 2.69 \text{ rad/sec}.$$

For resonance to occur,

$$\omega_v = \omega \quad \text{and} \quad \mathbf{S} = \frac{\omega D}{V_\infty} = \frac{\left(2.69 \dfrac{\text{rad}}{\text{sec}}\right)(12.75 \text{ in.})\left(\dfrac{\text{ft}}{12 \text{ in.}}\right)}{V_\infty} = \frac{\left(2.86 \dfrac{\text{ft}}{\text{sec}}\right)}{V_\infty}.$$

The wind velocity V_∞ is an unknown in both the Reynolds number and the Strouhal number and must be found by trial and error. Assuming a Strouhal number of 0.20 gives

$$V_\infty = \frac{\left(2.86 \dfrac{\text{ft}}{\text{sec}}\right)}{0.20} = 14.3 \text{ ft/sec}.$$

The Reynolds number for this wind velocity is

$$\mathbf{R} = \left(6250 \frac{\text{sec}}{\text{ft}}\right)\left(14.3 \frac{\text{ft}}{\text{sec}}\right) = 8.94 \times 10^4.$$

Figure 8.46 shows that the Strouhal number equals 0.20 for a Reynolds number of 8.94 × 10⁴, so our initial guess of the Strouhal number was correct. Therefore

$$V_\infty = 14.3 \frac{\text{ft}}{\text{sec}} = \left(14.3 \frac{\text{ft}}{\text{sec}}\right)\left(\frac{\text{mi}}{5280 \text{ ft}}\right)\left(\frac{3600 \text{ sec}}{\text{hr}}\right),$$

Answer ◀

$$| \ V_\infty = 9.75 \text{ mph}. \ |$$

Discussion

In a problem of this nature, convergence is usually faster by assuming a value for the Strouhal number rather than a value for the Reynolds number because the Strouhal number is relatively constant over a wide range of Reynolds numbers.

8.4.3 Aerodynamics in Sports

The science of aerodynamics is not all work and no play. Various flow phenomena play an important part in many sports. Although seldom coming under the heading of engineering, application of aerodynamic principles helps us understand some observed phenomena in the sports world. The application of aerodynamics to sports such as hang gliding, kite flying, and sailing is rather obvious. Aerodynamics has always played an important role in auto racing, and race cars have always been more aerodynamically styled than ordinary cars.

Golf and baseball provide several interesting applications of aerodynamic principles. Did you ever wonder why a golf ball has dimples? A golf ball in flight has a Reynolds number of about 10^5. Check the drag coefficient of a smooth sphere at this Reynolds number (Fig. 8.10b). The dimples on a golf ball promote transition to a turbulent boundary layer on the ball, delaying separation and lowering the drag coefficient. "Carefully controlled experiments" (i.e., people hitting several drives with dimpled and smooth golf balls) indicate that a smooth ball can be driven only about 60% as far as a dimpled ball [13].

A well-hit golf ball has underspin in addition to its linear motion. This spin, together with the dimpled surface, is responsible for lift on the ball (Fig. 8.31), which helps it carry farther. Figure 8.48 illustrates various effects on golf ball trajectories. Spin-generated "lift" (in this case, side force) is also responsible for hooking and slicing of balls that are not squarely hit. Ping-Pong and tennis players also use spin to control the trajectory of the ball.

Baseball pitchers are among the most clever practical aerodynamicists. A baseball is a sphere about 3 in. (8 cm) in diameter; however, the stitches on the ball give it both roughness and a certain asymmetry. By varying the spin on the ball, a pitcher can throw a wide variety of pitches. A fastball has a spin that generates vertical lift, allowing the

Figure 8.48 Various golf ball trajectories.

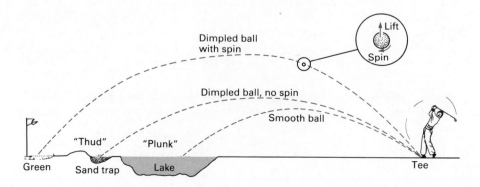

ball to follow a flatter trajectory without drop. A curveball is the result of spin that causes a side force. A knuckleball pitcher throws the ball with very little spin and at a relatively slow speed. The irregular pattern of the stitches causes irregular transition back and forth between laminar and turbulent boundary layers. The fluctuating forces thus generated cause the ball to wobble on its path to the plate, confusing the batter. The illegal spitball or "greaseball" allows the same effect at a faster speed. The batter usually prefers a fastball because if he is able to hit the ball, the lift-generating spin will help the ball travel farther. For the batter's sake, we hope that there is no wind blowing in from the outfield!

EXAMPLE 8.9   Illustrates applications of aerodynamics in sports

A pitcher releases a baseball at a velocity of 100 mph parallel to the ground. The ball has an underspin of 40 rev/sec in a vertical plane, as shown in Fig. E8.9(a). Estimate the y-coordinate of the ball when it reaches home plate (54 ft away).

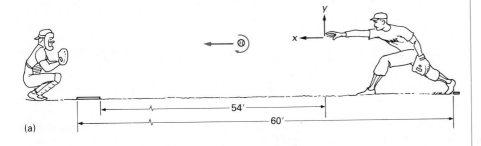

(a)

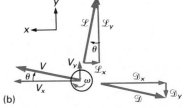

(b)

Figure E8.9 (a) Fastball with underspin; (b) free-body diagram of pitched baseball with underspin.

Solution

Given

Baseball thrown at 100 mph with an underspin of 40 rev/sec in a vertical plane
Figure E8.9(a)

Find

The y-coordinate of the ball after traveling 54 ft horizontally

Solution

Figure E8.9(a) shows the coordinate system; the origin is at the point where the ball is released by the pitcher at time $t = 0$, the x-axis is horizontal and directed toward the catcher, and the y-axis is directed vertically upward. The gravitational force on the ball will cause the ball to fall (move in the negative y-direction), while the lift due to the underspin will cause the ball to rise (move in the positive y-direction). The ball will remain in the x,y-plane. Although the ball is released parallel to the ground (or x-axis), its trajectory may not be parallel to the x-axis. The angle θ (Fig. E8.9b) measures the difference between the ball's trajectory and the x-axis.

We start with a free-body diagram of the ball as shown in Fig. E8.9(b). Applying Newton's second law in both the x- and y-directions gives

$$\sum F_x = m\left(\frac{dV_x}{dt}\right) \quad \text{and} \quad \sum F_y = m\left(\frac{dV_y}{dt}\right).$$

Substituting for the forces gives

$$-\mathcal{L}_x - \mathcal{D}_x = m\left(\frac{dV_x}{dt}\right) \quad \text{and} \quad -mg + \mathcal{L}_y - \mathcal{D}_y = m\left(\frac{dV_y}{dt}\right).$$

To simplify, we assume that the trajectory of the ball is relatively flat* ($\theta \approx 0$), so $\mathcal{L}_x \approx 0$ and $\mathcal{D}_y \approx 0$ and thus $\mathcal{L}_y \approx \mathcal{L}$ and $\mathcal{D}_x \approx \mathcal{D}$. The lift and drag are given by Eqs. (8.4) and (8.5):

$$\mathcal{L} = \frac{1}{2}\rho C_L V^2 S \quad \text{and} \quad \mathcal{D} = \frac{1}{2}\rho C_D V^2 S.$$

The square of the velocity is

$$V^2 = V_x^2 + V_y^2 = V_x^2\left[1 + \left(\frac{V_y}{V_x}\right)^2\right].$$

The assumption of a relatively flat trajectory implies

$$\left(\frac{V_y}{V_x}\right)^2 \ll 1,$$

so $V^2 \approx V_x^2$. The approximate equations of motion for the ball are then

$$-\frac{1}{2}\rho C_D V_x^2 S = m\left(\frac{dV_x}{dt}\right) \quad \text{and} \quad -mg + \frac{1}{2}\rho C_L V_x^2 S = m\left(\frac{dV_y}{dt}\right).$$

We now assume that the lift and drag coefficients of the baseball are approximated by the values for a spinning sphere given in Fig. 8.31. Figure 8.31 shows that both C_D and C_L vary with the velocity ratio U/V_∞ whose value at the time of release, $t = 0$, is

$$\left(\frac{U}{V_\infty}\right)_0 = \left(\frac{R\omega}{V_x}\right)_0.$$

A telephone call to a local sporting goods store gave the circumference of a baseball to be 9 in. and the mass to be 5 oz. The radius is

$$R = \frac{9 \text{ in.}}{2\pi}\left(\frac{\text{ft}}{12 \text{ in.}}\right) = 0.119 \text{ ft} \quad \text{and}$$

$$\left(\frac{U}{V_\infty}\right)_0 = \frac{(0.119 \text{ ft})\left(40\dfrac{\text{rev}}{\text{sec}}\right)\left(\dfrac{2\pi \text{ rad}}{\text{rev}}\right)}{\left(100\dfrac{\text{mi}}{\text{hr}}\right)\left(\dfrac{5280 \text{ ft}}{\text{mi}}\right)\left(\dfrac{\text{hr}}{3600 \text{ sec}}\right)} = 0.204.$$

Figure 8.31 shows the ball to be in a range where C_D and C_L change significantly with U/V_∞. We will assume U/V_∞ is constant (to be checked later), so C_D and C_L are constant. With this assumption, both equations of motion can be integrated. Separating the variables in the x-direction equation gives

$$\int_{V_{x0}}^{V_x} \frac{dV_x}{V_x^2} = -\frac{\rho C_D S}{2m}\int_0^t dt,$$

where V_{x0} is the x-component of velocity at time $t = 0$. Integrating gives

$$\left(\frac{1}{V_x}\right)_{V_{x0}}^{V_x} = \frac{\rho C_D S}{2m}(t - 0), \quad \text{or} \quad \frac{1}{V_x} = \frac{1}{V_{x0}} + \frac{\rho C_D S t}{2m}.$$

---

* We will check this assumption later.

Rearranging gives

$$V_x = \frac{1}{\dfrac{1}{V_{x_0}} + \dfrac{\rho C_D S t}{2m}} = \frac{V_{x_0}}{1 + \beta t},$$

where

$$\beta = \frac{\rho C_D S V_{x_0}}{2m}.$$

Substituting into the y-direction equation gives

$$-mg + \frac{\rho C_L S}{2}\left(\frac{V_{x_0}}{1 + \beta t}\right)^2 = m\left(\frac{dV_y}{dt}\right).$$

Rearranging and integrating, we have

$$\int_0^{V_y} dV_y = \int_0^t \left[-g + \frac{\left(\dfrac{\rho C_L S V_{x_0}^2}{2m}\right)}{(1 + \beta t)^2} \right] dt,$$

where $V_y = 0$ at $t = 0$ since the ball is released parallel to the ground. Integrating, we get

$$V_y = -gt - \frac{C_L}{C_D}\left(\frac{V_{x_0}}{1 + \beta t}\right)$$

$$= -gt - \frac{C_L V_{x_0}}{C_D}\left(\frac{1}{1 + \beta t} - 1\right)$$

$$= -gt + \frac{C_L}{C_D}(V_{x_0} - V_x).$$

After substituting this equation into

$$\frac{dy}{dt} = V_y,$$

multiplying by dt, and integrating, we obtain

$$\int_0^y dy = \int_0^t \left[-gt + \frac{C_L}{C_D} V_{x_0}\left(1 - \frac{1}{1 + \beta t}\right)\right] dt,$$

where $y = 0$ at $t = 0$. Performing the integration, we have

$$y = \frac{-gt^2}{2} + \frac{C_L}{C_D}(V_{x_0}t) - \frac{2mC_L}{\rho C_D^2 S} \ln(1 + \beta t)$$

$$= \frac{-gt^2}{2} + \frac{C_L}{C_D}(V_{x_0}t) - \frac{2mC_L}{\rho C_D^2 S} \ln\left(\frac{V_{x_0}}{V_x}\right).$$

Assuming 60°F air, Table A.4 gives $\rho = 0.077$ lbm/ft^3. For $U/V_\infty = 0.204$, Fig. 8.31 gives[*] $C_D \approx 0.35$ and $C_L \approx 0.070$. The only remaining unknown quantity needed to calculate the y-coordinate is the time t. It is found by substituting the equation for V_x into

$$\frac{dx}{dt} = V_x.$$

[*] These values are found by extrapolating the curves using $C_L = 0$ and $C_D = 0.30$ at $U/V_\infty = 0$ (see Fig. 8.10 and note $v = 1.7 \times 10^{-4}$ ft^2/sec and $R = 2.0 \times 10^5$).

and integrating. Thus

$$\int_0^x dx = \int_0^t \frac{V_{x_0} dt}{1 + \beta t},$$

where $x = 0$ at $t = 0$. Carrying out the indicated operations gives

$$x = \frac{2m}{\rho C_D S} \ln (1 + \beta t).$$

Solving for t gives

$$t = \frac{1}{\beta} \left[\exp \left(\frac{\rho C_D S x}{2m} - 1 \right) \right].$$

The numerical values of the various parameters are

$$\frac{\rho C_D S}{2m} = \frac{\rho C_D \pi R^2}{2m}$$

$$= \frac{\left(0.077 \dfrac{\text{lbm}}{\text{ft}^3} \right) (0.35) \pi (0.119 \text{ ft})^2}{2 (5 \text{ oz}) \left(\dfrac{\text{lbm}}{16 \text{ oz}} \right)} = 0.0019/\text{ft},$$

$$V_{x_0} = \left(100 \frac{\text{mi}}{\text{hr}} \right) \left(\frac{5280 \text{ ft}}{\text{mi}} \right) \left(\frac{\text{hr}}{3600 \text{ sec}} \right) = 147 \text{ ft/sec},$$

$$\frac{1}{\beta} = \frac{2m}{\rho C_D S V_{x_0}} = \frac{1}{\left(\dfrac{0.0019}{\text{ft}} \right) \left(147 \dfrac{\text{ft}}{\text{sec}} \right)} = 3.58 \text{ sec},$$

and

$$\frac{2m C_L}{\rho C_D^2 S} = \left(\frac{2m}{\rho C_D S} \right) \left(\frac{C_L}{C_D} \right) = \frac{1}{\left(\dfrac{0.0019}{\text{ft}} \right)} \left(\frac{0.070}{0.35} \right) = 105.3 \text{ ft}.$$

The time for the ball to travel 54 ft is

$$t = 3.58 \text{ sec} \, [\exp [(0.0019/\text{ft})(54 \text{ ft})] - 1]$$
$$= 0.387 \text{ sec}.$$

Using $g = 32.2$ ft/sec², the y-coordinate is

$$y = \frac{\left(32.2 \dfrac{\text{ft}}{\text{sec}^2} \right) (0.387 \text{ sec})^2}{2} + \left(\frac{0.070}{0.35} \right) \left(147 \frac{\text{ft}}{\text{sec}} \right) (0.387 \text{ sec})$$

$$- (105.3 \text{ ft}) \ln \left(1 + \frac{0.387 \text{ sec}}{3.58 \text{ sec}} \right)$$

$$= -2.41 \text{ ft} + 11.38 \text{ ft} - 10.81 \text{ ft},$$

Answer

$$| \; y = -1.84 \text{ ft}. \; |$$

Discussion

We must now check the assumptions used in developing our equations. The first assumption is that the trajectory is flat. Assuming a straight-line path for the ball, the actual ball path is at an angle of

$$\theta \approx \tan^{-1} \frac{y}{x} = \tan^{-1} \left(\frac{-1.84 \text{ ft}}{54 \text{ ft}} \right) \approx -2° \text{ (below horizontal)}.$$

The components of the drag force are thus

$$\mathscr{D}_x \approx \mathscr{D} \cos 2° = 0.9994\mathscr{D} \qquad \text{and} \qquad \mathscr{D}_y \approx \mathscr{D} \sin 2° = 0.035\mathscr{D}.$$

The components of the lift force are similarly

$$\mathscr{L}_y \approx 0.9994\mathscr{L} \qquad \text{and} \qquad \mathscr{L}_x \approx 0.035\mathscr{L}.$$

The second assumption is that

$$\left(\frac{V_y}{V_x}\right)^2 \ll 1.$$

The largest value of V_y and the smallest value of V_x occur when $x = x_p = 54$ ft. The value of V_y is

$$V_y = \left(-32.2\,\frac{\text{ft}}{\text{sec}^2}\right)(0.387\ \text{sec}) + \left(\frac{0.070}{0.35}\right)\left(147\,\frac{\text{ft}}{\text{sec}}\right)\left(1 - \frac{1}{1 + \dfrac{0.387\ \text{sec}}{3.58\ \text{sec}}}\right)$$

$$= -9.6\ \text{ft/sec}.$$

Also

$$V_x = \frac{\left(147\,\dfrac{\text{ft}}{\text{sec}}\right)}{1 + \dfrac{0.387\ \text{sec}}{3.58\ \text{sec}}} = 133\ \text{ft/sec}.$$

The maximum value of $(V_y/V_x)^2$ is

$$\left(\frac{V_y}{V_x}\right)^2_{\text{max}} = \left(\frac{9.6}{133}\right)^2 = 0.0052.$$

The ratio is considerably less than 1.

The third assumption is that U/V_∞ is constant. A complicated analysis involving the surface shear stresses is necessary to determine how much the ball's angular velocity decreases as the ball travels the 54 ft. Since both U and V_x decrease, a limiting case can be evaluated by assuming U remains constant and using the maximum and minimum values of V_x. These give

$$U = R\omega = (0.119\ \text{ft})\left(40\,\frac{\text{rev}}{\text{sec}}\right)\left(\frac{2\pi\ \text{rad}}{\text{rev}}\right)$$

$$= 29.9\ \text{ft/sec},$$

$$\left(\frac{U}{V_{x\text{max}}}\right) = \left(\frac{29.9\,\dfrac{\text{ft}}{\text{sec}}}{147\,\dfrac{\text{ft}}{\text{sec}}}\right) = 0.204,$$

and

$$\left(\frac{U}{V_{x\text{min}}}\right) = \left(\frac{29.9\,\dfrac{\text{ft}}{\text{sec}}}{133\,\dfrac{\text{ft}}{\text{sec}}}\right) = 0.225.$$

The difference is roughly 10%. The values of $\mathbf{C_D}$ and $\mathbf{C_L}$ may differ from the values used by about 10%. A more accurate analysis might use the values of $\mathbf{C_D}$ and $\mathbf{C_L}$ for the average value of U/V_x; however, the accuracy with which values from $\mathbf{C_D}$ and $\mathbf{C_L}$ can be read from Fig. 8.31 does not justify the more refined solution.

The final assumption to consider is use of Fig. 8.31 to obtain values for C_L and C_D. This figure is for a smooth sphere; however, a baseball is rather rough due to the stitches. The lift coefficient for a baseball may be as much as three times larger than the value for a smooth sphere. The drag coefficient for a baseball will also be somewhat larger. You might repeat the calculations using a three-times-larger value of C_L and a (say) two-times-larger value of C_D and compare your answers to ours.

For the interest of baseball fans, the fastball released by our pitcher at 100 mph arrives at the plate at 90.7 mph (133 ft/sec).

The quantity $\rho C_D S/2m$ is called the *ballistic drag parameter*. "Heavier" objects fall faster than "lighter" objects because heavier objects have lower values of this parameter.

8.5 Accelerating Bodies and Added Mass

So far we have considered forces on bodies only in steady flow. Suppose that a body immersed in a fluid is accelerated from rest to some finite velocity. In addition to the ordinary viscous drag force that resists the body's motion, an extra fluid dynamic force also resists acceleration of the body. This extra resisting force exists because the accelerating body must set fluid in motion, pushing it out of the way. This resistance force is usually modeled by an *added mass* or *hydrodynamic mass*. To accelerate the body, it is necessary to apply a propulsive force. Because of the extra resistance due to fluid inertia, a given propulsive force produces a smaller acceleration than we would expect from consideration of the drag alone. The same effect would occur if a body of larger mass were acted on by the net force (propulsive force minus drag). We write Newton's second law for the accelerating body as

$$(m_b + m_h)\frac{dV}{dt} = F_p - \mathcal{D}, \tag{8.38}$$

where m_b is the actual mass of the body, F_p is the propulsive force, $\mathcal{D}$ is the resisting viscous drag, and m_h is the added mass. This model, in effect, lumps the fluid inertia with the inertia of the body. The sum of the actual body mass and the added mass is called the *virtual mass*.

The added mass is a function of the fluid density and the shape and size of the body being accelerated. Added mass can be evaluated using a model based on the kinetic energy imparted to the fluid by the accelerating body. Suppose that a body is accelerated from rest to velocity V_∞. After steady conditions have been attained (with respect to an observer on the body), the total kinetic energy of the flow field can be calculated by

$$E_K = \int \frac{1}{2}\rho V^2 \, d\mathcal{V},$$

where the integral is taken over all of the fluid influenced by the body (effectively extending from the body surface to infinity). This kinetic energy is equivalent to the kinetic energy of a fluid mass equal to added mass moving with velocity V_∞. Thus

$$\frac{1}{2} m_h V_\infty^2 = E_K,$$

or

$$m_h = \frac{2E_K}{V_\infty^2}. \qquad\qquad (8.39)$$

Since the added mass is a measure of fluid inertia rather than viscous effects, a velocity field calculated for an imaginary inviscid fluid is usually adequate for determining added mass. Tables 8.7 and 8.8 give values of added mass for a few simple two-dimensional, axisymmetric, and three-dimensional bodies. References [14] and [15] contain more extensive information. You should note that these values of added mass apply only if the body is immersed in an infinite mass of fluid. If the object is near a solid surface or a free surface, the added mass will be different. Consult reference [15] for details.

Added masses apply only to translational acceleration. If a body experiences rotational acceleration, the inertial resistance of the accelerated fluid is described by an added or *hydrodynamic moment of inertia* according to

$$\sum \mathcal{M}_{0,x} = (I_{b,x} + I_{h,x}) \frac{d\omega_{b,x}}{dt},$$

Table 8.7 Added (hydrodynamic) mass for two-dimensional bodies [15]

Shape		Hydrodynamic Mass per Unit Length
Circular cylinder		$\frac{1}{4}\pi\rho D^2$
Flat plate		$\frac{1}{4}\pi\rho h^2$
Square cylinder		$0.3775\pi\rho s^2$
		$0.38\pi\rho s^2$
Elliptical cylinder		$\frac{1}{4}\pi\rho t^2$

Note: ρ is fluid density.

Table 8.8 Added (hydrodynamic) mass for axisymmetric and three dimensional bodies [14]

Shape		Hydrodynamic Mass
Sphere		$\frac{\pi}{12}\rho D^3$
Circular disk		$\frac{1}{3}\rho D^3$
Square plate		$0.1195\pi\rho s^3$
Cube		$2.32\rho s^3$
Ellipsoidal body		
$(\ell/D) = 2$		$0.0697\pi\rho D^3$
$= 4$		$0.0547\pi\rho D^3$
$= 8$		$0.0387\pi\rho D^3$

Note: σ is fluid density.

where $\sum \mathcal{M}_{0,x}$ is the sum of moments acting about an x-axis through the center of mass of the body, $I_{b,x}$ is the moment of inertia of the body about its centroidal x-axis, $\omega_{b,x}$ is the angular velocity of the body about its centroidal x-axis, and $I_{h,x}$ is the added moment of inertia about the centroidal x-axis. Added moments of inertia depend on fluid density and body geometry and can be calculated by an inviscid fluid–kinetic energy model similar to that for added mass. Table 8.9 gives a few values for added moments of inertia. Note that the added moment of inertia for rotation about an axis of symmetry is zero.

Table 8.9 Added (hydrodynamic) moments of inertia for a few bodies [14]

Shape	Hydrodynamic Moment of Inertia
Cylinder of elliptical cross section	$0.125\pi\rho(b^2 - a^2)^2$
Square cylinder	$0.014625\pi\rho s^4$
Circular cylinder	0
Thin plate	$0.0078125\pi\rho\ell^4$
Sphere	0
Disk of diameter D rotating about a diameter	$\frac{1}{90}\rho D^5$

Note: ρ is fluid density.

EXAMPLE 8.10 Illustrates hydrodynamic or added mass

Consider the 1.0-mm-diameter steel ball falling in light machine oil as shown in Fig. E8.5. Find the distance required for the ball to accelerate from zero velocity to the constant velocity $V_\infty = 0.0379$ m/s.

Solution

Given
1.0-mm-diameter steel ball
Oil viscosity 0.10 N·s/m²
Oil specific weight 8340 N/m³
Steel specific gravity 7.8

Find
Time for ball to accelerate from zero velocity to $V_\infty = 0.0379$ m/s

Solution
We first apply Eq. (8.38) to the ball in the vertical direction, with the positive direction taken as downward. This gives

$$(m_b + m_h)\frac{dV}{dt} = W - F_B - \mathcal{D},$$

where W is the weight of the steel ball and F_B is the buoyant force on the ball. Since the ball is known to fall very slowly, Stokes' law, Eq. (8.14), gives

$$\mathcal{D} = 3\pi\mu DV,$$

where μ is the fluid absolute viscosity, D the ball diameter, and V the ball velocity. The weight W is

$$W = \gamma_s \mathcal{V},$$

where γ_s is the ball specific weight and $\mathcal{V}$ its volume. The buoyant force is

$$F_B = \gamma_0 \mathcal{V},$$

where γ_0 is the oil specific weight. The ball's mass is

$$m_b = \rho_s \mathcal{V},$$

where ρ_s is the steel density. Table 8.8 gives the hydrodynamic mass as

$$m_h = \frac{\pi}{12}\rho_0 D^3 = \frac{1}{2}\rho_0 \mathcal{V},$$

where ρ_0 is the oil density. Substituting into Eq. (8.38) gives

$$\left(\rho_s + \frac{1}{2}\rho_0\right)\mathcal{V}\left(\frac{dV}{dt}\right) = \gamma_s \mathcal{V} - \gamma_0 \mathcal{V} - 3\pi\mu DV.$$

Rearranging gives

$$\frac{dV}{dt} + \frac{3\pi\mu DV}{\left(\rho_s + \frac{1}{2}\rho_0\right)\mathcal{V}} = \frac{\gamma_s - \gamma_0}{\rho_s + \frac{1}{2}\rho_0}.$$

Defining

$$a = \frac{3\pi\mu D}{\left(\rho_s + \frac{1}{2}\rho_0\right)\mathcal{V}} \quad \text{and} \quad b = \frac{\gamma_s - \gamma_0}{\rho_s + \frac{1}{2}\rho_0},$$

we have

$$\frac{dV}{dt} + aV = b.$$

The solution of this differential equation is

$$V = Ce^{-at} + \frac{b}{a},$$

where C is a constant. Setting time $t = 0$ when the ball is released ($V = 0$) gives

$$C = -\frac{b}{a}.$$

Then

$$V = \frac{b}{a}(1 - e^{-at}).$$

However,

$$\frac{b}{a} = \left(\frac{\gamma_s - \gamma_0}{\rho_s + \frac{1}{2}\rho_0}\right)\left(\frac{\left(\rho_s + \frac{1}{2}\rho_0\right)\cancel{V}}{3\pi\mu D}\right)$$

$$= \frac{(\gamma_s - \gamma_0)\cancel{V}}{3\pi\mu D} = \frac{(\gamma_s - \gamma_0)\left(\frac{\pi}{6}\right)D^3}{3\pi\mu D} = \frac{(\gamma_s - \gamma_0)D^2}{18\mu}.$$

Example 8.5 shows that this quantity is equal to the final velocity V_∞:

$$\frac{(\gamma_s - \gamma_0)D^2}{18\mu} = V_\infty.$$

Then

$$V = V_\infty(1 - e^{-at}).$$

This equation shows that the velocity V_∞ is never reached; however, V approaches V_∞ as time t increases. The mathematically "correct" answer is

Answer

$$\boxed{t \to \infty.}\ \blacktriangleleft$$

However, this is not a practical answer, so we will find the time for which

$$V = 0.99V_\infty.$$

Taking the logarithm of the velocity equation and solving for t gives

$$t = -\frac{1}{a}\ln\left(1 - \frac{V}{V_\infty}\right).$$

Referring to Example 8.5, we can find the numerical values of the parameters. Assuming $g = 9.81$ m/s^2,

$$\rho_0 = \frac{\gamma_0}{g} = \frac{\left(8340\ \dfrac{\text{N}}{\text{m}^3}\right)}{9.81\ \dfrac{\text{m}}{\text{s}^2}} = 850\ \text{kg/m}^3,$$

$$\rho_s = \rho_{w\,\text{at}\atop 4^\circ\text{C}}\, S_s = \left(1000\ \frac{\text{kg}}{\text{m}^3}\right)(7.8) = 7800\ \text{kg/m}^3,$$

and

$$a = \frac{3\pi\mu D}{\left(\rho_s + \frac{1}{2}\rho_0\right)V} = \frac{3\pi\mu D}{\left(\rho_s + \frac{1}{2}\rho_0\right)\left(\frac{\pi}{6}D^3\right)} = \frac{18\mu}{\left(\rho_s + \frac{1}{2}\rho_0\right)D^2}$$

$$= \frac{18\left(0.10\,\frac{N\cdot s}{m^2}\right)}{\left[7800\,\frac{kg}{m^3} + \frac{1}{2}\left(850\,\frac{kg}{m^3}\right)\right](1.0\,mm)^2\left(\frac{m}{1000\,mm}\right)^2} = 218.8\,m/s.$$

The time required to achieve a velocity equal to 99% of the final velocity is

$$t_{99} = -\frac{1}{\left(\dfrac{218.8}{s}\right)}\ln(1 - 0.99),$$

**Practical
Answer**
◀

$$|\ t_{99} = 0.021\ s.\ |$$

Discussion

It is instructive to find the distance the ball falls before it reaches $0.99\,V_\infty$. This is found by substituting the expression for V into

$$\frac{dy}{dt} = V.$$

Substituting, multiplying by dt, and integrating give

$$\int_0^{y_{99}} dy = \int_0^{t_{99}} V_\infty(1 - e^{-at})\,dt,$$

or

$$y_{99} = \left[V_\infty t + \frac{V_\infty}{a}e^{-at}\right]_0^{t_{99}}.$$

$$= V_\infty t_{99} - \frac{V_\infty}{a}(1 - e^{-at_{99}}).$$

The numerical values give

$$y_{99} = \left(0.0379\,\frac{m}{s}\right)(0.021\,s) - \frac{\left(0.0379\,\frac{m}{s}\right)}{\left(\dfrac{218.8}{s}\right)}\left[1 - \exp\left(-\left(\frac{218.8}{s}\right)(0.021\,s)\right)\right],$$

$$y_{99} = 0.000624\,m = 0.624\,mm,$$

a very short distance.

Problems

Extension and Generalization

1. A rotameter, a device used to measure flow rate, is illustrated in Fig. P8.1. Develop an expression for the volume flow rate Q of a liquid as a function of the position h of the float, which has a drag coeffi-cient C_D, and the velocity V. The float has weight W and rises in the glass cone as the flow rate increases. Why must the float be in the vertical position to operate properly?

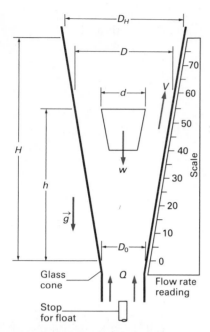

Glass cone

Stop for float

Figure P8.1

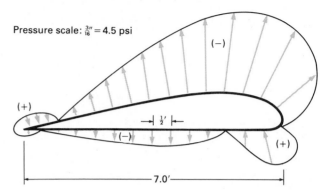

Pressure scale: $\frac{3''}{16} = 4.5$ psi

Figure P8.3

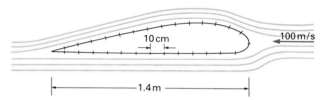

Figure P8.4

2. Apply the linear momentum equation to a control volume enclosing an airfoil (similar to the derivation of the Kutta-Joukowsky theorem) and obtain an expression for the drag in terms of measured upstream and downstream velocity profiles.

Lower-Order Application

3. The pressure distribution over the object in Fig. P8.2 is as shown. Find the drag force. The object is 3.0 m high and 10.0 m wide.

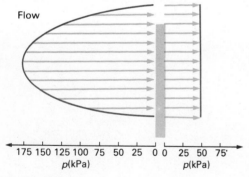

Figure P8.2

4. When warm water is running in a shower, the shower curtain is pulled inward. Explain why this happens. Will it happen if cold water is used? Explain.

5. Find the lift coefficient due to pressure for the airfoil shown in Fig. P8.3. $V_\infty = 350$ mph, $p_\infty = 29.4$ psia, and $T_\infty = 60°F$. Assume constant density.

6. The streamlines around an airfoil are shown in Fig. P8.4. Estimate the lift $\mathscr{L}$.

7. A 1.2-m-diameter spherical balloon weighs 1.0 N and contains helium at 10°C and 150 kPa. The ambient air is at 10°C and 101 kPa. Air blows horizontally over the balloon at 20 km/hr. Find the angle that a 10-m weightless string restraining the balloon makes with the ground; also find the tension in the string.

8. A flat movie screen in an outdoor theatre is 15 m square. Find the bending moment on the base of the structure that supports the bottom of the screen 15 m above the ground. The wind velocity is 60 km/hr and the atmospheric conditions are 15°C and 100 kPa.

9. A smokestack has an outside diameter of 10 m and a height of 120 m. Find the overturning moment due to a uniform velocity of 70 km/hr. The atmospheric conditions are 10°C and 101 kPa.

10. Repeat Problem 9 using the velocity distribution $V = Ay^2$, where $A = 0.00139$ km/m²·hr with y measured vertically above the ground.

11. A car-top carrier measuring 1 ft high, 4 ft wide, and 4 ft long is used for a trip. How much extra power (in hp) is required at 55 mph on a level road due to the addition of the carrier? Assume air flows over the entire carrier (i.e., there is an air space between the car top and the carrier).

12. The addition of the air deflector on top of a tractor has increased the gasoline mileage of a tractor-

trailer from 4 mi/gal to 4.25 mi/gal at 55 mph. The engine provides 22,000,000 ft·lb of energy to the drive wheels from every gallon of fuel. Half of this energy is used to overcome the drag resistance of the air on the tractor-trailer. Find the drag coefficient for the tractor-trailer with and without the air deflector. The tractor-trailer has a frontal area of 82 ft^2.

13. The railroad boxcar in Fig. P8.5 weighs 200 kN and rides on a track whose rails are 1.44 m apart. What wind velocity normal to the side of the boxcar will topple it?

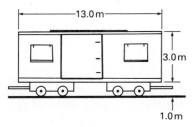

Figure P8.5

14. A man riding his bicycle is traveling along a horizontal road at 10 mph. The air is still and is at 85°F and 14.67 psia. Estimate the drag coefficient of this bicyclist if his frontal area is 4.5 ft^2 and his body burns up 150 calories per second with one-fourth of it going to work in pedaling the bicycle.

15. A woman riding her bicycle is sitting upright and coasting down a hill making an angle of 5° with the horizontal, with a velocity of 25 km/hr into an oncoming 25-km/hr wind. The air is at 15°C and 101 kPa. Estimate the drag coefficient of this bicyclist if her frontal area is 0.40 m^2. The woman and her bicycle weigh 520 N.

16. Repeat Example 8.6 if the wind velocity is given by

$$V = V_\infty \left(\frac{y}{\delta}\right)^{1/7},$$

where $V_\infty = 50$ mph and y and δ are both measured in ft, with $\delta = 850$ ft. Compare your answer with that of Example 8.6.

17. An 180-lb man is parachuting toward the ground using a hemispherical parachute in air at 20°F and 14.60 psia. Find the parachute diameter so the man's terminal velocity does not exceed 20 ft/sec. Neglect the weight of the parachute.

18. A dragster has accelerated to 180 mph and loses its brakes as the parachute opens. Assuming no friction between the wheels and the road, a drag coefficient of 0.24, and a frontal area of 7.5 ft^2, find

the time for the dragster to slow to 90 mph. The parachute is hemispherical with a 20-ft diameter. The atmospheric conditions are 70°F and 14.60 psia.

19. A cannon shoots a 5.5-cm-diameter spherical ball at a 45° angle with the horizontal and a muzzle velocity of 700 m/s in still air. The cannon is located at the edge of a plateau 100 m above a level plain. Find the horizontal distance traveled and the elapsed time when the ball hits the plain. Compare your answers with those related to a ball having no air drag.

20. Repeat Problem 19 and allow the initial inclination angle to be such that the spherical ball will travel the longest horizontal distance. Compare this angle with that of a ball having no air drag.

21. A vertical radio antenna on an automobile is 3.0 ft long and has a diameter of 0.125 in. Find the drag on the antenna at a car speed of 55 mph in still air.

22. A rocket weighs 10,000 N and has a diameter of 2.5 m and a length of 8.0 m. The rocket leaves the launching pad at an angle of 70° with the horizontal and burns fuel at the rate of 10.4 kg/s for 50 s. The exhaust gases leave the rocket at 1000 m/s relative to the rocket. The rocket has a constant drag coefficient of 0.70. Will the rocket travel 10 km before it uses all its fuel?

23. A 9000-N vehicle traveling at 11.4 m/s has a constant drag coefficient of 0.7 and a frontal area of 1.2 m^2. All four wheels skid on the ice when the brakes are applied. The coefficient of sliding friction between the tires and the ice is 0.10. Find the time for the vehicle to slow down to 8 m/s.

24. An automobile engine has a maximum power output of 70 hp, which occurs at an engine speed of 2200 rpm. A 10% power loss occurs through the transmission and differential. The rear wheels have a radius of 15.0 in. and the automobile is to have a maximum speed of 75 mph along a level road. At this speed the power absorbed by the tires due to their continuous deformation amounts to 27 hp. Find the maximum permissible drag coefficient for the automobile at this speed.

25. An airplane has a weight of 7300 N and a wing area of 15 m^2 and is cruising at 175 km/hr at an altitude of 1500 m in still air. Its lift and drag coefficients are given by curves C_{DB} and C_{LB} in Fig. P8.6 and are based on the wing planform area. The engine stops suddenly and cannot be restarted. Find the maximum horizontal distance the airplane can travel before landing.

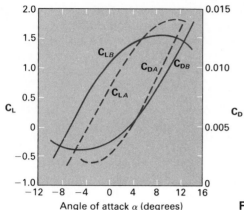

Angle of attack α (degrees)

Figure P8.6

26. A small airplane has a velocity of 120 mph traveling in 70°F still air at sea level. The propeller develops a forward thrust of 350 lb. Find the percent of the power developed by the propeller that is used to overcome the air resistance of 40 ft of 0.25-in.-diameter guy wires. Find the reduction in this power requirement if the guy wires are replaced by streamlined struts having the same frontal area.

27. A copper telephone wire has a 0.25-in. diameter, a 200-ft span, and a maximum drop of 5 ft (catenary shape). Find the forces on the support at one end of the telephone wire for a 50-mph wind normal to the plane of the wire.

28. A water tower is a spherical shell with a diameter of 5.9 m. The shell is supported by four vertical steel pipes that are 7 m in length and have a 7.5-cm outside diameter. A wind with a uniform velocity of 11 km/hr blows over the water tower. Find the total force and bending moment at the base of the tower.

29. Air flows at 10,000 cfm through a horizontal, 24-in.-square duct made of galvanized sheet metal. A 0.125-in.-diameter rod spanning the duct holds a 6-in.-diameter spherical ball in the center of the duct. Find the pressure drop over a length of the duct enclosing the sphere. Include both the pressure drop contribution due to wall friction between the air and the wall and the resisting force by the sphere and the wire.

30. Neglecting tire rolling resistance, does it require more energy to increase a car's speed from 15 to 20 mph or from 40 to 45 mph? Assume constant drag coefficient and the same rate of acceleration.

31. The two hydrofoils on a watercraft have a planform area of 50 ft^2 and a cross-sectional profile of a 2415 airfoil. How great a load W can the hydrofoil support at a speed of 20 mph in still water if the angle of attack is 12°?

32. Find the horsepower required to propel the watercraft of Problem 31 if it has a frontal area of 50 ft^2 and a drag coefficient of 0.30. Neglect the drag on the hydrofoils moving through the water.

33. The dirigible *Akron* had a length of 785 ft, a maximum diameter of 132 ft, and a maximum speed of 84 mph. The lifting force was 182,000 lb. Estimate the power required to overcome skin friction in the Standard Atmosphere at 5000 ft.

34. An automobile has a drag coefficient of 0.45 and a frontal area of 20 ft^2 and weighs 3000 lb. Assume that the frictional resistance between the road and a nondriving (free-rolling) wheel is a constant value of 2.5 lb. The automobile is traveling at 55 mph on a level road. Find the distance for the automobile to slow down to 15 mph after the clutch is disengaged.

35. A speed limit sign along a highway is shown in Fig. P8.7. Find the bending moment at the base of the 2-in.-outside-diameter support poles for a 60-mph wind.

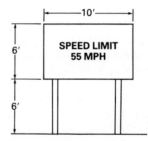

Figure P8.7

36. Why do airplanes try to take off and land while headed into the wind?

37. The next time a large truck overtakes and passes you while you are driving a car, observe the relative positions of the car and the truck where the car is pulled toward the truck. Why does this occur?

38. Wind tunnel tests have given the drag coefficient information in Fig. P8.8 for the missile in Fig. 7.32. When all the fuel has been depleted, the missile has a velocity of 3000 ft/sec and weighs 1500 lb. If the missile is travelling in atmospheric air

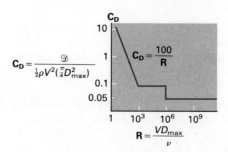

$$C_D = \frac{\mathcal{D}}{\frac{1}{2}\rho V^2 (\frac{\pi}{4}D^2_{max})}$$

$$C_D = \frac{100}{R}$$

$$R = \frac{V D_{max}}{\nu}$$

Figure P8.8

$(\rho = 0.075 \text{ lbm/ft}^3)$, calculate the time required for the velocity of the missile to decrease to 1 ft/sec (neglect gravitational effects).

39. The supports for Phil's Pizza Parlor sign in Example 8.4 consist of two steel pipes 25 cm long and 3.34 cm outside diameter. Estimate the engine horsepower required to overcome the air drag on these two supports for an advertising sign facing forward.

40. The cross section of an airplane wing is an NACA 2415 airfoil. The wing has a constant chord length of 3.5 m and a length of 15 m. Find the lift and drag on this wing at an airplane speed of 200 km/hr in air at 98 kPa and 0°C and angle of attack of 12°.

41. An airplane weighs 320,000 lb and has a wing area of 2800 ft² and a wing length of 140 ft. The atmospheric pressure and temperature are 14.67 psia and 60°F respectively. The airplane is moving along a runway at 200 mph and the wing has the lift and drag coefficients given in Fig. 8.39. Find the lift and drag on the airplane wings for an angle of attack of 12°.

42. An airplane flying at 10,000 m in the Standard Atmosphere has a weight of 400 kN and a drag of 32 kN while traveling at 450 km/hr. Find the lift and drag coefficients for a wing area of 120 m² and a wing length of 20 m if the wing provides 90% of the lift and 60% of the drag.

43. A race car has an airfoil installed above the rear wheel for downthrust. The airfoil makes an angle of 10° with the track and has a chord length of 1.0 ft and a length of 3.0 ft. The race car is traveling at 180 mph in air at 90°F and 14.72 psia. Find the downthrust and drag on the airfoil if it has the lift and drag coefficient characteristics of Fig. 8.39.

44. Assume the race car in Problem 43, which weighs 2000 lb, is rounding a curve of radius 200 ft. The coefficient of sliding friction between the track and the rear wheels is 0.6. Will the rear wheels of the car slide sideways? The curve makes an angle of 20° with the level ground and is slanted inward.

45. Leading edge slats on an airplane wing create a very high lift if the wing angle of attack is increased. The slat is retracted at cruise conditions. This reduces the drag at cruising conditions and permits the installation of a smaller wing. Figure P8.9 shows this increase in the lift coefficient for a particular wing design with and without slats. Consider an airplane that weighs 120,000 lb, has a wingspan planform area of 1035 ft², and cruises at 520 mph in the Standard Atmosphere at 30,000 ft. The takeoff and landing speeds are 160 mph and 140 mph respectively. Find the increase in the required wing plan-

form area if the slats are not used during takeoff and landing. Assume the wing and slats provide all of the lift and that the wing accounts for 25% and the slats account for 2% of the airplane weight. The lift coefficients in Fig. P8.9 are based on the wing area.

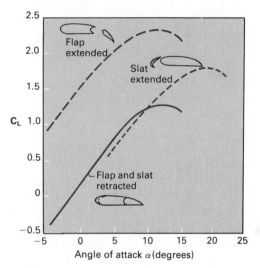

Figure P8.9

46. Trailing edge tail flaps increase the lift coefficient of an airplane wing without changing the wing's angle of attack. Figure P8.9 shows this increase in the lift coefficient for a particular wing design with and without a trailing edge wing flap. Consider the airplane in Problem 45 that cruises at 30,000 ft and has takeoff and landing speeds of 160 mph and 140 mph respectively. Find the increase in the wing planform area if the flaps are not used. Assume the wing and flaps provide all of the lift and that the wings account for 37% and the flaps account for 2% of the airplane weight.

47. The lift and drag coefficients for a flat plate inclined at angle of attack α are shown in Fig. P8.10.

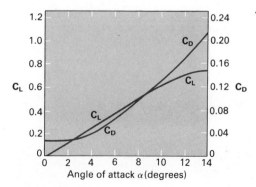

Figure P8.10

A ceiling fan with a vertical shaft rotates at 60 Hz with four flat blades that make an angle of 12° with the horizontal. Each blade is 50 cm long and 20 cm wide. The hub radius is 10 cm. Find the vertical force on the bearings in the fan.

48. Find the velocity as a function of time of a 1.0-mm-diameter steel ball falling in glycerine. What is the terminal velocity?

49. Find the velocity as a function of time of a 1.0-mm-diameter spherical air bubble rising in water. What is the terminal velocity?

50. A particle of sand is assumed to be a sphere with 1-mm diameter. What is the maximum water velocity so that a sand particle will fall a vertical distance of 1.0 m before it is carried a horizontal distance of 3.0 m by a flowing stream? Sand has a specific gravity of 1.60.

51. Find the terminal velocity of a spherical bubble rising in a bottle of cold soft drink.

52. Consider two identical aluminum rods, each having a diameter of 0.01 mm and a length of 1.0 mm. Both are placed in 20°C kerosene. One rod is vertical, and the second is horizontal. Find the terminal velocity of both rods, compare the velocities, and make any worthwhile observations.

53. Consider an aluminum rod having a diameter of 0.01 mm and a length of 1.0 mm. It is falling in 20°C kerosene with the axis of the rod making an angle θ with the horizontal. Find the terminal velocity (magnitude and direction) of the aluminum rod for $\theta = 30°$ and $\theta = 45°$.

54. Consider the model with the two spiral wires in Fig. P8.11(a). One spiral wire is clockwise, and the second is counterclockwise. The rubber band is wound so the clockwise spiral would rotate in the clockwise direction and the counterclockwise spiral would rotate in the counterclockwise direction. The model is then placed in a thick syrup so

the motion of each segment of each spiral constitutes low Reynolds number flow. Figures P8.11(b) and P8.11(c) show the velocity of a typical segment of each spiral and the drag components on each segment. Determine the direction of travel of the model.

55. The spirals in Fig. P8.11(a) have a wire diameter of 0.125 cm, a radius $R = 2.0$ cm, and a spiral angle $\theta = 30°$ and are to rotate at 0.25 rev/s when placed in a thick syrup ($v = 1.0 \times 10^{-2}$ m²/s). Find the torque to which the rubber band must be wound. Next find the propulsive force developed by the two spirals.

56. A domestic engineer dusts off a small table 24 in. high. The dust particles from the table are spherical with diameters ranging from 0.001 in. to 0.005 in. Dust has a density of 90 lbm/ft³. How long will it take for the dust to settle to the floor?

57. Show that the Kutta-Joukowsky theorem, Eq. (8.31), is valid for a free vortex superimposed on a uniform flow.

58. Estimate the lift of the airfoil in Problem 5 using the Kutta-Joukowsky theorem.

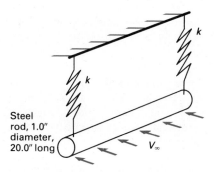

Steel rod, 1.0" diameter, 20.0" long

Figure P8.12

59. The solid steel, circular rod in Fig. P8.12 is supported by two springs, each having a spring constant k of 50 lb/in. (i.e., a force of 50 lb will compress or stretch each spring a distance of 1.0 in.). The natural frequency of vibration of the rod-spring system is given by

$$\omega = \frac{1}{2\pi}\sqrt{\frac{k}{m\ell}},$$

where m is mass and ℓ is length of the rod. Find the wind velocity V_∞ that causes vortex-shedding resonance at the natural frequency of the rod.

60. A solid aluminum rod has a length of 4.0 ft and a diameter of 2.0 in. It is pinned at its ends and has a spring constant k for a center deflection given by

$$k = \frac{384\,EI}{5\ell^3},$$

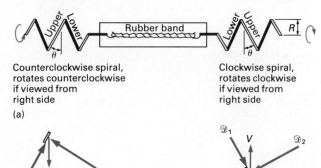

Counterclockwise spiral, rotates counterclockwise if viewed from right side

(a)

Clockwise spiral, rotates clockwise if viewed from right side

Rubber band

R

(b)

$\mathcal{D}_1$ V $\mathcal{D}_2$

(c)

$\mathcal{D}_1$ V $\mathcal{D}_2$

Figure P8.11

where E is the modulus of elasticity of the rod, I the second moment of area about the diameter of its circular cross section, and ℓ its length. Find the wind velocity V_∞ that causes vortex-shedding resonance at the natural frequency of vibration of the rod. Refer to Problem 59 for a description of how to calculate the natural frequency of the rod.

61. Find the rotational spin required for a baseball to break 2 ft (i.e., to move a distance of 2 ft perpendicular to the direction the ball is thrown) after it leaves the pitcher's hand at 90 mph and travels a distance of 54 ft. Neglect air drag and gravity. Assume the ball leaves the pitcher's hand with a velocity parallel to the ground and travels in a horizontal plane. Assume the lift coefficient on a baseball may be approximated by three times the values given in Fig. 8.31.

62. Estimate the energy required for a 6-ft-tall person to run a mile in 4 min in still air.

63. A baseball has a constant underspin of 100 rev/sec with an initial speed of 150 mph. How much farther will the ball travel than if it had no spin?

64. A tennis ball traveling in a straight line and hit 10 ft above the ground at the baseline must be served between an angle of 9.5° and 10.2° below the horizontal to land in the opponent's forecourt. See Fig. P8.13. Assume the ball is served at a speed of 25 mph with a top spin of 40 rev/sec. Will the difference of these two angles be increased? If so, by how much? Include the effect of gravity. What is the advantage of the top spin?

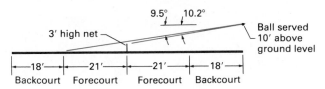

Figure P8.13

65. A baseball has a mass of about 0.25 lbm and leaves a bat with no spin, a speed of 150 mph (220 ft/sec), and an angle of 45° above the horizontal. Will the baseball clear a 10-ft-high fence 450 ft away? If not, what initial speed is needed?

66. Find the optimum angle for the baseball in the Problem 65 to be hit for it to travel the longest horizontal distance.

67. Find the initial speed of a nonspinning golf ball hit at an angle of 20° above the horizontal to travel 250 yards. A golf ball has a 1.68-in. diameter and a mass of 1.62 oz.

68. Find the optimum angle for the golf ball in Problem 67 to travel the longest horizontal distance.

69. Assume that the golf ball in Problem 67 has a constant underspin of 100 rev/sec with an initial speed of 220 ft/sec. How much farther will the spinning ball travel than the nonspinning ball?

70. It is desired to build a vertical wind tunnel with an upward flow to keep people in suspension to learn sky-diving techniques. What should be the velocity of the air past the sky divers?

71. A sky diver having a mass of 60 kg jumps out of an airplane at 3000 m above the ground. Find how far she has fallen and her velocity in 45 s if she has not yet opened her parachute and has provided the largest possible drag coefficient.

72. Find the velocity of the sky diver in Problem 71 after 45 sec if she opens her 25-ft-diameter, hemispherical parachute 25 sec after jumping out of the airplane. How far has she fallen?

73. To sail into the wind a sailboat tacks or zigzags into the wind as shown in Fig. P8.14. Find the velocity V_{SB} and the angle θ for the sailboat to advance into the wind at an angle of $\phi = 45°$. Assume the sail has the lift coefficient and drag coefficient characteristics of the NACA 2415 airfoil. The sail planform area is 15 m². The resistance of the water to the sailboat is expressed in terms of a frontal drag coefficient $C_{Df} = 0.50$ and a sideways drag coefficient $C_{Ds} = 1.15$. The frontal submerged projected area is 6 m² and the sideways submerged projected area is 12 m².

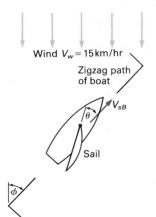

Figure P8.14

74. Consider a long cylinder of radius R moving with velocity V_∞ through a still fluid. Calculate the kinetic energy of the fluid set in motion for a unit length of the cylinder.

75. Consider a sphere of radius R moving with velocity V_∞ through a still fluid. Calculate the kinetic energy of the fluid set in motion.

76. Tabulate the drag coefficient for large Reynolds numbers versus the hydrodynamic mass per unit volume of the object. Do this for a sphere and a cube whose sides are normal and parallel to the flow. Do you expect any correlation to exist? If so, why?

Higher-Order Application

77. A design group has two possible wing designs (*A* and *B*) for an airplane wing. The planform area of either wing design is 130 m^2 and each must provide a lift of 1,550,000 N. The airplane is to fly at 700 km/hr at an altitude of 10,000 m in the Standard Atmosphere. The lift and drag coefficients are shown in Fig. P8.6. Both sets of lift and drag coefficients give the total lift $\mathscr{L}$ and the total drag $\mathscr{D}$ on the airplane per unit area of the wing so that

$$\mathscr{L}_{\text{airplane}} = \frac{1}{2} C_L \rho A_{\text{wing}} V^2 .$$

and

$$\mathscr{D}_{\text{airplane}} = \frac{1}{2} C_D \rho A_{\text{wing}} V^2,$$

where *V* is the airplane velocity. Which wing design would you recommend? Support your recommendation.

78. Find the angle ϕ, the angle θ, and the velocity *V* in Fig. P8.14 so the sailboat will travel into the wind with the maximum possible velocity. Refer to Problem 73 for approximate values of drag and lift coefficients and areas. [*Hint:* Choose several values of ϕ and find the corresponding values of *V* and θ. Plot *V* cos ϕ versus ϕ to find the maximum value of *V* cos ϕ.]

79. A spherical balloon is filled with helium at 109 kPa and 15°C and has a diameter of 5 m at sea level where the atmospheric conditions are 101 kPa and 15°C. The surface tension in the balloon may be assumed constant as the balloon rises to an altitude of 4000 m in the Standard Atmosphere. The balloon carries an internal payload of 150 N. Find the altitude of the balloon as a function of time.

80. A windmill consists of four flat plates measuring 5.0 m long and 1.0 m wide and connected to a 1.0-m-diameter hub. The blades rotate at 0.1 Hz. The blades have a twist so each segment of each blade makes an angle of 15° with the relative velocity of the wind and the blade segment. Find the torque developed by the four blades for a wind velocity of 15 mph parallel to the windmill's axis of rotation. Refer to Fig. P8.10 for the lift and drag coefficients of a flat plate at angles of attack.

81. You are the center fielder on a major league baseball team. You are 200 ft from home plate and a batter hits a knuckleball directly at you with a velocity of 170 mph at an angle of 20° above the level ground. How fast must you run in a straight line at a constant speed to catch the ball? You start running the instant the ball is hit.

82. A regulation football has a circumference of approximately 20.8 in. at its widest point, a length of 11 in., and a weight of 14.5 oz. How far will the football travel if it is thrown with a velocity of 45 ft/sec at an angle of 30° with level ground in still air? Assume the longitudinal axis is always aligned with the velocity vector, the effect of the spin about this axis is negligible, and the ball is released 6.0 ft above the ground. The drag coefficient based on the minor diameter is given below:

R	4×10^4	1×10^5	2×10^5	4×10^5	8×10^5
C$_D$	0.30	0.10	0.055	0.070	0.080

83. Consider the football in Problem 82. Find the optimum angle θ above level ground so that the football will travel the farthest horizontal distance.

84. The sweep-oar rowing boat in Fig. P8.15 is 18 m long, 0.60 m wide at the top, and 0.50 m high and weighs 1250 N. It carries eight crewmen and a coxswain. Estimate the number of strokes per minute by each crewman so the boat will travel at a speed of 15 km/hr. Figure P8.15 shows the oar depth during each stroke.

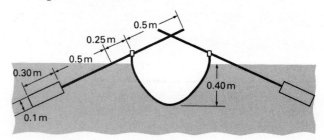

Figure P8.15

85. Boy Scouts are taught to "feather" oars on the backstroke when rowing a boat. Estimate the energy saved in 100 oar strokes.

Comparison

86. Determine the extra power required for an automobile traveling at 55 mph to travel up a 10° grade rather than along a level road. There is no wind. The automobile may be assumed to have a constant drag coefficient of 0.40 and a frontal area of 24 ft^2.

87. An airplane pilot has the option to fly either at 3000 m or at 10,000 m in the Standard Atmosphere

in still air. The airplane velocity is to be 800 km/hr. Assuming a constant drag coefficient, find the percent decrease in the power requirement to fly at 10,000 m.

88. A woman and her son Eddie have identical sports cars, leave home at the same time, and drive 4 mi along a level road to a family picnic. The speed patterns of each are shown in Fig. P8.16. Both automobiles have a drag coefficient of 0.35 and a frontal area of 13 ft^2. Each gallon of fuel releases 1.0×10^8 ft·lb of energy, and 30% of this energy is applied to the drive wheels. Neglect the energy used in continually deforming the tires and find the amount of fuel the woman saved on this trip.

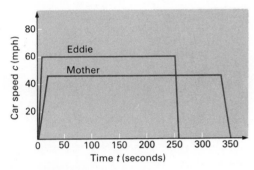

Figure P8.16

89. A full-size automobile has a frontal area of 24 ft^2, and a compact car has a frontal area of 13 ft^2. Both have a drag coefficient of 0.5 based on the frontal area. Find the horsepower required to move each automobile along a level road in still air at 55 mph. Assume the power required to continuously deform the tires at this speed (called rolling resistance) is equal to the power to overcome the air resistance. Estimate the gas mileage of both automobiles if the energy supplied to the drive wheels is one-fourth that available in the fuel. A gallon of fuel has 1.0×10^8 ft·lb of available energy.

90. Estimate the distances traveled by smooth and dimpled golf balls when driven from the tee at 100 mph at 20° above the horizontal with an underspin of 1200 rpm.

91. The oscillating-tail model in Fig. P8.17 is placed in a very viscous syrup so that low Reynolds number flow exists. The tail is made to oscillate by the rubber band and the crank mechanism. Will the model move forward as indicated? Will the motion be different if the model is placed in water so that high Reynolds number flow exists?

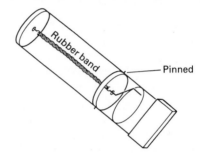

Figure P8.17

References

1. Shapiro, A. H. (Principal), *The Fluid Dynamics of Drag* (4 parts), NCFMF Film, available from Encyclopaedia Britannica Films, Chicago, IL.
2. White, F. M., *Viscous Fluid Flow,* McGraw-Hill, New York, 1974.
3. Hoerner, S. F., *Fluid Dynamic Drag,* published by Author, Midland Park, NJ, 1965.
4. Whitford, D. H., and J. E. Minardi, "Utility-Sizes Madaras Wind Plants," *The International Journal of Ambient Energy,* vol. 2, no. 1, 1981.
5. Abbott, I. H., and A. E von Doenhoff, *Theory of Wing Sections,* Dover Books, New York, 1959.
6. Lissaman and Shollenberger, "Formation Flight of Birds," *Science,* vol. 168, no. 3934, May 1970.
7. *Design for Fuel Economy: The General Motors X Cars,* SAE/SP-80/452, Society of Automotive Engineers, Warrendale, PA, 1980.
8. Scibor-Rylski, A. J., *Road Vehicle Aerodynamics,* Halsted Press, New York, 1975.
9. *Automotive Aerodynamics,* SAE/PT-78/16, Society of Automotive Engineers, Warrendale, PA, 1978.
10. Morel, T., and C. Dalton (Eds.), *Aerodynamics of Transportation,* proceedings of a symposium held at the Joint ASME-CSME Applied Mechanics, Fluids Engineering and Bioengineering Conference, Niagara Falls, NY, June 1979.

11. "Tacoma Narrows Bridge Collapse" (a film), available from Ohio State University, Motion Picture Division, Columbus, OH.

12. Blevins, R. D., *Flow Induced Vibration,* Van Nostrand–Reinhold, New York, 1977.

13. Fox, R. W., and A. T. MacDonald, *Introduction to Fluid Mechanics* (2nd ed.), Wiley, New York, 1978.

14. Wendel, K., "Hydrodynamic Masses and Hydrodynamic Moments of Inertia," U.S. Navy, David Taylor Model Basin Transactions 260, 1956.

15. Patton, K. T., "Tables of Hydrodynamic Mass Factors for Translating Motion," ASME Paper 65-WA/UNT-2, 1965.

Applications of Flow Analysis: The Potential Flow and Boundary Layer Model of a High Reynolds Number Flow Field

9

We are now able to solve the following types of flow problems using the methods presented in Chapters 4–8 (primarily for incompressible flow):

- Any one-dimensional, one-directional inviscid flow
- Fully developed (i.e., one-dimensional and one-directional) laminar or turbulent flow in a pipe or duct with constant cross-sectional area
- Flow and energy loss problems in piping systems provided that reliable information is available on local loss coefficients
- Forces on bodies in external flow if drag and/or lift coefficient data are available (we can also calculate lift if circulation is known)

This list contains many problems of practical importance, but it omits a number of situations of engineering interest. If we consider the last two items, which involve multidimensional, multidirectional viscous flow, we find that *we are incapable of dealing with any type of flow device that has not been previously built and extensively tested.* If our fluid mechanics remained limited to these four items, the only "new" devices and systems that we could analyze would be made up of "old" components simply rearranged.*

If we limit consideration to incompressible Newtonian fluids, the differential continuity equation (Eq. 3.19) and the Navier-Stokes equations (Eqs. 4.57, 4.58, and 4.59) are capable of predicting any laminar flow. If the flow is turbulent, the apparent stresses have to be included (Eq. 7.15) and an ad hoc model of these stresses must be devised, a difficult but not impossible task [1]. Unfortunately, these equations have defied solution for all but a few simple cases (such as fully developed pipe and duct flows); therefore, engineering calculations of multidimensional flows are often based on simplified models. In this chapter, we will consider the most widely used multidimensional flow model, the potential flow and boundary layer model.

---

* Actually, we could not even analyze all systems of this type because our information deals mainly with components in isolation, neglecting interference effects.

9.1   The Potential Flow and Boundary Layer Model

The motion of any fluid particle is determined by the balance of forces (including inertia) on the particle. If, in a particular flow, either inertia force or shear force is negligible, flow prediction becomes relatively easy. We are already familiar with Bernoulli's equation for the case of negligible shear and with fully developed pipe flow solutions for the case of zero inertia.

We have learned that the Reynolds number expresses the ratio of inertia and shear forces for a typical fluid particle. Since many engineering flows have high Reynolds number, it might seem that we would be justified in neglecting shear force completely; however, our discussions in Section 8.2.2 pointed out that this assumption usually leads to completely erroneous results. We found that the answer to this perplexing problem was the existence of the boundary layer. In Chapter 8, we used the boundary layer concept to explain the behavior of flow over an object at high Reynolds number, but we did not introduce any quantitative consideration of the boundary layer itself or of the flow outside of the boundary layer. We will now do just that.

The beginning point is the complete boundary layer hypothesis, first put forth by L. Prandtl in 1904 [2]:

> In a high Reynolds number flow, the effects of fluid viscosity and rotation (i.e., shear stress and turbulence) are confined to thin regions near solid boundaries or surfaces of discontinuity (like the boundary of a jet issuing into a still fluid). Outside of this thin region, the flow behaves as if it were ideal (i.e., inviscid and irrotational).

Figure 9.1 illustrates these ideas for both external and internal flows. We will show later that, for laminar flow, the boundary layer thickness is estimated by

$$\frac{\delta}{x} \sim \frac{1}{\sqrt{\mathbf{R}_x}}, \quad \text{where} \quad \mathbf{R}_x = \frac{Vx}{\nu}.$$

Note that the boundary layer grows in the downstream direction, so if

Figure 9.1 Division of high Reynolds number flow into potential flow and boundary layer regions: (a) Boundary layer and wake in external flow; (b) boundary layer and ideal flow regions in developing flow.

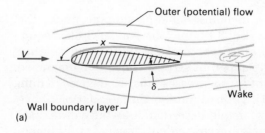

(a)

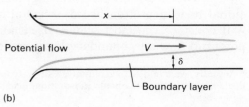

(b)

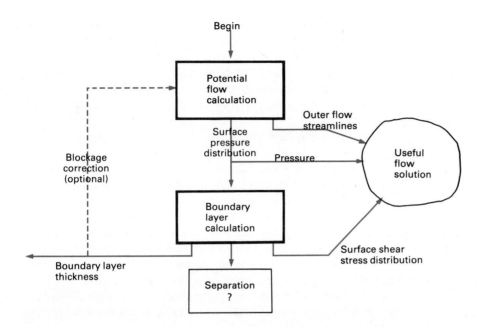

Figure 9.2 Schematic of calculations in the high **R** potential flow/boundary layer model.

the flow field is of limited extent (an internal flow), the boundary layer may grow to occupy the entire field, as in fully developed pipe flow.

Using the boundary layer hypothesis, we divide the flow field into two different regions and analyze each separately. A complete flow analysis is obtained by "patching" the two regions together. The outer region is analyzed using the assumption of zero shear stress. For this analysis, the boundary layer is usually assumed to be so thin that it can be neglected.

It can be shown that the condition of zero shear stress corresponds to the condition of irrotational flow (Sections 3.5.1 and 3.5.2). Irrotationality guarantees that the fluid velocity field can be obtained from the gradient of a scalar potential function; hence this type of flow (inviscid, irrotational) is often called *potential flow*. The most useful results of a potential flow calculation are the pressure distribution on the solid surface and the streamline pattern "near" the surface (but outside of the boundary layer).

The boundary layer is analyzed by assuming that it is so thin that the pressure is constant *across* the layer and equal to the surface pressure calculated by the (outer) potential flow analysis. Viscous, pressure, and inertia forces are retained in the boundary layer analysis, but the pressure is known and the viscous forces are simplified by neglecting the normal viscous stresses. The condition of constant pressure across the boundary layer replaces the force-momentum balance in the direction perpendicular to the surface, so only the momentum equation in the direction parallel to the surface and the continuity equation are considered. The results of a boundary layer calculation are the surface shear stress, the thickness of the boundary layer, and information on the possible occurrence of boundary layer separation.

The interaction between potential flow and boundary layer calculations is illustrated in Fig. 9.2. Calculations begin with the potential flow, neglecting the boundary layer. After the surface pressure is calculated, the boundary layer calculation is made. Often the calculations end here. Sometimes, for greater accuracy, the effects of the slower-moving boundary layer fluid are fed back into the potential flow calculation to obtain a more accurate pressure distribution. This step is especially important in calculation of internal flows such as in the entrance region of a pipe.

It is important to note how the physical boundary conditions are handled in the potential flow and boundary layer model. The potential flow calculation handles the uniform flow at "infinity" condition for external flow or the specified mass flow condition for internal flows. The boundary layer calculation handles the no-slip condition at a solid surface. Both calculations use the zero normal (relative) velocity condition at a solid surface. The potential flow calculation produces a finite tangential (slip) velocity at a solid surface (it does not satisfy the no-slip condition). This slip velocity is taken to be the velocity at the *outer* edge of the boundary layer in the boundary layer calculation.*

The potential flow and boundary layer model of a high Reynolds number flow allows us to consider its two components separately. To make a potential flow calculation, we don't have to know anything about the boundary layer. To make a boundary layer calculation, we need to know only the surface pressure distribution. The velocity at the outer edge of the layer can be obtained from the pressure via Bernoulli's equation since the flow is ideal just outside the layer. A great deal of development work on boundary layer calculation has been done using *measured* pressure distributions, bypassing potential flow calculations entirely. The remainder of this chapter will be devoted to separate consideration of potential flow and boundary layer calculation. Our treatment of both subjects will be such that they can be studied in any order, or only one of them can be considered. Since entire books have been written on each subject ([3, 4] and [5, 6]), our consideration here will be only a brief introduction. If you are interested in further study, also see [7, 8, 9].

9.2 Potential Flow

According to the procedure outlined in Fig. 9.2, a complete calculation begins with potential flow, so it seems reasonable that we consider it first. Historically, potential flow theory is the oldest branch of analytical fluid mechanics, dating from the mid-1700s. In fact, by the time Prandtl introduced the boundary layer hypothesis in 1904, a great many potential flow solutions had been worked out by mathematicians, only to be discarded by engineers as unrealistic.[†] Potential flow theory became important in practical fluid mechanics largely because boundary layer theory was developed to complement it.

---

* Note that the inner boundary of the outer region is matched to the outer boundary of the inner region.

† Remember d'Alembert's paradox (Section 8.2). This was first studied by d'Alembert in 1752.

9.2.1 The Basics of Potential Flow Theory: Irrotationality, Velocity Potential, and Stream Function

Figure 9.3 illustrates the type of flows we wish to calculate. We neglect boundary layers and assume that the fluid is inviscid and incompressible. The governing equations are the continuity equation,

$$\frac{\partial u}{\partial x} + \frac{\partial v}{\partial y} + \frac{\partial w}{\partial z} = 0, \tag{3.19}$$

and Euler's equations,

$$\frac{\partial u}{\partial t} + u\left(\frac{\partial u}{\partial x}\right) + v\left(\frac{\partial u}{\partial y}\right) + w\left(\frac{\partial u}{\partial z}\right) = -\frac{1}{\rho}\left(\frac{\partial p}{\partial x}\right) + g_x. \tag{4.32}$$

$$\frac{\partial v}{\partial t} + u\left(\frac{\partial v}{\partial x}\right) + v\left(\frac{\partial v}{\partial y}\right) + w\left(\frac{\partial v}{\partial z}\right) = -\frac{1}{\rho}\left(\frac{\partial p}{\partial y}\right) + g_y, \tag{4.33}$$

$$\frac{\partial w}{\partial t} + u\left(\frac{\partial w}{\partial x}\right) + v\left(\frac{\partial w}{\partial y}\right) + w\left(\frac{\partial w}{\partial z}\right) = -\frac{1}{\rho}\left(\frac{\partial p}{\partial z}\right) + g_z. \tag{4.34}$$

We can simplify these equations slightly by defining

$$p' = p + \rho g(\text{Elevation}), \tag{9.1}$$

so

$$\frac{\partial p'}{\partial x} = \frac{\partial p}{\partial x} + \rho g\left(\frac{\partial(\text{Elevation})}{\partial x}\right) = \frac{\partial p}{\partial x} - \rho g_x.$$

See Fig. 4.11. Similar relations hold for $\partial p'/\partial y$ and $\partial p'/\partial z$, so the Euler equations can be written as (using the Eulerian derivative operator D/Dt)

$$\frac{Du}{Dt} = -\frac{1}{\rho}\left(\frac{\partial p'}{\partial x}\right), \tag{9.2}$$

$$\frac{Dv}{Dt} = -\frac{1}{\rho}\left(\frac{\partial p'}{\partial y}\right), \tag{9.3}$$

$$\frac{Dw}{Dt} = -\frac{1}{\rho}\left(\frac{\partial p'}{\partial z}\right). \tag{9.4}$$

Since depth (h) is the negative of elevation, p' is the difference between the fluid static pressure (p) and the hydrostatic pressure (γh). For incompressible flow, we can solve for p' and then *add* the hydrostatic pressure to obtain p.

Equations (3.19), (9.2), (9.3), and (9.4) are four equations with four unknowns, u, v, w, and p'. We have already seen that the Euler equations can be integrated to yield Bernoulli's equation for steady, irrotational flow (Section 4.1.6). Our assumption of an inviscid fluid usually implies the condition of irrotationality. We will demonstrate the connection between the inviscid (or zero-viscosity) condition and the kinematic condition of irrotational flow both by physical arguments and by mathematical manipulation.

Recall that an irrotational flow is one in which the average angular velocity of all fluid particles is zero. Consider a fluid at rest or in uniform, parallel (one-directional, zero-dimensional) motion. Obviously all

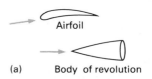

Airfoil

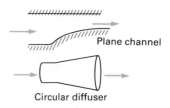

(a)   Body of revolution

Plane channel

Circular diffuser

(b)

Turbomachine blade passage

Figure 9.3 Typical flow geometries for potential flow problems: (a) External flows; (b) internal flows.

the fluid particles have zero angular velocity. The only way that a fluid particle can obtain any angular velocity is from the action of a torque on it. If there are no shear stresses (inviscid fluid), the only forces on the particle are normal forces (pressure, which acts through the geometric center) and gravity (which acts through the center of mass), neither of which is capable of exerting a net torque on a fluid particle.* Even if the still fluid is disturbed by a solid moving through it or if an object is placed in the moving stream, there will be no rotation if there are no shear stresses. We conclude that in the absence of viscosity, a fluid particle that initially has zero angular velocity will maintain that condition.

The mathematical demonstration that zero viscosity implies irrotational flow is more involved. The basic idea is to convert the Navier-Stokes equations for a viscous fluid (Eqs. 4.57–4.59) into an equation for the fluid vorticity. This is done by cross-differentiating the equations and subtracting them. The resulting equation is called the *vorticity transport equation* and, in vector notation, is

$$\frac{D\vec{\zeta}}{Dt} = (\vec{\zeta} \cdot \nabla)\vec{V} + \frac{\mu}{\rho}\nabla^2\vec{\zeta},$$

where $\vec{\zeta}$ is the vorticity vector, defined for two-dimensional and two-directional flow in Section 3.5.1. For three-dimensional, three-directional flow, the vorticity is

$$\vec{\zeta} = \zeta_x\hat{i} + \zeta_y\hat{j} + \zeta_z\hat{k}, \tag{9.5}$$

where

$$\zeta_x = \frac{\partial w}{\partial y} - \frac{\partial v}{\partial z}, \tag{9.6}$$

$$\zeta_y = \frac{\partial u}{\partial z} - \frac{\partial w}{\partial x}, \tag{9.7}$$

and

$$\zeta_z = \frac{\partial v}{\partial x} - \frac{\partial u}{\partial y}. \tag{9.8}$$

Notice that the vorticity transport equation does not contain the pressure; it was eliminated by cross differentiation and subtraction. If the fluid is inviscid ($\mu = 0$), the second term on the right in the vorticity equation vanishes, and we get

$$\frac{D\vec{\zeta}}{Dt} = (\vec{\zeta} \cdot \nabla)\vec{V}.$$

Recall that D/Dt of any quantity is the time rate of change of that quantity for a specific fluid particle. If a fluid particle initially has $\vec{\zeta} = 0$, then $D\vec{\zeta}/Dt$ is zero and the fluid particle never gains vorticity. Since the vorticity equation applies to all fluid particles, an initially irrotational flow field will remain irrotational.

---

* In a fluid of nonuniform density, the center of mass and the geometric center of a fluid particle do not coincide, so a torque about the center of mass of the particle may be generated. We will not consider such cases here.

All of this discussion about vorticity and rotation is often very confusing to the beginner. A summary of the results is as follows:

- If we neglect fluid viscosity, an initially irrotational flow will remain irrotational.
- If the flow is irrotational and steady, Euler's (differential) equations can be replaced by Bernoulli's (algebraic) equation,

$$\frac{p}{\rho} + \frac{V^2}{2} + gz = \text{Constant},$$

for the entire flow field. This was shown in Section 4.1.6.

The condition of irrotationality simplifies the process of finding the velocity distribution. The irrotationality condition,

$$\vec{\zeta} = \nabla \times \vec{V} = 0,$$

is a *necessary and sufficient condition* for the existence of a scalar function $\Phi(x, y, z)$ such that

$$\vec{V} = \text{Gradient} (\Phi) = \nabla\Phi. \tag{9.9}$$

In component form,

$$u = \frac{\partial\Phi}{\partial x}, \tag{9.10}$$

$$v = \frac{\partial\Phi}{\partial y}, \tag{9.11}$$

and

$$w = \frac{\partial\Phi}{\partial z}. \tag{9.12}$$

If we assume incompressible flow, a differential equation for Φ can be obtained by substituting Eqs. (9.10)–(9.12) into the continuity equation, Eq. (3.19), giving

$$\frac{\partial^2\Phi}{\partial x^2} + \frac{\partial^2\Phi}{\partial y^2} + \frac{\partial^2\Phi}{\partial z^2} = \nabla^2\Phi = 0. \tag{9.13}$$

Equation (9.13) is called *Laplace's equation* and is probably the best-known partial differential equation in all mathematics.*

The function Φ is called the *velocity potential*. A velocity potential can be defined only for irrotational flow, and any velocities derived from a potential function will always constitute an irrotational flow. It is usually simpler to solve for a single scalar potential function than for two or three velocity components, so inviscid-irrotational flow problems are almost always analyzed in terms of velocity potential. The entire field of study is often called *potential flow theory*. A velocity potential can even be defined for a compressible fluid. Potential flow theory

---

* Laplace's equation describes many phenomena in addition to inviscid fluid flow. Among these are deflection of a thin elastic membrane, electric potential in a uniform conducting material, and distribution of temperature in steady heat conduction in a solid. These analogies are sometimes useful since the (mathematical) solution to one type of problem can be adapted to the other types.

limited to incompressible fluids is sometimes called *ideal flow theory* (ideal = incompressible + inviscid) or *hydrodynamics*.

Calculation of three-dimensional, three-directional potential flow is difficult. Two-dimensional, two-directional flows exhibit most of the important characteristics of potential flow and are easier to solve. If we limit consideration to two-dimensional, two-directional flows and choose rectangular coordinates (x, y), the velocities (u, v) are given by Eqs. (9.10) and (9.11) and Φ satisfies the two-dimensional Laplace equation

$$\frac{\partial^2 \Phi}{\partial x^2} + \frac{\partial^2 \Phi}{\partial y^2} = 0. \tag{9.14}$$

Rectangular coordinates are not always the most convenient for analyzing potential flow. For two-dimensional flow in a plane, it is sometimes convenient to use polar coordinates (Fig. 9.4a). In plane polar coordinates, the velocity potential equation is

$$\frac{\partial}{\partial r}\left(r\frac{\partial \Phi}{\partial r}\right) + \frac{1}{r}\left(\frac{\partial^2 \Phi}{\partial \theta^2}\right) = 0, \tag{9.15}$$

and the velocity components are

$$V_r = \frac{\partial \Phi}{\partial r} \tag{9.16}$$

and

$$V_\theta = \frac{1}{r}\left(\frac{\partial \Phi}{\partial \theta}\right). \tag{9.17}$$

Flow over an axisymmetric body at zero angle of attack or through a passage of circular cross section is conveniently analyzed in cylindri-

Figure 9.4 Alternative coordinate systems for potential flow analysis: (a) Plane polar coordinates; (b) cylindrical coordinates.

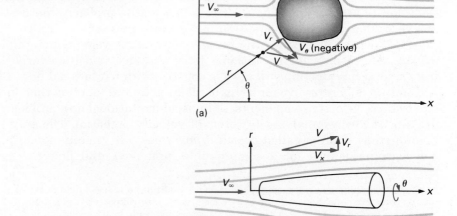

cal coordinates (Fig. 9.4b). The equations are

$$\frac{1}{r}\left(\frac{\partial}{\partial r}\right)\left(r\frac{\partial \Phi}{\partial r}\right) + \frac{\partial^2 \Phi}{\partial x^2} = 0, \qquad \textbf{(9.18)}$$

$$V_x = \frac{\partial \Phi}{\partial x}, \qquad \textbf{(9.19)}$$

and

$$V_r = \frac{\partial \Phi}{\partial r}. \qquad \textbf{(9.20)}$$

In a two-dimensional flow (either plane or axisymmetric), it is possible to define a second scalar function from which the velocities can be derived. First consider two-dimensional flow in a plane. The continuity equation can be written

$$\frac{\partial u}{\partial x} - \frac{\partial(-v)}{\partial y} = 0.$$

This equation is the necessary and sufficient condition for the existence of a function $\Psi(x, y)$ such that

$$u = \frac{\partial \Psi}{\partial y} \qquad \textbf{(9.21)}$$

and

$$v = -\frac{\partial \Psi}{\partial x}. \qquad \textbf{(9.22)}$$

The function Ψ, called the *stream function,* satisfies the continuity equation identically.

The stream function definition can be put into polar coordinates. The continuity equation is

$$\frac{1}{r}\left(\frac{\partial(rV_r)}{\partial r}\right) + \frac{1}{r}\left(\frac{\partial V_\theta}{\partial \theta}\right) = 0. \qquad \textbf{(9.23)}$$

Defining

$$V_r = \frac{1}{r}\left(\frac{\partial \Psi}{\partial \theta}\right) \qquad \textbf{(9.24)}$$

and

$$V_\theta = -\frac{\partial \Psi}{\partial r} \qquad \textbf{(9.25)}$$

gives

$$\frac{1}{r}\left(\frac{\partial}{\partial r}\right)\left(\frac{\partial \Psi}{\partial \theta}\right) + \frac{1}{r}\left(\frac{\partial}{\partial \theta}\right)\left(\frac{\partial \Psi}{\partial r}\right) \equiv 0.$$

We can also define a stream function for axisymmetric flow, but the stream function must have a slightly different form. For axisymmetric flow, the continuity equation is

$$\frac{1}{r}\left(\frac{\partial}{\partial r}\right)(rV_r) + \frac{\partial V_x}{\partial x} = 0 \qquad \textbf{(9.26)}$$

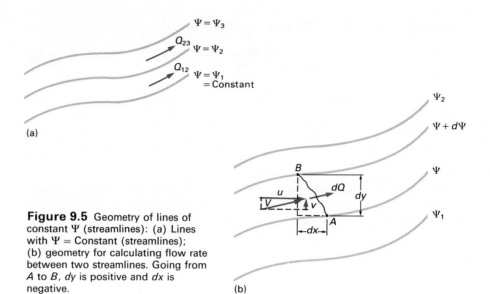

Figure 9.5 Geometry of lines of constant Ψ (streamlines): (a) Lines with Ψ = Constant (streamlines); (b) geometry for calculating flow rate between two streamlines. Going from A to B, dy is positive and dx is negative.

or, multiplying by r and rearranging,*

$$-\frac{\partial}{\partial r}(-rV_r) + \frac{\partial}{\partial x}(rV_x) = 0.$$

This allows us to define a function Ψ_s such that

$$rV_r = -\frac{\partial \Psi_s}{\partial x} \quad \text{and} \quad rV_x = +\frac{\partial \Psi_s}{\partial r},$$

or

$$V_r = -\frac{1}{r}\left(\frac{\partial \Psi_s}{\partial x}\right) \tag{9.27}$$

and

$$V_x = \frac{1}{r}\left(\frac{\partial \Psi_s}{\partial r}\right); \tag{9.28}$$

Ψ_s is called *Stokes' stream function*.

The stream function has more physical significance than the velocity potential. Consider a curve along which Ψ = Constant (Fig. 9.5a). Along such a curve,

$$d\Psi = \frac{\partial \Psi}{\partial x}\,dx + \frac{\partial \Psi}{\partial y}\,dy = 0.$$

The slope of this curve is

$$\left.\frac{dy}{dx}\right)_{\Psi=\text{constant}} = -\frac{\dfrac{\partial \Psi}{\partial x}}{\dfrac{\partial \Psi}{\partial y}} = \frac{v}{u},$$

* Note that $\partial r/\partial x = 0$.

which is the slope of a streamline. Thus *lines of constant* Ψ *are stream-lines.**

Next consider the change of Ψ between two streamlines (Fig. 9.5a):

$$\Psi_2 - \Psi_1 = \int_1^2 d\Psi = \int_1^2 \frac{\partial \Psi}{\partial x}\, dx + \int_1^2 \frac{\partial \Psi}{\partial y}\, dy.$$

Substituting from Eqs. (9.21) and (9.22) gives

$$\Psi_2 - \Psi_1 = \int_1^2 (-v\, dx + u\, dy)$$

$$= \int_2^1 v\, dx + \int_1^2 u\, dy.$$

From Fig. 9.5(b) the integral represents the volume flow rate (per unit depth) between the two curves; thus

$$\Psi_2 - \Psi_1 = \int_2^1 v\, dx + \int_1^2 u\, dy = \int_1^2 dQ$$

$$= Q_{1-2}.$$

Therefore:

> The difference in Ψ between two streamlines is the volume flow rate per unit depth between the two streamlines.

The stream function automatically satisfies the continuity equation for two-dimensional flow. To obtain the stream function for a particular flow, we need a second equation. For irrotational flow, an equation for Ψ is easily obtained from the zero-vorticity condition. In terms of Ψ, the vorticity for two-dimensional, two-directional flow in a plane is

$$\zeta_z = \frac{\partial v}{\partial x} - \frac{\partial u}{\partial y} = \frac{\partial}{\partial x}\left(-\frac{\partial \Psi}{\partial x}\right) - \frac{\partial}{\partial y}\left(\frac{\partial \Psi}{\partial y}\right) = -\nabla^2 \Psi.$$

For irrotational flow, ζ_z is zero and Ψ must satisfy

$$\frac{\partial^2 \Psi}{\partial x^2} + \frac{\partial^2 \Psi}{\partial y^2} = 0. \tag{9.29}$$

This is the same form of equation (Laplace's) that governs Φ, but of course this does not mean that Φ and Ψ are the same function because each depends on different boundary conditions. In plane polar coordinates, the differential equation for Ψ is

$$\frac{\partial}{\partial r}\left(r\frac{\partial \Psi}{\partial r}\right) + \frac{1}{r}\left(\frac{\partial^2 \Psi}{\partial \theta^2}\right) = 0. \tag{9.30}$$

Stokes' stream function (Ψ_s) has all the physical characteristics of the plane two-dimensional stream function (Ψ). The proof of this is left as a problem at the end of the chapter. The differential equation for Stokes' stream function is obtained from the irrotationality condition

$$-\zeta_\theta = \frac{\partial^2 \Psi_s}{\partial r^2} - \frac{1}{r}\left(\frac{\partial \Psi_s}{\partial r}\right) + \frac{\partial^2 \Psi_s}{\partial x^2} = 0. \tag{9.31}$$

This is *almost* Laplace's equation in axisymmetric cylindrical coordinates, but not quite; the sign of the second term is wrong.

---

* Now you know the reason for the choice of notation in Section 3.3.2.

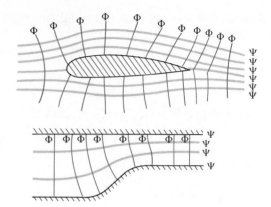

Figure 9.6 Equipotential (Φ) lines and streamlines (Ψ) for external and internal flow.

An extremely important property of potential and stream functions is as follows:

> Lines of Φ = Constant (called *equipotential lines*) are perpendicular to lines of Ψ = Constant (*streamlines*).*

This can be demonstrated by noting that

$$\left. \frac{dy}{dx} \right)_{\Phi=\text{constant}} = -\frac{\dfrac{\partial \Phi}{\partial x}}{\dfrac{\partial \Phi}{\partial y}} = -\frac{u}{v}$$

and

$$\left. \frac{dy}{dx} \right)_{\Psi=\text{constant}} = -\frac{\dfrac{\partial \Psi}{\partial y}}{\dfrac{\partial \Psi}{\partial x}} = \frac{v}{u}.$$

Since equipotential lines and streamlines have slopes that are negative reciprocals, they are perpendicular. This relation fails only at stagnation points, where both u and v are zero.

Figure 9.6 shows equipotential lines and streamlines for external and internal flows. Determining this network of lines for a particular boundary shape is tantamount to solving the corresponding potential flow problem. Solving potential flow problems is essentially a geometric exercise that might even be done (approximately) by graphical construction of a flow net like those shown in Fig. 9.6. It is particularly interesting to note that the primary equations of potential flow, such as Eqs. (9.13) and (9.29), do not contain any fluid properties (not even density). There are no dynamic similarity parameters in (incompressible) potential flow; geometric similarity alone is sufficient to guarantee kinematic similarity. The only thing that distinguishes one potential flow from another is the shape of the boundaries.

---

* This is also true for Ψ_s. The proof is left as an end-of-chapter problem.

Table 9.1 Boundary conditions for potential flow

Boundary	Physical Boundary Condition	Condition on Φ	Condition on Ψ	Condition on Ψ_s
Solid surface	Velocity normal to surface = 0	$\dfrac{\partial \Phi}{\partial n} = 0$	$\dfrac{\partial \Psi}{\partial s} = 0,$ $\Psi = \text{Constant}$	$\dfrac{\partial \Psi_s}{\partial s} = 0,$ $\Psi_s = \text{Constant}$
Inlet or outlet	Velocity a known function of x, y	$\dfrac{\partial \Phi}{\partial x}$ and $\dfrac{\partial \Phi}{\partial y}$ known	$\dfrac{\partial \Psi}{\partial x}$ and $\dfrac{\partial \Psi}{\partial y}$ known	$\dfrac{1}{r}\left(\dfrac{\partial \Psi_s}{\partial x}\right)$ and $\dfrac{1}{r}\left(\dfrac{\partial \Psi_s}{\partial r}\right)$ known
Inlet or outlet	Uniform parallel flow in x-direction	$\Phi = V_\infty x + C$	$\Psi = V_\infty y + C$	$\Psi_s = \dfrac{1}{2} V_\infty r^2 + C$
Outer boundary far from surface	Uniform parallel flow in x-direction	$\Phi = V_\infty x + C$	$\Psi = V_\infty y + C$	$\Psi_s = \dfrac{1}{2} V_\infty r^2 + C$

Notes: n = Coordinate normal to surface; s = Coordinate parallel to surface; C = Constant of integration.

To solve for Φ, Ψ, or Ψ_s, it is necessary to specify boundary conditions. These are derived from boundary conditions on velocity by noting that the derivative of Φ in a given direction is equal to the velocity in that direction and that the derivative of Ψ (or Ψ_s) in a given direction is the velocity perpendicular to that direction. Various boundary conditions are listed in Table 9.1. Also see Fig. 9.7.

Note that the stream function is constant on a solid surface, that is, a solid surface is a streamline. This can be put to great use; we can

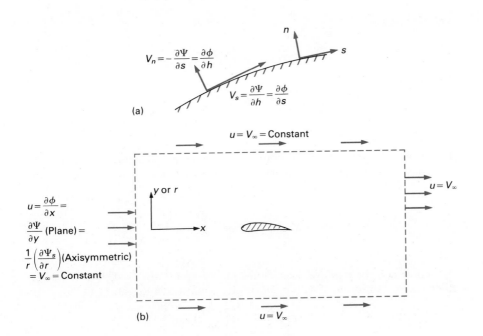

Figure 9.7 Boundary conditions for potential flow: (a) Velocities near solid surface; (b) velocities at faraway boundaries.

replace any streamline in a potential flow with a solid surface. Many useful potential flow solutions are obtained by finding an interesting streamline pattern and then replacing one or more of the streamlines by a solid surface. The remaining streamlines represent the potential flow over that surface.

EXAMPLE 9.1   Illustrates the velocity potential and stream function for flow over an object in a uniform stream

Show that the velocity potential Φ and the stream function Ψ given by

$$\Phi = V_\infty \left(r + \frac{R^2}{r} \right) \cos \theta, \qquad r \geq R,$$

and

$$\Psi = V_\infty \left(r - \frac{R^2}{r} \right) \sin \theta, \qquad r \geq R,$$

satisfy the continuity equation and the irrotationality condition and yield a velocity field that satisfies the boundary conditions:

$$V_r = 0 \qquad\qquad \text{at } r = R,$$
$$V_r = V_\infty \cos \theta \qquad \text{at } r \to \infty,$$
$$V_\theta = -V_\infty \sin \theta \qquad \text{at } r \to \infty.$$

Therefore, demonstrate that the given potential and stream functions describe the flow over an infinite cylinder of radius R.

Solution

Given
Velocity potential:

$$\Phi = V_\infty \left(r + \frac{R^2}{r} \right) \cos \theta, \qquad r \geq R$$

Stream function:

$$\Psi = V_\infty \left(r - \frac{R^2}{r} \right) \sin \theta, \qquad r \geq R$$

Show
That Φ and Ψ both satisfy the continuity equation and the irrotationality condition and yield a velocity field that satisfies the boundary conditions:

$$V_r = 0 \qquad\qquad \text{at } r = R,$$
$$V_r = V_\infty \cos \theta \qquad \text{at } r \to \infty,$$
$$V_\theta = -V_\infty \sin \theta \qquad \text{at } r \to \infty$$

Solution
Polar coordinates are used. Since only r and θ are used, the flow is two-dimensional. In terms of Φ, the continuity equation is Eq. (9.15),

$$\frac{\partial}{\partial r}\left(r \frac{\partial \Phi}{\partial r} \right) + \frac{1}{r}\left(\frac{\partial^2 \Phi}{\partial \theta^2} \right) = 0.$$

Substituting for Φ gives

$$\frac{\partial}{\partial r}\left[rV_\infty\left(1-\frac{R^2}{r^2}\right)\cos\theta\right]+\frac{1}{r}\left[V_\infty\left(r+\frac{R^2}{r}\right)(-\cos\theta)\right]=0,$$

$$V_\infty\left(1+\frac{R^2}{r^2}\right)\cos\theta-V_\infty\left(1+\frac{R^2}{r^2}\right)\cos\theta=0,$$

or

$$0=0,$$

and   **Answer** ◀

| Φ satisfies the continuity equation. |

In terms of Ψ, the continuity equation is Eq. (9.30),

$$\frac{\partial}{\partial r}\left(r\frac{\partial\Psi}{\partial r}\right)+\frac{1}{r}\left(\frac{\partial^2\Psi}{\partial\theta^2}\right)=0.$$

Substituting for Ψ gives

$$\frac{\partial}{\partial r}\left[rV_\infty\left(1+\frac{R^2}{r^2}\right)\sin\theta\right]+\frac{1}{r}\left[V_\infty\left(r-\frac{R^2}{r}\right)(-\sin\theta)\right]=0,$$

$$V_\infty\left(1-\frac{R^2}{r^2}\right)\sin\theta+V_\infty\left(1-\frac{R^2}{r^2}\right)(-\sin\theta)=0,$$

or

$$0=0,$$

and   **Answer** ◀

| Ψ satisfies the continuity equation. |

The irrotationality condition for plane flow in polar coordinates is

$$\zeta_z=\frac{1}{r}\left[\frac{\partial}{\partial r}(rV_\theta)-\frac{\partial}{\partial\theta}(V_r)\right]=0.$$

Now

$$V_r=\frac{\partial\Phi}{\partial r}=V_\infty\left(1-\frac{R^2}{r^2}\right)\cos\theta\quad\text{and}\quad V_\theta=\frac{1}{r}\left(\frac{\partial\Phi}{\partial\theta}\right)=-\frac{V_\infty}{r}\left(r+\frac{R^2}{r}\right)\sin\theta.$$

Substituting gives

$$\zeta_z=\frac{1}{r}\left(\frac{\partial}{\partial r}\right)\left[-V_\infty\left(r+\frac{R^2}{r}\right)\sin\theta\right]-\frac{1}{r}\left(\frac{\partial}{\partial\theta}\right)\left[V_\infty\left(1-\frac{R^2}{r^2}\right)\cos\theta\right]$$

$$=\frac{1}{r}\left[-V_\infty\left(1-\frac{R^2}{r^2}\right)\sin\theta\right]+\frac{1}{r}\left[V_\infty\left(1-\frac{R^2}{r^2}\right)\sin\theta\right],$$

or

$$\zeta_z=0.$$

Therefore   **Answer** ◀

| Φ satisfies the irrotationality condition. |

Noting that

$$V_r=\frac{1}{r}\left(\frac{\partial\Psi}{\partial\theta}\right)=\frac{1}{r}\left[V_\infty\left(r-\frac{R^2}{r}\right)\cos\theta\right]$$

$$=V_\infty\left(1-\frac{R^2}{r^2}\right)\cos\theta$$

and

$$V_\theta = -\frac{\partial \Psi}{\partial r} = -V_\infty \left(1 + \frac{R^2}{r^2}\right) \sin \theta$$

are the same expressions obtained for V_r and V_θ from Φ, we conclude that

Answer

$$\boxed{\Psi \text{ satisfies the irrotationality condition.}}$$

Last we consider the boundary conditions. The velocity components are given by Eqs. (9.16), (9.17), (9.24), and (9.25):

$$V_r = \frac{\partial \Phi}{\partial r} \qquad V_\theta = \frac{1}{r}\left(\frac{\partial \Phi}{\partial \theta}\right)$$

$$V_r = \frac{1}{r}\left(\frac{\partial \Psi}{\partial \theta}\right) \qquad V_\theta = -\frac{\partial \Psi}{\partial r}.$$

At $r = R$,

$$V_r = \frac{\partial \Phi}{\partial r}\bigg|_{r=R} = \left[V_\infty\left(1 - \frac{R^2}{r^2}\right)\cos\theta\right]_{r=R},$$

Answer

$$\boxed{V_r = 0 \quad \text{for } \Phi \text{ at } r = R.}$$

$$V_r = \frac{1}{r}\left(\frac{\partial \Psi}{\partial \theta}\right)\bigg|_{r=R} = \left[\frac{V_\infty}{r}\left(r - \frac{R^2}{r}\right)\cos\theta\right]_{r=R},$$

Answer

$$\boxed{V_r = 0 \quad \text{for } \Psi \text{ at } r = R.}$$

At $r \to \infty$,

$$V_r = \frac{\partial \Phi}{\partial r}\bigg|_{r\to\infty} = \left[V_\infty\left(1 - \frac{R^2}{r^2}\right)\cos\theta\right]_{r\to\infty},$$

Answer

$$\boxed{V_r = V_\infty \cos\theta \quad \text{for } \Phi \text{ at } r \to \infty,}$$

$$V_r = \frac{1}{r}\left(\frac{\partial \Psi}{\partial \theta}\right)\bigg|_{r\to\infty} = \left[\frac{V_\infty}{r}\left(r - \frac{R^2}{r}\right)\cos\theta\right]_{r\to\infty},$$

Answer

$$\boxed{V_r = V_\infty \cos\theta \quad \text{for } \Psi \text{ at } r \to \infty,}$$

$$V_\theta = \frac{1}{r}\left(\frac{\partial \Phi}{\partial \theta}\right)\bigg|_{r\to\infty} = \left[\frac{-V_\infty}{r}\left(r + \frac{R^2}{r}\right)\sin\theta\right]_{r\to\infty},$$

Answer

$$\boxed{V_\theta = -V_\infty \sin\theta \quad \text{for } \Phi \text{ at } r \to \infty,}$$

and

$$V_\theta = -\frac{\partial \Psi}{\partial r} = -\left[V_\infty\left(1 + \frac{R^2}{r^2}\right)\sin\theta\right]_{r\to\infty},$$

Answer

$$\boxed{V_\theta = -V_\infty \sin\theta \quad \text{for } \Psi \text{ at } r \to \infty.}$$

Discussion
Since $V_r = 0$ at $r = R$, no fluid crosses the circle given by $r = R$. The velocity field thus represents flow over a circle in the r,θ-plane or flow over an infinitely long cylinder perpendicular to the plane.

Solutions to potential flow problems can be obtained by purely mathematical means, without any need for experimental information. There are three solution methods commonly used:

- The method of singularities, in which "interesting" flow fields are built up by combining the flows generated by isolated singularities.
- The method of complex variables (applicable only to two-dimensional flows in a plane). In this method, a few basic flows are geometrically transformed into other flows by using the theory of complex variables.
- Direct, but approximate, methods. This extremely wide category runs from graphical construction of a net of equipotentials and streamlines to computer solution of the potential or stream function equations using finite-difference or finite-element approximations. There is some overlap with the method of singularities because one of the most popular computer methods is based on simulating an arbitrary flow by a *distribution* of singularities.

We will consider each of these three methods; however, we will devote the most effort to the method of singularities. Many students lack the mathematical background to effectively use the method of complex variables, so our discussion of that method will be brief. The numerical methods that make up the majority of the material for the approximate methods are mostly beyond the scope of this book (but *not* beyond the capabilities of most engineering students). We will concentrate on two-dimensional flows in a plane. Methods for axisymmetric flows are similar in principle and can be found in advanced texts [4, 7].

9.2.2 Plane Potential Flows from Singularities

The elementary plane potential flows are those due to *singularities*. A singularity is a point* where the governing equations are *violated*. At these singular points, mass is created or destroyed, or the fluid has infinite vorticity. It is paradoxical that mass-conserving irrotational flows can be built up from singularities. This is due to the fact that the singularities that we consider *do* generate flows that obey the governing equations everywhere *except* at the singularity itself. Singularities are just mathematical "tricks" to generate realistic flows. The basic singularities are the *source* (and *sink*) and the *line vortex*.

Figure 9.8 shows a source located at the origin. The source introduces fluid into the field at a constant volume flow rate, λ, called the *strength* of the source. The fluid flows radially outward from the source, uniformly in all directions. The streamlines are straight radial lines. Because of symmetry, the velocity is purely radial and varies only with radius. This flow most easily described in plane polar coordinates. To calculate the velocity, consider a circular control volume of radius r enclosing the source. The integral continuity equation gives

$$Q = \int_0^{2\pi} V_r r \, d\theta = \lambda.$$

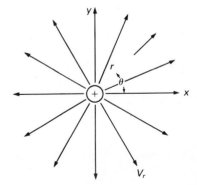

Figure 9.8 Source located at origin.

---

* In two-dimensional flow in a plane, this point is the piercing point of a line perpendicular to the plane.

Since V_r is independent of θ,

$$2\pi r V_r = \lambda,$$

or

$$V_r = \frac{\lambda}{2\pi r}; \left.\begin{array}{c} \\ \\ \\ \end{array}\right\} \text{Plane source} \tag{9.32}$$

by symmetry,

$$V_\theta = 0. \tag{9.33}$$

To demonstrate that this flow does indeed satisfy the governing equations, we first note that the continuity equation requires

$$\frac{1}{r}\left(\frac{\partial}{\partial r}\right)(rV_r) + \frac{1}{r}\left(\frac{\partial}{\partial \theta}\right)(V_\theta) = 0, \quad \text{or} \quad \frac{1}{r}\left(\frac{\partial}{\partial r}\right)\left(\frac{\lambda}{2\pi}\right) = 0,$$

which is obviously true except possibly at $r = 0$. In addition, the irrotationality condition requires

$$\zeta_z = \frac{1}{r}\left[\frac{\partial(rV_\theta)}{\partial r}\right] - \frac{1}{r}\left(\frac{\partial V_r}{\partial \theta}\right) = 0,$$

which is satisfied identically everywhere.

The velocity potential and stream functions for a plane source are obtained from

$$V_r = \frac{\partial \Phi}{\partial r} = \frac{1}{r}\left(\frac{\partial \Psi}{\partial \theta}\right) = \frac{\lambda}{2\pi r}.$$

Integrating gives

$$\Phi = \frac{\lambda}{2\pi}\ln r + C \left.\begin{array}{c} \\ \\ \end{array}\right\} \tag{9.34}$$

and

$$\left.\begin{array}{c} \\ \end{array}\right\} \text{Plane source}$$

$$\Psi = \frac{\lambda}{2\pi}\theta + C. \tag{9.35}$$

The constants of integration are customarily assigned values of zero. The streamlines are straight radial lines ($\theta =$ Constant) and the equipotential lines are circles ($r =$ Constant).

A second singularity, easily related to the source, is the *sink*. A sink is simply a negative source, ingesting fluid at a constant rate. The velocity, velocity potential, and stream function for a sink are simply those for a source with a negative quantity substituted for the strength:

$$\lambda_{\text{sink}} = -\lambda_{\text{source}}.$$

The velocity, velocity potential, and stream functions for a source or sink can be transformed into rectangular coordinates. The results are

$$u = V_r \cos\theta = \frac{\lambda}{2\pi}\left(\frac{x}{\sqrt{x^2 + y^2}}\right), \tag{9.36}$$

$$v = V_r \sin\theta = \frac{\lambda}{2\pi}\left(\frac{y}{\sqrt{x^2 + y^2}}\right), \tag{9.37}$$

$$\Phi = \frac{\lambda}{4\pi}\ln(x^2 + y^2), \tag{9.38}$$

and

$$\Psi = \frac{\lambda}{2\pi} \tan^{-1} \frac{y}{x}. \tag{9.39}$$

These results are for a source/sink located at the origin. If the singularity is located at point (a, b), the quantity $(x - a)$ replaces x, and $(y - b)$ replaces y. For example, the x-component of velocity generated by a source at point (a, b) is

$$u = \frac{\lambda}{2\pi} \left(\frac{(x - a)}{\sqrt{(x - a)^2 + (y - b)^2}} \right).$$

Another basic singularity is the *line vortex*. This singularity can be obtained by noting that, since Φ and Ψ are a set of orthogonal lines, a "new" potential flow can be obtained from a known one by interchanging Φ and Ψ. In this case, we simply switch the roles of Φ and Ψ that we obtained from the source, letting

and*

$$\Phi = \frac{\Gamma}{2\pi} \theta \tag{9.40}$$

$$\Psi = -\frac{\Gamma}{2\pi} \ln r. \tag{9.41}$$

Line vortex

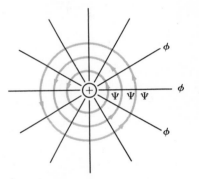

Figure 9.9 Equipotential and streamlines for free vortex.

Streamlines are circles, and equipotential lines are radii (Fig. 9.9). The velocities are

and

$$V_r = \frac{\partial \Phi}{\partial r} = \frac{1}{r} \left(\frac{\partial \Psi}{\partial \theta} \right) = 0 \tag{9.42}$$

$$V_\theta = \frac{1}{r} \left(\frac{\partial \Phi}{\partial \theta} \right) = -\frac{\partial \Psi}{\partial r} = \frac{\Gamma}{2\pi r}. \tag{9.43}$$

Line vortex

Substituting these velocities into the continuity equation, we have

$$\frac{1}{r} \left(\frac{\partial}{\partial r} \right) (r V_r) + \frac{1}{r} \left(\frac{\partial V_\theta}{\partial \theta} \right) = 0.$$

The vorticity equation,

$$\zeta_z = \frac{1}{r} \left[\frac{\partial (r V_\theta)}{\partial r} \right] - \frac{1}{r} \left(\frac{\partial V_r}{\partial \theta} \right) = \frac{1}{r} \left(\frac{\partial}{\partial r} \right) \left(\frac{\Gamma}{2\pi} \right),$$

shows that the flow is irrotational everywhere except at the origin, where the vorticity is infinite. The parameter Γ is the circulation, encountered in Chapter 8. If we choose a circular path about the origin and calculate the circulation, we get

$$\Gamma = \oint \vec{V} \cdot d\hat{s} = \int_0^{2\pi} V_\theta r \, d\theta = \int_0^{2\pi} \frac{\Gamma}{2\pi} \, d\theta = \Gamma.$$

* The minus sign is inserted as a result of hindsight, to make later results come out right. The reason for writing the constant as $\Gamma/2\pi$ will also become apparent soon.

It is possible to show that the circulation about any path that encloses the origin is Γ and that the circulation about any path that does not enclose the origin is zero. We have encountered this type of flow before (Example 4.8), where we called it a *free vortex*. The vortex may be positive or negative, depending on the sign of Γ and the corresponding direction of V_θ.

The equations for the free vortex can be written in rectangular coordinates for a vortex located at an arbitrary point (a, b). The results are

$$u = -\frac{\Gamma}{2\pi} \frac{y - b}{\sqrt{(x - a)^2 + (y - b)^2}}, \tag{9.44}$$

$$v = \frac{\Gamma}{2\pi} \frac{x - a}{\sqrt{(x - a)^2 + (y - b)^2}}, \tag{9.45}$$

$$\Phi = \frac{\Gamma}{2\pi} \tan^{-1}\left(\frac{y - b}{x - a}\right), \tag{9.46}$$

and

$$\Psi = -\frac{\Gamma}{4\pi} \ln\left[(x - a)^2 + (y - b)^2\right]. \tag{9.47}$$

An important basic potential flow (but not a singularity) is *uniform, parallel flow*. For flow aligned at an arbitrary angle α with the x-axis, the velocities are

$$u = V_\infty \cos \alpha \tag{9.48}$$

and

$$v = V_\infty \sin \alpha. \tag{9.49}$$

The velocity potential is

$$\Phi = V_\infty x \cos \alpha + V_\infty y \sin \alpha, \tag{9.50}$$

and the stream function is

$$\Psi = V_\infty y \cos \alpha - V_\infty x \sin \alpha. \tag{9.51}$$

9.2.3 Simple Body Flows by Superposition of Singularities

By themselves, the flows generated by singularities are not very interesting. We can, however, generate some interesting flows by combining several singularities in the same plane. This is called the method of *superposition* and works only because the potential and stream function equations (Eqs. 9.14 and 9.29) are linear. A linear equation has the property that if Φ_1 and Φ_2 are both solutions to the equations, then $(\Phi_1 + \Phi_2)$ is also a solution. If, for example, we put a source and a vortex in the same plane, then the potential and stream functions for the resulting flow are

$$\Phi = \Phi_{\text{source}} + \Phi_{\text{vortex}} \quad \text{and} \quad \Psi = \Psi_{\text{source}} + \Psi_{\text{vortex}}.$$

We can plot the streamlines for the resulting flow by adding the two (algebraic) stream functions, setting the result equal to various con-

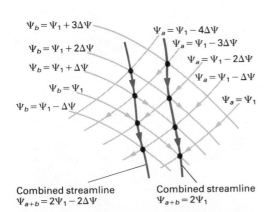

Combined streamline
$\Psi_{a+b} = 2\Psi_1 - 2\Delta\Psi$

Combined streamline
$\Psi_{a+b} = 2\Psi_1$

Figure 9.10 Graphical combination of streamlines.

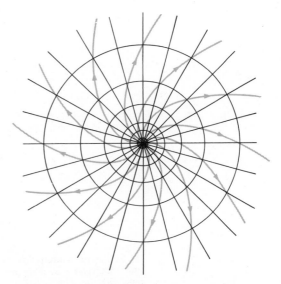

Figure 9.11 Streamline pattern for combined source and vortex.

stants, and plotting the resulting x,y-curves. A much easier procedure is to simply draw the streamlines of the basic singularities and then combine them graphically as done in Fig. 9.10.

Figure 9.11 shows the flow due to a combined source and vortex, both located at the origin. The potential and stream functions are

$$\Phi = \Phi_{\text{source}} + \Phi_{\text{vortex}} = \frac{\lambda}{2\pi} \ln r + \frac{\Gamma}{2\pi} \theta \qquad \textbf{(9.52)}$$

and

$$\Psi = \Psi_{\text{source}} + \Psi_{\text{vortex}} = \frac{\lambda}{2\pi} \theta - \frac{\Gamma}{2\pi} \ln r. \qquad \textbf{(9.53)}$$

The streamlines are logarithmic spirals, given by

$$r = \exp\left(\frac{2\pi\Psi}{\Gamma}\right) \exp\left(\frac{\lambda\theta}{\Gamma}\right).$$

Recall that λ and Γ are constants and that Ψ is constant for a particular streamline.*

The most interesting flows of this type are obtained by combining a uniform flow with one or more singularities. Suppose we combine a source and a uniform flow parallel to the x-axis ($\alpha = 0$). The stream function (in polar coordinates) is

$$\Psi = V_\infty r \sin\theta + \frac{\lambda}{2\pi} \theta. \qquad \textbf{(9.54)}$$

A few streamlines are sketched in Fig. 9.12(a). This flow looks relatively uninteresting until we examine the streamline marked A. This streamline separates the fluid introduced by the source from the uniform

* Can you think of any practical examples of this flow pattern?

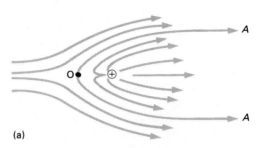

(a)

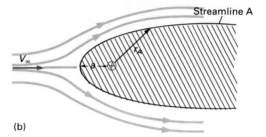

Streamline A

V_∞

(b)

Figure 9.12 Flow over half-body: (a) Streamline pattern for source in uniform stream; (b) flow over blunt half-body.

stream. The point 0 is a stagnation point. If we use the fact that any streamline can be replaced by a solid surface, the flow outside of streamline A represents flow over a blunt half-body (Fig. 9.12b). The half-body is open in the downstream direction. We might use the resulting flow pattern to model the flow over the front of a blunt object. It does not bother us at all that the flow pattern inside the body looks very unrealistic and, in fact, even contains a point where conservation of mass is violated. We have hidden this point inside the body.

The contour of streamline A can be obtained by setting $\Psi = 0$ along the centerline and noting that exactly one-half of the source flow lies between the centerline and streamline A; thus

$$\Psi_A = \frac{\lambda}{2},$$

and the streamline contour is given by

$$r_A = \frac{\lambda(\pi - \theta)}{2\pi V_\infty \sin\theta}. \tag{9.55}$$

We find the location of the front stagnation point, $x = -a$, by setting

$$u = \frac{\partial \Psi}{\partial x} = 0 \qquad \text{at } r = a, \quad \theta = \pi,$$

with the result

$$a = \frac{\lambda}{2\pi V_\infty}. \tag{9.56}$$

The velocity on the body surface is found from the general expression

$$V^2 = V_r^2 + V_\theta^2 = \left[\frac{1}{r}\left(\frac{\partial \Psi}{\partial \theta}\right)\right]^2 + \left(\frac{\partial \Psi}{\partial r}\right)^2,$$

with the result

$$V^2 = V_\infty^2 \left(1 + \frac{a^2}{r^2} + 2\frac{a}{r}\cos\theta\right). \qquad (9.57)$$

Although the flow over the half-body is interesting, the flow over a closed body would be more interesting. We can obtain a closed body by introducing a sink with strength equal to that of the source downstream of the source. The sink will ingest all of the output from the source. The resulting streamline pattern is sketched in Fig. 9.13. The most convenient placement of the origin is midway between the source and sink, with the source located at $(-a, 0)$ and the sink located at $(+a, 0)$. The stream function becomes

$$\Psi = \Psi_{\substack{\text{uniform} \\ \text{flow}}} + \Psi_{\substack{\text{source at} \\ -a}} + \Psi_{\substack{\text{sink at} \\ +a}}.$$

Substituting the various stream functions, we have

$$\Psi = V_\infty y + \frac{\lambda}{2\pi}\tan^{-1}\left(\frac{y}{x+a}\right) - \frac{\lambda}{2\pi}\tan^{-1}\left(\frac{y}{x-a}\right).$$

This can be rearranged to yield

$$\Psi = V_\infty y - \frac{\lambda}{2\pi}\tan^{-1}\left(\frac{2ay}{x^2+y^2-a^2}\right). \qquad (9.58)$$

The streamline marked Ψ_A in Fig. 9.13 separates the fluid introduced and ingested by the source-sink pair from the main stream fluid. Since the net flow from the source-sink pair is zero,

$$\Psi_A = 0.$$

Streamline A is a closed curve and can be replaced by a solid surface. This surface is called a *Rankine oval* and is somewhat blunter than an ellipse. The streamlines lying outside of Ψ_A (that is, streamlines with $\Psi > 0$) form the flow pattern over the oval. The internal streamlines ($\Psi < 0$) are unrealistic but are hidden inside the body contour. The geometric parameters of the body, t and ℓ, are related to the source-sink spacing $2a$, which is in turn related to the strength of the source and

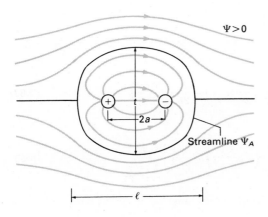

Figure 9.13 Flow over Rankine oval produced by source-sink pair in uniform stream.

sink and the free-stream velocity. The relations are

$$\frac{t}{a} = \cot\left(\frac{t/a}{\dfrac{\lambda}{\pi V_\infty a}}\right) \tag{9.59}$$

and

$$\frac{\ell}{a} = \left(1 + \frac{\lambda}{\pi V_\infty a}\right)^{1/2}. \tag{9.60}$$

The maximum velocity, which occurs at the point of maximum thickness on the body, is

$$\frac{V_{max}}{V_\infty} = 1 + \frac{\dfrac{\lambda}{\pi V_\infty a}}{1 + (t/a)^2}. \tag{9.61}$$

Now consider what happens if we vary the source-sink spacing. If $a \to \infty$, $t \to 0$, and $\ell \to \infty$, we obtain a uniform, parallel flow over a very thin flat plate. If $a \to 0$ (λ and V_∞ remaining constant), both t and ℓ approach zero, but, more important, their ratio becomes unity. As we bring the source and sink together, the body becomes a vanishingly small circular cylinder. The flow over a finite circular cylinder would be very interesting. If we note that increasing λ makes the body larger for a given source-sink spacing, it appears that we can generate the flow over a circular cylinder by letting λ increase without bound as a decreases to zero; in other words we keep the *product* λa constant as $a \to 0$. What we are actually doing is generating a new type of singularity called a *doublet*. A doublet is the direct superposition of a source and sink, both of infinite strength. The fact that this type of singularity has almost no physical significance should not bother us. We will be very careful to "hide" any doublets inside a body contour.

The potential and stream functions for a doublet are obtained by purely mathematical procedures:

$$\Phi_{doublet} = \lim_{\substack{a \to 0 \\ \lambda a = \text{constant}}} \left(\Phi_{\substack{source \\ at\ -a}} + \Phi_{\substack{sink \\ at\ +a}}\right)$$

and

$$\Psi_{doublet} = \lim_{\substack{a \to 0 \\ \lambda a = \text{constant}}} \left(\Psi_{\substack{source \\ at\ -a}} + \Psi_{\substack{sink \\ at\ +a}}\right).$$

Substituting the appropriate source and sink singularities, defining μ as the constant value of λa, and taking the limits,[*] we obtain

$$\Phi_{doublet} = \frac{\mu}{2\pi}\left(\frac{x}{x^2 + y^2}\right) = \frac{\mu}{2\pi}\left(\frac{\cos\theta}{r}\right) \tag{9.62}$$

and

$$\Psi_{doublet} = \frac{\mu}{2\pi}\left(\frac{y}{x^2 + y^2}\right) = \frac{\mu}{2\pi}\left(\frac{\sin\theta}{r}\right). \tag{9.63}$$

---

[*] Taking the limit in the Φ-equation involves expanding the logarithm function in an infinite series [4].

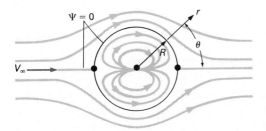

Figure 9.14 Flow resulting from combination of doublet and uniform stream.

If we combine a doublet with a uniform stream, we expect to obtain the flow over a circular cylinder. The stream function is

$$\Psi = \Psi_{\substack{uniform \\ stream}} + \Psi_{doublet},$$

or

$$\Psi = V_\infty y - \frac{\mu}{2\pi}\left(\frac{y}{r^2}\right). \tag{9.64}$$

Setting Ψ equal to zero to find a closed contour, we get

$$r)_{\Psi=0} = \sqrt{\frac{\mu}{2\pi V_\infty}} \equiv R, \quad \text{a constant,}$$

which is of course a circle. Figure 9.14 illustrates some other streamlines of the flow. As usual, the streamlines inside the contour are quite unrealistic, but the flow outside is just as we would expect. The velocity components are given by* (see Example 9.1):

$$V_r = V_\infty \cos\theta\left(1 - \frac{R^2}{r^2}\right) \tag{9.65}$$

and

$$V_\theta = -V_\infty \sin\theta\left(1 + \frac{R^2}{r^2}\right). \tag{9.66}$$

On the surface of the cylinder, $r = R$,

$$V_r)_{r=R} = 0, \quad \text{and} \tag{9.67a}$$
$$V_\theta)_{r=R} = -2V_\infty \sin\theta. \tag{9.67b}$$

The maximum velocity occurs at $\theta = \pi/2$ and is

$$V_{max} = 2V_\infty.$$

The pressure distribution on the surface of the cylinder can be obtained from Bernoulli's equation,

$$p = p_\infty + \frac{1}{2}\rho(V_\infty^2 - V_\theta^2),$$

or, in terms of pressure coefficient,

$$C_p \equiv \frac{p - p_\infty}{\frac{1}{2}\rho V_\infty^2} = 1 - \left(\frac{V_\theta}{V_\infty}\right)^2. \tag{9.68}$$

---

* Note that θ in these equations is measured from the *rear* stagnation point.

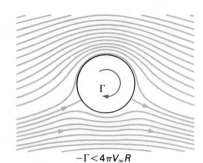

$-\Gamma < 4\pi V_\infty R$

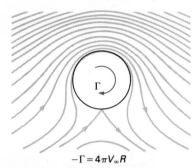

$-\Gamma = 4\pi V_\infty R$

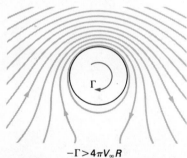

$-\Gamma > 4\pi V_\infty R$

Figure 9.15 Streamlines for flow over circular cylinder with circulation. Circulation is negative so as to generate positive lift.

Substituting Eq. (9.67b), we have

$$C_p = 1 - 4\sin^2\theta. \tag{9.69}$$

This pressure distribution was plotted in Fig. 8.12(b).*

An item of considerable interest is the force on the cylinder. Since the pressure distribution is symmetrical both front to back and top to bottom, the net force is zero (d'Alembert's paradox).

In Section 8.3.3 we learned that a *spinning* cylinder generates lift. We can simulate this type of flow by placing a line vortex at the origin, buried inside the circular cylinder. In Section 8.3.1, we learned that the lift is proportional to the circulation of this vortex. To obtain positive lift, we must introduce negative circulation (see Eq. 8.31). The resulting stream function is

$$\Psi = V_\infty r \sin\theta - R^2 V_\infty \left(\frac{\sin\theta}{r}\right) + \frac{\Gamma}{2\pi}\ln r.$$

Some streamlines are sketched in Fig. 9.15. The surface velocity is[†]

$$V_\theta)_{r=R} = -\frac{\partial\Psi}{\partial r}\bigg)_{r=R} = -2V_\infty \sin\theta - \frac{\Gamma}{2\pi R}. \tag{9.70}$$

Stagnation points are located on the surface at points where $V_\theta = 0$. These points are located at

$$\theta = \sin^{-1}\left(-\frac{\Gamma}{4\pi V_\infty R}\right).$$

If $\Gamma > 4\pi V_\infty R$, no stagnation points occur on the surface, but a single stagnation point occurs in the fluid below the cylinder.

The lift on the cylinder is obtained from the Kutta-Joukowsky theorem (Section 8.3.2) and is

$$\mathscr{L} = -\rho V_\infty \Gamma.$$

If we had a real flow in which the cylinder were spinning, rather than our artificial flow with a fixed cylinder and a vortex, the surface velocity of the cylinder (U) would be given by the second term of Eq. (9.70),

$$U = \frac{\Gamma}{2\pi R}.$$

The circulation necessary to model the spin is

$$\Gamma = -2\pi RU, \quad \text{so} \quad \mathscr{L} = 2\pi R V_\infty U,$$

and

$$C_L = \frac{\mathscr{L}}{\frac{1}{2}\rho V_\infty^2 (2R)} = 2\pi\left(\frac{U}{V_\infty}\right), \tag{9.71}$$

the relation plotted in Fig. 8.30.

---

* Refer to Section 8.2.2 for a discussion of the realism of these potential flow results.
[†] Recall that θ is positive *counterclockwise*.

EXAMPLE 9.2 Illustrates the addition of potential flow streamlines to produce an interesting flow

Investigate the addition of the stream function Ψ_U for a uniform flow parallel to the x-axis and the stream function Ψ_S for a sink. Show that this simulates the flow over the blunt tail of a body.

Solution

Given

$$\Psi = \Psi_U + \Psi_S$$

$$= V_\infty r \sin \theta + \frac{\lambda}{2\pi} \theta,$$

where $\lambda < 0$ and is the strength of the sink per unit depth (perpendicular to the page)

Show

That the superposition produces the flow over the blunt tail of a body

Solution
We plot lines of

$$\Psi_U = V_\infty r \sin \theta = \text{Constant}$$

and lines of

$$\Psi_S = \frac{\lambda}{2\pi} \theta = \text{Constant.}$$

For illustration purposes, we choose $\lambda = -32.0 \text{ m}^2/\text{s}$, $V_\infty = -1 \text{ m/s}$, and measure θ in radians. Then

$$r \sin \theta = -\Psi_U \qquad \text{and} \qquad \theta = \frac{2\pi\Psi_S}{\lambda} = -\frac{\pi}{16}\Psi_S.$$

We choose values of $\Psi_U = 0$, ± 2, ± 4, ± 6, ± 8, ± 10, ± 12, ± 14, and ± 16 and identical values for Ψ_S. The corresponding streamlines are plotted in Fig. E9.2. Lines of constant Ψ_U are parallel to the x-axis and lines of constant Ψ_S are radial lines. Lines for which $\Psi_U + \Psi_S = \pm 16$ and ± 18 are also plotted in the figure. Lines for values of ± 16 form the rearward surface of a blunt body. Lines for values of ± 18 represent two streamlines of the flow over the rearward surface or tail.

Answer

◀

| The answer is Fig. E9.2. |

Discussion
Note that the surface of the body is formed by two streamlines, one by $\Psi = +16$ and one by $\Psi = -16$.

We have chosen convenient values for λ and V_∞ to produce an easy numerical superposition. Other values of λ and V_∞ would be slightly more cumbersome; however, the procedure would be identical to that used here, but the size and shape of the body would be somewhat different.

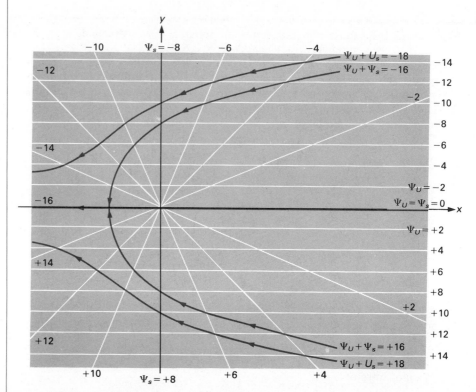

Figure E9.2 Graphical combination of streamlines.

EXAMPLE 9.3 Illustrates calculation of pressure and force in a potential flow

The following stream function represents the flow of a uniform stream over a cylinder of radius R with a line vortex at the center of the cylinder:

$$\Psi = V_\infty r \sin \theta - \frac{\mu}{2\pi}\left(\frac{\sin \theta}{r}\right) - \frac{\Gamma}{2\pi} \ln \frac{r}{R},$$

with

$$\Gamma = +2\pi R V_\infty \qquad \text{and} \qquad \mu = 2\pi R^2 V_\infty.$$

Determine the pressure distribution over the cylinder and calculate the lift.

Solution

Given
Stream function

$$\Psi = V_\infty r \sin \theta - R^2 V_\infty \left(\frac{\sin \theta}{r}\right) - R V_\infty \ln \frac{r}{R}$$

$$= V_\infty \left[\left(r - \frac{R^2}{r}\right)\sin \theta - R \ln \frac{r}{R}\right]$$

Find
Pressure distribution over the cylinder and the lift

Solution
A sketch of the flow field is shown in Fig. E9.3(a). Equations (9.42) and (9.43) give

$$V_r = \frac{1}{r}\left(\frac{\partial \Psi}{\partial \theta}\right) = \frac{V_\infty}{r}\left(\frac{\partial}{\partial \theta}\right)\left[\left(r - \frac{R^2}{r}\right)\sin\theta - R\ln\frac{r}{R}\right]$$

$$= V_\infty\left(1 - \frac{R^2}{r^2}\right)\cos\theta$$

and

$$V_\theta = -\frac{\partial \Psi}{\partial r} = -V_\infty\left(\frac{\partial}{\partial r}\right)\left[\left(r - \frac{R^2}{r}\right)\sin\theta - R\ln\frac{r}{R}\right]$$

$$= V_\infty\left[\frac{R}{r} - \left(1 + \frac{R^2}{r^2}\right)\sin\theta\right].$$

We now apply Bernoulli's equation to streamlines immediately adjacent to the surface of the cylinder. See Fig. E9.3(a). Since $V_r = 0$ at the cylinder surface, Bernoulli's equation gives

$$\frac{p_\infty}{\rho} + \frac{1}{2}V_\infty^2 = \frac{p_s}{\rho} + \frac{1}{2}V_\theta^2,$$

where p_s is the pressure on the cylinder surface. Substituting for V_θ, we have

$$\frac{p_\infty}{\rho} + \frac{1}{2}V_\infty^2 = \frac{p_s}{\rho} + \frac{V_\infty^2}{2}\left[\frac{R}{R} - \left(1 + \frac{R^2}{R^2}\right)\sin\theta\right],$$

or

$$\frac{p_\infty}{\rho} + \frac{1}{2}V_\infty^2 = \frac{p_s}{\rho} + \frac{V_\infty^2}{2}[1 - 2\sin\theta]^2,$$

or

$$\frac{p_s - p_\infty}{\rho} = \frac{V_\infty^2}{2}[1 - (1 - 4\sin\theta + 4\sin^2\theta)].$$

Since ρ, V_∞, and p_∞ are not known, the pressure coefficient is used:

$$\left| C_\mathbf{p} = \frac{p_s - p_\infty}{\frac{1}{2}\rho V_\infty^2} = 4\sin\theta - 4\sin^2\theta \right| \qquad \textbf{Answer} \blacktriangleleft$$

This result is plotted in Fig. E9.3(b).

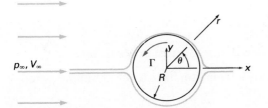

Figure E9.3 (a) Uniform flow over a cylinder.

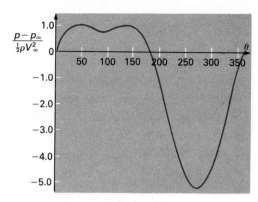

Figure E9.3 (b) Pressure coefficient for a uniform stream flowing over a rotating circular cylinder.

The lift $\mathscr{L}$ is found using Eq. (8.3), with $p = p_s$ and $\tau_w = 0$:

$$\mathscr{L} = \oint (-p_s \sin \theta)\, dA = \int_0^{2\pi} (-p_s \sin \theta)(2R\ell)\, d\theta,$$

where ℓ is the cylinder length. Substituting for p_s and dA gives

$$\mathscr{L} = -2R\ell \int_0^{2\pi} \left[p_\infty + \frac{1}{2}\rho V_\infty^2 (4 \sin \theta - 4 \sin^2 \theta) \right] \sin \theta\, d\theta.$$

Integrating gives

$$\mathscr{L} = -2R\ell \left[-p_\infty \cos \theta + \frac{4}{2}\rho V_\infty^2 \left(\frac{\theta}{2} - \frac{\sin 2\theta}{4} \right) \right.$$

$$\left. + \frac{4}{2}\rho V_\infty^2 \left(\frac{\cos \theta}{3} (\sin^2 \theta + 2) \right) \right]_0^{2\pi}$$

$$= -2R\ell \left[-p_\infty (1 - 1) + \frac{4}{2}\rho V_\infty^2 \left(\frac{2\pi - 0}{2} - \frac{0 - 0}{4} \right) \right.$$

$$\left. + \frac{4}{2}\rho V_\infty^2 \left(\frac{1(0 + 2) - 1(0 + 2)}{3} \right) \right],$$

$$\boxed{\mathscr{L} = -2\pi\rho V_\infty^2 R\ell.}$$

Answer ◀

Discussion

This result is equal to that of the Kutta-Joukowsky theorem, Eq. (8.31),

$$\mathscr{L} = -\rho V_\infty \Gamma \ell,$$

since $\Gamma = 2\pi R V_\infty$. Note that the length ℓ is absent from Eq. (8.31), which gives the lift per unit length.

The negative (downward) lift is due to the counterclockwise (positive) circulation. A negative circulation produces a positive (upward) lift. Compare with Fig. 9.15.

9.2.4 Plane Potential Flows from Complex Variables

Construction of potential flows from superposition of simple singularities has two limitations:

- Only a limited number of flows can be investigated.
- It is impossible to generate a surface with a sharp corner (such as an airfoil trailing edge).

It is possible to generate some new types of plane potential flows and to extend the flows generated by singularities to new geometries by application of the methods of complex variables. This method is strictly limited to two-dimensional geometries in a plane.

A complex variable (z) is composed of a real and an imaginary part:

$$z = x + iy,$$

where $i^2 = -1$ and x and y are "components" of the complex variable. A complex variable represents a point in a plane and can also be expressed in polar form:

$$z = x + iy = r(\cos \theta + i \sin \theta) = re^{i\theta}. \tag{9.72}$$

A *function* of a complex variable also has real and imaginary parts:

$$f(z) = f(x + iy) = \xi + i\eta. \tag{9.73}$$

To illustrate, the simple function

$$f(z) = z^2 = (x + iy)^2 = (x^2 - y^2) + i(2xy)$$

has

$$\xi = x^2 - y^2 \qquad \text{and} \qquad \eta = 2xy.$$

The most important type of complex functions are *analytic functions,* which have a unique derivative at every point. Two important theorems about analytic functions are the following:*

- For a function $f(z)$ to be analytic, it is necessary that

$$\frac{\partial \xi}{\partial x} = \frac{\partial \eta}{\partial y} \tag{9.74}$$

and

$$\frac{\partial \xi}{\partial y} = \frac{\partial \eta}{\partial x}. \tag{9.75}$$

These equations are called the *Cauchy-Riemann conditions.*
- Both components of an analytic function satisfy Laplace's equation, that is,

$$\frac{\partial^2 \xi}{\partial x^2} + \frac{\partial^2 \xi}{\partial y^2} = 0 \qquad \text{and} \qquad \frac{\partial^2 \eta}{\partial x^2} + \frac{\partial^2 \eta}{\partial y^2} = 0.$$

Since the velocity potential and stream function satisfy the equations

$$\frac{\partial \Phi}{\partial x} = u = \frac{\partial \Psi}{\partial y}, \qquad \frac{\partial \Phi}{\partial y} = v = -\frac{\partial \Psi}{\partial x}, \qquad \text{and}$$

$$\frac{\partial^2 \Phi}{\partial x^2} + \frac{\partial^2 \Phi}{\partial y^2} = 0 = \frac{\partial^2 \Psi}{\partial x^2} + \frac{\partial^2 \Psi}{\partial y^2},$$

they can be combined into an analytic *complex potential* function,

$$\Omega \equiv \Phi + i\Psi. \tag{9.76}$$

---

* You can verify that $f(z) = z^2$ satisfies these theorems.

The derivative of Ω, given by

$$w = \frac{d\Omega}{dz} = \frac{\partial}{\partial x}(\Phi + i\Psi) = u - iv, \tag{9.77}$$

is called the *complex velocity*.

We can obtain complex potentials for the singularities. First consider the uniform stream. The complex potential is

$$\Omega = \Phi + i\Psi = V_\infty x \cos \alpha + V_\infty y \sin \alpha + i(V_\infty y \cos \alpha - V_\infty x \sin \alpha).$$

Rearranging, we have

$$\Omega = V_\infty(x + iy)(\cos \alpha - i \sin \alpha).$$

Using Eq. (9.72), we can write this as

$$\Omega_{\substack{\text{uniform} \\ \text{stream}}} = V_\infty z e^{-i\alpha}. \tag{9.78}$$

Similarly,

$$\Omega_{\text{source}} = \frac{\lambda}{2\pi} \ln z, \tag{9.79}$$

$$\Omega_{\text{vortex}} = -\frac{\Gamma}{2\pi} \ln z, \tag{9.80}$$

and

$$\Omega_{\text{doublet}} = \frac{\mu}{2\pi z}. \tag{9.81}$$

Since we already have the solution for the singularity flows, we don't gain anything by recasting them in complex variables. We can find some new flows, however, when we recognize that *any* analytic function of a complex variable can be thought of as a complex potential. If the resulting curves of $\Psi = $ Constant are interesting, we have a useful flow solution. Consider the power function

$$\Omega = Az^n, \tag{9.82}$$

where A and n are real constants. The velocity potential and stream functions are

$$\Phi = Ar^n \cos(n\theta) \quad \text{and} \quad \Psi = Ar^n \sin(n\theta).$$

Lines of $\Psi = $ Constant (streamlines) for some values of n are plotted in Fig. 9.16. These flows are rather interesting, representing flow over exterior and interior corners. The corner angle is related to n by

$$\beta = \pi/n. \tag{9.83}$$

Note that we have generated flow over surfaces with sharp corners. We could postulate other functions of z and investigate them, but we leave that for you to try.

The most powerful application of complex variable techniques in potential flow theory involves *conformal mapping*. We noted earlier that a specific value of z represents a point in a plane. Any function of z produces a new variable (say ζ), defined by

$$\zeta = f(z) = \xi + i\eta,$$

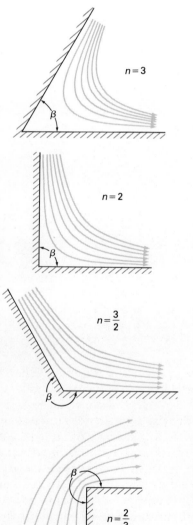

Figure 9.16 Complex potential $\Omega = Z^n$ represents flow in a corner.

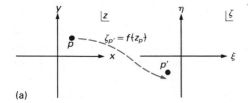

(a)

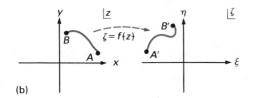

(b)

Figure 9.17 Mapping of points and lines by functions of complex variable: (a) Point mapping; (b) line mapping.

that represents a point in a *new* complex plane. The function $f(z)$ transforms or *maps* the z-plane, point by point, into the ζ-plane (see Fig. 9.17). A line in the z-plane will be mapped into a corresponding line in the ζ-plane. Maybe we can use this trick to transform known flows (that is, known equipotential and streamline patterns) into new flows. The geometric process of transforming old lines into new lines will always work; the key question is whether the new pattern will represent a real flow. The answer is yes—*if* the mapping function $f(z)$ is analytic. It can be shown that

- functions that satisfy Laplace's equation transform into functions that still satisfy Laplace's equation under mapping by an analytic function; and
- the angle of intersection of any pair of lines (say equipotential lines and streamlines) in the z-plane is preserved under mapping by an analytic function.

This latter property gives rise to the term *conformal mapping*.

Using the technique of conformal mapping, we can transform a familar flow (usually the flow over a circular cylinder) to a new and interesting flow. The transformation changes both the body shape and streamline pattern.

Probably the most well known example of conformal mapping in fluid mechanics is the transformation of the flow over a circular cylinder into the flow over an airfoil. This is done by the *Joukowsky transformation*,

$$z = \zeta + \frac{C^2}{\zeta}.$$

Note that we put the circle flow in the ζ-plane and transform it to the z-plane.

The transformation has *singular points* at $\zeta = \pm C$. Any pair of lines passing through a singular point will transform into a sharp corner, so we can obtain corners by locating a singular point on a surface. The mathematical details of the transformation are too lengthy to give here [7]. Figure 9.18 illustrates various mappings of the flow over a circle into an airfoil flow. If both singular points are located on the circle, it transforms into a flat plate (Fig. 9.18a). If the circle is offset along the ζ-axis, with one singular point remaining on the circle, a shape called the *symmetrical Joukowsky airfoil* is generated (Fig. 9.18b). Figure 9.18(c) shows this airfoil with an angle of attack. The unrealistic flow near the trailing edge is removed by adding circulation to rotate the rear stagnation point on the cylinder to the singular point so that, in the airfoil plane, the flow leaves the sharp trailing edge smoothly (Fig. 9.18d). This is called the *Kutta condition*. The necessary circulation is

$$\Gamma \approx -\pi V_\infty \ell \sin \alpha.$$

The lift on the airfoil is given by the Kutta-Joukowsky theorem,

$$\mathscr{L} = -\rho V_\infty \, \Gamma = \pi \rho V_\infty^2 \ell \sin \alpha. \tag{9.84}$$

Figure 9.18 Transformation of various circle flows into airfoil flows by Joukowsky transformation: (a) Flat plate airfoil; (b) symmetrical Joukowsky airfoil; (c) symmetrical airfoil at angle of attack; (d) symmetrical airfoil with circulation; cambered airfoil with angle of attack and circulation.

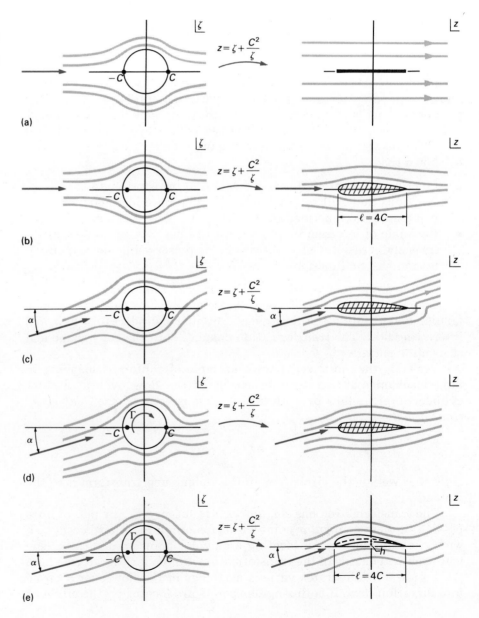

Also, the lift coefficient is

$$\mathbf{C_L} = \frac{\mathscr{L}}{\frac{1}{2}\rho V_\infty^2 \ell} = 2\pi \sin \alpha \approx 2\pi\alpha, \tag{9.85}$$

since α is small for unstalled flow.

A cambered airfoil can be generated by shifting the circle upward, still fixing one singular point on the surface (Fig. 9.18e). It is again necessary to add circulation to the cylinder to obtain smooth streamlines at the trailing edge. The necessary circulation is

$$\Gamma \approx -\pi V_\infty \ell \sin\left(\alpha + \frac{2h}{\ell}\right), \tag{9.86}$$

so

$$\mathbf{C_L} \approx 2 \sin\left(\alpha + \frac{2h}{\ell}\right) \approx 2\pi\left(\alpha + \frac{2h}{\ell}\right). \tag{9.87}$$

These predictions of lift coefficient are reasonably close to actual values for small camber and angle of attack. Note also that to predict lift (as well as to obtain a realistic streamline pattern), we were forced to arbitrarily add circulation to both the cylinder and airfoil flows. This was accomplished by placing an imaginary line vortex with the proper circulation inside the cylinder and, hence, inside the airfoil. The fact that the vortex is not really there* does not detract from the realism of the flow outside the airfoil. Note also that, because we placed a free vortex inside the cylinder and airfoil, the circulation for any path enclosing the cylinder (airfoil) is the same as the circulation for any other path enclosing the cylinder (airfoil), so a unique value of $\mathbf{C_L}$ is obtained by application of the Kutta-Joukowsky theorem.

Several other interesting plane two-dimensional flow patterns can be obtained by the conformal mapping method. One of these is the potential flow over a flat plate normal to the flow (see Fig. 8.13). If you are interested in the details of the mappings used to obtain this flow or other applications of the conformal mapping method, see [3, 4, 7, 9].

9.2.5 Direct Methods for Calculation of Potential Flow in a Plane

The method of singularities and the method of complex variables for plane potential flow problems are indirect methods in which a mathematical solution to the governing equations is postulated and then the resulting streamline pattern is examined to see if it represents an interesting flow. In most engineering fluid mechanics problems, we know the shape of the body or internal flow passage and we want to find the streamlines for flow over or internal to that specific configuration. It is obvious that we could waste a lot of time investigating possible mathematical functions and mappings to find one that fits a particular problem. In fact, most of the possible elementary solutions

---

* The real source of vorticity is the airfoil boundary layers.

and useful mappings have already been found. To calculate the potential flow for an arbitrary geometry, it is necessary to turn to approximate methods. There are many such methods [4]. We will outline only the three most popular ones:

- Graphical construction of a *flow net*
- Numerical solution of *finite-difference approximations* of the differential equations
- Simulation of arbitrary geometry by *distributed singularities*

Our treatment of these topics will be only a brief introduction.

Graphical Flow Net. The flow net method makes use of the fact that equipotential lines and streamlines always intersect at right angles. In this method, a grid of lines is laid out by hand in a trial and error procedure to conform to the rule of intersection at right angles and to the rule that, in the limit of an infinitely fine flow net, the lines must reduce to a grid of squares. Note that this amounts to construction of a "steamline and normal" grid of coordinates like that illustrated in Fig. 4.1. Figure 9.19 shows a flow net for flow through a bend. After a flow net is constructed, the velocity can be found by noting that it is parallel to the streamlines and at any point is given by (see Fig. 9.19)

$$V_s \equiv \frac{\partial \Phi}{\partial s} \approx \frac{\Delta \Phi}{\Delta s},$$

or, alternatively,

$$V_s \equiv \frac{\partial \Psi}{\partial n} \approx \frac{\Delta \Psi}{\Delta n}.$$

It is necessary to know the velocity at at least one point in the field (say the inlet) to obtain the proper dimensional values. The graphical flow net method is used only for rough calculations, for obvious reasons.

Figure 9.19 Flow net for flow through a plane bend.

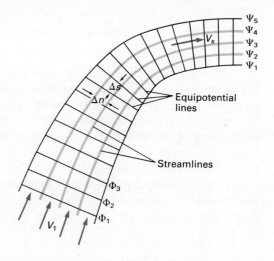

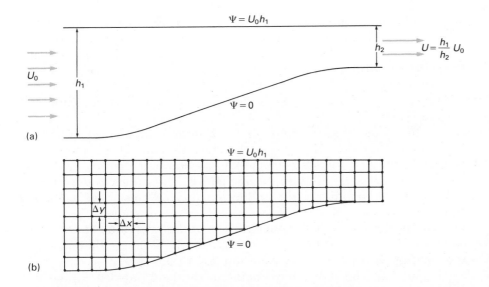

Figure 9.20 Finite-difference setup for flow in a converging channel.

Finite-Difference Approximation. The finite-difference approach has become very important in all types of flow problems because the digital computer has made handling of extremely large numbers of repetitive calculations commonplace. This approach has been successfully applied to all types of flow problems including potential flows, boundary layer flows, and even solution of the full Navier-Stokes equations. Several books, typified by [10] and [11], many conferences, and quite a few journals are devoted exclusively to computational fluid mechanics, of which the finite-difference approach constitutes about 90% of the effort. Literally hundreds of new papers on this subject appear each year. Luckily for us, potential flow is about the simplest application of the finite-difference approach.

Consider the problem of calculating the stream function for potential flow in a converging passage (Fig. 9.20). If the stream function is found, then the velocity can be obtained by differentiation and the streamlines can be plotted by setting Ψ equal to various constants. To calculate the stream function in the passage, we must find Ψ such that

$$\frac{\partial^2\Psi}{\partial x^2} + \frac{\partial^2\Psi}{\partial y^2} = 0 \qquad \textbf{(9.29)}$$

everywhere in the passage. We must also satisfy the boundary conditions

$$\Psi = 0 \qquad \text{on the lower wall,*} \qquad \textbf{(9.88a)}$$

$$\Psi = U_0 h_1 \qquad \text{on the upper wall,*} \qquad \textbf{(9.88b)}$$

$$\Psi = U_0 y \qquad \text{at the inlet, and} \qquad \textbf{(9.88c)}$$

$$\Psi = U_0\left(\frac{h_1}{h_2}\right)[y - (h_1 - h_2)] \qquad \text{at the outlet.} \qquad \textbf{(9.88d)}$$

* Remember that the change in Ψ between two streamlines is equal to the volume flow between the streamlines and that solid surfaces are streamlines.

The outlet condition is obtained from the continuity equation and the inlet condition. We have formulated the problem in terms of Ψ rather than Φ because the boundary conditions are easier. For a channel, boundary conditions for Ψ are simply known values on the boundaries, but boundary conditions for Φ involve the derivative normal to the boundary. Consideration of methods for calculating Φ can be found in advanced texts [4].

In the finite-difference method, we select a finite number of points in the flow field (usually lying at the intersections of a rectangular grid of lines) and calculate the unknown function (in this case the stream function) *only* at these specific points. The values of the stream function at the points are the unknowns, and we must have as many equations to calculate the unknowns as there are unknowns. Note that any points that lie on the boundaries have known values of Ψ. The equations for the unknown Ψ's are obtained by a finite-difference approximation to the governing equation, Eq. (9.29).

Figure 9.21 shows a typical point in the field, labeled P, and its neighboring points, labeled N, E, S, and W.* We seek approximations for the second derivatives that appear in Eq. (9.29). We can approximate the first derivatives by

$$\left.\frac{\partial \Psi}{\partial x}\right)_P \approx \frac{\Psi_E - \Psi_P}{\Delta x} \qquad \textbf{(9.89a)}$$

and

$$\left.\frac{\partial \Psi}{\partial x}\right)_P \approx \frac{\Psi_P - \Psi_W}{\Delta x}. \qquad \textbf{(9.89b)}$$

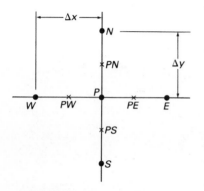

Figure 9.21 Typical points in finite-difference grid.

These two approximations, called a *forward difference* and a *backward difference,* become more accurate as Δx becomes smaller. It is possible to show that the error in these approximations is proportional to $(\Delta x)^2$. Although these are intended as approximations of the derivative at P, they are better approximations for the derivative at point PE midway between P and E and point PW, midway between P and W. At PE and PW, the error in the approximation is smaller, being proportional to $(\Delta x)^3$. We can obtain an approximation to the second derivative by

$$\frac{\partial^2 \Psi}{\partial x^2} = \frac{\partial}{\partial x}\left(\frac{\partial \Psi}{\partial x}\right) \approx \frac{\left.\dfrac{\partial \Psi}{\partial x}\right)_{PE} - \left.\dfrac{\partial \Psi}{\partial x}\right)_{PW}}{\Delta x}.$$

Equations (9.89a) and (9.89b) are excellent approximations for the derivative at points PE and PW, so

$$\frac{\partial^2 \Psi}{\partial x^2} \approx \frac{\dfrac{\Psi_E - \Psi_P}{\Delta x} - \dfrac{\Psi_P - \Psi_W}{\Delta x}}{\Delta x} = \frac{\Psi_E + \Psi_W - 2\Psi_P}{(\Delta x)^2}. \qquad \textbf{(9.90)}$$

This approximation for the second derivative is most accurate at P, since it is "centered" there, and has error proportional to $(\Delta x)^3$.

---

* The labeling scheme is adapted from the points on a compass.

By similar reasoning, the y-derivative for Eq. (9.29) is approximated by

$$\frac{\partial^2 \Psi}{\partial y^2} \approx \frac{\Psi_N + \Psi_S - 2\Psi_P}{(\Delta y)^2}, \tag{9.91}$$

and Eq. (9.29) is approximated by

$$\left(\frac{1}{\Delta x}\right)^2 (\Psi_E + \Psi_W - 2\Psi_P) + \left(\frac{1}{\Delta y}\right)^2 (\Psi_N + \Psi_S - 2\Psi_P) = 0.$$

If $\Delta x = \Delta y$ (a square grid), we get

$$\left(\frac{1}{\Delta x}\right)^2 (\Psi_E + \Psi_W + \Psi_N + \Psi_S - 4\Psi_P) = 0,$$

or, since Δx is finite,

$$\Psi_P = \frac{1}{4} (\Psi_E + \Psi_W + \Psi_N + \Psi_S). \tag{9.92}$$

Equation (9.92) can be written for every point in the grid (except points on the boundaries). Of course we will have to change the *PNESW*-notation to a global point-numbering system. For points adjacent to the boundaries, at least one of the surrounding points will be on a boundary and the value of Ψ there will be known. Sometimes the point located a distance Δx away may lie *outside* the boundary (Fig. 9.22), in which case a value for Ψ at the point can be found by extrapolation:

$$\Psi_S = \Psi_P + \frac{\Delta x}{\Delta x_B} (\Psi_B - \Psi_P). \tag{9.93}$$

After equations are written for each internal point, we find the following:

- There are exactly as many equations as there are unknowns.
- The known boundary values of Ψ appear in the equations.
- The equation for Ψ at a point contains the values for Ψ at the *four* surrounding points.

There are many ways to solve the resulting set of equations for the unknown values of Ψ. Probably the simplest is the *Gauss-Sidel iteration* method, which proceeds as follows:

- Guess values for all unknowns Ψ's.
- "Visit" each point, calculating a new value of Ψ_P from

$$\Psi_P = \frac{1}{4} (\Psi_N + \Psi_E + \Psi_S + \Psi_W),$$

using the best available value of each Ψ. If we move through the grid from bottom to top and from left to right, the values of Ψ_S and Ψ_W will be "new" values and the values of Ψ_N and Ψ_E will be "old" values.
- Continue the process until the largest change in Ψ_P is as small as you wish.

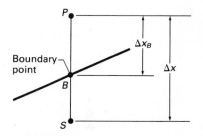

Figure 9.22 Regular point (s) lies outside a boundary.

This process can be easily programmed for a computer or programmable calculator.

After the Ψ's have been calculated, streamlines can be plotted by connecting points with the same value of Ψ. Interpolation will usually be required because Ψ is known only at discrete points. Velocities in the field can be calculated by

$$u = \frac{\partial \Psi}{\partial y} \approx \frac{\Delta \Psi}{\Delta y} \qquad (9.94)$$

and

$$v = -\frac{\partial \Psi}{\partial x} \approx -\frac{\Delta \Psi}{\Delta x}. \qquad (9.95)$$

These velocities will be most accurate at points midway between those at which Ψ is known; that is, u is most accurate at points PW and PE, and v is most accurate at points PN and PS (see Fig. 9.21).

This introduction has been rather brief, but we have covered the important ideas. For further discussion, consult [4]. An undergraduate heat transfer text would also be helpful because exactly the same numerical method is used to calculate temperatures in steady heat conduction in solids [12, 13, 14].

EXAMPLE 9.4 Illustrates how a network of streamlines and equipotential lines are used to determine velocity field

Consider the lower half of a symmetric enlargement, shown in Fig. E9.4. The stream function Ψ and the velocity potential Φ have units of cm²/s. The channel's length scale is shown in the figure. Find the velocity components along the centerline and along the outer wall for $x = 0$, $x = 4$ cm, and $x = 8$ cm. Then find the volume flow rate per unit depth of the channel (perpendicular to the paper).

Figure E9.4 Flow net for flow through a two-dimensional expansion.

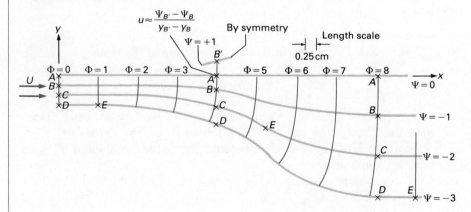

Solution

Given
Figure E9.4

Find
Velocity components along centerline and along outer wall at $x = 0$,
    $x = 4$ cm, and $x = 8$ cm
Volume flow rate per unit depth of channel

Solution
We will use Eqs. (9.94) and (9.95) to estimate the velocity components u and v:

$$u \approx \frac{\Delta \Psi}{\Delta y}, \qquad v \approx -\frac{\Delta \Psi}{\Delta x}.$$

Along the centerline, Ψ is constant and $v = 0$. Referring to Fig. E9.4, we see*

$$x = 0 \text{ cm:} \qquad u \approx \frac{\Psi_{B'} - \Psi_B}{y_{B'} - y_B} = \frac{1 - (-1)}{0.25 - (-0.25)} \frac{\text{cm}^2/\text{s}}{\text{cm}},$$

$$\left| \begin{array}{l} u \approx 4.0 \text{ cm/s}, \\ v = 0. \end{array} \right|$$

Answer ◀

In a similar manner,

$$x = 4 \text{ cm:} \qquad u \approx \frac{\Psi_{B'} - \Psi_B}{y_{B'} - y_B} = \frac{1 - (-1)}{0.4 - (-0.4)} \frac{\text{cm}^2/\text{s}}{\text{cm}},$$

$$\left| \begin{array}{l} u \approx 2.5 \text{ cm/s}, \\ v = 0, \end{array} \right|$$

Answer ◀

and

$$x = 8 \text{ cm:} \qquad u \approx \frac{\Psi_{B'} - \Psi_B}{y_{B'} - y_B} = \frac{1 - (-1)}{1.0 - (-1.0)} \frac{\text{cm}^2/\text{s}}{\text{cm}},$$

$$\left| \begin{array}{l} u \approx 1.0 \text{ cm/s}, \\ v = 0. \end{array} \right|$$

Answer ◀

Along the outer wall,

$$x = 0: \qquad u \approx \frac{\Psi_C - \Psi_D}{y_C - y_D} = \frac{(-2) - (-3)}{(-0.50) - (0.75)} \frac{\text{cm}^2/\text{s}}{\text{cm}},$$

$$\left| \; u \approx 4 \text{ cm/s}, \; \right|$$

Answer ◀

and

$$v \approx -\left(\frac{\Psi_E - \Psi_D}{x_E - x_D} \right) = -\left[\frac{(-3) - (-3)}{1.0 - 0} \right] \frac{\text{cm}^2/\text{s}}{\text{cm}},$$

$$\left| \; v \approx 0 \text{ cm/s}. \; \right|$$

Answer ◀

* Note that B' is the mirror image of B.

In a similar manner,

$$x = 4 \text{ cm}: \qquad u \approx \frac{\Psi_C - \Psi_D}{y_C - y_D} = \frac{(-2) - (-3)}{(-0.80) - (-1.20)} \frac{\text{cm}^2/\text{s}}{\text{cm}},$$

$$\left| \; u \approx 2.5 \text{ cm/s}, \; \right| \qquad \qquad \textbf{Answer} \; \blacktriangleleft$$

$$v \approx -\left(\frac{\Psi_E - \Psi_D}{y_E - y_D} \right) = \frac{(-2) - (-3)}{5.1 - 4.0} \frac{\text{cm}^2/\text{s}}{\text{cm}},$$

$$\left| \; v \approx 0.91 \text{ cm/s}, \; \right| \qquad \qquad \textbf{Answer} \; \blacktriangleleft$$

$$x = 8 \text{ cm}: \qquad u \approx \frac{\Psi_C - \Psi_D}{y_C - y_D} = \frac{(-2) - (3)}{(-2) - (-3)} \frac{\text{cm}^2/\text{s}}{\text{cm}},$$

$$\left| \; u \approx 1.0 \text{ cm/s}, \; \right| \qquad \qquad \textbf{Answer} \; \blacktriangleleft$$

and

$$v \approx -\left(\frac{\Psi_E - \Psi_D}{x_E - x_D} \right) = -\left[\frac{(-3) - (-3)}{9 - 8} \right] \frac{\text{cm}^2/\text{s}}{\text{cm}},$$

$$\left| \; v \approx 0. \; \right| \qquad \qquad \textbf{Answer} \; \blacktriangleleft$$

Assuming constant density, we see that the volume flow rate is the same at any cross section of the channel. This volume flow rate for a one-unit depth is

$$Q = \int_{-0.75 \text{ cm}}^{+0.75 \text{ cm}} u(1 \text{ cm}) \, dy.$$

Recognizing that u is uniform at the inlet, we have

$$Q = u(1 \text{ cm}) \int_{-0.75 \text{ cm}}^{0.75 \text{ cm}} dy = 1.5 \text{ cm}^2 \, u = (1.5 \text{ cm}^2)(4 \text{ cm/s}),$$

$$\left| \; Q = 6.0 \text{ cm}^3/\text{s}. \; \right| \qquad \qquad \textbf{Answer} \; \blacktriangleleft$$

Discussion
The velocity components u and v may also be evaluated by

$$u \approx \frac{\Delta\Phi}{\Delta x} \qquad \text{and} \qquad v \approx \frac{\Delta\Phi}{\Delta y},$$

It would be beneficial to compare the results of this method with the above answers. Another interesting exercise would be calculation of the pressure distribution along the centerline and the lower wall.

Distributed Singularities. Both the flow net method and the finite-difference method are better suited to internal flows than to external flows. They work best when all of the flow boundaries are clearly defined. The method based on distributing singularities along a line or over a surface in the flow field is more suited to external flows.

Recall that in the consideration of simple singularities, we had some slight control over body shape by positioning of point singularities. We could change a half-body into an oval by adding a sink, and we could change the shape of the oval by changing the distance between the singularities or altering their strength. This suggests that it may be possible to simulate flow over an arbitrary body by the proper combination of a uniform stream with sources and sinks (and vortexes, if a lifting body is to be simulated). This can be done, although it requires the continuous distribution of singularities.

Consider a line distributed source* (somewhat like a slit with fluid flowing out of it all along its length) rather than a single point source (Fig. 9.23). The strength of a differential strip of the source is

$$\lambda'(\xi)\,d\xi,$$

where λ' is the strength per unit length. The stream function at a point $P(x, y)$ due to a tiny strip of the source is

$$d\Psi_P = \frac{\lambda'(\xi)\,d\xi}{2\pi}\,\theta_\xi = \frac{\lambda'(\xi)\,d\xi}{2\pi}\,\tan^{-1}\left(\frac{y}{x-\xi}\right).$$

The *total* stream function at point P due to the entire strip of source, lying between ξ_1 and ξ_2, is[†]

$$\Psi_P = \int_{\xi_1}^{\xi_2} \frac{\lambda'(\xi)\,d\xi}{2\pi}\,\tan^{-1}\left(\frac{y}{x-\xi}\right). \tag{9.96}$$

If we combine a uniform stream with the distributed singularity, we get

$$\Psi = V_\infty y + \frac{1}{2\pi}\int_{\xi_1}^{\xi_2} \lambda'(\xi)\,\tan^{-1}\left(\frac{y}{x-\xi}\right)d\xi. \tag{9.97}$$

If we select values for ξ_1 and ξ_2 and a form for the singularity distribution $\lambda'(\xi)$ (say $\lambda'(\xi) = A\xi^n$), we can obtain the flow over various bodies.

Suppose we wish to calculate the flow over a specific body with surface given by $y = Y_B(x)$. The problem is to find the form of $\lambda'(\xi)$ that, when substituted into Eq. (9.97), will generate the flow over the given body. For a closed body, $\Psi_B = 0$. If we evaluate Eq. (9.97) on the known body surface and set Ψ_B equal to zero, we get

$$0 = V_\infty Y_B(x) + \frac{1}{2\pi}\int_{\xi_2}^{\xi_1} \lambda'(\xi)\,\tan^{-1}\left(\frac{Y_B(x)}{x-\xi}\right)d\xi. \tag{9.98}$$

Equation (9.98) is an integral equation for λ' and is quite difficult to solve. In practice, it is possible to *approximate* the continuous distribution by a series of strips of uniformly distributed source (Fig. 9.24), so

$$\int_{\xi_2}^{\xi_1} \lambda'(\xi)\,\tan^{-1}\left(\frac{Y_B(x)}{x-\xi}\right)d\xi \approx \lambda'_1\int_{\xi_1}^{\xi_2}\tan^{-1}\left(\frac{Y_B(x)}{x-\xi}\right)d\xi + \cdots$$
$$+ \lambda'_n\int_{\xi_n}^{\xi_2}\tan^{-1}\left(\frac{Y_B(x)}{x-\xi}\right)d\xi.$$

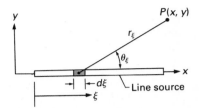

Figure 9.23 Line source.

* The line source in a plane is actually the intersection of a sheet normal to the plane of the flow with the plane of the flow.
† Note that ξ, not x or y, is the variable of integration.

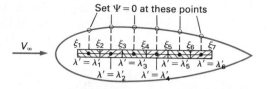

Figure 9.24 Simulation of flow over arbitrary body by strips of uniformly distributed source.

After integration, *one* value of x is selected for each strip at which Eq. (9.98) is required to hold. This produces n equations for the n unknown λ's.

In principle, this method should work for any body but, in practice, it doesn't. The set of equations produced by this technique does not always have a solution. The remedy for this difficulty is to distribute the singularities directly on the surface. This always produces a solution. The mathematics of the surface singularity method are too involved to take up here but have been completely treated elsewhere [4, 15]. Computer codes capable of simulating the flow over any two-dimensional or axisymmetric body using this method have been in existence for quite some time [15].

9.3 Boundary Layers

According to Prandtl's boundary layer hypothesis, at high Reynolds numbers viscous effects are confined to a thin region near a solid surface or line of discontinuity (see Fig. 9.1). Because this boundary layer is thin, certain simplifications can be introduced into the equations of motion; however, it is necessary to retain both inertia (acceleration) terms and stress (viscous) terms. Pressure terms may or may not be present, depending on the nature of the flow outside the boundary layer. Since the flow is viscous, it is rotational; that is, the vorticity of the boundary layer fluid is not zero. This means that it is not possible to find a velocity potential function for boundary layer flow and so the equation of motion must be attacked directly. This equation, even with the boundary layer simplifications included, is much more difficult to solve than the equations of potential flow.* Further complications are introduced by the fact that the boundary layer flow may be either laminar or turbulent (see Section 3.5.2 and Figs. 3.28, 3.29, 3.32, and 3.33).

Only a few exact solutions of the equations of motion for boundary layers are known, and these are all for laminar flow. The mathematics involved in obtaining these solutions is too complicated and lengthy to consider in this text, although we will derive the boundary layer equations. There are excellent approximate methods for calculating boundary layer flow, which we will consider, and digital computer–based numerical methods for calculating boundary layer flow, which we will not consider. Our aim is to help you establish a feel for boundary layer flow

---

* It is significant that, although Prandtl introduced the boundary layer concept and derived the appropriate equations of motion in 1904, the first solution to these equations did not appear until 1908 [16]. This solution was found by one of Prandtl's students.

and to examine a few simple cases in detail. To keep things as simple as possible, we will limit the discussion to steady, two-dimensional, two-directional, incompressible flow in a plane.

9.3.1 Dimensional Analysis and Qualitative Estimates of Boundary Layer Parameters

Figure 9.25 shows a boundary layer on a flat plate. The following boundary layer characteristics are of primary importance:

- The boundary layer is thin ($\delta \ll x$).
- The thickness of the boundary layer increases in the downstream direction, but δ/x is always small.
- The boundary layer velocity profile satisfies the no-slip condition at the wall and merges smoothly into the free-stream velocity at the edge of the layer.
- There is a shear stress at the wall.
- The streamlines of the boundary layer flow are *approximately* parallel to the surface; that is, the velocity parallel to the surface is considerably larger than the velocity normal to the surface. Note that the locus of points that defines the outer edge of the boundary layer is not a streamline.

In this section, we will demonstrate the validity of these characteristics and obtain *estimates* of key boundary layer parameters. We will show that most of the listed characteristics depend on the fact that the Reynolds number is large. Most boundary layer flows are at least qualitatively similar to flow over a flat plate, especially if we consider the flow using coordinates parallel and normal to the body surface. The flat plate flow has a uniform free-stream velocity (outside the boundary layer) and hence constant pressure. Most of our qualitative results for flat plate flow can be extended to flows with variable free-stream velocity by replacing V_∞, the constant free-stream velocity, with $U(x)$, the local free-stream velocity.*

Figure 9.25 Details of boundary layer near a thin flat plate.

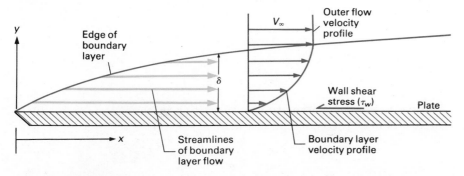

* Recall that this free-stream velocity can be obtained from a potential flow analysis that neglects the boundary layer.

As usual in complicated flow problems, it is helpful to begin with a dimensional analysis. The primary independent dimensionless parameter is the Reynolds number. If we adopt the local coordinate x as the reference length, we find

$$\frac{\delta}{x} = f_1(\mathbf{R}_x),$$

(9.99)

$$\mathbf{C_f} = \frac{\tau_w}{\frac{1}{2} \rho V_\infty^2} = f_2(\mathbf{R}_x),$$

(9.100)

and

$$\frac{v_e}{V_\infty} = f_3(\mathbf{R}_x),$$

(9.101)

where

$$\mathbf{R}_x = \frac{\rho V_\infty x}{\mu} = \frac{V_\infty x}{v}$$

(9.102)

and v_e is the y-velocity at the outer edge of the boundary layer.

We can estimate the form of the functions f_1, f_2, and f_3 by some clever, if not exactly rigorous, reasoning. The key is to recognize that, by definition, the boundary layer is the region of flow where viscous forces and inertia forces are the same order of magnitude. We can relate the flow parameters by examining this order of magnitude requirement. We will use the symbol $\sim$ to represent "of the same order of magnitude." Thus, in the boundary layer,

Inertia force $(\delta F_I) \sim$ Viscous force (δF_v).

We first consider the x (streamwise) components of the forces.

For a typical fluid particle within the boundary layer* (Fig. 9.26), the respective forces are

$$\delta F_{I,x} = \rho(\delta V)a_x = \rho u \left(\frac{\partial u}{\partial x}\right)(\delta V)$$

and

$$\delta F_{v,x} = \frac{\partial \tau}{\partial y} \delta y \, \delta A_y = \mu \left(\frac{\partial^2 u}{\partial y^2}\right)(\delta V).$$

Thus

$$\rho u \left(\frac{\partial u}{\partial x}\right) \sim \mu \left(\frac{\partial^2 u}{\partial y^2}\right).$$

(9.103)

We now proceed to estimate the quantities in this "equation." If the particle is near the center of the layer,

$$u \sim \frac{1}{2} V_\infty.$$

---

* The particle is located somewhere between the wall and the edge of the layer. At the wall the inertia force is zero, and at the edge the viscous force is zero.

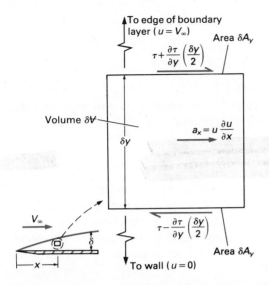

Figure 9.26 Stress and acceleration on fluid particle from "center" of boundary layer.

To estimate $\partial u/\partial x$, we note that, at the leading edge of the plate, $x = 0$; the velocity at the center of the layer is V_∞; and at x, the velocity at the center of the layer is approximately $V_\infty/2$, so

$$\frac{\partial u}{\partial x} \approx \frac{\Delta u}{\Delta x} \sim \frac{\frac{1}{2}V_\infty - V_\infty}{x - 0} = -\frac{V_\infty}{2x},$$

and thus

$$\rho u \left(\frac{\partial u}{\partial x} \right) \sim -\frac{1}{4} \rho \frac{V_\infty^2}{x}.$$

To estimate $\partial^2 u/\partial y^2$, we first estimate the derivative $\partial u/\partial y$. The velocity changes from $u = 0$ to $u = V_\infty$ in a distance δ, so

$$\frac{\partial u}{\partial y} \sim \frac{\Delta u}{\Delta y} = \frac{V_\infty}{\delta}.$$

The velocity gradient is greater than this at the wall and less than this (namely zero) at the boundary layer edge. The second derivative is estimated by

$$\frac{\partial^2 u}{\partial y^2} = \frac{\partial}{\partial y}\left(\frac{\partial u}{\partial y} \right) \sim \frac{\Delta\left(\frac{\partial u}{\partial y} \right)}{\Delta y} = \frac{\left. \frac{\partial u}{\partial y} \right)_{\text{edge of layer}} - \left. \frac{\partial u}{\partial y} \right)_{\text{wall}}}{\delta}.$$

If we assume that the velocity gradient at the wall is twice as large as the "typical" value (V_∞/δ), we find

$$\mu \frac{\partial^2 u}{\partial y^2} \sim \mu \frac{0 - 2\left(\frac{V_\infty}{\delta} \right)}{\delta} = -\frac{2\mu V_\infty}{\delta^2}.$$

Substituting our estimates into "Eq." (9.103), we get

$$\frac{1}{4} \rho \left(\frac{V_\infty^2}{x} \right) \sim 2\mu \left(\frac{V_\infty}{\delta^2} \right),$$

or, solving for δ,

$$\delta \sim \sqrt{\frac{8\mu x}{\rho V_\infty}}.$$

Of course we can't put very much faith in the 8 in this expression. Recognizing that we are dealing with an order of magnitude estimate, we drop it. Upon dividing by x and replacing μ/ρ with ν, we get

$$\frac{\delta}{x} \sim \sqrt{\frac{\nu}{V_\infty x}} = \frac{1}{\sqrt{\mathbf{R}_x}}, \tag{9.104}$$

which not only verifies the relation implied by Eq. (9.99) but also verifies that the boundary layer is thin for large Reynolds number. Equation (9.104) is limited to laminar boundary layers because only viscous stress was included in the estimate. Attempts at making a corresponding estimate for turbulent flow fail because a universal relation for turbulent stress is not available.

We continue our investigation by estimating the wall shear stress and skin friction coefficient. The wall shear stress is

$$\tau_w = \mu \left(\frac{\partial u}{\partial y} \right)_{y=0}.$$

An order of magnitude estimate gives*

$$\tau_w \sim \mu \left(\frac{V_\infty}{\delta} \right).$$

Using Eq. (9.104) for δ gives

$$\tau_w \sim \sqrt{\frac{\rho \mu V_\infty^3}{x}}.$$

Thus

$$C_f = \frac{2\tau_w}{\rho V_\infty^2} \sim \frac{1}{\sqrt{\mathbf{R}_x}}, \tag{9.105a}$$

which verifies Eq. (9.100). It is interesting to note that, with Eq. (9.104),

$$\frac{\mathbf{R}_\delta}{\mathbf{R}_x} = \frac{\dfrac{V_\infty \delta}{\nu}}{\dfrac{V_\infty x}{\nu}} = \frac{\delta}{x} \sim \frac{1}{\sqrt{\mathbf{R}_x}}, \qquad \text{so} \qquad \mathbf{R}_\delta \sim \sqrt{\mathbf{R}_x}.$$

Combining this with Eq. (9.105a), we obtain

$$C_f \sim \frac{1}{\mathbf{R}_\delta}, \tag{9.105b}$$

that is, the friction coefficient is inversely proportional to Reynolds number based on a *transverse* length dimension. We can compare this to the Hagen-Poiseuille formula for laminar pipe flow (Eq. 7.40) by noting that δ corresponds to $D/2$.

---

* No numerical "constants" will be included in our estimates henceforth.

We can estimate the transverse velocity at the edge of the boundary layer from the differential continuity equation,

$$\frac{\partial u}{\partial x} + \frac{\partial v}{\partial y} = 0, \qquad\qquad \textbf{(3.19a)}$$

or

$$\frac{\partial v}{\partial y} = -\frac{\partial u}{\partial x}.$$

Estimating orders of magnitude, we have

$$\frac{\partial v}{\partial y} \approx \frac{\Delta v}{\Delta y} = \frac{v_e - v_w}{\delta} = \frac{v_e - 0}{\delta} = \frac{v_e}{\delta} \qquad \text{and} \qquad -\frac{\partial u}{\partial x} \approx -\frac{\Delta u}{\Delta x} \sim \frac{V_\infty}{x},$$

so

$$\frac{v_e}{\delta} \sim \frac{V_\infty}{x}$$

and

$$\frac{v_e}{V_\infty} \sim \frac{\delta}{x} \sim \frac{1}{\sqrt{R_x}}. \qquad\qquad \textbf{(9.106)}$$

This not only verifies Eq. (9.101) but, more important, shows that the streamlines in the boundary layer are approximately parallel to the surface for large Reynolds number.

You might have noticed that none of our considerations have involved pressure, although in Section 9.1 we made much of the fact that it is possible to assume that the pressure does not vary across the boundary layer. We can now verify this statement by estimating the variation of pressure across a typical boundary layer. The complete force balance for a fluid particle considers pressure, inertia, and stress forces:

$$\delta F_{\text{pressure}} + \delta F_{\text{inertia}} + \delta F_{\text{stress}} = 0.$$

In a boundary layer, both inertia and stress forces are important. We can estimate the largest possible pressure force in a boundary layer by assuming that both inertia and stress contribute to the pressure variation:

$$\delta F_{\text{pressure}} = -(\delta F_{\text{inertia}} + \delta F_{\text{stress}}).$$

We are interested in the pressure variation across the boundary layer, so we apply this equation in a direction normal to the surface (the y-direction). For a typical fluid particle, we get

$$\frac{\partial p}{\partial y}(\delta V) = -\rho a_y \, \delta V + \delta F_{\text{stress}}.$$

In a boundary layer, the inertia force and stress force are the same order of magnitude, so

$$\frac{\partial p}{\partial y} \sim 2\rho a_y.$$

Now recognize that y is essentially normal to the streamlines, so, from Eq. (4.4), we have

$$a_y \approx \frac{V^2}{R} = \frac{u^2 + v^2}{R},$$

where R is the radius of curvature of the streamlines. In the boundary layer,

$$v \ll u,$$

and the streamlines are approximately parallel to the body, so

$$a_y \approx \frac{u^2}{R_B},$$

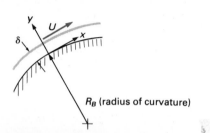

Figure 9.27 Boundary layer on curved surface.

where R_B is the radius of curvature of the body profile (Fig. 9.27). With these results, our estimate of the transverse pressure gradient becomes

$$\frac{\partial p}{\partial y} \sim 2\rho \left(\frac{u^2}{R_B} \right).$$

For a typical point in the boundary layer,

$$u^2 \sim \frac{V_\infty^2}{2},$$

so

$$\frac{\partial p}{\partial y} \sim \rho \left(\frac{V_\infty^2}{R_B} \right). \tag{9.107}$$

The change of pressure across the boundary layer is approximated by

$$\Delta p = \int_0^\delta \frac{\partial p}{\partial y}\, dy \approx \left(\frac{\partial p}{\partial y} \right) \delta, \qquad \text{so} \qquad \Delta p \sim \rho V_\infty^2 \left(\frac{\delta}{R_B} \right),$$

or, in terms of a pressure coefficient,

$$\mathbf{C_{\Delta p}} = \frac{2 \Delta p}{\rho V_\infty^2} \sim \frac{\delta}{R_B}. \tag{9.108}$$

Equation (9.108) verifies the following important statement:

> If the boundary layer thickness is small compared to the radius of curvature of the body surface, the pressure can be considered constant across the boundary layer.

This is certainly true for a flat plate because $R_B \to \infty$.

In this section, we have used the rules of mathematics a bit loosely. You should realize that we were interested only in estimates and general conclusions, not in producing equations for boundary layer calculation. We have verified all of the important assumptions of boundary layer theory. In the next section, we will use these assumptions to derive a simplified equation of motion for the boundary layer region. Our later solutions of this equation will verify our order of magnitude estimates as well as fill in the gaps. Although they *are not* computational equations, you should remember the principles of boundary layer theory that are embodied in Eqs. (9.104), (9.105), (9.106), and (9.107).

9.3.2 The Differential Equations of Boundary Layer Analysis

The important parameters of a boundary layer are the thickness, which is a function of distance along the surface, and the shear stress at the surface. The blockage that the boundary layer presents to the outer inviscid flow and information concerning the possibility of boundary layer separation are also important (see Fig. 9.2). All of these quantities can be determined from the boundary layer velocity field $u(x, y)$ and $v(x, y)$. The velocities are the solution to differential equations expressing the law of conservation of mass and the linear momentum theorem.

Figure 9.28 shows a control volume located in a boundary layer. Our analysis can be extended to curved surfaces (with $R_B \gg \delta$) by choosing coordinates parallel and perpendicular to the surface. To formulate the equations for the control volume, we use the following results from the previous sections:

- The pressure is constant across the boundary layer.
- The pressure is a known function of x and is related to the velocity immediately outside the boundary layer by Bernoulli's equation.
- The velocity changes much more rapidly across the layer than along it $[(\partial u/\partial x \sim V_\infty/x) \ll (V_\infty/\delta \sim \partial u/\partial y)]$.

The final point allows us to neglect the normal viscous stresses in the boundary layer.

We can make short work of the principle of conservation of mass by noting that, as far as mass flow is concerned, Fig. 9.28 is identical to Fig. 3.22 (except that ρ is constant), and so the continuity equation is

$$\frac{\partial u}{\partial x} + \frac{\partial v}{\partial y} = 0. \tag{9.109}$$

This is the continuity equation for two-dimensional, two-directional, incompressible flow, Eq. (3.19a). We have renumbered the equation here for convenience.

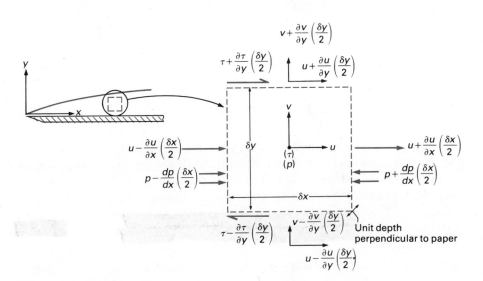

Figure 9.28 Control volume for boundary layer analysis. Only velocities and stresses relevant to continuity and x-momentum equations are shown.

Next we apply the linear momentum theorem in the x-direction. This derivation is similar to that of Section 5.3.3, except that the shear stress force must be included.* The x-momentum flux terms are

$$\delta \dot{M}_{x,\text{out}} - \delta \dot{M}_{x,\text{in}} = \delta \dot{M}_{x,\text{right}} + \delta \dot{M}_{x,\text{top}} - \delta \dot{M}_{x,\text{left}} - \delta \dot{M}_{x,\text{bottom}}$$

$$= \left[u + \frac{\partial u}{\partial x}\left(\frac{\delta x}{2}\right) \right] \rho \left[u + \frac{\partial u}{\partial x}\left(\frac{\delta x}{2}\right) \right] \delta y$$

$$+ \left[u + \frac{\partial u}{\partial y}\left(\frac{\delta y}{2}\right) \right] \rho \left[v + \frac{\partial v}{\partial y}\left(\frac{\delta y}{2}\right) \right] \delta x$$

$$- \left[u - \frac{\partial u}{\partial x}\left(\frac{\delta x}{2}\right) \right] \rho \left[u - \frac{\partial u}{\partial x}\left(\frac{\delta x}{2}\right) \right] \delta y$$

$$- \left[u - \frac{\partial u}{\partial y}\left(\frac{\delta y}{2}\right) \right] \rho \left[v - \frac{\partial v}{\partial y}\left(\frac{\delta y}{2}\right) \right] \delta x.$$

Carrying out the multiplications, canceling terms where possible, and dropping higher-order terms, we obtain

$$\delta \dot{M}_{x,\text{out}} - \delta \dot{M}_{x,\text{in}} = \rho \left[u\left(\frac{\partial u}{\partial x}\right) + v\left(\frac{\partial u}{\partial y}\right) + u\left(\frac{\partial u}{\partial x} + \frac{\partial v}{\partial y}\right) \right] \delta x \, \delta y.$$

The term $(\partial u/\partial x + \partial v/\partial y)$ is zero by Eq. (9.109), so

$$\delta \dot{M}_{x,\text{out}} - \delta \dot{M}_{x,\text{in}} = \rho \left[u\left(\frac{\partial u}{\partial x}\right) + v\left(\frac{\partial u}{\partial y}\right) \right] \delta x \, \delta y.$$

The net x-force is

$$\delta F_x = \delta F_{p,\text{left}} - \delta F_{p,\text{right}} + \delta F_{\tau,\text{top}} - \delta F_{\tau,\text{bottom}}.$$

Substituting the various pressures and stresses, we have

$$\delta F_x = \left[p - \frac{dp}{dx}\left(\frac{\delta x}{2}\right) \right] \delta y - \left[p + \frac{dp}{dx}\left(\frac{\delta x}{2}\right) \right] \delta y$$

$$+ \left[\tau + \frac{\partial \tau}{\partial y}\left(\frac{\delta y}{2}\right) \right] \delta x - \left[\tau - \frac{\partial \tau}{\partial y}\left(\frac{\delta y}{2}\right) \right] \delta x.$$

Simplifying gives

$$\delta F_x = \left(-\frac{dp}{dx} + \frac{\partial \tau}{\partial y} \right) \delta x \, \delta y.$$

Substituting into the linear momentum equation and canceling $\delta x \, \delta y$, we obtain

$$\rho \left[u\left(\frac{\partial u}{\partial x}\right) + v\left(\frac{\partial u}{\partial y}\right) \right] = -\frac{dp}{dx} + \frac{\partial \tau}{\partial y}. \tag{9.110}$$

This equation can be applied to either laminar or turbulent flow since we have not yet specified how τ is obtained. If we limit consideration to laminar flow of a Newtonian fluid with constant viscosity, the boundary layer momentum equation becomes (on dividing by ρ):

$$u\left(\frac{\partial u}{\partial x}\right) + v\left(\frac{\partial u}{\partial y}\right) = -\frac{1}{\rho}\left(\frac{dp}{dx}\right) + \nu\left(\frac{\partial^2 u}{\partial y^2}\right). \tag{9.111}$$

---

* Gravity forces can be combined with pressure in the same way as in Eqs. (9.1)–(9.4).

Equations (9.109) and (9.111) are two equations for the two unknowns u and v. These are the *Prandtl boundary layer equations*, first proposed in 1904 [2]. The pressure is assumed to be known. The condition

$$\frac{\partial p}{\partial y} \approx 0$$

replaces the entire y-momentum equation. If we wish, we can write Bernoulli's equation for the flow just outside the boundary layer*

$$p + \frac{\rho U^2}{2} = \text{Constant},$$

which, on differentiating, gives

$$-\frac{1}{\rho}\left(\frac{dp}{dx}\right) = U\left(\frac{dU}{dx}\right). \tag{9.112}$$

This equation can be used to eliminate the pressure from Eq. (9.111).

To solve the equations, we must specify boundary conditions. Two of the conditions are quite obvious. At $y = 0$ (the wall),

$$u = 0 \quad \text{at } y = 0, \tag{9.113a}$$

$$v = 0 \quad \text{at } y = 0. \tag{9.113b}$$

A third condition seems obvious when we recognize that the boundary layer must join with the outer inviscid flow at its edge. We would like to say

$$u = U \quad \text{at } y = \delta;$$

however, we do not know where the edge of the boundary layer is (δ is an unknown function of x). We have to fudge a bit and specify that

$$u \to U \quad \text{as} \quad y \to \infty. \tag{9.113c}$$

This condition is quite realistic since the boundary layer merges smoothly into the outer flow. We do have a problem calculating a value for the boundary layer thickness δ when condition (9.113c) is used to obtain a velocity profile. Conventionally, δ is defined as *the value of y at which u is 0.99U.*

A fourth condition is that the velocity profile must be known at some upstream or inlet position,

$$u = u_0(y) \quad \text{at } x = x_0. \tag{9.113d}$$

Equations (9.113a) and (9.113b) are all the boundary conditions that are needed. The boundary layer equations do not need and *cannot use* conditions on the velocity at an outlet (downstream) station or conditions on v at the outer edge of the layer.

9.3.3 Evaluation of Key Boundary Layer Parameters

A complete boundary layer calculation requires solution of Eqs. (9.109) and (9.111) to obtain the fluid velocities $u(x, y)$ and $v(x, y)$. This

---

* We write U for the velocity outside the boundary layer rather than V_∞ to emphasize that the equations are not restricted to a flat plate.

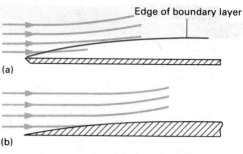

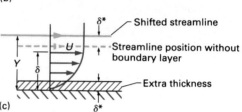

Figure 9.29 Streamline shift and blockage effects of a boundary layer: (a) Deflection of outer flow streamlines; (b) flow over thicker body; (c) equivalence of extra thickness and streamline shift.

is a rather tall order since the momentum equation is nonlinear.[†] For now, we will suppose that we have such a solution and consider how we evaluate some key boundary layer parameters.

The parameters of primary interest are the boundary layer thickness and the wall shear stress or skin friction coefficient. The boundary layer thickness is obtained from the velocity profile by the definition

$$\delta \equiv y)_{u/U=0.99}. \tag{9.114}$$

The wall shear stress is calculated by

$$\tau_w = \mu\left(\frac{\partial u}{\partial y}\right)_{\text{wall}} = \mu\left(\frac{\partial u}{\partial y}\right)_{y=0}, \tag{9.115a}$$

or, in terms of the skin friction coefficient,

$$C_f = \frac{2\tau_w}{\rho U^2} = \frac{2\nu}{U^2}\left(\frac{\partial u}{\partial y}\right)_{y=0}. \tag{9.115b}$$

We may also be interested in the boundary layer effects on the outer flow. There are two such effects. The first is streamline deflection. If there were no boundary layer, the streamlines near the surface would be parallel to it. The boundary layer produces a transverse velocity at its edge that deflects the outer flow streamlines outward slightly (Fig. 9.29a). The second effect is blockage. The mass flow in the boundary layer is reduced because the velocity is lowered. The streamline deflection and blockage effects are related because a thicker body (more blockage) would also cause more deflection of the outer flow (Fig. 9.29b). Both effects can be quantified by introducing the *displacement thickness* (δ^*), defined as either the distance that the outer flow

---

[†] The continuity equation can be eliminated by introduction of a *stream function* $\Psi(x, y)$ such that $u = \partial\Psi/\partial y$ and $v = -\partial\Psi/\partial x$, so that

$$\frac{\partial u}{\partial y} + \frac{\partial v}{\partial y} = \frac{\partial^2\Psi}{\partial x\,\partial y} - \frac{\partial^2\Psi}{\partial x\,\partial y} \equiv 0.$$

See Section 9.2.1.

streamlines are shifted (displaced) by the boundary layer or, equivalently, as the extra thickness that would be added to a body or surface to account for the extra blockage. Both concepts are illustrated in Fig. 9.29(c).

The displacement thickness can be calculated from a mass balance:

$$\int_0^Y \rho u\, dy = \int_{\delta^*}^Y \rho U dy.$$

The integral on the left is the actual mass flow between the wall and a point in the free stream. The integral on the right is the mass that would be carried by a uniform velocity $U(x)$ over an imaginary surface that is thicker by amount δ^* than the real surface. Since ρ and U are constants, we can cancel ρ and evaluate the right integral

$$\int_{\delta^*}^Y U dy = U(Y - \delta^*).$$

Solving for δ^*, we have

$$\delta^* = Y - \int_0^Y \frac{u}{U}\, dy = \int_0^Y dy - \int_0^Y \frac{u}{U}\, dy = \int_0^Y \left(1 - \frac{u}{U}\right) dy.$$

The actual value of Y is arbitrary so long as $Y > \delta$; however, note that

$$u = U \qquad \text{if } Y \geq \delta,$$

so the upper limit can be changed to δ (it could just as easily be changed to ∞):

$$\delta^* = \int_0^\delta \left(1 - \frac{u}{U}\right) dy. \tag{9.116}$$

The boundary layer blockage and streamline displacement effects are obtained by calculating δ^* from the velocity profile. The effects of the boundary layer on the outer flow can be obtained by adding the displacement thickness to the surface and recalculating the inviscid flow over the resulting body. This effect is especially important in internal flows such as the development region of a pipe (Fig. 7.6) where the displacement effect causes an acceleration of the inviscid core flow.

Besides accounting for blockage, δ^* provides a more rigorous and repeatable measurement of boundary layer thickness than δ does. This is especially true when we consider experimentally measured boundary layer velocity profiles. The point at which $u/U = 0.99$ is a matter of judgment in fitting a curve to the data; however, the integral in Eq. (9.116) gives almost the same value of δ^* no matter where the outer limit is located.

9.3.4 Momentum Integral Equation for Boundary Layers

Boundary layer flow is completely described by the Prandtl boundary layer equations, which are difficult to solve. Solutions have been found for only a few special (but nevertheless important) cases. This may be discouraging; however, if we consider the topics discussed in the last section, we note that a complete solution (that is, u and v as functions of x and y) is not really very important anyway. All that we really need for most practical applications are values for the boundary layer thickness (both δ and δ^*) and the skin friction coefficient. For most engineering purposes, reliable estimates of these parameters are sufficient. This situation is tailor-made for application of the control volume approach. Since we already know that δ and C_f are functions of x (Eqs. 9.104 and 9.105), we will use a "hybrid" control volume that is finite in the y-direction but differentially small in the x-direction. This will give us an ordinary (as opposed to partial) differential equation that is easier to solve. The catch is that the solution (although not the governing equation) is only an approximation, albeit a rather good one.

Figure 9.30 shows a control volume that spans the entire boundary layer, extending from the wall to an arbitrary height Y, outside the boundary layer. The length of the control volume is δx. The control volume has unit depth perpendicular to the paper. The velocity distributions at the inlet (left) and outlet (right) faces of the control volume are not uniform. The pressure is constant over each x-face. We first write the continuity equation for the control volume. Noting that no fluid crosses the lower surface, we get

$$\dot{m}_{\text{left}} = \dot{m}_{\text{right}} + \dot{m}_{\text{top}}. \tag{9.117}$$

The mass flow across the left (inlet) face is

$$\dot{m}_{\text{left}} = \int_0^Y \rho u \, dy \Big)_x. \tag{9.118}$$

The mass flow rate through the right face is

$$\dot{m}_{\text{right}} = \dot{m}_{x+\delta x} = \int_0^Y \rho u \, dy \Big)_x + \frac{d}{dx} \left(\int_0^Y \rho u \, dy \right) \delta x. \tag{9.119}$$

Substituting into Eq. (9.117) and solving for the mass flow across the top of the control volume give

$$\dot{m}_{\text{top}} = -\frac{d}{dx} \left(\int_0^Y \rho u \, dy \right) \delta x. \tag{9.120}$$

Now we consider the momentum equation. Momentum flows across the control surface at the right and left faces and at the top of the control

Figure 9.30 "Hybrid" control volume for integral analysis of boundary layer.

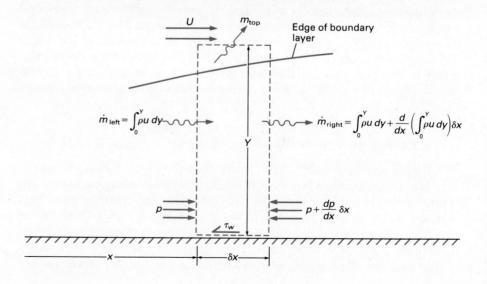

volume. The forces on the fluid are due to pressure and shear stress. The x-direction momentum equation is

$$\dot{M}_{x,\text{right}} + \dot{M}_{x,\text{top}} - \dot{M}_{x,\text{left}} = F_{x,\text{pressure}} + F_{x,\text{stress}}. \qquad (9.121)$$

The x-momentum flux at the left (inlet) face is

$$\dot{M}_{x,\text{left}} = \int_0^Y \rho u V_n \, dA = \left(\int_0^Y \rho u^2 \, dy \right)_x. \qquad (9.122)$$

The x-momentum flux at the face right is

$$\dot{M}_{x,\text{right}} = \dot{M}_x)_{x+\delta x} = \left(\int_0^Y \rho u^2 \, dy \right)_x + \frac{d}{dx}\left(\int_0^Y \rho u^2 \, dy \right)\delta x, \qquad (9.123)$$

and the x-momentum flux at the top is

$$\dot{M}_{x,\text{top}} = \dot{m}_{\text{top}} u_{\text{top}} = \dot{m}_{\text{top}} U.$$

Using Eq. (9.120), we have

$$\dot{M}_{x,\text{top}} = -U\left(\frac{d}{dx}\right)\left(\int_0^Y \rho u \, dy\right)\delta x. \qquad (9.124)$$

The forces are

$$F_{x,\text{pressure}} = pY - \left(p + \frac{dp}{dx}\delta x\right)Y = -\frac{dp}{dx}Y\delta x \qquad (9.125)$$

and

$$F_{x,\text{stress}} = -\tau_w \delta x. \qquad (9.126)$$

Substituting Eqs. (9.122)–(9.126) into Eq. (9.121), canceling terms, and dividing by δx, we obtain

$$\frac{d}{dx}\left(\int_0^Y \rho u^2 \, dy\right) - U\left(\frac{d}{dx}\right)\left(\int_0^Y \rho u \, dy\right) = -Y\left(\frac{dp}{dx}\right) - \tau_w. \qquad (9.127)$$

At the beginning of this section we implied that we would obtain an equation for the boundary layer parameters that is easier to solve than the Prandtl boundary layer equations. It certainly does not seem that Eq. (9.127) is it! To get this equation into useful form, we have to work with it a bit. Our first priority is elimination of the arbitrary height Y. To accomplish this, we first eliminate the pressure gradient using Eq. (9.112). Noting that

$$Y = \int_0^Y dy,$$

multiplying by -1, dividing by ρ, and rearranging, we get

$$U\left(\frac{d}{dx}\right)\int_0^Y u \, dy + U\left(\frac{dU}{dx}\right)\int_0^Y dy - \frac{d}{dx}\int_0^Y u^2 \, dy = \frac{\tau_w}{\rho}. \qquad (9.128)$$

Next consider the quantity[†]

$$\frac{d}{dx}\left(U\int_0^Y u \, dy\right) = \frac{dU}{dx}\left(\int_0^Y u \, dy\right) + U\left(\frac{d}{dx}\right)\int_0^Y u \, dy.$$

[†] This quantity does not appear in Eq. (9.128). It is not equal to the first integral in Eq. (9.128) because U is a function of x.

The second term on the right appears in Eq. (9.128). We solve for it in terms of the other two and substitute the result into Eq. (9.128) to get

$$\frac{d}{dx}\left(U\int_0^Y u\,dy\right) - \frac{dU}{dx}\left(\int_0^Y u\,dy\right) + U\left(\frac{dU}{dx}\right)\int_0^Y dy - \frac{d}{dx}\int_0^Y u^2\,dy = \frac{\tau_w}{\rho}.$$

Since U is not a function of y, it can be moved inside the first and third integrals. Performing this operation and grouping terms, we have

$$\frac{d}{dx}\int_0^Y u(U - u)\,dy + \frac{dU}{dx}\int_0^Y (U - u)\,dy = \frac{\tau_w}{\rho}.$$

Now note that if $y > \delta$, the integrands are zero, so we can change the upper limit of the integrals to δ to get

$$\frac{d}{dx}\int_0^\delta u(U - u)\,dy + \frac{dU}{dx}\int_0^\delta (U - u)\,dy = \frac{\tau_w}{\rho}. \qquad \textbf{(9.129)}$$

This equation is *von Karman's momentum integral equation.* It probably does not seem any simpler to you than the Prandtl boundary layer equations, but notice that it *is* an ordinary differential equation and that two key boundary layer parameters (δ and τ_w) appear explicitly. Also, the second integral reminds us strongly of the displacement thickness. The equation has the additional advantage that it can be applied to either laminar or turbulent flow since no specific expression for τ_w has been used.

To discover the real advantages of Eq. (9.129), we must manipulate it a bit more. We multiply the first integral by U^2/U^2 and the second integral by U/U to get

$$\frac{d}{dx}\left[U^2\int_0^\delta \frac{u}{U}\left(1 - \frac{u}{U}\right)dy\right] + U\left(\frac{dU}{dx}\right)\int_0^\delta \left(1 - \frac{u}{U}\right)dy = \frac{\tau_w}{\rho}.$$

Next we multiply each integral by δ/δ (noting that δ is independent of y) to get

$$\frac{d}{dx}\left[\delta U^2\left(\int_0^1 \frac{u}{U}\left(1 - \frac{u}{U}\right)d\left(\frac{y}{\delta}\right)\right)\right] + \delta U\left(\frac{dU}{dx}\right)\int_0^1 \left(1 - \frac{u}{U}\right)d\left(\frac{y}{\delta}\right) = \frac{\tau_w}{\rho}. \quad \textbf{(9.130)}$$

The integrals are complicated, but in this form they involve only the dimensionless shape of the velocity profile. Since the variable of integration has been changed from y to y/δ it does not matter how thick or thin the boundary layer is, as long as we know the shape of the velocity profile. It should not be too difficult for us to *assume* a reasonable approximate shape for the boundary layer velocity profile. With an assumed function[†] for the dimensionless velocity profile

$$\frac{u}{U} \approx f\left(\frac{y}{\delta}\right),$$

the integrals can be evaluated, and Eq. (9.130) becomes a differential equation involving U (which is known from potential flow analysis), δ, and τ_w. We can obtain some guidance in assuming a velocity profile function from the boundary conditions that it must satisfy (Eqs. 9.113a–

[†] Say a polynominal profile $u/U = a + b(y/\delta) + c(y/\delta)^2 + \cdots.$

9.113c). It is not necessary to know the exact shape because the velocity profile is integrated.

We can write Eq. (9.130) in a more abbreviated form. Notice that the second integral is the displacement thickness. The first integral also has the dimension of length and can be considered as a type of boundary layer thickness. This integral is called the *momentum thickness* and is assigned the symbol Θ:

$$\Theta \equiv \int_0^\delta \frac{u}{U}\left(1 - \frac{u}{U}\right)dy. \qquad (9.131)$$

In terms of δ^* and Θ, we can write von Karman's momentum integral equation as

$$\frac{d}{dx}[U^2\Theta] + \delta^* U\left(\frac{dU}{dx}\right) = \frac{\tau_w}{\rho}. \qquad (9.132)$$

The equation can also be written

$$\frac{d\Theta}{dx} + (2 + H)\frac{\Theta}{U}\left(\frac{dU}{dx}\right) = \frac{C_f}{2}, \qquad (9.133)$$

where

$$H \equiv \frac{\delta^*}{\Theta} \qquad (9.134)$$

is called the *shape factor*.

EXAMPLE 9.5 Illustrates boundary layer parameters

A 4-in.-nominal-diameter schedule 40 steel pipe has a water flow rate of 50 gal/min. Determine the values of the displacement thickness δ^*, momentum thickness Θ, and the shape factor H, for the fully developed flow velocity profile. The water temperature is 60°F.

Solution

  Given
4-in.-nominal-diameter schedule 40 steel pipe
Flow rate 50 gal/min of 60°F water
Fully developed flow

  Find
Displacement thickness δ^*
Momentum thickness Θ
Shape factor H

  Solution
Appendix D shows that the pipe inside diameter is 4.026 in. For 60°F water, Table A.6 gives $\nu = 1.22 \times 10^{-5}$ ft²/sec. The average velocity is

$$\bar{V} = \frac{Q}{A} = \frac{4Q}{\pi D^2}$$

$$= \frac{4\left(50\ \dfrac{\text{gal}}{\text{min}}\right)\left(\dfrac{\text{ft}^3}{7.48\ \text{gal}}\right)\left(\dfrac{\text{min}}{60\ \text{sec}}\right)}{\pi(4.026\ \text{in.})^2\left(\dfrac{\text{ft}}{12\ \text{in.}}\right)^2} = 1.26\ \text{ft/sec},$$

and the Reynolds number is

$$\mathbf{R} = \frac{VD}{\nu} = \frac{\left(1.26\,\dfrac{\text{ft}}{\text{sec}}\right)(4.026 \text{ in.})}{\left(1.22 \times 10^{-5}\,\dfrac{\text{ft}^2}{\text{sec}}\right)\left(\dfrac{12 \text{ in.}}{\text{ft}}\right)} = 3.5 \times 10^4.$$

For 4-in. commercial steel pipe, Fig. 7.12 gives $\varepsilon/D = 0.00043$. The Moody chart gives $f = 0.025$. Equations (7.49) and (7.51) give

$$\frac{u}{u_{\max}} = \left(\frac{y}{R}\right)^{1/n},$$

with

$$n \approx \frac{1}{\sqrt{f}} = \frac{1}{\sqrt{0.025}} = 6.3.$$

We are now able to calculate the boundary layer parameters. The displacement thickness δ^* is

$$\delta^* = \int_0^R \left(1 - \frac{u}{u_{\max}}\right) dy,$$

where the boundary layer thickness δ is equivalent to the pipe radius R. Substituting the velocity profile, we have

$$\delta^* = \int_0^R \left[1 - \left(\frac{y}{R}\right)^{1/n}\right] dy = \left[y - \frac{n}{n+1}\left(\frac{y^{(n+1)/n}}{R^{1/n}}\right)\right]_0^R$$

$$= \left[R - \frac{n}{n+1}R\right] = \frac{R}{n+1} = \frac{2.013 \text{ in.}}{7.3},$$

$$\boxed{\delta^* = 0.276 \text{ in.}}$$

Answer ◀

The momentum thickness Θ is

$$\Theta = \int_0^R \frac{u}{u_{\max}}\left(1 - \frac{u}{u_{\max}}\right) dy = \int_0^R \left[\frac{u}{u_{\max}} - \left(\frac{u}{u_{\max}}\right)^2\right] dy,$$

or

$$\Theta = \int_0^R \left[\left(\frac{y}{R}\right)^{1/n} - \left(\frac{y}{R}\right)^{2/n}\right] dy = \left[\frac{n}{n+1}\left(\frac{y^{(n+1)/n}}{R^{1/n}}\right) - \frac{n}{n+2}\left(\frac{y^{(n+2)/n}}{R^{2/n}}\right)\right]_0^R,$$

$$= \left[\left(\frac{n}{n+1}\right)R - \left(\frac{n}{n+2}\right)R\right] = \left[\frac{n(n+2) - n(n+1)}{(n+1)(n+2)}\right]R = \left(\frac{n}{(n+1)(n+2)}\right)R,$$

$$\Theta = \frac{6.3}{(7.3)(8.3)}(2.013 \text{ in.}),$$

Answer ◀

$$\boxed{\Theta = 0.209 \text{ in.}}$$

The shape factor is found from Eq. (9.134):

$$H = \frac{\delta^*}{\Theta} = \frac{0.276 \text{ in.}}{0.209 \text{ in.}},$$

Answer ◀

$$\boxed{H = 1.32.}$$

Discussion

Since the problem statement specifies fully developed flow, the "boundary layer thickness" is taken as the pipe radius, 2.013 in. The traditional boundary

layer thickness, δ, is given by

$$\delta = y_{u=0.99u_{\max}},$$

or

$$\delta = R\left(\frac{u}{u_{\max}}\right)^n\Bigg)_{u=0.99u_{\max}} = 2.013\,(0.99)^{6.3}\ \text{in.}$$

$$= 1.89\ \text{in.}$$

9.3.5 Calculation of Boundary Layer Parameters for Flow over a Flat Plate

Consider a flat plate aligned parallel to a uniform flow with velocity V_∞ (see Fig. 9.31). If the plate is very thin and we neglect the boundary layer displacement effect, then the inviscid outer flow solution is

$$U = V_\infty = \text{Constant} \quad \text{and} \quad p = p_\infty = \text{Constant.} \qquad \textbf{(9.135)}$$

The boundary layer on a flat plate is the simplest boundary layer to calculate because there is no free-stream pressure or velocity gradient.

Near the leading edge of the plate, the Reynolds number is small and the flow is laminar. If the plate is long enough, eventually the boundary layer becomes thick enough and the Reynolds number becomes large enough that transition occurs and the flow becomes turbulent. The laminar and turbulent portions of the boundary layer must be considered separately.

Laminar Boundary Layer. First we will use the momentum integral equation to obtain an estimate of the laminar boundary layer parameters. Since the free-stream velocity is constant, Eq. (9.130) reduces to

$$U^2\frac{d}{dx}\left[\delta\int_0^1 \frac{u}{U}\left(1-\frac{u}{U}\right)d\left(\frac{y}{\delta}\right)\right] = \frac{\tau_w}{\rho}. \qquad \textbf{(9.136)}$$

Since the flow is laminar, the wall shear stress is

$$\tau_w = \mu\left(\frac{\partial u}{\partial y}\right)_{y=0}.$$

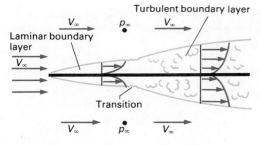

Laminar boundary layer

Turbulent boundary layer

Transition

Figure 9.31 Boundary layers on a thin flat plate parallel to the main flow.

Noting that U and δ are not functions of y, we can write this as

$$\tau_w = \frac{\mu U}{\delta}\left[\frac{\partial\left(\dfrac{u}{U}\right)}{\partial\left(\dfrac{y}{\delta}\right)}\right]_{y/\delta = 0}.\tag{9.137}$$

Substituting into Eq. (9.136) and dividing by U^2, we get

$$\frac{d}{dx}\left[\delta\int_0^1 \frac{u}{U}\left(1 - \frac{u}{U}\right)d\left(\frac{y}{\delta}\right)\right] = \frac{\nu}{U\delta}\left[\frac{\partial\left(\dfrac{u}{U}\right)}{\partial\left(\dfrac{y}{\delta}\right)}\right]_{y/\delta = 0}.\tag{9.138}$$

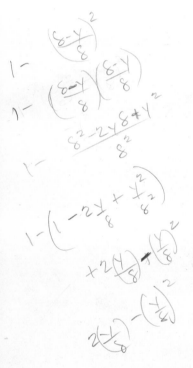

Now if we knew how u/U depends on y/δ, it would be possible to evaluate the integral on the left and the derivative on the right. All we need is a velocity profile that looks like a boundary layer. What is available? We might try various assumptions; however, if we reflect a bit we will remember that we already know at least one velocity profile for laminar flow—that for fully developed flow in a pipe. Combining Eqs. (7.33) and (7.34), we obtain the velocity profile for fully developed laminar *pipe* flow:

$$\frac{u}{u_{max}} = 1 - \left(\frac{r}{R}\right)^2.$$

To use this profile for a boundary layer analysis, we must replace u_{max} by U and R by δ and switch coordinates to measure outward from the wall. This latter transformation is accomplished by

$$y = R - r.$$

With these adjustments, our approximate boundary layer velocity profile becomes

$$\frac{u}{U} \approx 2\left(\frac{y}{\delta}\right) - \left(\frac{y}{\delta}\right)^2 = 2\eta - \eta^2, \qquad \text{where} \qquad \eta = \frac{y}{\delta}.$$

Note that this profile satisfies the conditions

$$u = 0 \quad \text{at } y = 0 \qquad \text{and} \qquad u = U \quad \text{at } y = \delta.$$

It is important to note that we are not assuming that the boundary layer is in a pipe and we definitely are not assuming that U (which corresponds to u_{max}) is in any way related to the parameter $(R^2/4\mu)(dp/dx)$. We use the dimensionless pipe flow velocity profile only because it looks reasonably like a laminar boundary layer velocity profile.

The dimensionless slope of the parabolic profile, evaluated at the wall, is

$$\left.\frac{\partial\left(\dfrac{u}{U}\right)}{\partial\left(\dfrac{y}{\delta}\right)}\right|_0 = \left.\frac{d\left(\dfrac{u}{U}\right)}{d\eta}\right|_{\eta = 0} = 2.$$

Substituting this result and the equation for u/U into the momentum integral equation, Eq. (9.138), gives

$$\frac{d}{dx}\left[\delta \int_0^1 (2\eta - \eta^2)(1 - 2\eta + \eta^2)\, d\eta\right] = 2\frac{v}{U\delta}.$$

Integrating, we obtain

$$\frac{2}{15}\left(\frac{d\delta}{dx}\right) = \frac{2v}{U\delta}.$$

Noting that v is constant and that the velocity at the boundary layer edge (U) is constant and equal to V_∞, we separate the variables,

$$\delta\, d\delta = 15\left(\frac{v}{V_\infty}\right) dx,$$

and integrate from $x = 0$, where $\delta = 0$, to $x = x$ to get

$$\frac{\delta^2}{2} = 15\left(\frac{vx}{V_\infty}\right).$$

Solving for δ, we obtain

$$\delta = 5.48\sqrt{\frac{vx}{V_\infty}}. \qquad (9.139a)$$

Dividing both sides by x, we get

$$\frac{\delta}{x} = 5.48\sqrt{\frac{v}{V_\infty x}} = \frac{5.48}{\sqrt{R_x}}. \qquad (9.139b)$$

This expression not only verifies Eq. (9.104) but also provides a value for the proportionality constant.

The wall shear stress is calculated from Eq. (9.137):

$$\tau_w = \frac{2\mu V_\infty}{5.48\sqrt{\dfrac{vx}{V_\infty}}} = 0.365\sqrt{\frac{\rho\mu V_\infty^3}{x}}. \qquad (9.140)$$

The skin friction coefficient is

$$C_f = \frac{2\tau_w}{\rho V_\infty^2} = 0.730\sqrt{\frac{v}{V_\infty x}} = \frac{0.730}{\sqrt{R_x}}, \qquad (9.141)$$

verifying Eq. (9.105a). We can calculate the displacement thickness and momentum thickness from

$$\delta^* = \delta \int_0^1 \left(1 - \frac{u}{U}\right) d\left(\frac{y}{\delta}\right) \approx \delta \int_0^1 (1 - 2\eta + \eta^2)\, d\eta \qquad \text{disp. thickness}$$

and

$$\Theta = \delta \int_0^1 \frac{u}{U}\left(1 - \frac{u}{U}\right) d\left(\frac{y}{\delta}\right) \approx \delta \int_0^1 (2\eta - \eta^2)(1 - 2\eta + \eta^2)\, d\eta. \qquad \text{mom. thickness}$$

The results are

$$\frac{\delta^*}{x} = \frac{1.83}{\sqrt{R_x}} \qquad (9.142)$$

Table 9.2 Effects of various velocity profile assumptions on momentum integral calculation of flat plate laminar boundary layer parameters

	$\dfrac{\delta}{x}\sqrt{\mathbf{R}_x}$	$C_f\sqrt{\mathbf{R}_x}$	$\dfrac{\delta^*}{x}\sqrt{\mathbf{R}_x}$
Linear profile $\dfrac{u}{U}\approx\dfrac{y}{\delta}$	3.46	0.578	1.73
Parabolic profile $\dfrac{u}{U}\approx 2\left(\dfrac{y}{\delta}\right)-\left(\dfrac{y}{\delta}\right)^2$	5.48	0.730	1.83
Cubic profile $\dfrac{u}{U}\approx\dfrac{3}{2}\left(\dfrac{y}{\delta}\right)-\left(\dfrac{y}{\delta}\right)^3$	4.64	0.646	1.74
Sine wave profile $\dfrac{u}{U}\approx\sin\left[\dfrac{\pi}{2}\left(\dfrac{y}{\delta}\right)\right]$	4.79	0.656	1.74
Exact solution [16]	5.0	0.664	1.721

and

$$\frac{\Theta}{x}\approx\frac{0.73}{\sqrt{\mathbf{R}_x}}. \tag{9.143}$$

These results are nice, but how accurate are they? Also, how much do they depend on the form of the function selected to approximate the velocity profile? To answer the second question, we note that changing the assumed velocity profile changes the numerical values of the definite integral and the dimensionless wall slope in Eq. (9.138).[†] This in turn causes slight changes in the numerical constants in Eqs. (9.139)–(9.143). Table 9.2 lists various approximations for the velocity profile and their effect on the results.

The accuracy of our momentum integral estimate can be evaluated by comparing the results to an exact solution of the Prandtl boundary layer equations. The first boundary layer calculation ever made was the solution of the Prandtl boundary layer equations for laminar flow over a flat plate, carried out by H. Blasius, one of Prandtl's students [16]. Blasius assumed that the velocity profile in the boundary layer at any location along the plate is similar to the velocity profile at any other location; that is, the velocity function is of the special form

$$\frac{u(x,y)}{V_\infty}=f\left(\frac{y}{\delta(x)}\right)=f\left(\frac{y}{\sqrt{\dfrac{vx}{V_\infty}}}\right).$$

[†] Note that the fundamental nature of δ-versus x-dependence indicated by $d\delta/dx\propto v/V_\infty\delta$ *is not* changed.

Using this assumption, Blasius was able to reduce the Prandtl boundary layer equations to a single ordinary differential equation. The details of the reduction and the solution of the differential equation are available elsewhere [5, 6, 8, 16]. Figure 9.32 is a plot of the resulting velocity profile in Blasius's coordinates. The boundary layer parameters can be evaluated from this velocity profile. They are

$$\delta = 5.0 \sqrt{\frac{vx}{V_\infty}}, \tag{9.144}$$

$$C_f = \frac{0.664}{\sqrt{R_x}}, \tag{9.145}$$

$$\delta^* = 1.721 \sqrt{\frac{vx}{V_\infty}}, \tag{9.146}$$

and

$$\Theta = 0.664 \sqrt{\frac{vx}{V_\infty}}. \tag{9.147}$$

Comparing Blasius's "exact" results with our momentum integral estimates, we find that we are about 10% too high—not bad in boundary layer theory. Blasius's results are also included in Table 9.2 so that you can evaluate the accuracy of other velocity profile approximations in the momentum integral equation.

With a boundary layer solution available, it is possible to calculate the drag coefficient for a plate of length L and unit width if the flow is laminar over the entire length. The drag on a flat plate is due entirely to shear stress, so

$$\mathscr{D} = \int_0^L \tau_w \, dx.$$

This drag is for only one side of the plate. The drag coefficient is

$$C_D = \frac{\mathscr{D}}{\frac{1}{2}\rho V_\infty^2 L} = \frac{1}{L} \int_0^L \frac{\tau_w}{\frac{1}{2}\rho V_\infty^2} \, dx = \frac{1}{L} \int_0^L C_f \, dx.$$

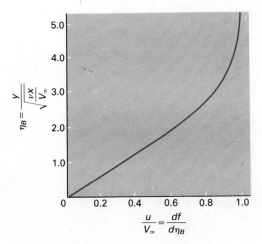

Figure 9.32 Velocity profile in laminar boundary layer on a flat plate from "exact" calculation.

Choosing Blasius's equation for C_f, Eq. (9.145), as the most accurate gives

$$C_D = \frac{1.328\sqrt{v}}{\sqrt{V_\infty L}} = \frac{1.328}{\sqrt{R_L}}.$$ **(9.148)**

It is interesting to note that

$$C_D = 2C_f(L) = 2\frac{\Theta(L)}{L}.$$ **(9.149)**

EXAMPLE 9.6   Illustrates estimation of boundary layer parameters

Equation (7.8) gives the entrance length for laminar flow in a circular pipe. See Fig. 7.6. Use Eq. (9.139b) to verify the form of Eq. (7.8).

Solution

Given

Equation (9.139b):

$$\frac{\delta}{x} = 5.48\sqrt{\frac{v}{V_\infty x}},$$

where δ is the boundary layer thickness

Show

That the form of Eq. (7.8) is

$$\frac{L_e}{D} = (\text{Constant})\left(\frac{V_\infty D}{v}\right),$$

where D is the pipe diameter and V is the average pipe velocity

Solution

Since $\delta = D/2$ at the end of the entrance length L_e, Eq. (9.139b) gives

$$\frac{D}{2L_e} \approx 5.48\sqrt{\frac{v}{V_\infty L_e}}.$$

Rearranging and squaring give

$$\frac{1}{L_e} = \frac{(10.96)^2}{D^2}\left(\frac{v}{V_\infty}\right).$$

Then

$$\frac{D}{L_e} = (10.96)^2\left(\frac{v}{V_\infty D}\right), \quad \text{or} \quad \frac{L_e}{D} = \frac{1}{(10.96)^2}\left(\frac{V_\infty D}{v}\right) \approx 0.01\left(\frac{V_\infty D}{v}\right).$$

The suggested form is then

$$\left| \frac{L_e}{D} = (\text{Constant})\left(\frac{V_\infty D}{v}\right). \right|$$ **Answer**

Discussion

Equation (9.139b) was developed for a laminar flow over a flat plate, and Eq. (7.8) describes laminar flow in a circular pipe. Therefore, Eq. (9.139b) should not be expected to provide an exact quantitative result. In the actual pipe flow, the acceleration of the inviscid core is accompanied by a pressure drop. This pressure drop is superimposed on the boundary layer to accelerate the boundary layer flow and reduce the growth rate of the boundary layer. The result is that the actual development length is about six times larger than our simple estimate predicts.

Turbulent Boundary Layer. If the laminar boundary layer is allowed to develop along the plate, it eventually becomes unstable and the process of transition to turbulent flow begins. The transition process is not completely understood; however, at some location, the boundary layer becomes fully turbulent. Although transition actually occurs over a finite distance, it is often sufficiently accurate to assume that it occurs instantaneously at a point. The location of the transition point depends on the Reynolds number and other parameters such as surface roughness (see Fig. 8.17) and the amount of turbulence in the flow adjacent to the boundary layer. For flow over a smooth plate, the transition Reynolds number is often taken to be

$$\mathbf{R}_{x,\text{tr}} = \frac{V_\infty x_{\text{tr}}}{\nu} \approx 5 \times 10^5. \tag{9.150}$$

The boundary layer downstream of the transition point is turbulent.

There is no "exact" solution of the Prandtl boundary layer equations for turbulent flow, even for a flat plate. This is because there is no rigorous expression for shear stress in turbulent flow. There are several numerical (computer) solutions of turbulent boundary layer flow. These solutions are based on semi-empirical models of the turbulent shear stress. The mixing length model (Eq. 7.7) is typical. These stress models are usually constructed from data on flat plate boundary layers, so any subsequent flat plate boundary layer calculation actually has the answer built in.

We can estimate the key parameters for a turbulent boundary layer on a flat plate by using the momentum integral approach; however, we will also have to use some empirical information. The turbulent boundary layer on a flat plate is described by the momentum integral equation with $dU/dx = 0$:

$$\frac{d}{dx}\left[\delta \int_0^1 \frac{u}{U}\left(1 - \frac{u}{U}\right)d\left(\frac{y}{\delta}\right)\right] = \frac{C_f}{2}. \tag{9.151}$$

This equation is identical to that for the laminar boundary layer, Eq. (9.136), except that the equation has been divided by U^2. All that we needed to do for the laminar boundary layer was to select an appropriate velocity profile. Turbulent boundary layer velocity profiles do not look like laminar ones, so we can't use the same profiles; however, borrowing the idea that pipe flow velocity profiles are reasonably accurate for flat plate boundary layers, we can use a power law velocity profile,

$$\frac{u}{U} \approx \left(\frac{y}{\delta}\right)^{1/n}. \tag{9.152}$$

The most common choice for n is 7.

For laminar flow, a velocity profile choice was all we needed. The skin friction coefficient could be calculated from Newton's law of viscosity by differentiating the assumed velocity profiles. This will not work for turbulent boundary layers because the slope of a power law profile is infinite at $y = 0$. To complete the problem, we must introduce independent information about $\mathbf{C_f}$. We can do this by again using the idea that the flat plate boundary layer is similar to fully developed pipe flow and by using a "friction law" for pipe flow to get $\mathbf{C_f}$. We might be tempted to use the Colebrook formula (Eq. 7.48) as a starting point, but this equation is implicit and would be difficult to use. If we limit consideration to smooth surfaces and relatively low Reynolds numbers, we can use the Blasius correlation,

$$f \approx 0.3164 \mathbf{R_D}^{-1/4}, \tag{7.45}$$

where

$$\mathbf{R_D} = \frac{\bar{V}D}{\nu}.$$

This can be converted to an equation for skin friction coefficient in a boundary layer by the following steps:

- Substituting $4\mathbf{C_f}$ for f (Eq. 7.31)
- Substituting 2δ for D
- Substituting

$$\frac{2n^2 V_\infty}{(n+1)(2n+1)}$$

for $\bar{V}$ (Eq. 7.52)

The resulting skin friction coefficient equation, for $n = 7$, is

$$\mathbf{C_f} \approx 0.045 \left(\frac{\nu}{V_\infty \delta}\right)^{1/4}. \tag{9.153}$$

We can now substitute Eqs. (9.152) and (9.153) into Eq. (9.151) to get

$$\frac{d}{dx}\left[\delta \int_0^1 \eta^{1/7}(1 - \eta^{1/7})\,d\eta\right] = 0.0225\left(\frac{\nu}{V_\infty}\right)^{1/4}\delta^{-1/4}.$$

The value of the definite integral is 7/72. The differential equation for δ becomes

$$\delta^{1/4}\left(\frac{d\delta}{dx}\right) = 0.231\left(\frac{\nu}{V_\infty}\right)^{1/4}.$$

Integrating gives

$$\frac{4}{5}\delta^{5/4} = 0.231\left(\frac{\nu}{V_\infty}\right)^{1/4}x + C.$$

To evaluate the constant of integration we specify that $\delta = 0$ at $x = 0$, neglecting the laminar portion of the boundary layer. This gives $C = 0$ and

$$\delta = 0.370\left(\frac{\nu}{V_\infty}\right)^{1/5} x^{4/5}, \tag{9.154}$$

or, on dividing by x,

$$\frac{\delta}{x} \approx 0.370\mathbf{R}_x^{-1/5}. \tag{9.155}$$

Note that Eq. (9.154) shows that the turbulent boundary layer grows at a faster rate ($\delta \sim x^{4/5}$) than a laminar boundary layer ($\delta \sim x^{1/2}$). Substituting Eq. (9.154) into Eq. (9.153) gives a skin friction equation

$$C_f \approx 0.0577\mathbf{R}_x^{-1/5}. \tag{9.156}$$

The displacement thickness is evaluated from

$$\delta^* = \delta \int_0^1 \left(1 - \frac{u}{U}\right) d\left(\frac{y}{\delta}\right) = \delta \int_0^1 (1 - \eta^{1/7}) \, d\eta,$$

which gives

$$\frac{\delta^*}{x} \approx \frac{1}{8}\left(\frac{\delta}{x}\right) = 0.046\mathbf{R}_x^{-1/5}. \tag{9.157}$$

As with the integral momentum solution for laminar boundary layers, we must ask two questions: How accurate are these results? and Can they be improved by different choices for the velocity profile and/or the skin friction equation? Unlike laminar flow, there are no exact boundary layer calculations for turbulent flow, so the question of accuracy can be answered only by comparison of predictions with experimental data. This comparison reveals that Eq. (9.156) is accurate to $\pm 3\%$ in the Reynolds number range $5 \times 10^5 < \mathbf{R}_x < 10^7$ (not bad for such a relatively crude analysis) but becomes progressively inaccurate at higher values of $\mathbf{R}_x$.

An improved momentum integral analysis can be made by using the logarithmic velocity profile

$$\frac{u}{u_\tau} = \frac{1}{0.41} \ln\left(\frac{yu_\tau}{\nu}\right) + 5.0, \tag{7.94}$$

which we discussed in connection with pipe flow. A review of Section 7.2.8 will reveal that this velocity profile was not specifically restricted to pipe flow. The logarithmic velocity profile equation contains its own friction law if we assume that it is valid all the way to the edge of the boundary layer. Recalling that

$$u_\tau = \sqrt{\frac{\tau_w}{\rho}} = \sqrt{\frac{C_f V_\infty^2}{2}}$$

and putting $u = V_\infty$ at $y = \delta$ in Eq. (7.94), we have

$$\left(\frac{C_f}{2}\right)^{-1/2} = 2.44 \ln\left[\mathbf{R}_\delta\left(\frac{C_f}{2}\right)^{1/2}\right] + 5.0, \tag{9.158}$$

an accurate but complicated implicit equation for C_f as a function of R_δ. White [8] used Eqs. (7.94) and (9.158) in a momentum integral analysis of flat plate boundary layers and obtained the following results:

$$\frac{\delta}{x} = 0.14R_x^{-1/7} \tag{9.159}$$

and

$$C_f = 0.027R_x^{-1/7}. \tag{9.160}$$

These equations are accurate to $\pm 5\%$ over the range $10^5 < R_x < 10^9$.

The drag coefficient of a flat plate with turbulent flow over its entire length can be obtained from

$$C_D = \frac{1}{L}\int_0^L C_f\,dx.$$

Using Eq. (9.156) for C_f gives

$$C_D = 0.072R_L^{-1/5}. \tag{9.161}$$

If we account for the laminar boundary layer on the front part of the plate, the drag coefficient is

$$C_D = \frac{1}{L}\left(\int_0^{x_{tr}} C_{f,lam}\,dx + \int_{x_{tr}}^L C_{f,turb}\,dx\right).$$

The transition location is calculated from

$$\frac{V_\infty x_{tr}}{\nu} \approx 5 \times 10^5.$$

Substituting Eqs. (9.145) and (9.156) and integrating, we obtain

$$C_D = 0.072R_L^{-1/5} - \frac{1700}{R_L}. \tag{9.162}$$

This equation provides a reasonable fit to the "smooth plate" curves of Fig. 8.8.

As lengthy as this development has been, we have considered only smooth plates. The analysis based on the logarithmic velocity profile can be extended to rough plates as well. Consult [8] for details.

9.3.6 Calculation of Boundary Layers with Pressure Gradient

The flat plate boundary layer solutions are easy to use, being a collection of algebraic equations for the key boundary layer parameters, but many practical flows occur with geometries other than a flat plate. In reality, it is not the flatness of the surface that is of primary importance but the absence of a streamwise pressure (or free-stream velocity) gradient. Many practical applications involve flow with variable pressure: flow over airfoils, flow in nozzles and diffusers, and even flow in the entrance region of a pipe or channel. If the streamwise pressure gradient is small (and especially if it is negative, with pressure decreasing in the flow direction), flat plate boundary layer results can be used to approximate the boundary layer parameters; however, such approximations are unsatisfactory if the pressure gradient is significant (Example 9.6). In this section we will briefly consider calculation of boundary layers in a flow with nonuniform pressure. We will concentrate on integral analysis of laminar boundary layers.

Figure 9.33 shows boundary layers developing on surfaces with a pressure gradient. The pressure and free-stream velocity are related by Eq. (9.112). The free-stream pressure and velocity are known functions of x. Although we talk about boundary layers with *pressure* gradient, the analysis deals with the free-stream *velocity* gradient. The boundary layer parameters are related by the momentum integral equation, which, for laminar flow, can be written as

$$\frac{d}{dx}\left[\delta U^2 \int_0^1 \frac{u}{U}\left(1-\frac{u}{U}\right)d\eta\right] + \delta U\left(\frac{dU}{dx}\right)\int_0^1\left(1-\frac{u}{U}\right)d\eta = \frac{\nu U}{\delta}\left(\frac{\partial\left(\frac{u}{U}\right)}{\partial\eta}\right)_{\eta=0}, \quad \textbf{(9.163)}$$

where

$$\eta = \frac{y}{\delta}.$$

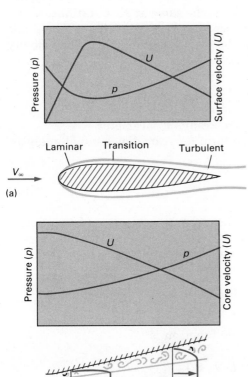

(a)

(b)

Figure 9.33 Examples of boundary layers in flow with pressure gradient: (a) Airfoil boundary layer; (b) diffuser wall boundary layer.

Unlike flat plate flow, U is a variable. This causes two immediate difficulties:

- U^2 cannot be factored out of the first term;
- dU/dx is nonzero, so the second term on the left does not vanish, as it did for flat plate flow.

We expect these difficulties to lead to increased mathematical complexity, but, since $U(x)$ and dU/dx are known, we should be able to handle it. *If* we assume

$$\frac{u}{U} = f(\eta),$$

as we did in flat plate flow, and select a "reasonable" profile (maybe the pipe flow profile again), we can evaluate the two definite integrals on the left and the derivative on the right and obtain an equation of the form

$$C_1 \frac{d}{dx}(\delta U^2) + C_2 \delta U\left(\frac{dU}{dx}\right) = C_3\left(\frac{\nu U}{\delta}\right),$$

where C_1, C_2, and C_3 are constants. Presumably this equation could be solved for $\delta(x)$ once $U(x)$ is specified. This approach seems straightforward enough; however, it will not work. The problem lies in the assumption that the velocity profile still depends on y/δ only. This means that the shape of the velocity profile is the same at all x-stations, which is simply not true for boundary layers in a pressure gradient. The shape of the velocity profile depends on the pressure gradient, so the velocity profile function must be of the form

$$\frac{u}{U} = f\left(\frac{y}{\delta}, \Lambda\right),$$

where Λ is a dimensionless pressure gradient parameter.

We can determine at least one form for Λ by dimensional analysis. We first assume

$$u = f\left(U, y, \delta, \frac{dU}{dx}, \nu\right). \tag{9.164}$$

Equation (9.164) contains six dimensional parameters and two fundamental dimensions ($[L]$ and $[T]$), so we expect four dimensionless parameters. Three obvious parameters are u/U, y/δ, and $\mathbf{R}_\delta = U\delta/\nu$. A fourth parameter is

$$\Lambda = \frac{\delta^2}{\nu}\left(\frac{dU}{dx}\right), \tag{9.165}$$

so the dimensionless velocity profile function is

$$\frac{u}{U} = F(\eta, \Lambda, \mathbf{R}_\delta).$$

Our experience with the flat plate boundary layer indicates that the dependence of velocity profile on Reynolds number is contained in the dependence of δ on $\mathbf{R}_\delta$ so, for boundary layers in a pressure gradient, we assume

$$\frac{u}{U} = f(\eta, \Lambda).$$ (9.166)

Similarly, we would expect

$$\mathbf{C_f} = \mathbf{C_f}(\Lambda, \mathbf{R}_\delta).$$ (9.167)

To calculate boundary layers with pressure gradient, we must find a velocity profile function of the form required by Eq. (9.166). When this profile is substituted into Eq. (9.163), a differential equation will result. Since we had such good luck with a second-degree polynomial profile for a flat plate, we will assume that the velocity profile can be represented by a polynomial,

$$\frac{u}{U} \approx a + b\eta + c\eta^2 + d\eta^3 + e\eta^4.$$ (9.168)

We might use more terms, but we will see that a fourth-order polynomial is adequate. We can determine the coefficients of the polynomial by making the velocity distribution satisfy certain conditions at the wall ($\eta = 0$) and at the edge of the boundary layer ($\eta = 1$). Two of these conditions are

$$\frac{u}{U} = 0 \quad \text{at } \eta = 0$$ (9.169a)

and

$$\frac{u}{U} = 1 \quad \text{at } \eta = 1.$$ (9.169b)

We can obtain a third condition by requiring that the boundary layer velocity profile join smoothly with the outer flow:

$$\frac{d}{d\eta}\left(\frac{u}{U}\right) = 0 \quad \text{at } \eta = 1.$$ (9.169c)

If we try to obtain a fourth condition by specifying a value for the first derivative of the velocity at the wall, we fail because the wall derivative is related to the wall shear stress ($\partial u/\partial y = \tau_w/\mu$), which is unknown. We can obtain a fourth condition by evaluating Prandtl's boundary layer momentum equation, Eq. (9.111), at the wall, where u and v are both zero:

$$-\frac{1}{\rho}\left(\frac{dp}{dx}\right) + v\left(\frac{\partial^2 u}{\partial y^2}\right) = 0 \quad \text{at } y = 0.$$

Substituting Eq. (9.112) and multiplying by δ^2/U, we have

$$\frac{d^2}{d\eta^2}\left(\frac{u}{U}\right) = -\frac{\delta^2}{v}\left(\frac{dU}{dx}\right) = -\Lambda \quad \text{at } \eta = 0.$$ (9.169d)

Notice that this fourth condition automatically introduces Λ into the velocity profile. The fifth condition is obtained by returning to the outer edge of the boundary layer and requiring a "very smooth" patching with

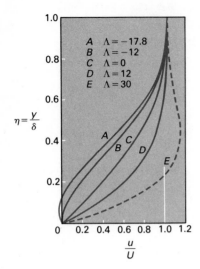

Figure 9.34 Fourth-order polynomial velocity profiles for various values of pressure gradient parameter Λ.

the outer flow by specifying

$$\frac{d^2\left(\dfrac{u}{U}\right)}{d\eta^2} = 0 \qquad \text{at } \eta = 1. \tag{9.169e}$$

Using Eqs. (9.169a)–(9.169e), we evaluate the coefficients in Eq. (9.168) and obtain,

$$\frac{u}{U} = 1 - (1 + \eta)(1 - \eta)^3 + \frac{\Lambda}{6}\eta(1 - \eta)^3. \tag{9.170}$$

Velocity profiles are plotted for various values of Λ in Fig. 9.34. In an actual boundary layer, Λ is a function of x that must be determined from the momentum equation. Once Λ is determined, the velocity profile shape is known.

Substituting Eq. (9.170) into Eq. (9.163) and carrying out the integration and differentiation yield an ordinary differential equation for δ (or Λ, since δ can be evaluated from Eq. 9.165 once Λ is known). The resulting equation is difficult to solve. In practice, it is rarely used to calculate boundary layers, so we do not include it here. You can consult [5] and [7] if you are interested.

In what follows, we will outline a much simpler procedure for calculation of laminar boundary layers with pressure gradient. This simple method, developed by Thwaites [17], is based on the ideas involved in the more complicated approach just developed. Suppose that we use the polynomial velocity profile (Eq. 9.170) to evaluate the boundary layer shape factor (H) and the skin friction coefficient. The results are

$$H = \frac{\delta^*}{\Theta} = \frac{\dfrac{3}{10} - \dfrac{\Lambda}{120}}{\left(\dfrac{37}{315} - \dfrac{\Lambda}{945} - \dfrac{\Lambda^2}{9072}\right)} \tag{9.171}$$

and

$$\frac{C_f}{2} = R_\delta^{-1}\left(2 + \frac{\Lambda}{6}\right). \tag{9.172}$$

The details of these equations are relatively unimportant; what is important is that they have the form

$$H = H(\Lambda) \tag{9.173}$$

and

$$\frac{C_f}{2} R_\delta = T_\delta(\Lambda). \tag{9.174}$$

Thwaites, as well as others before him, realized that the whole problem with the "polynomial profile" method of calculating boundary layers was that it was based on the wrong type of boundary layer thickness. Considering the integral momentum equation in the form

$$\frac{d\Theta}{dx} + (2 + H)\frac{\Theta}{U}\left(\frac{dU}{dx}\right) = \frac{C_f}{2}, \tag{9.133}$$

we see that Θ, not δ, is the most "natural" thickness parameter. This in turn means that a dimensionless pressure gradient parameter

$$\lambda \equiv \frac{\Theta^2}{v}\left(\frac{dU}{dx}\right) \tag{9.175}$$

should be used rather than Λ. Equation (9.133) can be converted into an equation for λ by multiplying by $\mathbf{R}_\Theta (= U\Theta/v)$ to give

$$\frac{U}{2}\left(\frac{d}{dx}\right)\left(\frac{\lambda}{U'}\right) + (2 + H)\lambda = T, \tag{9.176}$$

where

$$T \equiv \frac{\tau_w \Theta}{\mu U} \quad \left(= \frac{\mathbf{C_f R_\Theta}}{2}\right) \quad \text{and} \quad U' = \frac{dU}{dx}.$$

Equations (9.173) and (9.174) imply that H and T are functions of the pressure gradient parameter, so Eq. (9.176) is equivalent to

$$U\left(\frac{d}{dx}\right)\left(\frac{\lambda}{U'}\right) = 2[T(\lambda) - \lambda(2 + H(\lambda))] \equiv F(\lambda). \tag{9.177}$$

Thwaites realized that there is no real need to relate λ to the velocity profiles, deciding instead to seek the H and T functions more directly. Examining a great deal of boundary layer data, both from experiment and the few available exact solutions, Thwaites determined a relation between T and λ, between H and λ, and even between F (the entire right side of Eq. 9.177) and λ. These relations can be represented by the following curve fits [18, 17, 6]:

$$T(\lambda) \approx (\lambda + 0.09)^{0.62}, \tag{9.178}$$

$$H(\lambda) \approx \begin{cases} 2.61 - 3.75\lambda + 5.24\lambda^2, & \lambda > 0, \\[2mm] \dfrac{0.0731}{0.15 + \lambda} + 2.088, & \lambda < 0, \end{cases} \tag{9.179}$$

and

$$F(\lambda) \approx 0.45 - 6.0\lambda. \tag{9.180}$$

If we use Eq. (9.180), Eq. (9.177) becomes

$$U\left(\frac{d}{dx}\right)\left(\frac{\Theta^2}{v}\right) \approx 0.45 - 6.0\left(\frac{\Theta^2}{v}\right)\left(\frac{dU}{dx}\right),$$

which has the solution

$$\Theta^2 - \Theta_0^2 = \frac{0.45v}{U^6}\int_{x_0}^{x} U^5\, dx, \tag{9.181}$$

where Θ_0 is the value of Θ at $x = 0$. Using this equation, we have a simple calculation of the laminar boundary layer with a pressure gradient. The steps are as follows:

- $U(x)$, x_0, and Θ_0 must be known.
- At any x, calculate Θ from Eq. (9.181). This is the most difficult step because it is often necessary to evaluate the integral graphically or numerically.

- Calculate $\lambda \; [=(\Theta^2/\nu)(dU/dx)]$ and $\mathbf{R}_\Theta \; (=U\Theta/\nu)$.
- Calculate T from Eq. (9.178) and H from Eq. (9.179).
- Calculate $\mathbf{C_f}$ from

$$\mathbf{C_f} = \frac{2T}{\mathbf{R}_\Theta} \qquad (9.182)$$

or calculate τ_w from

$$\tau_w = \frac{\mu U}{\Theta} T \qquad (9.183)$$

- Calculate δ^* from

$$\delta^* = H\Theta. \qquad (9.184)$$

As a check, we can use Thwaites' method to calculate a flat plate boundary layer. Putting $U = V_\infty = \text{Constant}$ and $\Theta_0 = 0$, we obtain the following results:

$$\frac{\Theta}{x} = 0.671\mathbf{R}_x^{-1/2} \qquad \text{and} \qquad \mathbf{C_f} = 0.671\mathbf{R}_x^{-1/2}.$$

Compared to Blasius's values, these are quite satisfactory.

Unfortunately, no accurate, simple method like that of Thwaites is available for calculating turbulent boundary layers with pressure gradient. Many methods have been proposed over the years. In 1968, a great showdown between the various methods was held [19]. The best methods available were found to be equally distributed between momentum integral methods that used a logarithmic velocity profile, appropriately modified to account for pressure gradient, together with a consistent skin friction law and a third auxiliary equation based on a further integration of the boundary layer equations across the layer; and numerical finite-difference methods for solving the differential Prandtl boundary layer equations together with an appropriate model of the turbulent shear stresses. All currently available accurate methods of calculating turbulent boundary layers with pressure gradient require computer solution, even those that involve only ordinary differential equations.

EXAMPLE 9.7   Illustrates Thwaites' method and a boundary layer application

Use Thwaites' method to verify the shear stress distribution

$$\tau_w = [1.74 \times 10^{-3} \, (\pi - \theta) - 1.59 \times 10^{-5} \, (\pi - \theta)^2] \, \text{N/m}^2$$

used for the air flowing over the highway sign in Example 8.1. The air temperature is 20°C.

Solution

Given
20°C air flowing over the highway sign of Example 8.1 at 10 km/hr
The sign has the shape of a half-cylinder with diameter 2.0 m and length 25.0 m.

Show

That the shear stress can be approximated by

$$\tau_w = [1.74 \times 10^{-3} (\pi - \theta) - 1.59 \times 10^{-5} (\pi - \theta)^2] \text{ N/m}^2$$

Solution

We first calculate the Reynolds number to verify that the boundary layer is laminar. Table A.3 gives $\nu = 1.51 \times 10^{-5}$ m^2/s. Then

$$\mathbf{R} = \frac{V_\infty D}{\nu} = \frac{\left(10 \, \frac{\text{km}}{\text{hr}}\right)\left(\frac{1000 \text{ m}}{\text{km}}\right)\left(\frac{\text{hr}}{3600 \text{ sec}}\right)(2.0 \text{ m})}{\left(1.51 \times 10^{-5} \, \frac{\text{m}^2}{\text{s}}\right)} = 3.7 \times 10^5.$$

Our consideration of drag on a circular cylinder (Section 8.2.2) concluded that the boundary layer experiences transition somewhere on the cylinder if $\mathbf{R} > 5 \times 10^5$; therefore the cylinder boundary layer remains laminar,[*] and Thwaites' method may be used.

The shear stress is calculated using Eqs. (9.183), (9.178) and (9.181):

$$\tau_w = \frac{\mu U}{\Theta} T,$$

where

$$T \approx (\lambda + 0.09)^{0.62}, \qquad \Theta^2 = \Theta_0^2 + \frac{0.45\nu}{U^6} \int_{x_0}^{x} U^5 \, dx,$$

and

$$\lambda = \frac{\Theta^2}{\nu} \left(\frac{dU}{dx}\right).$$

For a circular cylinder, Eq. (9.66) gives

$$V_\theta = -V_\infty \left(1 + \frac{R^2}{r^2}\right) \sin \theta.$$

On the cylinder surface, $r = R$ and[†]

$$U = -V_\theta = 2V_\infty \sin \theta.$$

Then using $\theta = \pi - \phi$ and $x = \phi R$, we have

$$U = 2V_\infty \sin (\pi - \phi) = 2V_\infty \sin \phi$$

and

$$\Theta^2 = \Theta_0^2 + \frac{0.45(32 V_\infty^5)\nu}{64 V_\infty^6 \sin^6 \phi} \int_0^{\phi} \sin^5 \phi' R \, d\phi'.$$

Noting that $\Theta_0 = 0$, simplifying, and integrating,[**] we obtain

$$\Theta^2 = \frac{0.45\nu R}{2 V_\infty \sin^6 \phi} \int_0^{\phi} (1 - \cos^2 \phi')^2 \sin \phi' \, d\phi'$$

$$= \frac{0.45\nu R}{2 V_\infty \sin^6 \phi} \left[-\cos \phi' + \frac{2}{3} \cos^3 \phi' - \frac{\cos^5 \phi'}{5} + \frac{8}{15}\right].$$

---

[*] Example 9.8 gives an alternative procedure for estimating the state (laminar or turbulent) of the boundary layer.

[†] $U(x)$ is opposite V_θ for the direction of increasing x.

[**] Note that ϕ' is a dummy variable of integration.

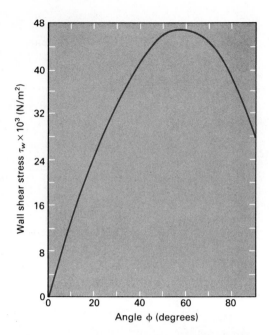

Figure E9.7 Wall shear stress over cylinder of Example 8.1.

Then

$$\Theta = \sqrt{\frac{0.45vR}{2V_\infty \sin^6 \phi} \left[-\cos \phi + \frac{2}{3} \cos^3 \phi - \frac{\cos^5 \phi}{5} + \frac{8}{15} \right]}.$$

To find the shear stress, note that

$$T = (\lambda + 0.09)^{0.62} = \left[\frac{\Theta^2}{v} \left(\frac{dU}{dx} \right) + 0.09 \right]^{0.62} = \left[\frac{\Theta^2}{vR} \left(\frac{dU}{d\phi} \right) + 0.09 \right]^{0.62}.$$

Substituting $U(\phi)$ gives

$$T = \left(\frac{2\Theta^2 V_\infty}{vR} \cos \phi + 0.09 \right)^{0.62}.$$

The wall shear stress is

$$\tau_w = \frac{\mu U}{\Theta} T = \frac{2\mu V_\infty \sin \phi \left(\dfrac{2\Theta^2 V_\infty}{vR} \cos \phi + 0.09 \right)^{0.62}}{\Theta},$$

where $\Theta(\phi)$ is given above. Now τ_w is plotted in Fig. E9.7 for the conditions of Example 8.1:

$$\mu = 1.81 \times 10^{-5} \text{ N/m}^2 \cdot \text{s}, \qquad v = 1.51 \times 10^{-5} \text{ m}^2/\text{s},$$
$$R = 1.0 \text{ m}, \qquad V_\infty = 10 \text{ km/hr} = 2.78 \text{ m/s}.$$

A second-order polynomial approximation of the curve is

$$\tau_w \approx (1.74 \times 10^{-3} \phi - 1.59 \times 10^{-5} \phi^2) \text{ N/m}^2,$$

or, since $\phi = (\pi - \theta)$,

Answer

$$\left| \; \tau_w = [1.74 \times 10^{-3} (\pi - \theta) - 1.59 \times 10^{-5} (\pi - \theta)^2] \text{ N/m}^2. \; \right| \; \blacktriangleleft$$

Discussion

We have made the simplifying assumption that the flow over half of a circular cylinder can be approximated by the flow over the front half of a complete cylinder. Our calculations are undoubtedly somewhat inaccurate near $\phi = 90°$. See Example 9.8.

9.3.7 Boundary Layer Separation and the Breakdown of the Potential Flow and Boundary Layer Model

The potential flow and boundary layer model of a high Reynolds number flow is based on the assumption that the boundary layer has only a minor effect on the external flow. This assumption is reasonably accurate, and the model produces acceptable results as long as the boundary layer is thin and remains attached to the surface. Under certain conditions, the boundary layer thickens rapidly and separates from the surface, forming a broad region of recirculating flow and grossly disturbing the outer flow streamlines (see Figs. 8.27, 8.43, and especially 8.15). In this case, although the concept of patching together a viscous-rotational flow and an inviscid-irrotational flow is still valid, simple boundary layer theory is no longer capable of calculating the viscous-rotational region, and the potential flow and boundary layer model fails.

Although the potential flow and boundary layer model is not capable of predicting the complete flow field, it is capable of yielding some very important information. Generally, the model is able to

- explain the conditions necessary for separation;
- indicate whether separation will occur;
- indicate the approximate location where separation will occur.

Figure 9.35 shows a boundary layer developing toward separation on a surface. As the point of separation is approached, the boundary layer thickens rapidly. Downstream of separation, there is a region of backflow near the surface. At the separation point, the velocity profile has a vertical slope at the wall and $\tau_w = 0$ (at separation). Physical

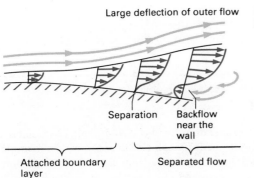

Large deflection of outer flow

Separation

Backflow near the wall

Attached boundary layer

Separated flow

Figure 9.35 Separating boundary layer.

reasoning (see Section 8.2.2) indicates that separation occurs in a rising pressure gradient, usually called an *adverse pressure gradient*. This fact can be verified with the aid of Prandtl's boundary layer momentum equation, Eq. (9.111), together with a consideration of the shape of the boundary layer velocity profiles. The velocity profiles in Fig. 9.35 indicate that for backflow to develop, a boundary layer must have a point of inflection where the second derivative $(\partial^2 u/\partial y^2)$ changes sign to obtain the S-shape characteristic of a separating profile. The second derivative of the velocity is always negative near the edge of the layer because the boundary layer must join smoothly with the free stream. The second derivative at the wall can be found from Eq. (9.111) evaluated at $y = 0$ where $u = v = 0$:

$$\left.\frac{\partial^2 u}{\partial y^2}\right)_{y=0} = \frac{1}{\mu}\left(\frac{dp}{dx}\right). \tag{9.185}$$

This relation is valid for laminar or turbulent boundary flow since the flow very near the wall is always laminar. According to this equation, the sign of the second derivative at the wall is the same as the sign of the pressure gradient. If the pressure gradient is positive, the second derivative must change sign between the wall and the edge of the layer, and the flow is susceptible to separation. If the pressure gradient is negative (falling pressure), there is no point of inflection, and the flow will not be susceptible to separation.

The location of the separation point (and hence whether separation will occur) can, in principle, be determined by calculating the boundary layer up to the point where

$$\mathbf{C_f} = \tau_w = 0,$$

which is the separation point. In practice, boundary layer calculations are often poorly behaved near separation, and location of the zero stress point is difficult. Monitoring the value of the shape factor H has been found to provide a reliable indication of separation. As a boundary layer approaches separation, the value of H increases rapidly. Separation is generally taken to occur at the point where

$$H > 3.5 \qquad \text{and} \qquad H > 2.4. \tag{9.186}$$

$$\text{Laminar flow} \qquad\qquad \text{Turbulent flow}$$

The lower value of H for turbulent separation (lower even than the laminar flat plate value of H) reflects the fact that the characteristic S-shape of separation occurs in the viscous sublayer near the wall while the outer turbulent portion of the profile remains somewhat fuller.

EXAMPLE 9.8 Illustrates how to determine the location of boundary layer separation

The critical wind velocity for vortex shedding for the 12.75-in.-outside-diameter sign pole in Example 8.8 was found to be 9.75 mph (14.3 ft/sec). Find the angle ϕ in Fig. E9.8 where the boundary layer separates from the pole for this flow condition. The free-stream velocity is given by $U(\phi) = 2V_\infty \sin \phi$ (see Example 9.7).

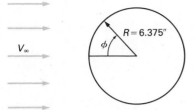

Figure E9.8 Cross section of sign pole in Example 8.8.

Solution

Given

Wind at 14.3 ft/sec flowing over a 12.75-in.-outside-diameter steel pipe
Air 60°F

Find

The angle ϕ where the boundary layer separates from the pole surface

Solution

The location of boundary layer separation is the point where $\tau_w = 0$. We must develop an expression for τ_w as a function of the angle ϕ for $U(\phi) = 2V_\infty \sin \phi$. An expression was developed in Example 9.7 for a laminar boundary layer. We therefore check that the boundary layer around the sign pole is laminar. The Reynolds number is

$$\mathbf{R}_x = \frac{Vx}{v} = \frac{V\phi R}{v}.$$

The air temperature in Example 8.8 is 60°F; Table A.4 gives $v = 1.7 \times 10^{-4}$ ft²/sec. Then

$$\mathbf{R}_x = \frac{\left(14.3 \, \dfrac{\text{ft}}{\text{sec}}\right)\left(\dfrac{1}{2}\right)(12.75 \text{ in.})\left(\dfrac{\text{ft}}{12 \text{ in.}}\right)\phi}{\left(1.59 \times 10^{-4} \, \dfrac{\text{ft}^2}{\text{sec}}\right)} = 4.78 \times 10^4 \, \phi.$$

The boundary layer becomes turbulent (transition occurs) at a Reynolds number (based on x) of about* 5×10^5. Therefore we expect a laminar boundary layer for

$$\phi < \frac{5 \times 10^5}{4.78 \times 10^4} = 10.5 \text{ rad}.$$

Since the entire cylinder is covered by $0 \le \phi \le \pi$ rad, the boundary layer will be laminar. Example 9.7 gives

$$\tau_w = \frac{2\mu V_\infty \sin \phi \left(\dfrac{2\Theta^2 V_\infty}{vR} \cos \phi + 0.09\right)^{0.62}}{\Theta},$$

where

$$\Theta = \sqrt{\frac{0.45 vR}{2V_\infty \sin^6 \phi}} \left[-\cos \phi + \frac{2}{3}\cos^3 \phi - \frac{\cos^5 \phi}{5} + \frac{8}{15}\right].$$

---

* This transition value is for flow over a flat plate. We assume that it also applies for flow over a cylinder.

The numerical values are

$$\mu = 3.77 \times 10^{-7} \text{ lb} \cdot \text{sec/ft}^3, \qquad \nu = 1.59 \times 10^{-4} \text{ ft}^2/\text{sec},$$

$$R = 6.375 \text{ in.} = 0.531 \text{ ft}, \qquad V_\infty = 14.3 \text{ ft/sec.}$$

Then

$$2\mu V_\infty = 2\left(3.77 \times 10^{-7} \frac{\text{lb} \cdot \text{sec}}{\text{ft}^2}\right)\left(14.3 \frac{\text{ft}}{\text{sec}}\right)$$

$$= 1.08 \times 10^{-5} \text{ lb/ft},$$

$$\frac{2V_\infty}{\nu R} = \frac{2\left(14.3 \dfrac{\text{ft}}{\text{sec}}\right)}{\left(1.59 \times 10^{-4} \dfrac{\text{ft}^2}{\text{sec}}\right)(0.531 \text{ ft})} = \frac{3.39 \times 10^5}{\text{ft}^2},$$

and

$$\frac{0.45 \nu R}{2 V_\infty} = \frac{0.45 \text{ ft}^2}{3.39 \times 10^5} = 1.33 \times 10^{-6} \text{ ft}^2.$$

Defining

$$\Omega = \frac{1}{\sin^3 \phi}\left[-\cos\phi + \frac{2}{3}\cos^3\phi - \frac{\cos^5\phi}{5} + \frac{8}{15}\right]^{1/2}$$

gives

$$\tau_w = \frac{\left(1.08 \times 10^{-5} \dfrac{\text{lb}}{\text{ft}}\right)}{(1.33 \times 10^{-6} \text{ ft}^2)^{1/2} \Omega}\left[\left(\frac{3.39 \times 10^5}{\text{ft}^2}\right)(1.33 \times 10^{-6} \text{ ft}^2)\Omega^2 \cos\phi + 0.09\right]^{0.62}.$$

Simplifying, we obtain

$$\tau_w = 0.00210\left[\frac{(1 + 5\Omega^2 \cos\phi)^{0.62}}{\Omega}\right] \text{ lb/ft}^2.$$

We must solve for the angle ϕ where $\tau_w = 0$. This can be accomplished graphically or by trial and error. The result is

$$\boxed{\phi_{\text{sep}} = \phi_{\tau_w = 0} = 104°.}$$

Answer

Discussion

In actuality, the separated boundary layer affects the pressure and free-stream velocity distribution, and the velocity distribution is approximately

$$\frac{U}{V_\infty} \approx 1.814\phi - 0.271\phi^3 - 0.0471\phi^5,$$

rather than $U/V_\infty = 2 \sin\phi$ [8]. Using this more realistic velocity distribution, we find that the calculated separation point is at about 80° (on the front of the cylinder). The 80° separation point is verified by experiments (see Figs. 8.16 and 8.17). The discrepancy between our calculated separation point and the experimental value is due not to inaccuracies in Thwaites' method but to the breakdown of the potential flow and boundary layer model near separation. Given the correct free-stream velocity distribution, Thwaites' method is reasonably accurate.

Problems

Extension and Generalization

1. Show that the slope for a constant-value Stokes stream function Ψ_s for axisymmetric flow is parallel to the velocity vector and is therefore a streamline.

2. Show that the difference $\Delta\Psi_s$ in the value of Stokes stream function between two streamlines for axisymmetric flow is the volume flow rate between the streamlines.

3. Use the irrotationality condition for axisymmetric flow to develop Eq. (9.31).

4. Show that lines of constant Φ for steady, axisymmetric, irrotational flow are normal to streamlines.

5. Consider the flow normal to a right circular cylinder. Use the realistic velocity distribution

$$\frac{U}{V_\infty} = 1.814\phi - 0.271\phi^3 - 0.0471\phi^5$$

and develop an expression for the angle ϕ where the boundary layer separates from the right circular cylinder. See Fig. E9.8.

6. The flow rate of current in an electrostatic field is given by

$$J_x = -\frac{1}{\rho}\left(\frac{\partial E}{\partial x}\right), \quad J_y = -\frac{1}{\rho}\left(\frac{\partial E}{\partial y}\right), \quad J_z = -\frac{1}{\rho}\left(\frac{\partial E}{\partial z}\right),$$

where J_x, J_y, and J_z are the x-, y-, and z-components of the total current density; ρ is the electrical resistivity and E the electric potential. Show that E satisfies Laplace's equation.

7. The flow rate of heat in a solid due to conduction is given by

$$q_x = -k\left(\frac{\partial T}{\partial x}\right), \quad q_y = -k\left(\frac{\partial T}{\partial y}\right), \quad q_z = -k\left(\frac{\partial T}{\partial z}\right),$$

where q_x, q_y, and q_z are the x-, y-, and z-components of the total heat flux; k is the thermal conductivity and T the temperature of the solid. Show that T satisfies Laplace's equation. Assume no heat is generated within the solid.

8. Turbulent boundary layer velocity profiles can be approximated by

$$\frac{u}{u_\tau} = 2.44\ln\left(\frac{yu_\tau}{\nu}\right) + 5.0 + 2.44\,\Pi\left[1 - \cos\left(\frac{\pi y}{\delta}\right)\right],$$

where Π is called the *wake parameter*. Find expressions for δ^*/δ, Θ/δ, and H as functions of u_τ/U and Π.

9. Using the velocity profile of Problem 8, find an (implicit) expression for the skin friction coefficient C_f as a function of R_δ ($\delta U/\nu$) and Π.

10. Consider the boundary layer on an axisymmetric body with $\delta \ll R$ (Fig. P9.1). Find the appropriate forms of the boundary layer partial differential equations.

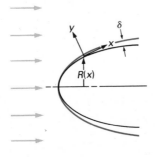

Figure P9.1

11. Derive the integral equations for a boundary layer by integrating the Prandtl boundary layer equations between the limits of zero and δ.

12. Derive the appropriate integral boundary layer equations for a thin boundary layer on an axisymmetric body (see Problem 10).

13. Suggest a modification of the Thwaites method that will permit the calculation of boundary layers on axisymmetric bodies (see Problems 10 and 12).

Lower-Order Application

14. Show that adding the stream functions for a uniform stream, a source of strength λ at $+a$, and a sink of strength λ at $-a$ gives Eq. (9.58).

15. Verify that the length ratios t/a and ℓ/a for a Rankine oval are given by Eqs. (9.59) and (9.60).

16. Show that the maximum velocity of the fluid around the Rankine oval described by Eq. (9.58) occurs at the point of maximum thickness of the Rankine oval and is given by Eq. (9.61).

17. Show that if Φ_1 and Φ_2 are both solutions of Laplace's equation, then the sum $(\Phi_1 + \Phi_2)$ is also a solution.

18. Show that the circulation of a free vortex for any closed path that does not enclose the origin is zero.

19. Show that the circulation of a free vortex for any closed path that encloses the origin is Γ.

20. Show that $\ln(ib) = \ln b + i\pi/2$, where b is a real positive constant.

21. Show that $e^{\pm\pi i} = \mp 1$.

22. Using $\Omega = \Phi + i\Psi$ and $z = x + iy$, show that

$$\Phi = \ln(x^2 + y^2) \quad \text{and} \quad \Psi = 2\tan^{-1}\frac{y}{x}$$

for $\Omega = \ln z^2$.

23. Using $\Omega = \Phi + i\Psi$ and $z = x + iy$, show that

$$\Phi = ax, \quad \Psi = ay, \quad \text{for } \Omega = az.$$

24. Using $\Omega = \Phi + i\Psi$, $z = x + iy$, and $r^2 = x^2 + y^2$, show that

$$\Phi = -V_\infty\left(r + \frac{a^2}{r}\right)\cos\theta$$

and

$$\Psi = -V_\infty\left(r - \frac{a^2}{r}\right)\sin\theta$$

for

$$\Omega = -V_\infty\left(z + \frac{a^2}{z}\right).$$

25. Find expressions for the velocity potential Φ and the stream function Ψ for the complex potential

$$\Omega = -V_\infty\left(z + \frac{a^2}{z}\right) - i\left(\frac{\Gamma}{2\pi}\right)\ln\left(\frac{z}{a}\right),$$

where V_∞, a, and Γ are constants.

26. A stream function is given by

$$\Psi = \sin\frac{x}{X}\sinh\frac{y}{Y},$$

where X and Y are constants, $0 \le x \le \pi X$, and $y \ge 0$. Can Ψ represent a potential flow? If so, locate any stagnation points and sketch several streamlines.

27. Find the corresponding stream function for which

$$\Phi = \frac{AV_\infty\cos\theta}{r}$$

if both A and V_∞ are positive constants.

28. A velocity potential is given by

$$\Phi = \frac{\cos\theta}{r} - V_\infty r\cos\theta.$$

Find $\Psi(r, \theta)$ and locate any stagnation points. Sketch several streamlines and determine a body shape about which Φ could represent the potential flow.

29. Plot lines of constant Ψ_1 for a source and lines of constant Ψ_2 for a free vortex. Graphically add Ψ_1 and Ψ_2 and show that the sum produces the streamlines in Fig. 9.11.

30. The stream function for a particular incompressible flow is

$$\Psi = Cxy,$$

where C is a constant. Is this flow irrotational? Verify your answer. Sketch a few streamlines for this flow. What practical situation does it represent? Calculate the pressure distribution.

31. Sketch a few streamlines in the z-plane for the complex potential

$$\Omega_\zeta = -\left(\zeta + \frac{a}{\zeta}\right)$$

under the transformation

$$z = \zeta + \frac{b^2}{\zeta},$$

where a and b are positive, real constants and $a > b$. Choose convenient values for the constants.

32. Sketch a few streamlines in the $(z = x + iy)$-plane for the complex potential

$$\Omega = -\left(\zeta + \frac{a^2}{\zeta}\right)$$

under the transformations

$$z_1 = \zeta + im \quad \text{and} \quad z = z_1 + \frac{b^2}{z_1},$$

where a, b, and m are positive, real constants and $a > b$. Choose convenient values for the constants.

33. Sketch a few streamlines in the $(z = x + iy)$-plane for the complex potential

$$\Omega = -\left(\zeta + \frac{a^2}{\zeta}\right)$$

under the transformations

$$z_1 = \zeta + me^{i\alpha} \quad \text{and} \quad z = z_1 + \frac{b^2}{z_1},$$

where a, m, α, and b are positive, real constants and $a > b$. Choose convenient values for the constants.

34. Sketch a few streamlines in the $(z = x + iy)$-plane for the complex potential

$$\Omega = -V_\infty\left(\zeta + \frac{a^3}{\zeta}\right) - i\left(\frac{\Gamma}{2\pi}\right)\ln\left(\frac{\zeta}{a}\right)$$

under the transformations

$$z_2 = \zeta e^{-i\alpha}, \quad z_1 = z_2 + im,$$

and

$$z = z_1 + \frac{b^2}{z_1},$$

where V_∞, a, Γ, α, m, and b are positive, real constants and $a > b$. Choose convenient values for the constants.

35. Consider the flow of a liquid of viscosity μ and density ρ down an inclined plate making an angle θ with the horizontal. The film thickness is t. The fluid velocity parallel to the plate is given by

$$V_x = \frac{\rho t^2 g \cos \theta}{2\mu}\left[1 - \left(\frac{y}{t}\right)^2\right],$$

where y is the coordinate normal to the plate. Calculate Φ and Ψ for this flow and show that neither satisfies Laplace's equation. Do you understand why not?

36. The velocity potential for a horizontal flow field is

$$\Phi = K \ln r.$$

Find the pressure $p(r, \theta)$ in terms of the stagnation pressure p_0.

37. A horizontal oil-bearing stratum is 3 m high and provides a volume flow rate of $Q = 1000$ m^3/hr. The flow moves radially inward and is collected by a vertical porous pipe having an outer radius of 1.0 m. The Laplace equation for the potential flow model of the oil is

$$\frac{1}{r}\left(\frac{\partial \Phi}{\partial r}\right) + \frac{\partial^2 \Phi}{\partial r^2} + \frac{1}{r^2}\left(\frac{\partial^2 \Phi}{\partial \theta^2}\right) = 0.$$

Find $\Phi(r, \theta)$.

38. The velocity potential for the combination of a source and a vortex is given by

$$\Phi = (3.0 \text{ m}^3/\text{s})(\ln r - \theta).$$

Calculate the pressure at $(4, \pi/4)$ if the pressure is 10^6 kPa at $(1, \pi/2)$. The fluid density is 1000 kg/m^3.

39. Air at 25°C flows normal to the axis of an infinitely long cylinder of 1.0-m radius. The approach velocity is 100 km/hr. Find the maximum and minimum pressures on the cylinder surface. Assume potential flow.

40. Consider the possibility of using two rotating cylinders to replace the conventional wings on an airplane for lift. Consider an airplane flying at 700 km/hr through the Standard Atmosphere at 10,000 m. Each "wing cylinder" has a 3.0-m radius. The surface velocity of each cylinder is 28 km/hr. Find the length ℓ of each wing to develop a lift of 1,550,000 N. Assume potential flow and neglect end effects.

41. Compare the drag of the two wing cylinders of Problem 40 with those of the wing designs in Problem 25 in Chapter 8. Is a potential flow model realistic?

42. Air at 25°C flows normal to the axis of an infinitely long cylinder of 1.0-m radius. The cylinder is rotating at 10 rad/s and the approach velocity is 100 km/hr. Find the maximum and minimum pressures on the cylinder surface. Assume potential flow.

43. Air at 25°C flows normal to the axis of an infinitely long cylinder of 1.0-m radius. The approach velocity is 100 km/hr. Find the rotational velocity of the cylinder so that a single stagnation point occurs on the lowermost part of the cylinder. Assume potential flow.

44. A 15-mph wind flows over a Quonset hut having a radius of 12 ft and a length of 60 ft. See Fig. P9.2. The upstream pressure and temperature are equal to those inside the Quonset hut, 14.696 psia and 70°F. Estimate the upward force on the Quonset hut. Find the location θ on the roof of the Quonset hut where the pressure is p_∞.

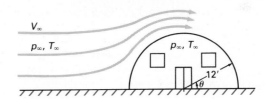

Figure P9.2

45. Consider an infinitely long cylinder of radius R in an air stream of approach velocity V_∞. Assume that separation occurs at angle β (defined in Fig. P9.3). Assume potential flow up to the separation line. Estimate the form (or pressure) drag component of a unit length of the cylinder for $R = 1.0$ in. and $V_\infty = 100$ mph. Calculate a drag coefficient based on the pressure profile. Do calculations for $\beta = 70°$ and $\beta = 85°$ and compare with published values. Draw any worthwhile conclusions.

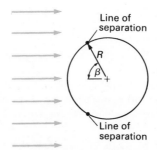

Figure P9.3

46. A telephone wire has a 1.5-cm diameter and is in a 10-m/s cross wind. Estimate the location on the wire where flow separation occurs.

47. Consider a small airplane whose wing cross section is shown in Fig. P9.4. The airplane is traveling at 100 mph in 20°F air. Experimentally observed streamlines are shown in the figure. Estimate the pressure drag and the shear drag on the airplane wing. Consider a unit depth of the wing.

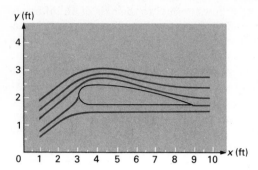

Figure P9.4

48. The open wind tunnel in Fig. P9.5 has a square cross section measuring 2 ft by 2 ft at the inlet. It is desired to flare out the cross section so the central core has a constant velocity along its 8-ft length. Determine the required exit area A_e for a 100-mph central core velocity.

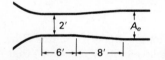

Figure P9.5

49. A coal barge is 1000 ft long and 100 ft wide and is submerged a depth of 12 ft. It is being towed at a speed of 12 mph. Estimate the friction drag on the barge.

50. Consider the laminar boundary layer for the flow over a flat plate. Plot the boundary layer thickness δ versus $\sqrt{vx/V_\infty}$ for each velocity profile listed in Table 9.2. Choose two specific values of $\sqrt{vx/V_\infty}$ and plot u/V_∞ versus y for each velocity profile. Make any significant observations.

51. Develop expressions for the wall shear stress and the displacement thickness by solving the momentum integral equation (9.151) for the laminar flow over a flat plate. Use the following expressions for the velocity u/V_∞:

a) $\dfrac{u}{V_\infty} = \dfrac{y}{\delta}$ b) $\dfrac{u}{V_\infty} = 2\left(\dfrac{y}{\delta}\right) - \left(\dfrac{y}{\delta}\right)^3$

c) $\dfrac{u}{V_\infty} = (1 - e^{-10y/\delta})$ d) $\dfrac{u}{V_\infty} = \tanh\left(\dfrac{5y}{\delta}\right)$

Do each of these expressions satisfy the boundary conditions for the flow over the flat plate? Which expression more closely represents the exact Blasius solution?

52. Consider a two-dimensional diverging passage that has straight sides, a constant width, and rectangular cross sections. Fluid enters with a uniform axial velocity U_0. The boundary layer is thin enough that it may be neglected. The inlet measures 1.0 m on a side, the outlet measures 1.0 m by 2.0 m, and the length is 4.0 m. Use finite differences to solve Laplace's equation for the stream function and find the velocity as a function of the distance from the inlet.

53. Consider a body whose free stream velocity $U(x)$ is given by

$$U(x) = U_\infty\left(1 - \frac{x}{\ell}\right)$$

for a laminar boundary layer, where x is the coordinate along the surface of the body in the direction of flow. Use Thwaites' method to calculate the value of x/ℓ where the laminar boundary layer separates from the wall.

54. Repeat Problem 53 using

$$U(x) = U_\infty\left(1 - \frac{x^2}{\ell^2}\right).$$

55. Consider the plane wall diffuser in Fig. P9.6. Use Thwaites' method to estimate the angle β at which separation of a laminar boundary layer occurs at the exit plane. The diffuser is very wide with $d = 0.25$ m, $\ell = 1.0$ m, and $U_\infty = 2.0$ m/s.

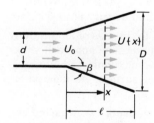

Figure P9.6

56. Calculate the numerical values for the quantities

$$\frac{\delta}{x}\sqrt{\mathbf{R}_x}, \qquad \mathbf{C}_f\sqrt{\mathbf{R}_x}, \qquad \text{and} \qquad \frac{\delta^*}{x}\sqrt{\mathbf{R}_x}$$

in the laminar boundary layer over a flat plate by assuming a sinusoidal profile,

$$\frac{u}{V_\infty} = \sin\left[\frac{\pi}{2}\left(\frac{y}{\delta}\right)\right].$$

Do your values agree with those in Table 9.2?

57. Calculate the numerical values for the quantities

$$\frac{\delta}{x}\sqrt{\mathbf{R}_x}, \qquad \mathbf{C}_f\sqrt{\mathbf{R}_x}, \qquad \text{and} \qquad \frac{\delta^*}{x}\sqrt{\mathbf{R}_x}$$

in the laminar boundary layer over a flat plate by assuming a linear profile, $u/V_\infty = y/\delta$.

58. Write a computer program to obtain a numerical solution of the equation for $f(\eta)$ in Problem 68. Compare your numerical results with Fig. 9.32.

59. Consider the flow of 20°C water flowing over a thin, very wide, smooth flat plate aligned with the flow. The approach velocity is 60 km/hr and the Reynolds number based on the plate length is 2×10^5. Calculate the drag per unit width of the plate for each of the velocity profiles listed in Table 9.2. Next calculate the drag coefficient $\mathbf{C_D}$ as used in Fig. 8.8. Which velocity profile(s) provide the best estimation of $\mathbf{C_D}$? Why?

60. A horizontal 100-km/hr wind blows over a 200-m-high, 10-m-diameter building. Separation occurs at an angle $\beta = 80°$ as shown in Fig. P9.3. Assume potential flow up to the separation point and a constant pressure from $\beta = 80°$ to $\beta = 180°$. Find the force tending to topple the building. Compare your result with that obtained by using the drag coefficient in Fig. 8.30.

61. Turbulent boundary layer velocity profiles are sometimes approximated by the equation

$$\frac{u}{u_\tau} = 2.44 \ln\left(\frac{yu_\tau}{\nu}\right) + 5.0 + 2.44\Pi\left[1 - \cos\left(\frac{\pi y}{\delta}\right)\right].$$

Determine an expression for u/U. Flat plate turbulent boundary layers have $\Pi \approx 0.5$. Plot turbulent velocity profiles for this value of Π and for a power law profile with $n = 7$. If necessary, use $\mathbf{R}_x = 10^6$.

Higher-Order Application

62. Write a general computer program to calculate laminar boundary layers for any arbitrary free-stream velocity distribution.

63. Using the logarithmic velocity profile

$$\frac{u}{u_\tau} = 2.44 \ln\left(\frac{yu_\tau}{\nu}\right) + 5.0,$$

calculate δ/x and $\mathbf{C}_f(x)$ for a turbulent boundary layer on a flat plate.

64. Calculate the development of the laminar boundary layer on a 4% thick symmetrical Joukowsky airfoil. Reynolds number based on chord length is 10^5.

65. Simulate the flow over a 10% thick Joukowsky airfoil by a distribution of sources and sinks along an axis. Use six strips, each with a constant length.

66. A cylinder on a stationary rotor ship has a diameter of 9 ft and a height of 50 ft and rotates at 125 rpm. Calculate the lift and drag on the cylinder in a 20-mph cross wind. Compare your result with that obtained using Fig. 8.31.

67. Repeat Problem 66 if the rotor ship is traveling at 10 mph and the wind is normal to the path of the ship.

68. Use the Blasius transformation

$$\Psi = \sqrt{\nu V_\infty x}\, f(\eta) \qquad \text{and} \qquad \eta = \frac{y}{\sqrt{\dfrac{\nu x}{V_\infty}}}$$

along with

$$u = \frac{\partial \Psi}{\partial y} \qquad \text{and} \qquad v = -\frac{\partial \Psi}{\partial x}$$

to show that the Prandtl boundary layer equations (9.109 and 9.111) reduce to

$$f''' + \frac{1}{2}ff'' = 0$$

69. Model an NACA 2415 airfoil as best you can and estimate the lift and drag coefficients at angles of attack of 0° and 4°. The Reynolds number, based on chord length, is 10^5. Compare your results to data from Fig. 8.34.

Comparison

70. Consider the flow over the very thin and very wide flat plate in Fig. P9.7. The flow on the lower side is turbulent over the entire plate, and the flow on the upper side is laminar on the front portion and then becomes turbulent. Compare the drag per unit width on the upper half with that on the lower half. The velocity $V_\infty = 10$ m/s; consider lengths x_{trans} and $2x_{trans}$.

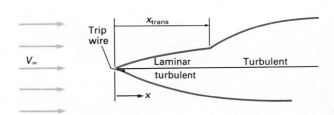

Figure P9.7

References

1. Launder, B. E., and D. B. Spalding, *Lectures on Mathematical Models of Turbulence,* Academic Press, London, 1972.
2. Prandtl, L., "Uber Flussigkeitsbewegung bei sehr kleiner Reibung," *Verhandlunger des III Internationalen, Mathematiker-Kongresses,* Liepzig, 1905.
3. Milne-Thompson, L. M., *Theoretical Hydrodynamics* (4th ed.), Macmillan, New York, 1960.
4. Robertson, J. M., *Hydrodynamics in Theory and Application,* Prentice-Hall, Englewood Cliffs, N.J., 1965.
5. Schlichting, H., *Boundary Layer Theory* (7th ed.), McGraw-Hill, New York, 1979.
6. Cebeci, T., and P. Bradshaw, *Momentum Transfer in Boundary Layers,* McGraw-Hill-Hemisphere, New York, 1977.
7. Curie, I. G., *Fundamental Mechanics of Fluids,* McGraw-Hill, New York, 1974.
8. White, F. M., *Viscous Fluid Flow,* McGraw-Hill, New York, 1974.
9. Pao, R. H. F., *Fluid Dynamics,* Merrill, Columbus, OH, 1967.
10. Roache, P. J., *Computational Fluid Dynamics,* Hermosa Publishers, Albuquerque, 1976.
11. Patankar, S. V., *Numerical Heat Transfer and Fluid Flow,* McGraw-Hill-Hemisphere, New York, 1980.
12. Holman, J. P., *Heat Transfer* (5th ed.), McGraw-Hill, New York, 1981.
13. Incoprea, F., and D. DeWitt, *Fundamentals of Heat Transfer,* Wiley, New York, 1980.
14. White, F. M., *Heat Transfer,* Addison-Wesley, Reading, MA, 1983.
15. Hess, J. L., and A. M. O. Smith, "Calculation of Potential Flow About Arbitrary Bodies" in *Progress in Aeronautical Sciences* (vol. 8), Pergamon, New York, 1966.
16. Blasius, H., "Grenzschechten in Flussigkerten mit kleiner Reibung," *Z. Mat. Physik, 56,* 1. (Available in English as NACA Tech Memo 1256).
17. Thwaites, B., "Approximate Calculation of the Laminar Boundary Layer," *Aeronautical Quarterly,* vol. 1, 1949.
18. White, F. M., *Fluid Mechanics,* McGraw-Hill, New York, 1979.
19. Kline, S. J., M. Morkovin, G. Sovran, and D. Cockrell, *Computation of Turbulent Boundary Layers—1968 AFOSR—IFP—Stanford Conference,* Department of Mechanical Engineering, Stanford University, Palo Alto, CA, 1968.

Applications of Flow Analysis: Compressible Flow

10

Most of our discussions in the previous chapters have concentrated on incompressible flow. The incompressibility condition permits two extremely important simplifications in flow analysis:

- The density of the fluid is known and can be treated as a constant in the continuity, momentum, and energy equations.
- The coupling between mechanical and thermal forms of energy is weak, permitting use of a special incompressible form of the energy equation. This special form concentrates on the fluid mechanical energy.

It is not always realistic to assume that fluids are incompressible, however, so it is not always possible to rely on results based on these two simplifications. All fluids, even liquids, are compressible. If the pressure changes that occur in a flow are sufficiently large to cause significant fluid density changes, it becomes necessary to abandon the incompressible flow assumption. This situation is much more likely to occur in a gas or vapor than a liquid. A pressure change of 500 kPa (72.6 lb/in^2) causes a density change of about 0.024% in water but a density change of about 250% in standard air!* Because compressibility is more likely to be significant in gas flow than in liquid flow, the study of compressible flow is sometimes called *gas dynamics*.

This chapter considers analysis of flows in which the effects of fluid compressibility are significant. This requires treatment of density as a variable and introduction of a relation between fluid density and pressure. The simplest such relation is the ideal gas law,

$$p = \rho RT. \tag{1.1}$$

This equation also involves temperature, which becomes an additional variable. The general energy equation and a relation between internal energy and temperature must also be included. Analysis of compressible flow requires simultaneous application of the continuity, momentum, and energy equations and a property relation to calculate fluid

* The process is assumed to occur without heat transfer.

velocity, pressure, density, and temperature. This is true even if the fluid is not an ideal gas.

There are two important elements in a study of compressible flow: consideration of the physical effects of fluid compressibility and development of analysis techniques for engineering calculations. The latter task is greatly simplified if we assume that the fluid is an ideal gas. Many of the equations that we will develop in this chapter are based on the ideal gas model. Generally, the information that we present about the physical behavior of compressible fluids is not limited to ideal gases. If you need to analyze flow of a fluid that is not adequately described by the ideal gas model (steam, for instance), be sure to recognize that results based on the equations and methods that we present in this chapter will be qualitatively correct, but numerical values should be treated with caution.

Both internal and external flows can be affected by fluid compressibility. This chapter will consider both types, but the major emphasis will be on internal flow. It is impossible to present anything but a brief introduction to compressible flow in a single chapter. Several books devoted to compressible flow (gas dynamics) are available [1–6].

10.1 Some Concepts from Thermodynamics

Analysis of *incompressible* flow is based on the continuity equation, the momentum equation, and a specialized form of the energy equation that considers only mechanical forms of energy. *Compressible* flow analysis also requires the continuity and momentum equations; however, the general energy equation must replace the mechanical energy equation because compressibility is associated with transformations between thermal and mechanical forms of energy. Compressible flow analysis also requires specification of a relation between fluid properties. Transformation between forms of energy and relations between fluid properties are the primary subject matter of thermodynamics. Concepts from thermodynamics and concepts from fluid mechanics are equally important in compressible flow analysis. Before we begin the study of compressible flow, we should consider a few basic concepts from thermodynamics. If you have already studied thermodynamics, this section should serve as a review. If you haven't studied thermodynamics yet, now is the time to start. This section presents all the concepts and equations that we need for compressible flow analysis, but the developments and explanations are brief. Consult [7–9] for more detail.

10.1.1 Properties, State, and Process

The following properties are important in thermodynamics:

- Density (ρ)
- Pressure (p)
- Temperature (T)
- Specific internal energy ($\tilde{u}$)
- Specific enthalphy ($\tilde{h} \equiv \tilde{u} + p/\rho$)
- Specific entropy ($\hat{s}$)

All of these properties except entropy should be familiar. Entropy will be defined later in this section.

Properties are used to describe the state of a system. Recall that a system is a collection of mass of fixed identity; a system can be as small as a single fluid particle. An important principle of thermodynamics is the *state postulate:*

> The state of a system is fixed if a certain minimum number of its properties are specified.

For a pure, compressible substance (gas or vapor), the minimum number of properties is two. For example, oxygen with $p = 14.70$ lb/in^2 and $T = 0°$F *always* has $\rho = 0.0954$ lbm/ft^3; no other density is possible at that pressure and temperature.

The state postulate tells us that the properties of any substance are related to each other. It is possible to quantify this relation using *equations of state* such as the ideal gas law, Eq. (1.1). Unfortunately, the ideal gas law is not complete because internal energy can't be determined from it. For an ideal gas, internal energy is related to temperature by the specific heat at constant volume (c_v):

$$\tilde{u}_2 - \tilde{u}_1 = \int_{T_1}^{T_2} c_v(T)\, dT.$$

In this chapter we assume that c_v is constant, so

$$\tilde{u}_2 - \tilde{u}_1 = c_v(T_2 - T_1). \tag{10.1}$$

The enthalpy of an ideal gas is obtained from the temperature and specific heat at constant pressure (c_p):*

$$\tilde{h}_2 - \tilde{h}_1 = \int_{T_1}^{T_2} c_p(T)\, dT.$$

For an ideal gas with constant c_p,

$$\tilde{h}_2 - \tilde{h}_1 = c_p(T_2 - T_1). \tag{10.2}$$

Since $\tilde{h} = \tilde{u} + p/\rho$,

$$\tilde{h}_2 - \tilde{h}_1 = \tilde{u}_2 - \tilde{u}_1 + \left(\frac{p}{\rho}\right)_2 - \left(\frac{p}{\rho}\right)_1 = c_v(T_2 - T_1) + R(T_2 - T_1).$$

Comparing this with Eq. (10.2), we note that

$$c_p = c_v + R. \tag{10.3}$$

Tables A.3 and A.4 in Appendix A give values of the gas constants R, c_v, c_p, and $k(\equiv c_p/c_v)$ for several gases. These tables show that c_p, c_v, and k vary slightly with temperature; but we will assume that they are constant. Although we will normally use values at standard sea-level temperature, accuracy can be improved slightly by using values appropriate for the actual temperatures in a particular problem.

Thermodynamics and fluid mechanics deal with changes of state. When a gas flows through a nozzle, the pressure and temperature of the

---

* The terms "constant volume" and "constant pressure" do not mean that these equations apply only if volume (density) or pressure is constant [7].

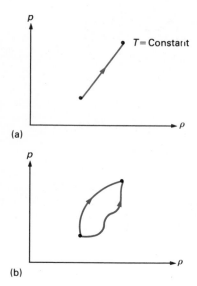

Figure 10.1 Processes on a property diagram: (a) Isothermal process on a p-ρ diagram; (b) two processes that have the same end states.

gas at the nozzle outlet are, in general, different from the pressure and temperature at the nozzle inlet, so the state of the gas will be different at the inlet and the outlet. If we follow a typical gas particle from the inlet to the outlet, we would observe that its state changes continuously. This sequence of different states is called a *process:*

> A process describes a sequence of states of a system.

There are many different types of processes. A process that occurs in such a way that the temperature remains constant is called an *isothermal* process. In an isothermal process, the pressure and density may change but the temperature does not. A particularly important type of process is the *adiabatic* process, in which no heat is transferred.

Processes can be represented as lines on a *property diagram*. Figure 10.1 illustrates a pressure-density (p-ρ) diagram. Since only two properties are required to specify the state of a substance, any point on a property diagram represents a specific state. Since, for an ideal gas,

$$p = (RT)\rho,$$

an isothermal process plots as a straight line on a p-ρ diagram. An adiabatic process would be represented by a different line.

An important characteristic of properties and processes is the following:

> Changes of properties in a process depend only on the end points of the process and not on the process itself.

Figure 10.1(b) illustrates two different processes that have the same end points. Obviously, the changes of p and ρ for the two processes are identical since the beginning and end points are identical; however, it is also true that the changes of all other properties (such as $T_2 - T_1$, $\tilde{u}_2 - \tilde{u}_1$) are identical for the two processes.

10.1.2 The Laws of Thermodynamics

The laws of thermodynamics deal with energy transformations and relate energy transfers to or from a system to changes of the system's properties. We introduced the first and second laws of thermodynamics in Section 3.2.

The first law of thermodynamics was considered extensively in Section 5.4, and we developed several equations representing the first law for a control volume. Consider, for example, Eq. (5.46), the general energy equation for steady flow along a streamline:

$$q - w_s = \tilde{u}_2 - \tilde{u}_1 + \frac{p_2}{\rho_2} - \frac{p_1}{\rho_1} + \frac{V_2^2}{2} - \frac{V_1^2}{2} + gz_2 - gz_1. \tag{5.46}$$

This equation relates energy transfers to or from a fluid particle (the terms on the left) to changes of the particle's properties (the terms on the right). The energy transfers are associated with the process that occurs between points 1 and 2. Because both heat and work appear on the left, it is impossible to determine both if we know only the change of fluid properties. Heat and work depend on the exact nature of the process that occurs between the end states. The two processes illustrated in

Fig. 10.1(b) would each have different amounts of heat transfer and work although the sum $q - w_s$ would be the same for both. This means that when we use the general energy equation to analyze a flow process, we must make an explicit statement about either the heat transfer or the work. To illustrate, in Example 5.10 it was necessary to assume that there was no heat transferred to the air during compression.*

The second law of thermodynamics deals with conversions between heat and work. Consider a sequence of processes that eventually returns the fluid to its initial state (Fig. 10.2). Such a sequence of processes is called a *cycle*. Heat and/or work will be transferred to the fluid in some of the processes and will be transferred from the fluid in others. Applying the first law to the entire sequence of processes gives

$$\sum q - \sum w_s = 0.$$

The right side is zero because the sequence returns the fluid to its initial state. Recall the sign conventions for heat and work:

- Heat addition (q_{in}) is positive.
- Work production (w_{out}) is positive.

Using subscripts "in" and "out" rather than signs, we get

$$\sum q_{in} - \sum q_{out} = \sum w_{out} - \sum w_{in} = w_{out,net}.$$

The second law of thermodynamics states that it is impossible to devise a sequence of processes that has no effect other than the conversion of a given amount of heat *input* into an equal amount of work output. In a *single process,* it is possible that work can be produced that is equal to or even greater than the amount of heat input during the process, but this can occur only with a change of properties, which is an additional effect. A concise statement of the second law of thermodynamics is as follows:

> It is impossible to convert a given amount of heat completely into useful work (or other form of mechanical energy).

It is, of course, possible to convert a given amount of work (or other form of mechanical energy) completely into heat; this is the process of mechanical energy "loss" that is so important in viscous flow.

Closely associated with the second law of thermodynamics is the concept of the *reversible process.* A reversible process is one that can be reversed or "run backward" with no net effect on either the system or the surroundings. Suppose a process changed the fluid from state 1 to state 2 while absorbing heat q and producing work w. The reversed process must not only return the fluid from state 2 back to state 1 but must also reject an amount of heat *exactly* equal to q and absorb an amount of work *exactly* equal to w. Any process that converts work (or other form of mechanical energy) into heat (by friction, for example) cannot be reversible because, according to the second law, it is impossible to convert the reverse heat completely into work. Other occurrences

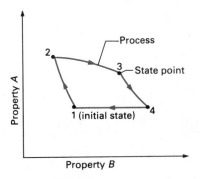

Figure 10.2 A cycle is a sequence of processes that returns the fluid to its initial state.

---

* This same type of problem carries over into incompressible flow. In incompressible flow, it is not possible to distinguish between work and mechanical energy loss by knowing only changes of pressure, velocity, and elevation.

that render a process irreversible are mixing of dissimilar substances and heat transfer across a finite temperature difference. Since all real-world processes involve friction, mixing, or heat transfer, they are irreversible.

10.1.3 Entropy and the Isentropic Process

The second law of thermodynamics implies the existence of a property called *entropy*. Consider two states of a substance, state 1 and state 2. The difference in specific entropy between the two states is defined by

$$\tilde{s}_2 - \tilde{s}_1 \equiv \int_1^2 \frac{dq_{\text{rev}}}{T}. \tag{10.4}$$

In this definition, q_{rev} is the heat transfer per unit mass that would occur in a *reversible* process between 1 and 2. It can be shown [7, 8] that

- $\int (dq_{\text{rev}}/T)$ is the same for any reversible path between states 1 and 2;
- $\tilde{s}_2 - \tilde{s}_1$ can be related to only the values of the other properties at states 1 and 2.

These facts tell us that entropy is truly a thermodynamic property, whose change between two state points depends only on the state points themselves and not on the process between them.

The second law of thermodynamics can be used to show that entropy changes in a *real* (irreversible) process can be related to actual heat transfer and temperature by

$$\tilde{s}_2 - \tilde{s}_1 \geq \int_1^2 \frac{dq}{T}. \tag{10.5}$$

The equality holds if the process is reversible and the inequality holds if the process is irreversible. The entropy gives us a way to evaluate the irreversibility of a process by comparing actual entropy change to the integral in Eq. (10.5).

Since entropy is a property, it can be related to other properties. The appropriate equations are

$$T\,d\tilde{s} = d\tilde{h} - \frac{dp}{\rho} \tag{10.6}$$

and

$$T\,d\tilde{s} = d\tilde{u} - \frac{p}{\rho}\left(\frac{d\rho}{\rho}\right). \tag{10.7}$$

Equations (10.6) and (10.7) are called the *Gibbs equations* or, sometimes, the *T-d$\tilde{s}$ equations*.

If we specify an ideal gas with constant specific heat, the Gibbs equations can be integrated to give

$$\tilde{s}_2 - \tilde{s}_1 = c_p \ln\left(\frac{T_2}{T_1}\right) - R \ln\left(\frac{p_2}{p_1}\right) \tag{10.8}$$

and

$$\tilde{s}_2 - \tilde{s}_1 = c_v \ln\left(\frac{T_2}{T_1}\right) - R \ln\left(\frac{\rho_2}{\rho_1}\right) \tag{10.9}$$

Note that Eqs. (10.8) and (10.9) allow calculation of entropy from other fluid properties. It is not necessary to know the heat transfer to calculate entropy change.

In analyzing compressible flow, we often assume that there is no heat transfer to or from the fluid during the flow; that is, we assume an adiabatic process. If we also assume that the process is reversible, then the entropy will remain constant. Such a process (reversible, adiabatic) is called an *isentropic process*. If the fluid is an ideal gas, Eqs. (10.8) and (10.9) yield simple and convenient process laws for an isentropic process. Considering Eq. (10.8) applied to an isentropic process, we have

$$\tilde{s}_2 - \tilde{s}_1 = 0 = c_p \ln\left(\frac{T_2}{T_1}\right) - R \ln\left(\frac{p_2}{p_1}\right).$$

Dividing by R and taking the antilogarithm give

$$\frac{p_2}{p_1} = \left(\frac{T_2}{T_1}\right)^{c_p/R}$$

Using Eq. (10.3), we have

$$\frac{c_p}{R} = \frac{k}{k-1}, \tag{10.10}$$

and we can write

$$\frac{p_2}{p_1} = \left(\frac{T_2}{T_1}\right)^{k/(k-1)} \tag{10.11}$$

Similarly,

$$\frac{\rho_2}{\rho_1} = \left(\frac{T_2}{T_1}\right)^{1/(k-1)} \tag{10.12}$$

and

$$\frac{p_2}{p_1} = \left(\frac{\rho_2}{\rho_1}\right)^{k} \tag{10.13}$$

Note carefully that Eqs. (10.11)–(10.13) are process laws relating changes of properties that occur in a specific type of process (isentropic). They *are not* equations of state and do not apply in a general sense.

10.1.4 Stagnation Properties

The thermodynamic state of a fluid particle is defined by its properties (p, $\tilde{u}$, $\tilde{s}$, and so on); but from the viewpoint of mechanics, it is also necessary to know the velocity of the particle and, possibly, its position in a gravitational field.

The thermodynamic properties are *static properties* and are the values that would be measured by instruments that are static with respect to the fluid.* The static properties represent the molecular structure of the fluid and obey all equations of state and other property-related laws and thermodynamic equations. The particle's velocity and elevation are specified separately.

It is often useful to combine the (static) thermodynamic properties with the velocity and elevation to obtain equivalent thermodynamic properties that represent the total (thermodynamic *and* mechanical) state. This is done by using *stagnation properties,* defined as follows:

> Stagnation properties are the properties that the fluid would obtain if it were brought to a condition of zero velocity and zero elevation in a reversible process with no heat transfer and no work.

We can obtain equations for stagnation properties with the aid of Fig. 10.3, which shows a (possibly imaginary) process that brings a fluid particle to rest and zero elevation. Applying the energy equation to the streamline between 1 and 0 gives

$$q - w_s = \tilde{h}_0 - \tilde{h}_1 + \frac{V_0^2}{2} - \frac{V_1^2}{2} + gz_0 - gz_1.$$

Since, by definition, q, w_s, V_0, and z_0 are all zero, we get

$$\tilde{h}_0 = \tilde{h}_1 + \frac{V_1^2}{2} + gz_1,$$

where $\tilde{h}_0$ is called the *stagnation enthalpy.*[†] In compressible flow analysis, potential energy and gravity force are almost always negligible, and the definition of stagnation enthalpy is usually simplified to

$$\tilde{h}_0 \equiv \tilde{h} + \frac{V^2}{2}. \tag{10.14}$$

In this equation, the subscript 1 has been dropped to obtain a general definition of stagnation enthalpy. Note that we did not need to use the specification of reversibility to obtain the stagnation enthalpy.

If the fluid is an ideal gas with constant specific heat, Eq. (10.14) can be written

$$c_p T_0 = c_p T + \frac{V^2}{2},$$

Figure 10.3 Process for defining stagnation properties.

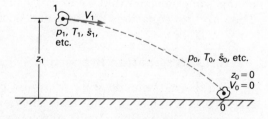

* To be static with respect to the fluid, the instruments have to move along with the fluid.

[†] The subscript zero on stagnation properties means "zero velocity (and elevation)."

and the *stagnation temperature* is given by

$$T_0 = T + \frac{V^2}{2c_p}. \tag{10.15}$$

We can obtain equations for the *stagnation pressure* and *stagnation density* by noting that, by definition, the stagnation process is isentropic. For an ideal gas,

$$\frac{p_0}{p} = \left(\frac{T_0}{T}\right)^{k(k-1)}. \tag{10.16}$$

and

$$\frac{\rho_0}{\rho} = \left(\frac{T_0}{T}\right)^{1/(k-1)}. \tag{10.17}$$

Dividing Eq. (10.15) by T, we obtain

$$\frac{T_0}{T} = 1 + \frac{V^2}{2c_p T}. \tag{10.18}$$

Substituting into Eqs. (10.16) and (10.17), we have

$$p_0 = p\left(1 + \frac{V^2}{2c_p T}\right)^{k/(k-1)}. \tag{10.19}$$

and

$$\rho_0 = \rho\left(1 + \frac{V^2}{2c_p T}\right)^{1/(k-1)}. \tag{10.20}$$

Equations (10.14)–(10.20) show that the stagnation properties are determined by the static properties *and* the fluid velocity. Since any fluid particle always has a pressure, a temperature, and a velocity (the velocity might be zero), any fluid particle can be described by both its static properties and its stagnation properties. It is perfectly proper to say "the fluid has a pressure of 100 kPa and a stagnation pressure of 125 kPa." The static properties actually are the properties of the fluid, and the stagnation properties are the properties that the fluid would have if it were brought to zero velocity in an isentropic process with no work. If (and only if) the fluid particle has zero velocity, its static and stagnation properties are equal.

Stagnation properties represent the maximum possible values of the fluid properties unless energy is added to the fluid by heat transfer or work. The postulated stagnation process represents the ideal conversion of the fluid's kinetic and potential energies into pressure and internal energy.

Stagnation properties are sometimes called *total properties*. Since the process of stagnation (that is, bringing the fluid to rest) can be accomplished by an irreversible process* as well as by a reversible process, there is some possible ambiguity in using the term *stagnation* in cases where viscous flow processes are considered. This ambiguity can also be remedied by use of the term "isentropic stagnation properties," an approach followed by some authors.

---

* An example is the deceleration that occurs in a viscous boundary layer
(see Chapter 9).

EXAMPLE 10.1 Illustrates basic concepts and calculations
in thermodynamics.

A turbine expands 10 lbm/sec of air from an inlet pressure and temperature of
40 psia* and 500°F to an outlet pressure of 14.9 psia. The turbine delivers 760 hp
to an electric generator. The air velocity is 500 ft/sec at the turbine inlet and
100 ft/sec at the turbine outlet. The turbine casing is well insulated. The flow
through the turbine is steady. Calculate the static temperature at the turbine
outlet, the stagnation temperature and pressure at the turbine inlet and outlet,
and the change of specific entropy between the inlet and outlet. Is the expansion a reversible process?

Solution

Given
A turbine, which expands hot air and produces power
Figure E10.1, a schematic of a turbine
Data on flow rate, air properties, and velocity given in the figure
Steady flow

Find
Static temperature at turbine outlet (T_2)
Stagnation pressure and temperature at turbine inlet (p_{0_1}, T_{0_1})
Stagnation pressure and temperature at turbine outlet (p_{0_2}, T_{0_2})
Entropy change ($\tilde{s}_2 - \tilde{s}_1$)
Whether expansion is reversible

Solution
The energy equation for a control volume enclosing the turbine is obtained
from Eq. (5.45):

$$\dot{Q} - \dot{W}_s = \dot{m}_2 \left[\bar{\tilde{u}}_2 + \overline{\left(\frac{p_2}{\rho_2}\right)} + \alpha_2 \overline{\left(\frac{\bar{V}_2^2}{2}\right)} + g\bar{z}_2 \right] - \dot{m}_1 \left[\bar{\tilde{u}}_1 + \overline{\left(\frac{p_1}{\rho_1}\right)} + \alpha_1 \overline{\left(\frac{\bar{V}_1^2}{2}\right)} + g\bar{z}_1 \right].$$

To simplify this equation, we assume that α_1 and α_2 are unity, that elevation
changes are negligible, and that the given data represent proper averages. The

Figure E10.1 Schematic of hot air
turbine.

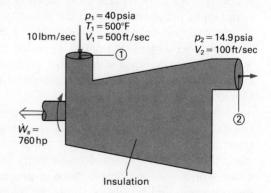

$p_1 = 40\,\text{psia}$
$T_1 = 500°F$
10 lbm/sec $V_1 = 500\,\text{ft/sec}$
①

$p_2 = 14.9\,\text{psia}$
$V_2 = 100\,\text{ft/sec}$

②

$\dot{W}_s =$
760 hp

Insulation

* In this chapter, we will follow the customary procedure in thermodynamics and write
"psia" for lb/in², abs.

continuity equation for the control volume is

$$\dot{m}_2 = \dot{m}_1 = \dot{m},$$

so the energy equation becomes

$$\dot{Q} - \dot{W}_s = \dot{m}\left[\left(\tilde{h}_2 + \frac{V_2^2}{2}\right) - \left(\tilde{h}_1 + \frac{V_1^2}{2}\right)\right].$$

Since the turbine is well insulated, $\dot{Q}$ is small and will be assumed to be zero. Solving for the enthalpy change of the air gives

$$\tilde{h}_2 - \tilde{h}_1 = -\frac{\dot{W}_s}{\dot{m}} + \frac{V_1^2 - V_2^2}{2}.$$

If we assume that the air is an ideal gas, Eq. (10.2) relates enthalpy to temperature. Substituting this relation and solving for T_2, we have

$$T_2 = T_1 - \frac{\dot{W}_s}{\dot{m}c_p} + \frac{V_1^2 - V_2^2}{2c_p}.$$

Substituting numbers gives (from Table A.4, $c_p = 186.7 \text{ ft·lb/lbm·°R}$)

$$T_2 = (500 + 459.7)°\text{R} - \frac{760 \text{ hp}\left(550 \dfrac{\text{ft·lb}}{\text{sec·hp}}\right)}{10 \dfrac{\text{lbm}}{\text{sec}}\left(186.7 \dfrac{\text{ft·lb}}{\text{lbm·°R}}\right)}$$

$$+ \frac{(500)^2 \dfrac{\text{ft}^2}{\text{sec}^2} - (100)^2 \dfrac{\text{ft}^2}{\text{sec}^2}}{2\left(186.7 \dfrac{\text{ft·lb}}{\text{lbm·°R}}\right)\left(32.2 \dfrac{\text{ft·lbm}}{\text{lb·sec}^2}\right)},$$

Answer

$$\left| \; T_2 = 755.8°\text{R} = 296.1°\text{F}. \; \right|$$

The stagnation temperature and pressure are calculated from Eqs. (10.15) and (10.16). At the turbine inlet,

$$T_{0_1} = T_1 + \frac{V_1^2}{2c_p} = 959.7°\text{R} + \frac{(500)^2 \dfrac{\text{ft}^2}{\text{sec}^2}}{2\left(186.7 \dfrac{\text{ft·lb}}{\text{lbm·°R}}\right)\left(32.2 \dfrac{\text{ft·lbm}}{\text{lb·sec}^2}\right)},$$

Answer

$$\left| \; T_{0_1} = 980.5°\text{R} = 520.8°\text{F}, \; \right|$$

and

$$p_{0_1} = p_1\left(\frac{T_{0_1}}{T_1}\right)^{k/(k-1)} = 40 \text{ psia}\left(\frac{980.5}{959.7}\right)^{1.4/0.4},$$

Answer

$$\left| \; p_{0_1} = 43.1 \text{ psia}. \; \right|$$

At the turbine outlet,

$$T_{0_2} = T_2 + \frac{V_2^2}{2c_p} = 755.8°\text{R} + \frac{(100)^2 \dfrac{\text{ft}^2}{\text{sec}^2}}{2\left(186.7 \dfrac{\text{lb·sec}}{\text{lbm·°R}}\right)\left(32.2 \dfrac{\text{ft·lbm}}{\text{lb·sec}^2}\right)},$$

Answer

$$\left| \; T_{0_2} = 756.6°\text{R} = 296.9°\text{F}, \; \right|$$

and

$$p_{0_2} = p_2 \left(\frac{T_{0_2}}{T_2}\right)^{k/(k-1)} = 14.9 \text{ psia} \left(\frac{756.6}{755.8}\right)^{3.5},$$

$$\left| \; p_{0_2} = 14.96 \text{ psia.} \; \right|$$

Answer ◀

The entropy change across the turbine is calculated by

$$\tilde{s}_2 - \tilde{s}_1 = c_p \ln\left(\frac{T_2}{T_1}\right) - R \ln\left(\frac{p_2}{p_1}\right).$$

Substituting numbers, we have

$$\tilde{s}_2 - \tilde{s}_1 = 186.7 \frac{\text{ft} \cdot \text{lb}}{\text{lbm} \cdot {}^\circ\text{R}} \ln\left(\frac{755.8}{959.7}\right) - 53.35 \frac{\text{ft} \cdot \text{lb}}{\text{lbm} \cdot {}^\circ\text{R}} \ln\left(\frac{14.9}{40}\right),$$

$$\left| \; \tilde{s}_2 - \tilde{s}_1 = 8.1 \text{ ft} \cdot \text{lb/lbm} \cdot {}^\circ\text{R.} \; \right|$$

Answer ◀

Since the expansion is adiabatic,

$$\int \frac{dq}{T} = 0 \quad \text{and} \quad \tilde{s}_2 - \tilde{s}_1 > \int \frac{dq}{T},$$

Thus:

Answer

$$\left| \; \text{The expansion is irreversible.} \; \right|$$

◀

Discussion

Several important points require emphasis. First note that the turbine power is treated as positive because the turbine delivers work. Second, note that temperatures can be expressed in either Farenheit or Rankine (absolute), but temperature ratios must be in terms of absolute temperature. Third, note that the difference between the static and stagnation properties is significant at the turbine inlet where the velocity is 500 ft/sec but nearly insignificant at the turbine outlet where the velocity is 100 ft/sec. Fourth, since the entropy at a stagnation state is equal to the entropy at the corresponding static state, we note that

$$\tilde{s}_{0_2} - \tilde{s}_{0_1} = c_p \ln \frac{T_{0_2}}{T_{0_1}} - R \ln \frac{p_{0_2}}{p_{0_2}} = \tilde{s}_2 - \tilde{s}_1.$$

You might substitute the appropriate numbers to verify this equality. Fifth, note that we used the room temperature values of c_p and k in the calculation. The accuracy could be improved slightly by using values appropriate to the "average" temperature, about 400°F.

Finally, note that Eq. (10.11) is not valid for relating the pressure and temperature between the turbine outlet:

$$\frac{p_2}{p_1} = \frac{14.9}{40} = 0.3725 \neq \left(\frac{T_2}{T_1}\right)^{k/(k-1)} = \left(\frac{755.9}{959.7}\right)^{3.5} = 0.4334.$$

This is because the expansion is not isentropic, even though it is adiabatic.

10.2 Waves in Compressible Fluids

It is a familiar fact that disturbances introduced in one part of a fluid are communicated to other parts of the fluid. If a valve in a pipeline is opened or closed, the flow in the entire pipeline is changed. Examination of Figs. 3.4 and 3.5 shows that fluid responds to the presence and motion of a surface well upstream of the surface itself.* The influence of a disturbance at a point is propagated to other points by waves that travel at a large, but finite, speed with respect to the fluid.

Consider the effect of suddenly closing a valve at the end of a long pipe. Before the valve is closed, fluid flows through the pipe at a steady rate. At the instant the valve is closed, the fluid at the valve is brought to rest. If the fluid continued to flow toward the valve, it would continue to "pile up" at the valve. A wave propagates from the closed valve back into the fluid, informing the fluid about the closed valve. The wave also causes a pressure increase as the kinetic energy of the fluid is rapidly changed. This pressure change can lead to a significant density change. The wave is called a compression wave. Experimental and theoretical investigations of such waves show them to be very thin (in a gas, wave thickness is a few molecular mean free paths). The speed at which such a wave travels in a particular fluid is a measure of the fluid's compressibility. The more compressible the fluid, the slower the wave. Because the existence and speed of such disturbance waves are intimately connected with fluid compressibility, analysis of compressible flow often involves consideration of wave speed and motion.

10.2.1 Wave Speed, Speed of Sound, and Mach Number

Consider a wave propagating into a still fluid (Fig. 10.4a). Ahead of the wave, the pressure and density of the fluid are p and ρ. The wave causes the pressure and density to change by Δp and $\Delta \rho$. The velocity of the wave is ω. The fluid is set in motion by the wave, and the velocity of the fluid behind the wave is ΔV; Δp, $\Delta \rho$, and ΔV are not necessarily small, and ω is finite. To analyze the wave, we choose a control volume that moves with the wave and so transform the flow into the situation shown in Fig. 10.4b. Since the wave is very thin, we pretend that it has no thickness at all so the areas of the front and back faces of the control volume are equal and the volume is zero. Applying the continuity equation to the control volume, we get

$$\dot{m} = \rho A \omega = (\rho + \Delta \rho) A (\omega - \Delta V).$$

Canceling A and solving for ΔV, we have

$$\Delta V = \omega \left(\frac{\Delta \rho}{\rho + \Delta \rho} \right). \tag{10.21a}$$

Note that if $\Delta \rho$ is small,

$$\Delta V \approx \omega \left(\frac{\Delta \rho}{\rho} \right), \tag{10.21b}$$

and ΔV will be much smaller than ω.

Next we apply the linear momentum equation to the control volume. Since the "sides" of the control volume are vanishingly small, the only force on the control volume is due to pressure on the inlet and

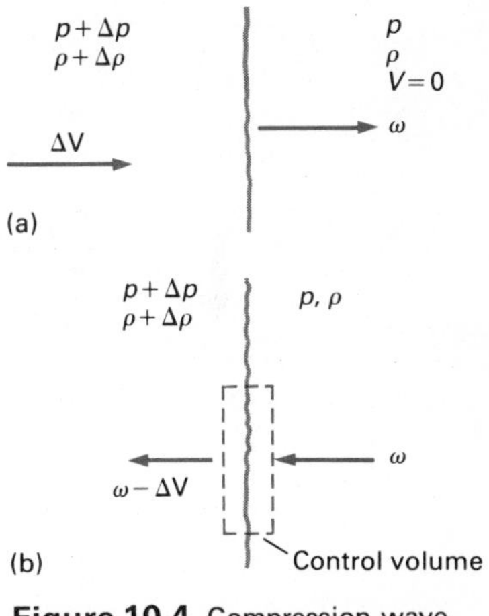

Figure 10.4 Compression wave in a fluid: (a) Wave moving relative to fluid; (b) wave motion removed by choosing a control volume fixed to the wave.

---

* Note the curvature of the streaklines (streamlines) upstream of the airfoil in Fig. 3.4.

outlet faces:

$$\sum F_x = pA - (p + \Delta p)A = \dot{m}[(\omega - \Delta V) - \omega].$$

Substituting for $\dot{m}$ and solving for Δp give

$$\Delta p = \rho \omega \Delta V. \tag{10.22}$$

This equation shows that waves can generate very large pressure changes if ω is large.

An expression for wave speed is obtained by substituting Eq. (10.21a) into Eq. (10.22):

$$\omega^2 = \frac{\Delta p}{\Delta \rho}\left(1 + \frac{\Delta \rho}{\rho}\right). \tag{10.23}$$

The *speed of sound* (c) is defined as the speed of a vanishingly weak wave,[*]

$$c^2 \equiv \lim_{\substack{\Delta p \to 0 \\ \Delta \rho \to 0}} \left[\frac{\Delta p}{\Delta \rho}\left(1 + \frac{\Delta \rho}{\rho}\right)\right] = \frac{\partial p}{\partial \rho}.$$

The limiting process produces a derivative that can be evaluated only if the thermodynamic process caused by the wave is specified. Two (relatively) obvious choices are an isothermal process and an adiabatic (isentropic) process. The correct choice is the isentropic process because there is minimal opportunity for heat transfer to or from the fluid as it passes rapidly through a very thin wave. The speed of sound is therefore defined by

$$c^2 \equiv \left.\frac{\partial p}{\partial \rho}\right)_s. \tag{10.24}$$

For an ideal gas, an isentropic process obeys the equation

$$\frac{p}{p_{\text{ref}}} = \left(\frac{\rho}{\rho_{\text{ref}}}\right)^k, \quad \text{and so} \quad \left.\frac{\partial p}{\partial \rho}\right)_s = k\left(\frac{p}{\rho}\right) = kRT.$$

The speed of sound in an ideal gas is calculated by

$$c = \sqrt{k\frac{p}{\rho}}, \tag{10.25}$$

or

$$c = \sqrt{kRT}. \tag{10.26}$$

The speed of sound will change from point to point in a flow if the temperature changes. Substituting gas constants for air gives the useful equations

$$c_{\text{air}} \approx 20.05\sqrt{T/(\text{K})} \text{ m/s} \quad \text{and} \quad c_{\text{air}} \approx 49.02\sqrt{T(^\circ\text{R})} \text{ ft/sec}.$$

Speed of sound in liquids and solids is related to the isentropic bulk modulus of elasticity

$$c = \sqrt{\frac{E_{v,s}}{\rho}}, \tag{10.27}$$

---

[*] Disturbances that the human ear perceives as sound are very weak, typically having a pressure change between 10^{-9} and 10^{-4} atm.

where

$$E_{v,s} \equiv \rho \left(\frac{\partial p}{\partial \rho} \right)_s.$$

For most liquids and solids the modulus of elasticity is nearly the same for any type of process, so the subscript s is dropped and c^2 depends on the general bulk modulus of elasticity E_v, defined in Section 1.3.4.

In Chapter 6 we stated that the primary dimensionless parameter of compressible flow is the Mach number, defined by

$$\mathbf{M} \equiv \frac{V}{\sqrt{E_v/\rho}}.$$

We now recognize that the term in the denominator of the Mach number is actually the speed of sound, so the Mach number is

$$\mathbf{M} = \frac{V}{c}. \tag{10.28}$$

If the Mach number is large, the effects of fluid compressibility are significant, but if it is small, compressibility effects are negligible. If we square both sides of Eq. (10.28) and assume an ideal gas, we get

$$\mathbf{M}^2 = \frac{V^2}{kRT} = \frac{\rho V^2}{kp}. \tag{10.29}$$

The criteria for neglecting fluid compressibility suggested on pages 7 and 301 are equivalent to specifying that the Mach number is small.

The importance of Mach number as a parameter in compressible flow can be illustrated by considering the (ideal gas) equations for stagnation properties. The stagnation-to-static-temperature ratio is

$$\frac{T_0}{T} = 1 + \frac{V^2}{2c_p T}. \tag{10.18}$$

Multiplying and dividing the second term on the right by kR and using Eqs. (10.3), (10.26), and (10.29), we obtain

$$\frac{V^2}{2c_p T} = \frac{k-1}{2} \left(\frac{V^2}{c^2} \right) = \left(\frac{k-1}{2} \right) \mathbf{M}^2,$$

so

$$\frac{T_0}{T} = 1 + \frac{k-1}{2} \mathbf{M}^2. \tag{10.30}$$

Using Eqs. (10.19) and (10.20), we get

$$\frac{p_0}{p} = \left(1 + \frac{k-1}{2} \mathbf{M}^2 \right)^{k/(k-1)} \tag{10.31}$$

and

$$\frac{\rho_0}{\rho} = \left(1 + \frac{k-1}{2} \mathbf{M}^2 \right)^{1/(k-1)} \tag{10.32}$$

Because of the exponential form of Eqs. (10.31) and (10.32), calculations involving these equations were quite tedious in precalculator (computer) days, and extensive tables of property ratios versus Mach number were prepared and used. Use of tables for such calculations is still quite common, so we have included a brief table of values for $k = 1.4$ (Table E.1 in Appendix E). Note that T/T_0 and p/p_0 (numbers less than 1.0) are tabulated with **M** as an independent parameter. There is no column for density ratio because

$$\frac{\rho}{\rho_0} = \left(\frac{p}{p_0}\right)\bigg/\left(\frac{T}{T_0}\right).$$

The remaining column of the table (A/A^*) will be explained later.

It is customary to define various regimes of flow based on the value of the Mach number. The regimes are as follows:

Incompressible flow	has	**M** $\ll 1$.
Subsonic flow	has	**M** < 1.
Transonic flow	has	**M** ≈ 1.
Supersonic flow	has	**M** > 1.
Hypersonic flow	has	**M** $\gg 1$.

These are essentially *local* definitions. It is possible for a flow field to contain regions of subsonic flow, regions of transonic flow, and regions of supersonic flow.

EXAMPLE 10.2 Illustrates Mach number, stagnation properties, and the use of Mach number tables

An airliner cruises at 250 m/s at an altitude of 10 km above sea level. Calculate the airliner's Mach number and the stagnation pressure and temperature relative to the airliner. Then compute the same quantities for an airliner speed of 500 m/s. Finally compute the same quantities for a speed of 250 m/s but change the altitude to 1 km above sea level.

Solution

Given
Airliner cruising at
250 m/s at 10 km
500 m/s at 10 km
250 m/s at 1 km

Find
Airliner Mach number
Stagnation pressure and temperature relative to the airliner

Solution
We first find the atmospheric pressure and temperature at the two altitudes. We assume that the atmospheric conditions are those of Standard Atmosphere. The atmospheric properties can be calculated from Eqs. (2.7)–(2.9) or read from Fig. 2.6 or Table A.1. At 10 km,

$$p_{10km} = 26.50 \text{ kPa} \quad \text{and} \quad T_{10km} = 223.3\text{K}.$$

At 1 km,

$$p_{1km} = 89.88 \text{ kPa} \quad \text{and} \quad T_{1km} = 281.7\text{K}.$$

The Mach number is

$$\mathbf{M} = \frac{V}{c}.$$

From Eq. (10.26),

$$c = \sqrt{kRT} = 20.05\sqrt{T/(\text{K})};$$

the gas is air. At 10 km,

$$c = 20.05\sqrt{223.3} \text{ m/s} = 299.6 \text{ m/s}.$$

Thus when $V = 250$ m/s,

$$\mathbf{M} = \frac{250 \text{ m/s}}{299.6 \text{ m/s}},$$

Answer

$$|\ \mathbf{M} = 0.834 \text{ at 10 km and 250 m/s.}\ |$$

The airliner is cruising at subsonic speed.
From Table E.1, the pressure and temperature ratios at this Mach number are

$$\frac{p}{p_0} = 0.633 \quad \text{and} \quad \frac{T}{T_0} = 0.878.$$

The static properties are those of the atmosphere. The stagnation properties relative to the airliner are

$$p_0 = \frac{p_{10km}}{\dfrac{p}{p_0}} \quad \text{and} \quad T_0 = \frac{T_{10km}}{\dfrac{T}{T_0}}.$$

Substituting, we have

$$p_0 = \frac{26.50 \text{ kPa}}{0.633},$$

Answer

$$|\ p_0 = 41.86 \text{ kPa at 10 km and 250 m/s;}\ |$$

$$T_0 = \frac{223.3\text{K}}{0.878},$$

Answer

$$|\ T_0 = 254.3\text{K at 10 km and 250 m/s.}\ |$$

Next consider 500 m/s at 10 km. The speed of sound and the static pressure and temperature are still the values appropriate to the 10-km altitude. The Mach number is

$$\mathbf{M} = \frac{V}{c} = \frac{500 \text{ m/s}}{299.6 \text{ m/s}},$$

Answer

$$|\ \mathbf{M} = 1.669 \text{ at 10 km and 500 m/s.}\ |$$

This is a supersonic cruise condition. At this Mach number, the pressure and temperature ratios are

$$\frac{p}{p_0} = 0.212 \quad \text{and} \quad \frac{T}{T_0} = 0.642.$$

Calculating p_0 and T_0, we have

$$p_0 = \frac{26.50 \text{ kPa}}{0.212},$$

$| p_0 = 125.0 \text{ kPa at 10 km and 500 m/s;} |$

Answer ◀

$$T_0 = \frac{223.3\text{K}}{0.642\text{K}},$$

$| T_0 = 347.8\text{K at 10 km and 500 m/s.} |$

Answer ◀

Finally, consider 250 m/s at 1 km. The speed of sound is now

$$c = \sqrt{kRT} = 20.05\sqrt{281.7\text{K}} \text{ m/s} = 336.5 \text{ m/s}.$$

The Mach number is

$$\mathbf{M} = \frac{V}{c} = \frac{250 \text{ m/s}}{336.5 \text{ m/s}},$$

$| \mathbf{M} = 0.743 \text{ at 1 km and 250 m/s.} |$

Answer ◀

Again a subsonic cruise condition. To calculate the stagnation pressure and temperature, we must use the atmospheric properties at 1 km. It is getting boring to use Table E.1, so we will use Eqs. (10.30) and (10.31) instead. They give

$$p_0 = p\left(1 + \frac{k-1}{2}\mathbf{M}^2\right)^{k/(k-1)} = 89.88 \text{ kPa}[1 + 0.2(0.743)^2]^{3.5},$$

$| p_0 = 129.7 \text{ kPa at 1 km and 250 m/s.} |$

Answer ◀

$$T_0 = T\left(1 + \frac{k-1}{2}\mathbf{M}^2\right) = 281.7\text{K}[1 + 0.2(0.743)^2],$$

$| T_0 = 312.8\text{K at 1 km and 250 m/s.} |$

Answer ◀

Discussion

Notice that the Mach number changed with altitude, even for the same velocity. This is because the temperature and, hence, the speed of sound changed. The stagnation properties changed with speed and altitude. Either Table E.1 (or a table from some other source) or equations can be used to calculate static-stagnation-property ratios. With a pocket calculator or computer it is often easier to make the calculation than to do table lookup and interpolation. It is usually best to calculate the temperature ratio first since this ratio is then raised to various powers to obtain pressure and density ratios.

10.2.2 The Radically Different Nature of Subsonic and Supersonic Flow

The response of a flowing fluid to a disturbance differs depending on whether the Mach number of the flow is less than or greater than unity. The nature of the difference can be demonstrated by the following discussion. Consider a large body of fluid that is influenced by a single point disturbance (Fig. 10.5). We assume that the point is fixed and that the fluid flows steadily past the point.* The point influences the flow

---

* This development can also be carried out by considering a moving point in a still fluid [1, 2]. Identical results are obtained; it is the relative motion between point and fluid that is important.

by changing its direction or speed or pressure; however, since we are considering only a single point, the influence is infinitesimally small. The disturbance is propagated into the fluid by waves that are generated at the point. Since the disturbance is infinitesimal, the waves are infinitesimal and travel at the speed of sound relative to the fluid.

Suppose that there is no fluid motion relative to the disturbance. Disturbance waves propagate uniformly from the point. Figure 10.5(a) shows the positions of a few waves. These waves were all generated at times earlier than the current time, t. If we select a uniform time interval Δt, the wave generated Δt seconds earlier (at time $t - \Delta t$) has spread into a circle of radius $c\,\Delta t$, the wave generated at $t - 2\,\Delta t$ has spread to a circle of radius $2c\,\Delta t$, and so on. The wave pattern is symmetrical about the point, and all points at a given distance from the disturbance are influenced equally.

Now suppose that the fluid is moving past the distrubance with a velocity very much smaller than the speed of sound ($\mathbf{M} \ll 1$). Figure 10.5(a) closely approximates the wave pattern because, in a time lapse (Δt) small enough to capture the wave motion, the fluid motion is barely perceptible.

Next suppose that the fluid is flowing past the disturbance at a speed that is the order of, but less than, the speed of sound. Suppose that $V = 0.5c$ (i.e., $\mathbf{M} = 0.5$). This situation is illustrated in Fig. 10.5(b). In addition to spreading into the fluid, the waves are swept downstream with the flow. As a result, the waves are bunched up on the upstream side of the disturbance and are spread out on the downstream side. The waves still propagate to all parts of the fluid; however, a fluid particle approaching the disturbance from upstream must get closer to the point to experience the same perturbation that it would experience farther away in a very small Mach number flow. If the waves actually represent

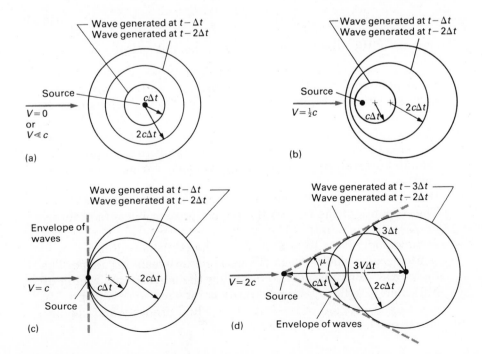

Figure 10.5 Wave pattern generated by compressible flow over an infinitesimal (point) disturbance.

sound, a listener moving toward the disturbance from upstream would hear a different frequency (pitch) than a downstream listener moving away from the disturbance. This is the well-known *Doppler effect*.

Now suppose that the fluid flows past the disturbance with a speed that is exactly equal to the sonic speed ($\mathbf{M} = 1$). In this case, the disturbance waves are swept downstream exactly as rapidly as they spread into the fluid, and the wave pattern appears as shown in Fig. 10.5(c). In this case, the waves do not propagate upstream and the fluid has no information about the disturbance until the instant it arrives! Note also that the waves form a straight envelope, perpendicular to the fluid velocity and passing through the disturbance. Since all of the waves are part of this envelope, the influence of the disturbance is concentrated and a fluid particle experiences it all at once on crossing the envelope.

Finally, suppose that the fluid flows past the point with a speed greater than the sonic speed ($\mathbf{M} > 1$). Figure 10.5(d) illustrates the resulting wave pattern for the specific case of $\mathbf{M} = 2$. At supersonic speeds, the waves are swept downstream faster than they spread. All of the waves are confined to a triangular (in two-dimensional flow in a plane) or conical (in three-dimensional or axisymmetric flow) region extending downstream from the point. The waves form an envelope of half-angle μ. Only the fluid inside the envelope "knows" about the disturbance. The effect of the disturbance is concentrated at the envelope, and the fluid adjusts suddenly on crossing the envelope. The envelope of infinitesimal waves is called a *Mach wave* in two-dimensional flow and a *Mach cone* in three-dimensional or axisymmetric flow. The half-angle of the envelope can be calculated from the geometry of Fig. 10.5(d):

$$\sin \mu = \frac{c\,\Delta t}{V\,\Delta t} = \frac{2c\,\Delta t}{2V\,\Delta t} = \cdots = \frac{c}{V} = \frac{1}{\mathbf{M}},$$

so the *Mach angle* (μ) is given by

$$\mu = \sin^{-1}\frac{1}{\mathbf{M}}. \tag{10.33}$$

The results of this discussion can be summarized as follows:

- In incompressible flow, the influence of any point is felt at all other points in the field. The influence is equally strong upstream and downstream. Fluid property and velocity variations must be continuous.
- In subsonic flow, the influence of any point is felt at all points. The effect is concentrated nearer to the point in the upstream region; however, fluid property and velocity variations must be continuous.
- In sonic flow, the influence of any point is not felt upstream. The response of the flow to the point's presence occurs abruptly on passing the point.
- In supersonic flow, the influence of any point is confined to a limited region downstream. There is no upstream influence. The higher the Mach number, the smaller the downstream region of influence. The flow responds to the point's presence abruptly on entering the region of the influence. Fluid property and velocity variations may be discontinuous.

The first two items tell us that compressible subsonic flow is qualitatively similar to incompressible flow, so we should expect subsonic flow in ducts or around bodies to behave similarly to the incompressible flows that we studied in Chapters 7–9. Transonic and supersonic flow is radically different from incompressible flow because of the removal of upstream influence. In almost all cases, supersonic flow responds differently to a given perturbation (such as area increase in a duct) than does incompressible or subsonic flow.

10.2.3 Shock Waves

Since we introduced the definition of the speed of sound, we have concentrated on infinitesimal disturbances and waves of infinitesimal strength. We will now consider waves that are associated with finite disturbances. In the previous section, we saw that introducing an infinitesimal disturbance into a supersonic flow generates a stationary (with respect to the disturbance) wave (the Mach wave). If a finite disturbance (say a sharp wedge or cone) is placed in a steady supersonic flow, a finite-strength wave will be generated (Fig. 10.6). This wave will turn the fluid as it passes over the wedge or cone. If a blunt object is placed in a supersonic flow, a finite-strength wave will be generated some distance ahead of the object (Fig. 10.7).

Finite-strength disturbance waves are called *shock waves*. Shock waves are extremely thin, and fluid properties and velocity change by large amounts (tens or even hundreds of percent) across a shock wave. Shock waves are actually rather common occurrences; examples of shock waves include explosion waves from dynamite (or maybe an extra-powerful Fourth of July firecracker), sonic "booms," which are actually the shock waves generated by supersonic aircraft, and water hammer, which can occur when a water faucet or valve is opened or closed rapidly.

Figure 10.6 Oblique shock in supersonic flow over a right circular cone. (Courtesy of Launch and Flight Division, Ballistic Research Laboratory/ARRADCOM, Aberdeen Proving Ground, MD.)

Figure 10.7 Curved shock standing ahead of a blunt body in supersonic flow. (Courtesy of Launch and Flight Division, Ballistic Research Laboratory/ARRADCOM, Aberdeen Proving Ground, MD.)

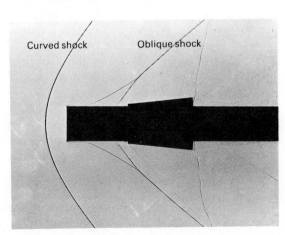

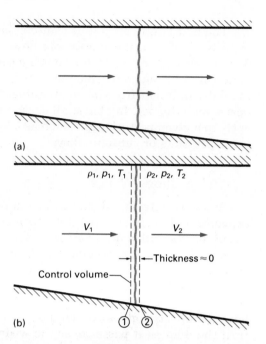

Figure 10.8 Normal shock in a duct; (a) Flow schematic; (b) flow view from control volume fixed to shock.

Shock waves may be moving, like the wave that results from rapidly closing a valve in a pipe, or stationary, as in flow over a stationary wedge or blunt body.* In either case, it is important to note that a shock wave adjusts the flow to some *downstream* condition. A *normal shock* is perpendicular to the inlet flow. An *oblique shock* is inclined at a constant angle to the inlet flow (Fig. 10.6). A *curved shock* has a variable angle between the wave and the inlet flow (Fig. 10.7). Flows that contain oblique or curved shocks are at least two-dimensional and two-directional because the shock changes the direction of the flow. One-dimensional, one-directional flow can contain only normal shocks. For the sake of simplicity, our consideration in this book will be limited to normal shocks. For the same reason, we will consider only shocks in an ideal gas.

Figure 10.8(a) shows a normal shock in a duct. The shock may be moving or stationary with respect to the duct. We can develop an analysis that applies to either case by using a control volume and coordinate system fixed to the shock. Velocities (and, therefore, stagnation properties) are defined relative to the shock. Figure 10.8(b) shows the flow into and out of the control volume. We assume that the velocity is uniform at the inlet and at the outlet. We assume that the shock has zero thickness, so

$$A_1 = A_2 = A,$$

and the volume of the control volume is zero.

---

* In reality, it's all relative. A wedge or blunt body moving in a still fluid would generate an identical shock to that generated by fluid flowing over a stationary body.

The continuity equation is

$$\rho_1 A V_1 = \rho_2 A V_2.$$ (10.34)

From the ideal gas law,

$$\rho = \frac{p}{RT}.$$ (10.35)

The velocity can be written as

$$V = \mathbf{M}c = \mathbf{M}\sqrt{kRT}.$$ (10.36)

Substituting into Eq. (10.34) and canceling A, k, and R, we obtain

$$\frac{p_1 \mathbf{M}_1}{\sqrt{T_1}} = \frac{p_2 \mathbf{M}_2}{\sqrt{T_2}},$$ (10.37a)

or

$$\frac{\mathbf{M}_2}{\mathbf{M}_1} = \frac{p_1}{p_2}\sqrt{\frac{T_2}{T_1}}.$$ (10.37b)

The linear momentum equation for the flow direction (x) is

$$\sum F_x = \rho_2 V_2^2 A - \rho_1 V_1^2 A.$$

Because the sides of the control volume in contact with the duct wall have zero thickness, there is no force on them. The only forces are due to pressure on the inlet and outlet planes, so the equation becomes

$$p_1 A - p_2 A = \rho_2 V_2^2 A - \rho_1 V_1^2 A.$$

From Eq. (10.29),

$$\rho V^2 = kp\mathbf{M}^2,$$

and the momentum equation can be reduced to

$$p_1(1 + k\mathbf{M}_1^2) = p_2(1 + k\mathbf{M}_2^2),$$ (10.38a)

or

$$\frac{p_2}{p_1} = \frac{1 + k\mathbf{M}_1^2}{1 + k\mathbf{M}_2^2}.$$ (10.38b)

Next we apply the general energy equation,

$$\dot{Q} - \dot{W}_s = \dot{m}\left(\tilde{h}_2 - \tilde{h}_1 + \frac{V_2^2}{2} - \frac{V_1^2}{2}\right).$$

Since the control volume has zero thickness, there can be neither heat transfer nor work, so the energy equation reduces to

$$\tilde{h}_2 + \frac{V_2^2}{2} = \tilde{h}_1 + \frac{V_1^2}{2},$$

or, alternatively, in terms of stagnation enthalpy,

$$\tilde{h}_{0_2} = \tilde{h}_{0_1}.$$ (10.39)

For an ideal gas, the energy equation gives

$$T_{0_2} = T_{0_1}.$$ (10.40)

Substituting Eq. (10.30) into Eq. (10.40), we have

$$T_2\left(1 + \frac{k-1}{2}\,\mathbf{M}_2^2\right) = T_1\left(1 + \frac{k-1}{2}\,\mathbf{M}_1^2\right), \tag{10.41a}$$

or

$$\frac{T_2}{T_1} = \frac{1 + \dfrac{k-1}{2}\,\mathbf{M}_1^2}{1 + \dfrac{k-1}{2}\,\mathbf{M}_2^2}. \tag{10.41b}$$

Equations (10.37b), (10.38b), and (10.41b) relate the ratio of temperature and pressure across a shock to the Mach numbers upstream and downstream of the shock. If we select a single parameter, say inlet Mach number $\mathbf{M}_1$, we can express the other three parameters in terms of the selected parameter. To do this, we square Eq. (10.37b):

$$\frac{\mathbf{M}_2^2}{\mathbf{M}_1^2} = \left(\frac{p_1}{p_2}\right)^2\left(\frac{T_2}{T_1}\right).$$

Then we substitute Eqs. (10.38b) and (10.41b) into the right side. The result is a quadratic equation relating $\mathbf{M}_2^2$ to $\mathbf{M}_1^2$; k appears in the equation as a parameter. Because the equation is quadratic, there are two solutions for $\mathbf{M}_2^2$:

$$\mathbf{M}_2^2 = \mathbf{M}_1^2 \tag{10.42a}$$

and

$$\mathbf{M}_2^2 = \frac{\mathbf{M}_1^2 + \dfrac{2}{k-1}}{\dfrac{2k}{k-1}\,\mathbf{M}_1^2 - 1}. \tag{10.42b}$$

The first solution seems trivial, but it expresses the important fact that it is not mandatory that the Mach number change discontinuously at a plane. This solution is valid if there is no shock in the control volume. If there is a shock in the control volume, the second solution, Eq. (10.42b), gives the downstream Mach number in terms of the upstream Mach number. For any realistic value of k (that is, $k > 1$), calculations with Eq. (10.42b) show that $\mathbf{M}_1$ and $\mathbf{M}_2$ always lie on opposite sides of $\mathbf{M} = 1.0$; that is, if $\mathbf{M}_1 > 1$, $\mathbf{M}_2 < 1$ and vice versa.

Equations (10.42a) and (10.42b) can now be substituted into the pressure ratio and temperature ratio equations (10.38b) and (10.41b). The constant Mach number solution gives pressure and temperature ratios equal to unity; that is, there can be no discontinuous change of pressure and temperature at a plane if there is no shock at the plane. If there is a shock, the pressure and temperature ratios are

$$\frac{p_2}{p_1} = \frac{2k}{k+1}\,\mathbf{M}_1^2 - \frac{k-1}{k+1} \tag{10.43}$$

and

$$\frac{T_2}{T_1} = \left(1 + \frac{k-1}{2}\,\mathbf{M}_1^2\right)\left(\frac{2k}{k-1}\,\mathbf{M}_1^2 - 1\right)\left(\frac{2(k-1)}{(k+1)^2\mathbf{M}_1^2}\right). \tag{10.44}$$

A shock may change the stagnation properties as well as the static properties. According to Eq. (10.40), the stagnation temperature ratio across a shock is*

$$\frac{T_{0_2}}{T_{0_1}} = 1. \tag{10.45}$$

The stagnation pressure ratio across a shock is obtained by first expressing stagnation pressure in terms of static pressure and Mach number. Using Eq. (10.31), we have

$$\frac{p_{0_2}}{p_{0_1}} = \frac{p_2 \left(1 + \dfrac{k-1}{2} \mathbf{M}_2^2\right)^{k/(k-1)}}{p_1 \left(1 + \dfrac{k-1}{2} \mathbf{M}_1^2\right)^{k/(k-1)}}.$$

We can reduce this to an equation involving only $\mathbf{M}_1^2$ and k by substituting Eq. (10.43) for the static pressure ratio and Eq. (10.42b) for $\mathbf{M}_2^2$. After some heavy algebra, we find

$$\frac{p_{0_2}}{p_{0_1}} = \left(\frac{k+1}{2} \mathbf{M}_1^2\right)^{k/(k-1)} \left(1 + \frac{k-1}{2} \mathbf{M}_1^2\right)^{-k/(k-1)} \left(\frac{2k}{k+1} \mathbf{M}_1^2 - \frac{k-1}{k+1}\right)^{-1/(k-1)} \tag{10.46}$$

Equations (10.42b)–(10.46) enable us to calculate properties on the downstream side of a shock if we know the properties on the upstream side and the inlet Mach number $\mathbf{M}_1$. Since the equations are algebraically complicated, it is difficult to determine the effects of a shock on the fluid by examining them. Table 10.1 indicates typical results of calculations made with the equations. Consider the second column, which

Table 10.1 Relative changes of properties as predicted by the normal shock equations

	$\mathbf{M}_1 > 1$ (Supersonic Inlet)	$\mathbf{M}_1 < 1$ (Subsonic Inlet)
M_2	<1 (subsonic)	>1 (supersonic)
$\dfrac{p_2}{p_1}$	>1	<1
$\dfrac{T_2}{T_1}$	>1	<1
$\dfrac{T_{0_2}}{T_{0_1}}$	$=1$	$=1$
$\dfrac{p_{0_2}}{p_{0_1}}$	<1	>1
$\tilde{s}_2 - \tilde{s}_1$	>0	<0

* Recall in both Eq. (10.45) and Eq. (10.46) that these are stagnation properties relative to the shock.

tabulates results for a (hypothetical) shock that occurs in a subsonic flow. Our discussion of the different nature of subsonic and supersonic flow indicates that discontinuous changes of fluid properties (like shock waves) do not occur in subsonic flow. Is there any way to decide, mathematically, whether the results in the right-hand column are realistic? Yes, we can use the second law of thermodynamics, in the mathematical form of Eq. (10.5):

$$\tilde{s}_2 - \tilde{s}_1 \geq \int_1^2 \frac{dq}{T}.$$

Since there is no heat transfer to or from the gas as it passes through a shock wave, it is necessary that

$$\tilde{s}_2 - \tilde{s}_1 \geq 0.$$

Entropy change across a shock wave can be calculated by using Eq. (10.8). It is convenient to work with a dimensionless entropy change. Dividing by R and using Eq. (10.10) give

$$\frac{\tilde{s}_2 - \tilde{s}_1}{R} = \frac{k}{k-1} \ln\left(\frac{T_2}{T_1}\right) - \ln\left(\frac{p_2}{p_1}\right).$$

Substituting Eqs. (10.43) and (10.44) for the pressure and temperature ratios yields an expression for entropy change in terms of Mach number. By selecting various values of $\mathbf{M}_1$, $(\tilde{s}_2 - \tilde{s}_1)$ can be determined. The entropy change is negative for subsonic inlet Mach numbers; therefore, no such "shock wave" can exist.* Shock waves can exist only if the inlet flow (relative to the shock) is supersonic. Real shock waves will always increase the static pressure and temperature and decrease the stagnation pressure (relative to the shock).

If you are analyzing a supersonic flow and need to calculate property changes across a shock wave, you can use Eqs. (10.42b)–(10.46); however, even with a calculator, these equations are quite tedious. This is especially true if you know, say, T_2/T_1 and are trying to calculate Mach number. Tables are helpful in such cases. Table E.2 in Appendix E presents numerical values of the normal shock functions for $k = 1.4$. Note that the tables are prepared with $\mathbf{M}_1$ as the independent parameter. If you want either density or velocity change across a shock, it is easily calculated from the ideal gas law and the continuity equation.

---

* It might be a good idea to cross out the right-hand column of Table 10.1.

EXAMPLE 10.3 Illustrates use of normal shock tables

A stream of nitrogen flowing in a duct passes through a normal shock. A Pitot tube and thermometer are placed downstream of the shock and a Bourdon tube pressure gage is attached to a wall tap upstream of the shock, as shown in Fig. E10.3. The pressure and temperature readings are $p_A = 2$ lb/in², $p_B = 40$ lb/in², and $T = 100°$F. Calculate the Mach number and velocity of the nitrogen upstream of the shock.

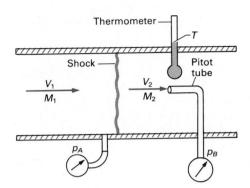

Figure E10.3

Solution

Given

Nitrogen flowing through a shock

Measurements made as indicated in Fig. E10.3

Find

Mach number and velocity upstream of shock

Solution

The first step is to convert all given data to absolute units. We assume that the atmospheric pressure in the vicinity of the pressure gages is sea-level standard, 14.7 psia. Then

$$p_A = 2 \text{ lb/in}^2 + 14.7 \text{ lb/in}^2 = 16.7 \text{ psia},$$
$$p_B = 40 \text{ lb/in}^2 + 14.7 \text{ lb/in}^2 = 54.7 \text{ psia},$$
$$T = 100° \text{ F} + 459.7°\text{F} = 559.7°\text{R}$$

For nitrogen,

$$R = 55.2 \text{ ft} \cdot \text{lb/lbm} \cdot °\text{R} \quad \text{and} \quad k = 1.4,$$

from Table A.4.

Next we interpret the data. Since the flow is parallel to the duct and the streamlines are straight, p_A is the *static* pressure upstream of the shock; p_B is the *stagnation* pressure downstream of the shock. Since the gas comes to rest at the thermometer bulb, T approximates the stagnation temperature. From the given data, for the shock

$$\frac{p_1}{p_{0_2}} = \frac{16.7 \text{ psia}}{54.7 \text{ psia}} = 0.3053.$$

We would like to obtain the corresponding Mach number $\mathbf{M}_1$ but a check of Eqs. (10.42)–(10.46) and Table E.2 reveals that we do not have a function or table of p_1/p_{0_2}. Note that we *cannot* use the isentropic table (Table E.1) or Eq. (10.31) because the static and stagnation pressures lie on different sides of the shock. We can, however, write

$$\frac{p_1}{p_{0_2}} = \left(\frac{p_1}{p_{0_1}}\right)\left(\frac{p_{0_1}}{p_{0_2}}\right) = \frac{\left(\dfrac{p_1}{p_{0_1}}\right)}{\left(\dfrac{p_{0_2}}{p_{0_1}}\right)}.$$

We do have functions and tables for p_1/p_{0_1} and p_{0_2}/p_{0_1}. To find $\mathbf{M}_1$ we must see iteration. A simple procedure is as follows:

- Guess a trial value of $\mathbf{M}_1$.
- Obtain a value of p_{0_2}/p_{0_1} from Table E.2.
- Calculate p_1/p_{0_1} from

$$\frac{p_1}{p_{0_1}} = \left(\frac{p_1}{p_{0_2}}\right)\left(\frac{p_{0_2}}{p_{0_1}}\right) = 0.3053\left(\frac{p_{0_2}}{p_{0_1}}\right).$$

- Obtain a new value for $\mathbf{M}_1$ from Table E.1.
- Continue the process until $\mathbf{M}_1$ doesn't change.

We first guess* $\mathbf{M}_1 = 1.5$, giving

$$\frac{p_{0_2}}{p_{0_1}} = 0.9298.$$

The trial value of p_1/p_{0_1} is

$$\frac{p_1}{p_{0_1}} = (0.9298)(0.3053) = 0.2837.$$

This implies a value

$$\mathbf{M}_1 = 1.473.$$

This Mach number gives

$$\frac{p_{0_2}}{p_{0_1}} = 0.9381 \quad \text{and} \quad \frac{p_1}{p_{0_1}} = (0.3053)(0.9381) = 0.2864.$$

From Table E.1, the corresponding Mach number is

$$\mathbf{M}_1 = 1.466.$$

Repeating the process until convergence, we have

$$| \ \mathbf{M}_1 = 1.463. \ |$$

Answer ◀

The nitrogen velocity is calculated by

$$V_1 = \mathbf{M}_1 c_1 = \mathbf{M}_1\sqrt{kRT_1}.$$

T_1 is computed by

$$T_1 = T_{0_1}\left(\frac{T_1}{T_{0_1}}\right),$$

where T_1/T_{0_1} is found for $\mathbf{M}_1 = 1.463$ and is

$$\frac{T_1}{T_{0_1}} = 0.7002.$$

The stagnation temperature is 559.7°R since a shock does not change T_0. Thus

$$T_1 = 559.7°\text{R}(0.7002) = 391.9°\text{R}.$$

The nitrogen velocity is

$$V_1 = 1.463 \sqrt{(1.4)\left(55.2\,\frac{\text{ft}\cdot\text{lb}}{\text{lbm}\cdot°\text{R}}\right)(391.9°\text{R})\left(32.2\,\frac{\text{ft}\cdot\text{lbm}}{\text{lb}\cdot\text{sec}^2}\right)},$$

Answer ◀

$$| \ V_1 = 1440 \text{ ft/sec.} \ |$$

* Of course we guess $\mathbf{M}_1 > 1.0$!

Discussion

Take careful note of the comments made during the solution. Do you understand why the isentropic table cannot be used with p_1/p_{0_2}? Note that we do use the isentropic table to relate $\mathbf{M}_1$ and p_1/p_{0_1} even though the flow across the shock itself is not isentropic. Of course you know that we couldn't use Bernoulli's equation to calculate the nitrogen velocity, but you might try it to see how close you come to the correct answer.

Substituting the appropriate Mach number functions allows us to find a new function

$$\frac{p_1}{p_{0_2}} = \left(\frac{k+1}{2}\,\mathbf{M}_1^2\right)^{-k/(k-1)}\left(\frac{2k\mathbf{M}_1^2}{k+1} - \frac{k-1}{k+1}\right)^{-1/(k-1)}$$

This is called the supersonic Pitot-tube formula, since a Pitot tube will always generate a shock upstream of itself in supersonic flow. If you had a lot of supersonic Pitot-tube data to reduce, you might make a table of values of this function.

The temperature of the thermometer bulb is actually slightly less than the stagnation temperature of the gas. See the discussion following Example 10.10 and Chapter 7 of [1].

10.3 Isentropic Flow and Bernoulli's Equation

Although real flows are often influenced by viscosity and heat transfer, it is sometimes convenient to ignore these effects. The resulting flow model, in addition to its (relative) simplicity, is reasonably accurate for external flow outside of viscous boundary layers and for flow in ducts if the boundary layers on the duct walls are thin and rates of heat transfer to or from the fluid are small. The usefulness of an inviscid flow model is immediately apparent when we recall that Bernoulli's equation (Eq. 4.20 or 4.21) describes steady flow of an *incompressible* inviscid fluid. In this section, we will consider the compressible flow equivalent(s) of Bernoulli's equation and demonstrate the inapplicability of Bernoulli's equation for high Mach number flow.

10.3.1 Steady Isentropic Flow of an Ideal Gas

Consider flow of a compressible fluid in a small streamtube. We make the following assumptions:

- The flow is steady.
- There are no shear stresses.
- There is no heat transfer.
- The streamtube does not pass through a shock wave.*

The last three assumptions taken together imply that the flow is adiabatic and reversible; therefore, the entropy of the fluid is constant:

$$\tilde{s} = \text{Constant};$$

---

* It is not necessary to assume that there are no shocks if the flow is subsonic everywhere along the streamtube because shocks are possible only if the inlet flow is supersonic.

or, in terms of two locations along the streamtube,

$$\tilde{s}_1 = \tilde{s}_2.$$

Flow with constant entropy is called *isentropic flow*.

For steady flow, the energy equation is

$$q - w_{\text{shaft}} - w_{\text{shear}} = \left(\tilde{h}_2 + \frac{V_2^2}{2}\right) - \left(\tilde{h}_1 + \frac{V_1^2}{2}\right),$$

where potential energy changes have been neglected, as is customary in compressible flow. According to our assumptions, q and w_{shear} are zero, but w_{shaft} is not necessarily zero and the energy equation becomes

$$- w_{\text{shaft}} = \left(\tilde{h}_2 + \frac{V_2^2}{2}\right) - \left(\tilde{h}_1 + \frac{V_1^2}{2}\right).$$

Using the definition of stagnation enthalpy,

$$- w_{\text{shaft}} = \tilde{h}_{0_2} - \tilde{h}_{0_1}.$$

If the fluid is an ideal gas,

$$- w_{\text{shaft}} = c_p(T_{0_2} - T_{0_1}). \tag{10.47}$$

Doing work on the fluid increases its *stagnation* temperature, even if there is no heat transfer.

Now we consider changes of stagnation pressure in isentropic flow. By definition of the stagnation state,

$$\tilde{s}_{0_1} = \tilde{s}_1 \qquad \text{and} \qquad \tilde{s}_{0_2} = \tilde{s}_2.$$

For isentropic flow,

$$\tilde{s}_{0_2} - \tilde{s}_{0_1} = \tilde{s}_2 - \tilde{s}_1 = 0;$$

thus

$$\frac{p_{0_2}}{p_{0_1}} = \left(\frac{T_{0_2}}{T_{0_1}}\right)^{c_p/R} = \left(\frac{T_{0_2}}{T_{0_1}}\right)^{k/(k-1)}. \tag{10.48}$$

Using Eq. (10.47), we can also write

$$\frac{p_{0_2}}{p_{0_1}} = \left(1 - \frac{w_{\text{shaft}}}{c_p T_{0_1}}\right)^{k/(k-1)}. \tag{10.49}$$

A situation of considerable interest in compressible flow is the case where no shaft work is transferred to or from the fluid. The following statement applies to this case:

In isentropic flow with no shaft work, all stagnation properties are constant.

Note that stagnation properties are constant in isentropic flow only if there is no shaft work. This special case ($w_{\text{shaft}} = 0$) is so important in compressible flow that some engineers and authors use the phrase "isentropic flow" to mean "isentropic flow with no work." We will use the phrase "isoenergetic-isentropic flow" to refer to isentropic flow with no work.*

---

* Isoenergetic means that there is no energy transfer (neither heat nor work) to the fluid from outside agencies.

Equations relating property and velocity changes in isoenergetic-isentropic flow of an ideal gas can easily be developed from the stagnation property definitions and the constancy of stagnation properties. Consider the calculation of the change in temperature between two points in an isoenergetic-isentropic flow. Since T_0 is constant, we can write

$$T_{0_1} = T_1 + \frac{V_1^2}{2c_p} = T_2 + \frac{V_2^2}{2c_p} = T_{0_2}. \tag{10.50a}$$

It is sometimes convenient to use Mach number instead of velocity. From Eq. (10.30),

$$\frac{T_2}{T_1} = \frac{T_{0_2}\left(1 + \dfrac{k-1}{2}\mathbf{M}_1^2\right)}{T_{0_1}\left(1 + \dfrac{k-1}{2}\mathbf{M}_2^2\right)},$$

but, since T_0 is constant,

$$\frac{T_2}{T_1} = \frac{1 + \dfrac{k-1}{2}\mathbf{M}_1^2}{1 + \dfrac{k-1}{2}\mathbf{M}_2^2}. \tag{10.50b}$$

A third approach uses the tables relating stagnation property ratios and Mach number. We can calculate the temperature ratio by

$$\frac{T_2}{T_1} = \frac{\dfrac{T}{T_0}(\mathbf{M}_2)}{\dfrac{T}{T_0}(\mathbf{M}_1)}, \tag{10.50c}$$

where

$$\frac{T}{T_0}(\mathbf{M})$$

is a numerical value read from Table E.1 at Mach number $\mathbf{M}$. A fourth method makes the calculation in two steps;

$$T_{0_2} = T_{0_1} = \frac{T_1}{\dfrac{T}{T_0}(\mathbf{M}_1)}$$

and

$$T_2 = T_{0_2}\left[\frac{T}{T_0}(\mathbf{M}_2)\right]. \tag{10.50d}$$

Any of Eqs. (10.50) could be used to relate temperature and velocities (or Mach numbers) at points 1 and 2. Which equation is most convenient depends on the exact information available. It is easy to fall into the trap of always using Mach number tables when direct calculations are sometimes easier. Similarly, pressure and velocity (or Mach number) in isoenergetic-isentropic flow can be related by any of the following equations:

$$p_1\left(1 + \frac{V_1^2}{2c_p T_1}\right)^{k/(k-1)} = p_2\left(1 + \frac{V_2^2}{2c_p T_2}\right)^{k/(k-1)}, \tag{10.51a}$$

$$\frac{p_2}{p_1} = \frac{\left(1 + \dfrac{k-1}{2}\mathbf{M}_1^2\right)^{k/(k-1)}}{\left(1 + \dfrac{k-1}{2}\mathbf{M}_1^2\right)^{k/(k-1)}}, \tag{10.51b}$$

$$\frac{p_2}{p_1} = \frac{\dfrac{p}{p_0}\,\{\mathbf{M}_2\}}{\dfrac{p}{p_0}\,\{\mathbf{M}_1\}}, \tag{10.51c}$$

and

$$p_2 = p_{0_2}\frac{p}{p_0}\,\{\mathbf{M}_2\}, \tag{10.51d}$$

where

$$p_{0_2} = p_{0_1} = \frac{p_1}{\dfrac{p}{p_0}\,\{\mathbf{M}_1\}}.$$

Don't forget that you can also use Eqs. (10.11)–(10.13) because the flow is isentropic.

EXAMPLE 10.4 Illustrates "routine" use of isentropic flow functions and tables for computation of isoenergetic-isentropic flow

Air flows in a certain duct. At one point in the flow, the pressure, temperature, and velocity are 50 psia, 100°F, and 500 ft/sec. At a downstream point, the Mach number is 1.80. The flow is steady, isoenergetic, and isentropic. Calculate the air pressure, temperature, and velocity at the downstream point. Do the calculation twice, once using isentropic flow equations and once using Table E.1.

Solution

Given
Air flow in a duct
At point 1, $p_1 = 50$ psia, $T_1 = 100°F$, and $V_1 = 500$ ft/sec
At a downstream point (point 2), $\mathbf{M}_2 = 1.80$
Figure E10.4

Figure E10.4 Flow between two points in a duct

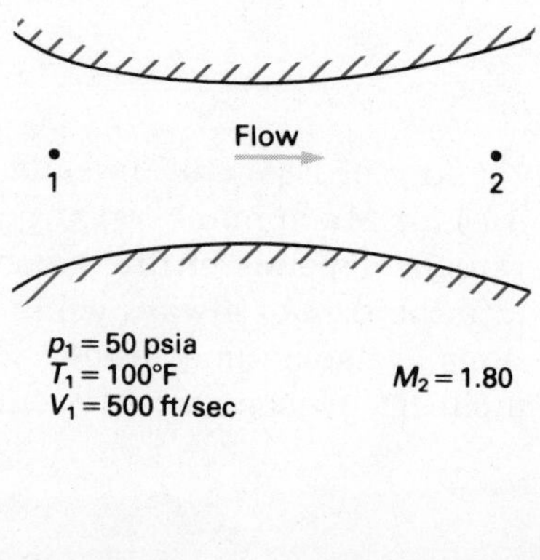

$p_1 = 50$ psia
$T_1 = 100°F$
$V_1 = 500$ ft/sec

$M_2 = 1.80$

Find

T_2, p_2, V_2 using both isentropic flow equations and Table E.1

Solution

The first step is to convert all given data into absolute units. The only conversion necessary is temperature;

$$T_1 = 100°F + 459.7°F = 559.7°R.$$

We first do the calculations with isentropic flow equations. The energy equation for adiabatic flow with no work between 1 and 2 is

$$c_p T_1 + \frac{V_1^2}{2} = c_p T_2 + \frac{V_2^2}{2}.$$

The velocity at 2 is

$$V_2 = M_2 c_2 = M_2 \sqrt{kRT_2}, \quad \text{so} \quad c_p T_1 + \frac{V_1^2}{2} = \left(c_p + \frac{kRM_2^2}{2} \right) T_2,$$

or

$$T_2 = \frac{c_p T_1 + \dfrac{V_1^2}{2}}{c_p + \dfrac{kRM_2^2}{2}}.$$

Substituting numbers, we have

$$T_2 = \frac{\left(186.7 \dfrac{ft \cdot lb}{lbm \cdot °R} \right)(559.7°R) + \dfrac{(500)^2 \dfrac{ft^2}{sec^2}}{2 \left(32.2 \dfrac{ft \cdot lbm}{lb \cdot sec^2} \right)}}{186.7 \dfrac{ft \cdot lb}{lbm \cdot °R} + \dfrac{1.4 \left(53.35 \dfrac{ft \cdot lb}{lbm \cdot R} \right)(1.80)^2}{2}},$$

Answer ◀

$$\boxed{T_2 = 352.2°R.}$$

The velocity at 2 is

$$V_2 = M_2 c_2 = [(1.80)(49.02)\sqrt{352.2}] \text{ ft/sec},$$

Answer ◀

$$\boxed{V_2 = 1660 \text{ ft/sec.}}$$

The pressure at 2 is computed from the isentropic process equation, Eq. (10.11):

$$p_2 = p_1 \left(\frac{T_2}{T_1} \right)^{k/(k-1)} = 50 \text{ psia} \left(\frac{352.2}{559.7} \right)^{3.5},$$

Answer ◀

$$\boxed{p_2 = 9.88 \text{ psia.}}$$

Now we repeat the calculation using Table E.1. The Mach number at 1 is

$$M_1 = \frac{V_1}{c_1} = \frac{V_1}{\sqrt{kRT_1}} = \frac{500 \text{ ft/sec}}{(49.02\sqrt{559.7}) \text{ ft/sec}} = 0.43.$$

Next we use Table E.1 to obtain the pressure and temperature ratio functions at 1 and at 2. At $M = M_1 = 0.43$,

$$\frac{p}{p_0} = 0.8806 \quad \text{and} \quad \frac{T}{T_0} = 0.9643.$$

At $\mathbf{M} = \mathbf{M}_2 = 1.80$,

$$\frac{p}{p_0} = 0.1740 \quad \text{and} \quad \frac{T}{T_0} = 0.6068.$$

Since the flow is isoenergetic and isentropic,

$$p_2 = p_1 \frac{\dfrac{p}{p_0}\,\cancel{(\mathbf{M}_2)}}{\dfrac{p}{p_0}\,\cancel{(\mathbf{M}_1)}} = 50 \text{ psia} \left(\frac{0.1740}{0.8806}\right),$$

$$\boxed{\; p_2 = 9.88 \text{ psia} \;}$$

Answer ◀

Now

$$T_2 = T_1 \frac{\dfrac{T}{T_0}\,\cancel{(\mathbf{M}_2)}}{\dfrac{T}{T_0}\,\cancel{(\mathbf{M}_1)}} = 559.7°\text{R}\left(\frac{0.6068}{0.9643}\right),$$

$$\boxed{\; T_2 = 352.2°\text{R.} \;}$$

Answer ◀

The velocity at 2 is

$$V_2 = \mathbf{M}_2\sqrt{kRT_2} = (1.80)(49.02)\sqrt{352.2} \text{ ft/sec.}$$

$$\boxed{\; V_2 = 1660 \text{ ft/sec.} \;}$$

Answer ◀

Discussion

Of course the answers determined by using Table E.1 are the same as those determined by using equations. We will leave the decision to you as to which method is easier.

Always convert pressures and temperatures to absolute units as the first step in solving a compressible flow problem.

10.3.2 Relation to Bernoulli's Equation and a Criterion for Incompressible Flow

One of the most widely used (and misused) equations in fluid mechanics is Bernoulli's equation,

$$\frac{p}{\rho} + \frac{V^2}{2} + gz = \text{Constant.} \tag{4.19}$$

The assumptions used in deriving Bernoulli's equation are similar to those we just used in deriving the isoenergetic-isentropic flow equations except that while Bernoulli's equation assumes incompressible flow, the isoenergetic-isentropic equations assume adiabatic flow of an ideal gas with no shocks. In this section, we will show that, in the limit of very small Mach number, the isoenergetic-isentropic equation for pressure becomes identical to Bernoulli's equation. As an added benefit we will develop a criterion that can be used to decide if a gas flow can be treated as incompressible.

Consider a steady flow with no shear stress, shaft work, or heat transfer. Under these conditions the stagnation pressure is constant.

We also assume that elevation changes are negligible. If the fluid is incompressible, the pressure at any location can be calculated from the "pressure form" of Bernoulli's equation:

$$p = p_0 - \frac{1}{2}\rho V^2. \qquad \text{Incompressible flow} \qquad \textbf{(10.52)}$$

If the fluid is compressible and an ideal gas, the static pressure and stagnation pressure are related by

$$p_0 = p\left(1 + \frac{k-1}{2}\mathbf{M}^2\right)^{k/(k-1)}. \qquad \text{Compressible flow} \qquad \textbf{(10.53)}$$

If we restrict consideration to "small" Mach numbers (say, less than 1), we can expand the Mach number term into an infinite series using the binomial theorem:

$$(1+x)^n = 1 + nx + \frac{n(n-1)}{2!}x^2 + \frac{n(n-1)(n-2)}{3!}x^3 + \cdots. \qquad \textbf{(10.54)}$$

If n is a positive integer, the series is finite. If n is not a positive integer, the series is infinite. If x is less than 1, the infinite series is convergent. Putting

$$x = \frac{k-1}{2}\mathbf{M}^2 \qquad \text{and} \qquad n = \frac{k}{k-1},$$

we get

$$\left(1 + \frac{k-1}{2}\mathbf{M}^2\right)^{k/(k-1)} = 1 + \frac{k}{2}\mathbf{M}^2 + \frac{k}{8}\mathbf{M}^4 + \text{H.O.T.,}$$

where H.O.T. involves higher *even* powers of $\mathbf{M}$. Substituting into Eq. (10.53) gives

$$p_0 = p\left(1 + \frac{k\mathbf{M}^2}{2} + \frac{k\mathbf{M}^4}{8} + \text{H.O.T.}\right).$$

Since we are interested in small values of $\mathbf{M}$, the higher-order terms are negligible, and the equation can be written

$$p_0 \approx p + \frac{kp\mathbf{M}^2}{2}\left(1 + \frac{\mathbf{M}^2}{4}\right). \qquad \textbf{(10-55)}$$

From Eq. (10.29), we have

$$kp\mathbf{M}^2 = \rho V^2.$$

Substituting and rearranging, we get the following equation for the pressure:

$$p \approx p_0 - \frac{1}{2}\rho V^2\left(1 + \frac{\mathbf{M}^2}{4}\right). \qquad \textbf{(10.56)}$$

If the Mach number is small (but nonzero), then $\mathbf{M}^2/4$ is small compared to 1, and we can write

$$p \approx p_0 - \frac{1}{2}\rho V^2; \qquad \text{Compressible flow, small } \mathbf{M} \qquad \textbf{(10.57)}$$

therefore Bernoulli's equation is an approximation to the isoenergetic-isentropic flow pressure relation for small Mach number. The accuracy of this approximation is dependent on the "smallness' of the Mach number. Equation (10.56) shows us that the error is proportional to $\mathbf{M}^2/4$ at low Mach numbers. If we wish to limit the error in using Bernoulli's equation for calculation of pressure to no more than, say, 2%, then

$$\mathbf{M} \leq \sqrt{4(0.02)} = 0.283.$$

There is nothing unique about 2% error. For rough estimates, 5% error may be acceptable, in which case the Mach number must be less than about 0.45. The most widely used criterion for the boundary between compressible and incompressible flow puts the threshold Mach number at 0.3:

Generally, a flow with $\mathbf{M} < 0.3$ everywhere can be assumed to be incompressible.

10.3.3 The Critical State

We have seen that the nature of a compressible flow changes drastically when the fluid velocity exceeds the speed of sound. The flow state that corresponds to exactly sonic flow is called the *critical state*. The fluid properties at the critical state are called the *critical properties*. Consider acceleration of a gas from zero velocity to the local sonic speed in an isoenergetic-isentropic process. As the fluid's velocity increases, its temperature and speed of sound decrease.[†] Since the stagnation temperature is constant for this process, we can write

$$T_0 = T\left(1 + \frac{k-1}{2}\mathbf{M}^2\right).$$

At the critical state, the Mach number is unity. Using the superscript * to denote conditions at the critical state, we have

$$T_0 = T^*\left(1 + \frac{k-1}{2}(1)^2\right),$$

or

$$\frac{T^*}{T_0} = \frac{2}{k+1}. \tag{10.58}$$

Since we have specified an isoenergetic-isentropic process, the pressure and density at the critical state are given by

$$\frac{p^*}{p_0} = \left(\frac{T^*}{T_0}\right)^{k/(k-1)} = \left(\frac{2}{k+1}\right)^{k/(k-1)} \tag{10.59}$$

and

$$\frac{\rho^*}{\rho_0} = \left(\frac{T^*}{T_0}\right)^{1/(k-1)} = \left(\frac{2}{k+1}\right)^{1/(k-1)}. \tag{10.60}$$

[†] Note that $c/c_0 = (T/T_0)^{1/2}$ for an ideal gas.

For an isoenergtic-isentropic flow, properties at the critical state are uniquely determined by the stagnation properties and k. The critical properties T^*, p^*, and ρ^* will be constant if the stagnation properties are constant.

10.4 Internal Flow: One-Dimensional Flow in a Variable Passage Area

When a fluid, compressible or otherwise, flows through a passage, the fluid motion is affected by changes in the passage cross-sectional area, friction at the walls, and heat transfer or work. Simultaneous consideration of all these effects is difficult. In this section we consider internal flow of a compressible fluid, and we assume that there is no friction, no work, and no heat transfer; we consider only the effects of area change.[†] We also assume that the flow is one-dimensional and one-directional.

The primary effect of area change is fluid acceleration or deceleration. A device for accomplishing fluid acceleration is called a *nozzle,* and a device for accomplishing fluid deceleration is called a *diffuser.* The assumptions of one-dimensional flow and negligible friction are much more realistic for nozzles than for diffusers, so we will concentrate on nozzle flow. We will assume steady or quasi-steady flow.

10.4.1 Isoenergetic-Isentropic Flow with Area Change: Preliminary Considerations

If there is no heat transfer or work, flow in a variable-area passage will be isoenergetic. If there are no shocks or friction, the flow will also be isentropic. Without limiting consideration to a specific type of fluid (such as an ideal gas), the fluid properties and velocity must satisfy the following equations:

■ Continuity equation:

$$\rho V A = \dot{m} = \text{Constant}$$

■ Energy equation:

$$\tilde{h} + \frac{V^2}{2} = \tilde{h}_0 = \text{Constant},$$

■ Property relation (Gibbs equation, Eq. 10.6, for isentropic flow):

$$d\tilde{h} - \frac{dp}{\rho} = T \, d\tilde{s} = 0.$$

Since the Gibbs equation is a differential equation, it is convenient to work with differential forms of the other equations. Taking the logarithm of the continuity equation and differentiating, we obtain

$$\frac{d\rho}{\rho} + \frac{dV}{V} + \frac{dA}{A} = 0. \tag{10.61}$$

[†] The film *Channel Flow of a Compressible Fluid* [10] provides an excellent introduction to this topic and is highly recommended.

Differentiating the energy equation gives

$$d\tilde{h} + V\,dV = 0.\tag{10.62}$$

With three equations, we can eliminate two variables. From the Gibbs equation,

$$d\tilde{h} = \frac{dp}{\rho}.$$

Substituting into Eq. (10.62), we have[†]

$$\frac{dp}{\rho} + V\,dV = 0.\tag{10.63}$$

Since the process is isentropic, we can write (using Eq. 10.24)

$$dp = \frac{\partial p}{\partial \rho}\bigg)_s d\rho = c^2\,d\rho,$$

and Eq. (10.63) becomes

$$c^2\left(\frac{d\rho}{\rho}\right) + V\,dV = 0.\tag{10.64}$$

We now solve the differential continuity equation for $d\rho/\rho$ and substitute into Eq. (10.64) to obtain

$$\left(\frac{V^2}{c^2} - 1\right)\frac{dV}{V} = \frac{dA}{A},$$

or, using the definition of Mach number,

$$\frac{dV}{V} = \left(\frac{1}{\mathbf{M}^2 - 1}\right)\frac{dA}{A}.\tag{10.65}$$

We can learn a great deal about the effect of area changes on the flow by examining the algebraic relation between dV and dA. The effect on fluid properties can be determined by noting that, according to Eqs. (10.62)–(10.64), changes of enthalpy, pressure, and density must have the opposite algebraic sign of velocity changes; when velocity increases, enthalpy, pressure, and density decrease.

Equation (10.65) reveals that the relation between velocity change and area change depends on the Mach number and in fact *changes sign* at $\mathbf{M} = 1$. Consider flow in a converging passage ($dA < 0$). If the flow is subsonic ($\mathbf{M} < 1$), the velocity increases, and hence the enthalpy, pressure, and density decrease. This behavior is familiar from consideration of incompressible flow. If the flow is supersonic ($\mathbf{M} > 1$), a converging passage causes the velocity to *decrease* (with corresponding increases in enthalpy, pressure, and density). A divergent passage ($dA > 0$) has the opposite effect, decelerating a subsonic flow and accelerating a supersonic flow. These observations are summarized in Table 10.2.

Since the speed of sound in a compressible fluid is determined by the fluid's thermodynamic properties, speed of sound decreases when

---

[†] Note that Eq. (10.63) is equivalent to the momentum equation (Euler's equation) for one-dimensional flow (Section 4.1.3)

Table 10.2 Effects of area change in isoenergetic-isentropic compressible flow

		M < 1	**M > 1**
Converging passage	Velocity (V)	Increases	Decreases
	Enthalpy ($\tilde{h}$)	Decreases	Increases
	Pressure (p)	Decreases	Increases
	Density (ρ)	Decreases	Increases
$dA < 0$	Passage acts as a	Nozzle	Diffuser
Diverging passage	Velocity (V)	Decreases	Increases
	Enthalpy ($\tilde{h}$)	Increases	Decreases
	Pressure (p)	Increases	Decreases
	Density (ρ)	Increases	Decreases
$dA > 0$	Passage acts as a	Diffuser	Nozzle

the thermodynamic properties decrease and increases when the thermo-dynamic properties increase. This means that in isoenergetic-isentropic flow, an increase in velocity always corresponds to a Mach number increase and vice versa. Combining this information with the information on velocity change, we conclude that a *converging* passage always drives the Mach number *toward* unity and a *diverging* passage always drives the Mach number *away* from unity. This implies that the Mach number cannot pass through unity (either accelerating or decelerating) in a passage that is only convergent or only divergent. If we consider the process of accelerating a gas from rest or a low subsonic Mach number in a simply converging passage, the *maximum* possible Mach number that can be achieved is unity. This maximum value can occur *only* at the end of the converging passage.

Suppose that the Mach number is unity at some point in a flow. Since infinite acceleration ($dV \to \infty$) is not possible unless there is a shock wave, which is specifically excluded by the assumptions, Eq. (10.65) implies that

$$dA = 0 \quad \text{when } \mathbf{M} = 1.$$

In isoenergetic-isentropic flow, sonic flow ($\mathbf{M} = 1$) can occur only at an area minimum (Fig. 10.9). The occurrence of sonic flow at an area maximum is excluded by the considerations in the preceding paragraph. To accelerate a fluid from rest (or a subsonic Mach number) to a supersonic Mach number, a *converging-diverging nozzle* is required.

Finally, suppose that a compressible fluid is flowing in a passage that actually has an area minimum (called a *throat*). Is it always possible to conclude that the Mach number is unity at the throat? To investigate this question, Eq. (10.65) is rearranged as follows:

$$\frac{dA}{A} = (\mathbf{M}^2 - 1)\frac{dV}{V}.$$

If dA is zero, then two options are possible:

$$\mathbf{M} = 1 \quad \text{or} \quad dV = 0.$$

Simply knowing that $dA = 0$ does not allow us to determine which

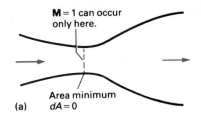

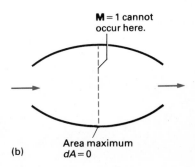

Figure 10.9 Passages with $dA = 0$.

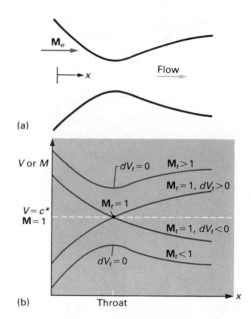

Figure 10.10 Possible velocity (or Mach number) distributions in passage with an area minimum.

option occurs. Figure 10.10 illustrates possible velocity (or Mach number) distributions in a passage with a minimum area. If the throat Mach number is not unity, then the velocity must have a local maximum or minimum at the throat ($dV_t = 0$). This behavior is observed in a Venturi in incompressible flow (Section 7.4.2 and Example 4.4). If the throat Mach number is 1, dV_t can be either positive or negative and the velocity can either increase or decrease downstream of the throat, depending on downstream conditions.

We summarize our findings about isoenergetic-isentropic flow:

- The response of the flow to a specific type of area change is exactly opposite for subsonic and supersonic flow. Details are given in Table 10.2.
- Sonic flow ($\mathbf{M} = 1$) can occur only at a minimum area. Minimum areas occur at the inlet of a simply diverging passage, the outlet of a simply converging passage, and the throat of a converging-diverging passage.
- It is possible, but not necessary, that the Mach number at the throat of a converging-diverging passage be equal to 1. If the Mach number at a throat is not 1, the velocity must pass through a maximum or minimum. If the throat Mach number is 1, the fluid may either accelerate or decelerate downstream of the throat.

Since we used a completely general property relation (Gibbs equation), these conclusions are valid for one-dimensional isoenergetic-isentropic compressible flow of any fluid (ideal gases included).

10.4.2 Isoenergetic-Isentropic Flow of an Ideal Gas

By assuming that the fluid is an ideal gas, we can obtain working equations for flow in variable-area ducts. Figure 10.11 illustrates the general flow geometry. The typical problem involves relating changes of fluid properties and velocity to the passage area. Calculation of mass flow rate or force on the duct walls is also important. Since we are assuming that the flow is isoenergetic and isentropic, Eqs. (10.50a)–(10.51d) and Table E.1 in Appendix E can be used to relate pressure, temperature, velocity, and Mach number for flow between any two planes in the duct. All we need for a complete analysis are equations relating, say, Mach number to passage area and mass flow rate to Mach number, flow properties, and passage area.

Area–Mach Number Relation.   To obtain an equation relating area and Mach number we use the continuity equation for a control volume with inflow at plane 1 and outflow at plane 2 (Fig. 10.11):

$$\rho_1 V_1 A_1 = \rho_2 V_2 A_2.$$

For an ideal gas,

$$\rho = \frac{p}{RT} \qquad \text{and} \qquad V = \mathbf{M}c = \mathbf{M}\sqrt{kRT}.$$

Substituting these equations into the continuity equation gives

$$\sqrt{\frac{k}{R}}\left(\frac{p_1 A_1 \mathbf{M}_1}{\sqrt{T_1}}\right) = \sqrt{\frac{k}{R}}\left(\frac{p_2 A_2 \mathbf{M}_2}{\sqrt{T_2}}\right).$$

Canceling k and R and solving for the area ratio, we have

$$\frac{A_2}{A_1} = \frac{\mathbf{M}_1}{\mathbf{M}_2}\left(\frac{p_2}{p_1}\right)^{-1}\left(\frac{T_2}{T_1}\right)^{1/2}.$$

Since the flow is isoenergetic and isentropic, Eqs. (10.50b) and (10.51b) can be substituted for the temperature and pressure ratios, giving

$$\frac{A_2}{A_1} = \frac{\mathbf{M}_1}{\mathbf{M}_2}\left[\frac{1 + \dfrac{k-1}{2}\mathbf{M}_2^2}{1 + \dfrac{k-1}{2}\mathbf{M}_1^2}\right]^{(k+1)/2(k-1)}. \tag{10.66}$$

If any three of the quantities A_1, $\mathbf{M}_1$, A_2, and $\mathbf{M}_2$ are specified, the fourth can be calculated. Usually A_1 and $\mathbf{M}_1$ are known and either $\mathbf{M}_2$ or A_2 is to be found. If $\mathbf{M}_2$ is given and A_2 is to be calculated, the calculation is simple; however, if A_2 is given and $\mathbf{M}_2$ must be calculated, that's not so easy. We can make the calculations a lot easier by making a table. To tabulate area ratio, we select a reference area at which the Mach number is known. The obvious[†] choice is the area corresponding

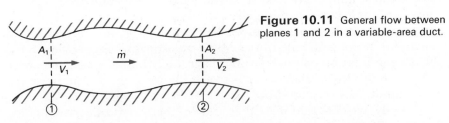

Figure 10.11 General flow between planes 1 and 2 in a variable-area duct.

[†] You might think that we would choose a "stagnation area" to correspond to the stagnation temperature and pressure. Do you know why this won't work?

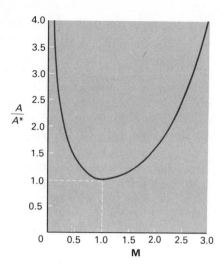

Figure 10.12 Area ratio function for $k = 1.4$.

to a Mach number of 1. We set $\mathbf{M}_1 = 1$, $A_1 = A^*$ and drop the subscript 2 to get

$$\frac{A}{A^*} = \frac{1}{\mathbf{M}} \left[\frac{2}{k+1} \left(1 + \frac{k-1}{2} \mathbf{M}^2 \right) \right]^{(k+1)/2(k-1)}. \tag{10.67}$$

In this equation, A is the area at the location where the Mach number is $\mathbf{M}$, and A^* (the critical area) is the area that would correspond to a Mach number of unity. Note that A^* is a reference area; it is not necessary for the actual area to be A^* anywhere in the flow. Values of A/A^* are included in the isentropic flow table (Table E.1). The area ratio function can be used to relate areas and Mach numbers at two planes in a duct by

$$\frac{A_2}{A_1} = \frac{\dfrac{A}{A^*} \langle \mathbf{M}_2 \rangle}{\dfrac{A}{A^*} \langle \mathbf{M}_1 \rangle}. \tag{10.68}$$

A plot of the area ratio function versus Mach number (Fig. 10.12) reveals some interesting information and confirms some of the observations that we made in the previous section. First, we note that A/A^* is always greater than unity except at $\mathbf{M} = 1$, where A/A^* is 1. This confirms that in isoenergetic-isentropic flow, the Mach number can be 1 only at the minimum area. Next, we note that if Mach number is given, there is a single corresponding area ratio so the problem of calculating the area required to obtain a given Mach number seems easy. We must be careful, however. If the initial Mach number ($\mathbf{M}_1$) is less than 1 and the desired final Mach number ($\mathbf{M}_2$) is greater than 1, a minimum area exactly equal to A^* must occur between planes 1 and 2; otherwise the flow cannot pass through $\mathbf{M} = 1$.

Finally, we note that if A/A^* is known, there are *two* corresponding Mach numbers. Choice of the proper Mach number for a known area ratio requires more information than just the area ratio itself. Generally, we must know the pressure or some other item of information besides the area ratio to obtain the Mach number. We can sometimes eliminate one of the possibilities. If we know that the flow is subsonic at plane 1 and that a minimum area exactly equal to A^* does not occur between planes 1 and 2, then the flow at 2 cannot be supersonic.

Before leaving this section on area ratio, we offer a few comments and warnings. The critical area A^* is the area necessary to accelerate or decelerate a given flow to sonic conditions. It is not necessary that A^* be identified with any *real* area in a flow passage. If we know that the flow is sonic somewhere in a passage, then the passage area equals the critical area at the sonic location. We also know that this must be the minimum area if the flow is isoenergetic-isentropic. The converse is not always true. If there is a minimum area in a flow passage, you *cannot* assume that the minimum area is equal to A^* without further evidence that the flow actually is sonic at the minimum area. Stated simply, in a passage with a throat, it is not true that the throat area (A_t) equals the critical area (A^*) unless other evidence indicates it.

EXAMPLE 10.5   Illustrates use of the area ratio function and the logic necessary for analysis of flow in a converging-diverging passage

The small laboratory wind tunnel shown in Fig. E10.5 draws air from the atmosphere. The pressure and temperature of the laboratory air are 100 kPa and 20°C. Schlieren photographs reveal a shock wave in the nozzle as shown in the figure. Calculate the air temperature and pressure at the throat and at planes 1, 2, and 3. Assume isoenergetic-isentropic flow except for the shock.

Solution

Given

Air flow in a laboratory wind tunnel
Shock wave at a specific location in the nozzle
Laboratory air at 100 kPa and 20°C
Areas given in Fig. E10.5
Isoenergetic-isentropic flow

Figure E10.5 Small laboratory wind tunnel.

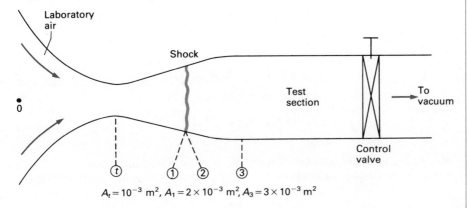

$A_t = 10^{-3} \ m^2, \ A_1 = 2 \times 10^{-3} \ m^2, \ A_3 = 3 \times 10^{-3} \ m^2$

Find

Air temperature and pressure at all areas shown

Solution

First convert the given air temperature to absolute units:

$$T_{\text{lab}} = 20°C + 273.2°C = 293.2K.$$

Since the passage is converging-diverging, we must be alert to the possibilities of sonic flow at the throat and supersonic flow in the diverging section. We note that there is a shock in the diverging section of the nozzle and conclude that the flow is supersonic at 1. Since the flow is subsonic ($\mathbf{M} \approx 0$) in the laboratory, the air must pass through $\mathbf{M} = 1$ somewhere upstream of plane 1. The *only* place this can occur is the throat, so

$$\mathbf{M}_t = 1.0.$$

Since the process is isentropic ahead of the shock, the stagnation properties are constant up to the shock and equal to the laboratory pressure and temperature (since $\mathbf{M} \approx 0$ in the ambient laboratory air). The critical area for the flow upstream of the shock is

$$A_{0 \to 1}^* = A_t.$$

The throat pressure and temperature are calculated with the aid of Table E.1:

$$p_t = p_0 \left(\frac{p}{p_0} (\mathbf{M}_t) \right) = 100 \text{ kPa}(0.5283),$$

$$| \ p_t = 52.8 \text{ kPa}; \ |$$ **Answer** ◀

$$T_t = T_0 \left(\frac{T}{T_0} (\mathbf{M}_t) \right) = 293.2K(0.8333),$$

$$| \ T_t = 244.3K. \ |$$ **Answer** ◀

The flow at plane 1, just ahead of the shock, has

$$A_1^* = A_t = 1.0 \times 10^{-3} \text{ m}^2.$$

The area ratio for plane 1 is

$$\left. \frac{A}{A^*} \right)_1 = \frac{2 \times 10^{-3} \text{ m}^2}{1 \times 10^{-3} \text{ m}^2} = 2.000.$$

From Table E.1, we find the *supersonic* Mach number corresponding to this area ratio:

$$\mathbf{M}_1 = 2.20.$$

The pressure and temperature ratios are (from Table E.1)

$$\frac{p_1}{p_{0_1}} = 0.0935 \quad \text{and} \quad \frac{T_1}{T_{0_1}} = 0.5081.$$

We now calculate

$$p_1 = 0.0935 p_{0_1} = 0.0935(100 \text{ kPa}),$$

$$| \ p_1 = 9.35 \text{ kPa}; \ |$$ **Answer** ◀

$$T_1 = 0.5081 T_{0_1} = 0.5081(293.2K)$$

$$| \ T_1 = 149.0K. \ |$$ **Answer** ◀

Plane 2 lies downstream of the shock. We must use the shock functions in Table E.2 to relate properties at plane 2 to those at plane 1. For $\mathbf{M}_1 = 2.20$, Table E.2

gives

$$\frac{p_2}{p_1} = 5.480, \qquad \frac{T_2}{T_1} = 1.857, \qquad \mathbf{M}_2 = 0.547.$$

Then

$$p_2 = \left(\frac{p_2}{p_1}\right) p_1 = (5.480)(9.35 \text{ kPa}),$$

Answer ◄

$$| \; p_2 = 51.2 \text{ kPa}; \; |$$

$$T_2 = \left(\frac{T_2}{T_1}\right) T_1 = (1.857)(149.0\text{K}),$$

Answer ◄

$$| \; T_2 = 276.7\text{K}. \; |$$

To calculate the properties at plane 3, we note that the flow from plane 2 to plane 3 is isentropic, although with different values of p_0 and A^* than the flow upstream of the shock. In particular, A_3^*, which is equal to A_2^*, *is not* equal to the throat area. We can calculate the area ratio function at plane 3 from

$$\left(\frac{A}{A^*}\right)_3 = \left(\frac{A_3}{A_2}\right)\left(\frac{A}{A^*}(\mathbf{M}_2)\right).$$

From Table E.1,

$$\frac{A}{A^*}(\mathbf{M}_2) = \frac{A}{A^*}(0.547) = 1.259, \qquad \text{so} \qquad \left(\frac{A}{A^*}\right)_3 = \left(\frac{3 \times 10^{-3} \text{ m}^2}{2 \times 10^{-3} \text{ m}^2}\right)(1.259) = 1.888.$$

This value of A/A^* corresponds to a supersonic Mach number and a subsonic Mach number. Since there is no throat between planes 2 and 3 the flow at plane 3 must be subsonic. From Table E.1,

$$\mathbf{M}_2 = 0.547, \qquad \frac{p_2}{p_{0_2}} = 0.8159,$$

$$\mathbf{M}_3 = 0.326, \qquad \frac{p_3}{p_{0_3}} = 0.9290, \qquad \frac{T_3}{T_{0_3}} = 0.9792.$$

We calculate p_3 and T_3:

$$p_3 = p_2 \frac{\dfrac{p}{p_0}(\mathbf{M}_3)}{\dfrac{p}{p_0}(\mathbf{M}_2)} = 51.2 \text{ kPa} \left(\frac{0.9290}{0.8159}\right),$$

Answer ◄

$$| \; p_3 = 58.3 \text{ kPa} \; |$$

and

$$T_3 = T_0 \left(\frac{T_3}{T_0}\right).$$

Since the shock does not affect T_0,

$$T_3 = (293.2\text{K})(0.9792),$$

Answer ◄

$$| \; T_3 = 287.1\text{K}. \; |$$

Discussion

It is important to note the logic used to decide that A^* upstream of the shock is equal to A_t, the geometric area of the throat. The value of A^* on the downstream side of the shock can be calculated as

$$A_2^* = \frac{A_2}{\frac{A}{A^*}\{M_2\}} = 1.59 \times 10^{-3}\ \text{m}^2.$$

Note that $A_2^* > A_1^*$.

There are alternative ways to manipulate the functions and tables to obtain the answers. For example, we could use

$$T_3 = T_2 \left(\frac{\frac{T}{T_0}\{M_3\}}{\frac{T}{T_0}\{M_2\}} \right) \quad \text{and} \quad p_3 = p_{0_1}\left(\frac{p_{0_2}}{p_{0_1}}\{M_1\}\right)\left(\frac{p}{p_0}\{M_3\}\right)$$

instead of the equations we used. It would be valuable for you to think of other possibilities. Suppose that only p_3 and T_3 were requested. Could they be calculated without calculating p_1, T_1, p_2, T_2, and so on?

This example demonstrates how to use the shock functions and tables to patch together different isentropic flows at a shock.

Mass Flow Relations and Choking. Several alternative equations can be used for calculation of mass flow rate in a duct. The "primitive variable" equation for mass flow rate is

$$\dot{m} = \rho V A. \tag{10.69}$$

If the fluid is an ideal gas, then

$$\rho = \frac{p}{RT} \quad \text{and} \quad V = \mathbf{M}c = \mathbf{M}\sqrt{kRT},$$

and an alternative equation for mass flow rate is

$$\dot{m} = \sqrt{\frac{k}{R}}\left(\frac{pA\mathbf{M}}{\sqrt{T}}\right). \tag{10.70}$$

Both Eqs. (10.69) and (10.70) evaluate the mass flow rate in terms of static properties. By substituting

$$p = p_0\left[1 + \left(\frac{k-1}{2}\right)\mathbf{M}^2\right]^{-k/(k-1)} \quad \text{and} \quad T = T_0\left[1 + \left(\frac{k-1}{2}\right)\mathbf{M}^2\right]^{-1},$$

we can obtain the mass flow rate in terms of stagnation properties:

$$\dot{m} = \sqrt{k}\left(\frac{p_0 A}{\sqrt{RT_0}}\right)\mathbf{M}\left[1 + \left(\frac{k-1}{2}\right)\mathbf{M}^2\right]^{-(k+1)/2(k-1)}. \tag{10.71}$$

Although this equation is complicated, it has certain advantages for isoenergetic-isentropic flow because p_0 and T_0 are constants. Any of Eqs. (10.69)–(10.71) can be used to calculate mass flow. The choice between them depends on what data are available.

By setting $\mathbf{M} = 1$ and $A = A^*$ in Eq. (10.71) we get

$$\dot{m} = \frac{p_0 A^*}{\sqrt{RT_0}}\sqrt{k}\left(\frac{2}{k+1}\right)^{(k+1)/2(k-1)}. \tag{10.72}$$

This equation shows us that A^* is constant in steady isoenergetic-isentropic flow with constant values of $\dot{m}$, p_0, T_0, R, and k.

We can obtain a dimensionless mass flow function from Eq. (10.71):

$$\frac{\dot{m}\sqrt{RT_0}}{p_0 A} = \sqrt{k}\mathbf{M}\left(1 + \frac{k-1}{2}\mathbf{M}^2\right)^{-(k+1)/2(k-1)}. \qquad \textbf{(10.73)}$$

A plot of this function is shown in Fig. 10.13. The key fact about Fig. 10.13 for this discussion is the occurrence of the maximum at $\mathbf{M} = 1$:

$$\frac{\dot{m}\sqrt{RT_0}}{p_0 A}\bigg)_{\max} = f(k \text{ only}).$$

We can draw several conclusions from this fact:

- Suppose that p_0 and T_0 are fixed (isoenergetic-isentropic flow). A specified mass flow rate can be forced through a limiting minimum area. No smaller area can pass the given flow.
- Suppose that p_0 and T_0 are fixed (isoenergetic-isentropic flow). Any particular fixed-geometry duct, with fixed minimum area, can pass only a certain maximum mass flow. Passage of this maximum mass flow occurs when the Mach number at the minimum area is 1. Once the Mach number at the minimum area becomes 1, it is not possible to increase the mass flow rate.
- It is possible to increase the limiting mass flow in a fixed geometry duct or to force a specified mass flow through a smaller area by *increasing* $p_0/\sqrt{T_0}$. Doing work on a gas (compression) increases both p_0 and T_0; however, the ratio $p_0/\sqrt{T_0}$ is usually increased by compression.

The preceding statements describe the phenomenon called *choking*. Simply stated:

> The choking phenomenon occurs in compressible duct flow when the local Mach number reaches 1 at the minimum area in the duct. When this occurs, the mass flow rate through the duct cannot be increased unless the ratio of stagnation pressure to square root of stagnation temperature is increased.

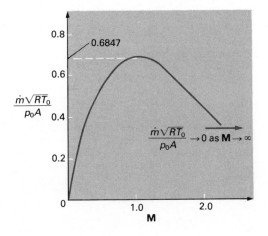

Figure 10.13 Dimensionless mass flow function for $k = 1.4$.

EXAMPLE 10.6 Illustrates use of compressible flow equations and tables in design of a flow passage

The duct between the turbine and exhaust nozzle of a jet engine is 0.60 m in diameter. The gas flowing in the duct has stagnation pressure and temperature of 275 kPa and 700K. The gas flow rate is 82 kg/s. Specify the relevant areas of a nozzle to expand the gas to standard atmospheric pressure. Calculate the force required to hold the nozzle to the duct. The gas has $k = 1.4$ and $R = 265$ N·m/kg·K.

Solution

Given

Schematic of the problem in Fig. E10.6(a)
Nozzle inlet diameter 0.6 m
Gas with $k = 1.4$, $R = 265$ N·m/kg·K, flowing at 82 kg/s
Gas with $p_0 = 275$ kPa, $T_0 = 700$K

Find

Relevant areas for nozzle to expand the gas to standard atmospheric
    pressure (101.33 kPa)
Force required to hold the nozzle to the duct

Solution

First verify that all given pressures and temperatures are absolute. Next establish what the nozzle geometry should be. The desired static-stagnation pressure ratio at the exit plane is

$$\frac{p_e}{p_{0_e}} = \frac{101.33 \text{ kPa}}{275.0 \text{ kPa}} = 0.3684.$$

From Table E.1, the desired exit Mach number is

$$\mathbf{M}_e = 1.285.$$

We assume that the inlet Mach number will be less than 1, so we need a minimum area (throat) between the entrance and the exit. The throat area is calculated from Eq. (10.72), with $k = 1.4$:

$$\dot{m} = 0.6847 \frac{p_0 A^*}{\sqrt{RT_0}}.$$

Figure E10.6 (a) Schematic of nozzle design problem; (b) control volume for nozzle force analysis.

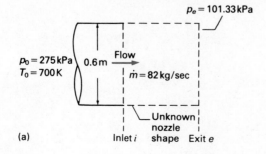

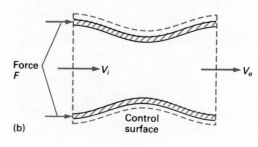

Since $A_t = A^*$,

$$A_t = \frac{\dot{m}\sqrt{RT_0}}{0.6847p_0} = \frac{82 \frac{\text{kg}}{\text{s}} \sqrt{\left(265 \frac{\text{N}\cdot\text{m}}{\text{kg}\cdot\text{K}}\right)(700 \text{ K})}}{0.6847\left(275{,}000 \frac{\text{N}}{\text{m}^2}\right)},$$

Answer

$$\boxed{A_t = 0.1876 \text{ m}^2.}$$

The throat diameter is

$$D_t = \sqrt{\frac{4A_t}{\pi}} = 0.4887 \text{ m}.$$

The inlet area of the nozzle is

$$A_i = \frac{\pi}{4} D_1^2 = \frac{\pi}{4}(0.6 \text{ m})^2,$$

Answer

$$\boxed{A_i = 0.2874 \text{ m}^2.}$$

The nozzle exit area is calculated as the value required to produce a Mach number of 1.285:

$$A_e = A_t\left[\frac{A}{A^*}(\mathbf{M}_e)\right].$$

From Table E.1,

$$\frac{A}{A^*} = 1.0601 \text{ at } \mathbf{M} = 1.285, \quad \text{so} \quad A_e = 0.1876 \text{ m}^2(1.0601),$$

Answer

$$\boxed{A_e = 0.1989 \text{ m}^2.}$$

The corresponding diameter is

$$D_e = \sqrt{\frac{4A_e}{\pi}} = 0.5032 \text{ m}.$$

To calculate the force, apply the linear momentum theorem to a control volume enclosing the nozzle, as shown in Fig. E10.6(b). The anchoring force, F, is assumed to act in the flow direction. The momentum equation for the flow direction is[†]

$$F + p_iA_i - p_eA_e - p_{\text{atm}}(A_i - A_e) = \dot{m}(V_e - V_i).$$

Solving for F, we have

$$F = (p_eA_e + \dot{m}V_e) - (p_iA_i + \dot{m}V_i) + p_{\text{atm}}(A_i - A_e).$$

It seems that we have a lot of work ahead to calculate pressures and velocities. We can avoid some of it by noting

$$\dot{m} = \rho_eA_eV_e = \rho_iA_iV_i,$$

so

$$F = (p_e + \rho_eV_e^2)A_e - (p_i + \rho_iV_i^2)A_i + p_{\text{atm}}(A_i - A_e).$$

From Eq. (10.29),

$$\rho V^2 = kp\mathbf{M}^2,$$

[†] The force due to uniform pressure on the outer surface of the control volume cancels except for the term involving p_{atm}.

so

$$F = p_e A_e (1 + k\mathbf{M}_e^2) - p_i A_i (1 + k\mathbf{M}_i^2) + p_{atm}(A_i - A_e),$$

where $\mathbf{M}_e$ and p_e are known. $\mathbf{M}_i$ and p_i are found from

$$\frac{A}{A^*}(\mathbf{M}_i) = \frac{A_i}{A_t} = \frac{0.2874 \text{ m}^2}{0.1876 \text{ m}^2} = 1.5319.$$

From Table E.1 (using subsonic Mach number),

$$\mathbf{M}_i = 0.419 \qquad \text{and} \qquad \frac{p_i}{p_0} = 0.8862.$$

Then

$$p_i = (0.8862)(275 \text{ kPa}) = 243.7 \text{ kPa}.$$

The force F is

$$F = \left(101{,}330 \, \frac{\text{N}}{\text{m}^2}\right)(0.1876 \text{ m}^2)[1 + 1.4(1.285)^2]$$

$$- \left(243{,}700 \, \frac{\text{N}}{\text{m}^2}\right)(2874 \text{ m}^2)[1 + 1.4(0.419)^2]$$

$$+ 101{,}330 \, \frac{\text{N}}{\text{m}^2}(0.2874 \text{ m}^2 - 0.1876 \text{ m}^2)$$

$$= -23{,}300 \text{ N}.$$

Since F is negative,

Answer

$$| \; F = 23{,}300 \qquad \text{Opposite the flow direction} \; |$$

Discussion
If you plan to do a lot of force calculations using the momentum theorem, you might investigate the possibility of tabulating

$$\frac{F}{F^*} \equiv \frac{pA(1 + k\mathbf{M}^2)}{p^* A^* (1 + k)}$$

as a function of $\mathbf{M}$ and k.

We treated p_{atm} and p_e as separate entities in the momentum equation even though they are equal for this problem. In some cases, the pressure at the exit of a supersonic nozzle is not equal to the pressure of the surrounding atmosphere (Section 10.4.4).

The nozzle flow was calculated as isoenergetic-isentropic because the exit Mach number was to be supersonic; therefore it is unlikely that any shocks occur in the nozzle.

The flow in the nozzle is choked. The given flow rate (82 kg/s) with the given stagnation properties (275 kPa, 700K) cannot pass through any area smaller than 0.1876 m².

10.4.3 Flow in a Converging Nozzle

We can see how the information about compressible flow in variable-area ducts fits together by examining the flow in a converging nozzle. Figure 10.14 shows a converging nozzle that exhausts gas from a large reservoir to a variable-pressure region. We assume that the

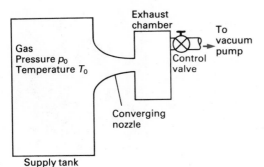

Figure 10.14 Apparatus for studying flow in a converging nozzle.

pressure and temperature in the reservoir are constant. We can vary the pressure in the exhaust region (called the *back pressure*) by means of the control valve that connects the exhaust region to a downstream vacuum pump. By opening or closing the control valve, we decrease or increase the back pressure and therefore expect to increase or decrease the nozzle flow rate. We will assume that there is no wall friction and no heat transfer or work.

We can make several preliminary observations about the flow in the nozzle:

- Since the nozzle is only convergent, it is not possible for the flow to pass *through* $M = 1$.
- Since the flow at the nozzle inlet (in the large reservoir) is obviously subsonic ($M \approx 0$), the flow in the entire nozzle will be subsonic with the possible exception of the nozzle exit.
- Since the flow cannot be supersonic in the nozzle, there can be no shocks and the flow will be isoenergetic-isentropic everywhere in the nozzle. The stagnation properties will be constant and equal to the gas properties in the reservoir.
- The maximum possible Mach number in the nozzle is 1.0. This value can occur only at the nozzle exit (minimum area).
- There is a maximum possible mass flow rate that can occur. This maximum is determined by the gas constants (k and R), the values of the reservoir (stagnation) properties, and the exit area.

The flow through the nozzle is described in terms of the pressure and Mach number distribution in the nozzle and the (dimensionless) mass flow rate ($\dot{m}\sqrt{RT_0}/p_0A_e$), all as functions of the ratio of back pressure to supply pressure and k. We also plot the ratio of pressure in the nozzle exit plane (p_e) to the back pressure (p_e/p_B). The various curves are shown in Fig. 10.15.

The curves and points labeled a correspond to a closed control valve. There is no flow. The pressure equals the reservoir pressure everywhere, and the Mach number is zero everywhere.

Case b corresponds to a slight opening of the control valve. The back pressure is less than the supply pressure and there is flow. The gas accelerates from the reservoir to the nozzle exit, and the pressure drops. Minimum pressure and maximum Mach number are at the nozzle exit. The pressure at the nozzle exit is equal to the back pressure. If we

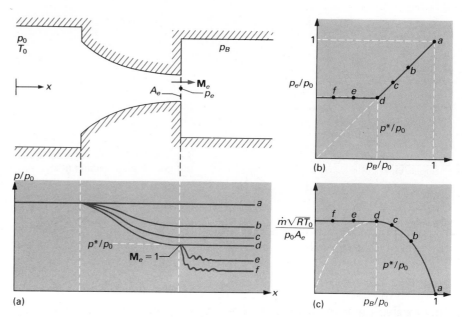

Figure 10.15 Performance of a converging nozzle at various values of pressure ratio.

know the ratio p_B/p_0, the equations

$$\frac{p_e}{p_0} = \frac{p_B}{p_0} \quad \text{and} \quad \mathbf{M}_e = \mathbf{M}\!\left(\frac{p_e}{p_0}\right)$$

enable us to determine the Mach number at the nozzle exit. The mass flow at the exit (and hence in the entire nozzle) can then be calculated from Eq. (10.73).

Case c is similar to case b except that a larger control valve opening permits a lower back pressure with correspondingly higher mass flow and exit Mach number and a lower exit pressure. Notice that the maximum Mach number still occurs at the nozzle exit and is less than 1.

In case d, the control valve has been opened a sufficient amount to bring the exit Mach number to a value of 1.0. The exit pressure now equals the critical pressure (p^*) and also equals the back pressure. Knowing $\mathbf{M}_e = 1.0$ allows calculation of the mass flow rate.

Case e corresponds to opening the control valve farther than in case d. When we try this, we find that no changes occur in the nozzle. In case d, we reached the limit of the nozzle's capability. The exit Mach number cannot exceed 1, the exit pressure cannot drop below the critical pressure, which is determined solely by the supply pressure and the specific heat ratio (Eq. 10.59), and the mass flow cannot exceed the choked value. The only difference between case d and case e is that the back pressure and exit pressure are no longer equal. The flow must adjust to the lower back pressure *after* leaving the nozzle. This is a multidimensional flow phenomenon, so the pressure curve is shown as a wavy line downstream of the nozzle exit. Opening the control valve with further lowering of the back pressure does not change the nozzle flow.

At first, these results may seem strange; however, they have a very simple physical explanation. Once the exit flow becomes sonic, pressure waves from the exhaust region cannot propagate upstream into the nozzle and supply reservoir to cause any flow adjustment.

There are two flow regimes in a simple converging nozzle: *unchoked flow* and *choked flow*. Which flow regime occurs in a given case depends on the relative values of the back pressure and critical pressure:

- If $p_B/p_0 > p^*/p_0$, the flow is unchoked.
- If $p_B/p_0 \leq p^*/p_0$, the flow is choked.

The critical pressure ratio (p^*/p_0) depends only on specific heat ratio. For a converging nozzle, the occurrence of choked flow does not depend on nozzle geometry.

You might be curious about the generality of these results. Although it appears that we have been discussing a very specific system, we did so only for the purpose of illustration. The large reservoir represents a supply of gas with specific stagnation conditions. The exhaust region/control valve/vacuum pump represents the resistance to flow downstream of the nozzle. In principle, it would be equally valid to consider a compressor and control valve upstream of the nozzle and a constant-pressure exhaust region. In this case, all of the *dimensionless* plots of Fig. 10.15 would still be valid; however, if we permit p_0 to increase, we can increase the *dimensional* mass flow rate $(\dot{m})$, even at choked conditions. Finally, we have shown a long, smoothly contoured nozzle. The principles apply to converging nozzles of any length, even orifices; however, the departure from one-dimensional, one-directional flow conditions becomes significant in short nozzles.

10.4.4 Flow in a Converging-Diverging Nozzle

We will complete our consideration of compressible flow in variable-area ducts by discussing the flow in a converging-diverging nozzle. Figure 10.16 shows a converging-diverging nozzle that exhausts gas from a large reservoir at constant pressure and temperature to a variable-pressure exhaust region. The back pressure can be varied by opening or closing a control valve (similar to Fig. 10.14). Possible generalizations of this arrangement are similar to those discussed for

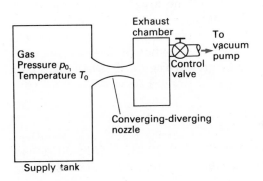

Figure 10.16 Apparatus for studying flow in a converging-diverging nozzle.

the converging nozzle. We assume that there is no wall friction, heat transfer, or work.

Preliminary observations about the flow in the converging-diverging nozzle are as follows:

- Since the nozzle is convergent-divergent, it is possible for the flow to pass through $\mathbf{M} = 1$. It is also possible for the flow to be subsonic everywhere.
- If $\mathbf{M} = 1$ anywhere, it must be at the throat.
- Since there may be supersonic flow in the divergent portion of the nozzle, there may be shocks in the flow. If there are shocks, the flow will not be completely isentropic, although it will be isoenergetic.
- If there are no shocks, the flow will be isentropic. If there are shocks, the flow from the reservoir up to the first shock will be isentropic. Flow downstream of a shock will also be isentropic but but with different values of entropy, stagnation properties, and critical area.
- The maximum possible Mach number that can occur anywhere in the passage corresponds to acceleration of the fluid in an isentropic process from the reservoir to the nozzle exit. The maximum possible Mach number can occur only at the exit and is determined by the ratio of exit area to throat area.
- The maximum possible mass flow rate in the nozzle is determined by the gas constants, the reservoir properties, and the minimum area, which occurs at the throat.

The flow in the converging-diverging nozzle will be described by plots of pressure and Mach number distributions in the nozzle, the dimensionless mass flow rate ($\dot{m}\sqrt{RT_0}/p_{0_1}A_t$), and the ratio of nozzle exit pressure to supply pressure (p_e/p_{0_1}), all as functions of the ratio of back pressure to supply pressure (p_B/p_{0_1}).

The nozzle performance curves are shown in Fig. 10.17. The case labeled a corresponds to complete closure of the control valve. There is no flow. The pressure is uniform, and the Mach number is zero everywhere.

Curves and points labeled b represent a slightly open control valve. The gas accelerates in the convergent portion of the nozzle, and the pressure drops. The throat Mach number is considerably less than 1 (subsonic flow). The flow decelerates in the divergent portion of the nozzle, and the pressure rises. The pressure and Mach number distributions are roughly symmetrical about the throat. There are no shocks because the flow is subsonic everywhere. The flow at the nozzle exit is subsonic, and the exit and back pressures are equal. The exhaust Mach number can be found from

$$\frac{p}{p_0}(\mathbf{M}_e) = \frac{p_B}{p_{0_1}},$$

and the mass flow can be calculated from Eq. (10.73). A value of A^* can be calculated from Eq. (10.67) or Eq. (10.72) applied at the exhaust plane. A^* will be less than A_t.

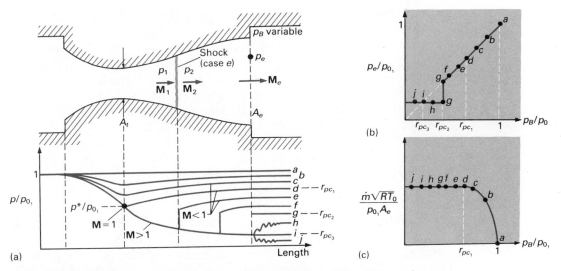

Figure 10.17 Performance of converging-diverging nozzle at various pressure ratios.

Curve c represents a slightly larger control valve opening. The situation is qualitatively similar to case b. Note that the mass flow is larger. The pressure and Mach number distributions are still roughly symmetrical about the throat. The maximum Mach number occurs at the throat.

In case d, the control valve has been opened just enough to bring the throat Mach number exactly to 1.0. The flow is still subsonic everywhere except exactly at the throat, where $\mathbf{M} = 1.0$. The flow is isentropic, but now

$$p_t = p^* \quad \text{and} \quad A_t = A^*.$$

The mass flow has reached its maximum possible value and the flow has just become choked. Since the pressure rises downstream of the throat, the back pressure at which the converging-diverging nozzle chokes is greater than p^*/p_0. The pressure ratio that causes choking to first occur in a converging-diverging nozzle is called the *first critical pressure ratio* (r_{pc_1}) and can be calculated by

$$r_{pc_1} = \frac{p}{p_0} \left(\frac{A}{A^*} = \frac{A_e}{A_t}, \mathbf{M} < 1 \right); \tag{10.74}$$

that is, by finding the pressure ratio corresponding to the subsonic value of A/A^*.

What happens when the valve is opened beyond the case d point? Since the throat is choked, the mass flow cannot be increased, and conditions upstream of the throat cannot be affected. There is a response in the flow downstream of the throat because, at case d, the downstream flow is subsonic. When the valve is opened and the back pressure lowered, the fluid begins to accelerate as it enters the divergent portion of the nozzle; that is, the flow becomes supersonic. If the control valve is opened only slightly past the case d point, the downstream resistance is too large for complete acceleration in the divergent portion of the nozzle, so a shock occurs in the divergent portion.

Curves and points labeled e represent the flow for a slightly more open control valve. The flow accelerates in the convergent portion of the nozzle, reaches sonic speed at the throat, and accelerates to supersonic speed downstream of the throat. The supersonic acceleration terminates in a shock wave. Downstream of the shock, the flow experiences subsonic deceleration and exits the passage with $\mathbf{M}_e < 1$. The exit pressure and back pressure are equal. The exact position of the shock depends on the back pressure and, for a given back pressure, is fixed. The mass flow rate for case e, as well as for all lower values of back pressure, is the same as for case d.

Opening the control valve and lowering the back pressure causes the shock to move downstream (case f). Note that once the shock passes a plane, the flow *up to* that plane is no longer affected by lowering the back pressure. As the valve is opened farther, the shock is eventually drawn to the nozzle exit. The flow accelerates isentropically all the way from the reservoir to the exit shock. Exactly at the exit, the pressure jumps to the back pressure as the fluid exits through the shock. This situation is shown as case g. Note that the exit pressure is double-valued at case g.

Lowering the back pressure further causes the shock to move out of the nozzle and become multidimensional (case h). The flow accelerates isentropically from the reservoir to the exit. The gas exits the nozzle supersonically with $\mathbf{M}_e$ corresponding to the nozzle's exit-to-throat area ratio. The exit pressure is determined only by the stagnation pressure (p_{0_1}) and the exit Mach number and is not equal to the back pressure. The gas adjusts to the back pressure externally. If we continue to lower the back pressure, it eventually reaches equality with the exit pressure (case i). At this condition there is no pressure adjustment in the exiting gas. Lowering the pressure further (case j) requires external expansion pressure adjustments but does not affect the nozzle flow.

We can summarize this information about converging-diverging nozzle flow as follows. There are four regimes of flow in a converging-diverging nozzle:

- *Venturi regime* (cases a–d). The flow is subsonic and isentropic everywhere. The flow accelerates in the convergent portion and decelerates in the divergent portion. Maximum Mach number and minimum pressure occur at the throat.
- *Shock regime* (cases d–g). The flow is subsonic in the convergent portion, sonic at the throat, and partly supersonic in the divergent portion. The acceleration terminates in a shock that stands in the divergent portion at a location determined by the exact value of the back pressure. The flow experiences subsonic deceleration from the shock to the exit. The flow is choked.
- *Overexpanded regime* (cases g–i). The flow accelerates throughout the nozzle. The throat flow is sonic, and the exit flow is supersonic. The pressure of the gas increases to the back pressure downstream of the nozzle exit.
- *Underexpanded regime* (cases i–j). This case is similar to the overexpanded regime except that the external pressure adjustments are expansive rather than compressive.

The boundaries between the four flow regimes are indicated by curves d, g, and i. The ratios of back pressure to supply pressure that correspond to these cases are called the *first critical pressure ratio* (r_{pc_1}), the *second critical pressure ratio* (r_{pc_2}), and the *third critical pressure ratio* (r_{pc_3}). These ratios are calculated as follows:

- First critical pressure ratio: See Eq. (10.74).
- Second critical pressure ratio: Assume that a shock stands at the nozzle exit and calculate the ratio of *downstream static* to *upstream stagnation* pressure:

$$r_{pc_2} = \frac{p_2}{p_1}(\mathbf{M}_e) \times \frac{p}{p_0}(\mathbf{M}_e), \qquad (10.75)$$

where $(p_2/p_1(\mathbf{M}))$ is the shock static pressure ratio function and $\mathbf{M}_e$ is given by
$$\mathbf{M}_e = \mathbf{M}(A/A^* = A_e/A_t).$$

- Third critical pressure ratio:

$$r_{pc_3} = \frac{p}{p_0}(A/A^* = A_e/A_t, \mathbf{M} > 1). \qquad (10.76)$$

Calculation of these ratios is easier than it looks, as the following example illustrates. Note that the three critical pressure ratios for a converging-diverging nozzle depend on the nozzle geometry as well as the specific heat ratio of the flowing gas. The first step in calculation of flow through a given converging-diverging nozzle is classification of the flow. This is accomplished by calculating the three critical pressure ratios and comparing them to the actual ratio of back pressure to supply pressure. After you have identified the proper flow regime, the rest of the problem falls into place easily.

EXAMPLE 10.7 Illustrates performance analysis for converging and converging-diverging nozzles

A thick-walled pressure vessel has a hole in its side. Figure E10.7 shows two possible configurations for the hole. The pressure and temperature of the air inside the vessel are 20 psia and 100.3°F. The pressure outside the vessel is 14.6 psia. Calculate the leak rate of air for each hole configuration.

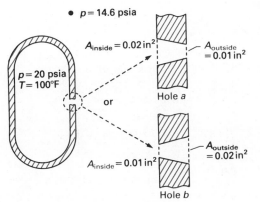

Figure E10.7 Thick wall pressure vessel with two possible hole configurations.

Solution

Given
Pressure vessel containing air at 20 psia and 100.3°F
Pressure of surrounding atmosphere 14.6 psia
Two possible configurations of a hole in the vessel wall, both with the same maximum and minimum areas

Find
Leak rate of air for each hole configuration

Solution
First convert given data to absolute units. The air temperature in the vessel is

$$T = 100.3°F + 459.7°F = 560°R.$$

Next, recognize that hole a acts like a simply converging nozzle and hole b acts like a converging-diverging nozzle. The "converging portion" of hole b will be formed by streamlines within the tank. Since both holes have the same minimum area and the supply temperature and pressure are the same, the maximum possible flow rate is the same for both holes. If both holes are choked, they will have the same flow rate. We first check both holes for choking.

Hole a
Hole a will be choked if

$$\frac{p_B}{p_{0_1}} < \frac{p^*}{p_0} = 0.5283.$$

For hole a,

$$\frac{p_B}{p_{0_1}} = \frac{14.6 \text{ psia}}{20 \text{ psia}} = 0.730.$$

Hole a will not be choked.

Hole b
Hole b will be choked if

$$\frac{p_B}{p_{0_1}} < r_{pc_1},$$

where

$$r_{pc_1} = \frac{p}{p_0}\left(\frac{A}{A^*} = \frac{0.02 \text{ in}^2}{0.01 \text{ in}^2} = 2, \mathbf{M} < 1\right).$$

From Table E.1,

$$r_{pc_1} = 0.9371.$$

The back pressure ratio is still 0.730, so hole b is choked. Next calculate the mass flow through hole b using Eq. (10.72), with $k = 1.4$ and $\mathbf{M} = 1$:

$$\dot{m}_b = 0.6847\left(\frac{p_0 A_{\min}}{\sqrt{RT_0}}\right) = 0.6847\left[\frac{\left(20 \dfrac{\text{lb}}{\text{in}^2}\right)(0.01 \text{ in}^2)\sqrt{32.2 \dfrac{\text{ft}\cdot\text{lbm}}{\text{lb}\cdot\text{sec}^2}}}{\sqrt{\left(53.35 \dfrac{\text{ft}\cdot\text{lb}}{\text{lbm}\cdot°R}\right)(560°R)}}\right],$$

$$\left| \dot{m}_b = 0.00450 \text{ lbm/sec.} \right|$$

Answer
◀

To calculate the leak rate in hole a, we need the Mach number at the exit. Since

$$\frac{p_e}{p_0} = \frac{p_b}{p_0} = 0.730,$$

Table E.1 gives

$$\mathbf{M}_e = 0.686.$$

Using Eq. (10.73), we have

$$\dot{m}_a = \frac{\left(20\,\frac{\text{lb}}{\text{in}^2}\right)(0.01\,\text{in}^2)\sqrt{32.2\,\dfrac{\text{ft}\cdot\text{lbm}}{\text{lb}\cdot\text{sec}^2}}}{\sqrt{\left(53.35\,\dfrac{\text{ft}\cdot\text{lb}}{\text{lbm}\cdot°\text{R}}\right)(560°\text{R})}}\sqrt{1.4}\,(0.686)[1 + (0.2)(0.686)^2]^{-3},$$

$$\left| \; \dot{m}_a = 0.00407\,\text{lbm/sec.} \; \right| \qquad\qquad \textbf{Answer} \;\; \blacktriangleleft$$

Discussion

These answers are only approximations since the flow through real holes in the wall of a pressure vessel would probably not be approximately one-dimensional. The third and second critical pressure ratios for hole b are

$$r_{pc_3} = \frac{p}{p_0}\left(\frac{A}{A^*} = 2,\, \mathbf{M}_e > 1\right) = 0.0935 \qquad (\text{note: } \mathbf{M}_e = 2.20)$$

and

$$r_{pc_2} = \frac{p_2}{p_1}\,(\mathbf{M} = 2.20) \times \frac{p}{p_0}\,(\mathbf{M} = 2.20).$$

Using Table E.2, we have

$$r_{pc_2} = (5.480)(0.0935) = 0.512.$$

Hole b is operating in the shock regime.

10.5 Internal Flow: Constant-Area Duct Flow with Friction

We will now examine the effects of wall friction in compressible internal flow. To simplify our discussion, we will limit consideration to flow in ducts[†] of constant cross-sectional area. We will assume that the flow is one-dimensional, with the velocity varying only with duct length (x) but *not* with distance across the duct. This is a somewhat artificial situation since a transverse velocity gradient is always associated with shear stress. The situation that we have assumed is most closely approximated by turbulent flow. The single value of fluid velocity associated with any duct cross section should be interpreted as an average velocity.

[†] In this section, the word *ducts* includes circular pipes as well as other cross-sectional shapes.

10.5.1 Preliminary Consideration: Comparison with Incompressible Duct Flow

Flow of an incompressible fluid in a constant-area duct is one of the most important practical problems in fluid mechanics. Incompressible flow in long ducts reaches a condition called fully developed flow, and the velocity profile does not change shape. Since the density is constant in incompressible flow, continuity requires that the average velocity remain constant, and so neither velocity profile shape nor magnitude vary in fully developed incompressible flow. Therefore, the fluid's kinetic energy and momentum flux are constant. The retarding effect of fluid friction is balanced by a pressure drop in the flow direction. This pressure drop is directly proportional to the irreversible loss of mechanical energy caused by fluid friction.

What changes does fluid compressibility introduce? It is possible for a compressible flow to become fully developed in the sense that the velocity profile does not change *shape,* so we could write

$$\frac{u}{u_{\max}} = f\left(\frac{r}{R} \text{ only}\right)$$

for flow in a circular pipe. If the density changes, then the average velocity must also change to satisfy the continuity equation. Changing the average velocity changes the magnitude of the entire velocity profile, even if its shape is fixed. More important, if the velocity changes, then the fluid's kinetic energy and momentum flux also change. Even if the flow is fully developed, the retarding effect of fluid friction is balanced by *both* pressure *and* momentum changes. Although an irreversible loss of mechanical energy is caused by fluid friction, it is not possible to interpret this loss in terms of a (static) pressure drop. Pressure changes in *compressible* flow *cannot* be calculated from the Darcy-Weisbach equation. In compressible flow, friction appears explicitly only in the momentum equation. The effects of friction on the flow can be determined only by simultaneous solution of the equations of continuity, momentum, energy, and state.

Since we must use the general energy equation, it is necessary to make an explicit assumption about work and heat transfer or to include equations for their calculation. Work transfer is usually "concentrated" in compressors or turbines and does not occur in long runs of duct. The question of heat transfer to or from the fluid is more complicated because the mechanisms of fluid friction and heat transfer are closely related. Three types of problems can be considered:

- Adiabatic flow, with no heat transfer
- Isothermal flow, in which sufficient heat is transferred to or from the fluid to maintain its static temperature constant
- Flow with simultaneous friction and heat transfer

If the duct is reasonably short, the adiabatic flow assumption is accurate. Since compressibility effects usually occur only if the fluid velocity is large and since, on dimensional grounds, mechanical energy loss increases as the square of the velocity, ducts in which friction and compressibility are both significant are usually kept short (duct length less than about 100 diameters) and heat transfer can be neglected.

Isothermal flow occurs in long pipelines that are exposed to a constant-temperature environment. A practical example is underground natural gas transmission lines. After traveling a few hundred diameters along the pipe, the gas attains the temperature of its surroundings. Sufficient heat is transferred to or from the gas to maintain constant temperature.

Our consideration in this book will be limited to adiabatic flow in a constant-area duct. This type of flow is sometimes called *Fanno flow*. Many of the qualitative features of isothermal flow are similar to those of Fanno flow; however, different working equations result from the $q = 0$ or $T = $ constant assumptions. Analysis of flow with simultaneous friction and heat transfer is not terribly difficult but is beyond the scope of a single chapter's introduction to compressible flow. Consult [1–4] for more information on isothermal flow and [1, 3] for analysis of flow with simultaneous friction and heat transfer.

10.5.2 Workless Adiabatic Flow (Fanno Flow) of an Ideal Gas

To obtain qualitative and quantitative information about Fanno flow, we assume that the flowing fluid is an ideal gas. The general conclusions that we will draw are valid for nonideal gas and vapor flows, but the equations will be restricted to an ideal gas.

Figure 10.18(a) illustrates the fundamental problem. A gas flows from plane 1 to plane 2 in a constant-area duct. There is no work or heat transfer, and elevation changes are negligible. There is shear

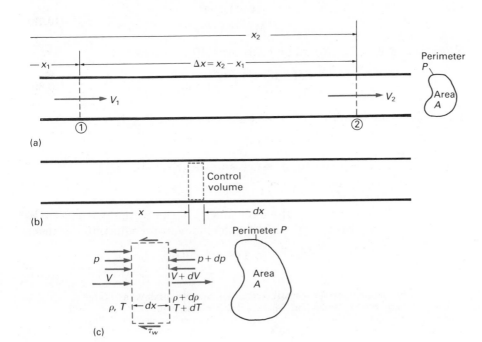

(a)

(b)

(c)

Figure 10.18 Analysis of flow in constant-area, duct: (a) Fundamental problem geometry; (b) differential control volume; (c) details of flow through control volume.

stress at the duct wall. We wish to relate changes of gas properties, velocity, and Mach number between planes 1 and 2 to the length, Δx, and other duct characteristics.

For an ideal gas, it is convenient to treat the pressure, temperature, and Mach number as dependent variables. Since the shear stress at the wall varies with location along the duct, it is necessary to apply the governing equations to a differential control volume lying somewhere between plane 1 and plane 2 (Fig. 10.18b). The resulting ordinary differential equations will then be integrated between planes 1 and 2. Figure 10.18(c) shows the details of the flow through the control volume.[†] The continuity equation is

$$\rho V A = (\rho + d\rho)(V + dV)(A),$$

which simplifies to

$$\rho \, dV + V \, d\rho = 0.$$

Dividing by ρV gives

$$\frac{dV}{V} + \frac{d\rho}{\rho} = 0. \tag{10.77}$$

From the equation of state,

$$\frac{d\rho}{\rho} = \frac{dp}{p} - \frac{dT}{T} \tag{10.78}$$

by logarithmic differentiation. The velocity can be written

$$V = \mathbf{M}c = \mathbf{M}\sqrt{kRT}.$$

Differentiating logarithmically gives

$$\frac{dV}{V} = \frac{d\mathbf{M}}{\mathbf{M}} + \frac{1}{2}\left(\frac{dT}{T}\right). \tag{10.79}$$

Substituting Eqs. (10.78) and (10.79) into (10.77) gives

$$\frac{d\mathbf{M}}{\mathbf{M}} + \frac{dp}{p} - \frac{1}{2}\left(\frac{dT}{T}\right) = 0. \tag{10.80}$$

The momentum equation for the control volume is

$$\sum dF_x = (\rho V A)[(V + dV) - V].$$

The forces are those due to pressure and shear stress,

$$\sum dF_x = pA - (p + dp)A - \tau_w P \, dx,$$

where P is the perimeter of the duct and τ_w is the wall shear stress. If we substitute the forces and simplify, the momentum equation becomes

$$\rho V A \, dV = -A \, dp - \tau_w P \, dx.$$

[†] Formally, we should consider a small control volume of length δx and fluid property changes δp, δV, and so on. The exact differential equations then result from taking the limit at the end of the derivation. This more exact procedure was used extensively in Chapters 2, 3, and 4.

We now divide the equation by $(\rho V^2 A)$ to get

$$\frac{dV}{V} + \frac{dp}{\rho V^2} + \frac{\tau_w}{\rho V^2}\left(\frac{P}{A}\right) dx = 0. \qquad (10.81)$$

Using Eqs. (10.29), (7.31), and (7.74), we reduce the momentum equation to

$$\frac{dV}{V} + \frac{1}{k\mathbf{M}^2}\left(\frac{dp}{p}\right) + \frac{1}{2}(4\mathbf{C_f})\frac{dx}{D_h} = 0,$$

where D_h is the hydraulic diameter

$$D_h = \frac{4A}{P}.$$

Substituting Eq. (10.79) and multiplying by $k\mathbf{M}^2$ give

$$k\mathbf{M}\, d\mathbf{M} + \frac{dp}{p} + \frac{k\mathbf{M}^2}{2}\left(\frac{dT}{T}\right) + \frac{k\mathbf{M}^2}{2}(4\mathbf{C_f})\frac{dx}{D_h} = 0. \qquad (10.82)$$

The energy equation for the control volume is

$$dq - dw = \left[(\tilde{h} + d\tilde{h}) + \frac{1}{2}(V + dV)^2\right] - \left[\tilde{h} + \frac{V^2}{2}\right].$$

Since the flow is adiabatic and there is no work, this equation gives

$$d\tilde{h} + V\, dV = 0.$$

For an ideal gas,

$$c_p\, dT + V\, dV = 0. \qquad (10.83)$$

Note that this equation is the same as

$$c_p\, dT_0 = c_p\left[dT + d\left(\frac{V^2}{2c_p}\right)\right] = 0;$$

in other words, the stagnation temperature of the gas is constant. Dividing Eq. (10.83) by $c_p T$ and using Eq. (10.79) and

$$\frac{V^2}{c_p T} = \frac{kR}{c_p}\left(\frac{V^2}{kRT}\right) = (k-1)\mathbf{M}^2,$$

we have

$$\left(1 + \frac{k-1}{2}\mathbf{M}^2\right)\frac{dT}{T} + (k-1)\mathbf{M}\, d\mathbf{M} = 0. \qquad (10.84)$$

Equations (10.80), (10.82), and (10.84) contain four (dimensionless) differentials, dp/p, dT/T, $d\mathbf{M}$, and dx/D_h. Since there are three equations, we can express three of the differentials in terms of one independent differential. The obvious choice for the independent variable is duct length dx/D_h. From Eq. (10.84),

$$\frac{dT}{T} = -\frac{(k-1)\mathbf{M}}{1 + \dfrac{k-1}{2}\mathbf{M}^2}\, d\mathbf{M}. \qquad (10.85)$$

Substituting this into Eq. (10.80) gives

$$\frac{dp}{p} = -\left(\frac{1}{\mathbf{M}} + \frac{\frac{k-1}{2}\mathbf{M}^2}{1 + \frac{k-1}{2}\mathbf{M}^2}\right)d\mathbf{M}. \tag{10.86}$$

Substituting Eqs. (10.85) and (10.86) into Eq. (10.82) gives, after simplification,

$$d\mathbf{M} = \frac{\mathbf{M}\left(1 + \frac{k-1}{2}\mathbf{M}^2\right)}{(1 - \mathbf{M}^2)}\left(\frac{k\mathbf{M}^2}{2}\right)\left(4\mathbf{C}_f\frac{dx}{D_h}\right). \tag{10.87}$$

Back substitution into Eqs. (10.85) and (10.86) gives

$$\frac{dT}{T} = -\frac{(k-1)\mathbf{M}^2}{(1 - \mathbf{M}^2)}\left(\frac{k\mathbf{M}^2}{2}\right)\left(4\mathbf{C}_f\frac{dx}{D_h}\right) \tag{10.88}$$

and

$$\frac{dp}{p} = -\frac{1 + (k-1)\mathbf{M}^2}{(1 - \mathbf{M}^2)}\left(\frac{k\mathbf{M}^2}{2}\right)\left(4\mathbf{C}_f\frac{dx}{D_h}\right). \tag{10.89}$$

We can draw several conclusions about the nature of adiabatic frictional flow from examination of Eqs. (10.87)–(10.89). We will consider how the pressure, temperature, and Mach number change as the flow proceeds downstream (dx is positive). Note that changes of pressure, temperature, and Mach number can be either positive or negative, depending on the sign of $(1 - \mathbf{M}^2)$. If the flow is subsonic, friction causes the Mach number to increase and the pressure and temperature to decrease. The *pressure* change in subsonic flow is qualitatively similar to that in incompressible flow; however, note the surprising fact that friction causes the fluid to accelerate. If the flow is supersonic, exactly the opposite happens! In supersonic flow, friction causes the Mach number to decrease while the pressure and temperature increase.

Consider the special case of sonic flow ($\mathbf{M} = 1$). Since infinite values of $d\mathbf{M}$, dp, and dT are not physically possible, the Mach number can be unity *only* if $4\mathbf{C}_f\,(dx/D_h)$ is zero. This can occur in two ways, $\mathbf{C}_f = 0$ or $dx = 0$. If $\mathbf{C}_f = 0$, then there is no friction and we are not considering a frictional flow but an isentropic flow. The situation $dx = 0$ means that the fluid cannot proceed down the duct any farther. This does not mean that the duct is closed but that the fluid is at the end of the duct. In frictional adiabatic flow, the Mach number can be 1 only at the end of the duct.

Frictional adiabatic flow in a constant-area duct is similar to isentropic flow in a converging nozzle in two respects:

- The Mach number is always driven toward a value of unity.
- The Mach number can reach unity only at the end of the passage.

Note that although we can say that if $\mathbf{M} = 1$, then $dx = 0$, we cannot always conclude that if $dx = 0$, then $\mathbf{M} = 1$. It is possible, but not necessary, that the flow is sonic at the end of a frictional duct.

It is instructive to consider the effect of friction on the stagnation properties. For adiabatic flow with no work, the stagnation temperature is constant, and

$$dT_0 = 0.$$

The stagnation pressure is related to the static pressure and Mach number by

$$p_0 = p\left(1 + \frac{k-1}{2}\mathbf{M}^2\right)^{k/(k-1)}.$$

Differentiating logarithmically, we get

$$\frac{dp_0}{p_0} = \frac{dp}{p} + \left(\frac{2k\mathbf{M}}{1 + \frac{k-1}{2}\mathbf{M}^2}\right)d\mathbf{M}.$$

Substituting Eqs. (10.87) and (10.89) and simplifying, we get

$$\frac{dp_0}{p_0} = -\frac{k\mathbf{M}^2}{2}\left(4\mathbf{C_f}\frac{dx}{D_h}\right), \tag{10.90}$$

which tells us that the stagnation pressure always decreases. The minimum value of stagnation pressure occurs at the end of the duct. The dimensionless mass flow rate

$$\frac{\dot{m}\sqrt{RT_0}}{p_0 A}$$

is always largest at the end of the duct because $\dot{m}$, T_0, and A are the same at all locations along the duct and p_0 is smallest at the end. Choking of the flow, if it occurs, can occur only at the *end* of the duct.

We can obtain equations for calculating pressure, temperature, and Mach number by integrating the differential equations from plane 1 to plane 2 in the duct. Equation (10.87) contains only Mach number and duct length as variables, so it can be rearranged and integrated

$$\int_{x_1}^{x_2} 4\mathbf{C_f}\frac{dx}{D_h} = \int_{\mathbf{M}_1}^{\mathbf{M}_2} \frac{2(1-\mathbf{M}^2)}{k\mathbf{M}^3\left(1 + \frac{k-1}{2}\mathbf{M}^2\right)}d\mathbf{M}.$$

If we assume that $\mathbf{C_f}$ is constant, the integral on the left will be simple enough; however, the integral on the right will obviously produce a rather complicated function of k, $\mathbf{M}_1$ and $\mathbf{M}_2$. We can simplify calculations by introducing a reference state at which the value of $\mathbf{M}$ is known and then constructing tables of the resulting function. Since we know that the flow approaches a limiting value of $\mathbf{M} = 1$, we write

$$\int_{x}^{x+\ell^*} 4\mathbf{C_f}\frac{dx}{D_h} = \int_{\mathbf{M}}^{1} \frac{2(1-\mathbf{M}^2)}{k\mathbf{M}^3\left(1 + \frac{k-1}{2}\mathbf{M}^2\right)}d\mathbf{M},$$

where ℓ^* is the critical length, the length of duct necessary to change the Mach number from $\mathbf{M}$ at x to 1 at $x + \ell^*$. Integrating gives

$$4\mathbf{C_f}\left(\frac{\ell^*}{D_h}\right) = \frac{1-\mathbf{M}^2}{k\mathbf{M}^2} + \frac{k+1}{2k}\ln\left[\frac{(k+1)\mathbf{M}^2}{2\left(1 + \frac{k-1}{2}\mathbf{M}^2\right)}\right]. \tag{10.91}$$

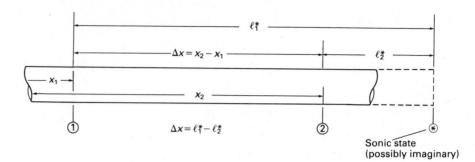

Figure 10.19 Relation between critical length and actual duct length.

This equation defines the *critical length function*. A table of values of $4C_f \ell^*/D_h$ versus $\mathbf{M}$ is provided in Appendix E, Table E.3.

Figure 10.19 illustrates how the choking length is used to relate Mach number to duct length. The Mach numbers at 1 and 2 are related by

$$4C_f \frac{\Delta x}{D_h} = 4C_f \frac{\ell^*}{D_h} (\mathbf{M}_1) - 4C_f \frac{\ell^*}{D_h} (\mathbf{M}_2). \tag{10.92}$$

The quantity on the left is calculated from the duct geometry and friction characteristics. The quantities on the right are calculated from Eq. (10.91) or, more frequently, obtained from Table E.3. Note that ℓ^* is purely a reference length; the actual duct length may be less than ℓ^*. Note also that the value of $4C_f \ell^*/D_h$ becomes very large for small Mach number and that small changes of $\mathbf{M}$ cause large changes of the critical length function, making calculations difficult. If the largest Mach number in the duct is less than 0.2, it is usually best to assume that the flow is incompressible and use the methods of Chapter 7.

Static and stagnation pressure and static temperature are most easily calculated by relating them to Mach number and so indirectly to duct length via the critical length function. To accomplish this, Eqs. (10.85), (10.86), and (10.90) must be integrated. To obtain an easily tabulated function, one limit of the integration is taken to be the sonic condition, so, for example, the pressure integration is

$$\int_{p^*}^{p} \frac{dp}{p} = -\int_{1}^{\mathbf{M}} \left(\frac{1}{\mathbf{M}} + \frac{\frac{k-1}{2}\mathbf{M}^2}{1 + \frac{k-1}{2}\mathbf{M}^2} \right) d\mathbf{M},$$

giving the pressure ratio function

$$\frac{p}{p^*} = \frac{1}{\mathbf{M}} \left(\frac{\frac{k+1}{2}}{1 + \frac{k-1}{2}\mathbf{M}^2} \right)^{1/2}. \tag{10.93}$$

In a similar manner, we find

$$\frac{T}{T^*} = \frac{\frac{k+1}{2}}{1 + \frac{k-1}{2}\mathbf{M}^2} \tag{10.94}$$

and

$$\frac{p_0}{p_0^*} = \frac{1}{\mathbf{M}} \left(\frac{1 + \dfrac{k-1}{2}\mathbf{M}^2}{\dfrac{k+1}{2}} \right)^{(k+1)/2(k-1)} . \qquad (10.95)$$

These functions are also tabulated in Table E.3. Note that the starred (*) properties are reference values, defined as the values the fluid properties would obtain if the fluid were brought to the sonic state in a frictional adiabatic flow. For a given frictional adiabatic flow, there is a unique limiting state, and starred properties are constant. The starred properties defined here are different from the critical properties defined in Section 10.3.3 because the earlier definition assumed an isentropic (frictionless) process.

Static and stagnation pressures and static temperatures at two points in a duct are related by

$$\frac{p_2}{p_1} = \frac{\dfrac{p}{p^*}\{\mathbf{M}_2\}}{\dfrac{p}{p^*}\{\mathbf{M}_1\}}, \qquad (10.96)$$

$$\frac{p_{0_2}}{p_{0_1}} = \frac{\dfrac{p_0}{p_0^*}\{\mathbf{M}_2\}}{\dfrac{p_0}{p_0^*}\{\mathbf{M}_1\}}, \qquad (10.97)$$

and

$$\frac{T_2}{T_1} = \frac{\dfrac{T}{T^*}\{\mathbf{M}_2\}}{\dfrac{T}{T^*}\{\mathbf{M}_1\}}. \qquad (10.98)$$

The terms on the right are calculated by Eqs. (10.93)–(10.95) or, more frequently, obtained from tables.

EXAMPLE 10.8   Illustrates computations in Fanno flow

A converging nozzle is connected to a large tank of air by a long pipe (see Fig. E10.8). The flow rate through the nozzle is 10 lbm/sec and the nozzle is choked. Calculate the static and stagnation pressures at planes 1 and 2.

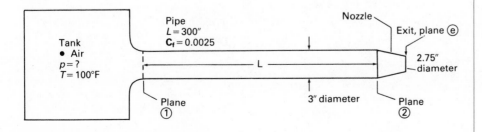

Figure E10.8 Tank-pipe-nozzle system.

Solution

Given

Converging nozzle connected to air tank by pipe with $L = 300$ in., $D = 3.0$ in., $C_f = 0.0025$
Nozzle exit diameter 2.75 in.
Nozzle passes 10 lbm/sec of air and is choked.
Air temperature in tank 100°F

Find

Static and stagnation pressures at pipe exit/nozzle entrance and at tank exit/pipe entrance

Solution

First convert given temperatures to absolute units:

$$T_{\text{tank}} = 100°F + 459.7°F = 559.7°R.$$

To solve the problem, we break the system into parts. We assume isoenergetic-isentropic flow in the nozzle and isoenergetic frictional (Fanno) flow in the pipe. Since we are given that the nozzle is choked, we start at the nozzle and work backward.

The nozzle flow rate is related to the nozzle stagnation pressure by Eq. (10.72), which, for $k = 1.4$, is

$$\dot{m} = 0.6847\left(\frac{p_{0_e}A_e}{\sqrt{RT_{0_e}}}\right),$$

where

$$A_e = \frac{\pi}{4}D_e^2 = 5.94 \text{ in}^2 \quad \text{and} \quad T_{0_e} = T_0 = T_{\text{tank}} = 559.7°R.$$

Solving for p_{0_e}, we have

$$p_{0_e} = \frac{\dot{m}\sqrt{RT_0}}{0.6847A_e}.$$

Since the nozzle flow is isentropic,

$$p_{0_2} = p_{0_e} = \frac{10 \dfrac{\text{lbm}}{\text{sec}} \sqrt{53.35 \dfrac{\text{ft} \cdot \text{lb}}{\text{lbm} \cdot °R} (559.7°R)}}{(0.6847)(5.94 \text{ in}^2)\sqrt{32.2 \dfrac{\text{ft} \cdot \text{lbm}}{\text{lb} \cdot \text{sec}^2}}},$$

$$\boxed{p_{0_2} = 74.9 \text{ psia.}}$$ **Answer** ◀

The static pressure at 2 is

$$p_2 = p_{0_2}\left(\frac{p_2}{p_{0_2}}\right).$$

To find p_2/p_{0_2}, we note that

$$\left.\frac{A}{A^*}\right)_2 = \frac{A_2}{A_e} = \left(\frac{D_2}{D_e}\right)^2 = \left(\frac{3.0 \text{ in.}}{2.75 \text{ in.}}\right)^2 = 1.190.$$

From Table E.1,

$$\mathbf{M}_2 = 0.599, \qquad \frac{p_2}{p_{0_2}} = 0.7846.$$

Thus

$$p_2 = 74.9 \text{ psia} (0.7846),$$

$$\left| \; p_2 = 58.8 \text{ psia}. \; \right|$$

Answer

To calculate the static and stagnation pressures at plane 1, we apply Fanno flow analysis between planes 1 and 2. From Eq. (10.92),

$$\left(\frac{4C_f \ell^*}{D_h} \right)_1 = 4C_f \left(\frac{\ell^*}{D_h} \right)_2 + 4C_f \frac{\Delta x}{D_h}.$$

From Table E.3, with $M_2 = 0.599$,

$$\left(\frac{4C_f \ell^*}{D_h} \right)_2 = 0.4948.$$

Also,

$$4C_f \frac{\Delta x}{D_h} = 4C_f \frac{L}{D} = 4(0.0025) \left(\frac{300 \text{ in.}}{3 \text{ in.}} \right) = 1.0,$$

so,

$$4C_f \left(\frac{\ell^*}{D_h} \right)_1 = 1.0 + 0.4948 = 1.4948.$$

From Table E.3

$$M_1 = 0.456.$$

The static pressure at plane 1 is given by Eq. (10.96):

$$p_1 = p_2 \left(\frac{\dfrac{p}{p^*} (M_1)}{\dfrac{p}{p^*} (M_2)} \right).$$

With the aid of Table E.3, we have

$$p_1 = 58.8 \text{ psia} \left(\frac{2.354}{1.767} \right),$$

$$\left| \; p_1 = 78.3 \text{ psia}. \; \right|$$

Answer

The stagnation pressure at plane 1 is

$$p_{0_1} = \frac{p_1}{\dfrac{p_1}{p_0} (M_1)} = \frac{78.3}{0.8671} \text{ psia},$$

$$\left| \; p_{01} = 90.3 \text{ psia}. \; \right|$$

Answer

Discussion

This complicated-looking problem became relatively simple when we broke it into pieces. Note that although we cannot use isentropic tables and functions to relate the flow at plane 1 to the flow at plane 2, we do use the isentropic tables to relate static and stagnation properties at the same location.

An important question for many practical applications is how to find C_f. From dimensional analysis, we expect

$$C_f = C_f\left(R, M, \frac{\varepsilon}{D}\right).$$

Based on a rather small amount of experimental evidence, it appears that the dependence of C_f on M is weak, at least for subsonic flow; accordingly, values of C_f for $M = 0$ (incompressible flow) are usually sufficient. For fully developed flow, we can obtain values for C_f from the Moody chart (Fig. 7.11), the Colebrook equation (7.48) or the Hagen-Poiseuille equation (7.38) by noting that the Darcy friction factor thus obtained is related to the skin friction coefficient by

$$C_f = \frac{f}{4}. \tag{10.99}$$

The factor of 4 is rather convenient, as we can write

$$4C_f \frac{\Delta x}{D_h} = f \frac{\Delta x}{D_h}.$$

Since f depends on Reynolds number, our assumption of a constant value of C_f along the duct seems questionable. The Reynolds number can be written

$$R = \frac{\rho V D_h}{\mu} = \frac{\dot{m}}{A} \frac{D_h}{\mu} = \frac{\dot{m}}{\mu} \frac{D_h}{A}. \tag{10.100}$$

The mass flow rate, hydraulic diameter, and cross-sectional area are all constant, so the variation of Reynolds number along the duct is due only to viscosity. For a gas, viscosity varies rather slowly with temperature ($\mu \sim T^{0.7}$). In addition, the dependence of f on R is weak in turbulent flow, which is the usual condition at velocities high enough for compressibility to be significant.

10.5.3 Frictional Flow in a Duct

We now consider the various possibilities for frictional flow in a duct. Figure 10.20 shows a duct that exhausts fluid from a large reservoir to a region with constant back pressure. We assume that the connection between the duct and the inlet reservoir is a converging nozzle with minimum area equal to the duct area.[†] The flow entering the duct must then be subsonic. The maximum possible Mach number anywhere in the flow is 1, and this can occur only at the *end* of the *duct*. The nozzle exit–duct entrance Mach number *cannot* be 1 unless the duct is frictionless. Figure 10.21 shows the pressure and Mach number distributions along the duct, the dimensionless mass flow rate, and the duct exit pressure, all as functions of the ratio of back pressure to supply pressure (p_B/p_{0_1}). In the case labeled a, the pressures are equal and there is no flow. In case b, the back pressure is lowered and there is flow. The gas accelerates in the inlet nozzle, enters the duct with a low

---

[†] It is not necessary that the inlet nozzle be smoothly contoured. The duct may be directly attached to a hole in the tank.

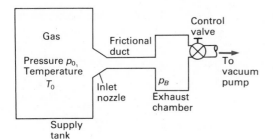

Figure 10.20 Apparatus for studying (subsonic) flow in a frictional duct.

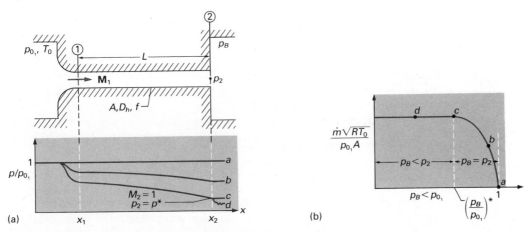

Figure 10.21 Performance of a frictional duct fed by a converging nozzle.

subsonic Mach number, and is accelerated in the duct. At the end of the duct, the flow is still subsonic and the exit pressure and back pressure are equal. The inlet Mach number, M_1, is small enough that

$$4C_f \frac{\ell^*}{D_h} (M_1) > 4C_f \frac{L}{D_h}.$$

In case c, the back pressure has been lowered to the point where the flow has just become sonic at the exit. In this case, the entering flow is still subsonic, but the gas accelerates to $M = 1$ at the duct exit. The exit pressure is equal to p^* and also equal to p_B. The mass flow rate is greater than in case b. The duct entrance Mach number satisfies the condition

$$4C_f \frac{\ell^*}{D_h} (M_1) = 4C_f \frac{L}{D_h}. \tag{10.101}$$

Case d represents a lower back pressure than case c. Since the flow was choked in case c, lowering the back pressure has no effect on the flow in the nozzle and duct. The entrance Mach number is still subsonic, and Eq. (10.101) still applies. The mass flow rate is the same as in case c. The exit plane pressure is equal to p^* and is higher than the back pressure.

The procedure for analysis of flow in a frictional duct depends on whether the flow is choked or unchoked. This question is decided by comparing the actual pressure ratio, p_B/p_{0_1}, to the value that corresponds to choking $(p_B/p_{0_1})^*$. The latter is calculated as follows:

- Calculate $4\mathbf{C}_f \ell^*/D_h$ from Eq. (10.101).
- Find $\mathbf{M}_1$ from Eq. (10.91) or Table E.3.
- Find p_1/p_{0_1} from Eq. (10.31) or Table E.1.
- Find p_1/p^* from Eq. (10.93) or Table E.3.
- Then

$$\left(\frac{p_B}{p_{0_1}}\right)^* = \frac{\left(\dfrac{p_1}{p_{0_1}}\right)}{\left(\dfrac{p_1}{p_B^*}\right)}. \tag{10.102}$$

If p_B/p_{0_1} is less than $(p_B/p_{0_1})^*$, the flow is choked, and the fact that $\mathbf{M}_{exit} = 1$ is the key to solving the problem. If p_B/p_{0_1} is greater than $(p_B/p_{0_1})^*$, then the flow is not choked, and the key to the problem is the equality of p_e and p_B.

As in incompressible duct flow, there are three types of problems:

- The pressure change problem
- The mass flow rate problem
- The duct sizing problem

The pressure change problem requires calculation of fluid property changes between two locations in the duct. Usually, one of these locations is the duct exit. In this problem, the duct dimensions, mass flow rate, and fluid properties at one location are given either explicitly or implicitly.[†] The Reynolds number and relative roughness can be calculated from the given data, and the friction coefficient is readily obtained. Fluid properties and velocity or Mach number at other locations in the duct can be calculated by direct application of Eqs. (10.91)–(10.98) or by using tables. Example 10.8 illustrated this type of problem.

Both the mass flow rate problem and the duct sizing problem require iterative solution. In both of these problems, duct length, inlet (static or stagnation) pressure, and exhaust region pressure are given. The fluid stagnation temperature is also given, either explicitly or implicitly. The friction factor depends on both $\dot{m}$ and D_h via the Reynolds number and relative roughness. The flow may be choked or unchoked, depending on the actual values of p_B, p_{0_1}, and $4\mathbf{C}_f(L/D_h)$. The mass flow and duct sizing problems are closely related because, at the duct inlet, the mass flow and duct area are related to each other and to p_{0_1}, T_0, and $\mathbf{M}_1$ by Eq. (10.71). Solution of either the mass flow rate problem or the duct sizing problem is carried out by iterating on the duct inlet Mach number $\mathbf{M}_1$. Figure 10.22 is a flow chart for a suggested iteration procedure. The procedure has the following characteristics:

- It will work for either the mass flow or duct sizing problem.
- It will work for either choked or unchoked flow.
- It is a successive approximation procedure that gives an updated value of $\mathbf{M}_1$ for use in the next cycle.

[†] If pressure, temperature, Mach number, and cross-sectional area are known, then mass flow rate can be immediately calculated.

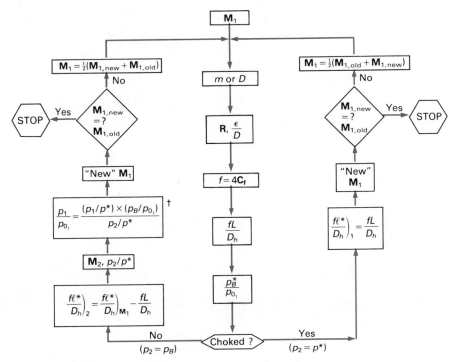

Figure 10.22 Flow chart for frictional duct flow calculation.
†This calculation is based on the requirement that $p_2 = p_B$ for unchoked flow.

EXAMPLE 10.9   Illustrates pipe sizing in compressible flow

An astronaut working in space breathes pure oxygen that is supplied to a spacesuit through an umbilical cord (Fig. E10.9a). The umbilical is 7 m long. The oxygen is supplied from a storage tank at 200 kPa and 9.8°C. The astronaut's suit pressure is 20 kPa. Compute the required inside diameter of the umbilical to supply 0.05 kg/s of oxygen to the astronaut. The inside of the umbilical is lined with a material with roughness of 0.01 mm.

Solution

Given
7-m-long umbilical (duct) with surface roughness of 0.01 mm ($= 10^{-5}$ m)
Flow rate 5×10^{-2} kg/s to be delivered from reservoir at 200 kPa and
9.8°C to region (space suit) at 20 kPa

Find
Required inside diameter of umbilical

Solution
Figure E10.9(b) is a schematic of the duct flow problem. First convert all given pressures and temperatures to absolute units:

$$T_{tank} = 9.8°C + 273.2°C = 283K.$$

The flow in the umbilical will obviously be subsonic (no converging-diverging nozzle at the inlet), so the problem can be solved by following the flow chart in

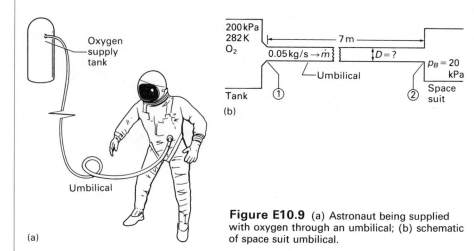

Figure E10.9 (a) Astronaut being supplied with oxygen through an umbilical; (b) schematic of space suit umbilical.

(a)

Fig. 10.22. We first obtain the properties of oxygen. They are $R = 260$ N·m/kg·K, $k = 1.4$, and $\mu = 1.97 \times 10^{-5}$ N·s/m² from Table A.3.

For a circular duct, from Eq. (10.100),

$$\mathbf{R} = \frac{\dot{m}}{\mu} \frac{D}{\dfrac{\pi D^2}{4}} = \frac{4\dot{m}}{\pi \mu D}.$$

Substituting numbers where possible, we have

$$\mathbf{R} = \frac{(4)\left(0.05 \,\dfrac{\text{kg}}{\text{s}}\right)}{(\pi)\left(1.97 \times 10^{-5} \,\dfrac{\text{N·s}}{\text{m}^2}\right)D} = \frac{3230}{D}, \tag{E10.1}$$

where D is in meters.

The equation for flow rate at plane 1 is Eq. (10.71). Since $\dot{m}$ is known and the area A is unknown, the equation can be rearranged to give the following by substituting $A = \pi D^2/4$:

$$D^2 = \frac{4}{\pi}\left(\frac{\dot{m}\sqrt{RT_0}}{p_{0_1}\sqrt{k}}\right)\left[\frac{1}{\mathbf{M}_1}\left(1 + \frac{k-1}{2}\mathbf{M}_1^2\right)^{(k+1)/2(k-1)}\right].$$

We can considerably speed up calculations if we recognize that the quantity in square brackets is (see Eq. 10.67)

$$\left(\frac{k+1}{2}\right)^{(k+1)/2(k-1)} \frac{A}{A^*}(\mathbf{M}_1) = 1.728 \frac{A}{A^*}(\mathbf{M}_1),$$

for $k = 1.4$. Thus

$$D^2 = \frac{4}{\pi}\left[\frac{\left(0.05\,\dfrac{\text{kg}}{\text{s}}\right)\sqrt{\left(260\,\dfrac{\text{N·m}}{\text{kg·K}}\right)(283\text{K})}}{\left(200{,}000\,\dfrac{\text{N}}{\text{m}^2}\right)\sqrt{1.4}}\right](1.728)\frac{A}{A^*}(\mathbf{M}_1)$$

and

$$D = \sqrt{D^2} = 0.0112\left(\sqrt{\frac{A}{A^*}(\mathbf{M}_1)}\right). \tag{E10.2}$$

The ratio of back pressure to supply pressure is

$$\frac{p_B}{p_{0_1}} = \frac{20 \text{ kPa}}{200 \text{ kPa}} = 0.1.$$

We now begin following the flow chart of Fig. 10.22. We guess

$$\mathbf{M}_1^{(1)} = 0.5.$$

The superscript (1) indicates that this is our first guess. From Table E.1,

$$\frac{A_1}{A^*} = 1.340.$$

From Eq. (E10.2),

$$D = 0.0112\sqrt{1.340} = 0.013 \text{ m}.$$

From Eq. (E10.1),

$$\mathbf{R} = \frac{3230}{0.013} = 2.48 \times 10^5.$$

Also

$$\frac{\varepsilon}{D} = \frac{1 \times 10^{-5}\,\text{m}}{1.3 \times 10^{-2}\,\text{m}} = 0.00077.$$

From Fig. 7.11 (the Moody chart),

$$f = 4\mathbf{C_f} = 0.020.$$

The duct length parameter is

$$4\mathbf{C_f}\frac{L}{D} = (0.020)\left(\frac{7\text{ m}}{0.013\text{ m}}\right) = 10.77.$$

Next we calculate the choking pressure ratio for this duct length parameter. Let $\mathbf{M}_{1_c}$ be the inlet Mach number corresponding to choking. From Table E.3, for $4\mathbf{C_f}\ell^*/D_h = 10.77$,

$$M_{1_c} = 0.227 \quad \text{and} \quad \frac{p}{p^*}(\mathbf{M}_{1_c}) = 4.801.$$

From Table E.1,

$$\frac{p}{p_0}(\mathbf{M}_{1_c}) = 0.9648.$$

From Eq. (10.102),

$$\frac{p_B^*}{p_{0_1}} = \frac{0.9648}{4.801} = 0.201.$$

Since

$$\frac{p_B}{p_{0_1}} = 0.10 < \frac{p_B^*}{p_{0_1}} = 0.201,$$

the duct seems to be choked and we follow the right branch of Fig. 10.22. Setting $4\mathbf{C_f}\ell^*/D_{h_1} = 4\mathbf{C_f}(L/D) = 10.77$, we find that the "new" Mach number is

$$\mathbf{M}_{1,\text{new}} = \mathbf{M}_{1_c}.$$

Since $\mathbf{M}_{1_c} = 0.227 \neq 0.5 = \mathbf{M}_1^{(1)}$, we now try

$$\mathbf{M}_1^{(2)} = \frac{0.227 + 0.5}{2} = 0.36.$$

From Table E.1,

$$\frac{A_1}{A^*} = 1.736.$$

From Eqs. (E10.2) and (E10.1), $D = 0.0148$ m and $\mathbf{R} = 2.18 \times 10^5$. Also

$$\frac{\varepsilon}{D} = \frac{10^{-5} \text{ m}}{1.48 \times 10^{-2} \text{ m}} = 6.76 \times 10^{-4}$$

and, from Fig. 7.11,

$$f = 4\mathbf{C_f} = 0.019.$$

This gives

$$4\mathbf{C_f}\frac{L}{D} = 0.019\left(\frac{7 \text{ m}}{0.0148 \text{ m}}\right) = 8.99.$$

From Tables E.3 and E.1,

$$\mathbf{M}_{1_c} = 0.244, \qquad \frac{p}{p^*}(\mathbf{M}_{1_c}) = 4.463, \qquad \frac{p}{p_0}(\mathbf{M}_{1_c}) = 0.9594,$$

and so

$$\frac{p_B^*}{p_{0_1}} = \frac{0.9594}{4.463} = 0.215.$$

The duct seems to be choked and so the "new" Mach number is $\mathbf{M}_{1_c}$. We now repeat the procedure with

$$\mathbf{M}_1^{(3)} = \frac{1}{2}(0.36 + 0.244) = 0.30.$$

The results of this calculation are

$$D = 0.016 \text{ m}, \qquad \mathbf{R} = 2.02 \times 10^5, \qquad \frac{\varepsilon}{D} = 6.25 \times 10^{-4},$$

$$f = 0.019, \qquad 4\mathbf{C_f}\frac{L}{D} = 8.31, \qquad \mathbf{M}_{1_c} = 0.252,$$

$$\frac{p_B^*}{p_{0_1}} = 0.221. \qquad \text{Flow choked}$$

This implies

$$\mathbf{M}_1^{(4)} = \frac{1}{2}(0.30 + 0.252) = 0.28.$$

The results for this trial are

$$D = 0.0165 \text{ m}, \qquad \mathbf{R} = 1.96 \times 10^5, \qquad \frac{\varepsilon}{D} = 6.07 \times 10^{-4},$$

$$f = 0.19, \qquad 4\mathbf{C_f}\frac{L}{D} = 8.060, \qquad \mathbf{M}_{1_c} = 0.255,$$

$$\frac{p_B^*}{p_{0_1}} = 0.224. \qquad \text{Choked}$$

Examining the calculations, we note that $\mathbf{M}_{1_c}$ has been changing very little. We make a calculation with

$$\mathbf{M}_1^{(5)} = 0.26.$$

The parameters of this calculation are

$$D = 0.017 \text{ m}, \qquad \mathbf{R} = 1.90 \times 10^5, \qquad \frac{\varepsilon}{D} = 5.88 \times 10^{-4},$$

$$f = 0.019, \qquad 4\mathbf{C}_f \frac{L}{D} = 7.82, \qquad \mathbf{M}_{1_c} = 0.258,$$

$$\frac{p_B^*}{p_0} = 0.226. \qquad \text{Choked}$$

Since $\mathbf{M}_{1_c} = 0.258 \approx 0.260 = \mathbf{M}_1^{(5)}$, the iteration has converged and we specify

Answer

$$\mid D = 0.017 \text{ m} = 1.7 \text{ cm.} \mid$$

Discussion

The iteration procedure converged very slowly. This is due to the fact that we averaged the "new" and "old" Mach numbers before repeating calculations. A table of values of "old" and "new" Mach numbers might be useful:

Trial	$\mathbf{M}_{old}$	$\mathbf{M}_{new}$
1	0.5	0.227
2	0.36	0.244
3	0.30	0.252
4	0.28	0.255
5	0.26	0.258

Since the "new" Mach numbers are changing very slowly, it is likely that they are close to the correct value. It might be a good idea to "weigh" them more heavily by an equation such as

$$\mathbf{M}_1^{(i+1)} = a\mathbf{M}_{1,old}^{(i)} + (1-a)\mathbf{M}_{1,new}^{(i)},$$

where a is selected between 0 and 1. Selecting $a = 0$ would produce more rapid convergence in the example since the "new" Mach numbers are always smaller than the "old" Mach numbers. If the "new" Mach numbers are alternately larger and smaller than the "old" ones, the iteration can become unstable if $a = 0$. Unstable iteration can occur on the left branch of the flow chart of Fig. 10.22 (nonchoked flow), so it is a good idea to stick with $a = 0.5$ in that case. Note that we used $a = 0.2$ in our last iteration.

Our discussion has concentrated on subsonic frictional flow. A discussion of supersonic frictional flow would be interesting but of limited practical value compared to the effort expended. There are two reasons for this. Equation (10.90) shows that loss of stagnation pressure is very large if the Mach number is greater than 1, so supersonic flow is usually avoided if friction is significant. Second, interaction between shocks, which can occur in supersonic flow, and boundary layers is a complex multidimensional process. The one-dimensional Fanno flow model is not sufficiently accurate to be of value. If you are interested in discussions of supersonic Fanno flow, see [1–4].

10.6 Effects of Compressibility in External Flow

Before ending our consideration of compressible flow, we will briefly consider the effects of compressibility in external flow. This discussion will be mainly qualitative. Suppose an object such as an airfoil is moving at high speed through a compressible fluid (Fig. 10.23). The nature of the flow over the airfoil will be different for subsonic, transonic, and supersonic motion. The key to understanding the differences is the discussion of Section 10.2.2. If the flow is subsonic everywhere, disturbances introduced by the airfoil motion are propagated to all parts of the fluid. The flow pattern (Fig. 10.23a) is similar to that which we would expect in a low-speed (incompressible) flow.

If the flow is supersonic (Fig. 10.23b), the upstream influence of the airfoil is limited, extending no farther than a leading shock wave. The disturbances caused by the airfoil will be concentrated in waves. In general, the fluid will contain both waves of finite strength (shocks) and waves of infinitesimal strength (Mach waves). Shock waves result from sudden compression of the fluid, and Mach waves result from expansion or gradual compression of the fluid. In external flow, these waves are multidimensional and cause multidimensional flow. Figure 10.24 illustrates these waves in an actual flow. Although the wave pattern and the resulting flow appear to be very complex, it is possible to analyze two- and three-dimensional supersonic flow by first analyzing the basic waves (oblique shocks and Mach waves) and then "patching" the waves together to obtain a complete flow field [1–6].

Figure 10.23(c) illustrates a transonic flow, in which some regions of the field are subsonic and some regions are supersonic. Shock waves exist in the supersonic portion of the flow. Analysis of transonic flow is very difficult because it is neither entirely like subsonic flow nor entirely like supersonic flow.

Figure 10.23 Airfoil in compressible flow: (a) Subsonic flow; (b) supersonic flow; (c) transonic flow.

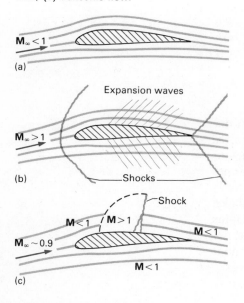

Figure 10.24 Supersonic flow over an object is influenced by a complex wave pattern. (Courtesy of Launch and Flight Division, Ballistic Research Laboratory/ARRADCOM, Aberdeen Proving Ground, MD.)

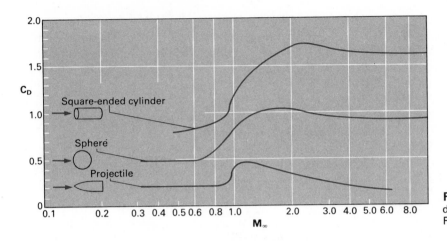

Figure 10.25 Effect of Mach number on drag of three different axisymmetric objects. Reynolds number effects are negligible.

Dimensional analysis of compressible external flow indicates that drag and lift coefficients depend on Mach number as well as Reynolds number:

$$C_D = C_D(\alpha, R, M_\infty) \quad \text{and} \quad C_L = C_L(\alpha, R, M_\infty).$$

It is very difficult to obtain test data in which both Reynolds number and Mach number are controlled (see Section 6.6). Most studies concentrate on the effects of Reynolds number alone (usually at low Mach number; see Chapter 8) or on the effects of Mach number alone. If the Reynolds number is high, then

$$C_D \approx C_D(\alpha, M_\infty \text{ only}) \quad \text{and} \quad C_L \approx C_L(\alpha, M_\infty \text{ only}).$$

Figure 10.25 illustrates the effect of Mach number on drag coefficient for three axisymmetric bodies. We note the following important trends from this figure:

- The drag coefficient increases with Mach number at subsonic speeds and decreases with Mach number at supersonic speeds.
- The drag coefficient at supersonic speeds is higher than the drag coefficient at low subsonic speeds.
- The drag coefficient rises sharply near $M_\infty = 1$.
- The value of the Mach number at which the "drag rise" occurs is pushed closer to 1 by making the body more "streamlined."

These trends can be explained as follows. Although the approaching fluid is warned of the presence of the object in subsonic flow, the strength of the upstream propagation decreases with increasing Mach number. Adjustments to the presence of the body occur closer to the body at higher Mach numbers, causing an increase in pressure drag. In supersonic flow, energy is required to maintain the wave pattern, so the drag is higher than in subsonic flow with no waves. The drag rise near $M_\infty = 1$ is associated with the sudden appearance of (nearly) normal shock waves in locally supersonic regions of the flow. The drag drop after $M_\infty = 1$ corresponds to a relative weakening of these shock waves as they become inclined to the flow. The drag rise delay for a more streamlined body is related to the fact that the first appearance of locally supersonic flow and shock waves is delayed by streamlining.

The value of $\mathbf{M}_\infty$ at which the highest local Mach number in the field first becomes 1 is called the *critical Mach number*. Note that since the Mach number near the maximum body thickness is always greater than the free-stream Mach number, the critical Mach number is always less than 1. The "drag rise" occurs at a Mach number slightly higher than the critical Mach number. The drag rise that occurs near "Mach One" is the true sound barrier that had to be overcome before supersonic flight was possible.

The effects of Mach number on lift and drag at subsonic and supersonic (but *not* transonic) speed are illustrated by the following equations [1, 3, 6].

- Prandtl-Glauert rules for subsonic flow:

$$C_D(M_\infty) \approx \frac{C_D(M_\infty = 0)}{\sqrt{1 - M_\infty^2}}, \tag{10.103}$$

$$C_L(M_\infty) \approx \frac{C_L(M_\infty = 0)}{\sqrt{1 - M_\infty^2}}, \tag{10.104}$$

where $\mathbf{M}_\infty = 0$ means incompressible flow.

- Ackeret rules for supersonic flow:

$$C_D(M_\infty) \approx \frac{\text{Constant}}{\sqrt{M_\infty^2 - 1}}, \tag{10.105}$$

$$C_L(M_\infty) \approx \frac{\text{Constant}}{\sqrt{M_\infty^2 - 1}}. \tag{10.106}$$

These rules apply only to *thin, two-dimensional* objects. The equations obviously cannot be valid when $\mathbf{M}_\infty$ is 1 and become less accurate as $\mathbf{M}_\infty$ approaches 1. The angle of attack must be constant and, strictly speaking, so must the Reynolds number. The "constant" in the Ackeret rules for supersonic flow is not the incompressible flow value of the corresponding coefficient because supersonic flow is not qualitatively or quantitatively similar to incompressible flow.[†] The "incompressible" flow coefficients and the "constants" in the Ackeret rules are functions of body shape and Reynolds number. It is best to consider Eqs. (10.103)–(10.106) as illustrative rather than directly useful for calculations.

[†] Obviously, the "constants" are the values of the coefficients when $\mathbf{M}_\infty = \sqrt{2}$.

EXAMPLE 10.10   Illustrates aerodynamic force calculations in compressible flow

The astronaut from Example 10.9 enters a spherical reentry capsule and returns to Earth. The reentry capsule is 2 m in diameter and has a mass of 1250 kg. At one point during the reentry, the capsule's altitude is 30 km, and it is traveling with a speed of 2300 m/s at an angle of 20° below the horizontal. Calculate the acceleration of the capsule in g's. Also estimate the temperature on the windward surface of the capsule.

Solution

Given

1250-kg spherical reentry capsule travelling at 2300 m/s at altitude of 30 km
Capsule velocity vector 20° below the horizontal
Capsule diameter 2 m
Figure E10.10(a)

Find

Capsule acceleration in g's
Temperature of windward surface

Solution

The capsule acceleration is computed from Newton's second law:

$$\vec{a} = \frac{\sum \vec{F}}{m} = \frac{\vec{F}_R}{m}.$$

Figure E10.10(b) shows a free body diagram of the capsule. The x-axis is aligned with the direction of travel. The mass m includes the mass of the capsule and the mass of the astronaut inside. Since the astronaut's mass is not given, we will have to assume a reasonable value, say 80 kg. Then

$$m = 1250 \text{ kg} + 80 \text{ kg} = 1330 \text{ kg}.$$

The forces on the capsule are the gravity force $\vec{W}$ and the aerodynamic drag, $\vec{\mathscr{D}}$. The magnitude of $\vec{W}$ is

$$W = mg.$$

At an altitude of 30 km, the value of g is found from Table A.1:

$$g = 9.72 \text{ m/s}^2.$$

This gives

$$W = (1330 \text{ kg})(9.72 \text{ m/s}^2) = 1.29 \times 10^4 \text{ N}.$$

The magnitude of the drag force is

$$\mathscr{D} = \mathbf{C_D}\left(\frac{1}{2}\rho V^2\right)\left(\frac{\pi}{4}D^2\right),$$

where D is the capsule diameter. From Eq. (10.29), we have

$$\rho V^2 = kp\mathbf{M}^2, \quad \text{so} \quad \mathscr{D} = \frac{\pi}{8}\mathbf{C_D}kp\mathbf{M}^2 D^2.$$

Figure E10.10 (a) Spherical reentry capsule in flight; (b) free-body diagram of reentry capsule. F_R is the resultant force.

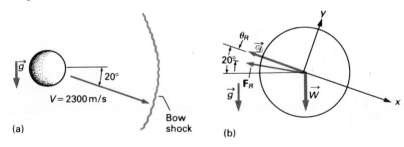

(a)

(b)

Figure 10.25 gives drag coefficients as a function of Mach number for spheres. To compute the Mach number, we need the temperature. From Eqs. (2.7)–(2.9), Fig. 2.6, or Table A.1, $T = 226.5$K and $p = 1.20$ kPa at 30-km altitude. Then

$$c = \sqrt{kRT} = (20.05\sqrt{226.5}) \text{ m/s} = 302 \text{ m/s}$$

and

$$\mathbf{M} = \frac{V}{c} = \frac{2300 \text{ m/s}}{302 \text{ m/s}} = 7.62.$$

From Fig. 10.25,

$$\mathbf{C_D} \approx 0.9,$$

so

$$\mathcal{D} = \frac{\pi}{8}(0.9)(1.4)\left(1.20 \times 10^3 \frac{\text{N}}{\text{m}^2}\right)(7.62)^2(2\text{m})^2$$

$$= 1.38 \times 10^5 \text{ N}.$$

Summing forces in the y-direction (see Fig. E10.10b), we have

$$\sum F_y = W \cos 20° = -1.21 \times 10^4 \text{ N}.$$

Summing forces in the x-direction, we have

$$\sum F_x = W \sin 20° - \mathcal{D}$$
$$= (1.29 \times 10^4 \text{ N}) \sin 20° - 13.8 \times 10^4 \text{ N} = -13.4 \times 10^4 \text{ N}.$$

The resultant force has magnitude

$$F_R = \sqrt{(1.21)^2 + (13.4)^2} \times 10^4 \text{ N} = 13.5 \times 10^4 \text{ N}.$$

The angle of the resultant force is

$$\theta_R = \tan^{-1}\left(\frac{F_y}{F_x}\right) = \tan^{-1}\left(\frac{-1.21}{-13.4}\right) = 5.2°.$$

The acceleration has the same direction as the resultant force, that is, it is directed away from the direction of travel and $(20° - 5.2°)$ 14.8° *above* the horizontal. The magnitude of the acceleration is

$$a = \frac{F_R}{m} = \frac{135,000 \text{ N}}{1330 \text{ kg}} = 101.5 \text{ m/s}^2.$$

Expressing the answer in g's (1 g = 9.807 m/s^2), we have

| $a = 10.35$ g's, opposite direction of travel and 14.8° above the horizontal. | **Answer** ◀ |

The temperature on the windward surface of the reentry capsule is approximately equal to the stagnation temperature relative to the capsule. Note that the bow shock does not affect the stagnation temperature. From Eq. (10.30),

$$T_0 = T\left(1 + \frac{k-1}{2}\mathbf{M}^2\right) = 226.5\text{K}[1 + 0.2(7.62)^2],$$

| $T_{\text{windward surface}} \approx T_0 = 2860\text{K }(4690°\text{F}).$ | **Answer** ◀ |

Discussion

Note that the strong bow shock ahead of the reentry capsule had no effect on our calculations. All of the objects in Fig. 10.25 will generate shocks when traveling at supersonic speed. The Mach number must be interpreted as the value far ahead of the object (i.e., upstream of the shock).

The deceleration of 10.35 g's is a little severe. A peak value of about 7 g's is generally taken as the safe upper limit for sustained exposure of humans. Note that the standard g, 9.807 m/s², is used.

Our surface temperature estimate is too high for two reasons. First, non-ideal gas effects are important at such high temperatures, so the true value of T_0 is about 10% lower than the calculated value. Second, the surface temperature will be somewhat lower than the stagnation temperature of the gas. The true temperature of the surface is called the *recovery temperature* and is calculated by [1]

$$T_{surf} = T\left[1 + r\left(\frac{k-1}{2}\right)\mathbf{M}^2\right],$$

where r, the recovery factor, is about 0.9. Combining recovery factor and non-ideal gas effects yields a surface temperature estimate of

$$T_{\substack{windward \\ surface}} \approx 2350\text{K} \ (3800°\text{F}).$$

Obviously some sort of heat shield is necessary to prevent an astronaut roast!

Problems

Extension and Generalization

1. Develop differential equations for the temperature and pressure changes for the flow of a gas through a constant-area duct with friction and specific heat $c_p = c_p(T)$. The gas still obeys the ideal gas law. This result would be useful for gas flows with a large temperature change along the duct.

2. Consider the frictional flow of an ideal gas through a very long pipe in a constant-temperature environment with heat transfer between the gas and the environment. The pipe is long enough that the fluid can be considered to have constant temperature throughout the entire length (i.e., the pipe entrance length along which the gas temperature changes is relatively short). Develop an expression for the differential pressure change along the pipe.

3. Repeat the analysis related to Fig. 10.19 for a frictional duct fed by a converging-diverging nozzle. Note the various regimes that can occur, specifically noting when the flow is choked, when shocks appear in either the nozzle or the duct, and when and where the flow is supersonic.

Lower-Order Application

4. A jet is flying at a local Mach number of 2.2 at 15,000 m in the Standard Atmosphere. Find the jet's velocity in km/hr.

5. Air at 25°C and 101 kPa is compressed adiabatically and reversibly to 404 kPa. Find the final density.

6. An airplane is flying at a flight (or local) Mach number of 0.70 at 10,000 m in the Standard Atmosphere. Find the ground speed (a) if the air is not moving relative to the ground and (b) if the air is moving at 30 km/hr in the opposite direction from the airplane.

7. A missile is moving parallel to the ground at Mach 4.0 at an altitude of 10,000 ft. How long after the missile is directly overhead will an observer on the ground hear any sound? Assume the Standard Atmosphere and still air.

8. An airplane flies at Mach 2.0 at an altitude where the static pressure and temperature are 5 psia and 0°F. A normal shock stands just ahead of the engine inlet. Calculate the stagnation pressure and air velocity (relative to the airplane) just downstream of the shock.

9. A jet-propelled airliner flies at an altitude of 15,000 m and a velocity of 265 m/s. What is the stagnation pressure and stagnation temperature of the air entering the engine? If the airliner is stationary on the ground, the engines are running with the same Mach number flow into the engines, and the air flow to each engine is 40 kg/s, what

is the airflow to each engine at the above flight conditions?

10. Consider the flow process in Fig. P10.1. Does the fluid flow from left to right or from right to left? Justify your answer.

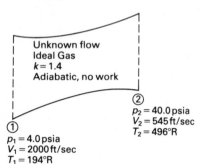

Unknown flow
Ideal Gas
$k = 1.4$
Adiabatic, no work

② $p_2 = 40.0\,psia$
$V_2 = 545\,ft/sec$
$T_2 = 496°R$

① $p_1 = 4.0\,psia$
$V_1 = 2000\,ft/sec$
$T_1 = 194°R$

Figure P10.1

11. The stagnation pressure in a Mach 2 wind tunnel operating with air is 900 kPa. A 1.0-cm-diameter sphere positioned in the wind tunnel has a drag coefficient of 0.95. Calculate the drag force on the sphere.

12. An airplane is flying at Mach 1.8 at 10,000 m, where the temperature is $-44°C$ and the pressure is 30.5 kPa.
    a) How fast is the airplane traveling in kilometers per hour?
    b) The stagnation temperature is an *estimate* of the surface temperature on the aircraft. What is the stagnation temperature under these conditions? Why is the stagnation temperature only an estimate of the surface temperature?
    c) What is the stagnation pressure?
    d) If the airplane slows down, at what speed (kilometers per hour) will it be traveling in the subsonic range?

13. Calculate the speed of sound in air, helium, and hydrogen. The temperature is 70°F.

14. Air flows through a constant-area, insulated passage. The inlet and exit Mach numbers are 0.80 and 1.00 respectively. Find T_2 and p_2, the entropy change $(\tilde{s}_2 - \tilde{s}_1)$, and the wall frictional force. The duct has 1.0-ft diameter.

15. A rocket is launched vertically upward in the Standard Atmosphere. The rocket's velocity and altitude are given by

$$V = V_e \ln\left(\frac{m_0}{m_0 - \dot{m}t}\right)$$

and

$$z = V_e t\left[1 - \left(\frac{m_0 - \dot{m}t}{\dot{m}t}\right)\ln\left(\frac{m_0}{m_0 - \dot{m}t}\right)\right],$$

where

$$V_e = 2500 \text{ m/s}, \qquad m_0 = 105{,}000 \text{ kg},$$
$$\dot{m} = 500 \text{ kg/s}, \qquad t = \text{time in seconds}$$
$$\text{(after liftoff)}$$

Calculate (relative to a coordinate system fixed on the rocket) the Mach number, stagnation pressure, and dynamic pressure at times of 10 s and 100 s after liftoff.

16. At some point for air flow in a duct, $p = 20$ psia, $T = 500°R$, and $V = 500$ ft/sec. Can a normal shock occur at this point?

17. A normal shock exists in a 400-m/s stream of nitrogen. The upstream static pressure and temperature are $-40°C$ and 70 kPa. Calculate the Mach number, pressure, and temperature downstream of the shock and the entropy change across the shock.

18. A Pitot tube is used to measure a supersonic Mach number. A detached normal shock appears just upstream of the tube, which measures a stagnation pressure of 20.0 psia. The static pressure upstream of the shock is 7.0 psia. Determine the Mach number just upstream of the normal shock.

19. A normal shock propagates at 2000 ft/sec into the still air in a tube. The temperature and pressure of the air are 80°F and 14.7 psia before "hit" by the shock. Calculate the air temperature, pressure, and velocity after the shock, the stagnation temperature and pressure *relative to the shock* ahead of and behind the shock, and the stagnation temperature and pressure *relative to the tube* ahead of and behind the shock.

20. Air at $V_1 = 800$ m/s, $p_1 = 100$ kPa, and $T_1 = 300$K passes through a normal shock. Calculate the velocity V_2, temperature T_2, and pressure p_2 after the shock. What would be the values of T_2 and p_2 if the same velocity change were accomplished isentropically?

21. Steam at 500 psia and 1000°F flows at 500 ft/sec in the reheat pipe of a power plant. The reheat stop valve is suddenly closed, generating a shock wave that travels upstream at a velocity of 2090 ft/sec. See Fig. P10.2(a). Figure P10.2(b) shows a transformation to a coordinate system fixed to the shock.

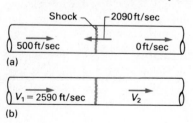

Shock — 2090 ft/sec

500 ft/sec — 0 ft/sec

(a)

$V_1 = 2590$ ft/sec — V_2

(b)

Figure P10.2

Assume steam is an ideal gas with $k = 1.3$ and find V_2 and M_1 for the stationary shock. Then find the steam temperature and pressure behind the moving shock.

22. The cylindrical rocket in Fig. P10.3 moves upward through the Standard Atmosphere at a constant velocity of 320 m/s. The drag coefficient for the rocket is also shown in the figure. Calculate the drag force at altitudes of 3500 m and 20,000 m, the altitude at which a shock wave first appears ahead of the rocket, and the pressure on the nose of the rocket at altitudes of 3500 m and 20,000 m.

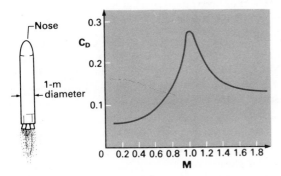

Figure P10.3

23. Air enters the primary nozzle of the jet pump in Fig. P10.4 at 300 kPa, 350K, and 20 m/s. The nozzle exit area and pressure are 0.5 m^2 and 50 kPa. The secondary flow rate is to be 3 times the primary flow rate. Find the total flow rate and the pressure and area at cross section 2.

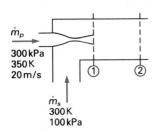

Figure P10.4

24. The nozzle of a rocket is to expand the gases from the combustion chamber to an exit pressure of 2.6×10^4 Pa. The combustion chamber temperature is 3000K, and the pressure is 1500 kPa. The exit area is 1.0 m^2. Find the flow rate for $k = 1.30$ and $R = 190$ N·m/kg·K.

25. An ideal supersonic diffuser reduces air from a Mach number of 3.5 to 0.5. The inlet temperature and pressure are 40°F and 12.0 psia. The exit area is to be 3.0 ft^2. Find the mass flow rate, inlet area, and throat area.

26. The gas entering a rocket nozzle has a stagnation pressure of 1500 kPa and a stagnation temperature of 3000°C. The rocket is traveling in the still Standard Atmosphere at 30,000 m. Find the throat and exit area for a flow rate of 10 kg/s. Assume $k = 1.35$, $R = 287.0$ N·m/kg·K. The gas is perfectly expanded to the ambient pressure.

27. Design a nozzle to isentropically expand steam from 150 psia and 1000°F to 15 psia at a flow rate of 1.0 lbm/sec. The steam may be considered an ideal gas.

28. A simplified schematic of a carburetor of a gasoline engine is shown in Fig. P10.5. The throat area is 0.5 in^2. The engine draws air downward through the carburetor Venturi and maintains a throat pressure of 10.0 psia. This low throat pressure draws fuel from the float chamber and into the air stream. The energy losses in the 0.06-in.-inside-diameter fuel line and valve are given by

$$h_L = \frac{KV^2}{2g},$$

with $K = 3.0$. The fuel specific gravity is 0.75. Assume an atmospheric pressure of 14.72 psia. Find the air-fuel ratio (that is, the ratio of the air mass flow rate to the fuel mass flow rate).

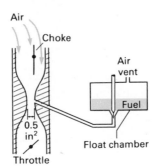

Figure P10.5

29. A new supersonic wing design is tested in a wind tunnel at Mach 1.80. The stagnation pressure and temperature of the wind tunnel are 200 psia and 500°F. The area of the model wing is 100 in^2. The lift and drag coefficients of the model wing are 1.50 and 0.20. Calculate the lift and drag.

30. A tank of oxygen has a hole of area 0.5 cm^2 in its wall. The temperature of the oxygen in the tank is 25°C. Calculate the rate (kg/s) at which oxygen leaks to the atmosphere for tank pressures of 135 kPa and 375 kPa. Assume frictionless flow.

31. Figure P10.6 shows a rocket engine on a test stand. The ambient pressure is p_a and the exit plane pressure is p_e. The exhaust gases have specific heat

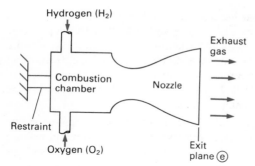

Hydrogen (H₂)

Exhaust gas

Combustion chamber

Nozzle

Restraint

Oxygen (O₂)

Exit plane ⓔ

Figure P10.6

ratio k and exit Mach number $\mathbf{M}_e$ through exit area A_e. Show that the thrust developed by the rocket engine (that is, the force on the restraint) is

$$F_T = kp_eA_e\mathbf{M}_e^2 + (p_e - p_a)A_e.$$

32. The combustion chamber reaction in Problem 31 is

$$H_2 + \frac{1}{2}O_2 \rightarrow H_2O.$$

The rocket engine is designed to produce 100 kN of thrust at an altitude of 10 km where the ambient pressure is 26.5 kPa (the water undergoes an isentropic expansion at this altitude). The combustion chamber temperature and pressure are 6.9 mPa and 3300K. Calculate the design throat and exit areas and the hydrogen and oxygen mass flow rates.

33. Air flows at 12.5 kg/s in a well-insulated nozzle. At one point in the nozzle, the cross-sectional area is 0.01 m², the air pressure is 120 kPa, and the air temperature is 0°C. At a point farther downstream, the air pressure is 90 kPa. The flow between the two points is reversible. Calculate the air velocity at both points and the temperature and area at the downstream point. Assuming that no work is done on the air, what is the *maximum* possible velocity that the air could achieve?

34. Steam enters the nozzles of a turbine stage with $T = 700°F$, $p = 100$ psia, and $V = 100$ ft/sec. At the nozzle outlet, $p = 30$ psia. Heat transfer is negligible and work is zero for the nozzle flow. Calculate the steam velocity and temperature at the nozzle outlet and the stagnation pressure and stagnation temperature at both the nozzle inlet and the nozzle outlet. Also calculate the ratio of the nozzle outlet area to the nozzle inlet area. Assume a reversible process and that the steam is an ideal gas with $k = 1.3$.

35. A simply converging nozzle with a 12.0-cm² exit area is supplied from an air reservoir. The temperature in the reservoir is 38°C and the back pressure is 100 kPa. Calculate the mass flow rate of air for reservoir pressures of 130 kPa and 350 kPa.

36. In a jet engine running on a test stand, the stagnation pressure and stagnation temperature just upstream of the convergent nozzle are measured to be 70 kPa gage and 700°C. The diameter of the nozzle exit is 0.5 m. Calculate the mass flow, jet exit velocity, and thrust of the nozzle.

37. A jet engine is to be designed for an altitude of 12,000 m, where the atmospheric pressure is 19.3 kPa. The jet nozzle has a supersonic exit Mach number and is perfectly expanded. The stagnation pressure and temperature of the gas are 100 kPa and 600°C. The flow rate of gas is 45 kg/s. Calculate the throat area, exit area, and exit velocity. Use $k = 1.3$ and $R = 260$ J/kg·K for the gas.

38. A convergent-divergent nozzle with a throat area of half the exit area is supplied with air at a stagnation pressure of 140 kN/m². It discharges into an atmosphere of static pressure 100 kPa. Find the stagnation pressure of the gas at the nozzle exit.

39. A rocket nozzle with an exit-to-throat-area ratio of 5:1 is operating at a stagnation pressure of 1.3 mPa and exhausting to an atmospheric pressure of 30 kPa. Determine if the nozzle is overexpanded, underexpanded, or ideally expanded if $k = 1.3$.

40. The pressure p and density ρ of the burning gas in the convergent-divergent nozzle of a rocket satisfy the approximate equation $p/\rho = C$ (constant). The gas accelerates throughout the nozzle. At the entry the gas is at a low velocity and pressure p_0. Find the pressure at the throat in terms of p_0. If the velocity at the exit is twice that at the throat, find the ratio (Exit area)/(Throat area). Neglect fluid friction.

41. An airliner is flying at 35,000 ft, where the ambient pressure and temperature are 3.468 psia and 395°R. The pressure and temperature in the aircraft cabin are 14.7 psia and 75°F. A terrorist fires a shot that makes a 1.0-in.-diameter hole in the aircraft. Assume that the hole behaves like a converging nozzle with an exit diameter of 1.0 in. and that the flow out of the hole is quasi-steady. Find the following:
a) The initial Mach number of the flow from the hole;
b) The initial mass flow rate from the hole;
c) The cabin pressure at which the Mach number at the hole reaches half the value of part (a).

42. At a certain point in a pipe, air flows steadily with a velocity of 150 m/s and has a static pressure of 70 kPa and a static temperature of 4°C. The flow is isoenergetic and reversible.
a) Calculate the maximum possible reduction in area and the following quantities for that

minimum area: stagnation pressure, stagnation temperature, static pressure, static temperature, velocity, and Mach number.

b) Calculate the quantities listed in part (a) at a point where the area is 15% smaller than the *initial* area.

43. Figure P10.7 shows a nozzle for the first (high-pressure) stage of a steam turbine. Steam enters the nozzle with $p = 500$ psia, $T = 800°F$, and negligible velocity. The nozzle exit pressure is 100 psia and the mass flow rate is to be 5.0 lb/sec. Assume that the steam is an ideal gas with $k = 1.3$ and find V_e, A_e, and A_t.

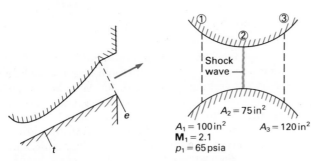

Figure P10.7   **Figure P10.8**

44. Air flows in the channel in Fig. P10.8. Determine the Mach number, static pressure, and stagnation pressures at station 3. Assume isentropic flow except for the normal shock wave.

45. A convergent-divergent nozzle has an exit throat area ratio of 3.0. It is to be supplied with air. Find:
a) the first, second, and third critical pressure ratios;
b) the exit plane Mach number in each case;
c) the throat Mach number in each case.
Find the location of a shock in the nozzle if the ratio of the exit pressure to the inlet stagnation pressure is 0.60.

46. Compute M_3, ρ_3, and p_{0_3} for the flow situation in Fig. P10.9.

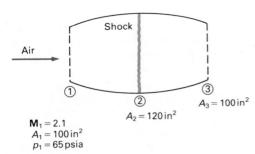

Figure P10.9

47. Air is supplied to a convergent-divergent nozzle from a reservoir where the pressure is 100 kPa. The air is then discharged through a short pipe into another reservoir where the pressure can be varied. The cross-sectional area of the pipe is twice the area of the throat of the nozzle. Friction and heat transfer may be neglected throughout the flow. If the discharge pipe has constant cross-sectional area, determine the range of static pressure in the pipe for which a normal shock will stand in the divergent section of the nozzle. If the discharge pipe tapers so that its cross-sectional area is reduced by 25%, show that a normal shock cannot be drawn to the end of the divergent section of the nozzle. Find the maximum strength of shock (as expressed by the upstream Mach number) that can be formed.

48. A converging-diverging nozzle exhausts air from a large tank to the atmosphere (see Fig. P10.10). The nozzle has a first critical pressure ratio (r_{pc_1}) of 0.90.
a) Calculate the second and third critical pressure ratios for the nozzle.
b) Calculate the tank pressure that will cause a normal shock to stand at the area halfway between the throat area and the exit area (i.e., $A_s = \frac{1}{2}(A_t + A_e)$).

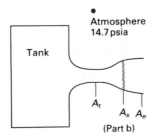

(Part b)   **Figure P10.10**

49. Hydrogen flows through a converging-diverging nozzle. The inlet stagnation conditions are 200 kPa and 310°K. The throat area is 12.5 cm² and the exit area is 25.0 cm². Find the exit pressures for normal shocks to occur at (a) the throat and (b) the exit.

50. The stagnation temperature and Mach number of air flowing out of a 1.0 in., type L copper pipe are 550°R and 0.65 respectively. The atmospheric pressure is 14.7 psia. Find the distance upstream of the exit where the Mach number is 0.25.

51. A compressor supplies air at 100 psia and 500°F to a 12-in.-diameter schedule 40 steel pipe. The pipe is 100 ft long, is thermally insulated, and discharges to an ambient at 40°F and 14.7 psia. Find the mass flow rate through the pipe.

52. A thermally insulated pipe has inside diameter of 0.50 m. The pipe is 100 m long and has a relative roughness of 0.001. The air at the inlet is at 275 kPa and 300K, and the pipe discharges to an ambient pressure of 100 kPa. Find the flow rate for choked conditions.

53. Air with $p_0 = 50$ psia, $T_0 = 150°F$ is to be transported through a 100-ft pipe. What is the minimum pipe diameter to transport the air without choking? The flow rate of air is 10 lbm/sec. The pipe is smooth and the friction coefficient is given by Blasius's formula:

$$4\mathbf{C_f} = \frac{0.3164}{(\mathbf{R})^{0.25}}.$$

54. Air enters a 4-cm-square galvanized steel duct with $p_0 = 150$ kPa, $T_0 = 400K$ and $V_1 = 120$ m/s. (*Note:* $\mu = 2.2 \times 10^{-5}$ N·s/m²).
   a) Compare the maximum possible duct length for these conditions.
   b) If the actual duct length is 0.75 times the maximum value, calculate the mass flow rate, the exit pressure, and the stagnation pressure.
   c) If the actual duct length is 1.3 times the maximum value for the stated conditions, compute the new mass flow rate and inlet velocity. Assume a low back pressure and use the same value of $\mathbf{C_f}$ as used in the maximum possible duct length case.

55. Assume the engine described in Problem 31 and Fig. P10.6 has a throat area of 0.321 m² and an exit area of 0.495 m². The engine is started on the ground, where the atmospheric pressure is 101 kPa. During the starting process, the stagnation pressure increases as shown in Fig. P10.11. Calculate the stagnation pressure when the nozzle first chokes and when the exit Mach number first becomes supersonic.

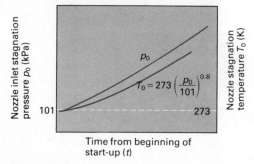

Figure P10.11

56. During the start-up of the engine in Problem 55 the stagnation temperature increases, as also shown in Fig. P10.11. Calculate the nozzle flow rate and gas velocity immediately downstream of the nozzle exit plane when the nozzle first chokes. Use $k = 1.3$ and $R = 260$ J/kg·K.

57. Waste gas (CO_2) is vented to outer space from a spacecraft through a circular pipe 0.2 m long. The pressure and temperature in the spacecraft are 35 kPa and 25°C. The gas must be vented at the rate of 0.01 kg/s. The friction factor for the flow in the pipe is given by

$$f = \frac{16}{\mathbf{R}}, \qquad \mathbf{R} < 5000,$$

$$f = 0.0032, \qquad \mathbf{R} > 5000, \qquad \mathbf{R} = \frac{4\dot{m}}{\pi\mu D}.$$

The viscosity (μ) is 4×10^{-4} N·s/m². Determine the required pipe diameter.

58. The jet pipe on a jet engine installation is 10 m long and 0.50 m in diameter and terminates in a short converging nozzle with an exit diameter of 0.46 m. The stagnation temperature and pressure at entry to the pipe are 800°C and 300 kPa respectively. The engine is run with the aircraft stationary on the ground, where the atmospheric pressure is 101.3 kPa. Calculate the net thrust of the engine when (a) friction in the jet pipe is neglected and (b) the friction coefficient in the jet pipe is 0.005. The gas may be assumed to be ideal with $k = 1.3$ and $c_p = 1.150$ kJ/kg·K.

59. A 9-m-long pipe is fabricated from brass and must carry an air flow of 0.18 kg/s. The stagnation temperature of the air is 373K, and the tolerable pressure drop is 130 kPa. The pipe discharges to a pressure of 100 kPa. Determine the diameter of the pipe.

60. A 10-ft-long hose is connected to the outlet of the regulator valve on a nitrogen bottle. The hose must deliver 0.065 lb/sec when the regulator pressure is 42 psia. The hose exhausts to a back pressure of 8.0 psia. The stagnation temperature is 120°F. The hose has a roughness equivalent to galvanized iron. Calculate the required hose diameter.

61. An air blower and pipe system is to be designed to convey pulverized coal through an 8-in. schedule 40 steel pipe. The pipe is to be 600 ft long and the outlet velocity is to be 150 ft/sec. If the pipe discharges to atmospheric pressure, what must be the pressure, velocity, and density of the air at the inlet end of the pipe? Assume isoenergetic flow with $T_0 = 50°F$ and neglect the effect of the coal particles.

62. The velocity of nitrogen in a 1-in.-diameter commercial steel pipe is 150 ft/sec. The static temperature and pressure are 67°F and 30 psia, respectively.

Calculate the length of pipe needed to achieve sonic flow. Calculate also the pressure at the end of the pipe.

63. Air at an absolute pressure of 500 kPa and temperature of 300K is contained in a pressure vessel. The vessel has a wall thickness of 6 mm and is found to be leaking to the atmosphere at a rate of 2×10^{-6} kg/s. Assuming that there is a small circular hole of length equal to the wall thickness, determine its diameter. The friction factor is given by $f = (4\pi\mu/\dot{m})D$, where μ is the air viscosity, $\dot{m}$ the mass flow rate, and D the hole diameter. The viscosity is constant and equal to 3×10^{-5} kg/m·s. Assume that the flow is choked at the hole outlet and the fluid acceleration up to the hole entrance is isentropic.

64. Estimate the maximum mass flow rate of air that can be passed by the duct in Fig. P10.12. $\mu = 4 \times 10^{-7}$ lb·sec/ft^2.

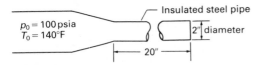

$p_0 = 100$ psia
$T_0 = 140°F$

Insulated steel pipe

2" diameter

20"

Figure P10.12

65. It is necessary to deliver air at the rate of 10 lbm/sec from tank A to tank B in Fig. P10.13. The two tanks are connected by 20 ft of insulated steel pipe. Find the required diameter to deliver the necessary flow rate.

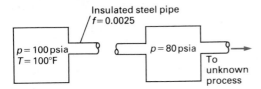

Insulated steel pipe
$f = 0.0025$

$p = 100$ psia
$T = 100°F$

$p = 80$ psia

To unknown process

Figure P10.13

66. A streamlined, two-dimensional body has a length-to-maximum-thickness ratio of 5:1 and a drag coefficient of 0.032 at a Mach number of $\mathbf{M}_\infty = 0.132$. Find the drag coefficient at $\mathbf{M}_\infty = 0.661$.

67. The constants in Eqs. (10.105) and (10.106) are $4\alpha^2$ and 4α, respectively, where α is the angle of attack. Find $\mathbf{C_D}$ and $\mathbf{C_L}$ for $\alpha = 4°$ for a two-dimensional airfoil if $\mathbf{C_D} = 0.012$ and $\mathbf{C_L} = 1.2$ at $\alpha = 12°$.

Higher-Order Application

68. Write a computer program to find the Mach number corresponding to a given area ratio A/A^* for an ideal gas with constant specific heat ratio k. Check your output against Table E.1.

69. Consider Problem 41, where a terrorist has put a 1.0-in.-diameter hole in an aircraft. Assume that flow is one-dimensional at the hole, that the pressure is uniform throughout the cabin at any instant, that the flow at the hole depends only on the instantaneous values of the cabin temperatures and pressure and the fixed atmospheric pressure, and that the expansion of the air in the cabin is reversible and adiabatic. Set up a differential equation for the cabin pressure as a function of time, the initial cabin temperature and pressure, the area of the hole, and appropriate gas constant(s).

70. Solve (numerically) the differential equation of Problem 69 and calculate the time for the cabin pressure to reduce to 5.0 psia.

71. Design a turbine nozzle to accelerate 1 lbm/sec of air from a stagnation pressure of 100 psia and temperature of 600°F to a pressure of 20 psia. Assume isentropic flow.

72. Consider the flow of air through the converging-diverging nozzle in Fig. P10.14. Determine the regime of the nozzle flow, whether the nozzle is choked, and whether there is a shock in the nozzle. Find the flow area at the shock if one does occur. Next calculate the air velocity at the nozzle exit.

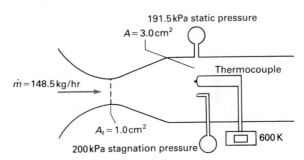

191.5 kPa static pressure

$A = 3.0$ cm^2

Thermocouple

$\dot{m} = 148.5$ kg/hr

$A_t = 1.0$ cm^2

200 kPa stagnation pressure

600 K

Figure P10.14

73. Figure P10.15 shows the converging-diverging nozzle for a supersonic wind tunnel. The nozzle width (perpendicular to the paper) is 2.0 in. During operation, there is a viscous boundary layer on the nozzle walls and the side walls. These boundary

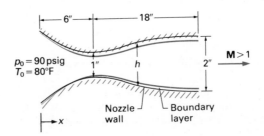

6"   18"

$p_0 = 90$ psig
$T_0 = 80°F$

1" h 2"

$M > 1$

x

Nozzle wall

Boundary layer

Figure P10.15

layers get thicker in the downstream direction. The flow blockage due to the boundary layers is given by the displacement thickness δ^*. The effective cross-sectional area is

$$A_{\text{eff}} = (h - 2\delta^*)(W - 2\delta^*).$$

Calculate the Mach number and velocity at both the nozzle throat and exit if the boundary layer is not present. Then calculate the exit Mach number and velocity by accounting for the "blockage" of the boundary layer. Use methods and equations from Chapter 9 to estimate δ^*. Is it true that $\mathbf{M} = 1$ at the geometric throat? Explain. See Problem 78(c) for a description of blockage.

74. An air compressor has the performance curve shown in Fig. P10.16. It discharges 700°F air through a 12-in. type L copper pipe. The pipe is 100 ft long and discharges to the 14.7-psia atmosphere. Find the mass flow rate through the copper pipe.

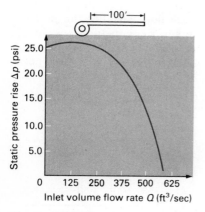

Figure P10.16

75. A fixed-geometry convergent-divergent jet engine intake is designed so that at maximum air flow a normal shock stands just upstream of the mouth in flight at a Mach number of 1.5. What is the ratio of the mouth area to the throat area? Neglecting friction, find the maximum possible mass flow rate into the engine, for 0.1-m² throat area, in flight at an altitude of 10,000 m at $\mathbf{M} = 1.5$, just before the shock is swallowed and at a slightly higher speed with the shock swallowed.

76. Consider the flow of air through the piping system of Fig. P10.17. If the system is choked, determine the location where the Mach number is 1. Is the flow in both pipes described by the same or different Fanno lines? Calculate the pressure ratio p_B/p_{0_1} that causes choking in this system.

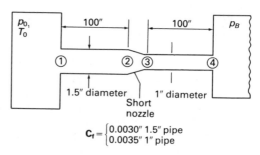

Figure P10.17

77. Consider the turbojet engine in Fig. P10.18. Assume isentropic flow in the diffuser and nozzle. Calculate the Mach number $\mathbf{M}_1$ and the mass flow rate through the engine. Next calculate p_2, p_{0_2}, the critical area for the diffuser, A_4 and A_5. Compare A_4 with the critical area for the diffuser. If they are not the same, explain why.

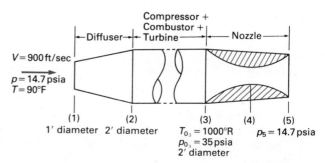

Figure P10.18

78. Models of high-speed aircraft are often tested in wind tunnels similar to that in Fig. P10.19. To test a certain model, a test section cross-sectional area of 2.0 ft² is required. The test section static properties are to be 14.6 psia and 60°F.

a) Sketch the nozzle shape necessary to obtain test section Mach numbers of (i) 0.60, (ii) 1.10, and (iii) 1.80. Indicate the necessary minimum nozzle area.

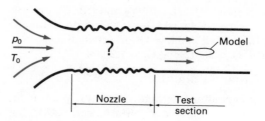

Figure P10.19

b) Calculate the stagnation properties and air mass flow rate for $M = 0.6$ and $M = 1.80$.

c) An important problem in wind tunnel testing is *blockage*. Since a model reduces the test section cross-sectional area, the flow properties at the model are different from those in the empty test section. Estimate the percentage error in air velocity at a model for $M = 0.6$ and $M = 1.10$. The model cross-sectional area is 1.5 in^2 (approximately 0.5% blockage).

References

1. Shapiro, A. H., *Dynamics and Thermodynamics of Compressible Fluid Flow,* vols. 1 and 2, Ronald Press, New York, 1953.
2. Zucker, R. D., *Fundamentals of Gas Dynamics,* Matrix Publishers, Portland, OR, 1977.
3. Zucrow, M., and J. D. Hoffman, *Gas Dynamics,* vols. 1 and 2, Wiley, New York, 1976.
4. John, J. E. A., *Gas Dynamics* (2nd ed.), Allyn-Bacon, Boston, 1984.
5. Anderson, J. D., *Modern Compressible Flow (with Historical Perspective)*, McGraw-Hill, New York, 1982.
6. Liepmann, H. W., and A. Roshko, *Elements of Gasdynamics,* Wiley, New York, 1957.
7. Van Wylen, G. J., and R. E. Sonntag, *Fundamentals of Classical Thermodynamics* (2nd ed.), Wiley, New York, 1973, 1976. Available in both English units and SI units versions.
8. Holman, J. P., *Thermodynamics* (3rd ed.), McGraw-Hill, New York, 1980.
9. Sears, F. W., M. W. Zemansky, and H. D. Young, *University Physics* (6th ed.), chaps. 14–19, Addison-Wesley, Reading MA, 1982.
10. Coles, D. (Principal), *Channel Flow of a Compressible Fluid,* National Committee for Fluid Mechanics Films, Distributed by Encyclopaedia Britannica Film Corporation.

Applications of Flow Analysis: Liquid Flow in Open Channels

11

An *open channel* is a conduit in which liquid flows with a free surface. Probably the most familiar examples of open channels are natural streams and rivers. Man-made or engineered open channels include irrigation and navigation canals, drainage ditches, and sewer and culvert pipes running partially full. Water running along the surface of a road or airport runway after a heavy rain is another example of open channel flow.

Since open channel flow involves the transport of fluid from one location to another and the flow occurs in a confined region, we might expect strong similarities between open channel flow and flow in closed pipes and ducts, which we studied in Chapter 7. In fact, the similarities are slight, extending only to the calculation of energy losses due to wall friction. In closed conduit flow, the extent of the flow stream is always known because the fluid fills the flow channel. There is usually a pressure change associated with the flow. In an open channel flow, the free surface is always at atmospheric pressure and the only pressure variation is hydrostatic. The driving force for open channel flows is *gravity:* there must be a net change in elevation of the channel to generate flow. The primary complicating factor in open channel flow analysis is that the location of the free surface is unknown a priori; in fact, the free surface rises and falls in response to perturbations to the flow (such as changes in channel slope or geometry). Thus the exact geometry of the flow stream itself is unknown and must be determined as part of the solution. This fact, plus the very large variety of types of open channels (ranging from natural streams to carefully designed flumes or spillways), makes analysis of open channel flow more difficult than analysis of pipe flow. In practice, open channel flow analysis makes more use of empirical data and is more approximate than pipe flow analysis.

Most practical open channels contain water, and almost all of the available experimental data were obtained with water. The general principles of open channel flow analysis are applicable to other fluids, but specific results may not be accurate. In this chapter, we will discuss the fundamentals of open channel flow analysis, usually concentrating on simple channel geometries. More complete treatments of the subject are contained in [1–3].

11.1 Introductory Concepts

In this section, we will examine some introductory concepts that provide the basis for open channel flow analysis.

11.1.1 Velocity Distributions and the One-Dimensional Flow Approximation

An open channel cross section generally has four boundaries: two sides, a bottom, and a free surface. The fluid in contact with the solid surfaces must satisfy the no-slip condition, while the fluid at the free surface has nonzero velocity. The velocity distribution in an open channel is generally three-directional and three-dimensional. If the channel is straight and has a regular and constant cross section, it is possible to have nearly one-directional flow, with the main velocity component parallel to the channel axis. Figure 11.1 illustrates velocity distributions for prismatic* channels of various shapes. Note that the maximum velocity always occurs at a point below the free surface, usually at 20–30% of the channel depth. The point of maximum velocity is nearer the surface in shallow channels.

If the channel cross section varies in the flow direction and/or the channel axis is curved, then the velocities are even more complicated. In these cases, significant velocity components in the cross-sectional plane exist and the flow is three-directional. These secondary motions are especially strong when a channel turns a sharp bend, due to centripetal acceleration of fluid particles, and can cause significant bottom erosion.

Figure 11.1 Typical velocity profiles in various open channels.

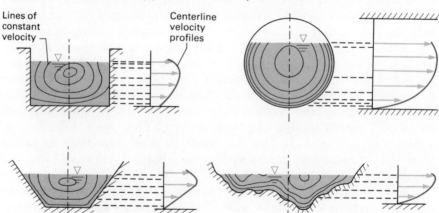

---

* A prismatic channel has constant cross-sectional shape and a straight axis.

There are few theories available for predicting velocity distributions in open channel flow. Engineering analysis methods are based on the assumption of one-dimensional, one-directional flow. Secondary motions in the channel cross-sectional plane are neglected, and the velocity is assumed uniform and parallel to the channel axis. Of course the uniform velocity should be taken as the average*

$$V = \frac{\int (\vec{V} \cdot \hat{n}) \, dA}{A} = \frac{Q}{A}, \qquad (11.1)$$

where Q is the volume flow rate in the channel and A is the cross-sectional area of the liquid in the channel (*not* including parts of the channel above the free surface). Since the free surface is at constant pressure, it is convenient to apply the energy equation between points on the free surface. When we do this, it is actually the surface velocity rather than the average velocity that we use to characterize the flow. Fortunately, the surface velocity in most open channel flows is not greatly different from the average velocity (Fig. 11.1).

When applying the mechanical energy and linear momentum equations to open channel flow, we can use the kinetic energy correction factor (α) and the momentum correction factor (β) to account for the nonuniformity of velocity. Measurements indicate that values of α range from 1.03 to 1.36 and values of β range from 1.01 to 1.12 for prismatic channels. These values are somewhat larger than typical values for fully developed turbulent pipe flow; nevertheless, most open channel flow analyses are based on the assumption that $\alpha = \beta = 1.0$. We will also follow this practice; however, if accurate estimates of α and β are available, flow calculations can be improved accordingly.

11.1.2 Dimensionless Parameters and Wave Speed

Water flowing in an open channel is influenced by forces due to

- gravity,
- friction at the channel walls, and
- surface tension at the free surface.

The effects of pressure and fluid compressibility are negligible. Table 6.1 gives the relevant dimensionless parameters:

- Froude number $\mathbf{F}$ ($\mathbf{F} = V/\sqrt{g\ell}$)
- Reynolds number $\mathbf{R}$ ($\mathbf{R} = V\ell/\nu$)
- Weber number $\mathbf{W}$ ($\mathbf{W} = \rho V^2 \ell/\sigma$).

The relevant dimension, ℓ, is either the depth or hydraulic diameter of the channel. In most open channel flows, ℓ is large and σ is small, so surface tension effects are negligible; therefore the important dimensionless parameters are the Froude and Reynolds numbers. By now the significance of Reynolds number should be familiar; however, the Froude number has not concerned us thus far. Since $\mathbf{F}$ is dimensionless,

---

* The usual overbar (̄) is omitted for convenience.

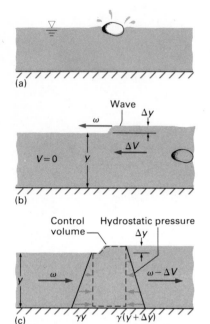

(a)

(b)

(c)

Figure 11.2 Free-surface gravity wave: (a) Disturbance in open channel; (b) disturbance produces surface wave; (c) control volume moving with wave.

the denominator, $\sqrt{g\ell}$, must have dimensions of velocity. According to the principles of similarity (Section 6.5), $\sqrt{g\ell}$ must represent a velocity significant to the flow. With proper selection of ℓ, $\sqrt{g\ell}$ represents the speed of propagation of surface waves.

Consider a still liquid in an open channel (Fig. 11.2a), and suppose a disturbance is introduced. Typical disturbances might be caused by a stone thrown into the liquid or a finger dipped into it. The depth of the liquid changes in the vicinity of the disturbance. Gravity forces act to level the surface, and so the depth change propagates as a surface wave. Ripples on a pond are familiar examples of this type of wave. We can determine the speed of free-surface waves by applying the continuity and momentum equations. Figure 11.2(b) shows a wave propagating in a still liquid. We attach a control volume to the wave and obtain the flow situation shown in Fig. 11.2(c), in which the wave appears fixed. Fluid approaches the control volume with speed ω and leaves the control volume with speed $\omega - \Delta V$, where ΔV is the velocity change induced by the wave. The disturbance increases the fluid depth an amount Δy; ΔV and Δy are not necessarily small. We assume that the geometry is uniform in all planes parallel to the paper. Since the control volume is very thin, there is no storage of mass or momentum. Applying the continuity equation gives

$$\dot{m} = \rho \omega y(1) = \rho(\omega - \Delta V)(y + \Delta y)(1),\qquad(11.2)$$

where (1) is the unit width perpendicular to the paper. Solving for ΔV, we have

$$\Delta V = \omega \left(\frac{\Delta y}{y + \Delta y} \right).\qquad(11.3)$$

Next we apply the linear momentum equation. Since the control volume is thin, the force due to shear stress on the channel bottom is negligible and the only force is due to hydrostatic pressure. The momentum equation is

$$\dot{m}(\omega - \Delta V) - \dot{m}\omega = \frac{1}{2}\gamma y(y)(1) - \frac{1}{2}\gamma(y + \Delta y)(y + \Delta y)(1).$$

Substituting the mass flow rate from Eq. (11.2), noting that $\gamma = \rho g$, and simplifying gives

$$\omega \Delta V = g \left(1 + \frac{\Delta y}{2y} \right) \Delta y.$$

Substituting ΔV from Eq. (11.3) and rearranging gives

$$\omega^2 = gy \left(1 + \frac{\Delta y}{2y} \right)\left(1 + \frac{\Delta y}{y} \right).\qquad(11.4)$$

We note that stronger waves, with larger Δy, travel faster. If we consider a weak wave with $\Delta y \to 0$, and use the symbol c to represent the speed of such a wave, we get

$$c^2 = \lim_{\Delta y \to 0} \omega^2 = gy.\qquad(11.5)$$

Therefore, if we choose the local water depth y as the length parameter in the Froude number, we see that

$$\mathbf{F} = \frac{V}{\sqrt{gy}}$$ (11.6)

represents the ratio of flow speed to the speed of an infinitesimal surface wave.

EXAMPLE 11.1 Illustrates surface wave speed

A child throws a stone into the center of a circular pool. The bottom of the pool is shaped as shown in Fig. E11.1(a). Compute the time for the first ripple to travel to the edge of the pool.

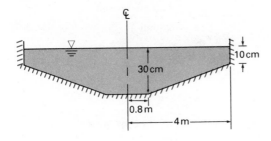

Figure E11.1 Cross section of a small circular pool.

Solution

   Given
Stone thrown into center of circular pool in Fig. E11.1(a)

   Find
Time for first ripple to travel to edge of pool

   Solution
   Assume that the wave is weak and may be approximated by a wave of infinitesimal strength. The wave speed, c, is then given by Eq. (11.5):

$$c = \sqrt{gy},$$

where y is the pool depth. Since the pool depth changes with the radial distance from the center, c varies with the radial coordinate r. The time, dt, for the wave to travel a short distance, dr, is

$$dt = \frac{dr}{c}.$$

The total time, t_0, for the wave to travel to the edge of the pool is

$$t_0 = \int_0^{t_0} dt = \int_0^{R_0} \frac{dr}{c},$$

where R_0 is the pool radius. We must evaluate the integral in two parts:

$$t_0 = \int_0^{R_i} \frac{dr}{c} + \int_{R_i}^{R_0} \frac{dr}{c},$$

where $R_i = 0.8$ m and $R_0 = 4$ m. The wave speeds are

$$c = \sqrt{gy_i}, \qquad 0 \le r \le R_i,$$

and

$$c = \sqrt{g\left[y_i - (y_i - y_0)\left(\frac{r - R_i}{R_0 - R_i}\right)\right]}, \qquad R_i \le r \le R_0,$$

where $y_i = 0.3$ m and $y_0 = 0.1$ m. Substituting and integrating give

$$t_0 = \frac{R_i}{\sqrt{gy_i}} + \frac{2}{\sqrt{g}}\left(\frac{R_0 - R_i}{y_i - y_0}\right)\left[\sqrt{y_i - (y_i - y_0)\left(\frac{r - R_i}{R_0 - R_i}\right)}\right]_{R_i}^{R_0}$$

$$= \frac{R_i}{\sqrt{gy_i}} + \frac{2}{\sqrt{g}}\left(\frac{R_0 - R_i}{\sqrt{y_i} + \sqrt{y_0}}\right).$$

For sea-level conditions, the numerical values give

$$t_0 = \frac{0.8 \text{ m}}{\sqrt{\left(9.81 \frac{\text{m}}{\text{s}^2}\right)(0.30 \text{ m})}} + \frac{2}{\sqrt{9.81 \frac{\text{m}}{\text{s}^2}}}\left(\frac{4.0 \text{ m} - 0.8 \text{ m}}{\sqrt{0.30 \text{ m}} + \sqrt{0.10 \text{ m}}}\right),$$

$$\left| \; t_0 = 4.17 \text{ s.} \; \right|$$

Answer ◀

Discussion
Note that a surface wave travels more slowly in shallow water.

11.1.3 Classification of Open Channel Flow

An orderly classification system for open channel flow helps engineers visualize what is likely to occur in a specific situation and select the proper methods for analysis. Some of the terms used to classify open channel flows should already be familiar.

An open channel flow may be *steady* or *unsteady,* depending on whether the flow rate is constant or varies with time. Although many open channel flows are unsteady (such as rainwater runoff in drainage ditches), we will limit our consideration to steady flows.

Open channel flow may be *laminar* or *turbulent.* Of course, we decide which flow state is likely to exist by examining the Reynolds number. Most open channel flows have rather large dimensions and involve water, which has a relatively low viscosity, so Reynolds numbers are quite large and the flow is usually turbulent. An exception to this generality is flow in very thin sheets such as rainwater runoff from streets and airport runways. Sometimes an engineer may be interested in the flow in a very short section (short *reach*) of a channel, in which case the flow may be assumed frictionless (i.e., neither laminar nor turbulent).

An important problem in open channel flow analysis involves predicting the change in liquid depth, and hence in the profile of the free surface, caused by disturbances in the channel. The most common disturbances are changes of bottom slope, changes of channel cross-

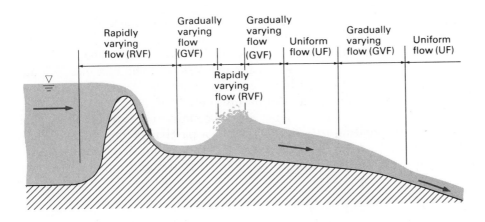

Figure 11.3 Illustration of varied flow and uniform flow in an open channel.

sectional shape or size, and the placing of obstructions in the stream. Flow with constant depth is called *uniform flow*. Flow with varying liquid depth is called *varied flow*. Varied flow is further subdivided into *gradually varying flow* (GVF), in which the rate of change of depth with distance along the channel is small so that the flow is nearly one-directional, and *rapidly varying flow* (RVF), in which the rate of change of depth is not small. Figure 11.3 illustrates the various types of flow. Obviously, a single channel can contain regions of all types, although the flow at each location is of a specific type.

The flow state at a particular location in an open channel is characterized by the local Froude number. The classifications are

$$\mathbf{F} < 1 \quad subcritical \text{ or } tranquil \text{ flow,}$$
$$\mathbf{F} = 1 \quad critical \text{ flow,}$$
$$\mathbf{F} > 1 \quad supercritical \text{ or } rapid \text{ flow.}$$

The response of the channel flow to disturbances is different for subcritical and supercritical conditions. If the flow past the disturbance is supercritical, then the waves generated by the disturbance cannot travel upstream since the wave speed is less than the flow speed. The waves will be swept downstream and only the downstream flow will be affected by the disturbance. If the flow is subcritical, waves generated by a disturbance travel both upstream and downstream and the upstream flow is "warned" of the disturbance.

The physical differences between subcritical and supercritical flow are reflected in methods of problem solution. In supercritical flow, we can start the solution at a point and proceed downstream; however, in subcritical flow it is usually necessary to begin at a point and calculate upstream.

11.2 Uniform Flow

Uniform flow is flow with a constant depth and constant velocity. It is the open channel equivalent of fully developed pipe flow. Uniform flow can only occur in a straight prismatic channel with a constant bottom slope. When liquid enters a reach of such a channel there is a development region of gradually varying flow called the *transitory zone*. The flow is determined by the balance between gravity and wall resistance forces and the inertia "force." In the transitory zone, the gravity force exceeds the wall force and the flow accelerates. The higher velocity results in an increase in the wall shear stress. If the channel is long enough, a condition of equilibrium between gravity force and wall force ultimately results, and the flow becomes uniform.

Many channels are designed for the condition of uniform flow. The depth corresponding to uniform flow in a particular channel is called the *normal depth* (y_n). The normal depth is an important design parameter, even if the flow in the channel is nonuniform.

11.2.1 Chezy and Manning Formulas

The fundamental problem of uniform flow is determining a relation between (average) fluid velocity, normal depth, and channel slope and geometry. Figure 11.4(a) illustrates uniform flow. The channel has arbitrary cross-sectional shape.* Since the uniform flow depth is constant, the liquid surface is parallel to the channel bottom. The energy equation for two points on the free surface is[†]

$$\frac{V_1^2}{2} + gz_1 = \frac{V_2^2}{2} + gz_2 + gh_L.$$

In uniform flow,

$$V_1 = V_2,$$

so

$$h_L = z_1 - z_2, \tag{11.7}$$

and we conclude that the energy grade line is also parallel to the bottom (and the free surface).

We now analyze the flow through the control volume, shown in Fig. 11.4(b). The continuity equation is

$$Q = V_1 A_1 = V_2 A_2,$$

Figure 11.4 Details of uniform flow in an open channel: (a) Geometry; (b) control volume for momentum analysis.

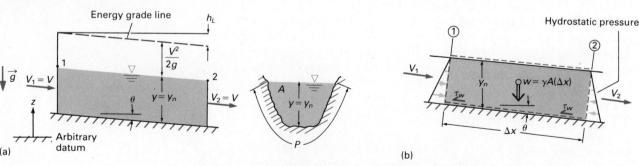

(a)   (b)

---

* In practice, only engineered channels with simple shapes have long enough runs of constant shape and slope to establish uniform flow.
[†] Kinetic energy correction factors are neglected (see Section 11.1.1).

where A is the area of the liquid cross section. For a given cross-sectional shape, the liquid area and depth are uniquely related, so if any two of velocity, flow rate, or depth are known, the remaining quantity can be calculated.

We next apply the momentum equation to the control volume. For uniform flow, the liquid velocity entering the control volume and exiting the control volume are equal and there is no momentum change. Since the depth is constant, the forces due to hydrostatic pressure at the control volume entrance and exit are equal and opposite. The momentum equation reduces to a balance between the gravity force and the resisting force due to shear stress at the walls:

$$\gamma(A\,\Delta x)\sin\theta = \bar{\tau}_w P\,\Delta x,$$

where P is the wetted perimeter of the channel cross section and $\bar{\tau}_w$ is the average shear stress. Channels with uniform flow generally have very small slopes, so

$$\sin\theta \approx \tan\theta = S_0,$$

where S_0 is the slope of the channel bottom. The momentum equation then gives

$$\bar{\tau}_w = \gamma S_0 R_h = \rho g S_0 R_h, \tag{11.8}$$

where R_h is the *hydraulic radius* defined by

$$R_h = \frac{A}{P}.$$

The hydraulic radius is one-fourth the hydraulic diameter D_h that was defined in Section 7.2.6. Note that the hydraulic radius depends on the liquid depth in the channel as well as the channel geometry.

The shear stress can be written in terms of the skin friction coefficient

$$\bar{\tau}_w = \mathbf{C_f}\left(\frac{1}{2}\rho V^2\right).$$

Since uniform open channel flow is somewhat similar to fully developed pipe flow, it is common practice to use a Darcy friction factor,

$$f = 4\mathbf{C_f},$$

giving

$$\bar{\tau}_w = \frac{f}{8}\rho V^2. \tag{11.9}$$

Substituting Eq. (11.9) into Eq. (11.8) and solving for the velocity, we have

$$V = \sqrt{\frac{8g}{f}}\sqrt{R_h S_0}. \tag{11.10}$$

The dependence of liquid velocity (and hence flow rate) on channel slope and hydraulic radius implied by Eq. (11.10) was first discovered by the French engineer A. Chezy in 1769. The equation is sometimes written as

$$V = C\sqrt{R_h S_0}. \tag{11.11}$$

Equation 11.11 is called the *Chezy formula*. The coefficient C is called the *Chezy coefficient* and is given by

$$C = \sqrt{\frac{8g}{f}}. \tag{11.12}$$

C is a dimensional coefficient, ranging in value from about 60 $\text{ft}^{1/2}/\text{sec}$ (33 $\text{m}^{1/2}/\text{s}$) for small, rough channels to about 160 $\text{ft}^{1/2}/\text{sec}$ (88 $\text{m}^{1/2}/\text{s}$) for large, smooth channels. Early research on open channel flow was involved with attempts to correlate C with channel geometry. Equation (11.12) indicates that C depends on the same parameters as f, namely Reynolds number ($\mathbf{R} \equiv VD_h/\nu = 4VR_h/\nu$) and channel surface roughness (ε/D_h). Many experiments have established that C can be evaluated by using the hydraulic diameter concept (see Section 7.2.6) together with the Colebrook formula or Moody chart to evaluate the friction factor f. Equation (11.12) then gives a value for C. We will call this the *friction-factor method*. By analogy with fully developed pipe flow, uniform open channel flow is laminar if $\mathbf{R} < 2300$ and turbulent if $\mathbf{R} > 4000$.

A more direct equation for C can be determined if we note that most open channel flows occur at very large Reynolds numbers and on moderately rough to very rough surfaces. In this case, the friction factor depends only on roughness and is given by (see Eq. 7.47)

$$\frac{1}{\sqrt{f}} \approx -2.0 \log\left(\frac{\varepsilon/D_h}{3.7}\right)$$

or, in terms of R_h ($= D_h/4$),

$$\frac{1}{\sqrt{f}} \approx -2.0 \log\left(\frac{\varepsilon/R_h}{14.8}\right).$$

Substituting into Eq. (11.12) gives

$$C \approx -\sqrt{32g} \log\left(\frac{\varepsilon/R_h}{14.8}\right). \tag{11.13}$$

This equation has one serious drawback. Values of surface roughness can be obtained for several materials from Fig. 7.12 or Table 7.1; however, these are engineered materials. Open channels are often lined with stones, rough earth, or vegetation. Natural channels can have extremely irregular bottom surfaces. Direct information on the absolute roughness (ε) of these types of channel surfaces is not easily available. Information about roughness of open channel surfaces is available in a form more suitable for use with *Manning's formula*. In 1891, Robert Manning, an Irish engineer, proposed the following form of Chezy's formula:*

$$V = \frac{1.0}{n} (R_h)^{2/3} (S_0)^{1/2} \tag{11.14a}$$

---

* Note that 1.0 has units of $\text{m}^{1/3}/\text{s}$ and 1.49 has units of $\text{ft}^{1/3}/\text{sec}$.

for SI units (R_h in meters) or

$$V = \frac{1.49}{n} (R_h)^{2/3} (S_0)^{1/2} \qquad \textbf{(11.14b)}$$

for BG or EE units (R_h in feet). In terms of the Chezy coefficient, Manning's proposal is equivalent to

$$C = \frac{1.0 \ (\text{m}^{1/3}/\text{s})}{n} [R_h \ (\text{m})]^{1/6} \qquad \text{SI units} \qquad \textbf{(11.15a)}$$

or

$$C = \frac{1.49 \ (\text{ft}^{1/3}/\text{sec})}{n} [R_h \ (\text{ft})]^{1/6} \qquad \text{BG units.} \qquad \textbf{(11.15b)}$$

The parameter n, called *Manning's n*, is a dimensionless measure of the roughness of the channel surface. Manning's research produced values of n for many surfaces; a partial list of values is presented in Table 11.1. Since n is a measure of surface roughness, there is some equivalence between n and the surface roughness ε. White [4] has shown that in the range $0.001 \le \varepsilon/D_h \le 0.1$, the relation can be represented by

$$n \approx 0.031[\varepsilon \ (\text{ft})]^{1/6}. \qquad \textbf{(11.16)}$$

Thus n varies much less rapidly than ε, with a thousandfold change in ε producing only a threefold change in n.

In practice, either Eq. (11.12), (11.13), or (11.15) can be used together with Eq. (11.11) to predict the flow velocity in an open channel, provided information on ε or n is available. You will not get identical answers, however, because of the approximations of the data correlations and the uncertainty in the data from which the correlations were derived.

Table 11.1 Values of Manning's n for various channel surfaces (uncertainty of values approximately 20%)

Channel Surface	n	Channel Surface	n
Artificial lined channels		Excavated earth channels	
Glass	0.010	Clean	0.022
Brass	0.011	Gravelly	0.025
Steel, smooth	0.012	Weedy	0.030
Painted	0.014	Stony, cobbles	0.035
Riveted	0.015		
Cast iron	0.013	Natural channels	
Cement, finished	0.012	Clean and straight	0.030
Unfinished	0.014	Sluggish, deep pools	0.040
Planed wood	0.012	Major rivers	0.035
Clay tile	0.014		
Brickwork	0.015	Floodplains	
Asphalt	0.016	Pasture, farmland	0.035
Corrugated metal	0.022	Light brush	0.05
Rubble masonry	0.025	Heavy brush	0.075
		Trees	0.15

EXAMPLE 11.2 Illustrates calculations in uniform flow and differences between equations for estimation of the Chezy coefficient

A 24-in.-diameter cast-iron storm sewer pipe is laid on a slope of 0.1°. The pipe runs half-full of water. Calculate the flow rate in the pipe using

a) Eqs. (11.11) and (11.12) with f evaluated from the Moody chart,
b) Eqs. (11.11) and (11.13),
c) Eq. (11.14b).

Solution

Given

24-in.-diameter cast-iron pipe
Slope 0.1°
Pipe half-full of water

Find

Flow rate using

a) Eqs. (11.11) and (11.12) and Moody chart,
b) Eqs. (11.11) and (11.13),
c) Eq. (11.14b)

Solution

a) The volume flow rate, Q, is given by

$$Q = VA = V\left(\frac{1}{2}\right)\left(\frac{\pi D^2}{4}\right) = \frac{\pi D^2 V}{8},$$

where D is the pipe inside diameter and V the average water velocity. Equations (11.11) and (11.12) give

$$V = C\sqrt{R_h S_0} = \sqrt{\frac{8g}{f}}\sqrt{R_h S_0}.$$

The hydraulic radius R_h is

$$R_h = \frac{A}{P} = \left(\frac{\pi D^2}{8}\right)\left(\frac{2}{\pi D}\right) = \frac{D}{4} = \frac{24 \text{ in.}}{4} = 6 \text{ in.}$$

The slope S_0 is

$$S_0 = \tan 0.1° = 0.00175.$$

With data from Table 7.1, the relative roughness is

$$\frac{\varepsilon}{D_h} = \frac{\varepsilon}{4R_h} = \frac{8.5 \times 10^{-4} \text{ ft}}{4(6.0 \text{ in.})\left(\dfrac{\text{ft}}{12 \text{ in.}}\right)} = 4.25 \times 10^{-4}.$$

Assuming 60°F water, Table A.6 gives $\nu = 1.22 \times 10^{-5}$ ft^2/sec. The Reynolds number is

$$\mathbf{R} = \frac{VD_h}{\nu} = \frac{4VR_h}{\nu} = \frac{4V(6 \text{ in.})\left(\dfrac{\text{ft}}{12 \text{ in.}}\right)}{\left(1.22 \times 10^{-5} \dfrac{\text{ft}^2}{\text{sec}}\right)} = 1.64 \times 10^5 V \text{ sec/ft.}$$

The velocity V must be found using an iterative procedure; the fastest convergence is obtained by guessing a value for f, calculating V from

$$V = \sqrt{\frac{8\left(32.2\,\frac{\text{ft}}{\text{sec}^2}\right)}{f}}\sqrt{(6\text{ in.})\left(\frac{\text{ft}}{12\text{ in.}}\right)(0.00175)} = \frac{0.475}{\sqrt{f}}\;\text{ft/sec,}$$

calculating the Reynolds number, finding f from the Moody chart, and comparing this value of f with the chosen one. Beginning with $f = 0.0162$ (the fully turbulent value), we have

$$V = \frac{0.475}{\sqrt{0.0162}}\;\text{ft/sec} = 3.73\;\text{ft/sec,}$$

$$\mathbf{R} = 1.64 \times 10^5 \left(3.73\,\frac{\text{ft}}{\text{sec}}\right)\left(\frac{\text{sec}}{\text{ft}}\right) = 6.11 \times 10^5, \qquad \text{and}$$

$$f = 0.0170.$$

The next iteration gives

$$V = \frac{0.475}{\sqrt{0.0170}}\;\text{ft/sec} = 3.64\;\text{ft/sec,}$$

$$\mathbf{R} = 1.64 \times 10^5 \left(3.64\,\frac{\text{ft}}{\text{sec}}\right)\left(\frac{\text{sec}}{\text{ft}}\right) = 6.0 \times 10^5, \qquad \text{and}$$

$$f = 0.0170.$$

Convergence has been obtained, and the flow rate is

$$Q = \frac{\pi D^2 V}{8} = \frac{\pi (24\text{ in.})^2 \left(\dfrac{\text{ft}}{12\text{ in.}}\right)^2 \left(3.64\,\dfrac{\text{ft}}{\text{sec}}\right)}{8},$$

$$\left|\; Q = 5.72\;\text{ft}^3/\text{sec.} \;\right|$$

Answer ◀

b) Combining Eqs. (11.11) and (11.13) gives

$$V = -\sqrt{32g}\,\log\left(\frac{\varepsilon/R_h}{14.8}\right)\sqrt{R_h S_0},$$

where

$$\frac{\varepsilon}{R_h} = \frac{8.5 \times 10^{-4}\text{ ft}}{6\text{ in.}\left(\dfrac{\text{ft}}{12\text{ in.}}\right)} = 17.0 \times 10^{-4}.$$

Then

$$V = -\sqrt{32\left(32.2\,\frac{\text{ft}}{\text{sec}^2}\right)}\,\log\left(\frac{17.0 \times 10^{-4}}{14.8}\right)\sqrt{(6\text{ in.})\left(\frac{\text{ft}}{12\text{ in.}}\right)(0.00175)}$$

$$= 3.74\;\text{ft/sec.}$$

The volume flow rate is

$$Q = \frac{\pi D^2 V}{8} = \frac{\pi (24\text{ in.})^2 \left(\dfrac{\text{ft}}{12\text{ in.}}\right)^2 \left(3.74\,\dfrac{\text{ft}}{\text{sec}}\right)}{8},$$

$$\left|\; Q = 5.87\;\text{ft}^3/\text{sec.} \;\right|$$

Answer ◀

c) Equation (11.14b) gives

$$V = \frac{1.49}{n} (R_h)^{2/3} (S_0)^{1/2}.$$

For cast-iron pipe, Table 11.1 gives $n = 0.013$, so

$$V = \frac{1.49}{0.013} \left(6.0 \text{ in.} \times \frac{\text{ft}}{12 \text{ in.}} \right)^{2/3} (0.00175)^{1/2} = 3.02 \text{ ft/sec}$$

and

$$Q = \frac{\pi D^2 V}{8} = \frac{\pi (24 \text{ in.})^2 \left(\dfrac{\text{ft}}{12 \text{ in.}} \right)^2 \left(3.02 \dfrac{\text{ft}}{\text{sec}} \right)}{8},$$

$$\boxed{Q = 4.74 \text{ ft}^3/\text{sec.}}$$

Answer ◀

Discussion
Note that Eq. (11.14b) gives the velocity in ft/sec if R_h is in ft. Although the values of Q calculated by the three methods differ by as much as 20%, it is not possible to decide which value is more accurate. The close agreement between the values calculated by the friction factor equations does not imply that they are more correct than the Manning's formula calculation.

11.2.2 Analysis and Design of Uniform Flow Channels

Uniform flow problems may involve either analysis or design of channels. In the analysis problem, channel cross section, surface type, and bottom slope are given and the objective is to determine liquid velocity, flow rate, or depth, given any one of them. The design problem involves selecting channel shape and/or slope to convey a given flow rate with a given depth. Often an engineer wishes to maximize the flow rate for a given slope or, conversely, to minimize the necessary slope to convey a given flow rate. We will first consider a few aspects of the analysis problem and then consider the design problem.

Analysis Problem. In the analysis problem, channel slope, shape, and surface condition are given. The value of n or ε can be estimated from tables. Given any one of flow rate (Q), velocity (V), or normal depth (y_n), the remaining two can be found from the continuity equation, the Chezy formula or Manning's formula, and geometric relations (relating A, P, and y_n). The calculations are straightforward if depth is given but require iteration if velocity or flow rate is given. Example 11.2 illustrated the procedure if the depth is given. The following example illustrates the procedure for calculating the depth for a specified flow rate.

EXAMPLE 11.3 Illustrates normal depth calculation in uniform open channel flow

Five of the storm sewer pipes described in Example 11.2 empty into a 48-in.-diameter unfinished cement pipe laid on a slope of 0.1°. Compute the depth of flow in the 48-in. pipe.

Solution

Given
Five storm sewer pipes of Example 11.2
Pipes emptying into 48-in.-diameter unfinished cement pipe
Pipe slope 0.1°

Find
Liquid depth in 48-in. pipe

Solution
Assuming that Manning's formula provides the most reliable estimate of the flow in the five smaller storm sewer pipes, the flow rate in the 48-in.-diameter pipe is

$$Q = 5(4.74 \text{ ft}^3/\text{sec}) = 23.7 \text{ ft}^3/\text{sec}.$$

The flow rate is given by

$$Q = VA,$$

where V is obtained from Eq. (11.14b) and Table B.1 gives

$$A = \frac{R^2}{2}(\theta - \sin \theta),$$

where R is the radius of the cement pipe. Also $P = R\theta$. Substituting for Q, V, and A gives

$$23.7 \text{ ft}^3/\text{sec} = \frac{1.49}{n}(R_h)^{2/3}(S_0)^{1/2}\frac{R^2}{2}(\theta - \sin \theta).$$

For the cement pipe,

$$R_h = \frac{A}{P} = \frac{\dfrac{R^2}{2}(\theta - \sin \theta)}{R\theta} = \frac{R(\theta - \sin \theta)}{2\theta}.$$

Substituting the hydraulic radius gives

$$23.7 \text{ ft}^3/\text{sec} = \frac{1.49}{n}\left[\frac{R(\theta - \sin \theta)}{2\theta}\right]^{2/3}(S_0)^{1/2}\frac{R^2}{2}(\theta - \sin \theta).$$

Using Table 11.1 and Example 11.2, we have for the remaining values

$$23.7 \text{ ft}^3/\text{sec} = \frac{1.49}{0.014}\left[\frac{(24 \text{ in.})\left(\dfrac{\text{ft}}{12 \text{ in.}}\right)}{2}\left(\frac{\theta - \sin \theta}{\theta}\right)\right]^{2/3}$$

$$\times (0.00175)^{1/2}\frac{(24 \text{ in.})^2\left(\dfrac{\text{ft}}{12 \text{ in.}}\right)^2}{2}(\theta - \sin \theta), \qquad \text{or}$$

$$2.66 = \frac{(\theta - \sin \theta)^{5/3}}{\theta^{2/3}}.$$

Solving for θ by iteration gives $\theta = 169.6°$. The liquid depth y_n is

$$y_n = R - R\cos\frac{\theta}{2} = R\left(1 - \cos\frac{\theta}{2}\right) = 24 \text{ in.}\left(1 - \cos\frac{169.6°}{2}\right),$$

$$\left|\; y_n = 21.8 \text{ in.} \;\right|$$

Answer
◀

Discussion

Solution of the equation for θ requires some care. You can easily verify that $\theta = 2.74°$ also satisfies

$$\frac{(\theta - \sin \theta)^{5/3}}{\theta^{2/3}} = 2.66.$$

This corresponds to a depth of only 0.007 in., which is, of course, quite unreasonable.

The problems that we have considered thus far have assumed that the entire channel surface can be characterized by a single value of ε or n. Consider flow in the channel shown in Fig. 11.5. This may represent a river flowing at flood stage. It is unlikely that the main channel has the same surface condition as the flow outside the riverbank. This flow can be analyzed by breaking the channel into sections as shown and writing, say, Manning's equation as

$$Q_{\text{total}} = \frac{1.49}{n_1} A_1 R_{h_1}^{2/3} S_0^{1/2} + \frac{1.49}{n_2} A_2 R_{h_2}^{2/3} S_0^{1/2} + \cdots. \qquad \textbf{(11.17)}$$

Note that the hydraulic radius of each section is based only on the perimeter of the channel surface and does not include the (imaginary) dividing line in the liquid.

Design Problem. When designing channels for uniform flow, an engineer must select the channel shape and slope to convey liquid at a given rate and with a given depth or velocity. There are many criteria that may be involved in the selection process. In one case, it might be that the channel must convey as large a flow rate as possible, with a large velocity desirable, while in another case the velocity must be kept small to avoid erosion. It is usually not possible to select all of the channel's geometric parameters based on specific criteria. For example, channel shape may be arbitrarily selected or may be dictated by the method of excavation.

If channel slope is to be determined, it is easily calculated from the Chezy or Manning formula, provided depth and flow rate are specified and channel shape and surface condition are known. An important practical problem involves selecting the channel shape and depth to most efficiently carry a given flow rate, that is, the shape that conveys a given flow with the smallest bottom slope. Combining Manning's formula with the continuity equation gives

$$Q = \frac{1.49}{n} (R_h)^{2/3} S_0^{1/2} A.$$

Substituting the definition of R_h, we have

$$Q = \frac{1.49}{n} S_0^{1/2} \frac{A^{5/3}}{P^{2/3}}.$$

For a given surface type (n), S_0 is minimized for a given Q by making $A^{5/3}/P^{2/3}$ as large as possible. This problem is purely geometric; the

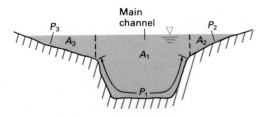

Figure 11.5 Flow in channel with sections with different characteristics.

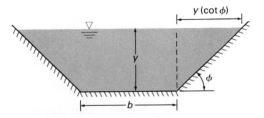

Figure 11.6 Trapezoidal channel.

most efficient channel section has the smallest perimeter for a given area. If the shape is totally unspecified, this is a very difficult problem; however, if a class of shapes is chosen, optimizing that class is not too difficult. Consider a trapezoidal channel filled with liquid to a depth y (Fig. 11.6). The area is*

$$A = by + y^2 \cot \phi \qquad (11.18)$$

and the wetted perimeter is

$$P = b + 2y(1 + \cot^2 \phi)^{1/2}. \qquad (11.19)$$

Eliminating b between these two equations gives

$$P = \frac{A}{y} - (\cot \phi)y + 2y(1 + \cot^2 \phi)^{1/2}. \qquad (11.20)$$

To find the optimum depth, we take $\partial P/\partial y$ and set the result equal to zero. This gives

$$y = \left[\frac{A}{2(1 + \cot^2 \phi)^{1/2} - \cot \phi}\right]^{1/2}. \qquad (11.21)$$

The optimum depth is a function of the area and channel side angle ϕ. A particularly interesting shape is a rectangle ($\phi = 90°$) for which the most efficient flow has depth

$$y = \sqrt{\frac{A}{2}}$$

or, since $A = by$ for a rectangle,

$$y = \frac{b}{2}.$$

We can find the most efficient trapezoidal channel shape by taking $\partial P/\partial \phi$ and setting the result to zero. This yields

$$\phi = 60°. \qquad (11.22)$$

If a trapezoidal channel is specified, with values of Q and S_0 given, Eqs. (11.21) and (11.22) can be combined with Manning's equation to select the optimum channel dimensions.

---

* Strictly speaking, the channel area is related to the distance between the channel bottom and the free surface, measured perpendicular to the bottom, while the depth is measured vertically. Since most channels with uniform or gradually varied flow have small slopes, it is adequate to use y to calculate the cross-sectional area.

This type of calculation can be repeated for other geometries. It will probably be no surprise (if you think about it) that the most efficient channel is one of semicircular shape.

11.3 Concepts for Analysis of Varied Flow

We will now consider some of the phenomena associated with varied flow and introduce the parameters that are commonly used to describe them.

11.3.1 Specific Energy and Critical Depth

Figure 11.7 illustrates varied flow in an open channel. For simplicity, we assume that the channel is rectangular and prismatic. The energy equation, in head form, for two points on the liquid surface is

$$\frac{V_1^2}{2g} + z_1 = \frac{V_2^2}{2g} + z_2 + h_L.$$

We divide the elevation into two parts,

$$z = z_B + y,$$

where z_B is the elevation of the channel bottom and y is the depth of liquid in the channel. The energy equation can then be written

$$\left(\frac{V_1^2}{2g} + y_1\right) - \left(\frac{V_2^2}{2g} + y_2\right) = z_{B_2} - z_{B_1} + h_L. \tag{11.23}$$

The quantity E, defined by

$$E \equiv \frac{V^2}{2g} + y, \tag{11.24}$$

is called the *specific energy** and is the energy per unit weight of the liquid, measured with the channel bottom as datum. At any cross section, the specific energy has a unique value, although the distribution of specific energy between kinetic and potential forms is not necessarily unique. The specific energy changes along the channel due to changes

Figure 11.7 Varied flow.

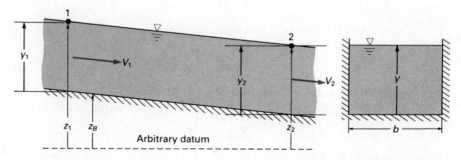

* A better name would be *specific head;* however, the term *specific energy* is widely used.

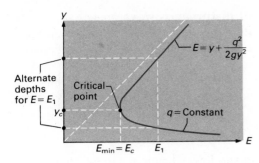

Figure 11.8 Specific energy curve for constant q.

of bottom elevation and mechanical energy loss. Specific energy is constant in a horizontal, frictionless channel, even if the flow is varied. In uniform flow, $\Delta z_B = -h_L$ and the specific energy is constant.

For a given value of specific energy and a given flow rate, there are *two* possible flow states. For a rectangular channel, the flow rate is

$$Q = Vby = qb,$$

where b is the width of the channel and q ($\equiv Q/b$) is the flow rate per unit width. The velocity is

$$V = q/y. \tag{11.25}$$

Substituting Eq. (11.25) into Eq. (11.24) gives

$$E = y + \frac{q^2}{2gy^2}. \tag{11.26}$$

For a given value of q, we can plot a specific energy curve, shown in Fig. 11.8. This curve shows that for given values of q and E, two depths, and hence two velocities, are possible. The two depths are called *alternate depths*.

We also note that there is a minimum possible value of E for a given value of q. At this minimum energy point, the alternate depths are equal. We can evaluate the flow state at the minimum energy point by noting that

$$\left.\frac{\partial E}{\partial y}\right)_q = 0$$

there. Differentiating Eq. (11.26) and setting the result to zero, we obtain

$$y_c = \left(\frac{q^2}{g}\right)^{1/3} \tag{11.27}$$

and hence

$$E = \frac{3}{2}\left(\frac{q^2}{g}\right)^{1/3} = \frac{3}{2}y_c. \tag{11.28}$$

The velocity at the minimum energy point is

$$V_c = \frac{q}{y_c} = (gy_c)^{1/2}. \tag{11.29}$$

The quantity $(gy_c)^{1/2}$ is the small disturbance wave speed, so, at the minimum energy point, the Froude number is 1 and the flow is critical.* The quantity y_c is called the *critical depth*. The critical depth depends on the flow rate and g only.

We can now label the upper and lower branches of the specific energy curve. Along the upper branch, the depth is large and the velocity is small, with

$$V < \sqrt{gy_c}$$

and the flow is subcritical. Along the lower branch, the depth is small and the velocity is large, with

$$V > \sqrt{gy_c}$$

and the flow is supercritical. For a given value of E, one of the alternate depths corresponds to subcritical flow (upper branch) and the other corresponds to supercritical flow (lower branch).

We can make several observations about varied flow by examining Fig. 11.8. The following are noted:

- When the flow depth is greater than the critical depth (subcritical flow), an increase in specific energy causes an increase in depth.
- When the flow depth is less than the critical depth (supercritical flow), an increase in specific energy causes a decrease in depth.
- If the flow is critical, small changes in specific energy cause large changes in depth. In practice, sustained critical flow (over a long reach of channel) is unstable, exhibiting a wavy surface profile.

The specific energy diagram of Fig. 11.8 is for a specific flow rate. By selecting various flow rates, we can generate a family of specific energy curves as shown in Fig. 11.9. Note that the larger the flow rate, the larger the minimum value of specific energy. For any value of E, there is a maximum flow rate, given by

$$q_{max} = q_c = \sqrt{g\left(\frac{2}{3}E_c\right)^3}. \tag{11.30}$$

This discussion of specific energy and critical depth has been limited to rectangular channels. It is not difficult to extend the discussion to nonrectangular channels. Figure 11.10 shows a channel of

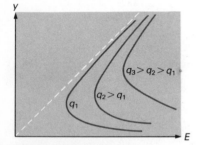

Figure 11.9 Specific energy curves for various flow rates.

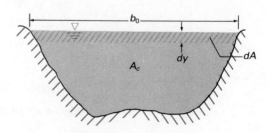

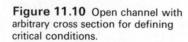

Figure 11.10 Open channel with arbitrary cross section for defining critical conditions.

---

* Hence the subscript c to identify parameters at the minimum energy point.

arbitrary cross section. The specific energy is

$$E = y + \frac{Q^2}{2gA^2},\tag{11.31}$$

where Q is the total flow rate and A is the cross-sectional area of the liquid. Since A is a function of y for a given cross section, it is possible to plot a specific energy curve similar to that in Fig. 11.8 for a given value of Q. The critical (minimum energy) point can be located by setting the derivative of E with respect to y equal to zero and noting that A is a function of y. This gives

$$\left.\frac{dA}{dy}\right)_c = \frac{gA_c^3}{Q^2}.$$

Recalling that the specific energy is defined for a point on the liquid surface, from Fig. 11.10 we have

$$\frac{dA}{dy} = b_0$$

at the channel surface. The critical area and velocity are given by

$$A_c = \left(\frac{b_0 Q^2}{g}\right)^{1/3}\tag{11.32}$$

and

$$V_c = \frac{Q}{A_c} = \left(\frac{gA_c}{b_0}\right)^{1/2}.\tag{11.33}$$

The Froude number for a nonrectangular channel is

$$\mathbf{F} = \frac{V}{V_c} = \frac{V}{(gA_c/b_0)^{1/2}}.\tag{11.34}$$

Uniform flow in an open channel can be subcritical, critical, or supercritical depending on whether the normal depth (y_n) is greater than, equal to, or less than the critical depth (y_c). The critical slope is the value of the bottom slope that yields uniform critical flow. The critical slope is evaluated by equating the critical velocity from Eq. (11.33) to the velocity from Manning's (or Chezy's) formula, Eq. (11.14):

$$\left(\frac{gA_c}{b_0}\right)^{1/2} = \frac{\xi}{n}\left(\frac{A_c}{P_c}\right)^{2/3}(S_c)^{1/2},$$

where $\xi = 1.0$ for SI units and $\xi = 1.49$ for EE or BG units. Solving for S_c, the critical slope, gives

$$S_c = \frac{gn^2}{b_0\xi^2}\frac{(P_c)^{4/3}}{(A_c)^{1/3}}.\tag{11.35}$$

For a wide rectangular channel ($b_0 \gg y$), Eq. 11.35 reduces to

$$S_c \approx \frac{gn^2}{\xi^2}(y_c)^{-1/3}.\tag{11.36}$$

Channel slopes less than the critical are called *mild* and channel slopes greater than the critical are called *steep*.

EXAMPLE 11.4 Illustrates the calculation of critical flow and critical slope

Figure E11.4 illustrates flow over a free overfall in a wide, rectangular channel. Measurements indicate that the depth at the brink is about 0.72 times the critical depth and that critical depth occurs about 5 times the critical depth upstream of the brink if the approaching flow is subcritical. In the channel shown, the depth at the brink is 0.5 m. Compute the flow rate per unit channel width. Calculate the critical slope if the channel is lined with (a) brick and (b) gravel. What is the Froude number at the overfall?

Solution

Given

Free overfall in Fig. E11.4
Brink depth of 0.5 m is 0.72 times critical depth.
Critical depth occurs upstream of the brink at a distance 5 times the
    critical depth.
Subcritical flow

Find

Flow rate per unit channel width
Critical slopes for brick and gravel channels
Froude number at overfall

Solution
The critical depth of the approaching flow is

$$y_c = \frac{y_{\text{brink}}}{0.72} = \frac{0.5 \text{ m}}{0.72} = 0.694 \text{ m}.$$

The flow rate per unit width, q, is found using Eq. (11.27):

$$q = \sqrt{gy_c^3} = \sqrt{\left(9.81 \frac{\text{m}}{\text{s}^2}\right)(0.694 \text{ m})^3},$$

Answer

$$|\ q = 1.81 \text{ m}^2/\text{s}.\ |$$

The critical slope is found using Eq. (11.36):

$$S_c = \frac{gn^2}{1.0}\,(y_c)^{-1/3},$$

for SI units with n found from Table 11.1. For a brick-lined channel, $n = 0.015$ and

$$S_c = \left(9.81 \frac{\text{m}}{\text{s}^2}\right)(0.015)^2(0.694 \text{ m})^{-1/3},$$

Answer

$$|\ S_c = 0.0025. \quad \text{brick channel}\ |$$

For a gravel-lined channel, $n = 0.025$ and

$$S_c = \left(9.81 \frac{\text{m}}{\text{s}^2}\right)(0.025)^2(0.694 \text{ m})^{-1/3},$$

Answer

$$|\ S_c = 0.0069. \quad \text{gravel channel}\ |$$

The Froude number at the overfall is found by applying Eq. (11.23) between the critical depth (c) and the brink (b). Neglecting the slight drop of the channel

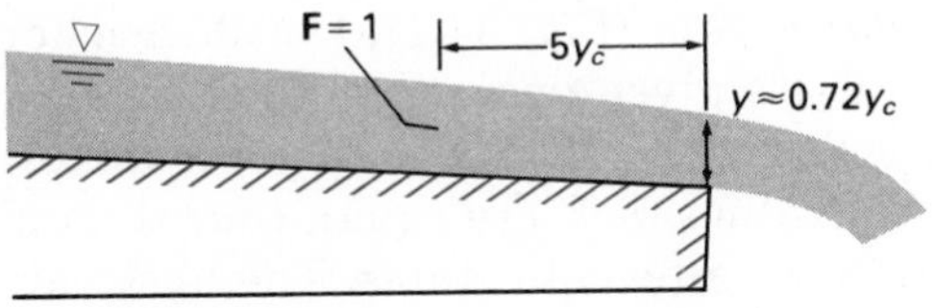

Figure E11.4 Water flow at free overfall.

bottom and assuming frictionless flow, we have

$$\frac{V_c^2}{2g} + y_c = \frac{V_b^2}{2g} + y_b.$$

At the critical depth, Eq. (11.29) gives

$$V_c^2 = gy_c, \quad \text{so} \quad \frac{y_c}{2} + y_c = \frac{V_b^2}{2g} + y_b.$$

The problem statement gives $y_b = 0.72y_c$, so

$$\frac{3}{2}y_c = \frac{V_b^2}{2g} + 0.72y_c \quad \text{and} \quad V_b = \sqrt{2g(0.78y_c)} = \sqrt{1.56gy_c}.$$

The Froude number at the brink is

$$\mathbf{F}_b = \frac{V_b}{\sqrt{gy_b}} = \sqrt{\frac{1.56gy_c}{gy_b}} = \sqrt{\frac{1.56y_c}{y_b}}.$$

The numerical values give

$$\mathbf{F}_b = \sqrt{\frac{1.56(0.694 \text{ m})}{(0.50 \text{ m})}},$$

$$|\ \mathbf{F}_b = 1.47.\ |$$

Answer ◀

Discussion

The approach flow is subcritical, and the brink flow is supercritical ($\mathbf{F} > 1$). The flow must pass through the critical depth for the transition to occur from subcritical to supercritical flow. The given values of brink depth ($0.72y_c$) and upstream distance from brink to critical depth ($\approx 5y_c$) are generally valid for subcritical flow.

11.3.2 Frictionless Flow in a Rectangular Channel

We can illustrate the usefulness of the specific energy concept by considering frictionless flow in a rectangular channel. For flow between any two locations 1 and 2, Eq. (11.23) can be written

$$E_2 - E_1 = -\Delta z_B - h_L, \tag{11.37}$$

where $\Delta z_B\ (= z_{B_2} - z_{B_1})$ is the increase in bottom elevation. The specific energy is changed by changes in bottom elevation (which may be positive or negative) or by energy loss, which must be positive. Uniform flow has $\Delta z_B = -h_L$, while frictionless flow has $h_L = 0$.

First consider a frictionless flow with constant specific energy. The channel bottom must be horizontal. These conditions are closely approximated by flow under a *sluice gate* (Fig. 11.11a), a device for regulating flow in open channels. For a given opening of the gate (y_G), a specific flow rate (q) exists. If we neglect friction, Eq. (11.37) indicates that the upstream and downstream values of E are equal; thus the upstream and downstream flow states are alternate states. The upstream state is subcritical and the downstream state is supercritical. Theoretically, the critical state occurs right at the opening, with $y_G = y_c$; however, the flow near the opening is not one-directional and one-dimensional.

If we specify that the flow rate remain constant and consider a larger gate opening (Fig. 11.11b), the upstream depth is less and the downstream depth is greater; the specific energy is smaller. For successively larger gate openings, the upstream and downstream states approach the critical state.

Next consider that the specific energy remains constant as the gate is opened (Fig. 11.11c and d). As the gate is opened, the flow rate increases, and the upstream and downstream state points are located on curves corresponding to larger flow rates. As the gate is opened and

Figure 11.11 Specific energy diagrams for flow under sluice gate: (a) and (b) q is constant; (c) and (d) E is constant.

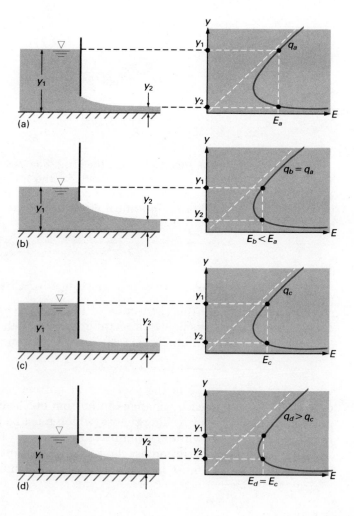

q is increased, the upstream depth decreases, the downstream depth increases, and the flow again approaches the critical condition. When the critical flow rate is reached, further opening of the gate will not increase the flow rate so long as the specific energy remains constant.

It would be interesting to consider the sequence of events as the gate is opened if we specify that the upstream depth remain fixed. We leave that as an exercise for you.

EXAMPLE 11.5   Illustrates flow with constant specific energy

A rectangular, concrete-lined irrigation canal is 80 ft wide and carries 1200 ft³/sec of water at a (normal) depth of 4 ft. A bridge is built over the canal. The bridge is supported by five concrete piers, each 3 ft wide. Calculate the water depth as it passes under the bridge. Assume frictionless flow in the vicinity of the bridge piers.

Solution

Given

Canal water flow rate of 1200 ft³/sec with a depth of 4 ft and a width of 8 ft
Canal width reduced by five 3-ft-wide piers
Frictionless flow

Find

Water depth in vicinity of piers

Solution

Combining Eqs. (11.23), (11.24), and (11.26) gives

$$\left(y_1 + \frac{q^2}{2gy_1^2}\right) - \left(y_2 + \frac{q^2}{2gy_2^2}\right) = z_{B_2} - z_{B_1} + h_L,$$

where points 1 and 2 are shown in Fig. E11.5(a). According to the problem statement, frictionless flow is assumed, so $h_L = 0$. The canal bed is assumed horizontal between points 1 and 2. Equations (11.26) and (11.37) give

$$y_2 + \frac{q_2^2}{2gy_2} = y_1 + \frac{q_1^2}{2gy_1} \qquad \text{or} \qquad E = \text{Constant.}$$

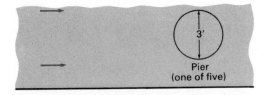

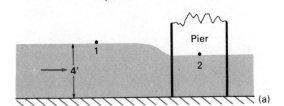

Fig E11.5 (a) Flow around bridge piers in a canal.

Top view

(a)

The numerical values in the problem statement give

$$q_1 = \frac{\left(1200 \, \dfrac{ft^3}{sec}\right)}{(80 \, ft)} = 15 \, \frac{ft^2}{sec},$$

$$E = E_1 = y_1 + \frac{q_1^2}{2gy_1^2} = 4 \, ft + \frac{\left(15 \, \dfrac{ft^2}{sec}\right)^2}{2\left(32.2 \, \dfrac{ft}{sec^2}\right)(4 \, ft)^2} = 4.22 \, ft,$$

and thus

$$E_2 = y_2 + \frac{q_2^2}{2gy_2} = 4.22 \, ft.$$

The continuity equation gives

$$b_2 q_2 = b_1 q_1, \qquad so \qquad y_2 + \frac{1}{2g}\left(\frac{b_1 q_1}{b_2 y_2}\right)^2 = 4.22 \, ft.$$

At the piers,

$$b_2 = 80 \, ft - 5(3 \, ft) = 65 \, ft$$

and

$$y_2 + \frac{1}{2g}\left(\frac{b_1 q_1}{b_2 y_2}\right)^2 = y_2 + \frac{1}{2\left(32.2 \, \dfrac{ft}{sec^2}\right)}\left(\frac{(80 \, ft)\left(15 \, \dfrac{ft^2}{sec}\right)}{(65 \, ft)y_2}\right)^2 = y_2 + \frac{5.29 \, ft^3}{y_2^2}.$$

Simplifying, we have

$$y_2^3 - 4.22 \, ft \, y_2^2 + 5.29 \, ft^3 = 0.$$

This equation has three roots. Solving for the first root, y_2, by trial and error gives

$$y_2 = 1.36.$$

The remaining roots are found by factoring out the first root, leaving the quadratic equation

$$y_2^2 - 2.86y_2 - 3.89 = 0.$$

Using the quadratic formula gives

$$y_2 = \frac{2.86}{2} \pm \sqrt{\left(\frac{2.86}{2}\right)^2 + 3.89} = 3.87, \, -1.01.$$

The two realistic roots are

$$y_2 = 1.36 \, ft \qquad and \qquad y_2 = 3.87 \, ft.$$

We must choose one of these as the correct answer. Noting that $q_1 = 15 \, ft^2/sec$ and

$$q_2 = \frac{1200 \, \dfrac{ft^3}{sec}}{65 \, ft} = 18.5 \, ft^2/sec,$$

we have as the corresponding critical depths

$$y_{c_1} = \left(\frac{q_1^2}{g}\right)^{1/3} = \left[\frac{\left(15\,\frac{ft^2}{sec}\right)^2}{32.2\,\frac{ft}{sec^2}}\right]^{1/3} = 1.91 \text{ ft}$$

and

$$y_{c_2} = \left(\frac{q_2^2}{g}\right)^{1/3} = \left[\frac{\left(18.5\,\frac{ft^2}{sec}\right)^2}{32.2\,\frac{ft}{sec^2}}\right]^{1/3} = 2.20 \text{ ft}.$$

Thus one of the two calculated depths at location 2 corresponds to subcritical flow and the other corresponds to supercritical flow. Perhaps this is the key to the correct root. The specific energy curves for q_1 and q_2 are shown in Fig. E11.5(b). Since the specific energy is constant from 1 to 2, the flow must pass through a critical depth y_c' corresponding to a flow rate q' to pass from subcritical at 1 to supercritical at 2. This critical depth y_c' corresponds to a critical width b_c' and critical velocity V_c'. At this critical condition,

$$E_c' = E_1 = 4.22 \text{ ft.}$$

Then Eqs. (11.24) and (11.29) give

$$E_c = \frac{V_c'^2}{2g} + y_c' = \frac{y_c'}{2} + y_c' = 1.5 y_c';$$

thus

$$y_c' = \frac{E_c'}{1.5} = \frac{4.22 \text{ ft}}{1.5} = 2.81 \text{ ft}$$

and

$$V_c' = (gy_c')^{1/2} = \left[\left(32.2\,\frac{ft}{sec^2}\right)(2.81 \text{ ft})\right]^{1/2} = 9.51 \text{ ft/sec.}$$

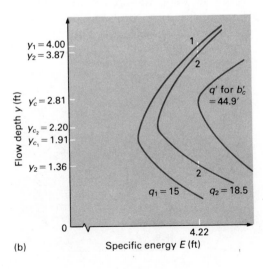

Figure E11.5 (b) Specific energy curve for flow around pier.

(b)

The corresponding channel width, b'_c, is obtained from the continuity equation:

$$b'_c = \frac{Q}{V'_c y'_c} = \frac{1200 \, \dfrac{ft^3}{sec}}{\left(9.51 \, \dfrac{ft}{sec}\right)(2.81 \, ft)} = 44.9 \, ft.$$

Since the irrigation canal does not narrow to 44.9 ft between locations 1 and 2, the flow will not pass from subcritical to supercritical, and the supercritical depth at 2 is not possible; therefore

Answer

$$|\ y_2 = 3.87 \, ft\ |$$

under the bridge.

Discussion

It is important to consider all possible solutions when a physical phenomenon is described by a higher-order equation. If we had stopped at the first realistic root ($y_c = 1.36$ ft), we would have had the wrong answer.

If you have studied Chapter 10 on compressible flow, you have probably noticed the similarity between this flow and isentropic flow in a converging-diverging nozzle. We will discuss this similarity in detail in a later section.

Note the surprising fact that the flow becomes shallower as the channel width is decreased.

Now we consider a frictionless flow in which the specific energy changes along the channel. According to Eq. (11.37), this can be accomplished by changing the channel bottom elevation. Figure 11.12(a) shows a horizontal channel with a bump on the bottom. Since the bump is higher than the upstream channel bottom, the specific energy will be less over the bump. If the upstream specific energy is known, the specific energy at any downstream point is

$$E_2 = E_1 - \Delta E = E_1 - \Delta z_{B_2}. \tag{11.38}$$

Figure 11.12(b) illustrates the process on a specific energy diagram.

The change in the liquid surface profile over the bump depends on whether the flow approaching the bump is subcritical (Fig. 11.13) or

Figure 11.12 Frictionless flow over a bump in a horizontal channel: (a) Flow schematic; (b) change of E.

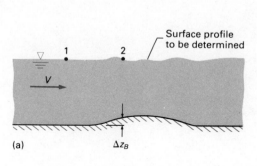
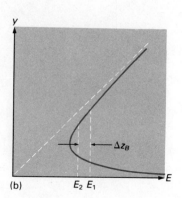

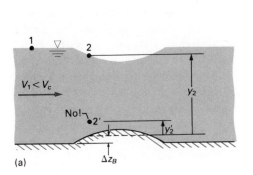

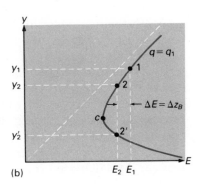

Figure 11.13 Subcritical flow over a bump.

supercritical (Fig. 11.14). First consider subcritical flow. The state point upstream of the bump is shown as point 1 in Fig. 11.13(b). When the specific energy is lowered to E_2, the flow state is either point 2 or point 2' on the specific energy diagram (Fig. 11.13b). Since the flow rate is constant, the only way to get to point 2' is to pass through the critical state, moving along the specific energy curve. This is impossible because the specific energy has not been lowered to E_c upstream of point 2 in the channel, so state point 2 describes the flow. The state is still subcritical and the depth is decreased. By choosing several points over the bump, we can work out the complete liquid surface profile.

Next assume that the flow approaching the bump is supercritical. The initial state is shown as point 1 in Fig. 11.14(b). Possible states with specific energy E_2 are shown as state points 2 and 2' on the specific energy curve. Since it is not possible for the flow to pass through the critical state, the downstream state must correspond to state point 2. The flow is still supercritical, and the depth increases over the bump. By selecting several points, we can determine the complete surface profile.

Now we consider what would happen if we made the bump higher. We note that the minimum specific energy always occurs at the crest of the bump. Suppose that the flow approaching the bump is subcritical. Raising the bump causes the depth at the crest to decrease and the velocity over the crest to increase. The state point at the crest is moved

Figure 11.14 Supercritical flow over a bump.

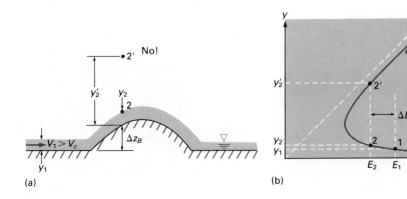

toward the critical state. If the flow state does not become critical over the crest, then on the downstream side of the bump the depth increases, the velocity decreases, and the flow returns to its initial state. If we make the bump high enough, it is possible that the flow over the crest can become critical. If this occurs, the flow on the downstream side of the crest either may return to the initial subcritical state or may accelerate to the alternate supercritical state, depending on conditions far downstream of the bump.

The height of the bump necessary to cause critical flow over the crest can be calculated from Eq. (11.38) by setting $E_2 = E_c$:

$$\Delta z_{B,\max} = E_1 - E_c.$$

Using Eqs. (11.26) and (11.28), we have

$$\Delta z_{B,\max} = y_1 + \frac{q^2}{2gy_1^2} - \frac{3}{2}\left(\frac{q^2}{g}\right)^{1/3}. \tag{11.39}$$

Now we ask what happens if the bump is made higher than $\Delta z_{B,\max}$. This would imply $E_2 < E_c$; however, this would put the downstream point off the specific energy curve for the given value of q. Physically, it would no longer be possible to maintain the given flow with the given upstream state 1. If the given flow were to be maintained, y_1 would have to increase so that E_1 would be greater. Alternatively, if y_1 were to remain fixed, q would be decreased. Assuming that q remains fixed, Fig. 11.15 illustrates the variation of the upstream and bump-crest depths with bump height. Note that once the flow over the crest becomes critical, it remains critical, with upstream depth adjusting to compensate for increases in bump height. This illustrates an important characteristic of subcritical flow: The upstream flow adjusts to compensate for downstream conditions.

Flow elements that cause critical flow such as sluice gates, bumps (dams), and changes in channel slope from mild to steep are often used as control sections for open channel flows. A *control section* controls the flow upstream if it is subcritical and controls the flow downstream if it is supercritical.

Figure 11.15 Variation of upstream depth and bump crest depth with bump height for constant q.

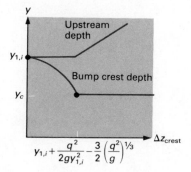

EXAMPLE 11.6   Illustrates frictionless channel flow with variable specific energy

The sides of the irrigation canal of Example 11.5 are 8 ft high. A new proposal calls for the canal to deliver water at a rate of 3000 ft³/sec. Can the canal handle the new flow rate? At a certain location, a telephone cable conduit spans the channel. The conduit is 5.5 ft above the canal bottom. Will the conduit be submerged at the new flow rate? If the conduit will be submerged, one proposal suggests lowering the water surface under the conduit either by excavating the canal bottom under the conduit or by filling in the bottom under the conduit. Which should be done? How much should the bottom be excavated or raised?

Solution

Given
Figure E11.6(a)
Canal 80 ft wide

Find
Water depth for 3000 ft³/sec flow rate
Whether the telephone conduit is submerged
The amount the canal bottom should be excavated or raised
    so the telephone conduit is not submerged

Solution
The slope of the canal bottom is found using Manning's formula, Eq. (11.14b), and the information in Example 11.5:

$$S_0 = \left(\frac{nV}{1.49R_h^{2/3}}\right)^2.$$

For an (assumed unfinished) concrete-lined canal, Table 11.1 gives $n = 0.014$. The hydraulic radius is

$$R_h = \frac{by}{b + 2y},$$

where b is the canal width (80 ft) and y the water depth (4 ft). Then

$$R_h = \frac{(80 \text{ ft})(4 \text{ ft})}{80 \text{ ft} + 2(4 \text{ ft})} = 3.64 \text{ ft}.$$

The velocity V is

$$V = \frac{Q}{by} = \frac{1200 \dfrac{\text{ft}^3}{\text{sec}}}{(80 \text{ ft})(4 \text{ ft})} = 3.75 \text{ ft/sec}$$

and

$$S_0^{1/2} = \frac{0.014\left(3.75 \dfrac{\text{ft}}{\text{sec}}\right)}{1.49(3.64 \text{ ft})^{2/3}} = 0.0149.$$

Now consider the 3000 ft³/sec flow rate. Manning's formula is

$$V = \frac{1.49}{n}(R_h^{2/3})S_0^{1/2}.$$

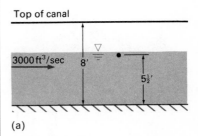

Top of canal

3000 ft³/sec   8′   5½′

(a)

Figure E11.6 (a) Conduit placed in irrigation canal.

Now

$$V = \frac{Q}{by} \qquad \text{and} \qquad R_h = \frac{by}{b + 2y}.$$

Substituting into Manning's formula gives

$$\frac{Q}{by} = \frac{1.49}{n} \left(\frac{by}{b + 2y} \right)^{2/3} S_0^{1/2}.$$

Solving for the depth, we have

$$\frac{1}{y} \left(\frac{b + 2y}{y} \right)^{2/3} = \frac{1.49 b^{5/3}}{nQ} (S_0^{1/2}).$$

The numerical values give

$$\frac{1}{y} \left(\frac{80 \text{ ft} + 2y}{y} \right)^{2/3} = \frac{1.49(80 \text{ ft})^{5/3}(0.0149)}{0.014 \left(3000 \dfrac{\text{ft}^3}{\text{sec}} \right)} = 0.785.$$

Solving by trial and error gives

$$y = y_n = 7.12 \text{ ft};$$

therefore

> | The canal can handle the new flow rate and the conduit is submerged. | **Answer** ◀

We next investigate whether raising or lowering the canal bottom will lower the water surface below the conduit. The first step is to determine whether the flow is subcritical or supercritical. Calculating the Froude number,

$$V = \frac{Q}{by} = \frac{3000 \dfrac{\text{ft}^3}{\text{sec}}}{(80 \text{ ft})(7.12 \text{ ft})} = 5.27 \text{ ft/sec},$$

$$\mathbf{F} = \frac{V}{\sqrt{gy}} = \frac{\left(5.27 \dfrac{\text{ft}}{\text{sec}} \right)}{\sqrt{\left(32.2 \dfrac{\text{ft}}{\text{sec}^2} \right)(7.12 \text{ ft})}} = 0.348.$$

The flow is subcritical. Assuming frictionless flow, Eqs. (11.23) and (11.25) give

$$\left(y_2 + \frac{q_2^2}{2gy_2^2} \right) = \left(y_1 + \frac{q_1^2}{2gy_1^2} \right) + z_{B_1} - z_{B_2}.$$

Choosing $z_{B_1} = 0$ and noting

$$q = \frac{3000 \dfrac{\text{ft}^2}{\text{sec}}}{80 \dfrac{\text{ft}^2}{\text{sec}}} = 37.5 \text{ ft}^2/\text{sec},$$

we have

$$y_2 + \frac{\left(37.5 \dfrac{\text{ft}^2}{\text{sec}} \right)^2}{\left(32.2 \dfrac{\text{ft}}{\text{sec}^2} \right) y_2^2} = 7.12 \text{ ft} + \frac{\left(37.5 \dfrac{\text{ft}^2}{\text{sec}} \right)^2}{2 \left(32.2 \dfrac{\text{ft}}{\text{sec}^2} \right)(7.12 \text{ ft})^2} - z_{B_2}$$

or

$$y_2 + \frac{21.84 \text{ ft}^3}{y_2^2} + z_{B_2} = 7.55 \text{ ft}.$$

Positive values of z_{B_2} represent raising the canal bottom, and negative values of z_{B_2} represent excavating the canal bottom. The water surface height above the $z_{B_1} = 0$ level is $(y_2 + z_{B_2})$. Using trial and error, we solve the above equation for y_2 as a function of z_{B_2}. We then plot the values of $(y_2 + z_{B_2})$ for both subcritical and supercritical flows versus z_{B_2} as the solid curve in Fig. E11.6b. The curve shows that the water level is lowered below the conduit (5.5 ft) if the canal bottom is raised less than 2.24 ft and the flow is forced to change from subcritical to supercritical. We find the bump height, z_{B_2}, required for transition from subcritical to supercritical flow by first using Eqs. (11.6) and (11.25) with $F = 1$ to give

$$1 = \frac{V_2}{\sqrt{gy_2}} = \frac{q_2}{y_2\sqrt{gy_2}}, \quad \text{or} \quad q_2 = y_2\sqrt{gy_2}.$$

Equations (11.23) and (11.25) then give

$$y_2 + \frac{(y_2)^2 g y_2}{2gy_2^2} = y_1 + \frac{q_1^2}{2gy_1^2} + z_{B_1} - z_{B_2}, \quad \text{or} \quad 1.5y_2 = y_1 + \frac{q_1^2}{2gy_1^2} + z_{B_1} - z_{B_2}.$$

The numerical values give

$$y_2 = \frac{7.55 - z_{B_2}}{1.5}, \quad \text{or} \quad (y_2 + z_{B_2}) = 5.03 + \frac{z_{B_2}}{3}.$$

This equation shows that the water surface height (above the $z_{B_1} = 0$ level) required for transition to supercritical flow depends on the value of the bump height z_{B_2}. In Fig. 11.6(b), $(y_2 + z_{B_2})$ is plotted versus z_{B_2} as the dashed line. The intersection of the dashed line and the solid curve shows that transition occurs at a $(y_2 + z_{B_2})$ value of 5.80 ft. The corresponding value of z_{B_2} is 2.25 ft. Noting that the value of z_{B_2} corresponding to $(y_2 + z_{B_2}) = 5.5$ ft is 2.24 ft, we see that

| the bump should look as sketched in Fig. E11.6(c). | **Answer** ◀ |

The maximum bump height is upstream of the conduit.

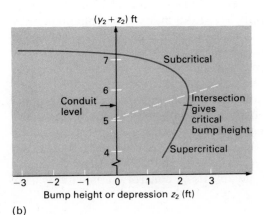

(b)

Figure E11.6 (b) Subcritical and supercritical water level heights above $z_1 = 0$ level.

Figure E11.6 (c) Bump schematic.

Canal wall height

• Conduit

2.25' <2.24'

(c)

Discussion

It no doubt seems strange that the channel bottom surface must be *raised* to lower the water surface below the conduit.

Could the water surface be lowered by changing the width of the canal at the conduit location? Should the canal be made wider or narrower to lower the water surface?

11.3.3 Hydraulic Jump

We have seen that open channel flow can be changed from subcritical to supercritical in a relatively efficient (low-loss) manner by sluice gates or changes in channel bottom geometry. In these cases, the flow state passes through critical with relatively small energy loss. We now consider transitions from supercritical to subcritical flow. Carrying out such transitions without significant energy loss is nearly impossible. This is largely due to the fact that disturbances cannot propagate upstream in supercritical flow. Consider trying to operate a sluice gate "backward," with the shallow supercritical flow entering and the deeper subcritical flow leaving. Intuition suggests that this is a highly unstable situation; any disturbance near the gate opening would cause the incoming liquid to "pile up" upstream of the gate.

Similarly, consider attempting to convert from supercritical flow to subcritical flow using a bump in the channel bottom. If the height of the bump is less than $\Delta z_{B,\max}$, then the flow will remain supercritical over the crest and will return to its original supercritical state downstream of the bump. If the height of the bump is greater than $\Delta z_{B,\max}$, then a physically impossible situation has been specified. In subcritical flow, the upstream depth adjusts to compensate; however, if the approach flow is supercritical, there is no mechanism by which the upstream flow can be "informed" about the too-high bump. Smooth passage through the critical state is possible only if the bump height is *exactly* equal to $\Delta z_{B,\max}$.

In practice, the transition between a supercritical flow and a subcritical flow is accomplished in a *hydraulic jump,* a reasonably short, extremely turbulent region of flow across which the state changes from supercritical to subcritical. Figures 6.15, 11.16, and 11.17 illustrate hydraulic jumps that occur in spillways. Like many other phenomena in fluid mechanics, the hydraulic jump is rather common. It is easy to observe a hydraulic jump in the kitchen sink as water from the faucet impinges on a flat surface and flows radially outward from the impingement point.* On a specific-energy diagram, a hydraulic jump is represented by a jump from the lower (supercritical) branch to the upper (subcritical) branch. Since a hydraulic jump is extremely turbulent, losses in the jump are not negligible, and the specific energy downstream of the jump is less than the specific energy upstream of the jump.

The flow in a hydraulic jump is extremely complicated but we usually do not need to consider its details. A control volume analysis of

---

* Try it in your sink.

Figure 11.16 Small hydraulic jump at outlet of a dam overflow. (Photograph by John Remington.)

Figure 11.17 Details of surface of the hydraulic jump shown in Fig. 11.16. (Photograph by John Remington.)

a hydraulic jump reveals most of what we need to know about it. We have already used this idea once, in Example 3.9. Figure 11.18 shows a control volume for analysis of a hydraulic jump. Since jumps are usually short, we assume that the channel bottom is horizontal and that the flow is identical in all planes parallel to the paper. The continuity equation is

$$V_1 y_1(1) = V_2(y_2)(1),$$

where (1) is the unit width perpendicular to the paper. The momentum equation is

$$\sum F_x = \dot{M}_{x,\text{out}} - \dot{M}_{x,\text{in}}.$$

The forces are due to hydrostatic pressure at the ends of the control volume and shear stress at the channel bottom. Since a hydraulic jump is short, the bottom force is negligible and the momentum equation becomes

$$\frac{\gamma y_1^2}{2}(1) - \frac{\gamma y_2^2}{2}(1) = \rho y_2(1) V_2^2 - \rho y_1(1) V_1^2.$$

Figure 11.18 Control volume for analysis of hydraulic jump.

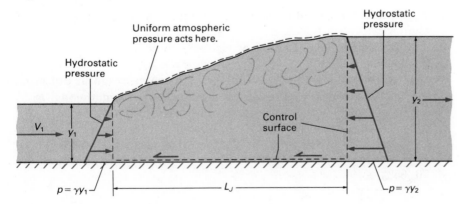

Solving for V_2 from the continuity equation, we have

$$V_2 = \frac{y_1}{y_2} V_1. \tag{11.40}$$

Substituting this into the momentum equation and solving for the depth ratio across the jump give

$$\frac{y_2}{y_1} = \frac{1}{2}\left[\sqrt{1 + 8\left(\frac{V_1^2}{gy_1}\right)} - 1\right],$$

or

$$\frac{y_2}{y_1} = \frac{1}{2}\left(\sqrt{1 + 8\mathbf{F}_1^2} - 1\right). \tag{11.41}$$

The depths y_1 and y_2 on opposite sides of a hydraulic jump are called *conjugate depths*.

The energy loss is an important parameter of a hydraulic jump. The energy equation across a jump is

$$\left(\frac{V_1^2}{2g} + y_1\right) - \left(\frac{V_2^2}{2g} + y_2\right) = h_L.$$

Substituting Eqs. (11.40) and (11.41) gives

$$h_L = \frac{(y_2 - y_1)^3}{4y_1y_2}. \tag{11.42}$$

Equation (11.41) works for $\mathbf{F}_1 > 1$ $(y_2/y_1 > 1)$ *and* $\mathbf{F}_1 < 1$ $(y_2/y_1 < 1)$; however, Eq. (11.42) shows that jumps with $\mathbf{F}_1 < 1$ have negative loss, which is a physical impossibility. Experiments have shown that Eqs. (11.41) and (11.42) are remarkably accurate for real jumps with $\mathbf{F}_1 > 1$ and $y_2 > y_1$.

Open channel designers often use hydraulic jumps as energy dissipators. A jump is usually located just downstream of a dam spillway (Figs. 6.15 and 11.16) to dissipate the kinetic energy of the flow. If this were not done, severe erosion of the downstream channel bed could result. Obviously, jumps must be located on a strong bed. Five classes of jumps, based on inlet Froude number, have been identified from experimental observation:

- $\mathbf{F}_1 = 1.0$ to 1.7: Standing wave, or undular, jump; dissipation less than 5%.
- $\mathbf{F}_1 = 1.7$ to 2.5: Smooth surface rise known as a weak jump; dissipation 5–15%.
- $\mathbf{F}_1 = 2.5$ to 4.5: Unstable, oscillating jump; each irregular pulsation creates a large wave that can travel far downstream, damaging earth banks and other structures; dissipation 15–45%.
- $\mathbf{F}_1 = 4.5$ to 9.0: Stable "steady" jump; best performance and action, insensitive to downstream conditions; dissipation 45–70%.
- $\mathbf{F}_1 > 9.0$: Rough, somewhat intermittent, strong jump; dissipation 70–85%.

Our control volume analysis is incapable of predicting the length of a hydraulic jump. Dimensional analysis indicates

$$\frac{L_J}{y_2} = f(\mathbf{F}_1).$$

Figure 11.19 shows experimentally determined jump length as a function of Froude number.

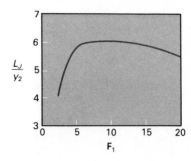

Figure 11.19 Length of hydraulic jump.

EXAMPLE 11.7   Illustrates hydraulic jump calculations

Figure E11.7(a) shows a hydraulic jump downstream of a sluice gate in a wide horizontal channel. Calculate y_2, y_3, and the percentage loss of mechanical energy. Sketch the flow to scale and include the energy grade line.

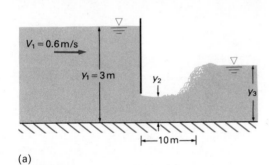

(a)

Figure E11.7 (a) Flow under sluice gate, with hydraulic jump.

Solution

    Given
Flow under sluice gate with hydraulic jump
Figure E11.7(a)

    Find
y_2, y_3, and percentage loss of mechanical energy in jump

    Solution
Assume the bottom of the channel is horizontal and the flow is frictionless. Equation (11.23) gives

$$\frac{V_1^2}{2g} + y_1 = \frac{V_2^2}{2g} + y_2.$$

Assuming the channel width is constant, we have

$$V_1 y_1 = V_2 y_2,$$

so the specific energy equation is

$$\frac{V_1^2}{2g} + y_1 = \frac{1}{2g}\left(\frac{y_1 V_1}{y_2}\right)^2 + y_2.$$

The numerical values give

$$\frac{\left(0.6\dfrac{m}{s}\right)^2}{2\left(9.807\dfrac{m}{s^2}\right)} + 3\text{ m} = \frac{1}{2\left(9.807\dfrac{m}{s^2}\right)}\left[\frac{(3\text{ m})\left(0.6\dfrac{m}{s}\right)}{y_2}\right]^2 + y_2$$

or

$$3.018\text{ m} = \frac{0.165\text{ m}^3}{y_2^2} + y_2.$$

Since the critical flow occurs at the sluice gate, we solve for the supercritical (smaller) value of y_2. Trial and error give

$$\boxed{y_2 = 0.244\text{ m.}}$$ **Answer** ◀

We find y_3 by first finding the Froude number $\mathbf{F}_2$ using

$$V_2 = \frac{y_1 V_1}{y_2} = \frac{(3\text{ m})\left(0.6\dfrac{m}{s}\right)}{(0.244\text{ m})} = 7.38\text{ m/s}$$

and

$$\mathbf{F}_2 = \frac{V_2}{\sqrt{gy_2}} = \frac{7.38\dfrac{m}{s}}{\sqrt{\left(9.807\dfrac{m}{s^2}\right)(0.244\text{ m})}} = 4.77.$$

Equation (11.41) then gives

$$y_3 = y_2\left[\frac{1}{2}\left(\sqrt{1 + 8\mathbf{F}_2^2} - 1\right)\right] = (0.244\text{ m})\left[\frac{1}{2}\left(\sqrt{1 + 8(4.77)^2} - 1\right)\right],$$

$$\boxed{y_3 = 1.528\text{ m.}}$$ **Answer** ◀

The energy loss in the jump, given by Eq. (11.42), is

$$h_L = \frac{(y_3 - y_2)^3}{4y_2 y_3}.$$

The percentage energy loss in the jump is

$$\Delta E = \frac{h_L}{\dfrac{V_2^2}{2g} + y_2}(100) = \frac{(y_3 - y_2)^3}{4y_2 y_3\left(\dfrac{V_2^2}{2g} + y_2\right)}(100).$$

The numerical values give

$$\Delta E = \frac{(1.528\text{ m} - 0.244\text{ m})^3(100)}{4(0.244\text{ m})(1.528\text{ m})\left(\dfrac{\left(7.38\dfrac{m}{s}\right)^2}{2\left(9.807\dfrac{m}{s^2}\right)} + 0.244\text{ m}\right)},$$

$$\boxed{\Delta E = 47.0\%.}$$ **Answer** ◀

The scaled sketch of the flow is shown in Fig. E11.7(b). The height of the sluice gate above the bottom of the channel is found using Eq. (11.23), assuming fric-

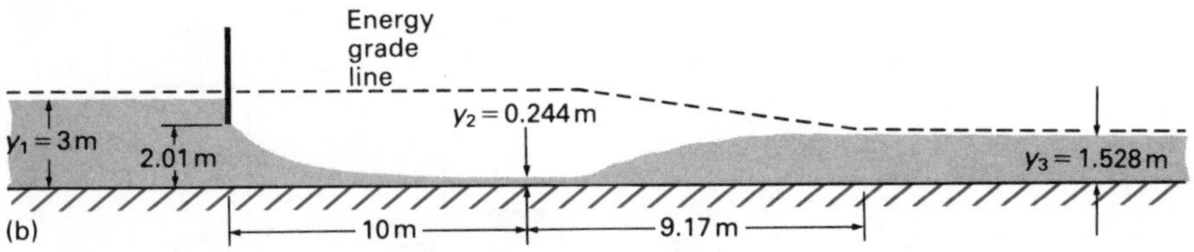

Figure E11.7 (b) Scaled sketch of flow and energy grade line.

tionless flow:

$$\frac{V_1^2}{2g} + y_1 = \frac{V_c^2}{2g} + y_c,$$

where Eq. (11.29) gives

$$V_c^2 = gy_c, \qquad \text{so} \qquad \frac{V_1^2}{2g} + y_1 = 1.5y_c.$$

The numerical values give

$$3.018 \text{ m} = 1.5y_c, \qquad \text{or} \qquad y_c = 2.01 \text{ m}.$$

The length of the hydraulic jump is found using $\mathbf{F}_2 = 4.77$ and Fig. 11.19 to give

$$L \approx 6y_3 = 6(1.528 \text{ m}) = 9.17 \text{ m}.$$

Discussion
Note that the energy grade line is level from point 1 to point 2 since frictionless flow was assumed and the floor of the channel is level.

11.4 Gradually Varied Flow

Thus far our discussion of open channel flow has been limited to simplified cases such as uniform flow (which requires uniform bottom slope, channel cross section, and surface roughness) and frictionless flow, which is a rather idealized case. In this section, we will discuss the analysis of varied flow with friction. We will permit changes in bottom slope, channel shape, and surface condition, and we will include friction losses; however, our analysis will not be completely general because we will restrict consideration to gradually varied flow (GVF). We assume that flow conditions vary slowly enough that the flow can be assumed one-dimensional (the independent variable is distance along the channel, x) and one-directional.

11.4.1 The Differential Equation of Gradually Varied Flow

The assumptions of one-dimensional and one-directional flow imply that the flow is described by an ordinary differential equation. Figure 11.20 illustrates the flow between locations x and $x + \delta x$ in a channel. The channel bottom slope, S_0, may vary with x. The liquid velocity V and depth y vary with x. We first write the energy equation between points x and $x + \delta x$ on the liquid surface:

$$\frac{V^2}{2g} + y + z_B = \frac{V^2}{2g} + \frac{d}{dx}\left(\frac{V^2}{2g}\right)\delta x + y + \left(\frac{dy}{dx}\right)\delta x + z_B + \left(\frac{dz_B}{dx}\right)\delta x + \delta h_L.$$

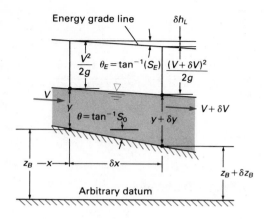

Figure 11.20 Gradually varied flow in small reach of channel.

The bottom slope (S_0) is

$$S_0 = -\frac{dz_B}{dx}.$$

We can write the head loss as

$$\delta h_L = \left(\frac{dh_L}{dx}\right)\delta x = S_E\,\delta x, \qquad (11.43)$$

where S_E is the slope of the energy grade line. Substituting S_0 and S_E into the energy equation, canceling terms where possible, and dividing by δx, we have

$$\frac{d}{dx}\left(\frac{V^2}{2g}\right) + \frac{dy}{dx} = S_0 - S_E. \qquad (11.44)$$

This equation involves derivatives of both velocity and depth. To eliminate the velocity derivative, we use the continuity equation for steady flow:

$$Q = VA = \text{Constant}.$$

Squaring and dividing by $2g$ give

$$\left(\frac{V^2}{2g}\right)A^2 = \frac{Q^2}{2g}.$$

Differentiating, we have

$$A^2\left(\frac{d}{dx}\right)\left(\frac{V^2}{2g}\right) + \left(\frac{V^2}{g}\right)A\left(\frac{dA}{dx}\right) = 0.$$

Substituting this equation into Eq. (11.44) gives

$$-\frac{V^2}{gA}\left(\frac{dA}{dx}\right) + \frac{dy}{dx} = S_0 - S_E. \qquad (11.45)$$

Recalling that this equation applies to points on the surface, from Fig. 11.10, we have

$$dA = b_0\,dy,$$

and the equation becomes

$$-\frac{V^2 b_0}{gA}\left(\frac{dy}{dx}\right) + \frac{dy}{dx} = S_0 - S_E.$$

Solving for dy/dx and using Eq. (11.34), we get

$$\frac{dy}{dx} = \frac{S_0 - S_E}{1 - \left(\dfrac{V^2 b_0}{gA}\right)} = \frac{S_0 - S_E}{1 - \mathbf{F}^2}. \tag{11.46}$$

Integration of this equation permits computation of liquid surface profiles; however, we can learn a great deal about gradually varied flow by examining the equation without integrating it. To simplify the discussion, we limit consideration to rectangular channels for which $b_0 = b = $ Constant and $A = by$.

Starting from a given initial point on the liquid surface, the fluid depth increases or decreases accordingly as the right side of Eq. (11.46) is positive or negative. The sign of the right side must be determined by examination of both the numerator and the denominator. The denominator changes sign at $\mathbf{F} = 1$ so, for a given sign of the numerator, the change of depth downstream of an initial point is opposite for subcritical or supercritical flow. We can easily decide if the flow at a given location is subcritical or supercritical by comparing the actual depth y to the critical depth y_c. If $y > y_c$, $\mathbf{F} < 1$ and the flow is subcritical, so the denominator in Eq. (11.46) is positive; if $y < y_c$, $\mathbf{F} > 1$ and the flow is supercritical, so the denominator is negative.

To examine the numerator, we first consider the friction slope S_E. From Eq. (11.43),

$$S_E = \frac{dh_L}{dx}.$$

In terms of friction factor and hydraulic radius,

$$S_E = \frac{dh_L}{dx} = \frac{f}{8g}\left(\frac{V^2}{R_h}\right). \tag{11.47}$$

Using the Chezy coefficient C, we have

$$S_E = \frac{V^2}{C^2 R_h}, \tag{11.48}$$

or, in terms of Manning's formula,

$$S_E = \frac{n^2}{\xi^2}\left(\frac{V^2}{R_h^{4/3}}\right), \tag{11.49}$$

where $\xi = 1.0$ for SI units and 1.49 for BG or EE units.

The comparison of S_E and S_0 is often most easily accomplished by comparing the actual depth (y) to the normal depth (y_n) for the same flow rate. Using Eq. (11.49) and assuming a wide channel $(R_h = y)$, we have for the flow rate

$$q = Vy = \left(\frac{\xi}{n}\right)y^{2/3}(S_E^{1/2})y. \tag{11.50}$$

Table 11.2 Possibilities for surface slope from Eq. (11.46)

	$y > y_c$ (Denominator Positive)	$y < y_c$ (Denominator Negative)	$y = y_c$ (Denominator Zero)
$y > y_n$ (numerator positive)	$\dfrac{dy}{dx} > 0$ Depth increases	$\dfrac{dy}{dx} < 0$ Depth decreases	$\dfrac{dy}{dx}$ infinite Not GVF
$y < y_n$ (numerator negative)	$\dfrac{dy}{dx} < 0$ Depth decreases	$\dfrac{dy}{dx} > 0$ Depth increases	$\dfrac{dy}{dx}$ infinite Not GVF
$y = y_n$ (numerator zero)	$\dfrac{dy}{dx} = 0$ Uniform flow, depth constant	$\dfrac{dy}{dx} = 0$ Uniform flow, depth constant	$\dfrac{dy}{dx} = \dfrac{0}{0}$ Uniform critical flow, unstable

The normal depth for the actual bottom slope (S_0) is calculated from Eq. (11.14):

$$q = \left(\frac{\xi}{n}\right) y_n^{2/3} (S_0^{1/2}) y_n. \tag{11.51}$$

Equating the two expressions for q gives

$$\frac{S_E}{S_0} = \left(\frac{y_n}{y}\right)^{10/3}$$

If the depth y is greater than the normal depth y_n, then S_E is less than S_0 and the numerator in Eq. (11.46) will be positive. Conversely if $y < y_n$, then $S_E > S_0$ and the numerator will be negative.

The sign of dy/dx in Eq. (11.46) and, therefore, the direction of depth change in gradually varied flow are determined by the actual depth relative to both the normal depth and the critical depth. Table 11.2 summarizes the various possibilities.

11.4.2 Classification of Surface Profiles

There are twelve different types of liquid surface profiles in gradually varied flow.* These types are identified according to the bottom slope and the actual depth relative to the normal and critical depths. The slope designations are based on comparison of the bottom slope S_0 with the critical slope S_c (Eqs. 11.35 and 11.36) and with zero slope. Table 11.3 summarizes slope names and symbols.

Numbers are associated with the slope symbols according to Table 11.4. All "1" profiles and all "3" profiles have depth increasing. All "2" profiles have depth decreasing. Since each slope symbol can be combined with a number, it would seem that there would be fifteen profile types rather than twelve; however, there is no $C2$ profile since $y_n = y_c$

* Excluding uniform flow, which would make thirteen types.

Table 11.3 Slope designations

Slope Name	Condition	Class Symbol
Mild	$S_0 < S_c$	M
Steep	$S_0 > S_c$	S
Critical	$S_0 = S_c$	C
Horizontal	$S_0 = 0$	H
Adverse	$S_0 < 0$ (uphill)	A

Table 11.4 Depth designations

Depth Condition	Number Designation
$y > y_n, y > y_c$	1
y between y_n and y_c	2
$y < y_n, y < y_c$	3

for critical slope, and there is no $H1$ or $A1$ since horizontal and adverse slope channels have no normal depth.

Figure 11.21 shows the twelve types of surface profiles. The horizontal (x) scale is greatly exaggerated. The sketches indicate the resulting profile shape in a reach of channel with constant slope, shape, and roughness. Note that all type 1 profiles approach a horizontal surface, and all type 2 and 3 profiles approach the *lower* of critical depth or normal depth. Profiles at critical depth are shown as dotted lines normal to the channel bottom due to the singularity in dy/dx at $y = y_c$, $\mathbf{F} = 1$.

Figure 11.22 illustrates how the various profile shapes might appear in the real world. Note that all profiles associated with subcritical flow ($M1$, $M2$, $S1$, $C1$, $H2$, $A2$) extend *upstream* from a control section (such as a dam, sluice gate, or overfall) while all profiles associated with supercritical flow ($M3$, $S2$, $S3$, $C3$, $H3$, $A3$) extend *downstream* from a control section.

Figure 11.21 The twelve types of surface profiles in gradually varied flow.

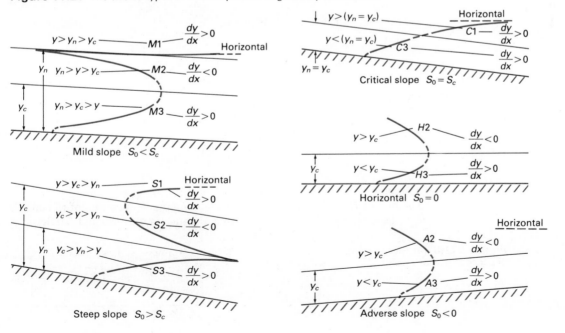

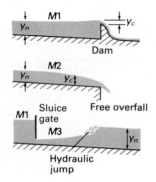

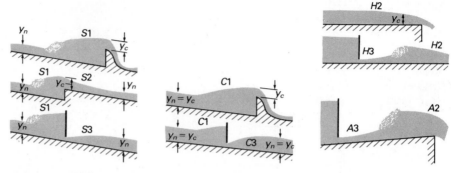

Figure 11.22 Occurrence of the twelve types of surface profiles in the real world.

EXAMPLE 11.8   Illustrates the classification of surface profiles

The channel shown in Fig. E11.8(a) is 6 m wide and carries 10 m³/s of water. Critical flow occurs over the dam. Indicate the missing surface profile shape from the initial point to the dam. Sketch and label the types of surface profiles.

Solution

   Given
Figure E11.8(a)
10 m³/s water flow rate
Critical flow over dam

   Find
The missing surface profiles
Sketch and label the various surface profiles.

   Solution
We first check the state of the approach flow. The approach velocity is

$$V_1 = \frac{Q}{by_1} = \frac{10\,\frac{m^3}{s}}{(6\ m)(0.30\ m)} = 5.56\ m/s.$$

The Froude number is

$$\mathbf{F}_1 = \frac{V_1}{\sqrt{gy_1}} = \frac{\left(5.56\,\frac{m}{s}\right)}{\sqrt{\left(9.807\,\frac{m}{s^2}\right)(0.30\ m)}} = 3.24.$$

The approach flow is supercritical.
   The height H of the water surface a short distance upstream of the dam is found by assuming that the flow over this distance is uniform and frictionless.

Figure E11.8 (a) Given conditions of a gradually varied flow.

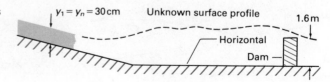

The specific energy equation, Eq. (11.23), gives

$$\frac{V_3^2}{2g} + y_3 = \frac{V_4^2}{2g} + y_4 + z_{B_4},$$

where point 3 is on the water surface a short distance upstream of the dam, $z_{B_3} = 0$, and point 4 is just above the dam. For critical flow over the dam, Eq. (11.6) gives

$$V_4^2 = c_4^2 = g y_4.$$

Then

$$Q = b y_4 V_4 = b y_4 \sqrt{g y_4}, \qquad \text{or} \qquad y_4 = \left(\frac{Q}{b\sqrt{g}}\right)^{2/3} = \left(\frac{10\,\dfrac{\text{m}^3}{\text{s}}}{6\text{ m}\sqrt{9.807\,\dfrac{\text{m}}{\text{s}^2}}}\right)^{2/3} = 0.657 \text{ m}.$$

The other numerical values give

$$\frac{V_3^2}{2g} = \frac{Q^2}{2gb^2 y_3^2} = \frac{\left(10\,\dfrac{\text{m}^3}{\text{s}}\right)^2}{2\left(9.807\,\dfrac{\text{m}}{\text{s}^2}\right)(6\text{ m})^2 y_3^2} = \frac{0.1416 \text{ m}^3}{y_3^2}$$

and

$$\frac{V_4^2}{2g} = \frac{Q^2}{2gb^2 y_4^2} = \frac{\left(10\,\dfrac{\text{m}^3}{\text{s}}\right)^2}{2\left(9.807\,\dfrac{\text{m}}{\text{s}^2}\right)(6\text{ m})^2(0.657\text{ m})^2} = 0.328 \text{ m}.$$

The specific energy equation becomes

$$\frac{0.1416 \text{ m}^3}{y_3^2} + y_3 = 0.328 \text{ m} + 0.657 \text{ m} + 1.6 \text{ m} = 2.59 \text{ m}.$$

Solving for y_3 by trial and error gives $y_3 = 0.2458$ m and $y_3 = 2.569$ m. The first value is not possible because the water would not flow over the dam. Therefore, the approaching flow (at 1) is supercritical, and the flow just upstream of the dam is subcritical with $y_3 = 2.569$; therefore, a hydraulic jump must occur upstream of the dam. Assuming that the hydraulic jump occurs over a relatively short distance, we have from Eq. (11.41)

$$y_2 = \frac{y_1}{2}\left(\sqrt{1 + 8\mathbf{F}_1^2} - 1\right) = \frac{0.30 \text{ m}}{2}\left(\sqrt{1 + 8(3.24)^2} - 1\right) = 1.23 \text{ m}.$$

Since

$$y_3 = 2.569 \text{ m} > 1.23 \text{ m} = y_2,$$

the hydraulic jump must occur on the incline, as shown in Fig. E11.8(b). This figure shows the water surface profile thus far. We must now classify the profile

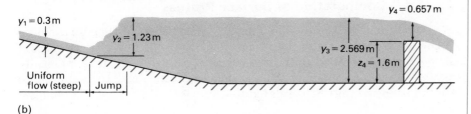

$y_4 = 0.657$ m

$y_1 = 0.3$ m

$y_2 = 1.23$ m

$y_3 = 2.569$ m

$z_4 = 1.6$ m

Uniform flow (steep)   Jump

(b)

Figure E11.8 (b) Initial sketch of surface profile segments.

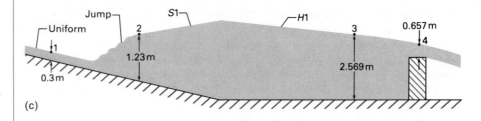

Figure E11.8 (c) Sketch of complete surface profile.

(c)

shapes in the reach from the hydraulic jump to the break in slope and from the break in slope to just upstream of the dam.

Downstream of the jump, the flow rate q, bottom slope S_0, and channel surface roughness (Manning's n) are the same as upstream of the jump; therefore, the normal depth, y_{n_2}, is

$$y_{n_2} = y_{n_1} = 0.3 \text{ m},$$

and the critical depth is

$$y_{c_2} = y_{c_1} = y_{c_4} = \left(\frac{q^2}{g}\right)^{1/3} = 0.657 \text{ m}.$$

Therefore, $y_{n_2} < y_{c_2}$ and the slope is steep (S). Also since

$$y_2 = 1.23 \text{ m} > y_{n_2} = 0.3 \text{ m},$$

the complete profile designation is $S1$.

Since the channel bottom is horizontal from the break in slope to the dam, the profile is an H profile. The critical depth at 3 is still

$$y_{c_3} = y_{c_4} = 0.657 \text{ m},$$

so, because

$$y_3 = 2.569 \text{ m} > y_{c_3} = 0.657 \text{ m},$$

the profile is type $H1$ ($dy/dx < 0$). Using this information,

| the complete profile is sketched in Fig. E11.8(c). |

Answer ◀

Discussion

The flow in the vicinity of the dam is not closely one-dimensional and one-directional, so the values by y_3 and y_4 are only approximate (note, however, that the value of y_c is accurate). If the dam has a sharp crest, it acts as a full-span weir, a flow measurement and control device that we will discuss in Section 11.5.1. A more accurate computation of y_3 can be based on the weir equations developed in that section.

Note that it was not necessary to actually calculate S_0 and S_c for the inclined reach of the channel since y_n could be calculated from the given data.

11.4.3 Computation of Surface Profiles

Examining Eq. (11.46) and classifying surface profiles give an engineer a general idea of the shape of a surface profile; however, the actual shape can be obtained only by integrating Eq. (11.46). It is generally assumed that Manning's formula is sufficiently accurate to evaluate the friction slope so that S_E is related to y and V by Eq. (11.49). The bottom slope is presumed known as a function of x. Since both V and y vary with x, S_E and $\mathbf{F}$ are variable. For an exact calculation, it is necessary to solve Eq. (11.46) simultaneously with the differential continuity equation. This solution must usually be done numerically.

A satisfactory approximate computation of a surface profile can be made by breaking the channel up into finite reaches and using average values of parameters in each reach to evaluate the right side of Eq. (11.46). This is most easily done by computing x as a function of y, that is, by computing the length of channel necessary to cause a certain depth change. The approximate equation is obtained from Eq. (11.44):

$$\Delta x_i \approx \frac{\left(\dfrac{V_{i+1}^2}{2g} + y_{i+1}\right) - \left(\dfrac{V_i^2}{2g} + y_i\right)}{(S_0 - S_E)_i}, \tag{11.52}$$

where Δx_i is the length required to change the depth from y_i to y_{i+1}, and V_i and V_{i+1} are related to y_i and y_{i+1} by

$$q = V_i y_i = V_{i+1} y_{i+1}.$$

Using Eq. (11.49), we have for the average value of $S_0 - S_E$

$$\overline{(S_0 - S_E)_i} \approx \left(\frac{S_{0,i} + S_{0,i+1}}{2}\right) - \frac{\bar{n}^2}{\xi^2} \frac{\left(\dfrac{V_i^2 + V_{i+1}^2}{2}\right)}{(\bar{R}_h)^{4/3}}, \tag{11.53}$$

where $\bar{n}$ is the average value of Manning's n for Δx_i, and $\bar{R}_h$ is computed from the average values of y and b:

$$\bar{y} = \frac{y_i + y_{i+1}}{2} \qquad \text{and} \qquad \bar{b} = \frac{b_i + b_{i+1}}{2}. \tag{11.54}$$

A typical computation step is as follows:

- V_i, y_i, and $S_{0,i}$ are known.
- A value of y_{i+1} is chosen (usually $|\Delta y| < 0.1 y_i$).
- V_{i+1} is calculated from $V_{i+1} = V_i y_i / y_{i+1}$.
- $\bar{R}_h$ is calculated and $\bar{n}$ is selected.
- $S_{0,i+1}$ is estimated.
- $\overline{(S_0 - S_E)_i}$ is calculated from Eq. (11.53).
- Δx_i is calculated from Eq. (11.52).
- $S_{0,i+1}$ is checked.
- If S_0 changes rapidly with x, it may be necessary to recompute $\overline{(S_0 - S_E)_i}$ and obtain a better value for Δx_i.

By repeating this process for each reach, we can work out a complete surface profile. It is not necessary that the entire surface profile be of a single type because the stepwise calculation allows us to patch together various profile types. It is even possible to patch in a hydraulic jump by computing the profile up to the jump, applying Eqs. (11.40) and (11.41) across the jump (possibly using Fig. 11.19 to account for jump length), and then continuing the profile computation on the other side of the jump. Note that the stepwise computation procedure works equally well in the upstream or downstream direction.

EXAMPLE 11.9 Illustrates computation of surface profiles in gradually varied flow

The unfinished concrete channel ($n = 0.014$) shown in Fig. E11.9 is 12 ft wide and is laid on a 0.06° slope. The reservoir water depth is 8 ft above the bottom of the channel inlet. Calculate the flow rate in the channel if the channel length is 1000 ft.

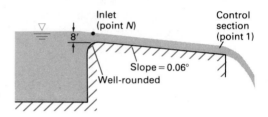

Figure E11.9 Gradually varied flow in long, rectangular channel.

Solution

Given
Figure E11.9
12-ft-wide channel having a 0.06° slope ($\tan 0.06° = 0.00105$)
$n = 0.014$

Find
Flow rate for channel length of 1000 ft

Solution
For a channel length of 1000 ft, it is improbable that uniform flow is established. We must, therefore, compute the gradually varied flow from the channel inlet to the free overfall. The *control section* for the flow is the critical point that occurs just upstream of the free overfall; we will have to begin calculations there and compute the profile back upstream toward the reservoir. This would be a straightforward application of the stepwise surface profile computation if the critical depth or the flow rate were known. Unfortunately, neither is known; only the elevation of the reservoir surface above the channel inlet is given. To solve this problem, we must combine an iteration with the stepwise calculation of the free-surface profile. The iteration proceeds as follows:

- A value for the critical depth y_c is assumed. From the information given in Example 11.4, critical depth occurs about $5y_c$ upstream of the brink.
- The flow rate q is calculated from the critical depth.
- The depth at the channel inlet is computed by applying the energy equation to the frictionless flow between the reservoir and the channel inlet (labeled point N in Fig. E11.9a). The equation is

$$\frac{q^2}{2gy_N^2} + y_N = 8 \text{ ft.}$$

- The distance from the critical point to the channel inlet is computed using Eq. (11.52) N times, with

$$\Delta y = y_{i+1} - y_i = \frac{(y_N - y_c)}{N}.$$

The number of steps is arbitrary, with larger N implying more work and more accuracy. Note that Δx_i will be negative.

■   The point where the depth equals y_N is computed from $x_N = 1000 - 5y_c + \sum_1^N \Delta x_i$. If the computed value of x_N does not equal zero, then another value of y_c is assumed and the process is repeated.

The computational equations are

$$q = V_c y_c = \sqrt{g y_c^3} = 5.675 y_c^{3/2}, \tag{E11.1}$$

$$\frac{q^2}{2g y_N^2} + y_N = \frac{y_c^3}{2 y_N^2} + y_N = 8, \tag{E11.2}$$

$$\Delta y = y_{i+1} - y_i = \frac{y_N - y_c}{N}, \tag{E11.3}$$

$$V_{i+1} = \left(\frac{y_i}{y_{i+1}}\right) V_i, \tag{E11.4}$$

$$\bar{y} = \frac{1}{2}(y_{i+1} + y_i), \tag{E11.5}$$

$$\bar{V} = \frac{1}{2}(V_{i+1} + V_i), \tag{E11.6}$$

$$\bar{R}_h = \frac{b\bar{y}}{b + 2\bar{y}} = \frac{12\bar{y}}{12 + 2\bar{y}}, \tag{E11.7}$$

$$\overline{(S_0 - S_E)_i} = S_0 - \frac{\bar{n}^2}{\xi^2} \frac{\bar{V}^2}{\bar{R}_h^{4/3}} = 0.00105 - 8.83 \times 10^{-5}\left(\frac{\bar{V}^2}{\bar{R}_h^{4/3}}\right), \tag{E11.8}$$

$$\Delta x_i = \frac{\dfrac{V_{i+1}^2 - V_i^2}{64.4} + y_{i+1} - y_i}{\overline{(S_0 - S_E)_i}}. \tag{E11.9}$$

Equations (E11.3)–(E11.9) are applied N times. The calculated value of x at the channel inlet is

$$x_N = 1000 \text{ ft} - 5y_c + \sum_1^N \Delta x_i.$$

If x_N is not zero, then it is necessary to assume a new value for y_c and return to Eq. (E11.1).

To illustrate the calculations, we choose $N = 5$ and guess

$$y_c^{(1)} = 5 \text{ ft}.$$

From Eq. (E11.1),

$$q^{(1)} = 63.44 \text{ ft}^2/\text{sec}.$$

Solving Eq. (E11.2) for y_N, we have

$$y_N = 6.538 \text{ ft}.$$

From Eq. (E11.3), we have

$$\Delta y = \frac{6.538 - 5}{5} = 0.308 \text{ ft}.$$

Table E11.9(a) summarizes the stepwise computation of the surface back to a depth of 6.538 ft. Since the calculated value of x_N is not zero, the assumed value of y_c is incorrect. Table E11.9(b) summarizes calculations of x_N for various values of y_c, all calculated with $N = 5$. The value

$$y_c = 4.803 \text{ ft}$$

yields satisfactory results for $N = 5$.

Table E11.9(a)

i	1 (control section)	2	3	4	5 (inlet)
		Channel Reach Number ($i = 1$ to $i = 5$)			
y_i	$5(=y_c)$	5.308	5.615	5.923	6.230
y_{i+1}	5.308	5.615	5.923	6.230	6.538
V_i	$12.69(=V_c)$	11.95	11.30	10.71	10.18
V_{i+1}	11.95	11.30	10.71	10.18	9.70
$\bar{y}$	5.154	5.461	5.769	6.076	6.384
$\bar{V}$	12.32	11.63	11.01	10.45	9.94
$\bar{R}_h$	2.772	2.859	2.941	3.019	3.093
$(S_0 - S_E)_i$	-2.39×10^{-3}	-1.89×10^{-3}	-1.49×10^{-3}	-1.16×10^{-3}	18.87×10^{-4}
Δx_i	-10.96	-37.63	-71.93	-117.33	-179.89
x_N	$-$	$-$	$-$	$-$	557.2

Table E11.9(b) Calculated value of channel inlet position for various values of y_c (channel divided into five reaches)

y_c	x_N
5.0	557.2
4.5	-3189.6
4.8	-12.6
4.81	26.60
4.803	-0.65

Table E11.9(c) Values of y_c for various numbers of reaches in channel

N	y_c (for $x_N \approx 0$)
2	4.772
5	4.803
10	4.807
20	4.808
50	4.809
100	4.809

To check the effect of the number of reaches, we repeat the calculations for $N = 2, 10, 20, 50,$ and 100. Table E11.9(c) summarizes the results of these calculations. Clearly, to the correct number of significant figures,

$$y_c = 4.81 \text{ ft,}$$

giving

$$q = 5.675(4.81)^{3/2} = 59.87 \text{ ft}^2/\text{sec} \quad \text{and} \quad Q = qb = \left(59.87 \frac{\text{ft}^2}{\text{sec}}\right)(12 \text{ ft}),$$

Answer

$$\boxed{Q = 718.4 \text{ ft}^3/\text{sec.}}$$

Discussion

The calculations in this example were quite laborious. These calculations can be easily programmed for a computer; in fact, all of our calculations were carried out on a small home computer. (You didn't really think we did 100 reaches by hand, did you?)

It is interesting to note that the calculated value of y_c is within about 1% of the correct value, with the channel broken into only two reaches.

Note that the profile computations started at the critical flow control section and proceeded back upstream.

11.5 Miscellaneous Topics in Open Channel Flow

We will close this chapter on open channel flow by discussing flow measurement methods for open channel flow and an analogy between open channel flow and compressible flow.

11.5.1 Measurement Methods for Open Channel Flow

Measurement of liquid flow rate in an open channel can be accomplished by the traverse method or by use of a weir. In the traverse method, liquid velocity is measured at several points in the channel

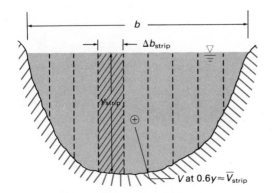

Figure 11.23 Flow rate can be measured by dividing channel into vertical strips and summing strip flows.

cross section.* The flow rate is the sum of each velocity multiplied by the appropriate area. Generally, the channel cross section is divided into vertical strips (Fig. 11.23), and the average velocity for each strip is determined. The flow rate for the strip is then

$$q_{strip} = \bar{V}_{strip} A_{strip} = \bar{V}_{strip} y_{strip} \Delta b_{strip},$$

and the total flow rate is the sum of the individual strip flow rates. Many investigations have shown that the liquid velocity measured at a point 60% depth below the free surface is a close approximation to the average velocity. An even better approximation to the average velocity is the average of velocities measured at 20% depth and 80% depth. Thus in practice, only one or two velocity measurements per strip need be made.

A *weir* is an obstruction built across the channel that causes the upstream liquid to rise. The height of the surface upstream of the weir is a measure of the flow rate. Figure 11.24 shows several types of weirs. The rectangular and triangular (or V-notch) weirs are called *sharp-crested weirs*. Note that all weirs create a free "jet" of liquid (called the *nappe*) on the downstream side of the weir. At low flow rates, the nappe may be absent, with the liquid adhering to the downstream surface of the weir. If this occurs, the weir is operating unsatisfactorily and accurate flow measurement is difficult. Figure 11.25 shows the flow over a sharp-crested weir and illustrates the nomenclature of weir flow measurement.

An analysis and discussion of a full-span rectangular weir was presented in Example 4.7. The result of that example was

$$Q = C_d \left(\frac{2\sqrt{2g}}{3} \right) L H^{3/2}, \tag{11.55}$$

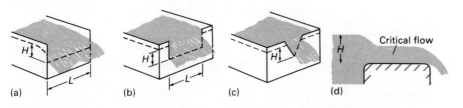

Figure 11.24 Various types of weirs for measuring flow rate in an open channel: (a) Full-span rectangular weir; (b) part-span rectangular weir with end contractions; (c) triangular or V-notch weir; (d) broad-crested weir.

* Velocities can be measured using a Pitot or Fechheimer probe (Figs. 4.7, E4.7a).

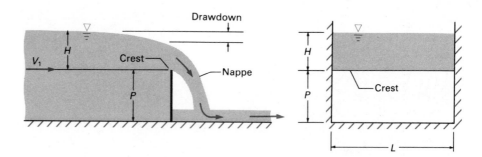

Figure 11.25 Flow over a full-span sharp-crested weir.

where $\mathbf{C_d}$ is an empirically determined discharge coefficient that accounts for certain assumptions in the derivation of the equation, such as neglect of the approach velocity and the drawdown, as well as for effects of viscosity and surface tension. Dimensional analysis indicates

$$\mathbf{C_d} = \mathbf{C_d}\!\left(\frac{H}{P}, \mathbf{R}, \mathbf{W}\right)\!.$$

In most practical cases, Reynolds number and Weber number effects are negligible and $\mathbf{C_d}$ depends on H/P only. A widely used correlation is

$$\mathbf{C_d} \approx 0.611 + 0.075\!\left(\frac{H}{P}\right)\!. \tag{11.56}$$

The flow through a rectangular weir with end contractions is similar to flow through a full-span weir except that the nappe contracts, making the effective flow area less than HL. Experiments have shown that if L is greater than $3H$, then the amount of contraction of the nappe is approximately $0.1H$.

A triangular weir is useful for measuring relatively small flow rates. Figure 11.26 shows a triangular weir with vertex angle ϕ. The velocity at any depth is proportional to $\sqrt{2gh}$. The flow through area dA is*

$$dQ = \mathbf{C_d}\sqrt{2gh}\; dA.$$

The differential area is

$$dA = b\, dh.$$

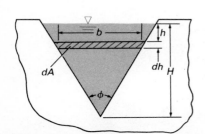

Figure 11.26 Illustration for analysis of triangular weir.

From geometry,

$$b = 2(H - h)\tan\!\left(\frac{\phi}{2}\right), \quad \text{and so} \quad Q = \int dQ = 2\sqrt{2g}\;\mathbf{C_d}\tan\!\left(\frac{\phi}{2}\right)\!\int_0^H (H - h)\sqrt{h}\; dh.$$

Integrating gives

$$Q = \frac{8}{15}\mathbf{C_d}\sqrt{2g}\;\tan\!\left(\frac{\phi}{2}\right)H^{5/2} \tag{11.57}$$

for a triangular weir.

The discharge coefficient is a function of vertex angle, Reynolds number, and Weber number. The two most commonly used vertex angles

---

* Anticipating the need for an empirical discharge coefficient, we insert $\mathbf{C_d}$ in the equation; however, $\mathbf{C_d}$ has meaning only for the integrated flow rate equation.

are 60° and 90°. For these angles and for H greater than about 0.2 m (0.6 ft) for $\phi = 90°$ and H greater than about 0.25 m (0.8 ft) for $\phi = 60°$, the value of $\mathbf{C_d}$ is approximately constant:

$$\mathbf{C_d} \approx 0.58. \tag{11.58}$$

A broad-crested weir is a large "bump" in the channel with a flat top surface (Fig. 11.27). A broad-crested weir should be high enough to cause critical flow to occur on the flat surface (see Section 11.3.2). The velocity over the weir crest is

$$V = V_c = \sqrt{gy_c},$$

and the flow rate is

$$Q = VA = \sqrt{g}y_c^{3/2}L. \tag{11.59}$$

The flow rate could be measured by measuring y_c; however, this is difficult to accomplish accurately since the flow is critical. It is easier to measure the upstream depth H. Applying the energy equation from a point on the upstream surface to a point on the crest and neglecting losses, we have

$$\frac{V_1^2}{2g} + H + P = \frac{V_c^2}{2g} + y_c + P.$$

Substituting $V_c = \sqrt{gy_c}$ gives

$$y_c = \frac{2}{3}H + \frac{V_1^2}{3g}.$$

If V_1 is small,

$$y_c \approx \frac{2}{3}H.$$

Substituting into Eq. (11.59) gives

$$Q \approx \frac{1}{\sqrt{3}}\left(\frac{2}{3}\right)\sqrt{2g}LH^{3/2}. \tag{11.60}$$

A more accurate equation replaces $1/\sqrt{3}$ with an experimentally determined discharge coefficient

$$Q = \mathbf{C_d}\left(\frac{2}{3}\right)\sqrt{2g}LH^{3/2}. \tag{11.61}$$

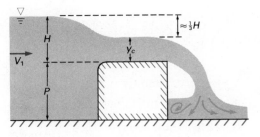

Figure 11.27 Flow over a broad-crested weir.

For flow with high Reynolds and Weber numbers, the discharge coefficient for a broad-crested weir is given by

$$C_d \approx \frac{0.65}{\left(1 + \dfrac{H}{P}\right)^{1/2}}. \tag{11.62}$$

EXAMPLE 11.10   Illustrates calculations for weirs and estimation of the uncertainty in a flow measurement

An engineering laboratory experiment uses a triangular weir in an open channel to measure flow rate. The nominal weir angle is 90°. In a certain test, the head of water above the weir was 4 in. The uncertainty in the weir angle is $\pm 2°$ and the uncertainty in the depth measurement is ± 0.05 in. Compute the flow rate and the percentage uncertainty in the flow rate.

Solution

   Given
Triangular weir with angle of 90° $\pm 2°$
Head of water H above weir is 4 in. ± 0.05 in.

   Find
Volume flow rate and uncertainty in the volume flow rate

   Solution
The volume flow rate through a triangular weir is found using Eq. (11.57):

$$Q = \frac{8}{15} C_d \sqrt{2g} \tan\left(\frac{\phi}{2}\right) H^{5/2}.$$

The numerical values give

$$Q = \frac{8}{15} (0.58) \sqrt{2\left(32.2 \frac{\text{ft}}{\text{sec}^2}\right)} \tan\left(\frac{90°}{2}\right)\left[(4 \text{ in.})\left(\frac{\text{ft}}{12 \text{ in.}}\right)\right]^{5/2},$$

$$\left| Q = 0.159 \text{ ft}^3/\text{sec.} \right| \qquad \textbf{Answer} \blacktriangleleft$$

The uncertainty in Q can be estimated in three different ways. The first method calculates both the maximum possible value and the minimum possible value of Q:

$$Q_{max} = \frac{8}{15} (0.58) \sqrt{2\left(32.2 \frac{\text{ft}}{\text{sec}^2}\right)} \tan\left(\frac{92°}{2}\right)\left[(4.05 \text{ in.})\left(\frac{\text{ft}}{12 \text{ in.}}\right)\right]^{5/2}$$

$$= 0.170 \text{ ft}^3/\text{sec},$$

and

$$Q_{min} = \frac{8}{15} (0.58) \sqrt{2\left(32.2 \frac{\text{ft}}{\text{sec}^2}\right)} \tan\left(\frac{88°}{2}\right)\left[(3.95 \text{ in.})\left(\frac{\text{ft}}{12 \text{ in.}}\right)\right]^{5/2}$$

$$= 0.149 \text{ ft}^3/\text{sec}.$$

The uncertainty U_Q in Q is

$$U_Q = (0.170 - 0.159) \text{ ft}^3/\text{sec} \qquad \text{and} \qquad U_Q = (0.149 - 0.159) \text{ ft}^3/\text{sec}, \qquad \text{or}$$

$$\left| \begin{array}{l} U_Q = +0.011 \text{ ft}^3/\text{sec} \\ -0.010 \text{ ft}^3/\text{sec}. \end{array} \right| \qquad \textbf{Answer} \blacktriangleleft$$

The second method is to estimate U_Q by taking the differential of Q as follows:

$$U_Q \approx dQ = \left(\frac{\partial Q}{\partial \phi}\right)_H d\phi + \left(\frac{\partial Q}{\partial H}\right)_\phi dH.$$

Now

$$\left(\frac{\partial Q}{\partial \phi}\right)_H = \frac{4}{15} C_d \sqrt{2g} \sec^2\left(\frac{\phi}{2}\right) H^{5/2} \quad \text{and} \quad \left(\frac{\partial Q}{\partial H}\right)_\phi = \frac{5}{2}\left(\frac{8}{15}\right) C_d \sqrt{2g} \tan\left(\frac{\phi}{2}\right) H^{3/2},$$

so

$$dQ = \frac{4}{3} C_d \sqrt{2g}\left[\frac{1}{5}\sec^2\left(\frac{\phi}{2}\right) H^{5/2}\, d\phi + \tan\left(\frac{\phi}{2}\right) H^{3/2}\, dH\right].$$

Using

$$d\phi = 2°\left(\frac{\pi\ \text{rad}}{180°}\right) = \frac{\pi}{90}\ \text{rad} \quad \text{and} \quad dH = 0.05\ \text{in.}\left(\frac{\text{ft}}{12\ \text{in.}}\right) = 0.00417\ \text{ft}$$

gives

$$dQ = \frac{4}{3}(0.58)\sqrt{2\left(32.2\,\frac{\text{ft}}{\text{sec}^2}\right)}\left[\frac{\sec^2 45°}{5}\left(\frac{4}{12}\,\text{ft}\right)^{5/2}\left(\frac{\pi}{90}\,\text{rad}\right)\right.$$

$$\left. + \tan 45°\left(\frac{4}{12}\,\text{ft}\right)^{3/2}(0.00417\ \text{ft})\right] = 0.0105\ \text{ft}^3/\text{sec}.$$

Using dQ as an estimate of the uncertainty in Q gives   **Answer**

$$\left|\ U_Q \approx \pm dQ = \pm 0.0105\ \text{ft}^3/\text{sec.}\ \right| \quad \blacktriangleleft$$

The third method provides a more realistic estimate of the uncertainty. It recognizes that the maximum possible deviation in the weir angle ($\pm 2°$) and the maximum possible deviation in the depth measurement (± 0.05 in.) will probably not occur simultaneously. Therefore, the above estimates give the maximum possible uncertainty. A more realistic estimate of the uncertainty is the Pythagorean summation [5, 6] of the individual uncertainties:

$$U_Q \approx \pm \sqrt{\left[\left(\frac{\partial Q}{\partial \phi}\right)_H d\phi\right]^2 + \left[\left(\frac{\partial Q}{\partial H}\right)_\phi dH\right]^2},$$

or

$$U_Q \approx \pm \frac{4}{3} C_d \sqrt{2g}\sqrt{\left[\frac{1}{5}\sec^2\left(\frac{\phi}{2}\right) H^{5/2}\, d\phi\right]^2 + \left[\tan\left(\frac{\phi}{2}\right) H^{3/2}\, dH\right]^2}.$$

The numerical values give

$$U_Q \approx \pm \frac{4}{3}(0.58)\sqrt{2\left(32.2\,\frac{\text{ft}}{\text{sec}^2}\right)}\left\{\left[\left(\frac{\sec^2 45°}{5}\left(\frac{4}{12}\,\text{ft}\right)^{5/2}\left(\frac{\pi}{90}\,\text{rad}\right)\right)\right]^2\right.$$

$$\left. + \left(\tan 45°\left(\frac{4}{12}\,\text{ft}\right)^{3/2}(0.00417\ \text{ft})\right)^2\right\}^{1/2},$$

Answer

$$\left|\ U_Q \approx \pm 0.0075\ \text{ft}^3/\text{sec.}\ \right| \quad \blacktriangleleft$$

Discussion
Note that the first two methods provide an estimate of the maximum uncertainty, and the third method provides a more reasonable estimate of the actual uncertainty.

11.5.2 Analogy Between Open Channel Flow and Compressible Flow

If you have studied Chapter 10, you have no doubt been struck by the strong similarities between certain phenomena in open channel flow and compressible flow. Probably the strongest similarity is that between the hydraulic jump in open channel flow and the normal shock wave in compressible flow. Both are strongly dissipative phenomena that occur over a relatively short distance. Both are extremely complicated in detail but can be satisfactorily analyzed by the control volume approach. Both convert the flow from one state (supercritical, supersonic) to another (subcritical, subsonic).

Another obvious similarity is the phenomena of wave motions and their relevant dimensionless parameters. Weak free-surface gravity waves in open channel flow are similar to sound waves in compressible flow. The flow state is classified by the Froude number in open channel flow and Mach number in compressible flow. For Froude or Mach numbers less than unity, disturbances at a point are propagated to all parts of the flow, but for Froude or Mach numbers greater than unity, disturbances propagate downstream only.

A final illustration of the similarities between open channel flow and compressible flow is the comparison between the response of a frictionless open channel flow to a change of bottom elevation (see Section 11.3.2) and the response of an isentropic gas flow to area change (Section 10.4.1). You may wish to study these sections in parallel and make a list of the similarities.

Are these similarities simply coincidence or is there a formal analogy between the two types of flow? If there is a formal analogy, we can interpret measurements or analysis in one type of flow in terms of the other type. We can demonstrate the formal analogy by considering the equations of continuity, momentum, and energy. Consider liquid flow in a rectangular open channel and gas flow in a closed passage. The continuity equations are

Channel flow:

$$Q = Vyb = \text{Constant},$$

Gas flow:

$$\dot{m} = V\rho A = \text{Constant}.$$

Differentiating each logarithmically, we have

$$\frac{dV}{V} + \frac{dy}{y} + \frac{db}{b} = 0, \qquad \text{channel flow} \tag{11.63}$$

$$\frac{dV}{V} + \frac{d\rho}{\rho} + \frac{dA}{A} = 0. \qquad \text{gas flow} \tag{11.64}$$

For isentropic gas flow, the differential energy and momentum equations are identical. The energy/momentum equation is Eq. (10.63):

$$V\,dV + \frac{dp}{\rho} = 0. \tag{10.63}$$

Using

$$dp = \frac{\partial p}{\partial \rho}\bigg)_s d\rho, \quad c^2 = \frac{\partial p}{\partial \rho}\bigg)_s, \quad \text{and} \quad \mathbf{M}^2 = \frac{V^2}{c^2},$$

we have for the energy/momentum equation for isentropic gas flow

$$\frac{d\rho}{\rho} = -\mathbf{M}^2 \frac{dV}{V}. \tag{11.65}$$

For open channel flow, the specific energy is

$$E = \frac{V^2}{2g} + y.$$

For frictionless flow in a horizontal channel, $E = \text{Constant}$, and

$$\frac{V\, dV}{g} + dy = 0.$$

Using

$$c^2 = gy \quad \text{and} \quad \mathbf{F}^2 = \frac{V^2}{c^2}$$

reduces the open channel energy equation to

$$\frac{dy}{y} = -\mathbf{F}^2\left(\frac{dV}{V}\right). \tag{11.66}$$

Comparing Eqs. (11.63) and (11.64) and Eqs. (11.65) and (11.66) suggests the following analogy between isentropic gas flow and frictionless flow in a horizontal rectangular channel:

- Mach number (gas flow) is analogous to Froude number (channel flow).
- Density (gas flow) is analogous to depth (channel flow).
- Area (gas flow) is analogous to channel width (channel flow).

Since the corresponding equations have identical form, the analogy is complete and formally correct. This analogy is put to use in a water table in which a free-surface flow of water is used to simulate compressible flow of a gas [7]. A water table can be used for flow visualization, with hydraulic jumps representing shock waves and depth representing density. Depth measurements in a water table can be used to deduce densities in the analogous compressible flow. A water table is much cheaper to build and operate than a wind tunnel; therefore, a water table may be called a "poor man's wind tunnel."

Problems

Extension and Generalization

1. Develop the vertical sluice gate equation,

$$\left(\frac{y_2}{y_1^2 - y_2^2}\right)\left(1 - \frac{y_2}{y_1}\right) = \frac{q^2}{2gy_1^3} = \frac{\mathbf{F}_1^2}{2y_1},$$

for frictionless flow, where $\mathbf{F}_1$ is the upstream Froude number.

2. Show that the maximum flow rate out of a partially open, vertical sluice gate and into a rectangular channel occurs when the gate is opened such that $y_1/y_2 = 1.5$. Show that the corresponding downstream Froude number is unity.

3. Consider laminar flow down a very wide rectangular channel making an angle θ with the hori-

zontal. The fluid has kinematic viscosity v and the volume flow rate per unit width is given by

$$q = \frac{gy^3 \sin \theta}{3v},$$

where y is the fluid depth perpendicular to the channel bottom. Find Manning's n for this flow.

4. Consider the flow down a prismatic channel having a rectangular cross section of width b. The channel bottom makes an angle θ with the horizontal. Show that

$$\frac{dy}{dx} = \frac{\sin \theta - \dfrac{n^2 Q^2}{A^2 R_h^{4/3}}}{1 - \dfrac{Q^2}{A^2 gy \cos^2 \theta}},$$

where y is the vertical fluid depth, $A = by \cos \theta$, and Q is the volume flow rate. Discuss the form of the equation for small values of θ ($\theta \sim 1°$).

5. Consider the flow down a prismatic channel having a trapezoidal cross section of base width b and top width $b + 2y \cos \theta \cot \phi$. The channel bottom makes an angle θ with the horizontal, and y is the vertical fluid depth. (See Fig. P11.1.) Show that

$$\frac{dy}{dx} = \frac{\sin \theta - \dfrac{n^2 Q^2}{A^2 R_h^{4/3}}}{1 - \left(\dfrac{Q^2}{A^2 gy \cos^2 \theta}\right)\left(\dfrac{b + 2y \cos \theta \cot \phi}{b + y \cos \theta \cot \phi}\right)}.$$

Note that the average fluid velocity is given by

$$V = \frac{Q}{A} = \frac{Q}{y \cos \theta (b + y \cos \theta \cot \phi)}.$$

Discuss the form of the equation for small values of θ ($\theta \sim 1°$).

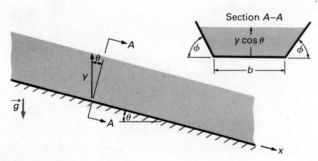

Figure P11.1

Lower-Order Application

6. Do shallow waves propagate at the same speed in all fluids? Explain why or why not.

7. Water flows down a concrete street having a slope of $10°$ at the rate of 0.01 ft^3/sec per foot of width. Find the water depth far down the street.

8. An unfinished concrete rectangular channel is 5 m wide and has a slope of $0.50°$. The water is 0.5 m deep. Find the discharge rate for uniform flow.

9. A rectangular brick-lined channel has a bottom slope of 0.0025 and is designed to carry a uniform water flow rate of 300 ft^3/sec. Would the channel need fewer bricks if the channel were 2 ft wide, 6 ft wide, or 10 ft wide? Explain.

10. Show that the portion of a circular channel that will most efficiently transport a given volume flow rate Q is a semicircle.

11. Determine the values of ϕ and h for the triangular channel in Fig. P11.2 for the most efficient triangular channel (i.e., the channel of the smallest perimeter for a given flow area, slope, and flow rate).

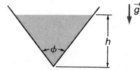

Figure P11.2

12. For $A = 1.0$ m^2 and $\phi = 90°$, plot P versus y using Eq. (11.20) and show that the optimum depth is $\sqrt{0.5}$ m.

13. Take the derivative of P in Eq. (11.20) with respect to ϕ and show that the optimum angle ϕ is $60°$ for a fixed area A.

14. A rectangular unfinished concrete channel with $b = 10.0$ m carries 100 m^3/s of water. The channel bottom slope is 0.0001. Find the normal depth.

15. A trapezoidal unfinished concrete channel with $b = 1.0$ m and $\phi = 30°$ carries 100 m^3/s of water. See Fig. 11.6. The channel bottom slope is 0.0001. Find the normal depth.

16. A prismatic unfinished concrete channel having a semicircular cross section is used as a water slide in an amusement park. The channel has a 1.0-m diameter, is 15 m long, and has a slope of $30°$. The water flow rate down the slide is 1.0 m^3/s. Find the normal depth.

17. A trapezoidal channel is made of wood planks and is used for a log slide. The bottom angle is $30°$ above the horizontal, the water flow rate is 5 ft^3/sec, the angle ϕ in Fig. P11.1 is $60°$, and the base b is 8.0 ft. Find the normal depth. Determine whether the flow is subcritical or supercritical.

18. A very wide river has a flow rate of 10 m^3/s·m. Find the critical depth and the critical slope.

19. A trapezoidal channel with $b = 10.0$ m and $\phi = 30°$ carries 100 m^3/s of water. See Fig. 11.6. Find the critical depth and the critical slope.

20. Water flows with an average velocity of 1.0 m/s and a normal depth of 0.5 m in a very wide rectangular channel. Is the flow subcritical or supercritical?

21. Water flows down a very wide rectangular channel having Manning's n 0.015 and bottom slope 0.0015. Find the rate of discharge and normal depth for critical flow conditions.

22. Find the depth at which water will flow at the rate of 6 ft^3/sec inside a 40-in.-inside-diameter concrete pipe having a slope of 0.0025.

23. A 90° triangular flume is built of planed wood and carries water at a rate of 0.25 m^3/s. Find the critical depth and the critical slope.

24. A 10-ft-wide rectangular channel is made of planed wood and carries 100 gal/min of water. The channel has a bottom slope of 0.001. Find the normal depth and the critical depth.

25. A very wide channel has a bottom slope $S_0 = 0.00015$, uniform laminar flow, and a normal depth $y_n = 0.125$ m. The fluid is water, a Newtonian fluid. Develop expressions for the water velocity parallel to the bottom and the discharge per unit width. Compare the form of this answer to that using Chezy's formula and solve for the Chezy coefficient.

26. A lubricating oil at 20°C flow down a wide inclined flat plate having a slope $S_0 = 0.00015$. Laminar flow exists and the Reynolds number is 200. Find the maximum flow rate per unit width of the plate.

27. Find the discharge per unit width for a very wide channel having a bottom slope of 0.00015. The normal depth is 0.003 m. Assume laminar flow and justify the assumption.

28. Figure P11.3 shows a cross section of an aqueduct that carries water at 50 m^3/s. The value of Manning's n is 0.015. Find the bottom slope.

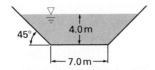

Figure P11.3

29. A 16-in.-inside-diameter concrete pipe is flowing half full of water. The bottom slope is 0.0001. Find the flow rate.

30. A 10.0-m-inside-diameter concrete pipe has Manning's n 0.015. What is the value of Manning's n for a 25.0-m-inside-diameter concrete pipe having the same value of absolute roughness?

31. The rate of discharge through the canal in Fig. P11.4 is to be 10 m^3/s. Find the width b and the bottom slope. The velocity will not exceed 50 m/s.

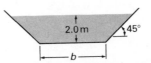

Figure P11.4

32. A construction engineer is designing a channel to carry a water flow rate of 100 ft^3/sec down an angle of 0.0025° with the horizontal. The construction material is planed wood. The engineer is debating whether to use a 60° triangular channel or a 90° rectangular channel. Would less wood be used for the triangular channel or the rectangular channel?

33. Consider a rectangular channel and a semicircular channel with the same bottom slope and Manning's n. Find the diameter of the semicircular channel that will have the same capacity as a rectangular channel measuring 3 m wide and 1.2 m deep.

34. Consider 100 ft^3/sec of water flowing down a rectangular channel measuring 10 ft wide. The normal depth is 3.00 ft. A 4.0-ft-diameter pier is located in the channel. Find the water depth as it flows past the pier.

35. A rectangular channel is made of planed wood, has a bottom slope of 0.0005, and carries 55 ft^3/sec of water. At section 1 the depth is 2.5 ft, and at section 2 the depth is 2.0 ft. Which section is located upstream of the other?

36. Consider the river and floodplains in Fig. P11.5. The river has a Manning's n of 0.030. The floodplains are to be planted with light brush. Find the width b of the floodplains for the river to handle 30,000 ft^3/sec of water. Bottom slope = 0.0001.

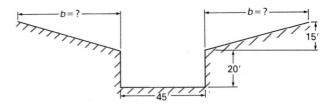

Figure P11.5

37. A river has a rectangular cross section and measures 50 ft wide and 20 ft deep. Recent spring thaws have flooded the banks of the river. The flood water is estimated to be 10% to 15% of the total river flow rate. The bank of one side of the river can be cut back 5 ft. Will this widening of the river alleviate the flooding? If not, what other modifications should be made? The river's bottom slope is 0.00020.

38. The channel in Fig. P11.6 has two floodplains as shown. Find the discharge if the center channel is lined with brick and the two floodplains are lined with cobblestones. The slope S_0 is 0.00025.

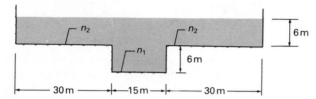

Figure P11.6

39. Subcritical, uniform flow of water occurs in a 5.0-m-wide, rectangular, horizontal channel. The flow has a depth of 2.0 m and a flow rate of 5.0 m³/s. The water flow encounters a 0.25-m rise in the channel bottom. Find the new normal depth. Is the flow above the rise subcritical, critical, or supercritical? Assume frictionless flow.

40. Subcritical uniform flow of water occurs in a 5.0-m-wide, rectangular, horizontal channel. The flow has a depth of 1.5 m and a flow rate of 15.0 m³/s. Find the rise in the channel bottom for the Froude number above the rise to be 1.5. Assume frictionless flow.

41. Repeat Problem 40 if the Froude number above the rise is to be 1.0.

42. Supercritical, uniform flow of water occurs in a 5.0-m-wide, rectangular, horizontal channel. The flow has a depth of 1.5 m and a flow rate of 45.0 m³/s. The water flow encounters a 0.25-m rise in the channel bottom. Find the normal depth after the rise in the channel bottom. Is the flow after the rise subcritical, critical, or supercritical? Assume frictionless flow.

43. The flow in Problem 42 encounters a 0.25-m drop in the channel rather than a rise. Find the new normal depth. Is the flow after the drop subcritical, critical, or supercritical?

44. Consider the flow in Example 11.5. Would your result be any different if the canal width were decreased by 15 ft? Explain why or why not. Is the assumption of frictionless flow important?

45. A channel has a rectangular cross section, a slope $S_0 = 0.005$, a width of 50 m, and a water flow rate of 460 m³/s. Find the uniform depth. A dam raises the water level 2.0 m. How far upstream of the high depth is the water level raised 0.5 m?

46. A major river is divided into three parts or courses—the upper course, the middle course, and the lower course. The slope is 70 ft per mile in the upper course, 10 ft per mile in the middle course, and 1.0 ft per mile in the lower course. All three courses have rectangular cross sections. The upper course has a normal depth of 40.0 ft. The river width is 120 ft. Find the normal depths of the middle and lower courses.

47. The higher velocities of the upper course of the river in Problem 46 carry sand and gravel along with the water and deposit them at the entrance of the middle course. Assume this buildup is 12.0 ft high. Will a hydraulic jump occur at the entrance of the middle course?

48. The lower velocities of the middle course of the river in Problem 46 do not wear away the sides of the channel as rapidly as do the velocities in the upper course. Assume the cross section of the middle course is that of a trapezoid with an angle ϕ (see Fig. 11.6) of 60°. Find the normal depth of the middle course if the base of the trapezoid b is 60 ft.

49. The lowest river velocities occur in the lower course of the river in Problem 46 and permit silt to be deposited in the bed of the lower course. This raises the bottom of the river, and the river can overflow into the floodplains. Assume that the cross section of the lower course is shown in Fig. P11.5. Find the width b.

50. A channel has a rectangular cross section, a width of 40 m, and a flow rate of 4000 m³/s. The normal water depth is 20 m. The flow then encounters a 4.0-m-high dam. Find the water depth directly above the dam.

51. A channel has a rectangular cross section, a width of 40 m, and a flow rate of 4000 m³/s. The water depth is 20 m. The channel bottom rises enough for the flow to become critical. The bottom then drops to the original level, and the flow encounters a 4.0-m-high dam. Find the water depth directly above the dam and the normal depth upstream of the dam.

52. What is the minimum water depth necessary for a 40-ft-wide stream to handle 4000 ft³/sec if the flow is not supercritical?

53. A 5.0-m-wide channel has a slope of 0.004, a 8.0-m³/s water flow rate, and a water depth 1.5 m after a hydraulic jump. Find the water depth before the jump.

54. A rectangular channel 3.0 m wide has a flow rate of 5.0 m³/s with a normal depth of 0.50 m. The flow then encounters a dam that rises 0.25 m above the channel bottom. Will a hydraulic jump occur? Justify your answer.

55. Find the water depth over the dam and classify the flow in Problem 54.

56. An unfinished concrete channel 10 ft wide has a slope of 0.005 and carries a water flow rate of 120 ft^3/sec. The flow then encounters an obstruction that rises 0.50 ft above the channel bottom. Will a hydraulic jump occur? Justify your answer.

57. A 100-ft-wide unfinished concrete rectangular channel has a bottom slope of 0.0025 and a flow rate of 5000 ft^3/sec. An upstream sluice gate produces a minimum water depth of 1.5 ft in the channel. Will a hydraulic jump occur in the channel? Justify your answer. Find the power loss in the jump.

58. A 90° triangular flume has sides 2.0 m wide, a water flow rate of 1.0 m^3/s, and a depth of 0.50 m. Find the depth after a hydraulic jump and the power loss in the jump.

59. Develop a dimensionless curve to find the normal liquid depth of water as a function of the diameter D of a circular pipe *or* of the base b and angle ϕ of a trapezoidal cross section and the quantity $(AR_h^{2/3}/b^{8/3})$ *or* $(AR_h^{2/3}/D^{8/3})$.

60. Develop a dimensionless curve to find the critical depth of water as a function of the diameter D of a circular cross section *or* of the base b and angle ϕ of a trapezoidal cross section and the quantity $(AR^{2/3}/b^{8/3})$ *or* $(AR_h^{2/3}/D^{8/3})$.

Higher-Order Application

61. Water flows out of a home faucet at the rate of 4.0 gal/min, hits a flat surface, and flows radially outward from the impingement point. The water depth is 0.10 in. at a radius of 3.0 in. Find the water depth after a hydraulic jump.

62. Consider an unfinished concrete drainage canal that has a trapezoidal cross section. The canal bottom is 15 ft wide and the angle ϕ is 60°. During a rainstorm the water in the canal rises to a level where a pump is activated to move water over a levee and into a lake. The canal has a bottom slope of 0.0008. The water flow rate is 500 ft^3/sec and the water depth at the pump inlet is 7.0 ft. Find the water depth $y(x)$ and classify the flow(s).

63. The river depth in most of the middle course in Problem 46 is 12.0 ft higher than at its entrance. Find the normal depth in the main section and the distance to attain 99% of the normal depth.

64. The spillway of a dam is 20.0 ft wide and has a flow rate of 5000 ft^3/sec. The spillway makes an angle of 70° with the horizontal. Find the vertical water depth down the spillway. See Problem 4.

65. A channel has a rectangular cross section, is 40 m wide, 20 m high, and has a flow rate of 4000 m^3/s. The channel bottom rises a distance of 5.0 m over a horizontal distance of 25 m. Find the water depth every 5.0 m from the start of the rise if the rise is linear.

66. A rectangular earth channel is 10 ft wide and 16,000 ft long and has a bottom slope of 0.0015. The water depth is 2.0 ft high directly above a 3.0-ft-high dam at the end of the channel. The water flow rate is 300 ft^3/sec. Find the water depths at every 1000-ft interval upstream of the dam.

67. A rectangular channel is 2.0 m wide and has a water flow rate of 3 m^3/s with a normal depth of 1.5 m. The channel makes a 90° bend in a horizontal plane. The radius of the inner side is 16.0 m and of the outer side 20.0 m. Find the difference in the water depths at the outer and inner sides.

68. A rectangular flume is made of unplaned wood, is 7.0 m wide and 300 m long, and is horizontal. It is fed by a reservoir whose surface is 4.0 m above the bottom of the flume at the entrance. Find the flow rate in the flume and the water depth 500 m from the entrance.

69. Write a FORTRAN computer program to find the water depth $y(x)$ for a fluid flowing down a prismatic channel with a trapezoidal cross section. The channel has a length L and a base width b, the sides make an angle ϕ with the horizontal, and the channel bottom makes an angle θ with the horizontal (see Fig. P11.1). The total volume flow rate Q and the water depth at the channel exit are known. Assume subcritical flow.

70. Write a BASIC computer program to find the fluid flow rate for a prismatic channel with a trapezoidal cross section. The channel has a length L and a base width b, the sides make an angle ϕ with the horizontal, and the channel bottom makes an angle θ with the horizontal (see Fig. P11.1). The channel is fed by a reservoir whose surface is a height H above the channel bottom at the channel entrance. Assume subcritical flow through the channel. Use Example 11.9 as a test case.

Comparison

71. Consider a circular channel and a rectangular channel whose height is one-half the width. Find the diameter of the semicircular channel that will have the same discharge as the rectangular channel. Both channels are full and have the same slope and the same value of Manning's n.

References

1. Chow, V. T., *Open Channel Hydraulics,* McGraw-Hill, New York, 1959.
2. Henderson, F. M., *Open Channel Flow,* Macmillan, New York, 1966.
3. Morris, H. M., and J. M. Wiggert, *Applied Hydraulics,* Ronald Press, New York, 1972.
4. White, F. M., *Fluid Mechanics,* McGraw-Hill, New York, 1979.
5. Kline, S. J. and S. McClintock, "Describing Uncertainties in Single Sample Experiments," *Mechanical Engineering,* January 1953.
6. "ASME Performance Test Code Supplement PTC 19.1—Evaluation of Uncertainties in Experimental Measurements," American Society of Mechanical Engineers, 1984.
7. Thompson, P. A., *Compressible Fluid Dynamics,* Chapter 11, McGraw-Hill, New York, 1972.

Appendixes

Appendix A Physical Properties

Table A.1 Properties of the U.S. Standard Atmosphere (SI units)

Geometric Altitude Z (m)	Temperature T (K)	Pressure p (Pa)	Density ρ (kg/m³)	Gravitational Acceleration g (m/s²)	Viscosity μ (N·s/m²)	Kinematic Viscosity ν (m²/s)
−5000	320.7	1.778 E5	1.931 E0	9.822	1.942 E-5	1.006 E-5
−4000	314.2	1.596	1.770	9.819	1.912	1.081
−3000	307.7	1.430	1.619	9.816	1.882	1.163
−2000	301.2	1.278	1.478	9.813	1.852	1.253
−1000	294.7	1.139	1.347	9.810	1.821	1.352
0	288.2	1.013	1.225	9.807	1.789	1.461
1000	281.7	8.988 E4	1.112	9.804	1.758	1.581
2000	275.2	7.950	1.007	9.801	1.726	1.715
3000	268.7	7.012	9.093 E-1	9.797	1.694	1.863
4000	262.2	6.166	8.194	9.794	1.661	2.028
5000	255.7	5.405	7.364	9.791	1.628	2.211
6000	249.2	4.722	6.601	9.788	1.595	2.416
7000	242.7	4.111	5.900	9.785	1.561	2.646
8000	236.2	3.565	5.258	9.782	1.527	2.904
9000	229.7	3.080	4.671	9.779	1.493	3.196
10000	223.3	2.650	4.135	9.776	1.458	3.525
15000	216.7	1.211	1.948	9.761	1.422	7.300 E-4
20000	216.7	5.529 E3	8.891 E-2	9.745	1.422	1.599
30000	226.5	1.197	1.841	9.715	1.475	8.013
40000	250.4	2.871 E2	3.996 E-3	9.684	1.601	4.007 E-3
50000	270.7	7.978 E1	1.027	9.654	1.704	1.659 E-2
60000	255.8	2.246	3.059 E-4	9.624	1.629	5.324
70000	219.7	5.520 E0	8.754 E-5	9.594	1.438	1.643 E-1
80000	180.7	1.037	1.999	9.564	1.216	6.058
90000	180.7	1.644 E-1	3.170 E-6	9.535	1.216	3.837 E0

Table A.2 Properties of the U.S. Standard Atmosphere (EE units)

Geometric Altitude Z (ft)	Temperature T (°R)	Pressure p (psia)	Density ρ (lbm/ft³)	Gravitational Acceleration g (ft/sec²)	Viscosity μ (lb·s/ft²)	Kinematic Viscosity ν (ft²/sec)
−15000	572.2	24.626	1.162 E-1	32.220	4.031 E-7	1.116 E-4
−10000	554.3	20.847	1.015	32.205	3.935	1.247
−5000	536.5	17.554	8.831 E-2	32.189	3.835	1.398
0	518.7	14.696	7.647	32.174	3.736	1.572
5000	500.8	12.054	6.590	32.159	3.636	1.776
10000	483.0	10.108	5.648	32.143	3.534	2.013
15000	465.2	8.297	4.814	32.128	3.431	2.293
20000	447.4	6.759	4.077	32.112	3.326	2.623
25000	429.6	5.461	3.431	32.097	3.217	3.017
30000	411.8	4.373	2.866	32.082	3.107	3.488
35000	394.1	3.468	2.375	32.066	2.995	4.057
40000	390.0	2.730	1.890	32.051	2.969	5.057
45000	390.0	2.149	1.487	32.036	2.969	6.423
50000	390.0	1.692	1.171	32.020	2.969	8.159
55000	390.0	1.332	9.219 E-3	32.005	2.969	1.036 E-3
60000	390.0	1.049	7.259	31.990	2.969	1.316
65000	390.0	0.826	5.716	31.974	2.969	1.671
70000	392.2	0.651	4.479	31.959	2.983	2.143
75000	395.0	0.514	3.511	31.944	3.001	2.750
80000	397.7	0.406	2.758	31.929	3.018	3.521
85000	400.4	0.322	2.170	31.913	3.035	4.501
90000	403.1	0.255	1.710	31.898	3.052	5.743
95000	405.8	0.203	1.350	31.883	3.070	7.316
100000	408.6	0.162	1.068	31.868	3.087	9.302
150000	479.1	0.020	1.112 E-4	31.716	3.512	1.016 E-1
200000	457.0	0.003	1.696 E-5	31.566	3.382	6.416
250000	351.8	0.000	2.263 E-6	31.42	2.721	3.868 E0
300000	332.9	0.000	1.488 E-7	31.27	2.593	5.608 E1

Table A.3 Physical properties of dry air and common gases at atmospheric pressure (SI units)

Gas	Temperature T (°C)	Density ρ (kg/m^3)	Absolute Viscosity μ (N·s/m^2)	Kinematic Viscosity v (m^2/s)	Gas Constant R (N·m/kg·K)	Constant Pressure Specific Heat c_p (N·m/kg·K)	Constant Volume Specific Heat c_v (N·m/kg·K)	Specific Heat Ratio k
Air	0	1.29	1.71 E-5	1.33 E-5	287.0	1003.6	716.4	1.40
	50	1.09	1.95	1.79	287.0	–	–	–
	100	0.946	2.17	2.30	287.0	1010.3	723.1	1.40
	150	0.835	2.38	2.85	287.0	–	–	–
	200	0.746	2.57	3.45	287.0	1024.5	737.3	1.39
	250	0.675	2.75	4.08	287.0	–	–	–
	300	0.616	2.93	4.75	287.0	1044.6	757.8	1.38
	400	0.525	3.25	6.20	287.0	1068.5	781.3	1.37
	500	0.457	3.55	7.77	287.0	1092.3	805.1	1.36
Hydrogen	0	–	8.4 E-6	–	4122.0	14194.9	10070.5	1.41
	100	–	1.03 E-5	–	4122.0	14448.2	10323.8	1.40
	200	–	1.21	–	4122.0	14504.3	10379.9	1.40
	300	–	1.39	–	4122.0	14533.2	10408.8	1.40
	400	–	1.54	–	4122.0	14508.9	10456.5	1.39
	500	–	1.69	–	4122.0	14662.2	10537.8	1.39
Oxygen	0	–	1.92 E-5	–	259.8	914.8	654.8	1.40
	100	–	2.44	–	259.8	933.7	673.7	1.39
	200	–	2.90	–	259.8	963.0	703.0	1.37
	400	–	3.69	–	259.8	1023.7	763.7	1.34
	600	–	4.35	–	259.8	1068.9	808.9	1.32
	800	–	4.93	–	259.8	1099.9	840.3	1.31
Nitrogen	0	–	1.66 E-5	–	296.7	1039.2	742.3	1.40
	100	–	2.08	–	296.7	1042.1	745.3	1.40
	200	–	2.46	–	296.7	1051.7	755.3	1.39
	400	–	3.11	–	296.7	1091.5	794.7	1.37
	600	–	3.66	–	296.7	1139.2	842.4	1.35
	800	–	4.13	–	296.7	1181.5	884.7	1.34
Helium	0	–	1.86 E-5	–	2079.0	5229.0	–	1.67
	100	–	2.29	–	2079.0	–	–	1.67
	200	–	2.70	–	2079.0	–	–	1.67
	300	–	3.07	–	2079.0	–	–	1.67
	400	–	3.42	–	2079.0	–	–	1.67
	600	–	4.07	–	2079.0	–	–	1.67
	800	–	4.65	–	2079.0	–	–	1.67
Carbon dioxide	0	–	1.38 E-5	–	188.8	814.8	625.9	1.30
	100	–	1.85	–	188.8	913.6	724.7	1.26
	200	–	2.68	–	188.8	992.7	803.9	1.23
	300	–	–	–	188.8	1056.7	867.9	1.22
	400	–	–	–	188.8	1110.3	921.1	1.21
	600	–	–	–	188.8	1192.0	1002.7	1.19

Table A.4 Physical properties of dry air and common gases at atmospheric pressure (EE units)

Gas	Temperature T (°F)	Density ρ (lbm/ft^3)	Absolute Viscosity μ (lb·s/ft^2)	Kinematic Viscosity v (ft^2/s)	Gas Constant R(ft·lb/lbm·°R)	Constant Pressure Specific Heat c_p (ft·lb/lbm·°R)	Constant Volume Specific Heat c_v (ft·lb/lbm·°R)	Specific Heat Ratio k
Air	0	0.0864	3.58 E-7	1.33 E-4	53.35	185.9	132.5	1.40
	32	0.081	3.61	1.44	53.35	186.7	133.7	1.40
	100	0.071	3.99	1.81	53.35	186.7	133.7	1.40
	200	0.0602	4.48	2.40	53.35	187.5	134.1	1.40
	400	0.0462	5.38	3.75	53.35	190.6	137.2	1.39
	600	0.0375	6.21	5.33	53.35	194.5	141.1	1.38
	800	0.0316	6.94	7.07	53.35	199.2	145.8	1.37
	1000	0.0272	7.63	9.07	53.35	203.8	150.4	1.36
Hydrogen	0	0.0059	1.76 E-7	9.61 E-4	766.0	2634.0	1868.0	1.41
	100	0.0049	1.93	1.27 E-3	766.0	2661.0	1895.0	1.40
	200	0.00415	2.14	1.66	766.0	2684.0	1918.0	1.40
	400	0.0032	2.56	2.58	766.0	2692.0	1926.0	1.40
	600	0.0026	2.95	3.65	766.0	2700.0	1934.0	1.40
	800	0.0021	3.40	5.21	766.0	2707.0	1941.0	1.39
	1000	0.0186	3.63	6.28	766.0	2707.0	1941.0	1.39
Oxygen	0	0.0955	3.77 E-7	1.27 E-4	48.3	170.0	121.7	1.40
	100	0.0785	4.41	1.81	48.3	171.2	122.9	1.39
	200	0.0666	5.00	2.42	48.3	173.3	125.0	1.39
	400	0.0511	6.07	3.82	48.3	179.3	131.0	1.37
	600	0.0415	7.02	5.45	48.3	185.9	137.6	1.35
	800	0.0349	7.86	7.20	48.3	191.8	143.5	1.34
	1000	0.0301	8.63	9.23	48.3	196.7	148.4	1.33
Nitrogen	0	0.0840	3.28 E-7	1.26 E-4	55.2	192.8	137.6	1.40
	100	0.0690	3.80	1.77	55.2	193.3	138.1	1.40
	200	0.0585	4.29	2.36	55.2	193.7	138.5	1.40
	400	0.0449	5.16	3.70	55.2	195.7	140.5	1.39
	600	0.0364	5.95	5.26	55.2	199.5	144.3	1.38
	800	0.0306	6.66	7.01	55.2	204.1	148.9	1.37
	1000	0.0264	7.31	8.91	55.2	209.2	154.0	1.36
Helium	0	0.012	3.74 E-7	1.00 E-3	386.0	972.5	586.5	1.67
	100	–	–	–	386.0	972.5	586.5	1.67
	200	0.083	4.70	1.82 E-4	386.0	972.5	586.5	1.67
	400	0.064	5.69	2.86	386.0	972.5	586.5	1.67
	600	0.0051	6.64	4.19 E-3	386.0	972.5	586.5	1.67
	800	0.0044	7.59	5.55	386.0	972.5	586.5	1.67
	1000	0.0037	8.54	7.42	386.0	972.5	586.5	1.67
Carbon dioxide	0	0.132	2.67 E-7	6.51 E-5	35.1	147.8	112.7	1.31
	100	0.108	3.26	9.72	35.1	157.9	122.8	1.29
	200	0.0915	3.74	1.32 E-4	35.1	169.6	134.5	1.26
	400	0.0702	4.73	2.17	35.1	185.2	150.1	1.23
	600	0.0570	5.62	3.17	35.1	198.7	163.6	1.21
	800	0.0480	6.38	4.28	35.1	208.8	173.7	1.20
	1000	0.0415	7.13	5.53	35.1	217.3	182.2	1.19

Table A.5 Physical properties of ordinary water and common liquids (SI units)

Liquid	Temperature T (°C)	Density ρ (kg/m³)	Specific Gravity S	Absolute Viscosity μ (N·s/m²)	Kinematic Viscosity ν (m²/s)	Surface Tension σ (N/m)	Isothermal Bulk Modulus of Elasticity E_v (N/m²)	Coefficient of Thermal Expansion α_T (K⁻¹)
Water	0	1000	1.000	1.79 E-3	1.79 E-6	7.56 E-2	1.99 E9	6.80 E-5
	3.98	1000	1.000	1.57	1.57	–	–	–
	10	1000	1.000	1.31	1.31	7.42	2.12	8.80
	20	998	0.998	1.00	1.00	7.28	2.21	2.07 E-4
	30	996	0.996	7.98 E-4	8.01 E-7	7.12	2.26	2.94
	40	992	0.992	6.53	6.58	6.96	2.29	3.85
	50	988	0.988	5.47	5.48	6.79	2.29	4.58
	60	983	0.983	4.67	4.75	6.62	2.28	5.23
	70	978	0.978	4.04	4.13	6.64	2.24	5.84
	80	972	0.972	3.55	3.65	6.26	2.20	6.41
	90	965	0.965	3.15	3.26	–	2.14	6.96
	100	958	0.958	2.82	2.94	5.89	2.07	7.50
Mercury	0	13600	13.60	1.68 E-3	1.24 E-7	–	2.50 E10	–
	4	13590	13.59	–	–	–	–	–
	20	13550	13.55	1.55	1.14	37.5	2.50 E10	1.82 E-4
	40	13500	13.50	1.45	1.07	–	–	1.82
	60	13450	13.45	1.37	1.02	–	–	1.82
	80	13400	13.40	1.30	9.70 E-8	–	–	1.83
	100	13350	13.35	1.24	9.29	–	–	–
Ethylene glycol	0	–	–	5.70 E-2	–	–	–	–
	20	1110	1.11	1.99	1.79 E-5	–	–	–
	40	1110	1.10	9.13 E-3	8.30 E-6	–	–	–
	60	1090	1.09	4.95	4.54	–	–	–
	80	1070	1.07	3.02	2.82	–	–	–
	100	1060	1.06	1.99	1.88	–	–	–
Methyl alcohol (methanol)	0	810	0.810	8.17 E-4	1.01 E-6	2.45 E-2	9.35 E8	–
	10	801	0.801	–	–	2.26	8.78	–
	20	792	0.792	5.84	7.37 E-7	–	8.23	–
	30	783	0.783	5.10	6.51	–	7.72	–
	40	774	0.774	4.50	5.81	–	7.23	–
	50	765	0.765	3.96	5.18	–	6.78	–
Ethyl alcohol (ethanol)	0	806	0.806	1.77 E-3	2.20 E-6	2.41 E-2	1.02 E9	–
	20	789	0.789	1.20	1.52	–	9.02 E8	–
	40	772	0.772	8.34 E-4	1.08	–	7.89	–
	60	754	0.754	5.92	7.85 E-7	–	6.78	–
Normal octane	0	718	0.718	7.06 E-4	9.83 E-7	–	1.00 E9	–
	16	–	–	5.74	–	–	–	–
	20	702	0.702	5.42	7.72	–	–	–
	25	–	–	–	–	–	8.35 E8	–
	40	686	0.686	4.33	6.31	–	7.48	–
Benzene	0	900	0.900	9.12 E-4	1.01 E-6	3.02 E-2	1.23 E9	–
	20	879	0.879	6.52	7.42 E-7	2.76	1.06	–
	40	858	0.857	5.03	5.86	–	9.10 E8	–
	60	836	0.836	3.92	4.69	–	7.78	–
	80	815	0.815	3.29	4.04	–	6.48	–
Kerosene	−18	841	0.841	7.06 E-3	8.40 E-6	–	–	–
	20	814	0.814	1.9	2.37	2.9 E-2	–	–
Lubricating oil	20	871	0.871	1.31 E-2	1.50 E-5	–	–	–
	40	858	0.858	6.81 E-3	7.94 E-6	–	–	–
	60	845	0.845	4.18	4.95	–	–	–
	80	832	0.832	2.83	3.40	–	–	–
	100	820	0.820	2.00	2.44	–	–	–
	120	809	0.809	1.54	1.90	–	–	–

Table A.6 Physical properties of ordinary water and common liquids (EE units)

Liquid	Temperature T (°F)	Density ρ (lbm/ft^3)	Specific Gravity S	Absolute Viscosity μ (lb·s/ft^2)	Kinematic Viscosity v (ft^2/s)	Surface Tension σ (lb/ft)	Isothermal Bulk Modulus of Elasticity E_v (lb/in^2)	Coefficient of Thermal Expansion α_T (°R^{-1})
Water	32	62.4	1.00	3.75 E-5	1.93 E-5	5.18 E-3	2.93 E5	2.03 E-3
	40	62.4	1.00	3.23	1.66	5.14	2.94	–
	60	62.4	0.999	2.36	1.22	5.04	3.11	–
	80	62.2	0.997	1.80	9.30 E-6	4.92	3.22	–
	100	62.0	0.993	1.42	7.39	4.80	3.27	1.7
	120	61.7	0.988	1.17	6.09	4.65	3.33	–
	140	61.4	0.983	9.81 E-6	5.14	4.54	3.30	–
	160	61.0	0.977	8.38	4.42	4.41	3.26	–
	180	60.6	0.970	7.26	3.85	4.26	3.13	–
	200	60.1	0.963	6.37	3.41	4.12	3.08	1.52
	212	59.8	0.958	5.93	3.19	4.04	3.00	–
Mercury	50	847	13.6	1.07 E-3	1.2 E-6	–	–	1.0 E-4
	200	834	13.4	8.4 E-4	1.0	–	–	1.0 E-4
	300	826	13.2	7.4	9.0 E-7	–	–	–
	400	817	13.1	6.7	8.0	–	–	–
	600	802	12.8	5.8	7.0	–	–	–
Ethylene glycol	68	69.3	1.11	4.16 E-4	1.93 E-4	–	–	–
	104	68.7	1.10	1.91	8.93 E-5	–	–	–
	140	68.0	1.09	1.03	4.89	–	–	–
	176	66.8	1.07	6.31 E-5	3.04	–	–	–
	212	66.2	1.06	4.12	2.02	–	–	–
Methyl alcohol (methanol)	32	50.6	0.810	1.71 E-5	1.09 E-5	1.68 E-3	1.36 E5	–
	68	50.0	0.801	–	–	1.55	1.19	–
	104	49.4	0.792	1.22	7.93 E-6	–	1.05	–
	140	48.9	0.783	1.07	7.01	–	–	–
	176	48.3	0.774	9.40 E-6	6.25	–	–	–
	212	47.8	0.765	8.27	5.58	–	–	–
Ethyl alcohol (ethanol)	32	50.3	0.806	3.70 E-5	2.37 E-5	1.65 E-3	1.48 E5	–
	68	49.8	0.798	3.03	1.96	–	1.31	–
	104	49.3	0.789	2.51	1.64	–	1.14	–
	140	48.2	0.772	1.74	1.16	–	9.83 E4	–
	176	47.7	0.754	1.24	8.45 E-6	–	–	–
	212	47.1	0.745	–	–	–	–	–
Normal octane	32	44.8	0.718	1.47 E-5	1.06 E-5	–	1.45 E5	–
	68	43.8	0.702	1.13	8.31 E-6	–	–	–
	104	42.8	0.686	9.04 E-6	6.79	–	1.08	–
Benzene	32	56.2	0.900	1.90 E-5	1.09 E-5	2.07 E-3	1.78 E5	–
	68	54.9	0.879	1.36	7.99 E-6	1.89	1.53	–
	104	53.6	0.858	1.05	6.31	–	1.32	–
	140	52.2	0.836	8.19 E-6	5.05	–	1.13	–
	176	50.9	0.815	6.87	4.35	–	9.40 E4	–
Kerosene	0	52.5	0.841	1.48 E-4	9.05 E-5	–	–	–
	77	50.8	0.814	3.97 E-5	2.55 E-5	–	–	–
Lubricating oil	68	54.4	0.871	2.74 E-4	1.61 E-4	–	–	–
	104	53.6	0.858	1.42	8.55 E-5	–	–	–
	140	52.6	0.845	8.73 E-5	5.33	–	–	–
	176	51.9	0.832	5.91	3.66	–	–	–
	212	51.2	0.820	4.18	2.63	–	–	–
	248	50.5	0.809	3.22	2.05	–	–	–

Table A.7 Physical properties of SAE oils and lubricants

Fluid	Temp. (°C)	SI Units			Temp. (°F)	EE Units		
		Specific Gravity	Kinematic Viscosity v (m²/s)			Specific Gravity	Kinematic Viscosity v (ft²/sec)	
			Minimum	Maximum			Minimum	Maximum
Oil								
SAE 50	99	–	1.68 E-5	2.27 E-5	210	–	1.81 E-4	2.44 E-4
40	99	–	1.29	1.68	210	–	1.08	1.81
30	99	–	9.6 E-4	1.29	210	–	1.03 E-2	1.08
20	99	–	–	5.7 E-4	210	–	–	6.14 E-3
20W	−18	0.92	2.60 E-3	1.05 E-2	0	0.92	2.80 E-2	1.13 E-1
10W	−18	0.92	1.30	2.60 E-2	0	0.92	1.40	2.80 E-2
5W	−18	0.92	–	1.30	0	0.92	–	1.40
Lubricants								
SAE 250	99	–	4.3 E-5	–	210	–	4.6 E-4	–
140	99	–	2.5	4.3 E-5	210	–	2.7	4.6 E-4
90	99	–	1.4	2.5	210	–	1.5	2.7
85W	99	–	1.1	–	210	–	1.2	–
80W	99	–	7.0 E-6	–	210	–	7.5 E-5	–
75W	99	–	4.2	–	210	–	4.5 E-5	–

Appendix B Properties of Common Geometric Areas

Shape	Area	x_c	y_c	I_{xc}	I_{yc}	I_{xyc}
 Rectangle	bh	$\dfrac{b}{2}$	$\dfrac{h}{2}$	$\dfrac{bh^3}{12}$ $I_x = \dfrac{bh^3}{3}$	$\dfrac{hb^3}{12}$ $I_y = \dfrac{hb^3}{3}$	0 $I_{xy} = \dfrac{bh}{3}(h^2 + b^2)$
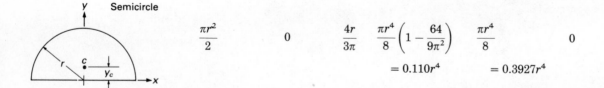 Any triangle, any side	$\dfrac{bh}{2}$	—	$\dfrac{h}{3}$	$\dfrac{bh^3}{36}$ $I_x = \dfrac{bh^3}{3}$	—	—
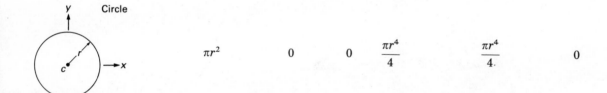 Circle	πr^2	0	0	$\dfrac{\pi r^4}{4}$	$\dfrac{\pi r^4}{4}$	0
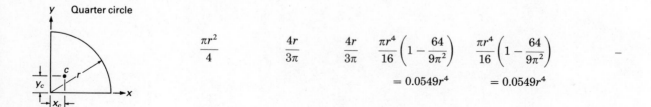 Semicircle	$\dfrac{\pi r^2}{2}$	0	$\dfrac{4r}{3\pi}$	$\dfrac{\pi r^4}{8}\left(1 - \dfrac{64}{9\pi^2}\right)$ $= 0.110 r^4$	$\dfrac{\pi r^4}{8}$ $= 0.3927 r^4$	0
Quarter circle	$\dfrac{\pi r^2}{4}$	$\dfrac{4r}{3\pi}$	$\dfrac{4r}{3\pi}$	$\dfrac{\pi r^4}{16}\left(1 - \dfrac{64}{9\pi^2}\right)$ $= 0.0549 r^4$	$\dfrac{\pi r^4}{16}\left(1 - \dfrac{64}{9\pi^2}\right)$ $= 0.0549 r^4$	—

Shape	Area	x_c	y_c	I_{xc}	I_{yc}	I_{xyc}
Circular segment	$\dfrac{r^2}{2}(\theta - \sin\theta)$	–	–	–	–	–
Circular sector	θr^2	$\dfrac{2r\sin\theta}{3\theta}$	0	$\dfrac{r^4}{16}\left(\theta - \dfrac{\sin 2\theta}{2}\right)$	–	0
Ellipse	πab	0	0	$\dfrac{\pi ba^3}{4}$	$\dfrac{\pi ab^3}{4}$	0
Semiellipse	$\dfrac{\pi ab}{2}$	0	–	–	$\dfrac{\pi ab^3}{8}$	0

Appendix C Conversion Factors

Table C.1 Time conversion factors

	Hour (hr)	Minute (min)	Second (sec)
Second (sec)			
Minute (min)	$60 \frac{\text{min}}{\text{hr}}$		
Hour (hr)	$3600 \frac{\text{sec}}{\text{hr}}$	$60 \frac{\text{sec}}{\text{min}}$	

Example: Convert 450.0 mils to centimeters.

Solution: Locate the *column* for the *smaller* unit (mil) and the *row* for the *larger* unit (cm). If unsure as to the smaller unit, try the second option if your first guess doesn't work. The intersection gives the conversion factor. In this case dividing by the conversion factor (equal to unity) cancels the mils, so

$$450.0 \text{ mil} \left(\frac{\text{cm}}{393.7 \text{ mil}} \right) = 1.143 \text{ cm.}$$

Box from diagram: $393.7 \frac{\text{mil}}{\text{cm}}$

Table C.2 Length conversion factors

	Angstrom (Å)	Micron (μ)	Mil	Millimeter (mm)	Centimeter (cm)	Inch (in.)	Foot (ft)	Yard (yd)	Meter (m)	Fathom (fath)	Rod (rd)	Statute mile (mi)
Micron (μ)	$10^4 \frac{Å}{μ}$											
Mil	$2.54E5 \frac{Å}{\text{mil}}$	$25.4 \frac{μ}{\text{mil}}$										
Millimeter (mm)	$1.0E7 \frac{Å}{\text{mm}}$	$1000 \frac{μ}{\text{mm}}$	$39.37 \frac{\text{mil}}{\text{mm}}$									
Centimeter (cm)	$1.0E8 \frac{Å}{\text{cm}}$	$1.0E4 \frac{μ}{\text{cm}}$	$393.7 \frac{\text{mil}}{\text{cm}}$	$10 \frac{\text{mm}}{\text{cm}}$								
Inch (in.)	$2.54E8 \frac{Å}{\text{in.}}$	$2.54E4 \frac{μ}{\text{in.}}$	$1000 \frac{\text{mil}}{\text{in.}}$	$25.4 \frac{\text{mm}}{\text{in.}}$	$2.54 \frac{\text{cm}}{\text{in.}}$							
Foot (ft)	$3.048E9 \frac{Å}{\text{ft}}$	$3.048E5 \frac{μ}{\text{ft}}$	$1.2E5 \frac{\text{mil}}{\text{ft}}$	$304.8 \frac{\text{mm}}{\text{ft}}$	$30.48 \frac{\text{cm}}{\text{ft}}$	$12 \frac{\text{in.}}{\text{ft}}$						
Yard (yd)	$9.144E9 \frac{Å}{\text{yd}}$	$9.144E5 \frac{μ}{\text{yd}}$	$3.6E5 \frac{\text{mil}}{\text{yd}}$	$914.4 \frac{\text{mm}}{\text{yd}}$	$9.144 \frac{\text{cm}}{\text{yd}}$	$36 \frac{\text{in.}}{\text{yd}}$	$3 \frac{\text{ft}}{\text{yd}}$					
Meter (m)	$1.0E10 \frac{Å}{\text{m}}$	$1.0E6 \frac{μ}{\text{m}}$	$2.54E5 \frac{\text{mil}}{\text{m}}$	$1000 \frac{\text{mm}}{\text{m}}$	$100 \frac{\text{cm}}{\text{m}}$	$39.37 \frac{\text{in.}}{\text{m}}$	$3.281 \frac{\text{ft}}{\text{m}}$	$1.094 \frac{\text{yd}}{\text{m}}$				
Fathom (fath)	$1.829E10 \frac{Å}{\text{fath}}$	$1.829E6 \frac{μ}{\text{fath}}$	$7.2E4 \frac{\text{mil}}{\text{fath}}$	$1.829E3 \frac{\text{mm}}{\text{fath}}$	$182.9 \frac{\text{cm}}{\text{fath}}$	$72 \frac{\text{in.}}{\text{fath}}$	$6 \frac{\text{ft}}{\text{fath}}$	$2 \frac{\text{yd}}{\text{fath}}$	$1.829 \frac{\text{m}}{\text{fath}}$			
Rod (rd)	$5.029E10 \frac{Å}{\text{rd}}$	$5.029E6 \frac{μ}{\text{rd}}$	$1.98E5 \frac{\text{mil}}{\text{rd}}$	$5029 \frac{\text{mm}}{\text{rd}}$	$502.9 \frac{\text{cm}}{\text{rd}}$	$198 \frac{\text{in.}}{\text{rd}}$	$16.5 \frac{\text{ft}}{\text{rd}}$	$5.5 \frac{\text{yd}}{\text{rd}}$	$5.029 \frac{\text{m}}{\text{rd}}$	$2.75 \frac{\text{fath}}{\text{rd}}$		
Statute mile (mi)	$1.609E13 \frac{Å}{\text{mi}}$	$1.609E9 \frac{μ}{\text{mi}}$	$6.336E7 \frac{\text{mil}}{\text{mi}}$	$1.609E6 \frac{\text{mm}}{\text{mi}}$	$1.609E5 \frac{\text{cm}}{\text{mi}}$	$6.336E4 \frac{\text{in.}}{\text{mi}}$	$5280 \frac{\text{ft}}{\text{mi}}$	$1760 \frac{\text{yd}}{\text{mi}}$	$1609 \frac{\text{m}}{\text{mi}}$	$879.7 \frac{\text{fath}}{\text{mi}}$	$320 \frac{\text{rd}}{\text{mi}}$	
Int'l nautical mile (mi)	$1.852E13 \frac{Å}{\text{mi}}$	$1.852E9 \frac{μ}{\text{mi}}$	$7.291E7 \frac{\text{mil}}{\text{mi}}$	$1.852E6 \frac{\text{mm}}{\text{mi}}$	$1.852E5 \frac{\text{cm}}{\text{mi}}$	$7.291E4 \frac{\text{in.}}{\text{mi}}$	$6076.12 \frac{\text{ft}}{\text{mi}}$	$2026 \frac{\text{yd}}{\text{mi}}$	$1852 \frac{\text{m}}{\text{mi}}$	$1012.7 \frac{\text{fath}}{\text{mi}}$	$368.3 \frac{\text{rd}}{\text{mi}}$	$1.151 \frac{\text{st mi}}{\text{I.N. mi}}$

Table C.3 Area conversion factors

	Square statute mile (mi²)	Acre	Square meter (m²)	Square yard (yd²)	Square foot (ft²)	Square inch (in²)
Acre	$640\ \frac{\text{acre}}{\text{mi}^2}$					
Square meter (m²)	$2.589\text{E}6\ \frac{\text{m}^2}{\text{mi}^2}$	$1233\ \frac{\text{m}^2}{\text{acre}}$				
Square yard (yd²)	$3.098\text{E}6\ \frac{\text{yd}^2}{\text{mi}^2}$	$4840\ \frac{\text{yd}^2}{\text{acre}}$	$1.196\ \frac{\text{yd}^2}{\text{m}^2}$			
Square foot (ft²)	$2.788\text{E}7\ \frac{\text{ft}^2}{\text{mi}^2}$	$4.356\text{E}4\ \frac{\text{ft}^2}{\text{acre}}$	$10.76\ \frac{\text{ft}^2}{\text{m}^2}$	$9\ \frac{\text{ft}^2}{\text{yd}^2}$		
Square inch (in²)	$4.014\text{E}9\ \frac{\text{in}^2}{\text{mi}^2}$	$6.272\text{E}6\ \frac{\text{in}^2}{\text{acre}}$	$1550\ \frac{\text{in}^2}{\text{m}^2}$	$1296\ \frac{\text{in}^2}{\text{yd}^2}$	$144\ \frac{\text{in}^2}{\text{ft}^2}$	
Square centimeter (cm²)	$2.589\text{E}10\ \frac{\text{cm}^2}{\text{mi}^2}$	$4.047\text{E}7\ \frac{\text{cm}^2}{\text{acre}}$	$1.0\text{E}4\ \frac{\text{cm}^2}{\text{m}^2}$	$8361\ \frac{\text{cm}^2}{\text{yd}^2}$	$929.0\ \frac{\text{cm}^2}{\text{ft}^2}$	$6.452\ \frac{\text{cm}^2}{\text{in}^2}$

Table C.4 Volume conversion factors

	Milliliter (ml) (cubic centimeter)	Cubic inch (in³)	U.S. fluid ounce (fl oz)	U.S. liquid quart (qt)	Liter (l)	U.S. liquid gallon (gal)	Cubic foot (ft³)	Cubic yard (yd³)	Cubic meter (m³)
Cubic inch (in³)	$16.39\ \frac{\text{ml}}{\text{in}^3}$								
U.S. fluid ounce (fl oz)	$29.574\ \frac{\text{ml}}{\text{fl oz}}$	$1.805\ \frac{\text{in}^3}{\text{fl oz}}$							
U.S. liquid quart (qt)	$946.4\ \frac{\text{ml}}{\text{qt}}$	$57.75\ \frac{\text{in}^3}{\text{qt}}$	$32\ \frac{\text{fl oz}}{\text{qt}}$						
Liter (l)	$1000\ \frac{\text{ml}}{\text{l}}$	$61.02\ \frac{\text{in}^3}{\text{l}}$	$33.81\ \frac{\text{fl oz}}{\text{l}}$	$1.057\ \frac{\text{qt}}{\text{l}}$					
U.S. liquid gallon (gal)	$3785.4\ \frac{\text{ml}}{\text{gal}}$	$231\ \frac{\text{in}^3}{\text{gal}}$	$128\ \frac{\text{fl oz}}{\text{gal}}$	$4\ \frac{\text{qt}}{\text{gal}}$	$3.7854\ \frac{\text{l}}{\text{gal}}$				
Cubic foot (ft³)	$2.83\text{E}4\ \frac{\text{ml}}{\text{ft}^3}$	$1728\ \frac{\text{in}^3}{\text{ft}^3}$	$957.5\ \frac{\text{fl oz}}{\text{ft}^3}$	$29.92\ \frac{\text{qt}}{\text{ft}^3}$	$28.3\ \frac{\text{l}}{\text{ft}^3}$	$7.48\ \frac{\text{gal}}{\text{ft}^3}$			
Cubic yard (yd³)	$7.646\text{E}5\ \frac{\text{ml}}{\text{yd}^3}$	$4.666\text{E}4\ \frac{\text{in}^3}{\text{yd}^3}$	$2.585\text{E}4\ \frac{\text{fl oz}}{\text{yd}^3}$	$807.9\ \frac{\text{qt}}{\text{yd}^3}$	$764.6\ \frac{\text{l}}{\text{yd}^3}$	$202.0\ \frac{\text{gal}}{\text{yd}^3}$	$27\ \frac{\text{ft}^3}{\text{yd}^3}$		
Cubic meter (m³)	$1.0\text{E}6\ \frac{\text{ml}}{\text{m}^3}$	$6.102\text{E}4\ \frac{\text{in}^3}{\text{m}^3}$	$3.381\text{E}4\ \frac{\text{fl oz}}{\text{m}^3}$	$1.057\text{E}3\ \frac{\text{qt}}{\text{m}^3}$	$1000\ \frac{\text{l}}{\text{m}^3}$	$264.2\ \frac{\text{gal}}{\text{m}^3}$	$35.31\ \frac{\text{ft}^3}{\text{m}^3}$	$1.308\ \frac{\text{yd}^3}{\text{m}^3}$	
Cubic statute mile (mi³)	$4.166\text{E}15\ \frac{\text{ml}}{\text{mi}^3}$	$2.544\text{E}14\ \frac{\text{in}^3}{\text{mi}^3}$	$1.409\text{E}14\ \frac{\text{fl oz}}{\text{mi}^3}$	$4.404\text{E}11\ \frac{\text{qt}}{\text{mi}^3}$	$4.166\text{E}12\ \frac{\text{l}}{\text{mi}^3}$	$1.101\text{E}12\ \frac{\text{gal}}{\text{mi}^3}$	$1.472\text{E}11\ \frac{\text{ft}^3}{\text{mi}^3}$	$5.452\text{E}9\ \frac{\text{yd}^3}{\text{mi}^3}$	$4.166\text{E}9\ \frac{\text{m}^3}{\text{mi}^3}$

Table C.5 Mass conversion factors

	Slug	Kilogram (kg)	Pound (lbm) (avdp.)	Ounce (oz) (avdp.)	Gram (gm)
Kilogram (kg)	$14.594\ \dfrac{kg}{slug}$				
Pound (lbm) (avdp.)	$32.174\ \dfrac{lbm}{slug}$	$2.2046\ \dfrac{lbm}{kg}$			
Ounce (oz) (avdp.)	$514.8\ \dfrac{oz}{slug}$	$35.27\ \dfrac{oz}{kg}$	$16\ \dfrac{oz}{lbm}$		
Gram (gm)	$1.459E4\ \dfrac{gm}{slug}$	$1000\ \dfrac{gm}{kg}$	$453.6\ \dfrac{gm}{lbm}$	$28.35\ \dfrac{gm}{oz}$	
Grain (gr)	$2.252E5\ \dfrac{gr}{slug}$	$1.543E4\ \dfrac{gr}{kg}$	$7000\ \dfrac{gr}{lbm}$	$437.5\ \dfrac{gr}{oz}$	$15.43\ \dfrac{gr}{gm}$

Table C.6 Force conversion factors

	Dyne	Poundal (pdl)	Newton (N)	Pound (lb) (avdp.)	Short ton	Metric ton	Long ton
Poundal (pdl)	$1.383E4\ \dfrac{dyne}{pdl}$						
Newton (N)	$E5\ \dfrac{dyne}{N}$	$7.233\ \dfrac{pdl}{N}$					
Pound (lb) (avdp.)	$4.448E5\ \dfrac{dyne}{lb}$	$32.174\ \dfrac{pdl}{lb}$	$4.448\ \dfrac{N}{lb}$				
Short ton	$8.896E3\ \dfrac{dyne}{s\ ton}$	$6.435E4\ \dfrac{pdl}{s\ ton}$	$8896\ \dfrac{N}{s\ ton}$	$2000\ \dfrac{lb}{s\ ton}$			
Metric ton	$9.807E3\ \dfrac{dyne}{m\ ton}$	$7.094E4\ \dfrac{pdl}{m\ ton}$	$9808\ \dfrac{N}{m\ ton}$	$2205\ \dfrac{lb}{m\ ton}$	$1.102\ \dfrac{s\ ton}{m\ ton}$		
Long ton	$9.964E3\ \dfrac{dyne}{l\ ton}$	$7.207E4\ \dfrac{pdl}{l\ ton}$	$9964\ \dfrac{N}{l\ ton}$	$2240\ \dfrac{lb}{l\ ton}$	$1.12\ \dfrac{s\ ton}{l\ ton}$	$1.016\ \dfrac{m\ ton}{l\ ton}$	

Table C.7 Pressure conversion factors

to ↓ / from →	atm	Bar	psi	in. Hg (32°F)	ft H₂O (39.2°F)	in. H₂O (39.2°F)	mm Hg (0°C)	Torr	Pascal $\left(\frac{N}{m^2}\right)$
Bar	$1.0132\ \frac{bar}{atm}$								
psi	$14.696\ \frac{psi}{atm}$	$14.504\ \frac{psi}{bar}$							
in. Hg (32°F)	$29.92\ \frac{in.\ Hg}{atm}$	$29.53\ \frac{in.\ Hg}{bar}$	$2.036\ \frac{in.\ Hg}{psi}$						
ft H₂O (39.2°F)	$33.90\ \frac{ft\ H_2O}{atm}$	$33.46\ \frac{ft\ H_2O}{bar}$	$2.307\ \frac{ft\ H_2O}{psi}$	$1.133\ \frac{ft\ H_2O}{in.\ Hg}$					
in. H₂O (39.2°F)	$406.8\ \frac{in.\ H_2O}{atm}$	$401.5\ \frac{in.\ H_2O}{bar}$	$27.68\ \frac{in.\ H_2O}{psi}$	$13.60\ \frac{in.\ H_2O}{in.\ Hg}$	$12\ \frac{in.\ H_2O}{ft\ H_2O}$				
mm Hg (0°C)	$760\ \frac{mm\ Hg}{atm}$	$750.1\ \frac{mm\ Hg}{bar}$	$51.71\ \frac{mm\ Hg}{psi}$	$25.4\ \frac{mm\ Hg}{in.\ Hg}$	$22.42\ \frac{mm\ Hg}{ft\ H_2O}$	$1.868\ \frac{mm\ Hg}{in.\ H_2O}$			
Torr	$760\ \frac{Torr}{atm}$	$750.1\ \frac{Torr}{bar}$	$51.71\ \frac{Torr}{psi}$	$25.4\ \frac{Torr}{in.\ Hg}$	$304.8\ \frac{Torr}{ft\ H_2O}$	$1868\ \frac{Torr}{in.\ H_2O}$	$1.00\ \frac{Torr}{mm\ Hg}$		
Pascal $\left(\frac{N}{m^2}\right)$	$1.013E5\ \frac{Pa}{atm}$	$1.0E5\ \frac{Pa}{bar}$	$6.895E3\ \frac{Pa}{psi}$	$3386\ \frac{Pa}{in.\ Hg}$	$2989\ \frac{Pa}{ft\ H_2O}$	$249.1\ \frac{Pa}{in.\ H_2O}$	$133.3\ \frac{Pa}{mm\ Hg}$	$100\ \frac{Pa}{Torr}$	
dyne/cm²	$1.032E6\ \frac{dyne/cm^2}{atm}$	$1.0E6\ \frac{dyne/cm^2}{bar}$	$6.895E4\ \frac{dyne/cm^2}{psi}$	$3.386E4\ \frac{dyne/cm^2}{in.\ Hg}$	$2.989E4\ \frac{dyne/cm^2}{ft\ H_2O}$	$2491\ \frac{dyne/cm^2}{in.\ H_2O}$	$1333\ \frac{dyne/cm^2}{mm\ Hg}$	$1000\ \frac{dyne/cm^2}{Torr}$	$10\ \frac{dyne/cm^2}{Pa}$

Table C.8 Velocity conversion factors

to ↓ / from →	$\frac{cm}{s}$	$\frac{m}{min}$	$\frac{in.}{sec}$	$\frac{km}{hr}$	$\frac{ft}{sec}$ (fps)	$\frac{statute\ mile}{hr}$ (mph)	Knots
$\frac{ft}{min}$ (fpm)	$1.969\ \frac{fpm}{cm/s}$	$3.281\ \frac{fpm}{m/min}$	$5\ \frac{fpm}{in./sec}$	$54.68\ \frac{fpm}{km/hr}$	$60\ \frac{fpm}{fps}$	$88\ \frac{fpm}{mph}$	$101.3\ \frac{fpm}{knot}$
$\frac{cm}{s}$		$1.667\ \frac{cm/s}{m/min}$	$2.54\ \frac{cm/s}{in./sec}$	$27.78\ \frac{cm/s}{km/hr}$	$30.48\ \frac{cm/s}{fps}$	$44.7\ \frac{cm/s}{mph}$	$51.44\ \frac{cm/s}{knot}$
$\frac{m}{min}$			$1.524\ \frac{m/min}{in./sec}$	$16.67\ \frac{m/min}{km/hr}$	$18.29\ \frac{m/min}{fps}$	$26.82\ \frac{m/min}{mph}$	$30.87\ \frac{m/min}{knot}$
$\frac{in.}{sec}$				$10.94\ \frac{in./sec}{km/hr}$	$12\ \frac{in./sec}{fps}$	$17.6\ \frac{in./sec}{mph}$	$20.26\ \frac{in./sec}{knot}$
$\frac{km}{hr}$					$1.097\ \frac{km/hr}{fps}$	$1.609\ \frac{km/hr}{mph}$	$1.852\ \frac{km/hr}{knot}$
$\frac{ft}{sec}$ (fps)						$1.47\ \frac{fps}{mph}$	$1.688\ \frac{fps}{knot}$
$\frac{statute\ mile}{hr}$ (mph)							$1.151\ \frac{mph}{knot}$
Knots							
$\frac{m}{s}$	$100\ \frac{cm/s}{m/s}$	$60\ \frac{m/min}{m/s}$	$39.37\ \frac{in./sec}{m/s}$	$3.6\ \frac{km/hr}{m/s}$	$3.281\ \frac{fps}{m/s}$	$2.237\ \frac{mph}{m/s}$	$1.944\ \frac{knot}{m/s}$

Table C.9 Absolute viscosity conversion factors

	$\dfrac{\text{lb}\cdot\text{sec}}{\text{ft}^2}$	$\dfrac{\text{lbm}}{\text{in}\cdot\text{sec}}$	$\dfrac{\text{lbm}}{\text{ft}\cdot\text{sec}}$	$\dfrac{\text{N}\cdot\text{s}}{\text{m}^2}$	$\dfrac{\text{kg}}{\text{m}\cdot\text{s}}$	Poise $\left(\dfrac{\text{gm}}{\text{cm}\cdot\text{s}}\right)$	Centipoise
$\dfrac{\text{lb}\cdot\text{sec}}{\text{in}^2}$	$144\ \dfrac{\text{lb}\cdot\text{sec/ft}^2}{\text{lb}\cdot\text{sec/in}^2}$	$386.1\ \dfrac{\text{lbm/in}\cdot\text{sec}}{\text{lb}\cdot\text{sec/in}^2}$	$4633\ \dfrac{\text{lbm/ft}\cdot\text{sec}}{\text{lb}\cdot\text{sec/in}^2}$	$6895\ \dfrac{\text{N}\cdot\text{s/m}^2}{\text{lb}\cdot\text{sec/in}^2}$	$6895\ \dfrac{\text{kg/m}\cdot\text{s}}{\text{lb}\cdot\text{sec/in}^2}$	$6.895\text{E}4\ \dfrac{\text{poise}}{\text{lb}\cdot\text{sec/in}^2}$	$6.895\text{E}6\ \dfrac{\text{centipoise}}{\text{lb}\cdot\text{sec/in}^2}$
$\dfrac{\text{lb}\cdot\text{sec}}{\text{ft}^2}$		$2.681\ \dfrac{\text{lbm/in}\cdot\text{sec}}{\text{lb}\cdot\text{sec/ft}^2}$	$32.174\ \dfrac{\text{lbm/ft}\cdot\text{sec}}{\text{lb}\cdot\text{sec/ft}^2}$	$47.88\ \dfrac{\text{N}\cdot\text{s/m}^2}{\text{lb}\cdot\text{sec/ft}^2}$	$47.88\ \dfrac{\text{kg/m}\cdot\text{s}}{\text{lb}\cdot\text{sec/ft}^2}$	$478.8\ \dfrac{\text{poise}}{\text{lb}\cdot\text{sec/ft}^2}$	$4.788\text{E}4\ \dfrac{\text{centipoise}}{\text{lb}\cdot\text{sec/ft}^2}$
$\dfrac{\text{lbm}}{\text{in}\cdot\text{sec}}$			$12\ \dfrac{\text{lbm/ft}\cdot\text{sec}}{\text{lbm/in}\cdot\text{sec}}$	$17.86\ \dfrac{\text{N}\cdot\text{s/m}^2}{\text{lbm/in}\cdot\text{sec}}$	$17.86\ \dfrac{\text{kg/m}\cdot\text{s}}{\text{lbm/in}\cdot\text{sec}}$	$178.6\ \dfrac{\text{poise}}{\text{lbm/in}\cdot\text{sec}}$	$1.786\text{E}4\ \dfrac{\text{centipoise}}{\text{lbm/in}\cdot\text{sec}}$
$\dfrac{\text{lbm}}{\text{ft}\cdot\text{sec}}$				$1.488\ \dfrac{\text{N}\cdot\text{s/m}^2}{\text{lbm/ft}\cdot\text{sec}}$	$1.488\ \dfrac{\text{kg/m}\cdot\text{s}}{\text{lbm/ft}\cdot\text{sec}}$	$14.88\ \dfrac{\text{poise}}{\text{lbm/ft}\cdot\text{sec}}$	$1488\ \dfrac{\text{centipoise}}{\text{lbm/ft}\cdot\text{sec}}$
$\dfrac{\text{N}\cdot\text{s}}{\text{m}^2}$					$1\ \dfrac{\text{kg/m}\cdot\text{s}}{\text{N}\cdot\text{s/m}^2}$	$10\ \dfrac{\text{poise}}{\text{N}\cdot\text{s/m}^2}$	$1000\ \dfrac{\text{centipoise}}{\text{N}\cdot\text{s/m}^2}$
$\dfrac{\text{kg}}{\text{m}\cdot\text{s}}$						$10\ \dfrac{\text{poise}}{\text{kg/m}\cdot\text{s}}$	$1000\ \dfrac{\text{centipoise}}{\text{kg/m}\cdot\text{s}}$
Poise $\left(\dfrac{\text{gm}}{\text{cm}\cdot\text{s}}\right)$							$100\ \dfrac{\text{centipoise}}{\text{poise}}$

Table C.10 Kinematic viscosity conversion factors

	Stoke (cm²/s)	$\dfrac{\text{in}^2}{\text{sec}}$	$\dfrac{\text{ft}^2}{\text{sec}}$	$\dfrac{\text{m}^2}{\text{s}}$
Centistoke	$100\ \dfrac{\text{centistoke}}{\text{stoke}}$	$645.2\ \dfrac{\text{centistoke}}{\text{in}^2/\text{sec}}$	$9.29\text{E}4\ \dfrac{\text{centistoke}}{\text{ft}^2/\text{sec}}$	$1.0\text{E}6\ \dfrac{\text{centistoke}}{\text{m}^2/\text{s}}$
Stoke (cm²/s)		$6.452\ \dfrac{\text{stoke}}{\text{in}^2/\text{sec}}$	$929\ \dfrac{\text{stoke}}{\text{ft}^2/\text{sec}}$	$10{,}000\ \dfrac{\text{stoke}}{\text{m}^2/\text{s}}$
$\dfrac{\text{in}^2}{\text{sec}}$			$144\ \dfrac{\text{in}^2/\text{sec}}{\text{ft}^2/\text{sec}}$	$929\ \dfrac{\text{in}^2/\text{sec}}{\text{m}^2/\text{s}}$
$\dfrac{\text{ft}^2}{\text{sec}}$				$10.76\ \dfrac{\text{ft}^2/\text{sec}}{\text{m}^2/\text{s}}$

Table C.11 Volume flow rate conversion factors

from \\ to	$\dfrac{m^3}{s}$	$\dfrac{ft^3}{sec}$	$\dfrac{liter}{s}$	$\dfrac{ft^3}{min}$	$\dfrac{gal}{min}$ (gpm)
$\dfrac{ft^3}{sec}$	$35.61\ \dfrac{ft^3/sec}{m^3/s}$				
$\dfrac{liter}{s}$	$1000\ \dfrac{l/s}{m^3/s}$	$28.32\ \dfrac{l/s}{ft^3/sec}$			
$\dfrac{ft^3}{min}$	$2119\ \dfrac{ft^3/min}{m^3/s}$	$60\ \dfrac{ft^3/min}{ft^3/sec}$	$2.119\ \dfrac{ft^3/min}{l/s}$		
$\dfrac{gal}{min}$ (gpm)	$4.403\ \dfrac{gpm}{m^3/s}$	$448.8\ \dfrac{gpm}{ft^3/sec}$	$15.85\ \dfrac{gpm}{l/s}$	$7.481\ \dfrac{gpm}{ft^3/min}$	
$\dfrac{cm^3}{s}$	$60E6\ \dfrac{cm^3/s}{m^3/s}$	$2.832E4\ \dfrac{cm^3/s}{ft^3/sec}$	$1000\ \dfrac{cm^3/s}{l/s}$	$471.9\ \dfrac{cm^3/s}{ft^3/min}$	$63.09\ \dfrac{cm^3/s}{gpm}$

Table C.12 Power conversion factors

from \\ to	$\dfrac{gm\cdot cal}{min}$	$\dfrac{BTU}{hr}$	Watt (W)	$\dfrac{ft\cdot lb}{sec}$	$\dfrac{gm\cdot cal}{s}$	$\dfrac{BTU}{min}$	Horsepower (hp)	Kilowatt (kW)	$\dfrac{BTU}{sec}$
$\dfrac{ft\cdot lb}{min}$	$3.086\ \dfrac{ft\cdot lb/min}{gm\cdot cal/min}$	$12.96\ \dfrac{ft\cdot lb/min}{BTU/hr}$	$44.25\ \dfrac{ft\cdot lb/min}{W}$	$60\ \dfrac{ft\cdot lb/min}{ft\cdot lb/sec}$	$185.2\ \dfrac{ft\cdot lb/min}{gm\cdot cal/s}$	$777.6\ \dfrac{ft\cdot lb/min}{BTU/min}$	$3.3E4\ \dfrac{ft\cdot lb/min}{hp}$	$4.425E4\ \dfrac{ft\cdot lb/min}{kW}$	$4.666E4\ \dfrac{ft\cdot lb/min}{BTU/sec}$
$\dfrac{gm\cdot cal}{min}$		$4.20\ \dfrac{gm\cdot cal/min}{BTU/hr}$	$14.34\ \dfrac{gm\cdot cal/min}{W}$	$19.44\ \dfrac{gm\cdot cal/min}{ft\cdot lb/sec}$	$60\ \dfrac{gm\cdot cal/min}{gm\cdot cal/s}$	$252.0\ \dfrac{gm\cdot cal/min}{BTU/min}$	$1.069E4\ \dfrac{gm\cdot cal/min}{hp}$	$1.432E4\ \dfrac{gm\cdot cal/min}{kW}$	$1.512E4\ \dfrac{gm\cdot cal/min}{BTU/sec}$
$\dfrac{BTU}{hr}$			$3.4134\ \dfrac{BTU/hr}{W}$	$4.629\ \dfrac{BTU/hr}{ft\cdot lb/sec}$	$14.29\ \dfrac{BTU/hr}{gm\cdot cal/s}$	$60\ \dfrac{BTU/hr}{BTU/min}$	$2542\ \dfrac{BTU/hr}{hp}$	$3413.4\ \dfrac{BTU/hr}{kW}$	$3600\ \dfrac{BTU/hr}{BTU/sec}$
Watt (W)				$1.356\ \dfrac{W}{ft\cdot lb/sec}$	$4.184\ \dfrac{W}{gm\cdot cal/s}$	$17.57\ \dfrac{W}{BTU/min}$	$745.7\ \dfrac{W}{hp}$	$1000\ \dfrac{W}{kW}$	$1054\ \dfrac{W}{BTU/sec}$
$\dfrac{ft\cdot lb}{sec}$					$3.086\ \dfrac{ft\cdot lb/sec}{gm\cdot cal/s}$	$12.96\ \dfrac{ft\cdot lb/sec}{BTU/min}$	$550\ \dfrac{ft\cdot lb/sec}{hp}$	$737.6\ \dfrac{ft\cdot lb/sec}{kW}$	$777.6\ \dfrac{ft\cdot lb/sec}{BTU/sec}$
$\dfrac{gm\cdot cal}{s}$						$4.200\ \dfrac{gm\cdot cal/s}{BTU/min}$	$6.416E5\ \dfrac{gm\cdot cal/s}{hp}$	$238.7\ \dfrac{gm\cdot cal/s}{kW}$	$252.0\ \dfrac{gm\cdot cal/s}{BTU/sec}$
$\dfrac{BTU}{min}$							$42.44\ \dfrac{BTU/min}{hp}$	$56.83\ \dfrac{BTU/min}{kW}$	$60\ \dfrac{BTU/min}{BTU/sec}$
Horsepower (hp)								$1.341\ \dfrac{hp}{kW}$	$1.414\ \dfrac{hp}{BTU/sec}$
Kilowatt (kW)									$1.054\ \dfrac{kW}{BTU/sec}$

Table C.13 Energy conversion factors

	hp·hr	kcal	ft³·atm	BTU	liter·atm	gm·cal	ft·lb	Joule (J) (N·m)	in·lb	ft·poundal	erg (dyne/cm²)
kWh	1.341 $\frac{hp·hr}{kWh}$	860.6 $\frac{kcal}{kWh}$	1255 $\frac{ft^3·atm}{kWh}$	3414 $\frac{BTU}{kWh}$	3.553E4 $\frac{liter·atm}{kWh}$	8.606E5 $\frac{gm·cal}{kWh}$	2.655E6 $\frac{ft·lb}{kWh}$	3.6E6 $\frac{J}{kWh}$	3.186E7 $\frac{in·lb}{kWh}$	8.543E7 $\frac{ft·poundal}{kWh}$	3.6E13 $\frac{erg}{kWh}$
hp·hr		641.6 $\frac{kcal}{hp·hr}$	935.9 $\frac{ft^3·atm}{hp·hr}$	2546 $\frac{BTU}{hp·hr}$	2.649E4 $\frac{liter·atm}{hp·hr}$	6.416E5 $\frac{gm·cal}{hp·hr}$	1.98E6 $\frac{ft·lb}{hp·hr}$	2.685E6 $\frac{J}{hp·hr}$	2.376E7 $\frac{in·lb}{hp·hr}$	6.370E7 $\frac{ft·poundal}{hp·hr}$	2.684E13 $\frac{erg}{hp·hr}$
kcal			1.459 $\frac{ft^3·atm}{kcal}$	3.966 $\frac{BTU}{kcal}$	41.29 $\frac{liter·atm}{kcal}$	1000 $\frac{gm·cal}{kcal}$	3086 $\frac{ft·lb}{kcal}$	4184 $\frac{J}{kcal}$	3.703E4 $\frac{in·lb}{kcal}$	9.929E4 $\frac{ft·poundal}{kcal}$	4.184E10 $\frac{erg}{kcal}$
ft³·atm				2.721 $\frac{BTU}{ft^3·atm}$	28.32 $\frac{liter·atm}{ft^3·atm}$	6855 $\frac{gm·cal}{ft^3·atm}$	2116 $\frac{ft·lb}{ft^3·atm}$	2869 $\frac{J}{ft^3·atm}$	2.539E4 $\frac{in·lb}{ft^3·atm}$	6.809E4 $\frac{ft·poundal}{ft^3·atm}$	1.0133E9 $\frac{erg}{ft^3·atm}$
BTU					10.41 $\frac{liter·atm}{BTU}$	252.0 $\frac{gm·cal}{BTU}$	777.6 $\frac{ft·lb}{BTU}$	1054 $\frac{J}{BTU}$	9332 $\frac{in·lb}{BTU}$	25020 $\frac{ft·poundal}{BTU}$	1.054E10 $\frac{erg}{BTU}$
liter·atm						24.2 $\frac{gm·cal}{liter·atm}$	74.74 $\frac{ft·lb}{liter·atm}$	101.3 $\frac{J}{liter·atm}$	896.8 $\frac{in·lb}{liter·atm}$	2405 $\frac{ft·poundal}{liter·atm}$	3.578E7 $\frac{erg}{liter·atm}$
gm·cal							3.086 $\frac{ft·lb}{gm·cal}$	4.184 $\frac{J}{gm·cal}$	37.03 $\frac{in·lb}{gm·cal}$	99.29 $\frac{ft·poundal}{gm·cal}$	4.184E7 $\frac{erg}{gm·cal}$
ft·lb								1.356 $\frac{J}{ft·lb}$	12 $\frac{in·lb}{ft·lb}$	32.174 $\frac{ft·poundal}{ft·lb}$	1.356E7 $\frac{erg}{ft·lb}$
Joule (J) (N·m)									8.850 $\frac{in·lb}{J}$	23.73 $\frac{ft·poundal}{J}$	1.0E7 $\frac{erg}{J}$
in·lb										2.681 $\frac{ft·poundal}{in·lb}$	1.130 $\frac{erg}{in·lb}$
ft·poundal											4.214E5 $\frac{erg}{ft·poundal}$

Table C.14 Specific energy conversion factors

	ft·lb/slug	Joule/kg	ft·lb/lbm	BTU/slug	BTU/lbm	gm·cal/gm	kcal/kg
erg/gm	929.4 $\frac{erg/gm}{ft·lb/slug}$	1.0E4 $\frac{erg/gm}{J/kg}$	2.989E4 $\frac{erg/gm}{ft·lb/lbm}$	7.224E5 $\frac{erg/gm}{BTU/slug}$	2.324E7 $\frac{erg/gm}{BTU/lbm}$	4.184E7 $\frac{erg/gm}{gm·cal/gm}$	4.184E7 $\frac{erg/gm}{kcal/kg}$
ft·lb/slug		10.76 $\frac{ft·lb/slug}{J/kg}$	32.174 $\frac{ft·lb/slug}{ft·lb/lbm}$	777.6 $\frac{ft·lb/slug}{BTU/slug}$	2.5020E4 $\frac{ft·lb/slug}{BTU/lbm}$	4.504E4 $\frac{ft·lb/slug}{gm·cal/gm}$	4.504E4 $\frac{ft·lb/slug}{kcal/kg}$
Joule/kg			2.990 $\frac{J/kg}{ft·lb/lbm}$	72.23 $\frac{J/kg}{BTU/slug}$	2324 $\frac{J/kg}{BTU/lbm}$	4184 $\frac{J/kg}{gm·cal/gm}$	4184 $\frac{J/kg}{kcal/kg}$
ft·lb/lbm				24.17 $\frac{ft·lb/lbm}{BTU/slug}$	777.6 $\frac{ft·lb/lbm}{BTU/lbm}$	1400 $\frac{ft·lb/lbm}{gm·cal/gm}$	1400 $\frac{ft·lb/lbm}{kcal/kg}$
BTU/slug					32.174 $\frac{BTU/slug}{BTU/lbm}$	57.90 $\frac{BTU/slug}{gm·cal/gm}$	57.90 $\frac{BTU/slug}{kcal/kg}$
BTU/lbm						1.800 $\frac{BTU/lbm}{gm·cal/gm}$	1.800 $\frac{BTU/lbm}{kcal/kg}$
gm·cal/gm							1.0 $\frac{gm·cal/gm}{kcal/kg}$

Appendix D Pipe, Tube, and Tubing Dimensions

Table D.1 Steel pipe (based on ASA Standards B36.10)

Nominal pipe size (in.)	Outside diameter (in.)	Schedule No.	Wall thickness (in.)	Inside Diameter (in.)
$\frac{1}{8}$	0.405	40*	0.068	0.269
		80†	0.095	0.215
$\frac{1}{4}$	0.540	40*	0.088	0.364
		80†	0.119	0.302
$\frac{3}{8}$	0.675	40*	0.091	0.493
		80†	0.126	0.423
$\frac{1}{2}$	0.840	40*	0.109	0.622
		80†	0.147	0.546
		160	0.187	0.466
$\frac{3}{4}$	1.050	40*	0.113	0.824
		80†	0.154	0.742
		160	0.218	0.614
1	1.315	40*	0.133	1.049
		80†	0.179	0.957
		160	0.250	0.815
$1\frac{1}{4}$	1.660	40*	0.140	1.380
		80†	0.191	1.278
		160	0.250	1.160
$1\frac{1}{2}$	1.900	40*	0.145	1.610
		80†	0.200	1.500
		160	0.281	1.338
2	2.375	40*	0.154	2.067
		80†	0.218	1.939
		160	0.343	1.689
$2\frac{1}{2}$	2.875	40*	0.203	2.469
		80†	0.276	2.323
		160	0.375	2.125
3	3.500	40*	0.216	3.068
		80†	0.300	2.900
		160	0.437	2.626
$3\frac{1}{2}$	4.000	40*	0.226	3.548
		80†	0.318	3.364
4	4.500	40*	0.237	4.026
		80†	0.337	3.826
		120	0.437	3.626
		160	0.531	3.438
5	5.563	40*	0.258	5.047
		80†	0.375	4.813
		120	0.500	4.563
		160	0.625	4.313

(continued)

Table D.1 (continued)

Nominal pipe size (in.)	Outside diameter (in.)	Schedule No.	Wall thickness (in.)	Inside Diameter (in.)
6	6.625	40*	0.280	6.065
		80†	0.432	5.761
		120	0.562	5.501
		160	0.718	5.189
8	8.625	20	0.250	8.125
		30*	0.277	8.071
		40*	0.322	7.981
		60	0.406	7.813
		80†	0.500	7.625
		100	0.593	7.439
		120	0.718	7.189
		140	0.812	7.001
		160	0.906	6.813
10	10.75	20	0.250	10.250
		30*	0.307	10.136
		40*	0.365	10.020
		60†	0.500	9.750
		80	0.593	9.564
		100	0.718	9.314
		120	0.843	9.064
		140	1.000	8.750
		160	1.125	8.500
12	12.75	20	0.250	12.250
		30*	0.330	12.090
		40	0.406	11.938
		60	0.562	11.626
		80	0.687	11.376
		100	0.843	11.064
		120	1.000	10.750
		140	1.125	10.500
		160	1.312	10.126
14	14.0	10	0.250	13.500
		20	0.312	13.376
		30	0.375	13.250
		40	0.437	13.126
		60	0.593	12.814
		80	0.750	12.500
		100	0.937	12.126
		120	1.062	11.876
		140	1.250	11.500
		160	1.406	11.188

* Designates former "standard" sizes.
† Former "extra strong."

Table D.2 Copper tube and tubing

Type K Tube			Type L Tube		
Nominal Size (in.)	Outside Diameter (in.)	Inside Diameter (in.)	Nominal Size (in.)	Outside Diameter (in.)	Inside Diameter (in.)
$\frac{1}{4}$	0.375	0.305	$\frac{1}{4}$	0.375	0.315
$\frac{3}{8}$	0.500	0.402	$\frac{3}{8}$	0.500	0.430
$\frac{1}{2}$	0.625	0.527	$\frac{1}{2}$	0.625	0.545
$\frac{5}{8}$	0.750	0.652	$\frac{5}{8}$	0.750	0.660
$\frac{3}{4}$	0.875	0.745	$\frac{3}{4}$	0.875	0.785
1	1.125	0.995	1	1.125	1.025
$1\frac{1}{4}$	1.375	1.245	$1\frac{1}{4}$	1.375	1.265
$1\frac{1}{2}$	1.625	1.481	$1\frac{1}{2}$	1.625	1.505
2	2.125	1.959	2	2.125	1.985
$2\frac{1}{2}$	2.625	2.435	$2\frac{1}{2}$	2.625	2.465
3	3.125	2.907	3	3.125	2.945
$3\frac{1}{2}$	3.625	3.385	$3\frac{1}{2}$	3.625	3.425
4	4.125	3.857	4	4.125	3.905
5	5.125	4.805	5	5.125	4.875
6	6.125	5.741	6	6.125	5.845
8	8.125	7.583	8	8.125	7.725
10	10.125	9.449	10	10.125	9.625
12	12.125	11.315	12	12.125	11.565

Type M Tube			Type DWV Tube			Tubing	
Nominal Size (in.)	Outside Diameter (in.)	Inside Diameter (in.)	Nominal Size (in.)	Outside Diameter (in.)	Inside Diameter (in.)	Outside Diameter (in.)	Inside Diameter (in.)
$\frac{3}{8}$	0.500	0.450	$1\frac{1}{4}$	1.375	1.295	$\frac{1}{8}$	0.065
$\frac{1}{2}$	0.625	0.569	$1\frac{1}{2}$	1.625	1.541	$\frac{3}{16}$	0.128
$\frac{3}{4}$	0.875	0.811	2	2.125	1.041	$\frac{1}{4}$	0.190
1	1.125	1.055	3	3.125	3.035	$\frac{5}{16}$	0.238
$1\frac{1}{4}$	1.375	1.291	4	4.125	4.009	$\frac{3}{8}$	0.311
$1\frac{1}{2}$	1.625	1.527	5	5.125	4.981	$\frac{1}{2}$	0.436
2	2.125	2.009	6	6.125	5.959	$\frac{5}{8}$	0.555
$2\frac{1}{2}$	2.625	2.495				$\frac{3}{4}$	0.680
3	3.125	2.981				$\frac{7}{8}$	0.785
$3\frac{1}{2}$	3.625	3.459				$1\frac{1}{8}$	1.025
4	4.125	3.935				$1\frac{3}{8}$	1.265
5	5.125	4.907					
6	6.125	5.881					
8	8.125	7.785					
10	10.125	9.701					
12	12.125	11.617					

Appendix E Tables of Compressible Flow Functions for Ideal Gas with $k = 1.4$

Table E.1 Isentropic flow functions [see Eqs. (10.30), (10.31), and (10.67)]

M	T/T_0	p/p_0	A/A^*	M	T/T_0	p/p_0	A/A^*
0.00	1.0000	1.0000	∞				
0.02	0.9999	0.9997	28.942	0.82	0.8815	0.6430	1.030
0.04	0.9997	0.9989	14.481	0.84	0.8763	0.6300	1.024
0.06	0.9993	0.9975	9.666	0.86	0.8711	0.6170	1.018
0.08	0.9987	0.9955	7.262	0.88	0.8659	0.6041	1.013
0.10	0.9980	0.9930	5.822	0.90	0.8606	0.5913	1.009
0.12	0.9971	0.9900	4.864	0.92	0.8552	0.5785	1.006
0.14	0.9961	0.9864	4.182	0.94	0.8498	0.5658	1.003
0.16	0.9949	0.9823	3.673	0.96	0.8444	0.5532	1.001
0.18	0.9936	0.9777	3.278	0.98	0.8389	0.5407	1.000
0.20	0.9921	0.9725	2.964	1.00	0.8333	0.5283	1.000
0.22	0.9904	0.9669	2.708	1.02	0.8278	0.5160	1.000
0.24	0.9886	0.9607	2.496	1.04	0.8222	0.5039	1.001
0.26	0.9867	0.9541	2.317	1.06	0.8165	0.4919	1.003
0.28	0.9846	0.9470	2.166	1.08	0.8108	0.4801	1.005
0.30	0.9823	0.9395	2.035	1.10	0.8052	0.4684	1.008
0.32	0.9799	0.9315	1.922	1.12	0.7994	0.4568	1.011
0.34	0.9774	0.9231	1.823	1.14	0.7937	0.4455	1.015
0.36	0.9747	0.9143	1.736	1.16	0.7880	0.4343	1.020
0.38	0.9719	0.9052	1.659	1.18	0.7822	0.4232	1.025
0.40	0.9690	0.8956	1.590	1.20	0.7764	0.4124	1.030
0.42	0.9659	0.8857	1.529	1.22	0.7706	0.4017	1.037
0.44	0.9627	0.8755	1.474	1.24	0.7648	0.3912	1.043
0.46	0.9594	0.8650	1.425	1.26	0.7590	0.3809	1.050
0.48	0.9560	0.8541	1.380	1.28	0.7532	0.3708	1.058
0.50	0.9524	0.8430	1.340	1.30	0.7474	0.3609	1.066
0.52	0.9487	0.8317	1.303	1.32	0.7416	0.3512	1.075
0.54	0.9449	0.8201	1.270	1.34	0.7358	0.3417	1.084
0.56	0.9410	0.8082	1.240	1.36	0.7300	0.3323	1.094
0.58	0.9370	0.7962	1.213	1.38	0.7242	0.3232	1.104
0.60	0.9328	0.7840	1.188	1.40	0.7184	0.3142	1.115
0.62	0.9286	0.7716	1.166	1.42	0.7126	0.3055	1.126
0.64	0.9243	0.7591	1.145	1.44	0.7069	0.2969	1.138
0.66	0.9199	0.7465	1.127	1.46	0.7011	0.2886	1.150
0.68	0.9154	0.7338	1.110	1.48	0.6954	0.2804	1.163
0.70	0.9108	0.7209	1.094	1.50	0.6897	0.2724	1.176
0.72	0.9061	0.7080	1.081	1.52	0.6840	0.2646	1.190
0.74	0.9013	0.6951	1.068	1.54	0.6783	0.2570	1.204
0.76	0.8964	0.6821	1.057	1.56	0.6726	0.2496	1.219
0.78	0.8915	0.6691	1.047	1.58	0.6670	0.2423	1.234
0.80	0.8865	0.6560	1.038	1.60	0.6614	0.2353	1.250

M	T/T_0	p/p_0	A/A^*	M	T/T_0	p/p_0	A/A^*
1.62	0.6558	0.2284	1.267	2.52	0.4405	0.05674	2.687
1.64	0.6502	0.2217	1.284	2.54	0.4366	0.05500	2.737
1.66	0.6447	0.2152	1.301	2.56	0.4328	0.05332	2.789
1.68	0.6392	0.2088	1.319	2.58	0.4289	0.05169	2.842
1.70	0.6337	0.2026	1.338	2.60	0.4252	0.05012	2.896
1.72	0.6283	0.1966	1.357	2.62	0.4214	0.04859	2.951
1.74	0.6229	0.1907	1.376	2.64	0.4177	0.04711	3.007
1.76	0.6175	0.1850	1.397	2.66	0.4141	0.04568	3.065
1.78	0.6121	0.1794	1.418	2.68	0.4104	0.04429	3.123
1.80	0.6068	0.1740	1.439	2.70	0.4068	0.04295	3.183
1.82	0.6015	0.1688	1.461	2.72	0.4033	0.04166	3.244
1.84	0.5963	0.1637	1.484	2.74	0.3998	0.04039	3.306
1.86	0.5911	0.1587	1.507	2.76	0.3963	0.03917	3.370
1.88	0.5859	0.1539	1.531	2.78	0.3928	0.03800	3.434
1.90	0.5807	0.1492	1.555	2.80	0.3894	0.03685	3.500
1.92	0.5756	0.1447	1.580	2.82	0.3860	0.03574	3.567
1.94	0.5705	0.1403	1.606	2.84	0.3827	0.03467	3.636
1.96	0.5655	0.1360	1.633	2.86	0.3794	0.03363	3.706
1.98	0.5605	0.1318	1.660	2.88	0.3761	0.03262	3.777
2.00	0.5556	0.1278	1.688	2.90	0.3729	0.03165	3.850
2.02	0.5506	0.1239	1.716	2.92	0.3697	0.03071	3.924
2.04	0.5458	0.1201	1.745	2.94	0.3665	0.02980	3.999
2.06	0.5409	0.1164	1.775	2.96	0.3633	0.02891	4.076
2.08	0.5361	0.1128	1.806	2.98	0.3602	0.02805	4.155
2.10	0.5314	0.1094	1.837	3.00	0.3571	0.02722	4.235
2.12	0.5266	0.1060	1.869	3.10	0.3422	0.02345	4.657
2.14	0.5219	0.1027	1.902	3.20	0.3281	0.02023	5.121
2.16	0.5173	0.09956	1.935	3.30	0.3147	0.01748	5.629
2.18	0.5127	0.09650	1.970	3.40	0.3019	0.01512	6.184
2.20	0.5081	0.09352	2.005	3.50	0.2899	0.01311	6.790
2.22	0.5063	0.09064	2.041	3.60	0.2784	0.01138	7.450
2.24	0.4991	0.08784	2.078	3.70	0.2675	0.009903	8.169
2.26	0.4947	0.08514	2.115	3.80	0.2572	0.008629	8.951
2.28	0.4903	0.08252	2.154	3.90	0.2474	0.007532	9.799
2.30	0.4859	0.07997	2.193	4.00	0.2381	0.006586	10.719
2.32	0.4816	0.07751	2.233	4.10	0.2293	0.005769	11.715
2.34	0.4773	0.07513	2.274	4.20	0.2208	0.005062	12.792
2.36	0.4731	0.07281	2.316	4.30	0.2129	0.004449	13.955
2.38	0.4689	0.07057	2.359	4.40	0.2053	0.003918	15.210
2.40	0.4647	0.06840	2.403	4.50	0.1980	0.003455	16.562
2.42	0.4606	0.06630	2.448	4.60	0.1911	0.003053	18.018
2.44	0.4565	0.06426	2.494	4.70	0.1846	0.002701	19.583
2.46	0.4524	0.06229	2.540	4.80	0.1783	0.002394	21.264
2.48	0.4484	0.06038	2.588	4.90	0.1724	0.002126	23.067
2.50	0.4444	0.05853	2.637	5.00	0.1667	0.001890	25.000

Table E.2 Normal shock functions [see Eqs. (10.42b), (10.43), (10.44), and (10.46)]

M_1	M_2	p_{0_2}/p_{0_1}	T_2/T_1	p_2/p_1	M_1	M_2	p_{0_2}/p_{0_1}	T_2/T_1	p_2/p_1
1.00	1.000	1.000	1.000	1.000					
1.02	0.9805	1.000	1.013	1.047	1.82	0.6121	0.8038	1.547	3.698
1.04	0.9620	0.9999	1.026	1.095	1.84	0.6078	0.7947	1.562	3.783
1.06	0.9444	0.9998	1.039	1.144	1.86	0.6036	0.7857	1.577	3.870
1.08	0.9277	0.9994	1.052	1.194	1.88	0.5996	0.7766	1.592	3.957
1.10	0.9118	0.9989	1.065	1.245	1.90	0.5956	0.7674	1.608	4.045
1.12	0.8966	0.9982	1.078	1.297	1.92	0.5918	0.7581	1.624	4.134
1.14	0.8820	0.9973	1.090	1.350	1.94	0.5880	0.7488	1.639	4.224
1.16	0.8682	0.9961	1.103	1.403	1.96	0.5844	0.7395	1.655	4.315
1.18	0.8549	0.9946	1.115	1.458	1.98	0.5808	0.7302	1.671	4.407
1.20	0.8422	0.9928	1.128	1.513	2.00	0.5774	0.7209	1.687	4.500
1.22	0.8300	0.9907	1.141	1.570	2.02	0.5740	0.7115	1.704	4.594
1.24	0.8183	0.9884	1.153	1.627	2.04	0.5707	0.7022	1.720	4.689
1.26	0.8071	0.9857	1.166	1.686	2.06	0.5675	0.6928	1.737	4.784
1.28	0.7963	0.9827	1.178	1.745	2.08	0.5643	0.6835	1.754	4.881
1.30	0.7860	0.9794	1.191	1.805	2.10	0.5613	0.6742	1.770	4.978
1.32	0.7760	0.9757	1.204	1.866	2.12	0.5583	0.6649	1.787	5.077
1.34	0.7664	0.9718	1.216	1.928	2.14	0.5554	0.6557	1.805	5.176
1.36	0.7572	0.9676	1.229	1.991	2.16	0.5525	0.6464	1.822	5.277
1.38	0.7483	0.9630	1.242	2.055	2.18	0.5498	0.6373	1.839	5.378
1.40	0.7397	0.9582	1.255	2.120	2.20	0.5471	0.6281	1.857	5.480
1.42	0.7314	0.9531	1.268	2.186	2.22	0.5444	0.6191	1.875	5.583
1.44	0.7235	0.9477	1.281	2.253	2.24	0.5418	0.6100	1.892	5.687
1.46	0.7157	0.9420	1.294	2.320	2.26	0.5393	0.6011	1.910	5.792
1.48	0.7083	0.9360	1.307	2.389	2.28	0.5368	0.5921	1.929	5.898
1.50	0.7011	0.9298	1.320	2.458	2.30	0.5344	0.5833	1.947	6.005
1.52	0.6941	0.9233	1.334	2.529	2.32	0.5321	0.5745	1.965	6.113
1.54	0.6874	0.9166	1.347	2.600	2.34	0.5297	0.5658	1.984	6.222
1.56	0.6809	0.9097	1.361	2.673	2.36	0.5275	0.5572	2.002	6.331
1.58	0.6746	0.9026	1.374	2.746	2.38	0.5253	0.5486	2.021	6.442
1.60	0.6684	0.8952	1.388	2.820	2.40	0.5231	0.5402	2.040	6.553
1.62	0.6625	0.8876	1.402	2.895	2.42	0.5210	0.5318	2.059	6.666
1.64	0.6568	0.8799	1.416	2.971	2.44	0.5189	0.5234	2.079	6.779
1.66	0.6512	0.8720	1.430	3.048	2.46	0.5169	0.5152	2.098	6.894
1.68	0.6458	0.8640	1.444	3.126	2.48	0.5149	0.5071	2.118	7.009
1.70	0.6406	0.8557	1.458	3.205	2.50	0.5130	0.4990	2.137	7.125
1.72	0.6355	0.8474	1.473	3.285	2.52	0.5111	0.4910	2.157	7.242
1.74	0.6305	0.8389	1.487	3.366	2.54	0.5092	0.4832	2.177	7.360
1.76	0.6257	0.8302	1.502	3.447	2.56	0.5074	0.4754	2.198	7.479
1.78	0.6210	0.8215	1.517	3.530	2.58	0.5056	0.4677	2.218	7.599
1.80	0.6165	0.8127	1.532	3.613	2.60	0.5039	0.4601	2.238	7.720

M_1	M_2	p_{0_2}/p_{0_1}	T_2/T_1	p_2/p_1	M_1	M_2	p_{0_2}/p_{0_1}	T_2/T_1	p_2/p_1
2.62	0.5022	0.4526	2.259	7.842	3.10	0.4695	0.3012	2.799	11.05
2.64	0.5005	0.4452	2.280	7.965	3.20	0.4644	0.2762	2.922	11.78
2.66	0.4988	0.4379	2.301	8.088	3.30	0.4596	0.2533	3.049	12.54
2.68	0.4972	0.4307	2.322	8.213	3.40	0.4552	0.2322	3.180	13.32
2.70	0.4956	0.4236	2.343	8.338	3.50	0.4512	0.2130	3.315	14.13
2.72	0.4941	0.4166	2.364	8.465	3.60	0.4474	0.1953	3.454	14.95
2.74	0.4926	0.4097	2.386	8.592	3.70	0.4440	0.1792	3.596	15.81
2.76	0.4911	0.4028	2.407	8.721	3.80	0.4407	0.1645	3.743	16.68
2.78	0.4897	0.3961	2.429	8.850	3.90	0.4377	0.1510	3.893	17.58
2.80	0.4882	0.3895	2.451	8.980	4.00	0.4350	0.1388	4.047	18.50
2.82	0.4868	0.3829	2.473	9.111	4.10	0.4324	0.1276	4.205	19.45
2.84	0.4854	0.3765	2.496	9.243	4.20	0.4299	0.1173	4.367	20.41
2.86	0.4840	0.3701	2.518	9.376	4.30	0.4277	0.1080	4.532	21.41
2.88	0.4827	0.3639	2.540	9.510	4.40	0.4255	0.09948	4.702	22.42
2.90	0.4814	0.3577	2.563	9.645	4.50	0.4236	0.09170	4.875	23.46
2.92	0.4801	0.3517	2.586	9.781	4.60	0.4217	0.08459	5.052	24.52
2.94	0.4788	0.3457	2.609	9.918	4.70	0.4199	0.07809	5.233	25.61
2.96	0.4776	0.3398	2.632	10.06	4.80	0.4183	0.07214	5.418	26.71
2.98	0.4764	0.3340	2.656	10.19	4.90	0.4167	0.06670	5.607	27.85
3.00	0.4752	0.3283	2.679	10.33	5.00	0.4152	0.06172	5.800	29.00

Table E.3 Fanno flow functions [see Eqs. (10.91), (10.93), (10.94), and (10.95)]

M	$4C_f \dfrac{\ell^*}{D_h}$	p_0/p_0^*	T/T^*	p/p^*	M	$4C_f \dfrac{\ell^*}{D_h}$	p_0/p_0^*	T/T^*	p/p^*
0.00	∞	∞	1.200	∞					
0.02	1778.450	28.94	1.200	54.77	0.82	0.05593	1.030	1.058	1.254
0.04	440.352	14.48	1.200	27.38	0.84	0.04226	1.024	1.052	1.221
0.06	193.031	9.666	1.199	18.25	0.86	0.03097	1.018	1.045	1.189
0.08	106.718	7.262	1.199	13.68	0.88	0.02180	1.013	1.039	1.158
0.10	66.922	5.822	1.198	10.94	0.90	0.01451	1.0089	1.033	1.129
0.12	45.408	4.864	1.197	9.116	0.92	0.00891	1.0056	1.026	1.101
0.14	32.511	4.182	1.195	7.809	0.94	0.00481	1.0031	1.020	1.074
0.16	24.198	3.673	1.194	6.829	0.96	0.00206	1.0014	1.013	1.049
0.18	18.543	3.278	1.192	6.066	0.98	0.00049	1.0003	1.007	1.024
0.20	14.533	2.964	1.191	5.456	1.00	0.0000	1.0000	1.000	1.000
0.22	11.596	2.708	1.189	4.955	1.02	0.00046	1.0003	0.9933	0.9771
0.24	9.387	2.496	1.186	4.538	1.04	0.00177	1.0013	0.9866	0.9551
0.26	7.688	2.317	1.184	4.185	1.06	0.00389	1.0029	0.9798	0.9338
0.28	6.357	2.166	1.182	3.882	1.08	0.00658	1.0051	0.9730	0.9134
0.30	5.299	2.035	1.179	3.619	1.10	0.00993	1.0079	0.9662	0.8936
0.32	4.447	1.922	1.176	3.389	1.12	0.01382	1.011	0.9593	0.8745
0.34	3.752	1.823	1.173	3.185	1.14	0.01819	1.015	0.9524	0.8561
0.36	3.180	1.736	1.170	3.004	1.16	0.02298	1.020	0.9455	0.8383
0.38	2.706	1.659	1.166	2.842	1.18	0.02814	1.025	0.9386	0.8210
0.40	2.309	1.590	1.163	2.696	1.20	0.03364	1.030	0.9317	0.8044
0.42	1.974	1.529	1.159	2.563	1.22	0.03942	1.037	0.9247	0.7882
0.44	1.692	1.474	1.155	2.443	1.24	0.04547	1.043	0.9178	0.7726
0.46	1.451	1.425	1.151	2.333	1.26	0.05174	1.050	0.9108	0.7574
0.48	1.245	1.380	1.147	2.231	1.28	0.05820	1.058	0.9038	0.7427
0.50	1.069	1.340	1.143	2.138	1.30	0.06483	1.066	0.8969	0.7285
0.52	0.9174	1.303	1.138	2.052	1.32	0.07161	1.075	0.8899	0.7147
0.54	0.7866	1.270	1.134	1.972	1.34	0.07850	1.084	0.8829	0.7012
0.56	0.6736	1.240	1.129	1.898	1.36	0.08550	1.094	0.8760	0.6882
0.58	0.5757	1.213	1.124	1.828	1.38	0.09259	1.104	0.8690	0.6755
0.60	0.4908	1.188	1.119	1.763	1.40	0.09974	1.115	0.8621	0.6632
0.62	0.4172	1.166	1.114	1.703	1.42	0.1069	1.126	0.8551	0.6512
0.64	0.3533	1.145	1.109	1.646	1.44	0.1142	1.138	0.8482	0.6396
0.66	0.2979	1.127	1.104	1.592	1.46	0.1215	1.150	0.8413	0.6282
0.68	0.2498	1.110	1.098	1.541	1.48	0.1288	1.163	0.8345	0.6172
0.70	0.2081	1.094	1.093	1.493	1.50	0.1361	1.176	0.8276	0.6065
0.72	0.1722	1.081	1.087	1.448	1.52	0.1434	1.190	0.8208	0.5960
0.74	0.1411	1.068	1.082	1.405	1.54	0.1506	1.204	0.8139	0.5858
0.76	0.1145	1.057	1.076	1.365	1.56	0.1579	1.219	0.8072	0.5759
0.78	0.09167	1.047	1.070	1.326	1.58	0.1651	1.234	0.8004	0.5662
0.80	0.07229	1.038	1.064	1.289	1.60	0.1724	1.250	0.7937	0.5568

M	$4C_f \dfrac{\ell^*}{D_h}$	p_0/p_0^*	T/T^*	p/p^*	M	$4C_f \dfrac{\ell^*}{D_h}$	p_0/p_0^*	T/T^*	p/p^*
1.62	0.1795	1.267	0.7870	0.5476	2.42	0.4144	2.448	0.5527	0.3072
1.64	0.1867	1.284	0.7803	0.5386	2.44	0.4189	2.494	0.5478	0.3033
1.66	0.1938	1.301	0.7736	0.5299	2.46	0.4233	2.540	0.5429	0.2995
1.68	0.2008	1.319	0.7670	0.5213	2.48	0.4277	2.588	0.5381	0.2958
1.70	0.2078	1.338	0.7605	0.5130	2.50	0.4320	2.637	0.5333	0.2921
1.72	0.2147	1.357	0.7539	0.5048	2.52	0.4362	2.687	0.5286	0.2885
1.74	0.2216	1.376	0.7474	0.4969	2.54	0.4404	2.737	0.5239	0.2850
1.76	0.2284	1.397	0.7410	0.4891	2.56	0.4445	2.789	0.5193	0.2815
1.78	0.2352	1.418	0.7345	0.4815	2.58	0.4486	2.842	0.5147	0.2781
1.80	0.2419	1.439	0.7282	0.4741	2.60	0.4526	2.896	0.5102	0.2747
1.82	0.2485	1.461	0.7218	0.4668	2.62	0.4565	2.951	0.5057	0.2714
1.84	0.2551	1.484	0.7155	0.4597	2.64	0.4604	3.007	0.5013	0.2682
1.86	0.2616	1.507	0.7093	0.4528	2.66	0.4643	3.065	0.4969	0.2650
1.88	0.2680	1.531	0.7030	0.4460	2.68	0.4681	3.123	0.4925	0.2619
1.90	0.2743	1.555	0.6969	0.4394	2.70	0.4718	3.183	0.4882	0.2588
1.92	0.2806	1.580	0.6907	0.4329	2.72	0.4755	3.244	0.4839	0.2558
1.94	0.2868	1.606	0.6847	0.4265	2.74	0.4792	3.306	0.4797	0.2528
1.96	0.2930	1.633	0.6786	0.4203	2.76	0.4827	3.370	0.4755	0.2499
1.98	0.2990	1.660	0.6726	0.4142	2.78	0.4863	3.434	0.4714	0.2470
2.00	0.3050	1.688	0.6667	0.4083	2.80	0.4898	3.500	0.4673	0.2441
2.02	0.3109	1.716	0.6608	0.4024	2.82	0.4932	3.567	0.4632	0.2414
2.04	0.3168	1.745	0.6549	0.3967	2.84	0.4966	3.636	0.4592	0.2386
2.06	0.3225	1.775	0.6491	0.3911	2.86	0.5000	3.706	0.4553	0.2359
2.08	0.3282	1.806	0.6433	0.3856	2.88	0.5033	3.777	0.4513	0.2333
2.10	0.3339	1.837	0.6376	0.3802	2.90	0.5065	3.850	0.4474	0.2307
2.12	0.3394	1.869	0.6320	0.3750	2.92	0.5097	3.924	0.4436	0.2281
2.14	0.3449	1.902	0.6263	0.3698	2.94	0.5129	3.999	0.4398	0.2256
2.16	0.3503	1.935	0.6208	0.3648	2.96	0.5160	4.076	0.4360	0.2231
2.18	0.3556	1.970	0.6152	0.3598	2.98	0.5191	4.155	0.4323	0.2206
2.20	0.3609	2.005	0.6098	0.3549	3.00	0.5222	4.235	0.4286	0.2182
2.22	0.3661	2.041	0.6043	0.3502	3.50	0.5864	6.79	0.3478	0.1685
2.24	0.3712	2.078	0.5990	0.3455	4.00	0.6331	10.72	0.2857	0.1336
2.26	0.3763	2.115	0.5936	0.3409	4.50	0.6676	16.56	0.2376	0.1083
2.28	0.3813	2.154	0.5883	0.3364	5.00	0.6938	25.00	0.2000	0.08944
2.30	0.3862	2.193	0.5831	0.3320					
2.32	0.3911	2.233	0.5779	0.3277					
2.34	0.3959	2.274	0.5728	0.3234					
2.36	0.4006	2.316	0.5677	0.3193					
2.38	0.4053	2.359	0.5626	0.3152					
2.40	0.4099	2.403	0.5576	0.3111					

Appendix F Fluid Mechanics Films

Fluid mechanics is a visual subject. Many films are available to help engineers and students visualize complex flow phenomena. The films listed in this Appendix provide excellent support for class lecture and laboratory work in teaching fluid mechanics. The films are available for a modest rental fee or sometimes free from the agencies listed. Several of the films can also be obtained from public libraries or university film services.

A. The National Committee for Fluid Mechanics Films (NCFMF). Available from: Encyclopaedia Britannica Educational Corp., 425 N. Michigan Ave., Chicago, IL 60611. The films are summarized in the book *Illustrated Experiments in Fluid Mechanics, the NCFMF Book of Film Notes,* available from MIT Press.

1. *The Fluid Dynamics of Drag* (119 min)
2. *Vorticity* (44 min)
3. *Flow Visualization* (31 min)
4. *Deformation of Continuous Media* (38 min)
5. *Pressure Fields and Fluid Acceleration* (30 min)
6. *Surface Tension in Fluid Mechanics* (29 min)
7. *Waves in Fluids* (33 min)
8. *Secondary Flow* (30 min)
9. *Boundary Layer Control* (25 min)
10. *Channel Flow of a Compressible Fluid* (29 min)
11. *Low Reynolds Number Flows* (33 min)
12. *Flow Instabilities* (27 min)
13. *Cavitation* (27 min)
14. *Fundamentals of Boundary Layers* (24 min)
15. *Turbulence* (29 min)

B. University of Iowa Films. Available from University of Iowa, Media Library, Audiovisual Center, Iowa City, IA 52242.

1. *Introduction to the Study of Fluid Motion* (25 min)
2. *Fundamental Principles of Flow* (23 min)
3. *Fluid Motion in a Gravitational Field* (24 min)
4. *Characteristics of Laminar and Turbulent Flow* (26 min)
5. *Form Drag, Lift, and Propulsion* (24 min)
6. *Effects of Fluid Compressibility* (17 min)

C. St. Anthony Falls Laboratory Films. Available from St. Anthony Falls Hydraulic Laboratory, Mississippi River at 3rd Ave. S.E., Minneapolis, MN 55414.

1. *Hydraulic Model Studies* (20 min)
2. *Some Phenomena of Open Channel Flow* (33 min)
3. *Flow in Culverts* (20 min)
4. *Spillway Design, Guayabo Dam, El Salvador* (15 min)
5. *Surface Waves* (24 min)
6. *Fluid Mechanics: The Boundary Layer* (30 min)

D. Films available from Engineering Societies Library, 345 E. 47th St., New York, NY 10017.

1. *Turbulent Flow (An Introductory Lecture)* (35 min)
2. *Flow Separation and Formation of Vortices* (16 min)
3. *Hydraulic Analogy for Flow Studies* (24 min)
4. *Visual Cavitation Studies of Mixed Flow Pump Impellers* (22 min)

Appendix G List of Symbols

This appendix contains a list of the major symbols used in this book. Since there are many more concepts than characters available in both the Roman and Greek alphabets, some duplication is unavoidable. We have generally tried to follow the American National Standards Institute (ANSI) standard symbols; however, there is no ANSI standard for fluid mechanics per se. We have considered ANSI-Y10.2 (*Hydraulics*), ANSI-Y10.7 (*Aeronautical Sciences*), and ANSI-Y10.4 (*Heat and Thermodynamics*). If a particular symbol is used for a specific concept in one chapter and for a different concept in another chapter, we have noted the chapters in the list. We also include a listing of the appropriate dimension(s) of each concept in the M, L, T, θ-dimension system. A 1 in the dimensions column indicates a dimensionless quantity and a $(-)$ in the dimensions column means that the idea of dimension is irrelevant to the particular concept.

Symbol	Concept	Dimensions
Roman Symbols		
A	Area (usually cross-sectional area)	L^2
A_{cv}	Surface area of control volume	L^2
A^*	Critical area	L^2
AR	Aspect ratio (Chapter 8)	1
a	Acceleration	LT^{-2}
a	Constant or exponent	Various
B	Atmospheric temperature lapse rate (Chapter 2)	θ^{-1}
B	Arbitrary system property (Chapter 5)	Various
B	Constant in logarithmic velocity profile (Chapter 7)	1
b	Specific (per unit mass) value of B (Chapter 5)	(Various) M^{-1}
b	Constant or exponent	Various
b	Width of open channel (Chapter 11)	L
C	Heat capacity (Chapter 1)	$ML^2T^{-2}\theta^{-1}$
C	Chezy coefficient (Chapter 11)	$L^{1/2}T^{-1}$
C	Constant	Various

(continued)

Symbol	Concept	Dimensions
Roman Symbols (continued)		
c	Airfoil chord (Chapters 8, 9)	L
c	Speed of sound (Chapter 10)	LT^{-1}
c	Speed of weak gravity wave (Chapter 11)	LT^{-1}
c_p	Specific heat at constant pressure	$L^2T^{-2}\theta^{-1}$
c_v	Specific heat at constant volume	$L^2T^{-2}\theta^{-1}$
C_c	Contraction coefficient	1
C_D	Drag coefficient	1
C_D'	Viscous drag coefficient	1
C_d	Discharge coefficient	1
C_f	Skin friction coefficient	1
C_L	Lift coefficient	1
C_p	Pressure coefficient	1
C_Q	Flow coefficient	1
C_v	Velocity coefficient	1
D	Diameter	L
D_{eff}	Effective diameter	L
D_{eq}	Equivalent diameter	L
D_h	Hydraulic diameter	L
$\mathscr{D}$	Drag force	MLT^{-2}
D/Dt	Eulerian (material, substantial) derivative operator	T^{-1}
E	Specific energy (Chapter 11)	L
E_K	Kinetic energy	ML^2T^{-2}
E_v	Bulk modulus of elasticity	$ML^{-1}T^{-2}$
$E_{v,T}$	Isothermal bulk modulus	$ML^{-1}T^{-2}$
F	Force	MLT^{-2}
F_B	Buoyant force	MLT^{-2}
f	Darcy friction factor	1
$F\{\ \}$	Function	—
$f\{\ \}$	Function	—
$\mathbf{F}$	Froude number	1
g	Acceleration of gravity	LT^{-2}
g_c	Force unit/mass unit conversion factor	1
gh_L	Mechanical energy loss	L^2T^{-2}
$G\{\ \}$	Function	1
H	Boundary layer shape factor (Chapter 9)	1
H	Depth of liquid over weir (Chapter 11)	L
H.O.T.	Higher-order terms (in a series)	Various
h	Height	L
h	Depth of liquid below a free surface	L
h	Airfoil chamber (Chapters 8, 9)	L
h	Head	L
h_L	Head loss	L
h_p	Pump head	L
h_T	Total head	L
h_t	Turbine head	L
$\tilde{h}$	Specific enthalpy	L^2T^{-2}
$\tilde{h}_0$	Stagnation enthalpy	L^2T^{-2}
I	Area moment of inertia (Chapter 2)	L^4
I	Mass moment of inertia (Chapter 8)	ML^2

Symbol	Concept	Dimensions

Roman Symbols (continued)

I_h	Hydrodynamic moment of inertia	ML^2
$\hat{i}$	Unit vector in x-direction	—
$\hat{j}$	Unit vector in y-direction	—
k	Number	—
k	Specific heat ratio	1
$\hat{k}$	Unit vector in z-direction	—
K	Loss coefficient	1
L	Length of pipe or plate	L
L_e	Entrance length	L
L_{eq}	Equivalent length	L
$\dot{L}$	Total rate of mechanical energy loss	ML^2T^{-2}
ℓ	Length (Chapter 8)	L
ℓ	Mixing length in turbulent flow (Chapter 7)	L
ℓ^*	Choking (critical) length	L
$\mathscr{L}$	Lift force	MLT^{-2}
M	Momentum	MLT^{-1}
$\dot{M}$	Momentum flux	MLT^{-2}
m	Mass	M
$\dot{m}$	Mass flow rate	MT^{-1}
m_h	Added mass	M
$\mathscr{M}$	Moment	ML^2T^{-2}
M	Mach number	1
N	Rotational speed (rpm)	T^{-1}
n	Coordinate normal to streamline (Chapter 4)	L
n	Number	—
n	Exponent in power law velocity profile (Chapters 5, 7, 9)	—
n	Manning's n (roughness factor) (Chapter 11)	1
$\hat{n}$	Unit vector normal (perpendicular) to area	—
p	Pressure	$ML^{-3}T^{-2}$
p_B	Back pressure	$ML^{-3}T^{-2}$
p_T	Total pressure	$ML^{-3}T^{-2}$
p_0	Stagnation pressure	$ML^{-3}T^{-2}$
p'	Modified pressure ($p' = p - \gamma h$) (Chapter 9)	$ML^{-3}T^{-2}$
Δp	Pressure difference	$ML^{-3}T^{-2}$
Δp_d	Pressure drop	$ML^{-3}T^{-2}$
Δp_L	Pressure loss	$ML^{-3}T^{-2}$
P	Perimeter	L
P	Height of weir	L
Q	Volume flow rate	L^3T^{-1}
$\dot{Q}$	Heat transfer rate	ML^2T^{-2}
q	Heat transfer per unit mass (Chapters 5, 10)	L^2T^{-2}
q	Volume flow per unit width of open channel (Chapter 11)	L^2T^{-1}
R	Radius of pipe, cylinder, or sphere	L
R	Radius of curvature (Chapters 4, 9)	L

(continued)

Symbol	**Concept**	**Dimensions**
Roman Symbols (continued)		
R	Specific gas constant (Chapters 1, 10)	$L^2T^{-2}\theta^{-1}$
R_0	Universal gas constant	$ML^2T^{-2}\theta^{-1}(\text{mole})^{-1}$
R_h	Hydraulic radius	L
r	Radial coordinate (cylindrical coordinates)	L
r'	Radial coordinate (spherical coordinates)	L
r_{pc}	Critical pressure ratio	1
$\mathbf{R}$	Reynolds number	1
S	Specific gravity	1
S	Scale factor (Chapter 6)	1
S	Projected area for force coefficients (Chapters 8, 9, 10)	L^2
S_E	Slope of energy grade line (Chapter 11)	1
S_0	Slope of channel bottom (Chapter 11)	1
s	Streamline coordinate (Chapter 4)	L
s	Arc length (Chapters 8, 9)	L
$\hat{s}$	Unit vector in s-direction	—
$\tilde{s}$	Specific entropy	$L^2T^{-2}\theta^{-1}$
$\mathbf{S}$	Strouhal number	1
T	Temperature	θ
T	Averaging time (Chapter 7)	T
T	Dimensionless wall shear stress (Chapter 9)	1 —
T_0	Stagnation temperature	θ
t	Time	T
$\mathscr{T}$	Torque	ML^2T^{-2}
U	Reference or given velocity	LT^{-1}
U	Velocity of coordinate system (Chapter 5)	LT^{-1}
U	Velocity at edge of boundary layer (Chapter 9)	LT^{-1}
u	Velocity in x-direction	LT^{-1}
u_τ	Friction velocity ($\sqrt{\tau_w/\rho}$)	LT^{-1}
$u_{\max}$	Velocity at pipe centerline	LT^{-1}
$\tilde{u}$	Specific internal energy	L^2T^{-2}
V	Velocity	LT^{-1}
V_r	Relative velocity	LT^{-1}
$\forall$	Volume	L^3
$\forall_{cv}$	Volume of control volume	L^3
v	Velocity in y-direction	LT^{-1}
W	Weight	MLT^{-2}
W	Width	L
W	Work	ML^2T^{-2}
$\dot{W}$	Power	ML^2T^{-3}
$\dot{W}_s$	Sum of shaft power and shear power	ML^2T^{-3}
w	Velocity in z-direction	LT^{-1}
w	Complex velocity (Chapter 9)	LT^{-1}
w_p	Pump work per unit mass	L^2T^{-2}
w_s	Sum of shaft work and shear work per unit mass	L^2T^2
w_t	Turbine work per unit mass	L^2T^{-2}

Symbol	Concept	Dimensions
Roman Symbols (continued)		
W	Weber number	1
x	Cartesian or cylindrical coordinate, usually parallel to main flow direction	L
x	Arbitrary dimensional parameter (Chapter 6)	Various
y	Cartesian coordinate, usually perpendicular to main flow	L
y	Coordinate parallel to plane of area (Chapter 2)	L
y	Depth in open channel flow (Chapter 11)	L
y_n	Normal depth	L
z	Cartesian coordinate, usually pointing upward	L
z	Elevation above specified datum (Chapter 4)	L
z	Complex variable (Chapter 9)	—
Greek Symbols		
α	Kinetic energy correction factor	1
α	Velocity diagram angle	1
α	Angle of attack (Chapters 8, 9)	1
α_T	Coefficient of thermal expansion	θ^{-1}
β	Momentum correction factor	1
β	Velocity diagram angle	1
β	Diameter ratio of a flow meter (Chapter 7)	1
Γ	Circulation	$L^2 T^{-1}$
γ	Specific weight	$ML^{-2} T^{-2}$
δ	Boundary layer thickness	L
δ^*	Boundary layer displacement thickness	L
$\Delta(\ )$	Change of ( )	Various
$\delta(\ )$	Small change of ( )	Various
ε	Surface roughness	L
ζ	Vorticity	T^{-1}
ζ	Transformed complex variable (Chapter 9)	—
η	Imaginary part of complex function	—
η	Dimensionless coordinate in boundary layer (y/δ)	1
η	Efficiency	1
Θ	Boundary layer momentum thickness	L
θ	Angle, angle coordinate in cylindrical coordinate system	1
θ'	Angle coordinate in spherical coordinate system	1
κ	Constant in logarithmic velocity profile	—
Λ	Boundary layer pressure gradient parameter $(\delta^2/v)(dU/dx)$	1

(continued)

Symbol	Concept	Dimensions
Greek Symbols (continued)		
λ	Strength of source in plane flow	L^2T^{-1}
λ	Boundary layer pressure gradient parameter $(\theta^2/v)(dU/dx)$	1
λ'	Strength of line distributed source	LT^{-1}
μ	Dynamic (absolute) viscosity	$ML^{-1}T^{-1}$
μ	Strength of plane doublet (Chapter 9)	L^3T^{-1}
v	Kinematic viscosity	LT^{-2}
ξ	Dummy variable of integration	—
ξ	Real part of complex function (Chapter 9)	—
ξ	Unit-specific constant in Manning's formula (Chapter 11)	$L^{1/3}T^{-1}$
Π	Dimensionless group	1
ρ	Density	ML^{-3}
ρ_0	Stagnation density	ML^{-3}
σ	Surface tension (Chapter 1)	MT^{-2}
σ	Normal stress (Chapter 4)	$ML^{-1}T^{-2}$
τ	Shear stress	$ML^{-1}T^{-2}$
τ_{app}	Apparent stress in turbulent flow	$ML^{-1}T^{-2}$
ϕ	Angle, usually angle of shear deformation (Chapter 1, 3, 4)	1
Φ	Velocity potential (Chapter 9)	L^2T^{-1}
$\dot{\phi}$	Shear strain rate	T^{-1}
Ψ	Plane flow stream function	L^2T^{-1}
Ψ_s	Stokes stream function	L^3T^{-1}
Ω	Complex potential	L^2T^{-1}
$\vec{\Omega}$	Angular velocity of coordinate system	T^{-1}
ω	Angular velocity	T^{-1}
ω	Wave speed (Chapter 10, 11)	LT^{-1}
$\dot{\omega}_v$	Vortex-shedding frequency	T^{-1}

Special Symbols

$\{\ \}$	Function	
$\langle\ \rangle$	Time average in turbulent flow	
∇	Gradient operator	
$\oint$	Closed or cyclic line integral	

Subscripts

Subscript	Meaning
abs	Absolute
atm	Atmospheric conditions
B	Bottom of open channel
c	Centroid (Chapter 2)
c	Critical (Chapter 11)
cv	Control volume
e	Exit
e	Edge of boundary layer (Chapter 9)
H	Horizontal
i	Initial (at $t = 0$)
i	Inlet
in	Entering control volume
m	Model

Subscript	Meaning
n	Normal or perpendicular, in n-direction
out	Leaving control volume
p	Center of pressure (Chapter 2)
p	Pertaining to specific fluid particle (Chapters 3, 4)
p	Prototype (Chapter 6)
p	At point p
r	In radial direction or evaluated at location r
ref	Reference quantity
rev	Reversible process
s	Isentropic process (Chapter 10)
s	In s (streamline) direction
sys	Pertaining to a system
t	Throat (Chapter 10)
V	Vertical
w	Wall
x	In x-direction or evaluated at location x
XYZ	Inertial reference frame
xyz	Noninertial reference frame
y	In y-direction or at location y
z	In z-direction or at location z
θ	In θ-direction
0	Stagnation
0	At point 0
1	At point 1 or in plane 1, and so on
∞	Far away, at infinity

Superscripts

Superscript	Meaning
$\rightarrow$	Vector
$\cdot$	Time derivative, rate
$-$	Average, usually over cross-sectional area
$'$	Alternate, similar quantity
$'$	Fluctuating component in turbulent flow (Chapter 7)
$+$	Dimensionless length or velocity in turbulent flow
$*$	Corresponding to sonic velocity, critical state
(1)	Iteration 1, and so on

Appendix H Differential Equations in Cylindrical (r, θ, x) and Spherical (r', ϕ, θ') Coordinates

Continuity Equation in Cylindrical Coordinates

$$\frac{\partial \rho}{\partial t} + \frac{1}{r}\left(\frac{\partial}{\partial r}\right)(\rho r V_r) + \frac{1}{r}\left(\frac{\partial}{\partial \theta}\right)(\rho V_\theta) + \frac{\partial}{\partial x}(\rho V_x) = 0$$

Continuity Equation in Spherical Coordinates

$$\frac{\partial \rho}{\partial t} + \frac{1}{(r')^2}\left(\frac{\partial}{\partial r'}\right)[\rho(r')^2 V_r] + \frac{1}{r' \sin \theta'}\left(\frac{\partial}{\partial \theta'}\right)(\rho V_0 \sin \theta')$$

$$+ \frac{1}{r' \sin \theta'}\left(\frac{\partial}{\partial \phi}\right)(\rho V_\phi)(\rho V_\phi) = 0$$

Navier-Stokes Equations in Cylindrical Coordinates for a Fluid with Constant Viscosity

r-component

$$\rho\left[\frac{\partial V_r}{\partial t} + V_r\left(\frac{\partial V_r}{\partial r}\right) + \frac{V_\theta}{r}\left(\frac{\partial V_r}{\partial \theta}\right) + V_x\left(\frac{\partial V_r}{\partial x}\right) - \frac{V_\theta^2}{r}\right]$$

$$= -\frac{\partial p}{\partial r} + \rho g_r + \mu\left[\frac{\partial^2 V_r}{\partial r^2} + \frac{1}{r}\left(\frac{\partial V_r}{\partial r}\right) + \frac{1}{r^2}\left(\frac{\partial^2 V_r}{\partial \theta^2}\right) + \frac{\partial^2 V_r}{\partial x^2} - \frac{V_r}{r^2} - \frac{2}{r^2}\left(\frac{\partial V_\theta}{\partial \theta}\right)\right]$$

θ-component

$$\rho\left[\frac{\partial V_\theta}{\partial t} + V_r\left(\frac{\partial V_\theta}{\partial r}\right) + \frac{V_\theta}{r}\left(\frac{\partial V_\theta}{\partial \theta}\right) + V_x\left(\frac{\partial V_\theta}{\partial x}\right) + \frac{V_r V_\theta}{r}\right]$$

$$= -\frac{1}{r}\left(\frac{\partial p}{\partial \theta}\right) + \rho g_\theta + \mu\left[\frac{\partial^2 V_\theta}{\partial r^2} + \frac{1}{r}\left(\frac{\partial V_\theta}{\partial r}\right) + \frac{1}{r^2}\left(\frac{\partial^2 V_\theta}{\partial \theta^2}\right) + \frac{\partial^2 V_\theta}{\partial x^2} + \frac{2}{r^2}\left(\frac{\partial V_r}{\partial \theta}\right) - \frac{V_\theta}{r^2}\right]$$

x-component

$$\rho\left[\frac{\partial V_x}{\partial t} + V_r\left(\frac{\partial V_x}{\partial r}\right) + \frac{V_0}{r}\left(\frac{\partial V_x}{\partial \theta}\right) + \frac{\partial V_x}{\partial x}\right]$$

$$= -\frac{\partial p}{\partial x} + \rho g_x + \mu\left[\frac{\partial^2 V_x}{\partial r^2} + \frac{1}{r}\left(\frac{\partial V_x}{\partial r}\right) + \frac{1}{r^2}\left(\frac{\partial^2 V_x}{\partial \theta^2}\right) + \frac{\partial^2 V_x}{\partial x^2}\right]$$

Navier-Stokes Equations in Spherical Coordinates for a Fluid with Constant Viscosity

r'-component

$$\rho\left[\frac{\partial V_{r'}}{\partial r'} + V_{r'}\left(\frac{\partial V_{r'}}{\partial r'}\right) + \frac{V_{\theta'}}{r'}\left(\frac{\partial V_{r'}}{\partial \theta'}\right) + \frac{V_\phi}{r'\sin\theta'}\left(\frac{\partial V_{r'}}{\partial \phi}\right) - \frac{V_{\theta'}^2 + V_\phi^2}{r'}\right]$$

$$= -\frac{\partial p}{\partial r'} + \rho g_{r'} + \mu\left[\frac{1}{r'^2}\left(\frac{\partial}{\partial r'}\right)r'^2\left(\frac{\partial V_{r'}}{\partial r'}\right) + \frac{1}{r'^2\sin\theta'}\left(\frac{\partial}{\partial \theta'}\right)\left(\sin\theta'\frac{\partial V_{r'}}{\partial \theta'}\right)\right.$$

$$\left. + \frac{1}{r'^2\sin^2\theta'}\left(\frac{\partial^2 V_{r'}}{\partial \phi^2}\right) - \frac{2V_{r'}}{r'^2} - \frac{2}{r'^2}\left(\frac{\partial V_{\theta'}}{\partial \theta'}\right) - \frac{2V_{\theta'}\cos\theta'}{r'^2} - \frac{2}{r'^2\sin\theta'}\left(\frac{\partial V_{\theta'}}{\partial \phi}\right)\right]$$

θ'-component

$$\rho\left[\frac{\partial V_{\theta'}}{\partial t} + V_{r'}\left(\frac{\partial V_{\theta'}}{\partial r'}\right) + \frac{V_{\theta'}}{r'}\left(\frac{\partial V_{\theta'}}{\partial \theta'}\right) + \frac{V_{\theta'}}{r'\sin\theta'}\left(\frac{\partial V_{\theta'}}{\partial \phi}\right)\right]$$

$$= -\frac{1}{r'}\left(\frac{\partial p}{\partial \theta'}\right) + \rho g_{\theta'} + \mu\left[\frac{1}{r'^2}\left(\frac{\partial}{\partial r'}\right)\left(r'^2\frac{\partial V_{\theta'}}{\partial r}\right) + \frac{1}{r'^2\sin\theta'}\left(\frac{\partial}{\partial \theta'}\right)\left(\sin\theta'\frac{\partial V_{\theta'}}{\partial \theta'}\right)\right.$$

$$\left. + \frac{1}{r'^2\sin^2\theta'}\left(\frac{\partial^2 V_{\theta'}}{\partial \phi^2}\right) + \frac{2}{r'^2}\left(\frac{\partial V_{r'}}{\partial \theta'}\right) - \frac{V_{\theta'}}{r'^2\sin^2\theta'} - \frac{2\cos\theta'}{r'^2\sin^2\theta'}\left(\frac{\partial V_{\theta'}}{\partial \theta'}\right)\right]$$

ϕ-component

$$\rho\left[\frac{\partial V_\phi}{\partial t} + V_{r'}\left(\frac{\partial V_\phi}{\partial r'}\right) + \frac{V_{\theta'}}{r'}\left(\frac{\partial V_\phi}{\partial \theta'}\right) + \frac{V_\phi}{r'\sin\theta'}\left(\frac{\partial V_\phi}{\partial \phi}\right)\right]$$

$$= -\frac{1}{r'\sin\theta'}\left(\frac{\partial p}{\partial \phi}\right) + \rho g_\phi + \mu\left[\frac{1}{r'^2}\left(\frac{\partial}{\partial r'}\right)\left(r'^2\frac{\partial V_\phi}{\partial r'^2}\right)\right.$$

$$+ \frac{1}{r'^2\sin\theta'}\left(\frac{\partial}{\partial \theta'}\right)\left(\sin\theta'\frac{\partial V_\phi}{\partial \theta'}\right) + \frac{1}{r'^2\sin^2\theta'}\left(\frac{\partial^2 V_\phi}{\partial \phi^2}\right)$$

$$\left. - \frac{V_\phi}{r'^2\sin\theta'} + \frac{2}{r'^2\sin^2\theta'}\left(\frac{\partial V_{r'}}{\partial \phi}\right) + \frac{2\cos\theta'}{r'^2\sin^2\theta'}\left(\frac{\partial V_{\theta'}}{\partial \phi}\right)\right]$$

Euler's Equations

Obtain from Navier-Stokes equations by putting $\mu = 0$.

Appendix I General Bibliography (Applicable to more than one chapter)

1. Allen, T. Jr., and R. L. Ditsworth, *Fluid Mechanics,* McGraw-Hill, New York, 1972.
2. Barry, R. G., and R. J. Chorley, *Atmosphere, Weather and Climate,* Methuen, London, 1968.
3. Berteaux, H. O., *Buoy Engineering,* Wiley, New York, 1976.
4. Daugherty, R. L., and J. B. Franzini, *Fluid Mechanics with Engineering Applications* (7th ed.), McGraw-Hill, New York, 1977.
5. DeNevers, N., *Fluid Mechanics,* Addison-Wesley, Reading, MA, 1970.
6. Fox, J. A., *An Introduction to Engineering Fluid Mechanics,* Ronald Press, New York, 1975.
7. Fox, R. T., and A. T. McDonald, *Introduction to Fluid Mechanics* (2nd ed.), Wiley, New York, 1973.
8. *Handbook of Chemistry and Physics* (35th ed.), Chemical Rubber Publishing Company, Cleveland, OH, 1953.
9. Hardenberg, W. A., and S. Baker, *Design of Dams: Irrigation,* International Textbook Company, Scranton, PA, 1933.
10. *Illustrated Experiments in Fluid Mechanics,* National Committee for Fluid Mechanics Films, MIT Press, 1972.
11. Janna, W. S., *Fundamentals of Fluid Mechanics,* Brooks/Cole Engineering Division, Monterey, CA, 1983.
12. John, J. E. A., and W. L. Haberman, *Introduction to Fluid Mechanics* (2nd ed.), Wiley, New York, 1973.
13. Kreith, F., *Principles of Heat Transfer* (3rd ed.), Harper & Row, New York, 1973.
14. Mironer, A., *Engineering Fluid Mechanics,* McGraw-Hill, New York, 1979.
15. Owczarek, J. A., *Introduction to Fluid Mechanics,* International Textbook Company, Scranton, PA, 1968.
16. Penzias, W., and M. W. Goodman, *Man Beneath the Sea,* Wiley-Interscience, New York, 1973.
17. Potter, M. C., and J. F. Foss, *Fluid Mechanics,* Ronald Press, New York, 1975.
18. Raznjevic, K., *Handbook of Thermodynamic Tables and Charts,* Hemisphere, New York, 1976.
19. Reynolds, A. J., *Turbulent Flows in Engineering,* Wiley, New York, 1974.
20. Roberson, J. A., and C. T. Crowe, *Engineering Fluid Mechanics* (2nd ed.), Houghton Mifflin, Boston, 1980.
21. *1971 SAE Handbook,* Society of Automotive Engineers, Warrendale, PA, 1971.
22. Schenck, H. Jr., *Introduction to Ocean Engineering,* McGraw-Hill, New York, 1975.
23. Shames, I. H., *Mechanics of Fluids* (2nd ed.), McGraw-Hill, New York, 1982.
24. Streeter, V. L., and E. B. Wylie, *Fluid Mechanics* (7th ed.), McGraw-Hill, New York, 1975.
25. Thomas, H. H., *The Engineering of Large Dams, Part I,* Wiley, New York, 1976.
26. *The U.S. Standard Atmosphere,* U.S. Government Printing Office, Washington, D.C., 1962.
27. Van Dyke, M., *An Album of Fluid Motion,* Parabolic Press, Stanford, CA, 1982.
28. White, F. M., *Fluid Mechanics,* McGraw-Hill, New York, 1979.
29. *The World Book Encyclopedia,* Field Enterprises Educational Corporation, Chicago.

Index